Lecture Notes in Computer Science 3044

Commenced Publication in 1973
Founding and Former Series Editors:
Gerhard Goos, Juris Hartmanis, and Jan van Leeuwen

Editorial Board

Takeo Kanade
　Carnegie Mellon University, Pittsburgh, PA, USA
Josef Kittler
　University of Surrey, Guildford, UK
Jon M. Kleinberg
　Cornell University, Ithaca, NY, USA
Friedemann Mattern
　ETH Zurich, Switzerland
John C. Mitchell
　Stanford University, CA, USA
Oscar Nierstrasz
　University of Bern, Switzerland
C. Pandu Rangan
　Indian Institute of Technology, Madras, India
Bernhard Steffen
　University of Dortmund, Germany
Madhu Sudan
　Massachusetts Institute of Technology, MA, USA
Demetri Terzopoulos
　New York University, NY, USA
Doug Tygar
　University of California, Berkeley, CA, USA
Moshe Y. Vardi
　Rice University, Houston, TX, USA
Gerhard Weikum
　Max-Planck Institute of Computer Science, Saarbruecken, Germany

Springer
Berlin
Heidelberg
New York
Hong Kong
London
Milan
Paris
Tokyo

Antonio Laganà   Marina L. Gavrilova
Vipin Kumar   Youngsong Mun
C.J. Kenneth Tan   Osvaldo Gervasi (Eds.)

Computational Science and Its Applications – ICCSA 2004

International Conference
Assisi, Italy, May 14-17, 2004
Proceedings, Part II

 Springer

Volume Editors

Antonio Laganà
University of Perugia, Department of Chemistry
Via Elce di Sotto, 8, 06123 Perugia, Italy
E-mail: lag@unipg.it

Marina L. Gavrilova
University of Calgary, Department of Computer Science
2500 University Dr. N.W., Calgary, AB, T2N 1N4, Canada
E-mail: marina@cpsc.ucalgary.ca

Vipin Kumar
University of Minnesota, Department of Computer Science and Engineering
4-192 EE/CSci Building, 200 Union Street SE, Minneapolis, MN 55455, USA
E-mail: kumar@cs.umn.edu

Youngsong Mun
SoongSil University, School of Computing, Computer Communication Laboratory
1-1 Sang-do 5 Dong, Dong-jak Ku, Seoul 156-743, Korea
E-mail: mun@computing.soongsil.ac.kr

C.J. Kenneth Tan
Queen's University Belfast, Heuchera Technologies Ltd.
Lanyon North, University Road, Belfast, Northern Ireland, BT7 1NN, UK
E-mail: cjtan@optimanumerics.com

Osvaldo Gervasi
University of Perugia, Department of Mathematics and Computer Science
Via Vanvitelli, 1, 06123 Perugia, Italy
E-mail: ogervasi@computer.org

Library of Congress Control Number: 2004105531

CR Subject Classification (1998): D, F, G, H, I, J, C.2-3

ISSN 0302-9743
ISBN 3-540-22056-9 Springer-Verlag Berlin Heidelberg New York

This work is subject to copyright. All rights are reserved, whether the whole or part of the material is concerned, specifically the rights of translation, reprinting, re-use of illustrations, recitation, broadcasting, reproduction on microfilms or in any other way, and storage in data banks. Duplication of this publication or parts thereof is permitted only under the provisions of the German Copyright Law of September 9, 1965, in its current version, and permission for use must always be obtained from Springer-Verlag. Violations are liable to prosecution under the German Copyright Law.

Springer-Verlag is a part of Springer Science+Business Media

springeronline.com

© Springer-Verlag Berlin Heidelberg 2004

Typesetting: Camera-ready by author, data conversion by PTP-Berlin, Protago-TeX-Production GmbH
Printed on acid-free paper   SPIN: 11010098   06/3142   5 4 3 2 1 0

Preface

The natural mission of Computational Science is to tackle all sorts of human problems and to work out *intelligent* automata aimed at alleviating the burden of working out suitable tools for solving complex problems. For this reason Computational Science, though originating from the need to solve the most challenging problems in science and engineering (computational science is the key player in the fight to gain fundamental advances in astronomy, biology, chemistry, environmental science, physics and several other scientific and engineering disciplines) is increasingly turning its attention to all fields of human activity.

In all activities, in fact, intensive computation, information handling, knowledge synthesis, the use of ad-hoc devices, etc. increasingly need to be exploited and coordinated regardless of the location of both the users and the (various and heterogeneous) computing platforms. As a result the key to understanding the explosive growth of this discipline lies in two adjectives that more and more appropriately refer to Computational Science and its applications: interoperable and ubiquitous. Numerous examples of ubiquitous and interoperable tools and applications are given in the present four LNCS volumes containing the contributions delivered at the 2004 International Conference on Computational Science and its Applications (ICCSA 2004) held in Assisi, Italy, May 14–17, 2004.

To emphasize this particular connotation of modern Computational Science the conference was preceded by a tutorial on Grid Computing (May 13–14) concertedly organized with the COST D23 Action (METACHEM: Metalaboratories for Complex Computational Applications in Chemistry) of the European Coordination Initiative COST in Chemistry and the Project *Enabling Platforms for High-Performance Computational Grids Oriented to Scalable Virtual Organization* of the Ministry of Science and Education of Italy.

The volumes consist of 460 peer reviewed papers given as oral contributions at the conference. The conference included 8 presentations from keynote speakers, 15 workshops and 3 technical sessions. Thanks are due to most of the workshop organizers and the Program Committee members, who took care of the unexpected exceptional load of reviewing work (either carrying it out by themselves or distributing it to experts in the various fields).

Special thanks are due to Noelia Faginas Lago for handling all the necessary secretarial work. Thanks are also due to the young collaborators of the High Performance Computing and the Computational Dynamics and Kinetics research groups of the Department of Mathematics and Computer Science and of the Department of Chemistry of the University of Perugia. Thanks are, obviously, due as well to the sponsors for supporting the conference with their financial and organizational help.

May 2004

Antonio Laganà
on behalf of the co-editors:
Marina L. Gavrilova
Vipin Kumar
Youngsong Mun
C.J. Kenneth Tan
Osvaldo Gervasi

Organization

ICCSA 2004 was organized by the University of Perugia, Italy; the University of Minnesota, Minneapolis (MN), USA and the University of Calgary, Calgary (Canada).

Conference Chairs

Osvaldo Gervasi (University of Perugia, Perugia, Italy), Conference Chair
Marina L. Gavrilova (University of Calgary, Calgary, Canada), Conference Co-chair
Vipin Kumar (University of Minnesota, Minneapolis, USA), Honorary Chair

International Steering Committee

J.A. Rod Blais (University of Calgary, Canada)
Alexander V. Bogdanov (Institute for High Performance Computing and Data Bases, Russia)
Marina L. Gavrilova (University of Calgary, Canada)
Andres Iglesias (University de Cantabria, Spain)
Antonio Laganà (University of Perugia, Italy)
Vipin Kumar (University of Minnesota, USA)
Youngsong Mun (Soongsil University, Korea)
Reneé S. Renner (California State University at Chico, USA)
C.J. Kenneth Tan (Heuchera Technologies, Canada and The Queen's University of Belfast, UK)

Local Organizing Committee

Osvaldo Gervasi (University of Perugia, Italy)
Antonio Laganà (University of Perugia, Italy)
Noelia Faginas Lago (University of Perugia, Italy)
Sergio Tasso (University of Perugia, Italy)
Antonio Riganelli (University of Perugia, Italy)
Stefano Crocchianti (University of Perugia, Italy)
Leonardo Pacifici (University of Perugia, Italy)
Cristian Dittamo (University of Perugia, Italy)
Matteo Lobbiani (University of Perugia, Italy)

Workshop Organizers

Information Systems and Information Technologies (ISIT)

Youngsong Mun (Soongsil University, Korea)

Approaches or Methods of Security Engineering

Haeng Kon Kim (Catholic University of Daegu, Daegu, Korea)
Tai-hoon Kim (Korea Information Security Agency, Korea)

Authentication Technology

Eui-Nam Huh (Seoul Women's University, Korea)
Ki-Young Mun (Seoul Women's University, Korea)
Taemyung Chung (Seoul Women's University, Korea)

Internet Communications Security

José Sierra-Camara (ITC Security Lab., University Carlos III of Madrid, Spain)
Julio Hernandez-Castro (ITC Security Lab., University Carlos III of Madrid, Spain)
Antonio Izquierdo (ITC Security Lab., University Carlos III of Madrid, Spain)

Location Management and Security in Next Generation Mobile Networks

Dong Chun Lee (Howon University, Chonbuk, Korea)
Kuinam J. Kim (Kyonggi University, Seoul, Korea)

Routing and Handoff

Hyunseung Choo (Sungkyunkwan University, Korea)
Frederick T. Sheldon (Sungkyunkwan University, Korea)
Alexey S. Rodionov (Sungkyunkwan University, Korea)

Grid Computing

Peter Kacsuk (MTA SZTAKI, Budapest, Hungary)
Robert Lovas (MTA SZTAKI, Budapest, Hungary)

Resource Management and Scheduling Techniques for Cluster and Grid Computing Systems

Jemal Abawajy (Carleton University, Ottawa, Canada)

Parallel and Distributed Computing

Jiawan Zhang (Tianjin University, Tianjin, China)
Qi Zhai (Tianjin University, Tianjin, China)
Wenxuan Fang (Tianjin University, Tianjin, China)

Molecular Processes Simulations

Antonio Laganà (University of Perugia, Perugia, Italy)

Numerical Models in Biomechanics

Jiri Nedoma (Academy of Sciences of the Czech Republic, Prague, Czech Republic)
Josef Danek (University of West Bohemia, Pilsen, Czech Republic)

Scientific Computing Environments (SCEs) for Imaging in Science

Almerico Murli (University of Naples Federico II and Institute for High Performance Computing and Networking, ICAR, Italian National Research Council, Naples, Italy)
Giuliano Laccetti (University of Naples Federico II, Naples, Italy)

Computer Graphics and Geometric Modeling (TSCG 2004)

Andres Iglesias (University of Cantabria, Santander, Spain)
Deok-Soo Kim (Hanyang University, Seoul, Korea)

Virtual Reality in Scientific Applications and Learning

Osvaldo Gervasi (University of Perugia, Perugia, Italy)

Web-Based Learning

Woochun Jun (Seoul National University of Education, Seoul, Korea)

Matrix Approximations with Applications to Science, Engineering and Computer Science

Nicoletta Del Buono (University of Bari, Bari, Italy)
Tiziano Politi (Politecnico di Bari, Bari, Italy)

Spatial Statistics and Geographic Information Systems: Algorithms and Applications

Stefania Bertazzon (University of Calgary, Calgary, Canada)
Borruso Giuseppe (University of Trieste, Trieste, Italy)

Computational Geometry and Applications (CGA 2004)

Marina L. Gavrilova (University of Calgary, Calgary, Canada)

Program Committee

Jemal Abawajy (Carleton University, Canada)
Kenny Adamson (University of Ulster, UK)
Stefania Bertazzon (University of Calgary, Canada)
Sergei Bespamyatnikh (Duke University, USA)
J.A. Rod Blais (University of Calgary, Canada)
Alexander V. Bogdanov (Institute for High Performance Computing and Data Bases, Russia)
Richard P. Brent(Oxford University, UK)
Martin Buecker (Aachen University, Germany)
Rajkumar Buyya (University of Melbourne, Australia)
Hyunseung Choo (Sungkyunkwan University, Korea)
Toni Cortes (Universidad de Catalunya, Barcelona, Spain)
Danny Crookes (The Queen's University of Belfast, (UK))
Brian J. d'Auriol (University of Texas at El Paso, USA)
Ivan Dimov (Bulgarian Academy of Sciences, Bulgaria)
Matthew F. Dixon (Heuchera Technologies, UK)
Marina L. Gavrilova (University of Calgary, Canada)
Osvaldo Gervasi (University of Perugia, Italy)
James Glimm (SUNY Stony Brook, USA)
Christopher Gold (Hong Kong Polytechnic University, Hong Kong, ROC)
Paul Hovland (Argonne National Laboratory, USA)
Andres Iglesias (University de Cantabria, Spain)
Elisabeth Jessup (University of Colorado, USA)
Chris Johnson (University of Utah, USA)
Peter Kacsuk (Hungarian Academy of Science, Hungary)
Deok-Soo Kim (Hanyang University, Korea)
Vipin Kumar (University of Minnesota, USA)
Antonio Laganà (University of Perugia, Italy)
Michael Mascagni (Florida State University, USA)
Graham Megson (University of Reading, UK)
Youngsong Mun (Soongsil University, Korea)
Jiri Nedoma (Academy of Sciences of the Czech Republic, Czech Republic)
Robert Panoff (Shodor Education Foundation, USA)
Reneé S. Renner (California State University at Chico, USA)
Heather J. Ruskin (Dublin City University, Ireland)
Muhammad Sarfraz (King Fahd University of Petroleum and Minerals, Saudi Arabia)
Edward Seidel (Louisiana State University, (USA) and Albert-Einstein-Institut, Potsdam, Germany)
Vaclav Skala (University of West Bohemia, Czech Republic)
Masha Sosonkina (University of Minnesota, (USA))
David Taniar (Monash University, Australia)
Ruppa K. Thulasiram (University of Manitoba, Canada)
Koichi Wada (University of Tsukuba, Japan)

Stephen Wismath (University of Lethbridge, Canada)
Chee Yap (New York University, USA)
Osman Yaşar (SUNY at Brockport, USA)

Sponsoring Organizations

University of Perugia, Perugia, Italy

University of Calgary, Calgary, Canada

University of Minnesota, Minneapolis, MN, USA

The Queen's University of Belfast, UK

Heuchera Technologies, UK

The project **GRID.IT**: *Enabling Platforms for High-Performance Computational Grids Oriented to Scalable Virtual Organizations*, of the Ministry of Science and Education of Italy

COST – European Cooperation in the Field of Scientific and Technical Research

Table of Contents – Part II

Grid Computing Workshop

Advanced Simulation Technique for Modeling Multiphase Fluid Flow
in Porous Media 1
  Jong G. Kim, Hyoung Woo Park

The P-GRADE Grid Portal 10
  Csaba Németh, Gábor Dózsa, Róbert Lovas, Péter Kacsuk

A Smart Agent-Based Grid Computing Platform 20
  *Kwang-Won Koh, Hie-Cheol Kim, Kyung-Lang Park, Hwang-Jik Lee,
  Shin-Dug Kim*

Publishing and Executing Parallel Legacy Code Using an OGSI
Grid Service... 30
  T. Delaitre, A. Goyeneche, T. Kiss, S.C. Winter

The PROVE Trace Visualisation Tool as a Grid Service 37
  Gergely Sipos, Péter Kacsuk

Privacy Protection in Ubiquitous Computing Based on Privacy Label
and Information Flow 46
  Seong Oun Hwang, Ki Song Yoon

Resource Management and Scheduling Techniques for Cluster and Grid Computing Systems Workshop

Application-Oriented Scheduling in the KNOWLEDGE GRID:
A Model and Architecture 55
  Andrea Pugliese, Domenico Talia

A Monitoring and Prediction Tool for Time-Constraint
Grid Application... 66
  Abdulla Othman, Karim Djemame, Iain Gourlay

Optimal Server Allocation in Reconfigurable Clusters with
Multiple Job Types 76
  J. Palmer, I. Mitrani

Design and Evaluation of an Agent-Based Communication Model for a
Parallel File System 87
  *María S. Pérez, Alberto Sánchez, Jemal Abawajy, Víctor Robles,
  José M. Peña*

Task Allocation for Minimizing Programs Completion Time in
Multicomputer Systems... 97
  Gamal Attiya, Yskandar Hamam

Fault Detection Service Architecture for Grid Computing Systems 107
  J.H. Abawajy

Adaptive Interval-Based Caching Management Scheme for Cluster
Video Server 116
  Qin Zhang, Hai Jin, Yufu Li, Shengli Li

A Scalable Streaming Proxy Server Based on
Cluster Architecture.. 126
  Hai Jin, Jie Chu, Kaiqin Fan, Zhi Dong, Zhiling Yang

The Measurement of an Optimum Load Balancing Algorithm in a
Master/Slave Architecture 136
  Finbarr O'Loughlin, Desmond Chambers

Data Discovery Mechanism for a Large Peer-to-Peer Based
Scientific Data Grid Environment 146
  Azizol Abdullah, Mohamed Othman, Md Nasir Sulaiman,
  Hamidah Ibrahim, Abu Talib Othman

A DAG-Based XCIGS Algorithm for Dependent Tasks in
Grid Environments... 158
  Changqin Huang, Deren Chen, Qinghuai Zeng, Hualiang Hu

Running Data Mining Applications on the Grid:
A Bag-of-Tasks Approach 168
  Fabrício A.B. da Silva, Sílvia Carvalho, Hermes Senger,
  Eduardo R. Hruschka, Cléver R.G. de Farias

Parallel and Distributed Computing Workshop

Application of Block Design to a Load Balancing Algorithm on
Distributed Networks.. 178
  Yeijin Lee, Okbin Lee, Taehoon Lee, Ilyong Chung

Maintenance Strategy for Efficient Communication at Data Warehouse .. 186
  Hyun Chang Lee, Sang Hyun Bae

Conflict Resolution of Data Synchronization in Mobile Environment 196
  YoungSeok Lee, YounSoo Kim, Hoon Choi

A Framework for Orthogonal Data and Control
Parallelism Exploitation 206
  S. Campa, M. Danelutto

Multiplier with Parallel CSA Using CRT's Specific Moduli
(2^k-1, 2^k, 2^k+1) 216
  Wu Woan Kim, Sang-Dong Jang

Unified Development Solution for Cluster and Grid Computing and
Its Application in Chemistry 226
  Róbert Lovas, Péter Kacsuk, István Lagzi, Tamás Turányi

Remote Visualization Based on Grid Computing 236
  Zhigeng Pan, Bailin Yang, Mingmin Zhang, Qizhi Yu, Hai Lin

Avenues for High Performance Computation on a PC 246
  Yu-Fai Fung, M. Fikret Ercan, Wai-Leung Cheung, Gujit Singh

A Modified Parallel Computation Model Based on Cluster 252
  Xiaotu Li, Jizhou Sun, Jiawan Zhang, Zhaohui Qi, Gang Li

Parallel Testing Method by Partitioning Circuit Based on the
Exhaustive Test 262
  Wu Woan Kim

A Parallel Volume Splatting Algorithm Based on PC-Clusters 272
  Jiawan Zhang, Jizhou Sun, Yi Zhang, Qianqian Han, Zhou Jin

Molecular Processes Simulation Workshop

Three-Center Nuclear Attraction Integrals for Density Functional
Theory and Nonlinear Transformations 280
  Hassan Safouhi

Parallelization of Reaction Dynamics Codes Using P-GRADE:
A Case Study 290
  Ákos Bencsura, György Lendvay

Numerical Implementation of Quantum Fluid Dynamics:
A Working Example 300
  Fabrizio Esposito

Numerical Revelation and Analysis of Critical Ignition Conditions
for Branch Chain Reactions by Hamiltonian Systematization Methods of
Kinetic Models 313
  Gagik A. Martoyan, Levon A. Tavadyan

Computer Simulations in Ion-Atom Collisions 321
  S.F.C. O'Rourke, R.T. Pedlow, D.S.F. Crothers

Bond Order Potentials for a priori Simulations of
Polyatomic Reactions 328
  Ernesto Garcia, Carlos Sánchez, Margarita Albertí, Antonio Laganà

Inorganic Phosphates Investigation by Support Vector Machine 338
  Cinzia Pierro, Francesco Capitelli

Characterization of Equilibrium Structure for N_2-N_2 Dimer in
1.2Å$\leq$R$\geq$2.5Å Region Using DFT Method 350
  Ajmal H. Hamdani, S. Shahdin

A Time Dependent Study of the Nitrogen Atom Nitrogen
Molecule Reaction 357
  Antonio Laganà, Leonardo Pacifici, Dimitris Skouteris

From DFT Cluster Calculations to Molecular Dynamics Simulation of
N_2 Formation on a Silica Model Surface 366
  M. Cacciatore, A. Pieretti, M. Rutigliano, N. Sanna

Molecular Mechanics and Dynamics Calculations to Bridge Molecular
Structure Information and Spectroscopic Measurements on Complexes
of Aromatic Compounds 374
  G. Pietraperzia, R. Chelli, M. Becucci, Antonio Riganelli,
  Margarita Alberti, Antonio Laganà

Direct Simulation Monte Carlo Modeling of Non Equilibrium
Reacting Flows. Issues for the Inclusion into a ab initio
Molecular Processes Simulator 383
  D. Bruno, M. Capitelli, S. Longo, P. Minelli

Molecular Simulation of Reaction and Adsorption in Nanochemical
Devices: Increase of Reaction Conversion by Separation of a Product
from the Reaction Mixture 392
  William R. Smith, Martin Lísal

Quantum Generalization of Molecular Dynamics Method.
Wigner Approach 402
  V. Filinov, M. Bonitz, V. Fortov, P. Levashov

$C_6NH_6^+$ Ions as Intermediates in the Reaction between
Benzene and N^+ Ions 412
  Marco Di Stefano, Marzio Rosi, Antonio Sgamellotti

Towards a Full Dimensional Exact Quantum Calculation
of the Li + HF Reactive Cross Section 422
  Antonio Laganà, Stefano Crocchianti, Valentina Piermarini

Conformations of 1,2,4,6-Tetrathiepane 432
  Issa Yavari, Arash Jabbari, Shahram Moradi

Fine Grain Parallelization of a Discrete Variable Wavepacket
Calculation Using ASSIST-CL 437
  Stefano Gregori, Sergio Tasso, Antonio Laganà

Numerical Models in Biomechanics Session

On the Solution of Contact Problems with Visco-Plastic Friction
in the Bingham Rheology: An Application in Biomechanics............. 445
  Jiří Nedoma

On the Stress-Strain Analysis of the Knee Replacement 456
  *J. Daněk, F. Denk, I. Hlaváček, Jiří Nedoma, J. Stehlík,
  P. Vavřík*

Musculoskeletal Modeling of Lumbar Spine under Follower Loads 467
  Yoon Hyuk Kim, Kyungsoo Kim

Computational Approach to Optimal Transport Network Construction
in Biomechanics 476
  Natalya Kizilova

Encoding Image Based on Retinal Ganglion Cell 486
  Sung-Kwan Je, Eui-Young Cha, Jae-Hyun Cho

Scientific Computing Environments (SCE's) for Imaging in Science Session

A Simple Data Analysis Method for Kinetic Parameters Estimation
from Renal Measurements with a Three-Headed SPECT System 495
  Eleonora Vanzi, Andreas Robert Formiconi

Integrating Medical Imaging into a Grid Based
Computing Infrastructure 505
  Paola Bonetto, Mario Guarracino, Fabrizio Inguglia

Integrating Scientific Software Libraries in Problem Solving
Environments: A Case Study with ScaLAPACK 515
  L. D'Amore, Mario R. Guarracino, G. Laccetti, A. Murli

Parallel/Distributed Film Line Scratch Restoration by
Fusion Techniques 525
  G. Laccetti, L. Maddalena, A. Petrosino

An Interactive Distributed Environment for Digital Film Restoration 536
  F. Collura, A. Machì, F. Nicotra

Computer Graphics and Geometric Modeling Workshop (TSCG 2004)

On Triangulations... 544
  Ivana Kolingerová

Probability Distribution of Op-Codes in Edgebreaker 554
  *Deok-Soo Kim, Cheol-Hyung Cho, Youngsong Cho, Chang Wook Kang,
  Hyun Chan Lee, Joon Young Park*

Polyhedron Splitting Algorithm for 3D Layer Generation............... 564
  Jaeho Lee, Joon Young Park, Deok-Soo Kim, Hyun Chan Lee

Synthesis of Mechanical Structures Using a Genetic Algorithm.......... 573
  In-Ho Lee, Joo-Heon Cha, Jay-Jung Kim, M.-W. Park

Optimal Direction for Monotone Chain Decomposition.................. 583
  Hayong Shin, Deok-Soo Kim

GTVIS: Fast and Efficient Rendering System for
Real-Time Terrain Visualization 592
  Russel A. Apu, Marina L. Gavrilova

Target Data Projection in Multivariate Visualization
– An Application to Mine Planning 603
  Leonardo Soto, Ricardo Sánchez, Jorge Amaya

Parametric Freehand Sketches 613
  Ferran Naya, Manuel Contero, Nuria Aleixos, Joaquim Jorge

Variable Level of Detail Strips 622
  J.F. Ramos, M. Chover

Bézier Solutions of the Wave Equation 631
  J.V. Beltran, J. Monterde

Matlab Toolbox for a First Computer Graphics Course for Engineers 641
  Akemi Gálvez, A. Iglesias, César Otero, Reinaldo Togores

A Differential Method for Parametric Surface Intersection............. 651
  A. Gálvez, J. Puig-Pey, A. Iglesias

A Comparison Study of Metaheuristic Techniques for Providing QoS
to Avatars in DVE Systems 661
  P. Morillo, J.M. Orduña, Marcos Fernández, J. Duato

Visualization of Large Terrain Using Non-restricted
Quadtree Triangulations 671
  Mariano Pérez, Ricardo Olanda, Marcos Fernández

Boundary Filtering in Surface Reconstruction 682
  Michal Varnuška, Ivana Kolingerová

Image Coherence Based Adaptive Sampling for Image Synthesis 693
  *Qing Xu, Roberto Brunelli, Stefano Messelodi, Jiawan Zhang,
  Mingchu Li*

A Comparison of Multiresolution Modelling in
Real-Time Terrain Visualisation 703
   C. Rebollo, I. Remolar, M. Chover, J.F. Ramos

Photo-realistic 3D Head Modeling Using Multi-view Images 713
   Tong-Yee Lee, Ping-Hsien Lin, Tz-Hsien Yang

Texture Mapping on Arbitrary 3D Surfaces 721
   Tong-Yee Lee, Shaur-Uei Yan

Segmentation-Based Interpolation of 3D Medical Images 731
   Zhigeng Pan, Xuesong Yin, Guohua Wu

A Bandwidth Reduction Scheme for 3D Texture-Based Volume
Rendering on Commodity Graphics Hardware 741
   *Won-Jong Lee, Woo-Chan Park, Jung-Woo Kim, Tack-Don Han,
   Sung-Bong Yang, Francis Neelamkavil*

An Efficient Image-Based 3D Reconstruction Algorithm for Plants 751
   Zhigeng Pan, Weixi Hu, Xinyu Guo, Chunjiang Zhao

Where the Truth Lies (in Automatic Theorem Proving in
Elementary Geometry) 761
   T. Recio, F. Botana

Helical Curves on Surfaces for Computer-Aided Geometric Design
and Manufacturing 771
   J. Puig-Pey, Akemi Gálvez, A. Iglesias

An Application of Computer Graphics for Landscape
Impact Assessment 779
   *César Otero, Viola Bruschi, Antonio Cendrero, Akemi Gálvez,
   Miguel Lázaro, Reinaldo Togores*

Fast Stereo Matching Using Block Similarity 789
   Han-Suh Koo, Chang-Sung Jeong

View Morphing Based on Auto-calibration for Generation of
In-between Views 799
   Jin-Young Song, Yong-Ho Hwang, Hyun-Ki Hong

Virtual Reality in Scientific Applications and Learning (VRSAL 2004) Workshop

Immersive Displays Based on a Multi-channel PC Clustered System 809
   Hunjoo Lee, Kijong Byun

Virtual Reality Technology Applied to Simulate
Construction Processes 817
   Alcínia Zita Sampaio, Pedro Gameiro Henriques, Pedro Studer

Virtual Reality Applied to Molecular Sciences 827
   Osvaldo Gervasi, Antonio Riganelli, Antonio Laganà

Design and Implementation of an Online 3D Game Engine 837
   Hunjoo Lee, Taejoon Park

Dynamically Changing Road Networks – Modelling and
Visualization in Real Time 843
   Christian Mark, Armin Kaußner, Martin Grein, Hartmut Noltemeier

EoL: A Web-Based Distance Assessment System 854
   Osvaldo Gervasi, Antonio Laganà

Discovery Knowledge of User Preferences: Ontologies in Fashion
Design Recommender Agent System.................................. 863
   Kyung-Yong Jung, Young-Joo Na, Dong-Hyun Park, Jung-Hyun Lee

When an Ivy League University Puts Its Courses Online, Who's
Going to Need a Local University? 873
   Matthew C.F. Lau, Rebecca B.N. Tan

Web-Based Learning Session

Threads in an Undergraduate Course: A Java Example Illuminating
Different Multithreading Approaches 882
   *H. Martin Bücker, Bruno Lang, Hans-Joachim Pflug,
   Andre Vehreschild*

A Comparison of Web Searching Strategies According to Cognitive
Styles of Elementary Students 892
   Hanil Kim, Miso Yun, Pankoo Kim

The Development and Application of a Web-Based Information
Communication Ethics Education System 902
   Suk-Ki Hong, Woochun Jun

An Interaction Model for Web-Based Learning: Cooperative Project 913
   Eunhee Choi, Woochun Jun, Suk-Ki Hong, Young-Cheol Bang

Observing Standards for Web-Based Learning from the Web............ 922
   Luis Anido, Judith Rodríguez, Manuel Caeiro, Juan Santos

Matrix Approximations with Applications to Science, Engineering, and Computer Science Workshop

On Computing the Spectral Decomposition of
Symmetric Arrowhead Matrices 932
   Fasma Diele, Nicola Mastronardi, Marc Van Barel, Ellen Van Camp

Relevance Feedback for Content-Based Image Retrieval Using
Proximal Support Vector Machine.................................. 942
  YoungSik Choi, JiSung Noh

Orthonormality-Constrained INDSCAL with Nonnegative Saliences 952
  Nickolay T. Trendafilov

Optical Flow Estimation via Neural Singular Value
Decomposition Learning 961
  Simone Fiori, Nicoletta Del Buono, Tiziano Politi

Numerical Methods Based on Gaussian Quadrature and Continuous
Runge-Kutta Integration for Optimal Control Problems 971
  Fasma Diele, Carmela Marangi, Stefania Ragni

Graph Adjacency Matrix Associated with a Data Partition 979
  Giuseppe Acciani, Girolamo Fornarelli, Luciano Liturri

A Continuous Technique for the Weighted Low-Rank
Approximation Problem 988
  Nicoletta Del Buono, Tiziano Politi

Spatial Statistics and Geographical Information Systems: Algorithms and Applications

A Spatial Multivariate Approach to the Analysis of Accessibility
to Health Care Facilities in Canada 998
  Stefania Bertazzon

Density Analysis on Large Geographical Databases. Search for an
Index of Centrality of Services at Urban Scale 1009
  Giuseppe Borruso, Gabriella Schoier

An Exploratory Spatial Data Analysis (ESDA) Toolkit for the
Analysis of Activity/Travel Data................................... 1016
  Ronald N. Buliung, Pavlos S. Kanaroglou

Using Formal Ontology for Integrated Spatial Data Mining........... 1026
  Sungsoon Hwang

G.I.S. and Fuzzy Sets for the Land Suitability Analysis 1036
  Beniamino Murgante, Giuseppe Las Casas

Intelligent Gis and Retail Location Dynamics: A Multi Agent
System Integrated with ArcGis 1046
  S. Lombardo, M. Petri, D. Zotta

ArcObjects Development in Zone Design Using
Visual Basic for Applications 1057
  Sergio Palladini

Searching for 2D Spatial Network Holes 1069
  Femke Reitsma, Shane Engel

Extension of Geography Markup Language (GML) for Mobile and
Location-Based Applications 1079
  Young Soo Ahn, Soon-Young Park, Sang Bong Yoo, Hae-Young Bae

A Clustering Method for Large Spatial Databases 1089
  Gabriella Schoier, Giuseppe Borruso

GeoSurveillance: Software for Monitoring Change in
Geographic Patterns 1096
  Peter Rogerson, Ikuho Yamada

From Axial Maps to Mark Point Parameter Analysis (Ma.P.P.A.) –
A GIS Implemented Method to Automate Configurational Analysis 1107
  V. Cutini, M. Petri, A. Santucci

Computing Foraging Paths for Shore-Birds Using Fractal Dimensions
and Pecking Success from Footprint Surveys on Mudflats:
An Application for Red-Necked Stints in the Moroshechnaya
River Estuary, Kamchatka-Russian Far East.......................... 1117
  Falk Huettmann

Author Index 1129

Table of Contents – Part I

Information Systems and Information Technologies (ISIT) Workshop, Multimedia Session

Face Detection by Facial Features with Color Images and Face
Recognition Using PCA... 1
   *Jin Ok Kim, Sung Jin Seo, Chin Hyun Chung, Jun Hwang,
Woongjae Lee*

A Shakable Snake for Estimation of Image Contours................... 9
   Jin-Sung Yoon, Joo-Chul Park, Seok-Woo Jang, Gye-Young Kim

A New Recurrent Fuzzy Associative Memory for Recognizing
Time-Series Patterns Contained Ambiguity........................... 17
   Joongjae Lee, Won Kim, Jeonghee Cha, Gyeyoung Kim, Hyungil Choi

A Novel Approach for Contents-Based E-catalogue Image Retrieval
Based on a Differential Color Edge Model........................... 25
   Junchul Chun, Goorack Park, Changho An

A Feature-Based Algorithm for Recognizing Gestures
on Portable Computers 33
   Mi Gyung Cho, Am Sok Oh, Byung Kwan Lee

Fingerprint Matching Based on Linking Information Structure
of Minutiae 41
   JeongHee Cha, HyoJong Jang, GyeYoung Kim, HyungIl Choi

Video Summarization Using Fuzzy One-Class Support Vector Machine... 49
   YoungSik Choi, KiJoo Kim

A Transcode and Prefetch Technique of Multimedia Presentations
for Mobile Terminals 57
   *Maria Hong, Euisun Kang, Sungmin Um, Dongho Kim,
Younghwan Lim*

Information Systems and Information Technologies (ISIT) Workshop, Algorithm Session

A Study on Generating an Efficient Bottom-up Tree Rewrite Machine
for JBurg 65
   KyungWoo Kang

A Study on Methodology for Enhancing Reliability of Datapath 73
   SunWoong Yang, MoonJoon Kim, JaeHeung Park, Hoon Chang

A Useful Method for Multiple Sequence Alignment
and Its Implementation 81
 Jin Kim, Dong-Hoi Kim, Saangyong Uhmn

A Research on the Stochastic Model
for Spoken Language Understanding................................. 89
 Yong-Wan Roh, Kwang-Seok Hong, Hyon-Gu Lee

The Association Rule Algorithm with Missing Data in Data Mining 97
 Bobby D. Gerardo, Jaewan Lee, Jungsik Lee, Mingi Park, Malrey Lee

Constructing Control Flow Graph for Java by Decoupling Exception
Flow from Normal Flow 106
 Jang-Wu Jo, Byeong-Mo Chang

On Negation-Based Conscious Agent 114
 Kang Soo Tae, Hee Yong Youn, Gyung-Leen Park

A Document Classification Algorithm Using the Fuzzy Set Theory
and Hierarchical Structure of Document 122
 Seok-Woo Han, Hye-Jue Eun, Yong-Sung Kim, László T. Kóczy

A Supervised Korean Verb Sense Disambiguation Algorithm
Based on Decision Lists of Syntactic Features....................... 134
 Kweon Yang Kim, Byong Gul Lee, Dong Kwon Hong

Information Systems and Information Technologies (ISIT) Workshop, Security Session

Network Security Management Using ARP Spoofing.................... 142
 Kyohyeok Kwon, Seongjin Ahn, Jin Wook Chung

A Secure and Practical CRT-Based RSA to Resist
Side Channel Attacks... 150
 ChangKyun Kim, JaeCheol Ha, Sung-Hyun Kim, Seokyu Kim,
 Sung-Ming Yen, SangJae Moon

A Digital Watermarking Scheme in JPEG-2000 Using the Properties
of Wavelet Coefficient Sign 159
 Han-Ki Lee, Geun-Sil Song, Mi-Ae Kim, Kil-Sang Yoo,
 Won-Hyung Lee

A Security Proxy Based Protocol for Authenticating
the Mobile IPv6 Binding Updates 167
 Il-Sun You, Kyungsan Cho

A Fuzzy Expert System for Network Forensics 175
 Jung-Sun Kim, Minsoo Kim, Bong-Nam Noh

A Design of Preventive Integrated Security Management System
Using Security Labels and a Brief Comparison with Existing
Models 183
  D.S. Kim, T.M. Chung

The Vulnerability Assessment for Active Networks;
Model, Policy, Procedures, and Performance Evaluations 191
  *Young J. Han, Jin S. Yang, Beom H. Chang, Jung C. Na,
  Tai M. Chung*

Authentication of Mobile Node Using AAA in Coexistence of VPN
and Mobile IP 199
  Miyoung Kim, Misun Kim, Youngsong Mun

Survivality Modeling for Quantitative Security Assessment
in Ubiquitous Computing Systems* 207
  Changyeol Choi, Sungsoo Kim, We-Duke Cho

New Approach for Secure and Efficient Metering
in the Web Advertising 215
  Soon Seok Kim, Sung Kwon Kim, Hong Jin Park

MLS/SDM: Multi-level Secure Spatial Data Model 222
  Young-Hwan Oh, Hae-Young Bae

Detection Techniques for ELF Executable File Using Assembly
Instruction Searching 230
  Jun-Hyung Park, Min-soo Kim, Bong-Nam Noh

Secure Communication Scheme Applying MX Resource Record
in DNSSEC Domain... 238
  Hyung-Jin Lim, Hak-Ju Kim, Tae-Kyung Kim, Tai-Myung Chung

Committing Secure Results with Replicated Servers 246
  Byoung Joon Min, Sung Ki Kim, Chaetae Im

Applied Research of Active Network to Control Network Traffic
in Virtual Battlefield 254
  Won Goo Lee, Jae Kwang Lee

Design and Implementation of the HoneyPot System with Focusing
on the Session Redirection 262
  Miyoung Kim, Misun Kim, Youngsong Mun

Information Systems and Information Technologies (ISIT) Workshop, Network Session

Analysis of Performance for MCVoD System 270
  SeokHoon Kang, IkSoo Kim, Yoseop Woo

A QoS Improvement Scheme for Real-Time Traffic
Using IPv6 Flow Labels.. 278
  In Hwa Lee, Sung Jo Kim

Energy-Efficient Message Management Algorithms in HMIPv6.......... 286
  Sun Ok Yang, SungSuk Kim, Chong-Sun Hwang, SangKeun Lee

A Queue Management Scheme for Alleviating the Impact
of Packet Size on the Achieved Throughput 294
  *Sungkeun Lee, Wongeun Oh, Myunghyun Song, Hyun Yoe,
  JinGwang Koh, Changryul Jung*

PTrace: Pushback/SVM Based ICMP Traceback Mechanism
against DDoS Attack 302
  Hyung-Woo Lee, Min-Goo Kang, Chang-Won Choi

Traffic Control Scheme of ABR Service Using NLMS in ATM Network... 310
  *Kwang-Ok Lee, Sang-Hyun Bae, Jin-Gwang Koh, Chang-Hee Kwon,
  Chong-Soo Cheung, In-Ho Ra*

Information Systems and Information Technologies (ISIT) Workshop, Grid Session

XML-Based Workflow Description Language
for Grid Applications 319
  *Yong-Won Kwon, So-Hyun Ryu, Chang-Sung Jeong,
  Hyoungwoo Park*

Placement Algorithm of Web Server Replicas 328
  Seonho Kim, Miyoun Yoon, Yongtae Shin

XML-OGL: UML-Based Graphical Web Query Language
for XML Documents 337
  Chang Yun Jeong, Yong-Sung Kim, Yan Ha

Layered Web-Caching Technique for VOD Services 345
  *Iksoo Kim, Yoseop Woo, Hyunchul Kang, Backhyun Kim,
  Jinsong Ouyang*

QoS-Constrained Resource Allocation for a Grid-Based Multiple
Source Electrocardiogram Application 352
  *Dong Su Nam, Chan-Hyun Youn, Bong Hwan Lee, Gari Clifford,
  Jennifer Healey*

Efficient Pre-fetch and Pre-release Based Buffer Cache Management
for Web Applications 360
  Younghun Ko, Jaehyoun Kim, Hyunseung Choo

A New Architecture Design for Differentiated Resource Sharing
on Grid Service 370
  Eui-Nam Huh

An Experiment and Design of Web-Based Instruction Model
for Collaboration Learning 378
  Duckki Kim, Youngsong Mun

Information Systems and Information Technologies (ISIT) Workshop, Mobile Session

Performance Limitation of STBC OFDM-CDMA Systems
in Mobile Fading Channels 386
  Young-Hwan You, Tae-Won Jang, Min-Goo Kang, Hyung-Woo Lee,
  Hwa-Seop Lim, Yong-Soo Choi, Hyoung-Kyu Song

PMEPR Reduction Algorithms for STBC-OFDM Signals 394
  Hyoung-Kyu Song, Min-Goo Kang, Ou-Seb Lee, Pan-Yuh Joo,
  We-Duke Cho, Mi-Jeong Kim, Young-Hwan You

An Efficient Image Transmission System Adopting OFDM Based
Sequence Reordering Method in Non-flat Fading Channel 402
  JaeMin Kwak, HeeGok Kang, SungEon Cho, Hyun Yoe, JinGwang Koh

The Efficient Web-Based Mobile GIS Service System
through Reduction of Digital Map 410
  Jong-Woo Kim, Seong-Seok Park, Chang-Soo Kim, Yugyung Lee

Reducing Link Loss in Ad Hoc Networks 418
  Sangjoon Park, Eunjoo Jeong, Byunggi Kim

A Web Based Model for Analyzing Compliance of Mobile Content 426
  Woojin Lee, Yongsun Cho, Kiwon Chong

Delay and Collision Reduction Mechanism for Distributed Fair
Scheduling in Wireless LANs 434
  Kee-Hyun Choi, Kyung-Soo Jang, Dong-Ryeol Shin

Approaches or Methods of Security Engineering Workshop

Bit-Serial Multipliers for Exponentiation and Division
in $GF(2^m)$ Using Irreducible AOP 442
  Yong Ho Hwang, Sang Gyoo Sim, Pil Joong Lee

Introduction and Evaluation of Development System Security
Process of ISO/IEC TR 15504 451
  Eun-ser Lee, Kyung Whan Lee, Tai-hoon Kim, Il-Hong Jung

Design on Mobile Secure Electronic Transaction Protocol
with Component Based Development 461
    Haeng-Kon Kim, Tai-Hoon Kim

A Distributed Online Certificate Status Protocol Based
on GQ Signature Scheme 471
    Dae Hyun Yum, Pil Joong Lee

A Design of Configuration Management Practices and CMPET
in Common Criteria Based on Software Process Improvement Activity ... 481
    Sun-Myung Hwang

The Design and Development for Risk Analysis Automatic Tool 491
    Young-Hwan Bang, Yoon-Jung Jung, Injung Kim, Namhoon Lee,
    Gang-Soo Lee

A Fault-Tolerant Mobile Agent Model in Replicated Secure Services 500
    Kyeongmo Park

Computation of Multiplicative Inverses in $GF(2^n)$
Using Palindromic Representation 510
    Hyeong Seon Yoo, Dongryeol Lee

A Study on Smart Card Security Evaluation Criteria
for Side Channel Attacks 517
    HoonJae Lee, ManKi Ahn, SeonGan Lim, SangJae Moon

User Authentication Protocol Based on Human Memorable Password
and Using RSA 527
    IkSu Park, SeungBae Park, ByeongKyun Oh

Supporting Adaptive Security Levels in Heterogeneous Environments 537
    Ghita Kouadri Mostéfaoui, Mansoo Kim, Mokdong Chung

Intrusion Detection Using Noisy Training Data 547
    Yongsu Park, Jaeheung Lee, Yookun Cho

A Study on Key Recovery Agent Protection Profile Having
Composition Function 557
    Dae-Hee Seo, Im-Yeong Lee, Hee-Un Park

Simulation-Based Security Testing for Continuity of Essential
Service 567
    Hyung-Jong Kim, JoonMo Kim, KangShin Lee, HongSub Lee,
    TaeHo Cho

NextPDM: Improving Productivity and Enhancing the Reusability
with a Customizing Framework Toolkit 577
    Ha Jin Hwang, Soung Won Kim

A Framework for Security Assurance in Component Based Development . 587
　　Hangkon Kim

An Information Engineering Methodology for the Security Strategy
Planning... 597
　　Sangkyun Kim, Choon Seong Leem

A Case Study in Applying Common Criteria to Development Process
of Virtual Private Network 608
　　Sang ho Kim, Choon Seong Leem

A Pointer Forwarding Scheme for Fault-Tolerant Location
Management in Mobile Networks..................................... 617
　　Ihn-Han Bae, Sun-Jin Oh

Architecture Environments for E-business Agent Based on Security...... 625
　　Ho-Jun Shin, Soo-Gi Lee

Authentication Authorization Accounting (AAA) Workshop

Multi-modal Biometrics System Using Face and Signature.............. 635
　　Dae Jong Lee, Keun Chang Kwak, Jun Oh Min, Myung Geun Chun

Simple and Efficient Group Key Agreement
Based on Factoring... 645
　　Junghyun Nam, Seokhyang Cho, Seungjoo Kim, Dongho Won

On Facial Expression Recognition Using the Virtual Image Masking
for a Security System 655
　　*Jin Ok Kim, Kyong Sok Seo, Chin Hyun Chung, Jun Hwang,
　　Woongjae Lee*

Secure Handoff Based on Dual Session Keys in Mobile IP
with AAA 663
　　Yumi Choi, Hyunseung Choo, Byong-Lyol Lee

Detection and Identification Mechanism against Spoofed Traffic
Using Distributed Agents 673
　　Mihui Kim, Kijoon Chae

DMKB : A Defense Mechanism Knowledge Base 683
　　Eun-Jung Choi, Hyung-Jong Kim, Myuhng-Joo Kim

A Fine-Grained Taxonomy of Security Vulnerability
in Active Network Environments 693
　　*Jin S. Yang, Young J. Han, Dong S. Kim, Beom H. Chang,
　　Tai M. Chung, Jung C. Na*

A New Role-Based Authorization Model
in a Corporate Workflow Systems 701
    HyungHyo Lee, SeungYong Lee, Bong-Nam Noh

A New Synchronization Protocol for Authentication
in Wireless LAN Environment 711
    Hea Suk Jo, Hee Yong Youn

A Robust Image Authentication Method Surviving Acceptable
Modifications 722
    Mi-Ae Kim, Geun-Sil Song, Won-Hyung Lee

Practical Digital Signature Generation Using Biometrics 728
    Taekyoung Kwon, Jae-il Lee

Performance Improvement in Mobile IPv6 Using AAA
and Fast Handoff 738
    Changnam Kim, Young-Sin Kim, Eui-Nam Huh, Youngsong Mun

An Efficient Key Agreement Protocol for Secure Authentication 746
    Young-Sin Kim, Eui-Nam Huh, Jun Hwang, Byung-Wook Lee

A Policy-Based Security Management Architecture Using XML
Encryption Mechanism for Improving SNMPv3 755
    Choong Seon Hong, Joon Heo

IDentification Key Based AAA Mechanism in Mobile IP Networks 765
    Hoseong Jeon, Hyunseung Choo, Jai-Ho Oh

An Integrated XML Security Mechanism for Mobile Grid Application.... 776
    Kiyoung Moon, Namje Park, Jongsu Jang, Sungwon Sohn,
    Jaecheol Ryou

Development of XKMS-Based Service Component for Using PKI
in XML Web Services Environment 784
    Namje Park, Kiyoung Moon, Jongsu Jang, Sungwon Sohn

A Scheme for Improving WEP Key Transmission between APs
in Wireless Environment 792
    Chi Hyung In, Choong Seon Hong, Il Gyu Song

Internet Communication Security Workshop

Generic Construction of Certificateless Encryption 802
    Dae Hyun Yum, Pil Joong Lee

Security Issues in Network File Systems 812
    Antonio Izquierdo, Jose María Sierra, Julio César Hernández,
    Arturo Ribagorda

A Content-Independent Scalable Encryption Model 821
  Stefan Lindskog, Johan Strandbergh, Mikael Hackman,
  Erland Jonsson

Fair Exchange to Achieve Atomicity in Payments
of High Amounts Using Electronic Cash 831
  Magdalena Payeras-Capella, Josep Lluís Ferrer-Gomila,
  Llorenç Huguet-Rotger

N3: A Geometrical Approach for Network Intrusion Detection
at the Application Layer 841
  Juan M. Estévez-Tapiador, Pedro García-Teodoro,
  Jesús E. Díaz-Verdejo

Validating the Use of BAN LOGIC 851
  José María Sierra, Julio César Hernández, Almudena Alcaide,
  Joaquín Torres

Use of Spectral Techniques in the Design
of Symmetrical Cryptosystems 859
  Luis Javier García Villalba

Load Balancing and Survivability for Network Services
Based on Intelligent Agents 868
  Robson de Oliveira Albuquerque, Rafael T. de Sousa Jr.,
  Tamer Américo da Silva, Ricardo S. Puttini,
  Clàudia Jacy Barenco Abbas, Luis Javier García Villalba

A Scalable PKI for Secure Routing in the Internet 882
  Francesco Palmieri

Cryptanalysis and Improvement of Password Authenticated Key
Exchange Scheme between Clients with Different Passwords 895
  Jeeyeon Kim, Seungjoo Kim, Jin Kwak, Dongho Won

Timeout Estimation Using a Simulation Model
for Non-repudiation Protocols 903
  Mildrey Carbonell, Jose A. Onieva, Javier Lopez, Deborah Galpert,
  Jianying Zhou

DDoS Attack Defense Architecture Using
Active Network Technology 915
  Choong Seon Hong, Yoshiaki Kasahara, Dea Hwan Lee

A Voting System with Trusted Verifiable Services 924
  Macià Mut Puigserver, Josep Lluís Ferrer Gomila,
  Llorenç Huguet i Rotger

Chaotic Protocols 938
  Mohamed Mejri

Security Consequences of Messaging Hubs in Many-to-Many
E-procurement Solutions 949
  Eva Ponce, Alfonso Durán, Teresa Sánchez

The SAC Test: A New Randomness Test, with Some Applications
to PRNG Analysis 960
  Julio César Hernandez, José María Sierra, Andre Seznec

A Survey of Web Services Security 968
  Carlos Gutiérrez, Eduardo Fernández-Medina, Mario Piattini

Fair Certified E-mail Protocols with Delivery Deadline Agreement....... 978
  Yongsu Park, Yookun Cho

Location Management and the Security in the Next Generation Mobile Networks Workshop

QS-Ware: The Middleware for Providing QoS and Secure Ability
to Web Server 988
  Seung-won Shin, Kwang-ho Baik, Ki-Young Kim, Jong-Soo Jang

Implementation and Performance Evaluation of High-Performance
Intrusion Detection and Response System 998
  Hyeong-Ju Kim, Byoung-Koo Kim, Ik-Kyun Kim

Efficient Key Distribution Protocol for Secure Multicast
Communication 1007
  Bonghan Kim, Hanjin Cho, Jae Kwang Lee

A Bayesian Approach for Estimating Link Travel Time
on Urban Arterial Road Network 1017
  Taehyung Park, Sangkeon Lee

Perimeter Defence Policy Model of Cascade MPLS VPN Networks 1026
  Won Shik Na, Jeom Goo Kim, Intae Ryoo

Design of Authentication and Key Exchange Protocol
in Ethernet Passive Optical Networks.............................. 1035
  Sun-Sik Roh, Su-Hyun Kim, Gwang-Hyun Kim

Detection of Moving Objects Edges to Implement Home Security
System in a Wireless Environment 1044
  Yonghak Ahn, Kiok Ahn, Oksam Chae

Reduction Method of Threat Phrases by Classifying Assets.......... 1052
  Tai-Hoon Kim, Dong Chun Lee

Anomaly Detection Using Sequential Properties of Packets
in Mobile Environment 1060
  Seong-sik Hong, Hwang-bin Ryou

A Case Study in Applying Common Criteria to Development Process
to Improve Security of Software Products 1069
  Sang Ho Kim, Choon Seong Leem

A New Recovery Scheme with Reverse Shared Risk Link Group
in GMPLS-Based WDM Networks 1078
  *Hyuncheol Kim, Seongjin Ahn, Daeho Kim, Sunghae Kim,
  Jin Wook Chung*

Real Time Estimation of Bus Arrival Time under Mobile Environment ... 1088
  Taehyung Park, Sangkeon Lee, Young-Jun Moon

Call Tracking and Location Updating Using DHS in Mobile Networks ... 1097
  Dong Chun Lee

Routing and Handoff Workshop

Improving TCP Performance over Mobile IPv6 1105
  Young-Chul Shim, Nam-Chang Kim, Ho-Seok Kang

Design of Mobile Network Route Optimization Based
on the Hierarchical Algorithm 1115
  Dongkeun Lee, Keecheon Kim, Sunyoung Han

On Algorithms for Minimum-Cost Quickest Paths
with Multiple Delay-Bounds 1125
  Young-Cheol Bang, Inki Hong, Sungchang Lee, Byungjun Ahn

A Fast Handover Protocol for Mobile IPv6 Using Mobility
Prediction Mechanism 1134
  Dae Sun Kim, Choong Seon Hong

The Layer 2 Handoff Scheme for Mobile IP
over IEEE 802.11 Wireless LAN 1144
  Jongjin Park, Youngsong Mun

Session Key Exchange Based on Dynamic Security Association for
Mobile IP Fast Handoff 1151
  Hyun Gon Kim, Doo Ho Choi

A Modified AODV Protocol with Multi-paths Considering Classes
of Services 1159
  *Min-Su Kim, Ki Jin Kwon, Min Young Chung, Tae-Jin Lee,
  Jaehyung Park*

Author Index 1169

Table of Contents – Part III

Workshop on Computational Geometry and Applications (CGA 04)

Geometric Graphs Realization as Coin Graphs 1
Manuel Abellanas, Carlos Moreno-Jiménez

Disc Covering Problem with Application to Digital Halftoning 11
Tetsuo Asano, Peter Brass, Shinji Sasahara

On Local Transformations in Plane Geometric Graphs Embedded
on Small Grids 22
Manuel Abellanas, Prosenjit Bose, Alfredo García, Ferran Hurtado, Pedro Ramos, Eduardo Rivera-Campo, Javier Tejel

Reducing the Time Complexity of Minkowski-Sum Based Similarity
Calculations by Using Geometric Inequalities 32
Henk Bekker, Axel Brink

A Practical Algorithm for Approximating Shortest Weighted Path
between a Pair of Points on Polyhedral Surface 42
Sasanka Roy, Sandip Das, Subhas C. Nandy

Plane-Sweep Algorithm of O(nlogn) for the Inclusion Hierarchy
among Circles 53
Deok-Soo Kim, Byunghoon Lee, Cheol-Hyung Cho, Kokichi Sugihara

Shortest Paths for Disc Obstacles 62
Deok-Soo Kim, Kwangseok Yu, Youngsong Cho, Donguk Kim, Chee Yap

Improving the Global Continuity of the Natural
Neighbor Interpolation 71
Hisamoto Hiyoshi, Kokichi Sugihara

Combinatories and Triangulations 81
Tomas Hlavaty, Václav Skala

Approximations for Two Decomposition-Based Geometric
Optimization Problems 90
Minghui Jiang, Brendan Mumey, Zhongping Qin, Andrew Tomascak, Binhai Zhu

Computing Largest Empty Slabs 99
Jose Miguel Díaz-Báñez, Mario Alberto López, Joan Antoni Sellarès

3D-Color-Structure-Code – A New Non-plainness Island Hierarchy 109
  Patrick Sturm

Quadratic-Time Linear-Space Algorithms for Generating Orthogonal
Polygons with a Given Number of Vertices 117
  Ana Paula Tomás, António Leslie Bajuelos

Partitioning Orthogonal Polygons by Extension of All Edges
Incident to Reflex Vertices: Lower and Upper Bounds
on the Number of Pieces 127
  António Leslie Bajuelos, Ana Paula Tomás, Fábio Marques

On the Time Complexity of Rectangular Covering Problems
in the Discrete Plane 137
  Stefan Porschen

Approximating Smallest Enclosing Balls 147
  Frank Nielsen, Richard Nock

Geometry Applied to Designing Spatial Structures:
Joining Two Worlds 158
  José Andrés Díaz, Reinaldo Togores, César Otero

A Robust and Fast Algorithm for Computing Exact and
Approximate Shortest Visiting Routes 168
  Håkan Jonsson

Automated Model Generation System Based on Freeform Deformation
and Genetic Algorithm 178
  Hyunpung Park, Kwan H. Lee

Speculative Parallelization of a Randomized Incremental Convex
Hull Algorithm 188
  Marcelo Cintra, Diego R. Llanos, Belén Palop

The Employment of Regular Triangulation for Constrained
Delaunay Triangulation 198
  Pavel Maur, Ivana Kolingerová

The Anchored Voronoi Diagram 207
  Jose Miguel Díaz-Báñez, Francisco Gómez, Immaculada Ventura

Implementation of the Voronoi-Delaunay Method for Analysis
of Intermolecular Voids 217
  *A.V. Anikeenko, M.G. Alinchenko, V.P. Voloshin, N.N. Medvedev,
  M.L. Gavrilova, P. Jedlovszky*

Approximation of the Boat-Sail Voronoi Diagram and Its Application.... 227
  Tetsushi Nishida, Kokichi Sugihara

Incremental Adaptive Loop Subdivision 237
  Hamid-Reza Pakdel, Faramarz F. Samavati

Reverse Subdivision Multiresolution for Polygonal Silhouette
Error Correction 247
  *Kevin Foster, Mario Costa Sousa, Faramarz F. Samavati,
  Brian Wyvill*

Cylindrical Approximation of a Neuron from
Reconstructed Polyhedron 257
  Wenhao Lin, Binhai Zhu, Gwen Jacobs, Gary Orser

Skeletizing 3D-Objects by Projections 267
  David Ménegaux, Dominique Faudot, Hamamache Kheddouci

Track on Computational Geometry

An Efficient Algorithm for Determining 3-D Bi-plane
Imaging Geometry 277
  Jinhui Xu, Guang Xu, Zhenming Chen, Kenneth R. Hoffmann

Error Concealment Method Using Three-Dimensional
Motion Estimation 288
  Dong-Hwan Choi, Sang-Hak Lee, Chan-Sik Hwang

Confidence Sets for the Aumann Mean of a Random Closed Set 298
  Raffaello Seri, Christine Choirat

An Algorithm of Mapping Additional Scalar Value in 2D Vector
Field Visualization 308
  Zhigeng Pan, Jianfeng Lu, Minming Zhang

Network Probabilistic Connectivity: Exact Calculation
with Use of Chains 315
  Olga K. Rodionova, Alexey S. Rodionov, Hyunseung Choo

Curvature Dependent Polygonization by the Edge Spinning 325
  Martin Čermák, Václav Skala

SOM: A Novel Model for Defining Topological Line-Region Relations 335
  Xiaolin Wang, Yingwei Luo, Zhuoqun Xu

Track on Adaptive Algorithms

On Automatic Global Error Control in Multistep Methods with
Polynomial Interpolation of Numerical Solution 345
  Gennady Yu. Kulikov, Sergey K. Shindin

Approximation Algorithms for k-Source Bottleneck Routing Cost
Spanning Tree Problems 355
  Yen Hung Chen, Bang Ye Wu, Chuan Yi Tang

Efficient Sequential and Parallel Algorithms for Popularity
Computation on the World Wide Web with Applications
against Spamming 367
  Sung-Ryul Kim

Decentralized Inter-agent Message Forwarding Protocols for
Mobile Agent Systems 376
  JinHo Ahn

Optimization of Usability on an Authentication System Built from
Voice and Neural Networks 386
  Tae-Seung Lee, Byong-Won Hwang

An Efficient Simple Cooling Schedule for Simulated Annealing 396
  Mir M. Atiqullah

A Problem-Specific Convergence Bound for Simulated
Annealing-Based Local Search 405
  Andreas A. Albrecht

Comparison and Selection of Exact and Heuristic Algorithms 415
  Joaquín Pérez O., Rodolfo A. Pazos R., Juan Frausto-Solís,
  Guillermo Rodríguez O., Laura Cruz R., Héctor Fraire H.

Adaptive Texture Recognition in Image Sequences with Prediction
through Features Interpolation 425
  Sung Baik, Ran Baik

Fuzzy Matching of User Profiles for a Banner Engine 433
  Alfredo Milani, Chiara Morici, Radoslaw Niewiadomski

Track on Biology, Biochemistry, Bioinformatics

Genome Database Integration 443
  Andrew Robinson, Wenny Rahayu

Protein Structure Prediction with Stochastic Optimization Methods:
Folding and Misfolding the Villin Headpiece 454
  Thomas Herges, Alexander Schug, Wolfgang Wenzel

High Throughput in-silico Screening against Flexible
Protein Receptors 465
  Holger Merlitz, Wolfgang Wenzel

A Sequence-Focused Parallelisation of EMBOSS on a Cluster
of Workstations.. 473
  Karl Podesta, Martin Crane, Heather J. Ruskin

A Parallel Solution to Reverse Engineering Genetic Networks........... 481
  Dorothy Bollman, Edusmildo Orozco, Oscar Moreno

Deformable Templates for Recognizing the Shape of
the Zebra Fish Egg Cell 489
  Ho-Dong Lee, Min-Soo Jang, Seok-Joo Lee, Yong-Guk Kim,
  Byungkyu Kim, Gwi-Tae Park

Multiple Parameterisation of Human Immune Response in HIV:
Many-Cell Models.. 498
  Yu Feng, Heather J. Ruskin, Yongle Liu

Track on Cluster Computing

Semantic Completeness in Sub-ontology Extraction Using
Distributed Methods 508
  Mehul Bhatt, Carlo Wouters, Andrew Flahive, Wenny Rahayu,
  David Taniar

Distributed Mutual Exclusion Algorithms on a Ring of Clusters......... 518
  Kayhan Erciyes

A Cluster Based Hierarchical Routing Protocol for Mobile Networks..... 528
  Kayhan Erciyes, Geoffrey Marshall

Distributed Optimization of Fiber Optic Network Layout
Using MATLAB 538
  Roman Pfarrhofer, Markus Kelz, Peter Bachhiesl, Herbert Stögner,
  Andreas Uhl

Cache Conscious Dynamic Transaction Routing in
a Shared Disks Cluster 548
  Kyungoh Ohn, Haengrae Cho

A Personalized Recommendation Agent System for
E-mail Document Classification 558
  Ok-Ran Jeong, Dong-Sub Cho

An Adaptive Prefetching Method for Web Caches 566
  Jaeeun Jeon, Gunhoon Lee, Ki Dong Lee, Byoungchul Ahn

Track on Computational Medicine

Image Processing and Retinopathy: A Novel Approach to
Computer Driven Tracing of Vessel Network 575
  *Annamaria Zaia, Pierluigi Maponi, Maria Marinelli,
  Anna Piantanelli, Roberto Giansanti, Roberto Murri*

Automatic Extension of Korean Predicate-Based Sub-categorization
Dictionary from Sense Tagged Corpora 585
  Kyonam Choo, Seokhoon Kang, Hongki Min, Yoseop Woo

Information Fusion for Probabilistic Reasoning and Its Application
to the Medical Decision Support Systems 593
  Michal Wozniak

Robust Contrast Enhancement for Microcalcification
in Mammography 602
  Ho-Kyung Kang, Nguyen N. Thanh, Sung-Min Kim, Yong Man Ro

Track on Computational Methods

Exact and Approximate Algorithms for Two–Criteria Topological
Design Problem of WAN with Budget and Delay Constraints 611
  Mariusz Gola, Andrzej Kasprzak

Data Management with Load Balancing in Distributed Computing 621
  Jong Sik Lee

High Performance Modeling with Quantized System 630
  Jong Sik Lee

New Digit-Serial Systolic Arrays for Power-Sum and Division
Operation in $GF(2^m)$ 638
  Won-Ho Lee, Keon-Jik Lee, Kee-Young Yoo

Generation of Unordered Binary Trees 648
  Brice Effantin

A New Systolic Array for Least Significant Digit First
Multiplication in $GF(2^m)$ 656
  Chang Hoon Kim, Soonhak Kwon, Chun Pyo Hong, Hiecheol Kim

Asymptotic Error Estimate of Iterative Newton-Type Methods and
Its Practical Application 667
  Gennady Yu. Kulikov, Arkadi I. Merkulov

Numerical Solution of Linear High-Index DAEs 676
  Mohammad Mahdi Hosseini

Fast Fourier Transform for Option Pricing: Improved Mathematical
Modeling and Design of Efficient Parallel Algorithm 686
   Sajib Barua, Ruppa K. Thulasiram, Parimala Thulasiraman

Global Concurrency Control Using Message Ordering of Group
Communication in Multidatabase Systems 696
   Aekyung Moon, Haengrae Cho

Applications of Fuzzy Data Mining Methods for Intrusion
DetectionSystems 706
   Jian Guan, Da-xin Liu, Tong Wang

Pseudo-Random Binary Sequences Synchronizer Based
on Neural Networks 715
   Jan Borgosz, Boguslaw Cyganek

Calculation of the Square Matrix Determinant:
Computational Aspects and Alternative Algorithms 722
   Antonio Annibali, Francesco Bellini

Differential Algebraic Method for Aberration Analysis
of Electron Optical Systems 729
   Min Cheng, Yilong Lu, Zhenhua Yao

Optimizing Symmetric FFTs with Prime Edge-Length 736
   Edusmildo Orozco, Dorothy Bollman

A Spectral Technique to Solve the Chromatic Number Problem
in Circulant Graphs 745
   Monia Discepoli, Ivan Gerace, Riccardo Mariani, Andrea Remigi

A Method to Establish the Cooling Scheme in Simulated
Annealing Like Algorithms 755
   Héctor Sanvicente-Sánchez, Juan Frausto-Solís

Packing: Scheduling, Embedding, and Approximating Metrics 764
   Hu Zhang

Track on Computational Science Education

Design Patterns in Scientific Software 776
   Henry Gardner

Task Modeling in Computer Supported Collaborative Learning
Environments to Adapt to Mobile Computing 786
   Ana I. Molina, Miguel A. Redondo, Manuel Ortega

Computational Science and Engineering (CSE) Education:
Faculty and Student Perspectives 795
   Hasan Dağ, Gürkan Soykan, Şenol Pişkin, Osman Yaşar

Computational Math, Science, and Technology: A New Pedagogical
Approach to Math and Science Education 807
  Osman Yaşar

Track on Computer Modeling and Simulation

Resonant Tunneling Heterostructure Devices – Dependencies
on Thickness and Number of Quantum Wells 817
  Nenad Radulovic, Morten Willatzen, Roderick V.N. Melnik

Teletraffic Generation of Self-Similar Processes with
Arbitrary Marginal Distributions for Simulation:
Analysis of Hurst Parameters....................................... 827
  Hae-Duck J. Jeong, Jong-Suk Ruth Lee, Hyoung-Woo Park

Design, Analysis, and Optimization of LCD Backlight Unit Using
Ray Tracing Simulation... 837
  Joonsoo Choi, Kwang-Soo Hahn, Heekyung Seo, Seong-Cheol Kim

An Efficient Parameter Estimation Technique for
a Solute Transport Equation in Porous Media 847
  Jaemin Ahn, Chung-Ki Cho, Sungkwon Kang, YongHoon Kwon

HierGen: A Computer Tool for the Generation of
Activity-on-the-Node Hierarchical Project Networks 857
  Miguel Gutiérrez, Alfonso Durán, David Alegre, Francisco Sastrón

Macroscopic Treatment to Polymorphic E-mail Based Viruses 867
  Cholmin Kim, Soung-uck Lee, Manpyo Hong

Making Discrete Games... 877
  Inmaculada García, Ramón Mollá

Speech Driven Facial Animation Using Chinese Mandarin
Pronunciation Rules... 886
  Mingyu You, Jiajun Bu, Chun Chen, Mingli Song

Autonomic Protection System Using Adaptive Security Policy 896
  Sihn-hye Park, Wonil Kim, Dong-kyoo Kim

A Novel Method to Support User's Consent in Usage Control for
Stable Trust in E-business 906
  Gunhee Lee, Wonil Kim, Dong-kyoo Kim

Track on Financial and Economical Modeling

No Trade under Rational Expectations in Economy
(A Multi-modal Logic Approach) 915
  Takashi Matsuhisa

A New Approach for Numerical Identification of
Optimal Exercise Curve... 926
   Chung-Ki Cho, Sunbu Kang, Taekkeun Kim, YongHoon Kwon

Forecasting the Volatility of Stock Index Returns:
A Stochastic Neural Network Approach............................. 935
   Chokri Slim

Track on Mobile Computing Systems

A New IP Paging Protocol for Hierarchical Mobile IPv6 945
   Myung-Kyu Yi, Chong-Sun Hwang

Security Enhanced WTLS Handshake Protocol 955
   Jin Kwak, Jongsu Han, Soohyun Oh, Dongho Won

An Adaptive Security Model for Heterogeneous Networks Using
MAUT and Simple Heuristics....................................... 965
   Jongwoo Chae, Ghita Kouadri Mostéfaoui, Mokdong Chung

A New Mechanism for SIP over Mobile IPv6 975
   Pyung Soo Kim, Myung Eui Lee, Soohong Park, Young Kuen Kim

A Study for Performance Improvement of Smooth Handoff Using
Mobility Management for Mobile IP 985
   Kyu-Tae Oh, Jung-Sun Kim

A Fault-Tolerant Protocol for Mobile Agent 993
   Guiyue Jin, Byoungchul Ahn, Ki Dong Lee

Performance Analysis of Multimedia Data Transmission with
PDA over an Infrastructure Network 1002
   Hye-Sun Hur, Youn-Sik Hong

A New Synchronization Protocol for Authentication in
Wireless LAN Environment 1010
   Hea Suk Jo, Hee Yong Youn

A Study on Secure and Efficient Sensor Network Management Scheme
Using PTD.. 1020
   Dae-Hee Seo, Im-Yeong Lee

Author Index 1029

Table of Contents – Part IV

Track on Numerical Methods and Algorithms

New Techniques in Designing Finite Difference Domain
Decomposition Algorithm for the Heat Equation 1
  Weidong Shen, Shulin Yang

A Fast Construction Algorithm for the Incidence Matrices
of a Class of Symmetric Balanced Incomplete Block Designs 11
  Ju-Hyun Lee, Sungkwon Kang, Hoo-Kyun Choi

ILUTP_Mem: A Space-Efficient Incomplete LU Preconditioner 20
  Tzu-Yi Chen

Optimal Gait Control for a Biped Locomotion
Using Genetic Algorithm 29
  Jin Geol Kim, SangHo Choi, Ki heon Park

A Bayes Algorithm for the Multitask Pattern Recognition Problem
– Direct and Decomposed Independent Approaches 39
  Edward Puchala

Energy Efficient Routing with Power Management
to Increase Network Lifetime in Sensor Networks 46
  Hyung-Wook Yoon, Bo-Hyeong Lee, Tae-Jin Lee, Min Young Chung

New Parameter for Balancing Two Independent Measures
in Routing Path 56
  Moonseong Kim, Young-Cheol Bang, Hyunseung Choo

A Study on Efficient Key Distribution and Renewal
in Broadcast Encryption 66
  Deok-Gyu Lee, Im-Yeong Lee

Track on Parallel and Distributed Computing

Self-Tuning Mechanism for Genetic Algorithms Parameters,
an Application to Data-Object Allocation in the Web 77
  *Joaquín Pérez, Rodolfo A. Pazos, Juan Frausto,
  Guillermo Rodríguez, Laura Cruz, Graciela Mora,
  Héctor Fraire*

Digit-Serial AB^2 Systolic Array for Division in $GF(2^m)$ 87
  Nam-Yeun Kim, Kee-Young Yoo

Design and Experiment of a Communication-Aware Parallel Quicksort
with Weighted Partition of Processors 97
  Sangman Moh, Chansu Yu, Dongsoo Han

A Linear Systolic Array for Multiplication in $GF(2^m)$
for High Speed Cryptographic Processors 106
  Soonhak Kwon, Chang Hoon Kim, Chun Pyo Hong

Price Driven Market Mechanism for Computational
Grid Resource Allocation 117
  Chunlin Li, Zhengding Lu, Layuan Li

A Novel LMS Method for Real-Time Network Traffic Prediction 127
  Yang Xinyu, Zeng Ming, Zhao Rui, Shi Yi

Dynamic Configuration between Proxy Caches within an Intranet 137
  *Víctor J. Sosa Sosa, Juan G. González Serna, Xochitl Landa Miguez,
  Francisco Verduzco Medina, Manuel A. Valdés Marrero*

A Market-Based Scheduler for JXTA-Based Peer-to-Peer
Computing System 147
  *Tan Tien Ping, Gian Chand Sodhy, Chan Huah Yong, Fazilah Haron,
  Rajkumar Buyya*

Reducing on the Number of Testing Items in the Branches
of Decision Trees 158
  Hyontai Sug

CORBA-Based, Multi-threaded Distributed Simulation of
Hierarchical DEVS Models: Transforming Model Structure
into a Non-hierarchical One 167
  Ki-Hyung Kim, Won-Seok Kang

The Effects of Network Topology on Epidemic Algorithms 177
  *Jesús Acosta-Elías, Ulises Pineda, Jose Martin Luna-Rivera,
  Enrique Stevens-Navarro, Isaac Campos-Canton,
  Leandro Navarro-Moldes*

A Systematic Database Summary Generation Using
the Distributed Query Discovery System 185
  Tae W. Ryu, Christoph F. Eick

Parallel Montgomery Multiplication and Squaring over $GF(2^m)$
Based on Cellular Automata 196
  Kyo Min Ku, Kyeoung Ju Ha, Wi Hyun Yoo, Kee Young Yoo

A Decision Tree Algorithm for Distributed Data Mining:
Towards Network Intrusion Detection 206
  Sung Baik, Jerzy Bala

Maximizing Parallelism for Nested Loops
with Non-uniform Dependences 213
   Sam Jin Jeong

Fair Exchange to Achieve Atomicity in Payments of High Amounts
Using Electronic Cash 223
   Magdalena Payeras-Capella, Josep Lluís Ferrer-Gomila,
   Llorenç Huguet-Rotger

Gossip Based Causal Order Broadcast Algorithm 233
   ChaYoung Kim, JinHo Ahn, ChongSun Hwang

Track on Signal Processing

Intermediate View Synthesis from Stereoscopic
Videoconference Images.. 243
   Chaohui Lu, Ping An, Zhaoyang Zhang

Extract Shape from Clipart Image Using Modified Chain Code –
Rectangle Representation 251
   Chang-Gyu Choi, Yongseok Chang, Jung-Hyun Cho, Sung-Ho Kim

Control Messaging Channel for Distributed Computer Systems 261
   Bogusław Cyganek, Jan Borgosz

Scene-Based Video Watermarking for Broadcasting Systems 271
   Uk-Chul Choi, Yoon-Hee Choi, Dae-Chul Kim, Tae-Sun Choi

Distortion-Free of General Information with Edge Enhanced Error
Diffusion Halftoning.. 281
   Byong-Won Hwang, Tae-Ha Kang, Tae-Seung Lee

Enhanced Video Coding with Error Resilience Based
on Macroblock Data Manipulation 291
   Tanzeem Muzaffar, Tae-Sun Choi

Filtering of Colored Noise for Signal Enhancement 301
   Myung Eui Lee, Pyung Soo Kim

Model-Based Human Motion Tracking and Behavior Recognition
Using Hierarchical Finite State Automata........................... 311
   Jihun Park, Sunghun Park, J.K. Aggarwal

Effective Digital Watermarking Algorithm by Contour Detection 321
   Won-Hyuck Choi, Hye-jin Shim, Jung-Sun Kim

New Packetization Method for Error Resilient Video Communications ... 329
   Kook-yeol Yoo

Maximizing Parallelism for Nested Loops
with Non-uniform Dependences 213
    Sam Jin Jeong

Fair Exchange to Achieve Atomicity in Payments of High Amounts
Using Electronic Cash 223
    Magdalena Payeras-Capella, Josep Lluís Ferrer-Gomila,
    Llorenç Huguet-Rotger

Gossip Based Causal Order Broadcast Algorithm 233
    ChaYoung Kim, JinHo Ahn, ChongSun Hwang

Track on Signal Processing

Intermediate View Synthesis from Stereoscopic
Videoconference Images... 243
    Chaohui Lu, Ping An, Zhaoyang Zhang

Extract Shape from Clipart Image Using Modified Chain Code –
Rectangle Representation 251
    Chang-Gyu Choi, Yongseok Chang, Jung-Hyun Cho, Sung-Ho Kim

Control Messaging Channel for Distributed Computer Systems 261
    Bogusław Cyganek, Jan Borgosz

Scene-Based Video Watermarking for Broadcasting Systems 271
    Uk-Chul Choi, Yoon-Hee Choi, Dae-Chul Kim, Tae-Sun Choi

Distortion-Free of General Information with Edge Enhanced Error
Diffusion Halftoning... 281
    Byong-Won Hwang, Tae-Ha Kang, Tae-Seung Lee

Enhanced Video Coding with Error Resilience Based
on Macroblock Data Manipulation 291
    Tanzeem Muzaffar, Tae-Sun Choi

Filtering of Colored Noise for Signal Enhancement 301
    Myung Eui Lee, Pyung Soo Kim

Model-Based Human Motion Tracking and Behavior Recognition
Using Hierarchical Finite State Automata........................... 311
    Jihun Park, Sunghun Park, J.K. Aggarwal

Effective Digital Watermarking Algorithm by Contour Detection 321
    Won-Hyuck Choi, Hye-jin Shim, Jung-Sun Kim

New Packetization Method for Error Resilient Video Communications . 329
    Kook-yeol Yoo

Achieving Fair New Call CAC for Heterogeneous Services
in Wireless Networks 460
    SungKee Noh, YoungHa Hwang, Kill Kim, SangHa Kim

Track on Visualization and Virtual and Augmented Reality

Application of MCDF Operations
in Digital Terrain Model Processing 471
    Zhiqiang Ma, Anthony Watson, Wanwu Guo

Visual Mining of Market Basket Association Rules 479
    Kesaraporn Techapichetvanich, Amitava Datta

Visualizing Predictive Models in Decision Tree Generation 489
    Sung Baik, Jerzy Bala, Sung Ahn

Track on Software Engineering

A Model for Use Case Priorization Using Criticality Analysis 496
    José Daniel García, Jesús Carretero, José María Pérez,
    Félix García

Using a Goal-Refinement Tree to Obtain and Refine
Organizational Requirements 506
    Hugo Estrada, Oscar Pastor, Alicia Martínez, Jose Torres-Jimenez

Using C++ Functors with Legacy C Libraries 514
    Jan Broeckhove, Kurt Vanmechelen

Debugging of Java Programs Using HDT with Program Slicing 524
    Hoon-Joon Kouh, Ki-Tae Kim, Sun-Moon Jo, Weon-Hee Yoo

Frameworks as Web Services 534
    Olivia G. Fragoso Diaz, René Santaolaya Salgado,
    Isaac M. Vásquez Mendez, Manuel A. Valdés Marrero

Exception Rules Mining Based on Negative Association Rules 543
    Olena Daly, David Taniar

A Reduced Codification for the Logical Representation
of Job Shop Scheduling Problems 553
    Juan Frausto-Solis, Marco Antonio Cruz-Chavez

Action Reasoning with Uncertain Resources 563
    Alfredo Milani, Valentina Poggioni

Track on Security Engineering

Software Rejuvenation Approach to Security Engineering 574
    Khin Mi Mi Aung, Jong Sou Park

A Rollback Recovery Algorithm for Intrusion Tolerant Intrusion
Detection System 584
    Myung-Kyu Yi, Chong-Sun Hwang

Design and Implementation of High-Performance
Intrusion Detection System... 594
    *Byoung-Koo Kim, Ik-Kyun Kim, Ki-Young Kim,
Jong-Soo Jang*

An Authenticated Key Agreement Protocol Resistant
to a Dictionary Attack... 603
    Eun-Kyung Ryu, Kee-Won Kim, Kee-Young Yoo

A Study on Marking Bit Size for Path Identification Method:
Deploying the Pi Filter at the End Host 611
    Soon-Dong Kim, Man-Pyo Hong, Dong-Kyoo Kim

Efficient Password-Based Authenticated Key Agreement Protocol 617
    *Sung-Woon Lee, Woo-Hun Kim, Hyun-Sung Kim,
Kee-Young Yoo*

A Two-Public Key Scheme Omitting Collision Problem
in Digital Signature 627
    Sung Keun Song, Hee Yong Youn, Chang Won Park

A Novel Data Encryption and Distribution Approach
for High Security and Availability Using LU Decomposition 637
    Sung Jin Choi, Hee Yong Youn

An Efficient Conference Key Distribution System Based
on Symmetric Balanced Incomplete Block Design 647
    Youngjoo Cho, Changkyun Chi, Ilyong Chung

Multiparty Key Agreement Protocol with Cheater Identification
Based on Shamir Secret Sharing 655
    Kee-Young Yoo, Eun-Kyung Ryu, Jae-Yuel Im

Security of Shen et al.'s Timestamp-Based Password
Authentication Scheme 665
    Eun-Jun Yoon, Eun-Kyung Ryu, Kee-Young Yoo

ID-Based Authenticated Multiple-Key Agreement Protocol
from Pairings.. 672
    Kee-Won Kim, Eun-Kyung Ryu, Kee-Young Yoo

A Fine-Grained Taxonomy of Security Vulnerability
in Active Network Environments 681
    *Jin S. Yang, Young J. Han, Dong S. Kim, Beom H. Chang,
Tai M. Chung, Jung C. Na*

A Secure and Flexible Multi-signcryption Scheme 689
  Seung-Hyun Seo, Sang-Ho Lee

User Authentication Protocol Based on Human Memorable Password
and Using RSA 698
  IkSu Park, SeungBae Park, ByeongKyun Oh

Effective Packet Marking Approach to Defend against DDoS Attack 708
  Heeran Lim, Manpyo Hong

A Relationship between Security Engineering and Security Evaluation ... 717
  Tai-hoon Kim, Haeng-kon Kim

A Relationship of Configuration Management Requirements
between KISEC and ISO/IEC 15408 725
  Hae-ki Lee, Jae-sun Shim, Seung Lee, Jong-bu Kim

Track on Information Systems and Information Technology

Term-Specific Language Modeling Approach to Text Categorization 735
  Seung-Shik Kang

Context-Based Proofreading of Structured Documents 743
  *Won-Sung Sohn, Teuk-Seob Song, Jae-Kyung Kim, Yoon-Chul Choy,
  Kyong-Ho Lee, Sung-Bong Yang, Francis Neelamkavil*

Implementation of New CTI Service Platform Using Voice XML 754
  Jeong-Hoon Shin, Kwang-Seok Hong, Sung-Kyun Eom

Storing Together the Structural Information of XML Documents
in Relational Databases 763
  Min Jin, Byung-Joo Shin

Annotation Repositioning Methods in the XML Documents:
Context-Based Approach... 772
  *Won-Sung Sohn, Myeong-Cheol Ko, Hak-Keun Kim, Soon-Bum Lim,
  Yoon-Chul Choy*

Isolating and Specifying the Relevant Information of an
Organizational Model: A Process Oriented
Towards Information System Generation 783
  Alicia Martínez, Oscar Pastor, Hugo Estrada

A Weighted Fuzzy Min-Max Neural Network for Pattern
Classification and Feature Extraction 791
  *Ho J. Kim, Tae W. Ryu, Thai T. Nguyen, Joon S. Lim,
  Sudhir Gupta*

The eSAIDA Stream Authentication Scheme 799
  Yongsu Park, Yookun Cho

An Object-Oriented Metric to Measure the Degree of Dependency
Due to Unused Interfaces 808
    René Santaolaya Salgado, Olivia G. Fragoso Diaz,
    Manuel A. Valdés Marrero, Isaac M. Vásquez Mendez,
    Sheila L. Delfín Lara

End-to-End QoS Management for VoIP Using DiffServ 818
    Eun-Ju Ha, Byeong-Soo Yun

Multi-modal Biometrics System Using Face and Signature.............. 828
    Dae Jong Lee, Keun Chang Kwak, Jun Oh Min,
    Myung Geun Chun

Track on Information Retrieval

Using 3D Spatial Relationships for Image Retrieval
by XML Annotation.. 838
    SooCheol Lee, EenJun Hwang, YangKyoo Lee

Association Inlining for Mapping XML DTDs to Relational Tables 849
    Byung-Joo Shin, Min Jin

XCRAB: A Content and Annotation-Based Multimedia Indexing
and Retrieval System 859
    SeungMin Rho, SooCheol Lee, EenJun Hwang, YangKyoo Lee

An Efficient Cache Conscious Multi-dimensional Index Structure 869
    Jeong Min Shim, Seok Il Song, Young Soo Min, Jae Soo Yoo

Track on Image Processing

Tracking of Moving Objects Using Morphological Segmentation,
Statistical Moments, and Radon Transform........................... 877
    Muhammad Bilal Ahmad, Min Hyuk Chang, Seung Jin Park,
    Jong An Park, Tae Sun Choi

Feature Extraction and Correlation for Time-to-Impact
Segmentation Using Log-Polar Images 887
    Fernando Pardo, Jose A. Boluda, Esther De Ves

Object Mark Segmentation Algorithm Using Dynamic Programming
for Poor Quality Images in Automated Inspection Process.............. 896
    Dong-Joong Kang, Jong-Eun Ha, In-Mo Ahn

A Line-Based Pose Estimation Algorithm for 3-D Polyhedral
Object Recognition... 906
    Tae-Jung Lho, Dong-Joong Kang, Jong-Eun Ha

Initialization Method for the Self-Calibration
Using Minimal Two Images 915
    Jong-Eun Ha, Dong-Joong Kang

Face Recognition for Expressive Face Images 924
    Hyoun-Joo Go, Keun Chang Kwak, Sung-Suk Kim,
    Myung-Geun Chun

Kolmogorov-Smirnov Test for Image Comparison..................... 933
    Eugene Demidenko

Modified Radius-Vector Function for Shape Contour Description 940
    Sung Kwan Kang, Muhammad Bilal Ahmad, Jong Hun Chun,
    Pan Koo Kim, Jong An Park

Image Corner Detection Using Radon Transform 948
    Seung Jin Park, Muhammad Bilal Ahmad, Rhee Seung-Hak,
    Seung Jo Han, Jong An Park

Analytical Comparison of Conventional and MCDF Operations
in Image Processing 956
    Yinghua Lu, Wanwu Guo

On Extraction of Facial Features from Color Images 964
    Jin Ok Kim, Jin Soo Kim, Young Ro Seo, Bum Ro Lee,
    Chin Hyun Chung, Key Seo Lee, Wha Young Yim, Sang Hyo Lee

Track on Networking

An Architecture for Mobility Management
in Mobile Computing Networks 974
    Dohyeon Kim, Beongku An

An Adaptive Security Model for Heterogeneous Networks
Using MAUT and Simple Heuristics 983
    Jongwoo Chae, Ghita Kouadri Mostéfaoui, Mokdong Chung

A Hybrid Restoration Scheme Based on Threshold Reaction Time
in Optical Burst-Switched Networks 994
    Hae-Joung Lee, Kyu-Yeop Song, Won-Ho So, Jing Zhang,
    Debasish Datta, Biswanath Mukherjee, Young-Chon Kim

Author Index 1005

Advanced Simulation Technique for Modeling Multiphase Fluid Flow in Porous Media

Jong G. Kim[1] and Hyoung W. Park[1]

Korea Institute of Science and Technology Information
P.O. Box 122, Yuseong, Daejeon 305-600, Korea
{jgkim, hwpark}@kisti.re.kr

Abstract. Modeling fluid flow in a porous medium is a challenging computational problem. It involves highly heterogeneous distributions of porous medium property, various fluid flow characteristics. For the design of better flow management scheme, high performance computing tools have been applied as a useful technique. In this paper, various parallel implementation approaches are introduced mainly based on high performance computing toolkits such as ADIFOR, PETSc, and Cactus. Provided by ADIFOR, accurate Jacobian matrix computation scheme was coupled with PETSc (Portable Extensible Toolkit for Scientific Computation), which allows data structures and routines for the scalable solution of scientific applications modeled by partial differential equations. In addition to the PETSc-ADIFOR-based approach, we also present a methodology to apply a grid-enabled parallel implementation toolkit, Cactus for the simulation of fluid flow phenomena in porous media. The preliminary result shows a significant parallel performance. On the basis of the current result, we further discuss the future plan for the design of multi-purpose porous media flow simulator.

1 Introduction

Parallel computing technology has been exploited in various fields of science and engineering areas. As addressed in many textbooks, the basic issues of parallel scientific computing are well identified. However, this does not mean that robust solutions are always available through the use of large computing power. Our particular motivation is to utilize advanced computational techniques to address the fundamental issues of solution approaches involved with modeling multiphase fluid flow in porous media.

Coupled with Darcy's law, the governing equations of the fluid flow in porous media are derived by mass and energy balances, which are resulted in a set of coupled partial differential equations (PDEs). Various numerical approaches such as finite difference, finite volume, and finite element methods are used to obtain the solution over the discretized PDE domain. The most popular method to solve such discretized PDE systems is the Newton's method. It requires a Jacobian matrix system to seek solution increments. The easiest and most widely used method for the Jacobian matrix computation is divided-differentiation method (or called finite difference scheme, FD), which obtains the derivatives by dividing the response perturbation of a functional calculation to the step-size variation of an independent variable. This method can lead to the breakdown of Newton iterations if high nonlinearity of a functional calculation generates a significant round-off and truncation errors [7]. Automatic differentiation (AD) method provides attractive alternative method for accurate and efficient Jacobian matrix computation. We explain how to couple the AD nethod in the parallel implementation procedure of a multiphase flow code. We used the SNES and SLES components of PETSc for the robust and flexible parallelization of the flow modeling code [2].

Additionally, we initialized an effort to implement a multiphase flow modeling system in a grid computing environment. Different from conventional computing frameworks, the grid computing technology enables large-scale aggregation and sharing of computing, data and other resources across geographical boundaries. This provides a new simulation approach to develop a next generation model of dynamically interactive, data-driven modeling strategies for application scientists. For an example, fine-resolution reservoir flow simulation models can be integrated with various observed data types derived from seismic/downhole sensors deployed in reservoir fields. This will eventually make it possible to establish an instrumented virtual operation systems providing more efficient, cost-effective, and user-friendly reservoir flow management schemes. In this paper, we explored to use this new grid-enabled approach to implement a multiphase flow modeling code using a grid middleware package, Cactus Computational Toolkit [4]. Through this demonstration, a prototype can be established to identify new grid application projects in computational fluid flow or other large scale simulation areas. Following the description of the model problems, we discuss the parallel implementation algorithms of both PETSc and Cactus and the test result of the implemented codes.

2 Governing Equations

Typically, a system of multiphase fluid flow in porous media can be described by three phases. The governing differential equations is given by

$$\nabla \cdot (T_l \nabla (P_l - \gamma_l Z)) = \frac{\partial}{\partial t}(\phi \frac{S_l}{B_l}) + \frac{q_l}{\rho_l} \tag{1}$$

In the equation, P_l (pressure of phase l) and S_l (saturation of phase l) are the primary variables to be solved for each phase. Volume formation factor is used to describe each phase behaviors. In solving the PDE system, finite difference, finite volume, or finite element methods are often used for a discretized domain. Explicit, implicit, or semi-implicit (implicit pressureand explicit saturation) formulations of the discretized difference or integral equations are derived for the time derivative terms. In the case of semi-implicit formulation, a hepta-diagonal system of pressure equation is implicitly derived for each time step. The detail procedure to obtain such finite difference and finite element formulations of the equations are found in the many publications [1,5,6]. In this work, typical constant initial and no flow boundary conditions were applied for pressure and saturation variables.

3 PETSc- and AD-Based Parallel Implemenetation

PETSc provides the parallel data structure and linear and nonlinear solvers necessary for high performance parallel computing problems. The implementations of Newton-like methods for nonlinear systems are collected in the SNES component. It includes both line search and trust region techniques with an interface to other linear solvers such as Krylov subspace methods. The SNES component of PETSc generally requires some derivative matrix such as Jacobian or Hessian matrices. By default, these derivatives are approximated with FD method. However, the simulation algorithm will be more efficient and accurate if the subroutines for analytical derivative computation are provided by users.

Based on the algebraic chain rule, AD method is a technique to generate a derivative-enhanced code for a function calculation given as a computer program. The core concept of the AD method is the fact that every computer program, no matter how complicated, is consisted of a sequence of basic arithmetic operations such as additions, subtraction, multiplication, division, or elementary functions such as *sin* and *cos*. If a chain rule for derivative calculus is repeatedly applied to a computer code for a function calculation, any form of arbitrary derivatives can be automatically computed with machine accuracy lever. In our study, ADIFOR package [3] is used to differentiate a multiphase flow modeling code. The procedure to provide the ADIFOR-generated derivatives to the SNES component is illustrated in Figure 1. In the figure, the user-defined routines to

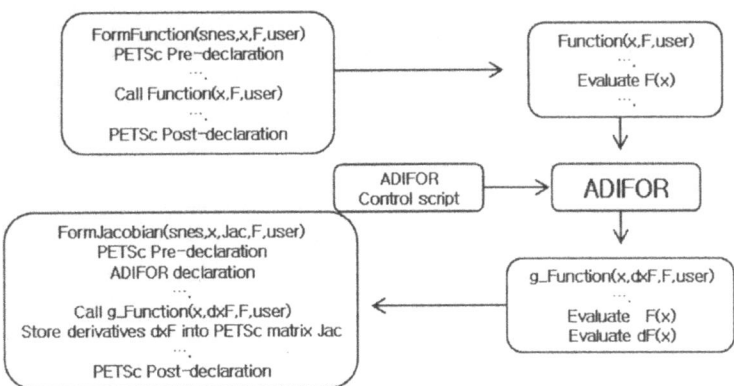

Fig. 1. Outline of procedure to couple PETSc with AD derivative code

compute the function and Jacobian matrix are referred as FormFunction and FormJacobian, which compute the nonlinear function $F(x)$ and the Jacobian matrix $dF(x, dx)$. Each routine includes several PETSc specific calls to set data structure and communication. In this process, local function computation code is isolated to generate the AD-derivative code, which is then combined back into the main program to assemble a global linear matrix system. The final linear matrix system is solved by the SNES and SLES components of PETSc. This AD-based numerical scheme was implemented to solve a finite element formulation of the two-dimensional, two-phase system described in the previous section.

4 Grid-Enabled Cactus Computational Toolkit

The Cactus Computational Toolkit is an open source PSE (problem solving environment) tool for the use in a grid computing environment. It allows efficient parallelism and portability in highly heterogeneous computing architectures. Originally, the development of the Cactus Computational Toolkit was begun by a group of computational physicists to solve the challenging astrophysics problem such as relativity equations. Internationally distributed, large amount of computational resources, various research groups, and large size of complicated legacy codes were involved in the begining of their project. In order to build efficient multidisciplinary collaboration in such a diverse working environment, they developed the framework of the Cactus Computational Toolkit. In its modular structure, the core flesh element provides a framework to attach additional application or utility modules called the thorns. This modular structure allows multidisciplinary collaborative environment where different application research groups co-develop computational codes/tools and interchanges the developed modules and expertise without learning the intricacies of MPI and other means of achieving parallelism. The basic structure of the Cactus is indicated in Figure 2 a).

Within the structure of the Cactus, user-supplied computing/utility modules

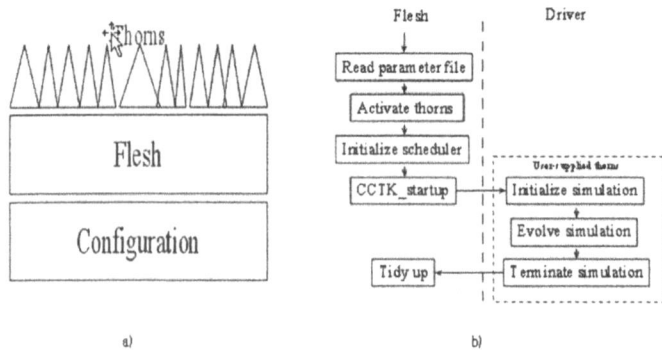

Fig. 2. a) Structural description of the Cactus Computational Toolkit, and b) the main program flow diagram

are collected into the thorns. These thorns are able to communicate with other thorns and to exchange data flows through the calls to the flesh. Independent from the thorns, the flesh provides the main function of program, which parses the involved parameters, activates the appropriate thorns, and passes required controls to involved thorns. The external appearance of thorns is generated in a configuration file during the compilation time, which describes the connection from a thorn to the flesh or to other thorns. Users need to write the computing or utility thorns for their simulation requirement. Figure 2 b) shows the diagram of main program flow that is initialized and controlled by the Cactus flesh and driver to run a Cactus code. In this work, we have implemented a thorn of the Cactus for the finite difference formulation of a three phase flow simulation scheme described in the previous section. Semi-implicit formulation was applied to implicitly solve a hepta-diagonal system of pressure equation. The implemenetation procedure is summarized as follows.

CCTK Startup
  Variable initialization
  Do loop for a given final time steps
    Pre − step computation
    Source/sink calculation
    Pressure equation : hepta − diagonal matrix system
    Linear matrix solver
    Update phase saturation
  Enddo
Termination

5 Model Problems and Numerical Experiment

We set up test problem sets with a simple configuration for the perforance test of the implementation approaches described in previous sections. The thermodynamic property data tables used in test runs were from Kim and Deo's work [6]. Constant sink/source terms were given to each diagonal corner of 2D and 3D test domains. As expected, symmetrical flow patterns were observed in each test run. The parallel performance of the PETSc-ADIFOR-based implementation approach was tested with typical two-dimensional finite element problem sets. In Table 1, the performance of AD-based Jacobain matrix computation scheme was compared with the one of FD-based computational approach. The table shows that the computational time to calculate five Jacobian matrices are about the same order; the AD method saves the computational time about 20%. However, a significant contribution in the iterations of linear matrix system was made by the AD method. It indicates that faster convergence behaviors in linear solution steps can be provided by the accurate Jacobian calculation. In this study, we used the GMRES method with the ASM preconditioner [2]. The observed parallel performance of the code on the IBM SP2 machine is shown in Table 2. With the problem size of 80,802 unknown variables, the scaled parallel efficiency of

Table 1. Computational time and convergence comparison of ADIFOR and FD mehods for a test problem with 13,122 unknown variables

	iterations		time (sec)		Residual norm
	nonlinear	linear	total	Jacobian	
ADIFOR	5	112	264	233	1.62E-6
FD	5	761	322	290	8.22E-7

about 78% was obtained. The efficiency was computed by the ratio of observed speedup to ideal speedup. Here, the speedup is defined as the time spent with one processor divided by the time spent N processors for a given problem size.

Table 2. Parallel performance for the test problem with 80,802 unknown variables

CPUs	iterations		time (sec)		
	nonlinear	linear	nonlinear	Jacobian	function
4	7	809	574	448	40
8	7	809	210	126	31
16	7	809	109	45	30
32	7	818	91	24	36

The parallel performance of the Cactus-based code was tested with a three dimensional problem with source/sink at each diagonal corner. A fast convergence behavior was observed since not much non-linearity was given to the test problem. Two different problem sizes from one to two million unknown variables were used to measure the parallel performance on a 16-node linux cluster system. In fact, the bigger problem shows a super-linear speedup performance. It is believed to be due to the memory cache problem (see Figure 4).

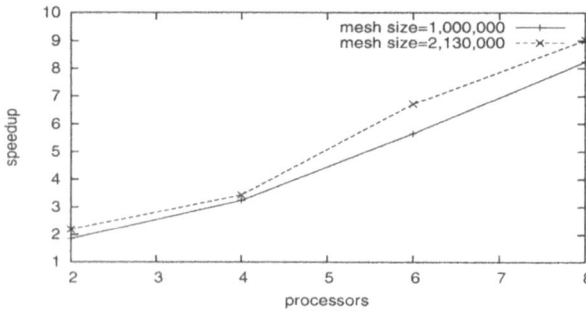

Fig. 3. Parallel performance of Cactus-based simulation code

Table 3 shows the distribution of the computational time for different tasks in the simulation code: source/sink, matrix coefficient, and linear solution with post-step variable update computation. Major computational time of about 62 71% was spent on the matrix coefficient calculation part. The problem set we solved in this test does not involve high non-linearity. Thus, linear solution step takes relatively less computational time with fast convergence behaviors. However, it is expected that the computational cost for linear solution step will be significantly increased with the increased complexity of the problem set. We

Table 3. Computational time distribution of each routine in the test simulation with 2,130,000 unknown variables

CPUs	source/sink	Matrix coeff.	Linear solver
1	3.71 (10%)	25.69 (71%)	6.77 (19%)
2	1.89.(11%)	10.13 (62%)	4.50 (27%)
8	0.59 (13%)	2.49 (62%)	1.01 (25%)
12	0.35 (13%)	1.71 (63%)	0.65 (24%)

also evaluated scaled speedup, which measures the speedup as the problem size increases linearly with the number of processes. Specifically, we fixed the problem size per process at 100×100×20, and we solved larger problems by adding more and more processes. The scaled efficiency of about more than 91% was obtained.

The Cactus provides a basic thorn to control and examine simulation through remote web-browsers, which allows an ideal ubiquitous way to interact with the running simulation. We used this HTTPD thorn to control and visualize our simulation. Figure 5 illustrates the computed saturation color map of non-aqueous phase after 600 time steps with the time step size of 0.01 day. As expected, we observed a symmetrical propagation of injected fluid from injection corner to the other corner.

6 Conclusions

We observed a significant parallel performance (scaled parallel efficiency of 78%) with the PETSc-ADIFOR-based parallel implementation approach. The AD method provided fast convergence behaviors in linear solution step with accurate Jacobian matrix computation. It is determined that the AD-based approach can make a further contribution to derive the complicated derivative code such as sensitivity enhanced simulation code for optimal flow management scheme. With the Cactus-based parallel implementation approach, we observed considerable speedup (factor of about more than 11 with twelve processors) on a linux cluster system. In the scaled efficiency measurement, average performance of about more than 91% was achieved with a three dimensional test problem. A

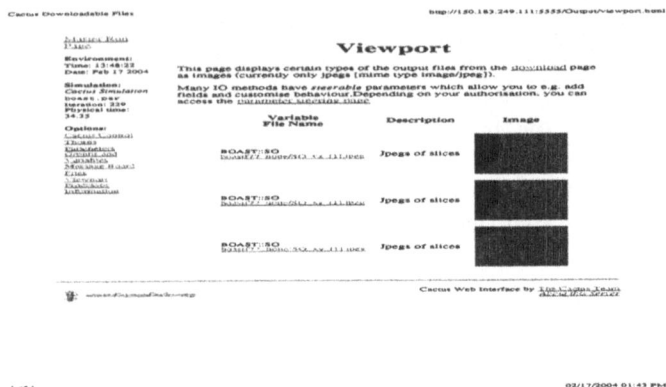

Fig. 4. Remote web-based visualization of a Cactus-based simulation: colormap of non-aqueous phase in a slice of the problem domain

significant portion of computational time was spent on matrix coefficient calculation. And linear solution step took about 25% of total computation time. The Cactus Computational Toolkit appeared to provide efficient parallel implementation scheme. Furthermore, we tested an ubiquitous feature of the Cactus package for the remote control and examination through web-browser. In our future plan, this work will be continued to develop a grid-based general purpose multiphase flow simulation code for modeling compositional and reactive fluid flow in porous media. This approach will provide prcatical computing environment for the optimal design of better flow management scheme, which requires reservoir flow simulation to be dynamically interacted with measured data set from the field.

Notation

∇ Difference operator in Cartesian coordinate
B_l Fluid formation volume factor of phase l
γ_l Density of phase l in terms of pressure/distance
ϕ Porosity
q_l Source and sink rate of phase l
ρ_l Density of phase l
T_l Transmissibility of phase l
Z Elevation

Acknowledgement. This work was supported in part by the Korea National Grid Implementation Program (K*Grid). The main sponsor of the K*Grid project is the Ministry of Information and Communication of Korea.

References

1. Aziz, K. and A. Settari, Petroleum Reservoir Simulation, Elsevier, London (1979)
2. Balay, S., K. Buschelman, W. D. Gropp, D. Kaushik, M. Knepley, L. C. McInnes, B. F. Smith, and H. Zang, PETSc User's Manual, ANL-95/11-Revision2.1.5, Argonne National Laboratory, (2002)
3. Bischof, C., A. Carle. P. Hovland, P. Khaderi, and A. Mauer, ADIFOR 2.0 User's Guide (Revision D), Report ANL/MCS-TM-192, Argonne National Laboratory, Argonne, Illinois, (1998)
4. The Cactus Team, Cactus 4.0 User Guide (www.cactuscode.org), Albert Einstein Institute, Max Planck Institute for Gravitational Physics (2002)
5. Chang, M. M., P. Sarathi, R. J. Heemstra, A. M. Cheng, and J. F. Pautz, User's Guide and Documentation manual for BOAST-VHS for PC, U.S. Dept. of Energy Report NIPER-542 Distribution Category UC-122 (1992)
6. Kim, J. G. and M. D. Deo, A Finite Element, Discrete-Fracture Model for Multiphase Flow in Porous Media, AIChE Journal, 46, 6, 1120-1130 (2000)
7. Kim, J. G. and S. Finsterle, Application of Automatic Differentiation for the Simulation of Nonisothermal, Multiphase Flow in Geothermal Reservoirs, Twenty-Seventh Workshop on Geothermal Reservoir Engineering Stanford University, Stanford, California, January 28-30, SGP-TR-171 (2002)

The P-GRADE Grid Portal*

Csaba Németh, Gábor Dózsa, Róbert Lovas, and Péter Kacsuk

MTA SZTAKI, Laboratory of Parallel and Distributed Systems,
H-1518 Budapest, Hungary,
{csnemeth,dozsa,rlovas,kacsuk}@sztaki.hu,
http://www.lpds.sztaki.hu

Abstract. Providing Grid users with a widely accessible, homogeneous and easy-to-use graphical interface is the foremost aim of Grid-portal development. These portals if designed and implemented in a proper and user-friendly way, might fuel the dissemination of Grid-technologies, hereby promoting the shift of Grid-usage from research into real life, industrial application, which is to happen in the foreseeable future, hopefully. This paper highlights the key issues in Grid-portal development and introduces P-GRADE Portal being developed at MTA SZTAKI. The portal allows users to manage the whole life-cycle of executing a parallel application in the Grid: editing workflows, submitting jobs relying on Grid-credentials and analyzing the monitored trace-data by means of visualization.

1 Introduction

Easy and convenient access of Grid systems is an utmost need for Grid end-users as well as for Grid application developers. Grid portals are the most promising environments to fulfill these requirements so we decided to create a Grid portal for our P-GRADE (Parallel Grid Run-time and Application Development Environment) system.

P-GRADE is an integrated graphical environment to develop and execute parallel applications on supercomputers, clusters and Grid-systems [11][12]. P-GRADE supports the generation of either PVM, or MPI code, creation and execution of Condor or Condor-G jobs [13]. More advanced features include support for migration, workflow definition and coordinated multi-job execution in the Grid. The aim of the portal is to make these functionalities available through the web. While P-GRADE in its original form requires the installation of the whole P-GRADE system on the client machines, the portal version needs only a web browser and all the P-GRADE portal services are provided by one or more P-GRADE portal servers that can be connected to various Grid systems.

The P-GRADE Grid portal currently provides the three most important Grid services needed by Grid end-users and application developers:

---

* The work presented in this paper was partially supported by the following grants: EU-GridLab IST-2001-32133, Hungarian SuperGrid (IKTA4-075), IHM 4671/1/2003, and Hungarian Scientific Research Fund (OTKA) No. T042459 projects.

- Grid certificate management,
- creation, modification and execution of workflow applications on grid resources available via Globus, and
- visualisation of workflow progress as well as each component job.

We developed the portal using GridSphere, a Grid portal development framework. The framework among other things saved us implementing user-management for the portal, by using its tag-library we could easily develop a coherent web interface, and enjoyed the way one can configure and manage portlets in GridSphere.

The paper describes the basic features of the P-GRADE portal and shows a meteorology application as a case study for using the portal. Section 2 overviews Grid portals in general and Section 3 introduces Grid portal development frameworks. Section 4 explains the security mechanism of the P-GRADE Grid portal. Section 5 describes the workflow system provided by the P-GRADE portal and finally, Section 6 gives details of the workflow visualisation facilities of the portal.

2   Grid-Portals

Web-portals can be considered as a single entry point to some underlying service, such as a web-store, a search-engine and alikes. Similarly, Grid-portals provide users with access to the underlying services. So portals for the Grid discussed in this paper are also Web-portals, they can be accessed from widely diverse platforms and clients, first of all from web-browsers. The need to support and integrate pretty heterogenous back-ends is in the nature of the Grid. Naturally, the criterion of system A not necessarily hold true for system B, still, for anyone attempting to develop a Grid-portal the following requirements may be of use. The list below is partly based on the list in [1].

1. Homogeneous: To let users easily manage the underlying diverse, distributed infrastructure the portal should be capable of hiding back-end heterogeneity by being homogeneous regarding it's operation and layout as well.
2. Integrated: All the related tasks can be performed.
3. Graphical: Visual elements might help the user to the greatest degree. This would not necessarily mean depriving the more experienced user of using the already available command-line tools, instead providing a new graphical user interface for the convenience of any user. Thinking of graphs and diagrams, one advantage of a central portal is its ability to process data that was collected from various sources.
4. Easy-to-use: Grid-portals should hide the complexity of the underlying infrastructure and have an easy-to-use interface.
5. Problem-oriented: The Grid-portal as a specific problem solving environment, should allow potential users to concentrate on their field without knowing the details of the things behind the scenes. As [1] cites: "A user communicates with the PSE in the language of the problem, not in the language of a particular operating system, programming language, or network

protocol." This implies that a portal should supply an appropriate abstraction layer on top of all the underlying components, and this layer's duty is also to make the portal more user-friendly and easy-to-use.

6. Widely accessible: There is no need for installed software on the client-side other than a browser. For enhanced graphical interfaces some plug-ins might come in handy, such as the Java plug-in for instance. Of course, further flexibility in this respect may be gained by portal-adaptations for less capable devices such as mobile phones.
7. Persistent: As operations in the Grid might take rather long times, support for persistency is needed.
8. Interactive: The user can initiate actions to intervene if he deems required. Apart from this, and maybe even unknown, the system too might intervene on the occurance of well-defined events. From the portal-side, interactivity may also take form in some notification services, such as sending an e-mail or SMS to the portal user on job-completion.

3 Grid-Portal Development Frameworks

3.1 Grid Portal Development Kit

Frameworks helping Grid-portal developers in concentrating on the particular needs of a specific portal are becoming mature enough soon. The first milestone on this way was the Grid Portal Development Kit [2], which attempted to provide a set of reusable components for accessing Globus based Grid services. The core concept of the GPDK is a three-tier architecture in which the client securely communicates with the web-server hosting the portal web-application. The dynamic content of the portal is generated by server-side components such as JSP pages, servlets and Java Beans. The diverse and mostly remote back-end components can be reached either by using the available Java APIs or calling native methods via JNI.

GPDK's functionality covered support for some core grid mechanisms, such as obtaining a delegated credential from a so called MyProxy server, managing user profiles, submitting jobs, transferring files and querying a simple LDAP-based information service.

3.2 GridSphere

Probably the most widespread, free Grid-portal development framework currently is being developed under the EU-funded GridLab project, and is called GridSphere [3]. The architecture and design is rather similar to that of GPDK, it is also a Java framework, however it's central building blocks, portlets are already conformant to a community supported API, the Portlet API [4]. Portlets have become a popular concept used to describe specific visual components which can be combined in order to achieve the desired functionality in a complex, still manageable and modular UI. All in all, the developer can focus almost exclusively on providing functionality, while layout questions can be left to the portlet itself. By leveraging this specification and providing portal developers with well designed, easy-to-use interfaces, together with using well known design patterns, this framework reasonably shortens the time required to develop a portal web-application.

The basic portlets in the framework are as follows: login/logout, account management, user management (eg. assign users to user groups), portlet subscription, which provides the ability for users to add or remove portlets from their workspace. Good news, the GridSphere developer group is expected to release a set of elaborate core Grid-portlets in the very near future. As the GridSphere community continues to grow and more and more knowledge and feedback is gathering on portal issues, we encourage anyone planning to develop a Grid-portal to start by evaluating this framework.

The overwhelming use of Java technologies first of all ends up in easily portable code, which is of utmost importance in the Grid, and due to the wide range of available free Java packages and tools, RAD may become reality at fairly low expenses. Java developers will find the Java CoG (Commodity Grid) Toolkit [5] a most useful API in programming for the Grid.

4 Security Mechanism of the P-GRADE Portal

Generally, security in the Grid is based upon using X.509 certificates. Users do not have accounts for all the resources instead they have public/private key pairs and identity certificates. Using certificates prevents the user from repeatedly typing username/password pairs. A critical issue is how to manage and store these certificates in the most practical way. One solution is to dedicate a server to store certificates securely at a single place, from where users can get them. To further diminish the chance that downloaded and then locally stored certificates get compromised, these credentials might be given a limited, reasonably short lifetime. This very service is implemented by MyProxy, the widely used online credential repository for the Grid [6]. These short term credentials are called proxy credentials.

In P-GRADE Portal the Credential Manager portlet allows the portal user to download proxy credentials from a MyProxy server (MPS). A user may have more than one proxy from different MPSs as well. The user can view the details of a proxy, including its lifetime and any proxy can be selected for usage. This credential will be passed to other portlets via the getUserCred(Object userId) interface, see Fig. 1. The proxies, that are no longer needed, can be deleted. Renewing an expired proxy requires connecting the MPS again. If there is no MPS holding appropriate credential for the user, he may upload his private key and certificate to an MPS, and later this MPS can create a proxy for him.

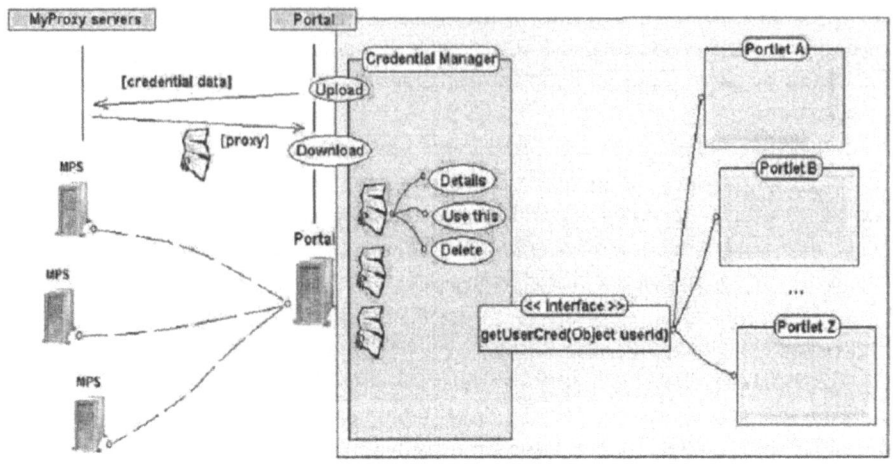

Fig. 1. Proxy-credential management

5 Workflow Editing and Job Submission in the P-GRADE Portal

A workflow is a set of consecutive and parallel jobs which are cooperating in the execution of a parallel program. One job's output may provide input for the next. In the literature workflow is also referenced as application flow. The workflow portlet helps to edit jobs, compose workflow-files, submit them, and visualize the trace-file describing each.

For illustration purpose we use a meteorological application [7] called MEANDER developed by the Hungarian Meteorological Service. The main aim of MEANDER is to analyse and predict in the ultra short-range (up to 6 hours) those weather phenomena, which might be dangerous for life and property. Typically such events are snow-storms, freezing rain, fog, convictive storms, wind gusts, hail storms and flash floods. The complete MEANDER package consists of more than ten different algorithms from which we have selected four to compose a workflow application for demonstration purpose. Each calculation algorithm is computation intensive and implemented as a parallel program containing C/C++ and FORTRAN sequential code.

Fig. 2 shows a workflow executing the MEANDER meteorology application. The portal contains a graphical editor by which the user can easily draw a workflow graph. The important aspect of creating a graph is the possibility of composing the workflow application from existing code components. These components can be sequential programs (e.g. Ready in Fig. 2), PVM programs (e.g. Delta in Fig. 2), or MPI programs (e.g. Satel in Fig 2). This flexibility

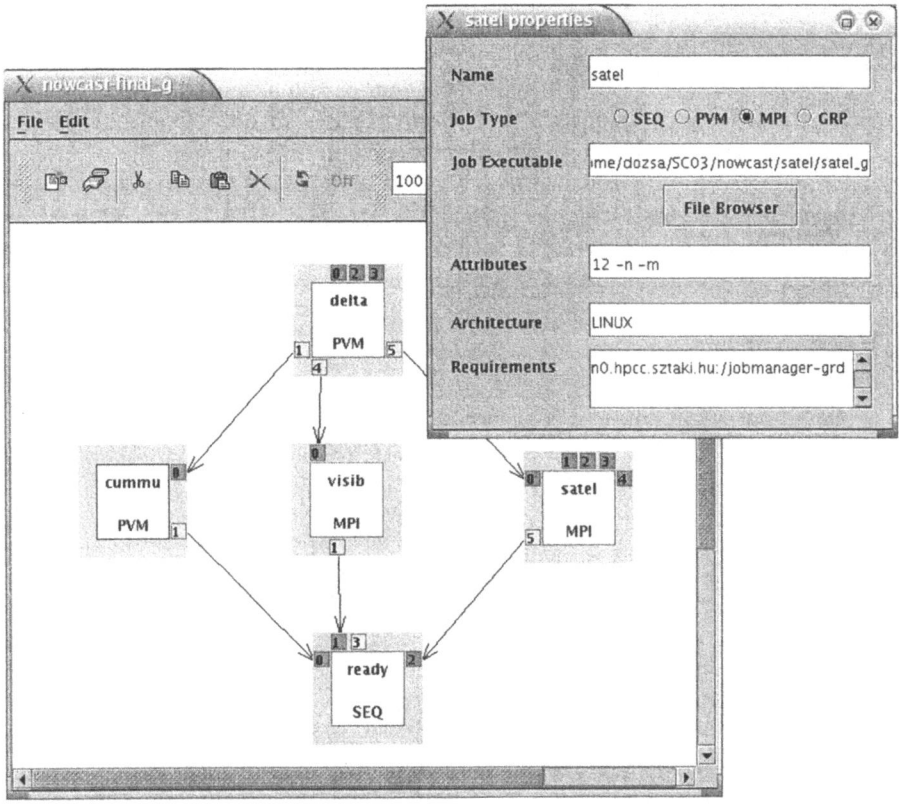

Fig. 2. Workflow editor of the P-GRADE portal

enables the application developer to create computation-intensive applications where several components are parallel programs to be run on supercomputers or clusters of the Grid. The P-GRADE workflow manager takes care of the parallel execution of these components provided that the end-user specifies a parallel resource for the component. The graph depicted in Fig. 2 consists of four jobs (nodes) corresponding four different parallel algorithms of the MEANDER ultra-short range weather prediction package and a sequential visualisation job that collects the final results and presents them to the user as a kind of meteorological map:

 – Delta: a PVM program
 – Cummu: a PVM program
 – Visib: an MPI program
 – Satel: an MPI program
 – Ready: a sequential C program

This distinction among job types is necessary because the job manager on the selected grid site should be able to support the corresponding parallel execution mode, and the workflow manager is responsible for the handling of various job types by generating the appropriate submit files.

Besides the type of the job and the name of the executable, the user can specify the necessary arguments and the hardware/software requirements (architecture, operating system, minimal memory and disk size, number of processors, etc.) for each job (see the dialog window in Fig. 2). To specify the resource requirements, the application developer can currently use either the Condor resource specification syntax and semantics for Condor based grids or the explicit declaration of grid site where the job is to be executed for Globus based grids (see the Requirement field of the dialog window in Fig. 2).

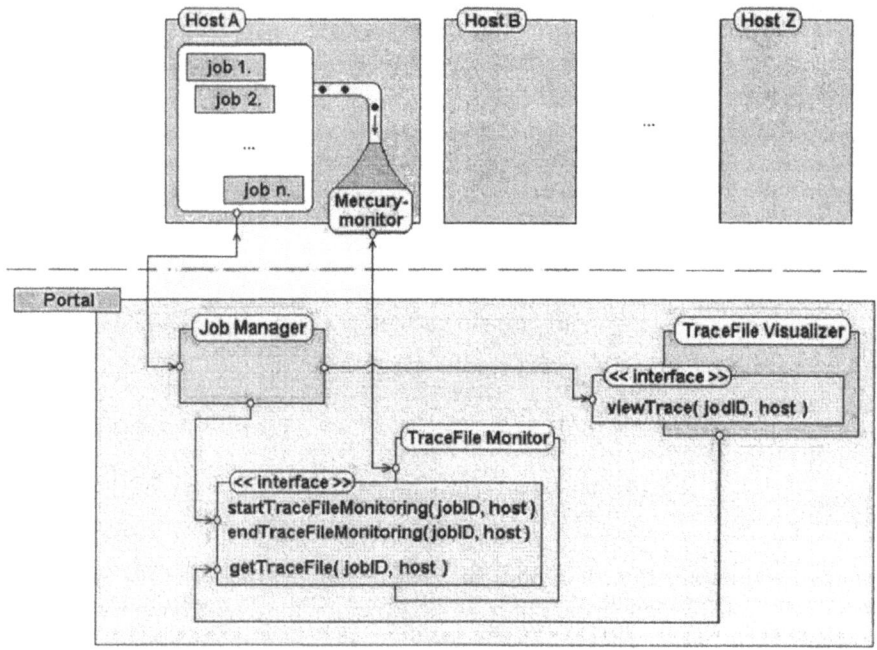

Fig. 3. Monitoring jobs

Once the workflow is specified, the necessary input files should be uploaded to the P-GRADE portal server and then the server will take care of the complete execution of the workflow and transfering results back to the client.

6 Monitoring and Execution Visualisation

Besides creating and managing workflows the P-GRADE portal enables on-line monitoring and visualisation of the workflow execution. For monitoring jobs we use the Mercury [8] monitoring tool. Its architecture complies with the Grid Monitoring Architecture proposed by the Global Grid Forum. Producers send measurement data to consumers upon request. For job monitoring, when submitting a job we subscribe to the corresponding metric. As progress unfolds, Mercury will send trace-data back to the TraceFile Monitor component in the portal, which collects trace-files and passes them to the visualiser (see Fig. 3).

The workflow level execution visualisation window shown in Fig. 4 graphically represents the progress of the workflow execution. In the workflow space-time diagram, the horizontal bar represents the progress of each component job in time (see the time axis at the bottom of the diagram) and the arrows among bars represents the file operations performed to make accessible the output file of a job as an input of another one. Finally, a similar window can be opened for any parallel component job. Interpretation of the same diagram elements is a bit different in this case (like job cummu, see the lower window in Fig. 4). Here the horizontal bars represent the progress of each process comprising the parallel job whereas arrows between bars represent (PVM or MPI) message transfers among the processes [14].

7 Conclusion and Future Work

Executing the meteorology application workflow through the P-GRADE Portal has already been successfully presented at Supercomputing Conference and Exhibition 2003 in Phoenix/AZ (U.S.A.) [9], and at IEEE International Conference on Cluster Computing in Hong Kong/China, 2003 [10].

The P-GRADE Grid portal currently supports managing proxy-credentials, editing workflows, uploading input files and executables, creating and submitting workflows to the Grid sites, and finally, monitored trace-data can be analyzed through space-time diagrams. The output of the jobs naturally can be downloaded, too.

Obviously, this set of functionalities is not complete and hence, some further work will be carried out to widen the services of the Grid portal. One important missing functionality is the connection to the Grid information service and providing that information to Grid users. Analyzing the visualized space-time diagrams the user may conclude that one resource is no longer available or the job should wait unacceptably long. In this case, he may reallocate the application. In a future version of the software the user could be dynamically offered alternate, more promising resources. He could even define some policy regarding priorities, costs and completion time, upon which a local scheduler would optimize execution.

Once these missing functionalities will be created and successfully integrated to the P-GRADE portal, this easy-to-manage interface may greatly promote Grid-computing on it's way towards an exciting breakthrough.

The P-GRADE portal is currently connected to the Hungarian SuperGrid [15] and soon will be connected to other Grid systems such as the Hungarian ClusterGrid [16], NorduGrid [17] and GridLab [18] testbed.

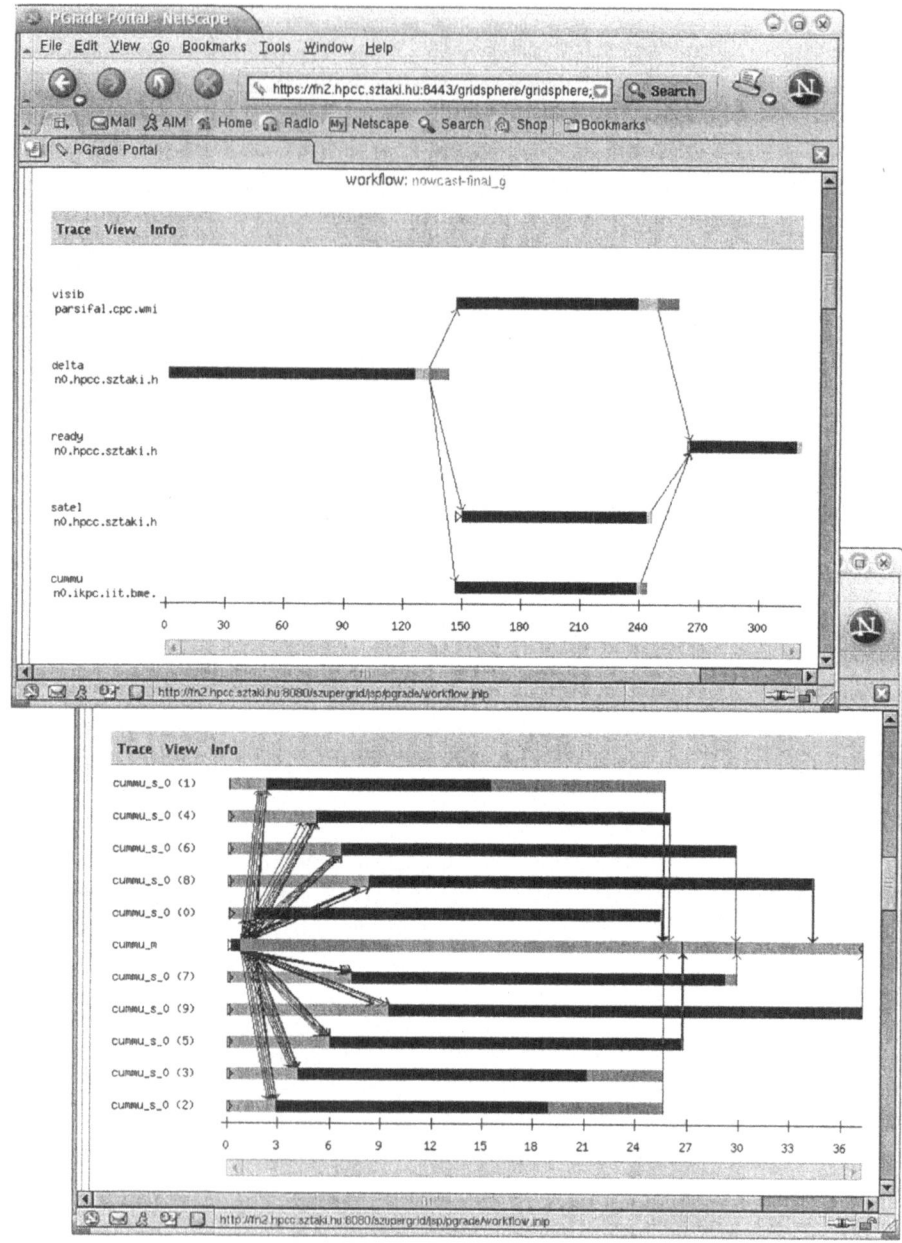

Fig. 4. Space-time diagrams of the MEANDER workflow (upper picture) and its parallel job 'cummu' (lower picture)

References

1. G. von Laszewski, I. Foster, J. Gawor, P. Lane, N. Rehn, and M. Russell: Designing Grid-based Problem Solving Environments and Portals. Proceedings of the 34th Hawaii International Conference on System Sciences, January 2001
2. J. Novotny: The Grid Portal Development Kit. IEEE Concurrency and Practice vol. 13, 2002
3. GridSphere homepage. http://www.gridsphere.org/
4. JSR-000168 Portlet Specification http://jcp.org/aboutJava/communityprocess/review/jsr168/
5. G. von Laszewski, I. Foster, J. Gawor, P. Lane: A Java Commodity Grid Kit. Concurrency and Computation: Practice and Experience, vol. 13, 2001
6. J. Novotny, S. Tuecke, V. Welch: An Online Credential Repository for the Grid: MyProxy. Proceedings of the 10th IEEE Intl. Symp. on High Performance Distributed Computing, 2001
7. R. Lovas, et al.: Application of P-GRADE Development Environment in Meteorology. Proc. of DAPSYS'2002, Linz, pp. 30-37, 2002
8. Z. Balaton, G. Gombs: Resource and Job Monitoring in the Grid., Proc. of EuroPar'2003 Conference, Klagenfurt, Austria, pp. 404-411, 2003
9. Supercomputing 2003 in Phoenix/AZ (U.S.A.). http://www.sc-conference.org/sc2003/
10. IEEE International Conference on Cluster Computing in Hongkong/China. http://www.hipc.org/
11. P-GRADE Graphical Parallel Program Development Environment. http://www.lpds.sztaki.hu/pgrade
12. P. Kacsuk, G. Dózsa, R. Lovas: The GRADE Graphical Parallel Programming Environment. In the book: Parallel Program Development for Cluster Computing: Methodology, Tools and Integrated Environments (Chapter 10), Editors: P. Kacsuk, J.C. Cunha and S.C. Winter, pp. 231-247, Nova Science Publishers New York, 2001
13. James Frey, Todd Tannenbaum, Ian Foster, Miron Livny, and Steven Tuecke: Condor-G: A Computation Management Agent for Multi-Institutional Grids. Journal of Cluster Computing volume 5, pages 237-246, 2002
14. Z. Balaton, P. Kacsuk, and N. Podhorszki: Application Monitoring in the Grid with GRM and PROVE. Proc. of the Int. Conf. on Computational Science ICCS 2001, San Francisco, pp. 253-262, 2001
15. P. Kacsuk: Hungarian Supercomputing Grid. Proc. of ICCS'2002, Amsterdam. Springer Verlag, Part II, pp. 671-678, 2002
16. P. Stefán: The Hungarian ClusterGrid Project. Proc. of MIPRO'2003, Opatija, 2003
17. O.Smirnova et al: The NorduGrid Architecture And Middleware for Scientific Applications. ICCS 2003, LNCS 2657, p. 264. P.M.A. Sloot et al. (Eds.) Springer-Verlag Berlin Heidelberg 2003
18. GridLab project, www.gridlab.org

A Smart Agent-Based Grid Computing Platform

Kwang-Won Koh[1], Hie-Cheol Kim[2], Kyung-Lang Park[1], Hwang-Jik Lee[1], and Shin-Dug Kim[1]*

[1] Dept. of Computer Science, Yonsei University, Seoul 120-749, Korea
[2] Dept. of Computer & Communication, Daegu University, Daegu 705-033, Korea

Abstract. Most of Grid computing platforms are usually configured with high performance computing servers, supercomputers, and cluster systems. The research is about to design an aggressive Grid computing platform by utilizing a large number of pervasive PCs as a Grid component. For this goal, an effective configuration method to construct a group of PCs is required as a single Grid component to be used in the same way as those cluster systems are designed and utilized as the same system technology. The configuration method is designed as a layered service architecture, i.e., physical information service, mobile management service, and point-based scheduling service, based on agent technology. The physical information service is to gather and maintain the status information of physical nodes, the mobile management service to guarantee the effective resource management by using the mobile agent, and the point-based scheduling service to provide a simple scheduling policy for the PC-based computing resource. Through this configuration method, the efficiency of the resource management and system throughput are expected to be increased. The experimental result shows that the system using this configuration method can support more than 90% of the expected performance given by a chosen set of running PC resources in general computing environment.

1 Introduction

Grid computing is an approach to provide high performance computing power by exploiting the unlimited computing resources connected to the Internet [1], [2]. Most of the Grid computing platforms such as Globus Toolkit [3] are usually constructed with high performance computing servers, supercomputers, and cluster systems. However, the most commonly used systems in the world, i.e., the personal computers (*PCs*), are not joined into recent Grid computing environments. For more aggressive and cost effective system configuration, those *PCs* need to be used as the Grid components. Also a large number of cluster systems are used at many organizations. Also both *PC* and cluster systems are generally used for the similar applications and designed based on the same system design technology. Thus, an aggressive method is required to utilize these pervasive *PCs* for a vast Grid computing environment.

---

* This work was supported by Korea Research Foundation Grant (KPF-2001-041-E00271)

The research is about to design an effective configuration method to utilize a group of adjacently located *PCs* as a single virtual cluster. A group of *PCs* is constructed as a single Grid component, where this system is managed in the same way as the usual cluster system. Thus, this type of system can be plugged into the Grid computing platform without any further modification. However, to build such a system, several factors caused by the operational characteristics of *PCs* in general computing environment should be considered. First, *PC* is a highly dynamic computing resource. Second, *PCs* in general computing environment are not well managed. Therefore, *PCs* cannot guarantee to provide any expected performance as Grid computing resources. Third, the complexity to predict the system usability is very high.

To solve problems, we define a virtual resource named a virtual cluster which is a framework to manage the unlimited *PCs* as Grid resources and to schedule corresponding processes simply. Many virtual clusters can be constructed to be joined to a specific Grid platform. The configuration mechanism for this virtual cluster is designed using agent technology. Agent is the software of user's procurator to perform tasks autonomously. Agent technology is applied to process the resource status information, to manage the information of all resources, and to schedule the request job to the appropriate resources. Especially, the mobility of smart agent technology is appropriate to manage these dynamic resources because of the low network load, autonomous operations of monitoring data analysis, and the integration of heterogeneous resources monitoring tools [5].

We simulate this configuration method by calculating how many resources can be managed, and how much performance can be achieved. The experimental result of our simulation shows that the system can manage more number of *PC* resources using this configuration method than the number of resources using traditional management method, in fully stressed network. Moreover the system using this configuration method can support more than 90% of the expected performance given by a chosen set of running *PC* resources in general computing environment.

In Section 2, we introduce previous related work. Section 3 provides all the required services according to the configuration method, their corresponding specific components, and operational model. In Section 4, this configuration method is explained by using a scenario, and Section 5 shows experimental results. Finally, we present the conclusion in Section 6.

2 Related Work

There are several Grid middlewares to manage the Grid resources [6], [7]. Representatively, the Globus toolkit is a set of components that can be used either independently or together to develop useful Grid applications and programming tools. It consists of several modules for resource location & allocation, communication, unified resource information service, authentication interface, process creation, and data access. The Globus toolkit is an effective method to construct Grid resource, but it does not support a large number of computing resources such as *PCs* in general computing environment. There is no effective method to manage the dynamic status of *PCs* in general computing environment. Therefore, a virtual resource constructed with many *PCs* proposed in this work can be used in the Globus toolkit as physical resource. The SETI@home is defined as the massively distributed computing for SETI (search for extraterrestrial intelligence) [8]. The SETI@home project is the first attempt to use large-scale distributed computing to perform a sensitive search for radio signals from extraterrestrial civilizations. So it uses many *PCs* to calculate an extreme amount of computations with a little communication, and the data to be calculated are independent from each other. computing performance is not important for each of its resource node. As a result, it provides a lot of computing resource nodes but cannot provide the function like the job submission. Thus it is quite different from our work.

3 System Configuration Method

The proposed Grid computing platform configuration is assumed to be constructed as supercomputers, clusters, meta-computers, and the proposed virtual resources as shown in Figure 1. Many virtual resources can be defined and joined

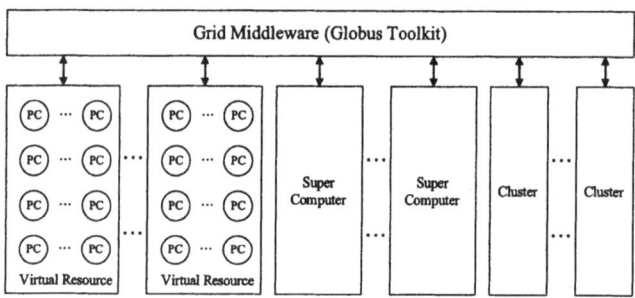

Fig. 1. The proposed Grid resource utilization.

to a specific Grid computing environment. In general, the utilization pattern of *PCs* is quite different from that of other Grid computing resource. Especially the availability of chosen *PCs* as a group is very dynamic. Thus to utilize *PCs* for the Grid computing, a mechanism to manage and maintain the dynamic state information for each *PC* is required to guarantee the expected performance. For this a hierarchically layered service architecture is designed as shown in Figure 2. Specifically three services are designed by using agent technology at each layer. To maintain accurate status information of each *PC*, a physical information service is required for real time status. Mobile management service and point-based scheduling service are required to manage *PCs* and schedule those processes onto the *PC* resource joined to the Grid environment.

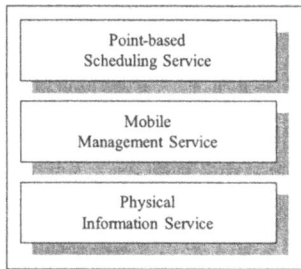

Fig. 2. Hierarchically layered service architecture.

To build these services, a hierarchical structure is considered to support interoperations between any two adjacent services as shown in Figure 2.

3.1 Physical Information Service

Physical information service (*PIS*) is to provide the status information of each physical resource to the upper level service and process the resource status information to the information structure which is consisted as the resource performance and network status as shown in table 1. As stated above, target re-

Table 1. Information provided by the *PIS*.

```
struct resource_status {
    bool_t resource_flag;
    int_t resource_point;
    resource_t min_latency_resource;
    resource_t max_latency_resource;
}
```

source is highly dynamic as thus inter-message should be minimized for effective management. Hence resource_status structure is leveraged to the system. A resource_status structure is constructed as a resource flag, a resource point and network information about latency.

Resource flag (RF): This resource flag represents that the computing resource can serve a user's request. It is an important factor to guarantee the success of user's request, to prevent the obstruction about the owner of resource, and to provide efficient performance. Each *PIS* of all resources supports the efficient operation of the upper layer service by providing its important status information with one bit flag. Therefore, the upper layer service can use information simply because the RF can be denoted by one bit flag as well as the upper layer service can maintain the information of *PIS* information simply on real-time.

The RF is determined by the formula as follows:

$$RF = S_{REQUEST} \&\& U_{CPU} \&\& T_{NOINPUT}, \tag{1}$$

where $S_{REQUEST}$ is specification of a request about architecture, free memory, and free storage. U_{CPU} is average CPU utilization in idle status and $T_{NOINPUT}$ means time without input by a user. $S_{REQUEST}$ is regarded as architecture, memory, and storage for the success of user's request. U_{CPU} is checked for user's background job. $T_{NOINPUT}$ reflects whether a resource is attended by its user. Consequently, it should be recognized by resource unavailable.

Resource point (RP): A resource point is to provide a capable performance of resources as a weight for an efficient scheduling. The RP can be obtained as follows:

$$RP = P \times n \times (1 - w/100) + t, \tag{2}$$

where P is the raw performance of a processor, n is the number of processors on the resource, w is average processor load, and t is the time duration of a resource unattended.

RP is the unit of weight for scheduling and is modelled to support the availability prediction of any resource. It shows high complexity to predict accurate availability of target resources. For decreasing the ratio of complexity, the hypothesis is formed by the computing resource, which is unattended by a user for a longer time or obtains a less chance to be attended by the user. Resources unattended can provide their full performance, and t reflects the hypothesis in equation (2).

Latency information: *PIS* passes information related to the network about each resource in the same sub-network. Especially, the rank information of the minimum latency resource and the maximum latency resource are provided for the mobile management service. The details of information are described in the following section.

3.2 Mobile Management Service

As stated above, the target *PCs* are highly dynamic and thus the management policy should recognize the target machine status and environment information continuously. But it is a big overhead because there are many communications to keep the accurate information about the resources, and then more the overhead affects the efficiency of the resource management when more the resources are participated. For solving the problem, mobile agent technology is applied to the mobile management service (*MMS*). The service searches an optimal place and moves to the optimal place automatically by itself, using the mobility of a smart agent in order to manage the resources effectively. In dynamic network environment, the reduction of the communication overhead makes effect of increasing the reliability of dynamic resource status information.

Optimal place (*OP*): Optimal place is defined as a resource place to manage the number of resources in the same sub-network effectively. In any environment that network status and resource status vary continuously, the mobile management service searches *OP* and places itself in the *OP* to gather the information from the *PIS* with little delay. Table 2 describes a simple algorithm for searching the *OP* with latency information provided by *PIS*. The algorithm is very simple, and it is enough to reduce the network load and to increase the effectiveness in high bandwidth network. The process of the *MMS* is shown in Figure 3. The

Table 2. Optimal place decision algorithm.

```
INPUT:
    n: the number of resources
    max_resource_net_point: better networked resource
    min_latency_resource
    max_latency_resource: network information from resource information
Algorithms:
    for count = 0 (to the size of resources) {
        get a resource information from count-th resource
        weight of min_latency_resource-th resource is increased
        weight of max_latency_resource-th resource is decreased
    }
    if max_resource_net_point-th resource
            can get information of all resources
            and can communicate the upper layer service
        return optimal place = max_resource_net_point
```

mobile agent, which is implemented as the *MMS*, is typically created at the step of a job submission, and transfers to a master resource of the allocated subnetwork, and it collects the information of each resource from the *PIS*, moves to the *OP* using an optimal place decision algorithm, divides the resource list of received information into two groups, by using the *RF*. Mobile agent sorts a resource list by the *RP* in each group, and transfers these information to the upper layer service. The *MMS* can reduce the communication delay time by using mobility of agent technology, and the reliability of resource information is increased. Resources can operate the computation job efficiently as a virtual Grid resource.

3.3 Point-Based Scheduling Service

Point-based scheduling service selects a specific number of resources and allocates processes onto them. It factorizes *RF* and *RP* and schedules the processes to the resources using preprocessed resource status information, so it can react rapidly for an unexpected event.

Point-based Scheduling (*PBS*): *PBS* is to schedule the processes in dynamic situation. In other words, it is a lightweight scheduling policy for the specific environment that resources operate actively, communicate several data, and are

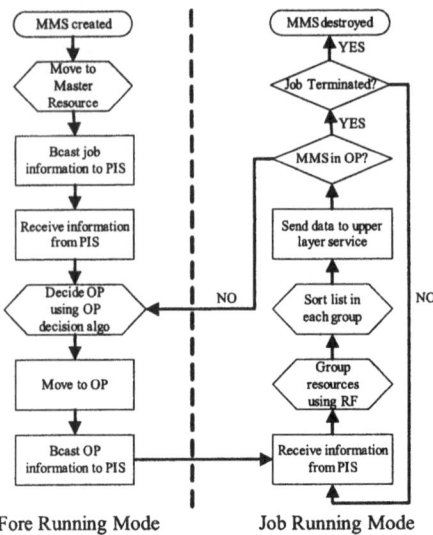

Fig. 3. Workflow of the *MMS*.

connected by high bandwidth network. In dynamic situation, it should schedule the processes to resources by several factors. These several factors are preprocessed to a resource_status structure, as stated in Table 1, and it is used as a scheduling factor. *RF* in resource_status structure is used as a primary factor. It should be considered to support the unobtrusiveness that local processes is prior to a remote job. The *PBS* divides the resources into the two groups; Guaranteed Group (G-Group) and Indistinct Group (I-Group). The G-Group consists of the resources for which *RF* value is true, and the I-Group consists of the resources whose *RF* value is false. The G-Group, basically, has priority over the I-Group. *RP* is used as a weight and is to improve the efficient operation of the system. In the same group, it is used to schedule processes to the resources directly. Moreover *RP* threshold is leveraged for the efficient operation of the system. *RP* threshold is the necessary condition of the remote job offered by the remote user. If the remote user offers *RP* threshold, the *PBS* can schedule the processes cost-efficiently.

By using these factors, the *PBS* is 3-step scheduling policy as follows:

First Select resources satisfying
    (resources.RF == true && resources.RP > RP threshold)
Second Select resources satisfying
    (resources.RF == true && resources.RP <= RP threshold)
Third Select resources satisfying
    (resources.RF == false && best RP)

The *PBS* uses the minimal-fit about *RP* in the first step. It chooses the resources satisfied with the *RP* threshold. It uses the best-fit about the *RP* in the second step and the third step. The minimal-fit means that the minimal resource satisfied with the request is preceded and the best-fit means that it selects the highest priority in accordance with the request of the user.

The point-based scheduling service is created at the step of a job submission. It checks that the number of resources is enough to provide the requested service at its initialization time. If it is available; we define this situation as an active condition. If it is not available, the point-based scheduling service notifies a user of an inactive condition which means that the system doesn't provide request service to a user. Then it allocates processes to the resources and receives information from the implementations of the *MMS* in each sub-network in order to collect resource information. The information is classified into the G-Group and the I-Group, and information is sorted by the *RP*. A point-based scheduling service typically merges this information grouped by the value of *RF*, and it goes its operation.

4 Experimental Result

Simulation is performed to evaluate the efficiency of the virtual resource attainable consistently at any given time. For this simulation, we typically make ten thousand environments to represent many different kinds of network status and resource status. Latency values between any two resources are set by random way by applying a weight of the distance according to the network size. As the size of a network grows larger, communication patterns between resources with a long distance become more complicated. The number of available resources is measured by latency information and an assumption that dynamic information like the average CPU utilization is valid in only 1000ms and only a resource having accurate information is available. As a weight for scheduling, *RP* of each resource is set in a random way. The proposed system means the system which adapts the proposed configuration method. Figure 4 shows the rate of the manageable resources using mobile management service. The traditional management service in a dynamic situation which resource status varies continuously and network is fully stressed. The proposed mobile management service increases the number

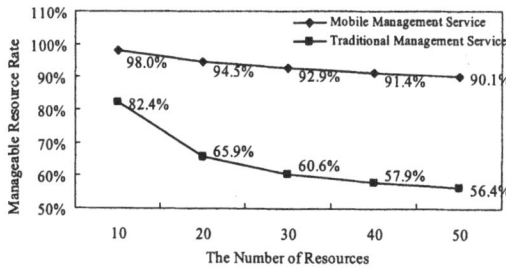

Fig. 4. Effectiveness of the mobsile management service.

of available resources in this situation. It is not only increasing the number of resources but also increasing the effectiveness of the system as a Grid resource because the gap between performance of the mobile management service and the traditional management service is increased continuously. In Figure 5, the

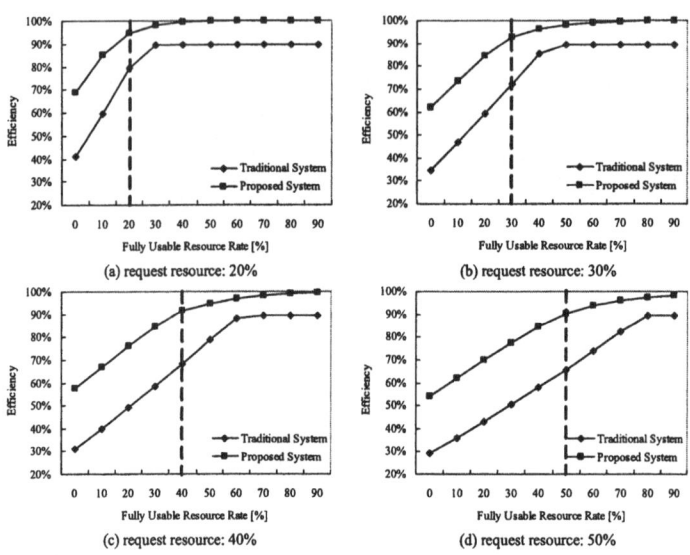

Fig. 5. Efficiency of the overall system utilization for various usable resources over total 200 resources.

proposed system is compared with the traditional systems with static information service and FIFO cluster scheduling algorithm [12]. The efficiency on the Y-axis and the fully usable resource rate on the X-axis are presented. Efficiency is calculated as following:

$$Efficiency = \frac{\sum_{i=0}^{n} P_i}{\sum P}, \qquad (3)$$

where P is raw performance of a processor and n is the number of resources participated in the system. As shown in Figure 5, as the rate of fully usable resources increases, the expected performance for the proposed method can increase higher than the traditional one. Therefore the proposed system shows more than 90% of expected computing performance in active condition though many resources are occupied.

5   Conclusion

The purpose of the research is to design an aggressive Grid computing platform by utilizing a large number of pervasive *PCs* as the Grid resources. This group of *PCs* can be plugged into the common Grid computing middleware, such as Globus Toolkit, like the usual cluster systems with compatibility.

This configuration method is designed by a layered service architecture, i.e., physical information service, mobile management service, and point-based scheduling service, based on the agent technology. Experimental result shows that the efficiency of any chosen system configuration can be achieved up to 90% of the overall system performance.

References

1. I. Foster and C. Kesselman, eds: The GRID : Blueprint for a New Computing Infrastructure, Morgan Kaufmann, (1988).
2. I. Foster, C. Kesselman, and S. Tuecke, The Anatomy of the Grid: Enabling Scalable Virtual Organizations, International J. Supercomputer Applications, 15 (3), (2001).
3. I. Foster and C. Kesselman, Globus: A metacomputing infrastructure toolkit, International Journal of Supercomputer Applications, 11 (2): 115-128, (1997)
4. Berman, F., Fox, G. and Hey, A. J. G., Eds. Grid Computing - Making the Global Infrastructure a Reality, John Wiley & Sons Ltd, (2003).
5. O. Tomarchio, L. Vita and A. Puliafito, Active monitoring in grid environments using mobile agent technology, In 2nd Int'l Workshop on Active Middleware Services, (2000).
6. de Roure, D., Baker, M., Jennings, N. R. and Shadbolt, N. The evolution of the Grid, in [4], pp. 65-100.
7. K. Krauter, R. Buyya, and M. Maheswaran, A taxonomy and survey of grid resource management systems for distributed computing, Software: Practice and Experience, 32(2):135-164, February (2002).
8. Korpela, E., Werthimer, D., Anderson, D., Cobb, J., and Leboisky, M., SETI@home-massively distributed computing for SETI, Computing in Science & Engineering [see also IEEE Computational Science and Engineering] , Volume: 3 Issue: 1 , Jan/Feb (2001).
9. O. F. Rana and D. W. Walker, The Agent Grid: Agent-based resource integration in PSEs, In Proceedings of the 16th IMACS World Congress on Scientific Computing, Applied Mathematics and Simulation, Lausanne, Switzerland, August (2000).
10. K. Czajkowski, I. Foster, N. Karonis, C. Kesselman, S. Martin, W. Smith, and S. Tuecke, A Resource Management Architecture for Metacomputing Systems, Proc. IPPS/SPDP '98 Workshop on Job Scheduling Strategies for Parallel Processing, pg. 62-82, (1998).
11. K. Czajkowski, S. Fitzgerald, I. Foster, and C. Kesselman, Grid Information Services for Distributed Resource Sharing, Proceedings of the Tenth IEEE International Symposium on High-Performance Distributed Computing (HPDC-10), IEEE Press, August (2001).
12. R. Henderson and D. Tweten, Portable Batch System: External reference specification, Technical report, NASA Ames Research Center, (1996).

Publishing and Executing Parallel Legacy Code Using an OGSI Grid Service*

T. Delaitre, A. Goyeneche, T. Kiss, and S.C. Winter

Centre for Parallel Computing,
Cavendish School of Computer Science,
University of Westminster
115 New Cavendish Street, London W1W 6UW
testbed-discuss@cpc.wmin.ac.uk

Abstract. This paper describes an architecture for publishing and executing parallel legacy code using an OGSI Grid service. A framework is presented that aids existing legacy applications to be deployed as OGSI Grid services and the concept is demonstrated by creating an OGSI/GT3 version of the Westminster MadCity traffic simulator application. This paper presents the Grid Execution Management for Legacy Code Architecture (GEMLCA), and describes the progress and achievements of its implementation.

1 Introduction

There are many different approaches realizing the computational Grid. However, basically all of these efforts are heading towards the implementation of a service-oriented Grid architecture. A service, in this context, is a software component that performs some kind of operations and that can be invoked through well-defined interfaces. One of the major advantages of using a service-based architecture is to aid the easy and relatively effortless re-engineering of existing legacy applications and to offer them as Grid services.

The Open Grid Services Infrastructure (OGSI), a specification by the Globus project together with IBM [1], describes a service-oriented Grid architecture based on Web services standards and protocols. GT3, the latest version of the Globus toolkit, which was first published in early 2003, is the first reference implementation of OGSI. There are several projects [2,3] aiming to install, test and further investigate GT3 and other middlewares based on OGSI.

The purpose of the UK OGSI Tesbed Project [3] is to test and evaluate the first implementation of OGSI core by deploying the GT3 toolkit in a service-based Grid testbed spanning organisational boundaries with differing institutional services, security policies and firewalls. The project deploys GT3 across several sites and creates a testbed that runs different OGSI compliant applications. One of the aims of the University of Westminster research group within

---

* The work presented in this paper is supported by an EPSRC funded project (Grant No.: GR/S77509/01): A proposal to evaluate OGSA/GT3 on a UK multisite testbed.

this project is to create a framework that aids existing legacy applications to be deployed as OGSI Grid services and to demonstrate this concept by creating a GT3 version of its MadCity traffic simulator.

This paper describes the general architecture designed to offer legacy applications as Grid services through the example of MadCity. It summarises the current results in implementing the architecture and outlines the future work ahead.

2 Traffic Simulation on Computer Clusters

Computational simulations are becoming increasingly important because they are the only way how some physical processes can be studied and interpreted. These simulations may require computational power not available even on most powerful supercomputers. A possible solution is to use computer clusters, multiple computers connected through a network, to run these complex tasks.

MadCity, a discrete time-based microscopic simulator, developed by the Centre of Parallel Computing at the University of Westminster, consists of two main components, the SIMulator (SIM) and the GRaphical Visualiser (GRV). The Simulator is organised around a compound data structure that represents the road network. The network file may contain several thousands of roads and hundreds of junctions. At each step of the simulation, each vehicle uses a set of simple localised rules to compute its new state and new position. In this, a vehicle must take into account its surrounding conditions, for example the proximity to a slower vehicle ahead will also influence the speed of the other vehicles. The Graphical Visualiser helps to design a possible road network generating the network file, and at the end of the simulation the created trace file is also loaded to the GRV in order to display the behaviour of cars.

The overall complex pattern of the urban traffic emerges from the simple local actions of individual vehicles. To simulate a city road network, a single processor is not able to perform such a computation intensive task within a limited period of time. Because of this reason parallel versions of MadCity have been developed [4,5,8] and the simulation is run on a computer cluster. To achieve parallelisation, the network file has to be distributed to all participating nodes and each node should work on a particular road area providing simulation locality.

The Parsifal cluster of the University of Westminster, that runs the simulation, is currently composed of 32 compute nodes and two master nodes. The primary master node is acting as the network gateway between the university public network and the isolated network of compute nodes. The second master node, namely node40, is the GT3 master node and it is configured as a submit host to the Parsifal Condor pool.

3 Grid Execution Management for Legacy Code: General Architecture

MadCity, as many other legacy code programs, has been developed and implemented to run on a computer cluster and does not offer a set of OGSI complying interfaces in order to be published and made available as a Grid Service. One approach to achieve this is to re-engineer the legacy code which in many cases implies significant efforts. Another solution to make available existing parallel legacy code programs as OGSI Grid services without having to re-engineer them is supported by the Grid Execution Management for Legacy Code Architecture (GEMLCA) proposed in this paper.

GEMLCA can be seen as a client front-end OGSI Grid service layer that offers a number of interfaces to submit and check the status of computational jobs, and get the results back. As any other Grid Service, GEMLCA has an interface described in Web Services Description Language (WSDL) [6] that can be invoked by any Grid services client to bind and use its functionality through Simple Object Access Protocol (SOAP) [7].

The general architecture to deploy existing legacy code as a service is presented in this paper based on OGSI and GT3 infrastructure. However, GEMLCA is a generic architecture and can also be applied to other service-oriented architectures. A similar solution is described in [9] for Jini, where Java RMI was used as communication protocol and the Jini lookup service to connect clients and services. Using different platforms the communication and the actual service implementation is different, but the concept of the architecture remains the same. This way the transition to new emerging standards like Web Services-Resource Framework (WSRF) [14] and GT4, will be straightforward.

As a general introduction, GEMLCA supports the following characteristics:

- Offers a set of OGSI interfaces, described in a WSDL file, in order to create, run and manage Grid service instances that offer all the legacy code program functionality.
- Interacts with job managers, such as Fork, Condor, PBS or Sun Grid Engine, allocates computing resources, manages input and output data and submits the legacy code program as a computational job.
- Administers and manages user data (input and output) related to each legacy code job providing a multi-user and multi-instance Grid service environment.
- Ensures that the parallel execution of the legacy code maps to the respective client Grid credential that requests the code to be executed.
- Presents a reliable file transfer service to upload or download data from the Grid service master node.
- Offers a single sign-on capability for submitting jobs, uploading and downloading data.
- A Grid service client can be off-line waiting for compute jobs to be completed and can request jobs status information and results any time before the GEMLCA instance termination time expires.

– Reduces complexity for application developers by adding a software layer to existing OGSI services and by supporting an integrated Grid execution life-cycle environment for multiple users/instances. The Grid execution life-cycle includes: upload of data, submission of job, check the status of computational jobs, and get the results back.

Figure 1 describes the GEMLCA implementation and its life-cycle. The Condor management system is used by the Westminster cluster as the job manager to execute legacy parallel programs.

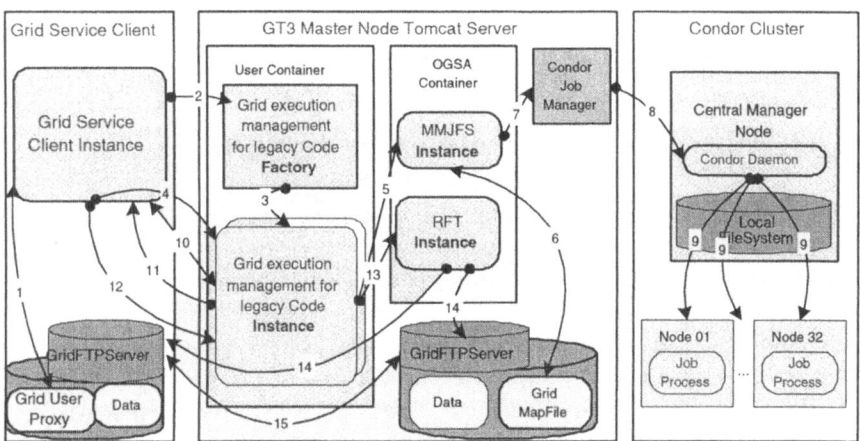

Fig. 1. GEMLCA life-cycle.

The scenario for submitting legacy code using the GEMLCA architecture is composed of the following steps:

- (1) The user signs his certificates to create a Grid proxy for authenticating to Grid services. The user Grid credential will later be delegated by the GEMLCA from the client to the Globus Master Managed Job Factory Service (MMJFS) for the allocation of resources.
- (2 & 3) A Grid service client, using the Grid Service Management for Legacy Code Factory, creates a number of Grid service Management for Legacy Code instances giving them a lifetime.
- (4) The Grid Client, using OGSI interfaces, calls a specific instance providing a set of input parameters needed by the legacy code program. The instance creates a Globus Resource Specification Language (RSL) file and a multi-user/instance environment to handle input and output data.
- (5 & 6) The instance, using the RSL file, contacts the Globus MMJFS on behalf of the client. The client credential is mapped to a specific local user using the Grid mapfile.

- (7, 8 & 9) If the client credential is successfully mapped, MMJFS contacts the Condor job manager which allocates resources and executes the parallel legacy code in a computer cluster.
- (10) At any time after the instance is created, the client can contact it in order to check the job status.
- (11) The Grid service instance can notify the client about any job status change using the OGSI notification framework if the client is on-line.
- (12 to 15) If the Grid Service instance time has not expired and the job status is finished, the Grid client contacts the Reliable File Transfer service instance to send results back to the client.

Finally, when the Grid Service instance is destroyed, the multi-user/instance environment is cleaned

4 GEMLCA Implementation and Results

The architecture presented in the previous section is implemented by deploying a secure Grid execution management service and tested by developing a secure Grid client. The implementation of the current development focuses on a subset of the architecture life-cycle and in particular on submitting jobs by the Grid client and executing them on the Westminster Condor pool cluster. The Grid client and service are both implementing GT3 security and in particular GT3 security proxy delegation [10] which allows the caller's Grid credential to be passed from the client via the Grid execution management service to the Master Managed Job Factory Service. MMJFS submits the job to the Condor pool via the Condor job manager and maps the execution of the MadCity parallel traffic simulator to the respective Unix user login name of the requestor's Grid credentials by using the GT3 system Grid-mapfile.

The Grid client is currently a Java program executed by the Java virtual machine from a Unix terminal window. The Grid execution management service and MMJFS are deployed in two separate Java servlet engine containers, and in particular Tomcat web application contexts, hosted by a single Tomcat server running on the Westminster GT3 master node. The reason to configure multiple containers hosted by the same Tomcat server is to ease the administration of a well-maintained GT3 server for Grid services deployments for multiple Grid developers on our GT3 master node. Another advantage is that each Grid developer can restart its own web application context using the Tomcat web application context manager without having to restart the whole Tomcat server.

The architecture presented in the previous section requires a specific job manager such as Condor, PBS or Sun Grid Engine to be configured for submitting computational jobs to clusters. Condor is selected as the job management facility for the Westminster cluster and it requires the Condor job manager interface to be installed and configured as well as the GT3 master node to be configured as a submit host to the Parsifal Condor pool. The default installation of GT3 only installs the Fork job manager and an additional step is required to install and configure the Condor job manager which is bundled with GT3. However, an updated file for the Condor job manager needs to be downloaded from the GT3 bugzilla facility (bugid 1425) as the native system condor.pm file contains a software fault which prevents Condor from getting the correct status of Condor jobs.

The parallel versions of the MadCity traffic simulator are implemented using PGRADE [12] and Spider [5] relying on PVM [13] as the message passing interface. This therefore requires the MadCity computational job to be submitted to the Condor pool using the PVM Universe. However, one of the issues of using Condor with the Globus Toolkit is that there is no native solution to set the PVM Universe in the Globus resource specification language (RSL). The solution to set the PVM Universe and specific Condor execution parameters is to use a Condor hash table in the RSL file to overcome the native RSL limitation.

The execution management Grid service uses the GT3 GramJob API to interface with MMJFS. The GramJob event notification does not work in a Grid service for interactive job submission because the GramJob API has been designed by Globus as a client API. Solutions include the use of a separate thread for the GramJob code, using the native OGSI APIs and extend the NotificationSink portType or use batch mode submission. Our approach is to submit jobs in batch mode to MMJFS which does not require an event listener to notify jobs completion.

Future development includes the implementation of the full execution management Grid service life-cycle which includes the upload and download of data prior and after the submission and completion of jobs, the handling of data transfer using the GT3 Reliable File Transfer (RFT) service for the upload and download of data between file systems located on different hosts. Future versions will also integrate with the Grid Application Monitoring Infrastructure (GAMI) [11] which includes the GRM trace collector and the Mercury monitor service developed by SZTAKI in the DataGrid and GridLab projects. It also envisaged that the Grid client could be embedded into a GridSphere portlet.

5 Conclusion

There is a clear need to deploy existing legacy code programs as OGSI Grid services. Two approaches can be identified to solve this problem. The first solution is to re-engineer the legacy code which in many cases implies significant efforts. The second one is to develop a front-end OGSI Grid service layer that contacts the target host environment. A solution for the second approach is addressed by the Grid Execution Management for Legacy Code Architecture (GEMLCA) which has been presented in this paper. An initial implementation of GEMLCA addressing a subset of the Grid service life-cycle has been developed and tested on the Westminster Condor pool cluster with the MadCity parallel traffic simulator application. Future work is planned to implement the full life-cycle of the Grid service and integrate with the Grid application monitoring infrastructure.

Acknowledgements. The authors wish to acknowledge the support and contributions of Damian Igbe, Agathocles Gourgoulis and Noam Weingarten in the traffic simulation aspects, and Kreeteeraj Sajadah and Alexandre Beaudouin in investigating and programming GT3.

References

1. S. Tuecke et al: Open Grid Services Infrastructure (OGSI) Version 1.0, June 2003, http://www.globus.org/research/papers/Final_OGSI_Specification_V1.0.pdf
2. The UK OGSA Evaluation Project Website, http://sse.cs.ucl.ac.uk/UK-OGSA/
3. UK OGSI Testbed Project Website, http://dsg.port.ac.uk/grid/testbed/
4. A. Gourgoulis, G. Terstyansky, P. Kacsuk, S.C. Winter, Creating Scalable Traffic Simulation on Clusters. PDP2004. Conf. Proc. of the 12-th Euromicro Conference on Parallel, Distributed and Network based Processing, La Coruna, Spain, 11-13th February, 2004,
5. D.Igbe, N.Kalantery, S.E Ijaha, S.C Winter, Parallel Traffic Simulation in Spider Programming Environment. In Distributed and Parallel Systems (Cluster and Grid Computing). Edited by Peter Kacsuk et al. Kluwer Academic Publishers, pp 165-172, 2002.
6. Web Services Description Language (WSDL) Version 1.2, http://www.w3.org/TR/wsdl12
7. Simple Object Access Protocol (SOAP) 1.1. W3C, Note 8, 2000 Center for telecommunication research, Columbia University.
8. D. Igbe, N. Kalantery, S.E. Ijaha, S.C. Winter, An open interface for the parallelisation of traffic simulators. In proceedings of the 7th IEEE international symposiums on distributed simulations and real time applications, pp158-163, 2003
9. G. Sipos, P. Kacsuk, Connecting Condor Pools into Computational Grids by Jini, 2nd European Across Grid Conference, Nicosia, Cyprus, January 2004,
10. Globus Team, Message Level Security, http://www-unix.globus.org/toolkit/3.0beta/ogsa/docs/message_security.html 2003.
11. Z. Balaton and G. Gombos: Resource and Job Monitoring in the Grid. Proc. of EuroPar 2003 Conference, Klagenfurt, pp. 404-411, 2003.
12. P. Kacsuk, G. Dozsa, R. Lovas: The GRADE Graphical Parallel Programming Environment, In the book: Parallel Program Development for Cluster Computing: Methodology, Tools and Integrated Environments (Chapter 10), Editors: P. Kacsuk, J.C. Cunha and S.C. Winter, pp. 231-247, Nova Science Publishers New York, 2001
13. A. Geist, et al. *PVM: Parallel Virtual Machine*, MIT Press, 1994.
14. Ian Foster, et al. Modeling Stateful Resources with Web Services, January 2004, http://www.globus.org/wsrf/

The PROVE Trace Visualisation Tool as a Grid Service*

Gergely Sipos and Péter Kacsuk

MTA SZTAKI Computer and Automation Research Institute, Hungarian Academy of Sciences
1618 Budapest P.O. Box 38.
{sipos, kacsuk}@sztaki.hu

Abstract. This paper introduces the way of separating the PROVE trace visualisation tool from the P-GRADE environment into a stand-alone Grid service. The separation resulted three PROVE implementations: a local service, a servlet based solution and an OGSA (Open Grid Services Architecture) enabled GT3 (Globus Toolkit) Grid service. The paper describes the problems and decisions during the development process of the different versions. Based on our experiences one can see how a local program can be relocated into the Grid as one of the several services the global infrastructure will one day offer.

1 Introduction

PROVE [5] is a visualisation tool integrated into the P-GRADE [12] program development environment. The goal of P-GRADE is to support users in every stage of the development and execution process of parallel programs and performance monitoring is an important functionality to achieve this. The application monitoring subsystem of P-GRADE consists of two parts: the GRM monitor infrastructure [10] and the PROVE visualisation tool. While GRM performs trace collection, PROVE is responsible for the data visualisation. PROVE can be separated into two logical parts: a data perser engine and a graphical engine that is tightly connected to the user interface of P-GRADE and performs the presentation itself.

In this paper we introduce how this two-layered PROVE could became an individual Grid service. Although the described results show the disintegration of one part of a complex software tool, it shows the general way of converting a well functioning local program into a Grid service. Therefore the conclusions we draw in this work can be generalised for every similar problems. In the next section we overview the role of the integrated PROVE inside P-GRADE. In Section 3 the new tasks of the grid-enabled PROVE service are introduced, while Section 4 gives a detailed overview about the different developed PROVE service implementations, namely the stand-alone application, the servlet based one and the GT3 based solution. Section 5 draws conclusions and outlines future works.

---

* The work presented in this paper was supported by the Ministry of Education under No. IKTA5-089/2002, the Hungarian Scientific Research Fund No. T042459 and IHM 4671/1/2003.

2 The Integrated Version of PROVE

As it was previously introduced GRM and PROVE are the main parts of the monitoring subsystem of P-GRADE. GRM uses source code instrumentation, hence an instrumentation API belongs to it [10]. Since P-GRADE provides a high level graphical interface that hides every low-level layers, users do not have to know anything about the GRM instrumentation API.

The first version of P-GRADE supported job execution on local clusters and supercomputers [6]. In such an environment it can easily set-up the GRM trace collector infrastructure before it starts the parallel program. The infrastructure contains local monitors – one for each node – and one main monitor has to be started on the machine where P-GRADE is located. The instrumented processes through the local monitors can send trace event messages to the main monitor, hence to P-GRADE. The main monitor controls the local monitors – when and where to send their locally buffered trace – and it creates a global trace file from the received data. Since the original PROVE is integrated into P-GRADE it runs on the host where the GRM main monitor does, thus they share the same file system. PROVE can open the global trace file any time and visualise its content from the client's needed aspect.

With the instrumentation API of GRM parallel applications can generate trace data in the Tape/PVM format [9], hence PROVE expects this type of trace files as well. Important to clarify that PROVE does not build onto the GRM infrastructure or its instrumentation API, solely onto the Tape/PVM file format. Someone could use PROVE to visualize trace data has been generated and collected with other tools than the discussed ones if they result a Tape/PVM formatted global trace file. Unfortunately the integrated PROVE version receives necessary starting and control parameters from its wrapper P-GRADE environment so users cannot exploit this independence. When we began the development of the PROVE service one of our motivations was to realise an independent tool.

3 The Role of PROVE in Grid Environments

In the past few years P-GRADE has overgrown the boundary of local resources and expanded its functionality into the Grid [7]. At the present P-GRADE supports the execution of PVM and MPI programs in Condor and Globus grids, but we are already working on the next version that will support JGrid [14], a Jini based Grid environment as well.

Since P-GRADE uses source code instrumentation based application monitoring in grids too, its trace collector infrastruecture had to be fundamentally changed [1]. While the job executor client can establish the trace collector infrastructure on local resources, in grids this task is impossible. The submitter entity does not know in advance where the processes will run and it cannot have necessary rights to login onto these machines anyway.

The new trace collector infrastructure has been already developed in the GridLab project [4]. It consists of two parts: the Mercury monitor service and a main monitor program. Mercury is a software service has to be installed on grid resources and its tasks are to collect trace from the jobs that run on that resource and to forward the trace to the interested clients. The main monitor is a software component that can register itself at remote Mercury services and create a global trace file from the received data. As it can be seen, in the new structure the local monitors (part of the Mercury service) are started by system administrators instead of job submitters.

Grid environments do not place new requirements on the PROVE tool unlike they did on the GRM infrastructure. If the main monitor cooperates with P-GRADE and registers itself at the appropriate remote Mercury services than even the "non-grid" version of PROVE can visualise the trace. Nevertheless, there are several reasons why we decided to develop a new, stand-alone PROVE version. The most important of them is to separate its function from P-GRADE into an individual service. One advantage of this step has been discussed in the previous section: it makes PROVE independent from P-GRADE in practice. Another important reason is that only the stand-alone PROVE can appear in the Grid as an OGSA [2] enabled Grid service. The following list contains all the aims we wanted to achieve with the stand-alone PROVE:

1. The service has to be able to visualise Tape/PVM formatted trace files.
2. The service must be able to parse trace files being situated anywhere in the network.
3. The result of a service call has to be a picture that illustrates the trace file from the requested point of view.

The first goal is inherited from the previous PROVE version. We did not want to change the supported trace file format, since Tape/PVM provides all the necessary data to generate space-time and statistical diagrams of parallel jobs. Based on the time fields of the Tape/PVM event entries the PROVE service is able to sort the data into chronological order and generate a picture. The sorting phase is sometimes very computation-intensive so performing it on a powerful PROVE service provider host can lower the load of client machines.

The second point of the previous list supposes that some entity collects the trace of a grid-job into a global file the PROVE service provider can access. This entity can be the previously discussed main monitor on the client machine or on a "trace collector service" provider host. The latter solution is more realistic, since trace files can easily overgrow the storage capacity of clients. Such a service has to provide similar functionality to the remote clients that the main monitor does locally.

The last point of the list also lowers the loads of grid clients: instead of large trace files they have to download rather pictures that graphically delineate the huge amount of data. Obviously, much less bandwidth is enough to download such a picture than to download the raw trace. Fig. 1. presents the role and the usage scenario of the PROVE service in grid environments. The scenario supposes that the client submitted the job, the processes already started trace production on the grid resources and there is a service provider host that collects the local traces into a global file.

When the client would like to check the actual state of the remotely running application he has to find a PROVE service provider. The detailed way in which this can be done is out of the scope of this paper, but generally it happens through the information system of the Grid. After the client sent the visualisation request to the PROVE provider (1), the provider finds the trace collector and instruct it to collect the actual trace (2). The trace collector host finds the job executors and gets the local traces (3). (The trace

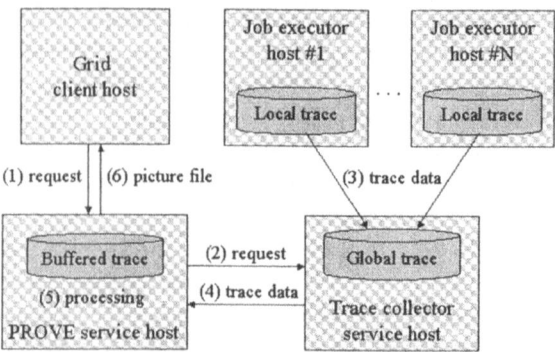

Fig. 1. The role of the PROVE service in the Grid

collector can find the job executors very similarly how the client found the PROVE provider and the PROVE provider found the trace collector.) The most obvious solution to generate the global trace file is to apply the Mercury service on the executor hosts and its corresponding client side main monitor program on the trace collector host. After the trace collector registered itself at the Mercury services that run on the executor machines these providers will automatically forward the local traces. The PROVE provider then can download the global trace and save it into a local buffer (4).

Since the visualisation could be requested already during the execution of a job, the results of two calls that request the space-time diagrams of the same job can be different. Because the first part of these trace files are totally the same (the newer file contains the older one), it seems quite logical to store job traces on the PROVE provider host more or less persistently. With this technique the network traffic can be radically reduced because when the second client request arrives the PROVE provider has to download only the locally missing tail of the global file but not the whole file again. After this data synchronisation phase the provider can sort the event entries of the file, generate the requested picture (5) and send it back using a popular image format (6).

4 Our Different PROVE Service Implementations

We implemented the introduced PROVE service in 3 different ways: as a client side local service, as a Java servlet and as a GT3 Grid service. While the first implementation provides local service the other two act as real services in grids. Although clients can use the three distinct versions in fundamentally different ways, their cores are the same. This service core can be seen in Fig. 2.

The most important part of the structure is the service program. It consists of two layers. The upper layer is written in Java while the bottom one in C. The double-layered architecture enables to exploit the advantages of both languages. While the Java layer provides an "easy to use" interface for clients and can be consumed from various platforms, the C layer allows more efficient memory allocation. Efficient memory usage is a key issue in our implementation because it uses the memory for the previously discussed

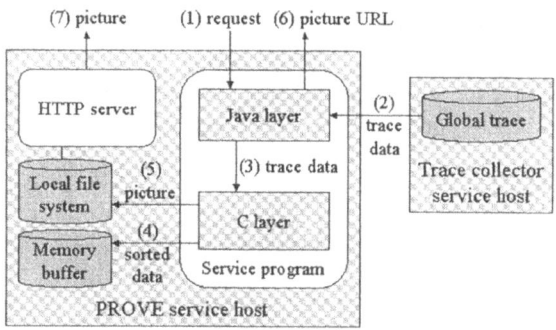

Fig. 2. The common structure of the PROVE service implementations

trace buffering purpose. Using memory buffers instead of file buffers we could achieve significant speed up in the data parsing and image generator phases. The detailed general scenario of the service usage presented in Fig. 2 is the following: after a received client request (1) the Java layer updates the local trace by downloading the missing part of the global trace from its collector host (2). It forwards the downloaded data with native calls to the C layer (3) which merges the new entries and saves them into a memory buffer (4). Then, the C library generates the requested image and saves it as a local file (5). The name of this file has been previously defined by the Java layer, so it can easily convert the file name to a URL and return the address to the client (6). The client then downloads the picture through the service side HTTP server and presents it with an appropriate tool (7). The introduced scenario is almost the same whichever implementation is being used. What mostly differs in them is the Java layer, thus how clients can contact the service.

4.1 PROVE as a Local Service

In the local version of PROVE the Java layer is a stand-alone Java application with a Swing GUI. A client can use this graphical interface directly to interact with the service. Through the graphical front-end one can set the parameters of the trace file to be analysed and can browse the result image. In this case the trace file is usually a local file, but it is not a restriction: even the local version of PROVE can visualise remote trace files that are accessible through web servers.

Although this version was developed to demonstrate the correctness of our new service core, we found this version a very useful tool for Tape/PVM formatted trace file visualisation. This stand-alone version can even replace the built in PROVE of P-GRADE, since it provides all of its functionalities. The only difference is that the new version presents the result as a static image file while the old one draws the picture onto the screen dynamically.

4.2 PROVE as a Java Servlet

The second version is the servlet based one. In this case the Java layer is a Java servlet that runs inside a servlet container. The structure of this implementation is presented in Fig. 3.

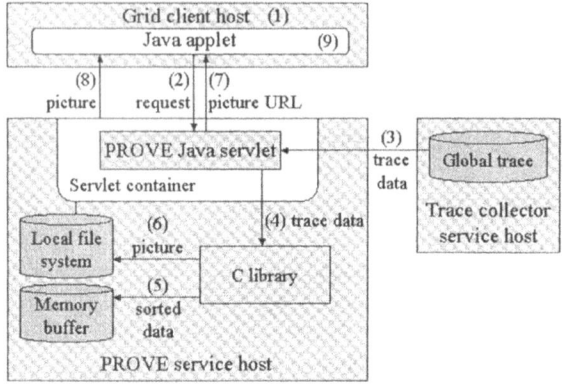

Fig. 3. The usage of the servlet based PROVE service

Since a servlet container is always contained in a web server, this servlet-based solution does not require the starting of a stand-alone HTTP server: the web server that hosts the servlet container can serve the image download requests as well.

In this case the service usage scenario is the following. The client first downloads an adequate web page that contains a Java applet (1). Such a web page could be linked for example into a Grid portal. The applet of the downloaded page acts like a stub during the service consumption: it knows how to communicate with the remote servlet. Besides, the applet provides local service for the client too, since it presents the result trace pictures.

When a client sends a visualisation request to the applet through its graphical elements it translates the event into an HTTP request the servlet can understand (2). The servlet receives the message and – together with the C library – updates the locally stored trace data (3, 4, 5) and generates the requested image file (6). (The data update and file generation processes happen in the previously discussed way.) After this the servlet wraps the URL of the generated image into an HTTP response message and sends it back to the applet (7). The applet automatically downloads the file (8) and presents it in the client side browser window (9).

As it can be seen in this description the servlet-based PROVE implementation has a big advantage: no special client side program has to be installed, a simple Web browser with a Java plug-in is enough. Because of this prosperous feature we built this PROVE service version into our P-GRADE Portal [12] has been presented during the Super Computing 2003 exhibition, and in the Grid Demo Session of the IEEE International Conference on Cluster Computing 2003.

4.3 PROVE as a Globus Toolkit 3 Grid Service

The third PROVE service version we developed uses the GT3 [13] framework. Although the OGSI [11] specification does not require the usage of the Java language for Grid service implementation we chose Java to realise the GT3 based PROVE. With this trick we could significantly reduce the development time since the core of the stand-alone PROVE implementation could be applied again.

Although the base structure remains the same, the GT3 framework grants it several extra features neither the stand-alone nor the servlet-based versions are able to provide. Globus services run inside containers that provide – among others – factory functionality for them. These GT3 containers do not automatically create a new service instance for every client call just when it is explicitly requested. In contrast, servlet containers provide stateless connection so they always initiate a new servlet instance for every incoming request [8]. In the present case it means that a new PROVE servlet is generated for every request while a single PROVE GT3 service can serve multiple calls. Since most of the visualisation requests that PROVE has to serve refers to the result of a previous one, this instantiation difference causes significant deviation in the service request protocols.

To illustrate it with an example imagine that one would like to zoom to a smaller part of a previously generated trace picture. In this case the new result does refer to the previous one – because the new image is part of the old one – so the request could be described with the old picture: obviously the relative coordinates of the new picture inside the old one are enough. Contrarily, the servlet-based PROVE service needs an exact description about the old picture besides these coordinates too, since the servlet that accepts the second visualisation request knows nothing about the previous image.

In case of the servlet-based version we could not eliminate this overhead so we had to develop a complicated protocol that can make the required applet-servlet communication possible. In case of the GT3 based PROVE the GT3 container starts one individual PROVE service per client and this instance serves every call of its owner. With this technique the previously discussed overhead can be saved since a service instance knows every previous result of its client. The usage of the GT3 based PROVE is presented in Fig. 4.

As it can be seen in Fig. 4 the Globus Toolkit 3 must be installed on both the client and the server sides. It is necessary since it provides several functions (API, server side container, engines for the communication) have to be used. The client application in this case is a stand-alone program because an applet cannot perform the appropriate usage of the GT3 API. This client application should have graphical interface otherwise it cannot present the result images to the user.

Since GT3 is a real grid infrastructure it contains information system as well, it is called Index service. One can get a reference to a factory of PROVE services, to a factory of trace collector services or to a factory of job executor services from an Index service. Fig. 4 supposes that the client already got those references, moreover he instantiated the executor service in the required number and created one instance from the trace collector service as well. After this instantiation the parallel job starts at the selected providers and the trace collector continously saves event data into a global trace file. The scenario of Fig. 4 begins at this point.

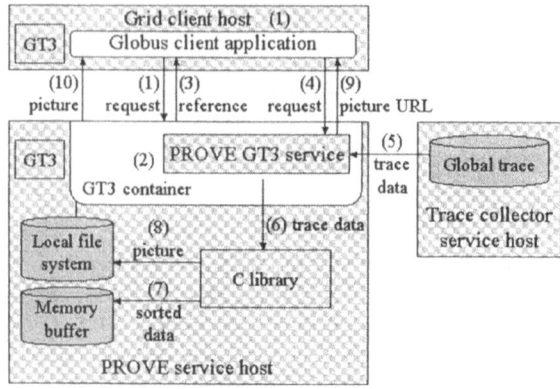

Fig. 4. PROVE as a GT3 Grid service

First the client application sends a service instantiation request to the GT3 container of the PROVE service (1). This request – as all others during this scenario – travels between the client and the PROVE provider as a SOAP message transmitted with HTTP protocol. As a response to the request the container creates an instance from the PROVE service (2) and sends back a reference on it to the client (3). The client application now can generate a stub to the newly instantiated service and send a visualisation request to it (4). The service, just like in case of the previous versions, downloads the trace file from its collector (5) and sends it for parsing to the C library (6). The C layer merges the data with the local trace (7) and generates the requested picture (8). The Java service than sends back the URL of the image file (9) what the client application finally downloads and presents in a client side graphical window (10).

Another feature of the GT3 framework is the notification infrastructure. In the GT3 version of PROVE this feature is used to make the job observation possible. The integrated version of PROVE can observe parallel applications, which means that it checks the global trace file on a regular basis and automatically performs visualisation when new entries appear. Since now the trace file is situated on the collector host and not on the one that runs PROVE it would be difficult to regularly check the content of this file through the network.

Instead of this obvious – but resource consuming – solution we applied the GT3 notification framework. The trace collector sends a notification message to the PROVE host every time when new data arrives from the executor sites. Based on these notification messages the PROVE service always knows whether the locally cached trace is up-to-date or not. If not it downloads the new data, so when a client requests a new visualisation the image can be generated at once. This solution for data actualisation can be applied only in the GT3 version, since using servlets solely the client side can initiate communication with the server the other way is impossible [3].

5 Conclusion and Future Works

This paper introduced the process has been applied to separate PROVE from its wrapper P-GRADE environment. The presented result is our first step toward the long-term goal to separate every function of P-GRADE into stand-alone Grid services. Since we would like to use fully qualified OGSA services to achieve this goal, the GT3 based version of PROVE perfectly fits into this schema. After finishing the full separation process P-GRADE will be able to appear in the Grid as a "super service" that can control every member of its underlying Grid service set. Another important purpose is to integrate a modified version of the PROVE service into a Jini based grid framework [14]. In such an environment PROVE can appear as a Jini service, and based on its functionality Jini clients can use a flexible application monitor infrastructure.

References

1. Z. Balaton, P. Kacsuk, N. Podhorszki and F. Vajda: From Cluster Monitoring to Grid Monitoring Based on GRM, Proc. of the 7th EuroPar'2001 Parallel Processing, Manchester, UK, 2001, pp. 874–881.
2. I. Foster, C. Kesselman, J. Nick and S. Tuecke: The Phisiology of the Grid: An Open Grid Services Architecture for Distributed Systems Integration, Globus Project, 2002, www.globus.org/research/papers/ogsa.pdf
3. M. Harding: Servlets: A Technical Discussion, presentation at the Software Forum Java Developers Special Interest Group meeting, Palo Alto, CA, USA, 2/3/1998.
4. GridLab Monitoring work package (WP11). Available at:
http://www.gridlab.org/WorkPackages/wp-11/index.html
5. P. Kacsuk: Performance Visualization in the GRADE Parallel Programming Environment, Proc. of the 5th international conference/exhibition on High Performance Computing in Asia-Pacific region (HPC'Asia 2000), Peking, 2000, pp. 446–450.
6. P. Kacsuk, G. Dózsa and R. Lovas: The GRADE Graphical Parallel Programming Environment, Parallel Program Development for Cluster Computing, Methodology, Tools and Integrated Environments, Nova Science Publishers, 2001, pp. 231–247.
7. P. Kacsuk: Parallel Program Development and Execution in the Grid, Proc. of PARELEC 2002, International conference on parallel computing in electrical engineering. Varsaw, Poland, 2002. pp. 131–138.
8. B. Kurniawan: How Servlet Container Work, available at:
http://java.sun.com/products/servlet/docs.html
9. É. Maillet: Tape/PVM: An Efficient Performance Monitor for PVM Applications. User's guide, LMAC-IMAG, Grenoble, France, 1995. Available at:
http://www-apache.imag.fr/software/tape/manual-tape.ps.gz
10. N. Podhorszki and P. Kacsuk: Design and Implementation of a Distributed Monitor for Semi-on-line Monitoring of VisualMP Applications, Distributed and Parallel Systems, From Instruction Parallelism to Cluster Computing Cluwer Academic Publishers, 2000, pp. 23–32.
11. S. Tuecke, K. Czajkowski, I. Foster, J. Frey, S. Graham, and C. Kesselman: Open Grid Service Infrastructure Version 1.0, Global Grid Forum, Draft 4/5/2003,
http://www.gridforum.org/ogsi-wg/drafts/draft-ggf-ogsi-gridservice-29_2003-04-05.pdf
12. P-GRADE Graphical Parallel Program Development Environment:
http://www.lpds.sztaki.hu/pgrade
13. The Globus Project: http://www.globus.org
14. *JGrid project: http://pds.irt.vein.hu/jgrid*

Privacy Protection in Ubiquitous Computing Based on Privacy Label and Information Flow

Seong Oun Hwang and Ki Song Yoon

Electronics and Telecommunications Research Institute
161 Gajeong-dong, Yuseong-gu, Daejeon, Republic of Korea
{sohwang, ksyoon}@etri.re.kr

Abstract. The vision of upcoming ubiquitous computing environment allows us to exchange and share information with any parties any time, anywhere if we want. However, the widespread use of computing devices and sensors networked together could be a considerable threat to the privacy of individuals living in the ubiquitous computing environment. In the paper, we investigate privacy issues in the ubiquitous computing environment and present a model of privacy protection and an example architecture to support it. The proposed architecture ensures that individuals not only get benefits and services from ubiquitous environment by freely exchanging and sharing information, but they also preserve their own privacy.

1 Introduction

With the rapid advancement of computing, communication technologies and convergence of those technologies, it is expected that in the near future we could live in the so-called ubiquitous computing environment. With ubiquitous computing environment, we refer to a scenario in which computing devices visible or invisible are present everywhere and those computing devices are networked to allow us to exchange and share information with any parties any time if we want. It seems that the coming ubiquitous computing environment plays a considerable part in blurring the distinction between our real world space and cyber space. The term of cyber space here means a space consisting of nodes which model our real world space as it is. In particular, ubiquitous sensing and the invisible form factor of embedded computing devices will have made it easier than ever to collect and use information about individuals without their knowledge [1]. The focus of this paper is on the discussion of informational privacy meaning "the right of people to determine for themselves when, how, and to what extent personal data about them is communicated to others." Personal data is data about an identified or identifiable person.

Here we list what implications the ubiquitous computing environment will have in the aspect of privacy protection [13, 14].

- **Control heterogeneity:** In the ubiquitous computing environment, there exist mixed both centrally controllable environment like the present time and one where central control is not available.

- **Task dynamism:** By virtue of being available everywhere at all times, it will have to adapt to the dynamism of users' environments and the resulting uncertainties.
- **Device heterogeneity:** The environment consists of various forms of devices, embedded or unembedded, information devices or appliances: computers, phones, cameras, TVs, sensors, refrigerators, media players, and so on.
- **Application mobility:** While moving with us, applications have to adapt to changing technological capabilities in their environment. Having application follows the user and move seamlessly between devices are very important. Totally new infrastructure are required on which applications are not pieces of software targeted to a particular device or environment but rather, high-level descriptions of the task a user needs to perform.
- **Context-aware:** a situation in which a mobile computer is aware of its user's state and surroundings, and modifies its behavior based on this information.
- **Various types of authentications** based on memorization or recognition

The objective of the paper is to model and design an architecture for privacy protection considering the above implications. Our approach to the privacy protection is a privacy-protecting framework based on information flow control [12, 15] from individuals to other communicating parties. We further develop this framework by introducing the concept of privacy label that consists of level and domain. We transform individual's private data into information set with appropriate privacy label attached. To structure the information flow under the control of user, we construct the privacy policy rule that allows information flow according to its privacy label and results in the protection of individual's privacy.

The remainder of the paper is organized as follows. In Section 2, we overview related works. Section 3 examines some examples of privacy protection and draws a formal model of privacy protection system. Section 4 presents the overall architecture of privacy protection system, whose components are explained in detail. Section 5 concludes the paper.

2 Related Work

Chaum introduced the model of anonymity and pseudonymity from which [2] defines a layered object that routes data through intermediate nodes, called mixes. These intermediate nodes may reorder, delay, and pad traffic to complicate traffic analysis. Even hostile observers who can monitor all the links in the mix network cannot trace a message from its source to its destination without the collusion of the mix nodes. Some work based on the concept of mix : Anonymous Remailers like [3] use mixes to provide anonymous e-mail services and also to invent an address through which mail can be forwarded back to the original sender. Mixes are also used to provide untraceable communication in an ISDN network [4], ATM network [5]. Goldschlag, et al. [6] presents an architecture that hide routing information on the Internet by establishing *anonymous socket connections by means of* proxy servers at initiator/responder sites and producing virtual circuits within link encrypted connections already running between routing nodes. Chaum also introduced the notion of anonymity sets from [7]. The anonymity set is "the set of all possible subjects who might cause an action." [8] It provides a measure of the unlinkability between the messages coming in and going out of the mix. Beresford et al. [9] develop two metrics for measuring location privacy, one based on anonymity sets and the other based on entropy [10]. Those two schemes [6, 7] only hide routing information and depend heavily on intermediate nodes' or proxy servers' trustworthiness. Those can be used as a component of our proposed system which results in protection of privacy. Jiang et al. [11] construct a model for privacy control by introducing decentralized information spaces and unified privacy tagging in pervasive computing environment. Jiang's scheme reflects more of access control rather than privacy itself. Our model focuses on the protection of individual's privacy based on the following observation that even on the same data, the sensitivity of privacy could be different depending on persons or its belonging (parsing) domain.

3 Modeling of Privacy Protection System

In the following, we take some examples to help readers understand our modeling of privacy protection system.

3.1 Some Examples

Example 1. Indirect information flow between parties (Fig. 1)
We say that there is indirect information flow when information flows among other parties except for the information owner. Figure 1 shows how information automatically flows between information generators A, B, C, D, E. Generators denote principals who deals with the information about some principal whom we call it Bob. We say that Bob is the owner of the information. It is because the exchanged information about him was generated through the interaction between Bob and generators. Bob, the owner could define additional constraints that should be followed by those generators during information exchanging or sharing. In this way, information about Bob flows among generators under the control of Bob. Note that information flow is allowed from A to D, but information flow from D to C is only allowed when there are de-personalization operations satisfying some properties which will be described later.
Example 2. Direct information flow
We say that there is direct information flow when information flows between the information owner and other parties. Let's suppose that Bob had a conference with Alice in some place, for example, in Alice's office. The conference result was stored at some place. The place might be a public place on the network or Bob's notebook or any other information storage. If the storage is Bob's notebook, let's suppose that Bob left Alice's office leaving his notebook behind him. Bob defined some constraints that only Alice could access only the conference result. In this case, Alice could access the

conference result only. She can not get any information that might include Bob's other private information.

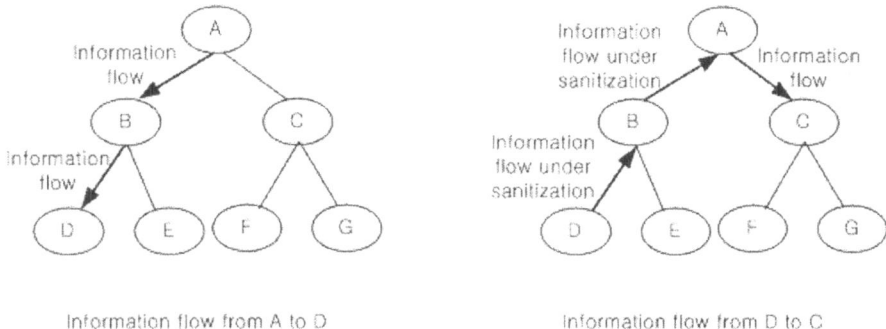

Fig. 1. Indirect Information Flow Diagram

Example 3. Protection of identity information
Individual could want their identity information be hided when requesting or receiving services (during service transactions). This is possible by employing an independent authentication service provider of the requesting service itself. Note that existing services mostly include authentication function within themselves. From services we separate the authentication function and assign it to some authentication service provider. Here we assume that the authentication service provider is a trusted third party in the sense that the authentication service provider should not reveal the identity to the service provider. This idea can be applied to any authentication scheme whether it is based on memorization, recognition, or any other type of authentication.

Next, we describe the theoretical model of privacy protection system that makes privacy control in ubiquitous computing environment.

3.2 Formal Descriptions

Definition 1. We define *privacy protection system* to be a set of 5-tuple <S,O,L,D,C> with some relationship among them, where S is a set of subjects, O is a set of information objects, L is a set of levels, D is a set of domains, and C is a set of constraints.
Definition 2. We define S, *a set of subjects (principals)*, to be a set of entities. Examples of subject could include a person, a set of persons, a particular (set of) device(s).
Definition 3. We define O, *a set of (information) objects* to which the privacy controls are applied by the owner entity. An information object represents every atomic object inside the system which is of interest from the point of view of privacy functionality and management. In other words, an information object is a target of privacy protection.
Definition 4. We define L, *a set of levels* which are assigned to O by the owner entity according to his or her own privacy policy. Levels denote the privacy sensitivity of the object. For example, *low, medium, high* can be used in increasing order of sensitivity.

The concept of level provides a good way of grading mechanism which allows users to structure O into levels of information. In this way, we distinguish between different levels of objects ranging.

Definition 5. We define D, *a set of domains* on which information about individual characteristics of each entity are generated, distributed and processed, that is, on which privacy levels are assigned and parsed. The concept of domain could represent/characterize activities of entities such as purposes. A domain could contain a number of attributes of sub-domain which could recursively contain sub-domains within itself. In this way, a set of domains can be hierarchically structured.

Definition 6. We define C, *a set of constraints*, to be a set of restrictions and conditions to be satisfied under which include ones such as information owner's location or network address or history of change of location. Constraints could specify under what conditions the privacy control is executed.

Definition 7. We define the *privacy label* of an object o to be the pair (f(o), g(o)), where f returns the level of o, and g returns the domain of o. Our observation is that the privacy level could be different on the same kind of data depending on persons. Furthermore, the privacy level of even the same data could be different depending on its belonging (parsing) domain.

Definition 8. We define a relation $l_i \le l_j$ holds when for any (each) element o_{ik} of o_i, there exists an element o_{jk} of o_j such that the level of o_{jk} is higher than that of o_{ik}. Note that l_i means the level of the object o_i.

Definition 9. We define a domain d_i *is covered by* d_j when the attribute set of d_i is contained in the attribute set of d_j.

Definition 10. We define a relation $(l_i, d_i) \le (l_j, d_j)$ holds among privacy labels when $l_i \le l_j$ and d_i is covered by d_j.

The privacy policy rule: Information flow is allowed from a subject having an object with label (l_i, d_i) to a subject having an object with label (l_j, d_j) only if

(1) $(l_i, d_i) \le (l_j, d_j)$ or
(2) there exist an integer k such that de-personalization operation s satisfies $s^k((l_i, d_i))$ $\le (l_j, d_j)$; where s^k means i times recursive application of s ,and de-personalization operation s: $O \times L \to P(O)$ gives information whose sensitive details has been purged.

Definition 11. We say that a privacy protection system is *safe* when all the level-assigned information at each domain exist on the intended privacy label by an object owner.

Property 1. The size of D per information object grows as information flow proceeds.

Theorem 1. A given privacy protection system {<S,O,L,D,C>} under the above privacy policy rule is safe.

Proof sketch. A formal proof of this theorem is proceeded by proof by contradiction. Suppose {<S,O,L,D,C>, the privacy policy rule} is not safe. Then there exists at least one information object of which one pair of consecutive state sequence violating the safety condition results in $f_i(o) < f_{i+1}(o) \land g_i(o) \subset g_{i+1}(o)$, or $f_{i+1}(o) > f_i(o) \land g_{i+1}(o) \subset g_{i+1}(o)$. The first result contradicts the privacy policy rule and the second result contradicts property 1. □

4 Design of Privacy Protection System

4.1 Design Objectives

Proposed system was designed to have the following specific objectives in mind:
- **User-centered approach:** The concept of privacy should be easy for users to understand and to manage. It also is to reflect users' different characteristics in regards to privacy from persons to persons, so we introduce the concept of agreement that can be achieved through negotiation with other entities reflecting each user's privacy characteristics and allows users to set privacy labels.
- **Autonomy:** A number of heterogeneous devices, networks, domains in ubiquitous computing environment could make users' privacy control complex. For efficient privacy control in each specific device/network/domain, we introduce the concept of intelligent agent that is responsible for executing its assigned privacy policy and coordinating or negotiating with other agent to result in consistent control of privacy.
- **Infrastructural approach:** To seamlessly apply consistent privacy policies in different device/network environment, we introduce a privacy layer between application layer and network layer that is responsible of taking privacy control.
- **Separation of authentication function from service itself:** Existing services are very difficult to be controlled in privacy aspects. It is because they include identification/authentication functions within themselves. To cope with this problem, we introduce the idea of separation of authentication function from service itself. In that way, we can support a wide range (spectrum) of services – from complete non-anonymous services, partly anonymous services, to complete anonymous services.

4.2 System Configuration

In the architectural viewpoint, proposed system includes the following (Fig. 2):
Privacy agreement is a relationship object that is the digital representation of a privacy agreement (agreed-upon terms and conditions of the relationship) between two or more parties. Agreements can be interpreted anywhere: it can be run by the content server (conventional access control), the client machine (for high interactivity or for the mobile case), or a third party. It can reside anywhere on the network, whether at a server(conventional access control), at a client(usage control), or with a third party(e.g., a rights clearing house)

We use a distributed team of cooperative autonomous agents to collectively user's privacy-management process. It is in favor of the agent approach because it is applied in larger pervasive forms of networks where centralized user management becomes increasingly difficult. Our agent approach's central concept is that by linking the sensory and intelligence capabilities of numerous agents distributed across a network, we can amplify the protection of user's privacy in the system. In addition, a central requirement for proposed architecture is to support privacy services not only within client-server environment, but also within peer-to-peer (P2P) and ad hoc networks, which require a distributive design approach. Some specific properties of a software *agent system that match the requirements of a distributed privacy protection platform* include autonomous goal-driven behavior, reasoning, and the capacity for proactive responses to potential threats. Client consists of a single master agent responsible for a set of sub-domains and a number of peer-slave agents responsible for actions in each sub-domain. Each agent maintains a separate user privacy profile for the local network domain that the agent manages. The master agent interacts with the users through user profile agent and ensures that all peer agents implement the centralized privacy policy by controlling and coordinating its subordinate slave agents. In real world, agents do privacy control in lieu of users in a network of information system. Each peer agent can maintain its own database of user behavior and it also can access and update user database under the control of master agent.

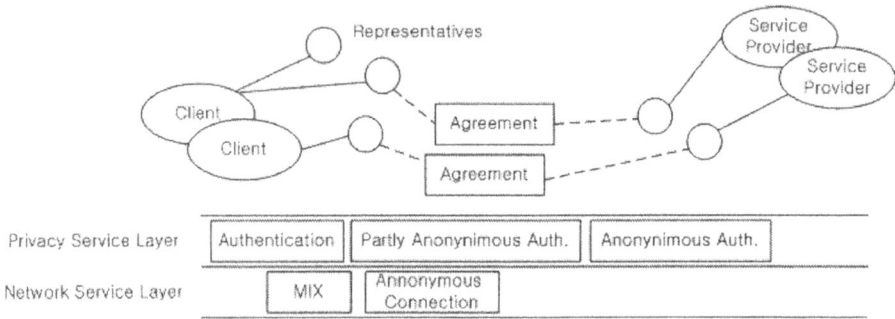

Fig. 2. Architecture of Proposed System

An Example Scenario

A user needs to access to a particular service from a device over a network link. The user first gets privacy profile by editing for himself. Based on the template profile, negotiation service of the agent completes the profile called privacy agreement below by negotiating with the service provider. The user assigns his service request to agents. According to the description of the agreement, the domain agents perform his user's service request with the service provider. On receiving the user's request, the agent parses it and fires its reasoning module to process the request. In this case, the system must validate that the communicating service provider has the necessary clearance to access the requested resource of the user. The agent then passes the user's details to its own privacy service module, which returns the user's valid status – that is, due label of user's data. If necessary, there can be de-personalization operations in the privacy service module.

4.3 Agent Architecture

This section shows what components the privacy agent consists of and how it works to achieve the privacy in lieu of the user. Privacy agent largely consists of two parts: Message Handler, which is in charge of connecting with other entities and Task Engine, which is in charge of doing services.

Task Engine consists of the following modules:

- **Coordination Engine** makes decisions concerning the agent's goals – for example, how to pursue the goals and when to abandon them. It also coordinates the agent's interactions with other agents using its known coordination protocols and strategies. This module infers the degree of privacy to process. This process is guided by the current privacy policy, which is communicated to the agent by its local master agent.
- **Resource Database** maintains a list of resources that are owned by and available to the agent.
- **Ontology Database and Editor** stores the logical definition of each privacy metadata – its legal attributes, the range of legal values for each attribute, any constraints between attribute values, and any relationships between the attributes of an agreement and other agreements.
- **Privacy Service**: Users cannot communicate with services directly – otherwise they would reveal their identity straight away. We therefore require an anonymizing proxy for all communication between users and services. The proxy lets services receive and reply to anonymous (or more correctly, pseudonymous) messages from the users. The middleware system is ideally placed to perform this function, passing user input and output between the service and the user. The privacy service checks the privacy labels of all outgoing traffic from the agent or user environment. Thus, privacy infringement through unauthorized information flow across different domains can be prevented. Our concept of privacy label is similar to what Xiaodong Jiang calls privacy tag [11].

5 Conclusion

This paper has investigated privacy issues in ubiquitous computing environment and has presented a model of privacy control and an architecture to support the model. The proposed design architecture is only one among many prospective architectures. In future research work, we plan to develop a prototype of privacy protection system based on both the model and the architecture.

References

1. Modeling Privacy Control in Context-Aware Systems, IEEE Pervasive Computing, Vol. 1, No. 3, July-September 2002.
2. D. Chaum, Untraceable Electronic Mail, Return Addresses, and Digital Pseudonyms, Communications of the ACM, V. 24, No. 2, Feb. 1981, pages 84-88.
3. C.Gulcu andG. Tsudik, Mixing Email with Babel, 1996 Symposium on Network and Distributed System Security, San Diego, February 1996.
4. A.Pfitzmann, B.Pfitzmann, and M.Waidner, ISDN-Mixes:Untraceable Communication with Very Small Bandwidth Overhead, GI/ITG Conference: Communication in Distributed Systems, Mannheim February 1991.
5. S. Chuang, Security Management of ATM Networks, Ph.D. thesis, Cambridge University.
6. David Goldschlag, Michael Reed, and Paul Syverson, Hiding Routing Information, Workshop on Information Hiding, Cambridge, UK, May 1996.

7. D. Chaum, "The Dining Cryptographers Problem: Unconditional Sender and Recipient Untraceablity, Jr. of Cryptography, Vol. 1, No. 1, 1988, pp. 66-75.
8. A. Pfitzmann and M. Köhntopp, Anonymity, Unobservability and Pseudonymity – A Proposal for Terminology, Designing Privacy Enhancing Technologies: Proc. Int'l Workshop Design Issues in Anonymity and Observability, LNCS, Vol. 2009, Springer-Verlag, 2000, pp.1-9.
9. A. Beresford and F. Stajano, Location Privacy in Pervasive Computing, January-March 2003.
10. C. Shannon, A Mathematical Theory of Communication, Bell System Tech. Jr., Vol. 27, July 1948, pp.379-423, and October 1948, pp.623-656.
11. Xiaodong Jiang and James Landay, Modeling Privacy Control in Context-Aware Systems, IEEE Pervasive Computing, July-September 2002.
12. D. E. Denning, Crptography and Security, Addison-Wesley, Reading, MA, 1982.
13. Special Issue: Issues and challenges in ubiquitous computing, Communications of the ACM, Vol. 45, No. 12, December 2002.
14. Special Issue: Technical and social components of peer-to-peer computing, Communications of the ACM, Vol. 46, No. 2, February 2003.
15. D.E.Bell, L.LaPadula, Secure Computer Systems: A Mathematical Model, Mitre Corporation, Bedford, Mass. 01730, January 1973.

Application-Oriented Scheduling in the KNOWLEDGE GRID: A Model and Architecture

Andrea Pugliese[1,2] and Domenico Talia[1]

[1] DEIS-Università della Calabria, Rende, Italy
[2] ICAR-CNR, Rende, Italy
{apugliese,talia}@deis.unical.it

Abstract. This paper describes a model and an architecture of an application-oriented scheduler designed in the context of the KNOWLEDGE GRID: an environment developed for supporting knowledge discovery on Grids. The KNOWLEDGE GRID scheduler allows abstracting from computational resources through the use of *abstract* hosts, i.e. hosts whose characteristics are only partially specified, and instantiates abstract hosts to concrete ones trying to improve applications' performances. The paper proposes a scheduling model, an open architecture for its support, and the specific scheduling functionalities used in the KNOWLEDGE GRID. The proposed model and architecture are general with respect to possible specific application domains and scheduling functionalities. Thus, they are potentially useful also in different Grid environments.

1 Introduction

Grids are geographically distributed computational platforms composed of heterogeneous machines that users can access via a single interface, providing common resource-access technology and operational services across widely distributed and dynamic *virtual organizations*, i.e. institutions or individuals that share resources [6,7]. Resources are generally meant as reusable entities employable to execute applications, and comprise processing units, secondary storage, network links, software, data, special-purpose devices, etc.

The term *Grid Resource Management(GRM)* [3,8,11] specifies a set of functionalities provided by Grid computing systems for managing a pool of available resources. As Grid resources often happen to be very heterogeneous with respect to their owners, providers, access policies, performances, and costs, resource management systems must take into account such heterogeneity (and possibly leverage it) to make resources exploitable at best, also providing support to performance optimization. The main requirements of GRM systems are adaptability to different domains, scalability and reliability, fault tolerance, and adoption of dynamic information monitoring and storing schemes. The *Global Grid Forum* is currently working on the definition of a general service-based resource management architecture for Grid environments, providing services for

(i) maintaining static and dynamic information about data and software components, network resources, and corresponding reservations; (ii) monitoring jobs during their execution; (iii) associating resource reservations with costs; (iv) assigning jobs to resources. Each of these services may have a different internal organization (centralized, cellular, hierarchical, peer-to-peer, etc.).

We refer to *jobs* as to the basic building blocks of Grid applications (Figure 1). Applications are composed of jobs, and jobs are coordinated by GRM systems by looking at applications' structures and available resources. A job, when executed on a certain host, may generate a number of local processes there, on which GRM systems have generally very little control (if any). The same control limitation exists over other resources, such as locally-managed networks, secondary storage devices, etc.

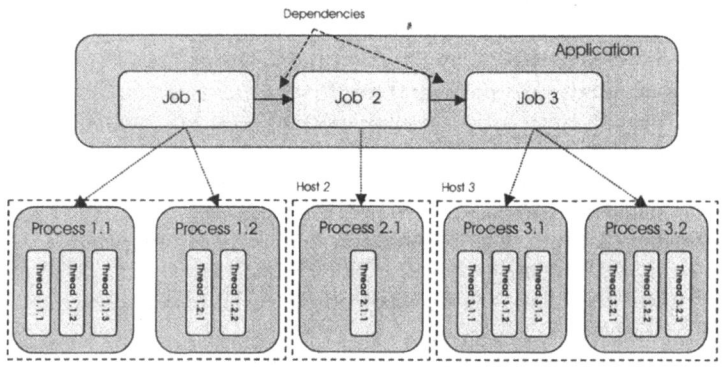

Fig. 1. Applications, jobs, and local processes

In order to provide sufficiently reliable information about resource availability and performance, *Service Level Agreements (SLAs)* may be made among the organizations participating in a Grid. Resource reservations and Quality-of-Service guarantees are being pointed out as a major requirement of modern GRM systems [6]; in the absence of SLAs, several approaches have been proposed so far [2,13,15] to the assessment of availability and performance of resources.

One of the most important features a GRM system should provide is the capability to "suitably" assign jobs to resources; in fact, in realistic Grid applications, it may generally be infeasible for users to find and specify all the needed resources during the composition of applications. This feature is typically provided by *schedulers*, i.e. sets of software components or services which, on the basis of knowledge or prediction about computational and I/O costs, generate *schedules*, i.e. assignments of jobs to resources along with their timing constraints, trying also to maximize a certain performance function.

In this paper we describe the scheduler developed in the context of the KNOWLEDGE GRID [4]. The main objectives of the scheduler are:

Abstraction from computational resources. By allowing the use of *abstract* hosts, i.e. hosts whose characteristics are only partially specified, and that can be matched to different concrete ones, users are allowed to disregard too low-level execution-related aspects, and to concentrate more on the structure of their applications.

Applications performance improvement. Given a set of available hosts, schedules are generated trying to minimize the application completion time.

The KNOWLEDGE GRID scheduler is *application-oriented*, as its performance function is related to individual applications. Furthermore, the scheduler is *application-level*, as it coordinates the execution of jobs looking at the entire structure of applications. To describe our scheduling approach under all the above-mentioned aspects, we first provide a *scheduling model* that comprises:

- A *resource model*, describing characteristics, performances, availability and cost of resources. Important requirements of this model are the timeframe-specificity of predictions, the use of dynamically-updated information, and the capability to adapt to changes.
- A *programming model* provided to the user for building applications [9]. This model may employ dataflow or workflow formalisms, mainly based on job dependency graphs, or domain-specific formalisms.
- A *performance metric* for measuring the performances to be improved.
- A *scheduling policy*, comprising (*i*) a *scheduling process*, i.e. the sequence of actions to be taken in coincidence with particular events, and (*ii*) a *scheduling algorithm*, defining the way in which jobs are assigned to resources. The main requirement of the scheduling policy is obviously a good effectiveness/efficiency tradeoff; the scheduling overhead must not overcome its performance benefits.

The rest of the paper is organized as follows. Section 2 briefly recalls the KNOWLEDGE GRID architecture, then details the scheduling model adopted. Section 3 describes the architecture of the scheduler. Section 4 provides a characterization of the scheduling problem to be dealt with, and discusses our approach to its solution. Finally, Section 5 outlines current and future work.

2 Scheduling Model

The KNOWLEDGE GRID (*K-Grid*) is an architecture built atop basic Grid middleware services that defines more specific knowledge discovery services [4] (Figure 2). The KNOWLEDGE GRID services are organized in two hierarchic levels: the *Core K-Grid layer* and the *High-level K-Grid layer*. The Core K-Grid layer offers the basic services for the definition, composition and execution of a distributed knowledge discovery computation over the Grid. The Core K-Grid layer comprises two main services: the *Knowledge Directory Service (KDS)* and the *Resource Allocation and Execution Management Service (RAEMS)*. The KDS is responsible for maintaining metadata describing KNOWLEDGE GRID resources. The RAEMS comprises three main modules: a *Scheduler*, an *Execution Manager* and a *Job Monitor*. The Scheduler is used to find a suitable mapping between an *abstract* execution plan, that is a plan comprising incompletely-specified jobs,

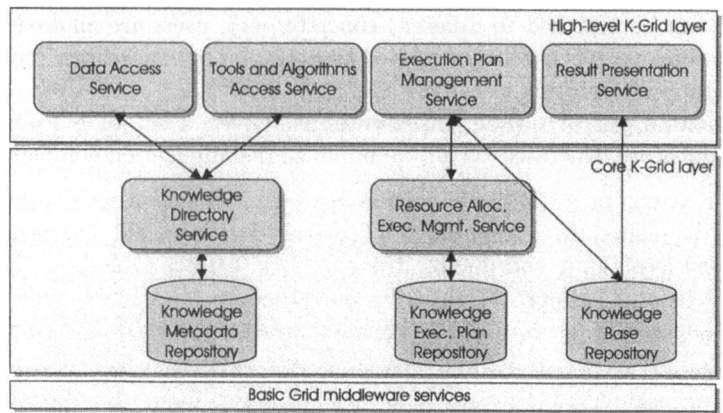

Fig. 2. The KNOWLEDGE GRID architecture

and available resources, with the goals of (*i*) satisfying the constraints (computing power, storage, memory, network performance) imposed by the execution plan, and (*ii*) minimizing its completion time. The output of this process is an *instantiated* execution plan, that is a plan with completely-specified jobs only. This plan is then managed by the Execution Manager, that translates it into requests to basic Grid services, submits these requests and, after their execution, stores knowledge results in the *Knowledge Base Repository (KBR)*. Finally, the Job Monitor follows the execution of submitted jobs, and notifies the Scheduler about significant events occurred. Note that, in the following, we shall mean equivalently the terms *application* and *execution plan*. The High-level K-Grid layer includes services used to compose, validate, and execute a parallel and distributed knowledge discovery computation. A detailed description of the High-level K-Grid layer can be found in [4]. In the following, the term KNOWLEDGE GRID *node* will denote a node implementing the KNOWLEDGE GRID services.

As seen above, the KNOWLEDGE GRID scheduling model adopts the application completion time as its performance metric. The scheduling process used is *dynamic with re-scheduling*, i.e. the Scheduler is invoked initially and then, during applications' executions, it is invoked again as a consequence of significant events occurred, to re-schedule unexecuted parts of the application. Scheduling algorithms are therefore allowed to produce *partial* outputs, i.e. comprising unassigned jobs (called *pending*), to be scheduled subsequently. No other assumption is made on the use of a particular algorithm.

2.1 Resource and Programming Models

The resource model provides simple abstractions for describing characteristics and performances of hosts, datasets, software components and network links. A *host* is modeled as a pair ⟨ *hostID, description* ⟩, where $description \subseteq \mathcal{P}$ is a set of attribute/value pairs describing e.g. the host's location, processor (e.g., model, clock, free percentage, count), operating system, memory (e.g., total RAM, available RAM, total secondary storage, free secondary storage). Obviously, $\mathcal{P}$ is the set of all possible (*attribute*, *value*) pairs. An *abstract* host is a partially-specified one, i.e. some of the values in its description are left null or indicate minimal requirements. A *dataset* is a triple ⟨*datasetID, description, hostID*⟩, where *description* $\subseteq \mathcal{P}$ provides information about the modalities for accessing the dataset (location, size etc.), its logical and/or physical structure, etc.; *hostID* is the identifier of the host at which the dataset resides. Finally, a *software component* is a triple ⟨*softwareID, description, hostID*⟩, where *description* $\subseteq \mathcal{P}$ includes the kind of data sources the software works on, the kind of knowledge that is to be discovered etc.; *hostID* identifies the host offering the software component.

Resource performances are modeled using three estimating functions (where $\mathcal{D}$, $\mathcal{S}$ and $\mathcal{H}$ are sets of available datasets, software components and hosts, respectively): (*i*) $\epsilon : 2^{\mathcal{D}} \times \mathcal{S} \times \mathcal{P} \times \mathcal{H} \times \mathbb{N} \to \mathbb{N}$ estimates computation times, i.e. associates to a quintuple ⟨*ds, s, p, h, t*⟩ the time needed to execute software *s* on host *h* with input datasets *ds* and parameters *p*, starting at time *t*; (*ii*) $\nu : \mathcal{H} \times \mathcal{H} \times \mathbb{N} \times \mathbb{N} \to \mathbb{N}$ estimates communication times, i.e. associates to a quadruple ⟨h_s, h_d, d_s, t⟩ the time needed to transfer a dataset of size d_s from host h_s to host h_d, starting at time *t*; (*iii*) $\omega : 2^{\mathcal{D}} \times \mathcal{S} \times \mathcal{P} \to \mathbb{N}$ estimates output sizes, i.e. associates to a triple ⟨*ds, s, p*⟩ the size of the output produced by software *s* when executed, with parameters *p*, on input datasets *ds*.

The programming model views *jobs* as a quintuples of the form ⟨*jobID, datasetIDs, softwareID, parameters, hostID*⟩, where *datasetIDs* is the set of input datasets' identifiers, *softwareID* identifies the software component to be executed, and *parameters* $\subseteq \mathcal{P}$ is a set of attribute-value pairs defining the parameters of the particular execution. *hostID* identifies (*i*) in the case of computational jobs, the host at which the job is to be executed, and (*ii*) in the case of data movement jobs, e.g. those using file transfer protocols, the host towards which a dataset (or software component) is to be moved. It should be noted that, with this definition, *hostID* always identifies the host at which an output is produced. Note also that in the case of data movement jobs, a *datasetID* can actually be a *softwareID*. A job is said to be *abstract* if its *hostID* is null or it refers to an abstract host.

Finally, an *application* is a job precedence DAG $G = (\mathcal{J}, \mathcal{A})$ where $\mathcal{J}$ is the set of jobs, and $\mathcal{A} \subseteq \mathcal{J} \times \mathcal{J}$ is the set of arcs, each arc $(j_i, j_k) \in \mathcal{A}$ dictating that job j_k uses the output of job j_i, so its start can not be scheduled before the completion of j_i (and after other possibly needed data movement jobs). An application is said to be *abstract* if it contains at least one abstract job; abstract applications need to be completely instantiated before being really executed, so they are the target of the scheduling activity.

In the remainder, we shall adopt simple array notations, denoting e.g. with $s[h]$ the host offering software *s*, with $j[d_k[h]]$ the host at which the *k*-th input dataset of job *j* resides, and so on.

3 The Knowledge Grid Scheduler

As seen above, the KNOWLEDGE GRID Scheduler is part of the Resource Allocation and Execution Management Service (RAEMS). Thus, each KNOWLEDGE GRID node has its own Scheduler which is responsible for responding to scheduling requests coming from the same or other nodes (Fig. 3). For each execution

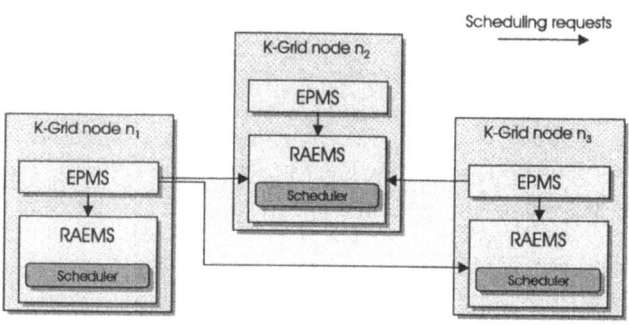

Fig. 3. Possible scheduling-related interactions among different KNOWLEDGE GRID nodes

plan to be scheduled, the Scheduler performs the following logical steps:

1. examines the structure of the execution plan and the jobs in it;
2. gathers information about features and performance of available resources;
3. selects resources usable to execute jobs;
4. evaluates a certain set of possible schedules and chooses the one which minimizes the completion time;
5. orders the execution of jobs;
6. adapts the schedule to new information about job status and available resources.

3.1 Scheduler Architecture

The architecture of the Scheduler, currently being implemented using Java, comprises three main components: a *Mapper*, a *Cost/Size Estimator (CSE)*, and a *Controller* (Fig. 4).

Mapper. The Mapper computes schedules from abstract execution plans employing some scheduling algorithm and making use of resource descriptions and computational and I/O cost evaluations for making its decisions.

Cost/Size Estimator. The Cost/Size Estimator builds the three estimation functions ϵ, ν, and ω, and makes them available to the Mapper. The CSE comprises *data gathering* modules that collect dynamic information about current and future availability and performance of resources and update the KNOWLEDGE GRID KDS accordingly. *Estimation* modules deal instead

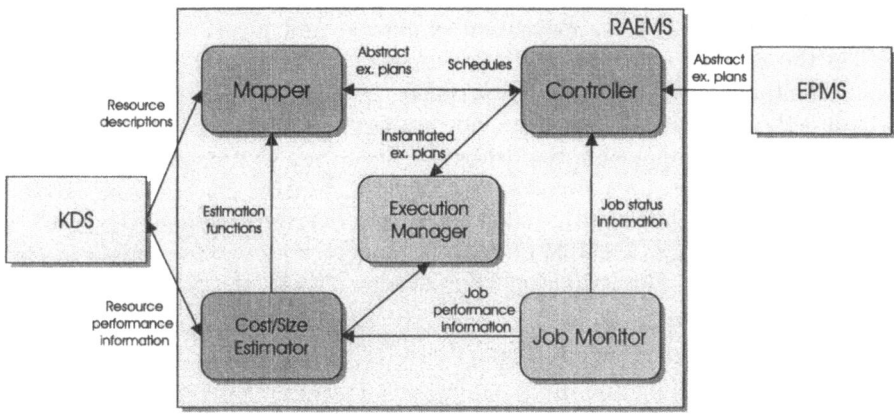

Fig. 4. Information flows among Scheduler modules (dark gray), other RAEMS modules (light gray), and other KNOWLEDGE GRID modules

with the actual construction of the estimation functions on the basis of the perceived status of resources w.r.t. time; this process is generally based on task profiling and analytical benchmarking [12].

Controller. The Controller guides the scheduling activity. It receives abstract execution plans from the EPMS, requests the corresponding schedules to the Mapper, and orders the execution of scheduled jobs to the Execution Manager. During job execution, the Controller also receives notifications about significant events, and re-schedules unexecuted parts of the application.

All of the Scheduler modules are extensible, as they provide an open Java interface allowing to "plug-in" user-defined functionalities and behaviors. The Scheduler can load Java classes implementing (i) scheduling algorithms (in the Mapper), (ii) scheduling processes (in the Controller), and (iii) data gathering and cost estimation activities (in the Cost/Size Estimator). Note that both the scheduling model and the architecture we propose are general with respect to possible specific application domains and scheduling functionalities; this makes them potentially useful also in Grid frameworks different from the KNOWLEDGE GRID.

3.2 Built-in Functionalities

The Scheduler currently under implementation provides the following built-in functionalities:

Scheduling algorithm. Several choosable algorithms are provided (described in the next section), all potentially producing partial schedules.

Scheduling process. The events that fire the re-scheduling activity are: (i) the completion of all jobs preceding a pending job in the current schedule; (ii) important performance variations; (iii) job/host failures.

Data gathering. For the evaluation of current and future CPU availability, in the absence of service level agreements, the Scheduler adopts as its information source the *Network Weather Service (NWS)* [15], if present, or directly the hosts' descriptions. For evaluating the processing requirements of software components, the Scheduler makes use of user-provided descriptions, if present, or a *sampling method* [12]. Finally, for assessing network performances (bandwidth and latency) the Scheduler adopts the NWS as well, if present, or a simple probing technique.

Cost estimation. For estimating computational costs, the Scheduler employs the formula $\epsilon(ds, s, p, h, t) = \frac{req(ds, s, p)}{perf(h) \cdot avail(h, t)}$, where $req(ds, s, p)$ represents the processing requirements (with respect to a reference host) of software s run with parameters p on datasets ds, $perf(h)$ is the no-load performance of host h, and $avail(h, t)$ is the fraction of processing cycles available on host h at time t. Finally, the Scheduler makes use of user-provided descriptions of the relationships between input and output sizes of software components.

4 Tackling the Scheduling Problem

In this section we characterize the scheduling problem the Mapper must deal with, then outline possible algorithms to build satisfactory solutions.

4.1 Problem Formalization

We model the constraints on the assignability of concrete resources to jobs, implicitly described by resource descriptions, through a *resource compatibility function* $\gamma : \mathcal{D} \times \mathcal{S} \times \mathcal{P} \times \mathcal{H} \to \{true, false\}$, where $\gamma(d, s, p, h) = true$ if software s can execute on host h with input dataset d and parameters p.

The Mapper's input consists of three sets $\mathcal{D}$, $\mathcal{S}$, and $\mathcal{H}$, containing datasets, software components, and hosts, three functions ϵ, ω and ν, a set $\mathcal{J} = \mathcal{J}_c \cup \mathcal{J}_m$ of jobs, where $\mathcal{J}_c$ is the set of computational jobs and $\mathcal{J}_m$ is the set of data movement jobs, and a job precedence DAG $G = (\mathcal{J}, \mathcal{A})$, as defined in Section 2.1. The Mapper's output consists of three new sets $\mathcal{D}'$, $\mathcal{S}'$, and $\mathcal{J}'$ of datasets, software components, and jobs and a *timing function* $\sigma_T : \mathcal{J}' \to \mathbb{N}$ associating each job with the time at which it must be started. Function σ_T must be such that (i) precedence constraints are satisfied, i.e., $(j_i, j_k) \in \mathcal{A} \Rightarrow \sigma_T(j'_i) + \hat{t}_{j'_i} + \hat{t}_{[j'_i, j'_k]} < \sigma_T(j'_k)$, where $\hat{t}_{[j'_i, j'_k]}$ is the duration of a data movement job between j'_i and j'_k (if needed); (ii) resource constraints are satisfied, i.e., $\forall j' \in \mathcal{J}', \gamma(j'[d], j'[s], j'[p], j'[h]) = true$; (iii) the overall completion time $\mathcal{T} = \max_{j' \in \mathcal{J}'}(\sigma_T(j') + \hat{t}_{j'})$ is as low as possible.

The Mapper deals with a very challenging problem. The search space, even disregarding possible variations in jobs' starting times, consists in a point for each possible assignment of hosts to abstract computational jobs that does not violate resource constraints, so it is exponential in the number of abstract computational jobs. Classical schedulers for distributed environments make strong assumptions about the underlying system. Typically, they assume to be in control of an entire completely-connected invariant resource pool, composed of processing units and network links having very similar (even identical) performances. Even under these assumptions, the resulting problem (called *precedence-constrained scheduling*) is $\mathcal{NP}$-hard if more than one host is considered. Moreover, as none of the above assumptions is realistic in the inherently more heterogeneous Grid environments, our problem is more general than classical precedence-constrained scheduling. Therefore, exact optimal techniques, such as integer-linear or constraint programming, are rarely usable as they incur in an exponential duration of the scheduling process.

More suitable approaches tackle the problem heuristically. The *NetSolve* system [1] employs a simple algorithm for scheduling independent jobs *online*, i.e. as soon as they are submitted to a system. Other approaches to scheduling independent jobs are presented in [12]. The main limitation of such approaches is that they disregard dependencies among jobs; in fact, looking only at the jobs that are ready to be run at scheduling time may produce performance losses for subsequent jobs. In the context of the *GrADS* project [5] an interesting approach is presented that groups together hosts belonging to the same organizational domain (therefore close to one another in terms of "network distance") and schedules entire applications to host groups. [14] discusses heuristics for scheduling job precedence DAGs in the so-called *online* and *batch* modes, i.e. looking at single jobs or entire subparts of the input DAG. Finally, in [10] natural heuristics are applied to workflow instances.

4.2 Algorithms

The scheduling algorithms currently provided in the Mapper share a preprocessing phase, whose objective is that of reducing the size of the search space. This phase comprises the following steps:

1. The input DAG is reduced by setting as pending the jobs whose input size is not known neither it can be evaluated using the ω function. The Mapper makes this reduction essentially because we deal with data-intensive applications, thus the choice of hosts on which to schedule jobs cannot leave aside the associated communication costs.
2. The input DAG is still reduced to comprise only *entry* jobs (i.e. jobs that have all inputs ready at scheduling time) and jobs that depend on them up to a certain depth p.
3. In the spirit of [5,10], by looking at the variations of ϵ w.r.t. its variable h, and of ν w.r.t. h_s and h_d (as these variations can be seen as indicators of machine and network heterogeneity, respectively), groups of similar hosts are formed through a simple clustering algorithm.
4. Finally, for each abstract job, the set of possible hosts on which that job can be scheduled is built, by looking at resource descriptions, in order to explore only feasible assignments.

After having performed the pre-processing phase, the Mapper can employ one of the following search procedures:
Exhaustive search. This procedure simply builds all possible solutions, then chooses the one minimizing the application completion time.
Simulated annealing. This procedure is based on the classical simulated annealing heuristic that, starting from random points in the search space, makes moves toward less costly points, sometimes also allowing moves toward more costly points with a probability that decreases for each iteration, thus mimicking the natural annealing phenomenon.
Min-Min, Max-Min, and Sufferage. This procedures are based on classical scheduling heuristics. *Min-Min*, at each step, evaluates the earliest completion time of each job over all compatible host, then chooses the job-host assignment having the minimum earliest completion time. The set of considered jobs comprises all the jobs ready to be run. *Max-Min* chooses instead the job-host assignment having the maximum earliest completion time. Finally, *Sufferage* assigns the host incurring in the minimum earliest completion time to the job that would "suffer" most of a suboptimal choice, that is the job having the largest difference between best and second best completion times.

5 Current and Future Works

We have planned a thorough experimental evaluation of the KNOWLEDGE GRID Scheduler. We will consider different approaches to cost estimation, and different scheduling algorithms, in order to assess their performance with respect to classical measures, such as distance from the optimal solution and *turnaround* time (i.e. total execution time, comprising the scheduling activity itself), but also in terms of other properties which are particularly desirable in Grid environments, such as robustness w.r.t. unpredictable changes and stability. Finally, we believe that several interesting extensions are worth investigating; these extensions comprise: (*i*) the possibility to define abstract software and datasets by partially specifying their description as well; (*ii*) the use of an ontological organization of resource descriptions; (*iii*) the definition of an *Open Grid Services Architecture (OGSA)*-compliant interface of the Scheduler, compatible with the Grid management architecture currently being defined by the Global Grid Forum.

Acknowledgements. This work has been partially supported by the "FIRB Grid.it" project funded by MIUR.

References

1. D.C. Arnold, H. Casanova, J. Dongarra. Innovations of the NetSolve Grid computing system. *Concurrency and computation: practice and experience*, 14, 2002.
2. Z. Balaton, P. Kacsuk, N. Podhorszki, F. Vajda. Comparison of Representative Grid Monitoring Tools. Tech. rep., Hungarian Academy of Sciences, 2000.

3. R. Buyya, S. Chapin, D. Di Nucci. Architectural models for resource management on the Grid. *First IEEE/ACM International Workshop on Grid Computing*, 2000.
4. M. Cannataro, D. Talia. The Knowledge Grid. *Communications of the ACM*, vol. 46, no. 1, pp. 89-93, 2003.
5. H. Dail, H. Casanova, F. Berman. A modular scheduling approach for Grid application development environments. Submitted manuscript.
6. Global Grid Forum. *http://www.ggf.org/*
7. The Globus Project. *http://www.globus.org/*.
8. K. Krauter, R. Buyya, M. Maheswaran. A taxonomy and survey of Grid resource management systems for distributed computing. *International Journal of Software: Practice and Experience*, Volume 32, Issue 2, 2002.
9. C. Lee, D. Talia. Grid programming models: current tools, issues, and directions. In: F. Berman, G.C. Fox, A.J.G. Hey, "Grid computing – making the global infrastructure a reality", 2003.
10. M. Mika, G. Waligora, J. Weglarz. A metaheuristic approach to scheduling workflow jobs on a Grid. In [11].
11. J. Nabrzyski, J. Schopf, J. Weglarz (Eds.). *Grid Resource Management*, Kluwer, 2003.
12. S. Orlando, P. Palmerini, R. Perego, F. Silvestri. Scheduling high-performance data mining tasks on a data Grid environment. *Europar Conf.*, 2002.
13. N. Sample, P. Keyani, G. Wiederhold. Scheduling under uncertainty: planning for the ubiquitous Grid. *COORDINATION 2002*, LNCS 2315, 2002.
14. H.J. Siegel, S. Ali. Techniques for mapping tasks to machines in heterogeneous computing systems. *Journal of Systems Architecture*, 46, 2000.
15. R. Wolski, N. Spring, J. Hayes. The network weather service: a distributed resource performance forecasting service for metacomputing. *Future Generation Computer Systems*, Vol. 15, 1999.

A Monitoring and Prediction Tool for Time-Constraint Grid Application

Abdulla Othman, Karim Djemame, and Iain Gourlay

School of Computing
University of Leeds
Leeds, LS2 9JT, UK
{othman, karim, iain}@comp.leeds.ac.uk

Abstract. A Grid system must integrate heterogeneous resources with varying quality and availability. For example, the load on any given resource may increase during execution of a time-constrained job. This places importance on the system's ability to recognise the state of these resources. This paper presents an approach used as a basis for applications and resources monitoring, in which Grid jobs are maintained at runtime. A reflective technique is used to simplify the monitoring of the Grid application. The monitoring tool is described and experimentally evaluated. Reflection is incorporated into the monitoring to separate functional and non-functional aspects of the system and facilitate the implementation of non-functional properties such as job migration. Results indicate that this approach enhances the likelihood of timely job completion in a dynamic Grid system.

1 Introduction

Grid computing infrastructures offer a wide range of distributed resources to applications [3]. Moreover a Grid system must integrate heterogeneous resources with varying quality and availability. This places importance on the system's ability to predict and monitor the state of these resources. The Grid is a dynamic system where resources are subjected to changes due to system performance degradation, new resources, system failure, etc.

To support application execution in the context of the Grid, a Grid Monitoring and Prediction Tool (MPT) is desirable. MPT detects when resource behaviour does not fall within an expected range (e.g., satisfying some user-specified requirements), helps identify the cause of the problem, and generates feedback so that other agents (e.g. broker) in the runtime system may take corrective action (e.g., rescheduling of the application on a different set of resources).

Traditional monitoring methods often assume that a dedicated set of resources will be available for the entire application run. Specifically, the resources are monitored rather than the application. For example, the Network Weather Service (NWS) [9] enables the user to obtain information about CPU percentage availability, rather than the CPU usage of a particular application. This information could be used to draw conclusions relating to application performance, if it is

1. Collected prior to run-time to allow the user to estimate likely resource availability, OR
2. The user knows that the resource is dedicated to their application.

However, in a Grid environment, resources are shared among many users, meaning that resource availability can change at any time during the application execution lifetime. The approach described in this paper differs from traditional methods in that it monitors the application during run-time and does not rely on a dedicated set of resources being available for the entire application run. It is often not possible to reserve resources or make performance guarantees because of the dynamic nature of the Grid.

This paper addresses the important issue of run-time monitoring and prediction (e.g. CPU usage of a particular job and predicted job completion time). A solution is presented, in which a reflective technique is used to simplify application monitoring in the Grid. The reflective technique aims to separate concerns for functional and non-functional behaviour of a program, in some code that is highly reusable and extensible [1]. In this paper, the aim is to address the concerns relating to application monitoring and prediction without affecting existing user applications.

An experiment was designed to enable the investigation of the monitoring tool using a reflective technique [2]. The experiment involves a time-constrained application with requirements specified by the user. An important issue is how information collected by the MPT should be used. The paper discusses a possible approach, to predicting application completion time, based on run-time CPU usage, for applications where the user is able to approximate the anticipated execution time. This is used to support the experiments carried out.

The paper is structured as follows. In Section 2, related work is discussed and its relevance to the work presented in this paper indicated. Section 3 discusses the design and implementation PMT, followed by experiments and results in section 4. The paper concludes with a summary and discussion of future work

2 Related Work

The Grid Monitoring Architecture working group at the Global Grid Forum is focused on producing a high-level architecture statement of the components and interfaces needed to promote interoperability between heterogeneous monitoring systems on the Grid [10].

The majority of existing monitoring systems, such as Gloperf [11] can collect data from distributed systems for analysing them through their specific tools. However, these monitoring systems cannot serve as data collection components for other tools and applications that may wish to use this information. Some of the existing monitoring systems do support external tools or applications for fault detection, resource scheduling, and QoS management. The NWS [9] measures available network bandwidth and system load to predict their future states. Future state predictions can be used by a scheduler to reserve and allocate these resources in an efficient manner. Several Grid monitoring systems are currently available. These systems include

Autopilot [13] and JAMM (Java Agents for Monitoring and Management) [12]. An infrastructure to monitor resources, services, and applications that make up the NASA IPG (Information Power Grid) and to detect resource usage patterns, faults, or deviations from required QoS levels is presented in.

Kapadia, et al [9] show that local learning algorithms can be successfully used to predict application performance and resource usage, given application input parameters and historical resource usage data.

Due to the heterogeneity of resources in the Grid, this paper seeks to separate the influences of the run-time system from the behaviour of the application. Similar to [9] this paper uses information about application input parameters (e.g. application requirements) when predicting application performance. However, rather than relying on application performance data from pervious runs on the same system, the approach combines the application fundamental knowledge with run-time prediction of resources availability from systems such as NWS [8] to derive acceptable application performance ranges.

3 Design of MPT

3.1 MPT through Reflection

MPT is implemented in such a way as to isolate the user from the complexities of the system. In particular, the user is not obliged to alter his/her code. This is achieved using reflection. A reflective system, as defined by [4], is one that is capable of reasoning about itself (besides reasoning about the application domain). The benefits of using reflection are [2]:
- Flexibility to customize policies dynamically to suit run-time environment.
- High-level transparency to applications.

Reflection is used to bind the MPT to the application object. This produces a reflected object and ensures a clean separation between the application and the MPT. Hence the MPT is transparent to the application.

3.2 MPT Architecture

The reflected object (consisting of user application bound to the MPT) can be submitted to an appropriate middleware component, e.g. a resource broker or scheduler, which then submits to appropriate Grid resources. For convenience, it is assumed here that a broker is used to perform this function. The MPT runs within the user application to monitor the application usage of resources required and predict when the application will finish. The MPT also ensures the application will be saved from where it stopped and signals the broker to initiate a migration.

The MPT consists of a Controller, Monitor, Check-pointer, Signaler and a Requirements Checker. The Controller links the other components in the MPT together. It is also responsible for deciding when to save the job and send a signal to the broker. Once the job begins execution, the Monitor collects dynamic data that varies during run-time from the resources. Specifically, the data of interest is that which describes the resource usage of the application (e.g. CPU usage of the application). This data is reported to the Controller. The Controller communicates with the Requirements Checker to ensure that the user requirements continue to be satisfied. If a violation of user requirements occurs, then the Controller is notified. The Controller then sends a request to the Check-pointer to save the job attributes, so that when the job is rescheduled to run on a different resource it starts from where it was stopped. The Controller then communicates with the broker to initiate a migration of the job to another resource. The MPT architecture is shown in Figure 1.

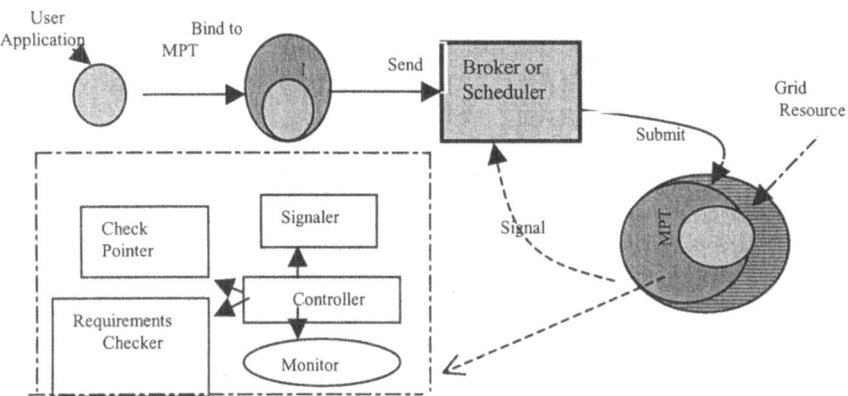

Fig. 1. MPT Architecture

4 System Implementation

4.1 Implementation of Reflection

The reflective technique is implemented using OpenJava [6, 7]. OpenJava enables programmers to handle source code as object oriented constructs through its class MetaObject API. OpenJava makes it easier to get information about methods, add methods and modify methods [5]. This performs the role of the binding which is performed at compile-time.

The OpenJava code creates a reflective object consisting of the application bound to the monitor tool components.

4.2 Prediction Modelling

This section discusses the prediction model that is used, during run-time to determine whether a pre-specified time constraint is expected to be satisfied. This model supports the experiments discussed in section 4.

Suppose the user can supply information, enabling an estimate of execution time to be obtained for any resource with a fixed CPU usage. In this case, a prediction model can be used to provide a dynamic estimate of the expected job completion time. The prediction is based on the pattern of CPU usage during run-time. The Run-Time Prediction (RTP) formula (see equation (1)) is used to determine whether a job is expected to complete execution within a time that is specified beforehand by the user.

Clearly, if the user has no knowledge of the job execution time, then the prediction model cannot ensure that the job runs within a specified time. In this case, the application monitoring is still useful. However the user requirements would need to be specified by means of different parameters, e.g. the user could require a minimum of 75% CPU usage for the entire duration of the job.

The equation below can best be described by considering the scenario depicted in Figure 2. The fractional CPU usage is measured periodically, at times T_0, T_1, etc. The fractional CPU usage at time T_i is labelled F_i, with 1 representing 100% CPU usage and 0 corresponding to no CPU usage. If n samples have been taken, then these can be used to predict the remaining time to completion of the job ($T_{remaining}$):

$$T_{remaining} = \frac{T_{100\%} - \sum_{i=1}^{n-1} T_i F_i}{F_{estimate}} \quad (1)$$

Here $T_{100\%}$ is the time the job would take to execute with 100% CPU usage. A value for $F_{estimate}$ can be obtained using

$$F_{estimate} = \overline{F_n} = \frac{1}{n}\sum_{i=1}^{n} F_i \quad (2)$$

Here, $F_{estimate}$ is the mean value of the fractional CPU usage over n samples. The estimate for $T_{remaining}$ obtained by using the RTP equation corresponds to the time the job would take to complete if the mean value of the CPU usage after the prediction (until the job is completed) is equal to the mean value calculated prior to the prediction. This assumption is inadequate if the CPU usage drops dramatically during the job execution and remains at a much lower value for a large proportion of the remaining running time. This could happen if, for example, another user submits a job with an execution time of the same order of magnitude or if there is a fault in the resource. In order to account for this, a worst-case prediction of the remaining time ($T_{worst-case}$) can be used. This estimate is obtained by replacing $\overline{F_n}$ with a worst-case estimate of the expected CPU usage. Details on the practical implementation of this are given in section 6.

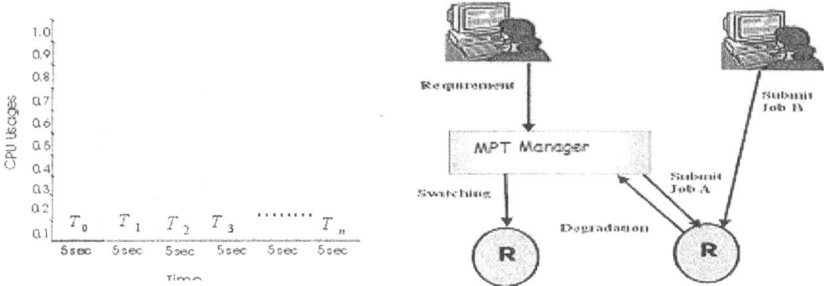

Fig. 2. Periodic Sampling of CPU Usage

Fig. 3. MPT Scenario

To illustrate this, consider the following scenario (Figure 3). The user specifies the job requirements (e.g. the time needed to execute the job). This will be referred to as job A. After the resources have been located and the job has been dispatched to the selected resources, the MPT monitors the resources and ensures the requirements are satisfied.

Suppose another user accesses a resource being used for job A and begins running another job (B). Since job B uses some portion of the resource, job A takes longer to finish. In this case, the monitor passes information to the MPT manager, based on the RTP formula and the MPT Manager initiates check pointing to ensure the job continues to run on resources that meet the minimum job requirements.

5 Experiments and Results

The experiments presented in this section involve the submission of jobs, with user requirements specified by the user, to a simple broker designed for this experiment. The function of this broker is to locate the resources according to user requirements and to interpret any signal from the MPT requesting that it locates alternative resources.

While a job is running, other jobs may be submitted to the same resource(s). The results obtained are compared to the case where the MPT is not used. In particular, the experiments address the following questions:

- When job requirements are not met, are jobs being successfully migrated?
- Does this result in shorter job execution time, compared to the case when the MPT is not used?

The experiments ran on a Grid test-bed consisting of 20 machines. The set of resources are labelled 1 to 20. The machines have different computing power, ranging from Pentium IV 1.2 GHZ to 2.4 GHZ with RAM from 256 to 512 MB. The operating system is Linux 2.4. All machines have Globus 2.4. Communication between resources is via a fast LAN Ethernet. Note that resources in a Grid environment are typically geographically distributed and communicate over a WAN. Hence the experiments discussed here do not address the issue of job migration times and their effect on the overall job execution time. This will be considered in future experiments.

Prior to running the experiments, the user's job is run and the CPU usage periodically measured. This is done so that a value for $T_{100\%}$ (see the RTP equation in section 4) can be obtained. Specifically, this is to enable the experiments to run with the assumption that the user knows how long the job would take to run with 100% (or some other percentage value, enabling $T_{100\%}$ to be calculated) CPU usage.

Two experiments are then run:
1. The job is executed from start to finish and the remaining job time predicted by the RTP formula during the course of the computation is compared to the actual execution time.
2. The job is executed and during execution, other jobs are submitted to the same resource. This initiates a migration. This is compared to case where MPT is not used.

Experiment 1 is used to assess the validity of the prediction formula. The second experiment is used to assess the effectiveness of MPT in ensuring timely job completion.

In experiment 2, the job is migrated under the following condition:

1. $T_{remaining} \geq T_{user} - T_{elapsed}$
   OR
2. $\overline{F}$ decreases for 10 successive samples and
   $T_{worst-case} \geq T_{user} - T_{elapsed}$

Here T_{user} is the user specified execution time and $T_{elapsed}$ is the time elapsed since the job began execution. The worst-case estimate of $\overline{F}$ is obtained by taking the mean value of the CPU usage over the 10 samples since it began to decrease. Thereafter, $\overline{F}$ is calculated from this point in the computation. Essentially, this is addressing the fact that the CPU usage has dropped and it may remain at its new level for the remainder of the computation. The choice of 10 samples in condition 2 is somewhat arbitrary: the larger this value, the less likely it is that a job is unnecessarily migrated due to a fluctuation in CPU usage, while the smaller this value, the faster the need for migration can be identified. Further investigation is planned to identify an *optimal* number of samples.

Prior to the experiments, the chosen job was run and took 4469 seconds, with a mean CPU usage of 0.98, resulting in a value of 4380 seconds (73 minutes) for $T_{100\%}$.

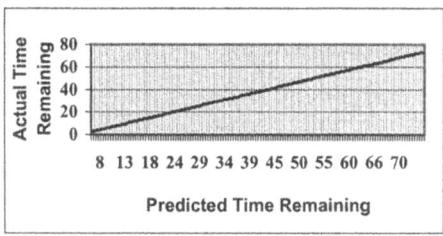

Fig. 4. Comparison of remaining execution time with predicted time remaining

Experiment 1 was carried out, as described above. No additional jobs were submitted to the chosen resource during execution. Figure 4 shows the remaining time predicted by the RTP formula plotted against the actual time remaining until the job completes. The latter is deduced retrospectively once the job has executed and the total run-time is known. The plot shows a good match between the predicted and the actual execution times, which enhances confidence in the validity of the RTP formula.

At the start of experiment 2, the user submits the job requirement. Specifically the job requirement is a time-constraint: the job must be executed within 4867 seconds (81 minutes 7 seconds). This corresponds to the length of time the job would take at 90% CPU usage. This is justified by the need to choose an execution time that is not less than the execution time obtained in the run prior to the experiments. Otherwise the likelihood the user's requirements would be met would be low, even if no other jobs were submitted to the system.

Figures 5 and 6 refer to the job running on resource 3. The job executes normally (i.e. with close to full CPU usage) for about 30 minutes.

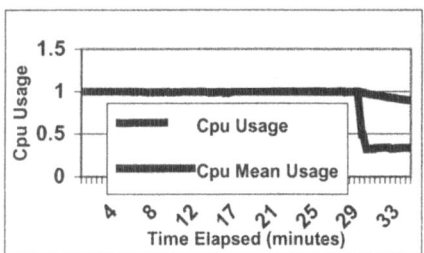 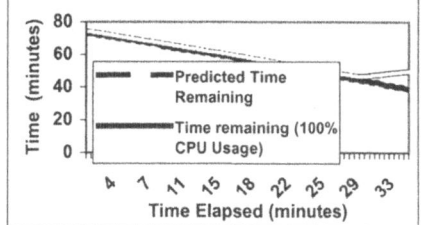

Fig. 5. CPU usage during run-time in resource 3

Fig. 6. Predicted time remaining during run-time in resource 3

As indicated by the dotted line in Figure 6, which shows the predicted remaining time, the remaining time begins to increase after this point. This indicates that the CPU usage has decreased sharply, as confirmed by the data shown in Figure 5. Hence the resource is no longer expected to finish the job within the specified time constraint. This means theis a need to take action and restart the job on another resource.

The broker has chosen resource 4 as shown in Figures 7 and 8. This shows that the job starts from where it stopped on resource 3. 34 minutes have elapsed since the job began execution. Hence, in order to meet the user requirement the job must complete within a further 47 minutes on resource 4.

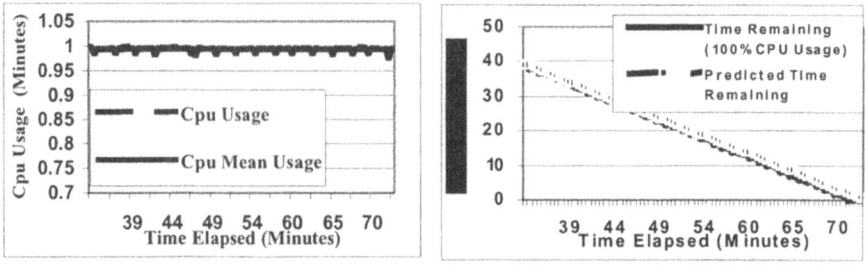

Fig. 7. Predicted time remaining during run-time in resource 4

Fig. 8. CPU usage during run-time in resource 4

Shortly after the job continues execution on resource 4, the RTP formula is used to predict the remaining time. Figure 8 show that the job is expected to finish within the specified time. The total execution time on both resources was 74 minutes. As a consequence of the use of MPT, the job is continues to meet the user requirements.

The same job was executed without the MPT, as shown in Figure 9; in this case the job took longer than the time it was assigned, i.e. it ran for more than 7200 (125 minutes). This results from the fact that there was another job running on the same resource during the job execution.

Referring to the questions posed at the beginning of this section, it has been shown through experiments that the MPT is successfully supporting job migration. This results in a reduction in the job execution time compared to the case where MPT is not used.

6 Conclusions

This paper describes a MPT that enables application monitoring and configuration based on resource characteristics and user preferences. A Reflective technique is proposed, controlling each aspect of an application in a different program. The reflective technique permits run-time mechanisms to automatically decide when and how application behaves in reaction to changes in resource conditions.

We have shown that our MPT is a viable contender for use in future Grid implementations. This is supported by the experimental results obtained on a Grid test-bed.

Future work will focus on developing an interactive job, where the user can change the attributes of the job during run time. In addition, more sophisticated methods for predicting job execution time are to be investigated.

References

1. G. Blair, F. Costa and G. Coulson. *Experiments with Reflective Middleware.* In Proceedings of ECOOP'98 Workshop on Reflective Object Oriented Programming and Systems, Springer Verlag, 1998. Lancaster University. 1999.
2. G. Blair B. and G. Coulson M.. *The Case for Reflective Middleware.* Internal report number MPG-98-38, Distributed Multimedia Research Group. Lancaster University. 1998.
3. I. Foster, C. Kesselman and S. Tuecke. *The Anatomy of the Grid: Enabling Scalable Virtual Organizations.* International Journal of Supercomputer Applications, *15(3), 2001.*
4. P. Meas. *Concepts and Experiments in Computational Reflection.* In Proceedings of OOPSLA'87, pp. 147-155. ACM. October 1987.
5. M. Tatsubori. *OpenJava Web Site.* August 2002;
http://www.csg.is.titech.ac.jp/openjava/
6. M. Tatsubori, S. Chiba, M-O. Killijian and K. Itano, *OpenJava: A Class-Based Macro System for Java.* Reflection and Software Engineering, W. Cazzola, R.J. Stroud, F. Tisato (Eds.), Lecture Notes in Computer Science 1826, pp.117-133, Springer-Verlag, 2000
7. M Tatsubori. *An Extension Mechanism* for the Java Language. Master of Engineering Dissertation, Graduate School of Engineering, University of Tsukuba, Ibaraki, Japan, Feb. 1999.
8. R. Wolski. Forecasting *Network Performance to Support Dynamic Scheduling Using the Network Weather Service.* In Proc. 6th IEEE Symp. on High Performance Distributed Computing. Portland, Oregon, 1997.
9. N. Kapadia, J. Fortes AND C. Brodely. *Predictive Application-Performance Modeling in a Computational Grid Environment.* In *Proceedings of the Eight IEEE Symposium on High-Performance Distributed Computing* (Redondo Beach, California, August 1999), pp. 47–54.
10. Grid Performance and Information Services;
http://www.gridforum.org/4_GP/Perf.htm
11. C. Lee, R. Wolski, I. Foster, C. Kesselman, J. StepanekA Network Performance Tool for Grid Computations.. *In* proceedings of Supercomputing '99, *1999*
12. B. Tierney, B. Crowley, D. Gunter, M. Holding, J. Lee and M Thompson. *A Monitoring Sensor Management System for Grid Environments..* Cluster Computing Journal, Vol 4-1, 2001, Baltzer Science Publications, LBNL-46847.
http://www-didc.lbl.gov/JAMM/
13. R. L. Ribler, J. S. Vetter, H. Simitci, and D. A. Reed. Autopilot: Adaptive control of distributed applications. In *Proc. 7th IEEE Symp. on High Performance Distributed Computing*, Aug 1998.

Optimal Server Allocation in Reconfigurable Clusters with Multiple Job Types

J. Palmer and I. Mitrani

School of Computing Science, University of Newcastle,
NE1 7RU, UK {jennie.palmer,isi.mitrani}@ncl.ac.uk

Abstract. We examine a system where the servers in a cluster may be switched dynamically and preemptively from one kind of work to another. The demand consists of M job types joining separate queues, with different arrival and service characteristics, and also different relative importance represented by appropriate holding costs. The switching of a server from queue i to queue j incurs a cost which may be monetary or may involve a period of unavailability. The optimal switching policy is obtained numerically by solving a dynamic programming equation. Two simple heuristic policies – one static and one dynamic – are evaluated by simulation and are compared to the optimal policy. The dynamic heuristic is shown to perform well over a range of parameters, including changes in demand.

Keywords: Optimal server allocation, Grid computing, Dynamic programming, Heuristic policies.

1 Introduction

This paper is motivated by recent developments in distributed processing, and in particular by the emerging concept of a *Computing Grid*. In a Grid environment, heterogeneous clusters of servers provide a variety of services to widely distributed user communities. Users submit jobs without necessarily knowing, or caring, where they will be executed. The system distributes those jobs among the servers, attempting to make the best possible use of the available resources and provide the best possible quality of service.

The random nature of user demand, and also changes of demand patterns over time, can lead to temporary oversubscription of some services, and underutilization of others. In such situations, it could be advantageous to reallocate servers from one type of provision to another, even at the cost of switching overheads. The question that arises in that context is how to decide whether, and if so when, to perform such reconfigurations.

We consider a system consisting of a pool of N machines, split into M heterogeneous clusters of sizes K_1, K_2,...,K_M, where $\sum_{i=1}^{M} K_i = N$. Cluster i is dedicated to a queue of jobs of type i ($i = 1, ..., M$). Job types may for example include short web accesses or long database searches. Different types of job have different response time requirements (e.g., some may be less tolerant of delays than others). It is possible to reassign any server from one queue to another, but the process is generally not instantaneous and during it the server becomes unavailable. In those circumstances, a reconfiguration policy would specify, for any given parameter set (including costs), and current state, whether to switch a server or not.

There is an extensive literature on dynamic optimization (some good general texts are [1,12,14]), but the problem described here does not appear to have been studied before. There is a body of work on optimal allocation in the context of polling systems, where a server visits several queues in a fixed or variable order, with or without switching overheads (see [4,5,8,9,10]). Even in those cases of a single server, it has been observed by both Duenyas and Van Oyen [4,5], and Koole [8,9], that the presence of non-zero switching times makes the optimal policy very difficult to characterize explicitly. This necessitates the consideration of heuristic policies. The only general result available for multiprocessor systems applies when the switching times and costs are zero: then the $c\mu$-rule is optimal, i.e. the best policy is to give absolute preemptive priority to the job type for which the product of holding cost and service rate is largest (Buyukkoc et al [3]).

A preliminary study of a model with just two job types was presented in [11]. A model similar to ours, also concerned with just two job types, was analyzed by Fayolle et al [7]. There the policy is fixed (servers are switched instantaneously, and only when idle), and the object is to evaluate the system performance. The solution is complex and rather difficult to implement.

Posed in its full generality, this is a complex problem which is most unlikely to yield an exact and explicit solution. Our approach is to formulate the problem as a Markov decision process, and to assume that there is a stationary optimal policy. This policy can be computed numerically by truncating the state space to make it finite. We then propose some heuristic policies which, while not optimal, perform well and are easily implementable. The quality of the heuristics, compared to the optimal policy, is evaluated by simulation.

The model assumptions are described in section 2. The dynamic programming formulation leading to the optimal policy is presented in section 3, while section 4 presents a number of numerical and simulation experiments, including comparisons between the optimal and heuristic policies. Section 5 summarizes the results.

2 The Model

Our model may be described as follows. Jobs of type i arrive according to an independent Poisson process with rate λ_i, and join a separate unbounded queue ($i = 1, 2, ..., M$). Their required service times are distributed exponentially with mean $1/\mu_i$. The cost of keeping a type i job in the system is c_i per unit time ($i = 1, 2, ..., M$). These 'holding' costs reflect the relative importance, or willingness to wait, of the M job types.

Any server currently allocated to queue i may be switched to queue j. Such a switch costs $c_{i,j}$ and takes an interval of time distributed exponentially with mean $1/\zeta_{i,j}$, during which the server cannot serve jobs. It is assumed that switches are initiated at job arrival or departure instants. Indeed, it is at those instants that switches may become advantageous, and if they do, they should be performed without delay. Also, it is assumed that the switching policy employed is stationary, i.e., switching decisions may depend on the current state but not on past history.

Any job whose service is interrupted by a switch returns to the appropriate queue and resumes service from the point of interruption when a server becomes available for it.

The system state at any time is described by the triple, $S = (\mathbf{j}, \mathbf{k}, \mathbf{m})$ where $\mathbf{j} = (j_1, j_2, ..., j_M)$ is the vector of current queue sizes (j_i is the number of jobs in queue i, including those being served), $\mathbf{k} = (k_1, k_2, ..., k_M)$ is the vector of current server allocations (k_i servers allocated to queue i) and $\mathbf{m} = (m_{i,j})_{i,j=1}^{M}$ is the matrix of switches currently in progress ($m_{i,j}$ servers being switched from queue i to queue j, $m_{i,i} = 0$). The valid states satisfy $\sum_{i=1}^{M} k_i + \sum_{i,j=1}^{M} m_{i,j} = N$.

Under the above assumptions, the system is modelled by a continuous time Markov process. The transition rates of that process depend on the switching policy, i.e. on the decisions (actions) taken in various states. Denote by $r_d(S, S')$ the transition rate from state S to state S' ($S \neq S'$), given that action d is taken. The possible actions are (a) do nothing, or (b) initiate a switch from queue i to queue j (if $k_i > 0$ and $i \neq j$). These actions are represented by $d = 0$ (do nothing) and $d = 1, 2, ..., M(M-1)/2$.

The values of $r_d(S, S')$, for $S = (\mathbf{j}, \mathbf{k}, \mathbf{m})$, $S' = (\mathbf{j}', \mathbf{k}', \mathbf{m}')$ and $d = 0$, are given by the following (where $i, j = 1, ..., M$):

$$r_0(S, S') = \begin{cases} \lambda_i & \text{if } \mathbf{j}' = \mathbf{j} + \mathbf{e}_i \\ \min(j_i, k_i)\mu_i & \text{if } \mathbf{j}' = \mathbf{j} - \mathbf{e}_i \\ m_{i,j}\zeta_{i,j} & \text{if } \mathbf{m}' = \mathbf{m} - \mathbf{e}_{i,j} \text{ and } \mathbf{k}' = \mathbf{k} + \mathbf{e}_j \\ 0 & \text{otherwise} \end{cases}$$

where $\mathbf{e}_i$ is the ith unit vector, and $\mathbf{e}_{i,j}$ is the matrix which has 1 in position (i, j) and zeros everywhere else.

The corresponding rates when $d \neq 0$ and the action taken is to switch a server from queue a to queue b ($a \neq b$) are obtained by replacing, in S', $\mathbf{k}'$ by $\mathbf{k}' - \mathbf{e}_a$ and $\mathbf{m}'$ by $\mathbf{m}' + \mathbf{e}_{a,b}$. Note that, in cases $d \neq 0$, there is a zero-time transition which changes k_a and $m_{a,b}$, and then an exponentially distributed interval with mean $1/r_d(S, S')$, after which the state jumps to S'.

The total transition rate out of state S, given that action d is taken, $r_d(S)$, is equal to:

$$r_d(S) = \sum_{S'} r_d(S, S').$$

3 Computation of the Optimal Policy

For the purposes of optimization, it is convenient to apply the technique of uniformization to the Markov process (e.g., see [13]). This entails the introduction of 'fictitious' transitions which do not change the system state, so that the average interval between consecutive transitions ceases to depend on the state, and then embedding a discrete-time Markov chain at transition instants. First, we find a constant, Λ, such that $r_d(S) \leq \Lambda$ for all S and d. A suitable value for Λ is

$$\Lambda = \sum_{i=1}^{M} \lambda_i + N\mu + N\zeta, \qquad (1)$$

where $\mu = max(\mu_i)$ is the largest service rate and $\zeta = max(\zeta_{i,j})$ is the largest switching rate.

Next, construct a Markov chain whose one-step transition probabilities when action d is taken, $q_d(S, S')$, are given by

$$q_d(S, S') = \begin{cases} r_d(S, S')/\Lambda & \text{if } S' \neq d(S) \\ 1 - r_d(S)/\Lambda & \text{if } S' = d(S) \end{cases},$$

where $d(S)$ is the state resulting from the immediate application of action d in state S. This Markov chain is, for all practical purposes, equivalent to the original Markov process.

Without loss of generality, the unit of time can be scaled so that the uniformization constant becomes $\Lambda = 1$.

The finite-horizon optimization problem can be formulated as follows. Denote by $V_n(S)$ the minimal expected total cost incurred during n consecutive steps of the Markov chain, given that the current system state is S. The cost incurred at step l in the future is discounted by a factor α^l ($l = 1, 2, \ldots, n-1$; $0 \leq \alpha \leq 1$). Setting $\alpha = 0$ implies that all future costs are disregarded; only the current step is important. When $\alpha = 1$, the cost of a future step, no matter how distant, carries the same weight as the current one.

Any sequence of actions which achieves the minimal cost $V_n(S)$, constitutes an 'optimal policy' with respect to the initial state S, cost parameters, event horizon n, and discount factor α.

Suppose that the action taken in state S is d. This incurs an immediate cost of $c(d)$, equal to $c_{i,j}$ if the action taken is to switch a server from queue i to queue j. In addition, since the average interval between transitions is 1, each type i job in the system incurs a holding cost c_i. The next state will be S', with probability $q_d(S, S')$, and the minimal cost of the subsequent $n-1$ steps will be $\alpha V_{n-1}(S')$. Hence, the quantities $V_n(S)$ satisfy the following recurrence relations:

$$V_n(S) = \sum_{i=1}^{M} j_i c_i + \min_d \left[c(d) + \alpha \sum_{S'} q_d(S, S') V_{n-1}(S') \right]. \qquad (2)$$

Thus, starting with the initial values $V_0(S) = 0$ for all S, one can compute $V_n(S)$ in n iterations. In order to make the state space finite, the queue sizes are bounded at some level, $j_i < J$ ($i = 1, ..., M$). Then, if $V_{n-1}(S)$ has already been computed for some n and for all S, the complexity of computing $V_n(S)$, for a particular state S, is roughly constant. There are no more than $2M + M(M-1)/2$ states S' reachable from state S, and $M(M-1)/2 + 1$ actions to be compared (corresponding to the $M(M-1)/2$ possible switches from queue i to queue j and action $d = 0$ to do nothing). The best action to take in that state, and for that n, is indicated by the value of d that achieves the minimum in the right-hand side of (2). Since there are on the order of $O(J^M N^{M-1+M(M-1)/2})$ states altogether, the computational complexity of one iteration is on the order of $O(J^M N^{M-1+M(M-1)/2})$, and hence the overall complexity of solving (2) and determining the optimal switching policy over a finite event horizon of size n, is on the order of $O(n J^M N^{M-1+M(M-1)/2})$.

If the discount factor α is strictly less than 1, it is reasonable to consider the infinite-horizon optimization, i.e. the total minimal expected cost, $V(S)$, of all future steps, given that the current state is S. That cost is of course infinite when $\alpha = 1$, but it is finite when $\alpha < 1$. Indeed, in the latter case it is known (see [2]), that under certain rather weak conditions, $V_n(S) \to V(S)$ when $n \to \infty$. When the optimal actions depend only on the current state, S, and not on n, the policy is said to be 'stationary'.

An argument similar to the one preceding (2) leads to the following equation for $V(S)$:

$$V(S) = \sum_{i=1}^{M} j_i c_i + \min_d \left[c(d) + \alpha \sum_{S'} q_d(S, S') V(S') \right]. \qquad (3)$$

The optimal policy (i.e. the best action in any given state) is specified by the value of d that achieves the minimum in the right-hand side of (3).

Equation (3) can be solved by applying the 'policy improvement' algorithm (see Dreyfus and Law [6]).

4 Experimental Results

We start with a simple system with $N = 2$ and $M = 2$, where switches cost money but do not take time. Although this case is not of great practical interest, it is included as an illustration. The system state is described by a triple, $S = (j_1, j_2, k_1)$. The number of servers allocated to type 2 is $k_2 = N - k_1$. The uniformization constant is now $\Lambda = \lambda_1 + \lambda_2 + 2(max(\mu_1, \mu_2))$. If action $d \neq 0$ is taken in state S, the value of k_i changes immediately as a switch initiated from queue i to queue j. Then a new state is entered after an exponentially distributed interval with mean $1/\Lambda$.

In this example, the arrival and service parameters of the two job types are the same, but waiting times for type 2 are twice as expensive as those for type 1. The discount factor is $\alpha = 0.95$. The stationary optimal policy for states where $k_1 = k_2 = 1$ is shown in table 1. The truncation level used in the computation was $J = 30$, but the table stops at $j_1 = j_2 = 10$; the actions do not change beyond that level. Actions d are numbered as follows:

$d=0$, do nothing;
$d=1$, switch a server from queue 1 to queue 2;
$d=2$, switch a server from queue 2 to queue 1.

Table 1. Optimal actions: zero switching times, $N = 2$, $M = 2$, $k_1 = 1$, $\lambda_1 = \lambda_2 = 0.086$, $\mu_1 = \mu_2 = 0.207$, $c_1 = 1$, $c_2 = 2$, $c_{1,2} = c_{2,1} = 10.0$

		\multicolumn{11}{c}{j_2}
		0	1	2	3	4	5	6	7	8	9	10
j_1	0	0	0	0	1	1	1	1	1	1	1	1
	1	0	0	0	0	0	1	1	1	1	1	1
	2	0	0	0	0	0	1	1	1	1	1	1
	3	0	0	0	0	0	1	1	1	1	1	1
	4	0	0	0	0	0	1	1	1	1	1	1
	5	0	0	0	0	0	1	1	1	1	1	1
	6	0	0	0	0	0	1	1	1	1	1	1
	7	2	0	0	0	0	1	1	1	1	1	1
	8	2	0	0	0	0	1	1	1	1	1	1
	9	2	0	0	0	0	1	1	1	1	1	1
	10	2	0	0	0	0	1	1	1	1	1	1

As expected, the presence of switching costs discourages switching; a server is sometimes left idle even when there is work to be done. Note that the $c\mu$-rule in this case would give preemptive priority to type 2: it would take action $d = 1$ whenever $j_2 \geq 2$, and action $d = 2$ when $j_2 = 0$, $j_1 \geq 2$.

The optimal policy for $k_1 = 0$ is to take action $d = 2$ when $j_1 = 1$ and $j_2 = 0$, or when $j_1 > 1$ and $j_2 < 2$. When $k_1 = 2$, it is optimal to take action $d = 1$ when $j_1 \leq 1$ and $j_2 > 0$, or when $j_1 > 1$ and $j_2 > 1$.

From now on, we examine models where switching takes non-zero time. To keep the number of parameters low, the monetary costs of switching will be assumed negligible, $c_{1,2} = c_{2,1} = 0$. The uniformization constant is given by (1), and the unit of time is chosen so that $\Lambda = 1$.

The next question to be addressed is "How can one use dynamic optimization in practice?" Ideally, the optimal policy would be characterized explicitly in terms of the parameters, providing a set of rules to be followed (like, for example, the $c\mu$-rule). Unfortunately, such a characterization does not appear feasible for this problem.

Another approach is to pre-compute the optimal policy for a wide range of parameter values, and store a collection of tables such as table 1. Then, having monitored the system and estimated its parameters, the optimal policy could be obtained by a table look-up. This is feasible, but will consume a lot of storage.

The third and most commonly used approach is to formulate a heuristic policy which (a) is simply characterized in terms of the parameters, and (b) performs acceptably well, compared with the optimal policy. That is what we propose to do.

4.1 Heuristic Policies

When the number of queues does not exceed the number of servers, it is possible to do no switching at all. Allocate the servers roughly in proportion to the offered load, $\rho_i = \lambda_i/\mu_i$, and to the holding cost, c_i, for each type. In other words, set

$$k_i = \left\lfloor N \frac{\rho_i c_i}{\sum_{j=1}^{M} \rho_j c_j} + 0.5 \right\rfloor \ (i = 1, ..., M-1) \ ; \ k_M = N - \sum_{j=1}^{M-1} k_j \ ,$$

if all k_i are non-zero. If any k_i is zero, replace k_i with 1 and the largest k_j ($i \neq j$) by $k_j - 1$. Repeat this process until all k_i are non-zero. Having made the allocation, leave it fixed as long as the offered loads and costs remain the same. This will be referred to as the 'static' policy. It certainly has the virtue of simplicity, and also provides a comparator by which the benefits of dynamic reconfiguration can be measured.

The idea behind our dynamic heuristic policy is to attempt to balance the total holding costs of the different job types. That is, the policy tries to prevent the quantities $j_i c_i$ ($i = 1, ..., M$) from diverging. The following rule is applied:

1. Calculate the following for each of the $M(M-1)/2$ possible switches from queue a to queue b ($a \neq b$ and $k_a > 0$):

$$c_b \left\{ j_b + \frac{1}{\zeta_{a,b}} [\lambda_b - \mu_b \min(k_b, j_b)] \right\}$$

$$- K c_a \left\{ j_a + \frac{1}{\zeta_{a,b}} [\lambda_a - \mu_a \min(k_a - 1, j_a)] \right\},$$

where K is a constant used to discourage too many switches from being initiated. The best value of K depends on the total load. For heavily loaded systems, $K = 5$ has been used.

2. Find the maximum of all quantities calculated in 1; if it is strictly positive, this will be the most advantageous switch to initiate. Take the action $d \neq 0$ corresponding to this switch. Otherwise, take action $d = 0$.

This rule is based on approximating the effects of a switch. If j_b jobs of type b are present and k_b servers are available for them, then the average queue b increment during an interval of length x may be estimated as $x[\lambda_b - \mu_b \min(k_b, j_b)]$. Similarly for queue a, except that if a server is switched from queue a then the available servers for this queue drops to $k_a - 1$. Thus, a server is switched if that switch would help to balance the holding costs, after taking account of its effect on the M queues. The above policy will be referred to as the 'heuristic'.

The optimal, static and heuristic policies are compared by simulation. In order to model changes in demand, the simulation includes a sequence of phases, with λ_i changing values from one phase to the next. There are two possible values for each λ_i: a high rate and a low rate. In any one phase, a particular λ_i is set at the high rate, while the remaining λ_i are at the low rate. This models demand peaking for a particular job type over a period of time. The performance measure in all cases is the total average holding cost, i.e. the simulation estimate of $E(\sum_{i=1}^{M} c_i j_i)$. The average phase duration is 100.

In all experiments, the parameters given below are renormalized to make the uniformization constant, Λ, equal to 1.

In figure 1, the average cost is plotted against total load. Here the number of servers is fixed at $N = 4$ and $M = 2$. The following parameters are used: $\mu_1 = \mu_2 = 1$, $\zeta_{1,2} = \zeta_{2,1} = 0.1$, $c_1 = 2$, $c_2 = 1$. The offered load increases, approaching saturation. The arrival rates are in the ratio 1:100 during phase 1 and 100:1 during phase 2, and are increased to produce the increase in total load.

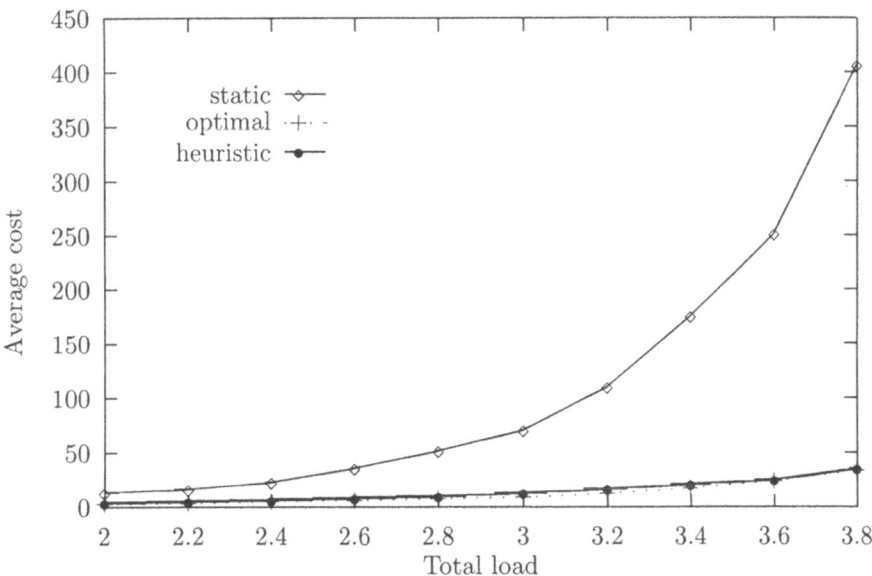

Fig. 1. Policy comparisons: $M = 2$ and increasing loads

This experiment shows emphatically that dynamic reconfiguration is advantageous. The cost of the static policy (which is not entirely static; it changes the allocation within each phase, as the arrival rates change) increases very quickly, while the heuristic, which is almost optimal, has much lower costs.

In figure 2, the average cost is plotted against the number of servers, N when $M = 3$. The following parameters are used: $1000\lambda_1 = \lambda_2 = \lambda_3$, $b_1 = 1000b_2 = 1000b_3$, $c_1 = 2$, $c_2 = 1$, $c_3 = 1$. Switching rates are equal and given by $\zeta_{i,j} = \mu_2/10 = \mu_3/10$. Arrival rates are increased with N so that the total offered load, $\rho_1 + \rho_2 + \rho_3$, is equal to $4N/5$ (i.e., the system is heavily loaded). There are no phase changes. This models a system where type 1 jobs are long and types 2 and 3 are much shorter. Requests of type 1 arrive at a much slower rate than for types 2 and 3, although the total load for each job type is the same.

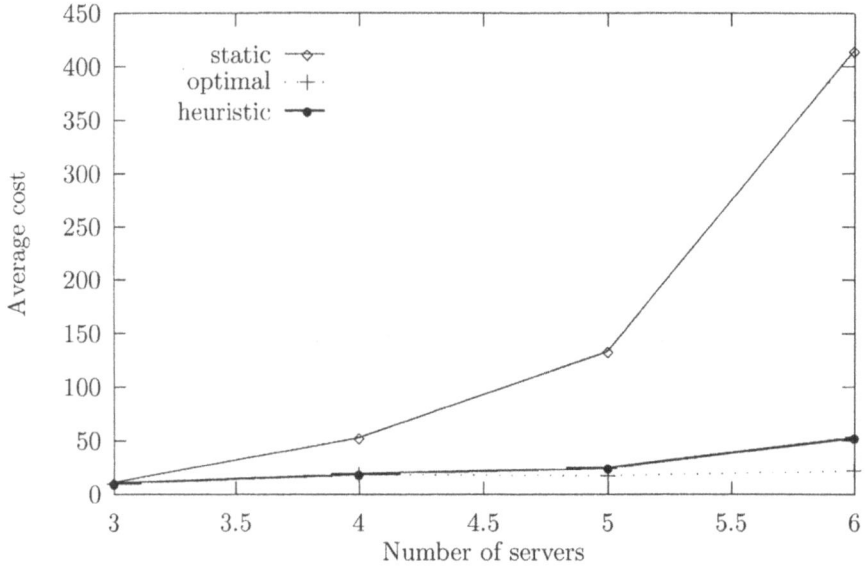

Fig. 2. Policy comparisons: $M = 3$ and increasing N

Another comparison when $M = 3$ is shown in figure 3. Here, the number of servers is fixed at $N = 4$ and the total load increases. Once again there are no phase changes, and all arrival rates are equal. The following parameters are used: $\mu_1 = \mu_2 = \mu_3 = 1$, $\zeta_{i,j} = 0.1$, $c_1 = 2$, $c_2 = 1$, $c_3 = 1$.

Figures 2 and 3 demonstrate clearly the benefit of dynamic reconfiguration of servers when the number of job types is increased to $M = 3$. The static policy performs poorly as the number of servers or the load is increased, while the heuristic policy performs almost as well as the optimal policy. In each case, choosing dynamic reconfiguration dramatically reduces the average holding costs.

In these experiments, 200000 job completions were simulated. Where phase changes were simulated, approximately 1000 phase changes occurred during the duration of the simulation. The longest simulation runs were for the optimal policy, because of the table look-ups.

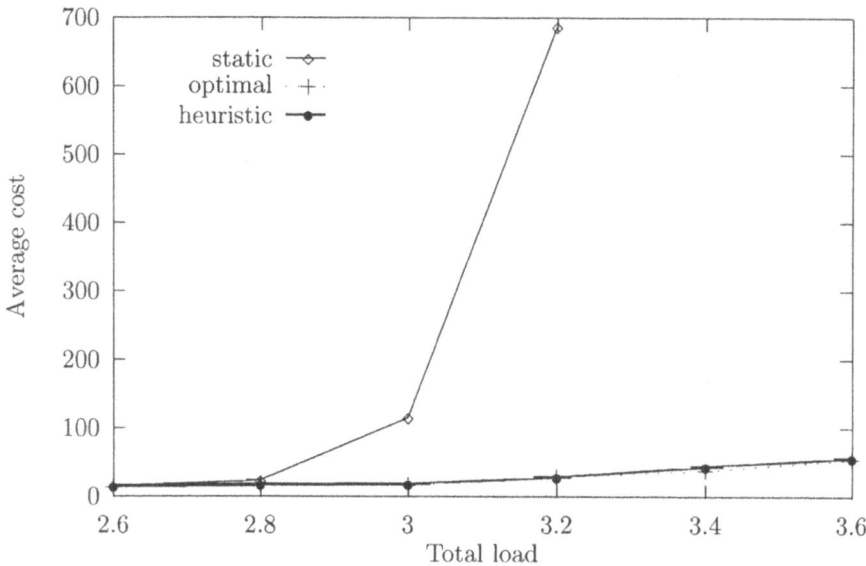

Fig. 3. Policy comparisons: $M = 3$ and increasing loads

Calculations of the table look-ups for the optimal policy have been executed on a Linux machine with an Intel Xeon 2.80GHz processor and 1GB RAM. To calculate the optimal policy for $N = 4$, $M = 3$ and a truncated queue size of $J = 30$ requires 223MB of available memory. As an illustration of the complexity of the calculations and the large size of the state space, if the number of servers is increased to $N = 6$, with $M = 3$ and $J = 30$, calculating the optimal policy now requires 1.564GB of available memory. Each table of decisions, when calculated using the Policy Improvement algorithm described in Section 3, took approximately 16 minutes. The Policy Improvement algorithm took 10 iterations to converge to the optimal policy, with the initial 'guess' of the policy set to the heuristic policy. Solving a large set of simultaneous equations using an iterative process is the most computationally expensive stage of the Policy Improvement algorithm. Initializing the cost matrix to the holding cost of the current state, this set of equations takes approximately 180 iterations to converge to within an accuracy of 0.01 in the first Policy Improvement iteration. This reduces upon each iteration.

5 Conclusions

A problem of interest in the area of distributed processing and dynamic Grid provision has been examined. The optimal reconfiguration policy can be computed and tabulated, subject to complexity constraints imposed by the size of the state space and the ranges of parameter values. However, for practical purposes, an easily implementable heuristic policy is available. The encouraging results of figures 1-3 suggest that its performance compares quite favourably with that of the optimal policy.

Acknowledgement. This work was carried out as part of the collaborative project GridSHED (Grid Scheduling and Hosting Environment Development), funded by British Telecom and the North-East Regional e-Science centre.

References

1. J. Bather, *Decision Theory - An Introduction to Dynamic Programming and Sequential Decisions*, Wiley, 2000
2. D. Blackwell, "Discounted dynamic programming", *Annals of Mathematical Statistics*, 26, pp 226-235, 1965
3. C. Buyukkoc, P. Varaiya and J. Walrand, "The $c\mu$-rule revisited", *Advances in Applied Probability*, 17, pp 237-238, 1985
4. I. Duenyas and M.P. Van Oyen, "Heuristic Scheduling of Parallel Heterogeneous Queues with Set-Ups", *Technical Report 92-60*, Department of Industrial and Operations Engineering, University of Michigan, 1992
5. I. Duenyas and M.P. Van Oyen, "Stochastic Scheduling of Parallel Queues with Set-Up Costs", *Queueing Systems Theory and Applications*, 19, pp 421-444, 1995
6. S.E. Dreyfus and A.M. Law, "The Art and Theory of Dynamic Programming", Academic Press, New York, 1977
7. G. Fayolle, P.J.B. King and I. Mitrani, "The Solution of Certain Two-Dimensional Markov Models", Procs., 7th International Conference on Modelling and Performance Evaluation, Toronto, 1980
8. G. Koole, "Assigning a Single Server to Inhomogeneous Queues with Switching Costs", *Theoretical Computer Science*, 182, pp 203-216, 1997
9. G. Koole, "Structural Results for the Control of Queueing Systems using Event-Based Dynamic Programming", *Queueing Systems Theory and Applications*, 30, pp 323-339, 1998
10. Z. Liu, P. Nain, and D. Towsley, "On Optimal Polling Policies", *Queueing Systems Theory and Applications*, 11, pp 59-83, 1992
11. J. Palmer and I. Mitrani, "Dynamic Server Allocation in Heterogeneous Clusters", Procs. of HETNETs '03 : First International Working Conference on Performance Modelling and Evaluation of Heterogeneous Networks, pp. 12/1-12/10, UK, July 2003.
12. S. M. Ross, *Introduction to Stochastic Dynamic Programming*, Academic Press, 1983
13. E. de Souza e Silva and H.R. Gail, "The Uniformization Method in Performability Analysis", in *Performability Modelling* (eds B.R. Haverkort, R. Marie, G. Rubino and K. Trivedi), Wiley, 2001
14. P. Whittle, *Optimisation over Time*, Vols 1 and 2, Wiley,

Design and Evaluation of an Agent-Based Communication Model for a Parallel File System

María S. Pérez[1], Alberto Sánchez[1], Jemal Abawajy[2], Víctor Robles[1], and José M. Peña[1]

[1] DATSI. FI. Universidad Politécnica de Madrid. Spain
[2] School of Computer. Carleton University. Ottawa, Canada

Abstract. Agent paradigm has become one of the most important topics appeared and widely developed in computing systems in the last decade. This paradigm is being sucessfully used in a large number of fields. MAPFS is a multiagent parallel file system, which takes advantage of the semantic concept of agents in order to increase its modularity and performance. This paper shows how to use agent theory as conceptual framework in the design and development of MAPFS. MAPFS implementation is based on nearer technologies to system programming, although its design makes usage of the abstraction of a multiagent system.

Keywords: Agent, multiagent system, parallel file system, prefetching.

1 Introduction

Agent technology constitutes a new computing paradigm. Despite agents are very related to the Distributed Artificial Intelligence (DIA) area [2], some works have demonstrated that the agent technology can be used in fields totally different to the DIA. In the last decade, a large number of applications have appeared in business [6], electric management [11], control [3], or industrial applications in general [10].

Agent technology provides several concepts, which allow programmers to analyze and design applications in a way close to the natural language. Furthermore, agents give applications a set of useful features for tackling complex and dynamic environments.

On the other hand, there are important differences between *system programming* and agent technology. Firstly, agent paradigm interacts with the system at a higher level than system programming. Furthremore, the efficiency is a very strict requirement in the case of the system programming. Agent technology introduces an abstraction layer and, thus, it involves a lost of efficiency. Nevertheless, these disadvantages can be avoided, since the agent paradigm differ clearly agent theory, which provides the concepts, and agents architectures, which provides concrete solutions and implementations.

This paper shows how to use agent theory as conceptual framework in the design and development of a parallel multiagent system, called MAPFS (MultiAgent Parallel File System), using nearer technologies to system programming, but taking advantage of the semantic concepts of agents.

The outline of this paper is as follows. Section 2 presents the overview of MAPFS, focusing on the multiagent subsystem of such file system, and describes the related work.

Section 3 describes our proposal for the generic structure of an agent in MAPFS. Section 4 shows the implementation and evaluation of MAPFS, in order to measure the influence of the agents in the parallel file system. Finally, Section 5 summarizes our conclusions and suggests further future work.

2 Problem Statement and Related Work

2.1 MAPFS Overview

MAPFS is a multiagent parallel file system for clusters, which provides a file system interface that includes traditional, advanced, collective, caching, and hints operations [14]. MAPFS consists of two subsystems with two clearly defined tasks: (i) MAPFS_FS, which provides the parallel file system functionality and (ii) MAPFS_MAS, responsible for the information retrieval. In order to provide data to MAPFS_FS, MAPFS_MAS is constituted by a set of agents which interact among them, that is, a *multiagent system* (MAS). The use of a MAS implies coordination among their agents. The cooperation model of MAPFS is defined in [15]. Agents must be reconfigured because of the dynamic and changing environment in which they coexist. These agents adapt their behavior depending on the response of the medium and their own learning. MAPFS uses an *agent hierarchy*, which solves the information retrieval problem in a transparent and efficient way. The taxonomy of agents used in MAPFS is composed of: (i) Extractor agents, responsible for information retrieval; (ii) distributor agents, which distribute the workload to extractor agents; (iii) caching and prefetching agents, associated with one or more extractor agents, caching or prefetching their data; and (iv) hints agents, which must study applications access patterns to build hints for improving data access.

Files are stored finally in several servers, which constitute the server-side of the underlying architecture. The grouping of servers from a logical point of view in MAPFS is named *storage group*.

2.2 Related Work

Nowadays, most of the frameworks are influenced by their environment, so that the environment conditions affect their performance in a dynamic way. For this reason, the usage of the agent technology is being widely used, since this paradigm adapts to changing and dynamic environments. The agent paradigm is usually implemented on distributed systems.

In a complex system, the interaction of several agents is required and, thus, a mechanism of communication between agents is necessary. For achieving agents communication and interoperability, it is necessary to use: (i) A common language; (ii) common ideas about the knowledge agents interchange; and (iii) capacity for interchanging this information. For standardizing this way of communication, a common or standard language is used. In this sense, KSE (*Knowledge Sharing Effort*) has several research lines [1]. This paper focuses on the ability of agents for communicating among them and developing a specific task within MAPFS.

As it is shown in [13], the usage of agents simplifies the distribution and optimizes the messages interchange among them. The idea of using agents to access data is not an innovating idea. Nowadays, a great number of agents platforms are widely deployed for accesing web databases. Different access methods are used, such as JDBC. The web popularity has created the need for developing Web Distributed Database Management Systems (DBMS), obtaining simple data distribution, concurrency control and reliability. However, DBMS offer limited flexibility, scalability, and robustness. Some suggestions propose the use of agents to solve this problem [16]. With respect to file accesses, several approaches have been made. Two paradigmatic approaches are described next.

MESSENGERS [4] is a system based on agents used for the development and deployment of distributed applications from mobile agents, called *messengers*. This system is composed of a set of daemons distributed in every node an used for managing received agents, supervising their execution and planning where agents must be sent. Several features are defined in [8] in order to measure the system performance. Some of them are load balancing, agent code optimization and availability and efficient sharing of available resources.

DIAMOnDS [17] stands for Distributed Agents for MObile and Dynamic Services, a system built under Java/Jini. This system is composed of a client module that accesses data of a remote file system, where an agent is responsible of managing this interaction.

Other research projects about agent systems for accesing files have been developed. Nevertheless, there are not agent systems focused on the development of parallel file systems features. MAPFS constitutes a new approach of this kind of systems.

3 Generic Structure of an Agent in MAPFS

Agents provide a set of very interesting properties. Some of these characteristics are autonomy, reactivity and proactivity, which makes the system flexible for adapting to changing environments. Furthermore, an aditional characteristic very related to agents and useful in the case of the MAPFS system is the intelligence. Intelligent agents usually take decisions in the system. In this context, MAPFS agents are responsible for building hints dynamically, modifying them according to the acquired knowledge in the process of analysis of data patterns.

On the other hand, as is described in the previous section, there are different kind of agents. Therefore, it is necessary to identify the role of every agent in the system. This method have been already identified and used in some agent architectures, such as MADKIT architecture [9]. This architecture defines the AGR model (*Agent-Group-Role*), in which the role or task of an agent constitutes one of the key concept. This role is the abstract representation of a function or service provided by the agent. Analogously, in MAPFS the role is used for setting the specific function of an agent.

Definition 1 *In MAPFS, an agent is defined in a formal way as the following tuple:*

$$< \mathtt{Ag\_Id}, \mathtt{Group}, \mathtt{Role}, \mathtt{Int\_Net} >$$

where:

- Ag_Id: *Agent identification, which is used in order to identify every agent of the system.*

- Group: *Storage group which the agent belongs to.*
- Role: *This field represents the kind of agent, taking values in the following domain:* [Cache, Distributor, Extractor, Hint]. *This domain can be increased with other values, if other kind of service must be implemented.*
- Int_Net: *This field represents the interaction network of an agent with other agents of its storage group. This network can be implemented as a vector or relations between the agent* Ag_Id *and the rest of agents of the same storage group.*

A key aspect of a multiagent system is the communication among agents. There are specific agent languages, oriented to communication of agents. KQML (Knowledge Query Manipulation Language) [7], is one of the most known agent communication languages. This language is composed of a set of messages, known as *performatives*, which are used for specifying agent communication elements. In [12], Labrou and Finin widely describe the KQML reserved performatives. Some of them are used in MAPFS.

According to the MAPFS cooperation model, a set of performatives has been defined. In order to define MAPFS performatives, several sets of elements are defined for a concrete storage group: (**DA**: Set of distributor agents, **EA**: Set of extractor agents, **CA**: Set of cache agents, **HA**: Set of hints agents). Next section defines KQML performatives for the interaction among agents.

3.1 MAPFS Performatives

When an element d is requested, a distributor agent is responsible of asking data to several extractor agents. Let x be a distributor agent of a storage group G_x. Figure 1(a) includes the KQML performative of the distributor agent. If the extractor agent has the element d, then such agent does the performative of Figure 1(b), indicating that the data item d is available in the storage group G_x.

On the other hand, if the extractor agent has not the element d, that is, the element is not in the cache structure, the extractor agent does the performative of Figure 2, asking required data to all the cache agents. The predicate ask(d,z) in the cache agent z involves the execution of the MAPFS function obtain(d) (read operation).

Next, the cache agent sends information about the completion of the operation to the distributor agent, through the extractor agent, indicating that the element d is available in the storage group G_x. This process corresponds to the performative of Figure 3. Thus, the cycle is closed. Nevertheless, the cache structure has a maximum number of entries, which must be replaced by other elements with a concrete replace policy. When the entry is invalidated, the cache agent z sends the performative represented in Figure 4.

Cache agents use metadata provided by hints agents, sending the performative of Figure 5(a). In this way, metainformation identified by h is required. A hint agent build the required metainformation, sending it to the cache agent by means of the performative of Figure 5(b). Figure 6 shows the control flow of system performatives.

3.2 Cache Agent. A Sample Agent

A cache agent is a sample MAPFS agent. In order to increase the efficiency of the I/O system, cache agents make prefetching and caching tasks. Prefetching is used for

```
Step 1
    x ∈ DA
    y ∈ EA
    (ask-if
        :sender      x
        :receiver    y
        :reply-with  id_da
        :language    Prolog
        :ontology    MAPFS
        :content     "exists(d,G_x)")
```

(a) Performative for the data request from a distributor agent to an extractor agent.

```
Step 2.1
    (tell
        :sender       y
        :receiver     x
        :in-reply-to  id_da
        :reply-with   id_ea
        :language     Prolog
        :ontology     MAPFS
        :content      "exists(d,G_x)")
```

(b) Response performative from an extractor agent to a distributor agent, if the agent has the required data.

Fig. 1. Performatives related to a distributor agent

```
Step 2.2
    z ∈ CA
    (achieve
        :sender       y
        :receiver     z
        :in-reply-to  id_da
        :reply-with   id_ea
        :language     Prolog
        :ontology     MAPFS
        :content      "ask(d,z)")
```

Fig. 2. Performative for the data request from an extractor agent to a cache agent

increasing the performance of read operations, since it is possible to read in advance data used in posterior operations. Caching is used for increasing the performance of read and write operations, due to the locality of data and the possibility of delayed write. The

```
Step 3.1
  (forward
      :from       z
      :to         x
      :sender     z
      :receiver   y
      :reply-with id_ca
      :language   KQML
      :ontology   kqml-ontology
      :content    (achieve
                      :sender      z
                      :receiver    x
                      :in-reply-to id_ea
                      :reply-with  id_ca
                      :language    Prolog
                      :ontology    MAPFS
                      :content     "exists(d,$G_x$)")
```

Fig. 3. Response performative from a cache agent to a distributor agent, once data are obtained

definition of a cache agent, according to the previous definition of a generic agent is the following:

Definition 2 *A cache agent of a storage group G_x is defined as the following tuple:*

$$< \texttt{C\_Ag\_Id}, G_x, \texttt{Cache}, \texttt{C\_Int\_Net} >$$

where C_Int_Net *represents the interaction network of the cache agent with other agents linked to it, as a vector of relation between cache agent and the rest of agents that belong to the same storage group. According to Figure 6, cache agents are related with other cache agents, with extractor and hint agents.*

The relation between a cache agent and an extractor agent is derived directly from the performative of Figure 2. Therefore, a cache agent is associated unless to an extractor agent. Furthermore, there may be a relation between different cache agents, with the aim of providing a caching service to a single extractor agent. Cache agents are related to hint agents, as we can see in the performative of Figure 5(a).

```
Invalidation
 (forward
    :from       z
    :to         x
    :sender     z
    :receiver   y
    :reply-with id_ca'
    :language   KQML
    :ontology   kqml-ontology
    :content    (unachieve
                    :sender     z
                    :receiver   x
                    :in-reply-to id_ea
                    :reply-with id_ca'
                    :language   Prolog
                    :ontology   MAPFS
                    :content    "exists(d,$G_x$)")
```

Fig. 4. Performative of invalidation of data in the cache

4 MAPFS Agents Implementation and Evaluation

Agents are useful in the design of a complex system, and, concretely in the design of a parallel file system, as we have shown in previous sections. Nevertheless, it is necessary to validate this paradigm within this field, evaluating the increase of the performance of the implementation of MAPFS and its multiagent subsystem.

The implementation of the multiagent subsystem is based on MPI technology. MPI provides a framework for deploying agents and their main features. MPI is able to create dynamically independent and autonomous processes with communication capacities. Additionally, agents can react to the environment or changes in other processes by means of a MPI message. In fact, KQML performatives are translated to MPI messages by MAPFS, as it is described below.

KQML defines an abstraction for transport for agent communication and can be implemented with different solutions. MPI is a good choice, since this technology fulfill the requirements of KQML performatives. The translation of the most relevant KQML performatives into MPI messages is shown in Table 1.

```
Step 3.2                              Step 4.1
  u ∈ HA                                (tell
  (achieve                                :sender     u
     :sender    z                         :receiver   z
     :receiver  u                         :in-reply-to  id_ca"
     :reply-with  id_ca"                  :reply-with  id_ha
     :language  Prolog                    :language  Prolog
     :ontology  MAPFS                     :ontology  MAPFS
     :content   "ask(h,v)")               :content   "exists(h,$G_x$)")
```

(a) Performative for the hint request from an cache agent to a hint agent.

(b) Response performative from a hint agent to a cache agent, once hints are obtained.

Fig. 5. Performatives related to a hint agent

Table 1. Translation of KQML performatives into MPI messages

KQML Performative	MPI Messages
ask-one	MPI_Send
	MPI_Recv a single basic element
ask-all	MPI_Send
	MPI_Recv a vector of elements
tell	MPI_Send
untell	MPI_Send
deny	MPI_Send
insert	MPI_Send
uninsert	MPI_Send
achieve	MPI_Send
unachieve	MPI_Send
error	MPI_Send
sorry	MPI_Send
forward	MPI_Send
	Interpretation of the included performative

It is important to emphasize that MPI only solves the communication problem. The semantic is provided by the messages content and the processing of such message by the receiver agent.

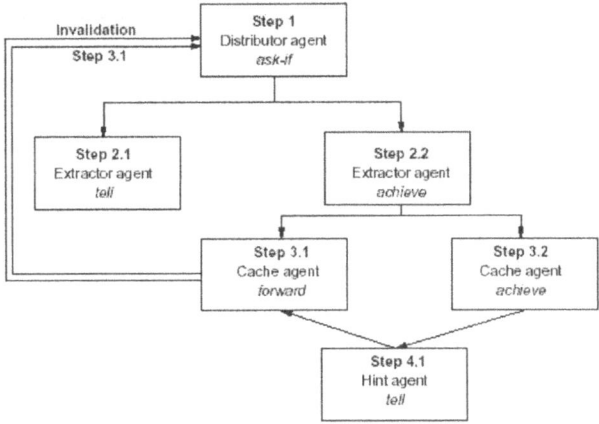

Fig. 6. Control flow of system performatives

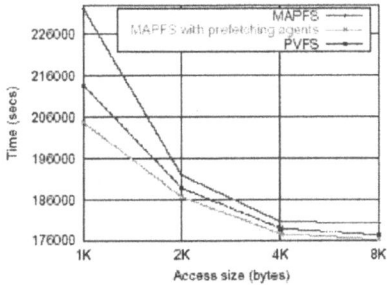

Fig. 7. Comparison of a scientific application in PVFS and MAPFS

A MAPFS multiagent subsystem responsible for prefetching has been implemented and evaluated. If we compare PVFS (Parallel Virtual File System) [5] and MAPFS (Figure 7), we conclude that the usage of agents is a flexible way of increasing the performance, improving the efficiency of PVFS by means of a multiagent subsystem oriented to prefetch probably used data in next executions.

5 Conclusions and Future Work

MAPFS is a multiagent parallel file system, whose design is based on agent theory. Several multiagent subsystems are implemented in MAPFS for tackling different aspects related to its performance. This paper shows the evaluation of a multiagent subsystem used for prefetching probably used data in next executions, concluding that MAPFS system can improve its efficiency in a flexible way by means of the usage of these multiagent subsystems. This paper also describes how to translate conceptual abstractions based on agents in implementations of close technologies to system programming. In the case of MAPFS, a implementation based on MPI has been used.

As future work, we are developing new multiagent subsystems for improving other aspects of the MAPFS system. Additionally, we have implemented MAPFS-Grid as a version of MAPFS for grid environments. We need to adapt these multiagent subsystems to this new version. Nevertheless, in this kind of system we have to face with different problems, related to the heterogeneity and geographical distribution of grids.

References

1. American National Standard. Knowledge Interchange Format. *Draft Proposed American National Standard (dpANS), NCITS.T2/98-004*, 1998.
2. N. M. Avouris and L. Gasser. *Distributed Artificial Intelligence: Theory and Praxis. Volume 5 of Computer and Information Science*. Kluwer Academic Publisher, Boston, MA, 1992.
3. C. P. Azevedo, B. Feiju, and M. Costa. Control centres evolve with agent technology. *IEEE Computer Applications in Power*, 13(3):48-53, 2000.
4. Lubomir Bic, Munehiro Fukuda, and Michael B. Dillencourt. Distributed programming using autonomous agents. *IEEE Computer*, 29(8):55-61, 1996.
5. P. H. Carns, W. B. Ligon III, R. B. Ross, and R. Thakur. PVFS: A parallel file system for linux clusters. In *Proceedings of the 4th Annual Linux Showcase and Conference*, pages 317-327, October 2000.
6. N. R. Jennings et al. ADEPT: Managing business processes using intelligent agents. In *Proceedings of the BCS Expert Systems 96 Conference, Cambridge, UK*, pages 5-23, 1996.
7. Tim Finin, Yannis Labrou, and James Mayfield. KQML as an agent communication language. *"Software Agents", MIT Press. Cambridge*, 1997.
8. Eugene Gendelman, Lubomir F. Bic, and Michael B. Dillencourt. Fast file access for fast agents. *Proceedings of the 5th International Conference, MA 2001.*, 2240:88-102, 2001.
9. Olivier Gutknecht and Jacques Ferber. The MADKIT agent platform architecture. *Infrastructure for Agents, Multi-Agent Systems, and Scalable Multi-Agent Systems*, March 2001.
10. Staffan Hägg. Agent technology in industrial applications. In *Proceedings of the Australia-Pacific Forum on Intelligent Processing and Manufacturing of Materials (IPMM'97)*, 1997.
11. N. R. Jennings, J. M. Corera, L. Laresgoiti, E. H. Mamdani, F. Perriollat, P. Skarek, and L. Z. Varga. Using ARCHON to develop real-world DAI applications for electricity transportation management and particle accelerator control. *IEEE Expert*, 1995.
12. Yannis Labrou and Tim Finin. A Proposal for a new KQML Specification. Technical Report TR CS-97-03, Baltimore, MD 21250, 1997.
13. E. Pitoura. Transaction-Based Coordination of Software Agents. In *Proceedings of the 9th International Conference on Database and Expert Systems Applications (DEXA)*, 1998.
14. María S. Pérez, Félix García, and Jesús Carretero. A new multiagent based architecture for high performance I/O in clusters. In *Proceedings of ICCP'01*, September 2001.
15. María S. Pérez, Félix García, and Jes´us Carretero. MAPFS MAS: A model of interaction among information retrieval agents. In *2nd IEEE/ACM CCGrid 2002*, May 2002.
16. K. Segun, A. Hurson, V. Desai, A. Spink, and L. Miller. Transaction management in a mobile data access system. *Annual Review of Scalable Computing*, 3:85-147, 2001.
17. Aamir et al. Shafi. DIAMOnDS - DIstributed Agents for MObile and Dynamic Services. In *Proceedings of the CHEP03*, March 24-28 2003.

Task Allocation for Minimizing Programs Completion Time in Multicomputer Systems

Gamal Attiya and Yskandar Hamam

Groupe ESIEE Paris, Lab. A^2SI
Cité Descartes, BP 99,
93162 Noisy-Le-Grand, FRANCE
{attiyag,hamamy}@esiee.fr

Abstract. Task allocation is one of the biggest issues in the area of parallel and distributed computing. Given a parallel program composed of M communicating modules (tasks) and a multicomputer system of N processors with a specific interconnection network, the problem is how to assign the program modules onto the available processors in the system so as to minimize the entire program completion time. This problem is known to be NP-complete and therefore untractable as soon as the number of tasks and/or processors exceeds a few units. This paper presents a heuristic algorithm, derived from Simulated Annealing, to this problem taking into account several kinds of constraints. The performance of the algorithm is evaluated through experimental study on randomly generated instances that being allocated into a multicomputer system of bus topology. Furthermore, the quality of solutions are compared with those derived using the Branch-and-Bound on the same sample problems.

1 Introduction

A fundamental issue affecting the performance of a parallel program running on a multicomputer system is the assignment of the program modules (tasks) into the available processors of the system. Modules of a program may be executed on the same or different processors. Because of different processors capabilities, the cost of executing a module may vary from one processor to another. The module execution cost depends on the work to be performed by the module and the processor attributes such as its clock rate (speeds), instruction set and cache memory. On the other hand, modules that are executed on different processors but communicate with one another do so using the interconnection network and so incur communication costs due to the overhead of the communication protocols and the transmission delays in the network. To realize performance potential, two goals need to be met: interprocessor communication cost has to be minimized and modules execution costs need to be balanced among processors. These two goals seem to conflict with one another. On one hand, having all modules on one processor will remove the interprocessors communication costs but result in poor balance of the processing load. On the other hand, an even distribution of modules among processors will maximize the processors utilization but might also increase the interprocessors communication costs. Thus, the purpose is to balance the two often conflicting objectives of minimizing the interprocessors communication costs and maximizing the processors utilization.

Several approaches, in both fields of computer science and operations research, have been suggested to solve the allocation problem. They may be roughly classified into four broad categories, namely, graph theory [1,2], state space search [3,4], mathematical programming [5,6] and heuristics [7]–[13]. However, most of the existing approaches deal with homogeneous systems. Furthermore, they do not consider many kinds of constraints that are related to the application requirements and the availability of system resources. The allocation problem may be further complicated when the system contains heterogeneous components such as different processors speeds and different resources availability (memory and/or communication capacities). This paper addresses the allocation problem in heterogeneous multicomputer systems taking into account both memory and communication capacity constraints. It first models the allocation problem as an optimization problem. It then proposes a heuristic algorithm, derived from Simulated Annealing (SA), to allocate modules of a parallel program into processors of a multicomputer system so as to minimize the entire program completion time. The performance of the proposed algorithm is evaluated through experimental studies on randomly generated instances. Furthermore, the quality of the resulting solutions are compared with those derived by applying the Branch-and Bound algorithm [14] on the same instances.

The remainder of this paper consists of five sections. Section 2 defines the task allocation problem. Section 3 describes how the allocation problem may be formulated as an optimization problem. Section 4 presents the Simulated Annealing (SA) technique and describes how it may be employed for solving the allocation problem. The simulation results are discussed in Section 5 while the paper conclusions are given in Section 6.

2 Problem Definition

The problem being addressed in this paper is concerned with allocating modules (tasks) of a parallel program into processors of a heterogeneous multicomputer system so as to minimize the program completion time. A multicomputer system consists of a set of N heterogeneous processors connected via an interconnection network. Each processor has some computation facilities and its own memory. Furthermore, the interconnection network has some communication capacities and cost of transferring a data unit from one computer sender to other computer receiver. An example of a multicomputer system is shown in Figure 1(b). A parallel program consists of a set of M communicating modules corresponding to nodes in a task interaction graph $G(V, E)$, as shown in Figure 1(a). Where, V represents set of M modules and E represents set of edges. In the graph, each module $i \in V$ is labeled by memory requirements and each arc $(i, j) \in E$ is labeled by communication requirements among modules. Furthermore, a vector is associated with each module representing the execution cost of the module at different processors in the system, as shown in Figure 1(c). The problem is how to map tasks (modules) of the task graph into processors of the system so as to minimize the program completion time under one or more constraints.

Briefly, we are given a set of M communicating modules representing a parallel program to be executed on a heterogeneous multicomputer system of N processors. Modules of the program require certain capacitated computer resources. They have computational and memory capacity requirements. Furthermore, they communicate with each other at a given rate. On the other hand, processors and communication resources are also capacitated. Thus, the purpose is to find an assignment of the program modules to the system processors such that the program completion time is minimized, the requirements of modules and edges are met, and the capacities of the system resources are not violated.

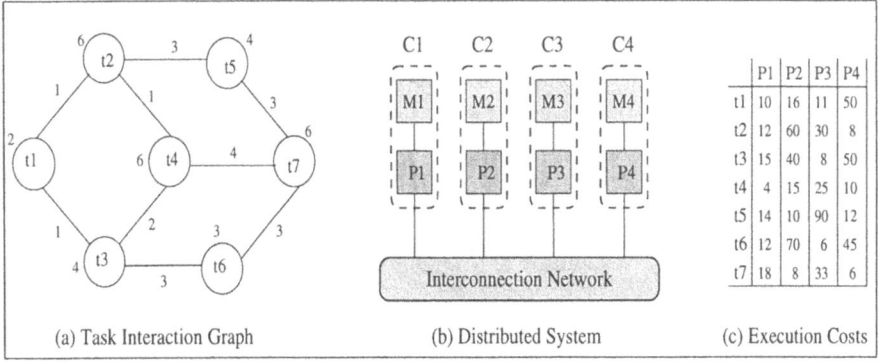

Fig. 1. An Example of a Task Interaction Graph and a Distributed System.

3 Model Description

Designing a model to the allocation problem involves two steps; (i) formulate a cost function to represent the objective of the task allocation, (ii) formulate set of constraints in terms of the modules requirements and the availability of the system resources. To describe the allocation model, let A be an assignment vector such that $A(i)=p$ if module i is assigned to processor p, and TC_p be a task cluster representing the set of tasks assigned to processor p.

3.1 Allocation Cost Development

For an assignment A, a processor load comprises all the execution and the communication costs associated with its assigned tasks. The time needed by the bottleneck processor (i.e., the heaviest loaded processor) will determine the program completion time. Therefore, minimizing the entire program completion time may be achieved by minimizing the cost at the maximum loaded processor.

- *Actual Execution Cost:*
  The actual execution load at a processor p is the cost of processing all tasks assigned to p for an assignment A. Define C_{ip} as the cost of processing task i on processor p, then the actual execution cost may be formulated as

  $$EXEC_p = \sum_{i \in TC_p} C_{ip}$$

- *Actual Communication Cost:*
  The actual communication load at a processor p is the cost of communicating data between tasks resident at p with other tasks resident at other processor q. Let cc_{avg} be the average communication cost of transferring a data unit through the network transmission media and d_{ij} be the data to flow between two communicating tasks i and j, then the actual communication cost may be formulated as

  $$COMM_p = \sum_{i \in TC_p} \sum_{j \neq i, (i,j) \in E, A(j) \neq p} d_{ij} * cc_{avg}$$

  It is worth noting that if two communicating tasks are assigned to different processors p and q, the communication cost contributes the load of the two processors. Furthermore, if two communicating tasks are assigned to the same processor, the communication cost is assumed to be zero as they use the local system memory for data exchange.

- *Bottleneck in the System:*
  For an assignment A, the workload at a processor p comprises all the execution and the communication costs associated with its assigned tasks. Hence, the workload at p may be formulated as

  $$L_p = EXEC_p + COMM_p$$

  The bottleneck in the system is the critical processor which has the maximum cost over all processors. Thus the maximum load at a bottleneck processor may be formulated as

  $$L_{max} = max\ \{L_p \mid 1 \leq p \leq N\}$$

- *Objective Function:*
  To minimize the entire program completion time, the cost at the maximum loaded processor must be minimized. Therefore the objective function of the allocation model may be formulated as

  $$min\ L_{max}$$

3.2 Assignment Constraints

The assignment constraints depend on the characteristics of both the application involved (such as memory and communication capacities requirements) and on the system resources including the computation speed of processors, the availability of memory and communication network capacities.

- *Memory Constraints:*
  Let m_i be the memory capacity requirements of a task i and M_p be the available memory capacity of a processor p. For an assignment A, the total memory required by all modules assigned to processor p must be less than or equal to the available memory capacity of p. That is, the following inequality must hold at each processor p.

$$\sum_{i \in TC_p} m_i \leq M_p$$

- *Network Capacity Constraints:*
  Let b_{ij} be the communication capacity requirements of edge (i,j) and R be the available communication capacity of the network transmission media. For an assignment A, the total communication capacity required by all arcs mapped to the transmission media must be less than or equal to the available communication capacity of the media. That is, the following inequality must hold at the network transmission media.

$$\sum_p \sum_{i \in TC_p} \sum_{j \neq i, (i,j) \in E, A(j) \neq p} \frac{b_{ij}}{2} \leq R$$

3.3 Allocation Model

This paper considers a model where a task must be allocated to exactly one processor and takes into account both the memory and the network capacity constraints. Let X be an $M \times N$ binary matrix whose element $X_{ip} = 1$ if module i is assigned to processor p and $X_{ip} = 0$ otherwise. Then the allocation problem may be formulated as follows:

$$\min \ L_{max} = max_{1 \leq p \leq N} \{EXEC_p + COMM_p\}$$

$$\text{subject to}$$

$$\sum_p X_{ip} = 1 \qquad \forall \ tasks \ i$$
$$\sum_{i \in TC_p} m_i \leq M_p \qquad \forall \ processors \ p$$
$$\sum_p \sum_{i \in TC_p} \sum_{j \neq i, (i,j) \in E, A(j) \neq p} \frac{b_{ij}}{2} \leq R \qquad \forall \ networks$$

4 Heuristic Algorithm

This section presents a heuristic algorithm derived from Simulated Annealing (SA) to the allocation problem. It first defines the basic concept of SA and then explains how it may be employed for solving the allocation problem.

4.1 Basic Concepts

Simulated Annealing (SA) is a global optimization technique which attempts to find the lowest point in an energy landscape [15,16]. The technique was derived from the observations of how slowly cooled molten metal can result in a regular crystalline structure. It emulates the physical concepts of *temperature* and *energy* to represent and solve the optimization problems. The procedure is as follows: the system is submitted to high temperature and is then slowly cooled through a series of temperature levels. At each level, the algorithm searches for the system equilibrium state through elementary transformations which will be accepted if they reduce the system energy. However, as the temperature decreases, smaller energy increments may be accepted and the system eventually settles into a low energy state close to the global minimum. Several functions have been proposed to determine the probability that an uphill move of size Δ may be accepted. The algorithm presented in this paper uses $exp(-\Delta/T)$, where T is the temperature.

4.2 Simulated Annealing Algorithm

To describe the SA algorithm, some definitions are needed. The set of all possible allocations of tasks into processors is called *problem space*. A *point* in the problem space is a mapping of tasks to processors. *Neighbors* of a point is the set of all points that are reachable by moving any single task from one processor to any other processor. The *energy* of a point is a measure of the suitability of the allocation represented by that point. The structure of the algorithm may be sketched as follows:

```
Randomly select an initial solution s;
Compute the cost at this solution E_s;
Select an initial temperature T;
Select a cooling factor α < 1;
Select an initial chain n_rep;
Select a chain increasing factor β > 1;
Repeat
  Repeat
    Select a neighbor solution n to s;
    Compute the cost at n, E_n;
    Δ = E_n - E_s;
    If Δ < 0,
      s = n; E_s = E_n;
    Else
      Generate a random value x in the range (0,1);
      If x < exp(-Δ/T),
        s = n; E_s = E_n;
    End
  End
  Until iteration = n_rep (equilibrium state at T)
  Set T = α * T;
  Set n_rep = β * n_rep;
Until stopping condition = true
E_s is the cost and s is the solution.
```

4.3 Applying the Algorithm

As can be seen from above, the SA algorithm requires an energy function, a neighbor function, a cooling function and some annealing parameters. The energy function is the heart of the algorithm. It shapes the energy landscape which affects how the annealing algorithm reaches a solution. For the allocation problem, the energy function represents the objective function to be optimized and has to penalize the following characteristics:
(i) Tasks duplication,
(ii) Processors with a memory utilization $\succ$ 100%,
(iii) Transmission media with capacity utilization $\succ$ 100%.
These characteristics should be penalized to achieve the application requirements and validate the resources availability.

In the solution development, the first property is penalized by constructing an allocation vector A whose element $A(i)$ represents the processor p where a task i may be assigned and therefore each task can not be allocated to more than one processor. For an assignment A, the second property is penalized by comparing the required memory capacity of all the tasks allocated to a processor p and the available memory capacity of p. An energy component E_{mem} is determined such that $E_{mem} = 1$ if the memory constraints are not satisfied and 0 otherwise. By the same strategy, the third property is penalized by testing and returning an energy component E_{cap} such that $E_{cap} = 1$ if the communication constraints are not satisfied and 0 otherwise. Let k be a penalty factor, then the energy function E may be formulated as:

$$E = L_{max} + k\left(E_{mem} + E_{cap}\right)$$

In this paper, a neighboring solution is obtained by choosing at random a task i from the current allocation vector A and assign it to another randomly selected processor p. For the cooling process, a geometric cooling schedule is used. An initial temperature T is set after executing a sufficiently large number of random moves such that the worst move would be allowed. The temperature is reduced in so that $T = \alpha \times T$, where $\alpha = 0.90$. At each temperature, the chain n_{rep} is updated in a similar manner: $n_{rep} = \beta \times n_{rep}$, where $\beta = 1.05$.

Figure 2 shows the behaviour of the SA algorithm for allocating a randomly generated task graph of 100 tasks into a distributed systems of 4 computers. The figure shows that the cost is unstable at the beginning but it rapidly improves and converges after a short latency time to the best cost.

5 Performance Evaluation

The proposed algorithm is coded in Matlab and tested for a large number of randomly generated graphs that being mapped into a distributed system of 4 computers with bus topology. Furthermore, the quality of the resulting solutions are compared with those derived using the Branch-and-Bound (BB) algorithm [14] which is also coded in Matlab and applied on the same instances.

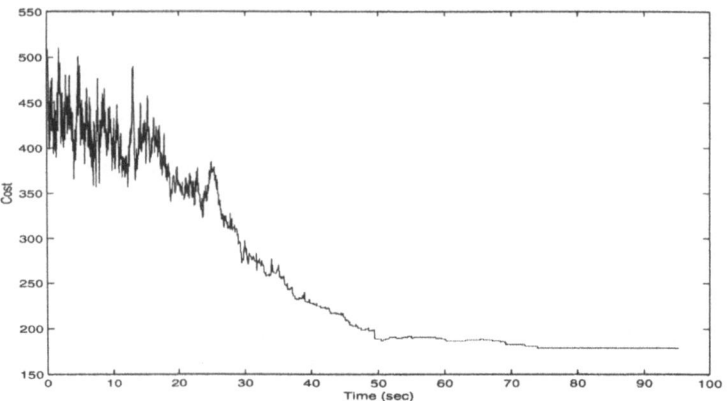

Fig. 2. Typical Behaviour of Simulated Annealing

The simulation results are shown in Figures 3, 4, 5 and 6. Figure 3 illustrates the computation time of the SA and the BB algorithms as a function of the number of tasks. The figure shows that the SA algorithm finds a solution very fast in comparing with the BB algorithm. Furthermore, the computation time of the SA algorithm slowly increases as the number of tasks increases. Figure 4 shows the quality of solutions resulting by SA with those derived by using the BB on the same instances. Figures 5 and 6 illustrate the workload distribution on the 4 processors of the system by applying SA and BB algorithms respectively. As shown in the figures, the workload distribution resulting by using the SA algorithm is very close to the optimal workload distribution.

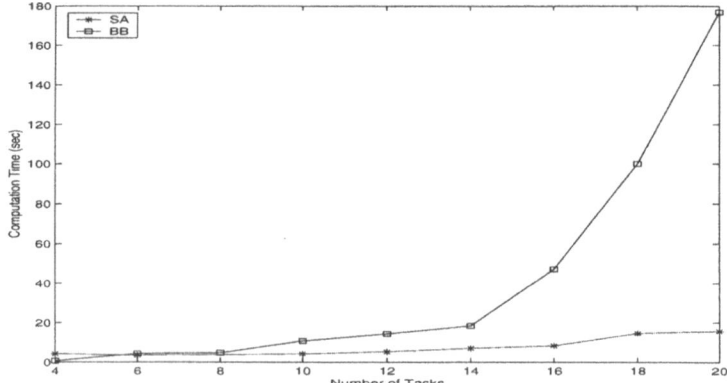

Fig. 3. Algorithms Computation Time

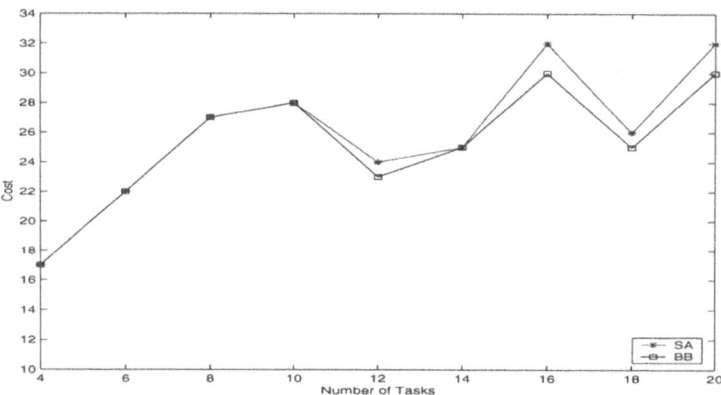

Fig. 4. Optimal and Suboptimal Completion Time

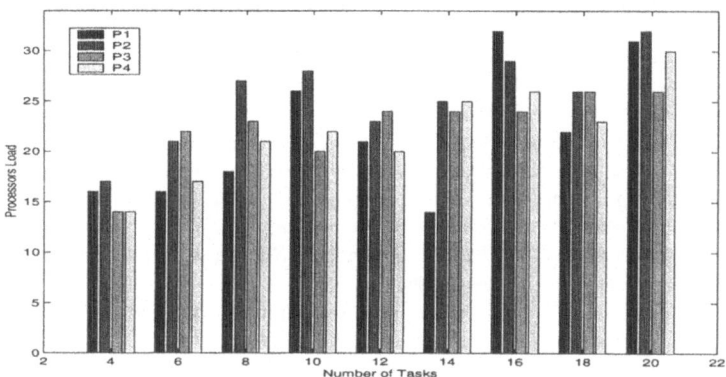

Fig. 5. Suboptimal Workload Distribution by SA

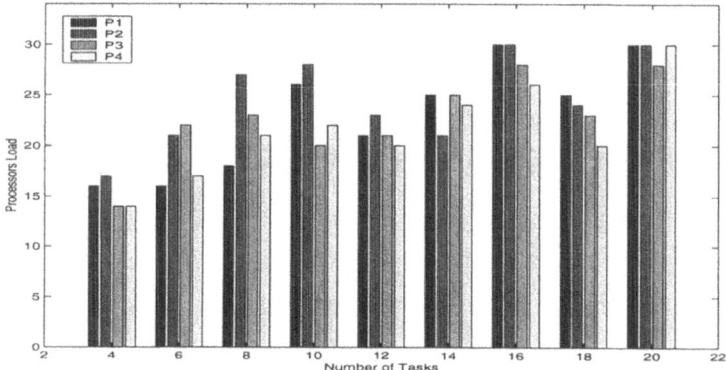

Fig. 6. Optimal Workload Distribution by BB

6 Conclusions

A heuristic algorithm for allocating a parallel program into a heterogeneous distributed computing system is presented in this paper. The goal is to minimize the entire program completion time. The algorithm derived from the well known Simulated Annealing (SA) and tested for a large number of randomly generated task graphs. The effectiveness of the algorithm is evaluated by comparing the quality of solutions with those derived using the Branch-and-Bound (BB) technique on the same sample problems. The simulation results show that, the SA is an efficient approach to the task allocation problem. The algorithm guarantees a near optimal allocation in acceptable amount of computation time.

References

1. C.-H. Lee, D. Lee, and M. Kim. Optimal Task Assignment in Linear Array Networks. *IEEE Trans. Computers*, 41(7):877-880, July 1992.
2. C.-H. Lee, and K. G. Shin. Optimal Task Assignment in Homogeneous Networks. *IEEE Trans. on Parallel and Dist. Systems*,8(2):119-129, Feb. 1997.
3. M. Kafil and I. Ahmed. Optimal Task Assignment in Heterogeneous Distributed Computing Systems. *IEEE Concurrency*, 42-51, July-Sept. 1998.
4. A. Tom and C. Murthy. Optimal Task Allocation in Distributed Systems by Graph Matching and State Space Search. *The J. of Systems and Software*, 46:59-75, 1999.
5. Y.-C.Ma and C.-P.Chung. A Dominance Relation Enhanced Branch-and-Bound Task Allocation. *J. of Systems and Software*, 58:125-134, 2001.
6. Gamal Attiya and Yskandar Hamam. Static Task Assignment in Distributed Computing Systems. 21st *IFIP TC7 Conference on System Modeling and Optimization*, Sophia Antipolis, Nice, France, 2003.
7. V. M. Lo. Heuristic Algorithms for Task Assignment in Distributed Systems. *IEEE Trans. on Computers*, 37(11):1384-1397, Nov. 1988.
8. P. Sadayappan, F. Ercal and J. Ramanujam. Cluster Partitioning Approaches to Mapping Parallel Programs Onto a Hypercube. *Parallel computing*, 13:1-16, 1990.
9. T. Chockalingam and S. Arunkumar. A Randomized Heuristics for the Mapping Problem: the Genetic Approach. *Parallel Computing*, 18(10):1157-1165, 1992.
10. T. Bultan and C. Aykanat. A New Mapping Heuristic Based on Mean Field Annealing. *J. of Parallel and Distributed Computing*, 10:292-305, 1992.
11. P. Bouvry, J. Chassin and D. Trystram. Efficient Solution for Mapping Parallel Programs. *Proceedings of EuroPar'95*, Vol. 966 of LNCS, pages 379-390, Springer-Verlag, August 1995.
12. J. Aguilar and E. Gelenbe. Task Assignment and Transaction Clustering Heuristics for Distributed Systems. *Information and Computer Sciences*, 97:199-219, 1997.
13. M. A. Senar, A. R. Ripoll, A. C. Cortes and E.Luque. Clustering and Reassignment-Based Mapping Strategy for Message-Passing Architectures. *J. of Systems Architecture*, 48:267-283, 2003.
14. Gamal Attiya and Yskandar Hamam. Optimal Allocation of Tasks onto Networked Heterogeneous Computers Using Minimax Criterion. *Proceedings of International Network Optimization Conference (INOC'03)*, pp. 25-30, Evry/Paris, France, 2003.
15. S. Kirkpatrick, C. D. Gelatt and J. M. P. Vecchi. Optimization by Simulated Annealing. *Science*, 220:671-680, May 1983
16. E. Aarts and J. Korst. Simulated Annealing and Boltzmann Machines. John Wiley and Sons, New York, *1989*.

Fault Detection Service Architecture for Grid Computing Systems

J.H. Abawajy

Deakin University
School of Information Technology
Geelong, Victoria, Australia.

Abstract. The ability to tolerate failures while effectively exploiting the grid computing resources in an scalable and transparent manner must be an integral part of grid computing infrastructure. Hence, fault-detection service is a necessary prerequisite to fault tolerance and fault recovery in grid computing. To this end, we present an scalable fault detection service architecture. The proposed fault-detection system provides services that monitors user applications, grid middlewares and the dynamically changing state of a collection of distributed resources. It reports summaries of this information to the appropriate agents on demand or instantaneously in the event of failures.

Keywords: Fault-tolerance, grid computing, fault-detection, grid scheduler, reconfigurable infrastructure.

1 Introduction

Grid computing [11] allows coupling of a wide variety of geographically distributed computational resources (e.g., supercomputers, compute clusters, storage systems, data sources, and instruments) in such a way that they are used as a single, unified resource for solving large-scale compute and data intensive applications. Although successful grid infrastructures [4], [2], [9] have simplified access and usage to grid systems, grid applications, tools, and systems have been either ignoring fault-tolerance issues, or different applications have been adopting ad hoc fault-tolerance mechanisms which cannot be reused, nor shared among them [5], [7].

Fault-detection scheme in grid computing is a necessary prerequisite to fault tolerance and fault recovery [10]. Faults can be generated from the communications links, grid system component (e.g., scheduling middleware), grid node (e.g., a node hosting the scheduling, middleware), grid task and grid application. Therefore, these system components and user applications have to be monitored and in case a failure occurs, actions have to be taken to correct it. Previous research on the fault detection service systems has listed scalability; low-overhead; ability to concurrently control of the monitored resources; flexible; timeliness;

accuracy and completeness; and fault-tolerant as main desired properties of the fault detection services [5], [10], [7].

In this paper, we present a generic failure detection service that can be used for early detection of failures in applications, grid middleware and grid resources. The common trend among existing FDS are designed for monitoring user application processes or system processes exclusively. In contrast, the proposed FDS is designed to allow simultaneous monitoring of both grid system processes (e.g., scheduling middleware) and user application processes. Another problem with existing systems is that they tend to be useless under network partitioning and host failure. For example, the Globus HBM [10] detects process failure by the received and missing heartbeats and is only effective when the host and network connections are functioning properly. Also, they are developed under the assumption that both the grid generic server and the heartbeat monitor run reliably [5]. Moreover, they scale badly in the number of members that are being monitored [6], require developers to implement fault tolerance at the application level [10]; difficult to implement [10] and have high-overhead [7]. We present an efficient and low-overhead multi-layered distributed failure detection service that release grid users and developers from the burden of grid fault detection and fault recovery. Our approach sets up a cascading hierarchy of monitoring domains to improve its scalability, which is the ability to handle large-scale monitoring tasks. The hierarchy allows us to retrieve and process data in parallel, restrict nodes to interesting sets of the resources, and cut down the communication to higher level domains. The proposed failure detection scheme is provided as a basic services designed to enable the construction of application specific fault recovery mechanisms, critical system and grid resources.

The rest of the paper is organized as follows: Section 2 presents the fault detection service architecture and its components. In Section 3, the details of the proposed fault detection service along how it enables the detection of the applications, scheduling middleware and so forth is illustrated. The conclusion and future directions is given in Section 4.

2 Fault Detection Service Architecture

Our goal here is to provide a simple, highly reliable mechanism failure detection and recovery services (FDS) for monitoring the state of user applications, grid middleware and grid resources. Figure 1 shows the overall architecture of the proposed failure detection and recovery services (FDS) system. In the figure, processors are represented by circles and health monitors (HMs) are represented by rectangles. The grid computing of interest is composed of n sites (i.e., $Site_1, ..., Site_n$) each site is managed by a resource management system (RMS) that performs several important functions such as scheduling tasks, controlling access to shared resources, providing for job submission and monitoring.

With respect to user applications, the FDS is designed to detect and report the failure of applications that have identified themselves to the FDS. It en-

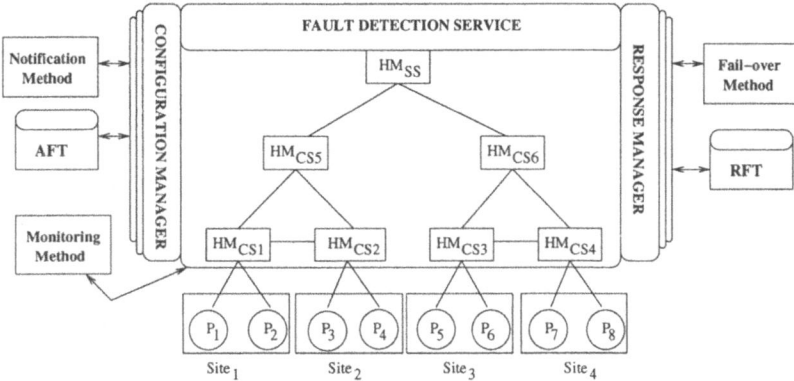

Fig. 1. Fault detection and recovery service architecture

ables the detection of generic task crash failures as well as allows users to define exceptions to handle application-specific failures. Also, it provides notification of application status exception events, so that recovery actions can be taken. By virtue of allowing users to control which entities are monitored, how often they are monitored, the criteria used to report failure, and where failures are reported, the proposed fault detection service provides the flexibility property. An application is made reliable by requesting fault-tolerance services from the FDS. Also, an application can register itself to FDS for its own failure detection and recovery. These services are realized with three main components of the FDS: (i) Configuration Manager (CM); (ii) Health Monitor; and (iii) Response Manager (RM). Figure 1 illustrates the relationships between these three components. These three fault management components collectively construct system configuration information by monitoring the entities and restarting them if need be or reconfigure the system to provide continuous availability of services. The following subsections explain the function of each of these components in detail.

2.1 Configuration Manager

The configuration manager (CM) provides methods for applications to register with the system in order to use the FDS service. It is also responsible for notifying all interested parties (e.g., applications) when a change in status of a particular application, host, link or grid middleware occurs (e.g., when a host crashes or when it recovers and rejoins the network). The CM maintains all the state associated with application. When an application registers with the CM, the information provided in the registration is added to the AFT table. The registration information includes notification-listener(s), the method of notification (polling or heartbeat), type of error to be monitored, fail-over strategy, and the failure recovery method.

A notification-listener is an entity that is interested in the aliveness of application and its components (i.e., processes). The notification-listener can be the scheduling middleware or the user who submitted the application. The main role of the notification listener is to receive the notification messages sent by the CM, and analyze those notification messages being delivered to determine the state of the entities of interest (e.g., whether they die or not, whether they have finished successfully or not etc.). The CM also provides two methods to communicate the status of the application to users: *on-demand notification* and *instant-notification*. In the former case, CM simply keeps track of the information and only make available to the user upon a request from users. This method is appropriate for the case where the user cannot specify the notification-listener. In the *instant-notification*, the user is notified as soon as the failure of a task is detected. This is appropriate for the case when there is a specific event listener is defined. An application may optionally register with CM a customized recovery actions. the RM (discussed blow) is responsible for invoking the recovery action after a process failure.

2.2 Response Manager

Response Manager (RM) provides methods for instantiating the appropriate recovery strategies for the registered applications. It also activates new replicas for the health monitors group during a fail-over to make the group size constant. In addition, reconfigures the resources such that the system will be able to deliver a continuous scheduling service to applications despite system component failures and repairs. The default recovery action is kill-and-restart. Hence, RM is also responsible for killing and restarting an application process during failure recovery. However, we allow an application to register its recovery actions and recovery objects. This allows RM to perform application customized recovery actions when the failure occurs.

A process migration (fail-over) is needed not only when a host failure is detected but also when a process fails too frequently on a host. In the latter case, it is useful to migrate a process to a more healthy host even if no host crash has occurred. FDS supports two fail-over strategies - Threshold-based and Failure-based. The Threshold-based strategy allows a process to be migrated to another host after the number of times that a process has failed on a given node exceeds a given threshold. In contrast, the Failure-based scheme migrates a process to another host each time a process failure occurs.

2.3 Health Monitor

The health monitor is responsible for monitoring the state of the registered applications, grid resources and middleware service. In case they deviate from their normal state, it immediately notifies the change to CM for the necessary actions to be taken. To address the scalability property without sacrificing its efficiency (i.e, low-overhead), the HMs are arranged in a hierarchy as shown in Figure 1. Each HM is responsible to monitor its immediate descendants (i.e., children) and provides following function:

1. Register or unregister to the parent.
2. Receive the monitoring instructions from parent and attribute the monitoring session to its children.
3. Merge/sort the monitoring data from its children and forward it to it parent.
4. Handle the fault events or forward them to the appropriate agent.

For example, HM_{CS1} monitors the health of P_1 and P_2 and periodically forwards the result to HM_{CS5}, which in turn merges the results from HM_{CS1} and HM_{CS2} and forewords it to HM_{SS}.

To address the fault-tolerance aspect of the HM itself, we explores the *parent-child* and *sibling* relationship between the HMs to implement fail-over strategy. Each parent HM is complemented with a configurable number of backup HM selected from its children HMs. These backup HMs are ranked as first backup, second back,..., (k-1) backup. The ranking of the backup HMs can be based on several factors such as the distance from the parent HM or the capacity of the child HM. The highest ranking backup HM is said to be a *hot backup HM (HBS)*. In other words, every state change in the primary HM is propagated to the HBS right away. The other backup HMs are *cold backup health monitor (CBHM)*, which means that there is no propagation of the primary HM state.

3 Fault Detection Service

The proposed FDS is built with mechanisms to tolerate a large variety of failures. The various types of failures dealt with are: process crash/hang, host crash, link crash, health monitor crash, configuration management crash and response manager crash. In this section, how the proposed FDS detects some of these possible failure scenarios.

3.1 User Application Failure Detection

Once an application has registered, the health monitor (HM) periodically monitors the application to detect failures. The HM monitors applications at job-level (between non-leaf nodes and their parents) and at task-level (between leaf nodes and their parents) as shown in Figure 2. A monitoring message exchanged between a parent and a leaf-level node is called a *report* while that between non-leaf nodes is called a *summary*. A report message contains status information of a particular task running on a particular node and sent every *REPORT-INTERVAL* time units. In contrast, the *summary* message contains a collection of many reports and sent every *SUMMARY-INTERVAL* time periods such that *REPORT-INTERVAL < SUMMARY-INTERVAL*.

HM supports two failure detection mechanisms: *polling* and *heartbeat*. In polling, HM periodically sends a ping message to an application. If the ping fails, it assumes that the process has crashed. In the heartbeat mechanism, an application actively sends heartbeats to HM either on a periodic basis or on a per request basis. If HM does not receive a heartbeat within certain duration,

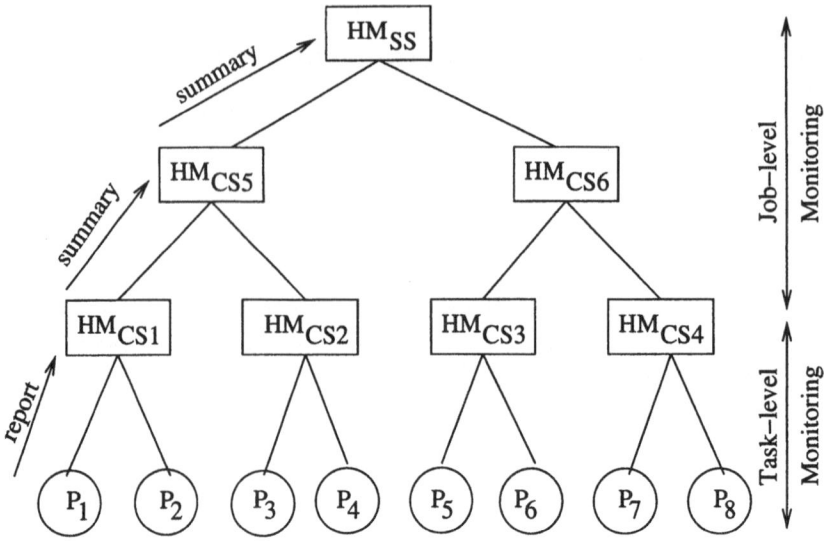

Fig. 2. Monitoring Jobs and Tasks.

the process is considered hung. The heartbeat mechanism is capable of detecting both crash and hang failures of a process or a host, whereas the polling mechanism is only capable of detecting crash failures. Applications may select one of these two approaches based on their reliability needs.

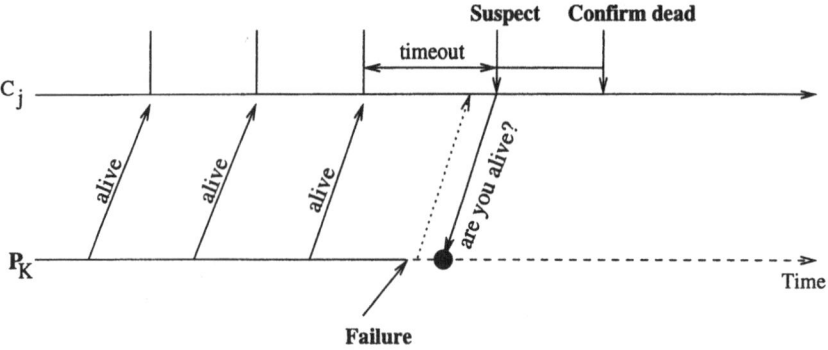

Fig. 3. Failure detection with health monitor.

Figure 3 illustrates how the failure detection works under heartbeats-based scheme. Node P_k periodically sends to the health monitor (HM) a small message indicating that it is alive. When the HM fails to receive the message from the node and after it has waited for timeout period it declares the node as failed. When HM detects a process crash or hang, it reports the failure to CM so that a possible fail-over actions can be taken. If a process fails too many times on a host or if the host on which the process is executing crashes, CM migrates (i.e. fails over) the process from one host to another host.

3.2 Detecting Link Faults

At regular intervals, the primary HM (i.e., parents) will send a *heartbeat* message to at least two children. The HMs on the two children use the *heartbeat* message from the primary HM for the purpose of confirming that the primary HM is alive and the link between them is healthy. The HMs on the children monitor the state of the network between them and their parents as well as the proper functionality of their parents. As shown in Figure 4, the HM running on the parent HM (i.e., P_k), at regular intervals sends an *alive* message to at least two children HMs (i.e., C_i and C_j) indicating that it is alive. In Figure 4, the dotted lines indicate a missing alive messages from P_k. Assume that C_i does not receive the alive message from P_k at the scheduled time. In this case, C_i consults C_j asking it to confirm whether or not it has received the *alive* message from P_k in the current monitoring period. We have two cases to consider:

1. C_j received the alive message: In this case, C_i assumes that the *link* between it and P_k may have failed or very slow. Therefore, C_i waits for another round of monitoring period for the *alive* message from P_k to arrive. If the *alive* message arrives before the waiting time expires, then the link and the P_k are presumed alive. Otherwise, C_i notifies the *CM* the fact that it has not received communication from P_k while C_j did.
2. C_j has not received the alive message: In this case, C_i and C_j send to P_k, the "*Are you alive?*" messages. If no response is received by both children (i.e., C_i and C_j) from P_k within a configurable time period, then P_k is declared dead. If one of the two children, say C_j, receives the response from P_k, then the link between C_i and P_k is declared dead.

If the HM on any of the two children does not receive the *heartbeat* message at scheduled time, the HMs send a *request for confirmation* message to each other. If one of the two HMs confirms that it has received the *heartbeat* message from the parent in the current monitoring period, then the *link* is suspected as being failed. In this case, the HM waits for another round of monitoring period for the *heartbeat* message to arrive before declaring the link as failed. If the *heartbeat* message arrives before the waiting time expires, then the link and the primary schedulers are presumed alive. Otherwise, the link is declared faulty. In case where both monitoring children have not received the *heartbeat* message from the primary scheduler, then either the parent or the link between them has failed. In this case, a *request for confirmation* message is sent to the primary HM at the next level of the tree. If no response message is received by the HMs within a configurable time period, the parent or the link between them is declared as failed.

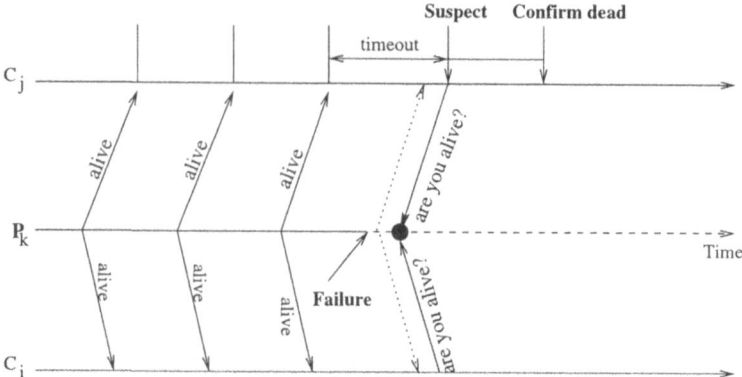

Fig. 4. Failure detection architecture. The dotted lines indicate the missing *alive* messages.

4 Conclusion and Future Direction

There is active research around the world to develop fundamental technologies that would provide pervasive, dependable, consistent and inexpensive access to advanced computational capabilities of the grid system allowing scientists to solve the most challenging computational problems. Failure detection service is one of the core components of a fault-tolerant grid computing systems, which we tackled in this paper. The proposed failure detection service scales well, relative to existing approaches, in the number of nodes. Also, it reduces the communication overhead as it requires the processes or processors to prove that they are alive instead of the usual two-way communication where the failure detection service sends a message to each process or processors and in turn each of these entities reply. In addition, the proposed service reduces the potential of false failures as it uses a two step process where the monitored entity is first suspected as failed and then a second phase to prove or disprove the suspicion is entered. Moreover, it provides timely detection as it uses only one way communication compared to similar approaches. For example in [10], every host in the system runs a single status detector process. The status detector receives heartbeats from the locally running processes in order to detect process failures and it exchanges the heartbeats with other detectors to detect processor failure. This can take a long time to detect failures. Last but not least, the proposed service can also be used to identify, in addition to processes and processors, network failures as will be discussed shortly. We are currently in the process of implementing the proposed approach to study its viability and induced overheads.

Acknowledgment. Financial help is provided by Deakin University. The help of Maliha Omar is also greatly appreciated.

References

1. Tierney, B. Crowley, D. Gunter, M. Holding, J. Lee, M. Thompson A.: Monitoring Sensor Management System for Grid Environments, In Proceedings of HPDC, (2000) 97-104.
2. Grimshaw, A. Ferrari, A. Knabe, F. and Humphrey, M.: Wide-Area Computing: Resource sharing on a large scale, IEEE Computer, 5 (1999) 29-37.
3. Namyoon, W. Soonho, C. hyungsoo, J. Jungwhan, M. and Heon, Y. Taesoon, P. and Hyungwoo, P.: MPICH-GF: Providing Fault Tolerance on Grid Environments, In Proceedings of CCGrid, (2003).
4. James, F. Todd, T. Foster, I. Livny, M. and Tuecke, S.: Condor-G: A Computation Management Agent for Multi-Institutional Grids, In proceedings of HPDC'10, (2001), Available at http://www.cs.wisc.edu/condor/condorg/.
5. Soonwook H.: A Generic Failure Detection Service for the Grid, Ph.D. thesis, institution = "University of Southern California, (2003).
6. Renesse, R. Minsky, Y. and Hayden, M.: A Gossip-Style Failure Detection Service, Technical Report, TR98-1687, (1998).
7. Abawajy, J. H. and Dandamudi,S.P.: A Reconfigurable Multi-Layered Grid Scheduling Infrastructure, In proceedings of PDPTA'03), (2003) 138-144.
8. Nguyen-Tuong, A.: Integrating Fault-Tolerance Techniques in Grid Applications, Ph.D. thesis, The University of Vergina, (2000).
9. Foster, I. and Kesselman,C.: The Globus Project: A Status Report, In proceedings of Heterogeneous Computing Workshop, (1998) 4-18.
10. Stelling, P. Foster, I. Kesselman, C. Lee, C. Laszewski, G.: A Fault Detection Service for Wide Area Distributed Computations, In proceedings of HPDC, (1998) 268-278.
11. Foster, I: The Grid: A New Infrastructure for 21st Century Science. Physics Today, 2 (2002) 42-47.
12. Li, M. Goldberg, D. Tao, W. Tamir, Y.: Fault-Tolerant Cluster Management For Reliable High-Performance Computing, In proceedings of Parallel and Distributed Computing and Systems, (2001) 480-485.

Adaptive Interval-Based Caching Management Scheme for Cluster Video Server[*]

Qin Zhang, Hai Jin, Yufu Li, and Shengli Li

Huazhong University of Science and Technology, Wuhan, 430074, China
hjin@hust.edu.cn

Abstract. In a video server, a good buffer management scheme can reduce the number of disk I/O. Traditional buffer replacement algorithms such as LRU (*least recently used*) and MRU (*most recently used*) only aim at increasing buffer hit ratio, can not guarantee the continuous and real-time needs for continuous media. Interval-based caching scheme can efficiently support cache share, which is a good candidate for video server cache management. In this paper, we propose a new interval-based caching scheme **PISR** (*adaptive dynamic Preemptive Interval-based caching with Size and Rate*), which can fit for heterogeneity environments. Our scheme is proven to outperform the original **PIS** (*preemptive interval-based caching with size*) and **PIR** (*preemptive interval-based caching with rate*) by simulation and performance evaluation.

1 Introduction

With the development of computer technology, the information quantity and quality needed have increased dramatically. Traditional video server cannot meet needs of most people due to its high price or low performance. A cluster architecture that connects some common PCs or workstations with network has high availability, high dependability, and high ratio of performance to price in nature. Therefore, modern video server also takes cluster as its architecture. Figure 1 gives the basic framework of a video server based on cluster architecture.

A video server not like a web server, it has its own characteristics. First, a video server has large amount of data. Usually one MPEG-1 coding video file can last 90 minutes with normal playing speed 1.5Mbps. Hence one file can take nearly 1GB disk capacity. Second, a video server has huge bandwidth demand that enlarge disk system and other computer subsystem. Disk I/O bandwidth does not match with the memory and CPU. It becomes the bottleneck of video server. Third, a video server has real-time constrains and continuous guarantee. A video server has to serve large amount of concurrent clients. All of those put forward new challenge to the video server performance. At the same time, a video server must adapt to multiple media type, different system configurations, various QoS requirements and various interactivities to support heterogeneous environment.

---

[*] This paper is supported by National 863 Hi-Tech R&D project under grant No. 2002AA1Z2102.

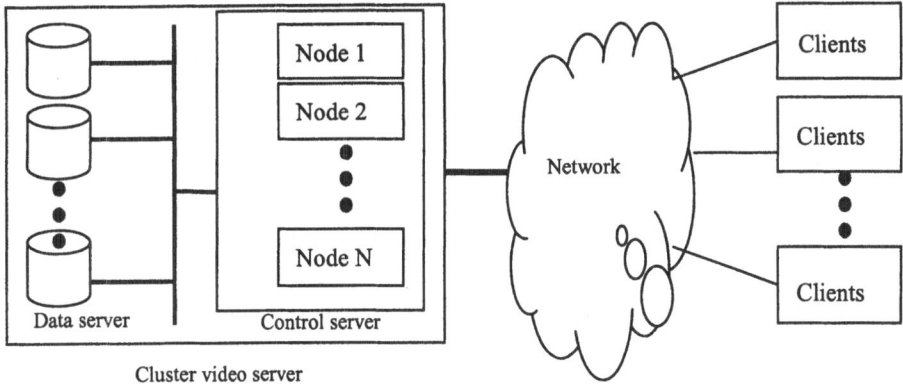

Fig. 1. Cluster Video Server Architecture

Caching has been extensively used in microprocessors and operating systems to speed up memory look-ups or improve the performance of the virtual memory system. If the requested data is already in the cache, a "hit" is scored and the data can be delivered quickly to the requesting client. If the data is not in the cache, a "miss" occurs and the data must be fetched from the disk at the cost of performance slowdown. The ratio of hits to the total number of requests is called the "cache hit-rate". But the cache capacity is limited when the requested data is not in the cache and it is fetched from disk. And then it is written into the cache, some items will then need to be removed from the cache. The item needs to be removed depends on the cache replacement policy of the cache. The video server can take good use the memory cache to provide data share so as to save some disk I/O bandwidth to serve more client requests.

The paper is organized as follows. Section 2 proposes a new cache management scheme that fit for modern video server environment based on carefully study of related work. Section 3 gives a new video stream management scheme that supports interactive control. The implementation that based on open source Linux kernel is given in Section 4. The simulation and performance evaluation is given in Section 5. Section 6 ends with conclusion and future work.

2 Related Work and PISR Cache Management Scheme

Cache management is important to system performance that has large amount of disk I/O. Most operating systems take LRU (*Least Recently Used*) page replacement algorithm as the cache management scheme by default. But for the video server, there are a large amount of sequence disk references; therefore, LRU is not suitable. Other page replacement algorithm such as LRU-K, MRU, LRFU [8][9][10], only aimed at page-hit ratio, not taking continuous and real-time guarantee into account. Hence, they do not fit for the video server, either. Some page replacement algorithms take page prefetching into account, such as BASIC and DISTANCE [7] algorithm. They only support CBR (*Const Bit Rate*), have high implementation load and the parameters are hard to tune.

Some other cache management schemes [1][4][6] take application characteristic into account. Based on different reference patterns, it would take different replacement scheme. For sequence or periodic reference pattern, MRU (*Most Recently Used*) replacement algorithm is used. For probability reference pattern, LFU (*Least Frequency Used*) replacement algorithm is used. For the temporal localization, LRU page replacement algorithm is used. But all of those schemes must know the application's reference pattern, either given by application directly or inspect by themselves online. The interval-based caching was proposed [5] to cache the interval dynamically between the two successive streams for the same video object. Therefore, the later one can be served from the cache that has been visited by the previous stream. Figure 2 illustrates the concept of interval-based caching.

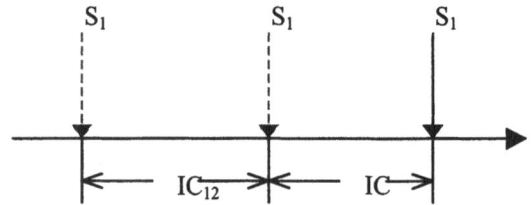

Fig. 2. Concept of Interval-based Caching

There are two kinds of interval-based caching scheme, preemptive and non-preemptive. The former means the interval cache established can be preempted by later stream, which can form another interval less than the preempted one. This only happens with some admission control to guarantee the preemption is safe. The later means that the interval established would not be preempted though there are some later streams can form new interval less than established ones. It is evident that the preemptive scheme outperforms the non-preemptive scheme. The former takes into account of system change and is adaptable dynamically.

There are some preempting algorithms. PIS (*preemptive interval caching-size*) is the original preemptive interval-based caching scheme. It preempts the interval with largest size, and makes the stream number served by maximum cache size. PIR (*preemptive interval caching-rate*) preempts the interval with smallest consuming rate, so as to decrease the I/O pressure on disk. For heterogeneous environment, different type of medium have different consuming rates, and at the same time different QoS for the same media object can have different consuming rates. Taking only one factor (size or rate) would not fit for all situations. In this paper, we propose a new scheme PISR (*preemptive interval based caching-size and rate*), it preempts the interval with largest ratio of size to rate. Therefore, we can adapt for different size and rate.

3 Video Stream Management

In interval-based caching management scheme, the stream served by interval cache is different from the stream served by disk I/O directly. They have different resource requirements, so it was given different state for a stream in different stage. Interval-based cache management only has five states: *Cached, Accepted, Disk Resident, Transient* and *Finished*. The corresponding state transition diagram is given in Fig.3.

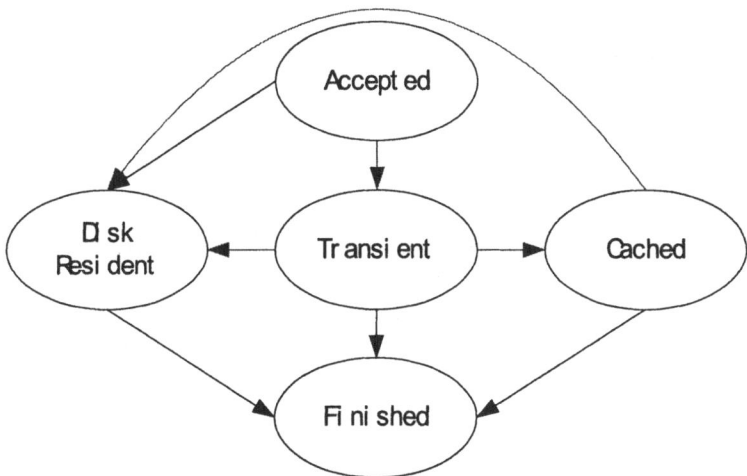

Fig. 3. Stream State Transition Diagram

Accepted state represents the new arrival stream just passed the admission control process. *Disk Resident* state represents the stream served from disk. *Transient* state represents the stream served from disk, at the same time it will consume an interval, which is being fed in the cache space. *Cached* state represents the stream served completely from cache. No disk bandwidth is consumed for this stream in this state. *Finished* state represents the stream finished. No resource is needed for the stream in this state. In Fig.3, the red lines stand for the state transition when preempt happens. The *Transient* or *Cached* state can transform to *Disk Resident* state. The blue lines stand for the process of interval cache comes into being.

Though interval-based caching scheme fits for video server, but original scheme did not support stream interactive such as *Seek, Pause, Restart*. Based on original state we add a new state *Paused*, which stands for a stream has been paused and now has no resource consuming. We define an event set E, which consists of the event that causes stream transform state, E = {E1, E2, E3, E4, E5, E6, E7, E8}. E1 indicates there is no interval but the disk bandwidth is enough. E2 indicates there is an interval but it has not formed. E3 indicates the interval has come into being. E4 indicates the stream is preempted. E5 indicates the stream received *Restart* command and passed admission control. E6 indicates the stream received *Pause* command. E7 indicates the stream received *Closed* command. E8 indicates the stream received *Seek* or other control command and passed admission control. The improved state transition diagram is showed in Figure 4. All the broken lines in Fig.4 are added for interactive control.

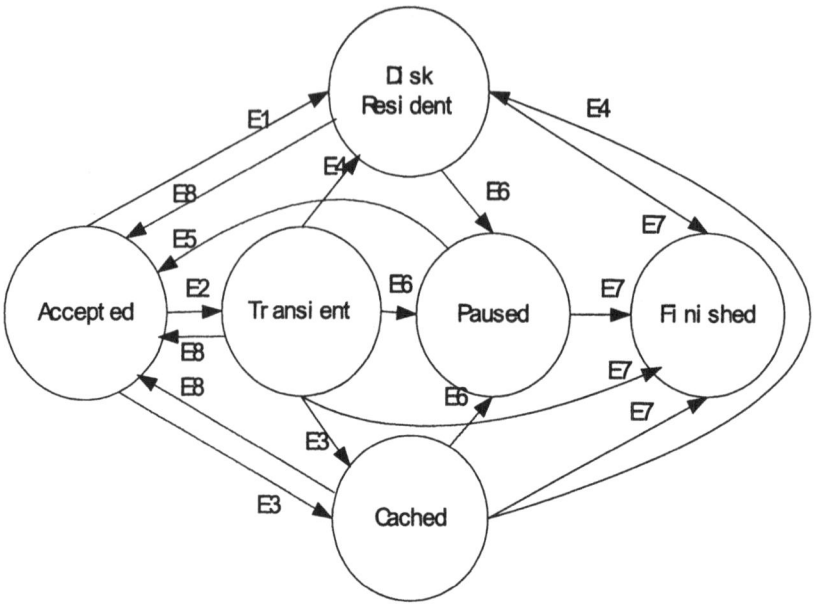

Fig. 4. Improved Stream State Transition Diagram

4 Implementation of PISR Scheme

In order to implement our PISR video server cache management scheme, we analyze open source GUN/Linux kernel cache management in detail. Based on kernel version 2.4.7, we implement PISR cache management and stream management. Figure 5 gives the relationship of stream management data structure. *File* structure is a file descriptor that stands for an opened file. *Inode* structure is a file object, which stands for an entity such as a file on a disk. An inode can have many files at the same time for different users, but all the files opening the same object refer to the same inode.

In order to decreasing memory footprint, we add *stream* and *stream group* data structure. *Stream group* manages multiple streams that visit the same object in succession and form interval cache. At the same time we modify the *file* and *inode* structure, add a pointer in *file* and *inode* that points to *stream* and *stream group* structure, respectively. Therefore, all the streams accessing the same video object are linked by *inode stream group* pointer, and the streams with different consuming rate are in different *stream group*s.

In order to support stream service in Linux kernel, we add some file system calls as user calling interface. We keep the traditional file service syntax, but with *stream_* prefix, such as *stream_open, stream_read, stream_ioctl,* and *stream_llseek*. We give a mixed service frame that maximum the resource utilization, supporting stream and ordinary file services at the same time. The resource of different service taken is dynamically changed based on system payload.

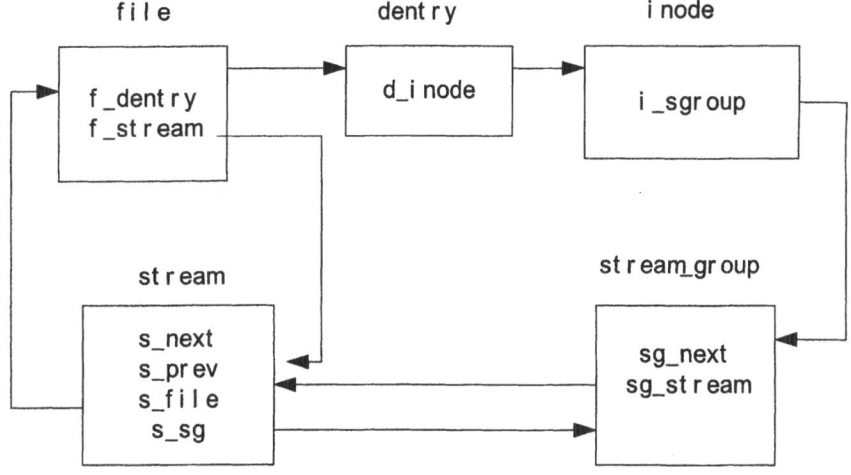

Fig. 5. Relationship of Stream Management Data Structure

The interval cache is managed by extending existing page cache management of Linux kernel. We add *Interval cached* page state to manage the pages belong to interval cache. When a page is used for interval cache, the *Interval cached* state is marked. At the same time the page reference count is increased by one so that it is not freed. When the page is not used for interval cache, the *Interval Cached* state is unmarked. The page reference count is decreased by one, so that it is freed immediately.

5 Performance Evaluation and Simulation

In a video server, the average client arrival interval follows poison distribution with value λ, and the client arrival ratio is $\frac{1}{\lambda}$. The popularity of a video file follows zipf [2][3] distribution with skew θ. The probability of a client visiting *i*th most popular video is P_i, $P_i = \frac{G}{i^{1-\theta}}, i=1,\cdots,I$, where $G = \frac{1}{1 + 1/2^{1-\theta} + \cdots + 1/I^{1-\theta}}$. For simplicity, we suppose a video file is CBR (*Const Bit Rate*) and the network bandwidth is adequate, set zipf distribution with skew 0.271 for all experiments. In order to get the result more accurate, we run large amount of testing for the same condition and give the average value as the final result.

We test multiple media types with normal consuming rate in first experiment. There are 30 files with 90 minutes playing time. They are encoded by MPEG-1 or MPEG-2 with uniform distribution. The request rate is 192KB/s and 400KB/s, respectively. The test result is showed in Figure 6 and Figure 7 to illustrate multiple media type impacts.

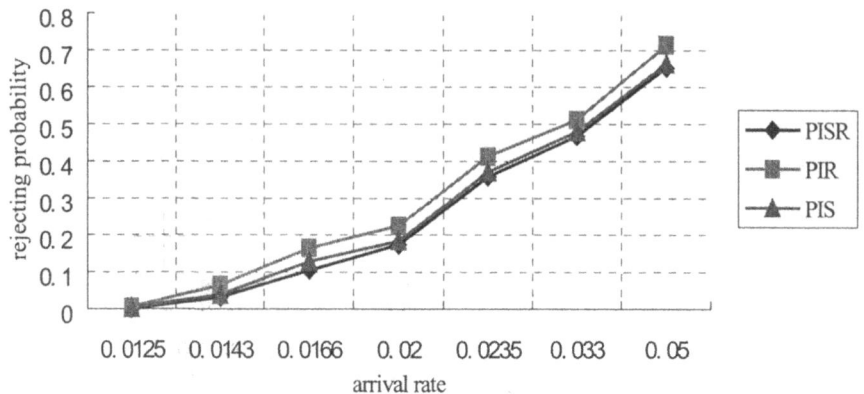

Fig. 6. Rejection Rate for Small Server (Memory 128MB, bandwidth 20MB/s)

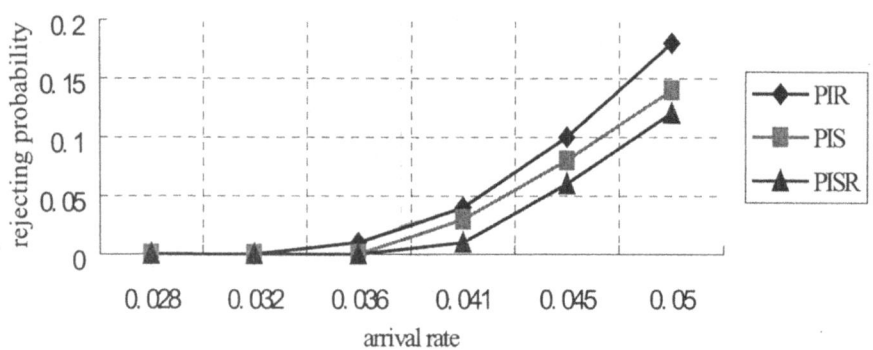

Fig. 7. Rejection Rate for Large Server (Memory 512MB, bandwidth 40MB/s)

We test different QoS requirements for single media type in second experiment. This time we use only MPEG-1 file but with different request rate as 96KB/s, 192KB/s and 384KB/s, respectively. The system configuration is the same as Fig.6. The requesting rate follows uniform distribution, and the play out length changes according to requesting rate. The test result is showed in Figure 8.

We test different system configurations in third experiment. This time the payload is 60 video files encoded in MPEG-1 or MPEG-2 format. The file type follows uniform distribution too. The system disk bandwidth is 20MB/s. The client arrival rate is 0.025. Figure 9 and Figure 10 show the cache capacity impacts for three schemes.

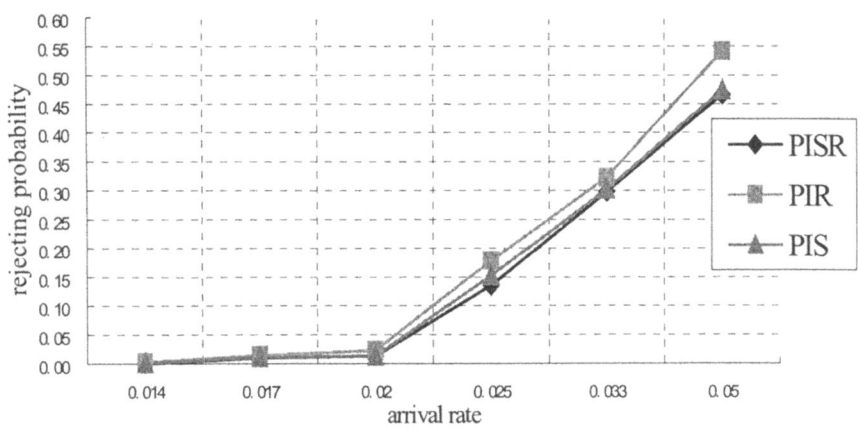

Fig. 8. Different QoS Requesting for Single Media Type

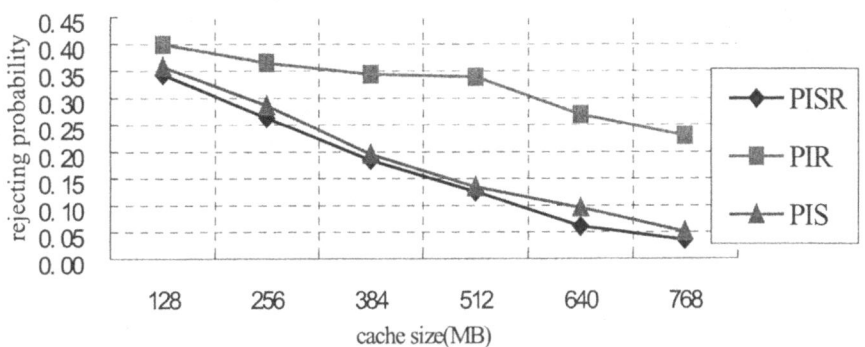

Fig. 9. The Impact of Cache Size for Three Schemes

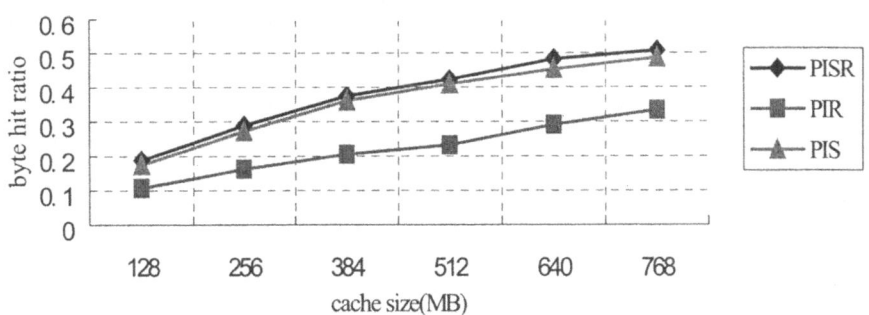

Fig. 10. The Byte Hit Ratio with Different Cache Size

We test effect of different payload on three schemes in fourth experiment. The system is configured with 128MB memory and 20MB/s disk bandwidth. The client arrival rate is 0.02. The payload is changed from 30 files to 80 files. The file distribution is the same as in first experiment. The test result is showed in Figure 11.

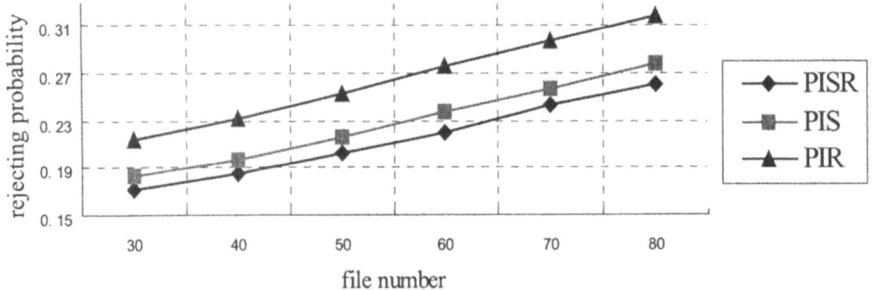

Fig. 11. The Payload Effect for Three Schemes

From above experiments, we find that our PISR scheme outperforms the other two schemes in different conditions. In Fig.6 and Fig.7, the rejection probability of three schemes increases with the client arrival rate, the performance of PIS scheme is close to our PISR scheme. In Fig.9 and Fig.10, with different cache size, rejection probabilities for three schemes decrease and the byte hit-rate increase. It turns out that cache size is important to the video server performance under interval-based caching schemes. We can make up disk bandwidth by adding memory capacity. From Fig.11 we find the performance of a video server is different with different payload.

6 Conclusions and Future Work

Cache management for the video server is the key point to enhance the video server performance. In this paper we summarize some cache management schemes that have been brought forward before. Based on characteristics of the cluster video server, we propose a new cache management scheme PISR. It can fit for heterogeneity environments effectively. We implement our scheme based on open source GNU/Linux kernel, providing video stream management and interactive control. Through simulation and performance evaluation, our PISR scheme outperforms both PIS and PIR scheme.

PISR scheme is implemented and verified in a single computer for the time being. For our cluster video server architecture, we will implement multiple computer cooperative interval-based caching based on Linux kernel to support NUMA (*Non-Uniform Memory Access*) architecture.

References

[1] V. Andronache, E. H.-M. Sha, and N. Passos, "Design of efficient application-specific page replacement techniques for distributed memory systems", *Proceedings of 2000 Conference on Parallel and Distributed Computing and Systems*, IASTED, ACTA Press, 2000, pp.551-556.

[2] M. F Arlitt and C. L Williamson, "Internet web server: workload characterization and performance implications", *IEEE/ACM Transactions on Networking*, 1997, Vol.5, No.5, pp.631-645.

[3] B. Brooks, "Quantitative analysis in the humanities: the advantage of ranking techniques", *Studies on Zipf's Law*, 1982, pp.65-115.

[4] J. Choi, D. Lee, and S. H. Noh, "Characterization and automatic detection of block reference patterns", *Journal of KISS (A) (Computer Systems and Theory)*, Sept. 1999, Vol.26, No.9, pp.1083-1095.

[5] A. Dan and D. Sitaram, "A generalized interval caching policy for mixed interactive and long video workloads", *Proceedings of the SPIE Multimedia Computing and Networking 1996*, SPIE, 1996, pp.344-351.

[6] J. M. Kim, J. Choi, and J. Kim, "A low-overhead, high-performance unified buffer management scheme that exploits sequential and looping reference", *Proceedings of 4th Symposium on Operating Systems Design and Implementation (OSDI 2000)*, USENIX Assoc, 2000, pp.119-134.

[7] T. Kimbrel, A. Tomkins, R. H. Patterson, B. Bershad, P. Cao, E. W. Felten, G. A. Gibson, A. R. Karlin, and K. Li, "A trace-driven comparison of algorithms for parallel prefetching and caching", *Proceedings of Second USENIX Symposium on Operating Systems Design and Implementation (OSDI), Operating System Review*, 1996, Vol.30, Special Issue, pp.19-34.

[8] J.-S. Ko, J.-W. Lee, and H.-Y. Yeom, "An alternative scheme to LRU for efficient page replacement", *Journal of KISS (A) (Computer Systems and Theory)*, May 1996, Vol.23, No.5, pp.478-486.

[9] D. Lee, S. H. Noh, and L. M. Sang, "LRFU: a block replacement policy that exploits infinite history of references", *Journal of KISS (A) (Computer Systems and Theory)*, July 1997, Vol.24, No.7, pp.632-641.

[10] E. J. O'Neil, P. E. O'Neil, and G. Weikum, "An optimality proof of the LRU-K page replacement algorithm", *Journal of the ACM*, 1999, Vol.46, No.1, pp.92-112.

A Scalable Streaming Proxy Server Based on Cluster Architecture*

Hai Jin, Jie Chu, Kaiqin Fan, Zhi Dong, and Zhiling Yang

Cluster and Grid Computing Lab
Huazhong University of Science and Technology, Wuhan, 430074, China
hjin@hust.edu.cn

Abstract. With the explosive growth of multimedia streaming service, streaming proxy server is deployed to reduce response time, server load and network traffic. Existing single node proxy server has limitation on delivering many simultaneously streams. To solve this problem, in this paper, we propose a scalable streaming proxy server based on cluster architecture. We conduct some simulation experiments, which exhibit high scalability and high performance with our design.

1 Introduction

Recent research has explored many streaming proxy caching techniques. For prefix caching [1][2][14], the proxy server caches an initial frames from the beginning of a clip or a movie instead of storing the entire contents, thus upon receiving a client request, the proxy server initiates transmission to the client from the prefix cache while simultaneously requesting the remainder of the stream from the origin server. For segment-based proxy caching [3], blocks of a media object received by the proxy server are grouped into variable-sized, distance-sensitive segments. For memory SRB [4], the proxy server fully utilizes the currently buffered data of streaming sessions by adopting a new memory-based caching algorithm for streaming media objects using *Shared Running Buffers* (SRB). For partial caching [5][6], the caching algorithm based on network-aware and streaming-aware allows partial caching of streaming media objects and joint delivery of content from caches and origin servers.

In this paper, we propose a cluster-based streaming proxy server architecture with support of segment-mass-movie data placement scheme, adaptive cost-based replacement policy and service and streaming coherence.

The rest of the paper is organized as follows. Section 2 lists some related works and issues. Section 3 describes our streaming proxy server architecture based on cluster. Section 4 discusses cache data stripping policy. Section 5 describes disk cache replacement policy. Section 6 presents our performance study, and finally section 7 gives our conclusion and the future work.

---

* This paper is supported by National 863 Hi-Tech R&D project under grant No. 2002AA1Z2102.

2 Related Works

There are much works on stream proxy caching techniques to improve video delivery, reduce the network congestion and origin server load over wide area networks, such as memory caching scheme [4][7][17], disk cache scheme [1][3][8][9][12][13], prefix caching scheme [1][2][14], caching admission policy [10][11] and batching and patching scheme [4][15].

There are also much works being focused on the proxy caching architecture. The "MiddleMan" [16] consists of two types of components: proxy servers and coordinators. It operates a collection of proxy servers as a scalable cluster cache coordinated by a centralized coordinator. *Self-Organizing Cooperative Caching Architecture* (SOCCER) [18] defines a unified streaming architecture with several novel techniques, including segmentation of streaming objects, combined use of dynamic and static caching. The Rcache and Silo project [19] proposes a cooperating distributed caches focusing on scalable fault tolerant data placement and replacement techniques for cached multimedia streams.

3 Streaming Proxy Server Architecture

Figure 1 illustrates our proposed *Cluster-based Streaming Proxy Server* (CSPS) architecture consisting of five types of nodes: *Client Request Nodes* (CRN), *Load Balancer Node* (LBN), a collection of *Arbitrator Handler Nodes* (AHN), a collection of interconnected *Streaming Cache Nodes* (SCN) and *Origin Server Nodes* (OSN).

According to the action of the user, CRN generates corresponding RTSP standard command (*Describe, Play, Pause, Teardown* and so on) and sends command packet to a VOD server. In an environment of configured proxy at client end, the packet is received by the proxy server LBN.

LBN is the single entry point for CGSC. It acts as a load balancer to forward RTSP request to AHNi. AHN accepts RTSP request from LBN, uses RTSP parser module to parse standard the RTSP command packet into internal communication packet, and forwards it to SCNi through UDP protocol. It maintains a global mass map table which entry is (*Moviename, Mass, SCN.IP*) pair.

SCN receives packet from AHNi, parses it and extracts some useful information, sends RTSP response packet to client directly, and begins to delivery stream data simultaneously in the case of cache hit, or sends request to the origin server for fetching streaming data in the case of cache miss.

OSN acts as a VOD server to response the client's request.

In Fig.1, when a user wants to play a movie, CRN sends standard RTSP command packet to LBN. LBN sends a request to AHNi. AHNi decides if the requested movie is in the global map table. If not, selects a SCNi based on SCN selection algorithm and forwards the request to it; otherwise forwards the request to SCNi. Here are two cases to be considered.

Cache hit: Since SCNi caches the partial or full part of the requested movie, SCNi directly sends RTSP response and stream data to the client.

Cache miss: SCNi originates the RTSP request and sends it to the origin server. Upon receiving the request, the origin server sends RTSP response and delivers stream data to SCNi. Once having "intercepted" these, the SCNi begins to cache stream data.

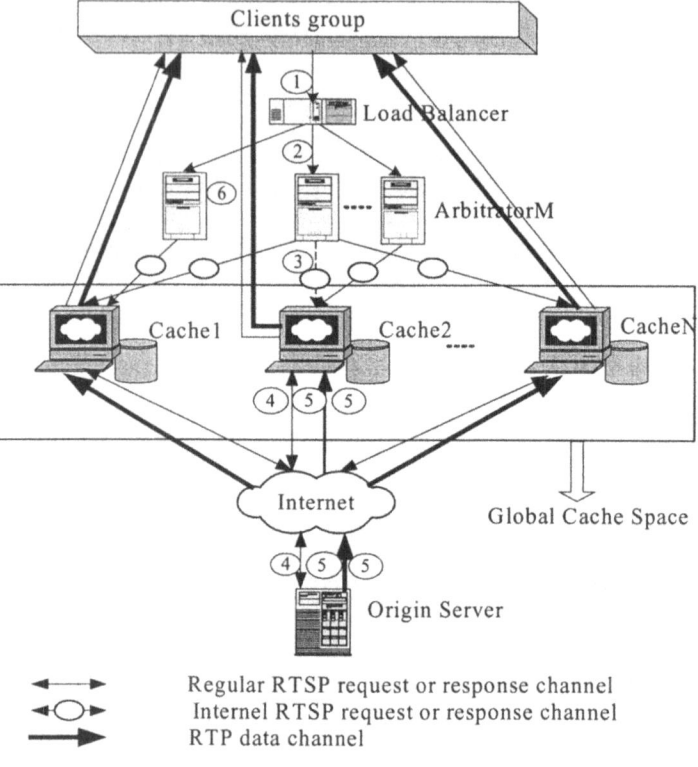

Regular RTSP request or response channel
Internel RTSP request or response channel
RTP data channel

Fig. 1. Streaming Cache Architecture

4 Cache Data Stripping Policy

4.1 Cache Storing and Load Balancing

In our caching system, we introduce a segment mechanism for cache storing. Each segment is a fixed time length (N seconds) composed of many cached RTP packets.

The caching load balancing scheme has a strong influence to the performance of cluster-based proxy server system. For the segment distribution policy, [16][18][19] propose their data stripping policies. But they did not solve the load balancing issue. At the proxy server end, the policy of small-size segment fully striped across the SCN can induce the following problems. As the dynamic distribution of the segments for a video, the AHN must maintain a global segment map table based on segment other than movie. This may exhaust ARN storage space with the increasing of total SCN number. On the other hand, to schedule the request to SCN accurately, when a segment is fetched from the origin server or deleted by replacement algorithm, SCN must inform the AHN to modify the global map table frequently. This introduces many internal communications. To address such problems mentioned above, we must provide an efficient mechanism for stripping segments across SCN, thus allowing for better load balancing, flexible and fine granularity disk allocation and replacement.

We propose a scheme called *Segment-Mass-Movie,* showed in Figure 2. In this scheme, the basic unit of cache storing and cache replacement is a segment. Each mass containing many continuous segments is only a virtual layer on the top of segment. The entire movie comprises N masses, where N is the number of proxy cache nodes configured in the system. When a segment within a mass is chosen by the cache replacement algorithm, if it is the only one in the mass, the cache replacement algorithm evicts it from the disk cache and sends *delete* message to the AHN, then AHN deletes corresponding mass entry. If not, it is evicted directly. When a segment whose mass does not exist is added into the system, a mass should first created in the SCN selected based on the SCN load. Then, SCN sends *register* message to the AHN, and AHN adds the mass to the global mass map table.

In Fig.2, the segments belonging to the same video locate in the same directory, namely *Mass*. movie.meta index file describes key information and SDP information of all cached movies. mass.meta index file describes the cached segments information. segM.meta index file describes cached RTP packets information. Thus the proxy server can access every RTP packet easily according to these index files. Therefore VCR operations can be well supported.

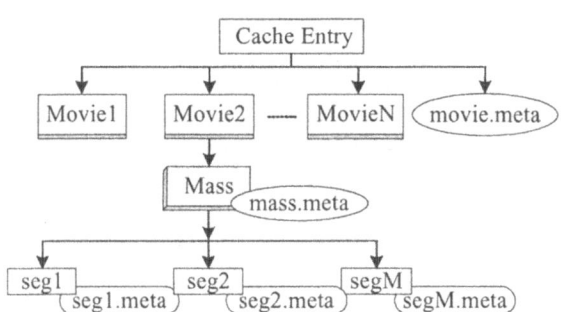

Fig. 2. The Layout of Cache Storing

4.2 Streaming Service Takeover

When the *Segment-Mass-Movie* cache storing and load balancing mechanism is deployed in the system, we must solve the streaming service takeover issue, which includes two issues: searching for the next mass, and RTSP request migration. For the former, the SCNi can send *query* message to AHN to get all masses information of a movie. If another SCN has the required mass, say SCNj, the request is migrated from SCNi to SCNj. If not, the SCNj must also fetch the required mass. For the latter, as long as the RTSP control channel based on TCP protocol between the LBN and the client remains connected, we can smoothly complete the request migration by regenerating the RTP data channel based on UDP protocol.

5 Disk Cache Replacement Policy

An efficient cache replacement policy is needed to make full use of limited disk cache space. To achieve this goal, network cost, media characteristic, user access pattern must be taken into account. Media-characteristic-weighted (MCW-n) cache replacement policy for mixed caching media is proposed in [8]. In [9], a cost-based replacement policy for multimedia proxy across wireless Internet is proposed. To our knowledge, there is no reported work integrating the cost replacement policy with LFU and LRU algorithm across the best-effort Internet.

In our *Adaptive Cost-based Replacement Policy* (ACRP), each segment is assigned a cost as a replacement criterion. When a new segment enters the proxy cache system, and if there is no available disk space, the cache replacement policy is called to evict the segments with the lowest cost. We define the cost value as follow:

$$C = \alpha \times (\beta \times Trend + \lambda \times Frequency + \delta \times Vpopularity), \text{ where } (\beta + \lambda + \delta = 1) \quad (1)$$

where α denotes the network condition between the proxy and the origin server, the poorer the network condition, the larger α. β, λ, and δ parameters reflect the preference for *Trend*, *Frequency*, and *VPopularity*. We can adjust the cache replacement policy to improve the replacement performance or to satisfy the need of users at runtime. *Trend* defines the function of interval since the last access time and reflects the tendency that the future requests may access it. *Frequency* represents the popularity of the segment to be accessed recently. *VPopularity* denotes the video popularity. In general, the condition of $\beta > \lambda > \delta$ stands. α is increased when the network condition between the proxy and the origin server is very poor, or when segments of a video does not want to be replaced. λ, β, and δ values can be tuned to adjust the cache replacement performance.

6 Performance Evaluation

In this section, we use simulations conducted in ns-2 [21] version 2.2.6 to evaluate the effectiveness of our proposed streaming cache techniques. The parameters of simulation system are showed in Table 1. The origin server contains 100 streaming media objects with size from 200MB to 500MB. Each media object is encoded in MPEG-1 format. Each media stream is a constant bit rate UDP stream with 1Kbyte packet size and 1.5 Mbps. We assume the request inter-arrival time follows the Possion distribution ($\lambda=5$) and request distribution conforms to Zipf distribution ($\theta=0.73$).

To evaluate the efficiency of cache load balancing scheme, we introduce the concept of variance and standard deviation to reflect load distribution between each proxy cache node. The definition is as follows:

Table 1. Simulation Parameters

PARAMETER	VALUE
Number of media object	100
Media object size	200MB~500MB
Media encode format	MPEG-1
Request distribution	Zipf(θ=0.73)
Request inter-arrival distribution	Possion(λ=5)
Segment size	10sec
The distribution of transfer delay between origin server and proxy cache	200+(n-1)*100ms (n is server number)
Max Caching Nodes	16
Simulated Caching Nodes	8
Simulated Disk Bandwidth	133Mbps

$$D_n = \sqrt{\frac{(L_1-\bar{L})^2+(L_2-\bar{L})^2+\cdots+(L_n-\bar{L})^2}{n}} \tag{2}$$

$$\bar{L} = \frac{(L_1+L_2+\cdots L_n)}{n} \tag{3}$$

where L_i denotes the proxy cache node load, n denotes the number of simulated cache nodes, $\bar{L}$ denotes the mean of all proxy cache nodes load. We assume that the proxy cache node load is proportional to the number of active streams being serviced. The result is showed in Fig. 3.

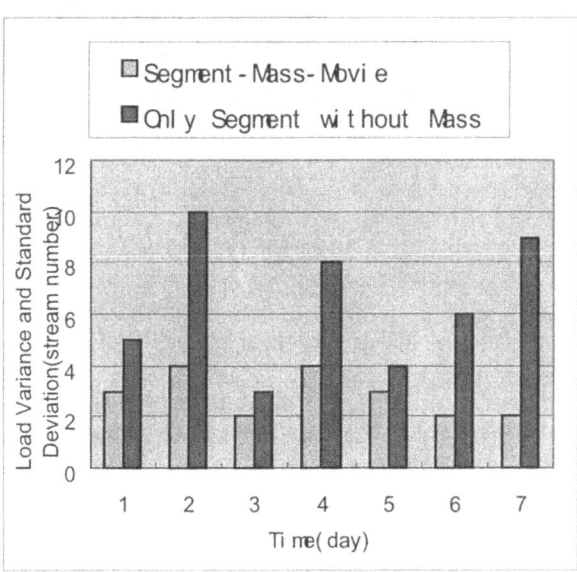

Fig. 3. Load Variance and Standard Deviation for Segment-Mass-Movie and Only Segment without Mass

In streaming proxy cache, *Byte Hit Ratio* (BHR) is introduced to measure the performance of streaming proxy cache replacement.

First, we explore the impact on BHR by varying the parameters of λ, β, and δ. In streaming proxy cache, under the same network conditions, the probability of accessed by requests in the future, *Trend*, contributes to the cost value mainly. *Frequency* takes the second replace. From Fig. 4 we demonstrate it from three aspects: (1) In the extreme case, λ, β, and δ values are all 0, which denotes ACRP does not take it into count in the cost function, when BHR is approximately 0.3, 0.4, and 0.48, respectively. Without considering the coefficient λ, the BHR is the lowest for all three cases. (2) In the other extreme case, λ, β, and δ values are all 1, which represents ACRP only takes it into count in the cost function, when BHR is approximately 0.47, 0.2, and 0.3, respectively. Apparently even if the sum of β and δ is 0, it does not have a great impact to the reduction of BHR. (3) BHR increases initially and subsequently decreases linearly with the increasing of value in horizontal axis, the extent of increasing and reduction also demonstrates this conclusion.

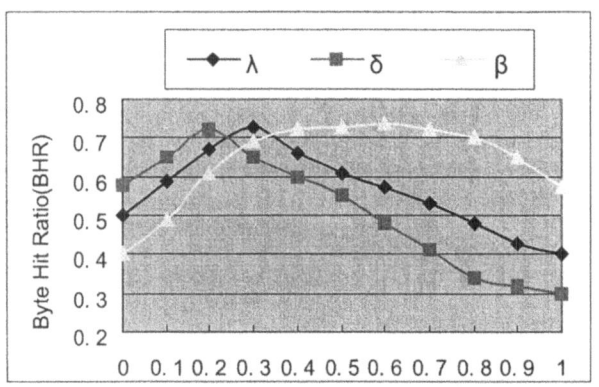

Fig. 4. λ, β, δ vs. Byte Hit Ratio

We assume that three control parameters (λ, β, δ) are used together to be configured as the following values (0.3, 0.5, 0.2). The impact on the BHR by varying the number of caching nodes and single node local cache size against traditional web caching algorithms such LFU, LRU, LRU-2 are described in Fig. 5 and Fig.6.

As showed in Fig.5, with the increasing of local cache size, the system wide cache size increases. It results in that BHR increases for all the algorithms. Furthermore, whatever local cache size is, the performance of ACRP always outperforms LFU, LRU, and LRU-2, which demonstrates the efficiency of our proposed ACRP algorithm. As we know, the scalability is the important issue of the design related to the cluster system. So we evaluate the scalability of our cluster-based streaming proxy cache server. In Fig.6, the BHR increasing behaviors similar to Fig.5, which can be explained that the increasing of caching nodes with constant local cache size equals the increasing of local cache size with const caching nodes. This linear increasing embodies the perfect scalability of our cache replacement policy.

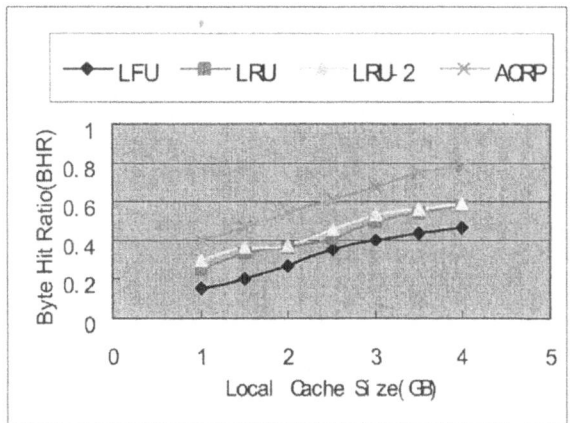

Fig. 5. Local Cache Size vs. Byte Hit Ratio

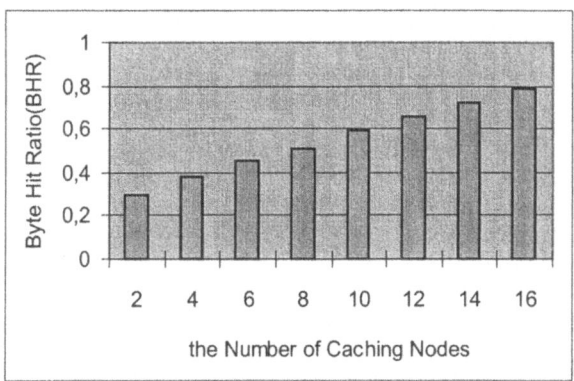

Fig. 6. The Number of Caching Nodes vs. Byte Hit Ratio

7 Conclusion and Future Work

In this paper, we propose a scalable streaming proxy server based on cluster architecture, and discuss techniques such as Segment-Mass-Movie storing and distribution policy, adaptive cost-based disk cache replacement algorithm. The simulation results show the benefits of them when they are integrated into our proxy cache system. More works needs to be done to evaluate the techniques discussed in the paper. In the future, we will explore the impact of the performance of the proxy system induced by the VCR operations. The policy of dynamic transmitting rate at *proxy server to reduce the user perceived latency* will be further studied.

References

[1.] S. Sen, J. Rexford, and D. Towsley, "Proxy prefix caching for multimedia streams", *Proc. of IEEE INFOCOM*, Mar 1999.

[2.] S. Gruber, J. Rexford, and A. Basso, "Protocol considerations for a prefix-caching proxy for multimedia streams", *Proc. 9th International World Wide Web Conference*, Amsterdam, 2000.

[3.] K. Wu, P. S. Yu, and J. L. Wolf, "Segment-based proxy caching of multimedia streams", *Proc. of the 10th International WWW Conference*, Hong Kong, 2001.

[4.] S. Chen, B. Sheng, Y. Yan, and S. Basu, "Shared running buffer based proxy caching of streaming session", *Technical Report HPL-2003-47*, Mobile and Media Systems Laboratory, HP Laboratories, Palo Alto, March, 2003.

[5.] S. Jin and A. Bestavros, "Accelerating Internet streaming media delivery using network-aware partial caching", *Technical Report BUCS-TR-2001-023*, Boston University, Computer Science Department, October 2001.

[6.] S. H. Park, E. Lim, and K. D. Chung, "Popularity-based partial caching for VOD systems using a proxy server", *Proc. of Int. Conf. Parallel and Distributed Processing Symposium*, 2001, pp.1164-1168.

[7.] E. Bommaiah, K. Guo, M. Hofmann, and S. Paul, "Design and implementation of a caching system for streaming media over the Internet", *Proceedings of IEEE Real Time Technology and Applications Symposium*, May 2002.

[8.] R. Rejaie, M. Handley, H. Yu, and D. Estrin, "Proxy caching mechanism for playback streams in the Internet", *Proceedings of Fourth International Web Caching Workshop*, March 1999.

[9.] F. Yu, Q. Zhang, W. Zhu, and Y.-Q. Zhang, "QoS-adaptive proxy caching for multimedia streaming over the Internet", *Proc. of First IEEE Pacific-Rim Conference on Multimedia*, Sydney, Australia, Dec 2000.

[10.] X. Jiang and P. Mohapatra, "Efficient Admission Control Algorithms for Multimedia Servers", *Multimedia Systems*, 1999, pp.294-304.

[11.] J. B. Kwon and H. Y. Yeom, "An admission control scheme for continuous media servers using caching", *Proceedings of International Performance, Computing and Communication Conference (IPCCC)*, 1999.

[12.] Y. W. Park, K. H. Baek, and K.-D. Chung, "Proxy caching based on patching scheme and prefetching", *Proc. of IEEE Pacific Rim Conference on Multimedia*, 2001, pp.558-565.

[13.] S. Chen, B. Shen, S. Wee, and X. Zhang, "Adaptive and lazy segmentation based proxy caching for streaming media delivery", *Proceedings of 13th ACM International Workshop on Network and Operating Systems Support for Design Audio and Video (NOSSDAV'03)*, Monterey, California, USA, June 2003.

[14.] Y. Yang, Z. Lang, D. Du, and D. Su, "A network conscious approach to end-to-end video delivery over wide area networks using proxy servers", *Proc. of IEEE Infocomm*, 1998.

[15.] O. Verscheure, C. Venkatramani, P. Frossard, and L. Amini, "Joint server scheduling and proxy caching for video delivery", *Proceedings of Sixth International Workshop on Web Caching and Content Distribution*, Boston, MA, May 2001.

[16.] S. Acharya and B. Smith, "MiddleMan: a video caching proxy server", *Proceedings of NOSSDAV*, June 2000.

[17.] Dan, D. Sitaram, "A generalized interval caching policy for mixed interactive and long video workloads", *Proceedings of ISWT/SPIE Multimedia Computing and Networking 1996*, San Jose, California, January 29-31, 1996.

[18.] M. Hofmann, E. Ng, K. Guo, S. Paul, and H. Zhang, "Caching techniques for streaming multimedia over the Internet", *Technical Report BL011345-990409-04TM*, Bell Laboratories, April 1999.

[19.] Y. Chae, K. Guo, M. Buddhikot, S. Suri, and E. Zegura, "Silo, rainbow, and caching token: schemes for scalable fault tolerant stream caching", *IEEE Journal on Selected Areas in Communications, Special Issue on Internet Proxy Services*, 2002.
[20.] K. O. Lee, J. B. Kwon, and H. Y. Yeom, "Exploiting caching for realtime multimedia systems", *Proc. of IEEE International Conference on Multimedia Computing and Systems*, Hiroshima, Japan, June 1996, pp.186-190.
[21.] http://www.isi.edu/nsnam/ns/.

The Measurement of an Optimum Load Balancing Algorithm in a Master/Slave Architecture

Finbarr O'Loughlin and Desmond Chambers

Department of Information Technology, National University of Ireland, Galway
boloughlin@ait.ie Des.Chambers@nuigalway.ie

Abstract. Identifying the optimum load balancing algorithm for a web site is a difficult and complex task. This paper examines a number of simulated algorithms based on a master/slave architecture. Three algorithms are used in order to have comparable results to discuss. The first algorithm is the use of a master/slave architecture and processing requests to the relevant servers as a batch of requests. The second algorithm investigated is the standard round robin algorithm used in a master/slave architecture. The final algorithm proposed in the paper is the use of a master/slave architecture that uses the round robin algorithm combined with a reverse proxy of requests. The use of this final combination of algorithms has showed a performance improvement of 19% over conventional master/slave round robin load balancing. The use of batch processing of request shows some interesting findings useful for very heavily loaded web sites with a constant high umber of requests.

1 Introduction

Load balancing as a concept has been around for quite a while. The growth in the popularity of the web in particular has increased the requirement for load balancing. It has been recently estimated that there are 605.6 million web sites in the world [13]. The increase of E-Commerce has lead many businesses to carry out the bulk of their day-to-day business online. As a result of the popularity of the web, providers of web sites want to ensure the availability of access to information for their users and the guarantee that requests are processed as quickly as possible.

Previous research at [12] has described load balancing as follows "In a distributed network of computing hosts, the performance of the system can depend crucially on dividing up work effectively across the participating nodes." Load balancing can also be described as anything from distributing computation and communication evenly among processors, or a system that divides many client requests among several servers. Another description of load balancing would be striving to spread the workload evenly so that all processors take roughly the same amount of time to execute and that no processor gets stuck with the bulk of the compute work, thus obtaining the highest possible execution speeds. Load Balancing can be tackled in a number of different ways depending on the particular system that you have in place and the type of requests that your system may be expected to handle. The handling of a static request and a dynamic request can be quite different depending on the load balancing methodology that is being used. Zhu, Smith and Yang have defined the distinction between static and dynamic requests at [7]. A static request in terms of a web server is a simple file fetch, which involves a simple read from disk. A dynamic request on the other hand is one which involves construction at the server before users requests are returned e.g. report production, database processing etc.

The next phase post examination of what request types are expected is to determine the architecture best suited for a particular system. This is followed by a selection of the optimum algorithm to balance the required load.

An architecture can be described as how it is intended to configure the network of nodes available as the load balancing pool. In the course of this paper an examination of three different architectures are conducted. These are the flat architecture, the master/slave architecture and the diffusive architecture.

While the list of potential load balancing algorithms is lengthy this research has concentrated on a comparison of three algorithms that fall within the master/slave architecture. The remainder of this paper is divided into 5 sections. Section 2 is a technical review of the current architectures and algorithms used in load balancing. Section 3 describes the simulator used to review the algorithms for this research. Section 4 details the experiments used to review and contrast the algorithms tested. Section 5 evaluates the results produced with observations noted. Section 6 concludes this paper.

2 Technical Review

The term flat architecture (Figure 1) has being assigned generally to load balancing that uses just a DNS (Domain Name Server) or switch approach.[7] The method treats all its servers with the same level of importance. The load balance is applied at the DNS or switch level where the decision on what server handles the next request is made. The main protocol used to load balance in this situation is round robin however some advances in DNS and switching has made it possible to introduce more advanced methods of effective load balancing. All servers within this architecture are considered to be at the same level, capable of handling all types of request, irrespective of the data type.

A more advanced architecture the master/slave architecture as in Figure 2 has been put forward by recent research. [5] This architecture is considered to be a more efficient option than the flat architecture. The architecture also offers much greater flexibility in relation to expanding the number of nodes in the server pool.

The methodology used for this architecture has been as a result of research into layered network service research [3] and the design of the Inktomi and AltaVista search engines. The system uses the principles of flat architecture in so far as a switch or a DNS is used for the clients to communicate through to the server pools.

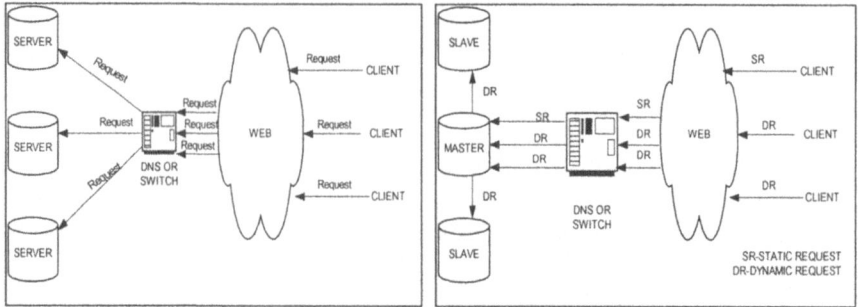

Fig. 1. Flat Architecture   **Fig. 2.** Master/Slave Architecture

The client request arrives from the DNS or switch to the master which handles all static requests. Dynamic requests may be sent to slave nodes or in a large system to other master nodes. Each master can have "n" slave nodes and in turn any slave can be a master with its' own slaves. The fact that the master/slave architecture separates dynamic and static content processing results in the masters servers being able to concentrate on the processing of static content requests leaving the slave servers dedicated to the processing of potentially long running, resource intensive CGI (Common Gateway Interface) scripts, etc.

Tests carried out by [7] showed a staggering improvement in the performance results achieved using a M/S architecture (using three distinct evaluation criteria with a simulation of the number of servers ranging from 32 to 128 and the number of requests ranging from 1000 to 4000 per second) over a flat-architecture.

Diffusive load balancing is a third member of the family of load balancing architectures. It has been researched widely in the past including [14][8]. This architecture works on the same principle as physical diffusion. The underlining principle of this diffusion is that the direction and speed of the diffusing items is determined by the items around them. In a computing environment all machines (nodes) side by side in a distributed manner for load balancing are considered to be neighbours. The decision on where to distribute the load and on how much to distribute depends on the amount of load the neighbouring nodes have. All nodes that act as neighbours for each other form part of a domain, see Figure 3.

The Diffusive architecture allows for four load balancing policies/algorithms to be used, firstly the direct neighbour policy, which works of the simple concept that each domain only uses two nodes therefore the decision on where to move the load is not required the only decision required is on when and what to move. The average neighbourhood policy uses all nodes in a domain. The master node determines what way to spread the load to ensure all nodes have an equal balance. The average extended neighbourhood policy works of the same principle as the average neighbourhood policy but extends the available nodes to a set of nodes that are a set figure or less nodes away from the master. The final policy is the direct neighbour repeated policy which check the load on the neighbour which in turn checks its neighbour and so on until it finds a node with a smaller load and passes the load to that node.

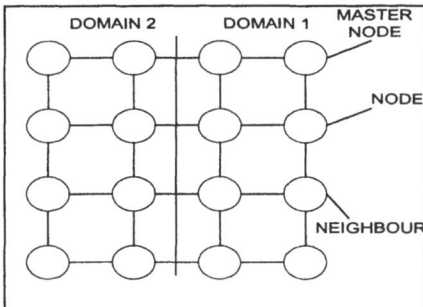

Fig. 3. Diffusive Architecture

The architecture used can dictate to a degree what algorithm is best suited to balance the websites load as has been identified for diffusive load balancing. The other two architectures described have a wide range of policies/algorithms that could be used. A list of the more common algorithms are provided at [11]. The simplest of these algorithms include load balancing with the DNS or switch dictating where each requests is sent.

The round robin algorithm which is another very common algorithm that has been used for some time allows requests to be passed onto each server in a consecutive manner. This algorithm has proved and continues to prove to be useful as a form of load balancing and is the main algorithm used with the flat architecture. This algorithm can also be used within the master/slave architecture as well as a number of adaptations and additions to it.

Round robin has been adapted to some degree to advance the effectiveness of the algorithm by including priority and weighting. Round robin with priority basically adds a new dimension to round robin processing by assigning a different priority to each server or server pool and passes requests to the servers with a highest priority in a consecutive manner. Promotion and demotion of server priorities continues throughout the processing.

Weighting can also be applied to round robin processing by weighting each server on response timings and/or processor power available. If weighting is applied then requests are passed to the server with the highest weight in a consecutive manner at that given time.

The least connections algorithm is simple load balancing processing where by requests are sent to the server with the least number of active sessions at the given time. This algorithm works best in environments where the servers or other equipment you are load balancing have similar capabilities.

Previous research such as [9][1] has investigated randomised load balancing. A randomised algorithm directs traffic randomly between servers. The only control used within the algorithm is that each server is assigned a maximum number of connections within a period. A weighted variety of this is also available which allocates traffic randomly but enables an administrator to configure a weight that requires the load balancing server to take server capacity into account.

Caching requests closer to the client instead of having to pass requests to a server further a field to be dealt with, has also been the focus of much previous research [6][4][2]. This research dealt with caching requests on the servers in the pool. This concept is referred to as DNS caching. Another, newer, variation of this is called the Reverse Proxy (Forward Proxy Caching). This approach can be used in conjunction with another algorithm. It is used by a number of web sites including Real Networks [10]. For this approach to work all requests that pass through a proxy or main server are processed in the normal manner the first time round, with the request farmed out to the relevant server to handle it, with the resultant pages cached in the proxy/main server on return to the client. The next time a user requests the same pages/data the proxy or main server has it available on it own file system. Therefore it does not have to forward the request to its subordinate servers, thus ensuring a faster reply for the client. This type of approach has particular attractions for web site that process the same request a large number of times.

This research has focused on the master/slave architecture with the algorithms listed below in Section 3. The master/slave architecture has been the main focus of this research as it is an expandable and efficient architecture as outlined by [5] in research into the web server clusters with intensive dynamic content processing.

3 Simulator Description

The simulator used for the results delivered in this paper was programmed in java. To simulate the master/slave architecture all static requests (page fetches) are handled by a master server and all dynamic requests (processing of results etc.) are handled by a slave server. The user specifies a number of settings including the number of masters and slaves, total number of requests (including the number of static and dynamic requests) for the test run. Each request is treated as a thread by the simulator allowing for comparable multi request runs.

All servers (master or slave) are separate instances of the server object. To make the simulation realistic a number of settings are applied for the runs. These settings are the CPU Quantum which is set to 10 milliseconds, the remote CGI latency which is set to 4 milliseconds, the priority update period which is set to 100 milliseconds, the input/output burst which is set to 2 milliseconds. These setting are based on previous research in this area [7].

The maximum number of requests a server can handle in a second is set to 1200. This figure may seem excessive but based on hardware available today and the fact that one of the worlds most popular web sites Google has an average of 2900 hits per second [15] it is felt that this is a realistic setting. Additional settings are also applied for the runs including database activity/processing of dynamic requests of 500 milliseconds, this setting is realistic for database processing such as Airline bookings etc.

Cached requests take on reduced settings of 100 milliseconds for dynamic requests as processing does not occur, only file searching and display. The quantum for static requests cached is also halved as the file search time is limited to the cache area on the main server instead of processing the request to the active server and carrying a file search on that server.

The simulator allows for the simulation of three algorithms as follows:

3.1 Batching Requests and Load Balancing in a Master/Slave Architecture

This research proposes the consideration and use of the above algorithm for web sites that have a guaranteed large number of hits per second. The switch or main server holds each request as they arrive and batches them into an even amount to be distributed among all the masters, and slaves, keeping the fundamental master/slave principal of master servers handling static requests and slave servers handling dynamic requests. This algorithm guarantees an even use of each server.

3.2 Load Balancing Using Round Robin in a Master/Slave Architecture

This algorithm has been put forward by a number of researchers including [7]. The simulation of this algorithm allows all the requests to be processed in a round-robin fashion among the masters and in turn each masters slaves. All static requests are handled by a master server, all dynamic requests are handled by a slave server. Each request is treated as a thread to represent the real world.

3.3 Load Balancing in a Master/Slave Round Robin Architecture with Reverse Proxy

This algorithm allows for the same functionality as described in 3.2 above with the added dimension of the reverse proxy. Each request is processed as normal on its first activation in any given run, and cached on the main master, thus when a user requests the same page or data processing there after a different method is run to simulate the processing of a cached request. This results in less file search time and file process time required to locate the required page or process the required data as it is cached on the main master.

4 Experiments/Tests

The flexibility of the simulator allowed for a number of variations in testing to be carried out. The tests concentrated on varying loads and request types to compare load balancing under the different algorithms described above namely: Batch Load Balancing against Round Robin Load Balancing and against Round Robin With Reverse Proxy Load Balancing.

The simulation tests were carried out with the settings of 12 Masters and 10 Slaves per master. The following were applied to the tests carried out:

Table 1. Test Settings

SETTINGS	TEST1	TEST2	TEST3
Number of requests	30000	30000	30000
Number of Static requests	15000	25000	5000
Number of Dynamic requests	15000	5000	25000
Number of possible different static transactions	100	100	100
Number of possible different dynamic transactions	100	100	100

During the course of the tests ten runs were carried out for each test to ensure the results were reflective of an average test. Other active processes were avoided from the testing machine to avoid non-reflective results.

5 Evaluations

5.1 Evaluation of Test 1

This test compared all three algorithms with an equal number of static and dynamic requests. As the graph in figure 4 shows there was a marginal difference between the batch algorithm and the round robin algorithm in relation to overall timing. The round robin with reverse proxy algorithm showed a saving of over 8% on both other algorithms. This obviously is as a result of caching requests following their first run and using the cache where possible there after to answer requests.

However there was a huge difference (in excessive of 90%) in relation to the time the average server and master spent active in the batch algorithm in comparison to both other algorithms. This is a result of having a predefined workload in the run to get through as opposed to both other algorithms that add jobs throughout the run.

5.2 Evaluation of Test 2

Test 2 compared all three algorithms once again, this time with the same number of requests, but with 83%(25000) of the request being static and 17% (5000) being dynamic. Again the findings identified similar results in relation to the batch algorithm for average master and slave activation time in comparison to both other algorithms.

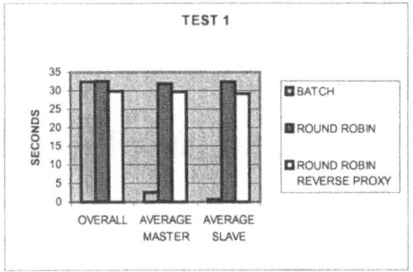
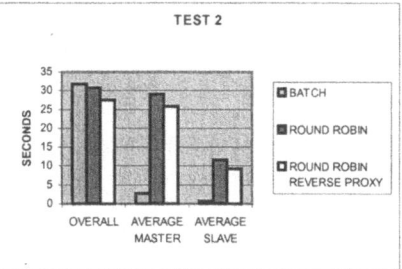

Fig. 4. Test 1 results   **Fig. 5.** Test 2 results

However the average slave time was much lower for the round robin and round robin with reverse proxy algorithms than test 1. This was as a result of 83% of the request being of a static nature, thus master intensive. The fastest algorithm proved to be the round robin with reverse proxy algorithm completing its run in excess of 10% faster than the basic round robin algorithm and some 13% faster than the batch algorithm. These results reiterate the advantage of caching results and not having to process the same request from fresh over and over again. The test results are outlined in figure 5.

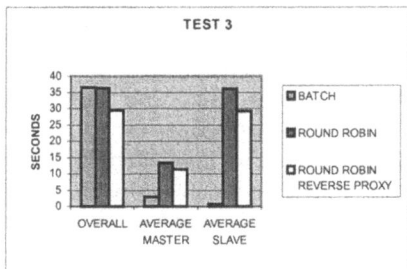

Fig. 6. Test 3 results

5.3 Evaluation of Test 3

Test 3 was similar to test 2 in that all three algorithms were tested under the same load. For this test however the type of requests were changed to reflect 83% of dynamic requests and 17% of static requests. Again, as expected the round robin algorithm with reverse proxy proved to be the most efficient from an overall time perspective. This algorithm was 19% faster than both other algorithms. This is due to the fact that there is a larger time saving in caching dynamic requests as they require significant processing to return values to the user. When a request is completed once and cached, the next time it is called it does not have to be processed, but instead just found in the cache and served. This algorithm is very much slave intensive for the round robin and round robin with reverse proxy algorithms, thus reducing the gap between the average master activation time for the Batch algorithm. The test results are outlined in figure 6.

6 Conclusions

This paper has reviewed load balancing architectures and algorithms. The bulk of the research has concentrated on comparing three distinct algorithms under the master/slave architecture. The algorithms were the conventional round robin algorithm, the round robin algorithm incorporating a reverse proxy for request caching and a batch algorithm that this researched proposed.

The main finding from this research was that processing requests using the round robin algorithm with a reverse proxy showed a 19% improvement in the time taken to process all requests in a given run over both other algorithms. This percentage is based on the overall time to process all requests, from the first to the last. This saving is a result of requests being cached following the normal processing occurring the first time a particular request is run.

An additional finding is the fact that although the batch and round robin algorithms are fundamentally different, both algorithms take a similar time to process all requests in a given run. However a staggering difference was identified in relation to the time that the individual servers spent processing requests for example in test 1, the batch algorithm resulted in constant server activity of in excess of 14 times less than the round robin algorithm. This difference varied with the other tests as was identified in figure 6 and figure 7. This saving is as a result of the algorithm design, which batches together all requests that any given server needs to process during a run. Therefore the servers have high intensive processing for a short period of time. The other two algorithms result in the individual servers being active for a longer period of time but to a much lower intensity.

This research proves that adapting existing algorithms to architectures can prove very beneficial in reducing the time required to process requests. Altering algorithms for load balancing will be an on going area of research for some time.

References

1. 1995 M.Adler, S. Chakrabarti, M. Mitzenmachner, L. Rasmussen. "Parallel Randomized Load Balancing". In Proceedings of the 27th Symposium on Theory of Computing, pages 238-247, New York, NY, USA, May29-June 1 1995. ACM Press.
2. 1995. Chankhunthod, P. Danzig, C. Neerdaels, M. Schwartz and K. Worrell, "A Hierarchical Internet Object Cache", Technical Report 95-611, Computer Science Department, University of Southern California, Los Angeles, California.

3. 1997 A Fox, S.D.Gribble, Y. Chawathe, E.A. Brewer, and P Gauthier. "Cluster-based scalable network services". In Proceedings of the Sixteenth ACM Symposium on Operating System principles, October 1997.
4. 1997 P Cao, S Irani, "Cost-Aware WWW Proxy Caching Algorithms". Proc of the USENIX Symposium on Internet Technologies and Systems, pp 193-206.
5. 1998 H. Zhu, B Smith and T Yang. "A Scheduling Framework for Web Server Clusters with Intensive Dynamic Content Processing". Technical Report TRCS-98-29, CS Dept, UCSB, November 1998. http://www.cs.ucsb.edu/research/rcgi
6. 1998 V. Holmedahl, B. Smith, T. Yang, "Cooperative Caching of Dynamic Content on a Distributed Web Server". In Proceedings of Seventh IEEE International Symposium on High performance Distributed Computing, pages 243-250.
7. 1999 H.Zhu, B.Smith, T. Yang. "Scheduling Optimization for Resource-Intensive Web Requests on Server Clusters". In Proceedings of the eleventh annual ACM symposium on Parallel algorithms and architectures.
8. 1999 A. Corradi, L. Leonardi, F. Zambonelli, "Diffusive Load Balancing Policies for Dynamic Applications". IEEE Concurrency, 7(1):22-31, 1999.
http://polaris.ing.unimo.it/Zambonelli/PDF/Concurrency.pdf
9. 1999 P. Berenbrink, T. Friedetsky, A. Steger. "Randomized and Adversarial Load Balancing". In proceedings of the eleventh annual ACM symposium on Parallel algorithms and architectures.
10. 2000 Realsystem Proxy 8 overview. RealSystem iQ Whitepaper, RealNetworks,
http://www.realnetworks.com/realsystems/
11. 2001 Kunal Dua, PC Quest, "Balance your Web Servers Load"
http://www.pcquest.com/content/technology/101071402.asp
12. 2002 E. Anshelevich, D. Kempe, J. Kleinberg. "Stability of Load Balancing Algorithms in Dynamic Adversarial Systems." In Proceedings of the thirty-fourth annual ACM symposium on Theory of computing.
13. 2002 Nua Internet www.nua.net/surveys/how_many_online/index.html
14. 2003 Jens-Christian Korth. Diffusive Load Balancing. Advanced Seminar "Load Balancing for Massive Parallel Systems". Department for Simulation of Large Systems. University of Stuttgart.
15. 2003 Danny Sullivan, Searches Per Day, Searchenginewatch.com
http://www.searchenginewatch.com/reports/article.php/2156461

Data Discovery Mechanism for a Large Peer-to-Peer Based Scientific Data Grid Environment

Azizol Abdullah[1], Mohamed Othman[1], Md Nasir Sulaiman[1], Hamidah Ibrahim[1], and Abu Talib Othman[2]

[1]Faculty of Computer Science and Information Technology
Universiti Putra Malaysia
43400 Serdang, Selangor
Malaysia
{azizol, mothman,nasir,hamidah }@fsktm.upm.edu.my

[2]Faculty of Information Technology and Communication
Universiti Pendidikan Sultan Idris
35900 Tanjong Malim, Perak
Malaysia
abutalib@upsi.edu.my

Abstract. Data Grid mostly deals with large computational problems and provide geographically distributed resources for large-scale data-intensive applications that generate large data sets. In a modern scientific computing communities, the scientists involves in managing massive amounts of a very large data collections in a geographically distributed environment. Research in the area of grid computing has given us various ideas and solutions to address these requirements. Recently, most of research groups working on the data distribution problems in Data Grids and they are investigating a number of data replication approaches on the data distribution. This leads to a new problem in discovery and access to data in Data Grids environment. Peer-to-peer networks also have become a major research topic over the last few years. In distributed peer-to-peer system, a discovery mechanism is required to locate specific information, applications, or users contained within the system. In this research work, we present our scientific data grid as a large peer-to-peer based distributed system model. By using this model, we study various discovery mechanisms based on peer-to-peer architecture and investigate these mechanisms for our Dynamic Scientific Data Grids Environment Model through our Grid Simulator. In this paper, we illustrate our model and our Grid Simulator. We then analyze the performance of the discovery mechanisms relative to their success rates and bandwidth consumption.

1 Introduction

The communities of researchers in a modern scientific computing involves in managing massive amounts of a very large data collections through a geographically distributed environment. Recently, research in the area of grid computing has given us various ideas and solutions to address these requirements. Peer-to-peer (hence P2P) networks also have become a major research topic over the last few years. In P2P distributed systems, object location is a major part in the operation. The grid and P2P computing represents the notion of sharing resources available at the edge of the Internet. There are two aspects of technology in grid computing: sharing of data, which is refers as Data Grid and sharing of resources such as processors, which is refers as Computational Grid. In P2P computing, it success was originally boosted by some very popular file sharing applications (e.g., Napster). In scientific communities, these resource sharing technology are required as to form a high-performance computing environment. However, in this paper, we are focusing only on Data Grid (data sharing) system.

A Data Grid system connects a collection of geographically distributed computer and storage resources that may be located in different parts of a country or even in different countries, and enables users to share data and other resources [1]. Recently, there are a few research projects directed towards the development of data grid such as Particle Physics Data Grid (PPDG) [2], Grid Physics Network (GriPhyN) [3], The China Clipper Project [4], and Storage Request Broker (SRB) [5]. All these projects aim to build scientific data grid that enable scientists sitting at various universities and research labs to collaborate with one another and share data sets and computational power. The size of the data that needs to be accessed on the Data Grid is on the order of Terabytes today and is soon expected to reach Petabytes. As an example, The Large Hadron Collider (LHC) experiment [12] is producing Terabytes of raw data a year, and by 2005, this experiment at CERN, will produce Petabytes of raw data a year that needs to be pre-processed, stored, and analyzed by teams comprising 1000s of physicists around the world. Through this process, more derived data will be produced and 100s of millions of files will need to be managed and stored at more than 100s of participate institutions. All data must always be available or guarantees that will always be located.

In scientific environments, even the data sets are created once and then remain reads only, but the data sets usage often leads to creation of new files, inserting a new dimension of dynamism into a system [13][14]. Ensuring efficient access to such huge and widely distributed data is a serious challenge to network and Data Grid designer. The major barrier to supporting fast data access in a Data Grid system are high latencies of Wide Area Network (WANs) and the Internet, which impact the scalability and fault tolerance of total Data Grid systems. There are a number of research groups working on the data distribution problems in Data Grids. They are still working and investigating a number of data replication approaches on the data distribution for Data Grid, in order to improve its ability to access data efficiently [17][18][19][20]. These lead to new problems in discovery and access to data in Data Grids environment since data will be replicated, and this replica set change dynamically as new replicas are being added and old one being deleted at runtime in the Data Grid system. To address these problems we have developed a model based on Peer-to-Peer (P2P) architecture to study various discovery mechanisms and investigate these mechanisms for our Dynamic Scientific Data Grid Environment. To evaluate our model and selected discovery mechanisms, we have developed our own

Data Grid Simulator using PARSEC to generate different network topologies and replication strategies to study the impact of discovery mechanisms on the cost of locating data and the data access on the overall Dynamic Scientific Data Grid.

The paper is organized as follows. Section 2 gives an overview of previous work on discovery mechanism. In Section 3, we describe our model. Section 4 describes the simulation framework and our testing results. Finally, we present brief conclusion and future directions in Section 5.

2 Discovery Mechanisms and Scientific Data Grids

Discovery mechanism is a basic functionality that enhances the accessibility of data in the distributed system context. Support for handling and discovering resource capabilities already exist in some metacomputing system such as Globus [21] and Legion [22]. In Globus, the Globus Resource Allocation Manager (GRAM) act as a Resource Broker is responsible for resource discovery within each administration domain, which works with an Information Service, and a Co-allocator for monitoring current state of resources, and managing an ensemble of resources respectively. Hence, application requirements in Resource Specification Language (RSL) are refined by Brokers into more specific requirement, until a set of resources can be identified. The Legion system, describes resources and tasks as a collection of interacting objects. A set of core objects, enable arbitrary naming of resources based on Legion Object Identitifers (LOIDs) and Legion Object Addresses (LOA). Specialized services such as Binding agents and Context objects are provided to translate between arbitrary resource name and its physical location – enabling the resource discovery to be abstracted as a translation between LOIDs and physical resource location. In CONDOR [23], resource describes their capabilities as an advertisement, which is subsequently matched with an advertisement describing the needs of an application. Most of these mechanisms based on centralized approach, which may not scale.

In a large and dynamic environment, network services such as discovery should be decentralized in order to avoid potential computation bottlenecks at a point and be more scale. In the last couple of years, Peer-to-Peer (P2P) system became fashionable [6][7][8][9][11]. They emphasize two specific attributes of resources sharing communities: scale and dynamism. Most of existing P2P systems such as Gnutella, Freenet, CAN [15] and Chord [16] provide their own discovery mechanisms and focus on specific data sharing environment and, therefore on specific requirements. As example, Gnutella emphasis on easy sharing and fast file retrieval, but with no guarantees that files will always be located. In Freenet, the emphasis is on ensuring anonymity. CAN and Chord guarantee that files are always located but accepting increased overhead for file insertion and removal.

According to researchers in [10], the typical uses of shared data in scientific collaboration have particular characteristics: *Group locality* – users tends to work in a group and uses the same set of resources (data sets). The newly produced data sets will be interest to all users in the group. *Time locality* – the same users may request the same data multiple times within short time intervals. However, file discovery mechanisms that proposed in existing P2P systems such as Gnutella, CAN, Chord, etc., do not attempt to exploit this behavior. In [10], they also have raised up a few questions. One of the questions that most interest to us is how do we translate the dynamics of scientific collaborations in self-configuring network protocols such as joining the network, finding the right group of interests, adapting to changes in user's interest, etc.? This is relevant and challenging in context of self-configuring P2P networks. We support this idea, and try to answer the question by developing a framework model which is discuss in the next section.

3 P2P Data Grids Framework Model

In this section, we illustrate our framework model that provides a generic infrastructure to deploy P2P services and applications. Figure 1 shows our proposed framework model for data grids environment that will support scientific collaboration. In our framework model, peers could be any network devices and in our implementation, the peers can include PCs, servers and even supercomputers. Each peer operates independently and asynchronously from all other peers and it can self-organized into a peer group. Peer group contains peers that have agreed upon a common set of services, and through this peer group, peer can discover each others on the network. Once peer join a group it uses all the services provide by the group. Peers can join or leave the group at anytime that they want. In our implementation, once peer join the group, all the data sets that are shared by other peers in the group will be available to him/her. Peers can also share any of their own data sets with others peers within the group. Peer may belong to more than one group simultaneously.

In designing this collaborative data-sharing model, we are following the ideas of power grid environment. In the case of power grid supply, the users are not concerned about which power generator delivers the electricity to their home appliances, but they are only concerned that the power is available to drive their appliances correctly. In our scientific communities data-sharing model, we try to provide the same concept as the analogy of electrical power grid, where the users or scientists (in our case is peers) can access to their required data sets without knowing which peers delivers the data sets. In other words, they can execute their applications, get the remote data sets (don't concern about the provider) and then wait for the results. This will be done all by the discovery mechanism. Our focus in this research is to propose a dynamic discovery mechanism for Dynamic Scientific Data Grid Environment.

4 Simulation Framework and Results

Simulation has many advantages over actual implementation/deployment since it offers a fully customizable and controllable environment, and it also gives us another alternative to conduct an experiment without deploying on the actual network infrastructure. With our limited resources, we couldn't deploy our experiment on the actual network infrastructure that we have. However, by developing and using the simulator we can conduct our experiments in a customizable and controllable network environment. The presented results were obtained with simulation of only read requests without any background traffic in network and the stream of requests for the Data Grid. The current simulation is described in the next subsection.

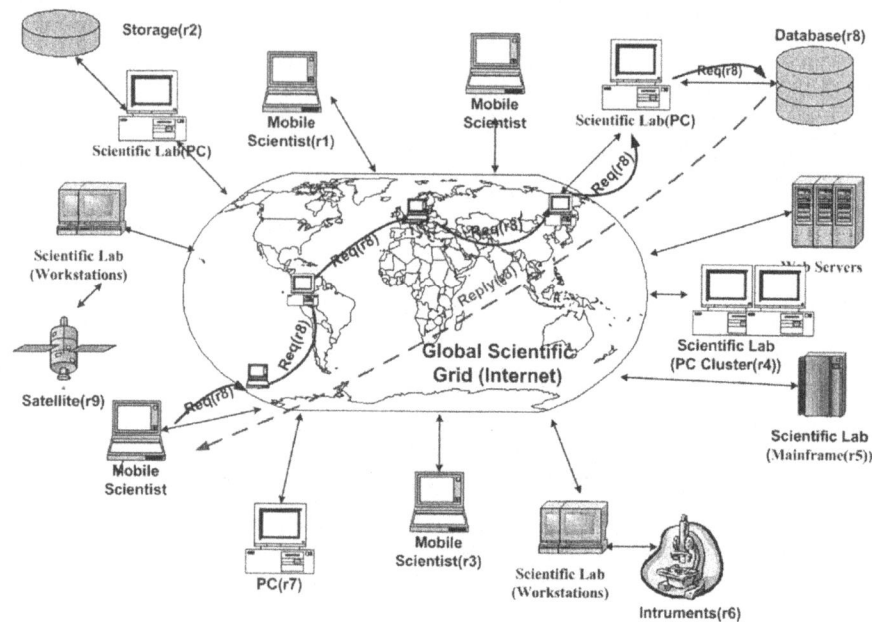

Fig. 1. P2P Framework Model for Data Grid

4.1 The Data Grid Simulator

Our simulator is written in PARSEC simulation language. PARSEC is a C-based discrete-event simulation. In PARSEC an object or set of objects is represented by a logical process. The simulator consists of three parts. The first part is the DRIVER entity that is responsible for creating the rest of entities: the entities that simulating the network nodes and network layer. It also reads all the inputs needed for the simulation. The second part is the network layer, which comprises of two entities: CHANNEL entity and DVMRP entity. The CHANNEL entity simulates a network forwarding protocol and the DVMRP entity simulates the Distance Vector Multicast Routing Protocol. The third part is the NETWORK_NODE entity that simulates the various nodes in the Data Grid.

4.2 Methodology of Simulation Study

The Data Grid Simulator is developed to identify a suitable resource discovery strategies and mechanism for Dynamic Scientific Data Grid Environment. The simulator is developed using a modular design approach that divided the simulator into various parts. An important benefit of the modular design is that it is relatively easy to replace any of the modules. Figure 2 shows our simulation architecture framework for data grid environment. Starting a simulation first involves specifying the topology of the grid, including the number of network nodes, how they are connected to each other and the location of the files across various nodes. The bandwidth of each link is not specified in our current simulation model in order to simplify the model.

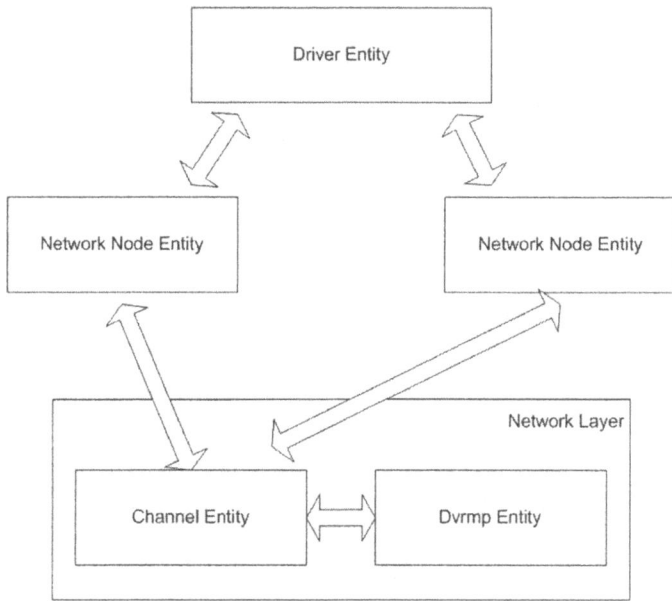

Fig. 2. The Data Grid Simulator Architecture

In this simulation model, the NETWORK_NODE will trigger various file requests originate according to file access patterns to be simulated. The CHANNEL will forward the request based on the information in the routing table provided by DVMRP. Before it forward the request to several neighbor nodes, it will check the file required and submit the request if the node has the requested file. In this case, the simulator will record the hop count for the successful request. If not, it will forward the request to other nodes until its time-to-live (TTL) for the request expired.

There are various proposed strategies and mechanism to discover or locate the resources in distributed system. We studied and compared different discovery strategies using this simulator. Through this simulation, various parameters such as hop count, success rates, response time and bandwidth consumption will be measured.

However, in this paper, we just show only the hop count measurement. The simulation was first run on the access patterns that were generated randomly. Table 1 is a sample access pattern file randomly generated during simulation. This being the worse case scenario, more realistic access pattern that contained varying amounts of temporal and geographical locality will be generated in the next version of our simulator.

Table 1. A Sample Access Pattern File

Time	File Name	Requestor Node
20	File1	1
39	File2	5
70	File2	6
75	File3	3
90	File1	2
92	File2	3
98	File3	5

4.3 Simulation Experiments and Results

In a Data Grid system, several factors influence the performance of discovery mechanism; the network topology, resource frequency and resource location. However, in our simulation studies, we will only focus on the last two factors that are related to the replication strategies. Replication strategies are incorporated into the simulation as to create the dynamic scenario of Data Grid system. In the real Data Grid system, the replica management system decides when to create a replica and where to place it.

Currently, in our simulation, we generated random flat networks topology consisting of 20 to 200 nodes. We placed the data file ourselves randomly, and measure data access cost in average number of hop, success number of request and bandwidth consumption. In our current simulation, we assume that the overlay network graph topology does not change during the simulation as to maintain the simplicity. In the simulation, we implemented a decentralized discovery mechanism based on flooding and request forwarding technique as to test our model and our data grid simulator. In a flooding protocol, nodes flood an overlay network with queries to discover a data file. In a request forwarding technique, nodes will forward the queries to the selected neighbor. In testing our model and our data grid simulator, we ran a number of simulations. In the first scenario of our simulations, we assume that, no replicas created at the beginning and during the simulation. In the second scenario, we created replicas at the beginning of simulation. We use average number of hop, success number of request and bandwidth consumption as our performance metric as it represents both the data discovery. Figure 3 shows a graph representing the data access cost in average number of hop for topologies with different number of nodes without replication using flooding discovery mechanism.

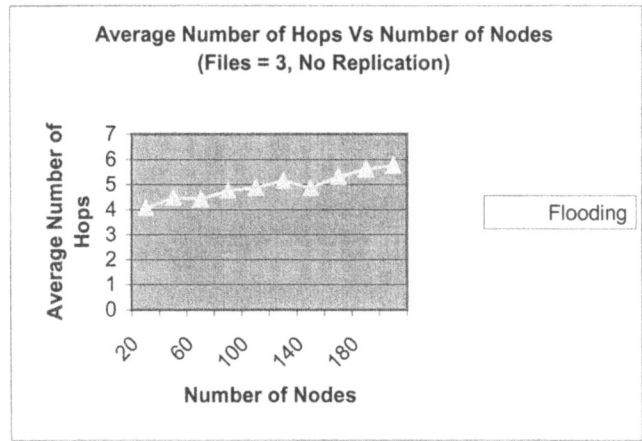

Fig. 3. Average Number of Hop in Different Number of Nodes

The results from figure 3 shows that, performance of the discovery mechanism based on flooding is worst when the number of nodes increases in the collaboration network. We expect that in real Dynamic Data Grid Environment, it will become worst as this mechanism may not provide the guarantees required by collaborators and it also doesn't scale well when the number of node increase. The mechanism might give optimal results in a network with a small to average number of nodes.

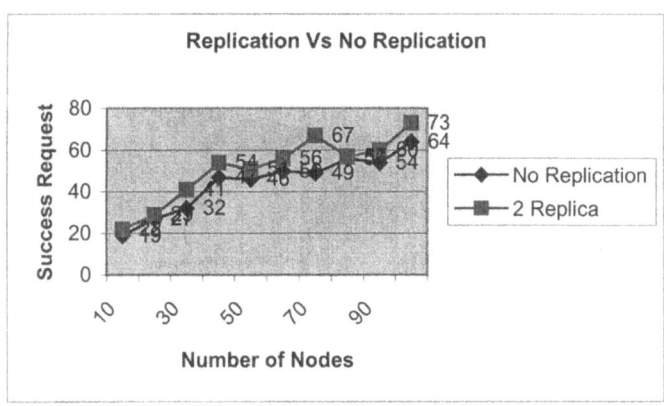

Fig. 4a. Number of Success Queries for Flooding Mechanism

Figure 4a shows a graph representing the data access cost in number of success queries for different topologies with different number of nodes with and without replication using flooding discovery mechanism. The result shows that, the number of success queries increase when the data file is replicated.

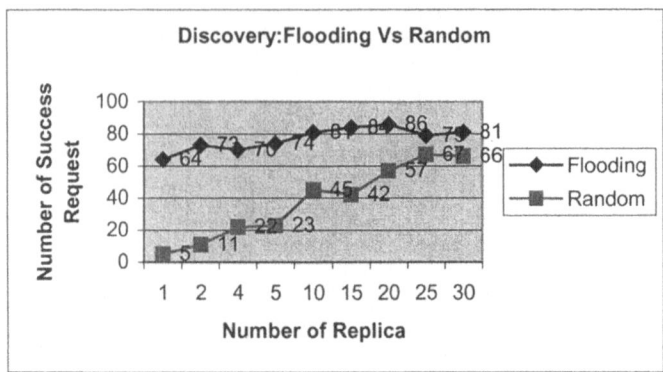

Fig. 4b. The 100 Nodes Network Topology with Various Number of Replicas

Figure 4b, shows a graph representing network topology with 100 nodes and the number of replicas created. The result shows that, when the number of replicas is increasing, the number of success queries also increasing accordingly.

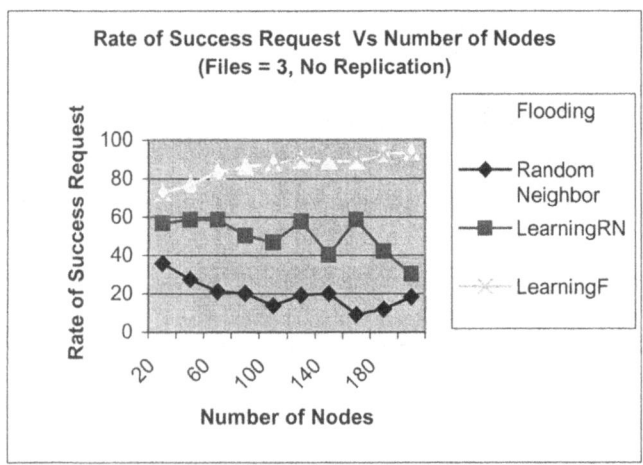

Fig. 4c. Number of Success Queries for four different Discovery Mechanism

Figure 4c shows a graph representing the data access cost in number of success queries for different topologies with different number of nodes without replication using four different discovery mechanisms. The result shows that, for flooding based discovery mechanism, the number of success queries increase accordingly when the number of node increase. It may be cause by the increasing number of different nodes making request when we increase the number of nodes. However, for forwarding based discovery mechanism (Random Neighbor and Learning Random Neighbor), the graph shows that, the number of success request is decreasing when the number of node is increasing.

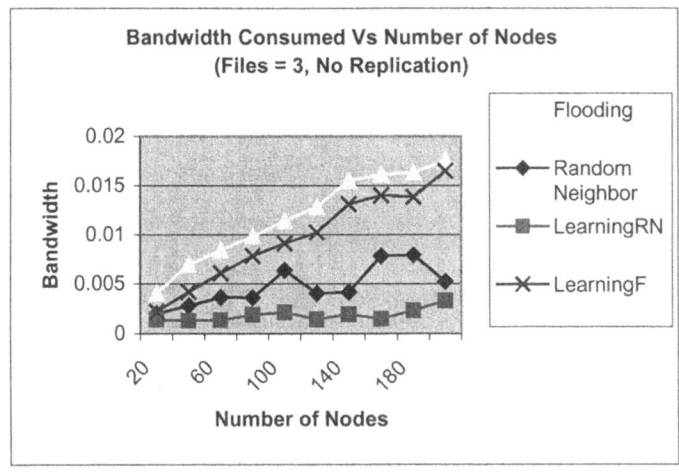

Fig. 5. Bandwidth Consumption for four different Discovery Mechanism

Figure 5 shows a graph representing the bandwidth consumption for different topologies with different number of nodes without replication using four different discovery mechanisms. The result shows that, for flooding based discovery mechanism, used more bandwidth to flood the queries. The bandwidth consumption increase accordingly when the number of node increase. However, for forwarding based discovery mechanism (Random Neighbor and Learning Random Neighbor), the graph shows that, there is only a small increase of bandwidth consumption.

5 Conclusions and Future Works

We have addressed the problem of discovery in Dynamic Data Grid Environment. We also have described our prototype implementation for Dynamic Data Grid Environment and our Data Grid Simulator for our study. Although the prototype is conceptually simple, it has provided a service that to be remarkably useful for a Large peer-to-peer based Data Grid Environment. We believe that, with a very simple simulation experiment that we have done on our model using our Data Grid Simulator, it shows that this approach can be used to support good services for data discovery in Dynamic Data Grid Environment. It also represents an interesting approach to data management in Data Grid.

Our results show that, in a Dynamic Data Grid Environment with large population of users, we need a discovery mechanism that could give a shortest time to discover and locate data resources, it also provide the guarantees required by collaborators and, with low bandwidth consumption. In our future work, we will study other strategies and mechanisms to discover or locate the resources in decentralized environment. Our simulator provides a modular framework within which optimization strategies for data discovery strategies can be studied under different data grid configuration. Various replication strategies will be incorporated as to create more dynamic data grid scenario. More realistic access pattern that contained varying amounts of temporal and geographical locality will be generated. The goal of our study, is to explore and propose a Decentralized Discovery Mechanism for Dynamic Data Grid Environment.

References

[1] Chervenak, A., Foster, .I., Kesselman, C., Salisbury, C., Tuecke, S. The Data Grid: Towards an Architecture for the Distributed Management and Analysis of Large Scientific Data Sets. Journal of Network and Computer Applications. (2000).
[2] PPDG: Particle Physics Data Grid, http://ppdg.net
[3] GriPhyn : Grid Physics Network, http://www.griphyn.org
[4] Johnston, W., Lee, J., Tierney, B., Tull, C.,Millsom, D. The China Clipper Project: A Data Intensive Grid Support for Dynamically Configured, Adaptive, Distributed, High-Performance Data and Computing Environments. In Proceeding of Computing in High Energy Physics 1998, Chicago (1998).
[5] SRB: Storage Request Broker, http://www.npaci.edu/DICE/SRB
[6] Gnutella website, http://www.gnutella.wego.com
[7] The Free Network Project website, http://freenetproject.org
[8] Napster website, http://www.napster.com
[9] Konspire website, http://konspire.sourceforge.net
[10] Iamnitchi, A., Ripeanu, M., Foster, I. Locating Data in (Small-World?) Peer-to-Peer Scientific Collaborations. 1st International Workshop on Peer-to-Peer Systems, Cambridge, Massachusetts, March (2002).
[11] Gong, L. Project JXTA: A Technology Overview. Sun Microsystems, Inc. April 25, (2001).
[12] LHC- The Large Hadron Collider website, http://lhc-new-homepage.web.cern.ch/lhc-new-homepage/
[13] Loebel-Carpenter, L., Lueking, L., Moore, C., Pordes, R., Trumbo, J., Veseli, S., Terekhov, I., Vranicar, M., White, S. and White, V. SAM and the Particle Physics Data Grid. In Proceedings of Computing in High-Energy and Nuclear Physics. Beijing, China. (2001)
[14] Lueking, L., Loebel-Carpenter, L., Merritt, W., Moore, C. The DO Experiment Data Grid – SAM. C.A. Lee(Ed.): GRID 2001, Lecture Notes in Computer Science, Vol 2242. Springer-Verlag Berlin Heidelberg (2001). 177-184
[15] Ratnasamy, S., Francis, P., Handley, M., Karp, R., Shenker, S. A Scalable Content-Addressable Network. In SIGCOMM 2001, San Diego, USA.(2001).
[16] Stoica, I., Morris, R., Karger, D., Kaashoek, M. F., Balakrishnan, H. Chord: A Scalable Peer-to-peer Lookup Service for Internet Applications. In SIGCOMM 2001, San Diego, USA. (2001).
[17] Ranganathan, K. and Foster, I. Identifying Dynamic Replication Strategies For a High Performance Data Grid. In Proceedings of the International Grid Computing Workshop. Denver. (2001)
[18] Ranganathan, K. and Foster, I. Design and Evaluation of Replication Strategies For a High Performance Data Grid. . In Proceedings of Computing in High-Energy and Nuclear Physics. Beijing, China. (2001)
[19] Stockinger, H., Samar, A., Allcock, B., Foster, I., Holtman, K., Tierney, B. File and Object Replication in Data Grids. In Proceedings of the Tenth International Symposium on High Performance Distributed Computing. IEEE Press. (2001).

[20] Vazhkudai, S., Tuecke, S., Foster, I. Replication Selection in the Globus Data Grid. In Proceedings of the First IEEE/ACM International Conference on Cluster Computing and the Grid. Pp. 106-113, IEEE Computer Society Press. (2001).
[21] Foster, I., Kesselman, C. Globus: A Metacomputing Infrastructure Toolkit. *Intl J. Supercomputer Applications*, 11(2) (1997)115-128
[22] Grimshaw, A., Lewis, M., Ferrari, A., Karpovich, J. Architectural Support for Extensibility and Autonomy in Wide-Area Distributed ObjectSystems. In Proceedings of the 2000 Network and Distributed System Security Symposium (NDSS2000). (2000)
[23] Frey, J., Tannenbaum, T., Foster, I., Livny, M., Tuecke, S. Condor-G: A Computation Management Agent for Multi-Institutional Grids. In Proceedings of the Tenth IEEE Symposium on High Performance Distributed Computing (HPDC10) San Francisco, California, August 7-9. (2001)

A DAG-Based XCIGS Algorithm for Dependent Tasks in Grid Environments[1]

Changqin Huang[1,2], Deren Chen[1], Qinghuai Zeng[3], and Hualiang Hu[1]

[1] College of Computer Science, Zhejiang University, Hangzhou, 310027, P. R. China
{cqhuang, drchen}@zju.edu.cn
[2] Department of Computer Science and Technology, Hunan University of Arts and Science, Changde, 415000, P. R. China
[3] Department of Education, Hunan University of Arts and Science, Changde, 415000, P. R. China

Abstract. Generating high quality schedules for scientific computation on a computational grid is a challenging problem. Many scheduling algorithms in grid computing are for independent tasks. However, communications commonly occur among tasks executed on different grid nodes. In this paper, an eXtended Communication-Inclusion Generational Scheduling (XCIGS) algorithm is proposed to schedule dependent tasks of an application with their DAG. During scheduling, those ineligible tasks are momentarily ignored, and a Buffer Set of Independent tasks (***BSI***) is conducted to leverage the utilization of grid resources. The predicted transferring time, the machine ready time and the expectation completion time of all predecessors are taken into consideration while an alternative auxiliary algorithm dynamically makes the schedule. Corresponding experimental results suggest that it betters resource utilization of grid experiments and improves execution performance.

1 Introduction

A computational grid is an emerging computing infrastructure that enables effective access to high performance computing resources to serve the needs of some Virtual Organization (VO) [1]. In order to utilize the dynamical and heterogeneous resources in gird environments, it is necessary to adopt an efficient task scheduling approach. With an objective of minimization of certain performance metrics and meeting some constraints, the task scheduling mainly allocates tasks to available resources both in time and in space in terms of a certain algorithm. The scheduling problems have been proved to be NP-Complete [2], and near-optimal heuristics are commonly studied and adopted in specific areas. However, most of the heuristics conduct exhaustive search scheduling and iterative optimization (e.g., simulated annealing and genetic algorithm), and they use static scheduling techniques, so these algorithms have large cost-

---

[1] The research reported in this paper is supported by the Science Research Foundation of Hunan University of Arts and Science, China (Grant No. JJ0231), and the National High-Tech. R&D Program for CIMS, China (Grant No. 2002AA414070).

time and are not suitable for dynamic grid environments in real time. This paper will adopt a dynamic approach to schedule tasks in computational grid.

In the area of engineering and scientific computations, there widely exists a type of application that consists of many dependent tasks. That is, the scheduling cannot violate precedence constraints among the tasks, and communications often occur between them. For example, computation in the areas of solid mechanics and fluid dynamics can be classified into that case. At present, independent task scheduling, rather than dependent task scheduling, is mainly concerned. A natural idea is that we should develop a heuristic algorithm that is suitable for scheduling dependent tasks and improving system efficiency as well. The directed acyclic graph (DAG) is a generic model that consists of a set of nodes connected by a set of edges, and it is well suitable to describe tasks and dependent relations of tasks. Most of precedence constrained scheduling make a schedule decision in the aid of DAG [5,6,7].

In this paper, we present a DAG-based extended algorithm of the Communication-Inclusion Generational Scheduling (CIGS) [8], termed XCIGS. During scheduling, a Buffer Set of Independent tasks (*BSI*) is proposed to adjust the utilization of grid resources like a task buffer, and the completion time of all predecessors is also considered for a better efficient decision.

This paper is organized as follows: Section 2 reviews related work in the arena of grid scheduling. The CIGS algorithm is presented in Section 3. Section 4 describes some assumptions and definitions. Details of our algorithm are proposed in Section 5. Case studies with experimental results are included in Section 6, and conclusions and future works are addressed in Section 7.

2 Related Works

For the development and deployment of computational grids, there are a few approaches to scheduling. Subramani et al [3] modify Metascheduler model to correspond to Multiple Simultaneous Requests, and Dogan et al [4] discuss how to schedule independent tasks with Quality of Service (QoS) requirements, but their scheduling approaches [3,4] are only suitable for independent tasks. Iverson et al. [5] present a framework for scheduling multiple DAG-structured applications and apt algorithms, but don't focus on novel algorithms for single application. Shang et al. [6] adopt a new list scheduling, but the approach is limited to homogeneous systems rather than heterogeneous grid environments. Wu et al. [7] propose a fast local search algorithm based on topological ordering to improve the scheduling quality and efficiency, whereas our work will adopt a different policy in which many heuristics can be utilized. Cheng et al. [9] study the feasibility problem of scheduling a set of start time dependent tasks with deadlines and identical initial processing times; however, they require strict constraints (e.g. a single machine). Beaumont et al. [11] apply a non-oriented graph to model the grid platform, and give the optimal steady-state scheduling strategy for each processor; their algorithm takes long execution time and is not dynamically appropriate for real-time environments. Gui et al. [10] present an optimal version of Generation Scheduling (GS) with a concern of both the completion time of task and ready time of task, The essential difference between their work and ours is that our heuristic takes data transferring into account and applies *BSI* to adjust the utilization of grid resources.

3 CIGS Algorithm

To develop a schedule of precedence constrained tasks on a set of heterogeneous machines/nodes that minimize the time required to execute the given tasks, Carter et al. [8] propose a fast, efficient scheduling algorithm, termed Generational Scheduling (GS), and also present an extended version concerning the overhead of data transferring among the tasks, i.e. a Communication-Inclusive extension of GS (CIGS). In CIGS, by analyzing the dependent relation of tasks, some "independent tasks", whose precedence constraints have been satisfied currently, are filtered out of the remaining tasks, and scheduling is only conducted for "independent tasks". The above process loops until all tasks are scheduled. During scheduling, the scheduled tasks may be being executed. CIGS maps a task to a particular host by an alternative heuristic algorithm (the literature [8] try using the "Fast-Greedy" algorithm), as well as, it considers the full amount of time required to transfer all of the associated data to the host and execute the task. Each loop in CIGS consists of the following four steps:

Step 1: It represents the scheduling problem. First let all tasks become ineligible for scheduling, then, by analyzing the dependent relation of all tasks, each of these tasks, whose precedence constraints have been satisfied, is found and set negligible.

Step 2: It establishes the set in aid of scheduling. All ineligible tasks are filtered out of the scheduling problem, the remainder forms a new Set of Independent tasks (*SI*) available to scheduling, which is only composed of the tasks immediately eligible for execution for there are no precedence constraints.

Step 3: It makes a scheduling by an auxiliary algorithm. The algorithm can be selected by the user, and applies a non-preemptive scheduling scheme to all eligible tasks including those that have not been started or completely executed.

Step 4: It initiates the scheduling problem. When a rescheduling event is detected, all associated tasks are collected and form a new scheduling problem for dependent tasks. The ready times of machines are refreshed, and it repeats the above steps until all tasks are scheduled.

In grid environments, the available machines dynamically vary both in the time-dimension and in the space-dimension. Under these circumstances, the CIGS algorithm is not best suitable for reducing the makespan and raising the resource utilization rate. We find that there exist two aspects to improve its efficiency:

a) In **Step 2** and **Step 3**, all ineligible tasks exist in one uniform set *SI*, and all tasks of *SI* will be scheduled during every single loop though they may have different priorities to enter into the scheduling. That may affect the makespan of the whole application and resource utilization rate. For example, a task T^1 on the critical path is out of *SI* during this cycle, however, it will exist in *SI* during the next cycle. On the basis of CIGS, one task T^2 of *SI*, which is not predecessor of any task not yet to be scheduled, will be mapped to one idle machine, and it is only one machine idle at this time for this loop. At this scenario, if T^2 is not scheduled, T^1 will be scheduled during the next loop. So, that will reduce the application's makespan, as well as, improve the resource utilization because these tasks like T^2 will enter into scheduling at arbitrary apt moment.

b) In **Step 4**, the CIGS algorithm only refreshes the ready times of machines and prepares for making a schedule decision in terms of these ready times. That does not benefit the system because the best basis of scheduling is not the ready time.

4 System Assumptions and Definitions

To solve the above-mentioned problems, we propose an eXtended CIGS (XCIGS) appropriate for the computational grid. During scheduling, a task with zero-out-degree is filtered out of the set ***SI*** and forms a new set ***BSI***. Each task of ***BSI*** may not be scheduled during the current scheduling cycle, i.e. the ***BSI*** works like a task buffer for scheduling. As well as, the predicted transferring time, the machine ready time and the expectation completion time of all predecessors are considered as the factors of scheduling to get a better efficient decision.

4.1 Assumptions

The assumptions of the XCIGS algorithm are presented as follows:

a) An application can be divided into many tasks. These tasks have dependent relations, and these precedent constraints are not time-variable, and they will be described by DAG.

b) One machine can only execute one task at the same time, and the tasks are non-preemptive.

c) During each cycle, the heuristic algorithm will be the same and takes the time of data transferring into account if data transferring happens.

d) Though the performance and status of machines and network connections dynamically vary in grid environments, they can be predicted more precisely at a certain moment during the scheduling. The execution time of tasks and the communication time among tasks can be predicted as well.

e) All scheduled tasks on the nodes can be completed in general. If a node fails at any time, the task on it will be enrolled into the new scheduling problem in the next scheduling cycle. The mechanism for checkpoint, migration and recover is not supported.

4.2 Definitions

To buffer the utilization of grid nodes (machines), We propose a special dynamic set of tasks (***BSI***). At the same time, we let this algorithm make dynamically a schedule decision in terms of the machine ready time, the data transferring time and the com-

pletion time of all predecessors. Detailed notations concerning XCIGS algorithm are described in the following.
- *N* is the total number of tasks which belong to an application scheduled in grid, *S* is a set of these tasks in the scheduling, $S=\{S[0], S[1], ..., S[N-1]\}$.
- *M* is the total number of machines/nodes participated in the scheduling in grid, *H* is a set of these machines/nodes, $H=\{H[0], H[1], ..., H[M-1]\}$.
- *US* is a set of these tasks unassigned a priority.
- *AS* is a set of these tasks assigned a priority.
- *SI* is a set of these tasks that can be executed for their predecessors have been completed.
- *BSI* is a set of these tasks which can be executed for their predecessors have been completed, and they are not the predecessors of any task not yet to be scheduled in a certain cycle during scheduling. It is formed by filtering all zero-out-degree tasks out of *SI*.
- *R* is a set of expectation ready time of machines/nodes, $R=\{R[0], R[1], ..., R[M-1]\}$, $R[j]$ is the expectation ready time of the machine/node $H[j]$.
- *T* is a matrix of the expectation speed of transferring data between machines/nodes, $T=\{T[0][0], T[0][1], ..., T[M-1][M-1]\}$, $T[k][j]$ denotes the expectation speed of transferring data between the machine $H[k]$ and $H[j]$, $T[x][x]$ is ∞, for the data transferring doesn't happen at the same machines/nodes $H[x]$.
- *E* is a matrix of expectation execution time of the tasks executed on machines/nodes, $E=\{E[0][0], E[0][1], ..., E[N-1][M-1]\}$, $E[k][j]$ is the expectation execution time of the tasks $S[k]$ executed on machines/nodes $H[j]$.
- *D* is a matrix of the size of the data necessarily transferred for the tasks executed on machines/nodes, $D=\{D[0][0], D[0][1], ..., D[N-1][M-1]\}$, $D[k][j]$ is the size of the data necessarily transferred for the tasks $S[k]$ executed on machines/nodes $H[j]$. If the task $S[k]$ is locally executed on the machine $H[j]$, $D[k][j]=0$.
- *TC* is a matrix of expectation transferring time of the tasks executed on machines/nodes, $TC=\{TC[0][0][0], TC[0][0][0], ..., TC[N-1][M-1][M-1]\}$, $TC[k][i][j]$ is the expectation transferring time of the dataset of the tasks $S[k]$ between $H[i]$ and $H[j]$.
- *C* is a matrix of expectation completion time of the tasks executed on machines/nodes, $C=\{C[0][0], C[0][1], ..., C[N-1][M-1]\}$, $C[k][j]$ is the expectation completion time of the tasks $S[k]$ executed on machines/nodes $H[j]$.
- *LC* and *LM* is the temporary set of the predicted completion time of tasks scheduled on the certain machine(s). $LC=\{LC[0], LC[1],...LC[N-1]\}$, and $LM=\{LM[0], LM[1],...LM[M-1]\}$. $LC[k]$ denotes the predicted completion time of a task $S[k]$ on the certain machine(s), similarly, the predicted completion times of a certain task on $H[j]$ is recorded in $LM[j]$.
- *Makespan* is the completion time of the application, and equal to the maximum completion time of all tasks, i.e. the maximum expectation ready time of all machines/nodes after all tasks have been scheduled.
- *TR* is a matrix of dependent relations among all tasks, $TR=\{TR[0][0], TR[0][1], ..., TR[k][k]\}$, $E[k][t]$ denotes the dependent relation between $S[k]$ and $S[t]$, $E[k][t]=1$

denotes that *S*[*t*] can not begin running until *S*[*k*] has completed execution. So *S*[*k*] is a predecessor of *S*[*t*], reversely, *S*[*t*] is a successor of *S*[*k*].

- **PS**[*k*] is a set of the predecessors of *S*[*k*], **SS**[*k*] is a set of successors of *S*[*k*], i.e. if a certain task *S*[*x*] is a predecessor of *S*[*k*], *S*[*x*]∈*PS*[*k*], *S*[*k*]∈*SS*[*x*]. If *PS*[*k*]={}, *S*[*k*] has a non-precedent constraints.
- **DAG** is a directed acyclic graph of the application executed in the grid environments, **DAG**=<*V,U*>, *V*[*k*] denotes the task *S*[*k*], the directed edge *U*[*k, t*] denotes the dependent relation *TR*[*k*][*t*].
- **ID** is a set of the in-degrees of all tasks, **OD** is a set of the out-degrees of all tasks. **ID** = {*ID*[*0*],*ID*[*1*],...,*ID*[*N-1*]}, and **OD** ={*OD*[*0*],*OD*[*1*],...,*OD*[*N-1*]}. *ID*[*k*] is the in-degree of *V*[*k*], and *OD*[*k*] is the out-degree of *V*[*k*].
- **TP** is a set of the priorities of all tasks, **TP**={*TP*[*0*], *TP*[*1*],..., *TP*[*N-1*]}. *TP*[*k*] denotes the priority of the task *S*[*k*], and $0 \leq TP[k] \leq N-1$.

5 XCIGS Algorithm

Before presenting the algorithm, we describe our scheduling strategy. During the scheduling, firstly, the system analyzes an application composed of many dependent tasks on the basis of DAG, and converts the DAG of this application into a matrix **TR** to compute the in-degree and out-degree of each task. Secondly, the algorithm assigns a fixed priority to each task in terms of the relation matrix. Finally, it selects apt tasks and performs scheduling: a) It filters all ineligible tasks out of the scheduling problem according to the same priority, and the remainder forms **SI**. b) It filters all tasks, which are not predecessors of any tasks not yet to be scheduled, out of **SI** during this loop, these tasks filtered out are put into the **BSI**. c) The scheduler maps the tasks of **SI** to all machines available by using a certain heuristic. d) By computing, if the number of remainder idle machines is more than the number of critical tasks during the next loop, the tasks in **BSI** will be scheduled, otherwise they will remain in the **BSI**. They will not be scheduled until there exists a suitable opportunity in the future.

Like CIGS, the system adopts an auxiliary heuristic algorithm, but the algorithm dynamically produces a schedule in terms of both the communication-inclusion completion times of all predecessors and the ready time of machines, not only the ready time of machines, in order to reduce the makespan.

5.1 In-Degree and Out-Degree Computations

The in-degree of *V*[*k*] is equal to the number of all predecessors of *V*[*k*]. Because of the relation between **TR** and **DAG**, it is equal to the number of "1" in the row *S*[*k*] in the matrix **TR**. For the same reason, The out-degree of *V*[*k*] is equal to the number of "1" in the column *S*[*k*] in the matrix **TR**. So all in-degrees and out-degrees can be acquired in the following two equations:

$$ID[k] = \sum_{i=0}^{N-1} TR[k][i], \qquad (1)$$

$$OD[k] = \sum_{i=0}^{N-1} TR[i][k]. \qquad (2)$$

5.2 Priority Computations

On the basis of dependent relations of tasks, these priorities of all tasks are determined with the method of task filtering, which can guarantee that there does not exist the dependent relation among the tasks with a same priority. Therefore, we can compute the priority according to the following method:

Because the set of the predecessors of $V[k]$ is $PS[k]$, so $PS[k]=\{PS[0], PS[1],…, PS[l]\}$, $l \geq 0$, first one initial value is given to the priorities of a "source task" (i.e. $\exists S[x]$ or $V[x]$, $PS[x]=\{\}$), then the priorities of the other tasks can be induced. The method in detail is presented in the following equation and steps:

$$TP[k] = \max_{S[i] \in PS[k]} (TP[i]) + 1. \qquad (3)$$

If one initial value is set 0, i.e. these "source tasks" will be scheduled and executed at first, all other priorities can be acquired in recursion as follows:

Step 1: Put all tasks of S into a set of task US.

Step 2: Find these tasks without any predecessor (if $\exists S[x]$ or $V[x]$, $PS[x]=\{\}$) from US, let their priorities 0.

Step 3: These tasks are removed out of US and are put into another set of task AS. If $US=\{\}$, it will go to **Step 6**, otherwise continues.

Step 4: Find one task from US, we assume T.

Step 5: If each of T's predecessors exists in AS, then the priorities of T are computed with equation (3) and the algorithm goes to **Step 3**, otherwise it goes to **Step 4**.

Step 6: The maximum priority of every task is its final priority.

5.3 Main Algorithm

The XCIGS algorithm is described below:

Step 1: Find all tasks of AS only if their priorities are equal to 0, and put these tasks into SI. Search each task $SI[k]$ of SI, and if $OD[k]=0$, this task is filtered out and put into BSI. The location of the datasets of tasks, the information of machines and network status are searched and predicted by NWS [12]. Both H and R are refreshed, and the predicted completion time of each task $S[k]$, which is scheduled on the certain machine(s) $H[j]$, is recorded in $LC[k]$ and in $LM[j]$ currently.

Step 2: On the basis of the available machines in *H*, a schedule is made for the set of *SI* by the heuristic algorithm (e.g. Fast-Greedy). All tasks of *BSI* are temporarily ignored.

Step 3: Filter the tasks, whose priorities (denoted *P*) are equal to 1, out of *AS*. These tasks filtered are put into *SI*. Filter each task *SI*[*k*] out of *SI* if *OD*[*k*]=0, and put these tasks into *BSI*. We assume each task *S*[*k*] is ready for being scheduled and located on the machines *H*[*i*], so the algorithm can compute the predicted completion time as follows. Both *H* and *R* are refreshed, and all tasks in *SI* are scheduled in terms of the *C* by the same algorithm in **Step 2**.

a) The expectation transferring time of *S*[*k*] in *SI* is computed by equation (4).

$$TC[k][i][j] = D[k][j] / T[i][j]. \qquad (4)$$

b) Because the execution occurs after all predecessors of *S*[*k*] have been finished and the machines mapped to the task *S*[*k*] are ready, so this task *S*[*k*] on the machine *H*[*j*] begins execution at this time *B*[*k*][*j*], and equation (5) exists.

$$B[k][j] = \max_{S[k] \in SS} (\max_{S[t] \in PS[k]} (LC[t]), R[j]). \qquad (5)$$

c) The predicted completion time of the task *S*[*k*] can be given as follows:

$$C[k][j]) \geq E[k][j] + B[k][j] + TC[k][i][j]. \qquad (6)$$

d) The above four steps are repeated until the predicted completion times of all tasks in *SI* are computed.

Step 4: Both *H* and *R* are refreshed, and the number (denoted N_1) of available machines in *H* is counted. The number (denoted N_2) of all the tasks, which are the critical tasks (denoted *S*[*x*]) in *AS* and *TP*[*x*])=*P*+1, is counted. If $N_1 \leq N_2$, the algorithm will go to Step 6.

Step 5: For each task of *BSI*, we compute the predicted completion time as Step 3, all or a part of the tasks in *BSI* are scheduled in terms of the *C* by the same algorithm in **Step 2**. These tasks scheduled are removed out of *BSI*.

Step 6: *P* = *P* +1, and the algorithm repeats from Step 3 to Step 5 until all tasks in *SI* and *BSI* are scheduled.

6 Experiments

In these experiments, we use our grid environment, named VGRID, which is an engineering computation oriented visual grid prototype system and consists of two LANs. In VGRID, the two LAN's bandwidths are 10M (LAN A) and 100M (LAN B), respectively, and they are connected by a 10M-bandwidth network. To test the efficiency of our algorithm, we use a special application in Computational Fluid Dynamics (CFD), all subtasks of which are dependent and we let the size of tasks different randomly. Case 1: we use 10 dependent tasks on 3 machines in LAN B; Case 2 :we use 140 tasks on 7 machines (5 machines in LAN B, the other in LAN A); Case 3 :we use 165 tasks on 9 machines (7 machines in LAN B, the other in LAN A). The scheduling algorithms adopt the CIGS–Greedy, CIGS–Min-Min, XCIGS–Greedy, XCIGS–Min-Min respectively, experimental results corresponding to the performance metric: makespans are illustrated in Fig. 1.

As shown in the figure, when the number of the dependent tasks increases, the makespan for XCIGS will reduce more obviously, reversely, there is only a little difference among their makespans. However, the makespans for XCIGS always reduce compared to those of CIGS by the same auxiliary algorithm. That is, the XCIGS algorithm can improve the system efficiency in grid environments.

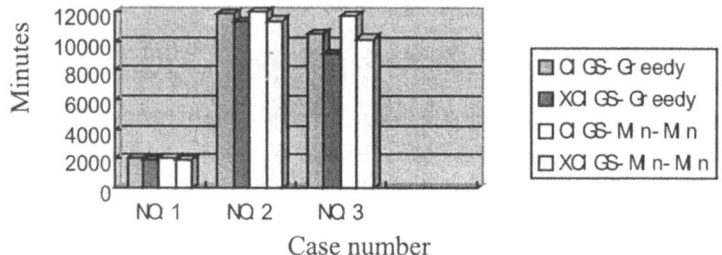

Fig. 1. Variation of the makespan.

7 Conclusions and Future Works

Computational grid is going to play an important role in scientific and engineering computing. To take full advantage of dynamic and heterogeneous grid resources, the scheduling algorithms are essential under many constraints. This paper focuses on the scheduling problem of a type of application composed of many dependent tasks. We have presented an extension version of CIGS suitable for the computational grid. To advance the execution of critical tasks, we propose a special dynamic set of tasks (***BSI***) to buffer the tasks suitable for scheduling. All tasks, which have a zero-out-degree in a certain phase, are put into this set, and these tasks are scheduled and removed out of this set while the system has adequate machines available in the other scheduling cycle. The predicted transferring time, the machine ready time and the expectation completion times of all predecessors are taken into consideration while an alternative auxiliary algorithm dynamically makes the schedule. Firstly, the algorithm assigns a fixed priority to each task. Secondly, it filters all ineligible tasks out of the scheduling problem according to the same priority, and the remainder forms a set ***SI***. Finally, the algorithm filters all tasks, which are not predecessors of any tasks not yet to be scheduled, out of ***SI*** at this time, and these tasks filtered out are put into ***BSI***. The scheduler maps the tasks of ***SI*** to the machines available by using a certain heuristic with a higher priority, after that, if there still exist more machines available, the tasks of ***BSI*** will be scheduled. Corresponding experimental results suggest that it improves execution performance.

In the future, we plan to use more cases and auxiliary algorithms to test the performance of our algorithms and better our algorithms in practice. We also have plans to study the impact of large dataset transferring on grid performance, and then to concern data-centric co-scheduling issues.

References

1. I. Foster, C. Kesselman et al.: The Anatomy of the Grid: Enabling Scalable Virtual Organizations. *International Journal of High Performance Computing Applications*, 2001, 15(3): 200-222
2. D. Fernandez-Baca: Allocating Modules to Processors in a Distributed System. *IEEE Transactions on Software Engineer*, 1989, 15 (11): 1427-1436
3. V. Subramani, R. Kettimuthu et al.: Distributed Job Scheduling on Computational Grids Using Multiple Simultaneous Requests. *Proc. of International Symposium on High Performance Distributed Computing*, 2002
4. A. Dogan and F. Özgüner: Scheduling Independent Tasks with QoS Requirements in Grid Computing with Time-Varying Resource Prices. *Proc. of Grid Computing (GRID 2002)*, 2002
5. M. Iverson and F. Özgüner: Dynamic, Competitive Scheduling of Multiple DAGs in a Distributed Heterogeneous Environment. *Proc. of the 7th Heterogeneous Computing Workshop (HCW'98)*, 1998
6. M. Shang, S. Sun, et al.: An Efficient Parallel Scheduling Algorithm of Dependent Task Graphs. *Proc. of the 4th International Conference on Parallel and Distributed Computing, Applications and Technologies*, 2003
7. M. Wu, W. Shu, et al.: Efficient Local Search for DAG Scheduling. *IEEE Transactions on Parallel and Distributed Systems*, 2001, 12(6): 617-627
8. B.R. Carter, D.W. Watson, et al.: Generational Scheduling for Dynamic Task Management in Heterogeneous Computing Systems. *Journal of Information sciences*, 1998, 106(3-4): 219-236
9. T.C.E. Cheng and Q. Ding: Scheduling Start Time Dependent Tasks with Deadlines and Identical Initial Processing Times on a Single Machine. *Computers & Operations Research*, 2003, 30: 51-62
10. X. Gui and D. Qian: OGS Algorithm for Mapping Dependent Tasks to Metacomputing Environment. *Chinese Journal of Computers*, 2002, 25(6): 584-588
11. O. Beaumont, A. Legrand, et al.: Scheduling Strategies for Mixed Data and Task Parallelism on Heterogeneous Clusters and Grids. *Proc. of the 11th Euromicro Conference on Parallel, Distributed and Network-Based Processing (Euro-PDP'03)*, 2003
12. R. Wolski, et al.: The Network Weather Service: a Distributed Resource Performance Forecasting Service for Metacomputing. *Future Generation Computing Systems*, 1999, (5-6): 757-768.

Running Data Mining Applications on the Grid: A Bag-of-Tasks Approach

Fabrício A.B. da Silva, Sílvia Carvalho, Hermes Senger,
Eduardo R. Hruschka, and Cléver R.G. de Farias

Universidade Católica de Santos (UniSantos)
R. Dr. Carvalho de Mendonça, 144, CEP 11030-906, Santos, SP, Brazil.
{fabricio,mygridgene,senger,erh,cleverfarias}@unisantos.br

Abstract. Data mining (DM) applications are composed of computing-intensive processing tasks working on huge datasets. Due to its computing-intensive nature, these applications are natural candidates for execution on high performance, high throughput platforms such as PC clusters and computational grids. Many data mining algorithms can be implemented as bag-of-tasks (BoT) applications, i.e., parallel applications composed of independent tasks. This paper discusses the use of computing grids for the execution of DM algorithms as BoT applications, investigates the scalability of the execution of an application and proposes an approach to improve its scalability.

1 Introduction

Knowledge discovery in databases is the non-trivial process of identifying valid, novel, potentially useful, and ultimately understandable patterns in data [1]. Data Mining (DM) is a step in this process that centers on the automated discovery of new facts and relationships in data, and it consists of three main steps: data preparation, information discovery and analysis of the mining algorithm output. All these steps exploit huge amounts of data and are computationally expensive. Therefore, several techniques have been proposed to improve the performance of DM applications, such as parallel processing [2] and implementations based on cluster of workstations [3].

A computational grid, or simply grid, provides access to heterogeneous resources in a geographically distributed area, allowing the integration of these resources into a unified computer resource. Computational grids are usually built on top of specially designed middleware platforms, the so-called grid platforms. Grid platforms enable the sharing, selection and aggregation of a variety of distributed resources, such as supercomputers, servers, workstations, storage systems, data sources and specialized devices, which are owned by different organizations [4].

In this work, we analyze the use of computational grids for processing data mining tasks that can be implemented as Bag-of-Tasks (BoT) applications, i.e., parallel applications whose tasks are independent of each other. BoT applications are one of the most suitable applications for running on a grid. However, depending on the size of the files to be mined and the amount of computation related to a task, BoT data mining applications can present scalability problems when executing on the grid. Therefore, this paper investigates a scheduling strategy to improve scalability by grouping sets of independent tasks into larger ones. For the experimental results, we used My-Grid [5], which is a grid platform specially designed to run BoT applications.

Our data mining application involves the Clustering Genetic Algorithm (CGA) [6], which was developed to find the *best* clustering in a dataset. In our simulations, we employ the *Congressional Voting Records Dataset,* which is a benchmark for data mining methods, available at the UCI Machine Learning Repository [13].

The remaining of this paper is organized as follows. Section 2 discusses the use of computational grids for the execution of DM applications. This section also presents both the grid platform and the CGA used in this work. Section 3 proposes a grouping strategy aimed at improving the scalability of BoT applications. Section 4 provides experimental results using the proposed strategy. Finally, section 5 presents conclusions and points out some future work.

2 Running Data Mining Algorithms as Bag of Tasks Applications

Some recent works suggest that grids are natural platforms for developing high performing data mining services [7,8,9]. More specifically, Orlando et al. [8] describe the application of two data mining algorithms in the *Knowledge Grid* [7]: the DCP Algorithm and the K-means Algorithm. The former enhances the popular Apriori Algorithm [10], which is an algorithm for frequent set counting, whereas the K-Means [12] is a popular clustering algorithm. However, other DM techniques can take advantage of a grid infrastructure. In general, several data-mining applications can be classified as parameter-sweep, BoT applications. Since those applications are composed of independent tasks, they are suitable for execution in a grid environment, where machine heterogeneity and significant network delays are common.

For example, the standard way of predicting the error rate of learning algorithms (like decision trees, neural networks, etc.) is to use stratified tenfold cross validation. To do so, the dataset is divided into ten equal parts. Then, the algorithm is trained in nine parts and tested in the other one. This procedure is repeated in ten different training sets and the estimated error rate is the average in the test sets. Clearly, these experiments can be performed in parallel, taking advantage of a grid platform.

Considering the use of neural networks (NNs) for data mining, it is usually necessary to train several different architectures to get the best possible one. In addition, NNs with specific architectures usually depend on random initializations of their parameters (weights). These parameters can have an important influence on the learning task. In a grid, several different trainings can be performed in parallel.

One of the most popular clustering algorithms is the K-means. Basically, this algorithm depends on the value of K (number of clusters), which is specified by the user. The determination of this value is very difficult and a practical approach consists of obtaining several different solutions (one for each value of K), and then choosing the *best* one. This approach can be performed in a parallel way, in which each machine runs the algorithm for a specific *K*.

Genetic algorithms can be used to solve problems where traditional search and optimization methods are less effective [11]. These algorithms are based on mechanisms of natural selection and genetics and they have been extensively used in data mining. Generally speaking, these algorithms use probabilistic rules to guide their search [11], usually providing non-deterministic results. Thus, several experiments are usually performed both to get the best possible solution and to obtain statistically significant results. Again, as each experiment is performed several times in each dataset, one can process them simultaneously, using different machines in a grid.

2.1 The Platform MyGrid

There has been considerable effort over the past few years in the development and implementation of platforms for grid computing, such as Condor [14], Legion [15], BOINC [16] and Globus [17], to name a few. All of these and other existing platforms address different objectives or implement different approaches to grid computing. In the context of this work, we adopted the platform MyGrid [5] for the implementation of data mining applications on the grid. MyGrid is a platform conceived to support the execution of BoT applications. BoT applications constitute a class of parallel applications that can be partitioned into several independent tasks. Usually, these tasks have an infrequent need for communication. The main benefits of MyGrid are twofold: minimal installation effort and ease of use. Most of the aforementioned grid platforms can only be installed and configured by system administrators. Moreover, installation procedures are usually manually repeated in a considerable number of machines. MyGrid enables non-privileged users to create their own grid to run applications on whatever resources they have access to, without the need for these users to get involved into grid details and administrative procedures for heterogeneous platforms.

In addition, a considerable effort is usually needed to enable parallel applications to run efficiently on loosely coupled environments such as the grid. BoT applications, however, can be easily adapted to run on MyGrid with *minimum* modifications. There is a clear trade-off in the design of MyGrid, by forfeiting the support for arbitrary applications to favor the support of BoT applications. The main benefit of this approach is simplicity. Since MyGrid focuses on BoT applications, its working environment consists of a small set of services which enable its users to manipulate their input and output files, distribute the application code, as well as spawn and terminate remote jobs on the grid. In spite of its reduced number of services, MyGrid supports the whole cycle of application deployment. These services allow the user to adapt her stand-alone application to run on the grid, regardless of the differences in configuration of multiple resources that comprise the grid. MyGrid has yet another advantage: it can be used in combination with other grid platforms. For example, one can combine MyGrid and Globus, by implementing a specific proxy which enables MyGrid applications to use resources available on a grid Globus.

2.2 Clustering Genetic Algorithm (CGA)

Clustering is a task in which one seeks to identify a finite set of categories (clusters) to describe a given data set, both maximizing homogeneity within each cluster and heterogeneity among different clusters. In this sense, objects that belong to the same cluster should be more similar to each other than objects that belong to different clusters. Thus, it is necessary to devise means of evaluating the similarities among objects. This problem is usually tackled indirectly, i.e. distance measures are used to quantify the dissimilarity between two objects, in such a way that more similar objects have smaller dissimilarity measures. Several dissimilarity measures can be employed for clustering tasks, such as the commonly used Euclidean distance.

There are three main types of clustering techniques: overlapping, hierarchical and partitioning. In this work, clustering involves the partitioning of a set **X** of objects into a collection of mutually disjoint subsets C_i of **X**. Formally, let us consider a set of N objects **X**={$x_1,x_2,...,x_N$} to be clustered, where each $x_i \in \Re^p$ is an attribute vector consisting of p real-valued measurements. The objects must be clustered into non-overlapping groups **C**={$C_1,C_2,...,C_k$} where k is the number of clusters, such that:

$$C_1 \cup C_2 \cup ... \cup C_k = X, \quad C_i \neq \varnothing, \quad \text{and} \quad C_i \cap C_j = \varnothing \text{ for } i \neq j. \qquad (1)$$

The problem of finding an optimal solution to the partition of N data into k clusters is NP-complete and, provided that the number of distinct partitions of N objects into k clusters increases approximately as $k^N/k!$, attempting to find a globally optimum solution is usually not computationally feasible. This difficulty has stimulated the search for efficient approximation algorithms. Genetic algorithms are widely believed to be effective on NP-complete global optimization problems and they can provide good sub-optimal solutions in reasonable time. Under this perspective, a genetic algorithm specially designed for clustering problems was introduced in [6]. This algorithm, called Clustering Genetic Algorithm (CGA), has provided encouraging results on several datasets and it is employed in our present work.

The CGA [6] is based on a simple encoding scheme. Let us consider a data set formed by N objects. Then, a genotype is an integer vector of (N+1) positions. Each position corresponds to an object, i.e. the i-th position (gene) represents the i-th object, whereas the last gene represents the number of clusters (k). Thus, each gene has a value over the alphabet {1,2,...,k}. For instance, in a dataset composed of 20 objects, a possible genotype is: 22345123453321454525. In this case, five objects {1,2,7,13,20} form the cluster whose label is 2. The cluster whose label is 1 has 2 objects {6,14}, and so on.

Generally speaking, the CGA works on a set of vectors (genotypes), in such a way that better solutions can be obtained. More specifically, the crossover operator combines clustering solutions coming from different genotypes, and it works in the following way: First, two genotypes (A and B) are selected. Then, considering that A represents k1 clusters, the CGA chooses randomly n ∈ {1,...,k1} clusters to copy in B. The unchanged clusters of B are maintained and the changed ones have their examples allocated to the cluster that has the nearest centroid. In this way, a child C is obtained. This same process is employed to get a child D, but now considering that the changed clusters of B are copied in A.

Two operators for mutation are used in the CGA. The first one works on genotypes that encode more than two clusters. It eliminates a randomly chosen cluster, placing its objects to the nearest remaining clusters according to their centroids (mean vectors). The second operator divides a randomly selected cluster into two new ones. The first cluster is formed by the objects closer to the original centroid, whereas the other one is formed by those objects closer to the farthest object from the centroid.

The objective function, whose role is to evaluate each genotype, is based on the silhouette [12]. To explain it, let us consider an object i belonging to cluster **A**. So, the average dissimilarity of i to all other objects of **A** is denoted by $a(i)$. Now let us take into account cluster **C**. The average dissimilarity of i to all objects of **C** will be called $d(i,C)$. After computing $d(i,C)$ for all clusters **C** ≠ **A**, the smallest one is selected, i.e. $b(i) = \min d(i,C)$, **C** ≠ **A**. This value represents the dissimilarity of i to its neighbor cluster, and the silhouette $s(i)$ is given by:

$$s(i) = \frac{b(i) - a(i)}{\max\{a(i), b(i)\}} \quad (2)$$

It is easy to verify that $-1 \leq s(i) \leq 1$. Thus, the higher $s(i)$ the better the assignment of object i to a given cluster. In addition, if cluster **A** is a singleton, then $s(i)$ is not defined and the most neutral choice is to set $s(i) = 0$ [12]. The objective function is the average of $s(i)$ over $i = 1, 2, ..., N$ and the best clustering is achieved when its value is as high as possible. In summary, the main steps of the CGA are described in Fig. 1 and its detailed description can be found in [6].

---
1. Initialize a population of random genotypes;
2. Evaluate each genotype according to its silhouette;
3. Apply a linear normalization;
4. Select genotypes by proportional selection;
5. Apply crossover and mutation;
6. Replace the old genotypes by the ones formed in 5);
7. If convergence is attained, stop; else, go to 2).

Fig. 1. Main steps of the Clustering Genetic Algorithm (CGA).

2.3 Running CGA as a Bag-of-Tasks Application

Our simulations were performed in a dataset that is a benchmark for data mining applications – Congressional Voting Records. This dataset is available at the UCI Machine Learning Repository [13] and it contains votes for each of the U.S. House of Representatives Congressmen on 16 key votes (attributes). There are 435 instances (267 democrats, 168 republicans) and each one of the 16 attributes is Boolean valued. Besides, there are 203 instances with missing values. These instances were removed and the CGA was employed in the 232 remaining ones.

As previously mentioned, the CGA employs probabilistic rules to optimize the objective function value described by equation (2). In other words, the employed data mining method provides non-deterministic outputs. Thus, in order to get statistically significant results, one runs the algorithm several times (usually 35 simulations provide good estimates). Besides, it must be emphasized that the probabilistic rules employed not only influence the final result (objective function value), but also the performed number of computations, and consequently, the execution time of each experiment. Thus, it is interesting to run several experiments (for example 10 of 35 simulations each) in order to obtain representative measures of execution time and throughput. Besides that, one should notice that the same final solution can be obtained by means of different ways of searching.

Under a data mining application view, the grid can provide more accurate and reliable results. For example, let us consider two experiments performed in a single machine. Initially, we run the CGA 10 times, using 20 genotypes and setting the maximum number of generations to 200. After, we run the CGA 35 times, using 50 genotypes in 500 generations. In the Congress dataset, the average objective function value improved from 0.5496 to 0.5497. On top of that, it is worth noting that better estimates of the CGA performance in these datasets were obtained. These results lead us to employ, in our grid experiments, populations formed by 50 genotypes (number of vectors in the search space), setting the maximum number of generations to 500 (number of iterations). Besides, in data mining applications, the program inputs are usually codified in plain text files. In this sense, the Congress dataset has approximately 8 Kb. When the output files are concerned, the Congress dataset provide 686 Kb of data in each simulation.

Additional simulations using this dataset were carried out to obtain estimates of performance gains. Initially, simulations were executed in a grid of 5, 10, 15, 20, 25, 30, 35, 40 and 45 machines with heterogeneous hardware characteristics. The grid used is composed of 45 machines, 22 machines located at UniSantos and the others located at Federal University of Campina Grande, Brazil (UFCG). A distance of about two and a half thousand kilometers separates these two sites. The basic scheduling strategy used in experiments was the standard workqueue algorithm [5].

For parallel execution, a Pentium IV home machine with 1GB of main memory is dedicated exclusively to dataset distribution as well as task dispatching to grid machines. Figure 2 illustrates the total execution time for 10 experiments comprised by 35 CGA simulations. One observes that the execution time of DM applications can be significantly reduced in a Grid environment, but the scalability of a grid with more than 35 machines is poor. It is due to both the application behavior and the way MyGrid interacts with grid machines to manage input and output files. Thus, a more detailed view of the MyGrid working environment becomes necessary.

Each MyGrid task is divided in three subtasks, named *init*, *execute* and *collect*, which is performed sequentially, in that order. Both *init* and *collect* subtasks are executed on the *home machine*, to send input files to grid machines and to collect the output files back to the home machine, respectively. The *execute* subtask runs on the grid machine, and is responsible to perform the computation itself. Depending on the size of input and output files as well as on the number of grid machines, the execution of *init* and *collect subtasks* on the *home machine* may become a bottleneck. The overhead related to file transfers in the home machine depends on both the size of the input and output files and the number of machines added to the grid. In addition, the data transfer rate observed in the home machine can also be affected by the running time of *grid* subtasks. The shorter is the execution time of *grid* subtasks, the more frequent is the transfer of input and output files, and consequently, the greater is the data transfer *rate demanded to the home machine*.

As evidenced by the previous tests, some applications can be critical because they cause higher data transfer ratios at the home machine. The CGA application in the Congressional dataset is such an example. Remember that home machine executes the processes that are responsible for managing the execution of application tasks on the grid machines. For each application task, a small number of processes are created to manage the transferring of input and output files, to and from grid machines. Such processes are eliminated as soon as their corresponding tasks have been completed. Thus, short application tasks can increase the rate of creation/destruction of such processes to critical levels. Moreover, these processes execute concurrently, competing for memory, I/O subsystem, network bandwidth, CPU cycles, and other resources. Our tests have shown that the scalability of the application can be severely impacted when we execute small tasks. This fact is evidenced by a new set of experiments, in which the Congress dataset was decreased from 500 to 150 generations, what reduce the computational time and move the bottleneck to from near 45 machines (Fig. 2) to around 20 machines in a dedicated cluster as illustrated in Figure 3.

For the results shown on figure 3, we run the set of tasks sequentially on a single, stand alone machine with a Pentium IV (1.8 GHz) processor, and 1 GB of main memory. Then, the same set of tasks were run on 2, 4, 8, 12, 16 and 20 dedicated machines with similar hardware characteristics, interconnected by a 100 Mbps Ethernet LAN.

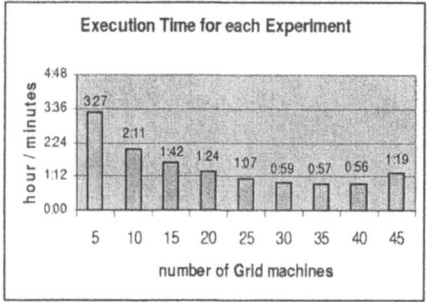

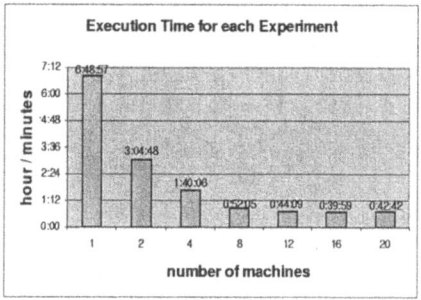

Fig. 2. Mean execution times in a Grid environment

Fig. 3. Mean execution times in a dedicated cluster environment

3 Grouping Independent Data Mining Tasks

The bottleneck represented by the home machine can be characterized by examining task execution times. For very short executions, like those in the Congressional dataset, the time of transferring data is almost the same as the running time. In these cases, there is a performance loss due to the way the home machine manages more and more processes. As an example, when executing 20 tasks in 20 machines, we obtained 52 minutes to transfer data and 58 minutes to perform the execution of the algorithm at the remote machine.

The variation in execution time is expected because the non-deterministic nature of the CGA. Despite the files transferred for each grid machine through the init are exactly the same, the amount of time needed for those transfers varies. We suspect that the bottleneck at the home machine is responsible for this variation.

As evidenced by the previous tests, some BoT applications do not scale partly because of the amount of data to be transferred to and from the home machine. The application of the CGA in the Congressional dataset is an example. Remember that the home machine is responsible for managing the execution of application tasks on the grid machines. For each application task, a small number of processes are created to manage the transfer of input and output files to and from grid machines. Such processes are terminated as soon as their tasks have been accomplished. Thus, short application tasks can increase the rate of creation/destruction of such processes to critical levels. Moreover, these processes execute concurrently, competing for memory, I/O subsystem, network bandwidth, CPU cycles, and other resources in the home machine.

One way of reducing the impact of the bottleneck represented by the home machine on the scalability of the platform is grouping sets of tasks in one larger task and then executing the "big" task in a grid machine. The main idea is grouping a set of tasks into one execution in a remote grid machine. This strategy is further investigated in the following section.

4 Experimental Results

We performed initial experiments using a cluster comprised of 20 machines at Unisantos in the same administrative domain. All machines have one Pentium IV (1.8 GHz) processor, and 128 MB of main memory. For those executions, we used a home machine with 1 GB of main memory.

In opposite to send a task per machine, we grouped sets of tasks in a single task. As the files transferred from the home machine to the grid machine should be the same for every task, it is possible to send the files just once. This, of course, reduces the amount of processing needed to manage remote tasks in home machine and minimizes the network traffic. On the other hand, execution time is increased at grid machines, but that fact does not increase the overall makespan.

As a first experiment we generate 350 tasks of the CGA using the same Congressional Voting dataset, sending one task per machine in a cluster with 20 machines with the same hardware. Afterwards we grouped the 350 tasks in groups of 5. The columns init and collect of Table 1 shows that the time for transferring data for individual tasks were larger than the corresponding times when tasks are grouped together.

Table 1. Mean times of grouped subtasks in a dedicated cluster environment

Group	init	execute	collect	total
1	00:22	00:50	00:09	01:25
5	00:16	04:30	00:05	04:56

To measure the improvement in scalability obtained when we group tasks, we executed experiments with 16 and 12 machines that groups 5 tasks to execute as a one larger task in a grid machine. Results are shown in Figure 4. The number of tasks to be grouped together was chosen after performing a set of empirical experiments. The number of tasks that have provided the best load balance among machines was 5.

The results in Figure 4 show that the total execution time decreases as more machines are added to the cluster. It is important to remember that we performed the same number of executions in all experiments, but the network traffic between the home machine and grid machines was reduced. The execution of each larger task is increased, as expected, because the remote machine runs 5 CGA tasks, instead of one, and sends packaged results back to the home machine.

We also run experiments in a grid environment comprised of 45 heterogeneous machines, 22 machines from Unisantos and the others from Federal University of Campina Grande. Comparing grouping and non-grouping executions we obtained an improvement when we clustered the tasks. Figure 5 shows the results of the experiment. When CGA tasks were executed individually the mean total execution time was 1 hour and 19 minutes (similar to the result shown in Figure 2) while when we grouped those tasks in sets of 5 the mean execution time was 48 minutes.

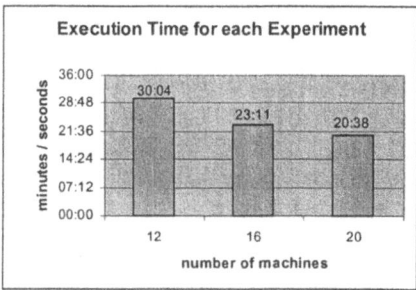 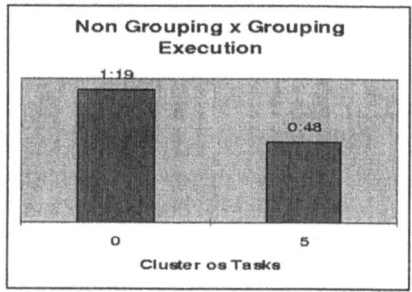

Fig. 4. Total execution times- groups of 4 tasks.

Fig. 5. Comparing Times of execution in a Grid environment

5 Conclusion

This paper has investigated the scalability of a data mining method (Clustering Genetic Algorithm) implemented as a BoT application. We proposed a scheduling strategy aimed at improved scalability, by grouping together sets of independent tasks to be executed in a grid machine. Our group is now focused in the development of a scheduling algorithm that will group tasks on the fly depending on both the characteristics of the applications and the grid platform.

Acknowledgements. Our group would like to acknowledge Walfredo Cirne and all the Mygrid team for insightful comments in earlier versions of this paper and for their support during the experiments. Fabrício Silva and Eduardo R. Hruschka acknowledge CNPq (proc. 55.2153/2002-8 and 301.353/03-4 respectively) for its financial support.

References

[1] Fayyad, U. M., Shapiro, G. P., Smyth, P. "From Data Mining to Knowledge Discovery : An Overview". In: Advances in Knowledge Discovery and Data Mining, Fayyad, U.M., Piatetsky-Shapiro, G., Smyth, P., Uthurusamy, R., Editors, MIT Press, pp. 1-37, 1996.

[2] Freitas, A.A., Lavington, S.H., *Mining Very Large Databases with Parallel Processing*, Kluwer Academic Publishers, 1998.

[3] Baraglia, R. et al., Implementation Issues in the Design of I/O Intensive Data Mining Applications on Clusters of Workstations, Proc. of the 3rd Workshop on High Performance Data Mining, International Parallel and Distributed Processing Symposium, Cancun, Mexico, 2000.

[4] Baker, M., Buyya, R., Laforenza, D. Grids and Grid Technologies for Wide-area Distributed Computing, Software, Pratice and Experience, v.32, pp.1437-1466, John Wiley and Sons, 2002.

[5] Cirne, W. et al. Running Bag-of_Tasks Applications on Ccmputational Grids: The MyGrid Approach. Proc. of the 2003 International Conference on Parallel Processing, October 2003.

[6] Hruschka, E. R., Ebecken, N.F.F. A genetic algorithm for cluster analysis, Intelligent Data Analysis (IDA), v.7, pp. 15-25, IOS Press, 2003.

[7] Canataro, M., Talia, D.The Knowledge Grid,Communications of the ACM, v.46, n.1, 2003.

[8] Orlando, S., Palmerini, P., Perego, R., Silvestri, F., *Scheduling High Performance Data Mining Tasks on a Data Grid Environment*, Proceedings of Int. Conf. Euro-Par 2002, 27-30 August 2002, Paderborn, Germany, LNCS 2400 - Springer-Verlag - Pag. 375-384.

[9] Hinke, H., Novotny, J., *Data Mining on NASA's Information Power Grid*, HPDC 2000, Pittsburgh, Pennsylvania, USA, pp.292-293, IEEE Computer Society.

[10] Agrawal, R. et al. *Fast Discovery of Association Rules* In: Advances in Knowledge Discovery and Data Mining, Fayyad, U.M., Piatetsky-Shapiro, G., Smyth, P., Uthurusamy, R., Editors, MIT Press, pp. 307-328, 1996.

[11] Goldberg, D.E., Genetic Algorithms in Search, Optimization and Machine Learning, USA, Addison Wesley Longman Inc, 1989.

[12] Kaufman, L., Rousseeuw, P. J., Finding Groups in Data, An Introduction to Cluster Analysis, Wiley Series in Probability and Mathematical Statistics, 1990.

[13] Merz, C.J., Murphy, P.M., UCI Repository of Machine Learning Databases, http://www.ics.uci.edu, Irvine, CA, University of California.

[14] Litzkow, M., Livny, M., Mutka, M. Condor – A Hunter of Idle Workstations, Proc. of the 8th International Conference of Distributed Computing Systems, pp. 104-111, June 1988.

[15] Grimshaw, A.,Wulf, W. Legion: The next logical step toward the world-wide virtual computer. Communications of the ACM, 40(1):39-45, January 1997.

[16] BOINC. Project homepage, available at http://boinc.berkeley.edu

[17] Foster, I., Kesselman, C. Globus: A Metacomputing Infrastructure Toolkit. Intl J. Supercomputer Applications, 11(2):115-128, 1997.

[18] Falkenauer, E., Genetic Algorithms and Grouping Problems, John Wiley & Sons, 1998.

Application of Block Design to a Load Balancing Algorithm on Distributed Networks

Yeijin Lee[1], Okbin Lee[1], Taehoon Lee[2], and Ilyong Chung[1]*

[1] Dept. of Computer Science, Chosun University, Kwangju, Korea
iyc@mail.chosun.ac.kr
[2] Dept. of Information Engineering, Kwangju University, Kwangju, Korea
thlee@hosim.gwangju.ac.kr

Abstract. In order to maintain load balancing in a distributed system, we should obtain workload information from all the nodes in the network. This processing requires $O(v^2)$ communication complexity, where v is the number of nodes. In this paper, we present a new synchronous dynamic distributed load balancing algorithm on a $(v, k+1, 1)$-configured network applying a symmetric balanced incomplete block design, where $v = k^2 + k + 1$. Our algorithm needs only $O(v\sqrt{v})$ communication complexity and each node receives workload information from all the nodes without redundancy. Later, for the arbitrary number of nodes, the algorithm will be designed.

1 Introduction

In a distributed system it is likely that some processors are heavily loaded while others are lightly loaded or idle. It is desirable that workload be balanced between these processors so that utilization of processors can be increased and response time can be reduced. A load balancing scheme[1]-[2] determines whether a task should be executed locally or by a remote processor. This decision can be made in a centralized or distributed manner. In a distributed system, distributed manner is recommended. In order to make this decision, each node can be informed about the workload information of other nodes. Also this information should be the latest because outdated information may cause an inconsistent view of the system state. So disseminating load information may incur a high link cost or a significant communication traffic overhead. For example, the ARPANET[3] routing algorithm is a distributed adaptive algorithm using estimated delay as the performance criterion and a version of the backward-search algorithm[4]. For this algorithm, each node maintains a delay vector and a successor node vector. Periodically, each node exchanges its delay vector with all of its neighbors. On the basis of all incoming delay vectors, a node updates both of its vectors.

In order to decrease communication overhead for obtaining workload information from all the nodes in the network, messages should be exchanged between

---

* Corresponding Author : Ilyong Chung iyc@chosun.ac.kr

adjacent nodes and then load-balancing process be performed periodically by using these local messages. So it can make the whole system be load-balancing[5]. CWA(Cube Walking Algorithm)[6] is employed for load balancing on hypercube network. It requires $O(v^2)$ communication complexity and a communication path is $O(log_2 v)$. To reduce communication cost, flooding scheme is applied. However, the overlap of workload information occurs[7]-[8]. Based on SBN(Symmetric Broadcast Networks), communication patterns between nodes are constructed. It also needs $O(v^2)$ communication complexity for collecting workload information from all the nodes and a communication path is $O(log_2 v)$[9]-[10].

In this paper we design the network topology consisting of v nodes and $v \times k$ links and each node of which is linked to $2k$ nodes, where $v = k^2 + k + 1$. On this network, each node receives information from k adjacent nodes and then sends these information to other k adjacent nodes periodically. So, each node receives workload information for $k^2 + k$ nodes with two-round message interchange. Our algorithm needs only $O(v\sqrt{v})$ communication complexity.

2 About (v, k, λ)-Configuration

Let $V = \{0, 1, ..., v-1\}$ be a set of v elements. Let $B = \{B_0, B_1, ..., B_{b-1}\}$ be a set of b blocks, where B_i is a subset of V and $|B_i| = k$. For a finite incidence structure $\sigma = \{V, B\}$, if σ satisfies following conditions, then it is a balanced incomplete block design(BIBD)[11], which is called a (b, v, r, k, λ)-configuration.

1. B is a collection of b k-subsets of V and these k-subsets are called the blocks.
2. Each element of V appears exactly r of the b blocks.
3. Every two elements of V appears simultaneously in exactly λ of the b blocks.
4. $k < v$.

For a (b, v, r, k, λ)-configuration, if it satisfies $k = r$ and $b = v$, then it is a symmetric balanced incomplete block design (SBIBD)[12] and it is called a (v, k, λ)-configuration. There are some relations among parameters b, v, r, k, λ that are necessary conditions for existence of this configuration, $bk = vr$ and $r(k-1) = \lambda(v-1)$.

3 Generation of a $(v, k+1, 1)$-Configuration

We now present an algorithm to generate an incidence structure $\sigma = \{V, B\}$ satisfying the condition for a $(v, k+1, 1)$-configuration in the case that k is a prime number. This $(v, k+1, 1)$-configuration is employed for constructing network topology below.

3.1 Design of an Algorithm to Construct $(v, k+1, 1)$-Configuration

Algorithm 1 for Generating an incidence structure

Incidence structure $T = \{V, B\}$, where $V = \{0, 1, ..., v-1\}$, $B = \{B_0, B_1, ..., B_{b-1}\}$, $|B_i| = k+1$. $B_{i,j}$ is the jth element of B_i

1. Select a prime number k and compute $v = k^2 + k + 1$.
2. Construct two incidence structures $X = \{V, C\}$ and $Y = \{V, D\}$.

   a) $C_{i,j} = \begin{bmatrix} 0 & if\ j=0 \\ t, t = i \times k + j & if\ j \geq 1 \end{bmatrix}$
   $0 \leq i, j \leq k$.

   b) $D_{i,j} = \begin{bmatrix} C_{0,t}, t = \lfloor i/k \rfloor + 1 & if\ j = 0 \\ C_{j,t}, t = 1 + (i + (j-1) \times \lfloor i/k \rfloor)\ mod\ k & if\ j \geq 1 \end{bmatrix}$
   $0 \leq i \leq (k^2 - 1),\ 0 \leq j \leq k$.

3. Generate $Z = \{V, B\}$ from X and Y.
   $B_i \leftarrow C_i$
   $B_{i+k+1} \leftarrow D_i$

The table below illustrates how to create $Z = \{V, B\}$, $V = \{0, 1, ..., 6\}$. We now prove that this structure satisfies the conditions of a $(v, k+1, 1)$-configuration.

Table 1. A set of blocks on Z generated from Algorithm 1

X	Y	Z
$C_0 = \{0, 1, 2\}$ $C_1 = \{0, 3, 4\}$ $C_2 = \{0, 5, 6\}$	$D_0 = \{1, 3, 5\}$ $D_1 = \{1, 4, 6\}$ $D_2 = \{2, 3, 6\}$ $D_3 = \{2, 4, 5\}$	$B_0 = \{0, 1, 2\}$ $B_1 = \{0, 3, 4\}$ $B_2 = \{0, 5, 6\}$ $B_3 = \{1, 3, 5\}$ $B_4 = \{1, 4, 6\}$ $B_5 = \{2, 3, 6\}$ $B_6 = \{2, 4, 5\}$

Definition 1 : On incidence structure Y, Sector S_i is the ith family of k blocks, $D_j \in S_i,\ i = \lfloor j/k \rfloor$.

For example, If k equals 3, then $\lfloor 0/k \rfloor = \lfloor 1/k \rfloor = \lfloor 2/k \rfloor = 0$. So, $S_0 = \{D_0, D_1, D_2\}$. There are k sectors in Y.

Lemma 1 : For two elements $D_{i1,j1}$ and $D_{i2,j2}$, $D_{i1,j1} \neq D_{i2,j2}$, if $j1 \neq j2$.
Proof: From Algorithm 1-2-(a), if $0 < j \leq k$, $0 \leq i \leq k$ then $C_{i,j} = i \times k + j$. This means if $j > 0$ then all the elements are distinct. And as shown in Algorithm 1-2-(b), an element of C_j is placed on the jth element of a certain block of Y if $D_{i,j} = C_{j,t}, t \neq 0$.

Lemma 2 : For a sector consisting of k blocks, the first element of each block has the same value and the other k^2 elements are equal to $V - C_0$.

Proof: In the case that $D_{i,0} = C_{0,\lfloor i/k \rfloor + 1}$, the first element of k blocks on a sector has the same value. According to Algorithm 1-2-(b), $D_{i,j} = C_{j,t}, t = 1 + (i + (j-1)\lfloor i/k \rfloor) \mod k$. Since k is a prime number, each element except the first element of each block is distinct and these distinct k^2 elements are equal to $V - C_0$.

Lemma 3 : For incidence structure Y, $D_{a,j} = D_{b,j}, j \geq 1$, if
$b = ((a - c(j-1)) \mod k + k(\lfloor a/k \rfloor + c)) \mod k^2$.
Proof: From Algorithm 1-2-(b), $D_{a,j} = C_{j,t}$. We now prove that $D_{b,j} = C_{j,t}$. t can be calculated from parameters b, j below. Then t obtained on this lemma is equal to that from Algorithm 1-2-(b). Therefore, $D_{a,j} = D_{b,j}$.

$t = 1 + (b + (j-1) \times \lfloor b/k \rfloor) \mod k$
$= 1 + (((a-c(j-1)) \mod k + k(\lfloor a/k \rfloor + c)) + (j-1)\lfloor ((a-c(j-1)) \mod k + k(\lfloor a/k \rfloor + c))/k \rfloor) \mod k$
$= 1 + (a - c(j-1)) + (j-1) \times (\lfloor a/k \rfloor + c) \mod k$
$= 1 + (a + (j-1)\lfloor a/k \rfloor) \mod k$

Here, if $D_{a,j}$ is in sector S_s then $D_{b,j}$ is in $S_{(s+c) \mod k}$. In case of $c \equiv 0 \pmod{k}$, then $a = b$.

Lemma 4 : Each element of V appears in exactly $k + 1$ times in Z.
Proof: According to Algorithm 1-2-(a), $C_{i,0} = 0$. Since $0 \leq i \leq k$, 0 appears $k + 1$ times. The other $v - 1$ elements, $V - \{0\}$, appear exactly once on X. From Lemma 3, each element of $C_{0,j}, 1 \leq j \leq k$, appears k times in a sector of Y and the rest k^2 elements appear once in every sector of Y. Therefore, each element appears $k + 1$ times in Z.

Lemma 5 : Any pair of elements of V appears in exactly only once in Z.
Proof: The first element of V makes a pair with all the other elements and this pair appears once by designing rule of incidence structure(see Algorithm 1-2-(a)). Each element of $C_{0,j}, 1 \leq j \leq k$ makes a pair with $V - C_0$ elements and it also appears once proven by Lemma 3. The rest k^2 elements are now considered. For an arbitrary pair $D_{a,j1} = D_{a,j2}, j1, j2 \geq 1$, in order to make the same pair on other block D_b, the two elements should be on the same block. According to Lemma 4, if $j1 = j2$, then they are located on D_b. However, this case does not occur since $j1 \neq j2$. Therefore, any pair of elements of V appears in exactly once in Z.

Therorem 1 : Z designed by Algorithm 1 satisfies the conditions of a $(v, k+1, 1)$-configuration.
Proof: Z satisfied the conditions of the SBIBD by Lemma 4 and Lemma 5.

3.2 Design of Network Configuration

In order to construct a network topology which have minimum link cost and traffic overhead, we imported $(v, k + 1, 1)$-configuration. An incidence structure $Z = \{V, B\}$ satifies the conditions for a $(v, k + 1, 1)$-configuration and M is a binary incidence matrix of Z. Then this matrix M can be transformed to an adjacent matrix of a graph $G = \{V, E\}$. Based on this idea, network topology can be designed as follows.

Algorithm 2 for Design of Network Configuration

1. Create an incidence structure $Z = \{V, B\}$ by Algoritm 1.
2. Generate $L = \{V, E\}$ from Z by exchanging blocks so that every block i includes object i.
   $E_0 \leftarrow B_0$
   for ($i = 1$; $i < v$; $i = i+1$) {
      if ($i \leq k$) { $j \leftarrow i \times k + 1$; $t \leftarrow B_{j,i}$ }
      else if ($(i \bmod k) = 1$){ $t \leftarrow B_{i,0}$ }
      else { $j \leftarrow \lceil i/k \rceil - 1$; $t \leftarrow B_{i,j}$ }
      $E_t \leftarrow B_i$
   }
3. Create an adjacent matrix $A = (a_{ij})$ for graph G from L, where G is a network topology containing v processors.

$$a_{ij} = \begin{bmatrix} 1 \text{ if } i \neq j, \text{ and if } i \in E_j \text{ or } j \in E_i \\ 0 \text{ otherwise} \end{bmatrix}$$

G has v nodes since G is created from $(v, k+1, 1)$-configuration. Each block $L[i]$ is composed of $k+1$ elements and i is the one of them. Each node obtains $2k$ links from Step 3 of Algorithm 2. So, G becomes a 2k-regular graph. Therefore there are $(2k \times v)/2 = vk$ links in G. Given $Z = \{V, B\}$ described on Fig.1, performance of Algorithm 2 is shown on Table 2 and Fig.1.

Table 2. Blocks of L generated from Z of Table 1

L
$E_0 = \{\,0, 1, 2\,\}$
$E_1 = \{\,1, 3, 5\,\}$
$E_2 = \{\,2, 3, 6\,\}$
$E_3 = \{\,0, 3, 4\,\}$
$E_4 = \{\,1, 4, 6\,\}$
$E_5 = \{\,2, 4, 5\,\}$
$E_6 = \{\,0, 5, 6\,\}$

4 Design of an Efficient Load Balancing Algorithm on (v,k+1,1)-Configured Networks

An efficient load balancing algorithm is now constructed on $(v, k+1, 1)$-configured networks generated by Algorithm 2.

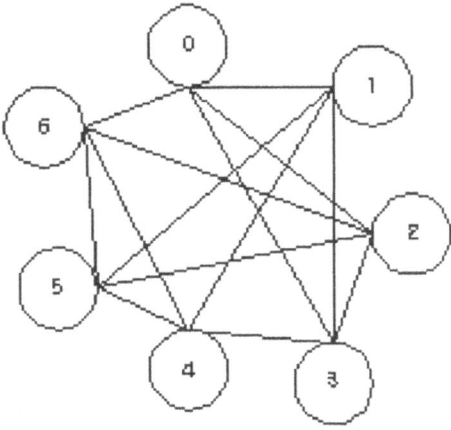

Fig. 1. (7,3,1)-configured network obtained from L

Definition 2 : Construct two sets S_i and R_i consisting of adjacent k nodes, where S_i is a set of nodes to which node i sends workload information and R_i is a set of nodes to receive i's workload information.
$S_i = \{v \mid v \in E_i - i\}$
$R_i = \{v \mid i \in E_v \text{ and } i \neq v\}$

Definition 3 : Generate two sets SF_i and RF_i, where $SF_i(j)$ is a set of workload information for i's adjacent nodes transmitted from node i to node j at time T_{2t} and $RF_i(j)$ is i's workload information transmitted from node i to node j at time T_{2t+1}.
$SF_i = \{SF_i(j) \mid j \in S_i, \ SF_i(j) = \{E_i - \{j\}\}\}.$
$RF_i = \{RF_i(j) \mid j \in R_i, \ RF_i(j) = i\}.$

Algorithm 3 for Construction of an Efficient Load Balancing Algorithm

1. Node i sends a set of workload information $SF_i(j)$ to node $j \in S_i$ at T_{2t} and renews a table of workload information.
2. Node i sends a set of workload information $RF_i(j)$ to node $j \in R_i$ at T_{2t+1} and renews a table of workload information.
3. Repeat the first step.

The following table indicates that node i sends workload information $SF_i(j)$ and $RF_i(j)$ to node j at times T_{2t} and T_{2t+1}, respectively. So every node can obtain workload information for all the nodes at T_{2t+2} and this fact is proven in Theorem 2.

Theorem 2 : According to Algorithm 3, every node obtains workload information for all the nodes at T_{2t+2}.
Proof: At T_{2t}, node i sends workload information for $SF_i(j)$ to node j. On an aritrary pair $(SF_{i1}(j), SF_{i2}(j))$, $i1 \neq i2$, intersection of these sets is empty

Table 3. Two steps for sending workload information from Node i

Node ID	T_{2t}	T_{2t+1}
0	$SF_0(1) = \{\ 0, 2\ \}\ SF_0(2) = \{\ 0, 1\ \}$	$RF_0(3) = \{\ 0\ \}\ RF_0(6) = \{\ 0\ \}$
1	$SF_1(3) = \{\ 1, 5\ \}\ SF_1(5) = \{\ 1, 3\ \}$	$RF_1(0) = \{\ 1\ \}\ RF_1(4) = \{\ 1\ \}$
2	$SF_2(3) = \{\ 2, 6\ \}\ SF_2(6) = \{\ 2, 3\ \}$	$RF_2(0) = \{\ 2\ \}\ RF_2(5) = \{\ 2\ \}$
3	$SF_3(0) = \{\ 3, 4\ \}\ SF_3(4) = \{\ 3, 0\ \}$	$RF_3(1) = \{\ 3\ \}\ RF_3(2) = \{\ 3\ \}$
4	$SF_4(1) = \{\ 4, 6\ \}\ SF_4(6) = \{\ 4, 1\ \}$	$RF_4(3) = \{\ 4\ \}\ RF_4(5) = \{\ 4\ \}$
5	$SF_5(2) = \{\ 5, 4\ \}\ SF_5(4) = \{\ 5, 2\ \}$	$RF_5(1) = \{\ 5\ \}\ RF_5(6) = \{\ 5\ \}$
6	$SF_6(0) = \{\ 6, 5\ \}\ SF_6(5) = \{\ 6, 0\ \}$	$RF_6(2) = \{\ 6\ \}\ RF_6(4) = \{\ 6\ \}$

since on $(v, k+1, 1)$-configuration, every two objects appears simultaneously in exactly one of v blocks and node j is an element of S_{i1} and S_{i2}, respectively. So node j obtains workload information for k^2 nodes. And at T_{2t+1}, node i transmits its workload information to node j by Algorithm 3-2. Then, node j receives k workload information. Therefore, node j receives workload information for $k^2 + k$ nodes at T_{2t+2}.

5 Conclusion

In order for the system to increase utilization and to reduce response time, workload should be balanced. In this paper, we present an efficient load balancing algorithm on $(v, k+1, 1)$-configured networks consisting of v nodes and vk links. Our algorithm needs only $O(v\sqrt{v})$ communication complexity and each node receives workload information from all the nodes without redundancy and load balancing is maintained so that every link has same amount of traffic for transferring workload information.

References

1. M. Willebeek-Lemair and A. P. Reeves, Strategies for dynamic load-balancing on highly parallel computers, IEEE Transactions on Parallel and Distributed Systems, vol. 4, no. 9, pp. 979-993, 1993.
2. B.A. Shirazi, Scheduling and load balancing in parallel and distributed systems, IEEE Computer Society Press, 1995.
3. M. Padlipsky, A perspective on the ARPANET Reference Model, Proc. of INFOCOM, IEEE, 1983.
4. L. Ford, D. Fulkerson, Flow in Network, Princeton University Press, 1962.
5. C.Hui, S.Chanson, Hydrodynamic Load Balancing, IEEE Transactions on Parallel and Distributed System, vol. 10, no. 11, pp. 1118-1137, 1999.
6. M. Wu, On Runtime Parallel Scheduling for processor Load balancing, IEEE Transactions on Parallel and Distributed System, vol. 8, no. 2, pp. 173-185, 1997.
7. K. Nam, J. Seo, Synchronous Load balancing in Hypercube Multicomputers with Faulty Nodes, Journal of Parallel and Distributed Computing, vol. 58, pp. 26-43, 1999.

8. H. Rim, J. Jang, S. Kim, Method for Maximal Utilization of Idle links for Fast Load Balancing, Journal of Korea Information Science Society, vol. 28, no. 12, pp. 632-641, 2001.
9. S. Das, D. Harvey, and R. Biswas, Adaptive Load-Balancing Algorithms Using Symmetric Broadcast Networks, Journal of parallel and Distributed Computing, vol. 62, no. 6, pp. 1042-1068, 2002.
10. S. Das, D. Harvey, and R. Biswas, Parallel Processing of Adaptive Meshes with Load Balancing, IEEE Transactions on Parallel and Distributed Systems, vol. 12, no. 12, 2001.
11. C.L.Liu, Block Designs in Introduction to Combinatorial Mathematics, McGraw-Hill, pp. 359-383, 1968.
12. I. Chung, W. Choi, Y. Kim, M. Lee, The Design of conference key distribution system employing a symmetric balanced incomplete block design, Information Processing Letters, vol. 81, no. 6, pp. 313-318, 2002.3.

Maintenance Strategy for Efficient Communication at Data Warehouse

Hyun Chang Lee[1] and Sang Hyun Bae[2]

[1] 604-5, Dangjung-Dong, Gunpo-Si, Kyunggi-Do, School of Information and Technology
Hansei University, Korea, 435-742
hclee@hansei.ac.kr
http://www.hansei.ac.kr
[2] 375, Seosuk-Dong, Dong-Gu, Kwangju, Computer Science & Statistics
Chosun University, Korea, 501-759
shbae@chosun.ac.kr

Abstract. Data warehousing is used for reducing the load of on-line transactional systems by extracting and storing the data needed for analytical purposes. Maintaining the consistency of warehouse data is challenging as views of the data warehouse span multiple sources, and user queries are processed using this view. The view has to be maintained to reflect the updates done against the base relations stored at the various data sources. In this paper, we present incremental and efficient view maintenance algorithm for a data warehouse derived from multiple distributed autonomous data sources and an enhanced algorithm that does not require the data warehouse be in a quiescent state for incorporating the new views. Also The proposed algorithm can be reduced server loads and the response time, by overlapping processing time and messages size between warehouse and sources. We show its performance result by comparing it to other conventional algorithms.

1 Introduction

A data warehouse collects information from distributed, autonomous, possibly heterogeneous, information sources and integrates it into a single source where it can be queried by clients of the warehouse. In short, a data warehouse is a repository of integrated information, available for queries and analysis (e.g., decision support, data mining)[4]. When relevant information is modified, the information is extracted from the source, translated into a common model, and integrated with existing data at the warehouse. Queries can be answered and analysis can be performed quickly and efficiently since the integrated information is available at the warehouse.

The relations stored at the warehouse represent materialized views over the data at the sources [2]. Because queries at the warehouse tend to be long and complex, a warehouse may contain materialized view in order to speed up query processing [3]. As changes are made to the data at the sources, the views at the warehouse become out of date. In order to make the views consistent again with the source data, changes to the source data in commercial warehousing systems are queued and propagated periodically to the warehouse view in a large batch update transaction. An important problem in data warehousing is how to execute queries and the periodic maintenance transaction so that they do not block one another, since both queries and maintenance transactions can be long running [5].

One of various approaches is to allow users to see an inconsistent database state. However, such inconsistency is not acceptable. During analysis it would be unacceptable to have the results change from query to query. Therefore, most of the recent research on materialized views has focused on techniques for incrementally updating materialized views in order to make the views consistent [1] and numerous methods have been developed for materialized view maintenance. Some representative approaches for view maintenance are the ECA, STROBE and SWEEP algorithms [7].

These algorithms have the following problems. First of all, the Eca algorithm is designed for a central database site whereas the Strobe and Sweep algorithms are based on multiple sites data model. The Eca and Strobe algorithms require key attributes and a quiescent state at the data warehouse that has to manage unanswered query set until it receives all answers. The Sweep algorithm requires heavy server loads and exponential message size with regard to the data sources. In this paper, we propose an incrementally efficient view maintenance algorithm in order to solve or reduce these problems. Our algorithm could reduce server loads by transferring the answer that would be compensated at the data warehouse to each data source, and message size by dividing a query to compensate at each source into two sub queries. It can reduce the query response time by overlapping the irrelevant part to each update as well. We illustrate the problem by using the relational model for data and relational algebra select-project-join queries for view.

The paper is organized as follows. In section 2, we briefly review related research, and in section 3, we provide a formal definition related to data warehouse. In section 4, we propose SFUIL algorithm using storage for update information list. Section 5 describes the performance of the proposed algorithm. In section 6, we conclude and discuss future directions of our work.

2 Related Research

Warehouse collects information from one or more data sources and integrates it into a single database, a large repository for analytical data and decision support, where it can be queried by clients of the warehouse. A general description of data warehousing may be found in [4] and the main problem that arises in the context of a data warehouse is to maintain the materialized view at the data warehouse in the presence of updates to the data sources. When the simple update information arrives at the warehouse, we may discover that some additional source data is necessary to update the views. Thus, the warehouse may have to issue queries to some of the sources as illustrated in figure 1.

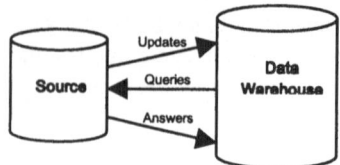

Fig. 1. Processing model between source site and data warehouse

A simple approach of recomputing the view as a result of each update is unrealistic. A more appropriate solution would be to update the data warehouse incrementally in response to the updates arriving from the data sources. Several approaches have focused on the problems associated with incremental view maintenance at data warehouse [7].

Many incremental view maintenance algorithms have been developed for centralized database systems [8] and an overview of materialized views and their maintenance can be found in [9]. Most of these solutions assume that a single system controls all of the base relations and understands the views, and hence can intelligently monitor activities and compute all of the information that is needed for updating the views.

Example of works in distributed database systems with multiple distributed sites includes Strobe algorithm [6] and Sweep algorithm [7,10]. Hull and Zhou [10] provide a framework for supporting distributed data integration using materialized views. However, their approach first materializes each base relation, and then computes the view from the materialized copies. Their algorithm needs an auxiliary data to store. The materialized view in Strobe is assumed to be an SPJ-expression where the projection list contains the key attributes from each relation. It also needs additional key information, and the algorithm has to wait to do the compensation until the entire query has been completely evaluated by querying all data sources. In Strobe and Sweep algorithms, an answer to a given update is evaluated by contacting all the sources in a regular order. This is responsible for the increase in the size of messages transferred between the warehouse and all data sources. The more the data sources increased, the size of messages is also increased. Also, in Sweep, to eliminate the effects of concurrent updates, it is handled at the data warehouse and this leads to heavier server loads. In this paper, we propose a SFUIL(storage for update information list) algorithm to solve or reduce the above mentioned problems in existing algorithms.

3 Data Warehouse Model

We mention the data model to be managed at the data warehouse environment. The data model in this paper is based on [6]. Updates occurring at the data sources are classified into three categories.
- Single update transactions where each update is executed at a single data source.
- Source local transactions where a sequence of updates are performed as a single transaction. However, all of the updates are directed to a single data source.
- Global transactions where the updates involve multiple data sources.

For the purpose of this paper, we assume that the updates being handled at the data warehouse are of type one and two. The approaches described in Strobe and Sweep can be used to the algorithm presented in this paper for type three updates. At each source, there are two types of message from the data source to the warehouse. One is reporting the update information and maintaining the reported update in a temporary storage to keep the order of update occurring at the source. We call the storage SFUIL(storage for update information list). The other is turning the answer to a query. There are also two types of messages from the data warehouse to the data sources. At this time, the data warehouse uses the SSFUIL, Server SFUIL, like the source to keep the order of updates occurring at each source. The warehouse may send queries to the data sources, and a command to delete the update information existed in SFUIL of that source, in which the update was originated at one source.

Figure 2 shows the structure of the data warehouse environment for the data warehouse and sources described above. It consists of n distributed sites for data sources and another site for storing and maintaining the materialized view of data warehouse. We assume that messages between each data source and data warehouse are not lost and are delivered in the order in which they are sent. Communication between the different data source sites may not be able to communicate with each other. The database model for each data source is assumed to be a relational data model. A data source may store any number of base relations, but in this paper we assume a single base relation Ri at data source i. Updates are executed atomically and transmitted asynchronously to the data warehouse as updates occur.

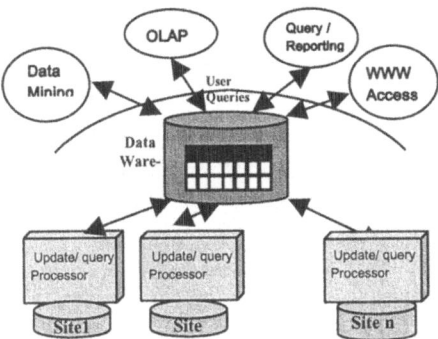

Fig. 2. Architecture of a data warehouse system

We assume that the materialized view used at the data warehouse is defined by the SPJ expression[6]. For instance, if there are n base relations at corresponding data sources, $R_1, .. R_i, .. R_n$, the materialized view denoted MV is as follows:

$$MV = \Pi_{ProAtrr} (\sigma_{selectcond} (R_1 \infty ... \infty R_i \infty ... \infty R_n)).$$

4 SFUIL Algorithm

We begin by describing the view maintenance algorithm in this section. Before understanding the algorithm, we investigate the notion of events.

4.1 Event at Data Source and Data Warehouse

We assume that each source has a SFUIL to keep the order of updates at each source, which the updates accumulate in order.
1) Events at the source
- $S_{\_req}$: the source executes an update U, and enrolls the update information into SFUIL and sends the update notification to the warehouse
- $S_{\_eva}$: if a query is a deletion operation, then the source deletes the first update information in SFUIL. Otherwise after the source evaluates a query Q using its current base relations, and the source evaluates the update information in SFUIL of the source with temporary relation to compensate at the source, and then sends the correct answer relation A back to the warehouse.

2) Events at the data warehouse
Suppose that the events of the data warehouse are processed in the following orders of a transaction.
- $W_{\_view}$: the warehouse receives an update U, appends U to SSFUIL(server side SFUIL) and generates a query Q_{11}, Q_{12}. After that, the warehouse sends a delete operation for removing the update information in the SFUIL to the source, which update has occurred, and Q_{11}, Q_{12} for evaluation to the different sources.
- $W_{\_ans}$: the warehouse receives the answer relations A_{11}, A_{12} to queries Q_{11}, Q_{12}, joins the two relations, and updates the materialized view based on the joined answer.

We will assume that events are atomic, and actions within an event follow the order described above. For example, within an event $S_{\_req}$, the source first executes the update operation, and then records the update information in SFUIL and sends the update notification to the warehouse.

4.2 SFUIL Algorithm

We present the details of a proposed algorithm, SFUIL termed storage for update information list, to reduce server loads and the message size transferred between data sources and the data warehouse. SFUIL algorithm is an incremental view maintenance algorithm based on distributed information sources. The algorithm calls two events for an update and processing at each source sites. Two events are the following. First is the SendUpates event for an update ΔR at relation R. Second is the ProcessQuery for a request to delete the update information in $SFUIL_i$ at the source i or to compute joined results at the sites except the original source.

We assume that each data source processes both requests concurrently. Hence requests can be processed simultaneously at different sites. The two events are shown in the following.

```
FUNCTION      CorrectAnswer      (TEMP:RELATION,ΔT:RELATION):
RELATION;
VAR
j: INTEGER
  BEGIN
    FOR(j=0; j < # of update information in SFUIL_i; j++)
      IF(jth update information in SFUILi is INSERTION)
      THEN TEMP=TEMP-ComputeJoin(jth update in SFUIL_i, ΔT);
      ELSE TEMP=TEMP+ComputeJoin(jth update in SFUIL_i, ΔT);
      ENDIF
    ENDFOR:
    RETURN(TEMP);
END CorrectAnswer
PROCESS SendUpdates;
  BEGIN
    LOOP
      RECEIVE ΔR FROM Relation R;
      APPEND ΔR TO SFUIL_i, SEND(ΔR,Index) TO DataWarehoue;
    FOREVER;
  END SendUpdates
PROCESS ProcessQuery;
  BEGIN
    LOOP
      RECIVE ΔT FROM DataWarehouse;
      IF ΔT is NOT DELETE Operation
      THEN Temporary = ComputeJoin(ΔT,R)
        SEND (ΔT = CorrectAnswer (Temporary, ΔT)) TO Data-
        Warehouse and Temporary = ∅;
      ELSE Delete ΔT in SFUIL_i;
      ENDIF;
    FOREVER;
  END ProcessQuery
```

The SendUpdates only append the update in the SFUIL and send the occurred update at that source to the data warehouse. After receiving a query from data warehouse, ProcessQuery identifies whether the query type is a deletion operation or a computing operation. The deletion only means that the source should remove the update information from SFUILi. The other case first computes the result of the join (the query and source relation). The computed result is not the correct answer if other updates have been already occurred at that source. Therefore, it needs to evaluate whether the result is an intentional state or not. If the result is not an intentional state, then the source calls the function CorrectAnswer to compensate locally. The function CorrectAnswer changes the temporary result into a final result in a way of compensating using the SFUIL of that source, i.e., the temporary result may be computed with each element in the SFUIL. After then, the source sends the final result to the data warehouse. It reduces server loads by transferring compensation done at the data warehouse into each data source.

In the data warehouse module, there are two parts. One is an event to process the UpdateSFUIL process for an update ΔR notification at the relation R. The process just records the notifications regarding which update has occurred at each source. The other part is the UpdateView event for a request to join two answer relations from sources, and to compute new incremental view, ΔMV. At the data warehouse, we assume that two requests are processed sequentially. The following are the algorithm for each event.

```
PROCESS UpdateSFUIL;
  BEGIN
      LOOP
         RECEIVE ΔR FROM Data Source i;
         APPEND (ΔR, i) TO SSFUIL(server side SFUIL);
      FOREVER;
END UpdateSFUIL
PROCESS UpdateView;
  BEGIN
   LOOP
    REMOVE first update information,(ΔR,i), FROM SSFUIL;
    MV = MV + ChangeView(ΔR, i);
   FOREVER;
END UpdateView
```

The event UpdateView includes a different meaning to improve the rate of concurrence. It is possible to improve the concurrency rate when the source of a next update in SSFUIL at the data warehouse is the same as the data source that has sent a temporary answer from one of sources. When the result of a currently processed query is sent back to the data warehouse, and the data source of a next update information in SSFUIL at the data warehouse is the same as the source that returns an answer relation. It means that the current query has nothing to do at that site. By overlapping the irrelevant part to each update, the next update information at data warehouse can be processed continuously. We call this method overlapping processing, which has not been mentioned in any literature to enhance concurrency.

5 Performance Evaluation

In the previous sections, we mentioned several strategies for view maintenance in a data warehouse environment. In this section, we analyze the performance of the view maintenance algorithm, SFUIL. We will briefly compare it with the Strobe and Sweep algorithms. The main properties of SFUIL are that it reduces the size of messages transferred between the warehouse and data sources, and also reduces server loads by transferring the answer that would be compensated at the warehouse to each source. The strategy also reduces the query response time by overlapping the irrelevant parts in each update.

Strobe is very restrictive in that it assumes that the view function includes the key attributes of every base relation in the view. The algorithm requires that the data warehouse be in a quiescent state for incorporating the new views, whereas there are no such problems in the Sweep algorithm. However, it increases the warehouse overhead because the data warehouse processes the operation to compensate temporary results generated that are sent from each source. In addition it increases the response time because a subsequent update has to wait until the preceding update is finished.

In SFUIL algorithm, all of the problems mentioned are reduced or solved because SFUIL handles an update as a local compensation at each source. As a result, it can reduce the server loads. It reduces the size of message transferred in half by sending in two directions on the basis of the site, where an update has occurred.

In general, a query at the warehouse is sent to the data sources for identifying whether an update can be incorporated with the materialized view or not. At this time, if the data warehouse has key attributes for each source, then data warehouse need not send a query to the data sources. Because an update information sent from a source includes key attributes, the data warehouse can use the key attributes to make the materialized view consistent: if an update from a data source is an insertion, then data warehouse has to send a query to the sources for identifying whether the update is incorporated with the materialized view. Whereas, if an update from a data source is a deletion, it is not necessary that the data warehouse should send a query.

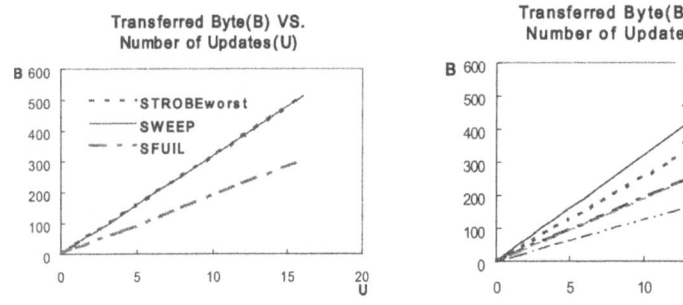

(a) transferred bytes of the three algorithms (b) transferred bytes, as the type of updates is different

Fig. 3. Bytes versus number of updates

That is why the data warehouse can make the materialized view a consistent state directly by removing the value of the materialized view corresponding to the key values.

Figure 3 shows the number B of bytes transferred as a function of the number of update. In figure 3-a, the Strobe algorithm presents the result of transferred bytes when all the types of update is insertion, whereas in figure 3-b, the Strobe shows the outstanding results under a various ratio of insertion to deletion operations. The Strobe algorithm based on figure 3-b shows enhanced results because of including key attributes. The Strobe has the advantage that the more there are deletion operations, the cost of the messages transferred between the data sources and warehouse is less.

However, the Sweep and Sfuil algorithms need not have any key attributes and require a quiescent state at the data warehouse. These are attractive points compared to the Strobe algorithm, where the key attribute is just included in each base relation and it is restrictive. The SFUIL algorithm is superior to the other algorithms as the rate of insert operation is more than half and also has the same advantages as the Strobe algorithm. .

6 Conclusion

A data warehouse is a repository of integrated information, available for queries and analysis. The relations stored at the warehouse represent materialized views to speed up query processing over the data at the sources. When relevant information is modified, the information is extracted from the source, translated into a common model, and integrated with existing data at the warehouse. Queries can be answered, and analysis can be performed, quickly and efficiently since the integrated information is available at the warehouse. In order to make the views consistent with the source data, changes to the source data are propagated incrementally to the warehouse view.

In this paper, we investigated the reason why the data warehouse environment has to propagate the updates occurred at each data source to the warehouse, and we considered algorithms related to supporting multiple distributed data model. We enumerated and considered the problems in these algorithms that are used in multiple distributed database sites, such as Strobe and Sweep algorithms, and proposed a new distributed algorithm, SFUIL. We also presented brief performance results for the SFUIL algorithm. The SFUIL algorithm has the following advantages. It does not require quiescence to incorporate the view changes into the materialized view and reduces the server load by transferring the answer that would be compensated from the data warehouse to each data source. It also reduces query response time by overlapping the irrelevant part of each update. A more detailed performance evaluation and analysis must be done in the future to evaluate the effectiveness of the proposed algorithm.

References

1. T. Griffin and L. Libkin. Incremental maintenance of views with duplicates. In M. Carey and D. Schneider, editors, Proceedings of ACM SIGMOD 1995 International Conference on Management of Data, pages 328-339, San Jose, CA, May 23-25 1995
2. D. Lomet and J. Widom, editors. Special Issue on Materialized Views and Data Warehousing, IEEE Data Engineering Bulletin 18(2), June 1995.
3. V. Harinarayan, A. Rajaraman, and J.D. Ullman. Implementing data cubes efficiently. In Proceedings of ACM SIGMOD 1996 International Conference on Management of Data, pages 205-216, 1996.
4. W.H. Inmon and C. Kelley. Rdb/VMS:Developing the Data Warehouse. QED Publishing Group, Boston, London, Toronto, 1993.
5. D. Quass and J. Widom. On-Line Warehouse View Maintenance. In Proceedings of ACM SIGMOD 1997 International Conference on Management of Data, pages 393-404, 1997.

6. Yue Zhuge, Hector Garcia-Molina, and Janet L. Wiener. The Strobe Algorithms for Multi-Source Warehouse Consistency. In Proceedings of the International Conference on Parallel and Distributed Information Systems, December 1996.
7. D. Agrawal, A. El Abbadi, A. Singh and T. Yurek. Efficient View Maintenance at Data Warehouses. In Proceedings of ACM SIGMOD 1997 International Conference on Management of Data, pages 417-427, 1997.
8. L. Colby, T. Griffin, L. Libkin, I. Mumick, and H. Trickey. Algorithms for deferred view maintenance. In SIGMOD, pages 469-480, June 1996.
9. A. Gupta and I. Mumick. Maintenance of materialized views: Problems, techniques, and applications. IEEE Data Engineering Bulletin, 18(2):3-18, June 1995.
10. R. Hull and G. Zhou. A framework for supporting data integration using the materialized and virtual approaches. In SIGMOD, pages 481-492, June 1996.

Conflict Resolution of Data Synchronization in Mobile Environment

YoungSeok Lee[1], YounSoo Kim[2], and Hoon Choi[2]

[1] Electronics and Telecommunications Research Institute
161 Gajeong-dong, Daejeon, 305-350, Korea
yslee@etri.re.kr
[2] Dept. of Computer Engineering, Chungnam National University
220 Gung-dong, Daejeon 305-764, Korea
{kimys,hchoi}@ce.cnu.ac.kr

Abstract. Proliferation of personal information devices results in a difficulty of maintaining replicated copies of the user data in data stores of the various devices. Data synchronization is an automated action to make the replicated data be consistent with each other and up-to-date. The traditional consistency models and the conflict resolution schemes do not fit to the synchronization of replicated data in the mobile computing environment due to the lacks of simultaneous update. This paper shows characteristics of data synchronization and conflict between replicated data in the mobile environment and defines the resolution rules for each conflict scenario under the recent-data-win policy. These rules have been implemented and verified in a data synchronization server which was developed based on the SyncML specification [1].

1 Introduction

Advance of mobile communication and device technologies has brought the mobile computing era. People can access Internet from anywhere, anytime by using personal information devices such as PDA(Personal Data Assistant)s, handheld PC(Personal Computer)s and cellular phones [2,3].

Proliferation of personal information devices has resulted in a difficulty of maintaining replicated copies of the user data in various devices' data storage. A user may have his electronic address book in his cellular phone and PDA also. If the user modifies a data item in the PDA, it causes the data being inconsistent with the replicated ones in other devices. Data synchronization is an automated action to make the replicated data be consistent and up-to-date with other copies. Data synchronization needs to resolve conflicts between multiple values or operations made to the different copies of the data due to independent updates on different devices.

There are numerous studies about data conflict and consistency models for the systems of the wired network, especially in the database and the file system context [4~11]. Balasubramaniam and Pierce [4], Ramsey and Csirmaz [6] studied about file synchronization in a replicated file system. While [4] showed the

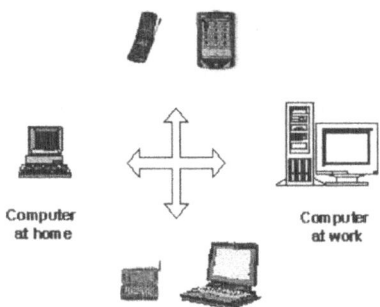

Fig. 1. Data Synchronization in Mobile Environment

conflict resolution rules based on a condition change of a file system before and after synchronization, [6] presented the rules based on the operations applied to each replica file system. Terry [11] and Page [7] studied synchronization of replicated database. Synchronization of the distributed replicated data store in the wired network, either it is the database or the file system, usually occurs in real-time, i.e., an update in one data store immediately propagates to the other replicated data stores. In this model, the conflict mainly concerns multiple simultaneous write operations on the same data item which is called the *write conflict*.

However synchronization of replicated data in the mobile computing environment has different characteristics. Personal information devices may not be always connected to the network and the updates made on a mobile device are propagated only after the user makes a network connection to the server and initiates synchronization. The update consistency model in this case classifies to the eventual consistency [13] that does not require simultaneous update property. Therefore the traditional conflict resolution scheme does not seem to fit for this case.

The purpose of this study is to clarify the resolution rules of data conflict that may occur in synchronization processing of replicated data in multiple devices in mobile computing environment for each conflict scenario. These rules have been implemented and verified in a data synchronization server which was developed based on the SyncML specification.

The paper is organized as follows. Section 2 describes the synchronization types and typical data consistency model of client-server systems. We describe the conflict resolution rules in Section 3 and mention briefly about implementation and test issue in Section 4.

2 Data Consistency Models

2.1 Data Consistency Model for Mobile Device

Update patterns of replicated database are generally categorized to a master-slave model and a peer-to-peer model. In the master-slave model of replicated database, the update on the slave database is temporary and it becomes valid only after committed by the master database. On the other hand, each replica database has the same competence as master database and update in any database is valid in peer-to-peer model.

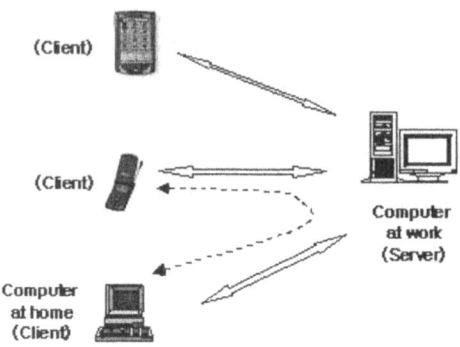

Fig. 2. Master-slave model:replicated data stores in mobile environment

Because replicated data store in mobile computing environment is usually contained in a small, portable device like a PDA or a cellular phone, the data store has limited capacity and is less stable compare with desktop computers. Thus, it is natural to have the master-slave model in which a desktop computer is used as the master data store (server) and portable devices are slaves(clients). When some data are modified in a client device, the update is available only in that device in this model. In order to propagate the update to other devices, it communicates indirectly through the server(Figure 2). The server stores the modified data and maintains the records of modification made either by the server or by one of the clients, the record is used when the server synchronizes with other client data store. This record, commonly called the *change-log* guarantees the consistency of replicated data in multiple devices.

Table 1 shows an example of the change-log implemented in a table form. In the table, the device ID is a sequence number assigned to each client devices of a user. Application ID identifies which application service the data is synchronized for.

Data stores in the server or client devices manage their data records by unique system-level identifier, not by the content of the data. For example, a data 'Tom' with system identifier 245 is different one from 'Tom' with system identifier 597. Commonly used identifiers are GUID (Globally Unique Identifier) and LUID (Locally Unique Identifier). LUID is used in the client devices and it is meaningful only in that local data store. GUID is used in the server and is meaningful in service system wide. When a data item is added in a client device, a new LUID is assigned to the data by the client device and used locally until the data is removed. When this new data is sent to the server along with the LUID for the synchronization, a new GUID is assigned to the data by the server. The (GUID,LUID) pair associated with the newly added data is kept in the server. Replicated data in the different data stores may have different LUIDs but their GUID is the same. If two data have different GUIDs, these data are different and they do not conflict each other.

Operation is the command that needs to be applied to the device, for instance, the first entry of Table 1 means that a data of vCard application was added by other john's device (for instance, device 1) and its GUID is 10001. This addition of the data also need to be performed at device 2 and device 3 of the user john.

Table 1. Information in the change-log

User ID	Device ID	GUID of data	Application ID	Operation $(op^{t-1}(o_i))$
john	2	10001	vCard	add
john	3	10001	vCard	add
dchamton	2	10035	vCalendar	replace
hsmith	1	10023	vCalendar	delete
hsmith	2	10087	vCard	add
hsmith	3	10087	vCard	add

Traditionally, replicated databases with a central master topology are regarded to avoid write conflicts whereas replicated databases with a peer-to-peer topology must detect and resolve the conflicts [12]. Therefore write conflict is no concern in such an environment of Figure 2. However propagation of updates often occurs in a different time zone. An update occurred in one mobile device may not be propagated to other devices of the user until he synchronizes the devices to the server. Consistency model for this case is the eventual consistency [13] that is characterized by the lack of simultaneous updates. Only if a user performs synchronization operations with the server before he accesses the local data store, this model guarantees any of the client-centric consistency [14] which is applied to ordinary replicated data stores in the networked distributed systems.

The major conflict type that may occur during the synchronization of mobile devices is between different operations that need to be performed to a data item of a device by either simultaneous access or in different time, which we define the *semantic conflict*. For example, suppose the device 1 has requested the server to 'delete' data A. After the server deletes the data A which is due to the device 1's request, it records in the change-log that 'delete data A' operation needs to be performed in other devices which belong to the same user. Other device, say the device 2, may later request the server to 'replace' the data A with some other value without knowing that the data has been already deleted, this is an example of the semantic conflict.

2.2 Conflict Resolution Policies

The synchronization between multiple replicated data inevitably encounters conflicts. A conflict resolution policy is the criteria of which data or operation to select as a valid one in case of write conflict or semantic conflict. Table 2 shows typical conflict resolution policies. Service systems or synchronization applications may support all or subset of these policies and the service administrator or the device-owner adopts one of them.

Table 2. Conflict resolution policies

Conflict resolution policy	Meaning
Originator Win	Take the data item of the originator
Recipient Win	Take the data item of the recipient
Client Win	Take the data item of the client device
Server Win	Take the data item of the server
Recent Data Win	Take the data item which has been updated recently in time
Duplication	Apply the requested modification on the duplicated data item while keeping the existing data

Due to the space limitation, this paper includes the resolution rules of the most generally used policy only, which is the recent-data-win policy.

3 Conflict Resolution Rules

In this section, we show the conflict resolution rules for recent-data-win policy to resolve the semantic conflict.

When a record exists in the change-log for a data item o_i to be synchronized, semantic conflict can occur between currently requested operation $op^t(o_i)$ and operation $op^{t-1}(o_i)$ recorded in the change-log. The resolution rules for such a conflict, i.e., actions to be taken to resolve the conflict should be defined at the server.

Denote Initiator($op^t(o_i)$) as the device identifier from which the operation $op^t()$ first applied, and let Target($op^{t-1}(o_i)$) denote the device identifier to which the operation $op^{t-1}(o_i)$ needs to be applied(propagated) to the data item o_i. There may be many synchronization operations but these operations can be classified into 'add', 'replace' and 'delete' type operations.

Conflict Resolution processing consists of the following two phases. Detailed actions in each phase depend on the conflict resolution policy.

Phase 1 decides which data wins among the server's value and the client's one.

Phase 2 updates the change-log information accordingly in the server.

We show the summary of the conflict resolution rules of recent-data-win policy in Table 3.

Table 3. Resolution Rules for the Recent-Data-Win Policy

$op^t(o_i)$	$op^{t-1}(o_i)$	Action	Rule
add	add	error	T.1
	delete		
	replace		
delete	add	delete/error	T.2
	delete	delete	T.3
	replace		
replace	add	replace/error	T.2
	delete	replace	T.4
	replace		

Rule T.1 :

$if(op^t(o_i) = \text{add}) \wedge (op^{t-1}(o_i) \neq null) \wedge (\text{Initiator}(op^t(o_i)) = \text{Target}(op^{t-1}(o_i)))$

action : error processing

This is the case that a client requests to add a new data item which already exists in the server, there has been operation $op^{t-1}(o_i)$ (either it is add, delete or replace) performed previously on this data item. Two data have been assigned the same GUID, this is error considering GUID is unique. The change-log information should not be modified in this case.

In case that $op^{t-1}(o_i)$ is null, then there is no semantic conflict and $op^t(o_i)$ can be performed.

Rule T.2 : It is the case of deleting (or replacing) the data item which has been added previously. This case consists of two sub cases.
T.2.1

$if(op^t(o_i) = \text{delete}) \wedge (op^{t-1}(o_i) = \text{add}) \wedge (\text{Initiator}(op^t(o_i)) = \text{Target}(op^{t-1}(o_i)))$

action : error processing

If there is a record in the change-log for o_i that tells the Initiator device to add a new data, then the Initiator device should not have data yet. But Initiator device already has some data with the same GUID and requests to delete it. This is an error by the same reasoning of Rule T.1.

T.2.2

$if(op^t(o_i) = \text{delete}) \wedge (op^{t-1}(o_i) = \text{add}) \wedge (\text{Initiator}(op^t(o_i)) \neq \text{Target}(op^{t-1}(o_i)))$

action:
    Phase 1 : the server performs the operation $op^t(o_i)$
    Phase 2 : the records in change-log can be silently deleted

The records in the change-log for o_i is not for the Initiator device but for other devices. This happens when the Initiator device wants to delete o_i after itself added o_i in the previous synchronization transaction. Devices, other than Initiator, that have not yet synchronized with the server for o_i will have an entry for themselves. Such records in change-log can be silently deleted. These devices will not be affected by the fact that the data was added and deleted.

Other than T.2.1 and T.2.2, it is not an error and performs $op^t(o_i)$.

In case that $op^t(o_i)$ is to replace, the rules T.2.1 and T.2.2 also apply except that Phase 2 of T.2.2 will be to change $op^{t-1}(o_i)$ from 'add' to 'replace'. Client devices will handle 'replace' in the change-log as 'add' if o_i does not exist in its local data store.

Rule T.3 : This is the rule to handle 'delete' request after o_i has been deleted or replaced previously.
T.3.1

$if(op^t(o_i) = \text{delete}) \wedge (op^{t-1}(o_i) = \text{delete}) \wedge (\text{Initiator}(op^t(o_i)) = \text{Target}(op^{t-1}(o_i)))$

action:
    Phase 1 : no action is required because it has been already deleted from the server
    Phase 2 : - delete the change-log entry which is for o_i and the Initiator's identifier
              - for $\forall o_i \in$ change-log s.t. $\text{Initiator}(op^t(o_i)) \neq \text{Target}(op^{t-1}(o_i))$, no action is required

T.3.2

$if(op^t(o_i) = \text{delete}) \wedge (op^{t-1}(o_i) = \text{replace}) \wedge (\text{Initiator}(op^t(o_i)) = \text{Target}(op^{t-1}(o_i)))$

action :
    Phase 1 : perform $op^t(o_i)$, i.e., delete o_i
    Phase 2 : - delete the change-log entry which is for o_i and the Initiator's identifier
              - for $\forall o_i \in$ change-log s.t. $\text{Initiator}(op^t(o_i)) \neq \text{Target}(op^{t-1}(o_i))$, modify the operation $op^{t-1}(o_i)$ of change-log entry to 'delete'

The devices that have not synchronized the previous operation 'replace' yet, they do not need to perform 'replace' followed by 'delete' because the replaced data will be deleted immediately. So the change-log is directly modified to have 'delete' as $op^{t-1}(o_i)$. The 'replace' operation will be hidden to the lazy devices.

Rule T.4 : This is the rule to handle 'replace' request after o_i has been deleted or replaced previously.

T.4.1

$if(op^t(o_i) = \text{replace}) \wedge (op^{t-1}(o_i) = \text{delete}) \wedge (\text{Initiator}(op^t(o_i)) = \text{Target}(op^{t-1}(o_i)))$

action :
  Phase 1 : add o_i with the previous GUID in the database
  Phase 2 : - delete the change-log entry which is for o_i and the Initiator's identifier
          - for $\forall o_i \in$ change-log s.t. Initiator$(op^t(o_i)) \neq$ Target$(op^{t-1}(o_i))$, modify the operation $op^{t-1}(o_i)$ of change-log entry to 'replace'

T.4.2

$if(op^t(o_i) = \text{replace}) \wedge (op^{t-1}(o_i) = \text{replace}) \wedge (\text{Initiator}(op^t(o_i)) = \text{Target}(op^{t-1}(o_i)))$

action :
  Phase 1 : perform $op^t(o_i)$
  Phase 2 : - delete target entry in the change-log
          - for $\forall o_i \in$ change-log s.t. Initiator$(op^t(o_i)) \neq$ Target$(op^{t-1}(o_i))$, no action is required

4 Implementation

We have developed a data synchronization server and implemented the aforementioned conflict resolution rules in the server in order to verify the rules [15]. We used the SyncML protocol suites from the SyncML Initiative [1,16,17,18] for the communication between client devices and the server and implemented vCard and vCalendar applications.

Figure 3 shows the functional architecture of the server. The Server Application module provides an interface for a user or the application administrator to modify contents data at the server. The SyncML Toolkit encodes and decodes SyncML messages. The Sync Agent implements the SyncML protocols, and it is application independent and can be used for all SyncML applications. The Sync Engine implements application dependent part such as a service policy, resolution rules in case of data conflict, etc. The Session Manager manages a temporary, session related information that needs to be maintained in the server during a session. Lastly, the Open DB Interface implements interfaces to access data in database. The server was implemented on the MS Windows 2000 and all the modules of the CNU SyncML Servers except the Session Manager were implemented into DLLs. We used HTTP(Hyper Text Transfer Protocol) to communicate with a SyncML client device and JNI(Java Native Interface) to load DLLs.

The server successfully accomplished synchronization according to various scenarios with multiple client devices [15]. We used intensive test cases having various conflict scenarios for the two applications. The server also passed the conformance test [19] of the SyncML Initiative.

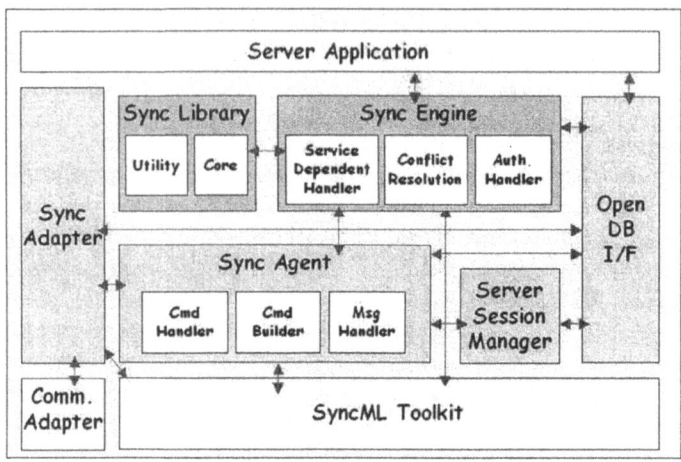

Fig. 3. Software Architecture of the CNU SyncML Server

5 Conclusions

As mobile communication technologies are being developed faster than ever, data processing using portable, mobile devices will be more popular. Importance of the data synchronization between mobile devices and a reliable server computer draws more attention than ever.

The semantic conflict may occur when we access the replicated data on multiple devices. This study showed the conflict resolution rules when multiple devices synchronize their data by the recent-data-win policy in the mobile computing environment. The conflict resolution rules and consistency model of the mobile computing has some different characteristics from traditional replicated databases in the wired network. Main characteristic is that the propagation of updates often occurs in a different time frame. An update occurred in one mobile device may not be propagated to other devices of the person until he synchronizes the devices to the server.

We hope that this study contributes to mobile, distributed computing by providing clear processing algorithm to correctly handle the semantic conflicts, therefore makes replicated data consistent and mobile applications reliable.

References

[1] SyncML Initiative, Building an Industry-Wide Mobile Data Synchronization Protocol, SyncML White Paper, Mar. 20, 2000, http://www.syncml.org
[2] Microsoft, ActiveSync Technology, http://www.microsoft.com/ mobile/pocketpc/ downloads/activesync/
[3] Palm Computing Inc., Palm HotSync Technology, http://www.palm.com/support/hotsync.html
[4] S. Balasubramaniam and B. C. Pierce, "What is a file synchronizer?" In Proc. of the 4th Annual ACM/IEEE International Conference on Mobile Computing and Networking (MOBICOM-98), pp. 98-108, Oct. 1998. (http://www.cis.upenn.edu/ bcpierce/unison)
[5] M.Satyanarayanan, J. Kistler, Kumar, et al. , "Coda: a highly available file system for a distributed workstation environment," IEEE Trans. On Computers, Vol. 39, No. 4, pp.447-459,1990.
[6] N. Ramsey and E. Csirmaz, "An Algebraic Approach to File Synchronization," ACM SIGSOFT Software Engi-neering Notes, Volume 26, Issue 5, pp.175-185, Sep. 2001.
[7] Thomas W. Page etal. "Perspectives on optimistically replicated, peer-to-peer filing," Practice and Experience, Vol. 27 (12), December 1997.
[8] P. J. Braam, "InterMezzo file system: Synchronizing folder collections," http://www.stelias.com
[9] Matteo Cavalleri, Rocco Prudentino "PostgreSQL Replicator: Conflict Resolution Algorithms," http://pgreplicator.sourceforge.net/
[10] M. Cavalleri, R.Prudentino, U. Pozzoli, G. Reni, "A set of tools for building PostgreSQL distributed databases in biomedical environment," Proc. of the 22th Annual International Conference of the IEEE Engineering in Medicine and Biology Society, Jul. 2000.
[11] D. B.Terry at al., "Managing update conflicts in Bayou, a weakly connected replicated storage system," Proc. of the 15th Symposium on Operating Systems Principle, 1995.
[12] M. Dahlin, A.Brooke, M. Narasimhan, B. Porter, "Data Synchronization for Distributed Simulation," European Simulation Interoperability Workshop, 2000.
[13] A. S. Tanenbaum and M. Steen, Distributed Systems, Principles and Paradigms, Prentice Hall, 2002.
[14] D. B. Terry, K. Petersen, M. Spreitzer, and M. Theimer, "The case for nontransparent replication: Examples from Bayou," IEEE Data Engineering, Vol.21, No. 4,pp.12-20, Dec. 1998.
[15] JiYeon Lee, ChangHoe Kim, Jong-Pil Yi, Hoon Choi, "Implementation of the Session Manager for a Stateful Server," 2002 IEEE TENCON, Beijing, pp. 387-390, Oct. 31, 2002.
[16] SyncML Initiative, SyncML Representation Protocol, version 1.0.1, June 15,2001.
[17] SyncML Initiative, SyncML Synchronization Protocol, version 1.0.1, June 15,2001.
[18] SyncML Initiative, SyncML HTTP Binding, version 1.0.1, June 15,2001.
[19] SyncML Initiative, SyncML Conformance Testing Process, version 0.2, Jan. 25, 2001.

A Framework for Orthogonal Data and Control Parallelism Exploitation

S. Campa and M. Danelutto

Dept. Computer Science – University of Pisa
Viale Buonarroti 2, I-56125 PISA – Italy
{campa,marcod}@di.unipi.it

Abstract. We propose an innovative approach to structured exploitation of both data and control parallelism. Our approach is based on the clear distinction between data and control parallelism exploitation mechanisms. This separation leads to a programming model where data and control parallelism exploitation is managed by means of independent, orthogonal mechanisms. By exploiting this orthogonal set of mechanisms, clear semantic transformation and verification tools can be developed. We show here a preliminary definition of the programming model, some sample skeleton applications and we discuss the basis for the development of a clear semantic framework that can be used to develop semantics preserving transformation rules as well as semantic based reasoning on parallel program properties.

1 Introduction

Several distinct parallel programming models recognized that parallelism exploitation may be better organized using compositions of instances of items belonging to a basic set of parallelism exploitation patterns, rather than explicitly using the mechanisms provided by the programming environment for parallel activities setup (processes, threads), for communications (point-to-point, collective), for mapping and scheduling of parallel activities. In particular, the algorithmical skeleton research track originated by Cole's works [13,14,15,17,18,19,35], the research track related to design patterns [10,22,24,31] and the track related to coordination languages [4,5,12,32], all individuated several standard methods (skeletons, design patterns, coordination patterns) that can be used to build a significant part of the interesting, scalable parallel applications currently required.

Some of the proposed programming models in this framework(s) explicitly assume that the parallelism exploitation methods can be arbitrarily composed [6,15]. Other models assume they can't be composed at all or they can be composed in rather limited forms [9,26,29,37]. Furthermore, some of the models provide users/programmers with both data and control (or task) parallel exploitation mechanisms (design patterns, coordination patterns, skeletons) [6,16,20,28,36,30] whereas other models just provide either control or data parallel exploitation mechanisms [1,21,25]. Some works concentrate on the need of integrating data and control parallelism into a well defined model [7,20,23,38]. The results of such attempts often led to the design of two-tier models for the development of libraries or frameworks but none of them has been able to define a clear, powerful, well defined separation of concerns.

This is a key point, indeed. Different classes of parallel applications can be exploited depending on the kind of parallelism exploitation patterns provided to the programmer: data parallel only, control parallel only, both. Furthermore, usually data and control parallel parallelism exploitation mechanisms are provided in quite different flavors. Control parallel mechanisms, such as pipelines and task farms, are usually provided by means of some kind of parallel control statement parametric in the pipeline stages or task farm worker computations. Data parallel mechanisms, such as independent or non independent `forall` or apply-to-all, reduce, prefix, etc. are defined in terms of some kind of parallel (array) data structure. As a consequence, task and data parallelism are rarely used in conjunction within the same application, but in some very simple, classical parallelism exploitation patterns. Furthermore, the semantics of joint data and control parallel applications is difficult to express and to use to derive program transformation or program verification techniques, which are usually very useful to support performance tuning as well as application optimizations.

In this work we will outline a structured parallel programming model that clearly separates data parallel and control/task parallel mechanisms in such a way they can be orthogonally composed to build more complex parallelism exploitation patterns while preserving the clear semantic properties deriving from the separation of concerns.

2 Coordinated Data and Control Parallelism Modeling

By keeping data and control parallelism separated, we achieve a perfect degree of orthogonalization that can be used both to express data and control parallelism exploitation in some kind of handy way and to allow programmers to design new parallelism exploitation patterns out of the existing, data and control parallel ones. In order to achieve such a separation, we must provide a formalism that allows programmers to describe both kind of parallelism exploitation in an independent way.

This goal is achieved by providing within the programming model clearly defined and semantically well founded *collective data access and control primitives* as well as by providing a suitable set of *abstract data types* describing the most common data structures used in the parallel programming frameworks. The collective operations will be defined on the abstract data types, in a sense, although all of them will be independent of, or better, parametric in the data set managed/transformed by the collective operation. The set of collective operations used onto the data set at hand defines an operator graph that actually defines the whole application too. Each one of the collective operations is associated with a clear and easily composable formal semantics. Exploiting such semantics, the operator graph defining the whole parallel application can be easily understood and, possibly, modified by applying source-to-source semantic preserving transformation rules just to enhance or tune application parallel behavior and performance.

A key role in our framework is played by *iterators* and abstract data set *views*. In classical data driven approaches, iterators provide a method to sequentially access to the elements of a data collection, without exposing the collection inner structure. Therefore, in classical data driven approaches, iterators allow to reason about data transformation in terms of collective operations, they can be defined ad hoc to solve particular access patterns, they can be concurrent, actually, rather that sequential, and, last but not least, their implementation is completely independent of the way we use them. The implementation only depends on the abstract data type representation.

Rather, in control driven approaches, iterators are used as accessories of the control structures to define dependencies between components of the parallel application. Such dependencies are relative to the component computations rather than to the type of data processed or to the kind of access performed to such data. Iterators actually implicitly define or explicitly use a *view* of the abstract data type, in this context. They implicitly define a view, as the iterator may define an access pattern which is general with respect to the abstract data type definition and therefore forces on the abstract data type a sort of "access model". They use a view, in that a general access pattern defined by an iterator may be specialized in subtle ways to follow a view of the abstract data type.

Let's exemplify the concept. Suppose to have an abstract data type implementing a generic set of values as well as a generic iterator enumerating all the elements of a given data structure. Suppose also that three views exist, specialized to the given abstract data type: a item view, regarding the array as a collection of $\langle i-th, j-th \rangle$ items, a row view and a column view, regarding the array as a collection of rows and columns, respectively. Depending on the different application at hand, we can use the item view (e.g. to compute the Mandelbrot set) or a row (column) view (e.g. to compute matrix multiplication). In this case, the iterator uses a view of the abstract data type. On the other side, in case we have two iterators, one accessing all the elements of a data collection sequentially and one accessing all them in parallel (concurrently), the usage of one of the two iterators define a view of the data structure at hand. In the former case, a set is viewed as a list while in the latter case it is viewed as a vector/array, in a sense.

Summarizing, our programming model is completely defined by a tuple $\mathcal{M} = \langle \mathcal{A}, \mathcal{V}, \mathcal{I}, \mathcal{C} \rangle$ where $\mathcal{A}$ represents the set of abstract data types (i.e. Arrays, Trees, Lists, Graphs,..), $\mathcal{V}$ represents the set of views that can be applied to a generic abstract data type (for instance, any data can be viewed as a vector or as a list of items), $\mathcal{I}$ represents the set of iterators that can be used to get items from an abstract data type with a given view and eventually $\mathcal{C}$ represents the set of collective operations that can be performed using iterators on abstract data type views.

For the sake of generality, and because here we just want to outline the methodology rather than describing a complete, working, perfectly assessed prototype, we do not detail the exact abstract data types nor the iterators and the views we want to take into account. Rather, in this paper we will provide different examples that show the kind of object populating the $\mathcal{A}$, $\mathcal{V}$, $\mathcal{I}$ and $\mathcal{C}$ sets. In other words, we assume that there are abstract data types modeling uni, bi, and multi-dimensional arrays, lists and trees, that there are views exposing an abstract data type as a vector, an array, a list or a tree of values, that there are iterators exposing view items and view item blocks and that there are collective operations that allow to concurrently apply a function to all the elements given by an iterator (ApplyToAll), to sequentially apply a function to all the elements given by an iterator (StreamApplyToAll), as well as to filter stream items (StreamFilter) and to sequentially compose two computations (Compose).

In our approach we provide AbstractDataType, View and Iterator whose objects can be used to support the exploitation of data and control parallelism using proper collective operations. As an example, suppose that the programmer wants to set up an application that processes in stages the data appearing on a file. Both stages operate exploiting data parallelism on one of the data items appearing in the file. The programmer can use a File abstract data type with a List or Vector view. He can Compose two operations both using an ApplyToAll collective to operate on the different items provided by an Item iterator. In case the programmer wants to experiment a different parallel solution, he can arrange to have an ArrayView over the File abstract data types and use an ApplyToAll of a Compose collective. In the former case, the parallelism exploited is a pipeline of two data parallel stages, in the latter it is plain data parallelism.

3 Implementation

In order to prove the feasibility of our approach, we started designing a prototype implementation of a programming framework providing the programmer the concepts outlined in Section 2. We chose Java as the implementation framework just to profit from the wide range of tools available there. The implementation of the prototype has been conceived in steps. The first step is to provide a multi-threaded implementation of a first, significant, subset of abstract data types, views and iterators. By programming the prototype, we have been forced to formalize the classes and, as a consequence, to model the features discussed above by means of the available Java OO mechanisms. This helped us understanding where exactly a given mechanism/property is to be implemented/confined, for instance.

The availability of a programming environment based on the concepts of abstract data types, iterators and views also allows a programmer to evaluate the expressive power of the approach, although the prototype only provides a small set of abstract data types, iterators and views. In the first prototype, parallel implementation of operators such as the ApplyToAll shown before happens to be concurrent, i.e. supported by threads. The idea is to evolve the prototype in such a way that this kind of parallel implementation is replaced by true parallel/distributed implementation when a network of processing elements (a cluster, a workstation network or a Java enabled grid, say) is (will be) available. This parallel implementation of our operators will rely on classical Java mechanisms, such as RMI or TCP/IP sockets, in order to achieve the maximum degree of portability possible.

The concurrent implementation of the framework actually demonstrates the expected speedup. We run some simple programs and the following table shows the results achieved, in terms of completion time and speedup in one of these experiments.

Thread#	MonoP	DualP	Speedup	Thread#	MonoP	DualP	Speedup
1	5837	5494	-	6	6457	2846	2.26
2	6808	2812	2.42	8	9705	2817	3.4
3	6480	3070	2.11	12	11506	2874	4.0
4	6527	2902	2.24	24	15706	2811	5.58

Data are relative to a simple data parallel program, using an `ApplyToAll` to compute a simple transformation of the original matrix. Columns show the number of threads used [1], the amount of milliseconds spent in evaluating the program on two different machines (a mono-processor, Pentium-III Mobile 800MHz and a dual processor, Pentium III 866MHz machine) and the speedup of the dual processor run vs. the mono-processor one. First of all, the efficiency measured when two processors are actually used on the dual processor machine (completion time 2812 msecs) w.r.t. the run using just one processor (completion time 5494 msecs) is near 97%. Second, when comparing single processor machine runs with dual processors ones, effective speedups are achieved. They are actually super-linear, as the single processor machine has a slower processor as well as different memory and disk configuration. The single processor machine also suffers from thread multiprogramming overhead much more than the dual processor machine, as shown in the table: the completion times increase with the number of threads used much more on the mono-processor than on the dual processor machine.

Just to give an idea of the expressive power of our prototype implementation, the parallel application transforming all the items of a two dimensional array by applying function f on them used to get the numbers shown in the table, can be expressed as follows:

```
AbstractDataType data = ...          // instantiate the data set
MatrixView matrix= data.getMatrixView(x,y); //set a matrix view
Iterator it = matrix.getIteratorRow();// set an iterator by row
// instantiate a parallel module
ApplyToAll application = new ApplyToAll(new SeqF(), it);
application.evaluate();              // run the application
```

In this case, the iterator `Iterator` is derived (in a specialized form) from the `MatrixView` view. The data parallel part of the application is hidden in the `ApplyToAll` application. The `ApplyToAll` is a simple parallel application pattern that computes a function (represented by its first parameter, actually) to all the items given by the iterator represented by its second parameter. Apply-to-all is a well known parallelism exploitation pattern, appearing in most of the

---

[1] the `Block` iterator used can be specialized to take as a parameter the block factor; this, in turn, can be used to force the parallelism degree of the `ApplyToAll` in terms of the number of threads used

structured parallel programming environments/models cited in Section 1 with different names: forall, map, apply-to-all, etc. In our approach, however, the apply-to-all happens to have a simple (w.r.t. other apply to all operators in data parallel frameworks, as an example), well defined behavior: it just processes, possibly concurrently, all the items provided by the parameter iterator. The SeqF code, i.e. the sequential code computing function f is something like:

```
public class SeqF extends Functor {
  public void run() {
    Iterator it = super.iterator();    // get iterator
      while (it.hasNext())             // while there are more items
        it.append(f(it.next()));       // process them
  }}
```

It inherits an iterator from the execution environment (the ApplyToAll, basically) and sequentially computes f on all the items of the row.

4 Assessment

Despite the fact that one has to reason about parallel programs in terms of views and iterators, we managed to design a framework that actually helps the programmer in developing good parallel applications in a reasonable amount of time. As the programmer is not concerned with all the details involved in concurrency/parallelism exploitation (concurrent activities setup, mapping and scheduling, communication handling, etc.) he may concentrate on the qualitative aspects of parallelism exploitation. This feature is common to other programming environments, typically all those derived in the skeleton and design pattern context. However, our approach adds the possibility to freely combine data access mechanisms (through iterators) and parallelism exploitation mechanisms (through collective operations).

The proposed framework allows common parallelism exploitation patterns to be covered. We've already shown plain data parallelism exploitation in Section 2. Plain control/task parallelism can be exploited using the StreamApplyToAll collective (basically getting a task farm) as follows:

```
AbstractDataType data = ... ;
ItemView stream = data.getItemSetView();
Iterator it = stream.getIteratorItem();
StreamApplyToAll application = new StreamApplyToAll(SeqF,it);
application.evaluate();
```

Pipelines can be programmed as follows:

```
AbstractDataType data = ... ;
XXXView stream = data.getXXXView();
Iterator it = stream.getXXXIterator();
Pipe pipe = new Pipe(it);
for (int i=0; i< NStages ; i++) pipe.add(new Sequential());
pipe.evaluate();
```

Client/server applications can be modeled by having clients (servers) producing service requests (getting service requests) and analyzing service answers (producing service answers): clients can be programmed as follows

```
AbstractDataType adt = ...;
FIFOBufferView view = adt.getFIFOBufferView();
Iterator it = ch.getIterator();
for (int i=0; i< Nrequest; i++) it.append(req);
```

The client uses an abstraction of the communication channel as if it was a FIFO buffer on which an iterator for appending messages is defined.
The server can be structured as a sequential or a parallel program where each request is evaluated by a Service function. The sequential version is given by:

```
AbstractDataType adt = ...;
FIFOBufferView view = adt.getFIFOBufferView();
Iterator it = ch.getIterator();
while(it.hasNext()) Service(it.next());
```

where each new service request is provided by `iterator.next()` associated to the `FIFOBufferView` view. The parallel version of the server looks like:

```
AbstractDataType adt = ...;
FIFOBufferView view = adt.getFIFOBufferView();
Iterator it = ch.getIterator();
StreamApplyToAll sat = new StreamApplyToAll(new Service(),it);
```

In this case, the server is programmed as a task farm application on the stream of requests generated by the Iterator created on a FIFOBuffererView data view. The Service class, in this case, implements a run method that gets the iterator on the *i-th* element of the stream passed by the `StreamApplyToAll` collective and processes the corresponding request. More complex parallelism exploitation patterns can be derived by composing the basic parallelism exploitation patterns.

We want also to point out another relevant difference with skeleton/design pattern based parallel programming frameworks. In those cases either there is no possibility to expand the skeleton/design pattern set or there is that possibility but adding a new parallelism exploitation pattern requires a huge effort as the programmer/system designer must provide code dealing with all the aspects related to parallelism exploitation in that pattern. In our case, instead, once a mechanism for data view, data access or collective operation has been recognized valid, then you can add code in the system enabling programmers to deal with the new feature and automatically the usage of that new feature propagates on the orthogonal combination of views, iterators and collective operations given a full range of new tools to the programmers.

5 Related Work

The whole framework design is based on the concept of iterators, views and collective operations that all represent, in different frameworks, assessed concepts.

By merging them into our framework we simply extended the possibilities they offer when kept alone. Iterators and views are common concepts in the OO world. C++ standard template library provides iterators, Java.util collection data types provide iterators, some STL implementations offer the view concept. Collective operations, on the other side, play a central role in the data parallel programming model world as well as in the whole skeleton framework.

In the field of parallel programming there are many examples of clever iterator usage. Several works are extension of the Standard Template Library as for example [27,34] and their design is heavily influenced by its predecessor. All of them offer some parallel implementation of iterators, containers and generic parallel algorithms. NESL[8] and some following works has reached a good degree of nested parallelism capability but the user intervention is still heavily needed.

STAPL (Standard Template Parallel Library) [3,33] inherits most of the Standard Template Library's principles. Programmers write parallel algorithms called pAlgorithms whose inputs are pRanges (sort of parallel iterators); pRanges are provided by pContainers (parallel STL containers), which guarantee access control to their encapsulated data through pRanges and/or conventional STL iterators. STAPL is very near to our idea of using iterator as access control mechanism, in particular for data parallel application. Nevertheless, it completely misses control flow features to exploit control/task parallelism.

Kuchen's skeleton library [28] provides type abstractions for the distribution and the partitioning of data structures; after the data has been statically distributed, on the basis of the selected distribution policy, the library provides operations to access it as a whole or to access a particular partition, regardless all communication details. This approach implies a sort of conceptual awareness about the distribution, explicit communication calls to manage overlapped partitions and the possibility to program parallel programs is strictly related to the availability of the data structures and policies provided by the library.

6 Conclusions

We discussed a framework aimed at separating the different concerns related to data and control joint parallelism exploitation. This framework is currently being developed and here we discussed preliminary results mainly concerning expressive power and preliminary performance evaluation. We are currently working to formalize the way the framework can be used to prove properties of parallel programs, in particular to develop a set of source-to-source transformation rules that can be used to perform performance tuning of parallel programs via parallelism exploitation restructuring.

Acknowledgments. This work has been partially supported by Italian National FIRB Project No. RBNE01KNFP "GRID.it" and by the Italian National strategic projects "legge 449/97" No. 02.00470.ST97 and 02.00640.ST97

References

1. S. Ahuja, N. Carriero, and D. Gelernter. Linda and friends. *Computer*, 19(8):26–34, August 1986.
2. M. Aldinucci, S. Campa, P. Ciullo, M. Coppola, M. Danelutto, P. Pesciullesi, R. Ravazzolo, M. Torquati, M. Vanneschi, and C. Zoccolo. ASSIST demo: an high level, high performance, portable, structured parallel programming environment at work. In *Proc. of Europar'2003*, Springer Verlag, LNCS, Vol. 2790, pages 1295–1300, August 2003.
3. P. An, A. Jula, S. Rus, S. Saunders, T. Smith, G. Tanase, N. Thomas, N. Amato, and L. Rauchwerger. STAPL: An Adaptive, Generic Parallel C++ Library. In *14th Workshop on LCPC*, August 2001. Cumberland Falls, KY.
4. F. Arbab, P. Ciancarini, and C. Hankin. Coordination languages for parallel programming. *Parallel Computing*, 24(7):989–1004, July 1998.
5. F. Arbab, I. Herman, and P. Spilling. An overview of Manifold and its implementation. *Concurrency: Practice and Experience*, 5(1):23–70, February 1993.
6. B. Bacci, M. Danelutto, S. Pelagatti, and M. Vanneschi. SkIE : A heterogeneous environment for HPC applications. *Parallel Computing*, 25(13–14):1827–1852, December 1999.
7. H.E. Bal and M. Haines. Approaches for integrating task and data parallelism. *IEEE Concurrency*, 6(3):74–84, July/September 1998.
8. G. E. Blelloch and J. Hardwick. A Library of Parallel Algorithms. http://www.cs.cmu.edu/~scandal/nesl/algorithms.html.
9. G. H. Botorog and H. Kuchen. Using algorithmic skeletons with dynamic data structures. *Lecture Notes in Computer Science*, 1117:263, 1996.
10. S. Bromling, S. MacDonald, J. Anvik, J. Schaeffer, D. Szafron, and K. Tan. Pattern-based parallel programming. subm. to ICPP-02, British Columbia
11. N. Carriero and D. Gelernter. Linda in context. *Communications of the ACM*, 32(4):444–458, April 1989.
12. P. Ciancarini and C. Hankin, editors. *Coordination Languages and Models*, volume 1061 of *LNCS*. Springer-Verlag, Berlin, Germany, April 1996.
13. Murray Cole. *Algorithmic Skeletons: structured management of parallel computation*. Monograms. Pitman/MIT Press, Cambridge, MA, 1989.
14. Murray Cole, 2004. The skeletal parallelism home page: http://homepages.inf.ed.ac.uk/mic/Skeletons/.
15. M. Danelutto, S. Orlando, and S. Pelagatti. P^3L: the pisa parallel programming language. Tech. Rep. HPL-PSC-91-27, HP Labs, Pisa Science Center (Italy), 1991.
16. M. Danelutto, F. Pasqualetti, and S. Pelagatti. Skeletons for data parallelism in P3L. *Lecture Notes in Computer Science*, 1300:619, 1997.
17. M. Danelutto and P. Teti. Lithium: A structured parallel programming environment in Java. *Lecture Notes in Computer Science*, 2330:844, 2002.
18. M. Danelutto and M. Stigliani. SKElib: Parallel programming with skeletons in C. *Lecture Notes in Computer Science*, 1900:1175, 2001.
19. J. Darlington, Y.-K. Guo, Hing Wing To, and J. Yang. Functional skeletons for parallel coordination. *Lecture Notes in Computer Science*, 966:55, 1995.
20. M. Díaz, B. Rubio, E. Soler, and J. M. Troya. Integrating task and data parallelism by means of coordination patterns. *LNCS*, 2026:16, 2001.
21. High Performance FORTRAN Forum. High performance fortran language specification. Tech Rep., Computer Science Dept., Rice University, Houston, Jan 1997.

22. Ian Foster. *Designing and building parallel programs: concepts and tools for parallel software engineering*. Addison-Wesley, Reading, MA, USA, 1995.
23. Ian Foster, David R. Kohr, Jr., Rakesh Krishnaiyer, and Alok Choudhary. A library-based approach to task parallelism in a data-parallel language. *Journal of Parallel and Distributed Computing*, 45(2):148–158, 15 September 1997.
24. E. Gamma, R. Helm, R. Johnson, and J. Vissides. *Design Patterns: Elements of Reusable Object-Oriented Software*. Addison Wesley, 1994.
25. David Gelernter and Nicholas Carriero. Coordination languages and their significance. *Communications of the ACM*, 35(2):97–107, February 1992.
26. Saniya Ben Hassen, Henri E. Bal, and Ceriel J. H. Jacobs. A task and data-parallel programming language based on shared objects. *ACM Transactions on Programming Languages and Systems*, 20(6):1131–1170, November 1998.
27. E. Johnson, D. Gannon, and P. Beckman. HPC++: Experiments with the parallel standard template library. In *Proc. of the 11th Intl. Conf. on Supercomputing (ICS-97)*, pages 124–131, New York, July 1997. ACM Press.
28. H. Kuchen. A skeleton library. *Lecture Notes in Computer Science*, 2400:620, 2002.
29. S. MacDonald, S. Bromling, D. Szafron, J. Schaeffer, J. Anvik, and K. Tan. From patterns to frameworks to programs. *Parallel Computing*, 28(12), Dec 2002.
30. S. MacDonald, D. Szafron, and J. Schaeffer. Object-oriented pattern-based parallel programming with automatically generated frameworks. *Proc of (COOTS '99), San Diego, California, USA*, Berkeley, CA, USA, May 1999.
31. B.L. Massingill, T.G. Mattson, and B.A. Sanders. A pattern language for parallel application programs (research note). *LNCS*, 1900:678, 2001.
32. G. A. Papadopoulos and F. Arbab. Control-driven coordination programming in shared dataspace. *Lecture Notes in Computer Science*, 1277:247, 1997.
33. L. Rauchwerger, F. Arzu, and K. Ouchi. Standard templates adaptive parallel library (STAPL). *Lecture Notes in Computer Science*, 1511:402–??, 1998.
34. J.V.W. Reynders, P.J. Hinker, J.C. C.ummings, S.R. Atlas, S. Banerjee, W.F. Humphrey, S.R. Karmesin, K. Keahey, M. Srikant, and M.Dell Tholburn. POOMA: A Framework for Scientific Simulations of Paralllel Architectures. In *Parallel Programming in C++*, MIT Press, 1996.
35. J. Serot, D. Ginhac, and J.P. Derutin. SKiPPER: A Skeleton-Based Parallel Programming Environment for Real-Time Image Processing Applications. In *Proc. of (PaCT-99)*, Sept 1999.
36. M. Vanneschi. The programming model of ASSIST, an environment for parallel and distributed portable applications. *Parallel Computing*, Dec 2002.
37. Marco Vanneschi. ASSIST: an environment for parallel and distributed portable applications. Tech. Rep. TR-02-07, Computer Science Dept.-Pisa, May 2002.
38. Emily A. West and Andrew S. Grimshaw. Braid: Integrating task and data parallelism. Technical Report CS-94-45, Department of Computer Science, University of Virginia, November 11 1994. Mon, 28 Aug 1995 21:06:39 GMT.

Multiplier with Parallel CSA Using CRT's Specific Moduli ($2^k-1, 2^k, 2^k+1$)*

Wu Woan Kim[1] and Sang-Dong Jang[2]

[1] Division of Information and Communication Engineering, Kyungnam University,
Masan, South Korea
wukim@zeus.kyungnam.ac.kr

[2] Dept. of Computer Engineering, Kyungnam University,
Masan, South Korea
angong@hawk.com.kyungnam.ac.kr

Abstract. Recently, RNS has received increased attention due to its ability to support high-speed concurrent arithmetic. Applications such as fast fourier transform, digital filtering, and image processing utilize the efficiencies of RNS arithmetics in addition and multiplication; they do not require the difficult RNS operations such as division and magnitude comparison of digital signal processor. RNS have computational advantages since operation on residue digits are performed independently and so these processes can be performed in parallel. There are basically two methods that are used for residue to binary conversion. The first approach uses the mixed radix conversion algorithm, and the second approach is based on the Chinese remainder theorem. In this paper, the new design of CRT conversion is presented. This is a derived method using an overlapped multiple-bit scanning method in the process of CRT conversion. This is achieved by a general moduli form($2^k-1, 2^k, 2^k+1$). Then, it simulates the implementation using an overlapped multiple-bit scanning method in the process of CRT conversion, In conclusion, the simulation shows that the CRT method which is adopted in this research, performs arithmetic operations faster that the traditional approaches, due to advantages of parallel processing and carry-free arithmetic operation.

1 Introduction

Digital signal processing hardware based in RNS is currently considered as an important method for high speed and low cost hardware realization[4][5][6]. However, it did not achieve widespread use in general purpose computing due to difficulties associated with division, magnitude comparison, sign and overflow detection. Various researches in order to overcome such problems, make RNS directly comprehensive attentions[1][2][3].

RNS have computational advantages since operation on residue digits are performed independently and so these processes can be performed in parallel. However, other

---

* This work was supported by Kyungnam University Research Fund.

operations such as input/output conversions are significantly more difficult. There are basically two methods that are used for residue to binary conversion. The first approach uses the mixed radix conversion algorithm, and the second approach is based on the Chinese remainder theorem[8][9][10].

Since special purpose processors are associated with general-purpose computers, binary-to-residue and residue-to-binary conversions become inherently important, and the conversion process should not offset the speed gain in RNS operations. While the binary-to-residue conversion does not pose a serious threat to the high-speed RNS operations, the residue-to-binary conversion can be a bottleneck. The Chinese Remainder Theorem (CRT) is considered the main algorithm for the residue conversion process. Several implementations of the residue decoder in [14] and [8] are based on using three moduli in the form $(2^n-1, 2^n, 2^n+1)$[12]; n is number of bit.

The moduli set, $2^n-1, 2^n, 2^n+1$ is of special interest because the numbers in this system can be scaled by one of the moduli efficient. Several publications are available in literature, which deal with algorithms and hardware implementations for efficient conversion of numbers.

In this paper, the new design of CRT conversion is presented. This is a derived method using an overlapped multiple-bit scanning method in the process of CRT conversion. This is achieved by a general moduli form $(2^k-1, 2^k, 2^k+1)$. In the following section, some backgrounds related the RNS theory and Overlapped Multiple-Bit Scanning Method are reviewed. Section III discuss how this method can be derived and designed and implemented. Section IV evaluates the speed performance.

2 Background

2.1 Preliminaries

A residue number system is characterized by a base that is not a single radix but an N-tuple of integers $(m_1, m_2, ..., m_n,)$. Each of these m_i ($i=1, 2, 3, ..., N$) is called a *modulus*. An integer x is represented in the residue number system by an N-tuple $(r_1, r_2, ..., r_n)$ where r_i is a nonnegative integer satisfying

$$x = q_i m_i + r_i \qquad i = 1,2,3,\cdots,N \qquad (1)$$

where q_i is the largest integer such that $0 \leq x_i < (m_i-1)$. x is known as the residue of x modulo r_i, and the notations $x \bmod m_i$ and $|x|_{m_i}$ are used. A commonly used form of (1) therefore is

$$x = m_i \left[\frac{x}{m_i}\right] + |x|_{m_i} \qquad (2)$$

It should be noted that x does not have to be a positive number but can be any integer. If x is negative, $[x/m_i]$ will also be negative. By definition, $|x|_{m_i}$ must be positive. In order to have a unique residue representation, the moduli must be pairwise relatively prime; that is,

$$GCD(m_i, m_j) = 1, \quad \text{for} \quad i \neq j \tag{3}$$

Then it is shown that there is a unique representation for each number in the range of

$$0 \leq x \langle \prod_{i=1}^{N} m_i = M$$

where N is the number of moduli.

2.2 Multiplicative Inverse

The multiplicative inverse of a number c modulo m is a number b, $0 \leq b \leq (m-1)$ satisfying $|cb|_m=1$, b is denoted by $|1/c|_m$. Any number c has an additive inverse $|-c|_m$, but the multiplicative inverse $|1/c|_m$ dose not always exist. The inverse $|1/c|_m$ exists if only if GCM$(c,m)=1$ and $|c|_m \neq 0$. If these conditions are satisfied then $|1/c|_m$ is unique.
If m is a prime number, then for every possible value c satisfying $1 \leq c \leq m-1$, GCM$(c, m)=1$ and the multiplicative inverse exist[14].

2.3 Chinese Remainder Theorem

If the moduli m_i are pairwise relatively prime we can use the Chinese Remainder Theorem in order to convert a number in the residue system to the conventional number system. This theorem states that

$$|x|_M = \left| \sum_{j=1}^{N} \hat{m}_j \left| \frac{r_j}{\hat{m}_j} \right|_{m_j} \right|_M \tag{4}$$

where $(\hat{m}_j = \dfrac{M}{m_j}, \ M = \prod_{j=1}^{N} m_j, \ \text{and} \ (m_j, m_k) = 1 \ (\text{for} \ j \neq k))$

2.4 Selecting the Moduli

We may have different objects when selecting the moduli. If our objective is to reduce the execution time of addition and multiplication, then a large number of small moduli is desirable, since the execution them of these operations is determined by the largest modulus. However, a large number of small moduli will lengthen the time for converting residue numbers to the associated mixed-radix system., since this conversion is a sequential procedure in which the number of steps is determined by the number of moduli. Such conversions are necessary for magnitude comparison, sign detection, or overflow detection

Another consideration when selecting moduli is the fact that the residues would be executed on their corresponding binary representation. We therefore have the following objectives:

1. Efficient binary representation to minimize the total number of bits
2. Convenient binary coding to simplify the execution of arithmetic operation

2.5 Sign Representation

In the residue system, sign detection is relatively difficult. The basic problem evolves from the fact that determining the relative magnitudes of two numbers in residue notation is not a simple, elementary operation. No equivalent to the simple inspection technique, based upon digit-by-digit comparison, is available. The question of relative magnitude is answered only by operations which involve all digits of a residue number.

Since, as noted, explicit sign representation does not eliminate the requirement for determining relative magnitude, the basic difference between the two representation methods appears in its effect on procedure. Specifically, the choice affects the placement of the sign-detection procedures in a series of arithmetic operations.

2.6 Overlapped Multiple-Bit Scanning

The idea is basic on the fact that the execution time can be reduced by shifting across a string of zeros in the multiplier. The greater the number of *zeros* in the multiplier, the faster the operation will be. Consider a string of k consecutive 1's in the multiplier as shown below:

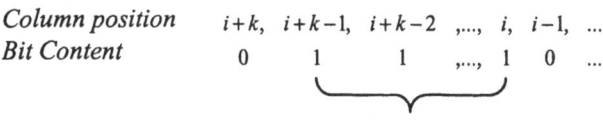

k consecutive 1's

By the following *string property*

$$2^{i+k} - 2^i = 2^{i+k-1} + 2^{i+k-2} + \ldots + 2^{i+1} + 2^i \quad (5)$$

We can replace the k consecutive 1's by the following string:

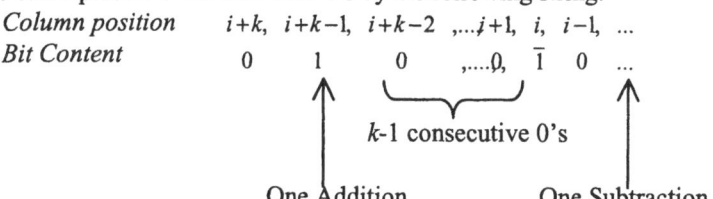

The low-order is $\overline{1}$ overbarred indicating $a-1$ corresponding to a subtraction to be performed. With this multiplier recording we can replace k consecutive additions by only one addition at the beginning and one subtraction at the ending of the string. This obviously saves a significant amount of add time, especially when the value of k is large. We are using this string property to explain why overlapped bit scanning is advantageous.

3 Derivation of the Proposed and Implementation

The arithmetic operation on two integers X and Y is equivalent to the arithmetic operation on its residue representation, that is,

$$|X \cdot Y|_M = \left(\left||X|_{m1} \cdot |Y|_{m1}\right|_{m1}, \left||X|_{m2} \cdot |Y|_{m2}\right|_{m2}, \left||X|_{m3} \cdot |Y|_{m3}\right|_{m3} \right) \quad (6)$$

Where "·" can be multiplication. Therefore, it is desired to convert binary arithmetic on large integers to residue arithmetic on smaller residue digits in which the operations can be parallely executed, and there is no carry chain between residue digit.

3.1 Derivation of the Proposed Method

Consider the moduli set, 2n-1, 2n, 2n+1. Let this moduli set be represented by {m1, m2, m3}. Let (x1, x2, x3) be the residue set for any number in this RNS system and M be the range of the moduli set. The range M is, therefore, m1 m2 m3. The legitimate range is [0, M], where M= m1 m2 m3[11].

$$m_1 = 2^k - 1 \quad m_2 = 2^k \quad m_3 = 2^k + 1$$

$$M = \prod_{i=1}^{3} m_i = (2^k - 1)2^k(2^k + 1) = 2^{3k} - 2^k$$

For this moduli set, the CRT procedure demonstrated by

$$|x|_M = \left| \sum_{j=1}^{N} \hat{m}_j \left| \frac{r_j}{\hat{m}_j} \right|_{m_j} \right|_M$$

reduces to:

$$|N|_M = \left| \hat{m}_1 \cdot \left| \frac{1}{\hat{m}_1} \right|_{m1} \cdot x + \hat{m}_2 \cdot \left| \frac{1}{\hat{m}_2} \right|_{m2} \cdot y + \hat{m}_3 \cdot \left| \frac{1}{\hat{m}_3} \right|_{m3} \cdot z \right|_M \quad (7)$$

where x, y, z is the residue digit of (6), respectively.
For each of the terms in (7) we can write:

$$|\hat{m}_1|_{m_1} = \left|2^k \times (2^k + 1)\right|_{2^k - 1} = |2|_{2^k - 1} = 2 \quad (if, k \geq 2)$$

$$|\hat{m}_2|_{m_2} = \left|(2^k - 1) \times (2^k + 1)\right|_{2^k} = |-1|_{2^k} = 2^k - 1 \quad (if, 2^k - 1 \langle 2^k)$$

$$|\hat{m}_3|_{m_3} = \left|(2^k - 1) \times (2^k)\right|_{2^k + 1} = |2|_{2^k + 1} = 2 \quad (if, k \geq 2)$$

$$|\hat{m}_1|_{m1} = 2, \quad |\hat{m}_2|_{m2} = 2^k - 1, \quad |\hat{m}_3|_{m3} = 2 \quad (8)$$

therefore the multiplicative inverses are derived:

$$\left|\frac{1}{\hat{m}_1}\right|_{2^k-1} = \left|\frac{1}{2}\right|_{2^k-1} = 2^{k-1}, \quad \left|\frac{1}{\hat{m}_2}\right|_{2^k} = \left|\frac{1}{2^k-1}\right|_{2^k} = 2^k - 1,$$

$$\left|\frac{1}{\hat{m}_3}\right|_{2^k+1} = \left|\frac{1}{2}\right|_{2^k+1} = 2^{k-1}+1$$

This allows a desirable simplication in the CRT equation where embedded multiplications can be replaced by simple shift left operations[6].

Table 1. Multiplicative inverse

	$\left|\frac{1}{\hat{m}_1}\right|_{2^k-1}$	$\left|\frac{1}{\hat{m}_2}\right|_{2^k}$	$\left|\frac{1}{\hat{m}_3}\right|_{2^k+1}$
K=2	1	3	3
k=3	4	7	5
k=4	8	15	9
k=5	16	31	17
k=6	32	63	33
⋮	⋮	⋮	⋮
Multiplicative inverse	2^{k-1}	2^k-1	$2^{k-1}+1$

We consider the conversion of residue numbers into binary format with respect to the {2k-1, 2k, 2k+1} moduli set. Substituting k≥2 in Table1, the multiplication inverses can be computed. Replacing the values in (9) result in:

$$\left|M\right|_{2^{3k}-2^k} = \left|2^k(2^k+1)\cdot\left|2^{k-1}\cdot x_1\right|_{2^k-1}\right.$$
$$+(2^k-1)(2^k+1)\cdot\left|(2^k-1)\cdot x_2\right|_{2^k}$$
$$+(2^k-1)(2^k)\cdot\left|(2^{k-1}+1)\cdot x_3\right|_{2^k+1}\Big|_{2^{3k}-2^k} \qquad (9)$$

Let $A = \left|2^{k-1}\cdot x\right|_{2^k-1}$, $B = \left|(2^k-1)\cdot y\right|_{2^k}$, $C = \left|(2^{k-1}+1)\cdot z\right|_{2^{k+1}}$
then we can write:

$$\left|2^k(2^k+1)\cdot A + (2^k-1)(2^k+1)\cdot B + (2^k-1)(2^k)\cdot C\right|_{2^{3k}-2^k}$$
$$= \left|(2^{2k}+2^k)\cdot A + (2^{2k}-1)\cdot B + (2^{2k}-2^k)\cdot C\right|_{2^{3k}-2^k}$$
$$= \left|(2^{2k+1}-2^{2k}+2^k)\cdot A + (2^{2k}-1)\cdot B + (2^{2k}-2^k)\cdot C\right|_{2^{3k}-2^k} \qquad (10)$$

We can find string property in (10). And it can be rearranged into:

$$\left|M\right|_{2^{3k}-2^k} = \left|\begin{array}{c} 2^{2k}\cdot B+(-2^k\cdot C)-B+2^{2k}\cdot C \\ +2^{2k+1}\cdot A+2^k\cdot A \\ +(-2^{2k}\cdot A) \end{array}\right|_{2^{3k}-2^k} \qquad (11)$$

The primary goal of this paper is to perform the computation of $|N|M$. That is, the implementation of this result(11) can be performed by multioperand addition and division(mod M) using A multilevel CSA tree consists of 2CSA's a carry propagate adder(CPA)[9] and two subtraction block(One is CHK block. The other is SUB block). CHK and SUB block uses string property and then it can be performed if:

$$\text{if} \quad K = 2^{2k} \cdot B + (-2^k \cdot C) - B + 2^{2k} \cdot C \geq 2^{3k} - 2^k$$
$$\text{then} \quad K = K - (2^{3k} - 2^k : CHK, CHK') \quad endif \quad (12)$$

Evidently, N is less than $3(2^{3k}-2^k)$ so it can be done CHK block(12) and SUB block(13) satisfying above condition.

$$\text{if} \quad K + 2^{2k+1} \cdot A + 2^k \cdot A + (-2^{2k} \cdot A) \geq 2^{3k} - 2^k$$
$$\text{then} \quad K = K) + 2^{2k+1} \cdot A + 2^k \cdot A + (-2^{2k} \cdot A)$$
$$- (2^{3k} - 2^k : SUB) \quad endif \quad (13)$$

3.2 Implementation

Example of Implementation for 8 bit Multiplier is shown in Fig. 1. We consider the conversion of residue numbers into binary format with respect to the $\{2^k-1, 2^k, 2^k+1\}$ moduli set. For moduli $m_1 = 31$, $m_2 = 32$, $m_3 = 33$ (substituting k=5), find M first.

$$M = \prod_{i=1}^{3} m_i = (2^k - 1) 2^k (2^k + 1) = 31 \times 32 \times 33 = 32736$$

We then obtain

$$|\hat{m}_1|_{m_1} = |32 \times 33|_{31} = 2, \quad |\hat{m}_2|_{m_2} = |31 \times 33|_{32} = 31, \quad |\hat{m}_3|_{m_3} = |31 \times 32|_{33} = 2$$

and the multiplication inverses can be computed and replacing the values in

$$\left|\frac{1}{\hat{m}_2}\right|_{32} = \left|\frac{1}{31}\right|_{32} = 31, \quad \left|\frac{1}{\hat{m}_1}\right|_{31} = \left|\frac{1}{2}\right|_{31} = 16, \quad \left|\frac{1}{\hat{m}_3}\right|_{33} = \left|\frac{1}{2}\right|_{33} = 17$$

result in:

$$|N|_{32736} = |(1056 \times |16 \times x|_{31}) + (1023 \times |31 \times y|_{32}) + (992 \times |17 \times z|_{33})|_{32736}$$

Let $A = |16 \cdot x|_{31}$, $B = |31 \cdot y|_{32}$, $C = |17 \cdot z|_{33}$ then we can write the calculation result in Table. 2. this result is calculated by string property:

Table 2. Calculation Result using String Property.

	String Property	Using String Property	Calculation Result
1056A	0 0 1 0 0 0 0 1 0 0 0 0 0 (strings: -2A, -4A, -2A, 0)	$2^{10} \times 2A = 2^{11}A$ $2^4 \times 2A = 2^5 A$ $2^8 \times -4A = -2^{10}A$	2048A − 1024A + 32A
1023B	0 1 1 1 1 1 1 1 1 1 1 (strings: -4B, 0, 0, -B)	$2^8 \times 4B = 2^{10}B - B$	1024B − B
992C	0 1 1 1 1 1 0 0 0 0 0 (strings: -4C, -2C, 0)	$2^4 \times -2C = -2^5 C$ $2^8 \times -4C = 2^{10}C$	1024C − 32C

The example of CHK block for 8 bit operation is shown in Table 3.

Table 3. Operation of CHK

$2^{2k} \cdot B + (-2^k \cdot C) - B + 2^{2k} \cdot C \geq 2^{3k} - 2^k$
32R1 and 33R2
32R2 and (33R31 or 33R32)
⋮
32R30 and (33R3 or 33R4 or…or 33R32)
32R31 and (33R2 or 33R3 or…or 33R32)

Table. 4. defines the number of bits which are used to A, B, C in (11). Since noneoverlapped among of them, it can be performed parallely.

The hardware realization of the CRT(10) is shown in Fig. 1. 2's complement operation will be performed after Fig. 1 is executed. The delay of a logic gate is assumed to be substituting k=5, a 8 by 8 multiplier has a delay of $61\Delta_T$. The multiplier is faster than Braun multiplier and the overlapped multiple bit scanning method that has a delay of $92\Delta_T$, $76\Delta_T$, respectively.

4 Conclusions

A multilevel CSA tree using overlapped multiple bit scanning method have computational advantages since operation on CSA tree are performed independently and so these processes can be performed in parallel. If k is more than 5, then a multilevel CSA tree has another advantages since tree level is fixed to three level.

In conclusion, the simulation shows that the CRT method which is adopted in this research, performs arithmetic operations faster than the traditional approaches, due to advantages of parallel processing and carry-free arithmetic operation

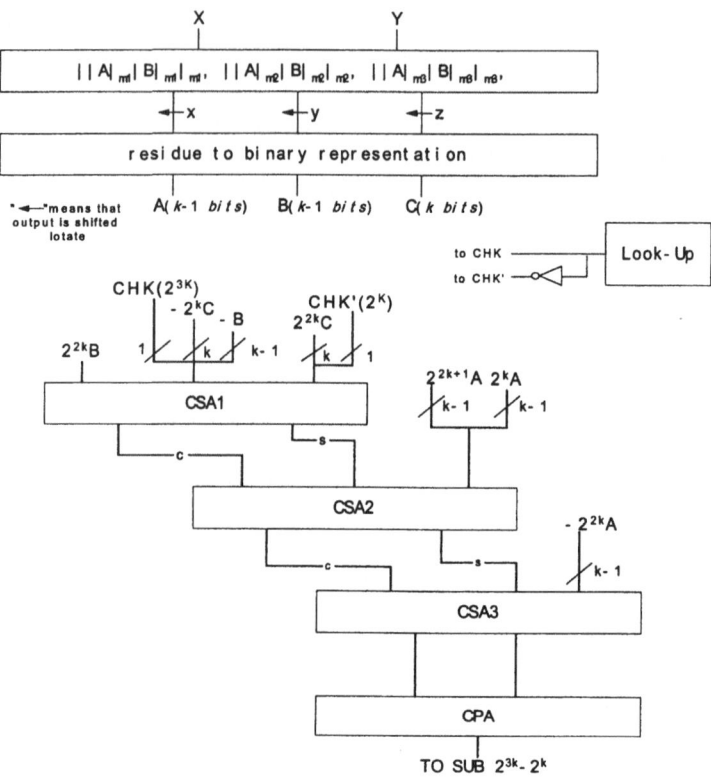

Fig. 1. Block Diagram of proposed Multiplier

Table 4. Number of bits for A, B, C

range of A, B, C	Number of bits
$0 \leq A = \left\vert (2^{k-1}) \cdot x \right\vert_{2^k-1} \leq 2^k - 1$	k bits
$0 \leq B = \left\vert (2^{k-1}-1) \cdot y \right\vert_{2^k} \leq 2^k - 1$	k bits
$0 \leq C = \left\vert (2^{k-1}+1) \cdot z \right\vert_{2^k+1} < 2^k$	k +1 bits

The result of this implementation is show in Table 5.

Table 5. Implementation result of Proposed Method

Area	Proposed Method	$4705\,A_g$
	Overlapped Multiple scanning method[13]	$1185\,A_g$
	Braun Multiplier[11]	$626\,A_g$
Delay	Proposed Method	$61\,\Delta_T$
	Overlapped Multiple scanning method	$76\,\Delta_T$
	Braun Multiplier	$92\,\Delta_T$

References

1. Ahmad A. Hiasat, "New Efficient Structure for a Modular Multiplier for RNS," *IEEE Trans Computers,* Vol. 49, pp. 170~174, Feb. 2000.
2. J. Bajard, L. Didier, and P. kornerup, "An RNS Montgomery Modular Multiplication Algorithm," *IEEE Trans Computers,* Vol. 47, pp. 766~776, July. 1998.
3. A. Wrzyszcz, D. Milford, and E. Dagless, "A New Approach to fixed-Coefficient Inner Product Computation over Finite Rings, " *IEEE Trans Computers,* Vol. 45, pp. no. 12, pp. 1,345~1,355, Dec. 1996.
4. K. Elleithy and M. Bayoumi, "*A Systolic Architecture for modulo Multiplication,* " *IEEE Trans Circuit and System-II: Analog and Digital Signal Processing,* Vol. 42, pp. 725~729, Nov. 1995.
5. W. K. Jenkins, "Techniques for residue to analog conversion for residue encoded digital filters, " *IEEE Trans, Circuits Syst.,* vol. CAS-25, pp 553-562, July 1978.
6. A. A Sawchuk and T. C Strand, "Digital optical computing" *Proc IEEE*, Vol. 72, pp. 758-779, July 1982.
7. H. M. Razavi, and J. Battelini., "Design of a residue arithmetic multiplier" *IEE Proceedings-G,* vol. 139, No. 5, pp. 581-585, October 1992
8. Khalid M. Ibrahim., and Salam N.Saloum, "An efficient residue to binary converter design," *IEEE Trans on Circuit and System,* Vol., 35, No. 9, pp. 1156-1158, September, 1988.
9. A. Hiasat "New designs for a sign detector and a residue to binary converter" *IEE Proceedings-G,* vol. 140, No. 4, pp. 247-252, August 1993.
10. F. Pourbigharaz, Memver, IEEE and H.M Yassine, "A Signed-Digit Architecture for Residue to Binary Transformation" *Algorithmica,* pp. 79-119, 3, 1988.
11. Braun, E. L., *Digital Computer Design,* Academic Press, New York, 1963
12. Khaled M. Elleithy, *Member, IEEE,* and Magdy A. Bayoumi, Senior *Member, IEEE,* "Fast and Flexible Architectures for RNS Arithmetic Decoding" *IEEE Transaction on Circuits and Systems,* Vol. 39, No. 4, pp. 226-235, April 1992.
13. K. Hwang, *Computer arithmetic*: *Principles, Architecture, and Design.* New York, Wiley, 1978.
14. Israel Koren, *Computer Arithmetic Algorithms*, Prentice-Hall. Inc, 1993.

Unified Development Solution for Cluster and Grid Computing and Its Application in Chemistry*

Róbert Lovas[1], Péter Kacsuk[1], István Lagzi[2], and Tamás Turányi[2]

[1] Computer and Automation Research Institute, Hungarian Academy of Sciences (MTA SZTAKI), 1518 Budapest, P.O. Box 63.
{rlovas,kacsuk}@sztaki.hu
[2] Department of Physical Chemistry, Eötvös University (ELTE), H-1518 Budapest, P.O. Box 32, Hungary
lagzi@vuk.chem.elte.hu, turanyi@garfield.chem.elte.hu

Abstract. P-GRADE programming environment provides high-level graphical support to develop parallel applications transparently for both the parallel systems and the Grid. This paper gives an overview on the parallelisation of a simulation algorithm for chemical reaction-diffusion systems applying P-GRADE environment at all stages of parallel program development cycle including the design, the debugging, the execution, and the performance analysis. The automatic checkpoint mechanism for parallel programs, which supports the migration of parallel jobs between different clusters, together with the application monitoring facilities of P-GRADE enable the long-running parallel jobs to run on various non-dedicated clusters in the Grid while their execution can be visualised on-line for the user. The presented research achievements will be deployed in a chemistry Grid environment for air pollution forecast.

1 Introduction

Computational Grid systems [1] are becoming more and more popular in natural science. In such systems, large number of heterogonous resources can be interconnected in order to solve complex problems. The aim of a joint national project, "Chemistry Grid and its application for air pollution forecast" is to investigate three important aspects of Grids and to find practical results in the areas as follows:

1. Establishment of a chemistry Grid, i.e. using Grid technologies for *supporting a specific scientific research area*; this new infrastructure provides access for chemists to both the Hungarian computational Grid resources

---

* The research described in this paper has been supported by the following projects and grants: Hungarian IHM 4671/1/2003 project, Hungarian OTKA T042459 and T043770 grants, OTKA Instrument Grant M042110, Hungarian IKTA OMFB-00580/2003, EU-GridLab IST-2001-32133, and COST Action D23/0003/01 project.

(e.g. Hungarian Cluster Grid [2]) and the European-wide chemistry Grid infrastructure being established as a result of the EU funded COST Action D23 project called SIMBEX.

2. The Grid as a *high performance computational infrastructure*; Computer and Automation Research Institute of the Hungarian Academy of Sciences (MTA SZTAKI) has elaborated an integrated parallel program development environment, called P-GRADE [3,4] that supports the parallelisation process of sequential applications in an efficient and clear way by means of its high level graphical approach [5,6] and special correctness and performance debugging and analysing tools [3]. In the chemistry Grid project, we are developing further P-GRADE to provide dedicated support for efficient execution of parallel programs in Grids, such as the migration of applications [4] across the Grid resources according to actual load and availability conditions. The P-GRADE system is currently used by all the participating chemists to parallelise their sequential simulations having high computational demands and afterwards to make them run on the chemistry Grid.

3. The Grid as *a computer system for supporting complex collaborative work* and its application for air pollution forecasting (elaboration of smog alarm plans); the partners will elaborate a collaborative application that will run on a supercomputer to forecast air pollution in Hungary in an operative manner. The same application will run on Grid as well to simulate earlier smog events and to analyse the efficiency of smog alarm plans and the prospective effects of various potential measurements against air pollution.

As a joint effort of MTA SZTAKI and Department of Physical Chemistry, Eötvös University (ELTE), P-GRADE environment has been applied to parallelise an existing simulator for chemical reactions and diffusions in the frame of the chemistry Grid project.
In this paper we introduce briefly the fundamentals of the basic problem of reaction-diffusion systems (see Section 2) and its parallelisation with P-GRADE programming environment in details through the design, debugging (see Section 3), and the performance analysis (see Section 4) phases of program development. Finally, we summarise the related and future works (see Section 5), and our achievements (Section 6).

2 Reaction-Diffusion Equations

A variety of spatiotemporal pattern formation arises from the interaction of chemical reaction and diffusion, such as chemical waves [7], autocatalytic fronts [8], Turing structures [9] and precipitation patterns (Liesegang phenomenon) [10]. Evolution of pattern formation can be described by second-order partial differential equations:

$$\frac{\partial c_i}{\partial t} = \nabla(D_i \nabla c_i) + R_i(c_1, c_2, \ldots, c_n), \quad i = 1, 2, \ldots, n, \tag{1}$$

where c_i is the concentration, D_i is the diffusion coefficient and R_i is the chemical reaction term, respectively, of the ith chemical species, and t is time. The chemical reaction term R_i may contain non-linear terms in c_i. For n chemical species, an n dimensional set of partial differential equations is formed describing the change of concentrations over time and space. These equations are coupled through the non-linear chemical reaction term. Assuming that the diffusion coefficient is constant in space, equation (1) can be rewritten as

$$\frac{\partial c_i}{\partial t} = D_i \nabla^2 c_i + R_i(c_1, c_2, \ldots, c_n), \quad i = 1, 2, \ldots, n, \tag{2}$$

Here

$$\nabla^2 \equiv \frac{\partial^2}{\partial x^2} + \frac{\partial^2}{\partial y^2},$$

where x and y are the spatial variables. The operator splitting approach is applied to equations (2), decoupling transport (diffusion) from chemistry, i.e.

$$c_{i,\hat{t}+\Delta t} = T_D^{\Delta t} T_C^{\Delta t} c_{i,\hat{t}}$$

where T_D and T_C are the diffusion and the chemistry operators, respectively, and $c_{i,\hat{t}+\Delta t}$ and $c_{i,\hat{t}}$ are the concentration of the ith species at time $\hat{t}$ and $\hat{t}+\Delta t$, where Δt is the time step.

The basis of the numerical method for the solution of the diffusion operator is the spatial discretisation of the partial differential equations on a two-dimensional rectangular grid. In these calculations, the grid spacing (h) is uniform in both spatial directions. This approach, known as the 'method of lines', reduces the set of partial differential equations (PDEs) of three independent variables (x, y, t) to a system of ordinary differential equations (ODEs) of one independent variable, time. A second order Runge-Kutta method is used to solve the system of ODEs arising from the discretisation of diffusion with no-flux boundary conditions on a 360×100 grid. The Laplacian is approximated as

$$\nabla^2 c_i^{j,k} \approx L_{c,i}^{j,k} = \frac{1}{6h^2} \sum_{p=-1,q=-1}^{p=1,q=1} a_{p,q} c_i^{i+p,j+q}.$$

The coefficients $a_{p,q}$ are taken according to the matrix below for a nine-point approximation:

$$a_{p,q} = \begin{pmatrix} 1 & 4 & 1 \\ 4 & -20 & 4 \\ 1 & 4 & 1 \end{pmatrix}$$

resulting in an error of $O(h^2)$ for the Laplacian.
The equations of the chemical term have the form

$$\frac{dc_i}{dt} = R_i(c_1, c_2, \ldots, c_n), \quad i = 1, 2, \ldots, n. \tag{3}$$

The time integration of system (3) is performed with the BDF method using the CVODE package [11,12], which solves stiff chemical kinetics equations. In the present study, a simple three variable, stiff chemical mechanism was used as proposed by S. C. Cohen and A. C. Hindmarsh [12]. The following parameter set was used in the simulation: $D_1 = D_2 = D_3 = 0.8$. The grid spacing and the time step were $h = 0.4$ and $\Delta t = 0.001$, respectively.

3 Parallel Implementation with P-GRADE

In order to parallelise the sequential code of the presented reaction-diffusion simulation the domain decomposition concept was followed; the two-dimensional grid is partitioned along the x space direction, so the domain is decomposed into horizontal columns. Therefore, the two-dimensional subdomains can be mapped onto a one-dimensional logical grid of processes. An equal partition of subdomains among the processes gives us a well balanced load during the solution of the reaction-diffusion equations. During the calculation of the diffusion of the chemical species communications are required to exchange information on the boundary concentrations between the nearest neighbour subdomains. In the rest of this chapter we illustrate how this idea can be implemented in P-GRADE.

The graphical language of P-GRADE consists of three hierarchical design layers [5]: (i) **Application Layer** is a graphical level, which is used to define the component processes, their communication ports as well as their connecting communication channels. Shortly, the Application Layer serves for describing the interconnection topology of the component processes or process groups (see Figure 1, *Application window*). (ii) **Process Layer** is also a graphical level where different types of graphical blocks are applied: loop construct (see Figure 1, in window labeled *Process: sim → sim_2*), conditional construct, sequential block, input/output activity block and macrograph block. The graphical blocks can be arranged in a flowchart-like graph to describe the internal structure (i.e. the behaviour) of individual processes (see Figure 1, *Process windows*). (iii) **Text Layer** is used to define those parts of the program that are inherently sequential and hence only pure textual languages like C/C++ or FORTRAN can be applied at the lowest design level. These textual codes are defined inside the sequential blocks of the Process layer (see Figure 1, at bottom of Process window labelled *Process: sim → sim_1*).

We defined a common process, called 'master' (see Figure 1, *Process: master*), which sets up the initial conditions in a sequential code block, *'init_cond'* and sends the necessary information, e.g. the initial concentrations (A, B and C matrices), the diffusion coefficient (*dc*), the time-step (*dt*) to each 'worker' process via the attached communication port with label '0' using collective communication operations (see the selected communication action icon in Figure 1, *Process: master*). The usage of predefined and scalable process communication templates enables the user to generate complex parallel programs from design patterns. A communication template describes a group of processes, which have a pre-defined regular interconnection topology. P-GRADE provides such communication templates for the most common regular process topologies like process farm, pipe, 2D mesh and tree, which are widely used among scientists. Ports of the member processes in a template are connected automatically based on the topology information. In our case the pipe communication template was selected (see Figure 1,

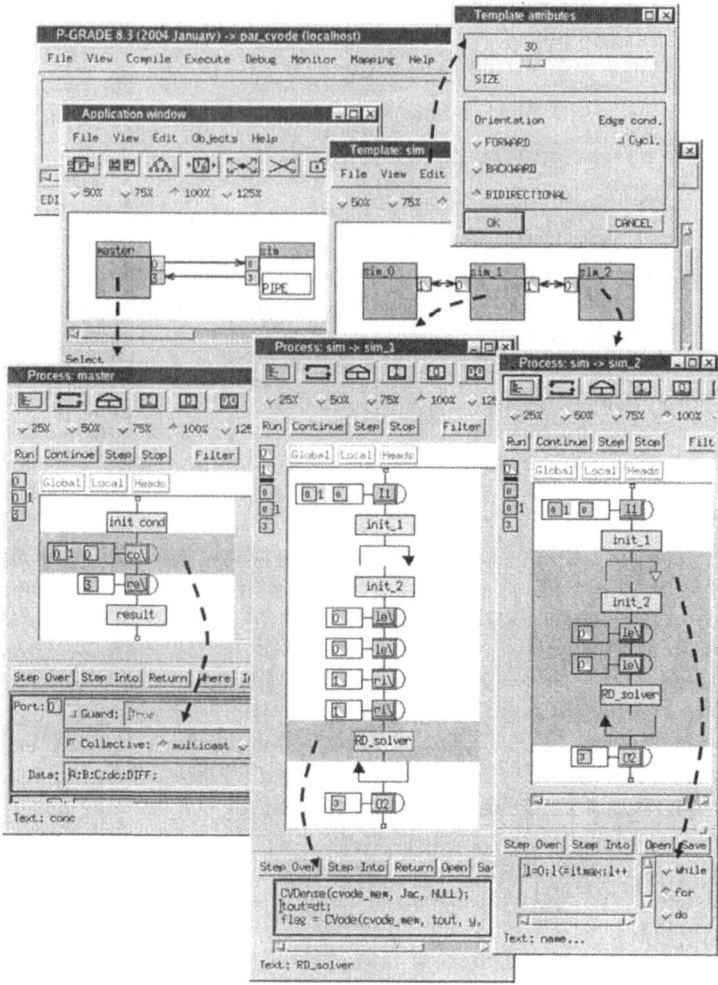

Fig. 1. Parallel code of reaction-diffusion simulation in P-GRADE

Template window) as the most suitable topology since the two-dimensional subdomains should be mapped onto a one-dimensional logical grid of processes and, during the calculation of the diffusion of the chemical species, communications are required to exchange information on the boundary concentrations between the nearest neighbour subdomains. A pipe communication topology consists of a linearly ordered set of processes where each process is interconnected only with its neighbours however, all processes in the pipe may communicate with outsider processes via group ports if such ports are defined for the template at Application level (see Figure 1, *Application window*).

The user must define only the code of the representative processes the number of which depends on the actual template attribute settings (see 'edge condition' below). In a separated dialog window (see Figure 1, *Template Attributes*) the significant attributes of the current template can be set by the user, e.g. in case of pipe:

- *Size*: Actual number of processes within the pipe at runtime.
- *Channel orientation*: Communication channels between neighbour processes can be directed forward, backward or both directions (i.e. bi-directional channels).
- *Edge condition*: Channel pattern can be cyclic if the last process is connected to the first (i.e. ring) one or acyclic otherwise (i.e. pipe).

Without applying that cyclic communication pattern, a pipe is defined by three representative processes since the communication interfaces (i.e. number and types of ports) of the first and last processes differ from that of the middle ones. For illustration purposes we describe only one inner process (see Figure 1, Process window labelled *Process: sim → sim_1*). First of all, the process receives the necessary input parameters for the calculation from the 'master' process. After the initialisation phase in each iteration step (applying loop construct) the process exchanges the boundary conditions (a vector of double precision numbers) with its neighbours (see Figure 1, communication actions in window labelled Process: *sim → sim_1*), and calculates the transportation and reaction of chemical species (see Figure 1, sequential code box of RD_solver in window labelled Process: *sim → sim_1*). For the calculation the process invokes external functions (see Figure 1, at the bottom of window labelled Process: *sim → sim_1*), which are available as sequential third-party code [11,12] written in C. Finally, the process sends back the results to the 'master' process, which is responsible for the collection of the results via collective (gather-type) communication. During the debugging stage we took all the advantages of DIWIDE [3] built-in distributed debugger of P-GRADE environment. DIWIDE debugger provides the following fundamental facilities of parallel debugging; the step-by-step execution on both graphical and textual levels, graphical user interface for variable/stack inspection, and for individual controlling of processes.

4 Performance Results

The parallel version of reaction-diffusion simulation with 1000 simulation steps has been tested on two clusters using Condor [13,14] job-mode of P-GRADE [4]: (1) a self-made Linux cluster of MTA SZTAKI containing 29 dual-processor nodes (Pentium III/500MHz) connected via Fast Ethernet, (2) dual mode cluster with 40 nodes (AMD Athlon/2GHz) located at ELTE and connected to the Hungarian Cluster Grid [2]. The simulation has been also tested with 10.000 iterations; the parallel application was able to migrate automatically to another friendly Condor pool when the actual pool had become overloaded, as well as to continue its execution from the stored checkpoint files [4].

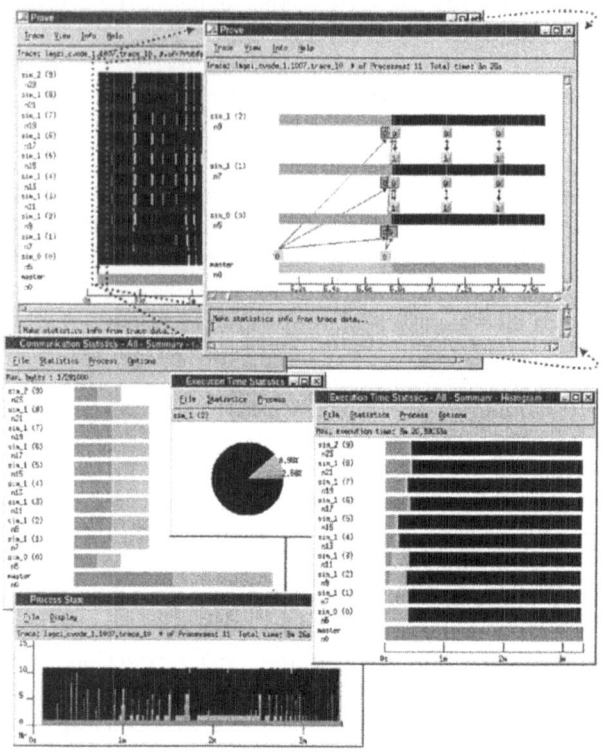

Fig. 2. Performance visualisation with PROVE on MTA SZTAKI's cluster

According to the available computational resources the actual size of the scalable pipe communication topology can be set by a dialog window in P-GRADE. To take an example, the calculation was executed on MTA SZTAKI's cluster within 3 min 26 sec (see Figure 2, PROVE window at the upper-right corner) with 10 worker processes, and it took 1 min 20 sec to calculate it with 40 processes. (The sequential execution time is approximately a half an hour).

In details, PROVE [3] performance analyser (based on the GRM/Mercury monitoring infrastructure [15]) as a built-in tool of P-GRADE system can visualise either event trace data, i.e. message sending/receiving operations, start/end of graphical blocks in a space-time diagram (see Figure 2, PROVE windows), or statistical information about the application behaviour (see Figure 2, Process State, Communication Statistics, and Execution Time Statistics windows). In all the diagrams of PROVE tool, the black colour represents the sequential calculations, and two different colours (green for incoming and grey for outgoing communication) used for marking the message exchanges. The PROVE space-time diagram presents a task bar for each process, and the arcs between the process bars are showing the message passing between the processes. We focused on some interesting parts of the trace (see Figure 2, PROVE windows) using zooming and filtering facilities of PROVE. The Process 'master' sends the input data to each 'worker' process, and they are starting the simulation and the message exchanges in each simulation steps. The *Process State* window (see Figure 2) is a Gantt chart of the application showing only the state of the processes, sorted by the different types of states. The horizontal axis represent the time while the vertical axis represents the three different states of the processes. For each state, the height of the colored column represents the number of processes in that state. Thus, this graph gives cumulative information about the state of the application. The *Execution Time Statistics* window offers another view; it shows the time of each process state independently for each process in bars or in a pie chart. The *Communication Statistics* (see Figure 2) provide information on the amount of exchanged messages in bytes for each process and for the entire application (see 'Max. bytes'). It is easy to recognise, the first process and the last one in the pipe communicates less (comparing to the other processes of the pipe) since they have only one neighbour. According to our measurements

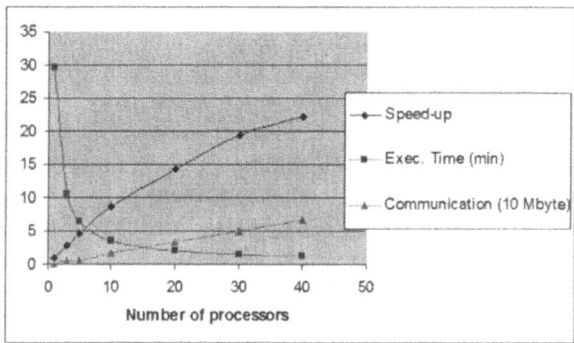

Fig. 3. Performance results on MTA SZTAKI's cluster in Condor job mode

and analysis with PROVE the communication overhead is showing nearly linear characteristics depending on the number of processors, and the curve of speed-up is getting closer and closer to the saturation when the number of processors reaches the 40 (see Figure 3).

5 Related and Future Works

P-GRADE has been applied for the parallelisation of the MEANDER ultra-short range weather prediction package [16] of Hungarian Meteorology Service and for the parallelisation of an urban traffic simulation system [17] by University of Westminster (UK). Institute of Chemistry, Chemical Research Centre of the Hungarian Academy of Sciences has recently parallelised a classical trajectory calculation written in FORTRAN [18] in the frame of joint chemistry

Grid project. Some other development systems, such as ASSIST [19], TRIANA [20], or CACTUS [21], target the same research community (biologist, chemists, etc.), and they can offer several useful facilities similarly to P-GRADE. On the other hand, P-GRADE is able to provide more transparent run-time support for parallel applications without major user interactions, such as code generation to different platforms (Condor [14] or Globus-2 [22] based Grids, PVM [23] or MPI [24] based clusters and supercomputers), migration of parallel jobs across grid sites based on automatic checkpointing facilities [4], or application monitoring of parallel jobs on various grid sites, clusters, or supercomputers [15]. As a new achievement, P-GRADE supports the creation of workflows to execute complex programs in the Grid [6]. In the frame of chemistry Grid project, the developed parallel applications, such as the presented simulation program, will be available via P-GRADE portal, where each component job will collaborate in the Grid based on our workflow concept to provide efficient air pollution forecasting (elaboration of smog alarm plans).

6 Summary

MTA SZTAKI developed a graphical programming environment, called P-GRADE that is able to support the entire life-cycle of parallel program development and the execution of parallel applications both for parallel systems and the Grid. One of the main advantages of P-GRADE is the transparency; P-GRADE users has not to learn the different programming methodologies for various parallel systems and the Grid, the same environment is applicable either for supercomputers, clusters or the Grid. As the presented work illustrates, P-GRADE enables fast parallelisation of sequential programs written in C, C++ or FORTRAN, and it provides an easy-to-use solution even for non-specialist programmers, like chemists, meteorologists, biologists, etc. P-GRADE can generate either PVM [23] or MPI [24] code that can be executed interactively on supercomputers and clusters, or as a job in Condor [14] or Globus-2 [22] based Grids. Moreover, remote monitoring, performance visualisation, fully automatic checkpointing, and migration mechanisms of Grid applications [4] are also supported by P-GRADE.

References

1. Foster, I., Kesselman, C.: Computational Grids, Chapter 2 of The Grid: Blueprint for a New Computing Infrastructure. Morgan-Kaufman, (1999)
2. Stefán, P.: The Hungarian ClusterGrid Project Proc. of MIPRO'2003, Opatija
3. P-GRADE Graphical Parallel Program Development Environment: http://www.lpds.sztaki.hu/projects/pgrade
4. Kacsuk, P., Dózsa, G., Kovács, J., Lovas, R., Podhorszki, N., Balaton, Z., Gombás, G.: P-GRADE: a Grid Programming Environment. Journal of Grid Computing (2004) (accepted for publication)

5. Kacsuk, P., Dózsa, G., Lovas, R.: The GRADE Graphical Parallel Programming Environment Parallel Program Development for Cluster Computing: Methodology, Tools and Integrated Environments (Chapter 10), Editors: P. Kacsuk, J.C. Cunha and S.C. Winter, pp. 231-247, Nova Science Publishers New York, 2001
6. Lovas, R., Dózsa, G., Kacsuk, P., Podhorszki, N., Drótos, D.: Workflow Support for Complex Grid Applications: Integrated and Portal Solutions. Proceedings of 2nd European Across Grids Conference, Nicosia, Cyprus, 2004 (accepted for publication)
7. Zaikin, A. N., Zhabotinsky, A. M.: Concentration wave propagation in two-dimensional liquid-phase self-oscillating system. Nature 225 (1970) 535-537
8. Luther, R.: Raumliche fortpflanzung chemischer reaktionen. Zeitschrift für Elektrochemie 12 (1906) 596-600
9. Turing, A. M.: The chemical basis of morphogenesis. Philosophical Transactions of the Royal Society of London series B 327 (1952) 37-72
10. Liesegang, R. E.: Üeber einige eigenschaften von gallerten. Naturwissenschaflichee Wochenschrift 11 (1896) 353-362
11. Brown, P. N., Byrne, G. D., Hindmarsh, A. C.: Vode: A variable coefficient ode solver. SIAM Journal of Scientific and Statistical Computing 10 (1989) 1038-1051
12. Cohen, S. C., Hindmarsh, A. C.: CVODE User Guide, Lawrence Livermore National Laboratory technical report UCRL-MA-118618 SIAM Journal of Scientific and Statistical Computing (1994) pp. 97
13. Litzkov, M. J., Livny, M., Mutka, M. W.: Condor – A hunter of idle workstations. Proceedings of the 8th IEEE International Conference on Distributed Computing Systems, pp. 104-111, 1988
14. Thain,D., Tannenbaum, T., Livny, M.: Condor and the Grid. in Fran Berman, Anthony J.G. Hey, Geoffrey Fox, editors, Grid Computing: Making The Global Infrastructure a Reality, John Wiley, 2003
15. Balaton, Z., Gombás, G.: Resource and Job Monitoring in the Grid Proceedings of EuroPar'2003 Conference, Klagenfurt, Austria, pp. 404-411, 2003
16. Lovas, et al.: Application of P-GRADE Development Environment in Meteorology. Proceedings of DAPSYS'2002, Linz, pp. 30-37, 2002
17. Gourgoulis, A., Kacsuk, P., Terstyanszky, G., Winter, S.: Using Clusters for Traffic Simulation. Proceedings of MIPRO'2003, Opatija, 2003
18. Bencsúra, Á., Lendvay, Gy.: Parallelization of reaction dynamics codes using P-GRADE: a case study. ICCSA/MPS 2004, Assisi, Italy (accepted for publication)
19. Vanneschi, M.: The programming model of ASSIST, an environment for parallel and distributed portable applications. Parallel Computing 28 (2002) 1709-1732
20. Taylor, I., et al.: Distributed P2P Computing within Triana: A Galaxy Visualisation Test Case. IPDPS'2003, April 2003
21. Goodale, T., et al.: The Cactus Framework and Toolkit: Design and Applications. 5th International Conference on Vector and Parallel Processin, 2002, pp. 197-227
22. www-unix.globus.org/toolkit/
23. Geist, A., Beguelin, A., Dongarra, J., Jiang, W., Manchek, B., Sunderam, V.: PVM: Parallel Virtual Machine – a User's Guide and Tutorial for Network Parallel Computing. MIT Press, Cambridge, MA, 1994.
24. Message Passing Interface Forum, MPI: A Message Passing Interface Standard, 1994

Remote Visualization Based on Grid Computing

Zhigeng Pan[1,3], Bailin Yang[2,3], Mingmin Zhang[3], Qizhi Yu[3], and Hai Lin[3]

[1] Institute of VR and Multimedia, HZIEE,Hangzhou,310032,P.R.China,
[2] Hangzhou University of Commerce, Hangzhou, 310035, P.R. China,
[3] State Key Lab of CAD&CG, Zhejiang University, Hangzhou, 310027, P.R. China,
(zgpan,ybl,zmm,yqz,lin)@cad.zju.edu.cn,
http://www.cad.zju.edu.cn/vrmm.html

Abstract. Remote visualization can not be well done in interactive way because of lacking of the expensive equipments and a large amount of resources in the past. With the development of Grid, it becomes possible. This paper discusses the pipeline of remote visualization technology based on Grid Computing. Moreover, we analyze and propose the basic framework of remote visualization technology for Grid. In addition, methods on improving the performance of visualization in Grid environment are discussed.

1 Introduction

Grid Computing has been proposed as a key technology for meeting the need of the rapid increase of computing resources and ability. It has been used in many important fields, such as geoscience, astrophysics, scienticfic visualization etc. It has evolved to provide scientists and engineers quick and transparent access to critical computing resources and advanced instrumentations. Thus, Grid users can directly obtain computers, software, data, and other critical resources with little concern for the physical location of those resources [1].

Visualization has emerged as an important tool for extracting meaning from large volumes of data that is provided by scientific instruments and simulations produce. But the result of advance visualization system can only be displayed in the local site in the past. With the development of network technology, we can see the visualization result of remote sites. But we can not manipulate the remote expensive devices and access the result in an interactive way due to the existed network infrastructure and the low network bandwidth. Grid can significantly increase distance visualization capabilities [2]. Grid users can achieve a real-time interaction with the remote visualization devices by only sending the visualization commands with parameters to the grid, which perform the visualization procedure with its high computing power.

Grid technologies set to surge in the past few years. The idea is to assemble a wide network of computers that can be seen from the user point of view as a single computational unit. A middleware layer provides facilities to submit jobs, store data, and get information from the grid while hiding the underlying grid architecture to the user and physical machines used for storage and computation. The most advantage of Grid is to provide the share of huge resources and computation ability[1]. The basic technology for exploiting computational grids is grid middleware, which is provided by toolkits such as Globus[3], Legion, and UNICORE. Such grid middleware toolkits offer information infrastructure services for security, communication, fault detection, resource and data management, and portability. We can utilize these grid middleware toolkits to support interactive remote visualization.

The exploration of grid computing includes the following key technologies: grid node, resource management, task management, task schedule tool, monitor tool and visualization tool. This paper focuses on the exploration of visualization tool. As we are all known, the major application field of grid computing is scientific computing, which may produce massive data. How to deal with these data efficiently is a difficult problem. By using the visualization technology, the researchers can avoid the puzzle of facing of massive meaningless data and draw a correct conclusion quickly by a friendly user interface.

A lot of researches have been done in the integration of Grid computing and remote visualization field. The Grid Visualization Kernel (GVK)[4] is a middleware extension for visualization. A new volume rendering framework based on three components in Grid environment is established[5]. Dv[6]is a framework for developing applications for distributed visualization of large scientific datasets on a computational Grid. OpenGL Vizserver[7], a product of SGI (http://www.sgi.com), is a client/server application that uses an SGI Onyx family system as a visualization server. Also,there are many other typical tools including Cactus Computational Toolkit and CAVERN etc.

Regarding the integration of grid computing environment and visualization, there are also some works in this field, which will be discussed in following section. The remainder of this paper is organized as follows. In Section 2, it first discusses the visualization technology and the traditional visualization pipeline, and further analyzes the visualization pipeline based on grid computing. In Section 3, we analyze visualization system architecture based on grid computing and propose a principal framework. Some improvements and performance optimization methods are also proposed in Section 3. In the end of this paper, we put forward the future work of the remote visualization based on grid computing.

2 Visualization Pipeline Based on Grid

2.1 Integration of Visualization and Grid

Visualization which is a kind of computing method converts raw data and information to certain kind of format amenable to understanding by the human perceptual system while maintaining the integrity of the information. Visualization provides a way of observing unseen objects which format the signs into geometry objects and help the researchers observe the simulation and computing result. Visualization has three features, which are interaction, multi-dimension and visibility.

Figure 1 describes the traditional visualization pipeline which converts raw simulation data into a displayable image [8]. There are three main transformations in the pipeline which are Data enrichment/enhancement, Visualization mapping and Rendering. A number of successful, contemporary visualize systems, such as VTK, AVS and OpenDX, are constructed using this visualization pipeline architecture. This environment is typically run on a machine with sufficient memory, processor and graphics capability to handle each stage of the pipeline. But it is unreachable for a common user who does not has so expensive machine to perform the visualization process. Consequently, the remote visualization technology is introduced for local user to perform visualizing process. It adopts some new system architectures, such as Client/Server and Web based system. We can do some computing operations on the server side and transmit the final result to the client side by Client/Server system. By this way we can fully explore the computing ability of server side and save some valuable resources on the client side. With the development of internet technology, users distributed over the world can also get the visualization result through exploiting remote visualization resources such as huge graphics capability.

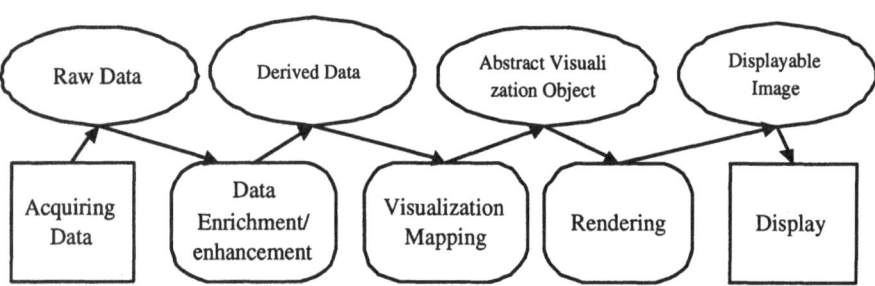

Fig. 1. Traditional visualization pipeline

These two approaches are certainly not optimal for several reasons. Firstly, they introduce a substantial delay when transmitting the data from the server to client. Secondly, they do not have proper resource management. They can not solve resource heterogeneity problem. Thirdly, it is difficult to use when lacking the flexibility and extensibility. For these reasons, remote visualization process requires a more suitable network system.

A new generation network, Grid computing, provides computing ability and resource management and store ability which are not possessed by a single server. At the same time, grid has a high speed network transmitting ability and various services involving resources management and information service etc. Therefore, Grid users can transparently access resources including the visualization resources distributed all over the gird. They have the ability to utilize remote expensive visualization equipments to practice the visualization process in a real-time interactive way.

2.2 Grid-Enabled Visualization Pipeline

Grid-enabled visualization pipeline based on traditional pipeline divides up the total visualization process into different sub-processes and spreads these sub-processes into the grid [9]. In this paper,it classifies the following different implementing approaches according to certain sub-processes into the grid. (1)Perform everything on the remote server within the grid. (2)Perform everything but rendering on the remote server under the grid. (3)Use a local proxy for the rendering, and other sub-processes are under the gird.

For these implementations, (1) and (2) are simple to utilize Grid enabled existing packages. But they can also incur high latency so that we need adopt some new ways to improve network performance. In addition, it removes some load from the remote site for the second implementation which requires every local site to have good rendering power. For the third implementation, it offloads work from the remote site and allows local sites to contribute additional resources. It leads to the application being more complex than others.

These pipeline implementations are static. So, these implementations are inflexible and can not cope with the varieties of new conditions under the grid. We can use dynamic pipeline implementation that adopts an agent to control the pipeline implementation to overcome this disadvantage. In detail, we can wrap the each part of the pipeline into certain service which is to be executed on different logical machines in different geographic locations. This separation also allows for each service to be scaled in the most appropriate manner for the computation that it is to perform, for example a render service can be hosted on a high performance graphics machine to produce high quality images. While the mapping service could be executing on a parallel computing array accessing data and filter services running at a high capacity storage farm. The agent can control all kinds of services to be executed in proper locations by means of the resources allocation, computing ability and user's suggestions. By this way, we can better utilize the grid resources and computing power.

3 Visualization Framework Based on Grid

Visualization framework based on Grid is different from the former systems such as MVEs and Client/Server because there are new conditions that are not having been anticipated during their initial design and implementation. What is needed is a new grid-enabled visualization framework, which is easy-to-use and requires considerations of both these new conditions and reusing our existing investments in visualization technology.

3.1 Grid Services and Component

It is simple for a program running on a desktop PC to find resources. But it is difficult for a system based on grid because the resources are distributed all over the grid. We can not gain some resources you wanted when the resources are not managed rightly. The grid provides the middleware that can help users access and manage distributed resources. Grid middleware such as Globus[3] that provides many kernel services. These services include Globus Resource Allocation Manager (GRAM) for resource management, the Grid Security Infrastructure (GSI) for authentication, the Meta-computing Directory Service (MDS) for information lookup and Nexus which is the Globus communication library for inter-component communication et al. It is transparent for users who do not care for who and where carry out the actual imple-mentation through some kinds of services. Any visualization system based on grid needs to make use of these services for high efficiency.

LC(liquid crystal) web portal [10] employs a multi-layer system architecture, which is constituted by client side as a web browser, middle layer provided some kernel services and the backend contained some Globus services which can manage distributed computing and data resources in safety.

Component technology is a flexible technology which is used by application researcher to develop certain function rapidly. But the visualization resources are limited, heterogeneous and dynamic distributed. These new features determine the old component used in single node unsuitable for grid-based visualization system. For satisfying these new features, new components need be developed. Firstly, these new components need satisfy distributed feature because resources are distributed in different locations. Secondly, these components can distinguish and use these heterogeneous resources. By these new features, users need not worry how to visit the heterogeneous resources. Norton et al[5] establish a new volume rendering framework based on three components , and these components can both be combined to do the remote visualization and be separated into the huge grid to be employed by others.

3.2 Grid-Enabled Visualization System Framework

Remote visualization system meets the following challenges: (1) Achieving real-time interaction between users and remote visualization servers. (2) Managing intensive data and heterogeneous resources properly. The visualization framework based on grid must cope with these challenges by certain ways. In general, we can accomplish it by both the system architecture improvement and some graphic and visualization optimizing. In this subsection, the architecture improvement will be discussed and the optimizing measures will be explained in the next section in detail.

From architecture point of view, the problem of heterogeneity is typically solved by well defined interfaces, which hides the complexity of the underlying architecture from the users. It is feasible to implement a grid-enabled visualization system such as GVK [4] as a kind of middleware extension, which can manage the visualization requests of the grid users. Inside the interface, we can implement following functions such as pipeline dynamically setting, resources adaptive management, visualization computing and analyzing user's request etc. The framework consists of different modules that have given function and collaboratively perform the remote visualization process.

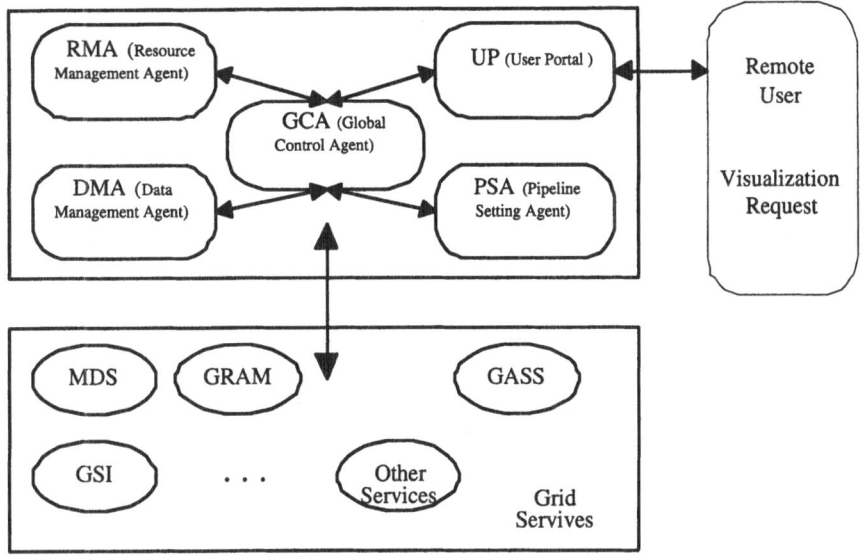

Fig. 2. Grid-enabled remote visualization framework

Figure 2 describes the framework of remote visualization system. It can be seen from the figure that there are three major parts, which are Remote users, Visualization Middleware and Grid services. For remote users, they only send the visualization requests to the Visualization Middleware and also get the result from it. Grid Services provided by some grid middleware, for example Globus services in this figure, offer some special services for the visualization middleware to manage the resources and data distributed all over the grid efficiently and conveniently through intercommunication each other. The kernel part is visualization middleware, which plays an important role through the whole remote visualization process. The middleware is composed of five main modules, which are UP (User Portal), RMA (Resources Management), DMA (Data Management Agent), PSA (Pipeline Setting Agent) and GCA (Global Control Agent), respectively.

UP is a portal to grid users. A session will be created between the user and the middleware. When the session created, remote grid users can make many visualization requests with some parameters to the grid through this session. UP receives the request and transmits the request to the GCA that will process it further after UP certificates it correctly. If the session is valid, the user can send requests continuously. The session will be ended after the result is forwarded to the user.

DMA manages the huge visualization data globally under the gird. Former Client/Server remote visualization system does not have data management. There are many kinds of data type for visualization system. For efficiently operating these different data, we need do some transforming, converting, filtering and sub-sample operations for these data. Administration of these operations into the DMA module can be centralized. Therefore, we can conveniently do some operations about data process by this DMA module.

Resource management is an important structural component of grid computing environments because they enable uniform access to the wide variety of resources necessary for scientific work. GMA can manage these heterogeneous resources efficiently by exploring the services provided by the grid middleware. We can use the Globus GRAM job-submission interface. GRAM provides a single mechanism to start jobs on a wide range of remote computer types, and permits higher level components to plan job placement and/or coordinate distributed tasks. MDS service is helpful on discovering resource wanted. GRAM and MDS collaborate to accomplish resources discovering and accessing. As discussed in Section 2.2, PSA can set the visualization pipeline dynamically. For achieving good visualization pipeline under the grid, PSA must inter-communicate with GMA for getting the information of what and which kinds of resources available. Also it must communicate with DMA for operating visualization data.

However how to communicate with the above modules is a big problem. We can implement a global control agent (GCA) which carries out centralized control of the communications between those modules. By this way, the middleware system can be integrated seamlessly and collaboratively as a whole unit. GCA can intercommunicate with these Grid Services and RMA module so that RMA can utilize both GRAM and MDS services. At the same time, PSA must also communicate with the GMA and DMA for its dynamic pipeline setting through the global collaboration of GCA. In a word, the GCA can help the whole system collaborates better than before each other. The system can also wrap these agents into different components for easy use.

3.3 Optimization for Grid Visualization

Remote visualization based on grid environments meets the real-time interactive challenge, and it can be improved by adopting optimization from different ways. In general, optimizations can be improved by the following aspects: graphic and image compressing technology [5], graphic and image optimization [4] and adaptive distributed technology [12].

Compressing technology being employed can reduce the magnitude of data transmitted over the network so that decrease the delay from the source to destination. The importance of visualization data compress is aroused by wide use of the internet and distributed application [13] [14]. Wavelet as a better theory is also used in graphic and visualization compress. Norton et al adopt D4 wavelet, D6 wavelet and Haar wavelet to compress the volume data and compares these compress efficiency.

Recently, visualization datasets are beyond the interactive rendering capabilities of the current graphics hardware. To bridge the increasing gap between hardware capabilities and graphics dataset size, the complexity of the graphics dataset is reduced such that its visual appearance is similar to the original. This reduction is achieved by level-of-detail rendering with multi-resolution hierarchies, occlusion culling, and image-based rendering. Meanwhile, these algorithms also can be considered in the grid enabled remote visualization. The main advantage of Multi-resolution technique deduce the data because the model can be represented by low resolution rather than the high resolution when low is enough. The processing procedure can be speeded up by transmitting information with different LODs [15,16,17].View-dependent simplifications have been introduced to enable various levels of detail to seamlessly coexist over different regions of the same surface[18,19].Occlusion-culling methods reduce the number of polygons for transmission [20].

It is inevitably to considere the network protocols, buffer, stream, and memory management when remote visualization is referred to Grid. The proper protocols must be chosen and better control mechanisms to be provided. Visapult project [21] takes the UDP protocol other than TCP when the application desires the low delay and high response. Buffer mechanism is employed both in local machine and network environment. DPSS (Distributed Parallel Storage System) [22] invented by LBL is an extensible, high performance and distributed parallel data storage system. Stream processing which utilizes parallel feature and hides the delay feature has widely used in many different places. For example, Chromium [23] coverts multi-stream to a single stream and switches their contexts between different streams.

Certainly, there are many other optimizations methods which can be used to improve the remote visualization performance. It is usual to integrate these optimization techiques into one system to improve the performance of the remote visualization.

4 Conclusion and Future Work

In recent years, research on Grid-based visualization is still just getting started. New methodologies and technologies that harness such resources for visualization applications are critical for the success of Grid environments. This paper analyzes the basic visualization pipeline, framework, and some optimization technologies of remote visualization based on Grid. With the development of Web technology, more researches on the combination of Web and Grid will be done[24][25].

Possible future work includes:

(1) Develop Grid-enabled visualization and VR toolkits.

(2) Support distributed collaborative visualization.

(3)Implement immersive VR steering for scientific visualization.

Acknowledgements. This research work is sponsored by National Fundamental Research Project (Grant No: 2002CB312100), Excellent Youth NSF in Zhejiang Province (RC00048) and Excellent Youth Teacher Awards of MOE in China.

References

1. Ian Foster, Carl Kesselman (eds.): The Grid: Blueprint for a Future Computing Infrastructure, Morgan Kaufmann Publishers, USA, 1999
2. John Shalf and, E.es Bethel: The Grid and Future Visualization System Architectures, Computer Graphics and Applications, IEEE Press, 2003, 23(2), pp. 6-9
3. I. Foster and C. Kesselman:Globus: A metacomputing infrastructure toolkit, International Journal of Supercomputer Applications, 1997, 11(2), pp. 4-18
4. D. Kranzlm¡§uller, G. Kurka, P. Heinzlreiter, J. Volkert: Optimizations in the Grid Visualization Kernel, Proc. PDIVM Workshop on Parallel and Distributed Computing in Image Processing, 2002.
5. Alan Norton,Alyn Rockwood:Enabling view-dependent progressive volume visualization on the grid, Computer Graphics and Applications, IEEE Press , 2003, 23(2), pp.22-31.
6. M. Aeschlimann, P. Dinda, J. Lopez, B. Lowekamp, L. Kallivokas, and D. O'Hallaron:Preliminary report on the design of a framework for distributed visualization, In Proceedings of the International Conference on Parallel and Distributed Processing Techniques and Applications (PDPTA'99), pp 1833-1839, Las Vegas, NV, June 1999.
7. SGI: SGI on the Grid: Unique Capablilities for Grid Compuing, SGI White Paper, http://www.sgi.com/Products/PDF/3307.pdf.
8. R. B. Haber, D. A. McNabb:Visualization Idioms: A Conceptual Model for Scientific Visualization Systems, Visualization in Scientific Computing, IEEE, 1996, pp. 74-93.
9. Wes Bethel:Visualization Viewpoints, Computer Graphics and Applications, IEEE Press, 2000, 20 (5), pp 17-20.
10. M.A. Baker , H.Ong:Grid Enabled Liquid Crystal Structure Modelling and Visualization Web Portal, DSG Technical Report 2002.02, January 2002.
11. Mark A. Duchaineau, Martin Bertram, Serban Porumbescu, Bernd Hamann, Kenneth I. Joy:Interactive Display Of Surfaces Using Subdivision Surfaces And Wavelets, Proceedings of 16th Spring Conference on Computer Graphics, Comenius University, Bratislava, Slovak Republic, 2001,pp 22-34.
12. Zhigeng Pan, Qizhi Yu:Visualization and Immersive VR Applications for Grid Environments, In Proceedings of the International Workshop on Grid and Cooperative Computing(GCC2002), 2002, pp:1148-1156.
13. L. Lippert, M. H. Gross, C. Kurmann:Compression domain volume rendering for distributed environments, In Proceedings of Eurographics '97, 1997, pp.95-107.
14. T. Chiueh, C. Yang, T. He, H. Pfister, and A. Kaufman: Integrated volume compression and visualization, In Proceedings of the IEEE Visualization 97, 1997, pp.329-336.
15. Zhigeng Pan, Mingmin Zhang, Kun Zhou, (et al):Level of Detail and Multi-resolution Modeling for Virtual Prototyping, International Journal of Image and Graphics, 2001, 1(2), pp.329-343.
16. Chiyi Cheng, Mingmin Zhang, Zhigeng Pan:Multi-resolution Modeling for Virtual Design, The International Journal of Virtual Reality, 2000,4(4),pp.52-56.
17. Zhigeng Pan, Xiaohu Ma, Jiaoying Shi, "An automatic generation algorithm of LOD models in virtual environment", Journal of Software, 1996,7 (9):pp 526-531
18. B. Gregorski, M. A. Duchaineau, P. Lindstrom, V. Pascucci, K. I. Joy:Interactive View-Dependent Rendering Of Large Isosurfaces, in Proceedings of the IEEE Visualization 2002, October/November 27-1, 2002.

19. Y. Livnat and C. Hansen, View dependent isosurface extraction, In Proceedings of IEEE Visualization '98, 1998, pp. 175-180.
20. S. Coorg and S. Teller: Real-time occlusion culling for models with large occluders, In Proc. of the ACM Symposium on Interactive 3D Graphics, 1997, pp 83-90.
21. J. Shalf , E. W. Bethel:Cactus and Visapult: An Ultra-High Performance Grid-Distributed Visualization Architecture Using Connectionless Protocols, IEEE Computer Graphics and Applications, 2003,23(2), pp.51-59.
22. B. Lee Tierney, B. Crowley, M. Holding, J. Hylton, F. Drake:A Network-Aware Distributed Storage Cache for Data Intensive Environments, Proceedings of IEEE High Performance Distributed Computting conference, August 1999.
23. G. Humphreys, M. Houston, R. Ng, R. Frank, S. Ahern, P. D. Kirchner, J. T. Klosowski:Chromium: A Stream-Processing Framework for Interactive Rendering on Clusters, ACM Transactions on Graphics (proc. Siggraph 2002), 21(3), pp.693-702.
24. J. Wood, K. Brodlie, and H. Wright:Visualization over the World Wide Web and Its Application to Environmental£-Data, Proc. IEEE Conf. Visualization 1996, IEEE CS Press, 1996, pp. 81-86.
25. C.Bajaj and S. Cutchin: Web-Based Collaborative Visualization of Distributed and Parallel Simulation, Proc. 1999 IEEE Parallel Visualization and Graphics Symposium, IEEE CS Press, 1999, pp. 47-54.

Avenues for High Performance Computation on a PC

Yu-Fai Fung[1], M. Fikret Ercan[2], Wai-Leung Cheung[1], and Gujit Singh[1]

[1] Department of Electrical Engineering
The Hong Kong Polytechnic University
Hung Hom, Kowloon, Hong Kong
eeyffung@polyu.edu.hk

[2] School of Electrical and Electronic Engineering, Singapore Polytechnic, Singapore
mfercan@sp.edu.sg

Abstract. Modern microprocessors used in personal computers have built in parallel computing features. In general, these microprocessors support Single Instruction Multiple Data (SIMD) type parallel computing mechanism. Recently, dual-CPU systems are becoming more common and therefore, it is possible to combine the SIMD mechanism and the computing power that comes with two microprocessors to form a two-level parallel computing system. In this paper, we present combined utilization of two types of parallelism and demonstrate its performance on a computationally intensive image processing algorithm. Our results show that utilizing the bi-level parallel mechanism, a far superior speed-up can be achieved than those by using only two CPUs.

1 Introduction

Modern microprocessors are becoming more and more powerful with the advances made in technology. Apparently, the very high operating speed of these devices is a clear indicator of this advancement. Currently, a typical personal computer is equipped with a microprocessor operating at frequencies above 2GHz. In addition, many advanced features are embedded to microprocessors in order to strengthen their computation capacity. The SIMD parallel mechanism is one such feature and is available in the two most popular microprocessor families in the PC market, namely Intel and AMD. For Intel, the SIMD mechanism is called the SSE (Streaming SIMD Extensions) [4], and in AMD microprocessors, it is the 3D Now! Technology [2]. Other examples include the SunSparc microprocessor [11] and the PowerPC [1].

The SSE mechanism allows 32-bit floating-point (as well as 8-bit, or 16-bit integer values) to be computed in parallel in its 128 bit registers. In the Pentium 4 processors, the SSE2 mechanism can also support 64-bit floating values. The support of floating-point operations broadens the application possibilities. For instance, Strey, A and Bange M. [10], reported implementation of neural network with SSE. In our earlier study, we demonstrated the application of SSE when developing a software electric railway simulator [5].

The gain in performance with SSE might be satisfactory when the problem size is small. However, to further improve the performance, a multiple-CPU platform to gether with the SSE mechanism is considered. As the cost of dual-CPU systems is becoming more affordable, therefore, it is feasible to combine the two different kinds of parallel mechanism to form a bi-level parallel system on a common personal computer. The term bi-level is used to express two forms of parallelism incorporated in this system. The SSE feature is embedded in the chip-level, whilst, the dual-CPU mechanism is at board level. In this paper, we will discuss how such a bi-level system can be implemented and we will present the performance achieved. Experiments presented in this paper are based on Intel microprocessors and its SSE/SSE2 features. It is also possible to perform similar application with the AMD processors.

2 Bi-level Parallel System

The first level of parallelism is the SIMD mechanism, which is embedded in the microprocessor and therefore regarded as the chip-level parallelism. The second level is based on the utilization of multiple CPUs and this is supported by the mother-board so it will be regarded as the board level parallelism.

The SIMD mechanisms supported by the Intel microprocessors include the SSE and SSE2 (SSE2 is only available in Pentium 4 series). The SSE mechanism is supported by the Xmm registers, which are 128-bit wide and there are eight Xmm registers. The registers can be directly addressed and utilized through a suitable programming tool, for example with the Intel compiler. The Xmm registers, or the SSE registers, can hold four 32-bit floating-point values. SIMD operations, such as add, subtract, multiply, and divide, can be applied to data (four 32-bit floating-point values) stored in two registers and a complete list of SIMD operations that are available can be found in [7]. Theoretically, a maximum speedup ratio of four can be achieved for floating-point operations. The SIMD feature can be invoked directly with assembly codes, which could be embedded in standard C/C++, or even Fortran, programs. On the other hand, the Intel compiler supports special data types that are devised to utilize the SSE mechanism and by employing the new data types no assembly language programming is needed and this is a simpler way to make use of the SIMD feature.

The new data type provided by the Intel compiler is called F32vec4, which represents a 128-bit storage that can store four 32-bit floating-point values; similarly, there is the F64vec2 type for storing two 64-bit double precision values. The F32vec4 is a C++ class and therefore can be used directly in any C/C++ program. In addition, there are functions provided to load, as well unload, data into, or from, the F32vec4 class. Data that are represented by a F32vec4 type can be processed in parallel, i.e. a set of four floating-point values can be manipulated in a single operation. For example, if data A, B, and C are all of the type F32vec4 then in our program, we can write down the statement C=A*B, which represents the multiplication of two sets (A and B) of four floating-point values and saving the results into C. From the example, we can observe that with the new data type, SIMD programming can be achieved in a very straight-forward manner. If a high-precision result is desired then we can apply the F64vec2 type and the steps required to carry out the calculation are the same as those in the F32vec4 case.

Board-level parallelism refers to systems that can support multiple CPUs. In the personal computer market, dual-CPU systems are becoming more common because its cost is reducing. Certainly, systems that can support four CPUs are also available, however, they are usually used as system server and are very expensive. Here, we concentrate on dual-CPU system due to its affordability though methods can easily be extended to four CPU systems.

The programming mechanism for a dual-CPU system is usually achieved by threading [9]. Multiple-thread programming is an efficient way to implement parallel programs on a SMP (Symmetrical Multi-Processor) system. A thread consists of a stream of control that can execute its instructions independently and therefore, a multi-threaded program can perform numerous tasks concurrently. In a SMP system, each thread can be executed by one processor so parallelism can be achieved. Multi-threaded programming provides a direct method for implementing software on a SMP system and function libraries, such as the POSIX library (pthread) [9], and the Win-Thread library [3], are available to support the implementation of software using thread. In our study, the POSIX Thread library is applied in our tests because our SMP computer is running the Linux operating system.

The combining of SSE and threading is a straight forward process. Each CPU will execute a thread and the thread will invoke a function, which will make use of the F32vec4 data type, as well as other SSE features. Therefore, we will have two functions executing simultaneously and both functions will be performing SIMD type parallel operations with SSE so a bi-level parallel system is formed.

3 Application with Hough Transform

Hough transform is widely used technique for detecting linear line segments or circular objects in an image. The computation for Hough transform has two phases. The first phase is a voting process where the result is accumulated in a parameter space based on the transform given in equation (1). In the second phase, the parameter space is elaborated and strong candidates are selected. The voting phase in line detection involves calculating the candidate lines which are represented in terms of parameters by using equation (1).

$$r = x\cos\theta + y\sin\theta \tag{1}$$

The parameter space is (r, θ) and for every point (x, y) in the original image space, a line is formed in the parameter space. With x and y being constants, different values of θ are put into equation (1) in order to obtain a series of r values in the parameter space.

We can utilize SSE registers to hold values of x, y, cosθ, and sinθ and carry out the calculation in parallel. Partitioning the processing tasks to different CPUs can be achieved by assigning different CPU to operate on different areas of the image and store the results in a local parameter space storage. When both CPUs terminate their processes then both local parameter spaces are combined to form a global space, from which the strong candidates are selected.

4 Experimental Results

We have conducted our tests on a system consists of two Pentium 4 microprocessors and running the Linux operating system. We observed speed-up obtained under eight different conditions and these include:

Float – 32-bit floating point is used in the calculations
Double – 64-bit floating point is used
SSE – SSE mechanism is used
SSE2 – SSE2 mechanism is used
Float with 2 CPU – using two CPUs but only 32-bit floating point is used
Double with 2 CPU – using two CPUs and 64-bit floating point
SSE and 2CPU – bi-level but only using 32-bit floating point with SSE
SSE2 and 2CPU – bi-level and using 64-bit floating point with SSE2

Hough transform is usually applied in image processing and the image size used in our tests is 1024x1024. As the complexity of Hough transform is related to the number of non-zero pixels in the image, therefore, in our tests, non-zero elements are located randomly in the image and different percentages of non-zero elements are tested.

The performance index is best represented by the speedup ratio under different conditions and it is given in Table 1. The speedup ratios for different cases are obtained by comparing the timing results to either to the float, or double, method.

The theoretical speedup ratio for the Bi-Level Parallel Algorithm (BLPA) should be 8 for cases involving SSE and 4 if SSE2 is applied. However, factors such as time to create threads, cache misses, and memory contention will hinder the performance. The performance of BLPA is reflected explicitly in the Hough transform algorithm, for which the speedup ratios obtained from BLPA are better than other approaches in all cases and the speedup ratios obtained are consistent among all cases as well. Using BLPA with SSE, the speedup is twice that of using SSE on itself and this demonstrates the advantage of combining the two types of parallel mechanisms. In Hough transform, the non-zero pixels are generated randomly and therefore, with each CPU processes half of the image, the processing steps executed by each CPU will be different. This will reduce the memory contention and improve the performance.

By using VTUNE [6], a performance monitor software from Intel, we can examine the L2 cache read misses frequency as well as the resource stalls, both are indicators for performance. When the size of the matrix is 400, or 800, then the frequency of the L2 cache misses (in term of percentage) increases sharply, the measured value is over 90%, as shown in Figure 1. For cases that involve two CPUs, we examined the resources stall events (in percentage) and the results are shown in Figure 2. As shown in Figure 2, the resource stalls events are more frequent for the two CPUs situation especially when the scale of the problem increases. Therefore, the problem size is a major factor affecting the overall performance and our results substantiate such findings.

Table 1. Speedup results for Hough transform for images with various percentage of non-zero pixel contents

	10%	20%	30%	40%
SSE vs. float	2.12	2.17	2.18	2.20
SSE2 vs. double	2.02	1.99	1.97	1.95
Float (2 CPUs) vs. float	1.98	1.99	1.97	1.99
Double (2 CPUs) vs. double	1.80	1.82	1.78	1.78
Bi-level (SSE) vs. float	3.83	3.80	3.80	3.82
Bi-level (SSE2) vs. double	2.60	2.60	2.56	2.54

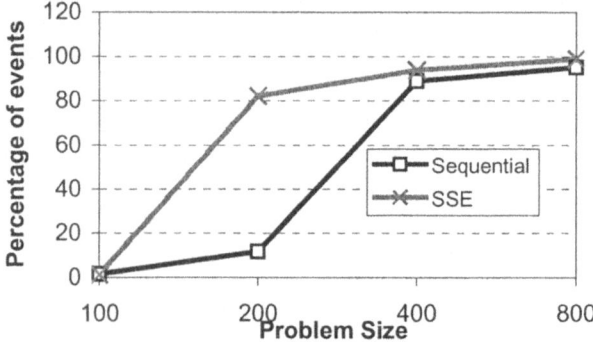

Fig. 1. L2 Comparing Cache Misses frequency for different conditions

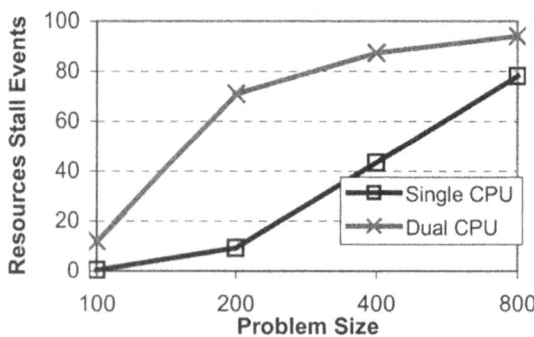

Fig. 2. Comparing Resource stalls events for different conditions

5 Conclusion

In this paper, we have investigated the bi-level parallel mechanism that involves chip-level, the SSE, or SSE2, mechanism, as well as board-level, multiple CPUs, parallel processing. The hardware requirement for implementing such a parallel system is a dual-CPU machine and this can now be easily obtain at an affordable price, therefore, the bi-level mechanism could become a standard approach for performance optimization. We implemented a computationally demanding image processing algorithm to experiment the speed-up that can be achieved. According to our results, we can conclude that the bi-level mechanism can improve the performance of the algorithm when comparing to performance obtained by utilizing only the board-level parallelism. In addition, the hyper-threading technology [8] that is now available in Intel's latest processors may further improve the performance and can be incorporated into the bi-level mechanism.

References

1. AltiVec Programming Environments Manual, Motorola, 2001.
2. AMD64 Architecture Programmer's Manual Volume 4: 128-bit Media Instructions, AMD Inc. (2003)
3. Beveridge J. E., Wiener R., Multithreading Applications in Win32, Addison-Wesley (1997)
4. Conte G., Tommesani S., Zanichelli F., The Long and Winding Road to High-performance Image Processing with MMX/SSE, IEEE Int'l Workshop for Computer Architectures for Machine(2000) 302-310
5. Ding Y., Ho T., Fung Y., Liu H., and Mao B., Applications of SSE on electrified railway train operation simulation, Journal of the China Railway Society, Vol. 25(2003)19-22
6. Gerber R., The Software Optimization Cookbook, Intel Press (2002)
7. Intel C/C++ Compiler Class Libraries for SIMD Operations User's Guide (2000)
8. Intel website: http://www.intel.com/technology/hyperthread/
9. Lewis B., Berg D.J., Multithreaded Programming with Pthreads, Sun Microsystems Press (1998)
10. Strey A., and Bange M., Performance Analysis of Intel's MMX and SSE: A Case Study, Lecture Notes in Computer Science, Vol. 2150(2001)142-147
11. VIS Instruction Set User Manual, Sun Microsystems Inc.(2001)

A Modified Parallel Computation Model Based on Cluster

Xiaotu Li, Jizhou Sun, Jiawan Zhang, Zhaohui Qi, and Gang Li

School of Electronic Information Engineering, Tianjin University, Tianjin 300072,
P.R.China,
tuteddy_lee@yahoo.com

Abstract. Based on cluster, a modified parallel computation model MSG (Master-Slave-Gleaner) model is presented. MSG model can be used as a parallel computation model to develop high performance volume rendering algorithms. A very complex parallel splatting algorithm was used to test the efficiency and the practicability of the MSG model in IBM Cluster 1350. The results of theoretical analysis and numerical testing showed that the modified computing model presented can decrease computing time, increase accelerator and improve computing efficiency greatly. In addition, the MSG model is suitable for a wide variety of volume rendering algorithms and has good scalability.

1 Introduction

In volume rendering, when the scene is complex, or when high-quality images or high frame rates are required, the rendering process becomes computationally demanding. To provide the necessary levels of performance, parallel computing techniques must be brought to bear [1].

The Master-Slave model is often used as parallel and distributed computation model to develop high performance volume rendering algorithms. However, in the parallel algorithm, the Master not only has to divide and allocate the tasks but also has to collect the results and rendering the graphics. As a result, the communications traffic is increased greatly and the bottleneck problem is brought. In this paper, the MSG (Master-Slave-Gleaner) model, a modified cluster computing model, was introduced by the authors for parallel volume rendering algorithms. Then a very complex parallel splatting algorithm was used to test the efficiency and the practicability of the MSG model in IBM Cluster 1350. By analyzing the results of the practical testing, the parallel computing accelerator was increased and computing time was decreased effectively.

2 Volume Rendering Algorithms Demand Parallel Computing

In computer graphics, rendering is the process by which an abstract description of a scene is converted to an image. For purposes of this discussion, a scene is a collection of geometrically-defined objects in three-dimensional object space, with associated lighting and viewing parameters. The rendering operation illuminates the objects and projects them into two-dimensional image space, where color intensities of individual pixels are computed to yield a final image.

For complex scenes or high-quality images, the rendering process is computationally intensive, requiring millions or billions of floating-point and integer operations for each image. The need for interactive or real-time response in many applications places additional demands on processing power. The only practical way to obtain the needed computational power is to exploit multiple processing units to speed up the rendering task, a concept which has become known as parallel rendering.

Parallel techniques are appropriate whenever rendering performance is an issue. Demanding applications such as real-time simulation, animation, virtual reality, photo-realistic imaging, and scientific visualization all benefit from the use of parallelism to increase rendering performance. Indeed, these applications have been primary motivators in the development of parallel rendering methods. Parallel rendering has been applied to virtually every image generation technique used in computer graphics, including surface and polygon rendering, terrain rendering, volume rendering, ray-tracing, and radiosity. Although the requirements and approaches vary for each of these cases, there are a number of concepts which are important in understanding how parallelism applies to the generic rendering problem. Instead of performing a sequence of rendering functions on a single data stream, it may be preferable to split the data into multiple streams and operate on several items simultaneously by replicating a number of identical rendering units.

The design of effective parallel rendering algorithms can be a challenging task. In some cases, existing sequential algorithms have straightforward parallel decompositions. In other cases, new algorithms must be developed from scratch. Whatever their origin, most parallel algorithms introduce overheads which are not present in their sequential counterparts. These overheads may result from some or all of the following:1)communication among tasks or processors; 2)delays due to uneven workloads; 3)additional or redundant computations; 4)increased storage requirements for replicated or auxiliary data structures.

3 Parallel Computing and Cluster Technology

The term cluster has become one of the most-used buzz words in the IT industry. Yet it is also one of the most misunderstood, or at least, misused terms as well. One of the reasons for this is that there are many types of clusters and clusters can be used for many different purposes.

In its simplest form, a cluster is two or more computers that work together to provide a solution. This should not be confused with a more common client server model of computing where an application may be logically divided such that one or more clients request services of one or more servers. The idea behind clusters is to join the computing powers of the nodes involved to provide higher scalability, more combined computing power, or to build in redundancy to provide higher availability. So rather than a simple client making requests of one or more servers, clusters utilize multiple machines to provide a more powerful computing environment through a single system image. Clusters of computers must be somewhat self-aware. That is, the work being done on a specific node often must be coordinated with the work being done on other nodes. This can result in complex connectivity configurations and sophisticated inter-process communications between the nodes of a cluster. In addition, the sharing of data between the nodes of a cluster through a common file system is almost always a requirement.

As mentioned, clusters are typically made up of a large number of computers (often called nodes). Those nodes are configured in a way that they can perform different logical functions in a cluster. Figure.1[2] presents the Beowulf logical architecture of cluster. A typical logical structure of a cluster provides the following parts: compute nodes with interconnect, operation system such as Linux, message passing library, parallel applications, management tools, manage node and local area network. Clusters come in many different forms. The three most common cluster types are high availability (HA), high performance computing (HPC), and horizontal scaling (HS). However, there are also obvious, but not so

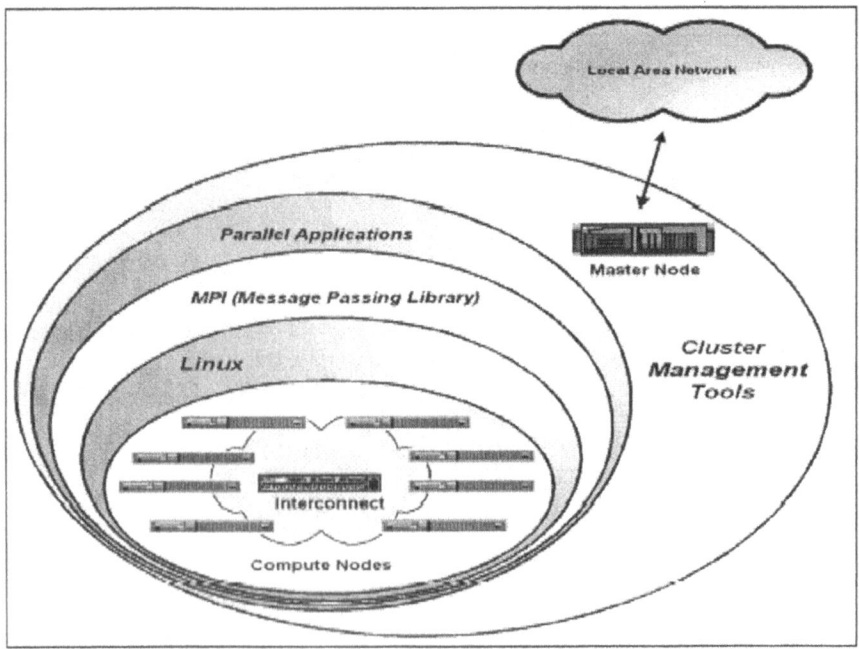

Fig. 1. The Beowulf logical architecture of cluster

obvious costs involved. In a large high performance computing cluster, the idea is to have as simple and efficient of an environment as possible to maximize the work that can be done on individual nodes. Operating systems such as Linux, which can be obtained virtually for free, provide a very economical solution to operating system licensing on large numbers of nodes.

In addition, with the popularity of Linux, there are many tools, utilities, and applications available to help build and manage a cluster. Many of these programs are available either for free or for a very reasonable cost. Using economical software such as Linux and Linux-based programs on Intel-based hardware such as IBM xSeries rack-mounted servers makes the overall cost of implementing a cluster much lower than it has been in the past.

Therefore, many enterprises are starting to make use of such clusters in environments where they have previously not been justified. Clusters (especially high performance clusters) typically have a large number of computers (often called nodes) and in general most of these nodes would be configured identically: The idea being that the individual tasks that make up a parallel application should run equally well on whatever node they are dispatched on.

4 Research and Implementation of Parallel Computation Model

When programmers develop parallel programs, the process must be used. Especially when the programs are tested on the platform which has some processing nodes, several processes may be mapped to one processor node. There are two ways to set up processes. One is static mode and the other is dynamic mode.

In static mode, all processes have to be assigned by programmers before being executed. And then these processes will be executed by the system. Programmers should sign them by using command line when the processes are being executed or before. In most applications, processes are neither identical nor different totally.

Generally, in a parallel program, there is a control process and we call it master process. The others are called workmen or slave processes. There are no essential differences between master process and slave process.

In the SPMD (Single program multiple data) mode, different processes are merged into one single program. The codes are customized by the control statements in the program. The control statements assign corresponding parts for each process. After the source file, in which the control statements have allocated the tasks for all processes, is developed, it will be compiled and changes to codes which can be executed by all processors (Fig.2[3]). Each processor will load the code in its local memory and execute. In the paper, IBM Cluster1350 which has a master node and 16 computation nodes and is interconnected by high speed Ethernet was used as the parallel environment. Considering that volume rendering algorithms depend much on computation resources and memory resources, the authors distributed these resource demands to the cluster node.

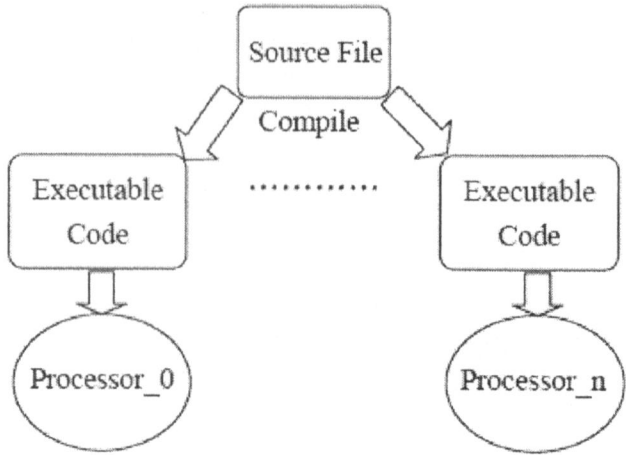

Fig. 2. Single program multiple data operation

4.1 The Master-Slave Model

In cluster environment, the Master-Slave model is often used as parallel and distributed computation model to develop high performance volume rendering algorithms. The architecture of the master-slave model is shown in Fig.3. Master-

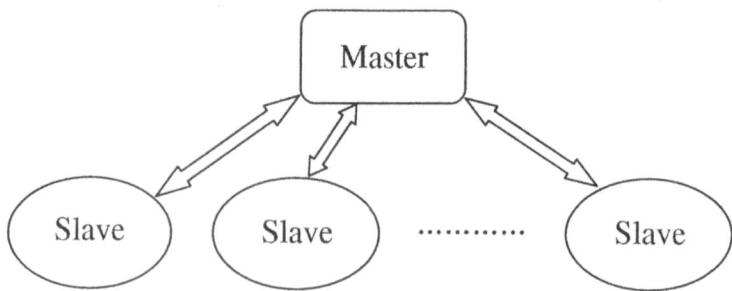

Fig. 3. The architecture of Master-Slave Model

slave tasking is a simple yet widely used technique to execute independent tasks under the centralized supervision of a control processor[4]. As one of parallel computation models, master-slave model is better[5]. The merits are:1)supports parallel in task level; 2)supports parallel in algorithm level; 3)has good stability and is easy to load; 4) has good flexibility and scalability.

However, it can be seen easily form the figure above, in the parallel algorithm, the Master not only has to divide and allocate the tasks but also has to collect the results and rendering the graphics. Further more the characteristics of low bandwidth and high communication delay exist also in cluster system. As a result, the communications traffic is increased greatly and the bottleneck problem is brought. To improve the performance of the model and parallel volume rendering algorithm, a modified model MSG (Master-Slave-Gleaner) model was designed and studied by authors.

4.2 The MSG Model

Considering the characteristics of parallel volume rendering algorithms, the authors modified the master-slave model. The MSG (Master-Slave-Gleaner) model was presented. Fig.4. shows the architecture of the master-slave-gleaner model. As shown in Fig.4, in the MSG model, the tasks are divided and assigned further.

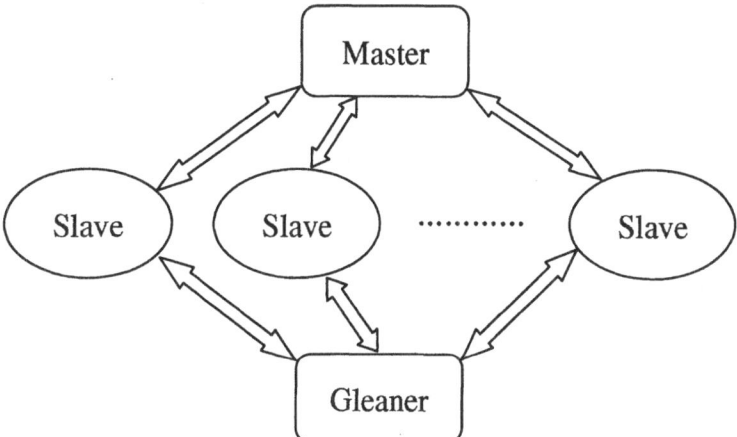

Fig. 4. The architecture of the Master-Slave-Gleaner Model

Master node divides and allocates the tasks to salve nodes. Slave nodes implement the volume rendering and computing tasks. Then the results processed are sent to the gleaner node. The gleaner node collects the results and rendering the graphics.

It is well known form the figure above, in the MSG model, allocating tasks and collecting processed results are independent. This can effectively alleviate the traffic load and decrease the communication traffic in master node. As a result, the communication bandwidth was expanded and the parallel computing efficiency was increased greatly. The MSG model has all the merits of the master-slave model and is simple and suitable for a wide variety of volume rendering algorithms. In addition the MSG model has good practicability and scalability.

In section 5, a very complex parallel splatting[6] algorithm will be used to test the efficiency and the practicability of the MSG model based on IBM Cluster 1350 system. By analyzing the result of the practical testing, the characteristics of the MSG model will be proved in the following.

5 Experiments and Analysis

The IBM Cluster 1350 is a new Linux cluster offering. It is a consolidation and a follow-on of the IBM Cluster 1300 and the IBM xSeries "custom-order" Linux cluster offering delivered by IGS[2]. This new offering provides greater flexibility, improved price/performance with Intel XeonTM processor-based servers (new xServer models x345 and x335). The Cluster 1350 is targeted at the High-Performance Computing field, with its main focus on a lot of industries. Also, with its high degree of scalability and centralized manageability, the Cluster 1350 is ideally suited for Grid solutions implementations.

There are 16 computing nodes (2.4GHz Intel Xeon X335, 512M*2DDR, 18.2G) and a manage node (2.4GHz Intel Xeon X345, 2*512M DDR, 18.2G) with 10/100Mbps Ethernet in our cluster system. The software platform is Linux (RedHat7.3). Computing nodes connect one another by two independent intercommunication switch hubs (100M, 16 ports) in TCP/IP Ethernet.

A splatting program, a kind of volume rendering algorithms, which has large complex volume data sets, was used to test the performance of the MSG model. Firstly the splatting program was paralleled using master-slave model and tested under the IBM Cluster 1350.Then the same program was paralleled using master-slave-gleaner model and tested under the same system.

In the experiment, there are several important parameters£(1) computing time of the parallel program£(2) total computing time of the system£(3) the absolute velocity of program using different model. It should be noticed that in this paper the absolute velocity of program was defined as the ratio of data size to computing time. The unit of time is second and data size is megabit. The absolute velocity is expressed by numerical part ofthe ratio directly. Each experiment had been done five times. The final result was the average value.

According to the time parameters and the amount of data sets, the authors got the absolute velocity of parallel program. Figure.5 shows graphics rendered by parallel spallting program. Figure.6 compares the absolute velocities of parallel program using master-slave model and master-slave-gleaner model from 1 to 16 computing nodes. In Figure.6, the horizontal axis measures the number of working nodes and vertical axis measures the absolute velocity. Curve1, 2 are the absolute velocity curve of parallel splatting program using master-slave-gleaner model and the curve of one using master-slave model respectively.

It has been observed from the graph that, there is a continual increase in the absolute velocity between 1 and 16 working nodes in these two curves. Furthermore the absolute velocity of MSG model parallel program is much higher than

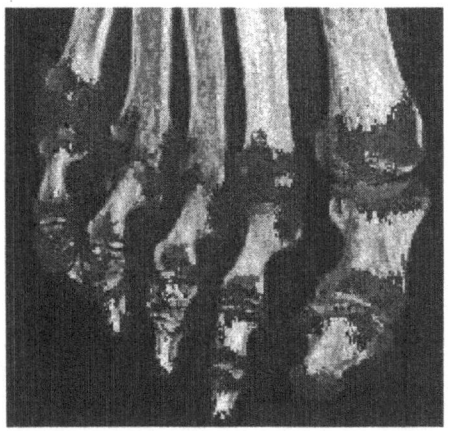

Fig. 5. The graphics rendered(MRI FOOT)

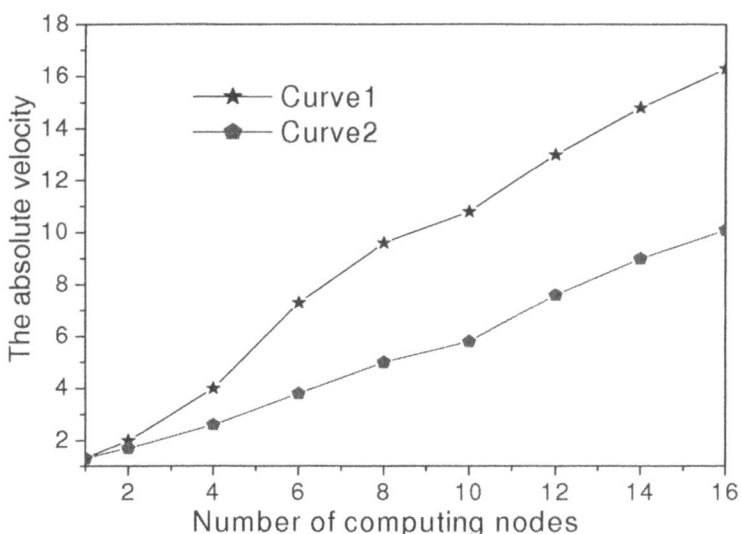

Fig. 6. The absolute velocity of parallel program using different model

that of master-slave one. For example, when there are 8 computing nodes in the cluster, the absolute velocity of MSG model parallel program is about 9.63 and the one of master-slave model parallel program about 5.01. The absolute velocity of MSG model parallel program is about 1.92 times than that of master-slave model. When there are 16 computing nodes in the system, the absolute velocity of MSG model parallel program and master-salve model parallel program are about 16.39 and 10.17 respectively. The absolute velocity of MSG model parallel program is 1.616 times than that of master-slave model.

This is due to that in master-slave model parallel program, the master node not only has to divide and allocate the data sets but also has to collect the results and then render the graphics. No doubt, the communications traffic between master and slave is increased greatly and the bottleneck problem is brought. However, in MSG model parallel program, the data sets are divided and assigned further. Master node divides and allocates the tasks to salve nodes. Slave nodes implement the volume rendering and computing tasks. Then the results processed are sent to the gleaner node. The gleaner node collects the results and rendering the graphics. Allocating tasks and collecting processed results are independent. This can effectively alleviate the traffic load and decrease the communication traffic in master node. As a result, the communication bandwidth was expanded and the parallel computing efficiency was increased effectively.

But in fig.6, it should not be neglected was that the slope of these two curves increased first and then reduced gradually. The pivotal point is the one when the number of working node is ten. This is because when the number of working nodes is increased excessively the communication traffic is aggravated. So in practice problems, we should study systematically to get optimum number of working node and optimum performance of parallel program.

The MSG model has all the merits of the master-slave model and is simple and suitable for a wide variety of volume rendering algorithms. Also the MSG model has good practicability and scalability. If needed, the number of gleaner node can be two or more. But this does not mean that more gleaner nodes can bring better performance of parallel volume rendering program. We ought to study this further in practical projects.

6 Conclusions

Based on cluster, a modified parallel computation model MSG model was designed by authors to support high performance volume rendering algorithms. A very complex parallel splatting algorithm was used to test the efficiency and the practicability of the MSG model in IBM Cluster 1350. The merits of the MSG model were proved by the results obtained. The parallel computing accelerator increased and computing time decreased greatly. And the communication bandwidth can be expanded and the parallel computing efficiency can be increased markedly. In addition the MSG model is suitable for a wide variety of volume rendering algorithms and has good scalability.

The characteristics of the MSG model are (1) a parallel computation model which can easily prompt the performance of volume rendering algorithms; (2) the Gleaner node was added to "glean" the results; (3) alleviation of traffic overload; (4) simple architecture and good scalability.

Acknowledgements. This work was supported by Natural Science Foundation of China under grant No. 60373061. The authors gratefully acknowledge the valuable comments and suggestions of their associates in Parallel Programming Research in CADC Group. Special thanks are expressed to Jiawan Zhang and Ce Yu for their interesting and rewarding comments.

References

1. Thomas W. Crockett,Parallel Rendering,Institute for Computer Applications in Science and Engineering NASA Langley Research Center, Pages: i-ii.
2. Ibm.com/redbooks, Building a Linux HPC Cluster with xCAT, Pages: 4
3. Addision-Wesley , Barry Wilkinson & Michael Allen,Parallel Programming:Techniques and Applications Using Networked Workstations and Parallel Computers,North Carolina, Pages:32
4. Olivier Beaumont, Arnaud Legrand, and Yves Robert,The Master-Slave Paradigm with Heterogeneous Processors,IEEE TRANSACTIONS ON PARALLEL AND DISTRIBUTED SYSTEMS, VOL. 14, NO. 9, SEPTEMBER (2003).
5. G.A. Geist, A. Beguelin, J. Dongorra, W.Jiang, R.Manchek, V.S.Sunderam, PVM:Parallel Virtual Machine - A User's Guide and Tutorial for Networked Parallel Computing, MIT Press, (1994)
6. Westover L.A. Footprint evaluation for volume rendering. Computer Graphics, (1990)14(4):367–376

Parallel Testing Method by Partitioning Circuit Based on the Exhaustive Test*

Wu Woan Kim[1]

[1] Division of Information and Communication Engineering,
Kyungnam University, Masan, South Korea
wukim@zeus.kyungnam.ac.kr

Abstract. This paper presents an approach, which is applicable to parallel testing, to the generation of test patterns using the partitioning technique based on the exhaustive testing scheme. The suggested method can discover faults faster than the exhaustive testing scheme. It also shows that testing can be performed in parallel using the functionally partitioned blocks, rather than in linear. In this paper, a functional level description as well as the Boolean differential is used to generate a test pattern that can be inserted into in parallel.

1 Introduction

In recent years, the complexity of digital systems has been increased dramatically. Although semiconductor manufacturers try to ensure that their products are reliable, it is almost impossible not to have faults somewhere in a system at any given period[1]. As complexity of VLSI circuits increases, more organized and automated methods for test generation become mandatory. Many approaches for test generation are based on search methods on a gate level description of the design[2].

Concern with the high complexity of these methods when applied to VLSI circuit, and the motivation to incorporate testing considerations early in the design cycle, have stimulated research in the area of functional level test generation[2]. Unlikely gate level techniques and the widespread stuck-at fault model, these methods for functional test generation lack a widely accepted fault model which accurately reflects physical defects. Furthermore, it has been observed that it is very difficult to project the results obtained from functional test generation and fault simulation, to the testability of a design's implementation[3].

In general, exhaustive testing schemes partition circuits in conformity to the partitioning technique and perform testing for each of the partitioned ones. Each of the partitioned circuit blocks is represented as a subspace for its input k bits. $2k$ binary patterns are required to test the whole blocks. Many researches into how to generate $2k$ test vectors for the entire k-bit patterns have extensively been performed. But, test-generating methods have some difficulty in being applied to complex LSI and VLSI

---

* This work was supported by Kyungnam University Research Fund.

circuit chips because time complexity required for generating test sets which are used to detect single Stuck-at faults and complex Stuck-at faults may inevitably be increased sharply as the circuits increase in complexity and size[4][6][7][8].

This paper presents an approach, which is applicable to parallel testing, to the generation of test patterns using the partitioning technique based on the exhaustive testing scheme. The suggested method can be performed in parallel using the functionally partitioned blocks, rather than in linear. Thus, testing can be done through analyzing the relationship among faults that are not dependent on other faults. In Section 2, various kinds of faults and how to detect them are presented. The suggested method for generating a test pattern is given in Section 3. The result of a simulation for the proposed method is presented in Section 4.

2 Testing Methods

A failure means an occurrence of deviations from prescribed actions, and a fault an occurrence of physical defects without having to do with whether any failure is occurred or not. Characteristics of faults can be pointed out in the viewpoint of nature, value, extent and duration. Faults, in their nature, are classified into logical and non-logical ones. A logical fault is one that generates an opposite value from a specified value, and a non-logical fault includes malfunction in clock signals, power failure and something like that. When any fault is referred in point of values, logical faulty values are fixedly yielded or continuously varied is concerned. The extent of a fault shows the influence of an occurred fault is limited to any local space or scattered around a whole circuit, and the duration of a fault represents how long the effect of an occurred fault lasts.

2.1 Boolean Differential

The basic principle of the Boolean difference is to derive two Boolean expressions- one of which represents normal fault-free behavior of the circuit, and the other represents the logical behavior under an assumed single s-a-1 or s-a-0 fault condition. These two expressions are then exclusive-Ored; if the result is 1 a fault is indicated[10][11].

Assume $F(X) = (x_1, \cdots, x_n)$, where $F(X)$ as a logic function of n variables. If there is one fault(x_i) among inputs of logic function, then the output is $F(x_1, \cdots, \overline{x_i}, \cdots, x_n)$. Boolean differential $F(X)$ with respect to x_i is defined as

$$\frac{dF\{x_1, \cdots, x_i, \cdots, x_n\}}{dx_i} = \frac{dF(X)}{dx_i} = F(x_1, \cdots, x_i, \cdots, x_n) \oplus F(x_1, \cdots, \overline{x_i}, \cdots, x_n)$$

The function $\dfrac{dF(X)}{dx_i}$ is called the Boolean difference of $F(X)$ with respect to x_i.

It is easy to see that when $F(x_1, \cdots, x_i, \cdots, x_n) \neq F(x_1, \cdots, \overline{x_i}, \cdots, x_n)$, $\dfrac{dF(X)}{dx_i} = 1$ and that when $F(x_1, \cdots, x_i, \cdots, x_n) = F(x_1, \cdots, \overline{x_i}, \cdots, x_n)$, $\dfrac{dF(X)}{dx_i} = 0$.

To detect a fault on x_i, it is necessary to find input combinations (tests) so that whenever x_i changes to $\overline{x_i}$ (due to a fault), $F(x_1, \cdots, x_i, \cdots, x_n)$ will be different from $F(x_1, \cdots, \overline{x_i}, \cdots, x_n)$. In other words, the aim to find input combinations for each fault occurring on x_i such that $\dfrac{dF(X)}{dx_i} = 1$.

Some useful properties of the Boolean difference are:

$$\overline{\dfrac{dF(X)}{dx_i}} = \dfrac{\overline{dF(X)}}{dx_i} \qquad (1)$$

$$\dfrac{dF(X)}{dx_i} = \dfrac{dF(X)}{d\overline{x_i}} \qquad (2)$$

$$\dfrac{d}{dx_i} \cdot \dfrac{dF(X)}{dx_j} = \dfrac{d}{dx_j} \cdot \dfrac{dF(X)}{dx_i} \qquad (3)$$

$$\dfrac{d|F(X)G(X)|}{dx_i} = F(X)\dfrac{dG(X)}{dx_i} \oplus G(X)\dfrac{dF(X)}{dx_i} \oplus \dfrac{dF(X)}{dx_i} \cdot \dfrac{dG(X)}{dx_i} \qquad (4)$$

$$\dfrac{d|F(X)+G(X)|}{dx_i} = \overline{F(X)}\dfrac{dG(X)}{dx_i} \oplus \overline{G(X)}\dfrac{dF(X)}{dx_i} \oplus \dfrac{dF(X)}{dx_i} \cdot \dfrac{dG(X)}{dx_i} \qquad (5)$$

A Boolean function $F(X)$ is said to be independent of x_i if and only if $F(X)$ is logically invariant under complementation of x_i, i.e. if

$$F(x_1, \cdots, x_i, \cdots, x_n) = F(x_1, \cdots, \overline{x_i}, \cdots, x_n)$$

This implies that a fault in x_i will not affect the final output $F(X)$ and $\dfrac{dF(X)}{dx_i} = 0$. Some additional properties can now be added to the original set (equations 1-5):

$$\dfrac{dF(X)}{dx_i} = 0 \text{ if } F(X) \text{ is independent of } x_i \qquad (6)$$

$$\frac{dF(X)}{dx_i} = 1 \text{ if } F(X) \text{ is depends only on } x_i \qquad (7)$$

$$\frac{d|F(X)G(X)|}{dx_i} = F(X)\frac{dG(X)}{dx_i} \text{ if } F(X) \text{ is independent of } x_i \qquad (8)$$

$$\frac{d|F(X)+G(X)|}{dx_i} = \overline{F(X)}\frac{dG(X)}{dx_i} \text{ if } F(X) \text{ is independent of } x_i \qquad (9)$$

So far, the Boolean difference method has been applied to derive tests for input line faults, but it can also be used for faults on lines which are internal to a circuit.

2.2 Exhaustive Testing

The exhaustive test-generating scheme is recognized to be an alternative to a pseudo random built-in pattern generator. This scheme has a coverage for all combined faults, and requires relatively a little information for the entire test sets[4][5].

A functional-level test generation has been considered for high-level fault models. Generally, it does not use any information on an internal structure of a circuit system.

3 Parallel Testing Using an Exhaustive Test Pattern

In this section, how to generate a pattern in accordance with a method proposed in this paper and how to test using this pattern are explained.

3.1 Partitioning for the Parallel Test

In general, the pseudo exhaustive scheme partitions circuits in conformity to the partitioning technique and performs testing for each of the partitioned ones.

The method suggested in this paper exclusively partitions circuits in the way that a certain fault is not influenced by any fault previously occurred so that parallel testing can be achieved. In this paper, primitive gate-level partitions for a circuit will be adopted. The sample circuit in <Fig. 1> is a 4-bit binary register that includes parallel loads, synchronous clears and increments. The dotted area will only be given a thought when generating test vectors and performing a testing because the dotted area that repeatedly occurs may yield a similar test pattern over and again. The following excerpted figure shows a repeated part of the entire register in <Fig. 1> from which a target test pattern would be generated.

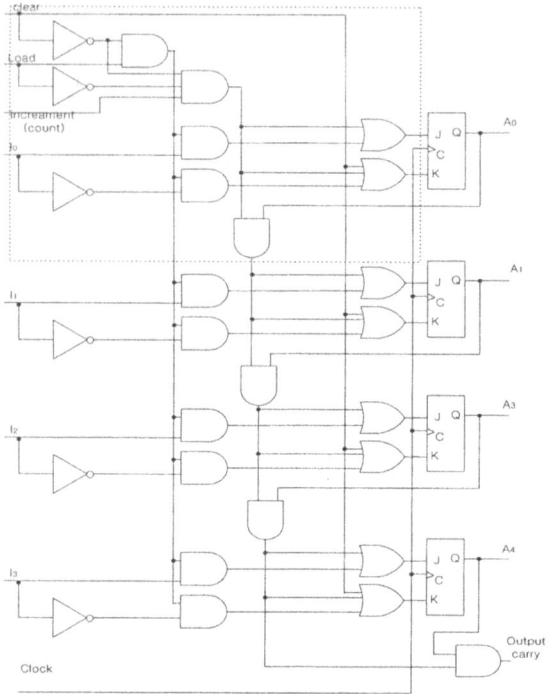

Fig. 1. 4-bit binary register

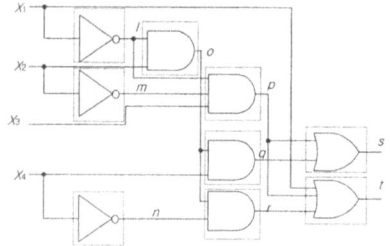

Fig. 2. An Example of Partitioning a Circuit

The more test points lead a complicated circuit, as the positions to be tested shall be increased. It means that deciding the size of partitioning is quite important. From <Fig. 2>, the fault in the entire circuit can be pointed out through 4 phases if it is partitioned on the gate level. *l*, *m* and *n* are the shortest paths on which each of x_1, x_2 and x_4 does not have an influence on the others. Let x_1 be *s-a-1*. This value never affects *m* and *n*. So, faults that may be occurred in the three points can be determined simultaneously. If there are no faults in the three points, the value can be used to find out any faults in the next phase *o* and *p* where faults in x_3, *l* and *m* can be detected. This is the way one or more faults can be found simultaneously.

3.2 Generating a Test Vector Using the Boolean Difference

The fundamental principles of the Boolean difference derive from two Boolean expressions. One represents the fault-free status of a circuit; the other does logical status based on the assumption of either *s-a-1* or *s-a-0* fault. On these two expressions, an exclusive OR (XOR) operation is applied to. If the result is 1, it means that there is a fault. In <Fig. 2>, assume that there is a *s-a-0* at x_1.

$$F(S) = \overline{x_1 x_2} x_3 + \overline{x_1} x_2 x_4 \tag{10}$$

$$\begin{aligned}
\frac{dF(S)}{dx_1} &= \frac{d(\overline{x_1 x_2} x_3 + \overline{x_1} x_2 x_4)}{dx_1} \\
&= \frac{d(\overline{x_1}(\overline{x_2} x_3 + x_2 x_4))}{dx_1} \\
&= \overline{x_1} \frac{d(\overline{x_2} x_3 + x_2 x_4)}{dx_1} \oplus (\overline{x_2} x_3 + x_2 x_4)\frac{d\overline{x_1}}{dx_1} \oplus \frac{d(\overline{x_2} x_3 + x_2 x_4)}{dx_1}\frac{d\overline{x_1}}{dx_1} \\
&= (\overline{x_2} x_3 + x_2 x_4)
\end{aligned} \tag{11}$$

By using the result, $x_1 \frac{dF(S)}{dx_1}$ must be 1. Therefore, if $x_1=1$ and $x_2=1$, x_3 does not affect to this vector, so both 0 and 1 can be the values, and x_4 should be 1, etc..

Then the test vector can be the following 4 sets; {1011}, {1011}, {1101}, {1111}. <Table 1> shows a vector table generated using the Boolean difference. In <Table 1>, there are two columns for *s-a-0* and *s-a-1*. A test pattern will be generated using these values.

Table 1. Stuck-at Test Vector for Each Gate

		s-a-0	s-a-1
Inverter	x_1	{1}	{0}
2-input AND	x_1	{1,1}	{0,1}
	x_2	{1,1}	{1,0}
3-input AND	x_1	{1,1,1}	{0,1,1}
	x_2	{1,1,1}	{1,0,1}
	x_3	{1,1,1}	{1,1,0}
2-input OR	x_1	{1,0}	{0,0}
	x_2	{0.1}	{0,0}
3-input OR	x_1	{1,0,0}	{0,0,0}
	x_2	{0,1,0}	{0,0,0}
	x_3	{0,0,1}	{0,0,0}

3.3 Generating an Exhaustive Test Pattern

As in <Fig. 2>, the circuit is partitioned on the gate level. *l, m, n, o, p, q, r, s* and *t* are detecting points. Multiple values can be input to faults, which are independent with one another. Consider the shortest path, *l, m* and *n*.

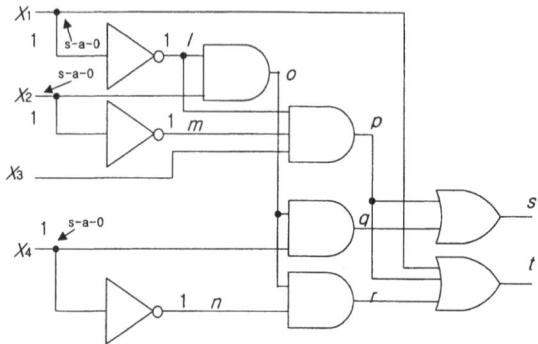

Fig. 3. Multiple Stuck-at Faults

x_1, x_2 and x_3 are input to this point. As these three inputs are independent with one another, three faults can be determined at the same time (<Fig. 3>). If vector is $x_1 x_2 x_3 x_4 = \{11\_1\}$, and *l, m* and *n* yield value 1 respectively, it can be inferred that all of them are s-a-0. This property may be applicable to multiple faults. And, values *o* and *p* in the next phase can be used to quickly identify which point in the previous phase (*l, m*) has produced a fault. Test vector is generated using the gate-level faults, and this result value is used to yield a possible fault vector in the form of combining gates in a nearest path with possibly faulty inputs. Fault testing is performed based on the fault vector.

3.4 Parallel Testing

Values at each point are detected first, and those values are compared with values from a fault-free circuit. If they are not equivalent, it is assumed that a fault has been occurred. A circuit for this comparison can simply be implemented through XOR gates. In this circuit, values in the nearest path are compared with those from fault-free circuit in XOR gate. If the result is normal, the next path is searched. This process is repeated over and again.

4 Implementation

The following <Fig. 4> shows one of interfaces of the implemented system. It represents test patterns, which is automatically generated, for respective faults from inverter, AND and OR gates. Optimized vectors, which are used for detecting faults for their corresponding patterns, are displayed in the left side of the interface. In order to find any faults in the circuit, the shortest path is set first, and then the input to this path is examined. A potential fault-producing pattern from fault patterns for each gate is input to this input path. An optimized pattern can then be determined.

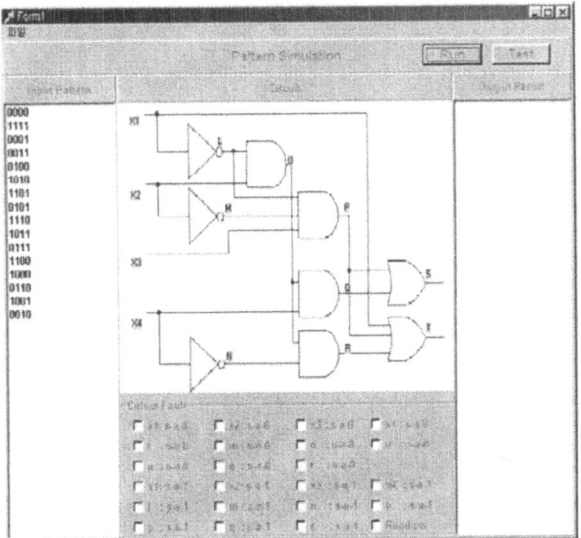

Fig 4. Generate Exhaustive input pattern

For instance, let's consider the situation where position *l*, *m* and *n* have their own input patterns, and try to find out a pattern that can be applied to the three positions at the same time. In this case, *1101* and *1111* would be selected for *s-a-0*. These patterns are practically used as input patterns. In this way, Next patterns that can be applied to the position *o* and *p* at the same time in the next shortest path are determined. Patterns that can be applied to each position at the same time are selected and supplied to input patterns. The <Fig. 5> shows one of interfaces of an implemented system. In the right side, any fault and its location are displayed.

5 Conclusions and Further Researches

An approach, which is applicable to parallel testing, to the generation of test patterns using the partitioning technique based on the exhaustive testing scheme is researched in this paper. The suggested method can discover faults faster than the exhaustive testing scheme because it has been approved to complement the weakness of the exhaustive testing and to be applicable to parallel testing. <Table 2> shows the comparison of the execution time between KSIM [17] and the proposed method in this paper applied to each circuits of ISCA '85. While on the other, the partitioning technique also has drawbacks in that checkpoints as well as inputs for test patterns may be increased when a circuit is partitioned to a slightly small size, and redundancy problems may be accompanied, too. In turn, when a circuit is partitioned to a relatively large size, test patterns may be increased. The optimum partitioning size is important for raising the efficiency of testing.

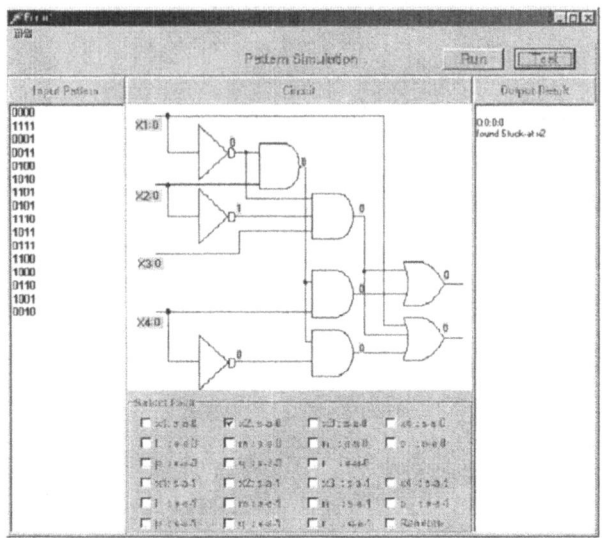

Fig 5. Interface of Implementation

Table 2. The Execution time of KSIM vs. the proposed method applied to ISCAS '85 Benchmarks

Netlist	KSIM(seconds)	Proposed Testing(seconds)
C17	0.03	0.015
C432	0.5	0.25
C499	1.3	0.65
C880	2.4	1.2
C1355	4.3	2.15
C1908	14.46	7.23
C2670	41.0	20.5
C3540	72.8	36.4
C5315	189.6	94.8
C6288	185.2	92.6
C7552	495.3	247.65

The method in this paper is based on gate-level partitioning. If it improves some insufficiency such as partitioning size, the overall performance will fairly be elevated.

References

1. Parag K. Lala, *Fault Tolerant and fault testable hardware design*, Prentice Hall International, New York, New York, 1985.
2. Grabriel M. Seiberman and Ilan Spillinger, "Functional Fault Simulation as a Guide for Biased-Random Test Pattern Generation," IEEE Transactions on Computer VOL. 40, NO. 1 January 1991.
3. S.P. Tomas and J.P. Shen, "A survey of: functional level testing and testability measures," Res. Rep. CMUCAD-83-18, CarnegiMellon Univ., Pittsburgh, PA, 1983.
4. Janusz Rajski and Jerzy Tyszer, "Recursive Pseudoexhaustive Test Pattern Generation," IEEE Transactions on Computer VOL. 42, NO. 12, December 1993.
5. A.K. Das, A. Sanyal, and P. Pal Chaudhuri, "On characterization of cellular automata with matrix algebra," Inform. Sci., vol. 65, pp 251-277, June 1992.
6. M. Abramovici, M.A. Breuer, and A.D. Friedman, *Digital Systems Testing and Testable Design*. Computer Science Press,New York,New York, 1990.
7. E.J. McCluskey "Verification testing - A pseudoexhaustive test technique," IEEE Trans Comput. vol. C-33, pp. 541-546, June 1984.
8. G.E. Sobelman and C.H. Chen, "An eficient approach to pseudoexhaustive test generation for BIST design," in Proc. ICCD, pp. 576-579, 1989.
9. Anderson, Tand P.Lee, *Fault-tolerance. principle and practice*, Prentice-Hall International, New York, New York, 1981.
10. Sellers, F.F, MoYo Hsiao and C.L. Bearnson, "Analyzing errors with the Boolean difference," IEEE Trans, Comput., pp. 676-683, July 1968.
11. Dhiraj K. Pradahan, *Fault-Tolerant Computing*, Prentice-Hall, Englewood Cliffs, New Jersey, 1986.
12. K.P. Parker and E.J. McCluskey, "Probablistic Treatment of General Combinational Networks," IEEE Trans. Computers, Vol. 24, No.6, pp. 668-670, June 1975.
13. Dhiraj K. Pradahan, *Fault-Tolerant Computer System Design*, Prentice-Hall International, New York, New York, 1993.
14. M.H. Konijnenburg, A.J. Van De Goor, and J.Th. Van der Linden, "Circuit Partitioned Automatic Test Pattern Generation Constrained by Three-State Buses and Ristrictors," 5th Asian Test Symposium (ATS '96), pp. 29-33, Nov. 1996.
15. Douglas Chang, and Malgorzata Marek-Sadowska, "Partitioning Sequential Circuits on Dynamically Reconfigurable FPGAs," IEEE Transactions on Computers, pp. 565-578, June 1999.
16. D.J.Huang, and A.B. Kahng, ``When Clusters Meet Partitions: New Density-Based Methods for Circuit Decomposition," Proc. of EDTC, pp. 60-64, March 1995.
17. Kristian Wiklund, "A gate Level Fault Simulation Toolkit," Charmers University of Technonology, Gothenburg, Sweden, Tech. Report 00-17, 2001.

A Parallel Volume Splatting Algorithm Based on PC-Clusters

Jiawan Zhang, Jizhou Sun, Yi Zhang, Qianqian Han, and Zhou Jin

IBM Computer Technology Center, Department of Computer Science, School of
Electronic and Information Engineering, Tianjin University, Tianjin 300072,
P.R.China
jwzhang@tju.edu.cn
http://ibm.tju.edu.cn/~zjw/

Abstract. Splatting is a widely used object-order volume rendering algorithm. By using a hybrid data-space and image-space partitioning scheme, a parallel volume Splatting algorithm based on PC-clusters is presented in this paper. Experiment results demonstrate that the proposed method can gain high rendering speed and excellent speedups on commercial PC-clusters.

1 Introduction

Direct volume rendering(DVR)[1][2][3] has gained great popularity in the visualization of three-dimensional data sets without generating geometry models. It has been widely used in medicine, computational fulid dynamics(CFD), finite element analysis(FEA), geology and other applications in recently years.

Generally, volume rendering algorithms can be divided into two categories. One is image-order method in which rays are cast into the volume through the screen pixels, for each ray, the volume is sampled at regular intervals along the ray. The optical properties of these samples are then composited to obtain the color of the corresponding screen pixel. The classical image-order volume rendering method is ray casting [2]. The other is object-order method. In this algorithm, the voxels are sorted slice by slice in the front-to-back or back-to-front order. Traversed in the order, each voxel is classified by using a proper transfer function,and shaded by a given illumination model. Then, the voxel is projected into the image plane, and its contribution is accumulated to an image buffer using a projected reconstruction kernel called footprint. In this way,successive slices are composited to produce the final image. And volume Splatting [4] is one of the most powerful and classical object-order volume rendering techniques.

However, as the development of visualization techniques and the rapid progress in the applications, the scale of data sets becomes even larger which leads to contradiction between the real-time interaction requirement and the low rendering speed. A lot of methods have been proposed by the research society to speedup the rendering, these methods include using interpolation algorithm with low complexity, low resample density, fewer rays projection, simplified illumination, simplified gradient computation. Another set of more scientific methods include efficient occlusion of the transparent voxels by using fine-tuned data structures such as octrees and run length encoding. Another important accelerating method is opacity termination. The methods mentioned above are most only fit for the visualization of small or medium scale data sets. When the scale of the data set becomes large or very large(such as the data set is large than 1024^3), it becomes impossible for a single PC or workstation to merely hold the data set into memory. Parallel volume rendering becomes the natural solution to this situation.

Parallel volume rendering algorithms can be generally divided into two categories based on what kind of data is distributed among the processors. One is the image-space partitioning scheme which subdivides the screen into small regions, and each region is assigned to each processor. The processor is responsible for the rendering of the portion of the screen, when all the processors finish there work, all the sub-screens are gathered together to produce the final rendering image. The other one is the object-space partitioning scheme which subdivides the volume into small subvolumes and each processor is responsible for rendering one or multiple subvolumes. Generally speaking, image-space partitioning techniques require less communication because each processor holds one copy of the entire volume. Image-space partitioning techniques are traditionally used in image-order volume rendering such as volume ray casting algorithms. However, the disadvantage of this method is the computationally expensive resampling that is required for generating high quality images, and it suffers from even more communications when the screen positions change abruptly. Compared to image-space partitioning techniques, object-space partitioning techniques do not require data redistribution or replication when the view is changed. And this method exhibit a better storage scalability when increasing the size of the data set and the number of processors. Object-space partitioning scheme is suitable for Splatting algorithm.

A new object-space volume and image partitioning strategy and a parallel volume Splatting algorithm based on PC-clusters are proposed in this paper.

2 Previous Work

Volume rendering parallelism has been exploited by many researchers in the last two decades.

Ma [5]presents a divide-and-conquer algorithm for parallel volume rendering. His method is a pure object-space partitioning scheme, where each processor receives a partition of the volume. Once rendering is done on the partition, neighboring processors repeatedly exchange half of their current images for compositing until all processors hold a small part of the final image. Scalability is a major advantage of his algorithm, but load imbalance is a drawback. Hsu [6]presents a segmented ray casting for data parallel volume rendering. In this method, the pixel values in the image plane are computed by casting rays through the volume data. The rays are segmented based on the intersection with the data subblocks in the processors. Each processor computes the color and opacity of the ray segments which pass through its subblock, which other segment values. This method does not require the transposition of volume data between processors at any time. M. Meißer presents an implementation of a parallel ray casting algorithm for orthogonal projections on a new single-chip SIMD architecture.

Under the context of Splatting, there are also much work has been done to realize parallelism. For example, Machiraju [7]demonstrates an efficient feed-forward volume rendering techniques for vector and parallel processors. Elvins [8] presents a volume rendering method based on a distributed memory parallel computer. Whitman[9] also implemented a parallel Splattting system for distributed visualization. Huang Jian [10]proposes a parallel Splatting algorithm with occlusion culling. In his paper, the volume data set is organized as object space bricks. The tight bounding box of the sparse voxel list within each brick facilitates fast and simple schemes for both brick level data culling and workload estimation. During the rendering process, the concept of image-aligned slab is leveraged to cull down the amount of data communication and data storage on the rendering nodes. Coupled with view coherence, the image-aligned slab representation reduces the data communication needed to two orders of magnitude lower than the storage of the raw volume itself. High load balance can be achieved by using the proposed method.

Recently, Antonio Garcia[11]presents a parallel volume rendering algorithm that combines the benefits of both image-space and object-space partition schemes based on the idea of pixel and volume interleaving. He first subdivides the processors into groups. Each group is responsible for rendering a portion of the volume. Inside of a group, every member interleaves the data samples of the volume and the pixels of the screen. Interleaving the data provides storage scalability and interleaving the pixels reduces communication overhead. This hybrid object and image-space partitioning scheme is able to reduce the image compositing cost, incur in low communication overhead and balance rendering workload at the expense of image quality. Experiments on a PC-cluster demonstrate encouraging results.

Because the rapid growing of computational ability, PC-clusters are now becoming widely used in many scientific computation areas. In this paper, we realize a new parallel volume Splatting method on a PC-cluster.

3 Method

Our method can be categorized as hybrid data-space and image-space partitioning scheme. There are two main features in this method. First, a new data distribution method is proposed. According to the data distribution strategy, the second feature, a new image compositing strategy is applied. In this Section, we discuss the main ideas behind,and at last the rendering results are given out to prove the proposed method.

3.1 Data Partitioning and Distribution Strategy

Because our work is aimed at solving volume rendering of large scale data sets, pure image-space partitioning scheme is discarded for one processor cannot hold the entire volume into memory.

In our method, first, we use data-space partitioning method to distribute the volume among nodes in the PC-clusters. If there are N nodes in the PC-clusters, the volume is first divided into N parallel slabs. Then each slab is subdivided into N voxel blocks. As demonstrated in Fig. 1.

Please note that the size of the voxel block can be different, however, all the slabs must be subdivided in the same way to ensure that the corresponding voxel blocks along the slab perpendicular.According to the partitioning strategy, the volume is totally divided into N^2 voxel blocks.

Now we introduce our data distribution strategy. The jth voxel block of the ith slab is assigned to $|j\text{-}i|^{th}$ nodes of the PC-clusters. For example, in the first slab, the 0th voxel block is assigned to the 0th node, and 1th voxel block to the 1th node, and so on. In the second slab, the 0th voxel block is assigned to the $(N\text{-}1)^{th}$ node, and 1th voxel block to 0th node, and so on.

The reason why we adopt these two strategies is to fully exploit better load balance by using a static load balance controlling strategy.

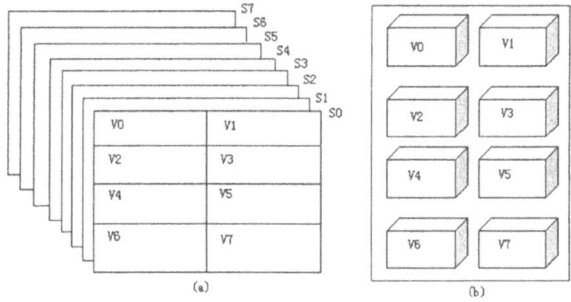

Fig. 1. Data distribution strategy: the volume is partitioned into N slabs first, and each slab is partitioned into N voxel blocks. Here N is the number of processors in the PC-clusters (a)Data partition in 3D format (b)Data partition in a single slab

3.2 Image Compositing Strategy

When the volume data is distributed by using the methods discussed above, each cluster node is responsible for the rendering of N corresponding intermediate images. When all the nodes finish their work, the image compositing process begins. Because each node holds N intermediate images, all the intermediate images along the slab perpendicular are sent to the same node, as shown in Fig.2. When this node receives the N intermediate images, it use **Under** or **Over**[12] operator in the compositing process to obtain the corresponding part of screen. Then at last, one node is responsible for the making up the final rendering image.

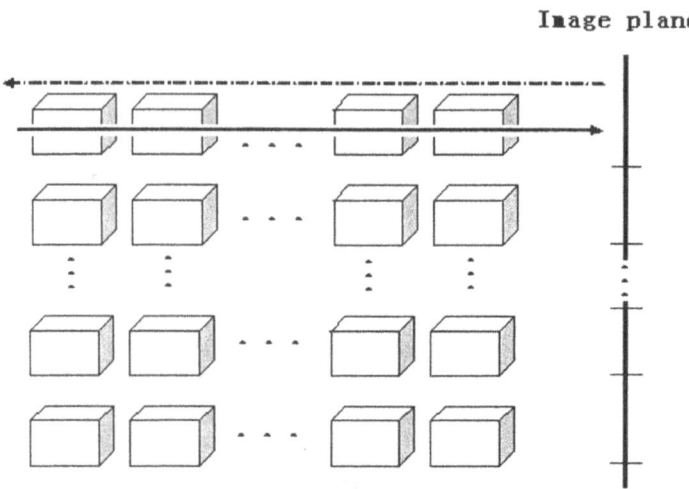

Fig. 2. The compositing process. The solid arrow indicates back to front(BTF) compositing, and dashed arrow indicates front to back(FTB) compositing

4 Experiments

Experiments have been conducted in our laboratory. A volume Splatting algorithm is implemented on a IBM eSever 1350 PC-clusters. There are 16 computing nodes (2.4GHz Intel Xeon X335, 2*512MDDR, 18.2G) and a manage node (2.4GHz Intel Xeon X345, 2*512M DDR, 18.2G) with 10/100Mbps Ethernet in our cluster system. The software platform is Linux (RedHat7.3),and PVM and MPI are equipped as parallel programming interfaces. Computing nodes connect one another by two independent intercommunication switch hubs (100M, 16 ports) in TCP/IP Ethernet. The experiment platform of this paper is shown in Fig.3.

The CT Visible Man (VM for short) data set is used in our experiments. We extract 512×1536× 1024 volume from the original data set and store each data is in one byte format. The data set size is 768M. The rendering results is shown in Fig.4.

Fig. 3. IBM eServer 1350 Cluster in our Lab. Left:front-end view, right:rear-end view.

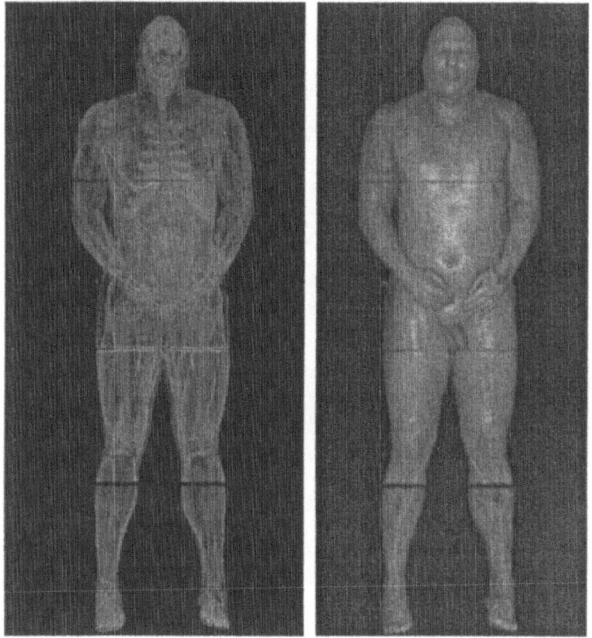

Fig. 4. Two rendering results of the VM data set. Left:bone and skin, right: skin

The rendering time in the experiment is shown in Fig.5. From the results, we can find that when only one node is used in the splatting process, the total rendering time is about 350 seconds. By adding nodes(2-12 nodes) to the rendering task, the rendering time reduces significantly. For example, when we use 12 computing nodes, the rendering time is only about 64 seconds. When all the

16 nodes of the cluster are added to the rendering task, the rendering time is only 36 seconds. The corresponding speedup gained by adding cluster nodes is shown in Fig.6. Through the experiments results, we can find that the proposed method in this paper has very excellent performance in the rendering of very large data sets.

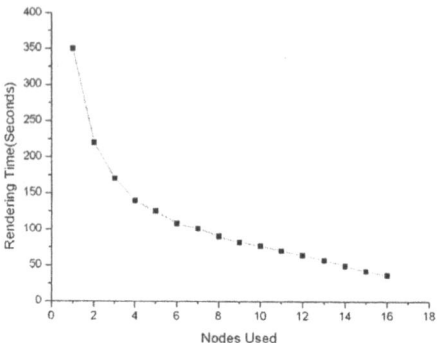

Fig. 5. The rendering time in the experiments

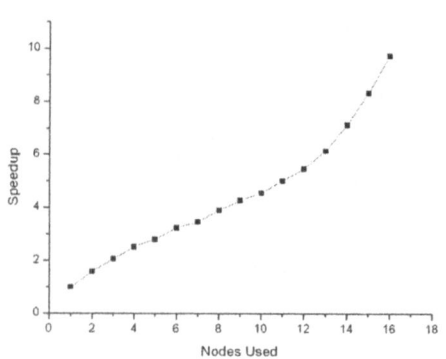

Fig. 6. The speedup gained in the experiments

5 Conclusion

A parallel volume Splatting algorithm based on PC-clusters which integrates hybrid data-space and image-space partitioning schemes are proposed in this paper. Experiments show that high rendering speed and excellent speedup can be obtained by adding nodes to the computation.

Acknowledgement. This paper is part work of Project – Research on Key Techniques and Supported Architectures for Real Time Volume Rendering, which is supported by the Natural Science Foundation of China(NSFC) under Grant No. 60373061. The authors thank NSFC for their support.

References

1. Drebin R. A., Carpenter L., and Hanrahan P. : Volume rendering. Computer Graphics. **22** (1988) 65–74
2. Levoy M. : Display of surfaces from volume data. IEEE Computer Graphics & Applications. **8**(1988) 29–37
3. Kajiya J. T., and Von Herzen B. P. : Ray tracing volume densities. Computer Graphics. **18** (1984) 165–174
4. Westover L. : Footprint evaluation for volume rendering. Computer Graphics. **24** (1990) 367–376
5. Ma K.L., Painter J.S., Hansen C.D., and Krogh M.F.: Parallel volume rendering using binary-swap image composition. IEEE Computer Graphics and Applications. **14** (1994) 59–68
6. Hsu W.M.: Segmented ray casting for data parallel volume rendering. Proceedings of 1993 Parallel Rendering Symposium. (1993) 77–83
7. Machiraju R., and Yagel R.: Efficient Feed-Forward Volume Rendering Techniques For Vector And Parallel Processors. SUPERCOMPUTING' 93. (1993) 699-708.
8. Elvins T.: Volume Rendering on a Distributed Memory Parallel Computer. Proceedings of Parallel Rendering Symposium. (1993) 93-98.
9. Whitman S., Li P. et al.: Parvox – A Parallel Splattting Volume Rendering System for Distributed Visualization. Proceedings of Parallel Rendering Symposium. (1997)
10. Jian Huang: A parallel splatting algorithm with occulsion culling . Proceedings of 2000 Eurographic Workshop on Parallel Graphics and Visualization. (2000)
11. Antonio Garcia, Han-wei Shen: An interleaved parallel volume renderer with PC-Clusters. Proceedings of 2002 Eurographic Workshop on Parallel Graphics and Visualization. (2002) 51–59
12. Porter T. and Duff T.: Compositing Digital Images. Computer Graphics. **8** (1984)

Three-Center Nuclear Attraction Integrals for Density Functional Theory and Nonlinear Transformations

Hassan Safouhi

Faculté Saint-Jean/University of Alberta
8406, 91 Street, Edmonton, Alberta T6C 4G9, Canada
hassan.safouhi@ualberta.ca

Abstract. In the present work, we present the progress realized in the fast and accurate numerical evaluation of three-center nuclear attraction integrals. These integrals are required for density functional theory and also for any accurate ab initio molecular structure calculations. They occur in many millions of terms, even for small molecules and require rapid and accurate evaluation, best obtained over Slater type functions. The Fourier transform method allowed analytic expressions to be developed for these integrals. These analytic expressions turned out to be extremely difficult to evaluate precisely and rapidly because of the presence of spherical Bessel integrals. Different approaches were used to develop efficient algorithms for the numerical evaluation of these spherical Bessel integrals. These approaches are discussed and comparisons in accuracy and rapidity are presented to demonstrate that the approaches based on nonlinear transformations present a valuable contribution to the literature on molecular integrals over Slater type functions.

Keywords: Nonlinear transformations. Numerical integration. Nuclear attraction integrals. Density functional theory. Exponential type functions.

1 Introduction

In numerical analysis, in applied mathematics and in physics, one has often to deal with infinite or semi-infinite integrals to represent solutions of many problems. In practice, these integrals have a very poor convergence and this presents severe numerical and computational difficulties. Therefore, nonlinear transformation methods for improving convergence of oscillatory integrals, have been studied for many years and applied to various situations. Their utility for enhancing and even inducing convergence has been amply demonstrated by Shanks [1]. These techniques form the basis of new methods for solving various problems and have many applications as well [2].

Three-center nuclear attraction integrals over Slater type functions (STFs) [3] are required for density functional theory (DFT) and for any accurate ab initio molecular structure calculations. These integrals are given by:

$$\mathcal{I}_{n_1,l_1,m_1}^{n_2,l_2,m_2} = \int_r \left[\chi_{n_1,l_1}^{m_1}\left(\zeta_1, \vec{R} - \overrightarrow{OA}\right)\right]^* \frac{1}{|\vec{R} - \overrightarrow{OC}|} \chi_{n_2,l_2}^{m_2}\left(\zeta_2, \vec{R} - \overrightarrow{OB}\right) d\vec{R}, \qquad (1)$$

where A, B and C are three arbitrary points of the euclidian space $\mathcal{E}_3$, O is the origin of the fixed coordinate system.

The above integral can be expressed as finite linear combination in terms of integrals over B functions [4,5]. Compared to STFs, the B functions have much more appealing properties applicable to multicenter integral problems [4,6,7]. Their Fourier transforms are exceptionally simple [8] and they are well adapted to the Fourier transform method [9,10], which led to analytic expressions for molecular integrals. These analytic expressions present severe numerical and computational difficulties due to the presence of two-dimensional integral representations. The inner semi-infinite integral is highly oscillatory due to the presence of spherical Bessel functions. Different approaches were used for the numerical evaluation of these spherical Bessel integrals, namely Gauss-Laguerre quadrature, the epsilon algorithm of Wynn [11] and Levin's u transform [12]. It is shown that These approaches require a large amount of CPU time for a sufficient pre-determined accuracy.

Nonlinear transformation methods are, without any doubt, the most successful approaches for accurate and fast numerical evaluation of the integrals under consideration. Algorithms based on the nonlinear $\overline{D}$ transformation of Sidi [13,14] are now developed. Numerical evaluation of the complete analytic expressions of the integrals under consideration for the H_2O molecule, were performed and they illustrate clearly the further improvement of the accuracy and the substantial gain in the calculation times. Convergence properties are discussed to show that from a numerical point of view, the nonlinear $\overline{D}$ transformation lead to the best ideal situation [13,15].

2 General Definitions and Basic Formulae

The Slater type functions (STFs) are defined in normalized form according to the following relationship [3]:

$$\chi_{n,l}^m(\zeta, r) = \zeta^{-n+1} \left[(2\zeta)^{2n+1}/(2n)!\right]^{\frac{1}{2}} r^{n-1} e^{-\zeta r} Y_l^m(\theta_r, \varphi_r), \qquad (2)$$

where n, l, m are the quantum numbers and they are such that $n = 1, 2, \ldots$, $l = 0, 1, \ldots, n-1$ and $m = -l, -l+1, \ldots, l-1, l$ and where $Y_l^m(\theta, \varphi)$ stands for the surface spherical harmonic [16].

STFs can be expressed as finite linear combinations of B functions, which are defined as follows [4,5]:

$$B_{n,l}^m(\zeta, r) = \frac{(\zeta r)^l}{2^{n+l}(n+l)!} \hat{k}_{n-\frac{1}{2}}(\zeta r) Y_l^m(\theta_r, \varphi_r), \qquad (3)$$

where $\hat{k}_{n+\frac{1}{2}}(z)$ stands for the reduced Bessel function [4].

The Fourier transform $\overline{B}_{n,l}^m(\zeta, p)$ of $B_{n,l}^m(\zeta, r)$ is given by [8]:

$$\overline{B}_{n,l}^m(\zeta, p) = \sqrt{\frac{2}{\pi}} \zeta^{2n+l-1} \frac{(-i|p|)^l}{(\zeta^2 + |p|^2)^{n+l+1}} Y_l^m(\theta_p, \varphi_p) \qquad (4)$$

The spherical Bessel function $j_l(x)$ is defined by [18] :

$$j_l(x) = (-1)^l x^l \left(\frac{d}{xdx}\right)^l \left(\frac{\sin(x)}{x}\right). \tag{5}$$

The spherical Bessel function $j_l(x)$ and its first derivative $j'_l(x)$ satisfy the following relation [18] :

$$x\, j_{l-1}(x) - (l+1)\, j_l(x) = x\, j'_l(x). \tag{6}$$

Three-center nuclear attraction integrals over STFs are expressed in terms of integrals over B functions. These integrals over B function are expressed, with the help of the Fourier transform method, in terms of two-dimensional integral representations [9, 10] :

$$\widetilde{\mathcal{I}}^{n_2,l_2,m_2}_{n_1,l_1,m_1} = 8\,(4\pi)^2\,(-1)^{l_1+l_2}\,(2l_1+1)!!\,(2l_2+1)!!$$

$$\times \frac{(n_1+l_1+n_2+l_2+1)!}{(n_1+l_1)!(n_2+l_2)!}\zeta_1^{2n_1+l_1-1}\zeta_2^{2n_2+l_2-1}$$

$$\times \sum_{l'_1=0}^{l_1}\sum_{m'_1=-l'_1}^{l'_1}(i)^{l_1+l'_1}\frac{<l_1 m_1|l'_1 m'_1|l_1-l'_1 m_1-m'_1>}{(2l'_1+1)!![2(l_1-l'_1)+1]!!}$$

$$\times \sum_{l'_2=0}^{l_2}\sum_{m'_2=-l'_2}^{l'_2}(i)^{l_2+l'_2}(-1)^{l'_2}\frac{<l_2 m_2|l'_2 m'_2|l_2-l'_2 m_2-m'_2>}{(2l'_2+1)!![2(l_2-l'_2)+1]!!}$$

$$\times \sum_{l=l'_{\min},2}^{l'_2+l'_1}<l'_2 m'_2|l'_1 m'_1|l m'_2-m'_1> R_2^l\, Y_l^{m'_2-m'_1}(\theta_{\vec{R}_2},\varphi_{\vec{R}_2})$$

$$\times \sum_{\lambda=l''_{\min},2}^{l_2-l'_2+l_1-l'_1}(-i)^\lambda<l_2-l'_2 m_2-m'_2|l_1-l'_1 m_1-m'_1|\lambda\mu>$$

$$\times \sum_{j=0}^{\Delta l}\binom{\Delta l}{j}\frac{(-1)^j}{2^{n_1+n_2+l_1+l_2-j+1}(n_1+n_2+l_1+l_2-j+1)!}$$

$$\times \int_{s=0}^{1} s^{n_2+l_2+l_1-l'_1}\,(1-s)^{n_1+l_1+l_2-l'_2}\,Y_\lambda^\mu(\theta_v,\varphi_v)$$

$$\times \left[\int_{x=0}^{+\inf} x^{n_x}\frac{\hat{k}_\nu[R_2\gamma(s,x)]}{[\gamma(s,x)]^{n_\gamma}} j_\lambda(vx)\,dx\right]\,ds, \tag{7}$$

where : $[\gamma(s,x)]^2 = (1-s)\zeta_1^2 + s\zeta_2^2 + s(1-s)x^2$
$\vec{v} = (1-s)\vec{R}_2 - \vec{R}_1$
$n_\gamma = 2(n_1+l_1+n_2+l_2) - (l'_1+l'_2) - l + 1$
$\nu = n_1+n_2+l_1+l_2-l-j+\frac{1}{2}$
$\mu = (m_2-m'_2) - (m_1-m'_1)$
$n_x = l_1-l'_1+l_2-l'_2$
$\Delta l = [(l'_1+l'_2-l)/2]$
v and R_2 stand for the modulus of $\vec{v}$ and $\vec{R}_2$ respectively.

3 The Numerical Evaluation of Three-Center Nuclear Attraction Integrals

The two-dimensional integral representations occurring in the analytic expression (7) turned out to be extremely difficult to evaluate rapidly and accurately because of the presence of semi-infinite spherical Bessel integral, which will be referred as $\widetilde{\mathcal{I}}(s)$, whose integrand will be referred to as $\mathcal{F}_s(x)$.

From equation (7), one can easily notice that the numerical evaluation of three-center nuclear attraction integral depends strongly on how fast and accurate, the spherical Bessel integrals are evaluated.

The evaluation of the above semi-infinite integral presents severe computational and numerical difficulties, particularly for large values of λ, which depends on the quantum numbers n, l and m. Note that when the value of v is large, the zeros of $j_\lambda(vx)$ become closer and the oscillations become strong. Note also that when the value of s is close to 0 or 1, the oscillation of the integrand become sharp. Indeed, if we make the substitution $s = 0$ or 1, in the integrand, the function $\gamma(s,x)$ becomes a constant ($\gamma(0,x) = \zeta_1$ and $\gamma(1,x) = \zeta_2$) and consequently the exponential decreasing part $\hat{k}_\nu$ of the integrand becomes a constant. From this it follows that the sharp oscillation of the spherical Bessel function could not be damped and suppressed by the exponential decreasing part of the integrand. The integrand could not de represented by a function of the form $e^{-\alpha x} j_\lambda(x)$ and this why Gauss-Laguerre quadrature cannot be efficient for the numerical evaluation of this kind of spherical Bessel integrals. We should also mention that the regions where s is close to 0 or 1 carry a very small weight due to factors $s^{i_2}(1-s)^{i_1}$ in the integrands [19, 20, 21, 22].

The semi-infinite integral $\widetilde{\mathcal{I}}(s)$ could be transformed into an infinite series:

$$\widetilde{\mathcal{I}}(s) = \sum_{n=0}^{+\inf} \int_{j_{\lambda,v}^n}^{j_{\lambda,v}^{n+1}} x^{n_x} \frac{\hat{k}_\nu[R_2 \gamma(s,x)]}{[\gamma(s,x)]^{n_\gamma}} j_\lambda(vx)\, dx, \tag{8}$$

where $j_{\lambda,v}^0 = 0$ and $j_{\lambda,v}^n = j_{\lambda+\frac{1}{2}}^n / v$ ($v \neq 0$) for $n = 1, 2, \ldots$ which are the successive positive zeros of $j_\lambda(vx)$.

The above infinite series is slowly convergent. To obtain 10 correct decimal in a certain regions where the value of v is large and s close to 0 or 1, one need to sum more than 8000 terms. This slowly convergence problem, prevented the use of this infinite series for the evaluation of $\widetilde{\mathcal{I}}(s)$. Note that, even for small molecules, millions of integrals are required for the calculations. Rapidity is a primordial criterion when a high accuracy could be reached.

The convergence of the above infinite series could be improved by the use of the epsilon algorithm of Wynn or Levin's u transform. It has been proved that these two convergence accelerators, are the best suited for the infinite series (8). Let I_k be the kth of the infinite series and S_k be the kth partial sum of the infinite series. The approximation of the infinite series using the epsilon algorithm of Wynn is given by [2]:

$$\epsilon_{-1}^{(n)} = 0, \quad \epsilon_0^{(n)} = S_n, \quad n = 0, 1, \ldots$$

$$\epsilon_{k+1}^{(n)} = \epsilon_{k-1}^{(n+1)} + \frac{1}{\epsilon_k^{(n+1)} - \epsilon_k^{(n)}}, \quad n = 0, 1, \ldots. \tag{9}$$

The approximation of the infinite series using Levin's u transform, can be obtained by using the following rules is given by [12] :

$$u_k(S_n) = \frac{\sum_{i=0}^{k}(-1)^i \binom{k}{i}(n+i+1)^{k-1} S_{n+i}/I_{n+i}}{\sum_{i=0}^{k}(-1)^i \binom{k}{i}(n+i+1)^{k-1}/I_{n+i}}, \quad n = 0, 1, \ldots, \tag{10}$$

The improvement of convergence of the infinite series was improved using the above convergence accelerators. Unfortunately, the calculation times are still prohibitively long for a high pre-determined accuracy.

In previous work [23, 24], on three-center nuclear attraction integrals over B functions, we used the nonlinear D (of Levin and Sidi) [25] and $\overline{D}$ (of Sidi) transformations for improving convergence of the semi-infinite integrals of interest. These two transformations improve convergence of oscillatory semi-infinite integrals, whose integrands satisfy linear differential equations having asymptotic expansions in inverse powers of x as $x \to +\infty$.

Let f be a function, satisfying a linear differential of order m and of the form required by the D transformation, such that $S = \int_0^{+\infty} f(t) \, dt < +\infty$. The D transformation, consists on developing an asymptotic expansion for the integral function $\int_x^{+\infty} f(t) \, dt$ as $x \to +\infty$ [25] :

$$\int_x^{+\inf} f(t) dt \sim \sum_{k=0}^{m-1} f^{(k)}(x) x^{j_k} \left(\beta_{0,k} + \frac{\beta_{1,k}}{x} + \frac{\beta_{2,k}}{x^2} + \ldots \right), \tag{11}$$

where $j_k \leq \max(i_{k+1}, i_{k+2} - 1, \ldots, i_m - m + k + 1)$, $k = 0, 1, \ldots, m-1$.

The above asymptotic expansion is transformed into a linear system by attributing values x_i, for $i = 0, 1, \ldots, n(m-1)$, for the variable x [25]. These values should be chosen efficiently to insure good convergence properties [13, 15, 25]. In the case of the $\overline{D}$ transformation x_l are chosen to be the successive positive zeros of the integrand. The approximation $\overline{D}_n^{(m)}$ for S is given by [13] :

$$\overline{D}_n^{(m)} = \int_0^{x_l} f(t) dt + \sum_{k=1}^{m-1} f^{(k)}(x_l) x_l^{\sigma_k} \sum_{i=0}^{n-1} \frac{\bar{\beta}_{k,i}}{x_l^i}, \quad l = 0, 1, \ldots, n(m-1). \tag{12}$$

In [13], Sidi showed that the functions of the form $f(x) = g(x) j_\lambda(x)$ satisfy second order linear differential equations of the form required to apply the $\overline{D}$ transformation. From this, it follows that $\overline{D}_n^{(2)}$ is a very good approximation of the semi-infinite integral. The x_l are chosen to be the leading positive zeros of $j_\lambda(x)$. Note that $f'(x_l) = g(x_l) j'_\lambda(x_l)$ and this is due to the fact that $j_\lambda(x_l) = 0$. In [24, 26], we demonstrated that the integrand occurring in the semi-infinite integral $\widetilde{\mathcal{I}}(s)$ satisfies the conditions required to apply $\overline{D}_n^{(2)}$.

The computation of $j'_\lambda(x_l)$ is not necessary. One can use the fact that $j_\lambda(x_l) = 0$ with equation (6) to obtain the following relation :

$$j'_\lambda(x_l) = j_{\lambda-1}(x_l). \tag{13}$$

To use the above expression, one should separate the case where $\lambda = 0$. Note also that when $\lambda = 0$, one can use the Cramer's rule for calculating the approximation $\overline{D}_n^{(2)}$, since the zeros of $j_0(x) = \dfrac{\sin(x)}{x}$ are equidistant [13]:

$$\overline{D}_n^{(2)} = \frac{\sum_{i=0}^{n+1} \binom{n+1}{i}(1+i)^n F(x_i) \big/ [x_i^2 g(x_i)]}{\sum_{i=0}^{n+1} \binom{n+1}{i}(1+i)^n \big/ [x_i^2 g(x_i)]}, \qquad (14)$$

where $F(x) = \int_0^{x_l} f(t)\,dt$.

Recently [27, 28], we introduced a highly efficient approach, which we called $S\overline{D}$, for a rapid and accurate evaluation of spherical Bessel integrals. This approach consists in transforming these semi-infinite integrals into a semi-infinite integrals involving the simple sine function and satisfying all the conditions required to apply the nonlinear $\overline{D}$ transformation with a second order differential equation. By applying the $S\overline{D}$ approach to the semi-infinite integral $\widetilde{\mathcal{I}}(s)$, we obtain [27] ($v \neq 0$):

$$\widetilde{\mathcal{I}}(s) = \frac{1}{v^{\lambda+1}} \int_0^{+\inf} \left[\left(\frac{d}{x\,dx}\right)^\lambda \left(x^{n_x+\lambda-1} \frac{\hat{k}_\nu[R_2\gamma(s,x)]}{[\gamma(s,x)]^{n_\gamma}} \right) \right] \sin(vx)\,dx. \qquad (15)$$

When $v = 0$, the semi-infinite integral $\widetilde{\mathcal{I}}(s)$ vanishes if $\lambda \neq 0$, and if $\lambda = 0$, we used the fact that $j_0(\alpha) = \dfrac{\sin(\alpha)}{\alpha} \to 1$ when $\alpha \to 0$ and the fact that the integrand is exponentially decreasing function, to obtain the following equation:

$$\widetilde{\mathcal{I}}(s) \approx \int_0^{+\inf} x^{n_x} \frac{\hat{k}_\nu[R_2\gamma(s,x)]}{[\gamma(s,x)]^{n_\gamma}}\,dx. \qquad (16)$$

The approximation of the semi-infinite integral (15) is obtained using the following expression:

$$S\overline{D}_n^{(2,j)} = \frac{1}{v^{\lambda+1}} \frac{\sum_{i=0}^{n+1} \binom{n+1}{i}(1+i+j)^n F(x_{i+j}) \big/ [x_{i+j}^2 G(x_{i+j})]}{\sum_{i=0}^{n+1} \binom{n+1}{i}(1+i+j)^n \big/ [x_{i+j}^2 G(x_{i+j})]}, \qquad (17)$$

where $j = 0, 1, 2, \ldots$ and $x_l = (l+1)\dfrac{\pi}{v}$ for $l = 0, 1, \ldots$ (the successive positive zeros of $\sin(v\,x)$) and $F(x) = \int_0^x \widetilde{\mathcal{F}}_s(t)\,dt$. The function $G(x)$ and $\widetilde{\mathcal{F}}_s(x)$ are given by:

$$G(x) = \left(\frac{d}{x\,dx}\right)^\lambda \left(x^{n_x+\lambda-1} \frac{\hat{k}_\nu[R_2\gamma(s,x)]}{[\gamma(s,x)]^{n_\gamma}} \right) \quad \text{and} \quad \widetilde{\mathcal{F}}_s(x) = G(x)\sin(vx),$$

The computation of the function $G(x)$ can be obtained easily by applying the Leibnitz formulae and by using the following relations:

$$\frac{d}{dx} = \frac{dz}{dx}\frac{d}{dz} \quad \text{and} \quad \left(\frac{d}{z\,dz}\right)^m \left[\frac{\hat{k}_{n+\frac{1}{2}}(z)}{z^{2n+1}}\right] = (-1)^m \frac{\hat{k}_{n+m+\frac{1}{2}}(z)}{z^{2(n+m)+1}}. \qquad (18)$$

Table 1. Values with 15 correct decimals of $\tilde{\mathcal{I}}(s)$ obtained using the infinite series with the sine function. $R_2 = 2.0$, $\zeta_1 = 1.5$ and $\zeta_2 = 1.0$.

s	ν	n_γ	n_x	λ	R_1	v	nmax	$\tilde{\mathcal{I}}(s)^{nmax}$
.24D-02	10	9	5	5	55.	53.0048	11106	.908143147723532D+00
.24D-02	7	6	4	4	55.	53.0048	9457	.335631573347817D-02
.24D-02	5	5	5	4	3.	3.4844	217	.607887949264881D+00
.24D-02	5	11	5	4	3.	4.4820	190	.446607335810847D-02
.998	7	6	4	4	55.	54.9960	10798	.377646226483030D-01
.998	8	9	5	4	55.	54.9960	10927	.243620544151150D-01
.998	5	9	4	3	3.	2.9890	183	.386472626382360D-02
.998	5	5	5	4	3.	2.9870	204	.429976963007154D-01

Note that the computation of the leading positive zeros of the spherical Bessel function and the computation of a method for solving linear systems are avoided by the introduction of the $\overline{SD}$ approach. Recurrence relations were developed for a fast computation of $\overline{SD}_n^{(2,j)}$ and for a better control of the degree of accuracy [29].

The convergence properties of the nonlinear $\overline{D}$ transformations were analyzed [13,15] and they showed that under certain conditions the approximations $\overline{D}_n^{(m)}$ transformation converge to the exact value of the semi-infinite integrals without any constraint. It is demonstrated [26,30], that the integrands of the semi-infinite integrals which occur in the analytic expression three-center nuclear attraction integral over STFs satisfy the conditions developed by Sidi for the convergence of the $\overline{D}$ transformation. The approximations $\overline{D}_n^{(2)}$ and $\overline{SD}_n^{(2,j)}$ converge to the exact values of the semi-infinite integrals $\tilde{\mathcal{I}}(s)$.

4 Numerical Discussion

The value with 15 correct decimals of the semi-infinite integral $\tilde{\mathcal{I}}(s)$ are obtained using the infinite series with the sine function, which converges faster than the series with the spherical Bessel function (8). To obtain these value with 15 correct decimals, we need to sum the infinite series until the order namx. Each terms of the infinite series were evaluated using Gauss-Legendre quadrature of order 20.

The finite integrals, which occur in (9), (10), (12), (14) and (17) are evaluated using Gauss-Legendre quadrature of order 20.

The linear system (12) was solved using the LU decomposition method.

Tables (1) contains value with 15 correct decimals of the semi-infinite integrals $\tilde{\mathcal{I}}(s)$, obtained using the infinite series with the sine function. In tables (2)we listed the errors obtained in the evaluation of the semi-infinite integrals using, $\tilde{\mathcal{I}}(s)$ obtained using, Levin's u transform (Error^u) of order n, the nonlinear $\overline{D}$ transformation of order n (Error$^{\overline{D}}$) and the $\overline{SD}$ approach (Error$^{\overline{SD}}$). The calculations are performed for s close to 0 and 1 (0.0024 and 0.998).

Three-Center Nuclear Attraction Integrals for Density Functional Theory   287

Table 2. Evaluation of $\widetilde{\mathcal{I}}(s)$. s, R_2, ζ_1, ζ_2, ν, n_γ, n_x, λ, R_1 and v are given in table (1). Time is in milliseconds.

n	Error^u	Time	n	Error$^{\overline{D}}$	Time	n	Error$^{\overline{SD}}$	Time
10	0.148D-06	0.141	8	0.784D-07	0.156	9	0.353D-13	0.266
10	0.362D-10	0.109	8	0.440D-11	0.141	7	0.580D-12	0.188
10	0.292D-04	0.125	10	0.151D-05	0.203	92	0.651D-10	2.000
10	0.217D-06	0.125	10	0.202D-07	0.188	81	0.751D-10	1.438
10	0.426D-09	0.141	8	0.419D-10	0.172	7	0.651D-11	0.188
10	0.399D-08	0.141	8	0.831D-09	0.172	3	0.426D-13	0.109
10	0.441D-07	0.125	10	0.587D-08	0.203	71	0.496D-10	1.391
10	0.157D-05	0.141	10	0.188D-06	0.203	92	0.721D-10	2.000

Table 3. Three-center nuclear attraction integrals over STFs (1). Values obtained with H_2O molecule (in atomic units), using the $\overline{SD}$ approach with an accuracy of 10 correct digits. The values are obtained with the geometrical parameters (in spherical coordinate) : $O(0., 0., 0.)$, $H_1(1.81, 52.5°, 0.0)$ and $H_2(1.81, 52.5°, 180°)$. Numbers in parentheses represent powers of 10.

$\mathcal{I}^{\bar{n}_2,l_2,m_2}_{\bar{n}_1,l_1,m_1}$	ζ_1	ζ_2	Values
$\langle 1s^O \vert (r_{OH_2})^{-1} \vert 1s^{H_1} \rangle$	7.67	1.25	0.3000060119(-1)
$\langle 2p_z^O \vert (r_{OH_2})^{-1} \vert 1s^{H_1} \rangle$	1.50	1.25	0.1700603538(0)
$\langle 2p_z^O \vert (r_{OH_2})^{-1} \vert 1s^{H_1} \rangle$	3.50	1.25	0.7739215284(-1)
$\langle 2p_{+1}^O \vert (r_{OH_2})^{-1} \vert 1s^{H_1} \rangle$	1.50	1.25	-0.7936139417(-1)
$\langle 2p_{+1}^O \vert (r_{OH_2})^{-1} \vert 1s^{H_1} \rangle$	3.50	1.25	-0.3124157389(-1)
$\langle 2p_{-1}^O \vert (r_{OH_2})^{-1} \vert 1s^{H_1} \rangle$	1.50	1.25	0.7936139417(-1)
$\langle 2p_{-1}^O \vert (r_{OH_2})^{-1} \vert 1s^{H_1} \rangle$	3.50	1.25	0.3124157389(-1)
$\langle 1s^O \vert (r_{OH_2})^{-1} \vert 1s^{H_1} \rangle$	7.67	1.21	0.3067870414(-1)
$\langle 2s^O \vert (r_{OH_2})^{-1} \vert 1s^{H_1} \rangle$	2.09	1.21	0.2313538730(0)
$\langle 2p_z^O \vert (r_{OH_2})^{-1} \vert 1s^{H_1} \rangle$	1.50	1.21	0.1710199961(0)
$\langle 2p_z^O \vert (r_{OH_2})^{-1} \vert 1s^{H_1} \rangle$	3.50	1.21	0.7740274814(-1)
$\langle 2p_{+1}^O \vert (r_{OH_2})^{-1} \vert 1s^{H_1} \rangle$	1.50	1.21	-0.7699898494(-1)
$\langle 2p_{+1}^O \vert (r_{OH_2})^{-1} \vert 1s^{H_1} \rangle$	3.50	1.21	-0.2997862990(-1)
$\langle 2p_{-1}^O \vert (r_{OH_2})^{-1} \vert 1s^{H_1} \rangle$	1.50	1.21	0.7699898494(-1)
$\langle 2p_{-1}^O \vert (r_{OH_2})^{-1} \vert 1s^{H_1} \rangle$	3.50	1.21	0.2997862990(-1)

Table (3) contain values of the complete analytic expression of the three-center nuclear attraction integrals over STFs, obtained with the H_2O molecule. The $\overline{SD}$ was used to evaluate all the semi-infinite integrals occurring in the analytic expression. The degree of accuracy was set to 10 correct decimals.

From Tables (1), one can notice that in certain regions, we need to sum more 9000 terms to obtain values with 15 correct decimals. There are two important advantages of using the infinite series with the sine functions, the first one is the fact that this infinite series converges faster than the infinite series with the spherical Bessel function, and the second one, is the fact that we do not need to compute the successive positive zeros of spherical Bessel function, in particular when we evaluate the semi-infinite integrals in these regions where the convergence is very slow and where we need to compute more than 9000 positive zeros of $j_\lambda(x)$.

From Tables (2) we can notice that the nonlinear $\overline{D}$ and the $\overline{SD}$ approach are highly efficient for evaluating these kind of spherical Bessel integrals. The accuracy obtained by using these approaches is higher than the accuracy that we obtained using Levin's u transform, which gave better results than the epsilon algorithm of Wynn. Note that the algorithm based on $\overline{SD}$ is very stable.

The gain in the calculation times realized by the introduction of the $\overline{SD}$ approach is clearly illustrated from the numerical tables.

The evaluation of the complete expression of three-center nuclear integrals over STFs with the H_2O molecule was performed to show that the $\overline{SD}$ approach is highly efficient and could lead to a rapid software package for accurate numerical evaluation of molecular integrals over STFs.

Extensive numerical results obtained with linear and nonlinear systems and comparisons with values form the literature can be found in [29]. A complete study of the stability of the algorithm based in the $\overline{SD}$ approach is presented in [29].

In Table (3), the abbreviation $2p_{+1}$ and $2p_{-1}$ refer to the Slater functions defined by the quantum numbers $(n = 2, l = 1, m = 1)$ and $(n = 2, l = 1, m = -1)$. The symbol $(r_{ab})^{-1}$ refers to the Coulomb operator.

All the calculations were performed on a Workstation with an Intel Xeon Processor with 2.4GHz.

5 Conclusion

With the help of nonlinear transformation, the improvement of convergence of highly oscillatory integrals can be remarkable. These methods were applied in various situations and led to highly satisfactory results. In the present work, we presented another application, that concerns a fast and accurate numerical evaluation of molecular integrals for DFT and for any accurate ab initio molecular calculation, where molecular orbitals are expressed as linear combinations of atomic orbitals. Slater type functions are used to represent these atomic orbitals.

STFs are expressed as finite linear combinations of B functions and the Fourier transform method was applied leading to analytic expressions for the molecular integrals under consideration. These analytic expressions involve highly oscillatory integrals. An efficient numerical evaluation of the molecular integrals depends strongly on the evaluation of these semi-infinite integrals. The approach based on Gauss-Laguerre quadrature is demonstrated to be inefficient for this kind of spherical Bessel integrals. The epsilon algorithm of Wynn or Levin's u transform could lead to a sufficient accuracy but in certain regions where the oscillations are strong, the algorithms are completely instable and the obtained results are inaccurate.

It is now shown, that nonlinear transformations constitute the key for a development of a highly efficient and fast algorithm for a numerical evaluation of molecular integrals over STFs.

References

1. D. Shanks, J.Math.Phys., 34, 1, (1955).
2. C. Brezinski and M. R. Zaglia, Extrapolation Methods: Theory and Practice, Edition North-Holland, Amsterdam, (1991).
3. J. C. Slater, Phys. Rev., 42, 33, (1932).
4. E. O. Steinborn and E. Filter, Theor.Chim.Acta. 38, 273, (1975).
5. E. Filter and E. O. Steinborn, Phys.Rev.A. 18, 1, (1978).
6. E. J. Weniger and E. O. Steinborn, J.Math.Phys., 30 (4), 774, (1989).
7. E. J. Weniger and E. O. Steinborn, Phys.Rev.A. 28, 2026, (1983).
8. E. J. Weniger and E. O. Steinborn, J.Chem.Phys. 78, 6121, (1983).
9. H. P. Trivedi and E. O. Steinborn, Phys.Rev.A. 27, 670, (1983).
10. J. Grotendorst and E. O. Steinborn, Phys.Rev.A. 38, 3857, (1988).
11. P. Wynn, Math.Tables Aids Comput., 10, 91, (1956).
12. D. Levin, Int.J.Comput.Math., B 3, 371, (1973).
13. A. Sidi, J. Inst. Maths. Applics., 26, 1, (1980).
14. A. Sidi, J. Comp. Appl. Math., 78, 125, (1997).
15. A. Sidi, J.Inst.Maths.Applics. 24, 327, (1979).
16. E. U. Condon and G. H. Shortley, The theory of atomic spectra, (Cambridge University Press, Cambridge, England), (1970).
17. E. J. Weniger and E. O. Steinborn, J.Math.Phys., 24, 2553, (1983).
18. G. B. Arfken and H. J. Weber, Mathematical methods for Physicists, Academic Press, Fourth edition (1995).
19. H. H. H. Homeier and E. O. Steinborn, Int.J.Quantum Chem. 41, 399, (1992).
20. E. O. Steinborn and H. H. H. Homeier, Int.J.Quantum.Chem. 24, 349, (1990).
21. H. H. H. Homeier and E. O. Steinborn, J.Comput.Phys. 87, 61, (1990).
22. H. H. H. Homeier and E. O. Steinborn, Int.J.Quantum.Chem. 39, 625, (1991).
23. H. Safouhi, D. Pinchon and P. E. Hoggan, Int.J.Quan.Chem., 70, 181, (1998).
24. H. Safouhi and P. E. Hoggan, J.Math.Chem. 25, 259, (1999).
25. D. Levin and A. Sidi, Appl.Math.Comput., 9, 175, (1981).
26. H. Safouhi, J.Comp.Phys., 165, 473, (2000).
27. H. Safouhi, J. Phys. A : Math. Gen., 34, 2801, (2001).
28. H. Safouhi, J. Comp. Phys., 176, 1-19, (2002).
29. L. Berlu and H. Safouhi, J. Phys. A : Math. Gen., 36, 11791, (2003).
30. H. Safouhi, J. Phys. A : Math. Gen., 35, 9685, (2002).

Parallelization of Reaction Dynamics Codes Using P-GRADE: A Case Study

Ákos Bencsura and György Lendvay

Institute of Chemistry, Chemical Research Center, Hungarian Academy of Sciences
P.O. Box 17, H-1525 Budapest, Hungary
{bencsura, lendvay}@chemres.hu
http://www.chemres.hu/eng/ic/depts/theoch.html

Abstract. P-GRADE, a graphical tool and programming environment was used to parallelize atomic level reaction dynamics codes. In the reported case study a classical trajectory code written in FORTRAN has been parallelized. The selected level was coarse grain parallelization. P-GRADE allowed us to use automatic schemes, out of which the task farm was selected. The FORTRAN code was separated into an input/output and a working section. The former, enhanced by a data transfer section operates on the master, the latter on the slaves. Small sections for data transfer were written in C language. The P-GRADE environment offers a user-friendly way of monitoring the efficiency of the parallelization. On a 20-processor NPACI Rocks cluster the speed-up is 99 percent proportional to the number of processors.

1 Introduction

Computer simulation of dynamical processes at the atomic level has been actively pursued for decades. Although the foundations of the methods were laid down long ago, modeling of polyatomic systems started only in the last few years when the speed and amount of computers achieved a critical level. The calculations can be sped up either by continuously purchasing faster and faster computers or by increasing the number of processors used for a single calculation. One of the recent achievements of computer technology is the rise of parallel and Grid computing, which further extends the resources available for a calculation [1,2,3]. In the following we briefly overview the issues arising when a user experienced in modeling molecular processes decides to switch to parallel programming. Our overview is strictly user-oriented. Doing this, we hope to help others familiar with scientific FORTRAN programming in finding efficient ways of parallelizing existing dynamics codes, without going through extensive education in computer science.

When a user decides to use a large number of computers, he needs to increase the number of processors that he can access. The possible ways of getting access to multiprocessor environments is a.) operating a large number of individual machines connected to a local network but operated and managed separately;

b.) setting up a cluster from a large number of single- or dual processor machines, c.) getting a supercomputer where an individual manufacturer produces a computer with multiple processors and the software that enables the available resources to be used efficiently, and d.) getting access to a computer Grid. From the point of view of an individual user who does have some resources and would like to develop them continuously, option b.), i.e. building a cluster from the existing resources seems to be a good starting point. This option has the advantage that the machines can be utilized all through their physical lifetime, and, by connecting them into a cluster, the management load is reduced spectacularly. An additional step may be if one gets access to a Grid by sharing his existing resources with those of other, geographically separated laboratories because then the resources can be utilized more efficiently. Once the resources are available, the codes that are used for modeling of molecular processes need to be altered so that they should be able to utilize the multiprocessor environment. Computer codes written for applications in the field of molecular modeling are generally prepared for sequential execution, i.e. the algorithm for the calculation is set up so that various steps are executed one by one, and the results of the computations are used in steps executed later in time. A different paradigm is needed when multiple processors are available for a calculation. The algorithm of the calculation has to be re-organized so that the possibility that several processors perform calculations at the same time could be allowed. The steps of the algorithm have to be set so that the computational tasks could be assigned to processors running simultaneously. Of course, the programmer still needs to take care of collecting data and for assigning tasks to processors so that the data should flow properly. The two major paradigms for executing jobs parallel are the task farm model and the pipeline model. In the first, a *master processor* organizes the data flow by assigning different tasks to the *slaves*, and collects data from them. The data that are produced simultaneously are equivalent in the sense that the slaves do not process data generated by any other slave in the same step. On the other hand, in the pipeline model the slaves process further the data produced by other slaves according to the same principle as vectorization. Once the original sequential code has been reorganized, which often can be easily performed by reorganizing subroutine calls, parallel codes can be created by inserting commands that take care of data flow and direct slaves to perform different tasks. There are "languages" to do this, which interface between the FORTRAN code and the multiprocessor system software. The most commonly accepted are PVM (Parallel Virtual Machine) [4] and MPI (Message Passing Interface) [5]. These commands have to be written carefully as they govern the sharing of tasks between processors. Much easier is the job of parallelization if one uses some object-oriented software. One of such software is P-GRADE (Parallel Grid Run-time and Application Development Environment) [6,7,8,9, 10]. This is a graphical program development tool and running environment. It provides users with clichés for different data flow models from which the user can build the graph of the code using graphical tools. P-GRADE then generates a code using PVM or MPI that enables parallel execution. The code can be debugged and the execution can be monitored using the "Prove" tool of P-GRADE, which graphically displays in real time how the processors are utilized, what time is spent on communication etc. The software prepares codes that can be used not only in a homogeneous environment that is available in a cluster of identical processors running under the same operating system but also on heterogeneous environments available on the Grid (i.e. a collection of different types of machines running under different operating systems). P-GRADE seems to be easy to learn and there are manuals and instructions available on the web [11]. Some case studies have also been presented in the literature [12,13].

In the present case study we describe how one can parallelize a sequential FORTRAN code using P-GRADE. We selected a simple application for this purpose, namely, a classical trajectory calculation. In the following we first describe how the algorithm was adapted for multiprocessor use in Section 2, then in Section 3 we detail how the code was modified using P-GRADE, and finally we present some data on the performance and some results.

2 Re-structuring the FORTRAN Code

The task selected for parallelization is quite simple and is appropriate for beginners: classical trajectory calculation for modeling bimolecular collisions. The basic idea of the procedure is that individual collisions are described by solving classical equations of motion for each atom in the colliding pair of molecules. Macroscopically observable properties are then calculated from the data on individual collisions by averaging over a statistical ensemble of initial conditions corresponding to the system to be modeled, using statistical mechanics. A classical trajectory is the solution of the equations of motion (a system of ordinary differential equations) in time corresponding to a given set of initial conditions, i.e. the path in phase space traced by the coordinates and momenta of the atoms in the system from a reactants-far-away-and-approaching to a products-far-away-and-departing arrangement. The initial conditions are selected by a Monte Carlo procedure that is set up to describe an ensemble that corresponds to the system to be modeled. From the point of view of parallelization it is important to note that calculation of individual trajectories is independent from each other, they can be performed in any sequence or even at different computational speed. This allows one to parallelize the code in the simplest way, i.e. "coarse grain" parallelization. This means that large units of the original code can be used without modification, and little care needs to be paid to the organization of efficient data flow. Classical trajectory codes generally consist of the following steps:

- input of static parameters: data on the potential surface
- input of dynamical parameters: data characterizing the system like atomic masses, temperatures, distance when a collision is considered to be finished
- input of data on the simulation (number of trajectories, random number seed etc.)
- initialization of a random number sequence

- in a loop of length equal to the number of trajectories to be run: generation of initial conditions, integration of equations of motion
- output of trajectory final data, generally in the routine that determines them (like final quantum state of a molecule, relative velocity etc.)
- optionally: calculation of cross sections or rate coefficients by calculating averages over the ensemble. This task is most efficiently done after completion of each trajectory, when the trajectory is put into the bin of final data that corresponds to those of that individual trajectory.

It can easily be seen that this sequence of steps offers very efficient parallelization. It is enough to arrange that working processors run individual trajectories simultaneously until the required number of trajectories is completed. There are many possible ways for doing this. We selected the following. We use the task farm model in which a master organizes the computation, and slaves perform the calculation of individual trajectories. This model is well adapted to our beowulf cluster which is controlled by a front-end machine and a number of computational nodes, a one-to-one mapping of front-end machine to master and nodes to slaves. The front-end machine is then not fully utilized by the parallel program, but this is not the purpose either: it has other tasks to do. The algorithm then is reorganized as follows:

- master reads all parameters
- master broadcasts all parameters to slaves
- master generates N seed numbers using a random number generator, where N is the number of slave processors, and distributes them among the slaves
- master instructs the slaves one by one to start a trajectory and watches for report
- each processor generates random numbers, and from them initial conditions for a trajectory, then solves equations of motion, and finally calculates final data and reports the latter to master. The record of the final data contains, in addition to the actual final data, the serial number of the trajectory, the serial number of the trajectory run on the given processor and the random number seed (this way individual trajectories can be reproduced, e.g. for visualization).
- master receives data from slave; monitors if there are more trajectories to run; if so, returns to the previous step. If not, then writes the collection of final data to file. Optionally performs the binning needed for cross sections.

In order to use the P-GRADE environment the code needs the following alterations at the FORTRAN level:

- the main program has to be made a subroutine
- generation of seed numbers for each node has to be performed in the "main" subroutine
- instructions for starting a trajectory have to be inserted
- the calculation of a trajectory needs to be separated into a distinct set of subroutines

- the section calculating the averages has to be moved to the main subroutine
- the output to disk file has to be moved to the main subroutine
- the FORTRAN segments have to be compiled on each operating system that will be involved in the computation

All the other steps can be performed using P-GRADE. The actual code that we parallelized is based on the 1988 version of VENUS [14] that was earlier heavily modified to match our needs.

3   Using P-GRADE to Organize the Parallel Code

P-GRADE is a high-level graphical environment providing a general tool to develop parallel applications. Central idea of the P-GRADE program is to provide an easy to use general set of programs for developing parallel applications without any knowledge of the commands of communication between processors.

The P-GRADE program provides a graphical programming environment and graphical tools for the whole program development. It has a graphical editor for program development, a debugger for interactive testing and monitoring of the performance.

The program can create both MPI or PVM jobs. It inserts the interface commands into the program so that when compiled, the whole code becomes an individual program. This way the user does not need to learn the language of the message passing software. The user can see a "coarse grain" graphical representation of the parallel code in the sense that a flow chart is visible and

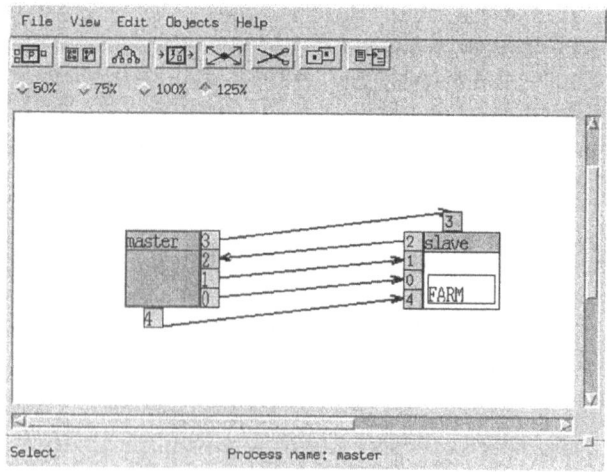

Fig. 1. P-GRADE representation of the classical trajectory program

the fine details are automatically taken care of by P-GRADE. P-GRADE has its own visual programming language, GRAPNEL, and an easy to use graphical

Parallelization of Reaction Dynamics Codes Using P-GRADE   295

editor. The visual editor helps the user to set up the graph of the code (Figs.1 – 3). The units of the graph can be sets of instructions written either in C or pre-compiled FORTRAN subroutines. Currently, FORTRAN commands can not be be used in these boxes so that a certain familiarity with C is necessary.
Fig. 1 shows the pre-built FARM architecture. Arrows between the boxes are graphical representations of parallel data transfer.

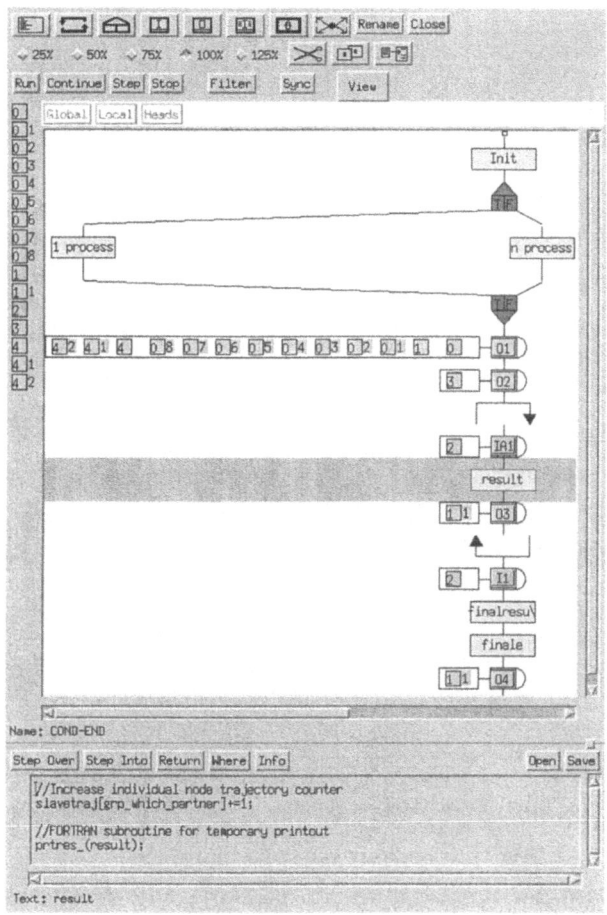

Fig. 2. P-GRADE representation of the master process

Fig. 2 displays the P-GRADE representation of the master program. The program starts with a condition depending on whether the program is going to run in parallel or in sequential mode. The next two lines are data broadcasting instructions where all the necessary input parameters are transferred to the slaves. When a slave receives the last piece of data, it starts a trajectory. The next is the main trajectory calculating loop. Here we collect results from the slave that sends a signal that it finished a trajectory, do some bookkeeping and send back a flag to the slave instructing it whether to continue the calculation or not. Then the master waits for signal from the next slave that finished its task. The content of the highlighted box called "result" can be seen at the bottom of Fig. 2. After collecting the required number of trajectories, each slave gets a "do not continue" signal, and the master analyzes and writes to disk the results and finally instructs all nodes to quit the program.

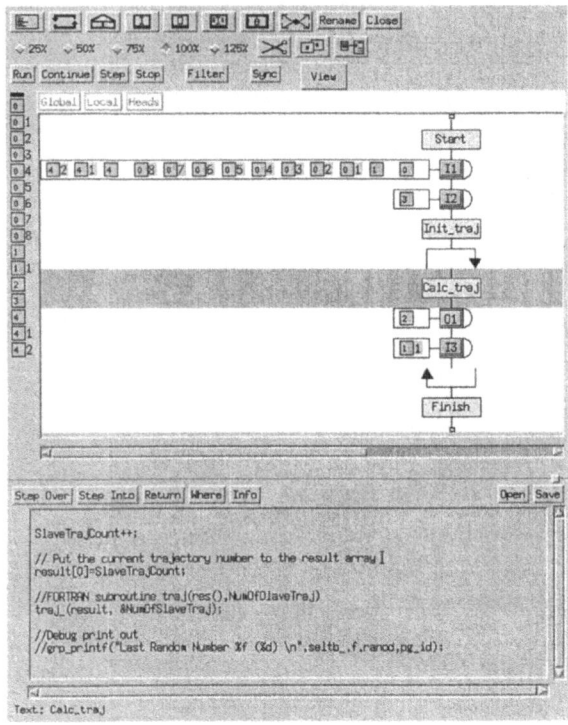

Fig. 3. P-GRADE representation of the slave process

Fig. 3 shows the representation of the slave code. After receiving the data from the master, the slave initializes it own random number generator, and goes into the slave loop. Here it calculates the next trajectory, and sends the results to master, receives the signal whether to continue or not; if so, it goes back to the beginning of the loop; if not, then quits.

4 Program Performance Analysis

Parallel execution of the program was tested on our 20 processor cluster running Rocks v3.0 [15]. The efficiency of the parallelization was examined with the

P-GRADE Monitor that collects trace information about the different events during the program execution. The collected data can be visualized with the PROVE visualization tool of P-GRADE in real time or later from a saved trace file. A result of a test run with 8 processors can be seen in Fig. 4.

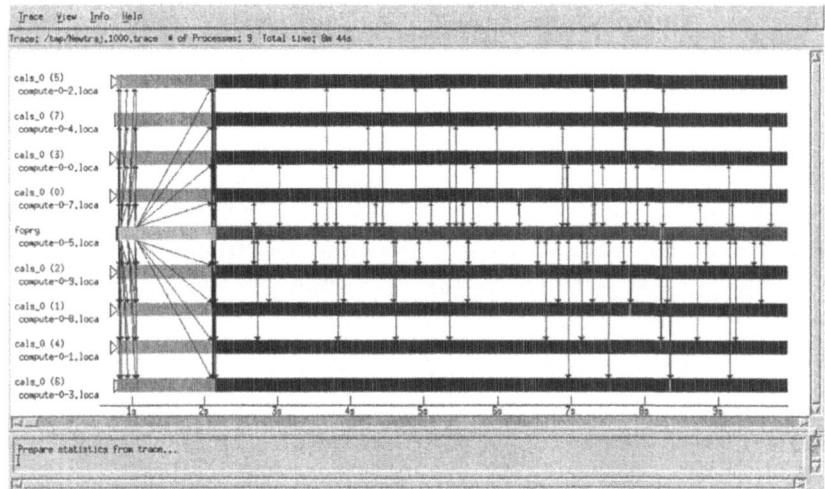

Fig. 4. PROVE visualization of the first 10 seconds of parallel execution of the classical trajectory program

In Fig. 4 different horizontal rows represent different processors with the master in the center row. Shades on each row indicate different process activity, black means calculation, gray (green in color print) communication, and lighter gray means idle. The arrows represent the direction of communication between processors. At the beginning of the run communication (three small and a large set of data) outweighs execution, but after the first couple of seconds all slaves

Table 1. Execution time and speed-up as a function of the number of processors (AMD Athlon 2400) for a calculation generating 10000 trajectories for the H+HF reaction

Proc. number	Execution time/s	Speed-up
1	8826	1.00
2	4444	1.99
4	2203	4.00
8	1116	7.91
16	552	15.99

start doing calculations. The rows corresponding to the slaves are all colored black indicating that there is no idling. We found that during the total execution the slave nodes spend more than 99 percent of the time by doing calculations. The communication time is negligible. This very efficient execution can be seen in Table 1. Data in this table show that the speed-up value is proportional to the number of processors meaning a perfect parallelization up to 16 processors.

5 A Quasiclassical Trajectory Calculation Using the Parallelized Code

As an example, we present data of a series of quasiclassical trajectory calculations of the reactive scattering cross section for the H + HF(v=4) reaction (Fig. 5). An entire curve was obtained within 90 minutes on 16 Athlon-2400 processors. The data indicate a switch between activated and capture-type behavior when the initial relative translational energy decreases as discussed in [16].

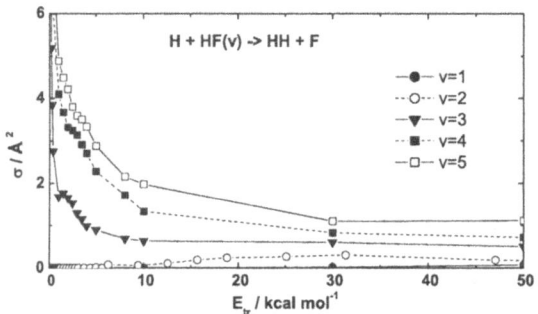

Fig. 5. Cross sections calculated using the parallelized classical trajectory code for the reaction of H atoms with vibrationally excited HF. Each point is obtained from 2000 trajectories.

6 Summary

A classical trajectory program written as a sequential code was parallelized using the P-GRADE graphical environment. From the case study we learned that a user not familiar with message passing can quickly parallelize his code according to the following algorithm: 1. design the parallel algorithm corresponding to the existing sequential code. 2. separate the FORTRAN or C code into units of the parallel algorithm and compile those subprograms 3. draw the graph of the algorithm using P-GRADE 4. fill in any missing boxes with the necessary program units (using C or C++ within the boxes in P-GRADE if no separate codes are available) 5. compile, check efficiency and run.

Acknowledgments. Financial support by the Hungarian Ministry of Education (IKTA00137/2002) and by the Hungarian National Scientific Research Fund (OTKA T29726) is gratefully acknowledged. This work is part of the workgroup METACHEM [3] supported by COST of EU. We thank Prof. P. Kacsuk, Drs. R. Lovas, G. Gombás and G. Hermann for their help with P-GRADE.

References

1. Laganà, A., Riganelli, A. (2000) *Lecture Notes in Chemistry*, **75**, 1.
2. A. Laganà, Towards a Grid-based Molecular Simulator, in: Theory of Chemical Reaction Dynamics, A. Laganà, G. Lendvay, Eds., Kluwer, New York, in press
3. COST Action No. D23, *METACHEM: Metalaboratories for complex computational applications in chemistry* http://costchemistry.epfl.ch
4. A. Geist, A. Beguelin, J. Dongarra, W. Jiang, B. Manchek and V. Sunderam, PVM: Parallel Virtual Machine. A User's Guide and Tutorial for Network Parallel Computing. MIT Press, Cambridge, MA, 1994.
5. Message Passing Interface Forum, MPI: A Message Passing Interface Standard, ftp.mpi-forum.org in pub/docs/.
6. P-GRADE Graphical Parallel Program Development Environment: http://www.lpds.sztaki.hu/index.php?menu=pgrade&load=pgrade.php
7. P. Kacsuk, G. Dózsa: From Supercomputing Programming to Grid Programming by P-GRADE, WESIC 2003, Lillafured, 2003, pp. 483-494
8. P. Kacsuk: Development and Execution of HPC Applications on Clusters and Grid by P-GRADE, European Simulation and Modelling Conference, Naples, Italy, 2003, pp. 6-13.
9. P. Kacsuk: Visual Parallel Programming on SGI Machines, Proc. of the SGI Users' Conference, Krakow, Poland, 2000, pp. 37-56
10. P. Kacsuk, G. Dózsa, R. Lovas: The GRADE Graphical Parallel Programming Environment, Parallel Program Development for Cluster Computing: Methodology, Tools and Integrated Environments (Chapter 10), Nova Science Publishers, New York, 2001, pp. 231-247
11. http://www.lpds.sztaki.hu/pgrade/p_grade/tutorial/tutorial.html
12. R. Lovas, et al., Application of P-GRADE Development Environment in Meteorology., Proc. of DAPSYS'2002, Linz, pp. 30-37, 2002
13. R. Lovas, P. Kacsuk, I. Lagzi, T. Turányi Unified development solution for cluster and grid computing and its application in chemistry ICCSA 2004, Assisi, Italy
14. Hase W.L., Duchovic R.J., Lu D.-H., Swamy K.N., Vande Linde S.R. and Wolf R.J. (1988) VENUS, A General Chemical Dynamics Computer Program.
15. P.M. Papadopulos, M,J. Katz, G. Bruno: NPACI Rocks: Tools and Techniques for Easily Deploying Manageble Linux Clusters, Cluster 2001 http://rocks.npaci.edu
16. E. Bene, G. Lendvay, and G. Póta, Quasiclassical trajectory studies of the dynamics of bimolecular reactions of vibrationally highly excited molecules, in: Theory of Chemical Reaction Dynamics, A. Laganà, G. Lendvay, Eds., Kluwer, New York, in press

Numerical Implementation of Quantum Fluid Dynamics: A Working Example

Fabrizio Esposito

Istituto Metodologie Inorganiche e Plasmi del C.N.R.
via Orabona 4, 70126 Bari Italy
f.esposito@area.ba.cnr.it

Abstract. In the last years new interest has been addressed to the fluid dynamical formulation of quantum mechanics of Madelung-De Broglie-Bohm. Various attempts to implement numerically this formulation have been presented recently concerning photodissociation and scattering, sharing the use of the lagrangian and semi-lagrangian method to discretize the spatial independent variables. The possibility of using moving numerical grids suited for wave packet propagation, in general computationally cheaper than fixed grids, is attractive but difficult when QFD is applied to molecular scattering. This aspect is discussed in the paper, and a particular solution of the problem is proposed. The result for transmission probability through the Eckart barrier as a function of momentum is obtained by Fourier transforming the final wave packet; the agreement with analytical results is good on a large range, the computational time is reasonable and the algorithm used is relatively simple in comparison to other ones proposed for solving this problem.

1 Introduction

The fluid-dynamic formulation of quantum dynamics, initially proposed by Madelung and De Broglie [1,2], is tightly linked to the name of Bohm, because of his well known interpretation of quantum theory [3], alternative [4,5] to the "standard" theory of Bohr, Heisenberg, Born and others. However, this interpretation is a step well beyond the starting point of the formulation - a mere manipulation of the equation of Schroedinger that, particularly in the last few years, has generated a growing interest from a computational point of view. In the original formulation, the wave function $\psi(x)$ is rewritten in the form:

$$\psi(x) = \tilde{R}(x) \cdot \exp(iS(x)/\hbar) \, . \qquad (1)$$

(in the text x and v are used for generally indicating vectors). After substitution into the time dependent Schroedinger equation, separating the real and imaginary parts, one obtains a system of two real equations:

$$\frac{\partial \rho}{\partial t} + \nabla \cdot \left(\rho \frac{\nabla S}{m}\right) = 0 \qquad (2)$$

$$\frac{\partial S}{\partial t} + \frac{1}{2m}(\nabla S)^2 + V + Q = 0 \qquad (3)$$

The first is a continuity equation for probability density, while the second is arranged in order to easily be interpreted as a Hamilton-Jacobi equation, typical of the classical evolution, with the additional term:

$$Q(x) = -\frac{\hbar^2}{2m}\frac{\nabla^2 \tilde{R}}{\tilde{R}} \qquad (4)$$

which is called "quantum potential" (QP). This formulation, totally equivalent to the one based on Schroedinger equation, has a formal beauty in its apparent logical simplicity (all the quantum effects arise from QP, the remainder has a classical-like behaviour) and in the "smooth" way of switching between classical and quantum dynamics; in fact, one could follow the system evolution classically, quantally or semiclassically by respectively neglecting, considering or approximating the quantum potential in eq.(3). It is worthwhile citing a semiclassical application of QFD [6], where the possibilities offered by this formalism are exploited to obtain an approximated model of nonadiabatic dynamics with quite good results, in spite of totally neglecting the quantum potential evaluation during the dynamical evolution (it is implicitly present in the initial conditions). From a stricter computational point of view, it is easy to see that, considering amplitude $\tilde{R}$ and phase S of wave function ψ should be better than using real and imaginary parts A and B of ψ, at least "very often" during time evolution, because of slow variation in time of both functions R and S in comparison to A and B. This is only *generally* true, but in some critical conditions, as shown in practically any recent work about scattering treated by QFD [7-10], being this one the major challenge to face, as will be clear also in this work.

The QFD formalism can be computationally exploited either in the Eulerian or in the Lagrangian description, that is respectively with fixed or "moving with the flow" spatial grids of computation of functions. The first type of spatial discretization is more commonly used, is simpler to be numerically implemented and lets the possibility of freely choosing the best way of evaluation of derivatives on the numerical grid. It has conversely the drawback of *fixing* the extension of the grid to the maximum value (generally the final one) from the beginning to the end of the propagation, although in some cases it is possible to reduce the initial waste of computational time in useless parts of the grid [11]. This problem is of minor importance when considering one-dimensional problems, but it becomes rapidly difficult or intractable for higher dimensions. It is in this computational context that the idea of implementing QFD on a Lagrangian or semi-Lagrangian grid has been developed. It will be briefly introduced in the next section.

2 QFD for Scattering Processes

2.1 Wave Function Scattering from an Eckart Barrier

Despite its more immediate physical justification, the continuity equation for probability density ρ is conveniently replaced, when QFD is computationally implemented, by an equation for the logarithm of ρ, obtained via direct substitution, as suggested by many authors. Here we adopt the form of the wave function and the equations as in [12]:

$$\psi = \exp\{(R+iS)/\hbar\}.$$

(note the difference between R and the preceding $\tilde{R}$), obtaining as a consequence the system of equations:

$$\frac{\partial R}{\partial t} + \frac{1}{m}\left\{\nabla R \cdot \nabla S + \frac{\hbar}{2}\nabla S^2\right\} = 0 \quad (5)$$

$$\frac{\partial S}{\partial t} + \frac{1}{2m}(\nabla S)^2 + V + Q = 0 \quad (6)$$

with m = mass of the system. In this case the QP is rewritten as:

$$Q(x) = -\frac{\hbar^2}{4m}\left\{\nabla^2\left(\frac{2R}{\hbar}\right) + \frac{1}{2}\left[\nabla\left(\frac{2R}{\hbar}\right)\right]^2\right\} \quad (7)$$

As numerical application of QFD, we have chosen the monodimensional scattering of a wave packet colliding with an Eckart barrier. The initial condition for scattering is, as usual, a gaussian wave packet, that in the Madelung form is described by:

$$S(x, t=0) = \hbar k_0 (x - x_0). \quad (8)$$

$$R(x, t=0) = \frac{\hbar}{2} \ln \rho. \quad (9)$$

where ρ is the normalized gaussian:

$$\rho(x) = \frac{1}{\sqrt{2\pi\sigma^2}} \exp\left[-\frac{(x-x_0)^2}{2\sigma^2}\right]. \quad (10)$$

Here x_0 is the initial wave packet center, σ the standard deviation of the gaussian, while $\hbar k_0$ is the initial average momentum (exactly the initial momentum in Bohmian mechanics [4]).

The potential used is the well known Eckart barrier:

$$V(x) = V_o / \cosh^2(\beta x) . \qquad (11)$$

Here V_o is the height of the barrier in energy units, while β determines the barrier width measure. The transmission probability as a function of k is analytically known to be:

$$P(k) = (\cosh(z)-1.0)/(\cosh(z)+\cosh(w)) .$$

with:

$$z = 2\pi k/\beta, \quad w = \pi\sqrt{\frac{8mV_o}{\beta^2 \hbar} - 1} . \qquad (12)$$

One can also define a total transmission probability P_{tot} as a function of initial k_0 by integrating P(k) weighted by the initial probability density ρ as a function of k, or by integrating the transmitted part of $\rho x, t = t_{final}$) [13]. Of course P_{tot} is a mean, so it is an indicator of the *global* agreement between analytical and numerical results. It numerically depends on the integration extrema on k domain, which are not univocal defined (wave packet propagation typically produces results on k domain of decreasing quality for increasing distance from k_0). P_{tot} is not be able to reveal if somewhere errors compensate each other.

Concerning the general time evolution of the wave packet with an Eckart barrier, one expects to see a separation in two parts, a smooth Gaussian-like transmitted part and a more complicated reflected packet, where interference effects should produce small fluctuations. These ripples are difficult to reproduce correctly, are produced by variations in the QP that can become very rapid and can eventually bring QFD calculations to a crash.

2.2 Semi-lagrangian Grids for QFD

The system in (5-6) is implicitly written in the Eulerian description. Switching to semi-Lagrangian (also indicated as arbitrary Lagrangian-Eulerian [9]) framework is readily obtained by simply taking into account that in this case the grid point x is moving at a velocity $v_{grid}(x)$, therefore variation in time of any function evaluated in x has to be calculated by total derivative:

$$d/dt = \partial/\partial t + v_{grid} \cdot \nabla . \qquad (13)$$

It is worthwhile to underline that v_{grid} is the velocity of the x grid point, *not necessarily* the flow velocity v (the velocity of a classical point subject to the sum of interaction and quantum potential V+Q; in this context it is given by $\nabla S/m$): if this is the case, the framework is Lagrangian, and the entire method of numerical time evolution can be called QTM (quantum trajectory method [7]). Despite the physical interpretation of these trajectories (these are just the trajectories of the Bohmian theory, see [4]), it seems that this is not an easy way of performing molecular scattering calculations [9]. In fact, Bohmian trajectories populate and depopulate some regions of wave packet propagation, depending on forces arising from interaction-quantum potential V+Q. This is desirable when one wants to "see" in a classical-like manner the wave packet time evolution [14,15], but can be inadequate if the wave function and its derivatives are evaluated exclusively on the points of this highly nonuniform grid. Taking into account (13) applied to R and S in (5-6), one easily obtains the semi-Lagrangian form of QFD:

$$\frac{dS}{dt} = v_{grid} \cdot \nabla S - \frac{1}{2m}(\nabla S)^2 - V(x) - Q(x) \quad (14)$$

$$\frac{dR}{dt} = v_{grid} \cdot \nabla R - \frac{1}{m}\left\{\nabla R \cdot \nabla S + \frac{\hbar}{2}\nabla^2 S\right\} \quad (15)$$

In this case, any prescription for moving the grid points can be put into the equations by simply choosing the appropriate vector field v_{grid}.

2.3 Calculating Derivatives on an Irregular Grid

If the grid points are irregularly spaced, the general strategy to calculate derivatives of a function g on the grid is to differentiate an interpolation function f fitted on the grid values of g. In QFD two methods have been principally used: the radial basis function interpolation (RBF [12]), which is a global method (but that can be used locally as in [12]), and the moving least square (MLS) method [7], a local method which has been used in this work. Essentially in MLS a least square minimization is calculated between the interpolating function f (a polynomial whose coefficients are to be determined) and the interpolated function g in the neighborhood of the point of interest x'. Each time the interpolation for a point x' on the grid is perfumed, one has to re-determine the neighborhood I(x') on the base of the points x_i nearest to x' in the grid. Each evaluation point x_i can be usefully weighted with a function of distance from x' as in [7], in order to give more importance to points nearest to x' in the neighborhood. In QFD for scattering, our experience is that choosing a correct way of defining this neighborhood is fundamental, because the time propagation of the numerical system is extremely sensitive to small variations in the parameters of the neighborhood. In this work, following [7] and many successive works, the criterion is to fix the number of points in I(x'), and then fitting the needed quantities taking as weight function a decaying exponential of the distance $|x_i-x'|$:

$$w(x) = \exp\left(-\frac{|x_i - x'|}{R_{max}}\right) \quad (16)$$

Following literature, the radius R_{max} is adjusted in order to give a weight 0.1 to the farthest point of $I(x')$. It could not be the best choice, especially when points are free to go away without restrictions. When the wave packet separates in the transmitted and reflected parts, in the middle the point density is rapidly decreasing and large errors could be generated by using points too far from x' (having fixed the number of points around x'). On the other end, the "crowding" of points is not better, because in that case R_{max} becomes too small and the fitting can amplify local errors up to numerical "explosion" of the code, generally through the cycle: excessive variation of the quantum potential, which generates a violently variating trend in $R(x)$, which in turn amplifies the initial error in QP, and so on. Of course the number np_{int} of points in $I(x')$ and the degree n_{pol} of the polynomial for interpolation is crucial, and much unsuccessful experimentation is necessary before arriving to some conclusion different from a crash. As will be clear in the result section, our preference is for a very low degree of polynomial and np_{int}. In this work a correction has been attempted to low R_{max} problems by simply imposing lower bound value $R^o{}_{max}$ for it, but it is really difficult to extract a general conclusion for the utility of this procedure. If a certain degree of stability is gained in this way, surely it is not determinant, unless $R^o{}_{max}$ is so large that the weight function becomes useless (while its importance in correctly smoothing interpolations is clear). Kendrick [16] in an ample and complex paper on the argument reports of a totally different way of determining R_{max}, depending on the maximum value of $R(x)$ in each time step and from two empirical parameters that are to be determined empirically, with very good behaviour. However, at this very "exploratory" stage of attempts this feature as not been implemented in the numerical code for this work, in order to have to determine the lowest possible number of parameters and to better understand what actually works and what should be changed.

2.4 Distributing the Grid Points

Strictly linked to function interpolation is "regridding". The point x' of the preceding paragraph is not necessarily a point of the grid, therefore it is possible to redistribute quite simply and inexpensively the whole grid while interpolating the function of interest in order to obtain its derivatives. It is only necessary to specify a way of making the new grid. Apparently it would be better to have many grid points where there is rapid function variation. The worst function to be evaluated is surely the QP, because of both its alternating trend and large variation near critical points and its direct influence on other functions. We have applied a simple equidistribution principle [17], trying to use as monitor function ρ and QP (separately), obtaining only more rapid crashes of the system than without equidistribution. We have also tried using equidistribution with "moving" mean values of these functions, in order to smooth too rapid variations. This is obtained by averaging function values in the same kind of neighborhood I(x') with the same weight function used for interpolation and the same R_{max}. Other possibilities have been explored using the same procedure but weighting with a function χ obtained by renormalizing the sum of ρ with a lower bound 0.1 added, in order to apply equidistribution only to higher peaks of ρ, with unsuccessful results. The underlying problem is that a redistribution good for QP function can be unsuited for another one. However, these smoothing procedures are interesting when applied to the grid velocity field, used to move the computational points, as explained in the next paragraph. The best use of regridding in QFD for scattering seems that operated on a uniform grid, the simplest one, as also noted in [16]. The uniform grid is constructed between the extrema of the grid obtained by dynamics. In this case the aim of the uniform redistribution is that, fundamental, of avoiding intersection of grid point trajectories.

2.5 Moving the Grid Points

At a first glance, the criterion for moving grid points should be: the *right* number of points should stay at the *right* place and time. Difficulties have been encountered in the application of this desirable criterion in our experiments as well as in the cited literature about scattering, because of lack of stability of the resulting dynamical code. The problem is that to guess the *right* values of the criterion in advance of at least one time step is relatively easy, *but* in some critical points, when all the useful functions are dramatically changing, and the application of the criterion generally accelerates the crash instead of preventing it. Many attempts have been done in this work in order to find a reasonable approximation of the criterion, exploiting the semi-Lagrangian implementation of QFD. As a first approximation, the use of center of mass velocity as grid velocity has been implemented, with a rigid translation of the grid. The center of mass position and velocity of the wave packet is calculated by weighting positions and flow velocities (∇S/m) in all grid points with the probability density ρ in those points. No numerical crash has been observed in this case (provided a reasonable interpolation is used, as illustrated in the next section), the grid follows the incident wave packet very well up to collision with the barrier, then it follows essentially the most massive of the two dividing parts of the wave packet. The result is encouraging, even if it is a first approximation: a "split" version of this procedure is in preparation, with two independent centers of mass and grids, one for each side of the barrier, and with expanding grids as the packets enlarge in time. This enlargement can be approximately followed numerically, by linking the dimensions of the grids to the standard deviations of density ρ on the left and right sides of the barrier independently.

Another partial success of application of semi-Lagrangian grids has been achieved by constructing a smoothed velocity field with the procedure explained in the preceding paragraph. In this case we tried with weights based on distance, on probability density and on both ones, as indicated here:

$$w(x) = \exp\left(-\frac{|x_i - x'|}{R_{max}}\right) \cdot \exp(-\chi) \tag{17}$$

where χ is defined in the preceding paragraph. If used with regridding, the simpler function of the distance $\exp(-|x_i-x'|/R_{max})$ has given some good results. The reason is in preventing crashes by very simply avoiding too sharp peaks in velocity field, independently on density. The action of this velocity smoothing is similar but much simpler than other solutions proposed in literature [9,16]. The ρ-depending weight has been formulated with the aim of reproducing *locally* a behaviour similar to that of center-of-mass-driven grids, because a moving mean value of flow velocity weighted with ρ in I(x') is essentially a local definition of the center of mass velocity. In this case the problem is the collisions among many different parts of the grid (a sort of small local grids), a problem that requires further studies.

A relevant problem during this experimentation is that the result (or the crash) depends on many different delicate parts of the numerical code: the parameters of the dynamics, the initial grid, the time step, the integration method, the interpolation of functions, the movement of the grid; all these parts are strictly interdependent, giving often unexpected results (for example a worse behaviour decreasing the time step, probably due to larger interpolation problems).

3 Numerical Results

After having discussed throughout the paper essentially of crashes of our system, the time of the good news has now arrived. In the numerical experiments of this work, the Eckart barrier is centered in the origin of the frame of reference, with $V_0 = 0.8$ eV and $\beta=0.5\text{Å}^{-1}$. The gaussian wave packet has a standard deviation $\sigma=0.5\text{Å}$, as a consequence the standard width σ_k of probability density as a function of k is 1Å^{-1}. The semi-extension γ of the initial packet, which is also the semi-extension of the initial grid, is 6σ, a detail often neglected in literature, where the typical extension (indirectly deduced) is one half of this value. Actually in QFD experiments of this work this "large" γ has been important, probably due to the smoother behaviour at the edges of all the interpolated functions. Good values for γ are in the range 5-7σ, independently of σ size. The position of packet center is $x_0= -\gamma\text{-}8/\beta$, chosen in order to minimize the initial interaction with the potential. The mass of the system is 54.9

a.u., $k_0 = 3.5\text{Å}^{-1}$. Three methods of integration have been used, a simple explicit Euler method, a second order Runge Kutta method and the central differencing scheme of integrator proposed in [12]. This last is well known as a stable integrator for the Schroedinger equation; however, in this case the problem is complicated by the need of expanding the grid increasing the point number. This means changing the grid both in a (semi-)lagrangian way (during an interval Δt) and by interpolation (at fixed time), and these operations introduce new possible causes of instabilities. The time step is $\Delta t_0 = 0.02$fs, with some experiments with values five times higher or lower. The initial number of grid points is 100, and the resulting spatial step Δx has been maintained as constant as possible in different ways:

1) by letting the system evolve in a Lagrangian way until any two points are too much near (typically $0.5\Delta x$) or far ($1.2\Delta x$): at that moment a uniform regridding is operated restoring the original point density; if central differencing integration is performed, the next step is integrated with the Euler method.

2) by regridding uniformly at each time step, before integrating the equations in the Lagrangian framework; this operation is made while interpolating for obtaining the derivatives of R and S, so no additional computational load is required. In this way one actually operates always on an almost uniform grid (the disomogeneity is due only to variations in one time step);

3) by integrating the equations for R and S in a semi-Lagrangian way, imposing a grid velocity field which has been:

3a) uniform, following the center of mass of the packet (but the reflected packet "slides away" at a certain point: this have to be corrected in future work);

3b) obtained by a moving mean of the the flow velocity; a spatial step check with possible regridding is necessary in this case;

3c) obtained by calculating velocities necessary to go in time Δt from initial positions known at time t to final positions at time $t+\Delta t$. Final positions can be determined by uniformly distributing points between two extrema that are calculated in a Lagrangian way from extrema at time t. This procedure is suggested in [16]. Of all these attempted paths, the best seems to be the second one with Runge-Kutta integrator, because for similar obtained results is, globally, the fastest one with good accuracy.

The interpolation is made with polynomials of degree n_{pol} from 2 to 5, with a number of points of interpolation np_{int} which cannot be smaller than 5 for $n_{pol}=2$, than 7 for $n_{pol}=3$, than 15 for $n_{pol}=5$, for avoiding severe instabilities. For any given n_{pol}, higher values of np_{int} increase stability but progressively degrade the final results, because the interpolation becomes less "flexible". The calculation of QP is the key problem, because its interpolation should be accurate everywhere but, possible, in critical points where the wave function has a node and the QP goes to infinity: in that case a rough but *finite* approximation is probably better than accurate result. In fig.1 the comparison between analytical and numerical transmission probabilities as a function of k is shown in the range $k_0 \pm 3\sigma_k$. The agreement is good in the whole interval, with maximum error less than 0.3% for probability values near 1 (around $k=3.8\text{Å}^{-1}$). Comparison of total probability reports an error of the order 10^{-3}, but this does not seems a good indicator of quality, because many times occurred that similar values appeared while not very good agreement was visible for small compensating oscillations of transmission probability around analytical curve. The result in fig.1 is interesting in the sense that it has been obtained by QFD formulation with semi-Lagrangian scheme in approximately 890 seconds of a 466 MHz personal computer. In this case second order polynomial with $np_{int}=5$ has been used. Many other successful runs have been made in order to be sure of reliability of the numerical code, in particular changing the initial k_0 to values $2,3,4,5\text{Å}^{-1}$, with results essentially identical in the whole range $k_0 \pm 3\sigma_k$. This is important because changing k_0 means changing drastically the fraction of the packet that should pass through the barrier.

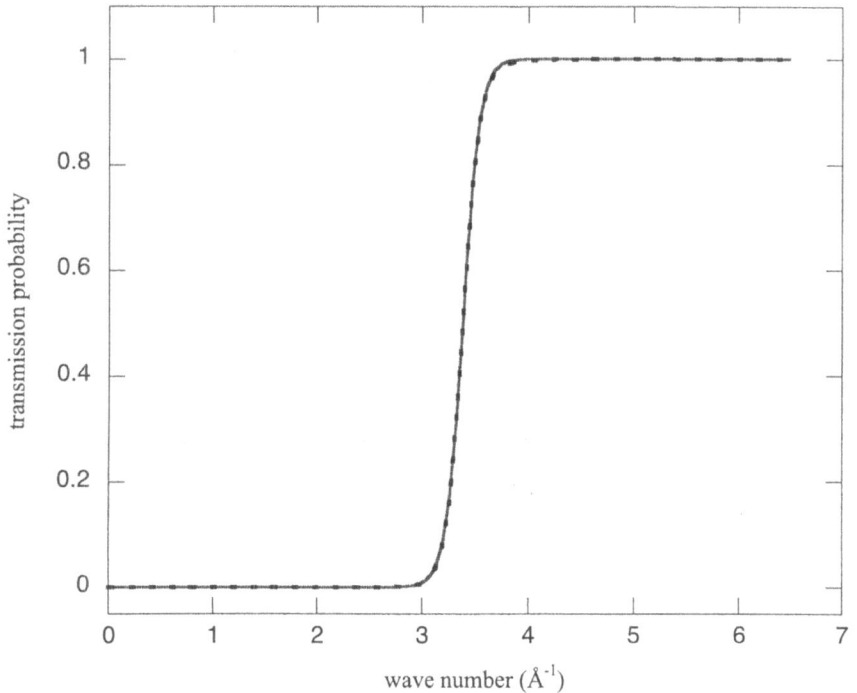

Fig. 1. Transmission probability as a function of k: analytical result (continuous line), numerical results (dotted line).

If higher accuracy is required, the best action is to increase interpolation quality by increasing polynomial degree and np_{int}. For example, for an agreement one order of magnitude better for total probability than that reported, it is possible to use $n_{pol}=5$, $np_{int}=15$, with a computational time 3.5 times larger.

Concerning the reflected part of the packet, it is highly dependent on QP calculation quality, because the expected ripples are due to dramatic variations in QP. Changing slightly interpolation features (e.g. np$_{int}$ from 5 to 6) can modify deeply the whole dynamics of reflected packet, after a sufficient amount of propagation time, changing results or determining a numerical crash. The drastic difference in behaviour between transmitted and reflected wave packet parts suggests the need for two separate numerical treatments as soon as the initial packet starts dividing.

It is worthwhile showing a semiclassical approximation that can be derived very simply from the QFD scheme. It is based on the fact that large part of the tunneling through the barrier is due in this formulation of quantum mechanics to passing or not *over* the barrier by the energy gained in the free propagation expansion of the wave packet, as suggested in [7]. Said in other words, the QP tends to "expand" the packet accelerating wave packet parts towards the barrier and decelerating the others: the first ones "tunnel" through the barrier, the others will be kicked back (the approximation here is in lack of consideration of QP action very near the barrier). Therefore it should be possible to have a semiclassical approximation by simply "switching off" the QP before a significant interaction with the barrier. The result is in fig.2, where QP has been used only for 10fs on about 80 fs of total collision time, essentially in conditions of free propagation, that are reproducible analytically [4]. It is clear the same ascending trend typical of a quantum interaction (classically there would be only the values 0 and 1 with a vertical transition, not a smooth curve). The plateau of the analytic curve for k>3.8Å^{-1} is not reproduced correctly: the trend of the curve is too high for low k values and decreasing for higher k values. This can be explained by the lacking of the "expansion" of the transmitted wave packet due to the lacking of QP in the final part of the semiclassical propagation. Different initial conditions for barrier and packet have given analogous results. Introducing a QP approximation during the whole collision can improve significantly the result, as will be shown in a future work on this topic [18]. However, it is of interest that by only introducing a sort of "quantum pre-propagation" of the packet (essentially a free analitycal propagation) deeply modifies the result in a semiclassical way.

4 Conclusions

Numerical implementation of QFD for scattering processes in the semi-Lagrangian framework is a quite difficult task, due to instabilities essentially introduced by the calculation of quantum potential. In this work the feasibility of a relatively simple calculation has been studied, in order to understand if the method can be used practically in a version as simple and fast as possible. The answer is surely affirmative for the transmitted (reactive) part of the problem, with a suggestion for a simple semiclassical model. For the reflected part much higher accuracy is required in the whole numerical code.

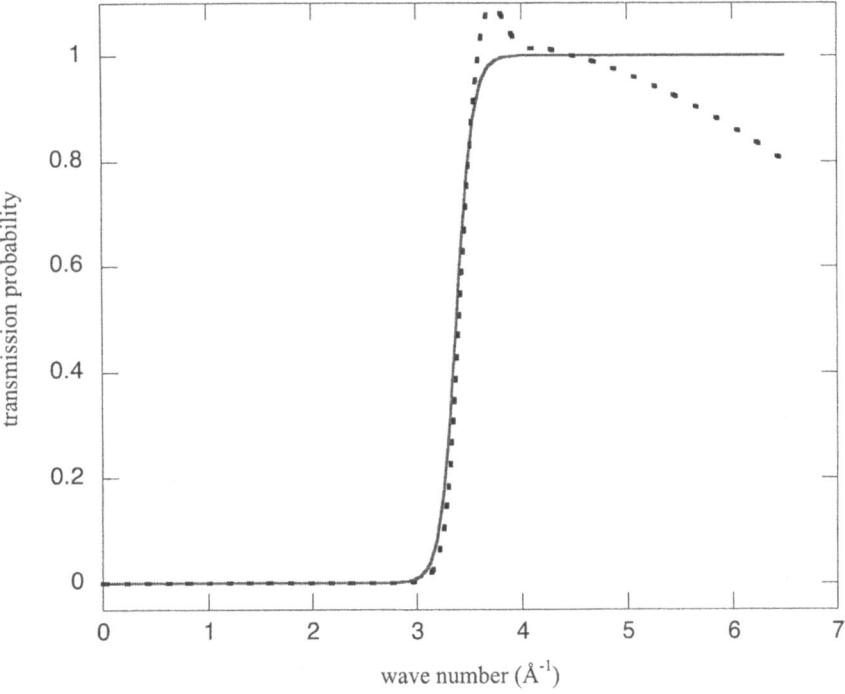

Fig. 2. Semiclassical approximation of transmission probability as a function of k: analytical result (continuous line), numerical results (dotted line).

The "working example" of the title is in this paper the collision with the Eckart barrier, which is a sort of minimal try for any dynamical method. A much more difficult and stringent test is the symmetric double maximum barrier in [19], where an accurate numerical P(k) is available, showing a sharp perfect transmission peak superposed to a smooth Eckart trend. Work is in progress to try reproducing quantitatively this result with QFD: at this moment a qualitative result has been obtained using numerical conditions very similar to those adopted for the Eckart barrier.

Acknowledgments. The author would like to thank Dr.Damiano Pagano for some discussions on QFD related problems. This work was supported by MIUR (cof.2003 Project 2003037912_010)

References

1. Madelung, V.E., Z.Phys.40 (1926) 332
2. De Broglie, L., C.R.Acad. Sci.Paris, 183 (1926) 447

3. Bohm, D.: A Suggested Interpretation of the Quantum Theory in Terms of "Hidden" Variables. I, Phys.Rev. 85 (1952) 166; A Suggested Interpretation of the Quantum Theory in Terms of "Hidden" Variables. II 85 (1952) 180; Bohm, D., and Hiley B.J., The Undivided Universe: An Ontological Interpretation of Quantum Theory. Routledge, London (1993)
4. Holland, P.R., The Quantum Theory of Motion. Cambridge University Press. New York (1993)
5. Selleri, F. The Wave-Particle Duality. New York/ London: Plenum, 1992
6. Burant,J.C. and Tully,J.C.: Nonadiabatic dynamics via the classical limit Schroedinger equation, Journal of Chemical Physics 112 (2000) 6097
7. Lopreore, C.L.,Wyatt, R.E.: Quantum Wave Packet Dynamics with Trajectories, Physical Review Letters 313 (1999) 189
8. Bittner, E.R., Wyatt, R.E.,: Integrating the quantum Hamilton–Jacobi equations by wavefront expansion and phase space analysis, Journal of Chemical Physics 113 (2000) 8888
9. Trahan, C.J.and Wyatt, R.E.: An arbitrary Lagrangian-Eulerian approach to solving the quantum hydrodynamic equations of motion: equidistribution with "smart" springs, Journal of Chemical Physics, 118 (2003) 4784-4790 and references therein.
10. Pagano, D. "Application of the Hydrodynamic Formulation of Quantum Mechanics to Atom-Surface Dynamics". Ph.D Thesis in Chimica dei Materiali Innovativi-XV Ciclo, University of Bari (2003)
11. Leforestier,C.,Bergeron,G. and Hiberty,P.C.: A quantum mechanical investigation of a collinear model for collision-induced dissociation, Chemical Physics Letters 84 (1981) 385-389
12. Hu, X., Ho, T. and Rabitz, H.: Multivariate Radial Basis Interpolation for Quantum Fluid Dynamical Equations, Computers and mathematics with applications 43 (2002) 525-537
13. Billing,G.D. and Mikkelsen,K.V., Advanced Molecular Dynamics and Chemical Kinetics, Chap.5
14. Kuppermann,A.: A Bohmian view of quantum reaction dynamics, oral presentation, Vth Workshop on Quantum Reaction Scattering, Dept.of Chemistry, University of Perugia (Italy), 25-27 June, 1999
15. Oriols,X., Martin,F. and Sune,J.: Oscillatory Bohm trajectories in resonant tunneling structures, Solid State Communications, 99(1996)123-128
16. Kendrick, B.K.: A new method for solving the quantum hydrodynamic equations of motion, Journal of Chemical Physics 119 (2003) 5805-5817
17. Baines, M.J.: Grid adaptation via node movement, Applied Numerical Mathematics 26 (1998) 77-96
18. Esposito.F, work in preparation
19. Manolopoulos,D.E. and J.C.Light: A log derivative formulation of reaction rate theory, Chemical Physics Letters 216 (1993) 18-26

Numerical Revelation and Analysis of Critical Ignition Conditions for Branch Chain Reactions by Hamiltonian Systematization Methods of Kinetic Models

Gagik A. Martoyan and Levon A. Tavadyan

Institute of Chemical Physics, National Academy of Sciences,
Republic of Armenia. 5/2 Sevak Street, Yerevan 375014, Armenia
Fax:(3742) 28-17-42, tavadyan@ichph.sci.am

Abstracts. A new numerical method to calculate explosion limits of gas mixtures based on Hamiltonian systematization of kinetic models is offered. As a criterion for the critical state of reaction system the extreme behaviour of functions describing total concentration of chemical species is selected.

The present approach enables numerical calculation of kinetic significance of the individual steps and species in the critical ignition conditions of the reaction. The abilities of this numerical method to calculate explosion limits are illustrated by means of the computer programme VALKIN for the reaction mechanism of hydrogen oxidation.

1 Introduction

The study of critical ignition conditions (explosion limits) is of considerable interest in both theory and the control of combustion processes [1, 2]. In critical ignition conditions for the branch chain reaction, at insignificant changes of reaction system parameters, qualitative transfer occurs from the slow process into the fast auto-accelerated one, leading to explosion of reaction mixtures. The task of determining explosion limits is gracefully solved in an monocentric approach [1, 2]. However, in case of kinetic models with a great number of individual steps the task of determination and analysis of critical ignition conditions is often limited by their strongly nonlinear dynamics [3-5].

The purpose of the present work is to offer an universal numerical method for the:
- revelation of explosion limits of reaction mixtures
- determination of kinetic significance of individual steps and species in the reaction model in these conditions.

2 Theoretical Fundamentals

The characteristic pecularities of the proposed approach is the critical ignition condition of chain reactions, which is considered to be an evolutionary state of the reaction mixture. As a criterion to determine the explosion limits the extreme behaviour of the function of particle total concentration is chosen. Based upon this,

calculus of variations is used for numerical calculations of the explosion limits, particularly using the Pontrjagin principle of maximum [6]. At the same time, by the method of Hamiltonian systematization, it is possible to reveal the role of steps and species in reaction systems [7-9].
The procedure of the calculation is performed steps for isothermal and non-isothermal regimes of the reaction.

2.1 Isothermal Case

The calculation includes the following stages:
a) Selection of the target control functional, $I(t)$:

$$I(t) = N(t) = \sum_{i=1}^{m} c_i(t) \tag{1}$$

where, in case of low expenditure of initial substances: $I(t)=N_{ac}(t)$ in which c_i is the concentration of the ith chemical specie of the reaction, N is the total concentration of the reaction species, N_{ac} is the total concentration of the chain carriers and m is the number of reaction species.

b) Recording of the criterion for determing the explosion limits.
In critical conditions, to which correspond definite initial conditions, the behaviour of the reaction system is extremal: $\delta N(t) = 0$ or $\delta N_{ac}(t) = 0$, which is equivalent to the condition:

$$I(t) = \int_0^t \frac{dN_{ac}}{dt} dt = \text{extremum} \tag{2}$$

c) Setting up the kinetic equations

$$\frac{dc_i}{dt} = f_i = f_i^+ - f_i^- + S_i \qquad i = 1, 2, ...m \tag{3}$$

f_i^+, f_i^- is the rate of formation and expenditure of ith specie; S_i is external source of the ith reaction specie; f_i is the rate of the change of the ith specie's concentration.

$$H = \psi_0 f_0 + \sum_{i=1}^{m} \psi_i f_i \qquad (k,c,\phi) \tag{4}$$

d) Setting up the Hamiltonian H with selection of the independent control parameter ϕ (e.g. initial temperature T, initial pressure P, rate constants, eta.):

where: $f_0 = \dfrac{dN}{dt}$, $\psi_0 = 1$ or -1 in case of solving for the maximum or minimum, respectively, c is the m-vector for the concentration of $c(t)$ species, $c(t_o)=c^o$, k is
the n vector of the rate constant and $\Psi(t)$ is the function conjucates to the concentrations c.

e) Setting up a system of differential equations for the conjugate functions $\Psi_i(t)$ (value of species, see equation (13))

$$\frac{d\psi_i}{dt} = -\frac{\partial H}{\partial c_i} \qquad i = 1, 2, \ldots m \tag{5}$$

f) Determination of the conjugate function $\Psi_i(t_o)$ for the initial moment of time [8-10]
g) Finding of the optimum of the parametre ϕ. In respect to the principle of maximum the conditions for the optimum are:

$$\sup H(\psi^*, c^*, \phi^*) = 0 \tag{6}$$

Soluting the system of kinetic equations (3) and the differential equations (5), and taking into account the condition (6) it can be obtained the optimal value of ϕ^*.
The latter, according to the determination (2), corresponds to the explosion limits.

2.2 Non-isothermal Case

When heat factors play a significant role, the special target variation functional is chosen.

$$I(t) = \int_0^t \frac{d(NT)}{dt} dt = \text{extremum} \tag{7}$$

In the functional (7) the magnitude NT correlates with the pressure of reactionary system and condition (7) is equivalent to demanding $\delta P = 0$.
Explosion limits are numerically determined according to condition (6) and (7) from the following systems of equations:

$$\frac{dc_i}{dt} = f_i$$

$$\rho c_v \frac{dT}{dt} = Q^+ - Q^- \tag{8}$$

$$\frac{d\psi_i}{dt} = -\frac{\partial H}{\partial c_i} \qquad i = 1, 2, \ldots, m$$

c_v, and ρ are the specific heat capacity and density of the reactionary system, respectively.
Q^+ and Q^- are the rates of chemical reaction heat generation and heat elimination from the reactionary system, respectively.
In this case, the corresponding Hamiltonian equals to:

$$H = \psi_o \frac{d(NT)}{dt} + \sum_{i=1}^{m+1} \psi_i f_i \tag{9}$$

The value $i = m + 1$ corresponds to the temperature of the system.

2.3 Determination of Kinetic Significance of Individual Steps and Species in Critical Ignition Conditions

To reveal the role of individual steps and species, the equation of the explosion limit $H=0$, looks as follows:

$$F\left(c_1^o, \ldots, c_m^o, T, P, v_1, \ldots, v_n, G_1, \ldots, G_n\right) = 0 \quad \text{or}$$

$$F\left(c_1^o, \ldots, c_m^o, T, P, f_1, \ldots, f_m, \psi_1, \ldots, \psi_m\right) = 0 \quad (10)$$

where c^o, T, P are the initial concentration of the species, the initial temperature and pressure, corresponding to the explosion limit, n is the number of the individual steps. The kinetic significance of individual steps and species is determined through the *value* parameters [7-9].
The *value* individual step is defined as

$$G_j(t) = \frac{\partial F\left[v_1(t), \ldots, v_n(t)\right]}{\partial v_j}\bigg|_{v_j = v_j(t_o)}, \quad j = 1, 2, \ldots, n \quad (11)$$

and enables to define the *value* contribution (h_j) to the step in critical ignition conditions.

$$h_j(t) = v_j(t) \cdot G_j(t) \quad (12)$$

The *value* of the ith reaction system species according to (10) is defined as

$$\psi_i(t) = \frac{\partial F\left[f_1(t), \ldots, f_m(t)\right]}{\partial f_i}\bigg|_{f_i = f_i(t_o)}, \quad i = 1, 2, \ldots, m \quad (13)$$

3 Illustrative Example

3.1 Numerical Calculations of Explosion Limits in a Hydrogen-Oxygen Mixture

Numerical calculation of explosion limits in hydrogen oxygen mixtures were carried out for the isothermal case, according to a kinetic model of the reaction [4] represented in Fig 1. Numerical calculations have been performed using the computer programme VALKIN, developed by us. The algorithm of the programme is developed on the basis of the Hamiltonian systematization of mathematical models of chemical reaction systems.
For the integration of differential equations the ROW4A algorithm developed by Gottwald [10] is used.

Table 1. Kinetic model and values of the rate constants for the $H_2 - O_2$ system [4] (units in mol, cm, s, degree Kelvin, cal).

no.	reaction	A	n	E/R[1]
1	$HO_2 + H \rightarrow H_2 + O_2$	2.50E13[2]	0	3.50E2
2	$H_2 + O_2 \rightarrow HO_2 + H$	3.10E13	0	2.87E4
3	$H + O_2 \rightarrow OH + O$	2.30E14	0	8.40E3
4	$OH + O \rightarrow H + O_2$	3.00E12	0.28	0
5	$O + H_2 \rightarrow OH + H$	1.80E10	1	4.48E3
6	$OH + H \rightarrow O + H_2$	8.30E09	1	5.50E3
7	$OH + H_2 \rightarrow H_2O + H$	2.50E13	0	2.60E3
8	$H_2O + H \rightarrow OH + H_2$	1.40E14	-0.03	1.02E4
9	$O + H_2O \rightarrow OH + OH$	5.80E13	0	9.07E3
10	$OH + OH \rightarrow O + H_2O$	6.30E12	0	5.50E2
11	$H + H + M^c \rightarrow H_2 + M$	2.57E18	-1	0
12	$H_2 + M \rightarrow H + H + M$	6.29E14	0	4.83E4
13	$H + OH + M \rightarrow H_2O + M$	2.40E22	-2	0
14	$H_2O + M \rightarrow H + OH + M$	1.14E24	-2.2	5.90E4
15	$O + O + M \rightarrow O_2 + M$	5.43E13	0	-9.0E2
16	$O_2 + M \rightarrow O + O + M$	5.14E18	-1	5.94E4
17	$H + O_2 + M \rightarrow HO_2 + M$	6.00E18	-1	0
18	$HO_2 + M \rightarrow H + O_2 + M$	6.00E15	0	2.30E4
19	$H + HO_2 \rightarrow OH + OH$	2.50E14	0	9.50E2
20	$OH + OH \rightarrow H + HO_2$	1.20E13	0	2.02E4
21	$H + HO_2 \rightarrow H_2O + O$	5.00E13	0	5.00E2
22	$H_2O + O \rightarrow H + HO_2$	4.80E14	0.45	2.87E4
23	$O + HO_2 \rightarrow OH + O_2$	5.00E13	0	5.00E2
24	$OH + O_2 \rightarrow O + HO_2$	1.30E13	0.18	2.80E4
25	$OH + HO_2 \rightarrow H_2O + O_2$	5.00E13	0	5.00E2
26	$H_2O + O_2 \rightarrow OH + HO_2$	5.60E13	0.17	3.66E4
27	$HO_2 + HO_2 \rightarrow H_2O_2 + O_2$	1.80E13	0	5.00E2
28	$H_2O_2 + O_2 \rightarrow HO_2 + HO_2$	3.00E13	0	2.16E4
29	$H_2O_2 + M \rightarrow 2OH + M$	1.71E17	0	2.29E4
30	$2OH + M \rightarrow H_2O_2 + M$	7.71E14	0	-2.65E3
31	$H + H_2O_2 \rightarrow H_2 + HO_2$	1.70E12	0	1.90E3
32	$H_2 + HO_2 \rightarrow H + H_2O_2$	6.00E11	0	9.30E3
33	$H + H_2O_2 \rightarrow H_2O + OH$	5.00E14	0	5.00E3
34	$H_2O + OH \rightarrow H + H_2O_2$	2.40E14	0	4.05E4
35	$O + H_2O_2 \rightarrow OH + HO_2$	2.00E13	0	2.95E3
36	$OH + HO_2 \rightarrow O + H_2O_2$	5.20E10	0.5	1.06E4
37	$OH + H_2O_2 \rightarrow H_2O + HO_2$	1.00E13	0	9.10E2
38	$H_2O + HO_2 \rightarrow OH + H_2O_2$	1.80E13	0	1.51E4
39	$O + OH + M \rightarrow HO_2 + M$	2.29E17	0	0
40	$HO_2 + M \rightarrow O + OH + M$	1.94E20	-0.43	3.22E4
41	$H_2 + O_2 \rightarrow OH + OH$	1.70E15	0	2.42E4
42	$OH + OH \rightarrow H_2 + O_2$	1.70E13	0	2.41E4
43	$H + O + M \rightarrow OH + M$	2.26E16	0	0
44	$OH + M \rightarrow H + O + M$	1.74E16	0	5.11E4
45	$H_2 + HO_2 \rightarrow H_2O + OH$	6.50E11	0	9.40E3
46	$H_2O + OH \rightarrow H_2 + HO_2$	7.20E09	0.43	3.61E4
47	$O + H_2O_2 \rightarrow H_2O + O_2$	8.40E11	0	2.13E3
48	$H_2O + O_2 \rightarrow O + H_2O_2$	3.40E10	0.52	4.48E4
49	$H \rightarrow$ wall destruction	1.20E01	0	0
50	$O \rightarrow$ wall destruction	9.20E01	0	0
51	$OH \rightarrow$ wall destruction	9.20E01	0	0
52	$HO_2 \rightarrow$ wall destruction	1.00E-1	0	0
53	$H_2O_2 \rightarrow$ wall destruction	1.00E-1	0	0

Note 1: the reaction rate constant is given by $k = AT^n \cdot \exp(-E/RT)$.
Note 2: the third body concentration has been evaluated as
$[M] = C_{H_2} + 0.4 C_{O_2} + 6.5 C_{H_2O}$, according to the efficiency of each molecule.

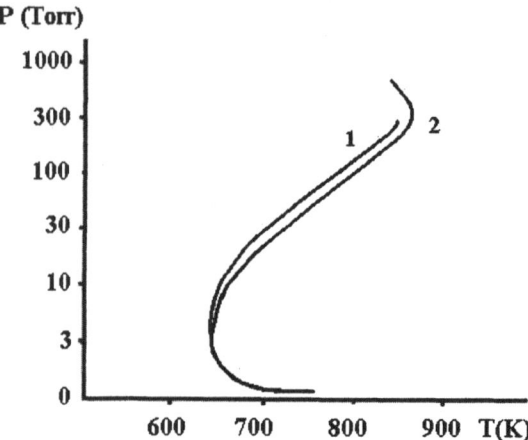

Fig. 1. Calculated explosion limits of the stoichiometric $H_2 - O_2$ reaction system.

(1) Calculation are made with the use of extremity criterion for the total concentration of reaction system species. (2) Calculations are made with the use of the criterion of the extreme behaviour, parametric sensitivity of the maximal concentrations of hydrogen atoms to the change of the initial pressure of the mixture $H_2 : O_2 = 2 : 1$ in $H_2 - O_2$ [4].

Curve (1) in Fig. 1 represents the explosion limits predicted using the explosion criteria based on the extreme behaviour of the total concentration of particles in hydrogen-oxygen reaction systems. For comparison, in the same figure are also shown the explosion limits calculated by Wu, Cao and Morbidelli[4]. In this work as a criterion of the explosion limit the extreme behaviour of parametric sensitivity of maximal concentration of hydrogen atoms to the initial pressure of the reaction mixture is chosen.

As the results on the Fig. 1 show, there is good agreement between the calculations for the first and second explosion limits by both methods. The small difference between the calculations may be connected with our consideration of isothermal condition. Moreover, in reference [4] agreement between experimental resalts and kinetic model is adequately shown.

3.2 Numerical Revelation of Kinetic Significance for Steps and Species in the Reaction of Hydrogen Oxidation

The revelation of kinetic significance for individual steps and species enables to understand the chemical essence of explosion limits.

In Fig. 2 the relative contributions for the most important steps of kinetic model of hydrogen oxidation reaction in critical ignition condition are represented.

In theoretical considerations [1] the branching step (5) in critical ignition condition play an essential role, as well as the steps of the chain termination (step (49) for the first explosion limits and step (11) for the second one).

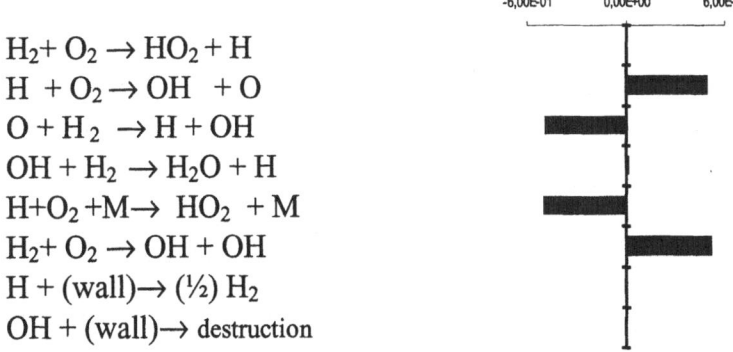

$H_2 + O_2 \rightarrow HO_2 + H$
$H + O_2 \rightarrow OH + O$
$O + H_2 \rightarrow H + OH$
$OH + H_2 \rightarrow H_2O + H$
$H + O_2 + M \rightarrow HO_2 + M$
$H_2 + O_2 \rightarrow OH + OH$
$H + (wall) \rightarrow (½) H_2$
$OH + (wall) \rightarrow$ destruction

Fig. 2. The relative value contributions (h_j) for the most important elementary reactions of gas stoichiometric hydrogen- oxygen mixture under the condition of the second explosion limit at 700K, $t=3 \bullet 10^{-2}$ s, $\bar{h}_j = h_j / (\sum h_j^2)^{½}$.

As it follows from the calculated *values* of the hydrogen - oxygen reaction mixture species shown in the Table.2, the role of H and O atoms as well as OH radical is essential. The *value* of HO_2 radical is relatively lower due to its comparative chemical passivity. Nevertheless, its role is ponderable, especially in the second explosion limit of hydrogen-oxygen mixture.

Table 2. The *value* of species of stoichometric hydrogen-oxygen mixture under the conditions of the first and second explosion limits at 700K, $t=3 \bullet 10^{-2}$ s.

$$\bar{\psi}_i = \psi_i / (\sum_{i=1}^{9} \psi_i^2)^{1/2}$$

Reaction condition	H_2	O_2	H_2O_2	H_2O	H	O	OH	HO_2	M
1st limit	0	-0.52E-14	0	0	0.33	0.33	0.33	-1.8E-15	3.6E-23
2nd limit	0	-0.26E-12	0	0.14E-12	0.43	0.31	0.30	-0.2E-12	2.3E-15

3.3 Reduction of the Kinetic Model

Hamiltonian systematization of multistep chemical reaction systems enables systematic reduction of their kinetic models [8]. Steps with low significance have been withdrawn from the model.

The calculation of the explosion limits by reduced kinetic model and by initial model from 59 steps are in agreement with a precision of 2%. The minimal kinetic model of the reaction includes the following 8 steps: 2, 3, 5, 7, 17, 41, 49, 51.

4 Conclusions

From the above results we conclude that:
Hamiltonian systematization of kinetic models using the Pontrjagin principle of maximum is an efficient approach for the numerical calculation of explosion limits, characterized by chain branch reactions.
The definition of kinetic significance of the steps and species by means of *value* magnitude can reveal the chemical essence of gas mixture explosion limits.
Our computer programme VALKIN worked out on the basis of *value* analysis for kinetic models is an effecient tool to be used for the study of multistep chemical reactions.

References

1. Semenov N.N.: Tsepniye Reaktsii (Chain Reactions) Nauka, Moscow (1986)
2. Lewis B., von Elbe G.: Combustion, Flames and Explosions of Gases, Academic, New York (1961)
3. Martoyan G.A, Gasparyan A.G., Arutunyan G.A.: The Calculate of Critical Ignition Conditions for Branch Chain Reaction. Chem.Phys. **5** (1986) 258-263(*in Russian*)
4. Wu H., Cao G., Morbidelli M.: Parametric Sensitivity and Ignition Phenomena in Hydrogen- Oxygen Mixtures. J. Phys. Chem. **97** (1993) 8422-8430
5. Ivanova A. N., Adrianova Z. S., Azatian V.V.: General Approach to Calculating Explosion Limits for the Reaction of Hydrogen Oxidation. Chem. Phys. Rep. **17** (1998) 1511-1524
6. Pontrjagin L.S., Boltyanski V.G., Gamkrelidze R.D., Mishenko I.F.: Mathematical Theory of Optimal Processes, Fizmatgiz, Moscow (1961)(*in Russian*)
7. Tavadyan L. A., Martoyan G.A.: Value Principle of Studying the Kinetics of Complex Kinetic Reactions. Chem. Phys. Rep. **13** (1994) 793-797
8. Martoyan G.A., Tavadyan L. A.: Numerical Revelation for Stepr. and Species in Complex Chemical Reaction Mechanisms by Hamiltonian Systematization Method. Lecture Notes in Computer Science. Vol. 2658. Springer- Verlag, Berlin Heidelberg New York (2003) 600-609
9. Tavadyan L. A., Martoyan G.A., Minasyan S. H.: Numerical Revelation of the Molecular Structure for Reaction Effective Stimulator or Inhibitor by Method of Hamiltonean Systematization of Chemical Reaction System Kinetic Models. Lecture Notes in Computer Science. Vol. 2658. Springer-Verlag, Berlin Heidelberg New York (2003) 593-599
10. Gottwald B. A.: Kiss – A Digital Simulation System for Coupled Chemical Reactions. Simulation. **33** (1981) 169-173

Computer Simulations in Ion-Atom Collisions

S.F.C. O'Rourke, R.T. Pedlow, and D.S.F. Crothers

Theoretical and Computational Physics Research Division, Queen's University
Belfast, Belfast BT7 1NN, UK

Abstract. One area in science that computer modelling and its applications are extensively used is ion-atom collisions. In this paper we present computer simulations for various theoretical models used in the study of single ionization of helium by multiply charged ions. We study several different numerical continuum-distorted-wave eikonal-initial-state (CDW-EIS) models and compare to recent experimental data for triply differential cross sections for the scattering plane for large perturbations. We observe that one type of CDW-EIS simulation based on a model potential with physically appropriate short-range and long-range behaviour leads to better agreement with experiment at high projectile charges, low electron energy and high momentum transfer than some of the earlier CDW-EIS models in the literature. It is hoped that further calculations using this CDW-EIS model to simulate triply differential cross sections, both qualitatively and quantitatively may lead to a fuller understanding of single ionization by highly charged ion impact.

1 Introduction

Computer simulation has become increasingly important over the last 50 years for modelling and solving different ideas within science.
One area in science that computer simulation and its applications is extensively used is ion-atom collisions. There are many different processes that may be modelled. In this paper we focus exclusively on the process of single ionization in ion-atom collisions.
In spite of decades of extensive research single ionization processes occurring in charged-particle atom collisions are still not fully understood. From a theoretical standpoint the main difficulty is that it is not possible to solve the Schrödinger equation in closed analytical form for three mutually interacting particles. A comprehensive formulation of the complete collision involving interactions between the projectile, electron and residual ion, thus necessitates a numerical approach. The validity of some of the models being employed is not always obvious and comparison with experimental data is therefore initally important.
A 'good' model will enable understanding of a complex situation, and this may be used to predict future events or behaviour of the particles. A commonly employed computer simulation in the study of ionization processes is the continuum-distorted-wave eikonal-initial-state(CDW-EIS) [1]. A write-up of the CDW-EIS package, which is very similar to the ones being considered, may be found in [2, 3]. This was thought to be adequate for single ionization as it takes into account

most of the post-collision interaction (PCI) that may occur, which is due to the final state CDW two-centre wavefunctions being simultaneously target and projectile based. Indeed, simulations of the model did provide a good approximation for many ion-atom single ionization collisions, [4,5,6]. It was therefore suprising to see the discrepancies between theory and experiment [7,8,9]. Upon discovery of the discrepancies it was thought that something, may have been omitted in the model or that the process had been over simplified. Hence, the aim was to update the simulation to a more quasi-realistic model, one that included as many influencing factors as possible.

In [7,8] the authors showed that especially at large perturbations the discrepancy between CDW-EIS and experimental results is more apparent.

Theoretical results, [7,8], have been presented, namely the CDW-EIS, CDW-EIS including projectile-target nucleus interaction (CDWNN-EIS) and First Born Approximation (FBA) models. Recently we have seen, [8], that the CDW-EIS model does not reproduce what was initially thought of as the recoil peak, see [7], which occurs in the forward scattering direction. The possible reason for the discrepancy was suggested to be that when the PCI is included in the model, the orthogonality between the initial state and the Coulomb-distorted wave, which describes the ionized electron, is lost. However, it was mentioned in [10] that orthogonality was not required as the interaction potential was already included in the wavefunctions. Therefore, it was concluded that the missing projectile-target internuclear potential might account for the lack of the recoil peak. We will discuss the effect which including the internuclear potential has on results and we will consider the effect of eliminating the Roothan-Hartree-Fock (RHF) wavefunction. The later was considered as the authors felt that care ought to be taken when considering RHF wavefunctions which can and do lead to unphysical components. Section 2 describes some of the equations used in the model, section 3 compares the simulation to experimental data and section 4 summarizes the results. Atomic units are used throughout except where stated and all theoretical results are absolute.

2 Theory

In this paper the various simulations used are to model the triply differential single-ionization cross sections for electrons emitted into the scattering plane in 3.6 MeV amu^{-1} Au$^{24+,53+}$+ He collisions. The perturbation parameter is given by Q/v, so the respective perturbations are 2.0 and 4.4, both of which are considered as quite large perturbations.

The triply differential cross sections (TDCS), $\sigma(\boldsymbol{k})$, is given by

$$\sigma(\boldsymbol{k}) = \int d\boldsymbol{\eta} \left| R_{if}(\boldsymbol{\eta}) \right|^2, \tag{1}$$

where $\boldsymbol{k}$ is the ejected-electron wave number, $\boldsymbol{\eta}$ is the transverse component of the change in the relative momentum of the nuclei and R_{if} is the scattering amplitude [1].

The scattering amplitude may be given by:

$$R'_{if}(\eta) = \frac{1}{2\pi} \int d\rho \exp(i\eta\cdot\rho)\tilde{a}_{if}(\rho), \qquad (2)$$

where ρ is the impact parameter and v is the impact velocity, with $\eta \cdot v = 0$ and $\rho \cdot v = 0$. $\tilde{a}_{if}(\rho)$ is the transition amplitude without the internuclear potential and the prime indicates that the internuclear potential has not been included. From [11], the transition amplitude including the internuclear potential is given by:

$$a_{if}(\rho) = (\rho v)^{\frac{2iZ_TZ_P}{v}} \tilde{a}_{if}(\rho), \qquad (3)$$

where Z_T and Z_P are respectively the target and projectile-nucleus charge. In all models $Z_T = 1.345$

This means that the 'new' scattering amplitude is given by:

$$R_{if}(\eta) = \int_0^\infty \rho d\rho J_0(\eta\rho)(\rho v)^{2i\bar\gamma} \int_0^\infty R'_{if}(\eta')J_0(\eta'\rho)\eta' d\eta', \qquad (4)$$

where $\bar\gamma = \frac{Z_T Z_P}{v}$, as in [12]. The authors call this method CDW-EIS+nn.
To eliminate the RHF wavefunctions the ground state wavefunction, $\varphi_i(r_T)$ of the outer electron orbital of the neutral target was obtained from [13], with He$^+$(1s) taken as the inner-electron orbital and with

$$\varphi(r_T, z_0) = z_0^{3/2} \pi^{-1/2} \exp(-z_0 r_T) \qquad (5)$$

where the value of z_0 is $(1.8072)^{1/2}$.

3 Comparison between Simulation and Experiment

The results in figures 1, 2 and 3 represent various computer simulations pertaining to triply differential single-ionization cross sections for electrons emitted into the scattering plane in 3.6 MeV amu^{-1} Au$^{24+,53+}$ + He collisions. In all figures the top left diagram (a) is FBA, top middle (b) is CDW-EIS, top right (c) CDW-EIS+nn. The bottom left (d) is CDW-EIS with RHF wavefunctions (CDW-EIS+RHF), bottom middle (e) is CDW-EIS with the internuclear potential by [14] (CDW-EIS Olivera) and bottom right (f) is CDW-EIS with the model potential of Bhattacharya, [15] (CDW-EIS Bhattacharya).

The model (e), uses the independent-electron approximation where one electron, the passive one, is assumed to be frozen during the reaction. The interaction between the projectile and the passive electron along with the internuclear potential is taken into account by a static potential. In [15] a slightly different approach is taken, with charge screening through the use of a model potential which includes short and long range effects. All the models except for (d), CDW-EIS+RHF, have used the modified wavefunction of (5) above, also the internuclear potential has been included to some degree in all models except for the FBA, (a), and CDW-EIS, (b), models.

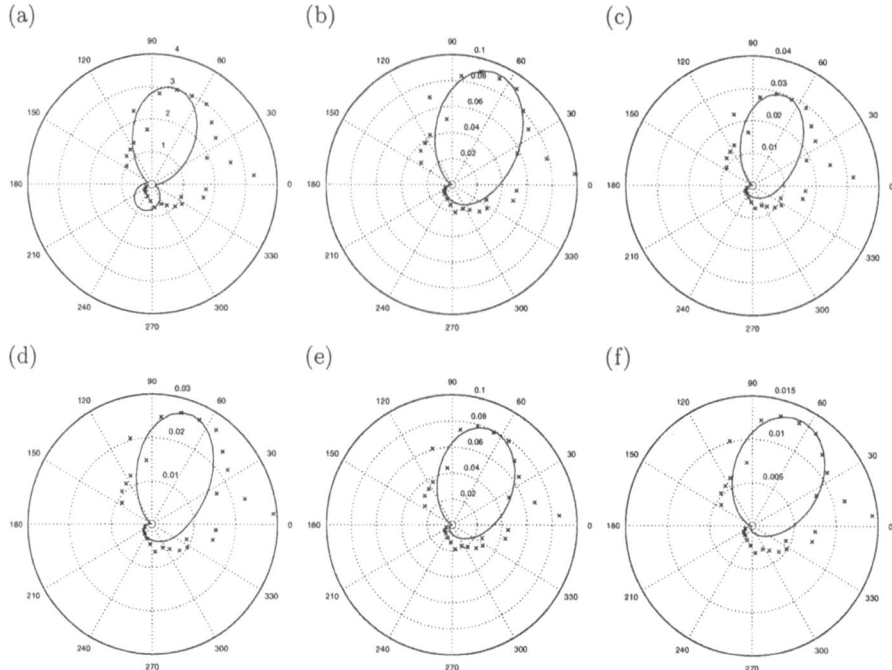

Fig. 1. TDCS for electrons emitted into the scattering plane for a fixed electron energy, $E_k = 4$ eV, and fixed magnitude of the momentum transfer, $q = 0.65$ a.u., as a function of the polar electron emission angle for 3.6 MeV amu^{-1} Au^{24+}+ He collisions. x experimental data [7], — theoretical results. (a) FBA, (b) CDW-EIS, (c) CDW-EIS+nn, (d) CDW-EIS+RHF, (e) CDW-EIS Olivera, (f) CDW-EIS Bhattacharya

In figure 1 we present the triply differential single-ionization cross sections (TDCS) for electrons emitted into the scattering plane in 3.6 MeV amu^{-1} Au^{24+}+ He collisions at momentum transfer of $q = 0.65$ a.u. and electron energy $E_k = 4$ eV, where the momentum transfer q is defined by $q^2 = \eta^2 + (\Delta\epsilon/v)^2$ given $\Delta\epsilon$ is the resonance defect. The perturbation parameter is Q/v which in this case is quite large at $(24/v) = 2.0$. The experimental data is taken from [7]. The computer simulations show that the FBA is a poor approximation for a high-perturbation regime, as expected, but the numerical simulations indicate that perhaps the best method for producing qualitative results is the CDW-EIS in this case. All the computer simulations reproduce the binary peak but do not reproduce the lobe in the forward direction, however, it was pleasing to see that all the CDW-EIS variations give a 'bottom' lobe, currently thought to be the recoil peak, [7]. It is a concern that the magnitude of the different theoretical models varies widely. The post-collision interaction ought to become more important for a stronger perturbation. To consider this we have looked at 3.6 MeV amu^{-1} Au^{53+}+ He collisions at momentum transfer of $q = 1.5$ a.u. and electron energy $E_k = 17.5$ eV. In this regime the perturbation parameter, $(53/v)$, is 4.4,

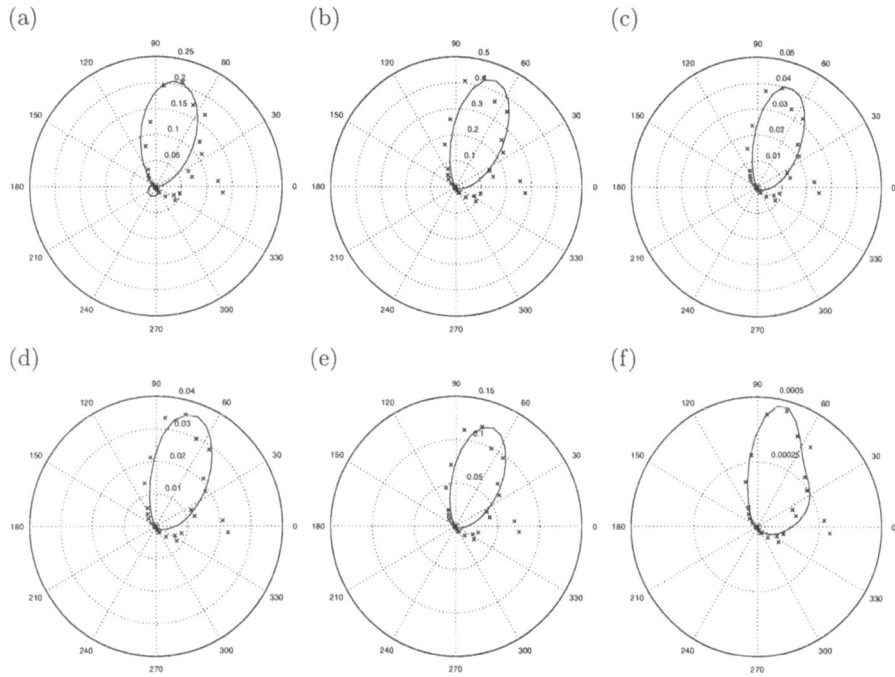

Fig. 2. Same as figure 1 but for 3.6 MeV amu^{-1} Au^{53+}+ He collisions with electron energy, $E_k = 17.5$ eV, and momentum transfer $q = 1.5$ a.u. and except for x experimental data [8]

which is considered as a large perturbation. Figure 2 shows the TDCS for the collision mentioned, with experimental data which appear in [8].

The FBA simulation does not give a good approximation as in the previous collision due to the large perturbation. The PCI is now seen to influence the models with the internuclear potential included and leads to slightly better results than models without the internuclear potential do, namely CDW-EIS. However, the model that gives the best qualitative agreement is the CDW-EIS Bhattacharya simulation. It can be seen that in this model the TDCS is starting to peak in the forward direction and is closer to the two results nearest 0° than any other model. It also gives good agreement for TDCS between 270° and 360° which is not accomplished in any of the other model simulations. It is interesting to note that the absolute value of the theoretical results is nearer to the experimental data than in any other model.

In figure 3 the TDCS for 3.6 MeV amu^{-1} Au^{53+}+ He collisions with electron energy, $E_k = 4.0$ eV, and momentum transfer $q = 1.0$ a.u., with experimental data from [7] has been presented. The high perturbation regime is once again considered. Once again the FBA simulation gives a poor approximation, however, it could be said that models (b) to (e) all appear to reproduce the binary peak, and the recoil peak to a lesser extent, but fail to reproduce the peak in the forward direction. Conversely, the CDW-EIS Bhattacharya simulation provides

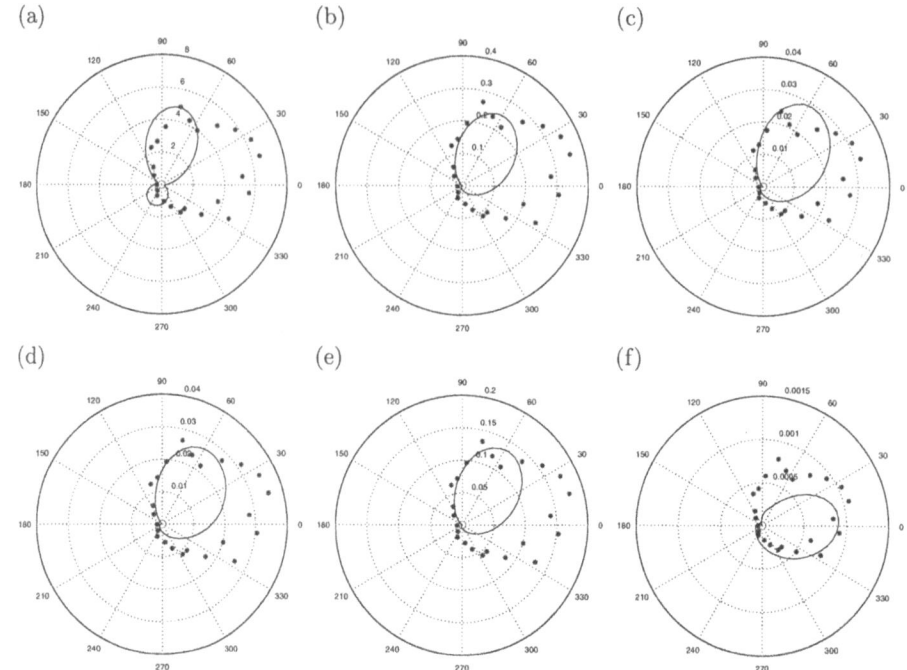

Fig. 3. Same as figure 1 but for 3.6 MeV amu^{-1} Au^{53+}+ He collisions with electron energy, $E_k = 4.0\ eV$, and momentum transfer $q = 1.0\ a.u.$ and except for x experimental data [7]

the best agreement with the peak in the forward direction but does not appear to have picked up the binary peak.

It appears that, at larger perturbation and with lower electron energy and higher momentum transfer the CDW-EIS Bhattacharya model will produce a peak in the forward direction.

4 Summary

In a recent paper [16], results are presented to illustrate post-prior discrepancies in the CDW-EIS simulation for ion-helium ionization. It suggests that using the prior version gives better results than the post version due to the prior version being less sensitive to the choice of the final state. This is of benefit as the final state cannot be computed exactly due to the 3-body problem and normally it is modelled with a hydrogenic continuum state for the electron-residual target state. This may account for some of the discrepancy between theory and experiment.

In summary, we have focused on different variations of the CDW-EIS model and found that, for 3.6 MeV amu^{-1} Au^{24+}+ He collisions, the standard CDW-EIS model provides a slightly better approximation than the variations of the CDW-

EIS approximations considered in this paper. It is apparent from our present study for 3.6 MeV amu^{-1} Au^{53+}+ He collisions, that the model which yields the best qualitative results is the CDW-EIS simulation with the model potential from [15]. We see that a 'bulge' in the forward direction is starting to appear for $E_k = 17.5\ eV$ but for $E_k = 4\ eV$ only the peak in the forward direction appears.

Acknowledgements. The authors thank M.Schulz for sending some experimental data and K.Roy for reference [15]. One (RTP) of the authors wishes to thank DEL for financial support.

References

1. Crothers DSF and McCann JF 1983 *J. Phys. B: At. Mol. Phys.* **16**, 3229.
2. O'Rourke SFC *et al* 2000 *Comput. Phys. Commum.* **131**, 129.
3. McSherry DM *et al* 2003 *Comput. Phys. Commum.* **155**, 144.
4. O'Rourke SFC *et al* 1997 *J. Phys. B: At. Mol. Opt. Phys.* **30**, 5281.
5. McGrath C *et al* 2000 *J. Phys. B: At. Mol. Opt. Phys.* **33**, 3693.
6. Schmitt W *et al* 1998 *Phys. Rev. Lett.* **81**, 4337.
7. Fischer D *et al* 2003 *J. Phys. B: At. Mol. Opt. Phys.* **36**, 3555.
8. Schulz M *et al* 2002 *J. Phys. B: At. Mol. Opt. Phys.* **35**, L161.
9. Schulz M *et al* 2001 *J. Phys. B: At. Mol. Opt. Phys.* **34**, L305.
10. Bubelev VE *et al* 1993 *J. Phys. B: At. Mol. Opt. Phys.* **26**, 3541.
11. McCann JF 1984 *PhD Thesis, QUB.*
12. McSherry DM 2001 *PhD Thesis, QUB.*
13. Crothers DSF 1986 *J. Phys. B: At. Mol. Phys.* **19**, 463.
14. Olivera GH *et al* 1993 *Phys. Rev. A* **47**, 1000.
15. Bhattacharya S *et al Private Communication*
16. Ciappina MF *et al* 2003 *J. Phys. B: At. Mol. Opt. Phys.* **36**, 3775.

Bond Order Potentials for a priori Simulations of Polyatomic Reactions

Ernesto Garcia[1], Carlos Sánchez[1], Margarita Albertí[2], and Antonio Laganà[3]

[1] Departamento de Química Física, Universidad del País Vasco, 01006 Vitoria, Spain
[2] Departamento de Química Física, Universidad de Barcelona, 08028 Barcelona, Spain
[3] Dipartimento di Chimica, Università di Perugia, 06123 Perugia, Italy

Abstract. A class of potential energy functionals based on bond order variables has been designed with the purpose of providing formulations of the interaction allowing a smooth switch between different arrangements and a proper description of the reaction channel. An application to the six atom $Cl + CH_4 \rightarrow HCl + CH_3$ reaction is considered.

1 Introduction

Accurate quantum and quasiclassical studies of the dynamics of few atom systems are a key ingredient for the a priori simulation of complex molecular processes. Recently, the possibility of building a priori simulators has been boosted by the rapid development of grid infrastructures and middleware that have allowed the assemblage of ubiquitous and distributed applications [1,2].

Molecular simulators are an important example of this type of applications. They are, in fact, increasingly needed nowadays as building blocks of many of the most advanced computational applications concerned with material, biology, life, energy, environment, pharmaceutics, food, etc. sciences. At the same time molecular simulators can only be built by gathering together the expertise of several laboratories, the infrastructures of many data repositories, and the specific visual rendering environments of different applications. This has focused the work of several laboratories on the production of grid based molecular simulators (GMS) for scientific and educational purposes. One of these GMS is SIMBEX that is discussed in detail in ref. [3]. To enhance the design of molecular simulators on the Grid the European Union has launched in the year 2000 an Action of the Chemistry domain of COST [4] aimed at developing Metalaboratories for complex computational applications in chemistry [5].

The general scheme of a GMS is structured into several modules like the calculation of the electronic energy, the integration of dynamics equations, the inclusion of kinetics and fluid dynamics treatments, the averaging over unobserved parameters. A key passage between the electronic structure calculation module and the dynamics integration one is given by the fitting of the calculated values using a suitable functional form to generate an analytic potential energy surface (PES) [6].

This guarantees that the resulting PESs are highly accurate (sometimes as accurate as hundredths of kcal mol^{-1}) and that they can be used for extended multiproperty comparisons with spectroscopic, scattering and kinetics data. Quite recently, however, dynamical calculations have been extended to large systems and to processes longer lived than nanoseconds. For these systems it is unthinkable to carry out high level *ab initio* calculations and molecular dynamics studies are usually carried out by formulating the interaction as a sum of model few body terms. These models are of empirical origin and the parameters are estimated without anchoring the search to a solid theoretical background. Accordingly, the estimates of the parameters of the adopted functional form are seldomly valid outside the boundaries of a specific family of molecules and processes. Moreover, most often the force fields constructed in this way are confined to conformation analyses and exclude reaction.

The goal of this study is to investigate the robustness of the empirical functional formulations of the interaction based on the bond order coordinates by investigating their behaviour in the limit of fairly small aggregates. Accordingly, the paper is articulated as follows: in Section 2 the polynomial formulation of the interaction in the bond order space is illustrated; in Section 3 the alternative formulation of the interaction as a collection of process potentials is discussed; in Section 4 the new minimum energy path approach for many processes expansions is described; in Section 5 the application of the new functional form to the six atom system reaction $Cl + CH_4$ is analyzed.

2 The Bond Order Polynomial Formulation of the Interaction

A key problem of the functional representation of the interaction is the fact that physical coordinates (like the internuclear distances in which most of the currently used potential energy functionals are expressed) cannot be used to formulate over the full range the interaction of neutral molecules neither as direct powers nor as inverse powers.

To overcome this difficulty it has been suggested to use bond order (BO) coordinates [7]. The BO coordinate of the generic $\kappa\lambda$ atom-atom pair is defined as $n_{\kappa\lambda} = \exp[-\beta_{\kappa\lambda}(r_{\kappa\lambda} - r_{e\kappa\lambda})]$ with $r_{\kappa\lambda}$ being the internuclear distance, $r_{e\kappa\lambda}$ being the corresponding equilibrium value and $\beta_{\kappa\lambda}$ being a parameter related to the bond strength of the $\kappa\lambda$ diatom [7,9]. These coordinates have built-in the metrics of molecular bonds since they become large at short distances, one at the atom-atom equilibrium distance and zero at infinity. The two body potential ($V^{(2)}$) of the diatom $\kappa\lambda$ can be expressed in the bond order formalism as a polynomial of the related BO variable [7]:

$$V^{(2)}(r_{\kappa\lambda}) = -D_{e\kappa\lambda} \sum_{j=1}^{G} a_{\kappa\lambda,j}\, n_{\kappa\lambda}^{j} \tag{1}$$

with G being the power of the polynomial and $D_{e\kappa\lambda}$ being the diatomic dissociation energy. For $G=2$ the BO $V^{(2)}$ reduces to a Morse potential (*i.e.* it is anharmonic in the physical space in spite of the harmonic-like formulation in the BO space). Then, the $\beta_{\kappa\lambda}$ parameter and the $a_{\kappa\lambda,j}$ coefficients are obtained by best fitting either the *ab initio* electronic energies or the spectroscopic data.

To adopt a polynomial formulation of the PES of a three atom system (say $\kappa\lambda\mu$) most often a Many Body Expansion (MBE) approach is followed [8]. In this case, in addition to the already discussed two-body terms, one needs a three-body one whose polynomial formulation [9] reads:

$$V^{(3)}(r_{\kappa\lambda}, r_{\lambda\mu}, r_{\mu\kappa}) = \sum_{\substack{i+j+k \neq i \neq j \neq k \\ i+j+k \leq H}}^{H} c_{\kappa\lambda\mu,ijk}\, n_{\kappa\lambda}^i\, n_{\lambda\mu}^j\, n_{\mu\kappa}^k \qquad (2)$$

where the $c_{\kappa\lambda\mu,ijk}$ coefficients are obtained by linear regression on the *ab initio* electronic energies (subtracted of the two-body contributions).

Sometimes, also its variant making use of the product of the BO coordinates for the related internuclear distances (RBO) [10,11] is used:

$$V^{(3)}(r_{\kappa\lambda}, r_{\lambda\mu}, r_{\mu\kappa}) = \sum_{\substack{i+j+k \neq i \neq j \neq k \\ i+j+k \leq H}}^{H} c_{\lambda\mu\nu ijk}(r_{\kappa\lambda}n_{\kappa\lambda})^i (r_{\lambda\mu}n_{\lambda\mu})^j (r_{\mu\kappa}n_{\mu\kappa})^k \qquad (3)$$

Both formulations have been used to fit the PES of three atom systems [9,10,11, 12,13,14,15]. A modified version of the RBO formalism has also been developed by Paniagua and collaborators [16,17,18,19,20].

A simple way of formulating the PES as a BO polynomial suitable for a straightforward generalization to polyatomic systems is the ALBO one in which the overall potential is formulated in the following pseudo pair additive form [21]:

$$V(\{n\}) = \sum_j D_j \left[1 + Q_j(\{n_k\}_{k \neq j})\right] P_j(n_j) \qquad (4)$$

where j runs over all diatomic pairs making the functional a sum of pseudo ("effective") diatomic model potentials having a shape depending on the vicinity of the other atoms. For this reason, Q_j and P_j are expressed as (low order) polynomials in the BO variables of the (other) pairs of atoms. In particular, Q_j depends on all the BO variables but n_j and makes the depth of the effective diatomic potential depend on the vicinity of the other atoms. Due to the nature of the BO variables these contributions vanish, as it should be, as the other atoms fly away. P_j depends explicitly only on n_j yet its coefficients depend parametrically on the other n_k variables.

Usually, in the BO polynomial approach the jth diatomic component (to which P_j tends as all the other atoms fly away) is expressed as a pure fourth (sometimes sixth) order polynomial. A simplified formulation of the ALBO functional in which the polynomial directly compares with the LEPS is obtained by truncating the polynomial to the second order. The dependence of P_j on the other BO variables is enforced via n_{0j} that for sake of simplicity here is taken to be linear ($n_{0j} = 1 + b_1 n_k + b_2 n_l$) which tends to 1 as the other atoms move to infinity. This makes the resulting ALBO asymptotic Morse like expression $P_j = (n_j/n_{0j})^2 - 2(n_j/n_{0j})$ coincide with that of the LEPS. In the strong interaction region n_{0j} deviates from 1 to account for the displacement of the minimum of the effective diatomic potential from 1 due to the action of multi- (larger than 2) body interaction when the other atoms gets closer.

The formulation adopted for Q_j ($Q_j = a_1 n_k + a_2 n_l$) is also linear and inspired to the same phylosophy. This formulation allows the dissociation energy to tend to the correct asymptotic value at both asymptotes and strong interaction limits. This leads to a functional representation of the interaction that is more flexible than the LEPS one (thanks to the larger number of parameters and to the incorporation of the asymptotic limits) and to a rms of 0.02 kcal mol^{-1} for the considered atom diatom test cases. Obviously the performance of these functional forms can be improved by increasing the order of the polynomials. However, in spite of the simplicity of the formulation and of the easy generalizability to larger molecules we have not yet adopted these formulations on a large scale since they do not allow a straightforward control of the features of the stationary points of the surface.

3 The Many Process Expansion of the Interaction

Alternative approaches allowing a direct calibration of the stationary points of the potential energy surface are those based on the Many Process Expansion (MPE) [22,23,24]. In the MPE scheme, given the set of atoms $\kappa\lambda\mu\nu...$ one can formulate the PES as a sum of all the possible processes ξ connecting reactants to products:

$$V_{\kappa\lambda\mu\nu...} = \sum_{\xi} W_\xi(\mathbf{s}_\xi) V_\xi(\mathbf{t}_\xi) \tag{5}$$

where $\mathbf{s}_\xi$ is the evolution coordinate (or reaction coordinate) of process ξ driving the transformation of the system from reactants to products while $\mathbf{t}_\xi$ is the set of coordinates describing the local deformation of the system. In eq. 5 $V_\xi(\mathbf{t}_\xi)$ is the functional describing the cut of the reaction channel at each point of the evolution coordinate while $W_\xi(\mathbf{s}_\xi)$ is a weight function that properly averages the contributions coming from different processes. In this way the permutational symmetry of the system is fully taken into account.

To define a suitable connection between the reactant and the product arrangement of process ξ one needs to properly specify the related evolution coordinate $\mathbf{s}_\xi$. In the case of three atom systems for the process $\kappa + \lambda\mu \to \kappa\lambda + \mu$ (let us assume for the moment that the related collision angle ϕ_λ, i.e. the angle formed by the $\kappa\lambda$ and the $\lambda\mu$ internuclear distances - is held fixed) one can adopt as evolution coordinate the angle α_λ. This angle is expressed in terms of the relevant BO coordinates $n_{\kappa\lambda}$ and $n_{\lambda\mu}$ as follows [23,24]:

$$\alpha_\lambda = \arctan\left(\frac{n_{\lambda\mu}}{n_{\kappa\lambda}}\right) \quad (6)$$

The angle α_λ can be seen as the rotation angle of a diatomic-like BO (ROBO) potential function of the associated coordinate ρ_λ. The rotation drives the systems from the $\kappa + \lambda\mu$ reactant arrangement to the $\kappa\lambda + \mu$ product arrangement. The asociated coordinate ρ_λ (the hyperradius) is defined as

$$\rho_\lambda = \sqrt{n_{\kappa\lambda}^2 + n_{\lambda\mu}^2}. \quad (7)$$

These polar internal variables have been called in the literature hyperspherical BO (HYBO) coordinates [26,27]. In three dimensions one has to consider not only that ϕ_ξ varies but also that all (three) possible reactive processes $\kappa + \lambda\mu \to \kappa\lambda + \mu$, $\lambda + \mu\kappa \to \lambda\mu + \kappa$ and $\mu + \kappa\lambda \to \mu\kappa + \lambda$ (each of which refers to a different set of HYBO process coordinates) can take place. Accordingly, the weight of eq. 5 has the normalized form:

$$W_\xi(\phi_\xi) = \frac{w_\xi(\phi_\xi)}{\sum_{\xi'} w_{\xi'}(\phi_{\xi'})} \quad (8)$$

where if the weight w_ξ depends on ϕ_ξ in a way that privileges the process closer to collinearity, the resulting functional form is called "largest angle generalization of ROBO" (LAGROBO). The LAGROBO functional has been successfully used to fit the potential energy surface of several three-atom systems [23,24,28,29,30].

The method has been also extended to four-atom systems [31,32,33] by properly defining the angles and by composing the fixed angle minimum energy paths as piecewise functions (all the quantities should be labeled after the particular process they refer to though in the followings we drop the label when not strictly necessary). Given in general the four atoms $\kappa\lambda\mu\nu$, the two plane angles can be ϕ (formed by the $\kappa - \lambda$ and $\lambda - \mu$ bonds) and ε (formed by the $\lambda - \mu$ and $\mu - \nu$ bonds) while the dihedral angle can be ζ (formed by the $\kappa\lambda\mu$ and $\lambda\mu\nu$ planes). Then one can still define as in eq. 6 (though in a different way) the internal angle α_ξ:

$$\alpha_\xi = \arctan\left(\frac{n_{\mu\nu}}{n_{\kappa\lambda}}\right) \quad (9)$$

(the angle α_ξ is again the continuity variable of the reactive process ξ), the pseudo diatomic like bond length ρ_ξ defined as

$$\rho_\xi = \sqrt{n_{\kappa\lambda}^2 + n_{\lambda\mu}^2 + n_{\mu\nu}^2} \quad (10)$$

and the additional angle σ_ξ defined as

$$\sigma_\xi = \arccos\left(\frac{n_{\lambda\mu}}{\rho_\xi}\right) \quad (11)$$

4 The MEP-MPE Method

When moving to larger systems it becomes increasingly more difficult to include explicitly all the possible processes. After all, in general, the key information for the calculation of the low energy dynamics of reactive systems is that linked to the portion(s) of the PES located in the proximity of the minimum energy path(s). For this reason one can construct the whole PES by limiting the expansion of eq. 5 to the processes (or parts of them) associated with the MEPs connecting the stationary points. Accordingly, for this MEP-MPE functional representation of the PES the sum of eq. 5 is limited to the N segments of the MEP (or MEPs) associated with the stationary points of the surface relevant to the investigated processes. In the functional form considered here each V_ξ term is formulated as a sum of two contributions: a contribution $V_{\xi,\text{mep}}$ associated with the minimum energy path closer to the considered arrangement and a contribution $V_{\xi,\text{mbe}}$ incorporating the correction associated with the deviation of the geometry considered from the nearest minimum energy configuration:

$$V_\xi = V_{\xi,\text{mep}} + V_{\xi,\text{mbe}} \tag{12}$$

$V_{\xi,\text{mbe}}$ is then articulated into two (2), three (3) and four (4) body components like in the Many Body Expansion approach [8]:

$$V_{\xi,\text{mbe}} = \sum_i V^{(2)}_{i\xi,\text{mbe}} + \sum_i V^{(3)}_{i\xi,\text{mbe}} + \sum_i V^{(4)}_{i\xi,\text{mbe}} \tag{13}$$

where the index i runs over all the two, three and four sets of atoms. The individual two body terms read:

$$V^{(2)}_{i\xi,\text{mbe}} = D^{(2)}_{i\xi,\text{mbe}} \left(n_i - n_{i\xi,\text{mep}}\right)^2 \tag{14}$$

to account for the difference between the related BO coordinate n_i and its reference value $n_{i\xi,\text{mep}}$ taken at the corresponding point on the nearest minimum energy path. The individual three body terms read:

$$V^{(3)}_{i\xi,\text{mbe}} = D^{(3)}_{i\xi,\text{mbe}} \left(\phi_{i\xi} - \phi_{i\xi,\text{mep}}\right)^2 \tag{15}$$

to account for the difference between the related actual bending angle $\phi_{i\xi}$ and that of the corresponding geometry on the nearest minimum energy path $\phi_{i\xi,\text{mep}}$. The individual four body terms read:

$$V^{(4)}_{i\xi,\text{mbe}} = D^{(4)}_{i\xi,\text{mbe}} \left(\zeta_{i\xi} - \zeta_{i\xi,\text{mep}}\right)^2 \tag{16}$$

to account for the difference between the related actual torsion angle $\zeta_{i\xi}$ and that of the corresponding geometry on the nearest minimum energy path $\zeta_{i\xi,\text{mep}}$. Finally the weight function w_ξ is defined as:

$$w_\xi = \prod_j \exp\left[-b\left(n_j - n_{j\text{mep}}\right)^2\right] \tag{17}$$

where the sum extends over all the j pairs of atoms of the system and b is an additional best fit parameter defining the proximity boundaries.

5 The MEP-MPE PES for the Cl + CH$_4$ Reaction

For illustrative purposes we have applied the MEP-MPE approach to the fitting of the Cl + CH$_4$ → HCl + CH$_3$ reaction. This reaction is important for the modeling of stratospheric ozone depletion and chorohydrocarbons flames. Three stationary states have been singled out from the available high level *ab initio* data [34,35,36,37,38,39,40,41]. The first stationary state is that of the reactants made by the isolated Cl atom and the CH$_4$ molecule. The second stationary state is that of the reactive saddle arrangement in which one H bond of CH$_4$ is weakened (the bond length is stretched) to allow the hydrogen atom to interact with Cl and form a collinear like C-H-Cl intermediate with the other three C-H bonds symmetrically arranged around the C-H-Cl axis. The third stationary state is that of the isolated HCl and CH$_3$ products. Accordingly, the minimum energy path of the reaction has been partitioned into two V_ξ segments (see eq. 5): the first goes from reactants to the saddle; the second goes from the saddle to products.

The best fit values of the parameters as well as the detailed description of the potential energy surface will be given elsewhere [42]. Here we sketch only some of the main features of the surface by showing a couple of contour maps. The first map is given in Figure 1 in which the isoenergetic contours of the fitted PES are plotted as a function of the cartesian projections x and y of the Cl-CH$_4$ vector giving the position of the Cl atom around the CH$_4$ molecule frozen at its transition state geometry. This implies an asymptotic energy of 22 kcal mol^{-1}. From this plot one can inspect the angular dependence of the potential as the Cl atom approaches the target molecule around the transition state region. As apparent from the Figure, the cone of acceptance (defined by the angle centred at the coordinate origin and covering the regions of the surface with potential lower than the 21 kcal mol^{-1}) is fairly wide (about 104°). In the Figure the saddle to reaction (7.9 kcal mol^{-1}) corresponds to the minimum placed at $x = -2.80$ Å and $y = 0$ (*i.e.*, along the collinear approach).

A detailed description of the collinear approach is given in Figure 2 using BO coordinates. In the figure reactants are located on the y axis (please note that the equilibrium value of n_{CH} differs from 1 since the CH bond in CH$_4$ is shorter than that of the isolated diatom taken for the construction of the BO variables). Product asymptotes (axis x) are more energetic than the reactants by 6.1 kcal mol^{-1} and the barrier to reaction is slightly higher than 7.5 kcal mol^{-1} above the reactant asymptote. It is worth to point out that, as in other reactions, fixed angle contour maps of the PES show an almost circular shape in BO coordinates representations.

6 Conclusions

The need for developing functional forms suitable for formulating the interaction of molecular systems with an increasing number of atoms has prompted the generalization of some few atom schemes. In this respect bond order coordinates

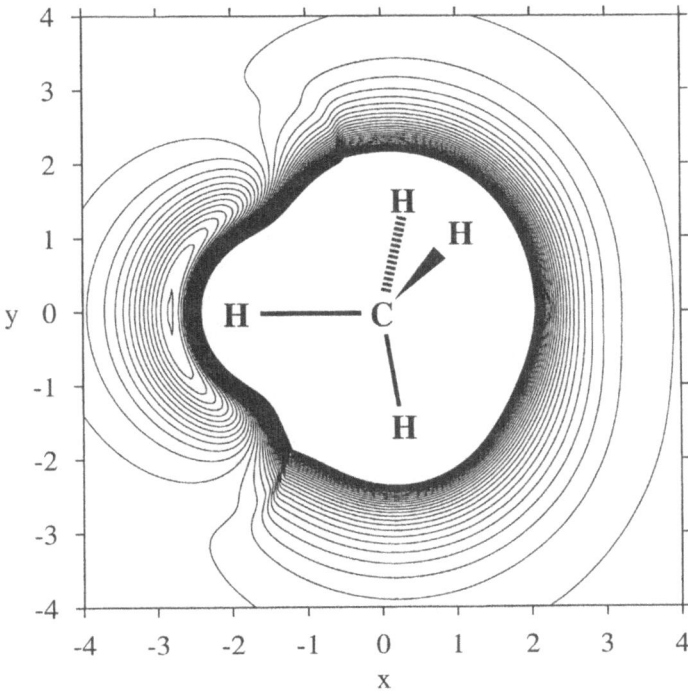

Fig. 1. Isoenergetic contour maps of the MPE-MEP PES (see ref. [42]) plotted as a function of the x and y cartesian projections of the Cl-CH$_4$ vector giving the position of Cl around CH$_4$, at the transition state geometry. Distances are given in Å. Contours are spaced by 1 kcal mol^{-1}. The zero of energy is fixed at the asymptotic reactants' equilibrium potental minimum.

have shown to be particularly suitable when using either a (pseudo) pairwise additive approach or a many process expansion one. The pairwise additive approaches are extremely promising to describe the interaction of large systems through a build up from individual (diatomic) properties. The many process expansion approaches are more appropriate when extended information on the interaction is available (including the definition of the main features of the stationary points and of the proximities of the minimum energy paths). The use of the latter approach in the present work has made it possible to construct a PES for the Cl + CH$_4$ flexible enough to allow an accurate reproduction of the room temperature thermal rate coefficient starting from the available extended *ab initio* information.

Acknowledgments. Financial support from MCyT, MIUR, ASI and COST in Chemistry is acknowledged

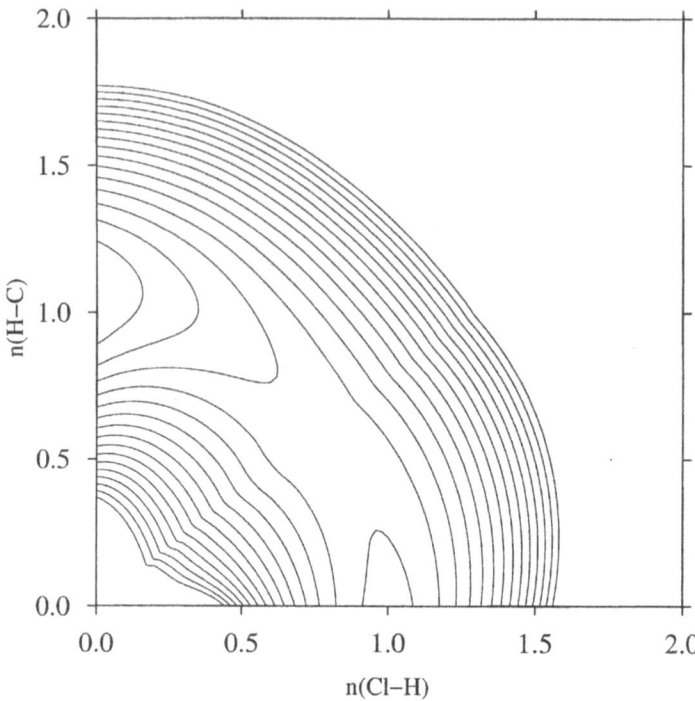

Fig. 2. Isoenergetic contour maps of the MEP-MPE PES (see ref. [42]) plotted as a function of Cl-H and H-C bond order coordinates. At each grid point the geometry of the CH_3 fragment has been optimized in the C_{3v} symmetry. Contours are spaced by 2.5 kcal mol^{-1}.

References

1. Foster, I., Kesselman, C. (eds.): The Grid Blueprint for a Future Computing Infrastructure. Kaufman Publisher, USA (1999)
2. The Globus project: http://www.globus.org
3. Gervasi, O., Laganà, A., Lobbiani, M.: Lecture Notes in Computer Science, **2331** (2002) 956-965
4. http://cost.cordis.lu/src/home.cfm
5. http://cost.cordis.lu/src/action_detail.cfm?action=D23
6. Schatz, G.C.: Lecture Notes in Chemistry, **75** (2000) 15-32
7. Garcia, E., Laganà, A.: Mol. Phys. **56** (1985) 621-627
8. Murrell, J.N., Carter, S., Farantos, S.C., Huxley, P., Varandas, A.J.C.: Molecular Potential Energy Surfaces. Wiley, New York (1984)
9. Garcia, E., Laganà, A.: Mol. Phys. **56** (1985) 629-639
10. Dini, M.: Tesi di Laurea. Università di Perugia, Perugia (1986)
11. Laganà, A., Dini, A., Garcia, E., Alvariño, J.M., Paniagua, M.: J. Phys. Chem. **95** (1991) 8379-8384
12. Alvariño, J.M., Hernández, M.L., Garcia, E., Laganà, A.: J. Chem. Phys. **95** (1986) 3059-3067

13. Palmieri, P., Garcia, E., Laganà, A.: J. Chem. Phys. **88** (1988) 181-190
14. Laganà, A., Hernández, M.L., Alvariño, J.M., Castro, L., Palmieri, P.: Chem. Phys. Lett. **202** (1993) 284-290
15. Laganà, A., Alvariño, J.M., Hernández, M.L., Palmieri, P., Martínez, T., Garcia, E.: J. Chem. Phys. **106** (1997) 10222-10229
16. Aguado, A., Paniagua, M.: J. Chem. Phys. **96** (1992) 1265-1275
17. Aguado, A., Tablero, C., Paniagua, M.: Comp. Phys. Comm. **108** (1998) 259-266
18. Aguado, A., Suárez, C., Paniagua, M.: J. Chem. Phys. **101** (1994) 4004-4010
19. Aguado, A., Tablero, C., Paniagua, M.: Comp. Phys. Comm. **134** (2001) 97-109
20. Tablero, C., Aguado, A., Paniagua, M.: Comp. Phys. Comm. **140** (2001) 412-417
21. Laganà, A.: A Molecular Dynamics Metalaboratory. In: Theory of the Dynamics of Elementary Chemical Reactions. NATO ARW, Laganà, A., Lendvay, G. Eds, Kluwer (2004) pp. 333-350
22. Laganà, A.: J. Chem. Phys. **95** (1991) 2216-2217
23. Garcia, E., Laganà, A.: J. Chem. Phys. **103** (1995) 5410-5416
24. Laganà, A., Ochoa de Aspuru, G., Garcia, E.: J. Chem. Phys. **108** (1998) 3886-3896
25. Laganà, A., Ferraro, G., Garcia, E., Gervasi, O., Ottavi, A.: Chem. Phys. **168** (1992) 341-348
26. Faginas Lago, N.: PhD Thesis. Università di Perugia, Perugia (2002)
27. Laganà, A., Crocchianti, S., Faginas Lago, N., Pacifici, L., Ferraro, G.: Collect. Czech. Chem. Commun. **68** (2003) 307-330
28. Laganà, A., Ochoa de Aspuru, G., Garcia, E.: J. Phys. Chem. **99** (1995) 17139-17244
29. Laganà, A., Garcia, E., Ochoa de Aspuru, G.: Faraday Discuss. Chem. Soc. **110** (1998) 211-213
30. Alagia, M., Balucani, N., Casavecchia P., Laganà, A., Ochoa de Aspuru, G., Van Kleef, E.H., Volpi, G.G., Lendvay, G.: Chem. Phys. Lett. **258** (1996) 323-329
31. Ochoa de Aspuru, G., Clary, D.C.: J. Phys. Chem. A **102** (1998) 9631-9637
32. Rodríguez, A., Garcia, E., Hernández, M.L., Laganà, A.: Chem. Phys. Lett. **360** (2002) 304-312
33. Garcia, E., Rodríguez, A., Hernández, M.L., Laganà, A.: J. Phys. Chem. A **107** (2003) 7248-7257
34. Truong, T.N., Truhlar, D.G., Baldridge, K.K., Gordon, M.S., Steckler, R.: J. Chem. Phys. **90** (1989) 7137-7142
35. Dobbs, K.D., Dixon, D.A.: J. Chem. Phys. **98** (1994) 12584-12589
36. Duncan, W.T., Truong, T.N.: J. Chem. Phys. **103** (1995) 9642-9652
37. Espinosa-Garcia, J., Corchado, J. C.: J. Chem. Phys. **105** (1996) 3517-3523
38. Roberto-Neto, O., Coitiño, E.L., Truhlar, D.G.: J. Phys. Chem. A **102** (1998) 4568-4572
39. Hu, H.-G., Nyman, G.: J. Chem. Phys. **111** (1999) 6693-6704
40. Corchado, J.C., Truhlar, D.G., Espinosa-Garcia, J.: J. Chem. Phys. **112** (2000) 9375-9389
41. Troya, D., Millán, J., Baños, I., González, M.: J. Chem. Phys. **117** (2002) 5730-5741
42. Garcia, E., Sanchez, C, Laganà, A.: A MEP MPE potential energy surface for the Cl + CH_4 reaction, J. Comp. Meth. Science Eng. (submitted)

Inorganic Phosphates Investigation by Support Vector Machine

Cinzia Pierro[1] and Francesco Capitelli[2]

[1] Center of Innovative Technologies for Signal Detection and Processing (TIRES), c/o Department of Physics, University of Bari, Via Orabona 4, 70125 Bari, Italy
cinzia.pierro@ba.infn.it
http://www.ba.infn.it/~tires/index.html
[2] Institute of Crystallography (IC-CNR), c/o Geomineralogic Department, Via Orabona 4, 70125 Bari, Italy
francesco.capitelli@ic.cnr.it
http://www.irmec.ba.cnr.it/

Abstract. We dealt the prediction of crystal chemical features of new sinthesized inorganic phosphates with a supervised-learning regression problem. Then, we analysed correlations between crystal chemical properties of phosphate crystals by a learning machine method, Support Vector Machine (SVM), to develop the detection algorithm. Using structural properties of phosphate crystal structures widely described in the literature, we developed several SVMs able to capture statistical relations between crystal chemical properties of the anhydrous phosphates from the available dataset. In this way, we demonstrated the suitability of SVM for the prediction of structural properties of crystals.

1 Introduction

Inorganic phosphates form an important family of materials which has been widely investigated in the last decades, since the first structure solutions of D.E.C. Corbridge [1], up to the nowadays more sistematic and complete works [2-5], which also reflect the great improvement of X-ray structural analysis techniques (most of all the coming of area detectors in single crystal and powder diffractometry). Parallel to the structure solutions, classification methods have been presented [6-8], with the aim to allow a better understanding of phosphate crystal chemistry; nevertheless, in presence of a new sinthesized compound, such classifications lack in the prediction of their main crystal chemical features, like stoichiometry, structural building units and cation coordination.

The prediction of crystal chemical features of new sinthesized compounds can be dealt with a supervised-learning regression problem, by a statistical analysis of available crystal structures of that given compound. Hence, in this paper, we describe an analysis of main phosphate crystal chemistry properties and an attempt to predict some structural features, both based on statistical analysis of phosphate crystal structures available in the current version of Inorganic Crystal Structure Database (ICSD) and in literature, using a learning machine system, the Support

Vector Machine (SVM) [9-11]. The crystal chemistry properties employed are symmetry center, degree of condensation of phosphate anions, anionic motive repetition, cation coordination number, cation coordination geometry, ionic radius, chemical formula, cation valence and cationic atomic number.

SVM is a powerful technique for learning relationships in data, successfully applied in different research fields for both classification and regression problems. Application of SVMs in crystal chemistry is new, even if they have been widely used in the last years in several classification problems, like separation of false signals in microcalcification in digital mammograms [12], damage identification [13], face authentication [14] and lesion detection ability in ultrasound images [15]. At the same time, recent applications of SVMs in regression problems, like estimation of power consumption [16], prediction of highway traffic flow [17], modelization of non-stationary financial time-series [18], encourage their application in prediction and forecasting problems. Such widespread repertory of applications of SVMs points out the feasibility of those learning machine methods in several application fields, like chemistry, medicine, signal processing, etc.

SVMs demonstrated their efficacy, being robust against incomplete and noisy data, and, most of all, able to work with low dimension datasets [9], unlike neural networks which tipically need large dimension ones [19-21]; nevertheless, SVMs presented high generalization ability on the test set.

At present, some thousands of phosphate crystal solutions are available in literature and in the ICSD: in this work, we somewhat narrowed the field, considering in our dataset only anhydrous monocationic phosphates, with general formula $A_xP_yO_z$, where A correspond to the cationic specie. This led to a number of 80 samples stored in our dataset, which is relatively sufficient in SVM modeling to get accurate predictions.

2 Support Vector Machine Regression

Let us consider a set of training data $T = \{(x_1, y_1), (x_2, y_2),...(x_l, y_l)\}$, where each $x_l \subset R^n$ denotes the input space of the sample and has a corresponding target value $y_l \subset R$ for $i = 1, ..., l$, where l corresponds to the size of training data [9]. The purpose of regression problem is to determine a function able to approximate future values accurately.

The generic SVM regression estimating function can be written as

$$f(x) = (w \cdot \Phi(x)) + b \qquad (1)$$

where $w \subset R^n$, $b \subset R$ and Φ denotes a non-linear transformation from R^n to high dimensional space. The aim of the method is to find the value of w and b so that the values of x can be determined by minimizing the regression risk:

$$R_{reg}(f) = C\sum_{i=0}^{l}\Gamma(f(x_i) - y_i) + \frac{1}{2}\|w\|^2 \qquad (2)$$

where C is a constant, Γ is the cost function and w can be written in terms of data points as:

$$w = \sum_{i=0}^{l} (\alpha_i - \alpha_i^*) \Phi(x_i) \qquad (3)$$

By substituting eq. (3) into eq. (1), the resulting equation is

$$f(x) = \sum_{i=0}^{l} (\alpha_i - \alpha_i^*)(\Phi(x_i) \cdot \Phi(x)) + b = \sum_{i=0}^{l} (\alpha_i - \alpha_i^*) k(x_i, x) + b \qquad (4)$$

In eq. (4) the dot product can be replaced with the kernel function $k(x_i, x)$. Such functions enable dot product to be performed in high-dimensional feature space using low dimensional space data input without knowing Φ; besides, they satisfy Mercer's condition that corresponds to the inner product of some feature space. The most common kernels are, over the linear one $k(x_i, x) = (x \cdot y)$,
the polynomial one

$$k(x_i, x) = ((x_i \cdot x_j) + 1)^d \qquad (5)$$

and the gaussian radial basis (rbf):

$$k(x_i, x) = \exp\left(-\frac{\|x_i - x_j\|^2}{2\sigma^2}\right) \qquad (6)$$

In the corse of our experiment, we used the polynomial and rbf kernels.
The most common cost function is the ε- intensive loss function, written as

$$\Gamma(f(x) - y) = \begin{cases} |f(x-y)| - \varepsilon, & \text{for } |f(x-y)| \geq \varepsilon \\ 0 & \text{otherwhise} \end{cases} \qquad (7)$$

The regression risk in eq. (2) and the ε - intensive loss function can be mimimized by solving the following quadratic optimization problem:

$$\frac{1}{2} \sum_{i,j=1}^{l} (\alpha_i^* - \alpha_i)(\alpha_j^* - \alpha_j) k(x_i, x_j) - \sum_{i=1}^{l} \alpha_i^* (y_i - \varepsilon) - \alpha_i (y_i - \varepsilon)$$

$$\text{subject to } \sum_{i=1}^{l} \alpha_i^* - \alpha_i = 0, \qquad \alpha_i, \alpha_i^* \in (0, C) \qquad (8)$$

The Lagrange multipliers, α_i and α_i^*, are solutions of the above quadratic problem that act as forces pushing predictions towards target value y_i. Only the Lagrange multipliers values different from zero in eq. (8) can be used in forecasting the regression line and are known as support vectors; if $|f(x)-y| \geq \varepsilon$, Lagrange multipliers can be non-zero values and used as support vectors.

The costant C assignes penalties to estimation errors: a large value of C determines higher penalties to errors, so that the regression is trained to minimize error with lower generalization; a small value determines fewer penalties to errors, allowing the minimization of margin with errors, and then higer generalization ability; if C becomes infinitely large, SVM regression would not allow the occurrence of any error and result in a complex model; last, when C reaches zero, the results would tolerate a large amount of errors and the model would be less complex.

On the basis of the obtained results, we can now introduce a SVM architectural scheme. Typically, it is composed by: an input layer; where a n-dimensional feature vector is introduced; a hidden layer which computes the kernel function on support- and input-vectors; an output layer which implements a decision function. The scheme is depicted in Fig. 1.

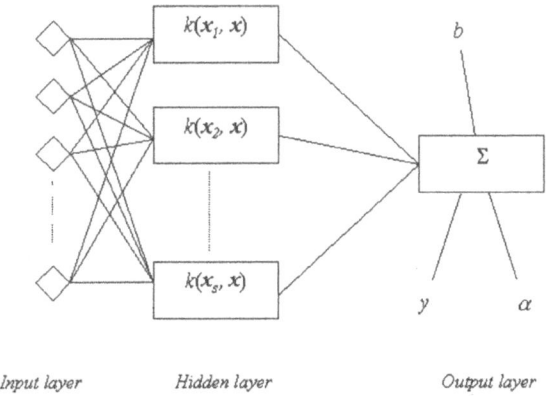

Fig. 1. SVM architectural scheme. $k(x_i, x)$ = kernel function avaluation on input vector and support vector x_i; b = known term of separating hyperplane; y = target vector corresponding to the support vectors $(x_1, \ldots x_s)$; α = Lagrange multipliers

The variable b can be computed by applying Karush-Kuhn-Tucker conditions which, in this case, implies that the product of the Lagrange multipliers and constraints is zero:

$$\alpha_i(\varepsilon + \zeta_i - y_i + (w, x_i) + b) = 0$$
$$\alpha_i * (\varepsilon + \zeta_i * + y_i - (w, x_i) - b) = 0 \qquad (9)$$

and

$$(C - \alpha_i)\zeta_i = 0$$
$$(C - \alpha_i *)\zeta_i * = 0 \qquad (10)$$

where ζ_i and ζ_i^* are slack variables used to measure errors. Since a_i, $a_i^* = 0$ and $\zeta_i^* = 0$, for $a_i^* \in (0, C)$ can be computed in the following way:

$$b = y_i - (w, x_i) - \varepsilon \quad \text{for } \alpha_i \in (0, C)$$
$$b = y_i - (w, x_i) + \varepsilon \quad \text{for } \alpha_i^* \in (0, C) \tag{11}$$

3 SVM Models of Phosphates

3.1 Crystal Properties

As previously introduced, in this paper we considered only anhydrous monocationic phosphates $A_xP_yO_z$. Every phosphate is characterized by a pentavalent phosphorous atom surrounded by a tetrahedron of four oxygen atoms. At the same time, we did not consider phosphates in which one or several oxygen atoms within the tetrahedron are substituted by other atoms (substituted phosphates). So, the compounds analyzed belong to two main families: *monophosphates*, characterized by an isolated $(PO_4)^{3-}$ tetrahedrical group, and *condensed phosphates*, which correspond to more anionic groups built from corner-sharing PO_4 tetrahedra. Condensed phosphates [2] can be shared, in their turn, in:

1) polyphosphates, with progressive linear linkage of n tetrahedra, characterized by general formula $(P_nO_{(3n+1)})^{(n+2)-}$: they are called oligophosphates, among which the commonly characterized ones are diphosphates ($n = 2$, $(P_2O_7)^{4-}$ anion), triphosphates ($n = 3$, $(P_3O_{10})^{5-}$), tetraphosphates ($n = 4$, $(P_4O_{13})^{6-}$) and pentaphosphates ($n = 5$, $(P_5O_{16})^{7-}$). When n becomes very large (>20), the formula is modified in $(PO_3)^{n-}$: they are called long-chain polyphosphates;

2) cyclophosphates, with cyclic condensation, with general formula $(P_nO_{3n})^{n-}$;

3) ultraphosphates: PO_4 tetrahedra share three of their oxygen atoms with the adjacent PO_4 groups: such condensation leads to various anion geometries: finite groups, infinite chains, infinite layers, three-dimensional frameworks: the simplest formula is $(P_{(n+2)}O_{(3n+5)})^{n-}$.

3.2 Data Representation

Before data processing, it was necessary to study the best way to embed these data in the input-output patterns. For this aim, we have encoded the set of phosphate crystal properties into a suitable form for support vector machines. The predicted properties have been extracted from the outputs of learning machines. The first data representation was suggested by the nature of properties of phosphates, even if with this natural coding the data presented numerical values of different order of magnitude, and then the training of the learning machines could be influenced by a particular property in input rather than by all properties. In order to avoid such difficulties, the data were before standardized, *i.e.* shifted to have a null average and, lastly, scaled to obtain the same unitary variance value.

The data were taken mostly from ICSD, and from crystallographic literature: all the most relevant crystal chemical features were analyzed and stored in the database employed for the SVMs. Incomplete or uncertain data were omitted, to avoid limitation in the SVM performance. Commonly, such features provide two kinds of input: the structural information is represented by the symmetry center, the degree of condensation of phosphate anions, the anionic motive repetition, the cation coordination number, the cation coordination geometry; the chemical information is represented by the ionic radius, the stoichiometry, the cation valence and the atomic number. Below, we report the investigated crystal chemical features of $A_xP_yO_z$ phosphates, conveniently associated with an appropriate shorthand name, and their coding:

(az) The presence or absence of inversion center in a crystal structure, represented by binary coding in a unit-length string.

(class) The degree of condensation of the phosphate anions, divided into six classes (isolate phosphate unit, finite group, ring, monodimensional infinite chain, bidimensional infinite sheet, tridimensional infinite framework), codified by natural numbers ranging from 1 up to 6.

(motive) The PO_4 tetrahedron structural unit of the crystal structures. We represented the type of phosphate as natural numbers in the interval {1-8}, which corresponds to the mostly characterized phosphates, *i.e.* monophosphates, diphosphates, triphosphates, tetraphosphates, pentaphosphates, long-chain poliphosphates, cyclophosphates and ultraphosphates.

(cn) The coordination number of cation A (coordination with oxygen): it ranges from 3 (tin monophosphate) up to 11 (barium diphosphate). We represented the coordination number as natural values in the interval {3-11}. Coordination number increases with ionic radius, as depicted in the experimental trend of Fig. 2.

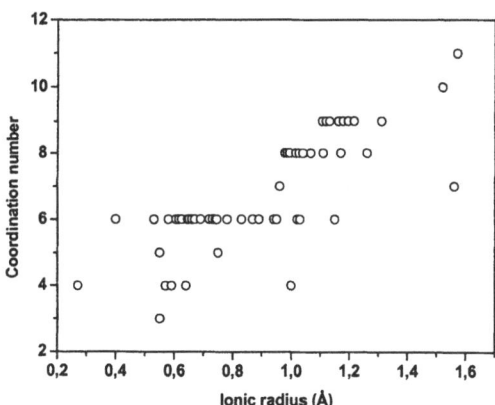

Fig. 2. Dependence of cation coordination geometry with ionic radius: the circles represent the employed dataset

(poly) Geometry of the polyhedron formed by coordinating cation A. The coordination polyhedra are characterized by eight fundamental types: octahedron, tetrahedron, tetragonal pyramid, trigonal prism, trigonal bipyramid, cube, pentagonal bipyramid and irregularly shaped coordination. We represented these eight coordination polyhedra as natural numbers in the interval {1-8}.

(ir) The ionic radius of coordinating cation A, as a function of the different coordinations [22]. We represented it as real numbers ranging from 0.270 Å (Be^{2+} in tetrahedrical coordination) up to 1.560 Å (Rb^+ in 7-fold coordination).

(formula) The chemical formula $A_xP_yO_z$ of the phosphate. We represented the numbers x, y and z by a string of natural numbers, each string element ranging in the interval $\{1\text{-}14\}$.

(val) Valence of the cation A. The valences of the cations found in the crystal structures range from 1 up to 4, and we represented them by natural numbers coding.

(Z) Atomic number Z of cation A. Since we classified phosphates with cation ranging from lithium up to uranium, we represented Z as a natural number in the interval $\{3\text{-}92\}$.

3.3 Training Results

We splitted the dataset formed by 80 available samples in two subsets, one used to set the learning machine parameters, the *training set*, and the other used to state the performance of learning machine, the *test set*. For this aim, we picked randomly the 65 phosphates who compose the training set, while the remaining 15 ones compose the test set. For the constant C a low value was used, $C = 20$, because the data were not affected by any uncertainties, while for the variable ε several investigation suggested a best value equal to 0.02. All learning machines were characterized by these constant values, whereas the kernel function changed case by case.

We extracted the feature values by output patterns rounding this value to the next integer value, because the coding of chemistry properties of phosphates is made by integer values (see Fig. 3).

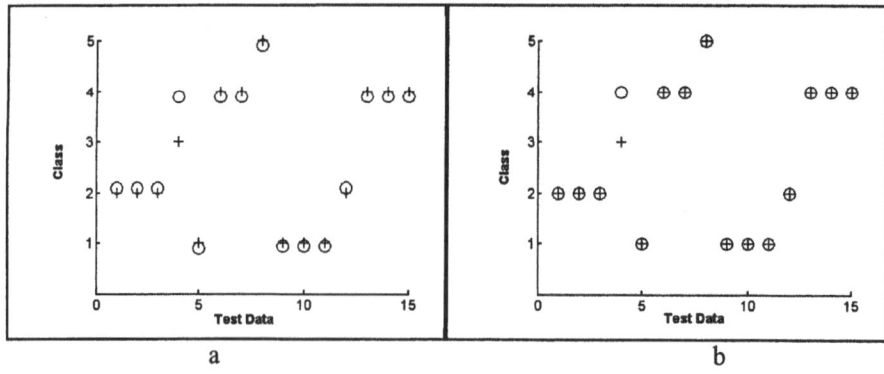

Fig. 3. a: Outputs of learning machine before rounding; b: Outputs of learning machine after rounding. The crosses indicate the target outputs; the circles indicate the prediction of learning machine

Tables 1, 2 and 3 show the results of training and testing sets. The measure chosen to evaluate the performance by SVM is the fraction of correctly classified phosphates. Fig. 4 shows the comparison between the output responses of some among the best trained learning machines from input data belonging to test data set and the true phosphate features.

Generally, the SVMs demonstrated a good ability to predict some features, like the degree of condensation (class), the anionic motive repetition (motive) and the cation coordination geometry (poly). The first important annotation which arises is the importance of the stoichiometric information (formula), necessary in this categorization problem, due to the fact that it is a triple input (formula $A_xP_yO_z$ of the phosphate); removing (formula) from the inputs, SVMs are not able to determine other phosphate features. Hence, we will not add the (formula) voice between the inputs.

In (class) prediction, the most important feature in input resulted (val), which alone or coupled with structural features like (poly) and (cn), and with (ir) performed best results, ranging respectively from 90.0% up to 93.3%. Lower percentages (80-86.7%) in (class) prediction were obtained with inversion centre (az) between the inputs, notwithstanding good results in training data. The best results with (az) were reached coupling it with (val), up to 86.7%. Remarkable results in (motive) prediction were obtained with (poly) in input (93.3%). The most used kernel function resulted the rbf; only few inputs ((val, poly), (val, cn) and (val, ir)) featured the polynomial one. (class) outputs are shown in Table 1. In Figs. 4a and 4b are depicted, respectively, results with (formula, val) and with (formula, val, ir) in input.

Table 1. Performance of SVM trained to predict the degree of condensation (class). % SV = employed support vectors on training data

Inputs	% SV	kernel function	Training data	Test data
formula, val	80.0	*rbf*	100	93.3
formula, val, poly	81.5	*poly*	95.4	93.0
formula, val, cn	81.5	*poly*	93.8	93.0
formula, val, ir	41.5	*poly*	93.8	90.0
formula, poly	78.5	*rbf*	98.5	86.7
formula, val, az	100	*rbf*	95.4	86.7
formula, az	86.2	*rbf*	95.4	86.7
formula, az, ir	75.4	*rbf*	92.3	86.7
formula, az, cn	100	*rbf*	93.8	80.0
formula, az, poly	86.2	*rbf*	93.8	80.0

rbf = gaussian radial basis; *poly* = polynomial

In (motive) prediction, good results were obtained with (val), alone or coupled with (az), and (poly), peaking 93.3% of test data. Quite good results were reached with (poly) coupled to (val), with (az) and with (az) coupled to (cn), everyone peaking 86.7%. Similarly to (class), most of the inputs featured the rbf kernel function: only one result (az, cn) featured the polynomial one. (motive) outputs are shown in Table 2. In Figs. 4c and 4d are depicted, respectively, results with (formula, az, val) and (formula, poly) in input.

Table 2. Performance of SVM trained to predict the anionic motive repetition (motive). % SV = employed support vectors on training data

Inputs	% SV	kernel function	Training data	Test data
formula, az, val	98.5	rbf	97.0	93.3
formula, val	98.5	rbf	95.5	93.3
formula, poly	83.1	rbf	95.5	93.3
formula, poly, val	92.3	rbf	98.5	86.7
formula, az	86.2	rbf	97.0	86.7
formula, az, cn	84.6	poly	95.4	86.7

rbf = gaussian radial basis; *poly* = polynomial

The determination of the cation coordination geometry (poly) resulted more difficult, but some results were satisfactory: on all, those with (val) and (cn) are significant, because we note that the test data show a percentage of 80.0 only with (cn), growing up to 86.7 adding (val) between the inputs, certifying the importance of (val) in the prediction of (poly). Minor results were obtained using (az) coupled with (val), (ir) and (cn): test data percentages range from 60% up to 73.3% (Table 3). Moreover, we note the importance of the cation coordination in (poly) prediction (50% of results featured (cn) in input), rationalizable with the close connection between (cn) and (poly): only the irregular coordination can display more than one coordination number. Also in this case, the most used kernel function was the rbf: only (az, cn) featured the polynomial one. (poly) outputs are shown in Table 3.

Table 3. Performance of SVM trained to predict the cation geometry coordination (poly). % SV = employed support vectors on training data

Inputs	% SV	kernel function	Training data	Test data
formula, val, cn	100	rbf	95.4	86.7
formula, cn	87.7	rbf	98.5	80.0
formula, az, cn	100	poly	98.5	73.3
formula, val, ir	75.4	rbf	92.3	73.3
formula, az, ir	69.2	rbf	98.5	66.7
formula, val, az	100	rbf	84.6	60.0

rbf = gaussian radial basis; *poly* = polynomial

Atomic number (Z) did not supply good information to get (class), (motive) or (poly): networks including such features did not fit accurately target data. It was not possible to use more than three inputs, because the support vector machines shown degraded performance on test data, in spite of promising results on training data, in some case better than those reported in Table 1, 2 and 3.

By the analysis of all the outputs, it resulted the importance of chemical inputs (formula, val) for degree of condensation and anionic motive repetition prediction, because same results (93%) can be obtained both only with (val) in input and with (val) coupled to a structural input, *e.g.* (poly) for (class) and (az) for (motive). In the cation geometry coordination the only chemical information is not sufficient, because best results are always obtained with (cn) in input.

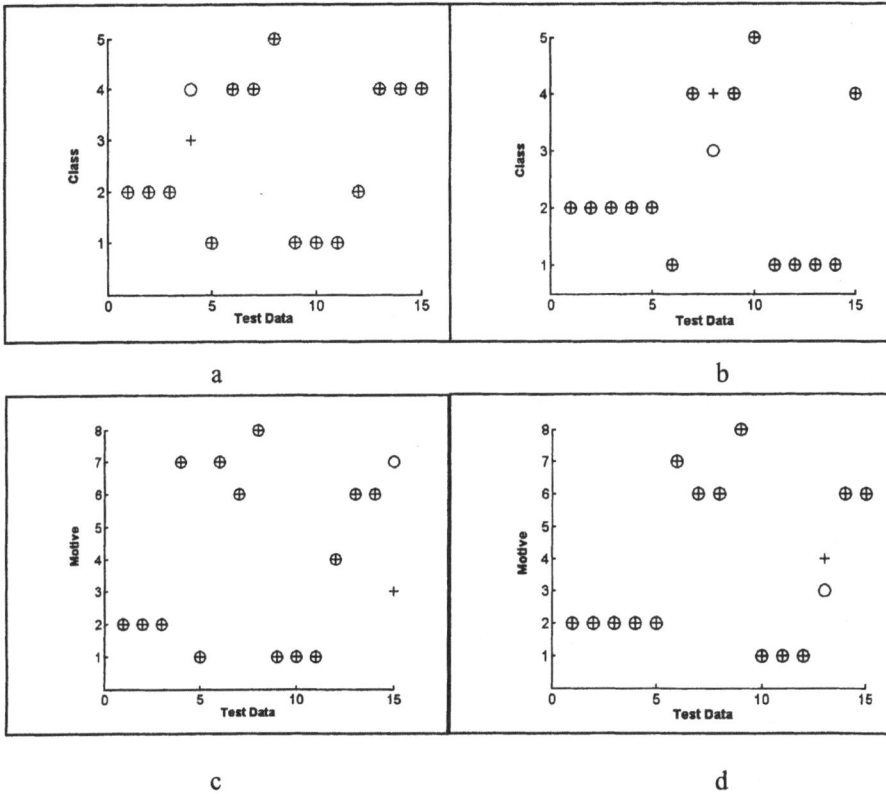

Fig. 4. Comparison between the prediction of learning machines (circles), being in input some features of 15 phosphates and the desired response (crosses). Fig. 4a shows the (class) prediction being in input (formula, val); Fig. 4b shows the (class) prediction being in input (formula, val, ir); Fig. 4c shows the (motive) prediction being in input (formula, az, val); Fig. 4d shows the (motive) prediction being in input (formula, poly)

4 Conclusions

We dealt the prediction of crystal chemical features of new sinthesized phosphates with a supervised-learning regression problem. So, we used Support Vector Machine to capture statistical relations between crystal chemical phosphate properties available from ICSD and from literature, and to predict some crystal chemical features. Such features are: symmetry center, degree of condensation of phosphate anions, anionic motive repetition, cation coordination number, cation coordination geometry, ionic radius, chemical formula, cation valence and cationic atomic number.

Unlike neural networks which tipically need large dimension datasets, SVMs were able to work with low dimension ones: therefore, we stored in our dataset 80 anhydrous monocationic phosphate structures, mostly taken from the current version of Inorganic Crystal Structure Database and in some cases from the existing literature.

SVMs performances resulted generally higher when a structural output was coupled to chemical inputs. In fact, most satisfactory results were reached in the prediction of degree of condensation of phosphate anions and of anionic motive repetition; worthy of consideration also results on cation coordination geometry. All these outputs required the stoichiometric input to work, and we recognized also the importance of the cationic valence in input and, for cation geometry coordination, also of the cation coordination number; besides, the gaussian radial basis resulted the most employed kernel function.

In the future, we plan to improve the method in the other structural features, like coordination number and symmeytry center. More, we plan to work with di- and tri-cationic phosphates, both anhydrous and hydrated, and also with acidic phosphate units; nevertheless, it is important the acquisition of new crystal structures, which will allow the improvement of SVMs performances.

References

1. Corbridge, D.E.C.: Crystallographic data on some hypophosphates and pyrophosphates. Acta Crystallogr. 10 (1957) 85-85
2. Durif, A.: Crystal Chemistry of Condensed Phosphates. Plenum Publ. Corp., New York (1995)
3. Ni, Y., Hughes, J.M., Mariano, A.N.: Crystal chemistry of the monazite and xenotime structures. Am. Mineral. 80 (1995) 21-26
4. Rao, C.N.R., Natarajan, S., Choudhury, A., Neeraj, S., Vaidhyanathan, R.: Synthons and design in metal phosphates and oxalates with open architectures. Acta Cryst. B 57 (2001) (2001) 1-12
5. Capitelli, F., Harcharras, M., Assaaoudi, H., Ennaciri, A., Moliterni, A. G. G., Bertolasi, V.: Crystal structure of new hexahydrate dicobalt pyrophosphate $Co_2P_2O_7 \cdot 6H_2O$: comparison with $Co_2P_2O_7 \cdot 2H_2O$, α-, β- and γ- $Co_2P_2O_7$. Z. Kristallogr. 218 (2003) 345-350
6. Van Wazer, J.R.: Phosphorous and Its Compounds. Interscience, New York (1966)
7. Corbridge, D.E.C.: The Structural Chemistry Of Phosphorous. Elsevier, Amsterdam (1974)
8. Averbuch-Pouchot, M.T., Durif, A.: Topics in Phosphate Chemistry. World Scientific Publ. Co., London-Singapore (1996)
9. Vapnik, V.: The Nature of Statistical Learning Theory. Springer, Berlin (1995)
10. Vapnik, V.: Statistical Learning Theory. Wiley, New York (1998)
11. Cristianini, N., Shawe-Taylor, J.: Support Vector Machines and other kernel-based learning methods. Cambridge University Press, Cambridge (2000)
12. Bazzani, A. Bevilacqua, A., Bollini, D., Brancaccio, R., Campanini, R., Lanconelli, N., Riccardi, A., Romani, D.: A SVM classifier to separate false signals from microcalcifications in digital mammograms. Phys. Med. Biol. 46 (2001)1651-1663
13. Worden, K., Lane, A.: Damage identification using support vector machines. J.: Smart Mater. Struct. 10 (2001) 540-547

14. Jonsson, K., Kittler, J., Li, Y.P., Matas, J.: Support Vector Machines for Face Authentication. Image Vision Comput. 20 (2002) 369-375
15. Kotropoulos, C., Pitas, I.: Segmentation of ultrasonic images using Support Vector Machines. Pattern Recogn. Lett. 24 (2003) 715-727
16. Chen, J., Chang, M.W., Lin, C.J.: Load forecasting using Support Vector Machines: a study on EUNITE Competition 2001. Report for EUNITE competition for smart adaptative system. Http://www.eunite.org
17. Ding, A., Zhao, X., Jiao, L.: Traffic flow time series prediction based on statistics learning theory. IEEE 5th International Conference on Intelligent Transportation System, Proceedings (2002) 727-730
18. Tay, F.E.H., Cao, L.J.: ε-Descending Support Vector Machines for Financial Time Series Forecasting. Neural Process. Lett. 15 (2002) 179-195
19. Zupan, J. Gasteiger, J.: Neural Networks for Chemists: An Introduction. VCH, Weinheim (1993)
20. Bishop, C.M.: Neural Netwoks for Pattern Recognition. Oxford University Press, Oxford (1995)
21. Barber, M.J., Becker, P.: Neural network investigation of borate crystal properties. Z. Kristallogr. 217 (2001) 205-211
22. Shannon, R.D.: Revised effective ionic radii and systematic studies of interatomic distances in halides and chalcogenides. Acta Crystallogr. A 32 (1976) 751-767

Characterization of Equilibrium Structure for N_2-N_2 Dimer in $1.2\text{Å} \leq R \geq 2.5\text{Å}$ Region Using DFT Method

Ajmal H. Hamdani and S. Shahdin

Pakistan Institute of Lasers ad Optics, P.O.Box 505, Rawalpindi, Pakistan

Abstract. Density Functional Theory (B3PW91 and B3LYP) has been used to investigate the equilibrium geometries and potential energy curves of singlet ground state of N_2-N_2 dimer for $1.20\text{Å} \leq R \leq 2.50\text{Å}$. D_{2h} symmetry proved to be the most stable. The effects of the BSSE upon the optimized geometries and potential energy curves of D_{2h} group have been studied. The BSSE is found to 1-2 % for Density Functional Theory with 6-311++G** basis set.

1 Introduction

The ab initio calculations [1-4] and experimental spectra [5-6] of N_2-N_2 dimer near van der Waals minimum ($3.00\text{Å} \leq R \leq 5.50\text{Å}$), have been studied for quite a long time. Recently Density Functional Theory (DFT) has been applied to compare the results of calculations with ab initio theory [7]. In these works, particular attention has been paid to characterize the most stable dimer geometry and calculation of the BSSE corrected potential energy curves for $3.00\text{Å} \leq R \leq 5.50\text{Å}$ using DFT. However nature of the ground state potential of N_2-N_2 dimer remains to be characterized by using DFT for $1.20\text{Å} \leq R \leq 2.50\text{Å}$ and then compared with prior results of [8].

In this paper we report DFT calculations on the singlet ground states of N_2-N_2 dimers for different symmetry groups with intermolecular distances in the range of $1.20\text{Å} \leq R \leq 2.50\text{Å}$. Particular attention has been paid to correct the optimized structures and potential energy curves for BSSE.

2 Methods

The N_2-N_2 dimer consists of two nitrogen molecules. The distance between the center of mass of the two atoms in the monomer is taken to be r, and that of between two molecules equal to R. The angles between the molecule one and the dimer axis and molecule two and dimer axis are θ_1 and θ_2 respectively. The dihedral angle is represented by θ_d as shown in Fig. 1.
The calculations were performed with Gaussian 94R program package [11] .The ground state geometries were optimized by Density Functional method with the Lee-Yang-Parr correlation function (B3LYP), and Perdew correlation function (B3PW91) using different basis sets.

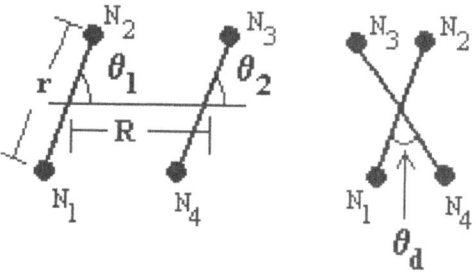

Fig. 1. The orientations of the coordinate system for $(N_2)_2$ dimer used for calculations

The intermolecular potentials in van der Waals N_2-N_2 dimer are generally calculated by fixing the inter atomic distance in the monomers to the natural bond length, i.e., equal to 1.10Å. However for calculating the intermolecular potentials of closely bound dimers it may not be correct approach, because the ground state of such dimers may be formed by the interaction between the vibrationaly excited monomers. In such a situation we may need a symmetry constrained optimization on the bond length of individual N_2 molecule. A total of six types of groups of N_2-N_2 dimer geometries have been studied. These include C_{2vz}, $D_{\infty h}$, C_{2vx}, C_{2h}, D_{2d} and D_{2h} symmetry groups. The Basis Set Superposition Error for N_2-N_2 dimer is calculated by using counter poise technique of Boys and Bernardi [13]. The BSSE is:

$$E_{BSSE} = E_{N_2^1}(R) + E_{N_2^2}(R) - 2E_{N_2}(\infty) \quad \ldots\ldots\ldots(1)$$

The interaction energy is defined as:

$$E_{int} = E_{(N_2)_2}(R) - E_{N_2^1}(R) - E_{N_2^2}(R) \quad \ldots\ldots(2)$$

Where $E_{(N_2)_2}(R)$ is the energy of the dimer for intermolecular distance equal to R.

3 Calculations and Results

The potential energy curves for C_{2vz} at B3PW91/6-311++G** and B3LYP/6-311++G** are shown in Fig.2. We can see that the ground state potential curves are steeply repulsive in the region of interest for us.

The results of the symmetry constrained optimization for r and R on the singlet ground states of D_{2d}, C_{2vx} and $D_{\infty h}$ and C_{2h} groups, after BBSE correction produced somewhat same results as those of C_{2vz}. These results are shown in Table.1 below.

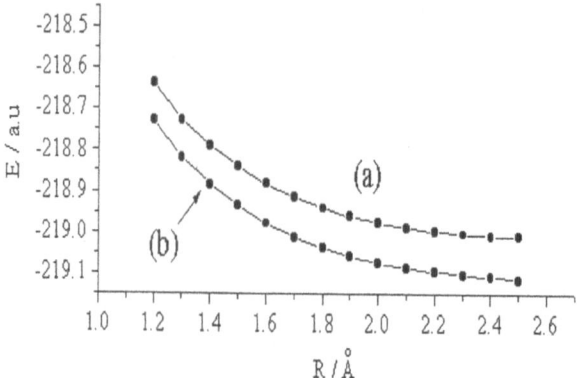

Fig. 2. The potential energy curves at (a) DFT -B3PW91/6-311++G** with r = 1.097A and (b) DFT-B3LYP/6-311++G** with r = 1.098 Å for singlet

Table. 1. Characterization of the BSSE corrected ground states of different symmetry groups (θ_1, θ_2 and θ_d in degrees, r and R_{min} in Å)

Group	configuration θ_1 θ_2 θ_d			DFT (B3PW91) nature of potential energy curve	DFT(B3LYP) nature of potential energy curve
C_{2vz}	45	135	0	Repulsive r = 1.096	Repulsive r = 1.096
	60	120	0	Repulsive r = 1.097	Repulsive r = 1.096
$D_{\infty h}$	0	0	0	Repulsive r = 1.098	Repulsive r = 1.096
C_{2h}	45	45	0	Repulsive r = 1.097	Repulsive r = 1.096
C_{2vx}	90	0	0	Repulsive r = 1.096	Repulsive r = 1.096
D_{2d}	90	90	90	Repulsive r = 1.098	Repulsive r = 1.096
D_{2h}	90	90	0	Bound r =1.2461 R_{min} = 1.53	Bound r = 1.2493 R_{min} = 1.54

D_{2h} Symmetry Group

It can be seen (Table.1) that N_2-N_2 dimer has the local minima which lies in the region of interest for us. Therefore this symmetry group will be the point of focus in the this section. The BSSE corrected geometries at B3PW91/6-311++G** and B3LYP/6-311++G** levels produced all positive values as shown in Table.2, confirming the existence of the real minimum at R = 1.52Å for DFT (B3PW91) and at R = 1.54Å for DFT (B3LYP) methods respectively.

Characterization of Equilibrium Structure for N_2-N_2 Dimer   353

Table 2. The BSSE corrected and uncorrected optimized structures and minimum energies of dimer for D_{2h} symmetry with different Basis sets for DFT theories.

Theory applied	r (Å)	R_{min} (Å) uncorrected	corrected	E_{min} (a.u) corrected	[a]NIF
B3PW91/6-31G*	1.2526	1.52	1.53	-218.67985	- 0 -
B3LYP/6-31G*	1.2558	1.53	1.53	-218.76751	- 0 -
B3PW91/6-31+G**	1.2514	1.52	1.53	-218.68907	- 0 -
B3LYP/6-31+G**	1.2550	1.53	1.56	-218.77871	- 0 -
B3PW91/6-311++G**	1.2461	1.52	1.52	-218.73560	- 0 -
B3LYP/6-311++G**	1.2493	1.53	1.54	-218.82704	- 0 -

[a]NIF: Indicates the number of imagionary frequencies

The calculations on the harmonic vibrational frequencies were carried out to characterize the nature of the minimum. All the frequencies turn out to be positive, indicating the exsitence of a real minimum in the potential energy curve and shown in Table.3.

Table 3. Vibrational harmonic frequencies(in cm^{-1}) calculated by using 6-311++G** basis set

Method	Harmonic frequency					
	$\omega_1 (A_U)$	$\omega_2 (B_{2U})$	$\omega_3 (A_G)$	$\omega_4 (B_{3G})$	$\omega_5 (B_{1U})$	$\omega_6 (A_G)$
DFT(B3PW91)	506.2288	587.7479	1018.7011	1054.8204	1483.3901	1708.4478
DFT(B3LYP)	513.9179	554.4521	989.2615	1053.2645	1450.1250	1680.2994

At this stage it is worth noting that the BSSE corrected potential energy curves for B3PW91/6-311++G** and B3LYP/6-311++G** levels also proved to have a local minima as shown in Fig 3 and Fig.4 respectively (also shown in Table.1).

As this is the only symmetry group, which could be optimized in the region of interest for us, hence BSSE and interaction energies are calculated on the ground state with different basis sets.

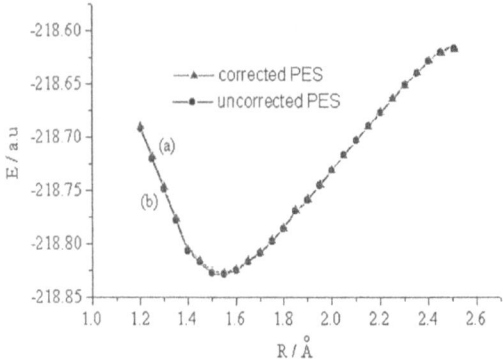

Fig. 3. The potential energy curve for B3LYP/6-311++G** with r = 1.098 Å for singlet ground state of D_{2h} group (a) BSSE corrected (b) BSSE uncorrected.

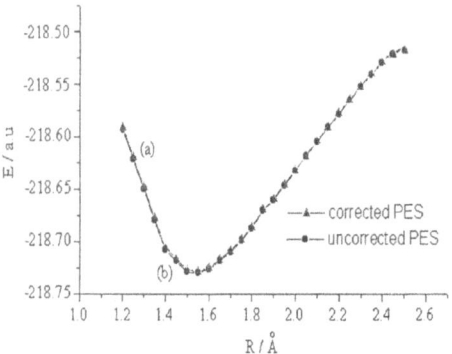

Fig. 4. The potential energy curve for B3PW91/6-311++G** with r = 1.097Å for singlet ground state of D_{2h} group (a) BSSE corrected (b) BSSE uncorrected.

4 Discussion

Following important results about the ground state interactions in N_2-N_2 dimer can be deduced from our theoretical calculations:

(1). Most of BSSE corrected geometries of N_2-N_2 dimer, with DFT approach, are not possible to be optimized in the region for 1.20Å≤R≤2.50Å, these results are consistent with our prior ones in [10]. This indicates that in most of the cases ground state of N_2-N_2 dimer is not stable in the investigated region.

(2). The ground state optimization and BSSE corrected potential energy scan, with DFT methods indicate that D_{2h} structure is the only and most stable structure in the investigated region.

(3). The BSSE corrected D_{2h} geometry, with B3PW91/6-311++G** has two minima, first at r = 1.2517Å and R = 1.52Å and second at 1.097Å and R = 4.92Å, and with B3LYP/6-311++G** level, at r = 1.2493Å and R = 1.54Å and r = 1.0966Å and R = 4.82Å. The second minima has already been investigated by [9] using DFT, however the first minima is reported for the first time in this paper.

Table 4. The BSSE, E_{int} and their comparion at the optimized structures by using different Basis sets with DFT

Method / Basis Set	$^{a}E_{int}$(cm^{-1}) uncorrected	$^{b}E_{BSSE}$(cm^{-1})	$^{c}\Delta E_{(int)}$ (cm^{-1}) corrected	E_{BSSE} / E_{int}(%)
B3PW91/6-31G*	40378.08	783.59	39594.44	1.94
B3LYP/6-31G*	41553.35	829.65	40723.70	1.99
B3PW91/6-31+G**	40523.08	553.75	39969.33	1.30
B3LYP/6-31+G**	41743.93	509.57	41234.36	1.22
B3PW91/6-311++G**	41930.51	395.01	41535.50	0.94
B3LYP/6-311++G**	43435.91	326.99	43108.92	0.75

$^{a}E_{int}$: Interaction energy before BSSE applied , $^{b}E_{BSSE}$: Basis Set Superposition Error
$^{c}\Delta E_{(int)}$: Interaction energy after BSSE applied

(4). The symmetry constrained optimization (for BSSE corrected geometry) on the ground state of D_{2h} group with DFT approaches produce all positive harmonic frequencies, in agreement to the results for ab initio [10]. Therefore we can say that the ground state of D_{2h} structure is a real minimum, not a transition state. The BSSE corrected potential energy well depth for the ground state at B3LYP/6-311++G** level is 3.53 e.v.

(5). The calculated BSSE and the E_{int} for DFT theories, at the optimized ground state of D_{2h}, by using different basis sets are shown in Table.4. The results indicate that BSSE is about 1-2% for DFT methods. Moreover the BSSE seems to be dependent on the selection of the basis sets, larger the basis set smaller the BSSE. A figure of BSSE around 1-2% for DFT method with 6-311++G** basis set is quite promising as compared to the results in vdW region where BSSE is about 20-40 % (for most stable geometry) of the total interaction energy for best achieved results [9]. This figure also suggests that BSSE is less effective in this region (1.20Å≤R≤2.50Å.), and means a strong interaction between two N_2 molecules in the dimer. The present values for BSSE are the lowest ever reported for N_2-N_2 dimer in any configuration or with any level of theory.

(6). The effect of the BSSE upon the variation in R for D_{2h} group at with different basis sets and at different levels of the theory has been presented in Table2. We can see that the effect of BSSE upon R is very small, in general the position of the minimum is shifted 0.01-0.03Å towards the longer distances. This small shift indicates the strong interactions and less effective BSSE. However in Vdw region this shift is found to be about 0.2-0.4Å [9].

(7). If we go back to the potential energy curves for a single ground state N_2 molecule, we note that for state $X^1\Sigma_g$ maximum allowed inter-atomic distance for $v \geq 3,4,....12$. are between $r = 1.237$Å-1.408 Å and even larger for $v = 13$-21. From our results of optimization for ground state X^1A_g of N_2-N_2 dimer, we see that the distance between the atoms in the N_2 is r = 1.2461Å for DFT (B3PW91) and r = 1.2493Å for DFT(B3LYP) method. These values of r are quite near to the above value of r = 1.260Å for v = 4 for N_2. Hence it seems to be quite possible that a vibrationaly excited ground state N_2 molecule with v ≥ 3,4,..12, when collides with a ground state nitrogen molecule N_2 ($X^1\Sigma_g$ v= 0,1,2,...) forms dimer of X^1A_g state. The production of the N_2-N_2 dimer can be illustrated by following path:

e + N_2 ($X^1\Sigma_g$ v= 0,1,2,...) → N_2 ($X^1\Sigma_g$ v≥4) + e

(N_2 vibrational excitation)

N_2 ($X^1\Sigma_g$ v≥4) + N_2 ($X^1\Sigma_g$ v= 0,1,2,...) → { N_2($X^1\Sigma_g$)-N_2($X^1\Sigma_g$)}*

(dimer formation)

{ N_2($X^1\Sigma_g$)-N_2($X^1\Sigma_g$)}* → {N_2 ($X^1\Sigma_g$)-N_2 ($X^1\Sigma_g$)} +hv (IR)

(transitions between vibrational levels of dimer)

{N_2 ($X^1\Sigma_g$)-N_2 ($X^1\Sigma_g$)} → N_2^* + N_2^*

(relaxation to lower vibrational levels)

N_2^* + N_2^* → N_2($X^1\Sigma_g$ v= 0,1,2,...) + N_2($X^1\Sigma_g$ v= 0,1,2,...)

(return to lower levels)

5 Conclusion

We have used DFT (B3PW91) and DFT (B3LYP) theory to study the singlet ground states of different symmetry groups of N_2-N_2 dimer. The effort is put on the characterization of the ground states in the short-range interaction region. On the basis of our calculations we found that except D_{2h} most of the structures could not be optimized in the region of interest for us. The BSSE has been found to be around 1-2% with DFT calculations with 6-311++G** basis set. According to best of our knowledge these BSSE values are the lowest ever reported for N_2-N_2 dimer.

References

[1]   J.Tennyson, A. vander Avoird, J. Chem. Phys. 77 (1982) 5664.
[2]   P.J.Hay, R.T.Pack, R.L.Martin, J. Chem. Phys. 81 (1984) 1360.
[3]   A.Wada, H.Kanamori, J. Chem. Phys. 109 (1998) 9434.
[4]   M.S.H.Ling, M.Rigby, Mol. Phys, 51 (1984) 855.
[5]   A.K.W.McKeller, J. Chem. Phys. 88 (1988) 4190.
[6]   A.Long et al, Chem. Phys. 2 (1973) 485.
[7]   O.Couronne, Y.Ellinger, Chem. Phys. Lett. 306 (1999) 71.
[8]   A.H.Hamdani et al Chem.Phys.Letts 325 (2000) 610
[9]   D.J.Graham et al, J. Chem. Phys. 98 (1993) 2564.
[10] D.S.Tinti and G.W.Robinson, J. Chem. Phys. 49 (1968) 3229.
[11] GUASSIAN 94, J.A.Pople et al. Gaussian, Inc., Pittsburgh. PA, 1995.
[12] W.J.Hehre, Ab initio molecular orbital theory, (John Wiley & Sons, New York ,1986).
[13] S.Boys and F.Bernardi, Mol. Phys 19 (1970) 533

A Time Dependent Study of the Nitrogen Atom Nitrogen Molecule Reaction

Antonio Laganà[1], Leonardo Pacifici[2], and Dimitris Skouteris[2]

[1] Department of Chemistry, University of Perugia,
via Elce di Sotto 8, I-06123 Perugia, Italy
lag@dyn.unipg.it

[2] Department of Mathematics and Computer Science, University of Perugia,
Via Vanvitelli, 1, I-06124 Perugia, Italy
{xleo,dimitris}@dyn.unipg.it

Abstract. In this paper a preliminary study of the $N + N_2$ reaction making use of quantum time-dependent calculations in Jacobi coordinates is presented.

1 Introduction

The interest in molecular processes involving the nitrogen atoms depends on the key role played by the nitrogen atom in processes taking place both in the lower and upper layers of the atmosphere. In particular, we have considered the reaction:

$$N(^4S) + N_2(^1\Sigma_g^+, v) \rightarrow N(^4S) + N_2(^1\Sigma_g^+, v') \tag{1}$$

that plays a key role in modeling the composition of the air both under equilibrium conditions [1] and typical non-equilibrium conditions [2] (like the ones occurring around reentering spacecrafts). Moreover, nitrogen-atom nitrogen-molecule reactions are important in high temperature environments [3], such as the Titan's (the largest moon of Saturn) atmosphere [4]. In fact, molecular nitrogen is the major component of the Titan's atmosphere that is quite similar to Earth's atmosphere. Since astrophysical and astronomic observations have revealed the presence of organic compounds containing C-N bonds, the importance of reaction 1 is also related to the understanding of the formation of organic molecules of fundamental importance for life.

The study of the $N + N_2$ reaction is also important from a theoretical point of view since $N + N_2$ is a suitable prototype for three identical heavy atom systems and an interesting case study for rationalizing bimolecular reactivity for heavy particles. In particular, its investigation is useful to assess the role played by internal and translational motions in promoting vibrational energy excitation and deexcitation when the reactive system does not contain hydrogen atoms.

A first difficulty in carrying out a theoretical study of this system lies in the capability of performing accurate *ab initio* calculations of the potential energy. A second difficulty lies in the capability of performing accurate quantum dynamics studies of the collisional process. In fact, in comparison with more celebrated systems, like H + H$_2$, little work has been carried out for the N + N$_2$ reaction both on experimental and theoretical side. We have tackled the study of this atom-diatom system as a part of our efforts to progress in the construction of a gas-phase molecular simulator developed in our laboratory called SIMBEX (Simulation of Molecular Beam Experiments) [5]. In fact, the above mentioned difficulties have motivated us to implement the quantum dynamics programs developed in our laboratory into the relevant blocks of SIMBEX.

Moreover, this has prompted also the need for implementing these sections of the molecular simulator on a recently constituted computing Grid for Chemistry [6]. In Section 2 details about the the theoretical and computational aspects of the work will be given. In Section 3 the results of the calculations will be discussed.

2 Theoretical and Computational Aspects

2.1 Previous Calculations

Extended theoretical studies of the N + N$_2$ reaction have been already carried out in the past in our laboratory. To this end a LEPS potential energy surface was assembled starting from the available theoretical and experimental information [7]. This LEPS PES shows a smooth monotonic approach to a barrier of 36 Kcal/mol corresponding to a collinear symmetric stretch of the external nitrogens and a clear collinear dominance. On this surface dynamics calculations of classical and quantum type were performed. In this way full dimensional classical and reduced dimensional quantum estimates of cross sections and rate coefficients were obtained over a whole range of energies and temperature (respectively) [8]. The resulting rate coefficients were used in modeling cold plasmas and spacecrafts reentry [9]. Also product vibrational distributions were extensively investigated and the influence of a well, located in the strong interaction region, was analyzed.

More recently, a three dimensional zero total angular momentum quantum study of this reaction has been performed on an *ab initio* potential energy surface (WSHDSP) [10]. The *ab initio* PES is based on the calculation of 3326 points using the CCSD(T) method (couple-cluster singles and doubles) in the large distance regions and a slightly different method for shorter distance regions (the difference is mainly concerned with the used basis set) where the interaction is stronger. *Ab initio* points were fitted using a non linear least square technique with 68 parameters.

The calculated PES has the interesting characteristic of showing a (symmetric) double barrier, i.e. two transition states connected by a shallow well. The two barriers are symmetric with respect to the interchange of two N atoms and are

located in corresponding positions of the entrance and the exit channel of the exchange reaction. The height of these two barriers is 47.2 Kcal/mol.
In order to compare with previous results and have the possibility of easily playing with the characteristics of the PES, we have carried out our calculations on the LEPS one.

The calculations reported in this paper are based on a new time dependent quantum reactive scattering program [11] using a product coordinates approach similar to that of Gray and Balint-Kurti [12] but propagating the whole complex wavepacket.

2.2 Present Time Dependent Treatment

In order to set up the initial wavepacket at R_0 in the asymptotic reactant region, we require the wavefunction of the overall system to be expressed in terms of the initial diatomic molecule BC wavefunction, $\varphi_{vj}^{BC}(r)$.

To start the calculations the initial wavepacket in the scattering coordinate R is built up by multiplying together a normalized Gaussian function [12] $Ne^{-\alpha(R-R_0)^2}$, a phase factor of asymptotic form $e^{-ik(R-R_0)}$, which gives the wavepacket a relative momentum towards the interaction region and the vibrational-rotational wavefunction of the diatomic reactant $\varphi_{vj}^{BC}(r)$. The phase factor is made up of appropriate incoming Ricatti-Hankel functions, thus avoiding the problem of having to start the wavepacket propagation sufficiently far away from the centrifugal barrier. The initial wavepacket may therefore be written as:

$$\Psi^{JK}(R,r,\Theta;t=0) \propto e^{-\alpha(R-R_0)^2} \cdot k(R-R_0) \cdot h_l^1(k(R-R_0)) \cdot \varphi_{vj}^{BC}(r) \cdot P_j^K(\Theta) \quad (2)$$

where K is the quantum number for the projection of the total angular momentum $\mathbf{J}$ on to the body-fixed z axis, $P_j^K(\Theta)$ is the normalized associated Legendre polynomial and k is the wavevector which determines the average relative momentum or kinetic energy, $E_{tr}^o = (k\hbar)^2/2\mu$, of the collision partners. A discrete representation is used to describe the wavepacket. The initial wavepacket is therefore set up in reactant coordinates and the potential and the wavefunction are represented by their values on a regular grid in the scattering coordinate, R, in the vibrational coordinate, r, and on a grid of Gauss-Legendre quadrature points in the Jacobi angle Θ. The initial wavepacket is transformed into the product Jacobi coordinates R', r' and Θ' of the channel of interest and the entire propagation is carried out in these coordinates. At the initial time ($t=0$) the wavepacket is placed far in the reactant channel and it is mapped into a product coordinate grid large enough to contain the initial wavepacket, the region where the analysis line is drawn, and the interaction region, and fine enough to accurately describe the structure of the wavefunction. As time progresses, it moves into the interaction region. At the grid edges an absorption region is introduced to prevent the wavepacket amplitude from reaching the edge of the grid and causing the problem known as aliasing in discrete Fourier transform theory. In this absorption region the wavepacket is multiplied by a damping function of the form:

$$\phi(R) = exp(-\frac{(R-R_0)^2}{2b^2}) \quad (3)$$

where b is a measure of the effective length of the potential and R_0 is the value of R beyond which it is different from unity. The wavepacket is then propagated in time until it has mainly been absorbed near the edge of the grid. The wavepacket is analysed at every time step along an analysis line in the asymptotic region of the product channel [12,13] so as to accumulate the data needed for the computation of the detailed state to state **S** matrix elements $S^J_{vjK,v'j'K'}(E_{tr})$ at the various values of the collision energy E_{tr} contained within the wavepacket.

By summing the square modulus of the detailed **S** matrix elements over K' and averaging over K one can evaluate state-to-state reaction probabilities $P^J_{vj,v'j'}(E_{tr})$. A further summation over v' and j' leads to initial state state-selected reaction probabilities $P^J_{vj}(E_{tr})$.

Before the propagation starts, suitable representations of the Hamiltonian to be used are calculated and stored in memory. The values of the potential energy surface are calculated for all points of the grid used, and are stored in the diagonal positions of the array *vpot*. In *vpot* are also stored the diagonal elements of the Hamiltonian which are local in the coordinate representation and therefore can be regarded as parts of the potential energy surface. One example of these contributions is the BJ^2 term for the overall rotation of the triatomic unit. Off-diagonal terms, both local and non-local in coordinate space (an example of the latter are the Coriolis terms decoupling internal rotation from the overall rotation) are stored in the appropriate off-diagonal positions in the *vpot* array (local ones) or in the *angc* array (non-local ones).

The radial kinetic energy terms are evaluated through a fast Fourier transform technique, and the imaginary exponentials required for this calculation are stored in the *akx, aky* arrays. The matrix elements of the j^2 operator (the rotation of the diatomic units) are calculated in the DVR representation used for the angles through expansion in a complete Legendre basis set, and are stored in the *dilj2l* array.

Finally, for the purposes of the propagation, the center and the width of the spectrum of the Hamiltonian in the basis set used is calculated using well-established formulae. The parameters of interest are E_0 (the center) and ΔE (the half-width) which are to be used to scale the Hamiltonian (with E being the total energy), according to the formula:

$$H_s = \frac{H-E_0}{\Delta E}$$

In order to calculate S - matrix elements from the overlaps computed during the propagation, the formula

$$S_{c\eta',a\eta}(E) = -\frac{f'(E)v_{c\eta'}^{1/2}e^{-ik_cR_\infty}\int_0^\infty <\phi_{c\eta'}|\psi(t)>e^{if(E)t}dt}{<\phi_{a\eta}(E)|\psi(0)>}$$

is used, where $f(E), f'(E)$ are respectively the f function at the required value of the energy and its derivative, $v_{c\eta'}$ is the speed of the recoiling products at the product channel, k_c is the corresponding wavevector, R_∞ is the Jacobi R coordinate at the analysis point, $\psi(t)$ is the wavepacket at time t and $\phi_{a\eta}, \phi_{c\eta'}$ are the channel wavefunctions for the reactants and the products set of vibration, rotation and helicity quantum numbers, respectively.

3 Calculations and Results

3.1 The Calculations Performed

The LEPS potential energy surface [7] used for the dynamical calculations of the present work was taken from the data base of the INTERACTION module of SIMBEX. This LEPS has a smooth monotonic approach to the barrier and a clear collinear dominance. On this surface a large batch of time dependent quantum mechanical calculations for a total angular momentum $\mathbf{J} = 0$ was carried out using product Jacobi coordinates. Due to the symmetry of the N + N_2 reaction only one product channel was considered. The reactants were set at their ground electronic state (^4S and $^1\Sigma_g^+$, respectively).

In a wavepacket approach, each calculation yields results for a range of translational energy, (E_{tr}). In the calculations reported here, different ranges of energies have been considered in order to be able to work out the reactive probabilities for different initial internal states of the reactants, at the same interval of collision energies. This allowed us to work out the fixed total collision energy product vibrational distributions (PVD), for various initial vibrational and rotational quantum numbers and to compare the energy dependence of the reactive probability of the system at different initial internal energies. In particular, the reactant rotational quantum number j has been varied from 0 to 3 and the reactant vibrational quantum number v has been varied from 0 to 5. Such a wide range of calculations poses heavy demands on computing resources. This led us into memory problems for the highest vibrational quantum numbers. Accordingly, since the local memory of the computer nodes of the clustered machine we used is of 1 Gbyte we could not accommodate the matrices needed to describe the system at $v = 6$ without introducing manipulations of the matrices so laborious to make the calculations practically unfeasible.

The energy range adopted for the present calculations was about 0.4 eV. Therefore, for the N_2 molecule in its ground state the range of total energies extends from 1.505 to 1.905 eV (the barrier to reaction is about 1.56 eV); this provided, for example, reaction probabilities over a translational energy range up to 1.759 eV. It is important to point out that in order to carry out calculations for higher vibrational states of the reactant at the same translational energy range

(1.359-1.759 eV) one has to modify the grid parameters. Accordingly, while for the vibrational ground state the grid interval is equal to 0.05 a_0 (with 240 grid points), for $v = 3$ the grid interval is set equal to 0.04 a_0 for a total number of grid points equal to 300. If this operation is not performed correctly, the absorption of the wavepacket at the grid edges is insufficient because of the recoil effect.

Each wavepacket calculation takes typically three days to carry out 1800 iterations on a single node of a Linux SSI cluster (the compiling and linking is performed using the Intel compiler specifically designed and optimized for the adopted platform). On the contrary, for higher initial vibrational quantum numbers of the reactant molecule, the time needed to complete a run exceeds five or six days.

3.2 State Specific Probabilities

We first consider the state specific reaction probabilities obtained by summing state-to-state probabilities over all product states open at the considered energy. The state specific probabilities for v varying from 0 to 5 and $j=0$ are given in Fig. 1. As clearly shown by the figure, in particular for the $v = 0$ curve, the probability is in general small and increases smoothly from threshold, in agreement with the fact that the system has a large barrier (about 1.56 eV). The vibrational excitation of the reactant N_2 molecule is quite efficient in enhancing the reactivity since, as is apparent from the figure, the higher v curves are shifted upwards by an almost constant quantity as the vibrational excitation of the reactant N_2 molecule increases. Some deviations from this behaviour start to be appreciated only at $v=5$. These results seem to indicate that the reaction mechanism is substantially the same for all the considered v states and that the reaction path remains basically collinear.

The calculated reactive probabilities indicate also that vibrational excitation is useful to lower the threshold energy. Plots of Fig. 1 when extended to lower translational energies show that vibrational excitation is almost quantitatively used to enhance reaction.

3.3 Product Vibrational Distributions

As already mentioned, the fact of carrying out the calculations using mass scaled product Jacobi coordinates allowed us to work out information on the product vibrational distributions. The PVDs calculated at the reactant vibrational states $v = 0, 1, 2, 3, 4$ and 5, at $E_{tr}=1.65$ eV and normalized to the maximum are given in Fig. 2.

An important indication obtained from the distributions plotted in the figure is that the reactive process is largely vibrationally adiabatic. In fact, the PVDs given in the figure always peak at $v = v'$. The plotted product vibrational

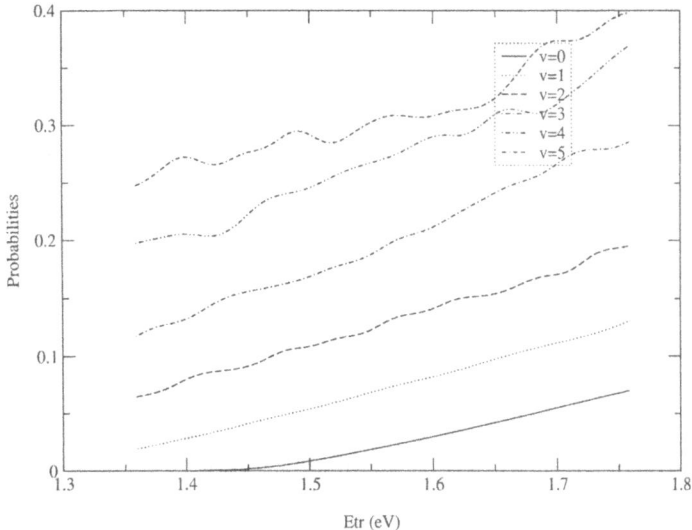

Fig. 1. State specific reactive probabilities calculated at zero total angular momentum, $j = 0$ and $v = 0$ (solid line), $v = 1$ (dotted line), $v = 2$ (dashed line), $v = 3$ (dotted-dashed line), $v = 4$ (double dotted-dashed line) and $v = 5$ (dotted-double dashed line), plotted against the translational energy E_{tr}.

distributions show also two other important features. The first of them is that the width of the PVD widens with v. This means that when increasing the reactant vibrational energy the spread of populated product vibrational states increases accordingly. The second is that a secondary peak shows up at $v > 2$. These peak may indicate that alternative paths leading to vibrationally more excited vibrational products open up at large v values. This agrees with the already found reactivity enhancement at threshold for $v > 2$.

A piece of evidence for the attribution of the secondary peak to alternative (non collinear) paths to reaction is offered by a comparison for the 3D quantum results with infinite order sudden calculations (shown as dotted lines in the same figure). Infinite order sudden results, in fact, while reproducing the primary (vibrationally adiabatic) maximum do not reproduce the secondary one. This implies that this is associated with reaction paths of non fixed angle nature (the collision angle is kept fixed in infinite order sudden calculations) and indicates that related contributions to reaction require rotations.

4 Conclusions

In this work we have reported on the preliminary study of the $N + N_2$ reaction by using the gas-phase molecular simulator SIMBEX. Full 3D quantum time

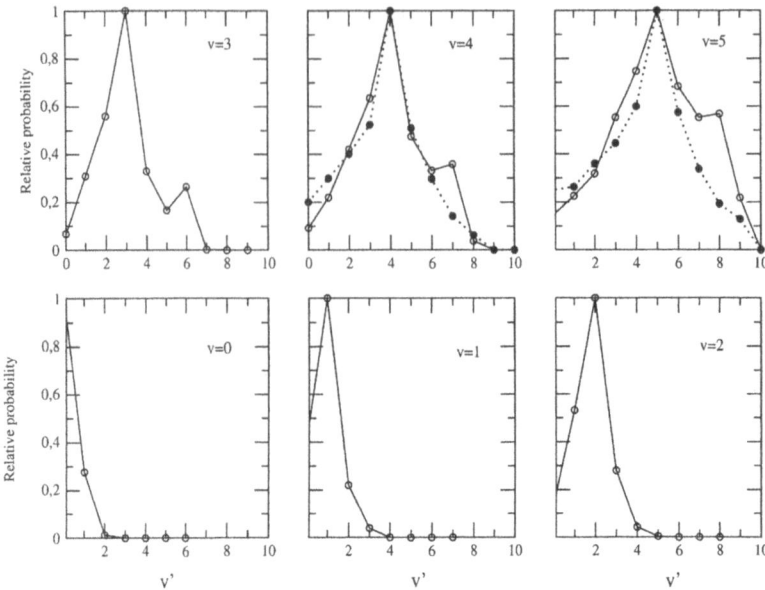

Fig. 2. Normalized product vibrational distributions for $v = 0$ (lower left panel), $v = 1$ (lower central panel), $v = 2$ (lower right panel), $v = 3$ (upper left panel), $v = 4$ (upper central panel) and $v = 5$ (upper right panel) at a collision energy of 1.65 eV. The infinite order sudden results for $v = 4$ and $v = 5$ are also shown (dotted lines).

dependent calculations have been carried out on a LEPS potential energy surface. The calculations have singled out some interesting features of the N + N_2 system. In particular, it has been pointed out that the system is scarcely reactive in the range of collision energy 1.4-1.9 eV. We have also investigated the effect of increasing the vibrational energy of the reactant N_2 molecule on the product vibrational distribution. We found that reactive processes are largely vibrationally adiabatic though non negligible contributions come from alternative reactive mechanisms.

Further work is in progress to assess whether the double barrier-intermediate minimum structure of the WSHDSP PES enhances, as expected, vibrational deexcitation.

References

1. Giordano, D., Maraffa, L.: Proceedings of the AGARD-CP-514 Symposium in Theoretical and Experimental Methods in Hypersonic Flows. **26**, (1992) 1.

2. Capitelli, M.: Non-equilibrium vibrational kinetics, Ed. Springer-Verlag, Berlin (1986); Capitelli, M. and Barsdley, J.N.: Non-equilibrium Processes in partially ionized gas. Plenum, New York (1990); Armenise, I., Capitelli, M., Garcia, E., Gorse, C., Laganà, A., Longo, S.: Deactivation dynamics of vibrationally excited nitrogen molecules by nitrogen atoms. Effects on non-equilibrium vibrational distribution and dissociation rates of nitrogen under electrical discharges. Chem. Phys. Letters, **200**, (1992) 597.
3. Esposito, F., Capitelli, M.: Quasiclassical molecular dynamic calculations of vibrationally and rotationally state selected dissociation cross-sections N + $N_2(v,j)\rightarrow$3N, Chem. Phys. Letters, **302**, (1999) 49-54. Esposito, F., Capitelli, M., Gorse, C.: Quasiclassical dynamics and vibrational kinetics of N + $N_2(v)$ system, Chem. Phys., **257**, (2000) 193-202.
4. Balucani, N., Alagia, M., Cartechini, L., Casavecchia, P., Volpi, G.G., Sato, K., Takayanagi, T. and Kurosaki, Y.: J. Am. Chem. Soc. **122**, (2000); Balucani, N., Cartechini, L., Alagia, M., Casavecchia, P., and Volpi, G.G., J. Phys. Chem. A **104**, (2000) 5655.
5. Gervasi, O., Laganà A. and Lobbiani, M.: Towards a GRID based portal for an a priori molecular simulation of chemical reactivity. Lecture Notes in Computer Science, eds. Sloot, P.M.A., Kennet Tan, C.J., Dongarra, J.J. and Hoekstra, A.G. (Springer, Berlin), **2331**, (2002) 956.
6. A. Laganà, A.: Piattaforme abilitanti per griglie computazionali ad elevate prestazioni orientate a organizzazioni virtuali saclabili. First deliverable of the workpackage 13, Firb project Prot. RBNE01JNFP.
7. Laganà, A., Garcia, E.: Temperature dependence of N + N_2 rate coefficients. J. Phys. Chem., **98**, (1994) 502-507.
8. Laganà, A., Ochoa de Aspuru, G. and Garcia, E.: Temperature dependence of quasiclassical and quantum rate coefficients for N + N_2. Technical report AIAA-94-1986, (1994) 1-6; Laganà, A., Ochoa de Aspuru, G. and Garcia, E.: Quasiclassical and quantum rate coefficients for the N + N_2 reaction, University of Perugia (1996).
9. Laganà, A.: Non equilibrium processes in partially ionized gases. Ed. Capitelli, M. and Barsdley, T. p. 105 (Plenum, New York, 1990).
10. Wang, D., Stallcop, J. R., Huo, W. M., Dateo, C. E., Schwenke, D. W. and Partridge, H.: Quantal study of the exchange reaction for N + N_2 using an *ab initio* potential energy surface. J. Chem. Phys., **118**, (2003) 2186-2189.
11. Skouteris, D., Laganà, A., Capecchi, G., Werner, H.J.: Int. J. Quant. Chem. **562**, (2004) 96.
12. Balint-Kurti, G.G. and Gray, S.K.: Quantum dynamics with real wavepackets, including application to three-dimensional (**J**=0) D + H_2 reactive scattering. J. Chem. Phys. **108**, (1998) 950-962.
13. Balint-Kurti, G.G.: Time dependent quantum approaches to chemical reactivity. Lecture Notes in Chemistry, Laganà, A. and Riganelli, A. Eds. Springer Verlag, **75**, (2000) 74-87.

From DFT Cluster Calculations to Molecular Dynamics Simulation of N_2 Formation on a Silica Model Surface

M. Cacciatore[1], A. Pieretti[2], M. Rutigliano[1], and N. Sanna[2]

[1] CNR-IMIP(Sez.Bari),c/o Chemistry Department of University,
Via Orabona 4, 70126 Bari – Italy
`{cacciatore,rutigliano}@area.ba.cnr.it`
[2] CASPUR Via dei Tizii 6b, 00185 Roma Italy
`{pieretti,sanna}@caspur.it`

Abstract.. B3LYP-DFT electronic structure cluster calculations have been performed to evaluate the adsorption properties of N and N_2 interacting with Si_xO_y clusters in a given adsorption site. To check the convergence of the calculated binding energy, clusters of different size were used in the calculations. As expected, the N atom is chemisorbed, $E_b \cong 2.75 eV$, while N_2 is weakly physisorbed. The *ab initio* results were used to build three PES of the LEPS-type having different activation barrier. The obtained PES have been used in the semiclassical scattering equations and the dynamics of the N_2 formation after atom recombination on a model silica surface was studied in great detail.

1 Introduction

The recombination reaction of N atoms catalyzed by silica surfaces is of great interest in many laboratory plasma processes and in aero-thermodynamics.
To get an insight into the N_2 formation on silica surface, molecular dynamics simulations can be of great interest, provided that a detailed collisional model able to describe the most fundamental behaviors of the interaction is available, and that a sufficiently accurate interaction potential for this system is known. This latter point, that is the knowledge of the potential energy surface where the reaction takes place, is the most critical problem to solve due to the nearly complete lack of both experimental and theoretical information on the properties of the chemisorbed species (binding energies, bond lengths and equilibrium geometry, reaction activation energies). In order to describe the N/N_2-surface interaction potential on firm physical basis, we have undertaken DFT electronic structure calculations as a preliminary but fundamental step for the scattering calculations. To this end, we have chosen a *size-scalable* cluster approach using Si_xO_y clusters of increasing size cleaved from the β-cristobalite unit cell. Then, from the 'nearly' convergent *ab initio* binding energies different PES of the LEPS-type potential, having different reaction barriers, have been considered and used in the scattering equations. These are obtained within the semiclassical collisional method [1],[2] developed for atom/molecule-surface

interactions. In the dynamics the rate determining step (2) for the two-step Eley-Rideal (E-R) reaction (1), (2), has been simulated:

$$N_{gas} + \text{silica} \rightarrow N_{ads}*\text{silica} \tag{1}$$

$$N_{ads}*\text{silica} + N_{gas} \rightarrow N_2(v, j) + \text{silica} + \Delta E_{exo} \tag{2}$$

where ΔE_{exo} is the reaction exothermicity.

As results of the trajectory calculations, the probability for each collisional surface process occurring at the silica surface, including the N_2 formation reaction, is calculated as a function of the kinetic energy of the impinging nitrogen atom. From the obtained reaction probabilities the recombination coefficient γ has also been obtained.

DFT calculations and PES determination for N_2/N interacting with silica surface models are given in Sec. 2. In Sec. 3 some details on the semiclassical collisional method used in the dynamics simulation are given, while in Sec. 4 the obtained results are presented and discussed.

2 Potential Energy Surface Determination

2.1 DFT Calculations

The sticking of atoms or molecules over a crystal surface is done by using either the wavefunction [3] or the density functional quantum chemistry methods [4],[5]. In this study the hybrid Hartree-Fock Self Consistent Field (HF-SCF) and Density Functional Theory (DFT) method of Becke [6] has been used. The DFT method of Lee/Yang/Parr [7] is considered in order to take into account the exchange term of the hamiltonian and the electron correlation contribution. According to the size scalable cluster approach, the size of the molecular cluster used to model the surface is increased until a convergence of the physico-chemical properties (binding energy and the R_{N-Si} distance in the equilibrium configuration) is reached.

Therefore, we first performed calculations of the interaction potential for the $N-SiO_2$ system along the C_2 symmetry axis by keeping the SiO_2 geometry (and the C_{2v} molecular symmetry) fixed at the experimental values of silica β-cristobalite unit cell. Then, with the same (N_2/N)-Si spatial approach, the Si_xO_y model was extended to the Si_3O_4 system and the corresponding potential calculated and later used to build the complete PES. To confirm the trends of the $N-Si_xO_y$ interaction energy versus the R_{N-Si} distance, we extended the model target to the Si_3O_6 and Si_7O_{14} cluster systems by performing a search of the minimum interaction energy and comparing the corresponding bond energies and N-Si distances with those obtained with the Si_3O_4 cluster. Figure 1 shows the Si_3O_4 cluster geometry.

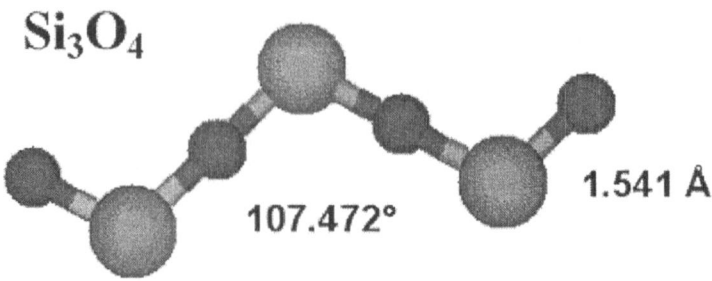

Fig. 1. Molecular geometry the of Si_3O_4 cluster (extracted from the β-cristobalite unitary crystallographic cell) assumed in DFT calculations.

The calculations were carried out with the Gaussian 98 package [8]. The basis set used in all scan calculations of the SiO_2 and Si_3O_4 models was the 6-311+G* as provided by the G98 program with the HF (RHF/UHF) and HF+DFT Hamiltonians including the Becke three parameters - Lee/Yang/Parr (exchange/correlation) - B3LYP [6],[7] functional. However, the PBE and PW91 functionals were also used in test cases and the superiority of the B3LYP functional was proved. For larger clusters Si_3O_6 and Si_7O_{14} we have used the HF+B3LYP method by limiting the basis set used over all atoms via the LANL2DZ [9] ECP (Effective Core Potential).

In Table 1 the binding energy E_b and the corresponding R_{N-Si} equilibrium distance obtained using B3LYP density functional calculations are reported.

Altough a complete convergence of the interaction energies has not been achieved in the considered clusters, the results in Table 1 shows a marked trend of the N-Si equilibrium distance toward 1.7 Å and the corresponding binding energy decreases (-2.84 eV) as the cluster size modelling the silica surface increases up to the Si_7O_{14}, system.

Table 1. Binding energy and bonding distance for different nitrogen/Si_xO_y cluster systems.

Molecular System	E_b (eV)	R_{N-Si} (Å)
N-SiO_2	-2.6	1.8
N-Si_3O_4	-2.74	1.65
N-Si_3O_6	-2.74	1.7
N-Si_7O_{14}	-2.84	1.7

Figure 2 shows the calculated interaction potentials for N_2 interacting in the perpendicular geometry with the SiO_2 and the Si_3O_4 clusters, respectively.

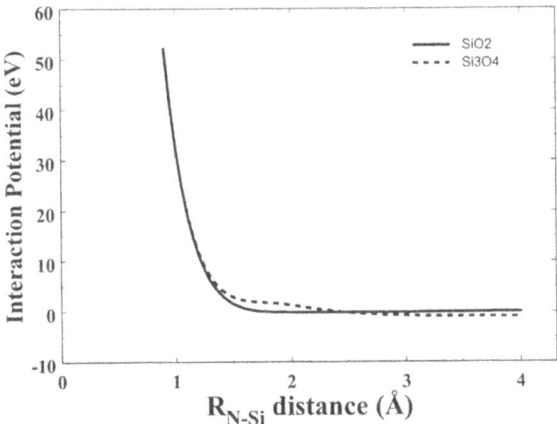

Fig. 2. The interaction potential for N_2 interacting perpendicularly with the silicon active atom of the SiO_2 (continue line) and Si_3O_4 (dotted line) clusters is shown as a function of the nearest N atom and the Si surface atom.

2.2 LEPS Potential

From the obtained *ab initio* data, the full PES of the LEPS-type has been built (in particular N-Si_3O_4 doublet spin state and N_2 -Si_3O_4). However, the determination of the activation barrier for the N_2 formation would imply massive multidimensional DFT calculations, not available due to computational time restrictions. Therefore, the two Sato parameters in the LEPS have been varied in order to obtain PESs with different barriers in the recombination reaction channel. This allowed us to study the sensitivity of the calculated collisional coefficients upon the reaction barrier height. The three considered PES have, respectively, E_{act}= 0.eV (PES-A), E_{act}= 0.1eV (PES-B) and E_{act}=0.3 eV (PES-C). In Figure 3 the potential contour map of the PES-B for N_2 interacting in the perpendicular geometry with the active Si surface atom is shown. The interaction potential is plotted as a function of the distance from the N atom closest to the surface, and the N-N bond distance. The energetic barrier to recombination is located at Z (N-silica) = 2.6Å.

3 Semiclassical Collisional Method

The molecular dynamics simulation is carried out using the semiclassical collisional method [1], [2] according to which the dynamics of the chemical particles, N and N_2, interacting with the silica surface is described classically by solving the relevant Hamilton's equations of motion while the phonons modes of the silica surface are

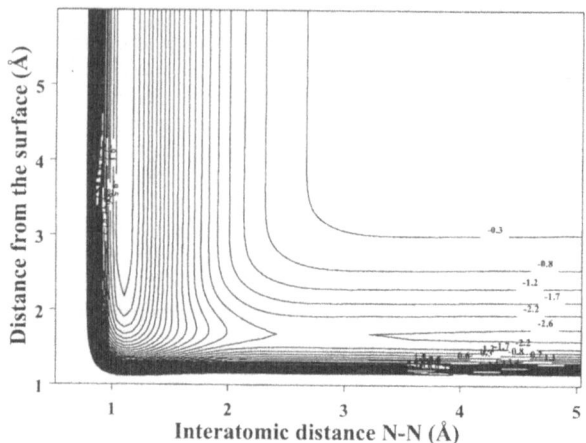

Fig. 3. Potential contour map of PES-B for N_2 interacting with a Si surface atom in the perpendicular geometry. The interaction potential is plotted as a function of the distance from the N atom closest to the surface, and the N-N bond distance. The contours energies are in eV.

treated quantum-mechanically. The coupling between the classical degrees of freedom with the phonons dynamics is made *via* a time and surface temperature (T_S) dependent effective potential H_{eff}, of the mean-field type. Thus the classical equations of motions are given by:

$$\frac{dR_i}{dt} = \frac{P_{R_i}}{m_i} \quad ; \quad \frac{dP_{R_i}}{dt} = -\frac{dH_{eff}(t, T_S)}{dR_i} \tag{3}$$

where m_i, P_{Ri} and R_i are mass, momentum and distance vector for i specie i=1,2 refers to the species in gas phase and

$$H_{eff}(t, T_S) = H_{cl} + V_{add}(t, T_S) \tag{4}$$

H_{cl} is the classical hamiltonian given by

$$H_{cl} = \frac{1}{2}\sum_{i,\gamma}\frac{1}{m_i}P^2_{i,\gamma} + V_{N_2}(r) + \Delta E_{ph} \tag{5}$$

where $P_{i\gamma}$ is the γ-th cartesian component of the momentum of atom i having mass m_i, $V_{N_2}(r)$ is the intramolecular potential of the free N_2 molecule, ΔE_{ph} is the energy exchanged with the surface phonons.

In Eq. (4) $V_{add}(t, T_S)$ is the additional term given as the expectation value of the interaction potential over the phonon wave functions:

$$V_{add}(r, R, t, T_S) = \langle \psi(t, T_S) | V(r, R) | \psi(t, T_S) \rangle \quad (6)$$

where t is the interaction time, V(r, R) is the interaction potential between the gas species and the surface atoms, $\Psi(t, T_S)$ is the total wave function of the phonon state at a given surface temperature T_S.

$$\psi(t, T_S) = U(t, t_0) \psi(t_0, T_S) \quad (7)$$

where $U(t,t_0)$ is the time evolution operator and $\Psi_0(t_0,T_S)$ the wave function at the surface phonon in the initial condition (at $t=t_0$ V(r,R)=0). The time evolution of the phonon wave function is then obtained by solving, in the LFHO approximation [10], the time dependent Schrödinger equations of motion self-consistently with the Hamilton equations of Eq.(3).

4 Collisonal Dynamics Simulation

The silica surface model assumed in the scattering calculations consists of 149 atoms arranged according to the β-cristobalite unitary cell [11]. The adsorbed nitrogen atom is placed at a distance of 1.65Å from an active Si atom, in thermal equilibrium at the surface temperature which is fixed at T_S=1000K. The kinetic energy of N_{gas} is varied between 0.01eV and 2.5 eV. The initial position coordinates for the gas-phase atom, impinging perpendicularly to the surface with a polar angle θ=0° to the surface normal, are chosen randomly within an aiming area equal to the β-cristobalite unit cell. For each kinetic energy a batch of 1200 trajectories is run so that a numerical convergence of about 15% is expected for the calculated collisional coefficients.

Some important collisional data relevant to the recombination reaction are calculated as a function of the kinetic energy of the impinging N atom: the state-selected recombination probability, P(v,j) and the energy distribution in the final product states. From the calculated recombination probability, the state-selected and the global recombination coefficient $\gamma(T_{gas},T_S)$ has been calculated at different temperature of the incident atomic N flux impinging the surface.

Figure 4 shows the recombination coefficient γ as a function of the gas temperature at T_S=1000K. At $T_{gas}=T_S$ we get $\gamma=3.85 \times 10^{-2}$, which is comparable with the experimental values $\gamma=9.0 \times 10^{-3}$ and $\gamma=4.0 \times 10^{-4}$, obtained in [12] and [13] but for N_2 formation on RGD and pure silica respectively.

In addition to the N_2 formation at surface, the N-silica interaction can lead to the activation of the following surface processes in competition with one another:
— adsorption processes (α):

$$N_{gas} + N_{ad}*silica \longrightarrow N_{ad}*silica + N_{ad}*silica \quad (8)$$

— adsorption/desorption processes with one atom adsorbed and the other reflected into the gas-phase (β):

$$N_{gas} + N_{ad}*silica \longrightarrow N_{ad}*silica + N_{gas} \quad (9a)$$

$$N_{gas} + N_{ad}*silica \longrightarrow N_{gas} + N_{ad}*silica \tag{9b}$$

– direct inelastic surface processes where both atoms are scattered in the gas-phase (γ)

$$N_{gas} + N_{ad}*silica \longrightarrow N_{gas} + N_{gas} + silica \tag{10}$$

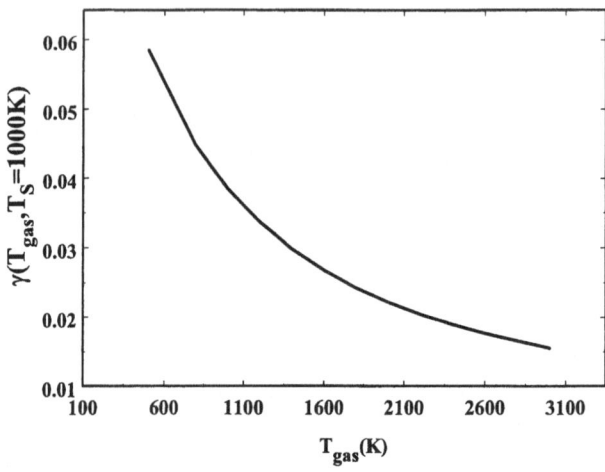

Fig. 4. Recombination coefficient γ as a function of gas temperature

In Table 2 the probability for the processes given above is reported as a function of the impact energy of the gas-phase nitrogen atom.

The obtained results show that at the higher collisional energies only the adsorption/desorption processes are the active ones, while the recombination process is very effective in a rather narrow range of impact energies, between 0.02 and 0.04 eV. This is due to the changing of the energy exchange mechanism between the silica substrate and the nitrogen atoms according to the collisional regime considered [14].

Table 2. Probabilities for different reaction channels (α)-(β)-(γ) of Eq. s (8)-(10)

E_{kin} (eV)	N_2 formation	(α)	(β)	(γ)
0.01	-	1		
0.02	6.84e-2	0.61	0.317	5.83e-2
0.03	0.115	0.72	0.154	1.25e-2
0.04	0.207	0.63	0.135	1.33e-2
0.055	-	1	-	-
0.5	-	1	-	-
1.0	-	1	-	-

Acknowledgements. This work was supported by the Italian Space Agency (ASI) in the framework of the project ASI/CNR IR/214/02: "Catalytic and Thermal Effects for Martian Atmospheric Entry".

References

1. G. D. Billing, "Dynamics of Molecule Surface Interactions", Wiley, 2000
2. M. Rutigliano, M. Cacciatore and G.D. Billing, Chem. Phys. Lett., 340 (2001) 13; M.Cacciatore and G.D.Billing, Surf.Science 232 (1990) 35
3. E.g., see, R. Dovesi, V.R. Saunders, C. Roetti, M. Causa, N.M. Harrison, E. Apra', CRYSTAL 95 User's Manual, University of Torino, Torino, 1996
4. E.g., see, K. G. Nakamura, Chem. Phys. Lett. 285 (1998) 21
5. G. P. Brivio, T. B. Grimley, Surf. Sci. Rep. 17 (1993)
6. A. D. Becke, J. Chem. Phys. *98* (1993) 5648
7. C. Lee, W. Yag and R. G. Parr, Phys. Rev., 1988, *37*, 785
8. Gaussian 98 (Revision A.7), M. J. Frisch, et al, Gaussian, Inc., Pittsburgh PA, (1998)
9. P. J. Hay and W. R. Wadt, J. Chem. Phys. 82 (1985) 270
10. P. Pechucas and J. C. Light, J. Chem. Phys. 44 (1966) 3897
11. R.W.G.Wickoff American Journal Science 5 (1925) 448
12. C. D. Scott, "Catalytic Recombination of Nitrogen and Oxygen on High Temperature Reusable Surface Insulation", in Aero and Planetary Entry, ed. A. L. Crosbie, Vol. 77 Progress in Astronautics and Aeronautics, AIAA , p.192, (1981)
13. Y.C. Kim and M.Boudart, Langmuir 7 (1991) 2999
14. M. Cacciatore, M. Rutigliano, A. Pieretti, N. Sanna, wotk in preparation

Molecular Mechanics and Dynamics Calculations to Bridge Molecular Structure Information and Spectroscopic Measurements on Complexes of Aromatic Compounds

G. Pietraperzia[1,2], R. Chelli[1,2], M. Becucci[1,2], Antonio Riganelli[3], M. Alberti[4], and Antonio Laganà[3]

[1] Dipartimento di Chimica, Università di Firenze, via della Lastruccia 3, I-50019, Sesto Fiorentino (FI), Italy
becucci@lens.unifi.it
[2] LENS, via Nello Carrara 1, I-50019, Sesto Fiorentino (FI), Italy
[3] Department of Chemistry, University of Perugia,
via Elce di Sotto, 8, I-06123 Perugia, Italy
{auto,lag}@dyn.unipg.it
[4] Departament de Química Física, ERQT
Parc Científic, Universitat de Barcelona, Martí i Franquès, 1. 08028 Barcelona (Spain)
m.alberti@ub.edu

Abstract. Molecular dynamics approaches are applied to the analysis of some properties of the anisole-water and benzene-argon complex. To this end, semiempirical pairwise additive potentials are used. The analysis of the potential energy surface allows to select the most probable conformation among the calculated ones. Then further calculations allow to estimate rotationally resolved spectra and interconversion ratios.

1 Introduction

The interpretation of high resolution spectroscopy experiments on molecular complexes is often a difficult task because of the lack of reliable hypotheses on the structure of the complexes. For several simple systems these hypotheses can be based on chemical intuition. However, as the complexity of the potential energy surface (PES) increases a more systematic approach is needed. To this end, when the system is strongly bound, *ab initio* methods can be used (especially when only a few electrons are involved). Unfortunately, when the complexity of the system under investigation increases and the forces coming into play are weak *ab initio* calculations may not be accurate enough. For example, Density Functional Theory (DFT) based approaches well account for cluster interactions when forces have a strong electrostatic character (*e.g.* systems with ions, highly dipolar molecules or H-bonding) whereas they strongly underestimate van der Waals interactions like those present in nonpolar molecular aggregates. On the other hand, other *ab initio* methods, such as the perturbative Møller-Plesset ones[1,2]. suffer exactly of the opposite problem than DFT (electrostatic forces are underestimated).

An example of this type is the anisole-water 1:1 complex. For this system accurate estimates of the rotational constants were obtained from high resolution laser induced fluorescence measurements[3]. There we showed that anisole and water form a H-bond and that the center of mass of water falls on the plane defined by the aromatic ring of anisole. Previous accurate *ab initio* calculations based on[4] high level of theory (MP2/aug-cc-pVDZ) had indicated, instead, that in the most stable complex geometry (called AW1) anisole and water are H-bonded and the water is displaced over the aromatic ring. At the same time, a geometry similar to that obtained in our experiment was found[4] to exist and to have an energy larger than AW1. Only subsequent calculations at the B3LYP/6-311++G(d,p) level were able to show that a theoretical evaluation of the OH frequencies is not accurate enough to allow a discrimination among different configurations of the H-bonded anisole-water complexes. A similar situation of unsatisfactory *ab initio* determination of the PES applies to benzene-Ar complexes for which experimental findings are not matched by theoretical conclusions[5].

Thus, to the aim of interpreting experimental measurements like the rovibronic bands obtained from high resolution spectroscopy, it could be adopted a hierarchical two step computational procedure. In the first instance, simple empirical potential models, such as the pairwise additive ones, or a slightly more sophisticated semiempirical potential model (though still computationally cheap) like the one we use for the benzene-argon complex, is used. Eventually, a second step, relying on more accurate methods, can be used to improve the agreement with the experimental data.

In the present paper the anisole-water and the benzene-argon cases are discussed using two different semiempirical potential models.

In section 2 the adopted potential models are discussed. In section 3 calculated and measured spectra are compared. In section 4 the benzene-Ar_3 cluster is studied.

2 Potential Energy Surfaces

2.1 Anisole – Water

All possible energy minima of the anisole-water 1:1 complex have been determined by quenching 3700 configurations of anisole and water[3] (see Ref.[6] and references therein for details on the computational protocol). These configurations were extracted randomly from a molecular dynamics simulation performed in the canonical thermodynamical ensemble (simulation box: cube of 15 Å sidelength; temperature: 300 K). The simulation conditions were sufficient to allow the complex to break and reform several times during the run. The anisole molecule has been modeled by the AMBER force field[7], and its atomic charges

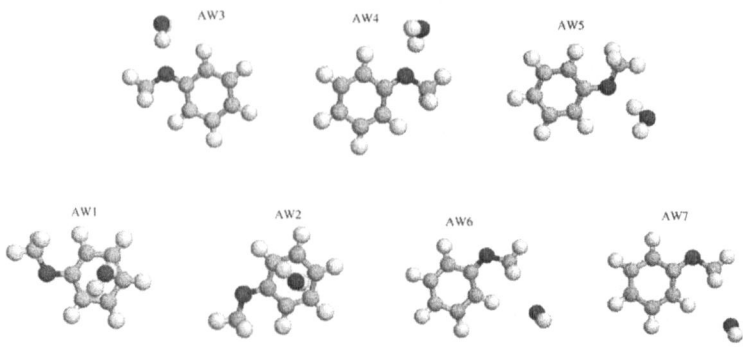

Fig. 1. Minimum energy structures of the anisole-water complex obtained by molecular mechanics calculations (structural class 1: AW1, AW2; structural class 2: AW3, AW4, AW5; structural class 3: AW6, AW7).

have been obtained by a fit of the *ab initio* (B3LYP/6-31G(d,p) level of theory) derived electrostatic potential[8]. The TIP3P model[9] has been utilized for water.

Seven minimum energy structures have been obtained (see Fig. 1) whose binding energy and rotational constants are reported in Tab. 1. The minimum energy configurations of Fig. 1 can be grouped in three structural classes (first class: AW1 and AW2; second class: AW3, AW4 and AW5; third class: AW6 and AW7). Each class is characterized by the fact of having structural configurations with similar anisole-water mutual arrangement. This geometrical similarity appears also evident from the differences in the binding energies of the complexes belonging to the same structural class (see Tab. 1) that are of the order of 0.1 kJ mol^{-1}. This could mean that the PES for the anisole-water 1:1 complex is quite flat around the class representative geometries. The structural geometries of the first and the second class seem to be physically reliable also on the basis of previous studies on complexes of water with aromatic compounds[4,10,11].

The complexes of the first class are characterized by a H-bond involving the π-system of anisole. In the complexes of the second class, anisole and water form instead a conventional nearly in plane σ H-bond, where water acts as hydrogen donor. From the calculations the complexes of the first class result more stable than those of the second one. We stress again that these data must be considered only in a semiquantitative fashion, since we are interested mainly in finding possible candidate structures for interpreting the experiments rather than the most stable of them. On the other hand, we notice incidentally that also for the p-cresol dimer, able to form both conventional and non conventional H-bonds, the binding energy of structures with conventional H-bond was found to be slightly lower than that of structures with non conventional H-bond[6].

Table 1. E_b: binding energy (in units of kJ mol^{-1}) for the anisole-water 1:1 complexes found during the energy minimization; A, B and C: rotational constants (in units of cm^{-1}).

complex	E_b	A	B	C
AW1	-17.93	0.0742	0.0369	0.0348
AW2	-17.67	0.0742	0.0366	0.0347
AW3	-15.69	0.0861	0.0329	0.0241
AW4	-15.63	0.0968	0.0300	0.0233
AW5	-15.56	0.0995	0.0294	0.0231
AW6	-9.97	0.0778	0.0299	0.0218
AW7	-9.60	0.0783	0.0297	0.0217

In the complexes of the third class the oxygen atom of water lies on the plane of the aromatic ring, while the hydrogen atoms are displaced out of the plane (C_s symmetry). For this class of complexes the binding energy is the lowest. Given that, contrarily to AW3, AW4 and AW5, H-bond between anisole and water is not established in the AW6 and AW7 configurations, we can estimate the additional energy stabilization due to H-bonding to be in the range of 5-10 kJ mol^{-1}. This low value is consistent with the quite large O$\cdots$H distance (1.88 Å) that agrees satisfactorily with the experimental value of 1.95Å.

2.2 Benzene – Argon

The Benzene-Ar complex is considered to be a suitable prototype of the van der Waals type interaction of aromatic molecules and rare gas atoms for both theory and experiment (see Ref.[12] and references therein). The system is sufficiently small to make it possible to use potential models more realistic than the atom-atom pair-additive ones. In previous works[13,14] we showed how this kind of potential is able to describe both the equilibrium geometries and the isomerization process of the benzene-Ar$_2$ complex. In that study the global potential (PES-H) of Pirani et al.[12] was used in the calculation. The same potential model is used in this work to calculate the minimum energy configuration. This minimum was obtained by a standard conjugate gradient searching technique. The corresponding structure has a C_{6v} symmetry, with the argon atom placed above the benzene ring at 3.53Å distance. The rotational constant used to reproduce the spectra are (A=0.09554 cm^{-1}, B=C=0.04037 cm^{-1}).

Calculations performed on PES-H, however, showed some critical features associated with the handling of atom-molecule rotations that caused large fluctuations in the total energy. Moreover, PES-H cannot be easily extended to systems in which more Ar atoms are clustered around the benzene molecule. For both these reasons, in view of extending the calculation to the simulation of more realistic clusters, a new functional form based on the atom-bond interactions (PES-AB) was derived[5] and used for the present work.

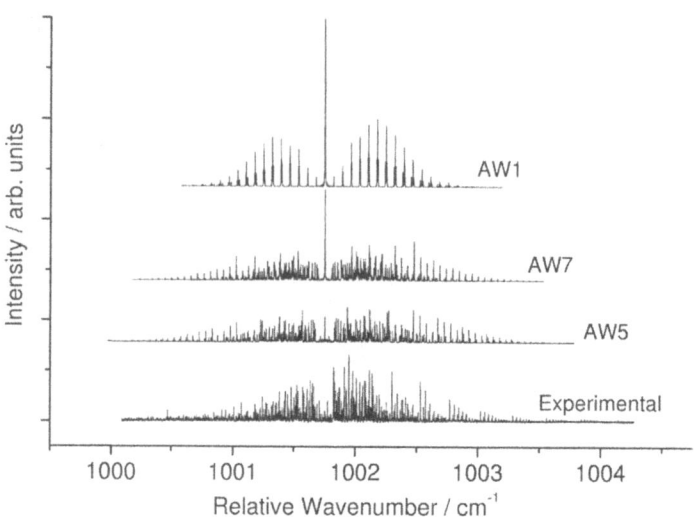

Fig. 2. Comparison between experimental and simulated spectra of the 0_0^0 band of the anisole-water complex.

3 Comparison between Calculated and Observed Spectra

3.1 Anisole – Water

The experimental spectrum of the origin band of the $S_1 \leftarrow S_0$ electronic transition of the anisole-water complex was recorded using a high resolution molecular beam spectrometer, described elsewhere in literature[15]. The complex is formed during the first stages of the supersonic expansion and lives long enough to perform the experiment thanks to the high vacuum kept in the molecular beam chamber.

The rotational structure of the spectra corresponding to the different complex geometries (see Fig. 2) was simulated using an asymmetric top, rigid rotor hamiltonian, with the rotational constants obtained from the calculation (one for each representative class). The other used information was the polarization of electronic transition and the orientation of the principal axis of inertia of the complex with respect to that of the anisole isolated molecule. No geometrical distortion of the complex in going from the S_0 to the S_1 electronic states was assumed.

From a comparison with the experimentally obtained spectrum, we were able to make a guess on the more probable structure of the complex and start the iterative procedure of fitting and simulating, that allows the complete assignment of the rotational fine structure of the spectra[3]. The almost planar AW3-5 struc-

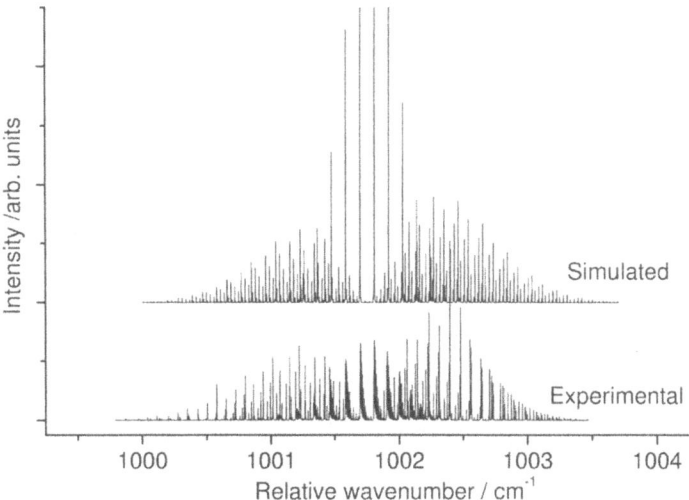

Fig. 3. Comparison between experimental and simulated spectrum of the 6_0^1 band of the benzene-argon complex.

tures are indicated as being the most probable by the simulation. The complete assignment of the rotational structure did also confirm this.

3.2 Benzene – Argon

The experimental spectra of several vibronic bands of the $S_1 \leftarrow S_0$ electronic transition for the benzene-argon complex was recorded under high resolution condition by Riedle et al.[16]. The structure of the benzene-argon complex was calculated according to the potential optimized as described in section 2.2. We compare the vibronic spectrum simulated according to the rotational and centrifugal distortion constants as obtained from the assignment of the rotational fine structure of the experimentally obtained spectrum[16], with the spectrum simulated using the rotational constants obtained from the calculation, using the same rotational constants both for the lower and the upper electronic state. The agreement between the two spectra, shown in Fig. 3, agree sufficiently well to take the structure as a good starting point for the iterative procedure.

4 Benzene – Ar_3 Simulation

As already mentioned, we extended the study to the clustering of additional Ar atoms around the benzene molecule. For this reason we started by considering the system benzene-Ar_3.

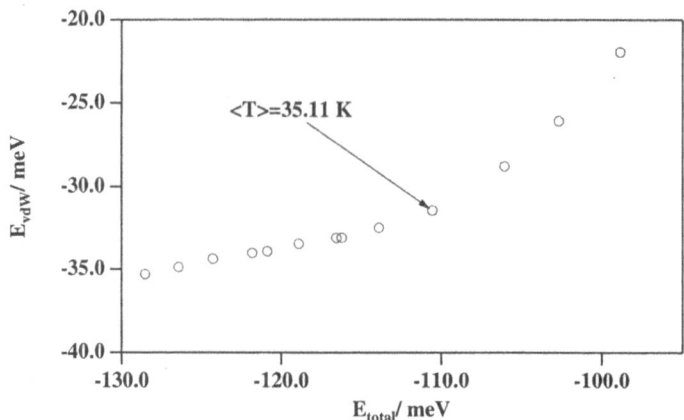

Fig. 4. Van der Waals energy due to the interaction between Ar atoms vs total energy

Simulations have been carried out using a microcanonical ensemble. As a first step a short simulation to equilibrate the temperature was carried out. A equilibration time of 75 ps with a temperature scaling of the velocities allowed to obtain the desired temperature. After obtaining the desired temperature, a long simulation of 100 ns is performed. To this end, as in our previous studies, the benzene molecule was taken as a rigid body. A time step of 0.0025 ps was proved to be sufficiently small to obtain converged values. Relative fluctuations of the total energy in the overall range are smaller than 10^{-6}.

Calculations have been performed for the interconversion between isomers having all the Ar atoms placed on the same side of the benzene plane (3—0 structures) to isomers having the Ar atoms placed in opposite sides of the plane (2—1 structures). A particular isomer was considered as formed when the time spent by the cluster in a particular configuration is equal or larger than 5 ps.

The calculations results show that isomerization processes begin to occur temperatures around 35 K. At this temperature isomeric interconversions are frequent though the probability of dissociation is rather small. When the temperature increases, the interconversion frequency and the dissociation probability increase. However, up to the highest investigated temperature (T=40 K), the cluster lifetime is always larger than 5 ns. The simulation time has been found to be large enough to ensure that the population ratio of the various isomers does not change when extending the simulation to longer times.

The beginning of the isomerization process is evidenced in Fig. 4 where the van der Waals energy (interaction energy of the Ar atoms, E_{vdW}) is plotted as a function of the total energy (E_{total}). At low values of total energy, only the most stable 3—0 isomer is found. By increasing total energy more configurations are explored leading to an increase of the van der Waals energy. Above a certain value of E_{total}, when interconversions between 3—0 and 2—1 isomers become possible, the E_{vdW} plot shows a clear change of slope.

The simulation also shows that the time spent by the benzene-Ar$_3$ cluster spends in the 2—1 configuration increases with E$_{total}$. The relative population of isomer 3—0 diminishes from a value of 92.8% at E$_{total}$=-106.6 meV (<T>=38.63 K) to a value of 61.6 % when the total energy is equal to -98.86 meV (<T>=42.18 K).

5 Conclusions

Molecular mechanics calculations have been used to start an iterative process to single out the most stable structures of aromatic molecular complexes and assign the rotational fine structure of the rovibronic spectra. The method has the advantage of allowing a complete exploration of the possible structures. It also estimates the relative energy of the various structures and the probability of their formation. This reflects the hybrid character of the control regime leading to the formation of the complex in supersonic expansions.

The method has been applied to two different systems. In the benzene-argon case, the molecular system is symmetrical and the interactions between the two partners in the complex are of pure dispersive nature; in the anisole-water case, the complex is scarcely symmetrical and the interaction responsible for the stabilization of the adduct is much stronger (hydrogen bond).

The quality of the calculated potential energy surfaces adopted in this work good enough to start a complete assignment of the rotationally resolved spectra and to support a rationalization of the experiments.

Acknowledgments. The Authors kindly acknowledge the support from Italian MIUR and EU (under Contract nr. HPRI-CT-1999-00111).

References

1. H.L. Williams and C. F. Chabalowski: Using kohn-sham orbitals in symmetry-adapted perturbation theory to investigate intermolecular interactions. J. Phys. Chem. A **105** (2001) 646–659
2. H.L. Williams and C. F. Chabalowski: State of the art in counterpoise theory. Chem. Rev. **94** (1994) 1873–1885
3. M. Becucci, G. Pietraperzia, M. Pasquini, G. Piani, A. Zoppi, R. Chelli, E. Castellucci, W. Demtroeder: A study on the anisole-water complex by molecular beam - electronic spectroscopy and molecular mechanics calculations. J. Chem. Phys. **in press** (2004)
4. B. Reimann, K. Buchold, H. D. Barth, B. Brutschy, P. Tarakeshwar, K. S. Kim: Anisole-(H$_2$O)$_n$ ($n = 1 - 3$) complexes: an experimental and theoretical investigation of the modulation of optimal structures, binding energies, and vibrational spectra in both the ground and first excited states. J. Chem. Phys. **117** (2002) 8805–8822
5. M. Alberti, A. Castro, A. Laganà, F. Pirani, M. Porrini, D. Cappelletti. Chem. Phys. Lett. (2003) submitter

6. F. L. Gervasio, R. Chelli, P. Procacci, V. Schettino: The nature of intermolecular interactions between aromatic amino acid residues. PROTEINS **48** (2002) 117–125
7. W. D. Cornell, P. Cieplak, C. I. Bayly, I. R. Gould, K. M. Merz, D. M. Ferguson, D. C. Spellmeyer, T. Fox, J. W. Caldwell, P. A. Kollmann: A second generation force field for the simulation of proteins, nucleic acids, and organic molecules. J. Am. Chem. Soc. **117** (1995) 5179–5197
8. M. J. Frisch, G. W. Trucks, H. B. Schlegel, G. E. Scuseria, M. A. Robb, J. R. Cheeseman, V. G. Zakrzewski, J. A. Montgomery, R. E. Stratmann, J. C. Burant, S. Dapprich, J. M. Millam, A. D. Daniels, K. N. Kudin, M. C. Strain, O. Farkas, J. Tomasi, V. Barone, M. Cossi, R. Cammi, B. Mennucci, C. Pomelli, C. Adamo, S. Clifford, J. Ochterski, G. A. Petersson, P. Y. Ayala, Q. Cui, K. Morokuma, D. K. Malick, A. D. Rabuck, K. Raghavachari, J. B. Foresman, J. Cioslowski, J. V. Ortiz, B. B. Stefanov, G. Liu, A. Liashenko, P. Piskorz, I. Komaromi, R. Gomperts, R. L. Martin, D. J. Fox, T. Keith, M. A. Al-Laham, C. Y. Peng, A. Nanayakkara, C. Gonzalez, M. Challacombe, P. M. W. Gill, B. Johnson, W. Chen, M. W. Wong, J. L. Andres, C. Gonzalez, M. Head-Gordon, E. S. Replogle, J. A. Pople: Gaussian 98, Revision A.5. Gaussian Inc., Pittsburgh, PA (1998)
9. W. L. Jorgensen, J. Chandrasekhar, J. D. Madura, R. W. Impey, M. L. Klein: Comparison of simple potential functions for simulating liquid water. J. Chem. Phys. **79** (1983) 926–935
10. P. Tarakeshwar, K. S. Kim, B. Brutschy: Fluorobenzene···water and difluorobenzene···water systems: an *ab initio* investigation. J. Chem. Phys. **110** (1999) 8501–8512
11. M. Raimondi, G. Calderoni, A. Famulari, L. Raimondi, F. Cozzi: The benzene/water/hexafluorobenzene complex: a computational study. J. Phys. Chem. A **107** (2003) 772–774
12. F. Pirani, M. Porrini, S. Cavalli, M. Bartolomei, D. Cappelletti: Potential energy surfaces for the benzene-rare gas systems. Chem. Phys. Lett. **367** (2003) 405–413
13. A. Riganelli, M. Memelli, A. Laganà: A molecular dynamics study of the benzene...ar$_2$ complexes. Lecture Notes in Computer Science **2331** (2002) 926–931
14. A. Zoppi, M. Becucci, G. Pietraperzia, E. Castellucci, A. Riganelli, M. Alberti: 16th International Symposium on Plasma Chemistry. Taormina, Italy (2003)
15. E. R. Th. Kerstel, M. Becucci, G. Pietraperzia, E. Castellucci: High-resolution absorption, excitation and microwave-UV double resonance spectroscopy on a molecular beam: S_1 aniline. Chem. Phys. **199** (1995) 263–273
16. E. Riedle, R. Sussmann, Th. Weber, J. Neusser: Rotationally resolved vibronic spectra of the van der Waals modes of benzene-Ar and benzene-Kr complexes. J. Chem. Phys. **104** (1996) 865–881

Direct Simulation Monte Carlo Modeling of Non Equilibrium Reacting Flows. Issues for the Inclusion into a *ab initio* Molecular Processes Simulator

D. Bruno[1], M. Capitelli[2], S. Longo[2], and P. Minelli[2]

[1]Istituto di Metodologie Inorganiche e dei Plasmi del C.N.R. - sez. Bari
Via Orabona 4, 70126 Bari, Italy
{cscpdb30, f.esposito}@area.ba.cnr.it
[2]Dipartimento di Chimica, Università di Bari
Via Orabona 4, 70126 Bari, Italy
{cscpmc05, cscpsl17, cscppm59}@area.ba.cnr.it

Direct Simulation Monte Carlo (DSMC) method for the modeling of non equilibrium reacting flows is presented. The is particularly suitable for the simulation of gas-phase systems with complex boundary conditions and with varying degrees of thermal and chemical non equilibrium. Since the description is done at the kinetic level, detailed information about the elementary processes, as derived from *ab initio* molecular dynamics calculations, can be used as input physical data for the simulation. These features make the DSMC method an ideal candidate for inclusion into a *ab initio* Molecular Processes Simulator for the gas phase.

1 Introduction

A large variety of gas flows can be modeled by means of the well-established Euler or Navier-Stokes equations. These include the study of reactive and/or hypersonic flows. Certain degrees of chemical and thermal non equilibrium can also be tackled by suitable extensions of the above equations. Nonetheless, the physical assumptions underlying the Navier-Stokes (and more strongly the Euler) equations restrict the applicability of such methods to systems that can be considered as continuum fluids, i.e. the mean free path of the particles, which is the physical lengthscale of the molecular granularity, must be negligible as compared the the lengthscale of the physically interesting phenomena. The dimensionless Knudsen number, defined as the ratio of the local mean free path to the macroscopic characteristic dimension of the flow, is usually used to set the limits of applicability of the continuum description. For Knudsen numbers less than 0.01 the continuum description applies; for Knudsen numbers greater than 10 the collisions between particles have a negligible effect as compared to the effect of the solid boundaries and the flow is called free molecular. For values between these two limits, the flow is called transitional or rarefied and the governing equation is the Boltzmann transport equation. Note that rarefaction is not dependent only on density. Rarefaction effects can appear in nano- or micro-scale flows where the characteristic dimensions of the flow are comparable to the mean free path even at moderate densities; in hypersonics, rarefaction effects appear since the characteristic macroscopic dimension is the spatial width of the gradients, which can become very steep. As a consequence there are several problems where rarefied gas effects play a crucial role. It is the case of the analysis of the rarefied environment of orbiting objects, be it a small satellite or a large space station; the study of the contamination produced by the firing of small control thrusters; hypersonic flow studies (reentry vehicle aerodynamics, thermal protection requirements, hypersonic propulsion) and fast reactive systems (detonations). The same issues also apply to the modeling of the vast class of problems dealing with low density plasmas for the surface treatment (etching, deposition). In particular, in the last three areas, matters are complicated by the presence of thermal and/or chemical non equilibrium in the flow. In these flows, many non equilibrium processes play an important role. Therefore, kinetic modeling is needed whose results crucially depend on the details with which each elementary process is described. This, in turn, calls for a careful description of many high temperature energy exchange and chemical reaction processes whose knowledge is far from complete. The numerical solution of the Boltzmann equation is a formidable task. Instead, a Monte Carlo method has been developed which gives an approximate solution to the same equation. The Direct Simulation Monte Carlo (DSMC) method [1] has been found in several previous works to be very effective to study gas phase kinetics, and it is now a standard simulation tool for the rarefied gas community. Being a convenient solution technique to the non linear Boltzmann equation [2], it allows the simulation of elementary processes at the phase-space kinetic level, i.e. the particle distribution f(r,v,t) is explicitly evaluated. The DSMC is therefore able, in principle, to consider any correlation between internal and translational kinetics, and the problem is changed to the one of availability of appropriate input data. It has been shown in [3, 4] that the method can be used to study the details of gas phase chemical physical processes. This work has been generalized by our group by showing that realistic models of the vibrational kinetics of gas molecules can be included [5], and that even the quantum coherent kinetics of fast reactions can be included by addressing any particle with a local density matrix [6]. Besides, case studied are the dissociation kinetics behind shock waves in nitrogen [7] and in oxygen [8] and the detonation dynamics in a model gas [9]. The non equilibrium kinetics in the former case involves a detailed modeling of several energy exchange and chemical processes. From this sketch the main features of the method are displayed: since it is a particle method it has no convergence or stability problems and it is easily applied to problems with complex geometry; further, a wide variety of boundary conditions can be implemented; the direct simulation of particle collisions ensures the strict fulfillment of the conservation laws; since the description of the system is done at the kinetic level, the physical input data are the cross sections for the elementary collision processes. This point, in particular, avoids the need to specify kinetic rate constants and/or relaxation times since the physical system can be described by the integration of the dynamics of the elementary processes. In the following we will describe these features in greater detail. The main point we wish to emphasize is that the DSMC method is a versatile and powerful tool for the simulation of rarefied gas flows that could be inserted as a module in the construction of a Molecular Processes Simulator for the gas phase.

2 The DSMC Method

DSMC is a particle simulation method that solves the nonlinear Boltzmann equation [2]. As such it can simulate flows in the rarefied and/or hypersonic regime that cannot be dealt with in the framework of a fluid dynamic treatment; besides, it can handle situations of strong thermal nonequilibrium where a clear hierarchy of relaxation times cannot be established and rate equation methods fail. The principle of the method is the decoupling, over a small timestep, of the processes of free flight and of collisional relaxation. A number of simulated particles are moved in the simulation domain according to their velocities and to prescribed boundary conditions. The collision step is based on the assumption that the velocity distribution function can be considered uniform in sufficiently small volumes. These volumes are generated by dividing the computational domain into cells. In a cell, the collision frequency for a couple of simulated particles is given by:

$$\frac{\sigma g w}{V}. \qquad (1)$$

where s is the cross section, g the relative speed, w the ratio of real to simulated particle number densities, V the volume of the cell. A Monte Carlo method is used to realise collision events with the appropriate frequency, eq. (1). The simulated particles have associated to them the position, the velocity vector and, if the case, labels to represent the chemical species and internal state. Notably, the collision event is a bimolecular collision that strictly conserves energy and momentum. The outcome of the collision event is sampled statistically from the probability distribution of the available exit channels as derived from the set of cross sections for the given collision. From this brief description some relevant features of the method are outlined. DSMC accounts for the coupling between degrees of freedom by treating on the same ground elastic, inelastic and chemically reactive collisions so that, in particular, the nonequilibrium velocity distribution function is explicitly taken into account. Although the assumptions underlying the Boltzmann equation (i.e.: binary, instantaneous collisions; molecular chaos) are retained, the simulation is done at the microscopic level, so that the description of the physical system enters the simulation in the form of cross sections for the elementary processes of interest. With respect to other particle simulation methods we should note the following. The conservation principles of momentum and energy are satisfied strictly. The simulation is always time dependent and a simple relation holds between real and simulated time. As an important drawback we mention that to give different statistical weights to different simulated species can have unpredictable behaviour so that, as a rule, trace species can only be handled at the expense of a high computational cost. The DSMC simulation thus proceeds through the following steps: the physical domain is mapped into a computational domain; the latter is divided into a network of cells for sampling the collision frequency; as such the dimension of the cells has to be less than the local mean free path. Then, the boundary conditions are specified. In general terms, these are specified by assigning the distribution function of particles reflected from the boundary; several different conditions can be implemented:

(a) the boundary is a free stream gas at equilibrium
(b) the boundary is a specularly reflecting solid boundary
(c) the boundary is a perfectly diffusing solid boundary at fixed temperature

(d) more complicated reflection kernels can be implemented based on the accommodation coefficients of the surface [10]
(e) detailed molecular dynamics data on the gas surface interaction can be inserted into the model, including surface absorption/desorption or etherogeneous chemical reactions [10]

The simulation is initialised by creating a number of simulated particles in the computational domain following a prescribed initial distribution of density, velocity, temperature, composition, internal state distribution. The number of simulated particles is usually much less than the number of real particles. This number can be chosen in order to reduce the statistical scatter within acceptable limits with the provision that at least 20 particles be present in each cell. The initialisation procedure is completed by choosing a suitable timestep: it should be less than the average collision time between particles. After the initialisation, the time cycle starts. A time counter, representing real time, is augmented by the timestep. New particles are introduced in the simulation as requested by the boundary conditions (free stream or outgassing surfaces) or by any source present in the domain (e.g.: rocket plumes). All particles are moved according to their velocities and the time step (a volume force could also be added) and boundary interactions are computed; then, in each cell, a number of collisions is sampled as to reproduce the local collision frequency. It is at this stage that all the information about the physical processes occurring in the gas enter the simulation. The detailed description of this kinetic modeling is deferred to the next section. Finally, bulk and surface properties are sampled by averaging on the particles quantities and the cycle starts again as shown in fig. 1.

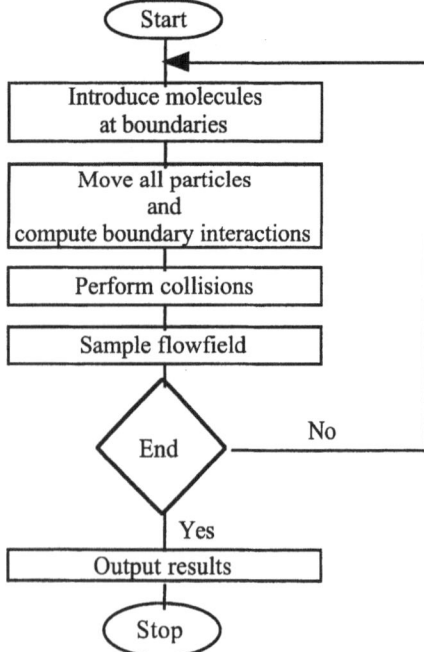

Fig. 1. Flowchart of DSMC simulation

3 Kinetic Model

During the collision step, in each cell, a number of collisions is computed between simulated particles, according to the timestep and the local collision frequency. There are several schemes to sample the correct number of collision pairs. One is the No Time Counter method of Bird [1], another, to which we adhere, in the majorant collision frequency technique of Ivanov [11]. Pairs of particles are selected randomly among those present in the cell. The particles are labeled by their velocities, internal state and chemical species. the outcome of the collision is selected at random from the possible exit channels (i.e. the possible elementary collision processes considered in the modeling) according to the following probability distribution:

$$p_{i \varnothing f} = \frac{\sigma_{i \varnothing f}}{\sigma_{Tot}} \qquad (2)$$

where $p_{i \varnothing f}$ is the probability that the collision pair in the initial state i, ends with the final state f; $\sigma_{i \varnothing f}$ is the cross section for the considered transition and σ_{Tot} is the total cross section for the pair in the given conditions. Therefore, the details of the molecular processes occurring in the gas system are specified by assigning the appropriate set of collision cross sections. In particular, the transport properties of the gas relate to the total cross section. Several phenomenological models exist that are able to reproduce the measured transport coefficients [1]; one such model is the Variable Soft Sphere model of Koura et al. [12]. It is assumed that the particles interact via an inverse power law potential, i.e. the total cross section is an inverse power law of the relative kinetic energy in collision. The exponent of this functional dependance reflects in the temperature dependance of the viscosity coefficient. Therefore, the reference diameter and the power law exponent can be adjusted to match the experimental values. Also the scattering angle functional dependance is modeled as to reproduce the behaviour of the diffusion coefficient. Obviously, these data could also be derived by more refined molecular dynamics calculations and inserted in the simulation in the form of tables, at the expense of computational cost. For the modeling of internal state and chemical kinetics, different approaches, with varying degrees of refinement are possible. we will discuss some of them, with reference to the work done on the simulation of strong shock waves in molecular gases [7, 8].

3.1 Rotation

Since we deal with high temperature systems, the rotational degrees of freedom can be considered classical. The rotational relaxation is described by a simple relaxation equation with a temperature dependent relaxation time. In the particle method this is implemented through the Larsen-Borgnakke method [13]. It accounts to assigning as an outcome to the collision event, the energies as prescribed by the equilibrium equipartition law. Such a relaxation occurs every Z collisions, Z being the collision relaxation number. The expression for the rotational relaxation number of ref. [14] with numerical parameters from ref. [15] is used. Note that the Larsen Borgnakke method had been extended to allow for discrete quantum levels [1]

3.2 Molecule-Molecule Processes

The diatom-diatom energy transfer processes are modeled by a simple statistical model [7]. Mono-quantum transitions are included along with quasi-resonant VV (vibration/vibration) energy transfers; i.e. the following processes are taken into account:

$$A_2(v) + A_2 \times A_2(v+1) + A_2$$
$$A_2(v) + A_2(w) \times A_2(v+1) + A_2(w-1)$$
$$; \quad A = N, O \quad (3)$$

The model allows for the dependence of the transition probabilities on the vibrational levels of the colliding particles (deduced from the harmonic oscillator model) and contains two adjustable parameters to scale the VT (vibration/translation) and VV transition rates with respect to the elastic collision frequency. The method was shown to reproduce satisfactorily the behaviour of the molecule-molecule relaxation kinetics. Here, we have adjusted the model parameters to reproduce the rate coefficients calculated by Billing and Fisher [16] in the temperature range of interest. Alternatively, the above rate coefficients can be directly inverted to get the cross sections [8]. Nonetheless the results can display oscillatory or other non physical behaviour. A method successfully tried is to assign the functional dependence of a given cross section with few parameters which are adjusted by a non linear fitting procedure so that the rate coefficients are reproduced. If any information about the desired rate coefficients is lacking, dedicated molecular dynamics calculations are necessary to obtain the desired data.

3.3 Atom-Molecule Processes

Atom-molecule vibrational energy exchange processes (including dissociation and exchange reactions) are modeled by using a set of cross sections obtained by QCT calculations [17, 18] based on classical trajectories on the adiabatic Potential Energy Surface (PES) which represents the interaction between 3 (nitrogen/oxygen) atoms. From these calculations a set of collision cross sections has been derived for the following processes that include exchange processes, multi-quantum energy exchanges and impact dissociation from every vibrational level:

$$A + A_2(v, j) \varnothing \; A + A_2(v', j' = all)$$
$$A + A_2(v, j) \varnothing \; 3A$$
$$; \quad A = N, O \quad (4)$$

Here, v and j are the vibrational and rotational quantum numbers, respectively. The resulting set is available in form of tables spanning systematically a wide range of v, j, and collision kinetic energy, and are therefore highly suitable for numerical implementation. The implementation of these data in the DSMC code is straightforward except for some caution that must be taken to ensure the fulfillment of detailed balance. In the present implementation we treat the rotational degrees of freedom in a phenomenological way and we do not consider any rotational-vibrational coupling: for the vibrational energy exchange processes we use rotationally averaged cross sections over the equilibrium rotational distribution at $T_{rot}=10^4$ K.

3.4 DSMC-QCT

Besides phenomenological modeling and cross section tables, a third option is viable for kinetic modeling: collisions in DSMC simulation are assumed to occur with random impact parameters and orientations, since the gas is supposed uniform inside each spatial cell. The outcome of each collision event could then be calculated by integrating the relevant dynamics. The high computational cost involved by this scheme prevents its widespread use, but the high degree of detail achieved by this kind of simulation makes it attractive for benchmark studies. The feasibility of DSMC-QCT calculations has been demonstrated for studying rotational relaxation in nitrogen [19] and for vibration-dissociation kinetics, again in nitrogen [20].

4 Case Study

In order to show the computational characteristics of a typical DSMC simulation a strong shock wave in oxygen has been taken as an example [8]. Details about the simulation can be found in the cited article. Here, emphasis is placed on the computational aspects. All tests have been performed on a PowerPC G4, 700 Mhz. The boundary conditions are set to obtain a Mach 11 shock wave into the unperturbed gas at T=300 K, P=100 Pa. The simulation region extends for about 500 (upstream) mean free paths downstream of the shock. This produces a computational domain made of 20000 cells. The number of simulated particles is 500000. The simulation has been run for 100000 timesteps. This is not sufficient for reaching steady state: depending on the initial distribution of particles, the transient regime can be much slower; however, it gives an order of magnitude estimate of the computational requirements for a meaningful sampling, once steady state has been attained. Under these conditions, the code spends 0.7 s in the move step, about 17 % of the total computational time, and 3.2 s (80 %) in the collision step. In the collision step, about $3\ 10^9$ collisions are performed and they account for 22 % of the time spent in this step. The remainder is used for sampling collision pairs. Even if the kinetic scheme is quite complex, and several processes are accounted for, each collision event takes about $2\ 10^{-5}$ s. This is to be compared to the time needed by a QCT method to integrate a single trajectory, in order to estimate the computational load of a DSMC-QCT implementation. As expected by any Monte Carlo method a huge number of random numbers are needed for each run. In this case about $3\ 10^{10}$ random numbers are generated, which, in this implementation occupy 15 % of the total computational time. The role played by the kinetic modeling in determining the computational load cab be better appreciated by comparison with a similar simulation with simplified kinetics. A shock wave in Argon, in fact. Now, all particles undergo only elastic collisions so that the number of operations performed *per collision event* is minimised. Results relate to the simulation of a Mach 9 shock wave into Argon gas at T=300 K, P=100 Pa. As before, the computational domain consists of 20000 cells; the number of simulated particles is 500000 and the simulation is run for 100000 timesteps. The time spent in the move step is comparable to the previous simulation, but now each collision step lasts about 0.5 s. Even though the total number of collisions computed in the simulation is comparable to the previous case, the computational time is strongly reduced. This is due to two factors. First, the collision dynamics is much simpler, so that, on the average, each collision needs only $8 \cdot 10^{-6}$ s to be computed; second, since a single chemical species is involved, the procedure for selection of collision pairs and the indexing of particles is simplified.

5 Conclusions

The DSMC method is a particle simulation method capable of modeling efficiently nonequilibrium kinetics in the gas phase. It can be a valuable tool for systems in which the continuum assumption breaks down: rarefied gas flows is a typical example, but also flows with strong spatial gradients as strong shock waves. It is also of valuable help in the kinetic modeling of systems for which a clear hierarchy of relaxation times does not apply so that different kinetic processes compete with the fluid dynamics. It has been shown that DSMC can be made to simulate realistic kinetics on a state-to-state basis and can incorporate in the modeling the physical details on elementary processes provided by molecular dynamics methods. In this way, however, it relies heavily on the knowledge of microscopic cross sections for the elementary processes being modeled. In this regard, the method is suitable to be inserted as the kinetic equations integrator into a *ab initio* Molecular Processes Simulator for the gas phase. In this environment it would be very easy to switch among different kinetic models with varying degrees of refinement. When, however, kinetic data are missing for the processes of interest, the DSMC simulation with QCT trajectories run 'on the fly' represents a valuable but computationally intensive alternative. Limitations can be traced to the particle nature of the method that introduces a statistical noise in the results. As a consequence, the method is able to simulate fine effects (e.g.: the tail of the distribution function, kinetics of trace species) only at the expense of high computational load. However, the inherent fulfillment of the conservation laws adds considerable value to the method.

Acknowledgments. This work was supported by M.I.U.R. under the F.I.R.B. project GRID.it.

References

[1] Bird, G.A.: Molecular Gas Dynamics and the Direct Simulation of Gas Flows. Clarendon Press, Oxford (1994)

[2] Wagner, W.: A Convergence Proof for Bird's Direct Simulation Monte Carlo Method for the Boltzmann Equation. J. Stat. Phys. **66** (1992) 1011-1044
[3] Dunn, S.M., Anderson, J.B.: Direct Monte Carlo simulation of chemical reaction systems: Internal energy transfer and an energy-dependent unimolecular reaction. J. Chem. Phys. **99** (1993) 6607-6612
[4] Dunn, S.M., Anderson, J.B.: Direct Monte Carlo simulation of chemical reaction systems: Dissociation and recombination. J. Chem. Phys. **102** (1995) 2812-2815
[5] Bruno, D., Capitelli, M., Longo, S.: DSMC modeling of vibrational and chemical kinetics for a reacting gas mixture. Chem. Phys. Lett. **289** (1998) 141-149
[6] Longo, S., Bruno, D., Minelli, P.: Direct simulation of non-linear interparticle collisional relaxation of ensembles of two-level systems. Chem. Phys. **256** (2000) 265-273
[7] Bruno, D., Capitelli, M., Esposito, F., Longo, S., Minelli, P.: Direct simulation of non-equilibrium kinetics under shock conditions in nitrogen. Chem. Phys. Lett. **360** (2002) 31-37
[8] Bruno, D., Capitelli, M., Esposito, F., Longo, S., Minelli, P.: Direct Monte Carlo simulation of oxygen dissociation behind shock waves. AIAA paper 2003-4059 (2003)
[9] Bruno, D., Capitelli, M., Longo, S.: Effect of translational kinetics on chemical rates in a Direct Simulation Monte Carlo model gas phase detonation. Chem Phys. Lett. **380** (2003) 383-390
[10] Bruno, D., Cacciatore, M., Longo, S., Rutigliano, M.: Gas-surface scattering models for Particle Fluid Dynamics: A comparison between analytical approximate models and Molecular Dynamics calculations. Chem. Phys. Lett. **320** (2000) 245-254
[11] Ivanov, M.S., Rogasinsky, S.V.: Analysis of numerical techniques of the direct simulation Monte Carlo method in the rarefied gas dynamics. Soviet J. Numer. Anal. Math. Modelling **3** (1988) 453-465
[12] Koura, K., Matsumoto, H.: Variable soft sphere molecular model for air species. Phys. Fluids A **4** (1992) 1083-1085
[13] Borgnakke, C., Larsen, P.S.: Statistical Collision Model for Monte Carlo Simulation of Polyatomic Gas Mixture. J. Comput. Phys. **18** (1975) 405-420
[14] Parker, J.G.: Rotational and vibrational relaxation in diatomic gases. Phys. Fluids **2** (1959) 449-462
[15] Boyd, I.D.: Rotational–translational energy transfer in rarefied nonequilibrium flows. Phys. Fluids A **2** (1990) 447-452
[16] Billing, G.D.: Vibration-Vibration and Vibration-Translation Energy Transfer, Including Multiquantum Transitions in Atom-Diatom and Diatom-Diatom Collisions. In: Capitelli, M. (ed.): Nonequilibrium vibrational kinetics. Springer-Verlag, Berlin Heidelberg New York (1986) 85-112
[17] Esposito, F., Capitelli, M.: Quasiclassical molecular dynamic calculations of vibrationally and rotationally state selected dissociation cross-sections: $N+N_2(v,j)->3N$. Chem. Phys. Lett. **302** (1999) 49-54; Esposito, F:, Capitelli, M., Gorse, C.: Quasi-classical dynamics and vibrational kinetics of $N+N_2(v)$ system. Chem.Phys. **257** (2000) 193-202
[18] Esposito, F:, Capitelli, M.: Quasiclassical trajectory calculations of vibrationally specific dissociation cross-sections and rate constants for the reaction $O+O_2(v)->3O$. Chem. Phys. Lett. **364** (2002) 180-187
[19] Koura, K.: Monte Carlo direct simulation of rotational relaxation of diatomic molecules using classical trajectory calculations: Nitrogen shock wave. Phys. Fluids **9** (1997) 3543-3549
[20] Fujita, K., Abe, T.: Coupled Rotation-Vibration-Dissociation Kinetics of Nitrogen using QCT Models. AIAA paper 2003-3779 (2003)

Molecular Simulation of Reaction and Adsorption in Nanochemical Devices: Increase of Reaction Conversion by Separation of a Product from the Reaction Mixture

William R. Smith[1] and Martin Lísal[2,3]

[1] Faculty of Science, University of Ontario Institute of Technology,
2000 Simcoe St. N., Oshawa ON L1H7K4, Canada
william.smith@uoit.ca
http://www.uoit.ca/schoolofscience

[2] E. Hála Laboratory of Thermodynamics, Institute of Chemical Process Fundamentals, Academy of Sciences of the Czech Republic,
165 02 Prague 6, Czech Republic
lisal@icpf.cas.cz
http://home.icpf.cas.cz/lisal/www/

[3] Department of Physics, J. E. Purkyně University,
400 96 Ústí n. Lab., Czech Republic

Abstract. We present a novel simulation tool to study fluid mixtures that are simultaneously chemically reacting and adsorbing within a molecularly porous material. The method is a combination of the Reaction Ensemble Monte Carlo method and the Dual Control Volume Grand Canonical Molecular Dynamics technique. The method, termed the Dual Control Cell Reaction Ensemble Molecular Dynamics (DCC-RxMD) method, allows for the calculation of both equilibrium and non-equilibrium transport properties in porous materials, such as diffusion coefficients, permeability and mass flux. Simulation control cells, which are in direct physical contact with the porous solid, are used to maintain the desired reaction and flow conditions for the system. The simulation setup closely mimics an actual experimental system in which the thermodynamic and flow parameters are precisely controlled. We present an application of the method to the dry reforming of methane within a nanoscale reactor in the presence of a semipermeable nanomembrane modelling silicalite. We studied the effects of the nanomembrane structure and porosity on the reaction species permeability by considering three different nanomembrane models. We also studied the effects of an imposed pressure gradient across the nanomembrane on the mass flux of the reaction species. Conversion of syngas (H_2/CO) increased significantly in all the nanoscale membrane reactor systems considered. The results of this work demonstrate that the DCC-RxMD method is an attractive computational tool in the design of nanoscale membrane reactors for industrial processes.

1 Introduction

With the rapid growth of nanotechnology and the development of various nanomaterials, some of these materials have been proposed as vehicles for nanochemical devices such as nanoscale reactors and nanoscale membrane reactors [1]. However, development of such applications is impossible without fundamental knowledge of reaction, adsorption and transport mechanisms in the nanoporous materials.

It is well established that molecular confinement brings about drastic changes in the thermodynamic properties of fluids, such as narrowing of the coexistence curve, lowering of the pore critical temperature or increasing the average pore densities; for a comprehensive review see Ref. [2]. Confinement also influences chemical reaction equilibrium. For example, generally the pore phase has a higher density than the corresponding bulk phase; this typically results in an increase in yield for reactions in which there is a decrease in the total number of moles (Le Chatellier's principle). Further, some components of the reaction mixture are selectively adsorbed on the solid surfaces, also affecting the reaction equilibrium. Finally, molecular orientations can be strongly influenced by proximity to a solid surface, which also can shift the reaction equilibrium with respect to that in the bulk phase.

Confinement also influences the transport properties of fluid particles inside the nanoporous materials. Physical space restrictions based on the fluid particle size or geometry may limit the flow of particles through pores in the material. Furthermore, attraction of the pore surface, $i.e.$, physisorption may play a critical role. Even further complicating matters, fluid particles may chemisorb. In addition to phenomena occurring between fluid particles and the pore surface, behavior in the porous material becomes increasingly more complex if chemical reactions occur between fluid particles.

In this work, we present a novel non-equilibrium molecular dynamics method, which we call the Dual Control Cell Reaction Ensemble Molecular Dynamics (DCC-RxMD) method, for the simulation of combined reaction and adsorption mechanisms in nanoporous materials. The novel method is a combination of the Reaction Ensemble Monte Carlo (REMC) [3,4,5] and Reaction Ensemble molecular dynamics (RxMD) [6] methods, and the Dual Control Volume GCMD (DCV-GCMD) technique [7,8]. We apply the DCC-RxMD method to study reactions and separations in nanoscale membrane reactors for the dry reforming of methane. This reaction underlies an important industrial process for producing syngas (H_2/CO) that utilizes membrane technology [9]. By application of the DCC-RxMD method, we demonstrate the possibility of downscaling the macroscopic membrane reactor to a nanoscale membrane reactor.

2 Simulation Methodology

The DCC-RxMD method uses a combination of stochastic and dynamic simulation steps, allowing for the simulation of both thermodynamic and transport properties. The method couples a MD system (dynamic cell) to reacting and non-reacting mixture reservoirs (control cells) that are formulated upon the REMC method [3,4,5] and the GCMC method [10], respectively; hence the term Dual Control Cell Reaction Ensemble Molecular Dynamics (DCC-RxMD) method. The control cells are in direct contact with the dynamic cell and the particles are able to move freely between the cells. Transport properties are calculated in the dynamic cell by using a MD simulation method [11]. REMC forward and reverse reaction steps, and GCMC particle insertion and deletion steps are performed in the control cells only, while MD steps are performed in both the dynamic cell and the control cells. The control cells, which act as sink and source reservoirs, are maintained at explicitly defined and controlled gradient conditions, e.g. a pressure gradient. The DCC-RxMD method is analogous to the DCV-GCMD method [7,8], both of which simulate conditions that directly relate to real, open systems. Computational details will be addressed below, but first we illustrate the overall setup of the method.

Consider a nanoporous material containing a system of *nanomembrane reactors*. Each nanomembrane reactor is made up of two adjacent voids separated by a semipermeable nanomembrane. One of the voids acts as a nanoreactor in which, for example, the model reaction

$$2A \rightleftharpoons B \qquad (1)$$

occurs. We term this void the *reaction* void. The accompanying void in the nanomembrane reactor system, termed the *transport* void, maintains a pressure gradient across the nanomembrane. We further assume that the nanomembrane separating the reaction and transport voids is not permeable to all molecular components. For example in the model system of Eq. (1), suppose the nanomembrane is permeable to component B only. For such a case, a difference in the partial pressures of component B in the reaction and transport voids results in a flux of component B through the nanomembrane. A schematic of the nanomembrane reactor system made up of the reaction and transport voids separated by a semipermeable nanomembrane is presented in Fig. 1. The nanomembrane reactor system is analogous to the slitpore model [12] which assumes that the individual nanomembrane reactors are taken as representative of all nanomembrane reactors in the porous material. The effect of pore connectivity can be readily invoked by considering a combination of connected nanomembrane reactors.

To mimic reaction and adsorption mechanisms in such an elementary nanoscale membrane reactor, we consider a DCC-RxMD simulation box as shown in the enlargement of Fig. 1. A nanomembrane of thickness $\delta = 2L_{1x}$ is placed at the center of the simulation box. The left-hand-side of the simulation box (from $-L_{1x}$ to $-L_{4x}$) with control cell (CC) A (from $-L_{2x}$ to $-L_{3x}$) corresponds to a reaction void, while the right-hand-side of the simulation box (from L_{1x} to L_{4x}) with CC B (from L_{2x} to L_{3x}) corresponds to a transport void. Box lengths in the y and z directions are denoted by L_y and L_z, respectively. For simplicity, we considered that both CC A and CC B are of the same size and are positioned symmetrically with respect to the yz plane at $x = 0$. We impose periodic boundary conditions in both the y and z directions, since we assume that the sizes of the reaction and transport voids in these directions are much larger than the nanomembrane thickness. To limit the size of the simulation setup in the x-

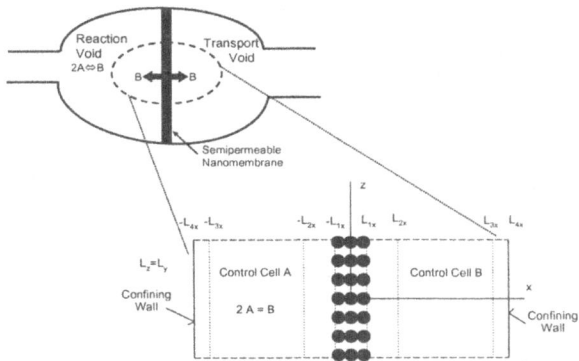

Fig. 1. Schematic of elementary nanoscale membrane reactor made up of a reaction void and a transport void separated by a semipermeable nanomembrane. A model reaction 2A⇌B takes place in the reaction void while component B is separated *via* the semipermeable nanomembrane. Enlargement shows schematic of the corresponding DCC-RxMD simulation setup. Periodic boundary conditions, applied in both the y and z directions, are omitted in the x direction due to the presence of repulsive confining walls (in the yz plane) at each end of the simulation setup.

direction, repulsive confining walls are placed at the ends of the simulation box. The size of the simulation box in the x-direction must be sufficiently large to ensure that these confining-wall boundaries have negligible effect on the calculated fluid properties. The alternative to including confining-wall boundaries is to allow the fluid particles to leave the simulation box at the boundaries. Use of either the confining-wall boundaries or open-end boundaries give the same simulation results [8,13]. An alternative scenario is possible here. For example, one can consider a DCC-RxMD simulation box with two confining walls located in the z direction at a distance L_z apart without confining walls in the x direction [13]. In this scenario, confinement effects in the z direction are also considered important.

In either scenario, the DCC-RxMD simulation proceeds as follows. After n_{MD} MD steps, the system is frozen, *i.e.* particle positions are held fixed, and we perform n_{REMC} forward and reverse reaction steps in CC A, and n_{GCMC} particle creation and destruction steps in CC B. Both the REMC and GCMC algorithms require insertion of particles into the CCs. The velocities for such particles are assigned from a Maxwell-Boltzmann distribution corresponding to the specified system temperature [6,14]. Values of n_{MD}, n_{REMC} and n_{GCMC} must be chosen appropriately to maintain reaction equilibrium in CC A, constant chemical potentials in CC B and reasonable transport rates at the boundaries between the CCs and the nanomembrane region (the dynamic cell).

3 APPLICATION: Nanoscale Membrane Reactor Model

We applied the DCC-RxMD method to simulate a multicomponent system in which both a chemical reaction and physisorption is occurring within a nanoporous solid. We consider the dry reforming of methane reaction, which is the basis for an important industrial process for producing syngas (H_2/CO) from CH_4 and CO_2 [9]. We focus our study on how the reaction conversion is affected by separating out H_2 from the reaction mixture *via* a semipermeable nanomembrane. Below we provide details of the reaction species models, nanomembrane models and various computational information for the systems considered in this work.

3.1 Reaction Species Potential Models

We consider a nanoscale membrane reactor system in which the dry reforming reaction

$$CH_4 + CO_2 \rightleftharpoons 2H_2 + 2CO \qquad (2)$$

occurs in the reaction void (CC A) and the nanomembrane is preferentially permeable to H_2. A difference in the H_2 partial pressures between the reaction and transport voids *i.e.* between CC A and CC B will result in a H_2 flux through the nanomembrane.

The species of the reaction system, $\{CH_4, CO_2, H_2, CO\}$, are modelled as Lennard-Jones (LJ) spheres. The fluid-fluid interactions are approximated with a truncated-and-shifted LJ (TS-LJ) potential with potential parameters listed in Table 1.

Table 1. Effective LJ energy (ε) and size (σ) parameters for CH_4, CO_2, H_2 and CO fluids, and nanomembrane particles.

Component	ε/k_B (K)	σ (nm)
CH_4	148.1	0.3810
CO_2	225.3	0.3794
H_2	38.0	0.290
CO	123.0	0.3662
Nanomembrane	82.0	0.270

The cross-term LJ parameters were evaluated using the Lorentz-Berthelot combining rules [11]. Fluid particles of component l interact with the repulsive confining walls (w) (see Fig. 1) *via* the TS-LJ potential with ε_{lw}, σ_{lw} and $r_{c,lw} = 2^{1/6}\sigma_{lw}$. For simplicity, we use $\varepsilon_{lw} \equiv \varepsilon_{ll}$ and $\sigma_{lw} \equiv \sigma_{ll}$, since the confining walls have no direct influence on the transport in the nanomembrane region and have only small effects on fluid properties in the portions of CCs adjacent to the confining walls.

3.2 Nanomembrane Models

In order to increase the reaction conversion in our nanomembrane reactor by separating out a particular product from the mixture, it is required that the nanomembrane separating the reaction and transport voids be permeable to that particular product only. For the dry reforming of methane reaction considered here, we define our nanomembrane model to be permeable to H_2 only while being impermeable to the remaining reaction species, namely CH_4, CO_2, and CO. Since the dry reforming reaction is typically carried out at quite high temperatures, in general, the separation of mixture components *via* a nanomembrane is based primarily on molecular sieving caused by the passage of smaller molecules of the mixture through the pores while the larger molecules are obstructed [15]. As evident in Table 1, H_2 molecules are smaller than the CH_4, CO_2 and CO molecules. Hence, a characteristic size for the nanomembrane pores should be greater than σ_{H_2} (to allow H_2 to permeate through the nanomembrane) but yet less than or comparable to $\sigma_{CH_4} \approx \sigma_{CO_2} \approx \sigma_{CO}$ (to obstruct CH_4, CO_2 and CO from permeating through the nanomembrane).

We utilized two types of nanomembrane models. The first model, proposed by Powles *et al.* [16,17], defines nanomembranes by several layers of LJ particles (characterized by the energy and size parameters ε_m and σ_m, respectively) with a distance between layers equal to $2^{1/6}\sigma_m$ (which corresponds to the LJ potential minimum). The particles in each layer are arranged in a face-centered cubic (fcc) structure. The layers are built by replicating a four-particle fcc primitive cell in the y and z directions. The chosen values for ε_m and σ_m correspond to silicalite molecules and are listed in Table 1. The nanomembrane particles interact with fluid particles *via* the TS-LJ potential and with the cross-term LJ parameters evaluated from the Lorentz-Berthelot combining rules. The Powles model generates nanomembranes with well-defined structures comprised of straight channels that are typical of zeolitic structures. We built two Powles nanomembranes, denoted as PNM1 and PNM2. Both models consist of seven fcc layers, and differ in their values of d_{pore}^{max}. PNM1 has $d_{pore}^{max} = 0.3253$ nm, which corresponds to a nanomembrane number density $\rho_m = 21.728$ nm^{-3}, while PNM2 is characterized by $d_{pore}^{max} = 0.4104$ nm and $\rho_m = 16.635$ nm^{-3}. Geometric pore size distributions (PSDs) [18] for PNM1 and PNM2 are displayed in Fig. 2. As expected, Fig. 2 shows PSDs that are quite narrow, with maximum peaks corresponding to d_{pore}^{max}.

The second type of nanomembrane model used in this study is comprised of random configurations of non-overlapping LJ spheres. This model, termed the random nanomembrane model (RNM), is generated from a canonical-ensemble Monte Carlo simulation of hard spheres of diameter of σ_m [19]. We built RNMs with various PSDs in the volume $2L_{1x} \times L_y \times L_z$, using the same value of L_{1x} as in PNM1 and PNM2. We then chose one RNM whose PSD peak is roughly at the same position as the PSD peak for PNM2. The chosen RNM has $\rho_m = 15.156$ nm^{-3} and its PSD is displayed in Fig. 2. This figure shows that the PSD for the RNM is substantially broader than that of PNM2, due to the random arrangement of the membrane particles.

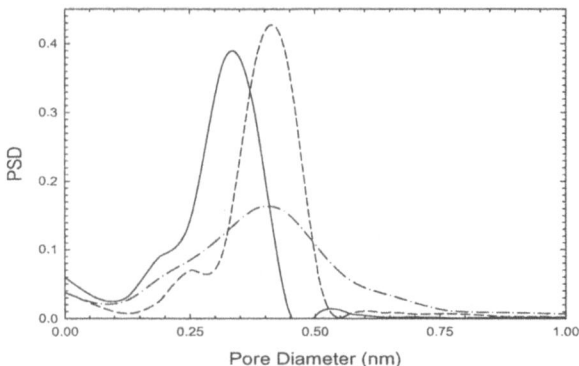

Fig. 2. Computed geometric pore-size distributions (PSD) for the three nanomembrane models: PNM1 (———), PNM2 (– – – –) and RNM (— · — · —).

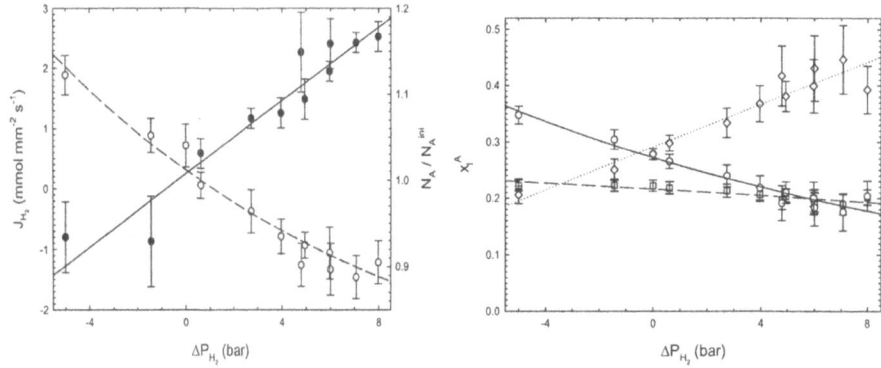

Fig. 3. (a) Hydrogen molar flux J_{H_2} (•), and N_A/N_A^{ini} (○) as a function of the hydrogen partial pressure difference ΔP_{H_2} and (b) the composition in the reaction void x_l^A (CH_4, □; H_2, ○; CO, ◊) as a function of ΔP_{H_2} in the case of PNM1 obtained from the DCC-RxMD simulations. Due to identical initial compositions of CH_4 and CO_2, $x_{CH_4}^A = x_{CO_2}^A$ within statistical uncertainties; therefore $x_{CO_2}^A$ is not plotted. The lines serve as a guide to the eye only.

4 Results

At $T = 1100$ K and a bulk volume $V \equiv V_A = (L_{3x} - L_{2x}) \times L_y \times L_z = 1576.71$ nm^3, both the REMC and RxMD methods predict the following bulk equilibrium properties of the reaction mixture $CH_4/CO_2/H_2/CO$: $u = -0.057_{26}$ kJ mol^{-1}, $P = 50.7_{11}$ bar, $\rho = 0.3291_{184}$ nm^{-3}, $x_{CH_4} = 0.222_{10}$, $x_{CO_2} = 0.222_{10}$, $x_{H_2} = 0.278_{10}$ and $x_{CO} = 0.278_{10}$; subscripts in numerical values denote standard deviations in the last digits. Simulations were initiated with $N_{CH_4}^{ini} = N_{CO_2}^{ini} = N_{H_2}^{ini} = N_{CO}^{ini} = 125$ molecules in the simulation box. At equilibrium, the total number of molecules in the bulk reaction system was $N = \sum_{l=1}^{c} N_l = 520_5$.

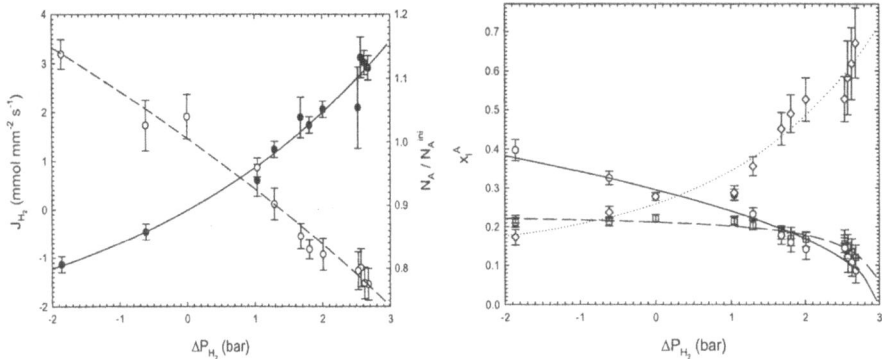

Fig. 4. (a) Hydrogen molar flux J_{H_2} (●), and N_A/N_A^{ini} (○) as a function of the hydrogen partial pressure difference ΔP_{H_2} and (b) the composition in the reaction void x_l^A (CH$_4$, □; H$_2$, ○; CO, ◊) as a function of ΔP_{H_2} in the case of PNM2 obtained from the DCC-RxMD simulations. Due to identical initial compositions of CH$_4$ and CO$_2$, $x_{CH_4}^A = x_{CO_2}^A$ within statistical uncertainties; therefore $x_{CO_2}^A$ is not plotted. The lines serve as a guide to the eye only.

We first calculated the permeabilities of the models for each component, obtaining values of $\{K_{CH_4}^*, K_{CO_2}^*, K_{H_2}^*, K_{CO}^*\} = \{0_1, 0_1, 46_6, 2_2\}$, $\{23_5, 9_6, 148_5, 17_1\}$, $\{4_1, 4_2, 66_7, 4_1\}$ for membranes PNM1, PNM2, and RNM, respectively (subscripts denote standard deviations in the final digits).

All DCC-RxMD simulations were started with $N_{CH_4}^{ini} = N_{CO_2}^{ini} = N_{H_2}^{ini} = N_{CO}^{ini} = 125$ in CC A, i.e. with $N_A^{ini} = 500$. Values of μ_l^B were chosen to obtain nearly pure H$_2$ in CC B. This was achieved by setting $\mu_{CH_4}^B/(RT) = \mu_{CO_2}^B/(RT) = \mu_{CO}^B/(RT) = -12.117$. Values of $\mu_{H_2}^B/(RT)$ were then varied from -4.712 to -8.078. This results in $\Delta P > 0$ but $\Delta P_{H_2} < 0$ for $\mu_{H_2}^B/(RT) > -5.25$ and $\Delta P_{H_2} > 0$ otherwise. Note that the pressure in the reaction void decreases with the removal of H$_2$, since the total number of particles decreases (see Figs. 3 to 5 below). For $\Delta P_{H_2} < 0$, H$_2$ flows from the transport void to the reaction void, while for $\Delta P_{H_2} > 0$ there is a H$_2$ flux from the reaction void to the transport void. Further, $\mu_{H_2}^B/(RT) = -8.078$ produces nearly vacuum conditions in CC B, i.e. the equilibrium number of particles in CC B becomes very small and thus $P_B \to 0$. Hence, the value of ΔP_{H_2} resulting from $\mu_{H_2}^B/(RT) = -8.078$ corresponds roughly to the maximum achievable ΔP_{H_2}.

Figs. 3 to 5 show J_{H_2} and N_A/N_A^{ini}, and x_l^A as a function of ΔP_{H_2} for all three nanomembrane models. Note that x_l^A at $\Delta P_{H_2} = 0$ corresponds to x_l^A of the bulk reaction system. We see from Figs. 3 to 5 that the maximal ΔP_{H_2} (corresponding to $\mu_{H_2}^B/(RT) = -8.078$) is directly related to value of the permeability K_{H_2}: a nanomembrane with a larger K_{H_2} has a lower ΔP_{H_2}. Also note that the upper portions of Figs. 3 to 5 show that the values of J_{H_2} corresponding to the maximal ΔP_{H_2} are approximately the same. In contrast to these values of J_{H_2}, the values of N_A/N_A^{ini} at the maximal ΔP_{H_2} differ significantly; they are lower for nanomembranes with a higher K_{H_2}. Next note that compositions x_l^A as a function of ΔP_{H_2} in the lower portions of Figs. 3 to 5 show that the

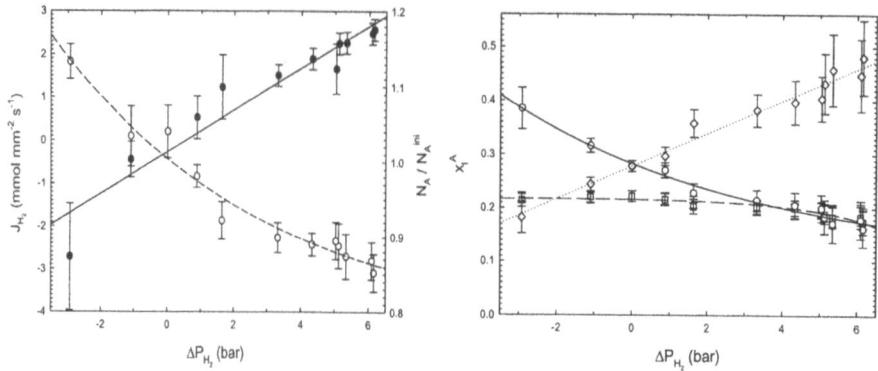

Fig. 5. (a) Hydrogen molar flux J_{H_2} (●), and N_A/N_A^{ini} (○) as a function of the hydrogen partial pressure difference ΔP_{H_2} and (b) the composition in the reaction void x_i^A (CH$_4$, □; H$_2$, ○; CO, ◊) as a function of ΔP_{H_2} in the case of RNM obtained from the DCC-RxMD simulations. Due to identical initial compositions of CH$_4$ and CO$_2$, $x_{CH_4}^A = x_{CO_2}^A$ within statistical uncertainties; therefore $x_{CO_2}^A$ is not plotted. The lines serve as a guide to the eye only.

compositions of reactants CH$_4$ and CO$_2$ decrease slowly with increasing ΔP_{H_2} (except $x_{CH_4}^A$ and $x_{CO_2}^A$ close to the maximal ΔP_{H_2} for PNM2). The composition of the product CO increases significantly with increasing ΔP_{H_2}. At the maximal ΔP_{H_2}, x_{CO}^A is larger for nanomembranes with a larger K_{H_2}. With respect to x_{CO}^A at $\Delta P_{H_2} = 0$, increases of x_{CO}^A at the maximal ΔP_{H_2} are $\sim 60\%$ for PNM1, $\sim 80\%$ for RNM and $\sim 150\%$ for PNM2. $x_{H_2}^A$ decreases with increasing ΔP_{H_2} due to hydrogen separation from the reaction void to the transport void. The decrease of $x_{H_2}^A$ is related to K_{H_2} and is larger for nanomembranes with larger K_{H_2}. The total yield of H$_2$ is a result of the H$_2$ amount in both the reaction and transport voids.

5 Conclusions

We have presented a novel simulation tool, termed the Dual Control Cell Reaction Ensemble Molecular Dynamics (DCC-RxMD) method, to study fluid mixtures that are simultaneously chemically reacting and adsorbing in a porous material. The DCC-RxMD method was developed by coupling a non-equilibrium molecular dynamics method with two Monte Carlo-based methods, namely Reaction Ensemble Monte Carlo and Grand Canonical Monte Carlo. Control cells, which are in direct physical contact with the porous solid, are used to maintain the desired reaction and flow conditions. The simulation setup closely mimics an actual experimental system in which the thermodynamic and flow parameters are precisely controlled. The method is akin to the Dual Control Volume Grand Canonical Molecular Dynamics method, which was developed to study the transport properties of fluid mixtures primarily in confined systems. The added feature of the DCC-RxMD method is the inclusion of chemical reactions, and thus its applicability is to a wider range of industrial processes beyond solely adsorption phenomena. The method presented here allows for the calculation of both equilibrium and non-equilibrium transport properties in porous materials such as diffusion coefficients, permeability and mass flux. Effects on these properties due to the characteristics of the porous material can be predicted; characteristics such as the pore size distribution, connectivity, porosity, and surface area.

This research was supported by the Grant Agency of the Czech Republic (Grant No. 203/02/0805) and by the National Research Council of Canada (Grant No. OGP 1041). Calculations were carried out on the SHARCNET (Shared Academic Hierarchical Computing Network), http://www.sharcnet.ca.

References

1. R. Breslow and M. V. Tirrell (Eds.), *Beyond the Molecular Frontier. Challenges for Chemistry and Chemical Engineering* (The National Academic Press, Washington, D.C., 2003).
2. L. D. Gelb, K. E. Gubbins, R. Radhakrishnan and M. Sliwinska-Bartkowiak, Rep. Prog. Phys. 62, 1573 (1999).
3. W. R. Smith and B. Tříska, J. Chem. Phys. 100, 3019 (1994).
4. J. K. Johnson, A. Z. Panagiotopoulos and K. E. Gubbins, Molec. Phys. 81, 717 (1994).
5. M. Lísal, I. Nezbeda and W. R. Smith, J. Chem. Phys. 110, 8597 (1999).
6. J. K. Brennan, M. Lísal, K. E. Gubbins and B. M. Rice, Phys. Rev. E submitted (2004).
7. G. S. Heffelfinger and F. van Swol, J. Chem. Phys. 100, 7548 (1994).
8. J. M. D. MacElroy, J. Chem. Phys. 101, 5274 (1994).
9. J. K. Dahl, A. W. Weimer, A. Lewandowski, C. Bingham, F. Bruetsch and A. Steinfeld, Ind. Eng. Chem. Res. in press (2004).
10. D. Frenkel and B. Smit, *Understanding Molecular Simulation: From Algorithms to Applications* (Academic Press, London, 2002).
11. M. P. Allen and D. J. Tildesley, *Computer Simulation of Liquids* (Clarendon Press, Oxford, 1987).
12. D. Nicholson and N. G. Parsonage, *Computer Simulation and the Statistical Mechanics of Adsorption* (Academic Press, New York, 1982).
13. R. F. Cracknell, D. Nicholson and N. Quirke, Phys. Rev. Lett. 74, 2463 (1995).
14. A. Papadopoulou, E. D. Becker, M. Lupkowski and F. van Swol, J. Chem. Phys. 98, 4897 (1993).
15. W. A. Steele, *The Interaction of Gases with Solid Surfaces* (Pergamon Press, Oxford, 1974).
16. E. Enciso, N. G. Almarza, S. Murad and M. A. Gonzalez, Molec. Phys. 100, 2337 (2002).
17. J. G. Powles, S. Murad and P. V. Ravi, Chem. Phys. Letts. 188, 21 (1992).
18. L. D. Gelb and K. E. Gubbins, Langmuir 15, 305 (1999).
19. I. A. Park and J. M. D. MacElroy, Molec. Simul. 2, 105 (1989).

Quantum Generalization of Molecular Dynamics Method. Wigner Approach

V. Filinov[1], M. Bonitz[2], V. Fortov[1], and P. Levashov[1]

[1] Institute for High Energy Density, Russian Academy of Sciences, Izhorskay 13/19, Moscow 125412, Russia
filinov@ok.ru

[2] Christian - Albrechts-Universitaet zu Kiel,
Institut fuer Theoretische Physik und Astrophysik,
Lehrstuhl Statistische Physik,
Leibnizstrasse 15, 24098 Kiel, Germany

Abstract. The new method for solving Wigner-Liouville's type equations and studying dynamics of quantum particles has been developed within the Wigner formulation of quantum statistical mechanics. This approach combines both molecular dynamics and Monte Carlo methods and computes traces and spectra of the relevant dynamical quantities. Considering, as an application, the quantum dynamics of an ensemble of interacting electrons in an array of random scatterers clearly demonstrates that the many-particle interaction between the electrons can lead to an enhancement of the electrical conductivity.

1 Introduction

It is well known that molecular dynamics method due to its highly efficiency is widely used in treatment of dynamic problems of classical statistical physics. The aim of this work is to develop the 'straight generalization' of classical molecular dynamics methods for rigorous consideration of quantum problems. The words 'straight generalization' mean that in classical limit the developed approach should exactly coincide with molecular dynamics method in the phase space. A generalization of molecular dynamics method is possible only in the phase space, so in our work it is naturally to use Wigner formulation of quantum mechanics. In 1932 Wigner proposed joint position and momentum (phase space) representation of quantum mechanics and derived quantum analog of the classical distribution function. This representation contains only the values common both for classical and quantum mechanics, which is especially convenient when one of two interacting subsystems is quantum and another - classical. Wigner's paper has given rise to an extensive literature on formal aspects of quantum theory in phase space.

Noninteracting electrons in an array of fixed random scatterers are known to experience Anderson localization at temperature $T = 0$ in one-dimensional systems [1]. However, it is expected that the many-particle interaction leads to delocalization tendencies which has been confirmed for simple models. To study the influence of the electron-electron Coulomb interaction on kinetic electron properties in a random environment we have simulated the quantum dynamics in a one-dimensional canonical ensemble at finite temperature for both interacting and noninteracting electrons using the developed Quantum-Dynamics-Monte-Carlo scheme. We discovered that the temporal momentum-momentum correlation functions and their frequency-domain Fourier transforms strongly depend on the electron-electron interaction, clearly demonstrating the delocalizing influence of the many-particle interaction at densities around $R_s = \bar{r}/a_0 = 5$ ($\bar{r}$ is the mean interparticle distance and a_0 the effective Bohr radius) even at finite temperatures. Our approach also treats the positions of the scattering centers as dynamical variables. We are, therefore, able to generate various initial conditions.

2 Wigner Representation of Quantum Mechanics

The basis of our consideration of quantum many particle system in the classical electromagnetic field is the gauge – invariant Wigner representation of the von Neumann equation – the Wigner-Liouville equation (WLE). To derive the WLE for the full density matrix of the N-particle system $\rho(x|x') = \psi(x)\psi^*(x')$, where $x = (ct, q_N)$ and $x' = (ct', q'_N)$, we introduce the Wigner variables $X = (cT, R) = (x + x')/2$ and $y = (c\eta, \xi) = (x' - x)$. The Wigner distribution function (WF) is defined by [2,3]

$$f(P, X) = \frac{1}{(2\pi\hbar)^\nu} \int dy \rho\left(X - \frac{y}{2}, X + \frac{y}{2}\right) \times$$
$$exp\left(-\frac{iy}{\hbar}(P - \frac{e}{c}\int_{-\frac{1}{2}}^{\frac{1}{2}} A(X + sy)ds)\right) \quad (1)$$

Here $\nu = 3N + 1$, $A(x) = (\varphi, \mathbf{A}(x))$ is the 4-vector potential of the applied electromagnetic field and the conjugate variable $P = (\epsilon, \mathbf{p})$ have the meaning of the kinetic 4-momentum. We assume that this system is described by the following Hamiltonian:

$$H = \frac{1}{2m}\left(i\hbar\nabla + \frac{e}{c}\mathbf{A}(x)\right)^2 + e\varphi(x) + V(q_N)$$

where potential energy $V(q_N)$ is the sum of the two particle interaction potentials.

The phase cofactor generated by the wave functions $\psi(x) \to \psi(x)\exp(\frac{ie}{ch}\Lambda(x))$ under the local guage transformatiom $A(x) \to A(x) - \nabla\Lambda(x)$ defined by means of an arbitrary real function $\Lambda(x)$, are canceled by the integral in the exponent of (1) and WF $f(P, X)$ remains inchanged. In the definition (1) the integration path is taken as a straight line betwee the points $X - y/2$ and $X + y/2$.

Using this definition it is straightforward to obtain the WLE for the *full* density matrix [2,4]. For simplicity let us consider system of charged particles in a static and homogeneous electrical $\mathbf{E}$ and magnetic $\mathbf{B}$ field:

$$\frac{\partial f}{\partial T} + \frac{p}{m}\frac{\partial f}{\partial R} + \left(eE + \frac{e}{mc}p\otimes B - \frac{\partial V(R)}{\partial R}\right)\frac{\partial f}{\partial p} + \frac{e}{m}pE\frac{\partial f}{\partial \epsilon} =$$
$$\int_{-\infty}^{\infty} ds\omega(s,R) f(\epsilon, p-s, T, R) \quad (2)$$

$$\epsilon f(\epsilon, p, T, R) = \frac{p^2}{2m} f(\epsilon, p, T, R) -$$
$$\frac{\hbar^2}{8m}\left(\nabla_R + \frac{e}{c}B\otimes\nabla_p + eE\frac{\partial}{\partial \epsilon}\right)^2 f(\epsilon, p, T, R) +$$
$$\frac{1}{2(2\pi\hbar)^\nu}\int d\tilde{\epsilon}d\tilde{p}d\eta d\xi \exp\left(\frac{i}{\hbar}[\xi(\tilde{p}-p) + \eta(\tilde{\epsilon}-\epsilon)]\right) \times$$
$$\{V\left(R-\frac{\xi}{2}\right) + V\left(R+\frac{\xi}{2}\right)\} f(\tilde{\epsilon}, \tilde{p}, T, R) \quad (3)$$

where

$$\omega(s, R) = F(R)\frac{d\delta(s)}{ds} + \frac{4\pi}{(2\pi\hbar)^\nu}\int d\xi V\left(R-\frac{\xi}{2}\right)\sin\left(\frac{s\xi}{\hbar}\right),$$

and $F(R) = -\partial V(R)/\partial R$ is the classical force. So WF should satisfy to two compatible equations. Interesting that for harmonic oscillator evolution equation Eq. (2) is fully classical (lhs Eq. (2) is equal to zero), while Plank's constant $\hbar$ is included in the quantum problem only due to existence of the Eq. (3). WF should satisfy to Eq. (3) identically for all time T.

Now let us consider evolution Eq. (2). Obviously, the force term in ω exactly cancels the term of the gradient of V on the lhs of Eq. (2). Retaining this term allows us to write the WLE as the classical Liouville equation [lhs of Eq. (2)] plus a quantum correction [all terms on the rhs of Eq. (2)] which vanish for $\hbar \to 0$. This form allows us identically to transform Eq. (2) into an integral equation [4]:

$$f(\epsilon, p, T, R) = f(\epsilon_0, p_0, T_0, R_0) +$$
$$\int_0^T d\tau \int_{-\infty}^{\infty} ds\, f(\bar{\epsilon}_\tau, \bar{p}_\tau - s, \tau, \bar{R}_\tau) \omega(s, \bar{R}_\tau) \quad (4)$$

The first contribution describes quantum dynamics and is given by the initial WF $f_0(\epsilon, p, R) \equiv f(\epsilon, p, 0, R)$, but taken at points $\epsilon_0 \equiv \bar{\epsilon}(0)$, $p_0 \equiv \bar{p}(0)$ and $R_0 \equiv \bar{R}(0)$ of the trajectories $\epsilon_\tau \equiv \bar{\epsilon}(\tau)$, $p_\tau \equiv \bar{p}(\tau)$ and $R_\tau \equiv \bar{R}(\tau)$ satisfying the Hamilton like equations associated to the WLE and connecting points (ϵ, p, R) at time T and points (ϵ_0, p_0, R_0) at time 0:

$$\frac{d\bar{R}}{dT} = \frac{\bar{p}}{m}$$
$$\frac{d\bar{p}}{dT} = eE + \frac{e}{mc}\bar{p}\otimes B + F(\bar{R})$$
$$\frac{d\bar{\epsilon}}{dT} = \frac{e}{m}\bar{p}E$$

Notice that even the first term in Eq. (4) may describe the evolution of a *quantum* many-body state if the initial WF $f_0(\epsilon, p, R)$ is chosen appropriately and contains the all powers of the Plank's constant. The integral term in Eq. (4) describes the perturbation of the classical trajectories due to quantum effects, for details we refer to Ref. [5]. Let us stress that only functions satisfying Eq. (3) or satisfying conditions obtained in [4] have the physical meaning and can be used as initial WF in the considered above evolution WL equation.

The structure of Eq. (4) suggests to construct its solution iteratively, starting with f_0. Let us rewrite Eq. (4) in the following compact form, $f^T = f_0^T + K_\tau^T f^\tau$, where the superscript on the WF denotes the time argument and $K_{\tau_1}^{\tau_2}$ denotes the time integral in Eq. (4). Then, the iteration series has the form:

$$f^T = f_0^T + K_{\tau_1}^T f_0^{\tau_1} + K_{\tau_2}^T K_{\tau_1}^{\tau_2} f_0^{\tau_1} + K_{\tau_3}^T K_{\tau_2}^{\tau_3} K_{\tau_1}^{\tau_2} f_0^{\tau_1} + \ldots\ldots, \quad (5)$$

where the first term describes the evolution of an initial (classical or quantum) WF f_0 (it may contain any order of Planck's constant). The remaining terms systematically take into account all dynamic quantum corrections [trajectories with momentum jumps arising from the shifted momentum arguments in the WF under the integral in (2)] including e.g. tunneling effects and correctly accounting for the Heisenberg uncertainty principle. Thus, the solution (5) can be understood as a properly weighted sum of classical and quantum phase space trajectories [5].

3 Wigner Representation of Time Correlation Functions

According to the Kubo formula the conductivity is the Fourier transform of the current–current correlation function. Our starting point is the general operator expression for the canonical ensemble-averaged time correlation function [6]:

$$C_{FA}(t) = Z^{-1} \text{Tr}\left\{ \hat{F} e^{i\hat{H}t_c^*/\hbar} \hat{A} e^{-i\hat{H}t_c/\hbar} \right\}, \quad (6)$$

where $\hat{H}$ is the Hamiltonian of the system expressed as a sum of the kinetic energy operator, $\hat{K}$, and the potential energy operator, $\hat{V}$. Time is taken to be a complex quantity, $t_c = t - i\hbar\beta/2$, where $\beta = 1/k_B T$ is the inverse temperature with k_B denoting the Boltzmann constant. The operators $\hat{F}$ and $\hat{A}$ are quantum operators of the dynamic quantities under consideration and $Z = \text{Tr}\left\{ e^{-\beta \hat{H}} \right\}$ is the partition function. The Wigner representation of the time correlation function in a v–dimensional space can be written as

$$C_{FA}(t) = (2\pi\hbar)^{-3N} \int\int d\mu_1 d\mu_2 \, F(\mu_1) A(\mu_2) W(\mu_1; \mu_2; t; i\hbar\beta), \quad (7)$$

where we introduce the short-hand notation for the phase space point, $\mu_i = (p_i, q_i), (i = 1, 2)$, and p and q comprise the momenta and coordinates, respectively, of all particles in the system. $W(\mu_1; \mu_2; t; i\hbar\beta)$ is the spectral density expressed as

$$W(\mu_1;\mu_2;t;i\hbar\beta) = Z^{-1} \int\int d\xi_1 d\xi_2\, e^{i\frac{p_1\xi_1}{\hbar}} e^{i\frac{p_2\xi_2}{\hbar}} \times$$

$$\left\langle q_1 + \frac{\xi_1}{2}\left|e^{i\hat{H}t_c^*/\hbar}\right|q_2 - \frac{\xi_2}{2}\right\rangle \left\langle q_2 + \frac{\xi_2}{2}\left|e^{-i\hat{H}t_c/\hbar}\right|q_1 - \frac{\xi_1}{2}\right\rangle, \quad (8)$$

and $A(\mu) = \int d\xi\, e^{-i\frac{p\xi}{\hbar}} \left\langle q - \frac{\xi}{2}\left|\hat{A}\right|q + \frac{\xi}{2}\right\rangle$ denotes Weyl's symbol [4] of operator $\hat{A}$. Similarly for the operator $\hat{F}$. Hence the problem of the numerical calculation of the canonically averaged time correlation function is reduced to the computation of the spectral density.

To obtain the integral equation for W let us introduce a pair of dynamic p, q-trajectories $\{\bar{q}_\tau(\tau;p_1,q_1,t), \bar{p}_\tau(\tau;p_1,q_1,t)\}$ and $\{\tilde{q}_\tau(\tau;p_2,q_2,t), \tilde{p}_\tau(\tau;p_2,q_2,t)\}$ starting at $\tau = t$ from the initial condition $\{q_1,p_1\}$ and $\{q_2,p_2\}$ propagating in 'negative' and 'positive' time direction:
$\frac{d\bar{p}_\tau}{d\tau} = \frac{1}{2}F[\bar{q}_\tau(\tau)]$; $\frac{d\bar{q}_\tau}{d\tau} = \frac{\bar{p}_\tau(\tau)}{2m}$,
with $\bar{p}_t(\tau=t;p_1,q_1,t) = p_1$; $\bar{q}_t(\tau=t;p_1,q_1,t) = q_1$,
$\frac{d\tilde{p}_\tau}{d\tau} = -\frac{1}{2}F[\tilde{q}_\tau(\tau)]$; $\frac{d\tilde{q}_\tau}{d\tau} = -\frac{\tilde{p}_\tau(\tau)}{2m}$,
with $\tilde{p}_t(\tau=t;p_2,q_2,t) = p_2$; $\tilde{q}_t(\tau=t;p_2,q_2,t) = q_2$,
where $F(q) \equiv -\nabla\tilde{V}$ with $\tilde{V}$ being the total potential, i.e. the sum of all pair interactions U_{ab}. Then, as has been proven in [7], W obeys the following integral equation

$$W(\mu_1;\mu_2;t;i\hbar\beta) = \bar{W}(\bar{p}_0,\bar{q}_0;\tilde{p}_0,\tilde{q}_0;i\hbar\beta) +$$

$$+\frac{1}{2}\int_0^t d\tau \int ds\, W(\bar{p}_\tau - s, \bar{q}_\tau; \tilde{p}_\tau, \tilde{q}_\tau; \tau; i\hbar\beta)\, \varpi(s,\bar{q}_\tau)$$

$$-\frac{1}{2}\int_0^t d\tau \int ds\, W(\bar{p}_\tau, \bar{q}_\tau; \tilde{p}_\tau - s, \tilde{q}_\tau; \tau; i\hbar\beta)\, \varpi(s,\tilde{q}_\tau), \quad (9)$$

where $\varpi(s,q) = \frac{4}{(2\pi\hbar)^\nu \hbar} \int dq'\, \tilde{V}(q-q')\sin\left(\frac{2sq'}{\hbar}\right) + F(q)\nabla\delta(s)$, and $\delta(s)$ is the Dirac delta function. Equation (9) has to be supplemented by an initial condition for the spectral density at $t=0$: $W(\mu_1;\mu_2;0;i\hbar\beta) = \bar{W}(\mu_1;\mu_2;i\hbar\beta) \equiv \bar{W}$. The τ-integrals connect the points $\bar{p}_\tau,\bar{q}_\tau;\tilde{p}_\tau,\tilde{q}_\tau$ at time τ of the mentioned above dynamic p,q-trajectories with the points $p_1,q_1;p_2,q_2$ at time t whereas in $\bar{W}$ the trajectories are to be taken at $\tau=0$. The function $\bar{W}$ can be expressed in the form of a finite difference approximation of the path integral [5,7,8]:

$$\bar{W}(\mu_1;\mu_2;i\hbar\beta) \approx$$

$$\int\int d\tilde{q}_1\ldots d\tilde{q}_n \int\int dq'_1\ldots dq'_n\, \Psi(\mu_1;\mu_2;\tilde{q}_1,\ldots,\tilde{q}_n;q'_1,\ldots,q'_n;i\hbar\beta), \quad (10)$$

with

$$\Psi(\mu_1;\mu_2;\tilde{q}_1,...,\tilde{q}_n;q'_1,...,q'_n;i\hbar\beta) \equiv$$

$$\frac{1}{Z}\left\langle q_1\left|e^{-\epsilon\hat{K}}\right|\tilde{q}_1\right\rangle e^{-\epsilon V(\tilde{q}_1)}\left\langle \tilde{q}_1\left|e^{-\epsilon\hat{K}}\right|\tilde{q}_2\right\rangle e^{-\epsilon V(\tilde{q}_2)}\ldots\ldots$$

$$e^{-\epsilon V(\tilde{q}_n)}\left\langle \tilde{q}_n\left|e^{-\epsilon\hat{K}}\right|q_2\right\rangle \varphi(p_2;\tilde{q}_n,q'_1) \times$$

$$\left\langle q_2 \left| e^{-\epsilon \hat{K}} \right| q_1' \right\rangle e^{-\epsilon V(q_1')} \left\langle q_1' \left| e^{-\epsilon \hat{K}} \right| q_2' \right\rangle e^{-\epsilon V(q_2')} \cdots \cdots$$
$$e^{-\epsilon V(q_n')} \left\langle q_n' \left| e^{-\epsilon \hat{K}} \right| q_1 \right\rangle \varphi(p_1; q_n', \tilde{q}_1) \tag{11}$$

where $\varphi(p; q', q'') \equiv (2\lambda^2)^{v/2} \exp\left[-\frac{1}{2\pi}\left\langle \frac{p\lambda}{\hbar} + i\pi\frac{q'-q''}{\lambda} \middle| \frac{p\lambda}{\hbar} + i\pi\frac{q'-q''}{\lambda} \right\rangle\right]$, and $\langle x|y \rangle$ denotes the scalar product of two vectors x, y. In this expression the original (unknown) density matrix of the correlated system $e^{-\beta(\hat{K}+\hat{V})}$ has been decomposed into $2n$ factors, each at a $2n$ times higher temperature, with the inverse $\epsilon = \beta/2n$ and the corresponding high temperature DeBroglie wave length squared $\lambda^2 \equiv 2\pi\hbar^2\epsilon/m$. This leads to a product of known high-temperature (weakly correlated) density matrices, however, at the price of $2n$ additional integrations over the intermediate coordinate vectors (over the "path"). This representation is exact in the limit $n \to \infty$, and, for finite n, an error of order $1/n$ occurs. The function Ψ has to be generalized to properly account for spin-statics effects. This gives rise to an additional spin part of the density matrix and antisymmetrization of one off-diagonal matrix element. To improve the accuracy of the obtained expression, we will replace $V_{ab} \to V_{ab}^{\text{eff}}$ where V_{ab}^{eff} is the proper effective quantum pair potential. For more details on the path integral concept, we refer to Refs. [9]–[11].

Let us now come back to the integral equation (9). For the discussion we note that the integral equation (9) can be exactly converted into an iteration series (which is obtained by successively replacing $W \to \bar{W}$ under the integrals). This series is, however, not a perturbative expansion in the interaction, neither in the electron-scatterer nor in the electron-electron interaction. It rather is an expansion in terms of corrections to classical trajectories of fully interacting electrons and electrons with scatterers. So multiple scattering effects are fully included. Physically the second order and other terms of the iteration series include corrections to the classical electron trajectories (momentum jumps related to the uncertainty principle between momentum–coordinate and energy–time). A detailed investigation of the conditions for which the contribution of the these terms of the iteration series should be taken into account is presented in [12]

The time correlation functions are linear functionals of the spectral density. Using the solution in the form of iteration series we can compute averages of arbitrary operators in standard way and obtain any dynamic macroscopic property of the correlated quantum particles without approximations on the potential interaction. Then for the same series representation holds,

$$C_{FA}(t) = (2\pi\hbar)^{-2v} \int\int d\mu_1 d\mu_2\, \phi(\mu_1;\mu_2)\, W(\mu_1;\mu_2;t;i\hbar\beta) \equiv (\phi|W^t) =$$
$$(\phi|\bar{W}^t) + (\phi|K_{\tau_1}^t \bar{W}^{\tau_1}) + (\phi|K_{\tau_2}^t K_{\tau_1}^{\tau_2} \bar{W}^{\tau_1}) + (\phi|K_{\tau_3}^t K_{\tau_2}^{\tau_3} K_{\tau_1}^{\tau_2} \bar{W}^{\tau_1}) + \ldots \tag{12}$$

where $\phi(\mu_1;\mu_2) \equiv F(\mu_1)A(\mu_2)$ and the parentheses $(\ldots|\ldots)$ denote integration over the phase spaces $\{\mu_1;\mu_2\}$, as indicated in the first line of the equation.

4 Quantum Dynamics

The possibility to convert a iteration series into a form convenient for probabilistic interpretation allows us to apply Monte Carlo methods to its evaluation. According to the general theory of Monte Carlo methods for solving linear integral equations, e.g. [13], one can simultaneously calculate all terms of the iteration series . Using the basic ideas of [13] we have developed a Monte Carlo scheme, which provides domain sampling of the terms giving the main contribution to the iteration series cf. [7]. The solution scheme is a combination of Quantum Monte Carlo and classical Molecular Dynamics methods: Quantum MC is used to generate the correlated initial state, MD generates the $p-q$ trajectories and Monte Carlo methods are applied to perform an importance sampling of the dominant terms of the iteration series (trajectories with momentum jumps in (5) [12] or in iteration series (12)). According to the structure of Kubo formula [6] our calculations include two different stages: (i) generation of the initial conditions (configuration of scatterers and electrons) in the canonical ensemble with probability proportional to the quantum density matrix $|\Psi|$(11) and (ii) generation of the dynamic trajectories on the time scale t' in phase space, starting from these initial configurations (phase cofactor of Ψ is included in trajectory's weight function). Naturally, the true particle number N is replaced by a greatly reduced number N_{sim} which is of the order $50-100$ in the MC cell with periodic boundary conditions.

Our numerical results below refer to finite temperature and moderate degeneracy $n_e \lambda_e = 0.2 \ldots\ldots 7$. We, therefore, will include in the following numerical analysis only the first term in iteration series of Eq. (12), , which is related to the propagation of the initial quantum distribution $\tilde{W}$ according to the Hamiltonian equation of motion. This term, containing all powers of Planck's constant, is the coherent sum of complex-valued contributions of a trajectory ensemble related to $\tilde{W}$. This term allows to describe quantum coherent effects such as Anderson localization, while other terms of iteration series describe deviations from the classical trajectories: the trajectories are perturbed by a finite momentum jump s occuring at arbitrary times τ, $0 \leq \tau \leq t$ [5]. These terms are essential for the recovery of tunneling effects, we expect that they do not give dominant contributions to coherence and localization phenomena. With increasing quantum degeneracy (i.e. decreasing temperature or/and increasing density) the magnitude of these terms will grow.

This approach allows us to generate, in a controlled way, various kinds of quantum dynamics and initial conditions of the many-body system, in particular (i) those which are characteristic of the fully interacting system [i.e.including scatterer-scatterer (s-s), electron-scatterer (e-s), and electron-electron (e-e)] and (ii) those which result if some aspects of these interactions are ignored.

5 Numerical Results

As an application, in this work we will consider a system composed of heavy particles (called scatterers) with mass m_s and negatively charged electrons with mass m_e. To avoid bound state effects due to attraction we consider in this case study only negatively charged scatterers, assuming a positive background for charge neutrality. The influence of electron-scatterer attraction will be studied in a further publication.

We now apply the numerical approach explained above to the problem of an interacting ensemble of electrons and disordered scatterers in one dimension. In all calculations times, frequencies and distances are measured in atomic units. The average distance between electrons, $R_s = 1/(n_e a_0)$, was varied between 12.0 and 0.55, with the densities of electrons and heavy scatterers taken to be equal. The results obtained were practically insensitive to the variation of the whole number of the particles in MC cell from 30 up to 50 and also of the number of high temperature density matrices (determined by the number of factors n), ranging from 10 to 20. Estimates of the average statistical error gave the value of the order 5–7%. We studied two different temperatures: $k_B T/|V_0^{es}| = 0.45$ and 0.28, corresponding to $\lambda_{ee}/a_0 \sim 2.2$ and $\lambda_{ee}/a_0 \sim 3.5$, respectively.

The results presented below are obtained from the first term of iteration series. This term, however, does not describe pure classical dynamics but accounts for quantum effects and, in fact, contains arbitrarily high powers of Planck's constant. The remaining terms of the iteration series describe momentum jumps [7,12] which account for higher-order corrections to the classical dynamics of the quantum distribution, which are expected to be relevant in the limit of high density. Obtained results related to two different cases: 1. *with* e-e interaction included in the dynamics ("interacting dynamics") and 2., *without* e-e interaction ("noninteracting dynamics"). In both cases, the initial state fully includes all interactions. Fig. 1 presents for our model the real part of the diagonal elements of the electrical conductivity tensor versus frequency (real part of the Fourier transform of the temporal momentum–momentum correlation functions) which characterizes the Ohmic absorption of electromagnetic energy and has the physical meaning of electron conductivity. To compare influence of electron interactions conductivities on Fig. 1 are given in the same arbitrary units. The first observation is that, in all cases, the conductivity for the non-interacting dynamics (2) has a maximum at some finite frequency related to the coherent oscillations in the time domain and vanishes at low frequency. The latter clearly indicates electron localization. The effect of the e-e interaction is, as shown by curves 1, a reduction of the maximum (damping of the coherent oscillations) and, in most cases, an increase of the zero-frequency conductivity. Thus, our calculations confirm the delocalizing effect of the interactions (Figs. a,b,d) at the considered densities. Interestingly, Fig. 1c) is an exception: even with interactions included, the localization behavior persists. The large oscillations in Fig. 1c) are not result of numerical noise, they exist inspite of very long simulation duration.

The reason for the observed behavior is an interplay of varying strength of the e-e-interaction (which is weakened with reducing R_s, i.e. from top to bottom figures) and of the magnitude of quantum effects (which grow with temperature reduction, i.e. from right to left figures). Thus, the delocalization tendency observed from Figs. c) to d) is due to thermal activation which, similarly as the interaction, destroys coherence phenomena.

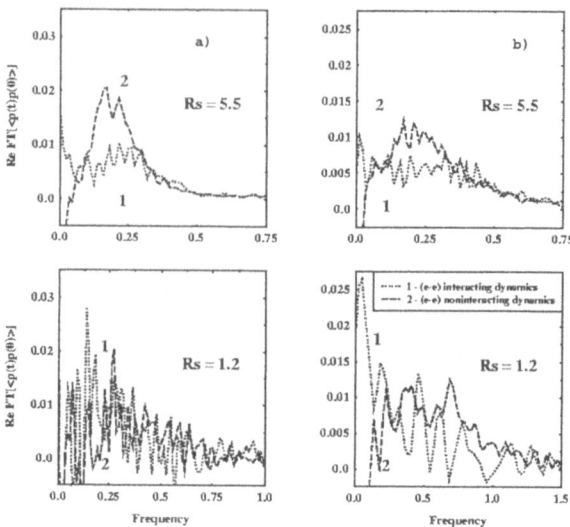

Fig. 1. Real part of the Fourier transform of the temporal momentum-momentum correlation functions for dynamics with (1) and without (2) e-e interaction. Figure parts are for two densities (a,b: $R_s = 5.5$; c,d: $R_s = 1.2$) and temperatures (a,c: $k_B T/|V_0^{es}| = 0.28$; b,d: $k_B T/|V_0^{es}| = 0.45$).

Our simulations qualitatively confirm analytical predictions for the low-frequency and zero temperature limit of the 1D conductivity. Yet our computer power allows us to generate dynamic trajectories up to times t' equal $100 \ldots \ldots 200$ in atomic units. Thus for small frequencies of the order 10^{-2}, large fluctuations of the conductivity appear[1], and the accuracy is not yet sufficient to extract an asymptotic frequency behavior. On the other hand, the advantage of our computational method is that it allows to study systematically the influence of finite temperature and of electron correlation effects on localization phenomena in a wide range of densities. We note that we have also performed simulations at lower densities and found that the delocalizing effect of the e-e-interaction has also been observed at lower density up to $R_s = 12$.

In summary, we have presented numerical results on the influence of Coulomb interaction on electron localization in a one-dimensional system. At low density ($R_s = 5.5$) the interaction is comparatively strong and localization is destroyed. With increasing density $R_s = 1.2$, localization is found to persist even in the presence of Coulomb interaction. For a full understanding of the physical pro-

---

[1] In fact, we observe negative values, although, the real part of the conductivity has to be positive. The reason are weakly damped oscillations with a period exceeding the scale t' used in the calculation of the dynamics. To overcome this deficiency of our model one has to increase the time t' and/or to take into account the slow motion of the heavy particles, which will destroy the coherent oscillations of the light electrons trapped by the heavy particles. Additional calculations with increased t' lead to decreasing negative contributions for low frequencies, as expected.

cesses additional investigations are needed which are presently under way. We are going also to study the influence of electrical and magnetic fields as well as of electron interaction with phonon bath.

Acknowledgments. The work is done under financial support of RF President Grant No. MK-1769.2003.08 and SS - 1953.2003.2, the RAS program No. 17 "Parallel calculations and multiprocessor computational systems" and the grant for talented young researchers of the Science support founda-tion. This work has been supported by a grant for CPU time at the NIC Jülich.

References

1. Belitz, D. and Kirkpatrik, T.R.: The Anderson – Mott transition. Rev. Mod. Phys. **64** 1994) 261 – 380
2. Levanda, M. and Fleurov, V.: Gauge – invariant quantum kinetic equations for electrons in classical electromagnetic fields. J.Phys.: Condens. Matter **6** (1994) 7889 –7908
3. Kremp, D., Bornath, Th., Bonitz M. and Schlanges M.: Quantum kinetic theory of plasmas in strong laser fields. Phys.Rev.E **60** (1999) 4725 – 4732
4. Tatarskii, V.: Wigner representation of quantum mechanics. Sov.Phys.Usp. **26** (1983) 311– 327
5. Filinov, V., Thomas, P., Varga, I., Meier, T., Bonitz, M., Fortov V. and Koch, S.: Interacting electrons in a one – dimensional random array of scatterers: A quantum dynamics and Monte Carlo study. Phys. Rev. **B65**, (2002) 165124-1 – 165124-11
6. Zubarev, D.N. *Nonequilibrium Statistical Thermodynamics*, Plenum Press, New York/London (1974)
7. Filinov, V.S.:Wigner approach to quantum statistical mechnics and quantum generalization molecular dynamics method. Part I. Part II. Mol. Phys. **88** (1996)1517 – 1528; 1529 – 1535
8. Feynman, R.P. and Hibbs, A.R. *Quantum Mechanics and Path Integrals*, McGraw-Hill, New York, Moscow (1965)
9. Zamalin, V.M., Norman, G.E. and Filinov, V.S., *The Monte Carlo Method in Statistical Thermodynamics*, Nauka, Moscow (1977) (in Russian)
10. Zelener, B.V., Norman, G.E. and Filinov, V.S., *Perturbation theory and Pseudopotential in Statistical Thermodynamics*, Nauka, Moscow (1981) (in Russian)
11. Details on our direct fermionic path integral Monte Carlo simulations are given in Filinov, V.S., Bonitz, M., Ebeling, W. and Fortov, V.E.: Thermodynanics of hot H-plasmas: path integral Monte Carlo simulatioms and analytical approximatioms. Plasma Phys. Contr. Fusion **43** (2001) 743 – 759
12. Filinov, V., Medvedev, Yu., Kamskyi, V.: Quntum dynamics and Wigner representation of quantum mechanics. J. Mol. Phys. **85** (1995) 717 – 726
13. Sobol, I.M., Messer, R. (Translator) *Monte Carlo Methods*, Univ. Chicago Publisher, Chicago (1975)

$C_6NH_6^+$ Ions as Intermediates in the Reaction between Benzene and N^+ Ions

Marco Di Stefano, Marzio Rosi, and Antonio Sgamellotti

Istituto di Scienze e Tecnologie Molecolari (ISTM) del CNR, c/o
Dipartimento di Chimica, Universitá degli Studi di Perugia,
Via Elce di Sotto 8, 06123 Perugia (Italy)
marco@thch.unipg.it

Abstract. The nitrenium ions of formula $C_6NH_6^+$ arising from the reaction between benzene molecules and atomic nitrogen ions, N^+ (^3P), are studied by means of theoretical methods so that their structures and relative stabilities are investigated. All calculations are carried out by using the DFT hybrid functional B3LYP in conjunction with the split valence 6-31G* basis set.

1 Introduction

Nitrenium ions are valid intermediates in several chemical reactions and the putative carcinogens formed in vitro from aromatic amines [1] because they are able to covalently modify DNA base sites. They are labelled $[R\text{-}N\text{-}R']^+$ and are isoelectronic with the carbenes. They exist in a singlet and triplet spin state and their relative stabilities depend on the substituents bound to the electron deficient atom. In the singlet state, the nonbonding electrons occupy an sp^2 orbital, leaving an empty p orbital on the nitrogen atom. In the triplet state, the nonbonding electrons have parallel spins and occupy an sp^2 orbital as well as an orbital with large p character and usually behave as diradicals [2]. Recent experiments indicate that singlet and triplet nitrenium ions show different reactivities: while the singlet ions tend to react with nucleophiles, on the triplet state we note the tendency to abstract H atoms [3].

Despite their well-known importance, most of the currently available information on the reactive nitrenium ions comes from quantum chemistry calculations because their inherent instability makes them rather unsuitable for X-ray crystallography or high-resolution NMR [4]. With regard to nitrenium ions with the heteroatom inside the aromatic ring, we mention that Cann supposed the formation of an aromatic nitrenium ion from the reaction of dibenz[b,f]azepine with silver(I), although he reported that he could not characterize this species [5]. Falvey and co-workers investigated by means of AM1 and MNDO methods the geometries and energies of a few arylnitrenium ions, among which five- and six-membered ring cations containing the nitrogen [6]. Smith and Bitar were the first to carry out ab initio molecular orbital studies on the aromatic nitrenium ions based on the azepine ring system [7].

Given their important role as reactive intermediates, in this paper we present a detailed theoretical study on the $C_6NH_6^+$ cations arising from the reaction between benzene and atomic nitrogen ions, N^+ (^3P). The interstellar relevance of this reaction has been given elsewhere [8] and is not repeated here. However we mention that N^+ (^3P) is isoelectronic with the carbon atom, whose addition into benzene has already been studied [9]. Even though much information is nowadays available on the C_7H_6 radicals, their isoelectronic $C_6NH_6^+$ ions have not been so carefully investigated. We concentrate on the structures and relative stabilities of a large number of $C_6NH_6^+$ cations, besides focussing on the various interactions between N^+ and benzene. Our theoretical study indicates the azepine nitrenium and the phenylnitrenium ions as the most stable intermediates and investigates possible ring contraction and ring expansion processes they can promote, in analogy with what previously deduced for the nitrenes [10]. The analysis of the transition states then allows a comprehensive view of the potential energy surface under study.

Since the formation of CN bond containing molecules from organic compounds is a subject of relevant interest [11], our study is aimed at further analyzing the role of this class of compounds as likely precursors in the synthesis of pre-biotic molecules in flames and interstellar clouds and might thus represent a worthy opening of further experimental and theoretical researches.

2 Theoretical Methods

Density functional theory, by using the hybrid [12] B3LYP functional [13] [14], is utilized in order to optimize the stationary points of the investigated systems. The DFT wavefunctions are implemented so that they result either restricted or unrestricted and their stabilities are tested. The further evaluation of the harmonic frequencies confirms that we have either a minimum or a saddle point. Thermochemical calculations are carried out at 298.15 K and 1 atm by adding the zero point correction and the thermal corrections to the calculated B3LYP energies. The absolute entropies are calculated by using standard statistical-mechanistic procedures from scaled harmonic frequencies and moments of inertia relative to the B3LYP/6-31G* optimized geometries. The 6-31G* basis set is used [15]. It contains a set of d functions on C and N and represents a good compromise between accuracy and saving in computational resources. All calculations are performed by using the Gaussian 98 computer codes [16].

3 Results and Discussion

The $C_6NH_6^+$ cations have got both a singlet and a triplet state spin configuration. The triplet ions are labelled **Tn**, where **n** is **1, 2,** ..., and the singlet **Sn**. The transition states linking the intermediates x and y are labelled **TS**$_{x-y}$. All thermochemical parameters are given in eV and calculated at 298.15 K. Finally, in Fig. 4 a schematic representation of the potential energy surface under study is given.

In agreement with the [C$_6$H$_6$ + C] reaction [9], the addition of N$^+$ ions into the benzene CC bond is a barrierless process. The electron deficient nitrogen ion, N$^+$ (^{3}P), attacks the benzene π electrons yielding the intermediate **T1**, whose B3LYP/6-31G* optimized geometry is given in Fig. 1. This addition is exothermic by 7.69 eV. The CC bond bridged by the nitrogen in **T1** is elongated (1.510 Å) if compared to that of benzene (1.395 Å, at our level of theory), revealing a single bond nature. The CN bonds (1.492 Å) are almost perpendicular to the ring plane. We are not able to locate the same stationary point on the singlet surface, because all attempts converge into the singlet azepine nitrenium ion, **S5** (see Fig. 2).

By passing through **TS$_{1-2}$**, **T1** yields **T2**. This isomerization is endothermic by 0.85 eV with an activation barrier of 0.87 eV. In **T2** the nitrogen interacts with three carbon atoms giving a symmetric structure. We observe the loss of the six-membered ring planarity and the nitrogen is more strongly bound to the carbon now lying out of the ring plane (1.404 Å). This interaction lengthens the non-planar CC bonds (1.533 Å). The same species also exists in the low-lying singlet state and is labelled **S2**. It lies 8.13 eV in energy below the reactants and might be reached from the triplet via InterSystem Crossing (ISC). The more elongated CN bonds (1.541 Å) indicate a weaker interaction between N$^+$ and benzene.

T3 lies 6.75 eV in energy below the separate reactants. The CN bonds are computed 1.519 Å and are perpendicular to the C$_6$H$_6$ ring, whose structure is rather bent with CC bonds showing either single (1.519 Å) or double (1.359 Å) character. We can note that the coordination with the "edge" of benzene, as in **T1**, yields stronger CN bonds than those of **T3**. The same species also exists in the low-lying singlet surface. It is labelled **S3** and lies 1.28 eV in energy below the triplet state isomer. The different spatial orientation of the nitrogen favours the formation of two additional CN bonds which are computed weaker than those of **T3**, 1.749 Å. Both **S3** and **T3** are isoelectronic with the 7-norbornalienylidene, or foiled methylene, a high-energy C$_7$H$_6$ isomer [17].

T4 lies 7.50 eV in energy below [C$_6$H$_6$ + N$^+$]. We note that the nitrogen atom lies out of the molecular plane. The CN bond is computed 1.461 Å and the positive charge is located on the heteroatom. At this level of theory, we are not able to optimize the same structure on the singlet potential energy surface.

T1 yields **T4** by passing through **TS$_{1-4}$**. At B3LYP/6-31G* level of theory this process is endothermic by 0.19 eV with an activation energy of 0.35 eV. Once formed, **T4** can evolve into **T3** via **TS$_{4-3}$**. This further isomerization is endothermic by 0.75 eV with a barrier of 1.22 eV. We point out that at our level of theory a transition structure for the pathway [**T1** → **T3**] is not located.

The definite insertion of N$^+$ into the CC bond of **T1** yields the azepine nitrenium cation, **T5**, via **TS$_{1-5}$** (see Fig. 2). An exothermicity of 1.09 eV and an activation barrier of 0.16 eV are computed at the B3LYP/6-31G* level of theory.

The azepine nitrenium cation was the object of HF, MP2 and CASSCF calculations [7]. Our optimized geometries are in good agreement with the available

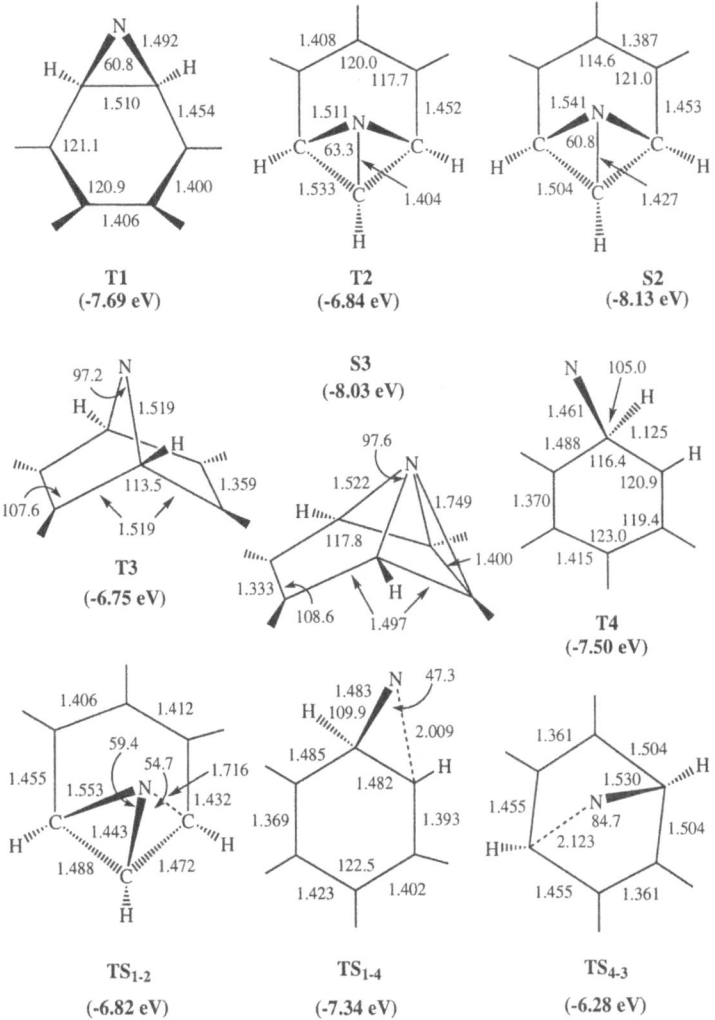

Fig. 1. Optimized geometries at B3LYP/6-31G* level of theory of the six-membered ring $C_6NH_6^+$ ions. Bond lengths are in Å and angles in degrees. In brackets, we report their relative stabilities (eV) with respect to the separate reactants at the B3LYP/6-31G* level

data, although a few slight differences are due to the choice of basis set and theoretical method. There are two atomic orbitals on the nitrogen which can accomodate the nonbonding electron pair, leading to either a singlet ground state or a triplet state. They are labelled **S5** and **T5**, respectively, and both own C_{2v} symmetry. As the former lies 10.63 eV in energy below the reactants, the latter 8.78 eV and the conversion between these two molecules occurs via InterSystem Crossing. Both species do not differ greatly and own a planar geometry. The

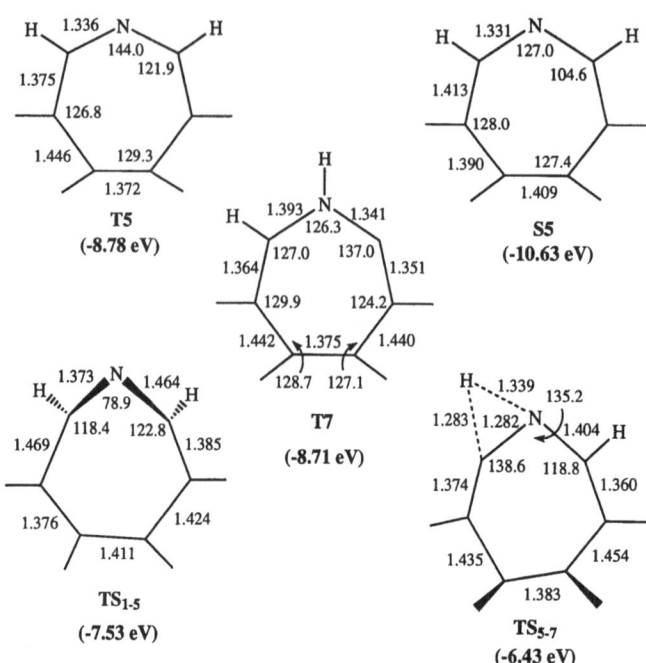

Fig. 2. Optimized geometries at B3LYP/6-31G* level of theory of the $C_6NH_6^+$ ions based on the azepine ring system. Bond lengths are in Å and angles in degrees. In brackets, we report their relative stabilities (eV) with respect to the separate reactants at the B3LYP/6-31G* level

positive charge is located on the nitrogen which results the electrophilic centre. We compute comparable CN bonds (1.336 Å in **T5** and 1.331 Å in **S5**), indicating a double bond nature. In **S5** we optimize a CNC angle of 127.0° which is in very good agreement with the value of the internal angles of a regular heptagon. It means that the empty p_z atomic orbital on the nitrogen is fully conjugated with the π orbitals around the ring, producing a planar cyclic fully conjugated system of six π electrons with an aromatic character. In **T5** the same angle is widened to 144.0°, as previously noted for other nitrenium ions [1]. Moreover, although other azaaromatics such as pyridine are known to become non planar upon excitation into the triplet state [18], the present system retains its planarity in both multiplicities. **T5** and **S5** are isoelectronic with the cycloheptatrienylidene radical, which has already been exhaustively studied [9,17,19]. The presence of the cyclic conjugated π-system adjacent to the carbene centre allows the existence of a singlet and a triplet state, the latter resulting less stable by nearly 0.87 eV. For the $C_6NH_6^+$ system, we calculate a larger difference (1.85 eV). This discrepancy is explained by the nonaromaticity of **T5** [7], whereas the cycloheptatrienylidene remains aromatic in both states [9].

T5 may shift a hydrogen atom and evolve into **T7** via TS_{5-7} (see Fig. 2). **T7** differs from **T5** for the presence of a NH bond. The carbon now without

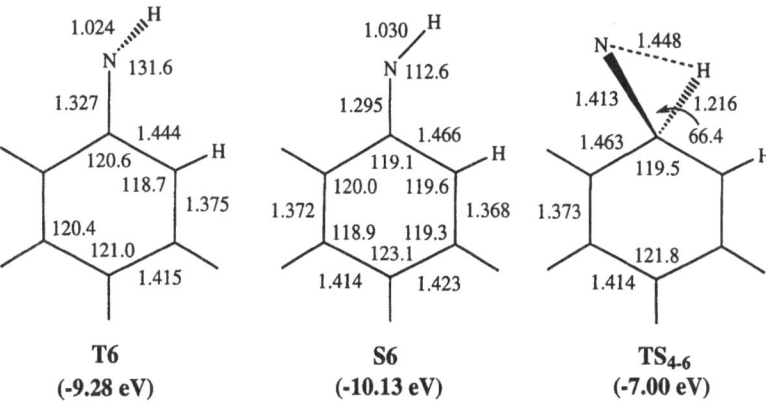

Fig. 3. Optimized geometries at B3LYP/6-31G* level of theory of the phenylnitrenium ions. Bond lengths are in Å and angles in degrees. In brackets, we report their relative stabilities (eV) with respect to the separate reactants at the B3LYP/6-31G* level

the hydrogen is the new electrophilic centre and it reduces the CNC angle to 126.3°. Both species are nearly isoenergetic, being the isomerization [**T5** → **T7**] endothermic by 0.07 eV. However, a high activation barrier is calculated, 2.35 eV, which makes this isomerization essentially interesting from a theoretical point of view.

In Fig. 3 we give the optimized geometries at B3LYP/6-31G* level of theory of the phenylnitrenium ion in its triplet, **T6**, and singlet, **S6**, state. Even though this intermediate is accessible from **T1**, we are not able to locate the direct route [**T1** → **T6**], as suggested for the [C_6H_6 + C] reaction [9]. On the $C_6NH_6^+$ potential energy surface, firstly **T1** has to evolve into **T4** and then **T4** yields **T6** via **TS$_{4-6}$**. This second pathway is exothermic by 1.78 eV with a barrier height of 0.50 eV. The phenylnitrenium cation has already been the object of theoretical studies in the past [6,20] which all indicate the singlet state lower in energy than the triplet. We point out that, whereas **S6** is a planar cation, in the triplet state a NH bond out of the C_6H_6 plane is calculated. Prior studies did not highlight this behaviour [20] which, on the contrary, is evident at the DFT level [6]. Both cations own the majority of the positive charge on the nitrogen, though it is more evident in the singlet state for the empty p orbital which serves as acceptor centre [2]. The hydrogen atom out of the plane opens the CNH angle (131.6° instead of 112.6°) and lengthens the CN bond (1.327 Å instead of 1.295 Å). At B3LYP/6-31G* level of theory, **T6** lies 9.28 eV in energy below the reactants, whereas the singlet 10.13 eV. As previously reported for the other $C_6NH_6^+$ ions, the process [**T6** → **S6**] may occur via InterSystem Crossing. Finally, in Fig. 4 we summarize the results reported so far.

It is known that ring expansion and ring contraction processes of arylcarbenes and arylnitrenes have been a subject of long-standing and continuing interest [10] [21]. However, it was pointed out that they lead to energetic species whose exper-

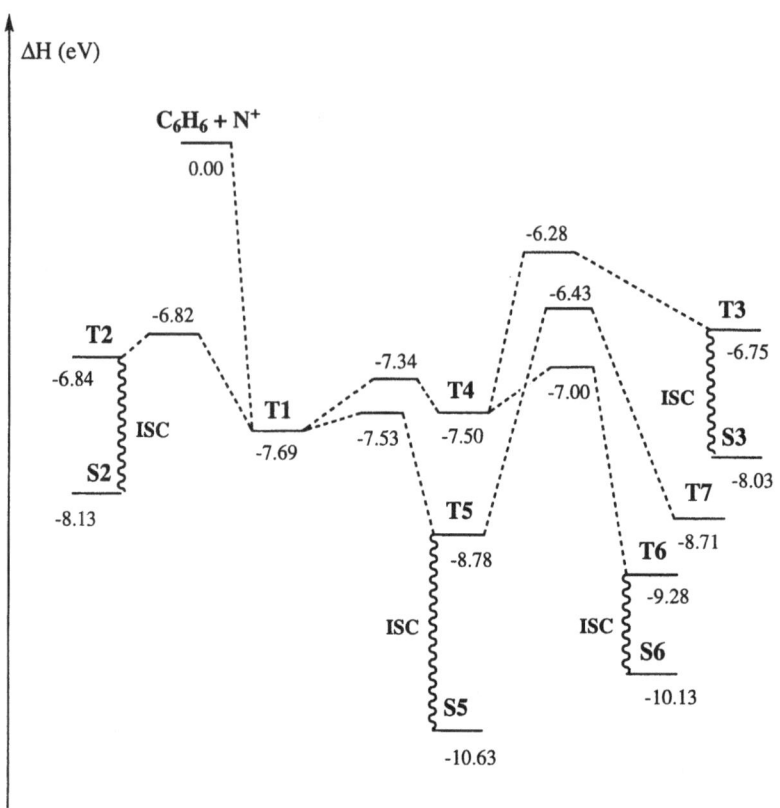

Fig. 4. Schematic representation of the $C_6NH_6^+$ potential energy surface at B3LYP/6-31G* level of theory. All thermochemical parameters are in eV at the temperature of 298.15 K

imental detection is rather difficult. On the $C_6NH_6^+$ potential energy surface we achieve comparable results which are schematically represented in Fig. 5, where we focus on such mechanisms promoted by both **T6** and **T7**.

T6 can either shift a H atom to yield **T8** or form a new CN bond to lead to the 7-azabicyclo[4.1.0]hepta cation, **T9**, with the nitrogen bridging a CC bond. While the first process is exothermic ($\Delta H = -0.19$ eV), the latter is endothermic ($\Delta H = 2.26$ eV). They both own comparable activation barriers of 2.24 eV and 2.33 eV, respectively, which indicate that the pathways under study might be rather improbable and, thus, essentially of interest from a theoretical point of view. **T9** can then shift a H atom and yield **T10**. This furter isomerization is endothermic by 0.91 eV with an activation barrier of 1.57 eV. The same intermediate may also be formed from **T8**, although an unfavourable thermochemistry is involved ($\Delta H = 3.37$ eV, $E_a = 3.39$ eV). A ring contraction process directly links **T9** and **T7**. At B3LYP/6-31G* level of theory, we calculate this mechanism endothermic by 1.73 eV with an activation energy of 1.91 eV. These values

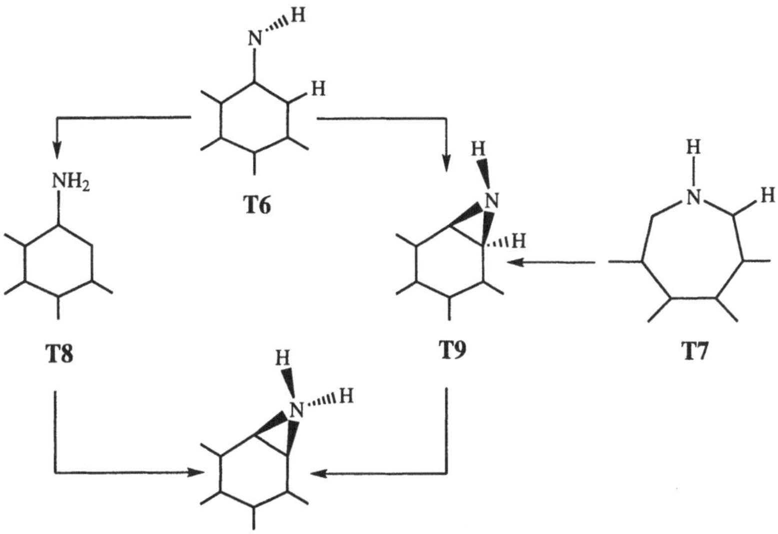

Fig. 5. Ring contraction and ring expansion processes promoted by **T7** and **T6**

ought to indicate that the formation of **T9** might be more likely from **T7** than from **T6**, repeating thus what previously calculated for the isoelectronic C_7H_6 radicals [9].

In Table 1 we list the thermochemical parameters at B3LYP/6-31G* level of theory of all the isomerization processes reported in this paper.

4 Conclusions

In this paper, we present B3LYP/6-31G* calculations on the structures and relative stabilities of nitrenium ions of formula $C_6NH_6^+$ arising from the reaction between benzene molecules and atomic nitrogen ions, N^+ (^3P). We compute the addition of N^+ into C_6H_6 a barrierless and strongly exothermic process which yields the intermediate labelled **T1**. Both the azepine nitrenium, **T5**, and the phenylnitrenium, **T6**, cations are calculated the most stable intermediates of the potential energy surface under study. They are both accessible from **T1** via the definite insertion of N^+ into the benzene ring. We study several isomerization pathways among the $C_6NH_6^+$ cations which lead to the characterization of numerous structures. However, the unfavourable thermochemistry involved makes many of these pathways essentially interesting from a theoretical point of view, at least at the current level of theory.

Table 1. Thermochemical parameters (eV) of the isomerization pathways among the investigated $C_6NH_6^+$ cations. All values are calculated at B3LYP/6-31G* level of theory at the temperature of 298.15 K.

Reaction	ΔH	E_a
$C_6H_6 + N^+ \to T1$	-7.69	-
T1 → T2	0.85	0.87
T1 → T4	0.19	0.35
T1 → T5	-1.09	0.16
T2 → S2	-1.29	-
T3 → S3	-1.28	-
T4 → T3	0.75	1.22
T4 → T6	-1.78	0.50
T5 → T7	0.07	2.35
T5 → S5	-1.85	-
T6 → T8	-0.19	2.24
T6 → T9	2.26	2.33
T6 → S6	-0.85	-
T7 → T9	1.73	1.91
T8 → T10	3.37	3.39
T9 → T10	0.91	1.57

References

1. Ford, G., Herman, P.S.,: Conformational Preferences and Energetics of N-O Heterolyses in Aryl Nitrenium Ions Precursors: *ab initio* and Semiempirical Molecular Orbital Calculations. J. Chem. Soc. Perkin Trans. **2** (1991) 607–616
2. Gonzalez, C., Restrepo-Cossio, A., Márquez, M., Wiberg, K.B., De Rosa, M.: Ab Initio Study of the Solvent Effects on the Singlet-Triplet Gap of Nitrenium Ions and Carbenes. J. Phys. Chem. A **102** (1998) 2732–2738
3. Falvey, D.E.: Singlet and Triplet State in the Reactions of Nitrenium Ions. J. Phys. Org. Chem. **12** (1999) 589–596
4. Boche, G., Andrews, P., Harms, K., Marsch, M., Rangappa, K.S., Schmiczek, M., Willecke, C.: Crystal and Electronic Structure of Stable Nitrenium Ions: A Comparison with Structurally Related Carbenes. J. Am. Chem. Soc. **118** (1996) 4925–4930
5. Cann, M.C.: Formation of Acridine from the Reaction of Dibenz[b,f]azepine with Silver(I): Formation of an Aromatic Nitrenium Ion?. J. Org. Chem. **53** (1988) 1112–1113
6. Falvey, D.E., Cramer, C.J.: Aryl- and Alkylnitrenium Ions: Singlet-triplet Gaps via Ab Initio and Semiempirical Methods. Tetrahedron Lett. **33** (1992) 1705–1708
7. Smith, D.A., Bitar, J.: Aromatic Nitrenium Ions Based on 1H-Azepine. 1. Ab Initio Studies at the RHF, UHF, and CASSCF Levels. J. Org. Chem. **58** (1993) 6–8
8. Di Stefano, M., Rosi, M., Sgamellotti, A., Ascenzi, D., Bassi, D., Franceschi, P., Tosi, P.: Experimental and Theoretical Investigation of the Production of Cations Containing C-N Bonds in the Reaction of Benzene with Atomic Nitrogen Ions. J. Chem. Phys. **119** (2003) 1978–1985 (references therein)

9. Hahndorf, I., Lee, Y.T., Kaiser, R.I., Vereecken, L., Peeters, J., Bettinger, H.F., Schreiner, P.R., Schleyer, P.v.R., Allen, W.D., Schaefer III, H.F.: A Combined Crossed-beam, *ab initio*, and Rice-Ramsperger-Kassel-Marcus Investigation of the Reaction of Carbon Atoms $C(^3P_j)$ with Benzene, C_6H_6 (X^1A_{1g}) and d_6-benzene, C_6D_6 (X^1A_{1g}). J. Chem. Phys. **116** (2002) 3248–3262 (references therein)
10. Kuhn, A., Vosswinkel, M., Wentrup, C.: Carbene and Nitrene Rearrangements: A Theoretical Study of Cyclic Allenes and Carbenes, Carbodiimides, and Azirines. J. Org. Chem. **67** (2002) 9023–9030
11. McEwan, M.J., Scott, G.B.I., Anicich, V.G.: Ion-molecule Reactions Relevant to Titan's Ionosphere. Int. J. Mass. Spectrom. Ion Processes **172** (1998) 209–219
12. Becke, A.D.: Density-functional Thermochemistry. III. The Role of Exact Exchange. J. Chem. Phys. **98** (1993) 5648–5652
13. Becke, A.D.: Density-functional Exchange-energy Approximation with Correct Asymptotic Behavior. Phys. Rev. A **38** (1988) 3098–3100
14. Lee, C., Young. W., Parr, G.G.: Development of the Colle-Salvetti Correlation-energy Formula into a Functional of the Electron Density. Phys. Rev. B **37** (1988) 785–789
15. Frisch, M.J., Pople, J.A., Binkley, J.S.: Self-consistent Molecular Orbital Methods 25. Supplementary Functions for Gaussian Basis Sets. J. Chem. Phys. **80** (1984) 3265–3269 (references therein)
16. Frisch, M.J. *et al.* GAUSSIAN 98, Revision A.7, Gaussian Inc., Pittsburg PA, (1998)
17. Wong, M.W., Wentrup,C.: Interconversions of Phenylcarbene, Cycloheptatetraene, Fulvenallene, and Benzocyclopropene. A Theoretical Study of the C_7H_6 Energy Surface. J. Org. Chem. **61** (1996) 7022–7029
18. Buma, W.J., Groenen, E.J.J., van Hemert, M.C.: Ab initio Calculations on the Structure of Pyridine in its Lowest Triplet State. J. Am. Chem. Soc. **112** (1990) 5447–5451
19. Matzinger, S., Bally, T., Patterson, E.V., McMahon, R.J.: The C_7H_6 Potential Energy Surface Revisited: Relative Energies and IR Assignment. J. Am. Chem. Soc. 118 (1996) 1535–1542
20. Cramer, C.J., Dulles, F.J., Falvey, D.E.: Ab Initio Characterization of Phenylnitrenium and Phenylcarbene: Remarkably Different Properties for Isoelectronic Species. J. Am. Chem. Soc. **116** (1994) 9787–9788
21. Galbraith, J.M., Gaspar, P.P., Borden, W.T.: What Accounts for the Difference between Singlet Phenylphosphinidene and Singlet Phenylnitrene in Reactivity toward Ring Expansion?. J. Am. Chem. Soc. **124** (2002) 11669–11674

Towards a Full Dimensional Exact Quantum Calculation of the Li + HF Reactive Cross Section

Antonio Laganà, Stefano Crocchianti, and Valentina Piermarini

Dipartimento di Chimica, Università di Perugia, 06123 Perugia, Italy

Abstract. Quantum mechanical calculations of the probabilities of the Li + HF$(v,j) \to$ LiF(v',j') + H elementary reaction have been performed by distributing the computations on two large scale computing facilities and using two different high level approaches. The calculations have been performed for several values of the total angular momentum quantum number, a large batch of energies, and the relevant reactant rotational states in order to progress towards an evaluation of the reactive cross section to compare with the experiment. A particular emphasis is put on the role played by internal energy.

1 Introduction

The key experimental observable of the Li + HF reaction still remains the reactive cross section measured by molecular beam experiments [1,2,3]. The energy dependence of the cross section of this reaction shows some interesting features which can be linked to the shape of the potential energy surface and the interplay among the various energy modes.

This problem can be nowadays tackled using rigorous theoretical means. From the theoretical point of view the

$$\text{Li} + \text{HF}(v,j) \longrightarrow \text{LiF}(v',j') + \text{H} \tag{1}$$

reaction can be considered a suitable prototype of atom diatom reactions that exhibits several interesting features. As a matter of fact, this reaction is endoergic (0.157 eV) and has a minimum energy path going through a barrier to products of 0.182 eV over the reactant asymptote (0.025 eV) [4,5]. Another important feature of this reaction is the large variation of the frequencies associated with the related bound states. In particular, on the surface used for present calculations, the zero point energy of reactants is higher than that at the saddle to reaction and at the product asymptote.

Quantum calculations have been already carried out for this reaction using a time independent infinite order sudden [6] method. Quantum calculations were also carried out using exact close coupling techniques for low values of the total angular momentum quantum number J and then extended to high J values using a J shifting model [7]. Preliminary quantum calculations were extended to much larger J values using an appropriate time dependent approach [8] and assuming that all the involved reactant rotational states would react in the same way.

The plot of the calculated cross section shows a maximum at very low energy followed by a small bump and a subsequent increase with energy. Such a structure is qualitatively confirmed by the measurements of Loesch [2] even if a quantitative agreement is still far from being reached.

In this paper we discuss the results obtained by using in a coordinated way the computing facilities of two supercomputer centers of EPCC (Edinburgh, UK) and CINECA (Casalecchio di Reno, IT) to figure out the main sources of error built-in into the used approximate treatments. In section 2 the computational procedures used for the calculations are sketched. In section 3 the dependence of the reactive probability on the value of J is discussed. In section 4 time dependent (TD) results are checked versus the time independent ones (TI). In section 5 we analyze the influence of the rotational excitation of the reactants on the efficiency of the reactive process.

2 The Computational Procedure

The first computational procedure used for the calculations is the time dependent (TD) one. The TD method is based on the wavepacket formalism described in refs. [9,10,11,12]. This has been already used to investigate the Li + HF reaction in two [13] and three dimensions [8]. In the related numerical procedure the calculation of the fixed total angular momentum quantum number J reactive probability is carried out using only the real part of the wavepacket [9,10] and a damped Chebyshev iteration [14,15]. In order to work out the state to state reactive probability, the initial wavepacket is prepared in Jacobi coordinates of the reactant arrangement (R, r and Θ). The wave packet is collocated on a matrix of the R and r coordinates, while for the angular part an expansion on a basis set of the collision angle is adopted. The collocated wavepacket is propagated in time for some thousand of steps using a helicity decoupling (CS) scheme. As time progresses, the wavepacket propagates into the interaction region and then also, eventually, into the product one. Near the edge of the grid an ad hoc potential ensures that aliasing effects are suppressed. The wavepacket is analysed at every time step along an analysis line placed in the asymptotic region of the product channel [9,10]. The coefficients of the analysis are stored so as to accumulate the data needed for the computation of the state specific **S** matrix elements. The square modulus of the state specific **S** matrix elements summed over Λ (the projection of J over the body fixed quantization axis) leads to the state selected reaction probability. The use of this method has allowed us to distribute fixed J calculations on various nodes for a given specific ($v = 0, j = 0$) initial state, obtaining results for a fine energy grid.

The second computational procedure we used is the time independent (TI) one. The TI quantum calculations are carried out at an exact full dimensionality (of the Coupled Channel hyperspherical type [16,17]) level using hyperspherical coordinates. To this end the reaction coordinate (ρ, the hyperspherical radius) is discretized and at each grid point the solution is expanded locally in the eigenfunctions of the hyperangular part of the hamiltonian (surface functions). When these surface functions are used to expand the global wavefunction, a large set of coupled differential equations in ρ are obtained. These are then integrated from small to large values of ρ (35 Å). In the asymptotic region the solution (carried out as an R matrix) is mappend onto the reactant and product states in the appropriate Jacobi coordinates and the S matrix is worked out. For the full dimensional method 230 sectors along the hyperspherical radius have been considered. At each grid point a set of 277 local basis functions have been calculated. The use of this method has allowed us to distribute fixed energy calculations on various nodes for all the open reactant (vj) and product ($v'j'$) vibrotational states. A set of 681 fixed energy runs were launched for each J value. A bottle-neck for the TI calculations is given by the capability of the node memory to accomodate the fixed J fixed Λ solution matrices that can be partitioned only at the price of a significant slowing down of the computing speed.

A key problem of the TI calculations is the availability (especially for larger J values) of a sufficient storage space for intermediate quantities (such as eigenvalues, eigenvectors, overlap matrices for each sector in which the hyperspherical radius was divided and the R matrices for each energy). The present extended calculations of fully converged exact quantum reactive probabilities have been made possible by an intensive use of the large scale computing facility of CINECA and EPCC. Calculations were distributed on the two large scale facility sites by assigning to each of them one of the two parities of the total wave function.

3 The J Dependence of the Time Dependent Probability

As already mentioned time dependent calculations, given a specific initial state (like the ground one $v = 0, j = 0$ considered here), produce results for a full batch of energies. The calculations were carried out for J values ranging from 0 to 40 in the $0 - 0.6$ eV translational energy interval. To cover the whole range of translational energy several runs were performed starting always from the $v = 0, j = 0$ state but using different values of the average relative initial kinetic energy E_{tr}^o. Then the results obtained for the various values of the average relative momentum were matched.

The probabilities calculated at different values of J are plotted in Fig. 1 as a function of the total energy E. As qualitatively evidenced by the dotted lines in the figure, an increase of value of J leads to a gradual shift of the curves towards higher energies. This is in line with the prediction of the basic J shifting model (already used for the Li + HF reaction [18]). That basic J shifting model mimes the J dependence of the reactive probability when plotted as a function of the translational energy E_{tr} by assuming that nonzero total angular momentum probabilities at a given J values can be derived from $J = 0$

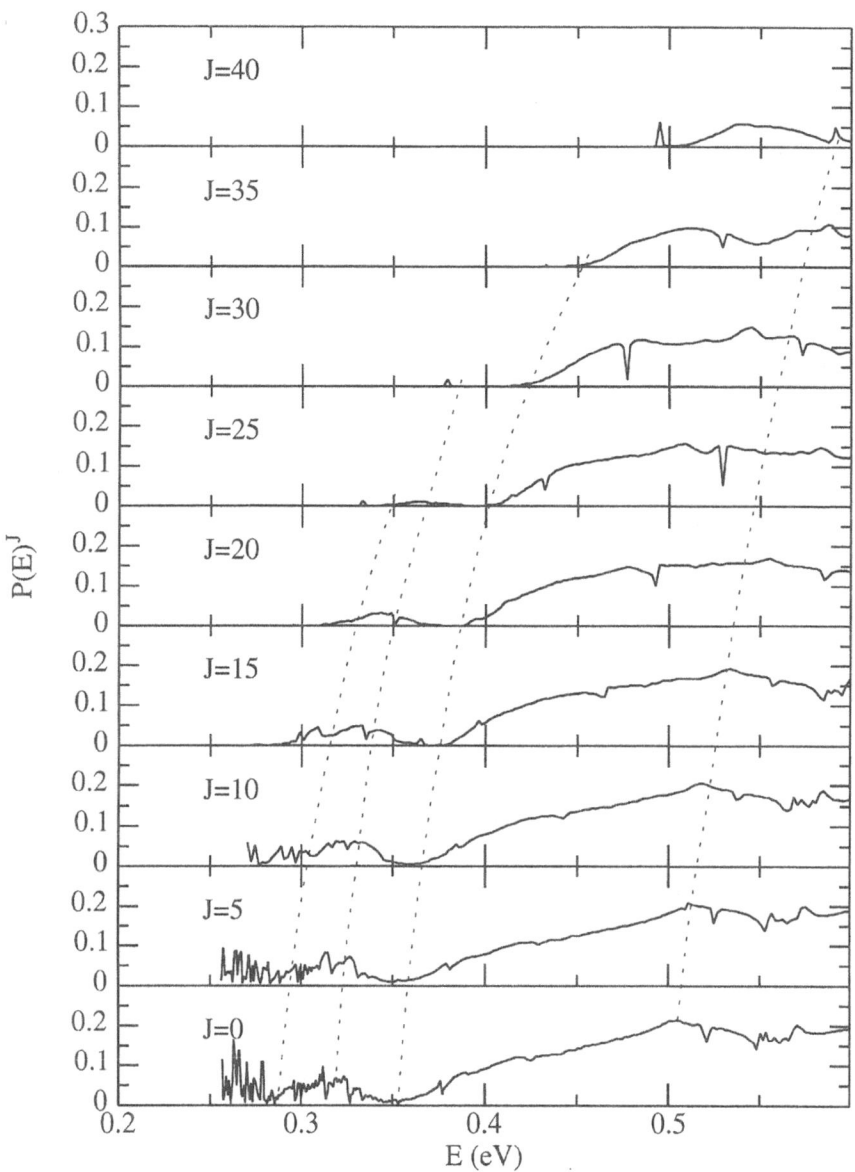

Fig. 1. Reaction probabilities calculated at $v = 0, j = 0$ using the TD method plotted as a function of the total energy E for J values ranging from 0 to 40 in steps of 5 (from the bottom panel up).

ones using the relationship $P^J_{vj,v'j'}(E_{tr}) \simeq P^{J=0}_{vj,v'j'}(E^J_{tr})$ where E^J_{tr} is defined by $E^J_{tr} = E_{tr} - BJ(J+1)$ and $B = \hbar^2/(2I_B) = \hbar^2/(2\mu r^2)$.

However, as suggested by the figure for J values up to 10, the threshold energy is characterized by a highly structured region dominated by resonant spikes. This resonant structure gradually vanishes (in addition to shifting to higher energies) as J increases. This behaviour is common to other parts of the probability curves. In particular, the wider range background oscillating structure masked at low energy by the resonant fine structure also gradually disappears as J increases. For this reason, the plot of the cross section given in Fig. 2 have been calculated using a modifed J shifting model.

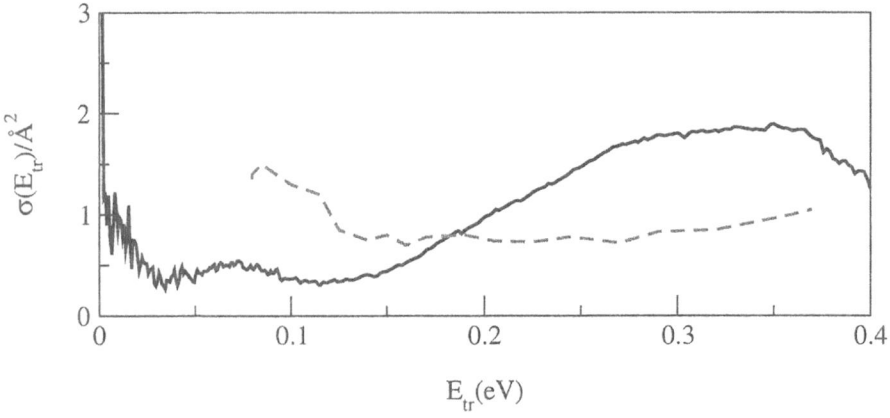

Fig. 2. Reactive cross section calculated using the modified J shifting model (solid line) plotted as a function of the translational energy E_{tr}. For comparison also experimental results [2] are plotted as dashed as a dashed line.

Using the modified J shifting model the value of the not calculated fixed J probability, $P^J_{v=0,j=0}(E_{tr})$, were evaluated by taking a weighted average of the corresponding probability at the nearest upper lying ($J_>$) and lower lying ($J_<$) calculated J values (related probabilities are $P^{J_>}(E_{tr} + E^{J_>J})$ and $P^{J_<}(E_{tr} - E^{JJ_<})$ respectively). The $E^{J_1 J_2}$ notation indicates that the shift in energy corresponds to the difference in the rotational energy between the J_1 and J_2 triatomic (LiFH) rotational states at the transition state geometry. The weight for the two contributions are $(E^{JJ_<})/(E^{J_>J_<})$ and $(E^{J_>J})/(E^{J_>J_<})$ respectively. Obviously, a further interpolation may be required when the energy value $E_{tr} + E^{J_>J}$ or $E_{tr} - E^{JJ_<}$ does not correspond exactly to a calculated one.

4 Time Dependent versus Time Independent Calculations

The cross section values discussed in the previous section were calculated by implicitly assuming that the CS probabilities calculated using the TD computational procedure are good approximations to the exact ones. This should be true for $J = 0$ since in this case the CS approximation is exact. To investigate this,

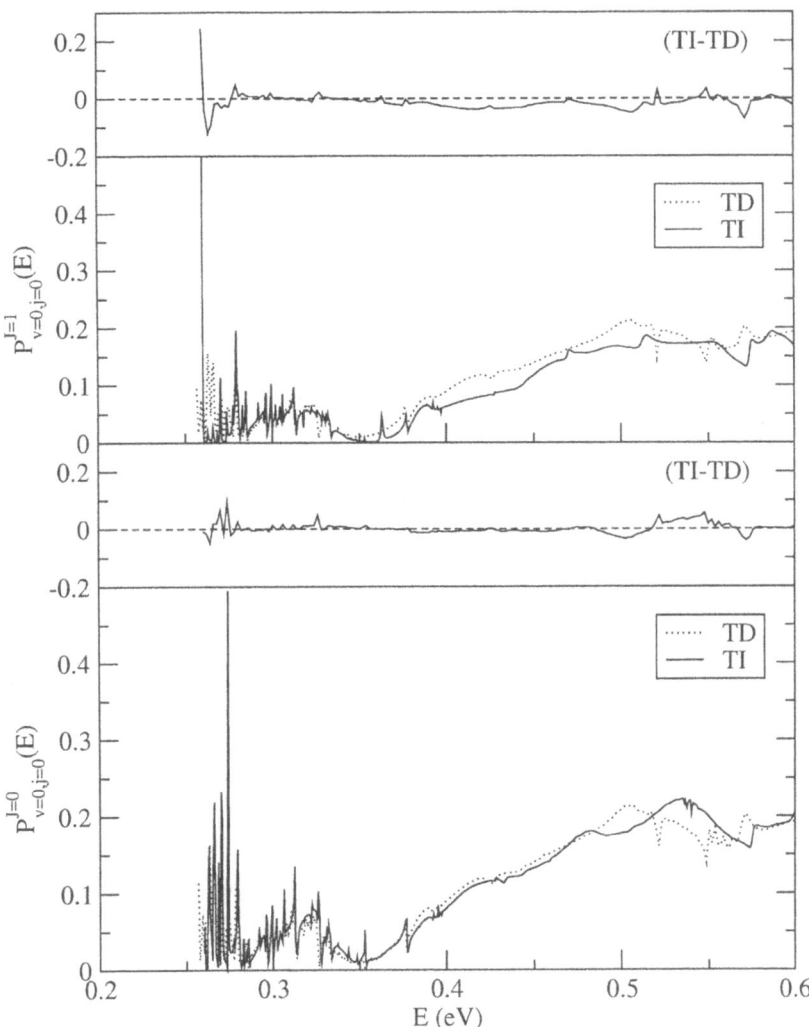

Fig. 3. Reaction probabilities calculated using the TI method (solid lines) at the vibrotational $v = 0, j = 0$ state of the reactants plotted as a function of the total energy E. Bottom panel: $J=0$; top panel: $J = 1$. For comparison also the TD results are given as dotted lines.

the TI calculations were carried out on a highly dense energy grid (coinciding with that obtained from the TD calculations) for $J=0$ and $J=1$.

The $J = 0$ $(P^{J=0}_{v=0,j=0}(E))$ TI reactive probabilities are plotted in the lower panel of Fig. 3 (solid line) as a function of the total energy E. The plot confirms that at low energy there is indeed, a dense resonant structure superimposed to a background oscillating structure having a larger period and that this is not an artifact of the computational procedure. In TI results too, the background structure becomes more evident at the high energy end of the plot, where the masking effect of the dense resonant structure vanishes. The corresponding TD CS values are plotted as a dotted line in the same figure. As apparent from the figure, thanks to the fine energy grid used (as already mentioned 681 points have been used) time independent calculations allow a detailed description of the resonance spikes. The results obtained from the wavepacket calculation coincide with the TI one in all but two major differences. The first difference, that has a dramatic effect on the calculated value of the low energy cross section, is a slightly lower value of the energy threshold and a mismatch of the first resonance peak. Such a discrepancy is not only associated with the fine structured shape of the probability that emphasizes even the smallest disagreement but is likely to be due to a difficulty of the time dependent approach to deal with the slow components of the wavepacket. The information on this discrepancy is plotted in the upper part of the bottom panel of the figure (the relevant root mean square deviation computed over all points is 0.06 for $P_{v=0,j=0}^{J=0}(E)$).

A similar calculation has been carried out at $J=1$. The resulting TI $J=1$ reactive probability for $v=0$ and $j=0$ ($P_{v=0,j=0}^{J=1}(E)$) is shown in the upper panel of Fig. 3 where also TD results are shown as a dotted line. In this case the discrepancy between TD and TI probabilities is larger (the relevant root mean square deviations computed over all points is 0.17). This is due not only to the numerical reasons already pointed out for $P_{v=0,j=0}^{J=0}(E)$, but also to the fact that TD calculations are carried out within an helicity decoupling scheme.

5   The Role Played by Rotational Energy

Another assumption made when estimating the cross section and that needs to be checked against rigorous calculations is the one based on the consideration that only a few lowest rotational states are populated at the temperature of the supersonic beam (T $\simeq$ 60° C) of the experiment and that relevant probabilities do not differ significantly. For this reason we examine here the probabilities calculated at $v=0$ and various j values for $J=0$. Since the temperature of the beam in the experiment is quite cold we need to consider only the lowest j values to investigate the impact of rotational excitation of the reactants on the measured cross section.

As apparent from Fig. 4, an increase of the rotational excitation of the target HF molecule has a clear effect on the threshold energy that moves to higher energies as j increases (such an effect becomes clearly visible expecially for $j \geq 2$). This causes a smaller contribution of related probabilities at threshold. However, from the figure is also apparent that there are other two effects leading to a substantial modification of the state specific probability plots. The first effect is concerned with the dense resonant structure of the threshold that is enhanced by some (low) excited rotational states. This is clearly the case of $j=2$ and partly also of $j=3$. This is, however, also the case of $j=1$ and $j=3$ for which two broad maxima become apparent just past the threshold region. The second

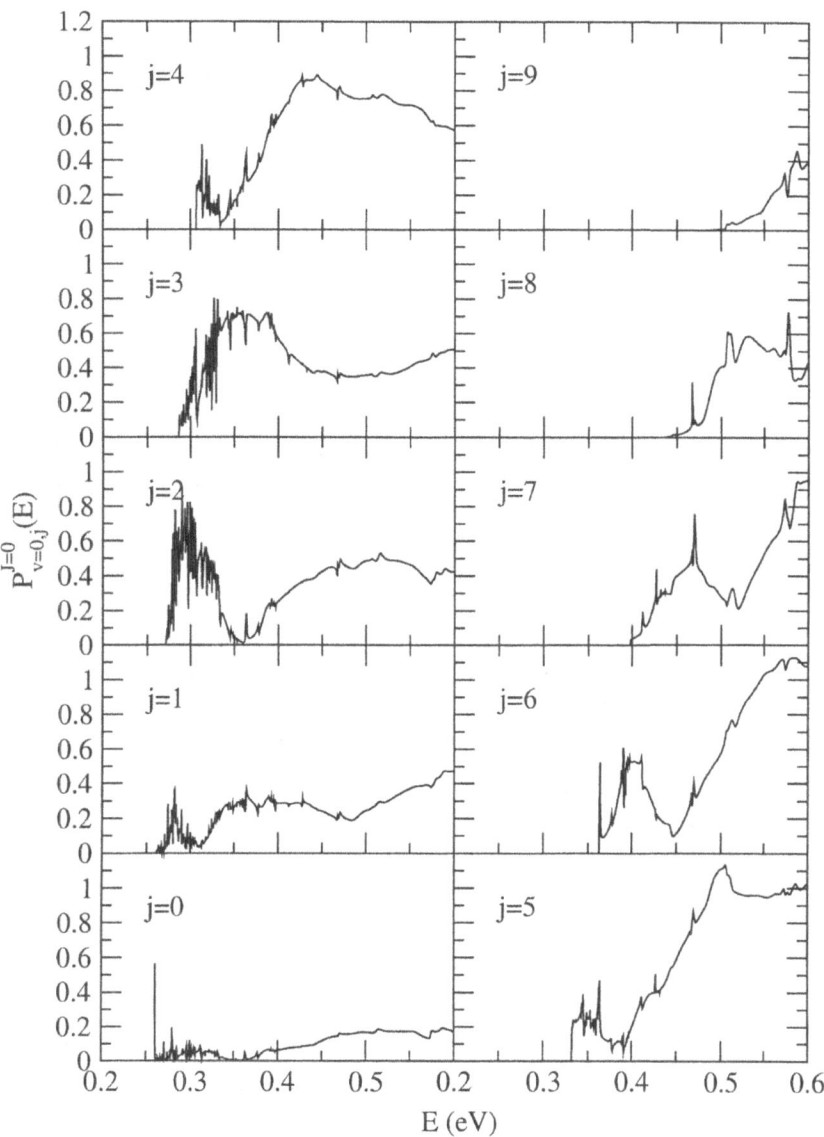

Fig. 4. Reaction probabilities calculated using the TI method at various vibrotational $v = 0, j$ states of the reactants plotted as a function of the total energy E. The rotational state increases from zero to 4 in going upward in the lhs column and from 5 to 9 in the rhs column.

effect is the clear increase of the wider range background contribution that is quite significantly enhanced by an increase of j.

A similar analysis applies to higher J value calculations. In this case, however, one has to take into account that for the different parities different contributions come into play. However, apart from minor effects, which still need to be understood, the j enhancement of the low j resonant structure and the increase of the background wide range probability structure are confirmed together with a clear shift to higher energies of the threshold.

6 Conclusions

Progress towards full dimensional quantum evaluations of reactive cross sections of three diffferent atom system can occur only by exploiting the potentialities of concurrent computing. For the calculations reported in this paper concurrency has been exploited at various levels. The coarsest grain level of concurrency has been exploited by distributing the calculations for different parities of the TI exact approach on two large scale computing facilities. Whereas the finest grain concurrency has been exploited at the level of the single program and in particular at the level of the propagation in time routine of the TD computational procedure. Thanks to this multilevel distribution it has been possible to investigate the effect of increasing both the total angular momentum quantum number and the diatomic rotational one on an extremely fine energy grid. Further work is in progress to calculate higher J contributions to the cross section using the exact TI method.

Acknowledgments. Thanks are due to the EU COST program in Chemistry. Financial support from MURST, ASI, CNR is acknowledged. Support by EC within the Research Training Network "Reaction Dynamics" (Contract No. HPRN-CT-1999-00007) is also acknowledged.

References

1. Becker, C.H., Casavecchia, P., Tiedmann, P.W., Valentini, J.J., Lee, Y.T., J. Chem. Phys. **73**, (1980) 2833-2850.
2. Hobel, O., Menendez, M., Loesch, H.J., Phys. Chem. Chem. Phys. **3**(17), (2001) 3633-3637.
3. Casavecchia, P., Rep. Progr. Phys., **63** (2000) 355-416.
4. Parker, G.A., Pack, R.T, and Laganà, A., Chem. Phys. Letters **202**, (1993) 75-81.
5. Parker, G.A., Laganà, A., Crocchianti, S., and Pack, R.T, J. Chem. Phys. **102**, (1995) 1238-1250.
6. Laganà, A., Aguilar, A., Gimenez, X., Lucas, J.M., J. Chem. Phys. **199**(30), (1995) 11696-11700.
7. Laganà, A., Bolloni, A., Crocchianti, S., and Parker, G.A., Chem. Phys. Letters **324**, (2000) 466-474.
8. Piermarini, V., Crocchianti, S., and Laganà, A., J. Comp. Meth. in Science and Engineering **2**, (2002) 361-367.
9. Gray, S.K., and Balint-Kurti, G.G., J. Chem. Phys. **108**, (1998) 950-962.

10. Meijer, A.J.H.M., Goldfield, E.M., Gray, S.K., and Balint-Kurti, G.G., Chem. Phys. Letters **293**, (1998) 270-276.
11. Balint-Kurti, G.G., Lecture Notes in Chemistry **75**, (2000) 74-87.
12. Hankel, M., and Piermarini, V., Lecture Notes in Chemistry, *75*, (2000) 209-221.
13. Balint-Kurti, G.G., Göğtas, F., Mort, S.P., Offer, A.R., Laganà, A., Gervasi, O., J. Chem. Phys. **99**, (1993) 9567-9584.
14. Mandelshtam, V.A., and Taylor, H.S., J. Chem. Phys. **102** (1995) 7390-7399.
15. Mandelshtam, V.A., and Taylor, H.S., J. Chem. Phys. **103** (1995) 2903-2907.
16. Pack R.T, and Parker, G.A., J. Chem. Phys. **87**, (1987) 3888.
17. Parker, G.A., Pack, R.T, Laganà, A., Archer, B.J., Kress J.D., and Bačić, Z., Supercomputer Algorithms for Reactivity, Dynamics and Kinetics of Small Molecules, Laganà, A., NATO ASI Series, 105-130.
18. Laganà, A., Pack, R.T, Parker, G.A., J. Chem. Phys. **99**(3) (1993) 2269.

Conformations of 1,2,4,6-Tetrathiepane

Issa Yavari*, Arash Jabbari, and Shahram Moradi

Department of Chemistry, Islamic Azad University, Science & Research Campus, Ponak,
Tehran, Iran and Department of Chemistry, Tarbiat Modarres University,
P. O. Box 14115-175, Tehran, Iran

Abstract. *Ab initio* calculations at HF/6-31+G*//HF/6-31+G* and B3LYP/6-31+G*//HF/6-31+G* level of theory for geometry optimization and MP2/6-31+G*//HF/6-31+G* level for a single point total energy calculation are reported for 1,2,4,6-tetrathiepane. This cyclic polysulfide is predicted to exist as a mixture of two unsymmetrical conformations, which interconvert *via* a low energy barrier of 11.1 kJ mol^{-1}. Conformational racemization of these forms can take place *via* the plane symmetrical boat geometry as a transition state and requires 48.4 kJ mol^{-1}; the C_s symmetric chair transition state is calculated to be slightly higher (51.1 kJ mol^{-1}).

Keywords: Cyclic polysulfide; Stereochemistry; Conformational analysis; *Ab initio* calculations

1,2,4,6-Tetrathiepane (**1**), the odorous principle from edible Shiitake mushroom (*Lenthinus edodes*) has shown promise as potential antibiotic [1]. This polysulfide has also been isolated from *Parkia speciosa* [2] and detected in cooked mutton [3]. Several methods have been reported for synthesis of this tetrathiepane [1, 4-6]. Although conformational studies of cycloheptane and seven-membered ring heterocycles based on dynamic NMR spectroscopy [7-10] can be found in the literature, no experimental or theoretical studies on the conformational properties of **1** are available.

We have recently reported [11] an *ab initio* study of conformational properties of lenthionine (1,2,3,5,6-pentathiepane) (**2**). In this study, we present the results of similar calculations for structural optimization and conformational interconversion pathways of 1,2,4,6-tetrathiepane by comparing the geometries (HF/6-31+G*) and conformational energies (MP2/6-31+G*//HF/6-31+G*). The results from MP2/6-31+G*//HF/6-31+G* calculations are used in the conformational energies discussions below.

---

* Corresponding author. Fax: +98-21-8006544; E-mail: isayavar@yahoo.com

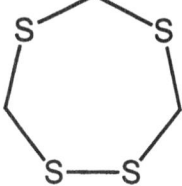
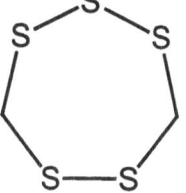

1,2,4,6-Tetrathiepane (**1**)   Lenthionine (**2**)

1 Results and Discussion

The conformational situation in cycloheptane [7, 8] was first unraveled by the classical study of *Hendrickson* [12], which represents the first application of the computer to molecular mechanics calculation. There are two families of conformations in cycloheptane, shown in Scheme 1. One comprises the chair and twist-chair, the other the boat and twist-boat. The situation is thus similar to that in cyclohexane [13, 14], except that the chair is part of a family of flexible conformations and, because of the severe eclipsing at the "flat" end, lies about 9.0 kJ mol^{-1} above the twist-chair into which it can be readily pseudorotated and which represents the most stable conformation of cycloheptane. In the second family, comprising the boat and twist-boat, the twist-boat is more stable (by 2.2 kJ mol^{-1}) than the true boat because of the more severe eclipsing in the latter, similarly as in cyclohexane [13, 14]. The difference between chair/twist-boat and the twist-chair is 10.4 kJ mol^{-1}, that is only about one-half the twist-chair difference in cyclohexane. The chair/twist-chair and boat/twist-boat families in cycloheptane, like their cyclohexane analogous, can be interconverted only by passing over a relatively high barrier of about 35.6 kJ mol^{-1}.

The conformational behavior of 1,2,4,6-tetrathiepane (**1**), with a disulfide (-S-S-) moiety, is expected to be quite different. The disulfide chromophore is inherently chiral [15], that is the chirality is built into the chromophore. The rotational strengths of inherently chiral chromophores tend to be very large. It has been known for a long time, on the basis of X-ray data and of theoretical considerations, that in the absence of geometrical constrains, the disulfide structural element (torsional barrier 21-63 kJ mol^{-1}) preferentially adopts a skewed conformation with $\varphi \approx 90°$. Thus, the energy difference between the chair and the twist-chair (boat and twist-boat) conformations is expected to be higher than that in cycloheptane.

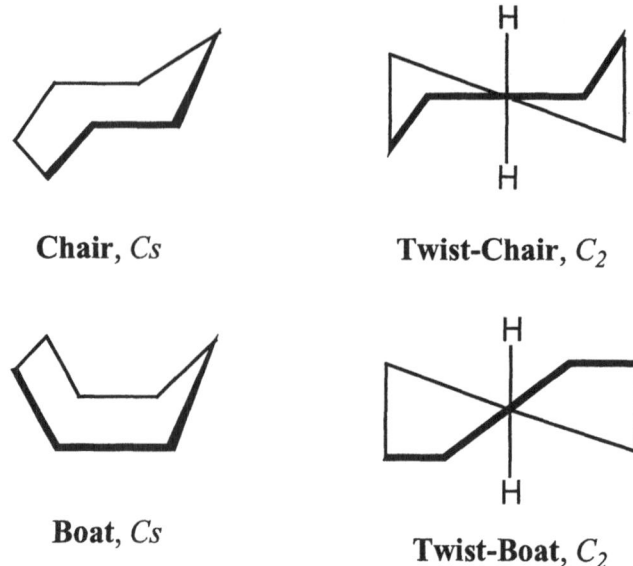

Scheme 1

The results of *ab initio* calculations for structure optimization and conformational interconversion pathways of **1** are shown in Table 1 and Figure 1. Altogether nine geometries were found to be important for description of the conformational features of this polysulfide. Three of these conformations belong to energy minima and six to one-dimensional energy maxima (saddle points or transition states) as shown in Figure 1 and Table 1.

All of the energy-minimum geometries are unsymmetrical. The twist-chair (**TC**) conformation is the global energy minimum for **1**. The twist-chair' (**TC'**) conformation is only 6.8 kJ mol^{-1} above **TC** [16]. These conformations are important, because they are expected to be significantly populated at ambient temperature. Interconversion of **TC** and **TC'** conformations can take place *via* a relatively low transition state, which is only 11.1 kJ mol^{-1} above **TC** form. The twist-boat (**TB**) conformation is calculated to be 14.7 kJ mol^{-1} less stable than **TC**, thus it is not expected to be significantly populated at room temperature, unless the calculations are seriously in error. As shown in Table 1, the **TC** and **TC'** forms have nearly skew S-S torsions and strongly expanded S-C-S bond angles. The simplest conformational process, and the one with the lowest barrier, involves degenerate interconversion of twist-chair with its mirror-image geometry. The transition state for this process has C_2 symmetry, and requires 22.5 kJ mol^{-1}. The plane symmetrical achiral boat geometry, with 48.4 kJ mol^{-1} is the transition state for conformational racemization of **TC**. If both processes just considered are fast, the time-averaged symmetry of the twist-chair conformation becomes C_{2v}, which is the maximum symmetry allowed by the chemical structure of **1**. Conformational racemization of **TC** conformation can also take place by chair process. The calculated strain energy (51.1 kJ mol^{-1}) barrier for this process is slightly higher than that of the chair process (see Figure 1).

Selected data for various geometries of **1** are given in Table 1. A comparison of bond lengths showed fairly small differences, however most of the bond angels are expanded in transition-state geometries. The calculated thermodynamic parameters (H^o, S^o, G^o) [17, 18] for various geometries of **1** at 0 and 298 K are shown in Table 1. The entropy differences for energy-minimum geometries of **1** are considered to be substantial.

In conclusion, *ab initio* calculations provide a picture of the conformations of 1,2,4,6-tetrathiepane from both structural and energetic points of view. The twist-chair (**TC**) and **TC'** conformations are the most stable forms for this polysulfide. According to these calculations, **1** is expected to exist as a mixture of **TC** and **TC'** conformations at ambient temperature. Interconversion or these conformations can take place *via* a fairly low energy barrier of 11.1 kJ mol^1.The calculated energy barrier (48.7 kJ mol^{-1}) for conformational enantiomerization of **TC** conformation *via* the boat process is lower than that (51.1 kJ mol^{-1}) calculated for the chair process.

2 Calculations

The *ab initio* molecular orbital calculations, were carried out using the GAUSSIAN 98 [19] program. Geometries for all structures were fully optimized by means of analytical energy gradients by *Berny* optimizer with no geometrical constraints [20, 21]. The restricted *Hartree-Fock* calculations with the split-valence 6-31+G* basis set which include a set of d-type polarization functions on all non-hydrogen atoms were used in these calculations [22]. Single point energy calculations at MP2/6-31+G*//HF/6-31+G* level were used to evaluate the electron correlation effect in the energies and order of stability of conformers.

Vibrational frequencies were calculated at the 6-31+G* level for all minimum energies and transition states, which were confirmed to have zero and one imaginary frequency, respectively. The frequencies were scaled by a factor of 0.9135 [23] and used to compute the zero-point vibrational energies.

References

[1] K. Morita and S. Kobayashi, *Chem. Pharm. Bull.*, **15**, 988 (1967); *Tetrahedron Lett.*, **7**, 573 (1966).
[2] R. Gemelin, R. Susilo and G. R. Fenwick, *Phytochemistry*, **20**, 2521 (1981).
[3] L. N. Nixon, E. Wong, C. B. Johnson and E. J. Birch, *J. Agric. Food Chem.*, **27**, 355 (1979).
[4] W. J. Still and G. W. Kutney, *Tetrahedron Lett.*, **22**, 1939 (1981).
[5] R. M. Bannister and C. W. Rees, *J. Chem. Soc., Perkin Trans. 1*, 509 (1990).
[6] H. C. Hawsen, A. Senning and R. G. Hazell, *Tetrahedron*, **41**, 5145 (1985).
[7] W. Tochtermann, *Top. Curr. Chem.*, **15**, 378 (1970).
[8] J. Diller and H. J. Geise, *J. Phys. Chem.*, **70**, 425 (1979).

[9] V. Elser and H. Strauss, *Chem. Phys. Lett.*, **96**, 276 (1983).
[10] D. F. Bocian and H. Strauss, *J. Am. Chem. Soc.*, **99**, 2866, 2876 (1977).
[11] Yavari, A. Jabbari and Sh. Moradi, *J. Mol. Struct. (Theochem)*, **631**, 225 (2003).
[12] J. B. Hendrickson, *J. Am. Chem. Soc.*, **83**, 4537 (1961).
[13] F. A. L. Anet and R. Anet, in L. M. Jackman and F. A. Cotton, Eds., *Dynamic Nuclear Magnetic Resonance Spectroscopy*, Academic Press (1975), New York, p. 543.
[14] J. Sandstrom, *Dynamic NMR Spectroscopy*, Academic Press, New York, 1982.
[15] E. L. Eliel and S. H. Wilen, *Stereochemistry of Organic Compounds*, Wiley, New York, 1994, p 1014.
[16] We are grateful to the referee of this paper for suggesting the **TC'** conformation.
[17] K. B. Wiberg, J. D. Hammer, H. Castejon, W. F. Bailey, E. L. Deleon, R. M. Jerret, *J. Org.Chem.*, **64**, 2085 (1999).
[18] K. B. Wiberg, H. Castejon, W. F. Bailey, J. Ochterski, *J. Org. Chem.*, **65**, 1181 (2000).
[19] Gaussian 98, Revision A. 6, M. J. Frisch, G. W. Trucks, H. B. Schlegel, G. E. Scuseria, M. A. Robb, J. R. Cheeseman, V. G. Zakrzewski, J. A. Montgomery, Jr., R. E. Startmann, J. C. Burant, S. Dapprich, J. M. Millam, A. D. Daniels, K. N Kudin, M. C. Strain, O. Farkas, J. Tomasi, V. Barone, M. Cossi, R. Cammi, B.Mennucci, C. Pomelli, C. Adamo, S. Clifford, J. Ochterski, G. A. Petersson, P Y. Ayala, Q. Cui, K. Morokuma, D. K. Malick, A. D. Rubuck, K. Raghavachari, J. B. Foresman, J. Cioslowski, J. V. Oritz, B. B. Stefanov, G. Liu, A. Liashenko, P. Piskorz, I. Komaromi, R. Comperts, R. L. Martin, D. J. Fox, T. Keith, M. A. Al- Laham, C. Y. Peng, A. Nanayakkara, C. Gonzalez, M. Challacombe, P. M. W. Gill, B. Johnson, W. Chen, M. W. Wong, J. L. Andres, C. Gonzalez, M. Head- Gordon, E. S. Replogle, and J. A. Pople, Gaussian. Inc., Pittsburgh PA, 1998.
[20] F. Jensen, *Introduction to Computational Chemistry*, Wiley, New York, 1999.
[21] W. J. Hehre, L. Radom, P. v. R. Schleyer, and J. A. Pople., *Ab Initio Molecular Orbital Theory*, Wiley, New York, 1986.
[22] P. C. Hariharan and J. A. Pople, *Theor. Chim. Acta*, **28**, 213 (1973).
[23] R. S. Grev, C. L. Janssen, H. F. Schaefer, *Chem. Phys.* **95**, 5128 (1991).

Fine Grain Parallelization of a Discrete Variable Wavepacket Calculation Using ASSIST-CL

Stefano Gregori[1], Sergio Tasso[1], and Antonio Laganà[2]

[1] Dipartimento di Matematica e Informatica, Università di Perugia,
06123 Perugia, Italy
[2] Dipartimento di Chimica, Università di Perugia,
06123 Perugia, Italy

Abstract. The fine grain parallelizzation of a code performing the time propagation of a wavepacket using a Discrete Variable approach is discussed. To this end a Task Farm model has been implemented. Performances obtained using MPI and the recently proposed skeleton based coordination language ASSIST-CL are compared.

1 Introduction

The scattering properties of reactive systems can be investigated using quantum mechanical approaches. To this end two main techniques have been adopted to deal with few atom reactive processes. In this paper we focus our attention onto a quantum wavepacket method integrating the time dependent Schrödinger equation of atom diatom systems[1]. For this purpose we used the numerical procedure TIDEP[2] that has been developed in our laboratory to study the time evolution of a simple atom-diatom quantum system. This technique collocates into a grid of the coordinate space of the reactants a wavepacket having the initial form:

$$\Phi^{J\Lambda}(R,r,\Theta,t=0) = \phi(R)Ne^{-\alpha(R-R_0)^2}\varphi_{vj}^{BC}(r)P_j^{\Lambda}(\Theta) \qquad (1)$$

where $Ne^{-\alpha(R-R_0)^2}$ is a normalized Gaussian function, $\phi(R)$ is the atom-diatom radial component, φ_{vj}^{BC} is the initial vibrational wavefunction of the diatomic molecule, $P_j^{\Lambda}(\Theta)$ is the angular wavefunction related to its rotational state j. The reaction space is given by a grid of appropriate dimensions able to contain the wavepacket not only in its initial form but also in its spread out form obtained after applying the time propagator long enough to almost completely absorb the wavepacket. At each step of the time propagation the wavepacket is analyzed in the asymptotic product channel region by mapping it over the product vibrational states. This allows to calculate the C coefficients

$$C_{vj\Lambda,v'j'\Lambda'}^{J}(t) = \int_{r'} dr' \int_{\Theta'} d\Theta' sin\Theta' P_{j'}^{\Lambda'}(\Theta')\phi_{v'j'}^{AB}(r')\Psi^{J\Lambda'}(R'=R_\infty,r',\Theta',t) \qquad (2)$$

(where primed quantities refer to products) from which it is possible to calculate the elements of the Scattering Matrix $S^J_{vj\Lambda,v'j'\Lambda'}$[2]. From the S matrix elements the reactive probabilities $P^J_{vj,v'j'}(E)$ and cross sections $\sigma_{vj,v'j'}(E)$ can be calculated as a function o fthe total energy E or translational energy E_{tr}.

Approaches to solve large problems usually require a parallelization of the related computational procedures. Hardware evolution to low cost platforms allows the assemblage of machines having hybrid configurations. Parallel languages and paradigms have also evolved to obtain high performances and friendly environments have been proposed to facilitate the structuring of programs aimed at solving complex problems. The most popular parallel programming tool is the MPI library whose routines can be called from Fortran and C/C++ codes. Using MPI the parallel structuring of the program is quite simple. However the use of MPI requires a deep knowledge of the algorithm and a detailed analysis of both the hardware architecture and the communication infrastructure in order to achieve high performances.

On the contrary, ASSIST-CL is an environment allowing the programmer to disregard the technicalities of the system (software and hardware) and offering a programming context in which only the data flow between the computational nodes needs to be worked out. This eventually grants a large extent of code reuse. ASSIST is the acronym of "A Software development system based on Skeleton Integrated Technology" and CL stands for "Coordination Language". The compiler of ASSIST is $astCC$ that is presently in version 1.1. ASSIST was developed as part of the project ASI-PQE2000 and derives from the predecessor skeleton programming language SkIE[3]. Both ASSIST and SkIE make use of an object oriented programming model. In this work we have tested ASSIST to restructure TIDEP.

2 The Av Routine

Our restructuring work has concentrated on the subroutine Av of TIDEP in which the time propagation step is iterated. In the Discrete Variable Representation (DVR) method this corresponds to iterating the following matrix operations

$$\boldsymbol{HPS} = \boldsymbol{HBR} \cdot \boldsymbol{C} + \boldsymbol{C} \cdot \boldsymbol{HSR}^T + \boldsymbol{V} \odot \boldsymbol{C} \tag{3}$$

where $\boldsymbol{HPS}$ is the matrix representation of the wavefunction propagated at time $t + \tau$, $\boldsymbol{HBR}$ and $\boldsymbol{HSR}$ are the R and r components of the DVR matrix representation of the kinetic part of the Hamiltonian operator $\hat{H}$, $\boldsymbol{C}$ is the current real part of the wavefunction representing the system at time t and $\boldsymbol{V}$ is the (diagonal) matrix representation of the potential. We refer here to a previously parallelized version of Av[4]. The used Task Farm model implements a domain decomposition of the matrices $\boldsymbol{HSR}, \boldsymbol{HBR}$ and $\boldsymbol{C}$ among the various nodes and the multiplication is carried out using an algorithm that allows a sequential access to the rows of the matrices so that the resulting $\boldsymbol{HPS}$ matrix is built row by row. The calculation is articulated as follows. First of all the master spreads the matrices $\boldsymbol{C}, \boldsymbol{HBR}, \boldsymbol{HSR}$ to the workers in blocks of rows ($\boldsymbol{HSR}, \boldsymbol{C}$) and

in blocks of columns (HBR). Second the time propagation of the wavefunction is performed. At each iteration of the propagation the master node reads matrix C row by row and computes the components of the $V \odot C$ direct product of the related line of matrix HPS. Then the master issues (via a broadcast) the row of matrix C that is used by workers to workout the remaining part of the line associated with the two matrix products. The algorithm used by the workers is:
Worker regime routine

```
receive by MPI_BROADCAST from master current Row(C);
initialize buffHPS where store partial calculated Row(HPS);
read from dataset file Row(HBR);
do cols = 1,num_cols_per_worker
   read from dataset file Row(C);
   calculate partial product HBR*C into buffHPS;
enddo
do cols = 1,num_cols_per_worker
   read from dataset file Row(HSR);
   calculate partial product C*HSRt and sum it into buffHPS;
enddo
send workout buffHPS to the master via MPI_REDUCE;
```

3 ASSIST Implementation

ASSIST programs are a direct representation of the computational graph of the algorithm[5] whose nodes are the computations carried out in a sequential or a parallel mode and branches represent the data flow through streams. Parallel computations are defined by particular structures called *parmod* (parallel module). The *parmods* manage the data distribution between input streams, Virtual Processors (VP) and output streams. The VP is a virtual abstraction of a computational node in which only sequential code runs. There is no static relationship between the VPs and the actual processors or nodes in the subsystem. ASSIST generates these relationships at runtime transparently to the programmer[6]. Therefore, the most important part of an ASSIST program is the *generic main* where the graph structure is defined using modules and streams. In Fig. 1 the Av computational graph is shown.

The ASSIST code necessary to generate this structure is[7]:

```
generic main(){
   //streams definition
   stream t_hbr_block[N] sHBR;
   stream t_block[N] sHSR;
   stream t_block[N] sC;
   stream double[nc] sRowC;
   stream double[nc] swRowHPS;
   stream double[nc] smRowHPS;
```

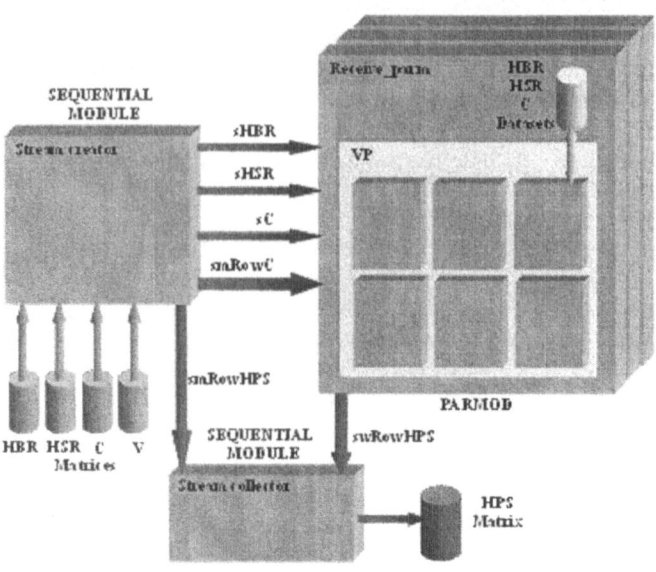

Fig. 1. Schematic representation of the computational graph

```
//modules definition
Stream_creator(output_stream sHSR,sHBR,sC,sRowC,smRowHPS);
Receive_parm(input_stream sHSR,sHBR,sC,sRowC
                                output_stream swRowHPS);
Stream_collector(input_stream swRowHPS, smRowHPS);
}
```

The sequential module Stream_creator plays the role of master in our task farm model. In this approach data input matrices are spread to the workers (all VPs in the *parmod* are workers!) making the domain decomposition and then the computational step begin with a sequential read from a file and a distribution through the stream *sRowC* of the current row of matrix C. At the same time the direct product of V and C related to the current row of the matrix HPS is carried out through the stream *smRowHPS*. The *parmod* Receive_parm defines the parallel computation. In ASSIST the input/output function between the various modules and the VPs is organized by a *parmod*. In particular, the *input section* of a *parmod* defines how to distribute the input data to the VPs. The VPs can be mapped using different *topologies*. In the present case a *topology array* of N elements has been used. A guard mechanism is adopted to handle the determinism in data manipulation. In the present application we have defined some guards that are verified when a data is useful in a particular input stream. As an example, we have defined a guard to activate an elaboration in all *parmods* when the matrix HSR is sent to the parmod over stream $sHSR$. Its structure reads

```
do input_section {
   ...
   guardHSR: on , , sHSR {
      //distribution of matrix HSR to VPs
      distribution sHSR[*s] scatter to Pvs[s];
   }
   ... other guards ...
}
```

In the input section it is also defined how to distribute incoming data to virtual processors using policies as *scatter and broadcast*. We have defined a particular data structure for streams that permits a simply isomorphus scattering of the matrices by blocks of rows (for matrix C and HSR) and by blocks of columns (for matrix HBR). When a guard is verified (there is a set of data over the stream) in the section *virtual processors* the corresponding elaboration is defined so as to invoke some procedures (*proc*) written in C++.
An example of this is:

```
virtual_processors{
   ...
   //elaboration
   HSR2disk(in guardHSR ){
      //VPs those participate to the elaboration
      VP i{
         //invoking the procedure which write to disk
         //matrix dataset for every VP
         proc_w2disk1(in sHSR[i],i );
         }
   }
   ... other elaborations ...
}
```

Similarly, when the module Stream_creator reads a row of the matrix C from the file and sends it over the stream $sRowC$ to all VPs, the VPs start the calculation to produce the related partial row of the output matrix HPS using data stored in local disk space. Then the *parmod* collects and sums all these rows from the VPs and puts the result into the stream $wsRowHPS$ (this is similar to the astion of MPI_REDUCE()). At the end the sequential module Stream_collector receives two lines issued by the master. The *parmod* sums them and then writes the calculated row of matrix HPS on disk.

4 Performance Analysis

When using the MPI library to implement a task farm model we identify the processors by their rank and distinguish between the master and the workers since they run the same program. The ASSIST approach is completely different: we define the parallelism degree of every *parmod* using different topology dimensions and this may not correspond to the actual number of processors. To carry out a meaningful performance test we have compiled the ASSIST code with a special compiler option (astCC -C -c source.ast). This generates some binary files to be run on the individual processor. The ASSIST compiler builds different binary files corresponding to various modules of the global program plus two other binary files (with suffix _ism and _osm) specifically handling the input and the output streams of the parmod. In the tests performed we run all the executables (excluded the binary ones with the parmod code and suffix _vpm) on a processor that acts as a master and one copy of the _vpm binary on all the slave processors. To work out an estimate of the efficiency and the speedup of the restructured code we have compared the execution time of the ASSIST version of *Av* with that of its serial version written in C++ and compiled using g++. In Fig. 2 the speedups pertinent to the complete test made using various values of the matrix dimension are shown. The speed up (defined as $T_{serial}/T_{parallel}$)

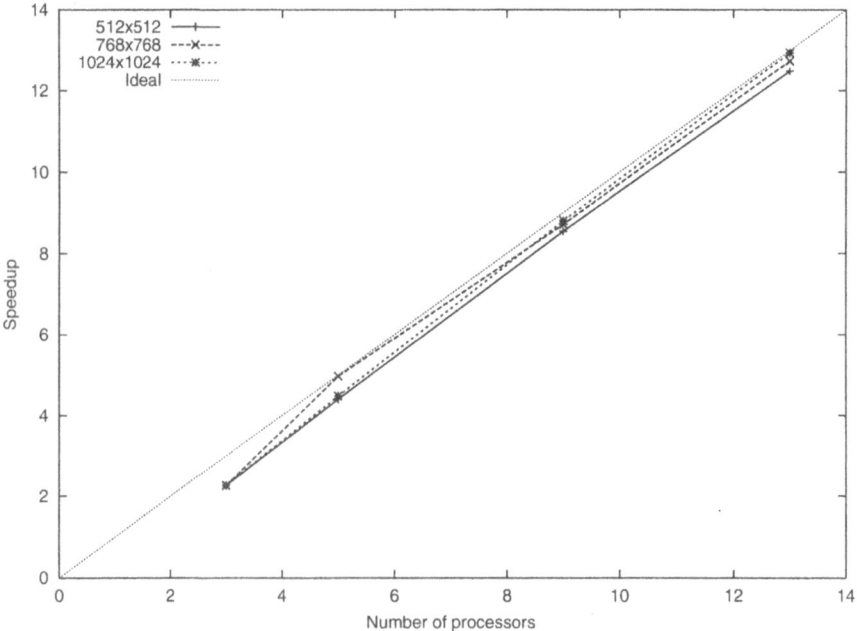

Fig. 2. Speedup calculated as T_s/T_p for different problem dimension.

does not exhibit the penalization shown by the previous MPI implementation as shown in Fig.3. This is likely to be due to a more sophisticated communication handling associated with the fact that various dedicated processes run on the same processor with the only scope of handling the data flow in the ASSIST implementation. This is important since, usually, more than one process run on a processor (system processes, other user programs, ...) and the operating system scheduler plays an important role in determining the performances. A particular

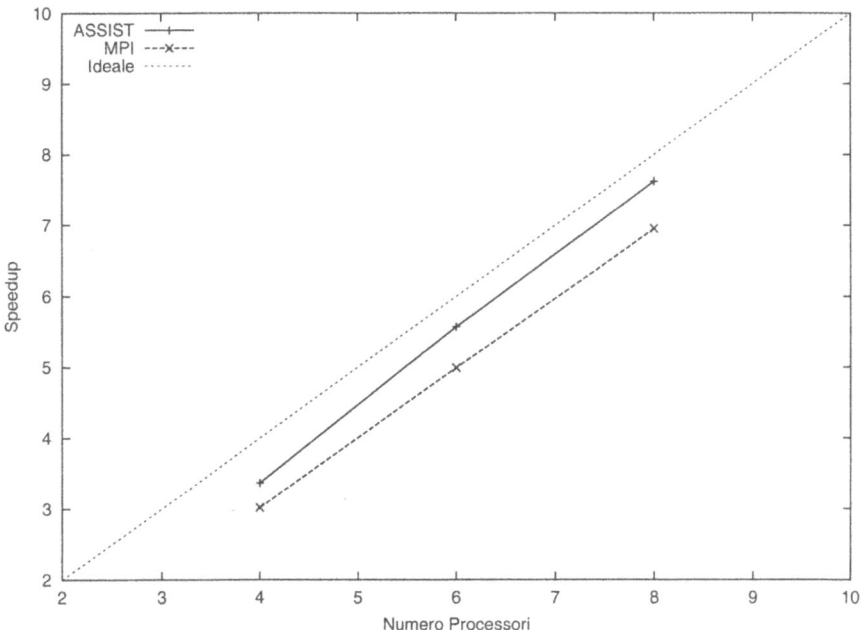

Fig. 3. Compared speedups from previous MPI and actual ASSIST implementation of the Av routine.

point to emphasize here is that the code has been implemented in a way that allows the use of large matrices that cannot be stored in the main memory so as the serial code needs to access data via the secondary memory. The same is true for the workers datasets stored in local disk space on all nodes. In these conditions the greatest limitation in speed is due the frequent accesses to hard disk and the initial spreading of the relevant matrices is not so important. A point that we have not yet investigated is the use of *Shared object* to share large matrices among the modules and the VPs. This is a further advantage for code reuse and may further optimize the performance of the code.

5 Conclusions

The good performances measured when using ASSIST to implement the Av routine of our wavepacket propagation program TIDEP confirms the strength of this coordination language and candidates the TIDEP program for further tests of other useful features of ASSIST. The measured figures indicate that ASSIST is a valid instrument when used also in heterogeneous platforms and paves the way for tests in GRID environments.

Acknowledgments. The authors are deeply grateful to Prof M. Vanneschi, M. Danelutto and Dr. P. Ciullo for providing a copy of ASSIST-CL and assistance for its use. Financial support from ASI, MIUR and COST is acknowledged.

References

1. Balint-Kurti, G.G.: Time dependent quantum approaches to chemical reactions, Lecture Notes in Chemistry **75** (2000) 74–88.
2. Laganà, A.: Innovative computing and detailed properties of elementary reactions using time dependent approaches, Computer Physics Communication, **116** (1999) 1–16.
3. Danelutto, M., Stigliani, M.: SKElib: parallel programming with skeletons in C, Lecture Notes in Computer Science **1900** (2000) 1175–1184.
4. Bellucci,D., Tasso, S., Laganà, A.: Fine Grain Parallelism for Discrete Variable Approaces to Wavepacket Calculation, Lecture Notes in Computer Science **2331** (2002) 918–925.
5. Vanneschi, M.: The programming model of ASSIST, an environment for parallel and distributed portable applications, *Parallel Computing*, **28(12)** 2002 1709–1732.
6. Baraglia, R., Danelutto, M., Laforenza, D., Orlando, S., Palmerini, P., Perego, R., Pesciullesi, P., Vanneschi, M.: AssistConf: A Grid Configuration Tool for The ASSIST Parallel Programming Environment, In *Proceedings of the Eleventh Euromicro conference on Parallel, Distributed and Network-based processing. IEEE, February 2003.*
7. Ciullo, P., Danelutto, M., Magini, S., Potiti, L.: ASSIST-CL User Manual, Technical report, Progetto ASI-PQE2000,July 2002.
8. Bellucci, D., Tasso, S., Laganà, A.: Parallel Models for a Discrete Variable Wavepacket Propagation, Lecture Notes in Computer Science **2658** (2003) 341–349.

On the Solution of Contact Problems with Visco-Plastic Friction in the Bingham Rheology: An Application in Biomechanics

Jiří Nedoma

Institute of Computer Science AS CR, Prague

Abstract. The paper deals with the solvability of model problems connected with a stress-strain rate analysis of loaded long human bones filled up by a marrow tissue. The problem is described by a contact problem with (or without) visco-plastic friction in the visco-plastic Bingham rheology. The variational and numerical solutions are discussed.

Keywords: visco-plasticity, Bingham rheology, contact problems with friction, variational inequalities, FEM, biomechanics

1 Introduction

Recent trends in total joint replacements and clinical failures have provided a strong impetus for the development of accurate stress analyses in normal and prosthetic joints. Theoretical stress analysis methods based on the finite element techniques are frequently used to evaluate the mechanical behaviour of laden bones with (or without) stems of protheses. Nevertheless, stress-strain analyses of laden bones filled up by a marrow tissue and analyses of stress propagation through a marrow tissue were not investigated up to the present.

This paper is addressed to a modelling of laden long bones, like the femur and the tibia, with (or without) the stem of the prothesis and filled up by a marrow tissue. It is evident that theoretical models using contact problems with (or without) friction in the Bingham rheology would be able to apply for structural stress analyses of laden bones in dependence on a transfer of loading also through a marrow tissue.

In the paper the constitutive relations of the Bingham rheology, the friction laws on the contact boundaries, an existence theorem and ideas how to solve the problem numerically, are presented.

2 Formulation of the Problem

Let $\mathbf{u} = (u_i)$ be the material velocity, let $D = (D_{ij})$, $D_{ij} = D_{ij}(\mathbf{u}) = \frac{1}{2}\left(\frac{\partial u_i}{\partial x_j} + \frac{\partial u_j}{\partial x_i}\right)$, denote the strain rate tensor and $D^D = (D_{ij}^D)$ the strain rate deviator defined by $D_{ij}^D = D_{ij} - \frac{1}{3}D_{kk}\delta_{ij}$, where δ_{ij} is the Kronecker symbol.

Denote by $\tau = (\tau_{ij})$ the Cauchy stress tensor and its deviator $\tau^D = (\tau_{ij}^D)$ by $\tau_{ij}^D = \tau_{ij} - \frac{1}{3}\tau_{kk}\delta_{ij}$, i.e. $\tau^D = \tau + pI_3$, where $-p$ denotes the spherical part of the stress tensor and has a meaning of the pressure; I_3 is the identity tensor. Besides the deviators D_{ij}^D and τ_{ij}^D another deviator $S^D = (S_{ij}^D)$, corresponding to plastic properties of biomaterials (see [3]), can be introduced. In the Bingham rheology for any process we find

$$\tau^D = S^D + 2\hat{\mu}D^D, \quad D^D = 2\lambda S^D, \tag{1}$$
$$f(S^D) = S_{II} - \hat{g}^2 \leq 0, \tag{2}$$

the so-called Mises relation or the yield condition, where $S_{II} = \frac{1}{2}|S^D|^2 = \frac{1}{2}S_{ij}^D S_{ij}^D$, $\hat{\mu} > 0$ is the threshold of viscosity, $\hat{g}$ the threshold of plasticity or the yield limit, $\hat{g}/\sqrt{2}$ is the yield stress in pure shear and λ is a scalar function defined as: (i) if $f(S^D) < 0$ or $(f(S^D) = 0$ and $\frac{\partial f(S^D(t))}{\partial t} < 0)$ then $\lambda(t) = 0$; (ii) if $f(S^D) = 0$ and $\frac{\partial f(S^D(t))}{\partial t} = 0$ then $\lambda(t) > 0$. The Mises relation yields that the invariant S_{II} cannot exceed the square of the yield limit $\hat{g}$.

The constitutive law in the Bingham rheology can be written as

$$\begin{aligned} D_{ij} &= (2\hat{\mu})^{-1}(1 - \hat{g}\tau_{II}^{-\frac{1}{2}})\tau_{ij}^D & \text{if } \tau_{II}^{\frac{1}{2}} > \hat{g}, \\ D_{ij} &= 0 & \text{if } \tau_{II}^{\frac{1}{2}} \leq \hat{g}, \end{aligned} \tag{3}$$

where $\tau_{II} = \frac{1}{2}\tau_{ij}^D\tau_{ij}^D = \frac{1}{2}|\tau^D|^2$ is an invariant of the stress tensor.

To invert relations (3) we have: let $|D| = 0$, then from (3) we find $|\tau^D| \leq \hat{g}$; let $|D| \neq 0$, then from (3) we obtain $|\tau^D| > \hat{g}$ and $|\tau^D| = 2\hat{\mu}|D| + \hat{g}$. Hence, the incompressibility condition and $\tau^D = \tau + pI_3$, then we find

$$\tau_{ij} = -p\delta_{ij} + \hat{g}D_{ij}D_{II}^{-1/2} + 2\hat{\mu}D_{ij} = -p\delta_{ij} + 2^{\frac{1}{2}}\hat{g}D_{ij}|D|^{-1} + 2\hat{\mu}D_{ij}, \tag{4}$$

where $D_{II} = \frac{1}{2}D_{ij}D_{ij}$, representing the constitutive stress-strain rate relation in the Bingham rheology. For $\hat{g} = 0$, we have a classical viscous incompressible (Newtonian) fluid; for small $\hat{g} > 0$, we have strongly visco-plastic materials close to a classical viscous fluid (e.g. haematoceles) and for $\hat{g} \to \infty$, we have absolutely rigid biomaterials, so that we can simulate all types of biomaterials.

Let $\Omega \subset \mathbb{R}^N$, $N = 2, 3$, $\Omega = \Omega_1 \cup \Omega_2$ be a union of bounded domains occupied by visco-plastic bodies with a smooth boundary $\partial\Omega = \Gamma_u \cup \Gamma_\tau \cup \Gamma_c$, where Γ_u represents one part of the boundary, where the velocity is prescribed; Γ_τ is the part of the boundary, where the loading is prescribed and $\Gamma_c \equiv \Gamma_c^{12}$, represents the contact boundary between the bone and the marrow tissue. As a spatial variable system the Eulerian coordinate system is assumed. Let $t \in I \equiv (0, t_p)$, $t_p > 0$, t_p be the duration of a loading, n the exterior unit normal of $\partial\Omega$, $u_n = \mathbf{u} \cdot \mathbf{n}$, $\mathbf{u}_t = \mathbf{u} - u_n\mathbf{n}$, $\boldsymbol{\tau} = (\tau_{ij}n_j)$, $\tau_n = \tau_{ij}n_jn_i$, $\boldsymbol{\tau}_t = \boldsymbol{\tau} - \tau_n\mathbf{n}$ be the normal and tangential components of the velocity and stress vectors. Any repeated index implies summation from 1 to N.

Next, we will solve the following problem:

Problem ($\mathcal{P}$): Find a function $\mathbf{u} : \Omega \times I \to \mathbb{R}^N \times I$, $N = 2, 3$, and a stress tensor $\tau_{ij} : \Omega \times I \to \mathbb{R}^{N \times N} \times I$ satisfying

$$\rho\left(\frac{\partial u_i}{\partial t} + u_k \frac{\partial u_i}{\partial x_k}\right) = \frac{\partial}{\partial x_j}\tau_{ij} + f_i \quad \text{in } \Omega \times I; \tag{5}$$

$$\text{div } \mathbf{u} = 0 \quad \text{in } \Omega \times I; \tag{6}$$

$$\tau_{ij} = -p\delta_{ij} + \hat{g} D_{ij} D_{II}^{-1/2} + 2\hat{\mu} D_{ij}; \tag{7}$$

provided $D_{II} \neq 0, |\tau^D| \leq 2^{\frac{1}{2}}\hat{g}$ if $D_{II} = 0$,

with the boundary value conditions

$$\tau_{ij} n_j = P_i \quad \text{on } \Gamma_\tau \times I, \tag{8}$$

$$\mathbf{u}(\mathbf{x}, t) = \mathbf{u}_1(\mathbf{x}, t) \quad \text{on } \Gamma_u \times I, \tag{9}$$

and with the bilateral contact condition with the local friction law on Γ_c^{12}

$$u_n^1 - u_n^2 = 0 \quad \text{and} \quad |\tau_t^{12}| \leq \mathcal{F}_c^{12}|S^{D12}|,$$
if $|\tau_t^{12}| < \mathcal{F}_c^{12}|S^{D12}|$ then $\mathbf{u}_t^1 - \mathbf{u}_t^2 = 0$,
if $|\tau_t^{12}| = \mathcal{F}_c^{12}|S^{D12}|$ then there exists $\lambda \geq 0$ such that $\mathbf{u}_t^1 - \mathbf{u}_t^2 = -\lambda \tau_t^{12}$,
$$\tag{10}$$

where $\mathcal{F}_c^{12}$ is a coefficient of friction, $u_n^1 = u_i^1 n_i^1$, $u_n^2 = u_i^2 n_i^2 = -u_i^2 n_i^1$, $\tau_n^{12} = \tau_n^1 = -\tau_n^2$, $\tau_t^{12} = \tau_t^1 = \tau_t^2$, and the initial condition

$$\mathbf{u}(\mathbf{x}, t_0) = 0. \tag{11}$$

Setting $S^D = |\tau^D|$ and if we determine $|\tau^D|$ from (4) we obtain $|\tau^D| = 2^{\frac{1}{2}}\hat{g} + 2\hat{\mu}|D(\mathbf{u})|$. Then the contact condition (10) depends on the solution of the investigated problem.

Remark 1. If $S^{D12} = |\tau_n^{12}|$ then (10) represents the classical Coulombian friction law. In some cases the bilateral contact condition $u_n = 0$, $\tau_t = 0$ can be also used.

3 Variational (Weak) Solution of the Problem

We introduce the Sobolev spaces as the spaces of all vector-functions having generalized derivatives of the (possibly fractional) order s of the type $[H^s(\Omega)]^k \equiv H^{s,k}(\Omega)$, where $H^s(\Omega) \equiv W_2^s(\Omega)$. The norm will be denoted by $\|\cdot\|_{s,k}$ and the scalar product by $(\cdot,\cdot)_s$ (for each integer k). We set $H^{0,k}(\Omega) \equiv L^{2,k}(\Omega)$. We introduce the space $C_0^\infty(\Omega)$ as a space of all functions in $C^\infty(\Omega)$ with a compact support in Ω. The space is equipped with the ordinary countable system of seminorms and as usual $C_0^\infty(\Omega, \mathbb{R}^N) = [C_0^\infty(\Omega)]^N$, $N = 2, 3$. We introduce the space $L^p(\Omega)$, $1 \leq p \leq \infty$, as the space of all measurable functions such that $\|f\|_{L^p(\Omega)} = \left(\int_\Omega |f(\mathbf{x})|^p d\mathbf{x}\right)^{1/p} < +\infty$. By $L^\infty(\Omega)$ we denote the set of all measurable functions almost everywhere on Ω such that $\|f\|_\infty = \operatorname{ess\,sup}_\Omega |f(\mathbf{x})|$ is finite. By $L^p(I;X)$, where X is a functional space, $1 \le p < +\infty$, we denote the space of functions $f : I \to X$ such that $\|f(\cdot)\|_X \in L^p(I)$. Moreover, we define for $s \ge 1$ the following spaces and sets:

$$^1H^{s,N}(\Omega) = \{\mathbf{v}|\mathbf{v} \in H^{s,N}(\Omega), \operatorname{div} \mathbf{v} = 0 \text{ in } \cup_{\iota=1}^2 \Omega^\iota,$$
$$v_n^1 - v_n^2 = 0 \text{ on } \Gamma_c^{12}\},\ V_s = \{\mathbf{v}|\mathbf{v} \in {}^1H^{s,N}(\Omega), \mathbf{v} = 0 \text{ on } \Gamma_u\},\ V_1 = V,\ V_0 = H.$$

Then the space V_s is a Hilbert space with the norm $\|\cdot\|_{s,N}$, $H^{1,N}(\Omega)$ with the norm $\|\cdot\|_{1,N} \equiv \|\cdot\|_1$ and $H^{0,N}(\Omega)$ with $\|\cdot\|_{0,N} \equiv \|\cdot\|_0$. Let us put $^1\mathcal{H}(\Omega) = {}^1H^{1,N}(\Omega) \cap C_0^\infty(\Omega, \mathbb{R}^N)$. We will denote the dual space of V_s by $(V_s)'$. Furthermore, we put $\mathbf{v}' = \frac{\partial \mathbf{v}}{\partial t}$,

$$^1H = \{\mathbf{v}|\mathbf{v} \in L^2(I;V_s), \mathbf{v}' \in L^2(I;H), \mathbf{v}(\mathbf{x}, t_0) = 0\}.$$

We will denote by $\mathbf{w}_j$ eigenfunctions of a canonical isomorphism $^1\Lambda_s : V_s \to (V_s)'$, i.e. $(\mathbf{w}_j, \mathbf{v})_s = \lambda_j (\mathbf{w}_j, \mathbf{v})_0$ $\forall \mathbf{v} \in V_s$, $\|\mathbf{w}_j\|_0 = 1$ (no summation over j).

Let $\mathbf{P}(\mathbf{x},t) \in L^2(I; L^{2,N}(\Gamma_\tau))$, $\mathbf{f}(\mathbf{x},t) \in L^2(I; V')$, ρ $\hat{g}$, $\hat{\mu}$ be piecewise constant and positive and let $\mathcal{F}_c^{12} \in L^\infty(\Gamma_c^{12})$, $\mathcal{F}_c^{12} \ge 0$ a.e. on Γ_c^{12} and $\mathbf{u}_1 \equiv 0$.

For $\mathbf{u}, \mathbf{v} \in H^{1,N}(\Omega)$ we put

$$a(\mathbf{u}, \mathbf{v}) = \sum_{\iota=1}^2 a^\iota(\mathbf{u}^\iota, \mathbf{v}^\iota) = 2\int_\Omega \hat{\mu} D_{ij}(\mathbf{u}) D_{ij}(\mathbf{v}) d\mathbf{x},$$

$$(\mathbf{u}', \mathbf{v}) = \sum_{\iota=1}^2 (\mathbf{u}^{\iota\prime}, \mathbf{v}^\iota) = \int_\Omega \rho \mathbf{u}' \mathbf{v} d\mathbf{x},$$

$$S(\mathbf{v}) = \sum_{\iota=1}^2 S^\iota(\mathbf{v}^\iota) = \int_\Omega f_i v_i d\mathbf{x} + \int_{\Gamma_\tau} P_i v_i ds \equiv (\mathbf{F}, \mathbf{v}),$$

$$b(\mathbf{u}, \mathbf{v}, \mathbf{w}) = \sum_{\iota=1}^2 b^\iota(\mathbf{u}^\iota, \mathbf{v}^\iota, \mathbf{w}^\iota) = \int_\Omega \rho u_i \frac{\partial v_j}{\partial x_i} w_j d\mathbf{x},$$

$$j(\mathbf{v}) = \sum_{\iota=1}^2 j^\iota(\mathbf{v}^\iota) = 2\int_\Omega \hat{g} (D_{II}(\mathbf{v}))^{\frac{1}{2}} d\mathbf{x},$$

$$j_g(\mathbf{v}) = \sum_{\iota=1}^2 j_g^\iota(\mathbf{v}^\iota) = \int_{\Gamma_c^{12}} \mathcal{F}_c^{12} |S^{D12}| |v_t^1 - v_t^2| ds,$$

Let us multiply (5) with (7) by $\mathbf{v} - \mathbf{u}(t)$, integrate over Ω and apply the Green theorem satisfying the boundary conditions. Then after some modification, including among other the integration in time over the interval I, we obtain the following variational (weak) formulation:

Problem $(\mathcal{P})_v$: Find a function $\mathbf{u}$ such that $\mathbf{u} \in {}^1H$ and

$$\int_I [(\mathbf{u}'(t), \mathbf{v} - \mathbf{u}(t)) + a(\mathbf{u}(t), \mathbf{v} - \mathbf{u}(t)) + b(\mathbf{u}(t), \mathbf{u}(t), \mathbf{v} - \mathbf{u}(t)) +$$
$$j(\mathbf{v}(t)) - j(\mathbf{u}(t)) + j_g(\mathbf{v}(t)) - j_g(\mathbf{u}(t))] dt \geq \int_I S(\mathbf{v} - \mathbf{u}(t)) dt \quad \forall \mathbf{v} \in {}^1H \tag{12}$$

holds for a.a. $t \in I$.

The bilinear form $a(\mathbf{u}, \mathbf{v})$ is symmetric, i.e. $a(\mathbf{u}, \mathbf{v}) = a(\mathbf{v}, \mathbf{u})$. Moreover, for $\mathbf{u} \in V$ there exists constant $c_B > 0$ such that $a(\mathbf{u}, \mathbf{u}) \geq c_B \|\mathbf{u}\|_{1,N}^2$ for all $\mathbf{u} \in V$. Furthermore, we have (see [9])

$$|b(\mathbf{u}, \mathbf{v}, \mathbf{w})| \leq c_4 \|\mathbf{u}\|_{1,N}^{1/2} \|\mathbf{u}\|_{0,N}^{1/2} \|\mathbf{w}\|_{1,N}^{1/2} \|\mathbf{w}\|_{0,N}^{1/2} \|\mathbf{v}\|_{s,N},$$

$$\mathbf{u} \in {}^1H^{1,N}(\Omega), \mathbf{w} \in H^{s,N}(\Omega), s = \frac{1}{2}N,$$

$b(\mathbf{u}, \mathbf{u}, \mathbf{u}) = 0$ and since $\mathcal{F}_c^{12} \in L^\infty(\Gamma_c^{12})$, $\mathcal{F}_c^{12} \geq 0$, then $j_g(\mathbf{u}(t)) \geq 0$

and, moreover, if $\mathbf{u}, \mathbf{v}, \mathbf{w} \in {}^1\mathcal{H}(\Omega)$ then $b(\mathbf{u}, \mathbf{v}, \mathbf{w}) + b(\mathbf{u}, \mathbf{w}, \mathbf{v}) = 0$, which is valid also for $\mathbf{u} \in L^{2,N}(\Omega)$, $\mathbf{v}, \mathbf{w} \in {}^1H^{1,N}(\Omega)$; $b(\mathbf{u}, \mathbf{u}, \mathbf{v}) = -b(\mathbf{u}, \mathbf{v}, \mathbf{u})$ for $\mathbf{u}, \mathbf{v} \in {}^1\mathcal{H}(\Omega)$.

The main result is represented by the following theorem:

Theorem 1. *Let $N \geq 2$, $s = \frac{N}{2}$. Let $\mathbf{f} \in L^2(I; V')$, $\mathbf{P} \in L^2(I; L^{2,N}(\Gamma_\tau))$, $\hat{g}, \hat{\mu}$ are piecewise constant, $\mathcal{F}_c^{12} \in L^\infty(\Gamma_c^{12})$, $\mathcal{F}_c^{12} \geq 0$ a.e. on Γ_c^{12}. Then there exists a function $\mathbf{u}$ such that*

$$\mathbf{u} \in L^2(I; V) \cap L^\infty(I; H), \quad \mathbf{u}' \in L^2(I; (V_s)'),$$
$$\mathbf{u}(\mathbf{x}, t_0) = 0$$

and satisfying the variational inequality (12).

Proof. To prove the theorem the triple regularizations will be used. Let us introduce the regularization of $j(\mathbf{v}(t))$ by

$$j_\varepsilon(\mathbf{v}(t)) = \frac{2}{1+\varepsilon} \int_\Omega \hat{g}(D_{II}(\mathbf{v}(t)))^{(1+\varepsilon)/2} d\mathbf{x}, \quad \varepsilon > 0 \text{ and } (j'_\varepsilon(\mathbf{v}), \mathbf{v}) \geq 0. \tag{13}$$

Since the functional $j_g(\mathbf{v})$ is not differentiable in the Gâteaux sense, therefore it can be regularized by its regularization $j_{g\varepsilon}(\mathbf{v})$. Let us introduce the function $\psi_\varepsilon : \mathbb{R} \to \mathbb{R}$ defined by $\psi_\varepsilon(y) = \sqrt{(y^2 + \varepsilon^2)} - \varepsilon$, regularizing the function $y \to |y|$. The function ψ_ε is differentiable and the following inequality

$$| |y| - \psi_\varepsilon(|y|)| < \varepsilon \quad \forall y \in \mathbb{R}, \varepsilon \geq 0$$

holds. Then the functional $j_g(\mathbf{v})$ will be regularized by its regularization $j_{g\varepsilon}(\mathbf{v})$, defined by

$$j_{g\varepsilon}(\mathbf{v}(t)) = \int_{\Gamma_c^{12}} \mathcal{F}_c^{12} S^{D12} \psi_\varepsilon(|\mathbf{v}^1 - \mathbf{v}^2|) ds \quad \text{and} \quad (j'_{g\varepsilon}(\mathbf{v}), \mathbf{v}) \geq 0, \qquad (14)$$

The third regularization will be introduced by adding the viscous terms $\eta((\mathbf{u}_{\varepsilon\eta}(t), \mathbf{v}))_s$ where $s = \frac{N}{2}$ and η is a positive number. For $N = 2$ we obtain $s = 1$ and $^1H^{s,N}(\Omega) = {}^1H^{1,N}(\Omega)$, $V_s = V$ and the added viscous term can be omitted. We remark that this term has a physical meaning as a viscosity.

Thus we will to solve the triple regularized problem:

Problem$(\mathcal{P}_r)_v$: Find a pair of functions $\mathbf{u}_{\varepsilon\eta} \in {}^1H$ such that

$$\int_I [(\mathbf{u}'_{\varepsilon\eta}(t), \mathbf{v} - \mathbf{u}_{\varepsilon\eta}(t)) + a(\mathbf{u}_{\varepsilon\eta}(t), \mathbf{v} - \mathbf{u}_{\varepsilon\eta}(t)) +$$
$$+ b(\mathbf{u}_{\varepsilon\eta}(t), \mathbf{u}_{\varepsilon\eta}(t), \mathbf{v} - \mathbf{u}_{\varepsilon\eta}(t)) + \eta((\mathbf{u}_{\varepsilon\eta}(t), \mathbf{v} - \mathbf{u}_{\varepsilon\eta}(t)))_s + j_\varepsilon(\mathbf{v}(t)) -$$
$$- j_\varepsilon(\mathbf{u}_{\varepsilon\eta}(t)) + j_{g\varepsilon}(\mathbf{v}(t)) - j_{g\varepsilon}(\mathbf{u}_{\varepsilon\eta}(t))] dt \geq \int_I S(\mathbf{v} - \mathbf{u}_{\varepsilon\eta}(t)) dt \quad \forall \mathbf{v} \in {}^1H. \qquad (15)$$

The method of the proof is similar to that of Theorem 3 in [4] (see also [5], [6], [7]) and it is as follows:

(1) the existence of the solution of (15) will be based on the Galerkin approximation technique;
(2) a priori estimates I and II independent of ε and η will be given;
(3) limitation processes for the Galerkin approximation (i.e. over m) and for $\varepsilon \to 0$, $\eta \to 0$ will be performed;
(4) the uniqueness of the solution of (12) for $N = 2$ can be proved only.

The existence of a function $\mathbf{u}_{\varepsilon\eta}$ will be proved by means of the finite-dimensional approximation. The proof is similar of that of Theorem 3 in [4]. We construct a countable base of the space $V_s(\Omega)$, i.e. each finite subsets are linearly independent and span $\{\mathbf{v}_i | i = 1, 2, \dots\}$ are dense in $V_s(\Omega)$ as $V_s(\Omega)$ is a separable space. Let us construct a space spanned by $\{\mathbf{v}_i | 1 \leq i \leq m\}$. Then the approximate solution $\mathbf{u}_m$ of the order m satisfies

$$(\mathbf{u}'_m(t), \mathbf{v}_j) + a(\mathbf{u}_m(t), \mathbf{v}_j) + b(\mathbf{u}_m(t), \mathbf{u}_m(t), \mathbf{v}_j) + \eta((\mathbf{u}_m(t), \mathbf{v}_j))_s$$
$$+ (j'_\varepsilon(\mathbf{u}_m(t)), \mathbf{v}_j) + (j'_{g\varepsilon}(\mathbf{u}_m(t)), \mathbf{v}_j) = S(\mathbf{v}_j), \quad 1 \leq j \leq m, \qquad (16)$$

$$\mathbf{u}_m(\mathbf{x}, t_0) = 0. \qquad (17)$$

Since $\{\mathbf{v}_j\}_{j=1}^m$ is linearly independent, the system (16), (17) is the regular system of ordinary differential equations of the first order, and therefore (16), (17) uniquely define $\mathbf{u}_m$ on the interval $\overline{I}_m = [t_0, t_m]$.

A priori estimate I:
It holds
$$b(\mathbf{u}_m(t), \mathbf{u}_m(t), \mathbf{u}_m(t)) = 0. \tag{18}$$

Via the integration of (16), with $\mathbf{v}_j(t) = \mathbf{u}_m(t)$ in time over $I_m = (t_0, t_m)$, and since $(j'_\varepsilon(\mathbf{v}), \mathbf{v}) \geq 0$, $(j'_{g\varepsilon}(\mathbf{v}), \mathbf{v}) \geq 0$, using the ellipticity of bilinear form $a(\mathbf{u}, \mathbf{u})$ and due to (18), then after some modifications as well as applying the Gronwall lemma and after some more algebra, we find the following estimates

$$\|\mathbf{u}_m(t)\|_{0,N} \leq 0,\ t \in I, \quad \int_I \|\mathbf{u}_m(\tau)\|_{1,N}^2 d\tau \leq c, \quad \eta \int_I \|\mathbf{u}_m(\tau)\|_{s,N}^2 d\tau \leq c. \tag{19}$$

From these estimates we obtain
$\{\mathbf{u}_m(t), m \in \mathbb{N}\}$ is bounded subset in $L^2(I; {}^1H)$, independent of m, ε, η;
$\{\eta^{1/2}\mathbf{u}_m(t), m \in \mathbb{N}\}$ is bounded subset in $L^2(I; V_s)$, independent of m, ε, η.

To prove a priori estimate II, then similarly as in [1], [4], [7] the system (16), (17) is equivalent to the following system

$$(\mathbf{u}'_m + A_B \mathbf{u}_m + \eta A_s \mathbf{u}_m + j'_\varepsilon(\mathbf{u}_m) + j'_{g\varepsilon}(\mathbf{u}_m) + h_m - \mathbf{F}, \mathbf{v}_j) = 0, \quad 1 \leq j \leq m, \tag{20}$$

with (17), where $a(\mathbf{u}, \mathbf{v}) = (A_B \mathbf{u}, \mathbf{v})$, $A_B \in \mathcal{L}(V_1, V'_1)$, $((\mathbf{u}, \mathbf{v}))_s = (A_s \mathbf{u}, \mathbf{v})$, $A_s \in \mathcal{L}(V_s, V'_s)$, $b(\mathbf{u}_m, \mathbf{u}_m, \mathbf{v}) = (h_m, \mathbf{v})$, $h_m \in {}^1K_m \subset L^2(I; V'_s)$. Then applying the technique of orthogonal projection and using the technique of [1] we find that

$$\mathbf{u}'_m \text{ is a bounded subset of } L^2(I; V'_s), \text{ independent of } m, \varepsilon, \eta. \tag{21}$$

The limitation over m (Galerkin), i.e. the convergence of the finite-dimensional approximation for ε, η being fixed and the limitation over $\varepsilon, \eta \to 0$ finish the existence of the solution $\mathbf{u}(t)$ satisfying (12). Uniqueness of the problem for $N = 2$ can be proved only.

4 Numerical Solution

(A) Dynamic case:
Let Ω_h be a polyhedral approximation to Ω in $\mathbb{R}^3$ and let its boundary be denoted as $\partial \Omega_h = \Gamma_{uh} \cup \Gamma_{\tau h} \cup \Gamma_{ch}$. Let $\mathfrak{T}_h$ be a partition of $\overline{\Omega}_h$ by tetrahedra $\mathcal{T}_h$. Let $h = h(\mathcal{T}_h)$ be the maximum diameter of tetrahedral elements $\mathcal{T}_h$. Let $\mathcal{T}_h \in \mathfrak{T}_h$ be a tetrahedron with vertices P_i, $i = 1, \ldots, 4$ and let R_i be the barycentres with respect to the points P_i, $i = 1, \ldots, 4$. Assume that $\{\mathfrak{T}_h\}$ is a regular family of partitioning $\mathfrak{T}_h$ of Ω_h such that $\overline{\Omega}_h = \cup_{\mathcal{T}_h \in \mathfrak{T}_h} \mathcal{T}_h$.

Now let us introduce the main idea how to solve the problem $(\mathcal{P})$. Let n be an integer and set $k = t_p/n$. Let

$$\mathbf{f}_k^{i+\Theta} = \frac{1}{k} \int_{ik}^{(i+\Theta)k} \mathbf{f}(t)dt, \quad i = 0, \ldots, n-1, \quad 0 \leq \Theta \leq 1, \quad \mathbf{f}_k^{i+\Theta} \in V'.$$

We start with the initial data $\mathbf{u}^0 = 0$. When $\mathbf{u}^0, \ldots, \mathbf{u}^i$ are known, we define $\mathbf{u}^{i+1}$ as an element of V. The existence of a function $\mathbf{u}^{i+1}$ for each fixed k and each $i \geq 0$ can be proved. As regards the spatial discretization, there are several classes of possibilities for approximation V_s by its finite element space V_h ([2], [9], [7], [8]). In the present paper the space V_s is approximated by a space of linear non-conforming functions V_h. Let $\mathcal{V}_h = \{\mathbf{v}_h | v_{hi} \in P_1^*, i = 1, \ldots, N, \forall \mathcal{T}_h \in \mathfrak{T}_h$, continuous in barycentres of tetrahedra B_j, $j = 1, \ldots, m$, and equal to zero in B_j laying on Γ_{uh}, $\sum_{i=1,\ldots,N} \partial v_{hi}/\partial x_i = 0$, $\mathbf{v}_h = 0$, $\mathbf{x} \in \Omega \setminus \overline{\Omega}_h\}$, where P_1^* denotes the space of all non-conforming linear polynomials, $V_h = \{\mathbf{v}_h | \mathbf{v}_h \in \mathcal{V}_h, \mathbf{v}_h = \mathbf{u}_1$ on $\Gamma_{uh}\}$.

Given a triangulation $\mathcal{T}_h$ of Ω_h, we assume $\mathbf{u}_h^0$ to be given in V_h, and, taken so that $\mathbf{u}_h^0$ tends to 0 in V_h when $h \to 0_+$. Moreover, we will assume that $\|\mathbf{u}_h^0\|$ are bounded. We define recursively for each k and h a family of elements $\mathbf{u}_h^0, \ldots, \mathbf{u}_h^i$ of V_h, which can be based on implicit, semi-implicit and/or explicit schemes. In our paper the semi-implicit scheme will be introduced and shortly discussed. The time derivatives will be approximated by the backward differences. Since $\frac{\mathbf{u}_h^{i+1} - \mathbf{u}_h^i}{k} = \frac{\mathbf{u}_h^{i+\Theta} - \mathbf{u}_h^i}{\Theta k}$, $0 \leq \Theta \leq 1$, then

$$\mathbf{u}_h^{i+\Theta} = \Theta \mathbf{u}_h^{i+1} + (1-\Theta)\mathbf{u}_h^i, \quad \mathbf{u}_h^{i+1} = \Theta^{-1}\mathbf{u}_h^{i+\Theta} - \frac{(1-\Theta)}{\Theta}\mathbf{u}_h^i, \quad 0 \leq \Theta \leq 1.$$

Since the contact conditions depend on the solution of the studied problem, we will assume that $j_h(\mathbf{v}_h) = j_h(\mathbf{u}_h^i, \mathbf{v}_h)$.

Scheme $(\mathcal{P}_{si})$: Let $\mathbf{u}_h^0, \ldots, \mathbf{u}_h^i$ be known, then $\mathbf{u}_h^{i+1}$ is a solution in V_h of

$$k^{-1}(\mathbf{u}_h^{i+1} - \mathbf{u}_h^i, \mathbf{v}_h - \mathbf{u}_h^{i+\Theta}) + a_h(\mathbf{u}_h^{i+\Theta}, \mathbf{v}_h - \mathbf{u}_h^{i+\Theta}) +$$
$$+ b_h(\mathbf{u}_h^i, \mathbf{u}_h^i, \mathbf{v}_h - \mathbf{u}_h^{i+\Theta}) + j_h(\mathbf{v}_h) - j_h(\mathbf{u}_h^{i+\Theta}) + j_{gh}(\mathbf{v}_h) - j_{gh}(\mathbf{u}_h^{i+\Theta}) \geq$$
$$\geq (\mathbf{F}_h^{i+\Theta}, \mathbf{v}_h - \mathbf{u}_h^{i+\Theta}) \quad \forall \mathbf{v}_h \in V_h. \tag{22}$$

For $\Theta = 1$ we have the semi-implicit scheme and for $\Theta = \frac{1}{2}$ we have the Crank-Nicholson scheme.

Let $|\cdot|_h$ be the norm in $L^2(\Omega)$ ($[L^2(\Omega)]^N$), $\|\cdot\|_h$ in $\mathcal{V}_h$ and V_h. According to [9] $|\mathbf{u}_h|_h \leq d_1 \|\mathbf{u}_h\|_h$, $\|\mathbf{u}_h\|_h \leq S(h)|\mathbf{u}_h|_h$ $\forall \mathbf{u}_h \in \mathcal{V}_h$, d_0, d_1 are independent of h. Furthermore,

$$|b_h(\mathbf{u}_h, \mathbf{v}_h, \mathbf{w}_h)| \leq d_1 \|\mathbf{u}_h\|_h \|\mathbf{v}_h\|_h \|\mathbf{w}_h\|_h \quad \forall \mathbf{u}_h, \mathbf{v}_h, \mathbf{w}_h \in \mathcal{V}_h,$$

where d_1 does not depend on h,

$$|b_h(\mathbf{u}_h, \mathbf{u}_h, \mathbf{v}_h)| \leq S_1(h)|\mathbf{u}_h|_h \|\mathbf{u}_h\|_h |\mathbf{v}_h|_h \quad \forall \mathbf{u}_h, \mathbf{v}_h \in \mathcal{V}_h,$$

where

$$S_1(h) \leq d_1 S^2(h), \quad b_h(\mathbf{u}_h, \mathbf{u}_h, \mathbf{u}_h) = 0 \quad \forall \mathbf{u}_h \in \mathcal{V}_h.$$

It can be shown that $\mathbf{u}_h^i$ and $\mathbf{u}_h^{i+1}$, defined by the scheme $(\mathcal{P}_{si})$, for $\Theta \geq \frac{1}{2}$ satisfy the following conditions:

Theorem 2. *Let the family of triangulations $\{\mathfrak{T}_h\}$ be uniformly regular, and let the angles in the tetrahedra be less or equal to $\frac{\pi}{2}$. Let k, h satisfy $kS_0(h) \leq d_0$, $kS(h) \leq d_1$, where d_0, d_1 are positive constants independent of k, h. Let $\Theta \geq \frac{1}{2}$. Then $\mathbf{u}_h^i$ are defined by the scheme $(\mathcal{P}_{si})$ and*

$$|\mathbf{u}_h^i|^2 \leq c, \ i=0,\ldots,n, \quad k\sum_{i=0}^{n-1}\|\mathbf{u}_h^{i+\Theta}\|^2 \leq c, \quad k\sum_{i=0}^{n-1}|\mathbf{u}_h^{i+1} - \mathbf{u}_h^i|^2 \leq c_0, \quad (23)$$

hold, where c, c_0 are constants independent of k, h.

The scheme $(\mathcal{P}_{si})$ is stable and convergent. The proofs are similar of that of [2], [7], [8]. The difficulty in practical computations is connected with the approximation of the constraint $\text{div } \mathbf{v}_h = 0$, i.e. the incompressibility condition of Bingham's fluid. For its approximation see e.g. [2], [9]. The studied problem can be also solved by using the penalty and the regularization techniques similarly as in the quasi-stationary case.

(B) Quasi-stationary case:
The analysis of the problem will be based on the penalization, regularization and finite element techniques. The algorithm is modification of that of [3]. We shall assume that $\mathbf{u}_1 \neq 0$.

The problem $(\mathcal{P})_v$ leads to the following problem:

Problem $(\mathcal{P}_{sf})_v$: Find a function $\mathbf{u} \in K = \mathbf{u}_1 + V_1$ satisfying for every $t \in I$

$$a(\mathbf{u}(t), \mathbf{v} - \mathbf{u}(t)) + b(\mathbf{u}(t), \mathbf{u}(t), \mathbf{v} - \mathbf{u}(t)) + j(\mathbf{v}(t)) - j(\mathbf{u}(t)) +$$
$$+ j_g(\mathbf{v}(t)) - j_g(\mathbf{u}(t)) \geq S(\mathbf{v} - \mathbf{u}(t)) \quad \forall \mathbf{v} \in K. \quad (24)$$

In connection with the given data, we suppose that $V_1 \neq \emptyset$ and that the physical data satisfy the similar conditions as above.

Let us introduce the space $\mathcal{W}$, a closed subspace of $H^{1,N}(\Omega)$, by

$$\mathcal{W} = \{\mathbf{v} | \mathbf{v} \in H^{1,N}(\Omega), \mathbf{v}|_{\Gamma_u} = 0, v_n^1 - v_n^2 = 0 \text{ on } \Gamma_c^{12}\}, \quad (25)$$

in which the incompressibility condition $\text{div } \mathbf{u} = 0$ is not introduced.

Since the linear space $V_1 = K - \mathbf{u}_1, \mathbf{u}_1 \in V_1$ on which the variational problem is formulated, contains the condition of incompressibility representing certain cumbersome for numerical solution, therefore we apply a penalty technique, similarly as in the case of incompressible Newtonian fluid (see [9]). The penalty term will be introduced by

$$P(\mathbf{u}_\varepsilon) = \frac{1}{\varepsilon} c(\mathbf{u}_\varepsilon, \mathbf{u}_\varepsilon), \ c > 0$$

where

$$c(\mathbf{u}, \mathbf{v}) = \int_\Omega (\text{div } \mathbf{u})(\text{div } \mathbf{v}) dx \quad \forall \mathbf{u}, \mathbf{v} \in H^{1,N}(\Omega).$$

It can be shown that for each $\varepsilon > 0$ the corresponding penalized variational inequality has a unique solution and that its corresponding solution converges strongly in $H^{1,N}(\Omega)$ to the solution of the initial problem when $\varepsilon \to 0$.

To solve the penalized problem numerically we set $\overline{\mathbf{u}} = \mathbf{u} - \mathbf{u}_1$ and then the finite element technique will be used. Let $\mathcal{W}_h \subset \mathcal{W}$ be a family of finite element subspaces with the property:

$$\forall \mathbf{v} \in \mathcal{W} \quad \text{there exists } \mathbf{v}_h \in \mathcal{W}_h \quad \text{such that } \mathbf{v}_h \to \mathbf{v} \text{ in } H^{1,N}(\Omega) \text{ for } h \to 0. \tag{26}$$

Then we will solve the following problem:

Problem $(\mathcal{P}_{sf})_h$: Find $\overline{\mathbf{u}}_{\varepsilon h} \in \mathcal{W}_h$ satisfying for every $t \in I$ the variational inequality

$$a_h(\overline{\mathbf{u}}_{\varepsilon h}, \mathbf{v}_h - \overline{\mathbf{u}}_{\varepsilon h}) + b_h(\overline{\mathbf{u}}_{\varepsilon h}, \overline{\mathbf{u}}_{\varepsilon h}, \mathbf{v}_h - \overline{\mathbf{u}}_{\varepsilon h}) + j_h(\mathbf{v}_h) - j_h(\overline{\mathbf{u}}_{\varepsilon h}) + j_{gh}(\mathbf{v}_h) -$$
$$- j_{gh}(\overline{\mathbf{u}}_{\varepsilon h}) + \frac{1}{\varepsilon} c_h(\overline{\mathbf{u}}_{\varepsilon h}, \mathbf{v}_h - \overline{\mathbf{u}}_{\varepsilon h}) \geq S_h(\mathbf{v}_h - \overline{\mathbf{u}}_{\varepsilon h}) \quad \forall \mathbf{v}_h \in \mathcal{W}_h. \tag{27}$$

Lemma 1. *Let $\overline{\mathbf{u}}_{\varepsilon h}$ be a solution of (27) for each $h > 0$ and let $\overline{\mathbf{u}}_\varepsilon$ be the solution of the penalized problem for a fixed $\varepsilon > 0$. Then*

$$\overline{\mathbf{u}}_{\varepsilon h} \to \overline{\mathbf{u}}_\varepsilon \quad \text{strongly in } H^{1,N}(\Omega) \text{ when } h \to 0.$$

The proof follows from Lemma A4.2 of [3].

Since the functionals $j(\mathbf{v})$ and $j_g(\mathbf{v})$ are not differentiable in the Gâteaux sense, they can be regularized similarly as above for the case. Then the functionals $j(\mathbf{v})$ and $j_g(\mathbf{v})$ will be regularized by their regularizations $j_\gamma(\mathbf{v})$ and $j_{g\gamma}(\mathbf{v})$, defined by

$$j_\gamma(\mathbf{v}) = \int_\Omega 2^{\frac{1}{2}} \hat{g} \psi_\gamma(|D(\mathbf{v} + \mathbf{u}_1)|) d\mathbf{x},$$

$$j_{g\gamma}(\mathbf{v}) = \int_{\Gamma_c^{12}} \mathcal{F}_c^{12} [2^{\frac{1}{2}} \hat{g} + 2\hat{\mu} \psi_\gamma(D|\mathbf{v} + \mathbf{u}_1|) \psi_\gamma(|\mathbf{v}^1 - \mathbf{v}^2 + (\mathbf{u}_1^1 - \mathbf{u}_1^2)|) ds.$$

Then we will solve the penalized-regularized problem:

find $\overline{\mathbf{u}}_{\varepsilon h \gamma} \in \mathcal{W}_h$ satisfying

$$a_h(\overline{\mathbf{u}}_{\varepsilon \gamma h}, \mathbf{v}_h - \overline{\mathbf{u}}_{\varepsilon \gamma h}) + b_h(\overline{\mathbf{u}}_{\varepsilon \gamma h}, \overline{\mathbf{u}}_{\varepsilon \gamma h}, \mathbf{v}_h - \overline{\mathbf{u}}_{\varepsilon \gamma h}) + j_{\gamma h}(\mathbf{v}) - j_{\gamma h}(\overline{\mathbf{u}}_{\varepsilon \gamma h}) +$$
$$+ j_{g\gamma h}(\mathbf{v}_h) - j_{g\gamma h}(\overline{\mathbf{u}}_{\varepsilon h}) + \frac{1}{\varepsilon} c_h(\overline{\mathbf{u}}_{\varepsilon \gamma h}, \mathbf{v}_h - \overline{\mathbf{u}}_{\varepsilon \gamma h}) \geq S_h(\mathbf{v}_h - \overline{\mathbf{u}}_{\varepsilon \gamma h}) \quad \forall \mathbf{v}_h \in \mathcal{W}_h.$$
$$\tag{28}$$

It can be shown that the functionals $j_{\gamma h}(\mathbf{v})$ and $j_{g\gamma h}(\mathbf{v})$ are convex and continuous, and therefore, the problem (28) has a unique solution $\overline{\mathbf{u}}_{\varepsilon \gamma h} \in \mathcal{W}_h$. As a result we have the following result:

Theorem 3. *Let $u_\varepsilon = \overline{u}_\varepsilon + u_1$, where $\overline{u}_\varepsilon$ is a solution of the penalized problem with homogenous condition on Γ_u, $u_{\varepsilon h} = \overline{u}_{\varepsilon h} + u_1$, $u_{\varepsilon \gamma h} = \overline{u}_{\varepsilon \gamma h} + u_1$ for all $\varepsilon, \gamma, h > 0$. Let u be the solution of the problem $(\mathcal{P}_{sf})_v$. Then*

(i) $u_\varepsilon \to u$ strongly in $H^{1,N}(\Omega)$ when $\varepsilon \to 0$,

(ii) $u_{\varepsilon h} \to u_\varepsilon$ strongly in $H^{1,N}(\Omega)$ when $h \to 0$, (29)

(iii) $u_{\varepsilon \gamma h} \to u_{\varepsilon h}$ strongly in $H^{1,N}(\Omega)$ when $\gamma \to 0$.

The problem leads to solving the non-linear algebraic system, which can be solved by e.g. the Newton iterative method (see [3]).

Remark 2. Analyses of model problems discussed in Remark 1 are modifications of that discussed above.

Acknowledgments. The author would like to thank to Prof. MUDr. Jiří Stehlík, Ph.D. for his information about this unsolved orthopaedic problem.

References

1. Duvaut, G., Lions, J.L.: Inequalities in Mechanics and Physics, Springer, Berlin (1976)
2. Glowinski, R., Lions, J.L., Trémolières, R.: Numerical Analysis of Variational Inequalities. North-Holland, Amsterdam (1976)
3. Ionescu, I.R., Sofonea, M.: Functional and Numerical Methods in Viscoplasticity. Oxford Univ. Press, Oxford (1993)
4. Nedoma, J.: Equations of Magnetodynamics in Incompressible Thermo-Bingham's Fluid under the Gravity Effect. J.Comput. Appl. Math. **59** (1995) 109-128
5. Nedoma, J.: Nonlinear Analysis of the Generalized Thermo-Magneto-Dynamic Problem. J.Comput. Appl. Math. **63** 1-3 (1995) 393-402
6. Nedoma, J.: On a Coupled Stefan-Like Problem in Thermo-Visco-Plastic Rheology. J.Comput. Appl. Math. **84** (1997) 45-80
7. Nedoma, J.: Numerical Modelling in Applied Geodynamics. J.Wiley & Sons, Chichester (1998)
8. Nedoma, J.: Numerical Solution of a Stefan-Like Problem in Bingham Rheology. Math. & Comput. in Simulation **61** (2003) 271-281
9. Temam, R.: Navier-Stokes Equations. Theory and Numerical Methods. North-Holland, Amsterdam (1979)

On the Stress-Strain Analysis of the Knee Replacement

J. Daněk[1], F. Denk[2], I. Hlaváček[3], J. Nedoma[4], J. Stehlík[5], and P. Vavřík[6]

[1] Centre of Applied Mathematics, University of West Bohemia,
Univerzitní 8, 306 14 Pilsen, Czech Republic
[2] Walter a.s., Jinonická 329, 158 01 Prague 5 - Jinonice, Czech Republic
[3] Mathematical Institute, Academy of Sciences of the Czech Republic,
Žitná 25, 115 67 Prague 1, Czech Republic
[4] Institute of Computer Science, Academy of Sciences of the Czech Republic,
Pod Vodárenskou věží 2, 182 07 Prague 8, Czech Republic
[5] Dept. of Orthopaedics, The Hospital České Budějovice,
B. Němcová Str. 585/54, 370 01 České Budějovice, Czech Republic
[6] 1st Orthopaedic Department, 1st Medical Faculty of the Charles University Prague,
V úvalu 84, 150 00 Prague 5 - Motol, Czech Republic

Abstract. The paper deals with the stress/strain analysis of an artificial knee joint. Three cases, where femoral part of the knee joint part is cut across under 3, 5 and 7 degrees, are analysed. Finite element method and the nonoverlapping decomposition technique for the contact problem in elasticity are applied. Numerical experiments are presented and discussed.

1 Introduction

The success of artificial replacements of human joints depends on many factors. The mechanical factor is an important one. The idea of a prothesis being a device that transfers the joint loads to the bone allows one to explain the mechanical factor in terms of the load transfer mechanism. A complex relation exists between this mechanism and the magnitude and direction of the loads, the geometry of the bone/joint prothesis configuration, the elastic properties of the materials and the physical connections at the material connections. Authors in [7,12,9,10] showed that the contact problems in a suitable rheology, and their finite element approximations [2,5] are a very useful tools for analyzing these relations for several types of great human joints and their artificial replacements. The aim of the paper is to analyze the total knee replacement in dependence on the femoral cut and the orientation of the anatomic joint line.

2 The Model

The model of the knee is based on the contact problem in elasticity, and the finite element approximation. The problem leads to solving variational inequalities, which describe physically the principle of virtual work in its inequality form.

In the present paper we deal with mathematical simulations of total knee joint replacements and simulations of mechanical processes taking place during static loadening. The model problem investigated was formulated as the primary semi-coercive contact problem with the given friction and for the numerical solution of the studied problem the nonoverlapping domain decomposition method is used ([1,2,3]).

Let the investigated part of the knee joint occupy a union Ω of bounded domains Ω^ι, $\iota = f, t$ in $\mathbb{R}^N$ ($N = 2$), denoting separate components of the knee joint - the femur (f) and the tibia together with the fibula (t), with Lipschitz boundaries $\partial \Omega^\iota$. Let the boundary $\partial \Omega = \partial \Omega^f \cup \partial \Omega^t$ consist of three disjoint parts such that $\partial \Omega = \Gamma_\tau \cup \Gamma_u \cup \Gamma_c$. Let $\Gamma_\tau = {}^1\Gamma_\tau \cup {}^2\Gamma_\tau$, where by ${}^1\Gamma_\tau$ we denote the loaded part of the femur and by ${}^2\Gamma_\tau$ the unloaded part of the boundary $\partial \Omega$. By Γ_u we denote the part of the tibia boundary, where we simulate its fixation. The common contact boundary between both joint components Ω^f and Ω^t before deformation we denote by $\Gamma_c = \partial \Omega^f \cap \partial \Omega^t$.

Let body forces $\mathbf{F}$, surface tractions $\mathbf{P}$ and slip limits g^{ft} be given.
We have the following problem: find the displacements $\mathbf{u}^\iota$ in all Ω^ι such that

$$\frac{\partial}{\partial x_j} \tau_{ij}(\mathbf{u}^\iota) + F_i^\iota = 0 \quad \text{in } \Omega^\iota, \ \iota = f, t, \ i = 1, ..., N, \tag{1}$$

where the stress tensor τ_{ij} is defined by

$$\tau_{ij}(\mathbf{u}^\iota) = c_{ijkl}^\iota e_{kl}(\mathbf{u}^\iota) \quad \text{in } \Omega^\iota, \ \iota = f, t, \ i = 1, ..., N, \tag{2}$$

with boundary conditions

$$\tau_{ij}(\mathbf{u}) n_j = P_i \quad \text{on } {}^1\Gamma_\tau, \ i = 1, ..., N, \tag{3}$$

$$\tau_{ij}(\mathbf{u}) n_j = 0 \quad \text{on } {}^2\Gamma_\tau, \ i = 1, ..., N, \tag{4}$$

$$\mathbf{u} = \mathbf{u}_0 \, (= 0) \quad \text{on } \Gamma_u, \tag{5}$$

$$u_n^f - u_n^t \leq 0, \ \tau_n^f \leq 0, \ (u_n^f - u_n^t)\tau_n^f = 0 \quad \text{on } \Gamma_c, \tag{6}$$

$$\begin{aligned} &|\tau_t^{ft}| \leq g^{ft} \quad \text{on } \Gamma_c, \\ &|\tau_t^{ft}| < g^{ft} \implies u_t^f - u_t^t = 0, \\ &|\tau_t^{ft}| = g^{ft} \implies \text{there exists } \vartheta \geq 0 \text{ such that } u_t^f - u_t^t = -\vartheta \tau_t^{ft}. \end{aligned} \tag{7}$$

Here $e_{ij}(\mathbf{u}) = \frac{1}{2}(\frac{\partial u_i}{\partial x_j} + \frac{\partial u_j}{\partial x_i})$ is the small strain tensor, normal and tangential components of displacement vector $\mathbf{u}$ ($\mathbf{u} = (u_i)$, $i = 1, 2$) and stress vector τ ($\tau = (\tau_i)$) $u_n^f = u_i^f n_i^f$, $u_n^t = u_i^t n_i^f$ (no sum over t or f), $\mathbf{u}_t^f = (u_{ti}^f)$, $u_{ti}^f = u_i^f - u_n^f n_i^f$, $\mathbf{u}_t^t = (u_{ti}^t)$, $u_{ti}^t = u_i^t + u_n^t n_i^t$, $i = 1, ..., N$, $\tau_n^f = \tau_{ij}^f n_i^f n_j^f$, $\tau_t^f = (\tau_{ti}^f)$, $\tau_{ti}^f = \tau_{ij}^f n_j^f - \tau_n^f n_i^f$, $\tau_n^t = \tau_{ij}^t n_i^t n_j^t$, $\tau_t^t = (\tau_{ti}^t)$, $\tau_{ti}^t = \tau_{ij}^t n_j^t - \tau_n^t n_i^t$, $\tau_t^{ft} \equiv \tau_t^f$.

Assume that c^ι_{ijkl} are positive definite symmetric matrices such that
$0 < c^\iota_0 \le c^\iota_{ijkl}\xi_{ij}\xi_{kl} \mid \xi \mid^{-2} \le c^\iota_1 < +\infty$ for a.a. $\mathbf{x} \in \Omega^\iota, \xi \in \mathbb{R}^{N^2}$, $\xi_{ij} = \xi_{ji}$,
where c^ι_0, c^ι_1 are constants independent of $\mathbf{x} \in \Omega^\iota$.

Let us introduce $W = \prod_{\iota=f,t}[H^1(\Omega^\iota)]^N$, $\|\mathbf{v}\|_W = (\sum_{\iota=f,t}\sum_{i \le N}\|v^\iota_i\|^2_{1,\Omega^\iota})^{\frac{1}{2}}$ and the sets of virtual and admissible displacements $V_0 = \{\mathbf{v} \in W \mid \mathbf{v} = 0 \text{ on } \Gamma_u\}$, $V = \mathbf{u}_0 + V_0$, $K = \{\mathbf{v} \in V \mid v^f_n - v^t_n \le 0 \text{ on } \Gamma_c\}$. Assume that $u^f_{0n} - u^t_{0n} = 0$ on Γ_c Let $c^\iota_{ijkl} \in L^\infty(\Omega^\iota)$, $F^\iota_i \in L^2(\Omega^\iota)$, $P_i \in L^2({}^1\Gamma_\tau)$, $\mathbf{u}^\iota_0 \in [H^1(\Omega^\iota)]^N$.

Then we have to solve the following variational problem $(\mathcal{P})$: find a function $\mathbf{u}$, $\mathbf{u} - \mathbf{u}_0 \in K$, such that

$$a(\mathbf{u}, \mathbf{v} - \mathbf{u}) + j(\mathbf{v}) - j(\mathbf{u}) \ge L(\mathbf{v} - \mathbf{u}) \quad \forall \mathbf{v} \in K \tag{8}$$

holds, where

$$\begin{aligned} a(\mathbf{u}, \mathbf{v}) &= \sum_{\iota=f,t}\int_{\Omega^\iota} c^\iota_{ijkl} e_{ij}(\mathbf{u}^\iota) e_{kl}(\mathbf{v}^\iota)\, dx,\\ j(\mathbf{v}) &= \int_{\Gamma_c} g^{ft} \mid v^f_t - v^t_t \mid ds,\\ L(\mathbf{v}) &= \sum_{\iota=f,t}\int_{\Omega^\iota} F^\iota_i v^\iota_i \, dx - \sum_{\iota=f,t}\int_{\Gamma^\iota_\tau} P^\iota_i v^\iota_i \, ds. \end{aligned} \tag{9}$$

Let us define the sets of rigid displacements and rotations
$P = P^f \times P^t$, $P^\iota = \{\mathbf{v}^\iota = (v^\iota_1, v^\iota_2) \mid v^\iota_1 = a^\iota_1 - b^\iota x_2, v^\iota_2 = a^\iota_2 + b^\iota x_1\}$ where $a^\iota_i, i = 1, 2$ and b^ι are arbitrary real constants and $\iota = f, t$.

It can be shown that the problem (8) has a unique solution, if (see [5]):

$$\begin{aligned} &L(\mathbf{v}) < j(\mathbf{v}) \quad \forall \mathbf{v} \in P \cap K - \{0\},\\ &\{\mathbf{v} \in P \cap V_0, v^f_n - v^t_n = 0 \text{ on } \Gamma_c\} \Rightarrow v = 0\\ &\text{and} \quad |L(\mathbf{v})| > j(\mathbf{v}) \quad \forall \mathbf{v} \in P \cap V_0 - \{0\}. \end{aligned}$$

3 Short Description of the Domain Decomposition Algorithm

Let every domain $\overline{\Omega}^\iota = \cup_{i=1}^{J(\iota)}\overline{\Omega}^\iota_i$, where $J(\iota)$ is a number of subdomains of Ω^ι. Let $\Gamma^\iota_i = \partial\Omega^\iota_i \setminus \partial\Omega^\iota$, $\iota = f,t, i \in \{1, ..., J(\iota)\}$, be a part of dividing line (boundary line) and let $\Gamma = \cup_{\iota=f,t}\cup_{i=1}^{J(\iota)}\Gamma^\iota_i$ be the whole interface boundary. Let

$$T^\iota = \{j \in \{1,..,J(\iota)\} : \overline{\Gamma}_c \cap \partial\overline{\Omega}^\iota_j = \emptyset\}, \iota = f, t, \tag{10}$$

be the set of all indices of subdomains of the domain Ω^ι which are not adjacent to a contact, and let

$$\Omega^{*j} = \cup_{[i,\iota]\in\vartheta}\Omega^\iota_i, \tag{11}$$

where $\vartheta = \{[i,\iota] : \partial\Omega^\iota_i \cap \Gamma_c \ne \emptyset\}$ represent subdomains in unilateral contact. Suppose that $\Gamma \cap \Gamma_c = \emptyset$. Then for the trace operator $\gamma : [H^1(\Omega^\iota_i)]^N \to [L^2(\partial\Omega^\iota_i)]^N$ we have

$$V_\Gamma = \gamma K|_\Gamma = \gamma V|_\Gamma \tag{12}$$

Let $\gamma^{-1} : V_\Gamma \in V$ be an arbitrary linear inverse mapping satisfying

$$\gamma^{-1}\overline{\mathbf{v}} = 0 \quad \text{on } \Gamma_c \quad \forall \overline{\mathbf{v}} \in V_\Gamma. \tag{13}$$

Let us introduce restrictions $\overline{R}_i^\iota : V_\Gamma \to \Gamma_i^\iota$; $L_i^\iota : L^\iota \to \Omega_i^\iota$; $j_i^\iota : j^\iota \to S$; $a_i^\iota(.,.) :$ $a(.,.) \to \Omega_i^\iota$; $V(\Omega_i^\iota) \to \Omega_i^\iota$ and let $V^0(\Omega_i^\iota) = \{\mathbf{v} \in V \mid \mathbf{v} = 0 \text{ on } \overline{(\cup_{\iota=f,t}\Omega^\iota)\setminus\Omega_i^\iota}\}$ be the space of functions with zero traces on Γ_i^ι. The algorithm is based on the next theorem and on the use of local and global Schur complements.

Theorem 3.1: A function $\mathbf{u}$ is a solution of a global problem $(\mathcal{P})$, if and only if its trace $\overline{\mathbf{u}} = \gamma\mathbf{u}|_\Gamma$ on the interface Γ satisfies the condition

$$\sum_{\iota=f,t}\sum_{i=1}^{J(\iota)}[a_i^\iota(\mathbf{u}_i^\iota(\overline{\mathbf{u}}), \gamma^{-1}\overline{\mathbf{w}}) - L_i^\iota(\gamma^{-1}\overline{\mathbf{w}})] = 0 \quad \forall \overline{\mathbf{w}} \in V_\Gamma, \overline{\mathbf{u}} \in V_\Gamma \tag{14}$$

and its restrictions $\mathbf{u}_i^\iota(\mathbf{u}) \equiv \mathbf{u}|_{\Omega_i^\iota}$ satisfy:
(i) the condition

$$\begin{aligned} a_i^\iota(\mathbf{u}_i^\iota(\overline{\mathbf{u}}), \varphi_i^\iota) - L_i^\iota(\varphi_i^\iota) \quad \forall \varphi_i^\iota \in V^0(\Omega_i^\iota), \; \mathbf{u}_i^\iota(\overline{\mathbf{u}}) \in V(\Omega_i^\iota), \\ \gamma\mathbf{u}_i^\iota(\overline{\mathbf{u}})|_{\Gamma_i^\iota} = \overline{R}_i^\iota\overline{\mathbf{u}}, \; i \in T^\iota, \iota = f, t, \end{aligned} \tag{15}$$

(ii) the condition

$$\begin{aligned} \sum_{[i,\iota]\in\vartheta} a_i^\iota(\mathbf{u}_i^\iota(\overline{\mathbf{u}}), \varphi_i^\iota) + j^\iota(\mathbf{u}_i^\iota(\overline{\mathbf{u}}) + \varphi_i^\iota) - j^\iota(\mathbf{u}_i^\iota(\overline{\mathbf{u}})) \geq \\ \geq \sum_{[i,\iota]\in\vartheta} L_i^\iota(\varphi_i^\iota) \quad \forall \varphi \in (\varphi_i^\iota, [i,\iota] \in \vartheta, \varphi_i^\iota \in V^0(\Omega_i^\iota), \end{aligned} \tag{16}$$

and such that

$$\mathbf{u} + \varphi \in K, \quad \gamma\mathbf{u}_i^\iota(\overline{\mathbf{u}})|_{\Gamma_i^\iota} = \overline{R}_i^\iota\overline{\mathbf{u}} \quad \text{for } [i,\iota] \in \vartheta. \tag{17}$$

For the proof see [2].

To analyze the condition (14) the **local and global Schur complements** are introduced. Let

$$V_i^\iota = \{\gamma\mathbf{v}|_{\Gamma_i^\iota} \mid \mathbf{v} \in K\} = \{\gamma\mathbf{v}|_{\Gamma_i^\iota} \mid \mathbf{v} \in V\}$$

and define a particular case of the restriction of the inverse mapping $\gamma^{-1}(.)|_{\Omega_i^\iota}$ by

$$\begin{aligned} & Tr_{i\iota}^{-1} : V_i^\iota \to V(\Omega_i^\iota), \; \gamma(Tr_{i\iota}^{-1}\overline{\mathbf{u}})|_{\Gamma_i^\iota} = \mathbf{u}_i^\iota, \; i = 1,..,J(\iota), \; \iota = f, t \\ & a_i^\iota(Tr_{i\iota}^{-1}\overline{\mathbf{u}}_i^\iota, \mathbf{v}_i^\iota) = 0 \; \forall \mathbf{v}_i^\iota \in V_0^0(\Omega_i^\iota), \\ & Tr_{i\iota}^{-1}\overline{\mathbf{u}}_i^\iota \in V(\Omega_i^\iota) \text{ for } i \in T^\iota, \iota = f, t, \end{aligned} \tag{18}$$

where $V_0^0(\Omega_i^\iota) = \{\mathbf{v} \in V_0 \mid \mathbf{v} = 0 \text{ on } (\cup\Omega_i^\iota)\setminus\Omega_i^\iota\}$. For $[i,\iota] \in \vartheta$ we complete the definition by the boundary condition (13), i.e.

$$Tr_{i\iota}^{-1}\overline{\mathbf{u}}_i^\iota = 0 \text{ on } \Gamma_c. \tag{19}$$

The local Schur complement for $i \in T^\iota$ is the operator $\mathcal{S}_i^\iota : V_i^\iota \to (V_i^\iota)^*$ defined by

$$\langle \mathcal{S}_i^\iota \overline{u}_i^\iota, \overline{v}_i^\iota \rangle = a_i^\iota(Tr_{i\iota}^{-1} \overline{u}_i^\iota, Tr_{i\iota}^{-1} \overline{v}_i^\iota) \quad \forall \overline{u}_i^\iota, \overline{v}_i^\iota \in V_i^\iota. \tag{20}$$

For subdomains which are in contact we define a **common local Schur complement** for the union $\Omega_i^f \cup \Omega_j^t$ (where $[i, f] \in \vartheta, [j, t] \in \vartheta$) as the operator $\mathcal{S}^{ft} : (V_i^f \times V_j^t) \to (V_i^f \times V_j^t)^* = (V_i^f)^* \times (V_j^t)^*$ defined by

$$\langle \mathcal{S}^{ft}(\overline{y}_i^f, \overline{y}_j^t), (\overline{v}_i^f, \overline{v}_j^t) \rangle = a_i^f(u_i^f(\overline{y}_i^f), Tr_{if}^{-1} \overline{v}_i^f) + a_j^t(u_j^t(\overline{y}_j^t), Tr_{jt}^{-1} \overline{v}_j^t)$$
$$\forall (\overline{v}_i^f, \overline{v}_j^t) \in V_i^f \times V_j^t, \tag{21}$$

where Tr_{if}^{-1} and Tr_{jt}^{-1} are defined by means of (18) and (19).

The condition (14) can be expressed by means of local Schur complements in the form

$$\sum_{\iota=f,t} \sum_{i \in T^\iota} \langle \mathcal{S}_i^\iota \overline{u}_i^\iota, \overline{v}_i^\iota \rangle + \sum_{\iota=f,t} \langle \mathcal{S}^{ft}(\overline{u}_i^f, \overline{u}_j^t), (\overline{v}_i^f, \overline{v}_j^t) \rangle =$$
$$= \sum_{\iota=f,t} \sum_{i=1}^{J(\iota)} L_i^\iota(Tr_{i\iota}^{-1} \overline{v}_i^\iota) \quad \forall \overline{v} \in V_\Gamma, [i, f] \in \vartheta, [j, t] \in \vartheta, \tag{22}$$

where $\overline{u} = \gamma u|_\Gamma$, $\overline{v}_i^\iota = \overline{R}_i^\iota \overline{v}$, $\overline{u}_i^\iota = \overline{R}_i^\iota \overline{u}$. Then we will solve the equation (22) on the interface Γ in the dual space $(V_\Gamma)^*$. We rewrite (22) into the following form

$$\mathcal{S}_0 \overline{U} + \mathcal{S}_{CON} \overline{U} = F, \tag{23}$$

where

$$\mathcal{S}_0 = \sum_{\iota=f,t} \sum_{i \in T^\iota} (\overline{R}_i^\iota)^T \mathcal{S}_i^\iota \overline{R}_i^\iota, \quad \mathcal{S}_{CON} = \sum_{\iota=f,t} \overline{R}_{ft}^T \mathcal{S}^{ft} \overline{R}_{ft},$$
$$F = \sum_{\iota=f,t} \sum_{i=1}^{J(\iota)} (\overline{R}_i^\iota)^T (Tr_{i\iota}^{-1})^T \mathcal{S}_i^\iota, \tag{24}$$

and $\overline{R}_{ft}(\overline{u}) = (\overline{R}_i^f(\overline{u}), \overline{R}_j^t(\overline{u}))^T$, $\overline{u} \in V_\Gamma, [i, f] \in \vartheta, [j, t] \in \vartheta$. Equation (23) will be solved by **successive approximations**, because the operators $\mathcal{S}^{ft}$ and therefore $\mathcal{S}_{CON}$ are nonlinear. As a initial approximation $\overline{U}^0$ we choose the solution of the global primal problem, where the boundary conditions on Γ_c are replaced by the linear bilateral conditions with $g^{ft} \equiv 0$ (i.e. $j(u) \equiv 0$)

$$u_n^f - u_n^t = 0, \ \tau_t^{ft} = 0 \ \text{on} \ \Gamma_{c0}. \tag{25}$$

On $\Gamma_c \backslash \Gamma_{c0}$ we consider homogeneous conditions of zero surface load $P_j^f = P_j^t = 0, j = 1, 2$.

Then we replace the set K by $K^0 = \{v \in V \mid v_n^f - v_n^t = 0 \ \text{on} \ \Gamma_{c0}\}$ and therefore, we solve the following problem

$$u^0 = \arg\min_{v \in K^0} \mathcal{L}(v) \tag{26}$$

where $\mathcal{L}(v) = \frac{1}{2} a(v, v) - L(v)$ and we set $\overline{U}^0 = \gamma u^0|_\Gamma$. The auxiliary problem (26) represents a linear elliptic boundary value problem with bilateral contact and it can be solved by the domain decomposition method again.

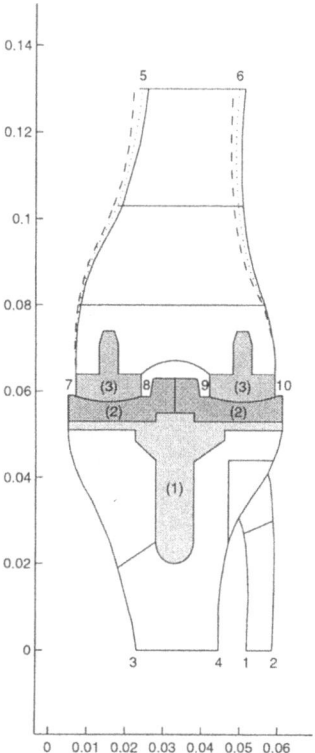

Fig. 1. The models.

The non-linear equation (23) will be solved by successive approximations. We will assume that the approximation $\overline{\mathbf{U}}^{k-1}$ is known and the next approximation $\overline{\mathbf{U}}^k$ we find as the solution of the following linear problem

$$\mathcal{S}_0\overline{\mathbf{U}}^k = \mathbf{F} - \mathcal{S}_{CON}\overline{\mathbf{U}}^{k-1}, k = 1, 2, \ldots \qquad (27)$$

In [2] the convergence of the method of successive approximation (27) to the solution of the original problem (23) in the space $(V_\Gamma)^*$ is proved.

Numerically (26) and (27) are solved by the finite element method.

4 Discussion of Numerical Results

The model of the knee joint replacement was derived from the X-ray image after application the total knee prothesis under the resulting femoral cuts 3, 5 and 7 degrees.

In the model the material parameters are as follows: Bone: Young's modulus $E = 1.71 \times 10^{10}$ [Pa], Poisson's ratio $\nu = 0.25$, (1) $Ti6Al4V$: $E = 1.15 \times 10^{11}$ [Pa], $\nu = 0.3$, (2) Chirulen: $E = 3.4 \times 10^8$ [Pa], $\nu = 0.4$, (3) $CoCrMo$:

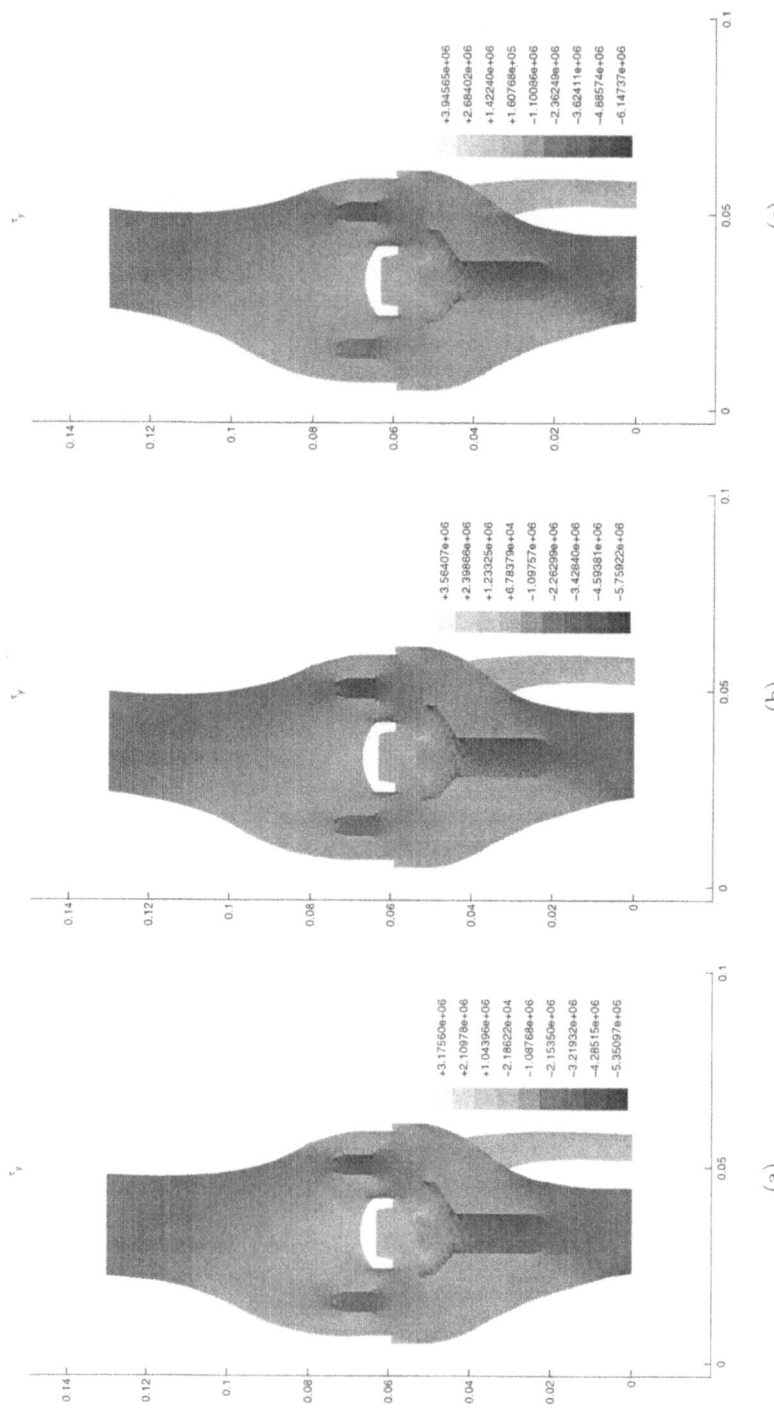

Fig. 2. The vertical stress tensor component - the frontal cross-section of the *CoCrMo* prothesis (a) the femoral cut under 3 degree, (b) the femoral cut under 5 degree, (c) the femoral cut under 7 degree.

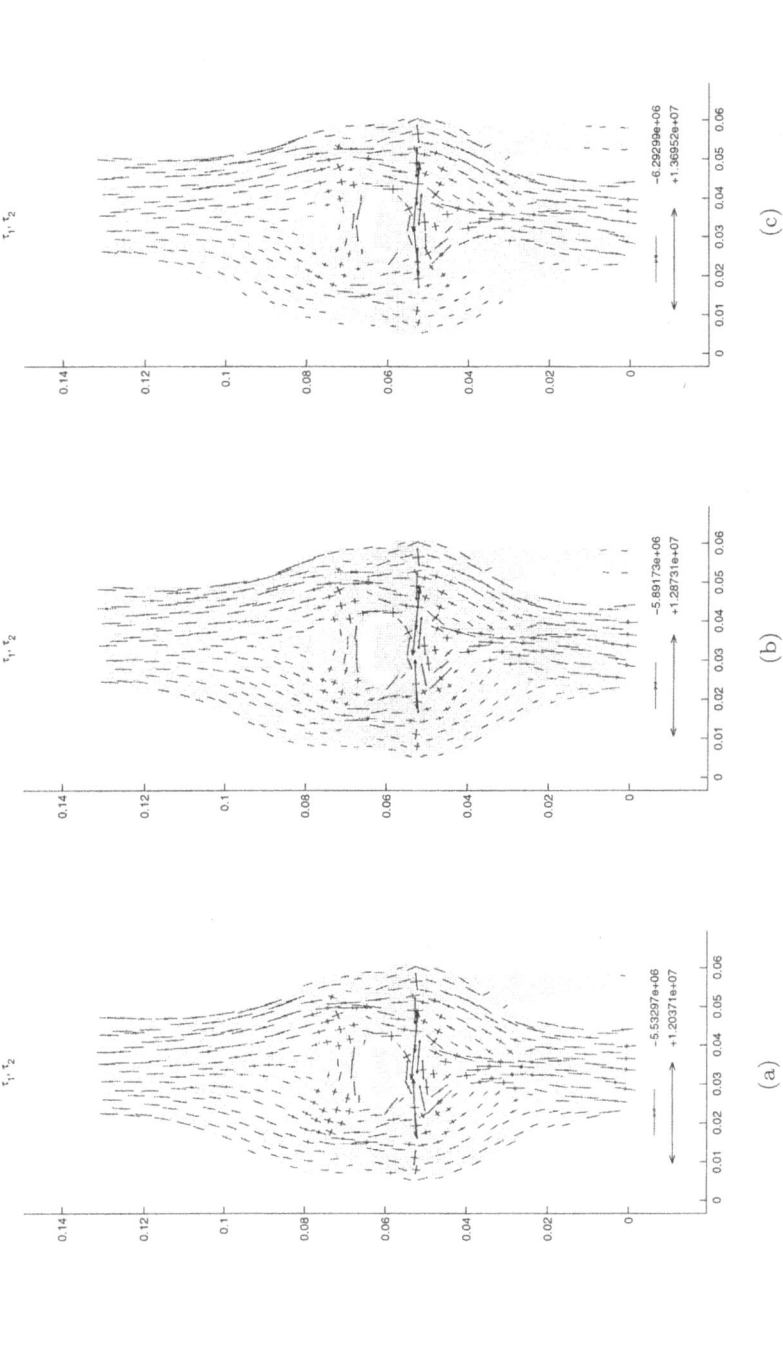

Fig. 3. The principal stresses - the frontal cross-section $CoCrMo$ prothesis ($\rightarrow\leftarrow$ represents compression and $\leftarrow\rightarrow$ extension) (a) the femoral cut under 3 degree, (b) the femoral cut under 5 degree, (c) the femoral cut under 7 degree.

464   J. Daněk et al.

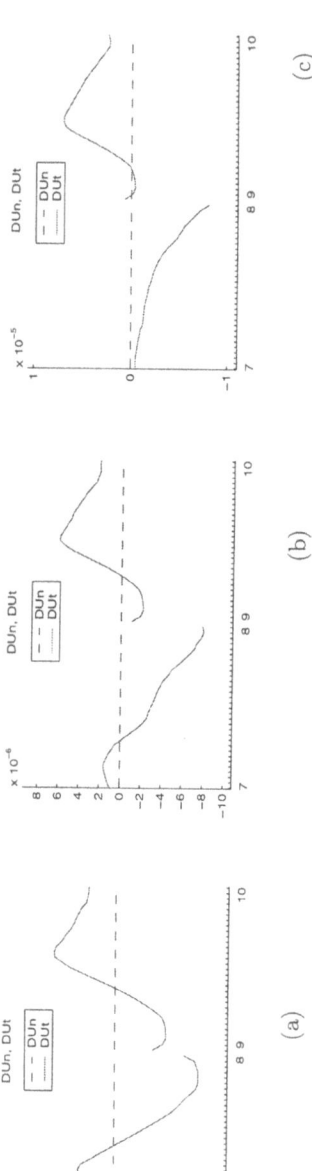

Fig. 4. Normal and tangential components of displacement - the frontal cross-section $CoCrMo$ prothesis with the cut under (a) 3 degree, (b) 5 degree, (c) 7 degree, where the normal and tangential components of displacement vector are denoted by $u_n \equiv DUn$, $u_t \equiv DUt$.

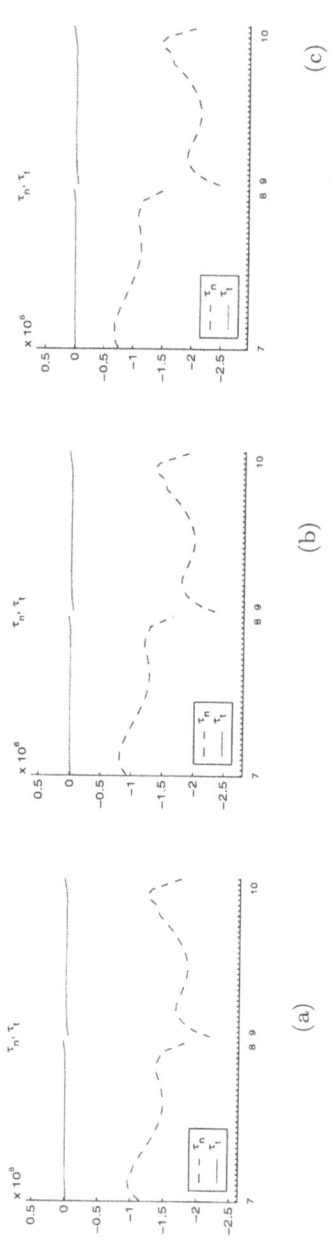

Fig. 5. Normal and tangential components of stress vectors - the frontal cross-section $CoCrMo$ prothesis with the cut under (a) 3 degree, (b) 5 degree, (c) 7 degree.

$E = 2.08 \times 10^{11}$ [Pa], $\nu = 0.3$. The femur is loaded between points 5 and 6 by a loading $0.215 \times 10^7 [Pa]$, the tibia and the fibula are fixed between points 1 and 2 (the tibia) and between 3 and 4 (the fibula) are fixed and the unilateral contact boundary are between points 7 and 8 as well as between 9 and 10. On the contact boundary we suppose that $g^{ft} = 0$. Discretization statistics are characterized by 13 subdomains of domain decoposition, 3800 nodes, 7200 elements, 62 unilateral contact nodes, 350 interface elements. The loadings evoked by muscular forces were neglected. The paper presents three models - the frontal cross-section prothesis with the cut under 3 degree - model (a), 5 degree - model (b) and 7 degree - model (c). All models are presented in Fig.1. In Fig.2 a,b,c the vertical stress tensor components for the

frontal cross-section are presented, while in Figs 3 a,b,c the principal stresses are presented. The presented graphical results represent distribution of stresses in the femur, in the total protheses and in the tibia as well as in the fibula. On Figs 4 a,b,c the normal and tangential components of displacement and on Figs 5 a,b,c stress vectors on the contact boundaries of both condyles (i.e. between points $7 - 8$ and $9 - 10$) are presented. We see that both parts of the prothesis are in a contact and that they mutually move in the tangential direction.

The obtained numerical results, in their graphical forms, corespond to the observed distribution of stress field in the bones and in the knee prothesis, and therefore, they are in a good agreement with the orthopaedic practice. The presented models facilitate to compare the protheses made from the different materials like the $CoCrMo$ alloy, the Al_2O_3 and ZrO_2 ceramics, respectively. The aim of the mathematical modelling of the knee prothesis is to determine the best version of the knee prothesis.

Acknowledgements. The authors have been partly supported by the Ministry of Education, Youth and Sport of the Czech Republic, MSM 235200001.

References

[1] Daněk, J. (2002). Domain decomposition algorithm for solving contact of elastic bodies. In: Sloot P.M.A. et al. (Eds), ICCS'2002, LNCS, Springer Vlg., Berlin, Heidelberg.

[2] Daněk, J., Hlaváček, I., Nedoma, J. (2003). Domain decomposition for generalized unilateral semi-coercive contact problem with friction in elasticity (to appear).

[3] Hlaváček, I. (1999). Domain decomposition applied to a unilateral contact of elastic bodies. Technical Report No 785, Institute of Computer Science AS CR, Prague.

[4] Hlaváček, I., Haslinger, J., Nečas, J., Lovíšek, J. (1988). Solution of Variational Inequalities in Mechanics Appl. Math. Sci 66. Springer Vlg., New York.

[5] Hlaváček, I., Nedoma, J. (2002). On a solution of a generalized semi-coercive contact problem in thermo-elasticity. Mathematics and Computers in Simulation, 60, pp. 1-17.

[6] LeTallec, P. (1994). Domain decomposition methods in computational mechanics. Comput. Mech. Advances 1, pp. 121-220.

[7] Nedoma, J. (1993). Mathematical Modelling in Bio-Mechanics. Bone - and Vascular Implant Systems. Habilitation Thesis. ICS AS CR, Prague.
[8] Nedoma, J., Bartoš, M., Hornátová, H., Kestřánek, Z., Stehlík, J. (2002). Numerical analysis of the loosened total hip replacement (THR). Mathematics and Computers in Simulation. 60, pp. 119-127.
[9] Nedoma, J., Klézl, Z., Fousek, J., Kestřánek, Z., Stehlík, J. (2003). Numerical simulation of some biomechanical problems. Mathematics and Computers in Simulation, 61, pp. 283-295.
[10] Nedoma, J., Hlaváček, I., Daněk, J., Vavřík, P., Stehlík, J., Denk, F. (2003). Some recent results on a domain decomposition method in biomechanics of human joints. Lecture Notes in Science, I-III, vol. 2667-2669, Springer Vlg.
[11] Pavarino, L.F., Toselli, A. (2002). Recent Developments in Domain Decomposition Methods. Lecture Notes in Comput. Sci Eng., vol. 23, Springer Vlg.
[12] Stehlík, J., Nedoma, J. (1989). Mathematical simulation of the function of great human joints and optimal designs of their artificial replacements. Technical Report. V-406/407. ICS AS CR, Prague (in Czech).

Musculoskeletal Modeling of Lumbar Spine under Follower Loads

Yoon Hyuk Kim[1] and Kyungsoo Kim[2]

[1] School of Advanced Technology, Kyung Hee University, Yongin-shi 449-701, Korea***
yoonhkim@khu.ac.kr
[2] IIRC, Kyung Hee University, Yongin-shi 449-701, Korea

Abstract. The lumbar spine can support a much larger compressive load if it is applied along a follower load path that approximates the tangent to the curve of the lumbar spine compared with the vertical load path. In order to investigate the quantitative role of multisegmental muscles to the generation of follower loading patterns, a mathematical model of the human lumbar spine subjected to the follower load was developed in the frontal plane using the finite element method.

1 Introduction

Compressive loads on the human lumbar spine reach 1000N during activities of daily such as standing and walking [1] and are increased up to several thousand Newtons in many lifting activities [2,3]. However, some experimental and computational studies showed that the intact ligamentous lumbar spine buckles at load levels far below those seen in vivo when the compressive loads were applied at the superior end in the vertical direction.

Patwardhan et al. [4] introduced a follower load concept that stabilized the spine and developed a new feasible experimental technique for applying physiologic compressive follower loads to a whole human lumbar spine. A follower load is a compressive load applied along the path that approximates the tangent to the curve of the lumbar spine, called a follower load path, thus subjecting the spine to nearly pure compression. They showed that the whole lumbar spine supported a compressive load of up to 1200 N with small angular changes in the sagittal and frontal planes under the follower load path. In contrast, it withstood only 120 N of vertical load without the follower load path [5,6].

Human trunk muscles are believed to act on the lumbar spine in order for the compressive load to be transmitted along the follower load path. However, since it is hard to investigate the role of the trunk muscles *in vitro* experiments, the computational modeling approach has been researched either through the classical Euler-beam theory to formulate a set of second-order differential equations [7], or the stiffness method with optimization techniques [8,9,10].

*** This study was partly supported by 2004 Kyung Hee University Research Grant, Korea.

The purpose of this study is to develop a mathematical model of the simplified lumbar spine including idealized muscle forces in the frontal plane in which compressive loads are applied along the follower load path to investigate the quantitative role of multisegmental muscles in the generation of follower loading patterns.

2 Materials and Methods

A two-dimensional elastic finite element model of a human lumbar spine was constructed to study the response of the whole lumbar spine (L1-Sacrum) in the frontal plane under external loads in the vertical direction and muscle forces acting on it. In this model, five node points of finite elements were identified with the vertebral body centers of L1-L5 using geometric data in [7] of the standing posture scaled to the length of the lumbar spine model (Fig. 1). Each finite element was assumed to have stiffness property given in [8,9], and the stiffness matrix data therein was used for motion of intervertebral segments in 3-dimensional space. This model was assumed to be fixed at the sacrum and free to translate and rotate at L1 in the frontal plane. In this study only three degrees of freedom, a translation in the lateral direction, a translation in the vertical direction, and a rotation around anterior-posterior axis, were used since the model was assumed to be a 2-dimensional problem. A global coordinate system in the frontal plane was defined, with the origin at the center of the first sacral vertebra, the x-axis pointing horizontally to the right, and the y-axis pointing upward.

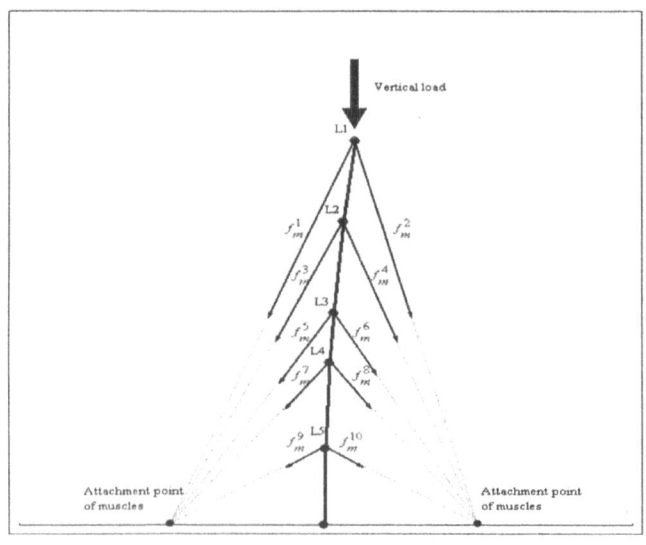

Fig. 1. The lumbar spine model in the frontal plane

In this paper, vertical loads were applied at the superior body center (L1) to simulate the external load, and only ten unknown muscle forces ($f_m^1 - f_m^{10}$) were applied at five nodes along the lines of action of muscles for an idealized muscle architecture that represented multisegmental muscles as shown in Fig.1. The muscles were assumed to be activated statically. In the global coordinate system, the initial locations of five node points were as follows: L1(10.00, 190.00), L2(6.23, 150.00), L3(3.05, 105.00), L4(1.77, 80.00), and L5(0.40, 38.00). The lateral distances between the origin and the attachment point of muscles were 50.00 (unit : millimeter).

A follower load path was defined as having a direction parallel to the lower segment at each node between two adjacent segments. A local coordinate system at each node in the frontal plane could be defined with the origin at the node that is the vertebral body center, the x-axis in the opposite direction to the follower load path called the follower direction, the y-axis in the perpendicular direction to x-axis as usual called the shear direction. At each node, the resultant compressive load could be decomposed into two force components, that in the follower direction and that in the shear direction. The former was called the follower force and the latter was called the shear force. In this study, the ideal follower load concept was defined so that all shear forces were zero at every node since all resultant moments at each node were automatically zero.

The lumbar spine model is in static equilibrium when all the forces and moments at each node equal zero. In order to analyze the force equilibrium equation, three kind of forces, motion segment forces, muscle forces, and external forces, are considered. Similarly, motion segment moments, moments produced by the muscle forces, and external moments are taken into account for the moment equilibrium equation. The stiffness property is a force-displacement relationship which leads directly to matrix equation relating element forces to their corresponding displacements in a mechanical model. Here, the motion segment forces and moments in a plane are related to two translations and one rotation at each node is formulated as

$$\mathbf{f} = \mathbf{k} \cdot \mathbf{d} \qquad (1)$$

$$\mathbf{f} = \begin{pmatrix} f_1 \\ f_2 \\ f_3 \end{pmatrix} = \begin{pmatrix} \text{force in } x - \text{direction} \\ \text{force in } y - \text{direction} \\ \text{moment} \end{pmatrix},$$

$$\mathbf{d} = \begin{pmatrix} d_1 \\ d_2 \\ d_3 \end{pmatrix} = \begin{pmatrix} \text{translation in } x - \text{direction} \\ \text{translation in } y - \text{direction} \\ \text{rotation} \end{pmatrix},$$

$$\mathbf{k} = \begin{pmatrix} k_{ij} \end{pmatrix}_{1 \leq i,j \leq 3},$$

where k_{ij} are the stiffness coefficients, and $\mathbf{k}$ is called the stiffness matrix. When a displacement d_i of unit value is imposed and all other degrees of freedom are zero, the force f_i is equal in value to k_{ij}.

Then we can simplify three equilibrium equations as follows:

$$f_m + f_{ext} = \mathbf{k} \cdot \mathbf{d} \tag{2}$$

where f_m denotes 3-dimensional vectors representing muscle forces in x-, y-direction and moments by muscle forces, f_{ext} external forces and moments.

By synthesizing these equilibrium equations for the five vertebrae, the static equilibrium for the whole lumbar spine model is given as the following vector form

$$F_m - \mathbf{K} \cdot \mathbf{D} + F_{ext} = 0 \tag{3}$$

where F_m represents the muscle forces and moment, F_{ext} the external forces and moments acting on the five lumbar vertebrae, $\mathbf{K}$ the 15 × 15 stiffness matrix, and $\mathbf{D}$ the translations and the rotations at each nodes. In this analysis, ten muscle forces and two translations and a rotation at each vertebra, i.e. fifteen displacements, were considered as variables. Thus the number of variables were quite larger than one of constraints. To solve this indeterminate problem, an optimization technique was indispensable, and the quadratic sequential programming method(MATLAB, MathWorks Inc., USA) was implemented in this paper.

Two kinds of cost functions with parameters were used in order to study and analyze the relation between the muscle forces and the resultant compressive loads, the follower forces and the shear forces, when the follower load concept was applied. Problem 1 was formulated as:

$$\text{Minimize } F = \sum_{i=1}^{n_e} (f_s^i)^2 \tag{4}$$

$$\text{s.t. } f_m^j \leq \alpha \cdot f_m^{max}, \quad j = 1, \cdots, n_m, \quad 0 \leq \alpha \leq 1,$$

where f_s^i denoted the shear force at the i-th node, $i = 1, \cdots, n_e$, f_m^j the j-th muscle force, $j = 1, \cdots, n_m$, f_m^{max} the maximal muscle force acting on the lumbar spine when no restriction concerning the direction of the resultant compressive load was given, n_e the number of elements and n_m the number of muscles. Here, $n_e = 5$ and $n_m = 10$.

Problem 2 was similarly formulated as:

$$\text{Minimize } F = \sum_{j=1}^{n_m} (f_m^j)^2 \tag{5}$$

$$\text{s.t. } f_s^i \leq (1-\beta) \cdot f_s^{max}, \quad i = 1, \cdots, n_e, \quad 0 \leq \beta \leq 1,$$

where f_s^{max} denoted the maximal shear force at all nodes when no restriction concerning the activation of muscles was given, and f_s^j, f_m^j, n_e and n_m are defined as in problem 1.

In problem 1, the shear forces were minimized for given muscle force constraints and in problem 2, the muscle forces were minimized for given shear force constraints. α and β were the penalty parameters of the constraints. In both cases, shear forces decreased as parameters α and β increased due to the expansion of the range of muscle forces (problem 1) and the diminution of the upper bound of the shear forces (problem 2). In the case that α and β were 1.0, it is supposed that the resultant compressive forces were applied along the follower load path.

In one simple case, a vertical load of 635 N applied to L1 in a slightly laterally flexed posture was tested so as to compare the result with that of Patwardhan et al. [7]. First, the responses of the lumbar spine model such as translations and a rotation at each node, and tilts of the superior element were investigated for the various penalty parameters α and β in problem 1 and problem 2, respectively and the deformed shapes of the lumbar spine dependent on the penalty parameters were examined. The resultant compressive forces, follower forces and shear forces, at each node were calculated and their tendencies to change in magnitude with the parameters were sought in both problems, respectively. Also, the muscle forces at every node were determined with varying α and β values in both problems, and the relations between the patterns of the muscle activation and the parameters were found.

3 Results

For the given external force condition, f_m^{max}=274.8N and f_s^{max}=113.0N. The deformed shape of the lumbar spine was obtained by calculating the two translations at each node L1-L5 for problem 1 and 2, respectively, when the penalty parameters α and β were 0.0, 0.2, 0.4, 0.6, 0.8, and 1.0. In problem 1, the spinal shape was deformed most when there was no muscle force ($\alpha = 0$), and it approached the initial shape as α increased (Fig. 2a). The tilt of L1 also approached that of the initial posture as α increased to 1.0. For β in problem 2, the shape of lumbar spine was deformed largest when β was zero, and it approached the initial shape as β increased (Fig. 2b). The tilt of L1 varied in the same way as in problem 1.

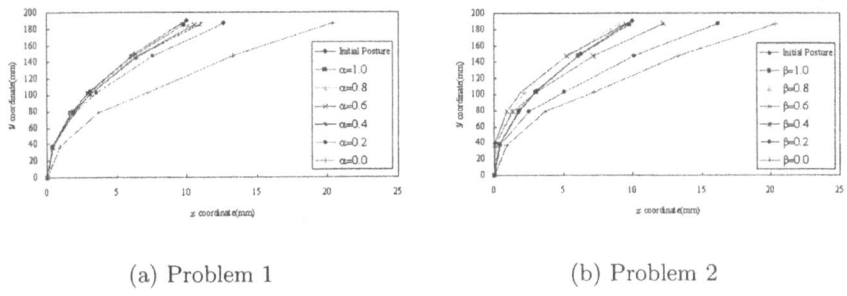

(a) Problem 1   (b) Problem 2

Fig. 2. Deformed shapes of the lumbar spine dependent on the penalty parameters

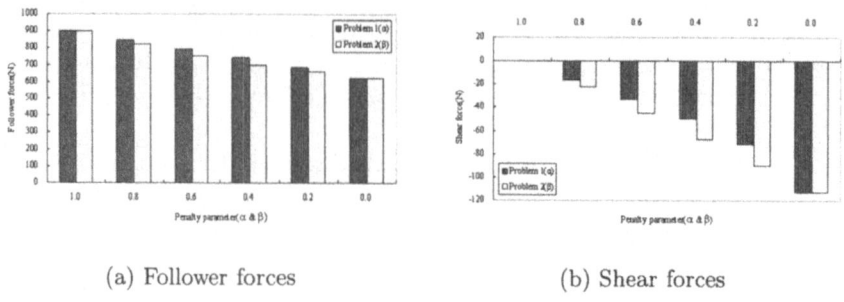

(a) Follower forces (b) Shear forces

Fig. 3. Comparison of follower forces and shear forces, applied at L1

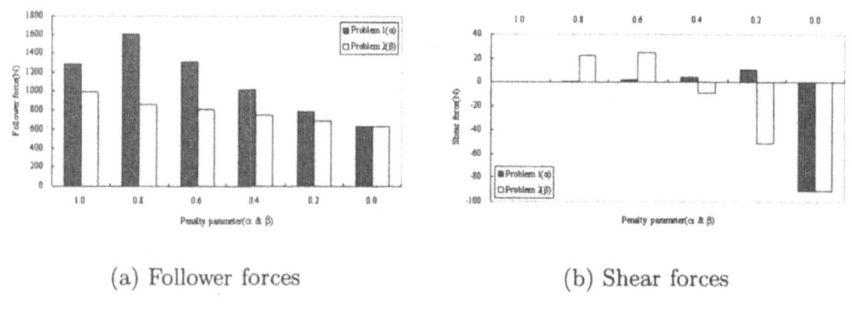

(a) Follower forces (b) Shear forces

Fig. 4. Comparison of follower forces and shear forces, applied at L3

In both problem 1 and problem 2, the magnitudes of the shear forces decreased, and those of the follower forces increased as α and β increased. At L1 the follower forces gradually increased from 624.9N to 900.4N in problem 1 and from 624.9N to 902.2N in problem 2 (Fig 3a). The shear forces at L1 gradually decreased from 112.9N to 0.0N in both problems (Fig. 3b). At L3 the follower force in problem 1 showed an increasing trend with increasing α, up to 0.8, which had a maximum force 1602.2N (Fig 4a). The magnitude of the shear force at L3 in problem 2 increased with increasing β except in the case $\beta = 0.6$ (Fig 4b). When the parameters were set to 1.0 the shear forces at all nodes equaled zero, therefore the resultant compressive forces were equal to the follower forces and were transmitted along the follower load path.

The global magnitude of muscle forces in both problems decreased as each parameter tended to zero (Fig. 5). For all values of the parameters, the largest muscle forces were stimulated on the left side at L1 in both problems. When the parameters values equaled 1.0, f_m^1 in problem 1 was 274.7N and f_m^1 in problem 2 was 276.4N. In addition, the muscle forces caused the resultant compressive

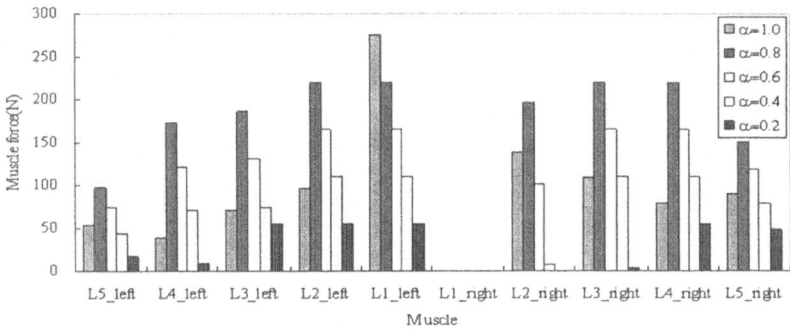

(a) Problem 1

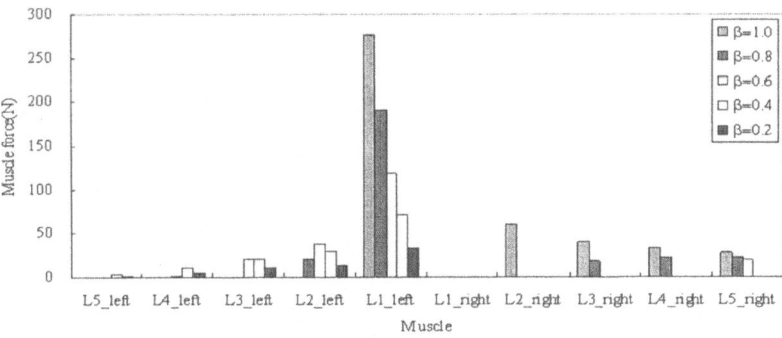

(b) Problem 2

Fig. 5. Distributions of muscle forces

forces to be directed along the follower load path in both problems. In problem 2, only one muscle was activated at each node in all case of β and thus while five muscle forces were applied to the lumbar spine model, almost all muscles were activated in problem 1.

4 Discussion and Conclusion

The lumbar spine can support a much larger compressive load if it is applied along a follower load path that approximates the tangent to the curve of the lumbar spine compared with the vertical load path, as shown by results of several experimental studies [5,6].

Only one result previously dealt with the mathematical model of the lumbar spine in the frontal plane subjected to a follower load using the differential equation theory by Patwardhan et al. [7]. However, the model in [7] was restricted in that it assumed that only one muscle was active in each muscle pair at all level. In order to analyze the model with many muscles, such as the lumbar spine model with its complex configurations and large number of muscles, the stiffness method was appropriate. In this paper, the lumbar spine model which had two muscles activated in each level was examined using the finite element method.

The magnitudes of the follower forces and the shear forces calculated under the vertical load of 635N in problem 1 and problem 2 with the follower load concept ($\alpha = \beta = 1.0$) were close to those in [7]. The pattern of muscle activation in the present paper, especially in problem 2 was similar to those in [7]. As the penalty parameters α and β increased, it was shown that the directions of the resultant compressive forces approached to the follower load path in problem 1 and problem 2 respectively. The magnitude of muscle forces activated was restricted in problem 1 while the amount of shear forces allowed was limited in problem 2. The muscle forces estimated in problem 1 were quite large compared with those in problem 2 because as much muscle forces as possible had to be stimulated to minimize the sum of shear forces in problem 1. In contrast, the activation of the muscle forces in problem 2 was as little as possible and the number of active muscles was only five since the sum of muscle forces was minimized within the range of the allowed shear forces in problem 2.

When the follower load concept was applied to the lumbar spine model ($\alpha = \beta = 1.0$), the pattern of the muscle activation in problem 1 was not equal to that in problem 2. This contradiction was caused by the indeterminateness of each optimization problem. To gain analytically the uniqueness of the solution would be much too complicated and almost impossible. The uniqueness of the solution was induced in problem 2 by testing various initial guesses with the optimization programming, but it was not true for problem 1. Additional physiological constraints had to be considered in order to obtain a reasonable solution.

So far, most experimental research has dealt with the follower load in the sagittal plane. Therefore it is also necessary that the lumbar spine model be formulated in the sagittal plane to guarantee the validity of the mathematical model. Moreover, in the end, the three-dimensional model must be made to physiologically investigate the human lumbar spine and the trunk muscles.

There are factors that decrease the accuracy of the developed mathematical model of the lumbar spine: the simplified and idealized geometric data, the linear stiffness matrix, and the static property of muscles. These drawbacks can be resolved by the development of a more complicated model.

In summary, the mathematical model of the human lumbar spine subjected to the follower load was developed in the frontal plane using the finite element method. The resultant compressive loads including the follower forces and the shear forces at each vertebral body center were investigated and the activations of the idealized trunk muscle forces were obtained.

References

1. Nachemson, A.: Lumbar Intradiscal Pressure. in The Lumbar Spine and Back Pain. Chap. 9 (1987) 191–203
2. Schultz, A.: Loads on the Lumbar Spine. in The Lumbar Spine and Back Pain. Chap. 10 (1987) 205–214
3. McGill, S.: Loads on the Lumbar Spine. in Biomechanics of the Spine: Clinical and Surgical Perspective, Chap. 5 (1990)
4. Patwardhan, A. G., Havey, R. M., Meade, K. P., Lee, B., Dunlap, B.: A Follower Load Increases the Load-Carrying Capacity of the Lumbar Spine in Compression. Spine **24** (1999) 1003–1009
5. Panjabi, M. M., Miura, T., Cripton, P. A., Wang, J.-L., Nain, A. S., DuBois, C.: Development of a System for In Vitro Neck Muscle Force Replication in Whole Cervical Spine Experiments. Spine **26** (2001) 2214–2219
6. Rohlmann, A., Neller, S., Claes, L, Bergman, G., Wilke, H.-J.: Influence of a Follower Load on Intradiscal Pressure and Intersegmental Rotation of the Lumbar Spine. Spine **26** (2001) E557–E561
7. Patwardhan, A. G., Meade, K. P., Lee, B.: A Frontal Plane Model of the Lumbar Spine Subjected to a Follower Load: Implications for the Role of Muscles. J. Biomech. Eng. **123** (2001) 212–217
8. Panjabi, M. M., Brand Jr., R. A., White III, A. A.: Three-Dimensional Flexibility and Stiffness Properties of the Human Thoracic Spine. J. Biomech. **9** (1976) 185–192
9. Gardner-Morse, M. G., Laible, J. P., Stokes, I. A. F.: Incorporation of Spinal Flexibility Measurements into Finite Element Analysis. J. Biomech. Eng. **112** (1990) 481–483
10. Stokes, I. A. F., Gardner-Morse, M. G.: Lumbar Spine Maximum Efforts and Muscle Recruitment Patterns Predicted by a Model with Multijoint Muscles and Joints with Stiffness. J. Biomech. **28** (1995) 173–186

Computational Approach to Optimal Transport Network Construction in Biomechanics

Natalya Kizilova

Department of Theoretical Mechanics,
V.N.Karazin Kharkiv National University,
Svobody Sq.,4, 61077 Kharkiv, Ukraine
knn@hall.nord.net.ua

Abstract. Long-distance liquid transport in biosystems is provided by special branching systems of tubes (arteries, veins, plant vessels). Geometry of the systems possesses similar patterns and can be investigated by computer methods of pattern recognition. Here some results on plant leaf venation investigation are presented. The lengths, diameters and branching angles are estimated for the leaves of different shape, size and venation type. The statistical distributions of the measured parameters are similar to the corresponding ones which have been obtained for arterial beds. The both correspond to the model of optimal branching pipeline which provide liquid delivering at minimum total energy consumptions. The biomechanical model of liquid motion in a system consisting of a long thin tube with permeable walls which is embedded into a biological porous medium is considered. The pressure distributions and velocity fields for different geometry of the system are obtained. The main result is when the delivering liquid is completely absorbed by the alive cells in the porous medium the relation between the diameter and the length of the tube and the total volume of the medium which correspond to the measured data is reached.

1 Introduction

Transport of liquids and dissolved substances on the distances comparable to the characteristic size of the biological system is provided by special conducting structures. In animals and higher plants the conducting systems are represented by branching networks with 5-9 branching orders in plant leaves and 20-30 branching orders in mammalian arterial and venous systems. Design principles of network geometry can be investigated on special plastic casts of blood vessels (3D-geometry), X-ray pictures of vessels filled with special substances (arteriography), cleared leaf preparations (leaves that have been bleached and stained to make their venation patterns more visible) by using computer methods of image analysis. A few general statistical relations between the diameters D_i, lengths L_i and branching angles φ_i of separate vessels (fig.1a) have been revealed in intra-and extraorgan arterial [1-4] and respiratory systems [5-6] of mammals, in fluid transport systems of sponges [1], vein branching in plant leaves [7-9], branching in trees and shoots [10-11]. Structure of arterial beds corresponds to the model of optimal branching pipeline. The appropriate biomechanical models of development of the optimal transport system in a growing biological system are based on theoretical problem of liquid motion through a branching system of tubes with special properties [2,3,12]. Corresponding models of liquid motion in the conducting systems of plants are not investigated yet.

2 Measurements and Principles of Construction of the Networks

Geometry of the conducting systems of plant leaves of different shape, size and venation type is investigated on high-resolution digital pictures of the fresh-cutted leaves (fig.1b). A few main stages have been provided:
- Contrast enhancement of an image and edges-finding (leaf blade perimeter, leaf veins) (fig.1c)
- Allocation of separate bifurcations, measurement of the diameters and branching angles (fig.1d)
- Skeletization of the vein system, measurement the lengths of veins (fig.1e)
- Allocation of influence domains of different veins (leaf areas S_i which are supplied by liquid through separate main veins) (fig.1f)

Using the procedure more then 327 images of dicotyledonous leaves have been investigated. In spite of the complicated topology of the conducting systems a few simple principles of their organization have been found out.

Principle 1. The diameters of the parent and daughter's vessels $d_{0,1,2}$ at each bifurcation obey the relation $d_0^\gamma = d_1^\gamma + d_2^\gamma$ which is called Murray's law. For mammalian vessels $\gamma \approx 3$ ($\gamma = 2.55 - 3.02$ for arterial, $\gamma = 2.76 - 3.02$ for venous, $\gamma = 2.61 - 2.91$ for respiratory systems) [1-6]. For the large blood vessels and bronchi $\gamma \sim 2.33$. For the small vessels when rheology of the fluid should be taken into account $\gamma \sim 2.92$. For the most part of investigated images of plant leaves $\gamma \approx 3$. The correspondent dependence is presented in fig.2a for one leaf (approximately 200-250 vein bifurcations). Validity of Murray's law for arterial systems can be explained by formation of an optimal vessel with $Q \sim d^3$ due to maintaining the shear stress τ_w at the vessel wall at a constant level $\tau_w \approx const$ [12]. In plant leaves the veins with $d \geq 0.08$ mm obeys Murray's law at $\gamma = 3$ (fig2a).

Principle 2. Branching angles $\varphi_{1,2}$ are defined by the diameters $d_{0,1,2}$ by the formula:

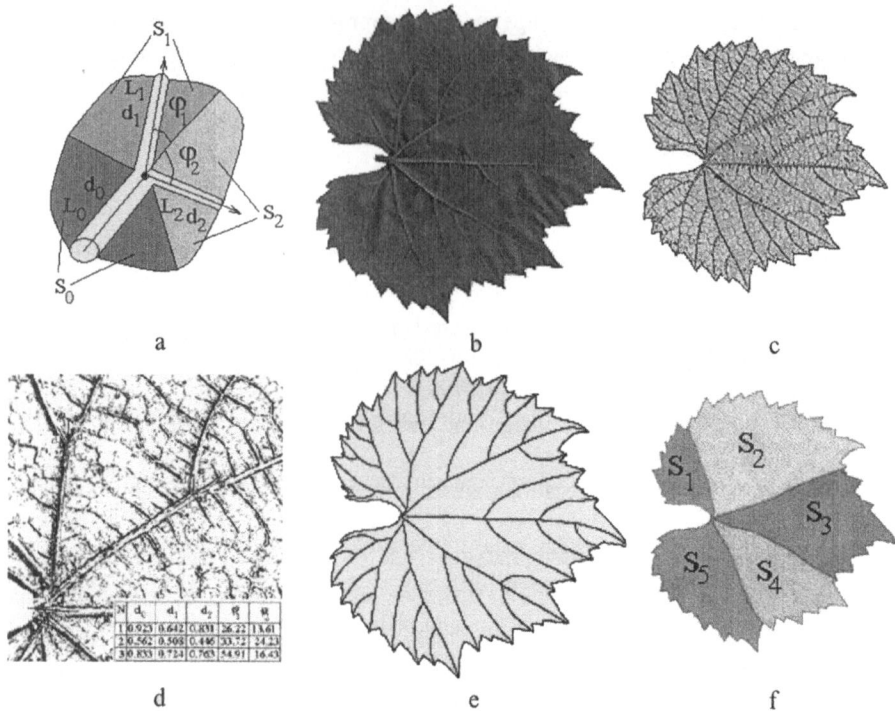

Fig. 1. Image analysis procedure: measured parameters of a bifurcation of the veins (a); digital picture of Vitis vinifera leaf (b); result of edge-finding procedure (c); measurement of the diameters $d_{0,1,2}$ and branching angles $\varphi_{1,2}$ in separate bifurcations 1-3 (d); result of the skeletization procedure, measurement the lengths of the veins (e); measurement of the areas S_{1-5} of influence domains of the first-order veins (f).

$$\cos(\varphi_1) = \frac{(1+\xi^3)^{4/3} + 1 - \xi^4}{2(1+\xi^3)^{2/3}}, \quad \cos(\varphi_j) = \frac{(1+\xi^3)^{4/3} + \xi^4 - 1}{2\xi^2(1+\xi^3)^{2/3}}$$

where $\xi = d_2/d_1 < 1$, $d_2 = \min\{d_{1,2}\}$ is asymmetry of the bifurcation. The relation between the diameters and angels in a bifurcation is valid for mammalian arterial beds and is derived from an optimality principle [12]. Both principles 1 and 2 represent an optimal branching that minimizes the costs of delivery of the liquid, the construction and maintenance of the transport system. For plant veins optimality of the branching angles is peculiar to the bifurcations with $K \in [1.5;2]$ where $K = (d_1^2 + d_2^2)/d_0^2$ (fig.2b).

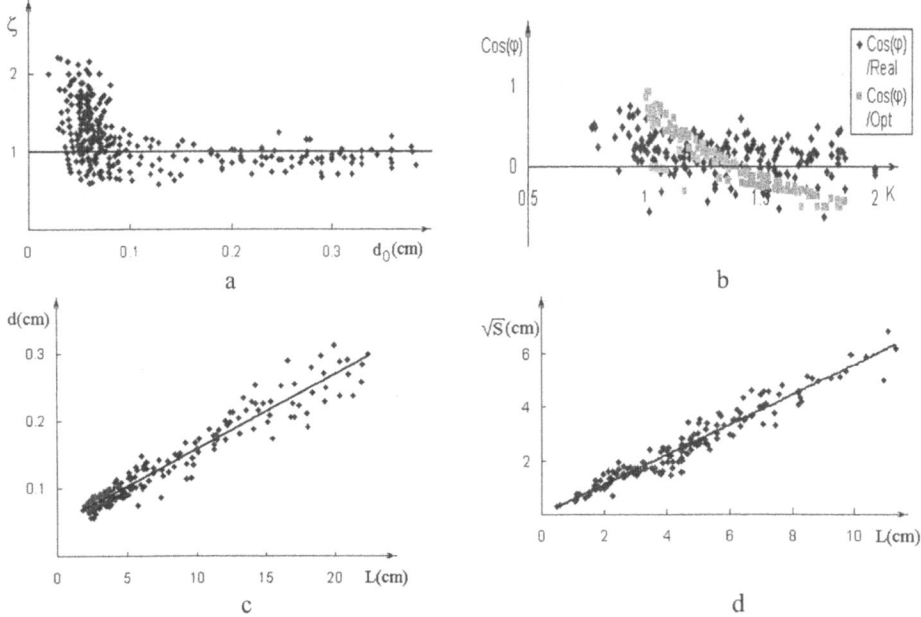

Fig. 2. Results of measurements by the image analysis procedure: a relation between the Murray's parameter $\zeta = (d_1^3 + d_2^3)/d_0^3$ and the diameter d_0 of the parent vein for 200 bifurcations of a leaf (a); dependence of the optimal (theoretical) and real (measured) angles on the branching coefficient K (b); dependence $d(L)$ for one Vitis vinifera leaf (200 segments of veins) (c); dependence $\sqrt{S}(L)$ for vein segments ($n = 1-3$) and their domains of influence (d).

Principle 3. The relation $L = \alpha d^\beta$ has been revealed for mammalian arterial systems [13] where $\alpha \in [2.6; 7.59]$, $\beta \in [0.84; 1.16]$ (approximately $\beta \sim 1$). The principle can be regarded as an allometric rule as applied to different kinds of mammals with different mass of the body. The dependence between the lengths and diameters of the leaf veins with orders $n = 1-3$ is presented in fig.2c. The linear dependence $L = \alpha d$ with $\alpha \in [82; 96]$ is valid to each leaf from our database.

Principle 4. The dependence $S = AL^2$ has been obtained for all leaves, $A \in [0.19; 0.32]$ varies insignificantly for leaves with different sizes (1-35 cm). Shapes and sizes of the domains of influence are quite different that indicate the nonallometric character of the relation. The same relation $L \sim \sqrt{S}$ is well-known for river systems and their drainage areas in geophysics as Hack's law. The geometrical similarity of the vein (fig2d) and river systems is deep and revealed the common design principles of network construction in Nature [14].

3 A Biomechanical Model of Liquid Motion in a Plant Leaf

A 2-D steady motion of an incompressible viscous liquid in a porous medium is considered. Domain of influence is modeled as a curvilinear polygon that bounded by thin channels (leaf veins) and impermeable border (edge of the leaf blade) taking into account symmetry of the region about the x_1-axis (fig.3a). The governing equations have been taken to be the following [15-16]:

$$\rho div(\bar{v}) = -q \tag{1}$$

$$v_i = -\frac{\Lambda_{ik}}{\mu}\left(\frac{\partial p}{\partial x_k} - \sigma\frac{\partial \pi}{\partial x_k}\right) \tag{2}$$

$$\frac{\partial C}{\partial t} + \frac{\partial}{\partial x_i}(Cv_i) = D_c \frac{\partial^2 C}{\partial x_i^2} + q_c \tag{3}$$

where $\bar{v}$ is velocity vector in the porous medium and the generalized Darcy law (2) is introduced, Λ_{ik} is permeability tensor, p, π are hydraulic and osmotic pressures, ρ, μ are the fluid density and viscosity, C is concentration of an osmotically active dissolved mineral component. As fluid moves along the channels and through their permeable walls into the porous medium it is adsorbed by alive cells of the leaf and evaporate into the atmosphere. The total absorption is modeled as distributes sinks of water $q(x, y)$ and of the mineral component $q_c(x, y)$, D_c, σ are diffusion coefficient and the so-called reflection coefficient of the mineral component. It was assumed the permeability tensor is defined by the directions of the secondary veins $i = 2 - 3$ which are thought to be parallel and form two orthogonal sets of directions. The directions of the secondary veins in different domains of influence can slightly differ (fig.3b). Here a Vitis vinifera leaf (fig.1b) is modeled by a cardioid (fig.3a) or by a circle with a notch at the base of the lamina (fig.3b).

The fluid moves from the inlet $x_{1,2} = 0$ of the transport system through the tubes 1-2, then through their permeable walls into the porous media I-II. The driving forces are gradients of hydrostatic and osmotic pressures. The osmotic pressure is defined by concentration of the solute which can be maintained by the alive cells at a constant level providing the propelling force for the flux. The van't Hoff equation for osmotic pressure can be written as

$$\pi = RTC/M_c \tag{4}$$

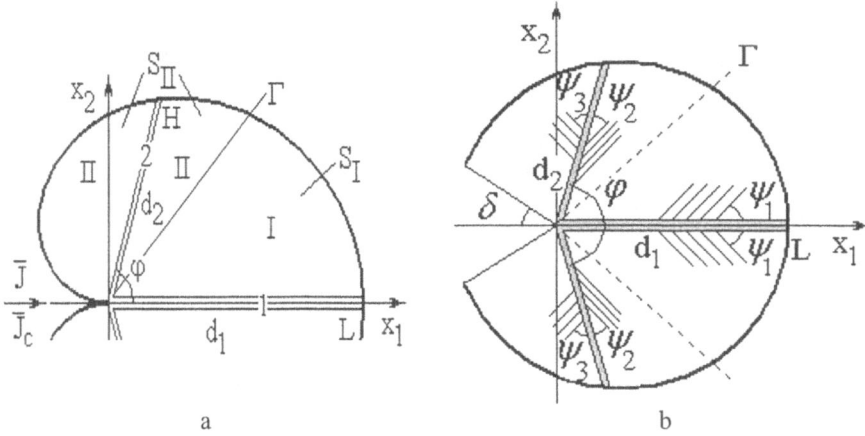

Fig. 3. Model of the leaf with 2 veins (1-2) and their domains of influence (I-II) (a); a round leaf with a notch at the base and venation pattern of the secondary veins (b).

where T is temperature that is taken to be constant, R is gas constant, M_c is the molar mass of the mineral component. Substituting (4) into (2) and into (1),(3) gives

$$\frac{\partial}{\partial x_i}\left(\frac{\Lambda_{ik}}{\mu}\left(\frac{\partial p}{\partial x_k}-\frac{RT\sigma}{M_c}\frac{\partial C}{\partial x_k}\right)\right)=\frac{q}{\rho} \quad (5)$$

$$\frac{\partial C}{\partial t}-\frac{\Lambda_{ik}}{\mu}\left(\frac{\partial p}{\partial x_k}-\frac{RT\sigma}{M_c}\frac{\partial C}{\partial x_k}\right)\frac{\partial C}{\partial x_i}-\frac{Cq}{\rho}=D_c\frac{\partial^2 C}{\partial x_i^2}+q_c \quad (6)$$

Numerical simulations have been used for solution of the system (5)-(6) and for illustration some key features of the water and the solute motion in regions with different geometry. Poiseulle flow in a rectangular channel is considered for the liquid motion in the veins. The pressure and flow continuity conditions at the walls of the channels and the constant water and solute fluxes $J = const$, $J_c = const$ at the inlet of the transport system $x_{1,2} = 0$ are given. The corresponding procedure of numerical calculations is described in [16]. Different distributions $q(x_1,x_2)$, $q_c(x_1,x_2)$ are introduced for simulations using the preliminary experimental data in form $q, q_c = \{const; q_0(1-x_1/L)(1-x_2/H); q_0 \exp(-a_1 x_1 - a_2 x_2)\}$.

The goal of the simulations was calculation an optimal branching angle φ and the asymmetry of the bifurcation $\zeta = d_2/d_1$ at given $H, L, d_1/L, q, q_c, J, J_c$ which define the optimal transport system with the optimization criterion [12]:

$$\Phi = \int_{S_I} \tau_{ik}^I v_{ik}^I dS_I + \int_{S_{II}} \tau_{ik}^{II} v_{ik}^{II} dS_{II} + \int_{S_1} \tau_{ik}^1 v_{ik}^1 dS_1 + \int_{S_2} \tau_{ik}^2 v_{ik}^2 dS_2 \to \min \quad (7)$$

$$V = \pi(d_1^2 L + d_2^2 H) + S_I + S_{II} = const \quad (8)$$

where $\tau_{ik} = 2\mu v_{ik}$, v_{ik} is rate of deformation tensor. Total value $S = S_I + S_{II}$ are given whereas the border between S_I and S_{II} are defined after the calculations on (5)-(6) as a border Γ with $J_\Gamma = \int_\Gamma v_n d\Gamma = 0$, where $\vec{n}$ is a normal vector to Γ. After calculation the pressure and concentration by (5)-(6) the velocity field was obtained by (2),(4). As an illustration the velocity field and the corresponding domains of influence are presented in fig.4 for $q, q_c = const$, $L = H$, $d_1 = d_2$, $d_1/L = 0.1$. The border Γ has been defined after the velocity (v_1, v_2) calculation. The iterative procedure is consisted of determining the direction $\vec{n}_\Gamma$ in each point starting with $x_{1,2} = 0$ so that $\vec{v}_I \parallel \vec{v}_{II} \parallel \vec{n}_\Gamma$. After that the areas $S_{I,II}$ were calculated.

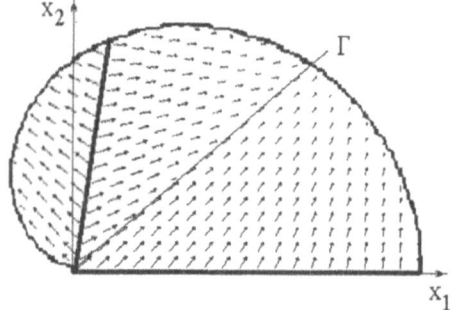

Fig. 4. Velocity field for a heart-shaped leaf with two first-order veins

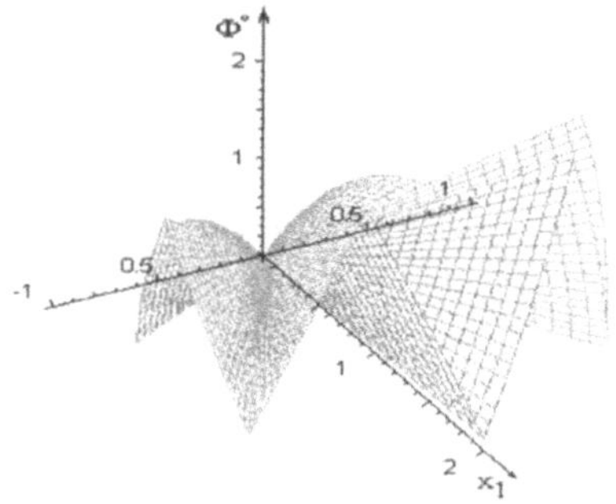

Fig. 5. Distribution $\Phi^\circ(x_1, x_2)$ for a round leaf, $L=2$, $d_1=d_2$, $d_1/L=0.1$.

4 Results and Discussions

Total energy dissipation strongly depends on geometry of the leaf: its shape, number and diameters of the veins. Hydraulic conductance of the main veins is significantly higher then the conductance of the porous medium where the most part of the total energy is expended (fig.5). During the computer simulations a few simple shapes (round, elliptic, cardioid's) as well as shapes of some real leaves have been investigated. All the angles of the secondary veins φ_j, ψ_j and of the notch of the blade δ, the radii r_j and lengths L_j of the veins have been obtained by measurements on the pictures of the leaves by image analysis procedure described in the previous chapter.

The optimal criterion (7)-(8) has been evaluated after the calculations on (5)-(6) and (2)-(4) by variation of separate geometrical parameters $\zeta, d_2/L, \varphi, \psi_{1-3}, \delta$ for a leaf with 3 main veins at a constant shape of the leaf blade (round with radius L, cardioid's $r = L(1 - \cos\theta)$, where (r, θ) are polar coordinates) or the parameters $\zeta_1 = d_2/d_1, \zeta_2 = d_3/d_1, d_2/L, d_3/L, \varphi_{1-2}, \psi_{1-5}, \delta$ for a leaf with 5 main veins. For real leaf blades the Lagrange function $\Theta = \Phi + \lambda V$, $\lambda = const$ has been evaluated after the calculations on (5)-(6) and (2)-(4) at small variations of the geometrical parameters of the corresponding models (figs 3a,b) at close range of the measured data. For the cases with simple geometry of the leaf blade and the veins the corresponding inverse problem (7)-(8) can be solved.

For a round leaf at a given number of the main veins the total volume of the system remains constant whereas the dimensionless function Φ° reaches its minima at certain branching angle between the main first-order veins (fig.6a). For wide variations of the branching angles ψ_{1-3} and the diameters of the veins $d_{1,2}$ at a constant L the energy dissipation decreases at increasing the conductivity of the veins. Nevertheless the optimal values of the branching angle φ vary insignificantly for all possible combinations of other geometrical parameters and close to $\varphi_{opt} = 35.1 - 35.5°$ (fig.6a). The corresponding results for a symmetrical cardioid's leaf with 3 main first-order veins (fig.3a) are presented in fig.6b. Here increasing of the diameters of the veins leads to decreasing of the total energy dissipation at a constant total volume of the system. The dependence $\Phi^\circ(\varphi)$ reaches its minimum at $\varphi = 68.5 - 75.8°$ for different pairs of the dimensionless geometrical parameters d_2/d_1, d_2/L of the model.

The optimal criterion (7)-(8) has been evaluated after the calculations on (5)-(6) and (2)-(4) by variation of separate geometrical parameters $\zeta, d_2/L, \varphi, \psi_{1-3}, \delta$ for a leaf with 3 main veins at a constant shape of the leaf blade (round with radius L, cardioid's $r = L(1 - \cos\theta)$, where (r, θ) are polar coordinates) or the parameters $\zeta_1 = d_2/d_1, \zeta_2 = d_3/d_1, d_2/L, d_3/L, \varphi_{1-2}, \psi_{1-5}, \delta$ for a leaf with 5 main veins. For real shapes of leaves the Lagrange function $\Theta = \Phi + \lambda V$, $\lambda = const$ has been evaluated after the calculations on (5)-(6) and (2)-(4) at small variations of the geometrical parameters $r_{1-4}, L, \delta, \varphi_{1-3}, \psi_{1-7}$ of the corresponding models (figs 3a,b) at close range of the measured data. For the cases with simple geometry of the leaf blade and the veins the corresponding inverse problem (7)-(8) can be solved.

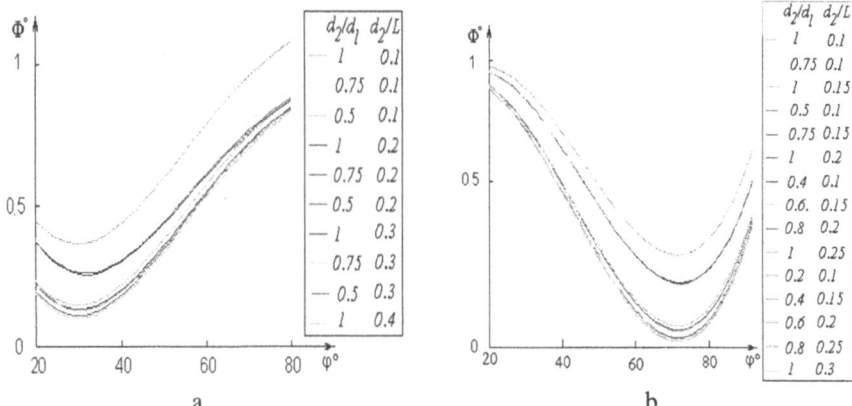

Fig. 6. Dependence $\Phi°(\varphi)$ at different pairs of the parameters d_2/d_1, d_2/L for the round leaf (a) and for the cardioid's leaf (b).

The main results of the computer simulations can be summarized as a number of regularities:
- For any given shape of the leaf blade the optimal branching angle of the first-order veins exists and depend on the number and diameters of the veins. Some increase/decrease in the branching angle comparable to the optimal value leads to significant increase in the energy loss due to the viscous dissipation in the I/II domain of influence. The corresponding mechanism of the optimal branching angle formation during the leaf growth and development can be defined by the balance of the liquid delivery by the vein and its absorption by the alive cells of the corresponding domain of influence. When the branching angle increases/decreases due to some growth fluctuations the area of I/II domain of influence will increase/decrease and the cells in I/II region receive relatively less/more amount of water and dissolves mineral and organic substances. It will lead to the corresponding decrease/increase of the growth rate in I/II and to decrease/increase the area I/II and the branching angle. The feedback system can underlay the mechanism of the optimal branching angle formation during the normal growth.
- Optimal branching pattern ψ_{1-3} of the secondary veins $i=2$ is defined by increasing all the angles up to $\psi_{1-3} \sim 90°$. The result corresponds to experimental observations and the statement of evolutionary biology concerning the increase the branching angle for the secondary veins during the evolutional transition from palmate to the pinnate (with $\psi \to 90°$) leaf venation pattern.
- When pressure at the ends of the main veins equals to the pressure in the surrounding porous medium so that the total volumetric rate of the liquid motion through the vein equals to its total absorption by the alive cells in the corresponding domain the relation $L_j \sim \sqrt{S_j}$ is found out for the main veins and their domains of

influence of all the investigated shapes of leaves. The relation can be regarded as a relation between form and function of a leaf.

The results of the dynamic computer simulations correspond to the results of measurements on the leaves of different types and to the experimental biological observations. The balance relations of the liquid motion, redistribution and adsorption can play an important role in formation the optimal transport systems in plant leaf venation.

References

1. La Barbera, M.: Principles of Design of Fluid Transport Systems in Zoology. Science. 249 (1990) 992-1000
2. Pries, A.R., Secomb, T.W., Gaehtgens, P.: Design Principles of Vascular Beds. Circulat.Res. 77 (1995) 1017-1023
3. Zamir, M., Bigelov, D.C.: (1984) Cost of Departure from Optimality in Arterial Branching. J.Theor.Biol. vol.109, pp. 401-409
4. Zamir, M.: Branching characteristics of human coronary arteries, Can.J.Physiol. 64 (1986) 661-668
5. Kitaoka, H., Suki, B.: Branching Design of the Bronchial Tree Based on a Diameter-Flow Relationship. J.Appl.Physiol. 82 (1997) 968-976
6. Kitaoka, H., Takak,i R., Suki, B.: A Three-Dimensional Model of the Human Airway Tree. J.Appl. Physiol. 87 (1999) 2207–2217
7. Kizilova, N.N.: Transport System and Growth of Leaves. In: Belousov, L., Stein, A. (eds): Mechanics of Growth and Morphogenesis. Moscow Univ.Press, Moscow (2001) 379-405
8. Kizilova, N.N.: Construction Principles and Control over Transport Systems Organization in Biological Tissues. In: Fradkov, A.L., Churilov A.N. (eds): Physics and Control, Vol.1. St.Petersburg. (2003) 303-308
9. McCulloh, K.A, Sperry, J.S., Adler, F.R.: Water Transport in Plants Obeys Murray's Law, Nature 421 (2003) 939-942
10. Honda, H.: Tree branch angle: maximizing effective leaf area. Science 199 (1978) 888-889
11. Zhi, W., Zhao, Ming, Yu, Qi-Xing: Modeling of Branching Structures of Plants. J.Theor.Biol. 209 (2001) 383-394
12. Rosen, R.: Optimality Principles in Biology. Plenum Press, New York (1967)
13. Dawson, C.A., Krenz, G.S., Karau, K.L. et al: Structure-Function Relationships in the Pulmonary Arterial Tree. J.Appl.Physiol. 86 (1999) 569-583
14. Pelletier, J.D., Turcotte, D.L.: Shapes of River Networks and Leaves: are they Statistically Similar? Phil.Trans.Royal Soc.London, Ser.B. 355 (1999) 307-311
15. Henton, S.M., Greaves, A.J., Piller, G.J., Minchin, P.E.H.: Revisiting the Münch Pressure-Flow Hypothesis for Long-Distance Transport of Carbonhydrates: Modeling the Dynamics of Solute Transport Inside a Semipermeable Tube. J.Experim.Bot. 53 (2002) 1411-1419
16. Kizilova, N.N.: Hydraulic Properties of Branching Pipelines with Permeable Walls. Appl.Hydromech. 38 (2003) 78-84

Encoding Image Based on Retinal Ganglion Cell

Sung-Kwan Je[1], Eui-Young Cha[1], and Jae-Hyun Cho[2]

[1]Dept. of Computer Science, Pusan National University,
[2]Dept. of Computer Information, Catholic University of Pusan
jimmy@harmony.cs.pusan.ac.kr

Abstract. Today's Computer vision technology has not been showing satisfying results and is far from a real application because currently developed computer vision theory has many assumptions, and it is difficult to apply most of them to real world. Therefore, we will come over the limit of current computer vision technology by developing image recognition model based on retinal ganglion cell. We have constructed the image recognition model based on retinal ganglion cell and had experiment upon recognition and compression processing of information such as retinal ganglion cell by handwritten character database of MNIST.

1 Introduction

It is important for human to detect changes of an environment, take a judgment, interpret a situation, and do an appropriate action to adapt to changes of an environment. As sensory organ provide information to human, human can recognize an environment, and control it actively. The information consists of sight 60%, hearing 20%, feeling 15%, and sense of taste 3%, sense of smell 2%. Human uses visual and hearing information very much. Even if these percentages exclude the precision, this is telling a magnitude of visual information [1-2].

A research of machine vision is to play the purpose of imitating function of a human visual system, and providing it to a machine. Algorithm and applications related to machine vision are developing gradually. But the machine vision technology used in real world is still applied restricted fields. And its information processing is developing into a different form with that of human visual system because of the limited application range.

Today's computer hardware technology can not deal with the information processing happening in a current human visual system. A human visual information processing composed with a lot of step. The primary information processing begins in the human retina. The human retina not only includes a process changing light energy into electricity chemical energy but also transmits visual information to visual path [1-3].

While information is transferred into ganglion in retina, total information is constricted with making loss of critical information minimized. That is, the information of 'what' related to recognition and information of 'where' related to part

is compressed with the minimum loss. Therefore, the human visual information about to 'what' is sensitive but information about' where is not sensitive [4].
Human retina consists of three layers constructed from six types of cells (photo-receptor, horizontal cell, amarcrine cell, interplexiform cell, bipolar cell, ganglion cell). The light is translated into the neural signal at photo - receptor. Then this signal is transferred to the brain through the many cells in retina [4]. We remember the number of photo-receptors, which transform light into neural signal transferring path in retina (the ratio is about 130:1) According to above fact, we can say that a few data are transferred to the brain, and then data are constricted in retina.

In this paper, we investigate the data compression property of retina through physiological evidence. We are going to implement a model about the visual information compression using wavelet transform that actual human gets up in a process of recognizing an image. We look into human visual system and wavelet in Chapter 2. We compose an image recognition model based on human visual information processing in Chapter 3. We analyze the experiment results in a Chapter 4. Finally, we will conclude our research in Chapter 5.

2 Human Visual Systems and Wavelet

2.1 Human Visual System

The research of human visual system has been preceded until 1960's by neurophysiology and psychophysics. It got a lot of achievements by an appearance of cognitive science in 1970's. Knowledge about visual information processing of a human visual system has been found with a medical science, engineering recently. And study to maximize performance of a current developed computer vision through modeling about this visual information processing is proceeded in the entire world. Also, there is active study to creature artificial vision for the blind.

A human visual system describes in figure 1. As for the human visual system, a light energy becomes a shadow with a reversed phase in the retina, and it transmits a displayed neural signal to ganglion cell and, as for this image, it is through a cognition process on delivery with primary visual cortex through optic nerve later by the retina. The retina located behind a pupil is complicatedly composed of ten vertical layers. However, since ganglion cell is composed with only about 1,000,000 one eye, a lot of visual information is compressed for moving from 125,000,000 receptors to ganglion cell. Actually it is turned into a neural signal by an operation of the retina, but the sampling image is transmitted to primary visual cortex of a brain through optic nerve of the retina which listened to this signal in 125,000,000 rod cells and 6,000,000 con cells. Because it is composed of about 1,000,000 medical charges of this ganglion cell, there is enable data compression about 130:1 information. That is, human recognizes an object with visual information transfer with primary visual cortex in 1,000,000 ganglion cell, but is not displaying a particular problem [5-6].

Also, one fovea sampling of non-uniform is performed in receptor level. This means that sampling is dense in a fovea region, but loose in peripheral region because output of a con cell is transmitted from peripheral region to ganglion cell in fovea by more limited dynamic range. This process means the important data compression occurs [4]. It is the place where a process is originated conversion of visual information becomes quit at the same time because the retina has both a receptor and neuron.

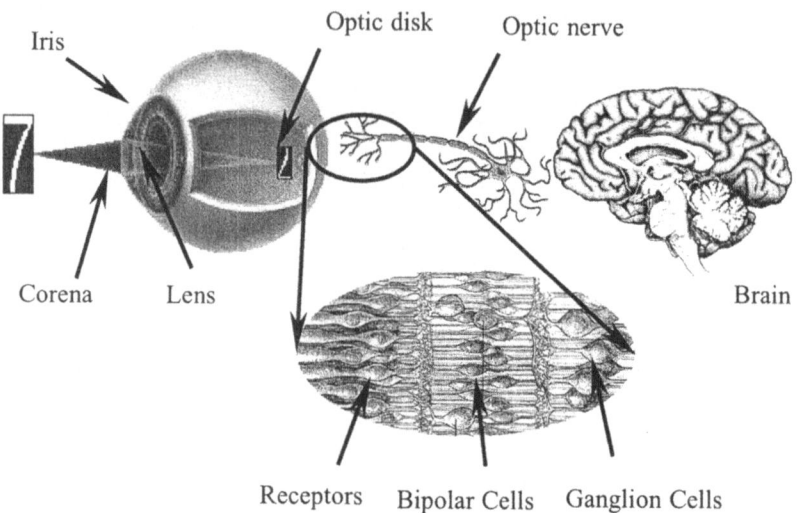

Fig. 1. Human visual system

2.2 Compression

Image compression technology considers an important characteristic of an image as well as provides a clear image in high compression. Because. DCT (Discrete Cosine Transform) mainly used in the conventional image compression technique divides an image into regular size if compression ratio rises, it displays blocking phenomenon, and quality of an image brings the results to become lower remarkably too [8][12]. The LOT (Lapped Orthogonal Transform) and wavelet transform which overlap each division has been used in order to overcome these disadvantages. Especially, wavelet transform which carry out decomposition and frequency about the image of whole surface and resolution removes a correlation among vectors while preserving correlation among coefficients unlike the existing division conversion.
Wavelet transform is adopted in JPEG2000; it has the following characteristics [11].

- Image quality to be more excellent than the existing standard proposal
- Excellent low bit-rate
- Loss and lossless compression are possible with one algorithm
- The progressive transmission that used resolution

The wavelet transform expresses well the local characteristic of spatial or time - frequency, as is MRA. So, it used image analysis and signal processing. The wavelet is a signal on which energy is concentrated on definite time (*t*); we can create a wavelet function with the scaling and translation of mother wavelet (φ) in equation (1).

$$\varphi^{a,b}(t) = |a|^{-1/2} \varphi((t-b)/a) \tag{1}$$

Because it needs so many calculations to calculate wavelet coefficients about every scale (*a*) and every location (*b*) in the wavelet transform, it is proved that if we do the wavelet transform in the moment the scale and location are at 2^n, we could get an effective and correct result. So, it called the DWT [9]. Mallat [10] suggested the scheme that we could effectively discrete transform with using filters. The Wavelet transform for the discrete signal can be accepted by using the FIR (Finite Impulse Response) which is digital filter originated from regular orthogonal based function. We will compare a characteristic which occurs during moving data from retina to ganglion cells of a human visual system in human visual system to that of general image recognition model in this paper.

3 Proposed Model

According to the current computer vision theory, the original image received goes though extracting feature of an object and a process of recognizing an object. A conventional computer vision process is same as figure 2. That is, it imitates human visual characteristics of extracting information in computer vision and is recognizing an object. However, the algorithm of current computer has been developing as a different form from human visual information processing in the actual application. Study about information processing of a human brain and a visual information processing are actively attempted recently. And a human visual information processing is gradually figured out. Currently study of these models is attempted actively each nation in the world. [1-4]. Computer vision is emulating a rough form of human visual information processing, and the information processing is showing them the form that is different from a human information processing [12].

In this paper, we propose a model about the visual information compression adapted to compression which human actually recognizes an image. The information handed down the human retina little loses critical information in ganglion cell and gets utmost compression.

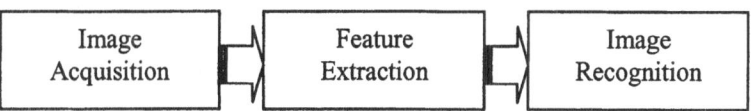

Fig. 2. Conventional computer vision for image recognition

These characteristics are very similar to wavelet transform. Information of low-band has very much delicate information to express original image in wavelet transform and much information of original image exists. As for theses, a mapping process of parvo-cellular of ganglion cell (P-cell) and the mechanism are very alike in receptor of the retina. Also, information of high-band of wavelet transform has edge information of original image compared to low-band which has a little information. These mechanisms are very similar to a mapping process of magno-cellular of ganglion cell (M-cell). Therefore, P-cell of ganglion cell deals with main information of an image like low-band, and M-cell is dealing with edge information of an image like the high-band. In this paper, we used wavelet transform in a compression process of this visual information and composed the model. We implemented the image recognition model in figure 3.

If a discrete signal is $c_0^M, c_1^M, \cdots, c_{N-1}^M$, $N = 2^M$ from sampled to original signal $f(t)$, we used equation (2) or (3) to show that wavelet transformed original image and calculated low-frequency, high-frequency filter of wavelet [11].

$$c_k^{J-1} = \sum_{m=0}^{N-1} h(m) c_{M+2k}^{J} \qquad (2)$$

$$d_k^{J-1} = \sum_{m=0}^{N-1} g(m) d_{M+2k}^{J} \qquad (3)$$

We applied low-frequency filter and high-frequency to a vertical direction and shared band with horizontal direction on original image and applied twice of filtering repeatedly again on a low-band section and use 3 level mallat tree to divide high-band section. Because 98% of high-band does division, we considered a diagonal, vertical, horizontal direction and did zigzag scan and did encoding in order to do encoding. We used the image that we passed through a decoding process and a reverse-quantization process again, and it was compressed encoding image on an image recognition process.

A lot of current developed neural network have so many algorithms. In each algorithm, application fields to meet in a characteristic exist. If it is necessary to classify data into similar characteristic, ART2 (Adaptive Resonance Theory) and a SOFM (Self-Organizing Feature Map) are frequently used.

In this paper, we did not give an answer about an input pattern having been given and adopted ART2 and the SOFM which were the non-supervised learning algorithm in which learning was possible. Especially, ART2 and SOFM the algorithm are similar to a biology recognition system. Specially, as for the SOFM, the practice speed is quite fast with neural network model proposed by Kohonen, and adaptability is excellent in a real-time with learning process and various changes about an input pattern. SOFM is the model in which application is possible on a binary input pattern and analog input pattern. And it is the same, and a connection weight change of SOFM reacts to a cluster generation having an average of all input patterns [7][10].

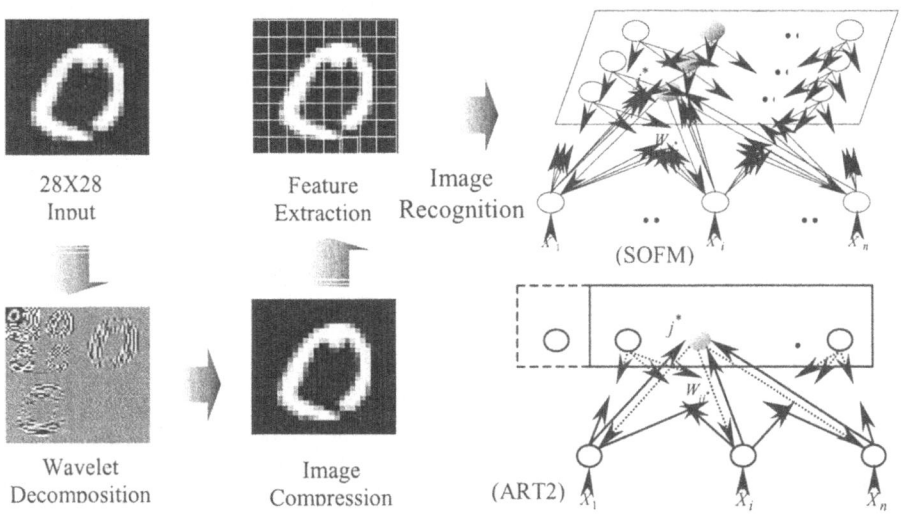

Fig. 3. Image recognition model based on retinal ganglion cell (X: image, j: cluster, W: connection weight)

4 Experimental Results

We used Visual C++ 6.0 in a Pentium 1.7GHz, 256MB memory, and windows XP environment and implemented the image recognition model based on human visual information processing that we proposed in this paper. We used MNIST (database of the AT&T Corp. which it was used in a lot of paper on handwritten off-line number in figure 4 at this paper and tested. It is MNIST database newly compounding database of original NIST (National Institute of Standards and Technology). This is database using handwritten digit 0-9 in learning and recognition. NIST database is the binary image which was normalized with 20X20 sizes while keeping horizontal vertical ratio. The MNIST fitted size to 28X28 with the 8-bit gray-scale image that anti-aliasing handled database of NIST. The training data set selected 5,000 in 60,000. And, as for the test data set, a random sampling selected 5,000 in 10,000. And the person who wrote a number is not same.

In this paper, we experimented on the recognition modeling which we used ART2, SOFM neural networks. We experimented with normal data and compressed data from retina to ganglion cell of a proposed model. The parameter which is necessary equally did learning rate and learning epoch of algorithm used in experiment. We experiment results is shown by following in table 1.

Fig. 4. MNIST off-line handwritten number database

Table 1. Experiment of off-line number recognition

bpp	MSE	PSNR	Recognition rate
0.75	0.9040	48.5689	93.45%
0.50	0.9040	48.5689	91.66%
0.25	5.7965	40.4992	88.69%
0.125	19.0545	35.3308	88.09%
0.0625	49.1714	31.2137	85.11%
0.0078	276.2008	23.7186	77.38%

In the table 1 ART2, SOFM algorithm was experimented with 5,000 training data. The experiment results, in ART2, the general recognition model recognized 5,000 characters showing 93.69% of recognition performance. But the machine vision technology used in real world is still applied restricted fields. And its information processing is developing into a different form with that of human visual system because of the limited application range. Compressed data of the proposed model showed similar performance in the visual information processing. Human's memory tends to degenerate with time. Human's recollection ability also fades with time, which makes their recognition less accurate. Meantime, due to the limited capacity of human brain, humans tend to compress old memories to recollect. Thus, this dissertation took the recollection ability into account, and conducted different experiments according to the compression ratio. Recognition rate decreased as compression ratio increased.

Human visual information processing system goes through the process in which a great deal of information is compressed by ganglion cell in the retina, but does not feel a particular problem on an actual recognition process. These reasons are because important information 'what' compresses with minimum loss in a compression process of information. And the information 'where' which is not important has utmost compression. Therefore, the results modeling did a compression process of a human visual information processing. And also general recognition model and performance difference were able to know null.

5 Conclusion

The proposed model did a change from photoreceptor to ganglion cell based on human visual information processing. Machine vision technology used in real world is still applied restricted fields. And ist information processing is developing into a different form with that of human visual system because of the limited application range.
In simulation results, there is not significant recognition result between conventional computer vision and proposed model. Future study, it is modeling from ganglion cell to primary visual cortex. And actual recognition performance will be improved more if modeling is finished from primary visual cortex to recognition step. The proposed model, It can be applied to image recognition, object tracking, artificial vision.

Acknowledgment. This work was supported by grant No.R01-2002-000-00202-0 from the Basic Research Program of the Korea Science & Engineering Foundation.

References

[1] M. fisxhler, O. Firschein, Intelligence: The eye, the brain and the computer, *Addison-Wesley*, 1987
[2] R. Arnheim, J. O. Kim, Translation, Visual thinking, *Ewha Woman Univ. Press*, 1982
[3] J. H. Cho, "Learning data composition and recognition using fractal parameter" *IJCNN'99*, Washington D.C., July 1999
[4] Brain Science Research Center, "A study on artificial vision and hearing based on brain information processing," *BSRC Research Report*, 2001
[5] I. S. Lee, Eye and Computer, *Computer-World Publishing Company*, 1989
[6] I. S. Lee, Human and Computer, *Gachi Publishing Company*, 1992
[7] D. S. Kim, Neural Network Theory and Application, *High-Tech Information*, 1992.
[8] J. M. Shapiro, "Embedded Image coding using zerotrees of wavelet coefficients," *IEEE Trans. on Signal Processing*, Vol. 41, No. 12, pp. 3445-3462, Dec. 1993.
[9] S. Mallat, "Multi-Frequency Channel Decomposition of Images Wavelets Models," *IEEE Trans. on Information Theory*, Vol. 11, No. 7, July 1992.
[10] S. Haykin, Neural Networks: A Comprehensive Foundation, *MacMillan*, 1994.
[11] S. H. Hong, "JPEG2000 Still-Image Encoding," *IDEC2002, IDEC*, 2002
[12] R. C. Gonzalez, Richard E. Woods, Digital image processing, Second edition, *Prentice Hall*, 2001

[13] Y. LeCun, L. D. Jackel, L. Bottou, C. Cortes, J.S. Denker, "Learning algorithms for Classification: A Comparison On Handwritten Digit Recognition," *Neural networks*: The Statistical Mechanics perspective, 1995
[14] D. Hubel, T. N. Wiesel, Brain Mechanisms of Vision, Neuro-Vision, *IEEE Press*, 1994.
[15] W. H. Dobelle, "Artificial Vision for the Blind by Connecting a Television Camera to the Visual Cortex," *ASAIO journal*, pp. 3-9, 2000
[16] C. S. Burrus, R. A. Gopinath and H. Guo, Introduction to Wavelets and Wavelet Transforms, *prentice hall*, 1998.
[17] S. Mallat, A Wavelet tour of Signal Processing, Academic Press, 1998.
[18] J. M. Zurada, Introduction to Artificial Neural Systems, *Boston: PWS Publishing Company*, 1992.
[19] M. L. Minsky, and S.A. Papert, Perceptrons, Cambridge, *MA: The MIT Press*, First edition, 1969, expanded edition, 1988.
[20] T. Kohonen, Self-Organizing Maps, *Berlin: Springer-Verlag*. First edition was 1995, second edition 1997.
[21] K. I. Diamantaras, and S.Y. Kung, Principal Component Neural Networks, Theory and Applications, *NY: Wiley*, 1996.

A Simple Data Analysis Method for Kinetic Parameters Estimation from Renal Measurements with a Three-Headed SPECT System

Eleonora Vanzi[1] and Andreas Robert Formiconi[1,2]

[1] Dep. of Phatophysiology, University of Florence, 50134 Florence, Italy
e.vanzi@dfc.unifi.it,
[2] INFM Genova, 16146 Genova, Italy

Abstract. We present here a new method for the determination of renal kinetic parameters from time-activity curves (TACs) obtained in dynamic tomographic Nuclear Medicine studies. The method allows to achieve subsecond sampling rates when using three-headed Single Photon Emission Computed Tomography (SPECT) systems.
We used three-projections data sets in order to directly reconstruct region-of-interest (ROI) contents and to achieve an accurate description of the TACs. By defining a kinetic model for the behavior of the tracer we were able to relate each sampling point of the renal TAC to the correspondent point of the plasmatic TAC (input function).
We analyzed simulated projections in 81 cases of region segmentation errors: the values of the renal clearances were reproduced within about -12% and +4%. The vascular fractions were reproduced within the measurement errors.

1 Introduction

The diagnostic potential of Nuclear Medicine investigations derives mainly from the functional information provided by the images. During the last couple of decades, a significant improvement has been brought by the introduction of the tomographic concept by means of both Single Photon Emission Computed Tomography (SPECT) and Positron Emission Tomography (PET). By eliminating the problem of background which is superimposed on the regions of interest in conventional scintigraphic planar imaging, these techniques enable the measurement of radiopharmaceutical concentration. The conventional use of SPECT requires the radiopharmaceutical distribution to be constant during the acquisition of data. Therefore, the application of compartmental modeling for the determination of useful physiological parameters is limited. Dynamic SPECT, by obtaining a series of images in time, allows one to follow the time evolution of the radiopharmaceutical distribution and, therefore, the scope of studies based on compartmental modeling may be broadened. However, even with dynamic SPECT, instrumental constraints may cause inconsistencies in the data since the acquisition speed is not fast enough with respect to the variation of radiopharmaceutical distribution in time.

In [1] we proposed a new method for acquiring and processing data when three-headed SPECT system are used in dynamical SPECT studies with certain tracers. Basically, the idea exploited the fact that these systems, during each sampling interval, acquire three projections simultaneously. The three projections provide a complete, even if very coarse, angular sampling, being placed at 120 degrees angular steps. In dynamic studies, the tracer may be homogeneously distributed in a small number of regions-of-interest (ROI) thus allowing us to reconstruct the content of the ROI from three-projections data sets for each time sample. Actually, for conventional imaging, in which thousand of voxels are reconstructed, three views represent a severe angular undersampling whereas in these kind of studies, activity concentrates in few large volumes and, by using a method of reconstructing ROI contents directly from projections [2], three views actually give a quite accurate description of the TACs.

In [1] and in this paper we focused on dynamic SPECT studies of the kidneys for the determination of renal clearance values, an important parameter (expressed in ml/min) describing the ability of kidneys to remove certain substances from blood. In this kind of studies basically 5 ROIs are involved and we will refer to them by using an index α: the two kidneys (α=1,2), two cardiac ROIs (the left ventricular blood pool, α=3, and "the rest of the heart", α=4), and the background (α=5).

The solution of the problem requires to define both a geometrical model of the acquisition process as well as a biological model of the tracer kinetic. The first one involves spatial relationships between ROIs and projection bins whereas the second one involves time relationships between ROI contents and the input function. The input function, which is represented by concentration of tracer in the blood in function of time, is obtained from the TAC of the left ventricular blood pool region.

In our first approach, the problem to be solved was posed by writing relationships between each projection bins and quantities derived from the input function according with both the geometrical and the biological models. A very large system of equations stems from this approach and the details relative to its solution by means of the Direct Least Squares Estimation (DLSE) ([3] - [6]) have been given in [1].

The idea of this paper is that, even if the results obtained with the DLSE method were encouraging, we could try to estimate the renal clearances in a more simple and consistent way in order to improve both the feasibility and the robustness of the procedure. Basically, the idea consisted in writing relationships directly between ROI contents and quantities related to the input function.

As far as details about study protocol and data collection are concerned, we refer here exactly to the same situation described in [1].

The paper is organized as follows. Section 2 outlines the new approach to parameter estimation and in Section 3 we recall how it is possible to obtain a subsecond sampling interval for the time activity curves by exploiting the sampling geometry of a three-headed SPECT system. In Section 4 we shortly rewrite the equations we had to manage in the DLSE approach, in order to compare the two methods in Section 5. In Section 6 results based on simulated data are shown. Conclusions are reported in Section 7.

2 TACs-Comparison Method

First of all, we need to consider the kinetic model for the behavior of the tracer.

The Hippuran mean renal transit time is about 2 minutes and since our analysis regards data acquired within the first two minutes from injection, kidneys can be considered as trapping compartments and the amount of tracer in a kidney at time t, $A_\alpha(t)$ (mg), can be described by an accumulation term and a vascular part:

$$A_\alpha(t) = K_\alpha \int_0^t C(\tau)\, d\tau + V_\alpha\, f_\alpha\, C(t) \qquad \alpha = 1, 2\ . \tag{1}$$

In this equation K_α is called renal clearance (ml/min), $C(t)$ is the input function (tracer concentration in the plasma, mg/ml), V_α is the volume of the kidney (ml) and f_α is an adimensional constant that includes both the hematocrit ratio (the ratio between plasma volume and blood volume, $c \approx 3.5/5$) as well as the vascular fraction, that is the fraction of volume occupied by blood vessels in the region. Multiplying each side of (1) by the activity of the radiopharmaceutical per unit of mass, a ($counts/min/mg$), and dividing by the number of voxel N_α that compose the kidney, we can write the number of counts acquired in each voxel of a kidney during the time interval Δt of acquisition of angle j, $j \in (1, \ldots, J)$, in rotation i, $i \in (1, \ldots, I)$:

$$X_{\alpha,ij} = \frac{1}{N_\alpha} \int_{t_{ij}-\Delta t}^{t_{ij}} A_\alpha(t)\, a\, dt =$$

$$= \frac{K_\alpha}{V_\alpha} \int_{t_{ij}-\Delta t}^{t_{ij}} \int_0^t a\, \beta\, C(\tau)\, d\tau\, dt + f_\alpha \int_{t_{ij}-\Delta t}^{t_{ij}} a\, \beta\, C(t)\, dt \qquad \alpha = 1,2 \tag{2}$$

where β is the conversion factor $ml/voxel$ ($\beta = V_\alpha/N_\alpha$).

Let us now consider the left ventricular blood pool ($\alpha = 3$): this region is simply full of blood and so we can suppose it to have a tracer concentration proportional to the plasma concentration:

$$A_3(t) = V_3\, c\, C(t) \tag{3}$$

where the proportionality constant is just the hematocrit ratio.

The number of counts acquired in each voxel of the left ventricular blood pool during the acquisition of angle j in rotation i will then be given by:

$$X_{3,ij} = C_{ij} = \int_{t_{ij}-\Delta t}^{t_{ij}} a\, \beta\, c\, C(t)\, dt\ . \tag{4}$$

We will call this amount C_{ij}, in order to remember that it is strictly related to the plasma input function $C(t)$, and W_{ij} its cumulative integral:

$$W_{ij} = \int_{t_{ij}-\Delta t}^{t_{ij}} [\int_0^t a\,\beta\,c\,C(\tau)\,d\tau]\,dt \ . \tag{5}$$

The basic equation (2) for the kidneys can then be written as:

$$X_{\alpha,ij} = \frac{K_\alpha}{c\,V_\alpha} W_{ij} + \frac{f_\alpha}{c} C_{ij} \qquad \alpha = 1,2 \ . \tag{6}$$

Since the volumes of the regions and the value of c are known, if we are able to measure the number of photons emitted in a voxel of the kidney and in a voxel of the left ventricular blood pool for each time interval $[t_{ij} - \Delta t,\ t_{ij}]$, we can easily determine the unknowns K_α and f_α. Let us observe that even if (6) is a system of IJ equations, just 1/3 of them are significant. In fact, a 3-head SPECT system acquires 3 projection angles simultaneously and it means that $X_{\alpha,ij}$, C_{ij} and W_{ij} must assume the same value for that 3 values of j that are the indices of 3 simultaneous views. We can translate this into equations by substituting the index j with a double index lm in which $l = 1,..,L = J/3$ identifies the specific set of 3 views and $m = 1,2,3$ enumerates the views: i and l are the indices involving time dependence. System (6) becomes:

$$X_{\alpha,il} = \frac{K_\alpha}{c\,V_\alpha} W_{il} + \frac{f_\alpha}{c} C_{il} \qquad \alpha = 1,2 \tag{7}$$

or, in matrix form,

$$\mathbf{X}_\alpha = \mathbf{M}\mathbf{U}_\alpha \qquad \alpha = 1,2 \tag{8}$$

in which $\mathbf{X}_\alpha$ is the vector of the sampling points of the renal TAC:

$$\mathbf{X}_\alpha = (X_{\alpha,11}, ..., X_{\alpha,1L}, ..., X_{\alpha,I1}, ..., X_{\alpha,IL})^T \ , \tag{9}$$

$\mathbf{U}_\alpha$ is the vector of the unknown kinetic parameters for kidney α:

$$\mathbf{U}_\alpha = (\frac{K_\alpha}{c\,V_\alpha}, \frac{f_\alpha}{c})^T \tag{10}$$

and $\mathbf{M}$ is the $[\,IL \times 2\,]$ matrix whose columns are, respectively,

$$\mathbf{W} = (W_{11}, ..., W_{1L}, ..., W_{I1}, ..., W_{IL})^T \tag{11}$$
$$\mathbf{C} = (C_{11}, ..., C_{1L}, ..., C_{I1}, ..., C_{IL})^T \ .$$

The least squares solution of (8), for an unweighted fit, is given by:

$$\mathbf{U}_\alpha = [\,\mathbf{M}^T\mathbf{M}\,]^{-1}\,\mathbf{M}^T\,\mathbf{X}_\alpha \tag{12}$$

and, since the matrix $[\,\mathbf{M}^T\mathbf{M}\,]$ we have to invert is just 2×2, we can calculate the solution directly from (12), without using any iterative method.

3 Estimation of the Time-Activity Curves

We begin with the statement of the tomographic problem in the following general way:

$$p_{ilmn} = \sum_k F^k_{lmn} Y_{k,il} \qquad (13)$$

where p_{ilmn} is the measured projection at revolution i, in view m at gantry position l, bin n; $Y_{k,il}$ is the number of emitted photons in voxel k during the acquisition of set l in rotation i and the elements F^k_{lmn} describe the acquisition process as well as the geometrical system response [7].

Next, let us remember that our image space has been divided into a number ($N_R = 5$) of regions with constant content such that $Y_{k,il} = X_{\alpha,il}$ for all values of k belonging to the region α ($k \in R_\alpha$). Equation (13) becomes

$$p_{ilmn} = \sum_{\alpha=1}^{N_R} \left(\sum_{k \in R_\alpha} F^k_{lmn} Y_{k,il} \right) = \sum_{\alpha=1}^{N_R} X_{\alpha,il} \sum_{k \in R_\alpha} F^k_{lmn} \,. \qquad (14)$$

We call "static sinogram" of a region α the amount:

$$G^\alpha_{lmn} = \sum_{k \in R_\alpha} F^k_{lmn} \qquad (15)$$

and, as we can see by rewriting (14) in the following way,

$$p_{ilmn} = \sum_{\alpha=1}^{N_R} G^\alpha_{lmn} X_{\alpha,il} \qquad (16)$$

the sinogram of a region weighs the contribution of that region to a generic bin of projection.

If we call $\mathbf{X}_{il} = \{X_{\alpha,il}\}_{\alpha=1}^{N_R}$ the $[N_R \times 1]$ concentration vector relative to gantry position l of revolution i, $\mathbf{p}_{il} = \{\{p_{ilmn}\}_{m=1}^3\}_{n=1}^N$ the $[3 \cdot 128^2 \times 1]$ projection data set relative to gantry position l of revolution i and finally $\mathbf{G}_l = \{\{\{G^\alpha_{lmn}\}_{\alpha=1}^{N_R}\}_{m=1}^3\}_{n=1}^N$ the $[3 \cdot 128^2 \times N_R]$ region projection matrix for the gantry position l, the least squares solution of (16) is

$$\mathbf{X}_{il} = [\mathbf{G}_l^T \mathbf{G}_l]^{-1} \mathbf{G}_l^T \mathbf{p}_{il} \qquad (17)$$

and, for all i and l values, it describes the time-activity curves in all the regions. Examples of plasmatic and renal time-activity curves reconstructed from noisy simulated data are shown in figure 1.

Starting from the ventricular TAC, coefficients W_{il} (5) can be numerically calculated by:

$$W_{il} = \Delta t \left(\sum_{i'=1}^{i-1} \sum_{l'=1}^{J/3} C_{i'l'} + \sum_{l'=1}^{l-1} C_{il'} + \frac{C_{il}}{2} \right). \qquad (18)$$

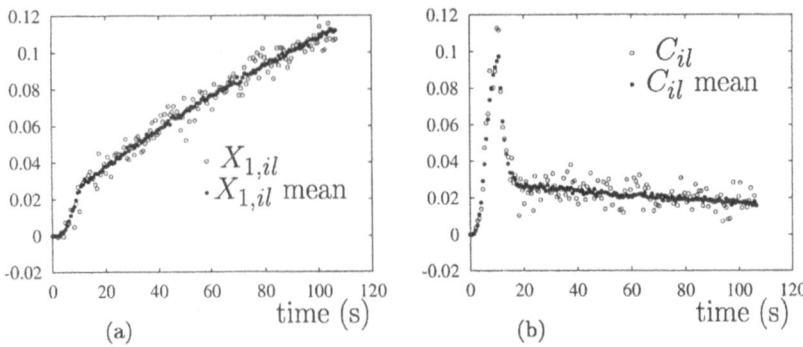

Fig. 1. (a) TACs for the right kidney. (b) TACs for the left ventricular blood pool. White circles have been obtained starting from a single noisy simulated data set; black circles are the mean values over 30 noisy data sets.

4 DLSE Method

In the DLSE approach a theoretical expression for the generic bin of projection p_{ilmn} has to be written in order to estimate the kinetic parameters by minimizing the difference between theoretical and experimental projections. The expression we wrote in [1] was:

$$p_{ijn} = G^1_{jn}\left[\frac{K_1}{c\,V_1}W_{ij} + \frac{f_1}{c}C_{ij}\right] + G^2_{jn}\left[\frac{K_2}{c\,V_2}W_{ij} + \frac{f_2}{c}C_{ij}\right] +$$
$$+ G^3_{jn}C_{ij} + G^4_{jn}\frac{f_4}{c}C_{ij} + G^5_{jn}\frac{f_5}{c}C_{ij}. \quad (19)$$

In order to write it, we considered all the 5 ROIs: the kinetic model we assumed for kidneys and left ventricular blood pool was the same we have described in Section 2. For the "remaining part" of the heart ($\alpha = 4$) and the background ($\alpha = 5$), we supposed activity to be proportional to the input function.

Equation (19) is a system of IJN equations in 6 unknowns, that we can rewrite in matrix form:

$$\mathbf{Rx} = \mathbf{b} \quad (20)$$

where

$$\mathbf{R} = \begin{pmatrix} G^1{}_1W_{11} & G^2{}_1W_{11} & G^1{}_1C_{11} & G^2{}_1C_{11} & G^4{}_1C_{11} & G^5{}_1C_{11} \\ \vdots & \vdots & \vdots & \vdots & \vdots & \vdots \\ G^1{}_JW_{1J} & G^2{}_JW_{1J} & G^1{}_JC_{1J} & G^2{}_JC_{1J} & G^4{}_JC_{1J} & G^5{}_JC_{1J} \\ \vdots & \vdots & \vdots & \vdots & \vdots & \vdots \\ G^1{}_1W_{I1} & G^2{}_1W_{I1} & G^1{}_1C_{I1} & G^2{}_1C_{I1} & G^4{}_1C_{I1} & G^5{}_1C_{I1} \\ \vdots & \vdots & \vdots & \vdots & \vdots & \vdots \\ G^1{}_JW_{IJ} & G^2{}_JW_{IJ} & G^1{}_JC_{IJ} & G^2{}_JC_{IJ} & G^4{}_JC_{IJ} & G^5{}_JC_{IJ} \end{pmatrix} \quad (21)$$

and
$$\mathbf{G}^\alpha{}_j = [G^\alpha_{j1}...G^\alpha_{jN}]^T, \quad \alpha = 1,..,5.$$

Vector **b** is given by

$$\mathbf{b} = \mathbf{p} - [\mathbf{G}^3{}_1{}^T C_{11}...\mathbf{G}^3{}_J{}^T C_{1J}...\mathbf{G}^3{}_1{}^T C_{I1}...\mathbf{G}^3{}_J{}^T C_{IJ}]^T$$

where $\mathbf{p} = \{\{\{p_{ijn}\}_{i=1}^I\}_{j=1}^J\}_{n=1}^N$ is the vector of projections.

The vector of unknowns is given by

$$\mathbf{x} = \left(\frac{K_1}{c\,V_1}, \frac{K_2}{c\,V_2}, \frac{f_1}{c}, \frac{f_2}{c}, \frac{f_4}{c}, \frac{f_5}{c}\right). \tag{22}$$

Once coefficients C_{ij} and W_{ij} had been estimated in the way we have described in Section 3, the linear system could be solved by some iterative method.

5 Comparison

An outline of the two methods has been given in the previous sections. Now we can compare the two approaches.

1) Computational task: in both the methods we have to determine the regions of interest by means of some segmentation procedure. Then we have to project them in order to produce their sinograms and reconstruct the time activity curves by eq. (17). The two approaches differ in the next step, in the way they estimate the unknown kinetic parameters. In the DLSE approach the estimates are obtained by solving system (20), a system of $[IJN]$ equations in 6 unknowns. For example, in the acquisition protocol we considered ($I = 8$, $J = 60$, $N = 128^2$), we had 7864320 equations, that could be solved only by an iterative algorithm (in our case LSQR, [8]). In the approach we are proposing in this paper, we have, instead, to solve a linear system (8) for each kidney, thus 2 systems of $I \times J/3$ equations in 2 unknowns (for a total of 320 equations in 4 unknowns) and we are able to calculate directly the least squares solution (12) without using any iterative approach.

In conclusion, in the new method the number of equations reduces of a factor $1.5N$ (about 25000 in our case!) since it does not use a relationship for each bin of projection but just a relationship for each sampling interval. The total number of unknowns also reduces, since (7) directly relates kidneys and plasma and there is no need to consider the distribution of activity in any other region.

2) Management of ROIs and of the geometrical model: in both the methods it is important to achieve a good geometrical identification of the ROIs, in order to produce the correct sinograms. But, while in the new method equation (17) is the unique step in which sinograms are involved, in the DLSE approach we use them again to write the theoretical expression for the projections (19).

3) Management of projection data: in both the methods the experimental data are used in equation (17) for the determination of the time activity curves. But, while in the new method equation (17) is the unique step in which

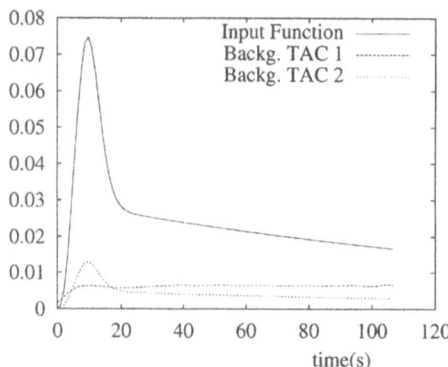

Fig. 2. Comparison between background Time-Activity Curves: in [1] we supposed proportionality between the background TAC (TAC 1) and the input function, but the real background TAC is quite different (TAC 2, smoothed patient data).

projections are involved, in the DLSE approach projections are used to determine the kinetic parameters by solving equation (20).

4) **Management of the kinetic model:** the tracer kinetic model we are assuming in the two methods is the same in what concerns kidneys and the left ventricular blood pool. But, while the description of the behavior in these regions ($\alpha = 1, 2, 3$) is sufficient to write equation (7), in the DLSE approach we also need to describe the kinetics in the "remaining part of the heart" ($\alpha = 4$) and in the background ($\alpha = 5$), in order to write equation (19): here we have assumed that the concentration in these regions is proportional to the input function. This hypothesis seems not to be true in patient studies for the first 20-30 seconds, since the background curve does not show the peak we can observe in the input function (fig.(2)): only after this time span, when both the background TAC and the input function become quite constant, the assumption of proportionality can be thought reasonable. In the new method we can drop all the assumptions for regions 4 and 5: it is sufficient the activity to be constant in each ROI.

6 Results

In order to make a general comparison between the two methods we applied the new approach to the same set of simulated data we processed in [1] with the DLSE method. The basic data needed to generate the software phantom and the input function were obtained by analyzing patient data. The phantom had the five three-dimensional regions listed above and the input function we used is shown in fig. (2). We obtained C_{il} and W_{il} values by means of numerical integration. For a more exhaustive description of the phantom and of the whole simulation set up see [1]. The model input function and the phantom ROIs were used to produce a simulated matrix **R** that we applied, see (20), to this parameter set ($K_1= 265$, $K_2=245$, $f_1=0.06$, $f_2=0.05$, $f_4=0.73$, $f_5=0.12$) in order to produce synthetic projections. Those data were successively perturbed with Poisson noise: noise level was adjusted in order to reproduce the situation of a typical dynamical SPECT study in which 100000 counts can be collected in a single rotation. Thirty different noise realizations were produced, thus giving rise to 30 noisy data sets. Synthetic data were then processed exactly in the same way we would have processed true patient data: first, we reconstructed images by Filtered Back Projection (FBP) and then we segmented them with a simple threshold method. Since resolution in images produced by emission data is very poor, it is very difficult to accurately delineate ROIs. That is why in our simulations we have devoted a fair amount of effort to assessing the robustness of parameter estimation with respect to region segmentation: we have considered three different cases of segmentation errors ($\pm 10 - 15\%$ and no-error) over the volumes of the regions $\alpha = 1, 2, 3, 4$ and we have reconstructed parameter values in all the 3^4 combinations of segmentation errors. For all the cases we analyzed 30 noise realizations in order to estimate mean and standard deviation of the distributions of kinetic parameters.

Lowest and highest parameter estimates are reported in table 1. For the clearance values, that are the most relevant kinetic parameters, quartiles are also reported. Comparing results with the ones in [1], we can observe that the new approach always gives a bit smaller (at most of about 1.5%) clearances estimates, but the results of the two methods are always consistent within the statistic uncertainty. In all the cases we studied we never underestimated K_1 and K_2 more than 12% and never overestimated them more than 4%. The estimates of the vascular fractions are instead a bit higher than the results obtained with the DLSE approach, that seemed to slightly underestimate these parameters.

In order to test how a wrong assumption about the kinetic behavior in the background region can affect the estimates in the DLSE approach, we also produced 30 new sets of noisy projections in which the background time-activity curve was not proportional to the input function. These data have been produced by substituting the C_{ij} coefficients in the sixth column of matrix **R** (21) with coefficients F_{ij} describing the new behavior of the background and then applying **R** to the same vector of parameters; F_{ij} have been obtained from patient data (TAC 2 in figure (2)). We analyzed these data in the hypothesis of

Table 1. Lowest and highest parameter estimates ($\pm\sigma$ relative to 30 sets of noisy projections) obtained in presence of segmentation errors. Quartiles for the clearance values are also reported. In parentheses results of [1] are shown.

	K_1	K_2	f_1	f_2
true values	265	245	0.06	0.05
lowest	233 ± 7	216 ± 7	0.05 ± 0.02	0.03 ± 0.02
	(235 ± 6)	(218 ± 7)	(0.04 ± 0.02)	(0.02 ± 0.02)
highest	273 ± 7	254 ± 6	0.07 ± 0.02	0.05 ± 0.02
	(276 ± 6)	(257 ± 6)	(0.06 ± 0.02)	(0.04 ± 0.02)

quartiles	K_1	K_2
25th	249	231
	(252)	(234)
50th	257	238
	(260)	(241)
75th	264	245
	(266)	(248)

no segmentation errors: in the DLSE approach we found $K_1 = 278 \pm 4$ and $K_2 = 260 \pm 4$ (+5% and +6% respect to the real values) and absurd values for the vascular fractions. As expected, the new method seemed to give better estimates for clearances ($K_1 = 263 \pm 4$ and $K_2 = 245 \pm 4$) and vascular fractions were well reproduced too ($f_1 = 0.06 \pm 0.01$ and $f_2 = 0.05 \pm 0.01$).

7 Conclusion

In conclusion, from a mathematical-computational point of view, the new method seems to be much more simple and economic than the DLSE and, moreover, it seems much more linear and consistent even under the theoretical aspect: the problem is posed by two different kind of models, a physical-geometrical model, and a biological-temporal one. In the new approach, the two models are kept separate: the first is used in conjunction to projection data in order to calculate the sampling points of the TACs; the second is used to determine the physiological parameters starting from these curves. Even the DLSE approach was a two-step procedure, but the two steps were much more interdependent: projection data were used to determine coefficients C_{ij} by the physical-geometrical model, then they were again used to estimate the physiological parameters in a procedure that involved both the physical-geometrical model and the biological-temporal model at the same time.

References

1. Vanzi E., Formiconi A. R., Bindi D., La Cava G., Pupi A.: Kinetic parameters estimation from renal measurements with a three headed SPECT system: a simulation study. IEEE Trans. Med. Imag. (in press)
2. Formiconi A. R.: Least squares algorithm for region of interest evaluation in emission tomography. IEEE Trans. Med. Imag. **12(1)** (1993) 90–100.
3. Zeng G. L., Gullberg G. T., Huesman R. H.: Using linear time-invariant system theory to estimate kinetic parameters directly from projection measurements. IEEE Trans. Nucl. Sci. **42(6)** (1995) 2339–2346.
4. Huesman R. H., Reutter B. W., Zeng G. L., Gullberg G. T.: Kinetic parameter estimation from SPECT cone-beam projection measurements. Phys. Med. Biol. **29(5)** (1998) 973–982.
5. Reutter B. W., Gullberg G. T., Huesman R. H.: Kinetic parameter estimation from attenuated SPECT projection measurements. IEEE Trans. Nucl. Sci. **45(6)** (1998) 3007–3013.
6. Reutter B. W., Gullberg G. T., Huesman R. H.: Direct least-squares estimation of spatiotemporal distributions from dynamic SPECT projections using a spatial segmentation and temporal B-splines. IEEE Trans. Med. Imag. **19(5)** (2000) 434–450.
7. Formiconi A. R., Passeri A., Calvini P.: Theoretical determination of the collimator geometrical transfer function for the reconstruction of SPECT data. IEEE Trans. Nucl. Sci. **46(4)** (1999) 1075–1080.
8. Paige C. C., Saunders M. A.: LSQR: an algorithm for sparse linear equations and sparse least squares. ACM Trans. Math. Softw. **8(1)** (1982) 43–71.

Integrating Medical Imaging into a Grid Based Computing Infrastructure

Paola Bonetto[1,2], Mario Guarracino[3], and Fabrizio Inguglia[1]

[1] Dipartimento di Informatica e Scienze dell'Informazione,
University at Genoa, Italy
[2] INFM, Istituto Nazionale di Fisica della Materia,
Unità di Genova, Italy
[3] Institute for High Performance Computing and Networking - Naples branch
National Research Council, Italy

Abstract. Since the capillary diffusion of powerful computing systems Medical Imaging has undergone a profound transformation providing new ways to physicians to conduct research and perform diagnosis and prevention. In this work a computational environment for medical imaging purposes based on a grid architecture is used to build a medical imaging application. It results from the interaction of scientists devoted to the design and deployment of new tomographic reconstruction techniques, researchers in the field of distributed and parallel architectures and physicians involved with tomographic imaging. The single software, hardware and medical components are described. Their integration into a distributed framework provides mutual benefits to both the scientific and the medical community, and represents the basis for the creation of a global, scientific, virtual community that reaches beyond the limits of space, remotely sharing not only physical resources but also experience and knowledge.

1 Introduction

In recent years medical diagnosis based on the imaging of the human body has more and more heavily relied on a digital approach. Huge amounts of data produced by modern medical instruments need to be processed, organized, and visualized in a suitable response time: the need for an efficient underlying computational infrastructure appears evident. Despite of the considerable results that have been obtained the increasing demand of high performance computing and storage facilities is still an open and unsolved problem. The elegant solution of such a complex, continuously growing problem may involve the coordinated availability of various resources and geographically distributed skills; it may require a shared access to facilities, interpersonal communication, and the joined efforts of multidisciplinary groups.
In the last decade the World Wide Web has been widely accepted as a tool for distributing and accessing any kind of information and service. With the increasing demand of applications that use distributed resources, new technologies are arising that offer

the means to solve more advanced difficulties related to coordination and sharing capabilities, such as authentication, resource discovery, and dynamic allocation. Such technologies are known as Computational Grids [1]. In short, they represent the middleware infrastructure for scientific computing that allows to transversely share information, knowledge and competence among disciplines, institutions and nations.

In this work we describe an application for the management, processing and visualization of biomedical images that integrates a set of software and hardware components, or, more specifically, a set of grid collaborative applications useful to nuclear doctors. It results from the interaction of scientists devoted to the design and deployment of new tomographic reconstruction techniques, researchers in the field of distributed and parallel architectures and physicians involved in neural medicine. The current configuration of the system builds upon the previously developed environment Medigrid [2], adding new features such as an easy to use graphical interface, a newly designed Repository for an efficient management of the data and an extended testbed that geographically ranges over the whole country. The enhanced system has now come to offer several benefits: it promotes the creation of a virtual organization that shares resources across a computational grid; it enables the use of remote resources, such as medical acquisition devices, data processing algorithms, or high performing computing resources, in a transparent way; it supports the distant collaboration between doctors; finally, it is an open architecture that can be easily tested, maintained and extended.

This paper is organized as follows. The next section describes the state of the art of the two areas involved in the project, namely imaging equipment and data processing applications in the medical reality on one side, and computational architectures on the other side. Thereafter, the adds on to MedIGrid are presented in detail, with particular attention to every single medical, hardware and software component and to the integration of the latter into a common framework. A typical case of use of the application is described to better illustrate the system. The last section discusses the benefits that derive from such integration and the advantages that both medical and scientific communities can take from a joined work and interaction.

2 State of the Art

In the last few decades, exciting advances in new imaging systems, such as computer tomography, ultrasound, digital radiography, magnetic resonance, and tomographic radioisotope imaging, have provided medical personnel a great insight into the patient's conditions. New and better ways to extract information about bodies have been provided, reflecting anatomy, dynamic body functions and internal biologic changes as they occur in vivo. Among the many imaging modalities, an expanding role in diagnosis, surgical and radio-therapy is played by tomographic techniques, which produce an image of the signal contained in a slice across the human body. The nature of the signal depends on the kind of imaging device. For instance, with CT one obtains a description related to the densities of tissues. With PET or SPECT the distribution of a radiopharmaceutical tracer is registered. All these devices collect data related to the spatial distribution of the pertinent quantity within the body and then transfer such data to a control workstation that processes it, mapping it into an image.

The most common techniques to reconstruct tomographic images involve the solution of an inverse problem that usually is an ill-posed one. Issues related to the ill-posedness are particularly relevant as far as the quality of the resulting images is concerned [3]. Hence, the use of some complex reconstruction method in replacement to the more naïve and rough ones that are usually directly provided along with the acquisition device itself, may represent a fundamental contribution to the final interpretability of the data. This aspect has been a matter of notable concern in the research community and significant progress has been made lately and proposed in the literature.

Recent research has also made step forward in the direction of developing and deploying suitable computational architectures and efficient systems to quickly archive, retrieve and exchange large amounts of medical data and information across institutions and hospitals. An interesting solution is offered by the so-called electronic Picture Archiving and Communications System (PACS) [4, 5], an integrated system of digital products and technologies that has been evolving since the 1980s. However, a large number of workstations is usually needed to satisfy the requirements of the hospital. These are expensive and run proprietary software, which makes it impossible to exchange data among devices, thus hampering diagnosis.

The difficulties left unsolved by PACSs may be overcome by grid computing technologies that allow geographically distributed resources to be coupled and offer consistent and inexpensive access to resources irrespective of their physical location. They enable sharing, selection, and aggregation of a wide variety of geographically distributed computational resources, such as supercomputers, computing clusters, storage systems, data sources, instruments, and, interestingly, also people. Grid architectures can be transparently accessed as a single, unified resource for solving large-scale computational and data intensive applications. Several Grid projects– such as the European DataGrid (http://datagrid.web.cern.ch) and CrossGrid (http://www.eu-crossgrid.org) - have already been successfully carried out in several categories of applications, ranging from biomedicine to high energy physics and earth observations.

3 The Application

This section is devoted to a description of MedIGrid and the new features that have lately been added. First, all single components are presented. Then a typical case of use is described, showing how all components interact and fit together and how they are integrated into a single framework which is the environment the nuclear doctor will be working in.

3.1 Data Acquisition Devices: The Medical Hardware Resources of the Grid

Although the final environment we developed is potentially usable in any kind of medical imaging reality, in order to answer to some concrete needs of the clinical situation in our country we have focused on tomographic nuclear imaging devices. Specifically, we have considered the **SPECT** and **PET** scanners that are located at the Department of Neurophysiopathology of the S. Martino Hospital at Genoa, Italy, and at the Department of Physiopathology at the Careggi Hospital at Florence, Italy, respectively. The SPECT devices are a CERASPECT (Digital Scintigraphics Incorporated) used for brain scans only, and featuring an annular collimator, and a Philips (ex Marconi), three heads IRIX model. The PET device is the whole body Advance tomograph produced by General Electric, which we used for brain studies as well as for cardio, lung and other studies.

3.2 Hardware Components: The Computing and Storage Elements of the Grid

From a physical point of view the application relies on the Internet. Hence, all of the units described in this section have an Internet access. Together with the ones described in above, these units make up the new testbed of the system that, compared to older versions, now geographically spans across the country.

A **PC** located next to each tomographic device is used to retrieve the data as soon as it is acquired by the scanner and "expose" it to the Grid. Such PCs can hence be seen within the framework as the front end of the acquisition resources. These PCs are also overloaded with two additional roles, namely, they can be used as computing element for a non parallel reconstruction procedure of the data, and they can be utilized by the user as a physical access point to the application itself. Although overloading a single hardware component with three logical roles may seem inelegant, it is a cheap solution to offer several services, optimizing the CPU utilization.

A third **PC** has been configured as a web and grid server and plays the role of a portal to which users have to connect in order to log into the system. Hence, all packages needed to configure it as a Web Server have been installed here, namely apache_1.3.23, mm-1.1.3, mod_ssl-2.8.7-1.3.23, and openssl-0.9.6c. Some of the software imaging tools that are described below and that run as Java applets are also stored on this server, along with the Java software development kit, in order to perform communication with the clients through the Globus layer via an applet-servlet scheme as described next.

Two **PC clusters** are located in Naples. One is configured as a high performing computing element, running Meditomo, a parallel version of the reconstruction procedures [11]. The other is just used and designed as a storage element in order to save significant scans and data collections for future reference.

3.3 Middleware Components: Globus and ImageJ

For the implementation of the system we relied on two existing software products: the Globus Toolkit and ImageJ. **Globus** (http://www.globus.org) is a project focused on enabling the application of Grid concepts to scientific and engineering computing [6]. The Globus Toolkit [7] is the set of services and software libraries that we have used as the middleware between the physical nodes above introduced and the additional software we developed. It includes modules for security, information infrastructure, resource management, data management, communication, fault detection, and portability. More specifically, we have equipped the PCs connected to the acquisition devices with the CoG Kit (cog-0.9.12), that provides access to the Grid services through a Java framework, and the clusters and the portal with the Globus Toolkit 2.

ImageJ (http://rsb.info.nih.gov/ij) is a public domain Java image processing program developed at the National Institutes of Mental Health. It consists of a collection of tools to read, display, edit, analyze, process, save and print images in various formats. Its main features are speed, platform independence, free availability of the source code, and an open architecture that provides extensibility via plugins. We have developed a set of plugins for our specific needs, as mentioned in the next section.

3.4 Software Components: MedItools and the ImageJ Plugins

Here, we separately describe each single component we have designed to run either upon the Globus Toolkit or plugged into the ImageJ framework, focusing on the most recent newborns such as the graphical user interface and the Repository. Other components have been inherited from initial version of MedIGrid, and have been accurately described in previous works [2, 8]. The next section will provide a deeper insight into how all software modules interact. On the basis of the previously mentioned Globus services we have designed the following MedIGrid components.

The **Repository** manages both raw data as acquired by the medical equipment and the data that have already been reconstructed. Its implementation is based on the Globus services for authentication and data transfer. Datasets are organized in the repository according to their date of acquisition and can be retrieved by the user for visualization or further processing by simply browsing in a directory tree as illustrated in Fig. 1.

MedIMan is a collection of services that coordinate basic grid operations such as authentication, file transfers, data consistency checks, and submissions of the reconstruction jobs. Some of them are invoked through a common graphical interface shown in Fig. 2. Others can not be explicitly called but are transparently executed as a side effect of other events: for instance data transfer is automatically performed when the user reconstructs a dataset, which needs to be copied from the repository to a computing element. So far, transfer security is implemented using SSL tunnelling [9], but a Globus oriented solution based on GASS and GridFTP is planned for a future implementation. Once data transfer is completed, data consistency is verified and the reconstruction script is submitted to the local queuing system. This script executes MedITomo and transfers the reconstructed data back to the repository.

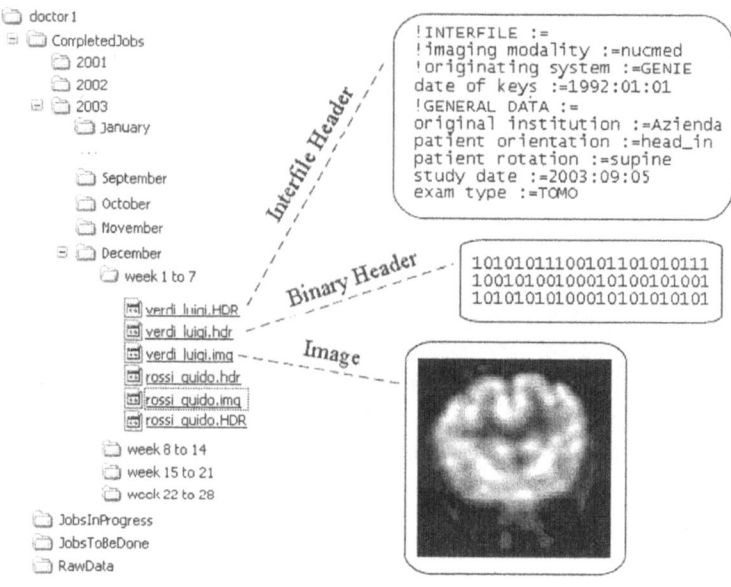

Fig. 1. Data organization in the Repository

According to the ImageJ philosophy we have organized several tools in terms of Java, platform independent plugins that can be freely downloaded from the ImageJ plugin page (http://rsb.info.nih.gov/ij/plugins). The ensemble of our tools leads the user through the main operations that can be applied to medical data: import and read, control the execution of tomographic reconstruction algorithms, explore, analyze and display the reconstructed volumes. None of these plugins is implemented using the Globus Toolkit. This means that they do not offer advantages of the grid technology such as transparent and secure data transfer, but, on the other hand, they can all be used as standalone tools, as well as within the MedIGrid environment. A detailed description of our tools can be found in [8] as well as on our web page (http://www.disi.unige.it/person/BerteroM/MedNuc), from which it is also possible to download the Java classes and the source files.

3.5 A Case of Use

In order to better understand how the above introduced components interact together a complete understanding of MedIGrid functioning may be helpful. Let us hence describe a typical case of use of the application. A doctor having a patient's data being acquired, needs to reconstruct them, and edit and analyze the resulting image. He may also want to store the results in a way that he or other colleagues, possibly from different working locations, may later be able to retrieve them. Since the reconstruction procedure may be time consuming, it is preferable to have it executed on a high performance computer: the necessary data transfers and remote reconstruction should be handled in a transparent way.

The system deals with the case in the following way. Once data are acquired they are transferred to the PC connected to the acquisition device. From any computer connected to the Internet the physician can authenticate himself with MedIGrid. This is done by visiting the URL https://grid.disi.unige.it:8443/GRID through a web browser, and logging in with one's username and password. On the server the user's encrypted certificate is stored, which is used for authentication: when the physician logs in, the certificate is decrypted and used to create a proxy credential with limited time validity, which will be transparently used by the application for all subsequent operations involved in the working session. Hereafter, a graphical user interface pops up into the user's browser, through which the latter can interact with the system.

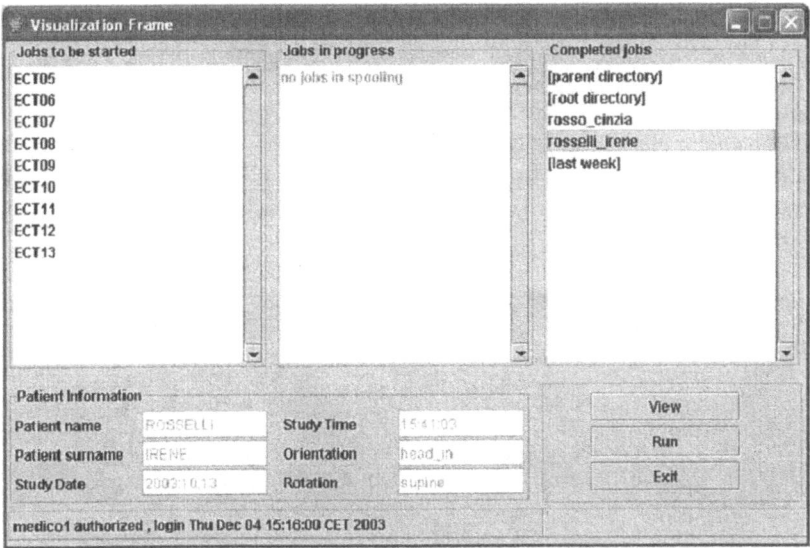

Fig. 2. The graphical interface through which the user can submit jobs and retrieve the reconstructions

As shown in Fig. 2 the interface is organized in three columns, holding the jobs to be started, the jobs in progress, and the completed jobs respectively. The columns are updated by the system in real time. By clicking on items in the left column new reconstructions jobs can be submitted, whereas by clicking on items in the right hand column, the physician can browse through the database of reconstructed images and select a specific one to be displayed. The lower part of the window is updated with the most relevant pieces of information for each currently selected item, such as the patient's identity and the study details.

When the user interacts with the application through the GUI shown in Fig. 2 a number of events transparently occur behind the scene. Most importantly, when a new reconstruction is submitted, a default configuration file is created with an optimal set of reconstruction parameters. These values can be freely modified by the user; however, the default configuration has been proven to lead to sufficiently accurate results of the reconstruction process and hence speeds up the working session and facilitates the physician's job, who is often not familiar enough with mathematical reconstruction settings. A series of data transfers transparently occurs: raw data are retrieved from the PC connected to the acquisition device and copied to the storage server and from here to the computational unit, i.e. the Beowulf in Naples, along with the configuration file. The file transfer uses SSL encryption to protect data privacy. Once data are copied, a checksum is performed to ensure no corruption occurred. A job is queued to the local scheduler: its purpose is to execute the reconstruction, transfer the data back to the storage server, and return to the user whether the job completed successfully or, in case an error occurred, what is the state of the computation. The reconstructed data can be accessed through the GUI and displayed on the user's monitor through an ImageJ plugin that runs as a Java applet. The latter is actually executed in combination with a Java servlet, thus providing the benefits of the applet approach for the design of the graphical user interface and of the servlets to overcome problems related to security, such as access to the remote File System where the results are stored. More specifically, the user accesses the results from his local PC through a GUI, which is an applet running on the portal. The applet connects to a servlet, which is also running on the portal. The Servlet performs its task (such as user authentication or data retrieval and transfer) using the API provided by the CoG Kit and the Globus Toolkit and finally returns to the applet.

4 Discussion and Future Work

MedIGrid represents the melting point of state of the art medical imaging techniques and the most recent advances in the field of high performance computing. It conjointly addresses problems of medical and computational/architectural nature thus providing cross-benefits to both the medical and the basic scientific community.

The most direct advantages offered to the medical community are facility of use and efficiency. The first is mostly due to the design that heavily relies on graphical user interfaces whose use is rather intuitive. The presetting and estimation of the optimal choices for reconstruction and analysis of the data, as well as the transparency offered by MedIGrid, by which dynamic resource allocation, data transferring, security and all other complex grid related problems are hidden to the final user, also significantly add to making our environment easy to approach and use.

Efficiency is guaranteed by the optimization of the mathematical models and algorithms that we have implemented, and by the underlying grid architecture, that dynamically gives access to distributed resources that may perform better than the locally available computing units. We have estimated that if the 3D reconstruction of a volume requires K time units, the 2D+1 reconstruction takes about K/N time units, where N is the number of transaxial planes of the volume (without any considerable loss of image quality, [10]). Under specific conditions, this value can be further reduced by a factor in the order of L if the reconstruction is run on a Beowulf PC cluster with L CPUs. Considering the processing power and the Internet bandwidth available at the time we are writing, by taking advantage of MedIGrid we are able to reconstruct a 128^3 volume from any geographical node of the grid in just only a few minutes, whereas the lowest computing time we were able to reach by locally running the corresponding 2D+1 reconstruction on a PC was in the order of half an hour. This apparently small difference has dramatic consequences on the way diagnosis and health care are led: indeed, it means that results on a patient's study can be provided almost immediately after the patient has been scanned, thus avoiding annoying waiting times, reservations and multiple hospital visits.

MedIGrid offers all the benefits of a virtual organization, in which members can share not only physical resources such as computing power and added storage capabilities, but also competences, experience and skills. Such an approach enables new ways for doctors to conduct research and diagnosis, and to benefit of resource sharing, distant collaboration and problem solving.

The benefits offered by a virtual organization to the medical community similarly apply to the basic scientific one: an open and widely accessible environment promotes the collaborative development of scientific software, stimulates in the direction of information sharing, and facilitates communication among different research groups. In this perspective, we focused on designing a product that could be easily and freely accessed and tested. We put special care on issues such as platform independence and remote execution and we tried as much as possible to rely on existing and emerging standards as far as choices such as image formats, programming language, and acquisition settings and protocols are concerned.

By being placed in the intersection of the medical and the basic scientific areas, MedIGrid simultaneously provides innovative solutions and advances to both fields. While the end user of the medical equipment can try out new reconstruction techniques, processing tools, and working environments and take advantage of new advances in imaging, the scientific community can benefit of an extended feedback from the end users of its products, receiving countless suggestions for future developments and valuable hints regarding the research direction to take.

In this regard we have been able to identify the main ways to improve and extend our project. A catalogue of both raw and reconstructed data produced by a variety of acquisition devices, as well as metadata regarding the patient, medical reports, diagnosis performed over the years, represents a rich database on which to carry out statistical studies and compare old and new reconstruction and processing techniques. The management of such data is a challenging task, in particular from the point of view of efficient distribution, retrieval and storage. Navigation and visualization tools need to be grid-enabled both with the use of techniques like data staging, caching and asynchronous striping, that help hiding variable latency and transfer speeds, and with algorithms capable of adapting to the dynamic behavior of the computing environment. Finally, in order to extend MedIGrid to a larger number of nodes and institutions, it is necessary to increase application scalability, dynamic resource discovery, and allocation, which allow to reserve and use resources in the most economic and efficient way.

5 Conclusion

MedIGrid is a grid enabled environment for medical diagnosis that integrates high performance computing platforms, last generation medical acquisition systems and recent advances in the field of tomographic imaging techniques. It has been designed with the aim of facilitating and optimizing reconstruction, display and analysis of medical images. Feedback from the medical community has proven this product to be useful and of undoubted interest.

Future work has to be oriented towards further stimulating information sharing and experience exchange in order to enable new ways for doctors to conduct research and diagnosis and to benefit of resource sharing and problem solving in a dynamic, multi-institutional organization.

References

1. I. Foster, C. Kesselman, "The Grid. Blueprint for a new computing infrastructure", Morgan Kaufmann, San Francisco 1999
2. M. Bertero, P. Bonetto, L. Carracciuolo, L. D'Amore, A. Formiconi, M.R. Guarracino, G. Laccetti, A. Murli, G. Oliva, MedIGrid: a Medical Imaging application for computational Grids, 17th International Parallel and Distributed Processing Symposium (IPDPS-2003), Nice Acropolis Convention Center, Nice, France, 22-26 April 2003, CD-ROM/Abstracts Proceedings, IEEE Computer Society 2003, ISBN 0-7695-1926-1
3. M. Bertero, P. Boccacci, "Introduction to Inverse Problems in imaging", Bristol: IOP Publishing, 1998
4. S. Bryan, G. C. Weatherburn, J. R. Watkins, et al. "The benefit of hospital-wide picture archiving and communication system: a survey of clinical users of radiology services", Br. J. Radiol., 72, 469-478, 1999
5. C. Creighton "A literature review on communication between picture archiving and communication systems and radiology information systems and/or hospital information systems", J. Digital Imaging, 12, 138-143, 1999
6. I. Foster, C. Kesselman, "The Globus Project: A Status Report", Proc. IPPS/SPDP '98 Heterogeneous Computing Workshop, pp. 4-18, 1998
7. I. Foster, C. Kesselman, "Globus: A Metacomputing Infrastructure Toolkit", Intl J. Supercomputer Applications, 1997
8. P. Bonetto, G. Comis, A.R. Formiconi, M. Guarracino, "A new approach to brain imaging, based on an open and distributed environment", 1st International IEEE EMBS Conference on Neural Engineering, 20-22 March 2003, Capri Island, Italy, CD-ROM Proceedings, IEEE Catalog Number 03EX606C, ISBN 0-7803-7819-9
9. A.O. Freier, P. Karlton, P.C. Kocher, "Secure Socket Layer 3.0", Internet Draft. IETF, 1996
10. P.Boccacci, P.Bonetto, P.Calvini, A.R.Formiconi, "A simple model for the efficient correction of collimator blur in 3D SPECT Imaging", Inverse Problems, 15, 907-930, 1999
11. L. Antonelli, M. Ceccarelli, L. Carracciuolo, L. D'Amore, A. Murli, "Total Variation Regularization for Edge Preserving 3D SPECT Imaging in High Performance Computing Environments", Lecture Notes in Computer Science, 2330:171-180, 2002

Integrating Scientific Software Libraries in Problem Solving Environments: A Case Study with ScaLAPACK

L. D'Amore[1], M.R. Guarracino[2], G. Laccetti[1], and A. Murli[1,2]

[1]Dipartimento di Matematica e Applicazioni,
University of Naples Federico II, Italy
{luisa.damore, giuliano.laccetti,
almerico.murli}@dma.unina.it
[2]Institute for High Performance Computing and Networking – Naples branch
National Research Council, Italy
{mario.guarracino}@na.icar.cnr.it

Abstract. We describe the reference software architecture of parIDL, a PSE for distributed memory parallel computers that integrates the parallel software library ScaLAPACK, needed to develop and test scientific applications related to adaptive optics simulations. It incorporates a run time support for the execution of the commercial problem solving environment IDL on parallel machines, whose development has begun in a project with astronomy researchers for adaptive optic simulation of terrestrial large binocular telescopes.

1 Introduction

Matlab (http://www.mathworks.com) and IDL (http://www.rsinc.com) are examples of Problem Solving Environments (PSE), which have in common a language with simple syntax, mathematical software libraries integrated with visualization tools, and capabilities to manage the overall computational process. PSEs meet scientists favors so well they have found widespread application, and user requests, in terms of computational power needed by handled simulations, have grown exponentially. To satisfy those needs the development of new tools for multicomputers begun, with a variety of approaches and related problems that will be detailed in the following section. What we provide here is not a general solution to the problem of integrating existing parallel libraries in PSE, but rather a methodology that can be applied to different environments.

In present work we describe the reference software architecture of *parIDL*, a runtime support for a PSE aimed to distributed memory parallel computers. The software provides a parallel virtual machine for the execution on parallel machines of the commercial problem solving environment IDL [20] and integrates the parallel software library *ScaLAPACK*. The development of *parIDL* has begun in a project with astronomy researchers for adaptive optic simulation of terrestrial large binocular telescopes [4], and has proved to significantly speed up computations involved in simulation needed to build the control mechanism of adaptive telescope lenses.

What we provide here are methods and mechanisms to incorporate tools for multicomputers, such as parallel scientific software libraries, visualization and navigation tools in such a way they can be used transparently and there is no need to collect and distribute data during computation. This methods can be applied to other PSE and provide a general framework to add functionalities. Indeed, *parIDL* provides support to run multiple instances of IDL on different computing nodes; each running process can call IDL sequential functions and exchange data through the run time support using message passing. In this way all sequential macros can still run, and it is possible to implement parallel macros adding calls to message passing primitives provided by *parIDL*. It will be shown how to incorporate tools for Beowulf multicomputers [14], and in particular how to use functions of existing parallel software libraries within the environment. The approach used permits to maintain data distributed in local memories during computations, in fact avoiding the need to gather and distribute the results, which can be a time and resource consuming task in case of long running computations.

The remaining of the paper is organized as follows: next section describes other contributions to the field and how present works relates to them. Section 3 describes functionalities of run time support, while section 4 gives details of how to incorporate an external parallel procedure in the framework. Section 5 provides a complete example regarding the solution of a linear system of equations. Then it follows a discussion on results, conclusions and plans for future work.

2 Related Work

Current concept of PSE has been established in an 1991 workshop founded by National Science Foundation (NFS) [7] and tracked in a second NFS workshop held in 1995. Findings can be found in white papers [13] written by the participants and summarized in [3].

Enabling to call functions of parallel software libraries within PSE language has the advantage of software reusability: the amount of work to port a user application depends on modifications made to the programming language and syntax. Therefore, it is important that such modifications leave the application programming interfaces and syntax as much as possible as they were in the original language. This strategy is possible when source code of PSE is available and all the efforts are focused on modifying the environment in order to use parallel software libraries. Examples are OpenDx (http://www.opendx.org), a commercial visualization software developed by IBM, and Plookat [9], an open source vrml browser, in which modules for distributed rendering can be transparently used, since the systems have been modified to call the parallel functions.

If the source code is not available, it is possible to use the remote procedure call (RPC) technique, to execute procedures on a remote computer; when the execution ends, results are transferred back to the calling process. Pros of such strategy are clear: users do not need to completely rewrite applications and computational environment is transparent to the user, who has not to worry about implementation details related to different implementations on different platforms. That is a particular case of a distributed PSE [18], in which different components take care of computational work, data management and user interface. Disadvantages are mainly related to the overhead time introduced by data transfers. Examples are Matpar [15] at JPL, Cactus and Netsolve [1] at University of Tennessee.

Then there is a way to solve this problem based on codes coupling, that means development of software components with their given interfaces executed on distributed resources. This is the approach of Pardis project [11], which extends the CORBA object model introducing SPMD objects.

Another methodology resides in the implementation of code-to-code compilers in which the PSE macros are automatically translated in a parallel program; in such a way calls to library functions can be translated into calls to parallel functions. Such strategy shares the same pros and cons of parallel compilers: although they provide a useful tool for automatic code generation, results are not always efficient. To this field belong the projects Falcon [12] and Conlab [10], which produce parallel programs from codes written in a pseudo language in Matlab style.

Finally there is the methodology we choose to adopt, that is to introduce a parallel virtual machine to provide message passing capabilities to the PSE. This strategy makes it possible to use the runtime support to incorporate parallel software libraries. Pros are the same of the first strategy, but cons are we have to implement ad hoc wrappers to enable the use of every routine. An example of implementation of a similar strategy for Matlab is MultiMATLAB [17].

3 The Parallel Virtual Machine

In this section we introduce and discuss the software architecture of *parIDL*, the parallel virtual machine that provides run-time support for the IDL environment. The architecture is built on top of a MPI standard message passing library and it is an efficient, portable and flexible tool for the development of a problem solving environment suitable for clusters of workstations. At the moment no other support is available for the execution of IDL on distributed memory parallel machines.

As seen in previous section, the chosen strategy is the one that provides the meanings to develop parallel algorithms in the language of the PSE, enabling the use of message passing primitives inside the programming language. Such choice well suited the aim of the project and avoided the obstacles posed by the unavailability of the PSE source code. Moreover it makes it possible to concurrently use IDL software tools and modules on each node to harness the power of a supercomputer. We will also describe how, using the runtime support, it is also possible to integrate parallel software libraries.

From a user point of view the computational environment is a MIMD (Multiple Instruction Multiple Data) multicomputer, in which different computers are connected via a network that allows both one to one and collective communications; moreover network topology is assumed to be a completely connected graph. The concurrent programming paradigm is *SPMD* (Single Program Multiple Data), in which multiple instances of the same program run on different computers and execute the same operation on different data, loosely synchronizing with message passing. For ease of explanation it is assumed that each node of the multicomputer is run by a single group of processes that communicates with other groups of processes.

3.1 parIDL

The run time support of the parallel virtual machine *parIDL* provides the mechanisms to spawn the group of processes on the single computing nodes of the multicomputer. It initializes the message passing environment, the interprocessor communication mechanisms and the shared memory segments that will be used to exchange data between IDL macros and *parIDL* itself. Finally it prompts the user for a macro. Once the name of a macro is given by the user, a copy of it is sent to each node, and a new process is created to execute the IDL parser and loader; when those operations are completed, each copy of the macro is executed. In the meantime *parIDL* remains waiting for a request instantiated by the IDL macro to communicate with another macro on a different node. An identification number identifies each and all communication primitives supported by runtime support: a request of communication is interpreted by the environment and executed. Let's suppose *parIDL* receives a request from one macro to send data to another macro: it reads (consumes) the data, written (produced) in the shared memory buffer by the sender macro, and sends them to the other macro. Race conditions are avoided serializing producer-consumer access to the shared memory segment and using a different buffer for send and receive operations; this means an IDL macro send request has to wait until a previous send has been fulfilled by the parallel virtual machine before it can be accepted, and that nonblocking operations may overwrite the MPI internal buffer and not the one used to exchange data between the runtime support and the IDL macro. The behavior of *parIDL* reflects the one of the message passing primitives involved: if we ask, for example, to execute a blocking send or receive, the runtime blocks on that operation, until the data buffers can be safely overwritten. On the other hand, if a nonblocking operation is required, the parallel machine does not block on call and it returns results of operation to the calling macro. Serialization guarantees the semantic of nonblocking operations: it is not possible that two consecutive calls overwrite the data buffer used for interprocess communication between the runtime support and the IDL macro.

3.2 Proxy Layer

A software layer, called *proxy*, is also needed to let the processes executing the IDL macro communicate with *parIDL*. This layer provides the functionality to initialize and finalize the parallel run time support and to communicate with another macro. In Figure 1, a layered view of the software architecture is given.

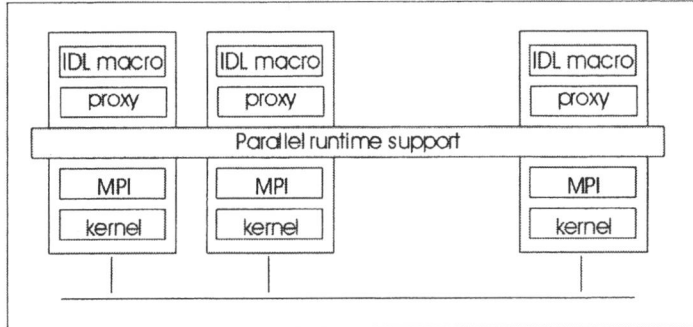

Fig. 1. Reference software architecture of parIDL

When a user wants to use the new functionality provided by the run time support, she has to add a call to the function *Initialize (menum, N)* at the beginning of the macro and *Finalize (menum)* at the end. The first calls an external C function whose aim is to inform the run time support a parallel macro is starting its execution and to query *parIDL* and return the number N of available concurrent processes and the identifier *menum* of the process. A call to *Finalize* acknowledges *parIDL* the macro has terminated its parallel execution.

Each API to be used within the environment has an associated IDL macro, which uses a call to an external function written in C. Such function casts the arguments to the correct data type and tries to locate the shared memory data segments provided by the run time support. If the operation is successful, it tries to attach it to its address space and tells *parIDL* what to execute using the identification number of the operation known by the runtime support. If an IDL macro need to send data, it copies data in the shared memory segments and informs the run time support the operation is completed. On the other side, a process willing to receive data, calls the macro to execute the proxy function. Such function, after initializing the shared memory segments, informs the run time support of its willingness to receive data and waits for data to be produced.

The possibility to use communication primitives is not restricted to the ones just described. Adding more message passing functionality to the software is an easy task. Let's take the example of a collective communication primitive, such as a broadcast. The first step is to have an IDL macro calling the appropriate C function in a dynamic library in the proxy layer. Such function will initialize the interprocess communication mechanisms and inform *parIDL*. Depending on the need to send o receive data, the virtual parallel machine will instantiate the call to the appropriate MPI broadcast function and it will exchange data with the proxy layer.

It is clear that the wall clock time is affected by an overhead due to the interprocess communication mechanism, used on each single computer to exchange data between the multiple layers of the run time software architecture, and the message passing system. Such additional time cannot be avoided, and, as it will be described later, performance results show reasonable performance can be gained.

The use of existing message passing libraries, such as the ones implementing MPI standard [8], makes it easier to develop robust and portable software. That reasoning prevented us from using the internal RPC mechanism provided by IDL, since, in that case, we had to build a robust message exchange mechanism and re-implement all collective and one-to-one communication primitives, that can be found in any MPI library.

If a message passing library is available and it is not MPI compliant, the changes needed will regard only the calls inside the parallel virtual machine to that library, leaving the overall structure still valid.

4 Implementation Details

In this section we describe the step by step procedure needed to add a single routine from a parallel library to *parIDL*. The ScaLAPACK software library [2] includes a subset of LAPACK routines [21] redesigned for distributed memory MIMD parallel computers. It is used to solve dense and band matrix problems, large sparse eigenvalue problems, sparse direct systems of equations and to compute preconditioners for large sparse iterative solvers.

In particular, we are going to detail the use of *PDGEMM* routine from ScaLAPACK. The operation consists in the implementation of three fragments of code and modification to library format for existing library. The first is the IDL layer, written in its script language, that provides a description of the function parameters to the parallel machine; the second, written in C language using standard POSIX system calls, in the proxy is used by the IDL process to locate and attach shared memory segments needed for interprocess communication with the run time support, which is the third layer of software, also written in C language, and that also has to be modified in order to provide the new function calls within IDL scripting language. Finally all libraries need to be compiled as shared object libraries.

4.1 IDL layer

In order to make it possible to use a parallel procedure, it is needed to instruct the runtime support on how parameters are called in the original routine. To do this a descriptor vector is used, which contains the size in bytes of each function argument. This information is passed by `mypdgemm.pro` to *parIDL*, together with the values of the parameters. This information will be used to correctly call the library function.

4.2 parIDL

The runtime support needs minor changes. When a new function is added, it is identified with a new identifier, which will be used in all calls to the routine, as it happens for communication primitives. The routine is called using as arguments the different parts of the shared memory segment that has been used to exchange data with the IDL layer and in which the parameters are stored, using the offset provided by the descriptor vector associated with the function. The arguments description array needs to be exchanged through the shared memory, since some arguments can assume variable size.

4.3 Proxy Layer

The proxy function *mypdgemm.c*, associated with the MYPDGEMM procedure of IDL, is needed to exchange data with the runtime support. It firstly informs *parIDL* of the operation to be executed via its identification number and after it locates and attaches to the shared memory segment used by *parIDL*; then it copies in the segment the input parameters needed by the external function. It waits until the computation is over and copies back data from the shared memory segment so results can be returned to *mypdgemm.pro*. On each node synchronization between proxy function and *parIDL* is obtained through a simple busy wait algorithm that uses a shared variable to serially and alternatively let the two processes access the shared data. The reason of this choice is the IDL macro cannot produce new input data if the results of PDGEMM have not been consumed, which precludes the use of other busy waiting algorithmic solutions; furthermore it allows the code to be run on a multicomputer with multiprocessors nodes. On the other hand, the use of techniques to avoid busy waiting, such as signal driven interrupt handlers and asynchronous callbacks needs to be justified by the cost of the operation with respect to the overall execution time, as it will be seen in next section.

Shared Libraries

IDL mechanism CALL_EXTERNAL [19], used to run external code, requires functions to be compiled as shared objects. To avoid the static linkage to the runtime support of all libraries called by ScaLAPACK, they have been recompiled as shared. Pros are the possibility to call all functions that do not need message passing directly from the script language. Cons are an overhead due to dynamic load of objects in memory and the fact that the libraries need to be accessible to all nodes on which parIDL has run.

5 Numerical Evaluation

In the following an example is provided of how to use a ScaLAPACK library function in *parIDL*. Let it start from a simple driver program for function PDGEMM. The routine is used to compute double precision matrix-matrix multiplication on a distributed memory multicomputer. It is possible to compute:

$$C = \alpha A * B + \beta C$$

where, A, B and C are matrices, α and β scalars. Tests have been performed on a Beowulf cluster [16] of 16 Pentium 4 1.5 GHz, with 512MB RAM, connected with a Fast Ethernet network. Each node runs a Linux kernel 2.4.20, gcc compiler 2.96, mpich 1.2.5, BLACS 1.1, ScaLAPACK 1.7, LAPACK 3.0, BLAS with ATLAS optimization; libraries have all been recompiled as shared objects.

Tests have been performed on idle workstations; the time refers to wall clock time of the slower executing node and it has been measured with function $SYSTIME()$ provided by IDL. The default maximum shared memory segment was 32MB, which led to the impossibility to run some test cases on a small number of processors.

Table 1. Matrix-matrix multiplication execution times (in seconds)

	1x1	2x1	2x2	4x2	4x4
1000	3.030344	2.962173	2.156209	2.038030	2.075710
2000	n.a.	n.a.	11.92654	8.338313	5.732544
3000	n.a.	n.a.	n.a.	23.18162	14.31242
4000	n.a.	n.a.	n.a.	n.a.	28.82498

Tests in Table 1 have been performed on logical 2D meshes of processors and show that, for an increasing number of processors, the execution time decreases, if the problem to be solved has sufficient computational complexity. Moreover, time reduction increases for larger problems, with a consistent gain in performance.

To better understand the overhead introduced with the call of *PDGEMM* routine from IDL, in Table 2 execution time on *16* processors is compared with total execution time. Last column shows the relative overhead with respect to total time.

Table 2. Execution time comparison

	PDGEMM	Total	Relative
1000	1.766357	2.075710	0.15
2000	5.528130	5.732544	0.03
3000	14.09291	14.31242	0.02
4000	28.43769	28.82498	0.01

The time spent in data exchange and synchronization on each node is the difference between total and *PDGEMM* execution time. It is only a small fraction of total time and, as the problem size increases, its impact on overall time decreases. If such overhead is not acceptable for a particular application, then different interprocess communication and synchronization mechanisms need to be investigated.

6 Discussion

The work involved in the integration requires a modification of both the proxy layer and the runtime support. Although it does not require a specialized knowledge, it needs to be repeated for all and each function to be added. This can become a time consuming task in case of libraries with a substantial number of functions. Nevertheless, the effort involved in the integration of the existing software is just a small fraction of that would be necessary to rewrite the whole routine in the programming language of PSE. Further investigation is required to automate the task.

Another issue concerns the use of the function API as it is in the software library. In fact, it is not necessary to give the user full details about the routine. For example, PDGEMM routine exposes as arguments parameters needed to choose the block factor for the matrix data distribution. Although they can be of much importance for performance, they are of no interest for the user of the PSE. Probably, it would be better to have a function with variable arguments, and to set defaults to that arguments that can be ignored by a user that has no interest in underlying details. It would be interesting to have a method to overload the primitive methods of the environment. For example it would be useful to overload the "*" matrix multiplication with a its parallel counterpart if asked by the user or detected by the environment. This feature unfortunately is not available in many commercial tools, although it is easy to implement in products, such as Octave (http://www.octave.org), for which the source code is available.

7 Conclusions and Future Work

The implemented software allows to run IDL on a low cost Beowulf class supercomputer, using external parallel software libraries. The obtained hardware/software architecture enables easy and fast applications development. With the adopted strategy, developers can harness the power of a supercomputing system with the ease of a PSE, obtaining a wall clock time reduction in their applications and a significant speed-up. Moreover it is possible to considerably reduce computing time and increase computing speed, while solving problems that cannot be solved on a single workstation.

Further work will be devoted to integrate the environment with middleware for distributed environment [6], to obtain a flexible tool for the development application that can benefit from the use of grid computing paradigm [5].

Acknowledgments. The work described has been partially supported by Italian Ministry for University and Research with FIRB project n. RBNE01KNFP Grid.it and PRIN 2003-2005 project n. 2003027003. Authors would like to thank reviews for their fruitful comments and suggestions.

References

[1] D. Arnold, S. Agrawal, S. Blackford, J. Dongarra, M. Miller and S. Vadhiyar. User's Guide to NetSolve V1.4. University of Tennessee, CS Department, Tech. Rep. UT-CS-01-467, 2001.

[2] L. S. Blackford, J. Choi, A. Cleary, E. D'Azevedo, J. Demmel, I. Dhillon, J. Dongarra, S. Hammarling, G. Henry, A. Petitet, K. Stanley, D. Walker, and R. C. Whaley. *ScaLAPACK Users' Guide*, SIAM, 1997.

[3] B. Boisvert and J.R. Rice. From scientific software libraries to problem solving environments. *IEEE Comp. Sci. Eng.*, 3, 1996.
[4] M. Carbillet, S. Correia, B. Femenia, and A. Riccardi. Adaptive Optics Simulations for Imaging with the Large Binocular Telescope Interferometer: a First Application. Proceedings of SPIE Vol. 4007, 2000.
[5] I. Foster and N. Karonis. A Grid-Enabled MPI: Message Passing in Heterogeneous Distributed Computing Systems. Proceedings of SC98, 1998.
[6] I. Foster, C. Kesselman and S. Tuecke. The Anatomy of the Grid: Enabling Scalable Virtual Organizations. *Intl J. Supercomputer Applications*, 15(3), 2001.
[7] E. Gallopoulos, E.N. Houstis and J.R. Rice. Workshop on problem-solving environments: Findings and recommendations. *Computing Surveys*, 27, 1995.
[8] W. Gropp, E. Lusk and A. Skjellum. *Using MPI - 2nd Edition*. The MIT Press, 1999.
[9] M.R. Guarracino, G. Laccetti and D. Romano. Performance evaluation of a graphics accelerators cluster. Proc. of the International Conference ParCo2001, Imperial College Press, pp. 449-456, 2002.
[10] P. Jacobson. The CONLAB environment. University of Umeå, S-901 87 Umeå (Sweden), Institute of Information Processing, Tech. Rep. UMINF-173.90, 1990.
[11] K. Keahey and D. Gannon. PARDIS: CORBA-based Architecture for Application-Level PARallel DIStributed Computation. Proceedings of SC97, 1997.
[12] B. Marsolf. Techniques for the Interactive Development of Numerical Linear Algebra Libraries for Scientific Computation. Ph.D. thesis, Department of Computer Science, University of Illinois at Urbana-Champaign. May 1997.
[13] J.R. Rice. Scalable Scientific Software Libraries and Problem Solving Environments. Purdue University, CS Department, TR-96-001, 1996.
[14] D. Ridge, D. Becker, P. Merkey, T. Sterling and P. Merkey. Beowulf: Harnessing the Power of Parallelism in a Pile-of-PCs. Proceedings of IEEE Aerospace, 1997.
[15] P.L. Springer. Matpar: Parallel Extensions for MATLAB.Proceedings of PDPTA, 1998.
[16] T. Sterling, J. Salmon, D. Becker and D. Savarese. How to build a Beowulf. The MIT Press, 1999.
[17] A. Trefethen, V. Menon, C. Chang, G. Czajkowski, C. Myers and L. Trefethen. MultiMATLAB: MATLAB on multiple processors. Cornell Theory Center, Ithaca, NY, Tech. Rep. 96-239, 1996.
[18] D. Walker, O.F. Rana, M. Li, M.S. Shields and Y. Huang. The Software Architecture of a Distributed Problem-Solving Environment. *Concurrency: Practice and Experience*, 12(15), 2000.
[19] External Development Guide. Research Systems, 1999.
[20] IDL Reference Guide. Research Systems, 1999.
[21] http://www.netlib.org/lapack/lug/lapack_lug.html, 1999.

Parallel/Distributed Film Line Scratch Restoration by Fusion Techniques

G. Laccetti[1], L. Maddalena[2], and A. Petrosino[2]

[1] "Federico II" University of Naples
giuliano.laccetti@dma.unina.it
[2] ICAR-CNR, Section of Naples
{lucia.maddalena,alfredo.petrosino}@na.icar.cnr.it

Abstract. Many algorithms have been proposed in literature for digital film restoration; unfortunately, none of them ensures a perfect restoration whichever is the image sequence to be restored. Here, we propose an approach to digital scratch restoration based on image fusion techniques for combining relatively well assested distinct techniques. The method has large memory requirements and is computationally intensive; due to this main reason, we propose parallel versions of the restoration approach, focusing on strategies based on data partition and pipelining to achieve good load balancing. The parallel approach well adapts also to be distributed.

Keywords: Scratch restoration, Parallel/distributed computing.

1 Introduction

With the recent advent of digital technologies, and the ever increasing need for speed and storage, occluded or missing parts in images and video is a more and more widespread problem. The problem can occur in several multimedia applications such as wireless communication and digital video restoration. Several classes of defects can be distinguished that affect movies, and many algorithms have been proposed in literature for their restoration; unfortunately, none of them ensures a perfect restoration whichever is the image sequence to be restored.

In the present paper, we focus on the class of scratch defects, intended as long and thin vertical scratches that affect several subsequent images of a sequence, due to the abrasion of the film by dust particles in the slippage mechanisms used for the development, projection and duplication of the film. A sufficiently general model of degraded video signal is the following for a pixel location $\mathbf{p} = (x, y)$:

$$I(\mathbf{p},t) = (1 - b(\mathbf{p},t))I^*(\mathbf{p},t) + b(\mathbf{p},t)c(\mathbf{p},t) + \eta(\mathbf{p},t) \quad (1)$$

where $I(\mathbf{p},t)$ is the corrupted signal at spatial position $\mathbf{p}$ in frame t, $b(\mathbf{p},t) \in \{0,1\}$ is a binary mask indicating the points belonging to missing parts of the degraded video, I^* is the ideal uncorrupted image. The (more or less constant) intensity values at the corrupted spatial locations are given by $c(\mathbf{p},t)$. Though noise is not considered to be the dominant degrading factor in the section, it is still included in (1) as the term $\eta(\mathbf{p},t)$. Commonly, the scratch restoration is a two-step procedure. In the first step the scratches need to be detected, i.e., an estimate for the mask $b(\mathbf{p},t)$ is made (*detection step*). In the second step the values of I^*, possibly starting from information about $c(\mathbf{p},t)$ inside the scratch, is estimated (*removal step*). As usual, we consider scratch reduction as a problem of detection and removal of missing part, i.e. we suppose that any information $c(\mathbf{p},t)$ has been lost within the scratch. Several attempts are reported in literature (see for instance [1,2,6,9,10,12,16]).

As expected, advantages and disadvantages characterize each technique, and any of them could be said to win the competition. A way to deal with this kind of problems is to adopt fusion techniques (see for instance [3,8]): input images which provide alternative and complementary "views" and "characteristics" of a given area are "fused" into a single image. Fusion techniques should ensure that all the important visual information found in input images is transferred into the fused output image, without the introduction of artifact distortion effects. In this sense machine vision systems can be organised as a set of separated visual modules that act as virtual sensors. In this paper, the term *visual module* indicates an algorithm that extracts information of some specific and descriptable kind from a numerical image. For what concerns digital scratch restoration, a fusion technique may be applied both at the detection stage and at the removal stage: in both cases it takes into account already existing promising algorithms and suitably combines the obtained results in order to provide a restored sequence as similar as possible to the original uncorrupted sequence. As we show, the results produced by the proposed approach upon different damaged movies greatly enhance those produced by each considered approach. The experimental results reported in the literature showed that the accuracy provided by the combination of an ensemble of visual modules can outperform the accuracy of the best single visual module.

On the other side, automatic batch processing is limited to low degraded video sections and is mainly aimed at the removal of speckle noise and brightness variations. The manual inpainting of high definition digital video is, of course, a very intensive and expensive activity which can be justified in cases of documents of particular historical and artistical meaningfulness. A great improvement in this field should be related to the development of high performance software for the automatic or semiautomatic restoration of degraded video; this could drastically reduce the costs and efforts for the recovery of whole movies. The proposed algorithm is specifically suited for (and indeed it has been devised for) *grid environments* [7], where several geographically distributed resources can be accessed and exploited for the solution of the problem under investigation. In the case of our image sequence restoration application, the resources we refer to are not only computational resources (such as high-performance architectures), but also data resources (data repositories where the huge amount of data necessary to store digitized image sequences can be accessed), and visual modules (to be taken into account by the approach). We analyze the performance of two proposed parallel algorithms and present several restoration examples on real image sequences, while remarking that one of them shows properties that makes it *grid*-aware.

The contents of this paper are as follows: Section 2 outlines the considered scratch detection modules and the proposed fusion strategy, while section 3 describes modules and fusion strategy relating to scratch removal. Section 4 describes the qualitative results achieved by the proposed fusion approach. Section 5 reports the parallelization strategies we propose and tested on real video sequences.

2 Detection: Methods and Fusion Strategies

For demonstrative purposes we choose three visual modules performing detection. In the following some hints will be provided about them.

2.1 Detection Algorithm 1

Following [10], the first detection algorithm taken into account can be described as follows, in the case of white scratches:

- pre-processing of each sequence frame, obtained by using the Radon projection in the vertical direction, with height A depending on the maximum inclination of the scratch to the vertical direction and on the image spatial resolution: $I_A(x,y) = \sum_{j=0}^{A-1} I(x, j + A * y)$ producing an image I_A with height M/A, M being the height of the original image.
- vertical linear structure detection in subsampled images I_A by a local maximum detection procedure based on an opening top-hat morphological transform followed by a thresholding process.

Based on the results obtained in the subsampled images I_A, a scratch mask I_B for each of the original sized images is created. The produced results could be enhanced, by retaining only the scratches that appear fixed in time; as suggested in [10] this can be made by tracking the scratches. Anyway to maintain comparable the computational complexities of the considered detection algorithms, we avoided to implement this part of the procedure.

2.2 Detection Algorithm 2

In [11] the detection scheme consists of:

- adaptive binarization of the sequence frame, producing B, by adopting a threshold, computed over a horizontal neighbourhood of any pixel at position (x, y) as: $\sigma = 1/w \sum_{i=-w/2}^{w/2} I(x+i, y)$
- construction of the scratch mask I_B by labelling as belonging to the scracth all the pixels of columns x of B that contain at least 70% of pixels with value $B(x, \cdot) = 1$.

2.3 Detection Algorithm 3

Following [12], the third detection algorithm taken into account consists of:

- low-pass filtering with a Gaussian vertical filter, followed by a vertical subsampling, producing I_G
- location of the horizontally impulsive noise in I_G by:
  - horizontal median filtering I_G, in order to eliminate straight lines with width lower than the filter window size, producing I_M;
  - thresholding the difference $I_G - I_M$, in order to obtain the binary mask I_B;
  - Hough transform in order to characterize the straight lines in I_B (the candidate scratches);

In [12] this process is followed by a Bayesian refinement strategy which allows to select among candidate scratches, those that can be best modelled by a horizontal scratch section model. As for detection algorithm, we avoided to implement this part of the procedure.

2.4 Detection Fusion Strategies

The main goal of using more than one detection visual module is to make up for deficiencies in the individual modules, while retaining their features, thus achieving a better overall detection result than each single module could provide. In this case the combination should be made among sets of pixels, representing those detected as corrupted and belonging to scratches. Let us indicate these sets as I_B^j, where $j = 1, \ldots, n$ (in our case $n = 3$). Here, two different combining methods or aggregation operators are adopted:

- Union aggregator operator: $\cup \{I_B^j(x,y) : j = 1, \ldots, n\} = \max_j \{I_B^j(x,y)\}$
- Maximum Covering (MC) aggregation operator, defined as:

$$MC\{I_B^j(x,y) : j = 1, \ldots, n\} = \\ \{(x,y) : x = x^m - W^-, \ldots, x^m + W^+ \text{ and } y = 0, \ldots, M-1\} \qquad (2)$$

where M represents the number of image rows, $x^m = \mathrm{median}_x\{(x,y) \in \cap_j I_B^j : I_B^j(x,y) = 1\}$, $W^+ = \max_x\{(x,y) \in I_B^j : I_B^j(x,y) = 1, j = 1, \ldots, n\}$, $W^- = \min\{(x,y) \in I_B^j : I_B^j(x,y) = 1, j = 1, \ldots, n\}$ and $\cap \{I_B^j(x,y) : j = 1, \ldots, n\} = \min_j \{I_B^j(x,y)\}$.

In section 4 the performance of these aggregation operators with respect to the problem at hand will be shown and commented.

3 Removal: Methods and Fusion Strategies

Two methods are adopted and compared: one belongs to the class of interpolating methods [13] and the other one to the class of inpainting algorithms [4]. Both algorithms attempt to reconstruct not only the structure of the image in the scratch domain (pixels where mask I_B gets value 1), but also its texture.

3.1 Removal Algorithm 1

In [4] a nonparametric Markovian model is adopted and adapted to our purposes. The probabilistic model is based on the assumption of spatial locality: the probability distribution for one pixel given the values of its neighbourhood is independent of the rest of the image. The neighbourhood $N(\mathbf{p})$ of a pixel $\mathbf{p}$ is chosen to be a square window around this pixel and is chosen to be adaptive. The model is non-parametric and it is approximated by a reference sample image, which must be large enough to capture the stationarity of the texture. During the synthesis process, the approximation of the probability distribution $P(I(\mathbf{p})|I(\mathbf{q}); \mathbf{q} \in N(\mathbf{p}))$ is made as follows:

- the sample image is first searched in order to find all pixels that have a neighbourhood "similar" to the one of the pixel being synthesized.
- one of these candidates is randomly drawn and its intensity value is assigned to the pixel being processed.

The similarity of two neighbourhoods is measured according to the normalized sum of square differences (L^2 distance) and it is weighted by a two-dimensional Gaussian, in order to give more importance to the pixels that are near the centre of the window than to those at the edge.

Specifically, from the binary mask I_B representing the pixels to be replaced, for each missing pixel p the number of its valid neighbours is enumerated. Pixels are then replaced starting from the ones having the most valid neighbours. All missing regions are thus simultaneously and progressively filled from the edges to the centre.

3.2 Removal Algorithm 2

In [13] a simple interpolating method is adopted and the interpolation result is corrected by adding to it the estimated displacement between the adopted model and the real model. Specifically, provided the scratch mask I_B, the procedure consists of:

- interpolation of the pixels pertaining to the scratch domain;
- estimate of the image texture in the scratch neighbourhood, obtained by computing the displacement between the least square fitting over an uncorrupted neighbourhood of the scratch and the same neighbourhood pixels.
- adding of the estimated texture to the pixels belonging to the scratch domain.

Different versions of the above described method can be obtained by adopting different interpolation methods and neighbourhood sizes.

3.3 Removal Fusion Strategies

The problem here can be stated as follows: given l images representing heterogeneous data on the observed phenomenon, take a decision D_i on an element

(x, y) where D_i belongs to a decision space D. In image fusion the information relating (x, y) to each possible decision D_i is represented as a number M_i^j, where j indexes the decision making module having different properties and different meanings depending on the mathematical framework. If we assume that, given I_R^j images obtained by different removal modules, $M_i^j(x,y) = f^j(x,y)$, with (x, y) in the scratch domain, represents the probability degree to which the pixel (x, y) could be seen as "restored" (i indexes the values of this appearance), we can claim all the advantages of the Bayesian framework relying in the variety of combination operators. Here we adopt the averaging operator, i.e. $f^c(x,y) = 1/2(f^1(x,y) + f^2(x,y))$, known in the Bayesian framework as the Basic Ensemble Method [14] for combining different classification modules for which it is demonstrated to significantly improve the classication performance of each single module. In the sequel, it will be shown that the same improvements hold for the restoration purposes. Alternative methods reside in the Lukasievic minimum and maximum operators, defined as $LAND(a,b) = \max\{a+b-1, 0\}$ and $LOR(a,b) = \min\{a+b, 1\}$, while it is under study the possibility of using more elaborate operators like the Ordered Weighted Aggregation (OWA) operators due to Yager [18] or the Nonliner Generalized Ensemble Method as introduced in [5].

4 Qualitative Results

The accuracy of the result of the detection algorithms taken into account is quite high, as it is shown by the scratch masks reported in Figs. 2(b-d) for the sequence frame of Fig. 2-(a). Nonetheless, we can observe that the union aggregation operator, whose results are reported in Fig. 2(e), seems more appropriate for the successive removal phase, compared with the MC aggregation operator (see Fig. 2(f)). Specifically, the union aggregation operator for scratch masks allows to consider all details captured by the different detection algorithms, while retaining the minimum support of the mask; the MC aggregation operator leads to a mask that appears unnatural for real images (being perfectly rectangular) but that allows to avoid several false positives detected by the algorithms taken into account. We tested the algorithm also on artificially corrupted real images, obtaining for all the detection algorithms considered (and therefore also for the compound algorithm) nearly 98% of correctly detected scratch pixels. Specifically, we created a white scratch (intensity value equal to 255) of width $w = 3$ and $w = 5$ on each sequence frame. Knowing a priori the scratch mask for such images, we applied our algorithm to the corrupted images, gaining an estimate of the error in terms of the following error measures: (a) the number N_{fp} false positives, i.e. the number of pixels in the computed scratch mask that do not belong to the scratch; (b) the number N_{fn} false negatives, i.e. the number of pixels of the scratch that are not in the computed scratch mask. For the considered images and image scratches, the values of N_{fp} and N_{fn} have been very low (nearly to 0.1). The results of the removal algorithm for the sequence frame of Fig. 2(a) are reported in Figs. 3(a) and 3(b) for the inpainting and interpolative algorithms respectively based on the mask produced by the union aggragation operator; Figs. 3(c) and 3(d) show the results over the mask produced by the MC aggregation operator. Fig. 3(e) and 3(f) report the obtained by adopting the averaging method over the images depicted in Fig. 3(a)-(b) and 3(c)-(d) respectively. From these results we can observe that, even though the removal algorithms taken into account perform quite well, their reconstruction accuracy can be enhanced; the aggregated results, instead, tend to smooth the inaccuracies, still retaining the good performance of the considered algorithms.

5 Parallel/Distributed Algorithms

Parallel algorithms. The first approach to the parallelization of the scratch restoration algorithm relies on **data partitioning**. Given an image sequence consisting of N frames and, assuming P available processors, the image sequence is decomposed into subsequences, each consisting of $\lfloor N/P \rfloor$ adjacent frames. Different subsequences are assigned to different processors, which perform the sequential scratch restoration algorithm on their own subsequence, in parallel with the remaining processors. This approach gives rise to an embarrasingly parallel algorithm; anyway the algorithm assumes the availability of data and the chance to distribute the sequence over the processors, that it is not always the case.

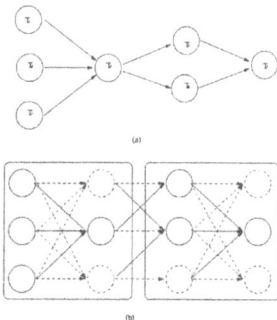

Fig. 1. The IATG (a) and an allocation strategy (b).

The second considered approach for the parallelization relies on **task partitioning**. In our case seven tasks, corresponding to the visual modules described in the previous sections, should be managed. As could be noted, detection and removal tasks are completely independent of each other, while fusion modules could be accomplished just only after the detection (or removal) modules have completed and released their results. The operations involved may be efficiently described as an Image processing Application Task Graph (IATG). Firstly, we recall that a task parallel program can be modeled by a Macro Dataflow communication Graph (MDG) [15] and an IATG could be seen as an MDG in which each node stands for a visual module and each edge stands for a precedence constraint between two adjacent operators. Let T_i denote the i-th task, $i = 1, \ldots, N_t$, with $N_t = 7$ in our case, the IATG corresponding to the fusion approach to the detection/removal of scratches is depicted in Fig. 1(a), also seen as a matrix of tasks/processing stage. Before describing the allocation task/processor, some considerations should be made about the computational load of each task. Since task T_5, that corresponding to the Removal algorithm 1, takes more than 50% of the total restoration time, any allocation made by letting rows or columns to each processor would result in an unbalanced scheme. We then propose to merge first and second columns, respectively third and fourth columns, in order to have a two-layered IATG graph; then we allocate each row to each processor. This simple scheme, that lets each processor to perform just detection or removal steps over a video frame, results in a better balanced scheme and, as it will be remarked in the following, well adapts to *grid* environments. Speedups and efficiencies obtained by both data and task parallelism approaches, over two different video sequences denote that both the approaches are quite similar, but even more significant, that they are both nearly optimal. Specifically Fig.s 4(a) and 4(b) report respectively the speedups obtained by both approaches over a B/W video sequence of 80 frames of dimension 256 × 256, and over a colour video sequence of 80 frames of dimension 720 × 576, by letting the number of processors of a Linux Beowulf to go up to 16. The same data are adopted for the efficiencies reported in Fig.s 5(a) and 5(b).

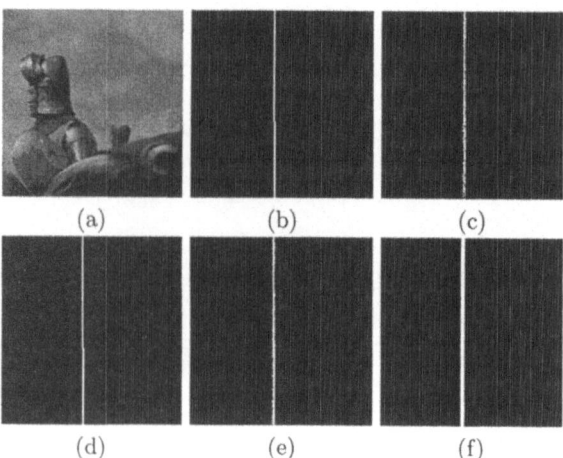

Fig. 2. Detection results over the original frame (a) by adopting (b) algorithm 1, (c) algorithm 2, (d) algorithm 3, (e) the union fused image and (f) the MC fused image.

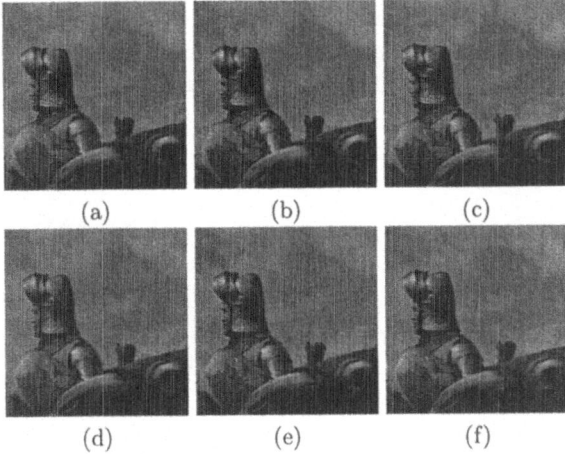

Fig. 3. Removal results by adopting (a) algorithm 1 and (b) algorithm 2 over the union mask, (c) algorithm 1 and (d) algorithm 2 over the MC mask. In (e) and (f) the results obtained by adopting the averaging method over the images (a)-(b) and (c)-(d) respectively.

Distributed algorithm. The main advantages in using *grids* are in an extremely high computing power, in a better use of idle resources, in a shared remote access of special purpose resources or data sources, in the support of collaborative work via a virtual shared space [7]. In the case of digital video restoration two considerations are in order: (a) each *grid* node typically cannot perform all the involved procedures, and we suppose that each one is enrolled to perform just detection or removal; (b) the data (video to be restored) should not be distributed over the *grid* nodes for copyright purposes, so the data distribution is made according to secret sharing schemes[17], according to which we divide data D into n pieces such that D could be easily reconstructable from any k pieces, but even complete knowledge of $k-1$ pieces reveals absolutely no information about D. From the above considerations it is clear that the data parallelism approach does not properly work in *grid* environments, while task parallelism approach, as described, would result in a more appropriate approach. The simplest architectural scheme can be seen as a two layered scheme, where a *grid* node, that holds the data, works as a master node, belonging to the first layer and the other nodes(the slaves) to the second layer. During the scratch detection phase, the master distributes the data to various slaves such that node i works on frames numbered by $i, i+P, \ldots$, where P is the number of *grid* nodes now. Each slave node operates the appropriate detection and maintains the data results in the secondary memory. As soon as a slave has completed the detection phase over a frame t, it sends the same to another *grid* node involving removal, while it signals its availability to the master that then sends another data set. This greedy approach that randomly partitions the video data and assigns them to the first available slaves achieves in practice a good balance, although it should be noted that this scheme in no way guarantees load balancing in the scratch restoration, since each *grid* node could be loaded in different manner, as for instance in our case those deputed to the detection and those to the removal. Experimental results will be reported in a more detailed paper.

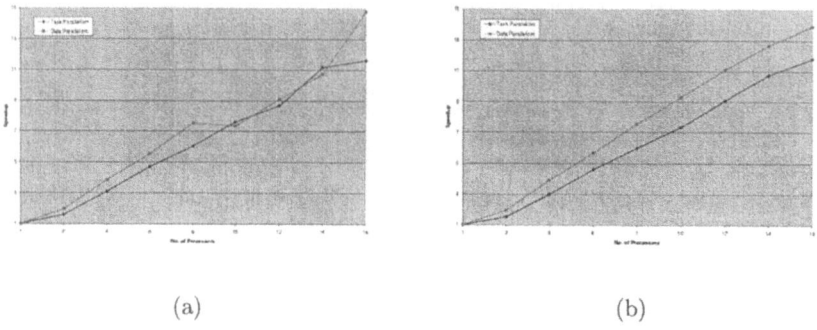

Fig. 4. The Speedups obtained by data parallelism and task parallelism over two real video sequences with (a) B/W frame dimension 256 × 256 (b) colour frame dimension 720 × 576.

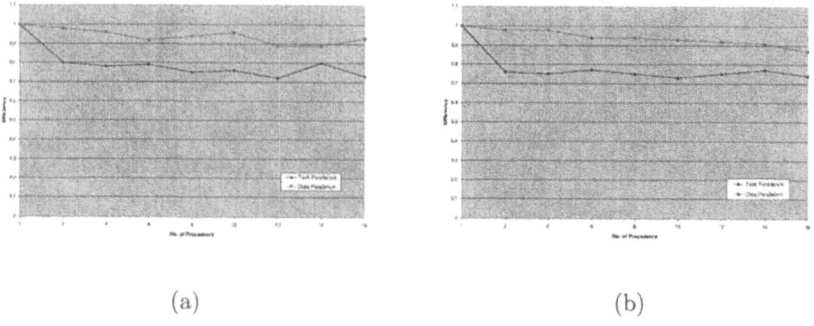

Fig. 5. The Efficiency obtained by data parallelism and task parallelism over two real video sequences with (a) B/W frame dimension 256 × 256 (b) colour frame dimension 720 × 576.

References

1. Acton S.T., Mukherjee D.P., Havlicek J.P., Bovik A. C., Oriented Texture Completion by AM-FM Reaction Diffution, *IEEE Transactions on Image Processing*, vol. 10, no. 6, pp. 885-896, 2001.
2. Beltramio M., Sapiro G., Caselles V. and Ballester C., Image Inpainting, *Computer Graphics*, pp. 417-424, 2000.
3. Bloch I., Information Combination Operators for Data Fusion: A Comparative Review with Classification, *IEEE Trans. Systems, Man, Cybernetics*, vol.26, no. 1, pp.52-67, 1996.
4. Bornard R., Lecan E., Laborelli L., Chenot J.-H., Missing Data Correction in Still Images and Image Sequences, *ACM Multimedia 2002*, Juan-les-Pins, France, December 2002.
5. Ceccarelli M., and Petrosino A., Multifeature adaptive classifiers for SAR image segmentation, *Neurocomputing*, vol. 14, pp. 345-363, 1997.
6. Chan T. and Shen J., Non-Texture Inpainting by Curvature-Driven Diffusions (CDD), *J. Visual Communication and Image Representation*, vol. 12, no. 4, pp. 436-449, 2001.
7. Foster I., Kesselman C. (Eds.), *The Grid: Blueprint for a New Computing Infrastracture*, Morgan Kauffmann Publisher, Los Altos, CA, 1998.
8. Ho T.K., Hull J.J. and Srihari S.N., Decision Combination in Multiple Classifier Systems, *IEEE Trans. on Pattern Analysis and Machine Intelligence*, vol. 18, pp. 66-75, 1994.
9. Isgró F. and Tegolo D., A distributed genetic algorithm for restoration of vertical line scratches, *Parallel Computing, in press*, 2004.
10. Joyeux L., Boukir S., Besserer B., and Buisson O., Reconstruction of Degraded Image Sequences. Application to Film Restoration, *Image and Vision Computing*, vol. 19, pp. 503-516, 2001.
11. Kao O. and Engehausen J., Scratch Removal in Digitised Film Sequences, *International Conference on Imaging Science, Systems, and Technology (CISST)*, pp. 171-179, 2000.
12. Kokaram A. C., Morris R., Fitzgerald W. and Rayner P., Detection of missing data in image sequences, *IEEE Transactions on Image Processing* Part I-II, pp. 1496-1519, 1995.
13. Maddalena L., Efficient Methods for Scratch Removal in Image Sequences, *Proc. IEEE Intern. Conf. Image Analysis and Processing*, pp. 547-552, 2001.
14. Perrone M. P., Cooper L. N., When networks disagree: Ensemble method for neural networks, in: R.J Mammone (Ed.), *Artificial Neural Networks for Speech and Vision*, Chapman & Hall, New York, pp. 126-142, 1993.
15. Ramaswamy S.S.S. and Banerjee P., A framework for exploiting task and data parallelism on distributed memory multicomputers, *IEEE transactions on parallel and distributed systems*, vol. 8, no. 11, pp. 1098-1115, 1997.
16. Rosenthaler L., Wittmann A., Gunzl A. and Gschwind R., Restoration of old movie films by digital image processing, *IMAGE'COM 96*, Bordeaux, France, pp. 1-6, May 1996.
17. Shamir A., How to share a secret, *Communications of the ACM*, vol. 22, pp. 612-613, 1979.
18. Yager R. R. and Kacprzyk J., *The Ordered Weighted Averaging Operation: Theory, Methodology and Applications*, Kluwer: Norwell, MA, 1997.

An Interactive Distributed Environment for Digital Film Restoration

F. Collura, A. Machì, and F. Nicotra

ICAR Consiglio Nazionale delle Ricerche, Sezione di Palermo
Viale delle Scienze 90128 Palermo Italy
colf@medialab.pa.icar.cnr.it,
{machi,nicotra}@pa.icar.cnr.it

Abstract. The paper presents FESR an interactive environment enabling collaborative digital film restoration over a digital network by connecting seamless the supervisor and operator's graphical frameworks, a parallel Image Processing server, a video stream server and a DBMS server. It allows a supervisor to remotely define the restoration protocol and to monitor job progress by using a metadata-driven navigation interface. The operators use a graphical desktop to interactively perform parameter steering of restoration filters and to activate and synchronize batch processing of defect detection and restoration filters on the IP server. The video stream server stages digital frames while the meta-data server stores defect masks and other film attributes by using MPEG7 formalism for video-segments representation. System prototype architecture, inspired to component technology is described.

1 Introduction

The animated documentation of a lot of historical events and of most part of cinema and TV production of XX th century is registered in movies on celluloid film ribbons. These films downgrade over time because of chemical changes due to film aging and to inappropriate ribbon handling and environment. Their recovery is essential for preservation of cultural heritage but restoration is highly costly process.

During the last ten years, several studies have been published concerning (semi) automatic digital restoration of degraded video and film image sequences. Algorithms and methods have been developed to detect and correct abnormal luminance fluctuations, line or frame misalignments, to detect and restore frame areas damaged because of dust, mould or dirt, excessive heating, improper drag, or just film aging [1-3].

A number of projects have been founded by the European Commission aimed to develop software systems supporting automated restoration [4-5], and to implement parallel algorithm libraries reducing time constraints of heavy computations [6]. Other projects have concentrated their efforts in developing tools and environments for minimizing the time spent by the operator to handle the huge quantity of temporary date produced [7].

In semi-automatic approaches to digital video restoration, the restorer browses the video and bounds, around damaged frames, video sections on which to apply appropriate sequences of spatial-temporal filters. Smoothness of change in scene properties around the damaged frames is essential for proper operation of filters. Of course, the reconstruction fails whenever scene changes abruptly or any scene component moves too quickly [8]. If a scratch spreads on more frames, the support area should be restricted inside the frame under restoration. If the defect area is large, its local neighborhood should not be included in the support area to obtain high quality reconstruction [9-10].

Fig. 1. Blotch restoration: defected input (left), defect mask (center), frame restored (right)

Adaptive algorithms need to be tuned for proper operation .The paper presents FESR a new interactive environment enabling collaborative digital film restoration over a digital network with express support of parameter steering.

In FESR approach the supervisor browses the video, defines the restoration protocol by remotely marking sections on which to apply appropriate series of spatial-temporal filters and monitors job progress by using a metadata-driven navigation interface. The operators use a graphical desktop to interactively perform parameter steering of restoration filters and to activate batch defect detection and restoration on a parallel Image Processing server. After the automatic processing step, he browses the restored sequences and identifies unsatisfactory results. Badly reconstructed frames are then restored by in painting or reprocessed with a different series of filters.

1.1 Digital Film Restoration: Supervised Process Description

Fig 2 outlines the phases composing a supervised digital motion picture restoration process. The damaged motion picture is firstly digitized from its original film tape support. Each picture frame is digitized in a high definition format (over 2k by 2k pixels) and stored in high-density digital tapes.

Four digital processing steps then follow and finally the restored version is saved on its final digital (DT, DVD) or analog (VHS tape, film) support. In the following, we will use term film to refer a motion picture even in its digital form.

Digital restoration includes a preprocessing indexing step aimed to partition movie into sections homogeneous for scene temporal and semantic continuity.

Defect detection follows indexing. It is an iterative process in which a number of Image Processing filters are applied on a small number of sample frames seeking for

best parameter tuning. After optimal configuration is obtained, automatic processing of an entire sequence is started.
The same steering procedure is applied before application of restoration filters and, in case of unsatisfactory reconstruction, reprocessing of difficult areas is worthwhile.

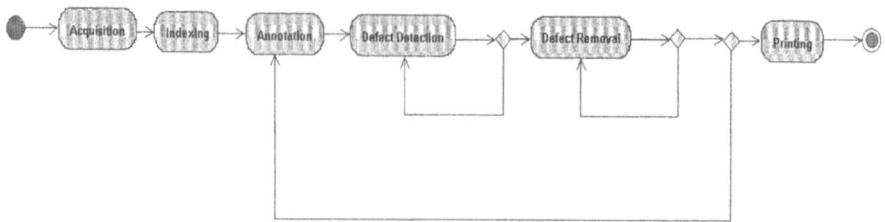

Fig. 2. Activities in the restoration process

In the FESR environment, a film is described by using an MPEG7 coding of its metadata [11]. Its sections are represented as Video Objects whose features are labels for defect typologies, identifiers of image processing filter chains and URI references to temporary masks produced.
The supervisor draws the restoration plan for the entire film by annotating defect typology and required operations in the meta-description record of each defected section. He can also perform trial processing on short sections and annotate preliminary values for parameter settings.
Operators are responsible for contemporary processing of different sections. They can perform detection trials and steering by restricting processing to small group of frames or selected frame areas, or by locally modifying the suggested parameter set. Two processing modalities support steering: procedure calls on the local engine or high-priority request on the remote parallel server. Batch processing submission is provided for final sequence processing. Each operator maintains and updates a work-session-scoped copy of the meta-descriptor of its file section. He updates meta-description fields for found defects list and annotations, parameters settings used, final defect masks, temporary and final results of restoration.
The supervisor reviews operator's meta-descriptions of its working session and confirms or refuses results. Accepted sessions meta-data and objects replace (or are merged to) corresponding ones in sections of the official movie meta-description.

2 The Interactive Distributed Environment

The process just described requires the environment to support patterns as event-driven programming, interactive graphics, quality-of-service processing, and session status management. FESR implements all such patterns though the coordination of several subsystems playing different roles and developed with component technology. Fig. 3 shows FESR subsystems cooperation diagram with component deployment on remote nodes interfaces and ports association. The application is developed as a three-tier client-server application with rendering interfaces (GUIs) on the user nodes, and computational, video-stream and meta-DBMS services deployed on the network.

The middle tier hosts the event subsystem, which coordinates user interaction from windows and adapters for local and remote procedure, calls (plug-ins and proxies).

It also hosts a memory engine able to store environment meta-data across procedure calls and to store and retrieve them from a DBMS service.

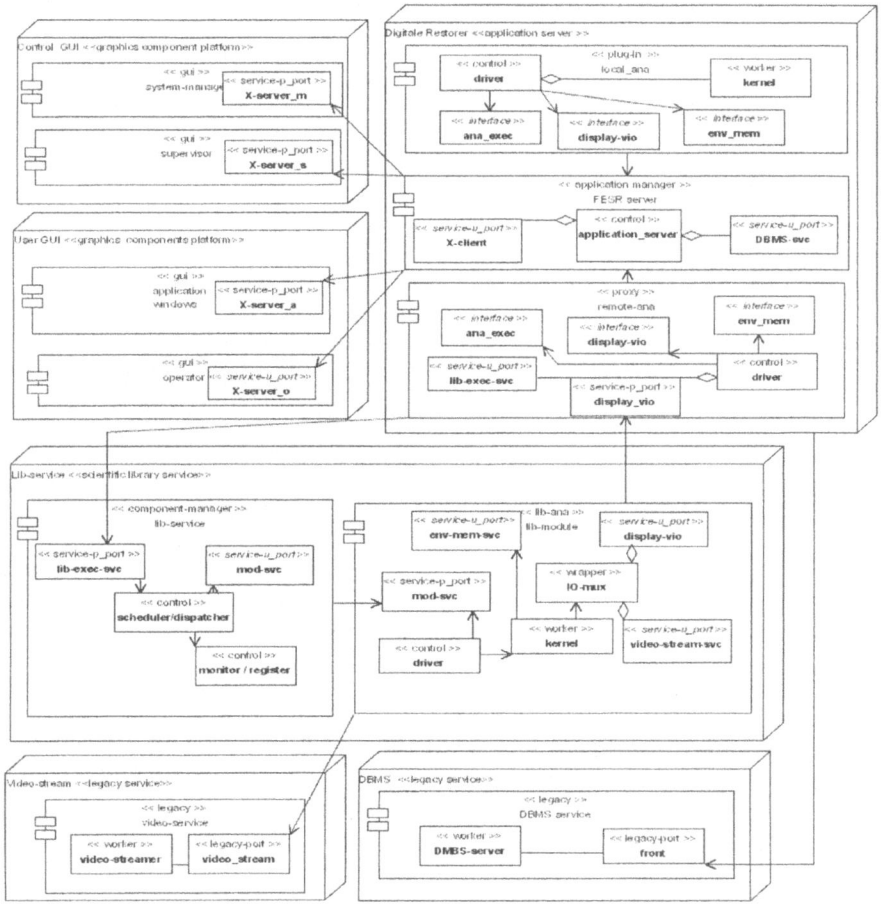

Fig. 3. FESR components cooperation diagram.

3 The Event Driven Application Server Subsystem

This subsystem is composed by the graphics framework and by the restoration server The windows of the interactive subsystem (GUI) may be made remote by using the intrinsic facilities of the underlying graphical X system.

Four windows arrangements are provided: *supervisor, operator, display and list*.

The *supervisor* window allows navigation of movie structure. Browsing and editing of Video Objects of type *Shot*, *Clip* and *Task* is supported.

The *operator* window manages parameter selection and activation of local or remote processing on a movie section. Three modalities are supported: *test*, *merge* results with previous run or replace session meta-data. The *display* windows may be connected to output streams from processing modules. Facilities for panning, zooming and snake-ROI selection are provided. Finally, the *list* window allows movie selection from movie database.

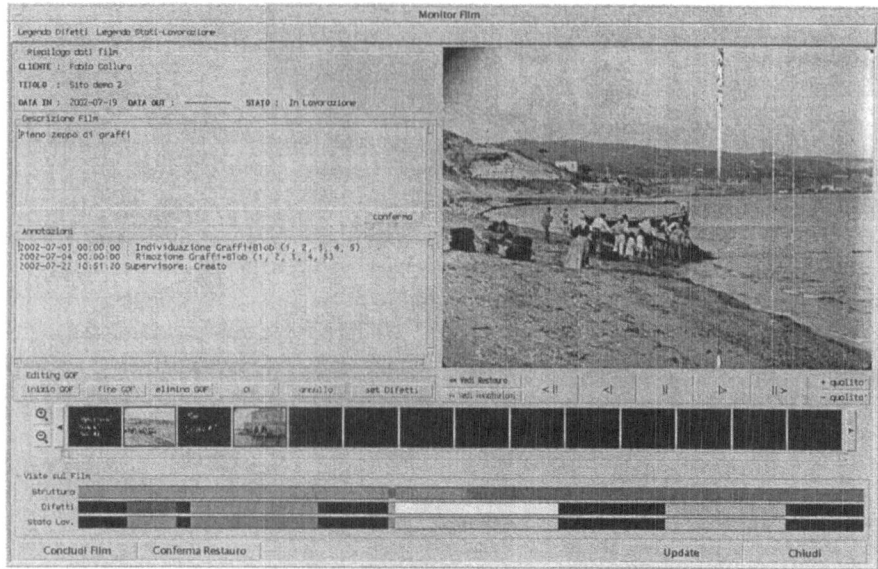

Fig. 4. Snapshot of a supervisor window showing movie structure, one frame and annotations.

The GUI subsystem is implemented by using the Trolltech Qt [13] Graphics Toolkit over the X-Window graphic system.

Graphics events generated by user interaction with a window through the mouse and keyboard are managed as signals, registered by an *event-engine* and routed to other platform components registered for receiving them.

An analysis engine manages processing. Adapters for local processing of restoration modules (plug-ins) or remote procedure-calls on the processing servers (proxies) may be loaded at run time as dynamically linked libraries.

At loading time, the analysis engine reads a description of graphics requirements for window control and a description of the module activation signature through the adapter introspection interface. At activation time, the adapter reads relevant data from execution environment, calls methods of the controlled module and updates environment, according to the test/merge/replace policy selected by the user.

For remote processing the environment, memory is reflected on the network through NFS. The analysis module gets direct access to video data streams from the Video-server.

An environment-engine manages persistence of meta-data throughout a session and their storage or retrieval from the DMBS server.

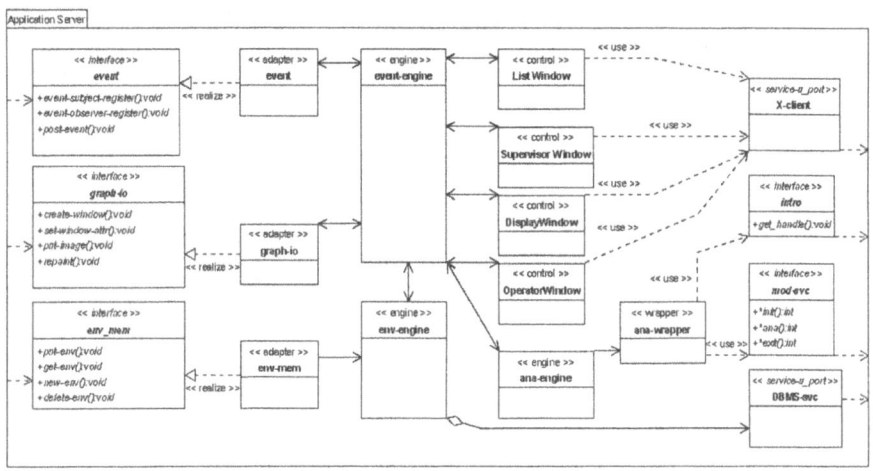

Fig. 5. Architecture of the application server.

4 The Image Processing Parallel Server

The parallel server offers Image Processing services according to two specific Quality of Service modalities: shortest-time and minimum-cost.

Shortest-time mode supports steering. Testing on short movie sections for parameter tuning needs near-real-time response and graphical rendering. Minimum-cost mode supports global optimization of the queue of long queue batch jobs during production. The parallel server is implemented as a container of parallel components. Each parallel component is an instance of a data-parallel or stream parallel version of a detection or restoration module and is identified by its URI. Modules are implemented using skeleton *map* and *ordering-farm* templates [12,14]. A Directed Acyclic Graph represents workflow among Virtual Processors implementing the skeleton. Actual mapping of DAG to physical nodes at run-time realizes skeleton configuration [15].

The server manages a set of 2*n virtual nodes. Half of them are devoted to execute tasks according to the minimum-cost contract (batch), the other half to serve tasks according to minimum-time contract (interactive). Each physical node hosts one interactive and one batch process. When active, the interactive process freezes its bat partner. A scheduler manages proper partitioning of virtual nodes among tasks for optimal DAG mapping.

Fig. 6 sketches the architecture of the Parallel server composed by a command *acceptor*, a task *scheduler*, a job *dispatcher*, a *control* engine and *configurable skeleton* modules. Service requests are accepted by the *acceptor* and inserted in two queues. The low-priority queue accepts requests for minimum-cost tasks, the high-priority one requests for minimum-time tasks.

The *scheduler* honors service requests according to a FIFO policy. It assigns to a ready-to-run task half of its interactive available nodes if it requires minimum-cost performance, otherwise an optimal number of batch virtual nodes. Optimality is based on evaluation of performance based on a cost model for the skeleton implementation. The *dispatcher* performs actual mapping of skeleton DAG on the node set assigned by the *scheduler* and the *control engine* manages component life cycle on the server-distributed platform.

Parallel version of restoration algorithms implemented through the data-parallel *map* skeleton are normally activated for interactive processing, parallel versions implemented through the stream-parallel *ordering-farm* skeleton are normally activated for batch processing

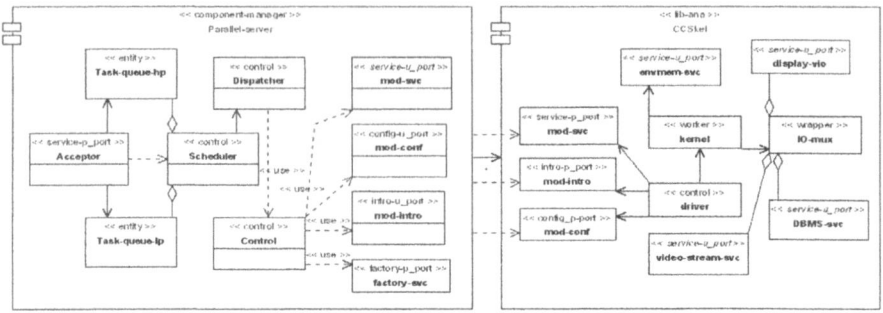

Fig. 6. Architecture of the Image Processing parallel server

5 The Video-Stream and the Metadata Servers

The Video-stream server provides a simple video-on-demand service optimized for video-frame streaming according to a GET and PLAY modes. A GET command provides a single frame or part of it selected, at original resolution or sub sampled. A play command provides a continuous video stream of a movie section with flow control. The server is implemented with a multi-tread architecture. An acceptor thread queues client requests and routes them to worker threads.

The MPEG7 meta-data server stores and retrieves movie Video Objects

6 Discussion

A process model for digital restoration has been discussed with reference to parameter steering. Support from to the FESR environment to interactivity has been described. An event driven platform supports user generation of workflow control in interactive execution and batch submission modalities from local and remote servers and application coordination. Support for meta-data storage is provided at three levels: temporary in processing modules heap during steering, session-persistent in the application server's memory component, and definitely persistent on a DBMS server. The parallel server provides support for single-frame high-priority data-parallel processing during steering and stream-parallel processing during batch task-queue optimization.

The FESR environment has been developed for supporting. digital film restoration but it is suited for supporting other applications with similar activity diagram as multimedia authoring or biomedical dynamic imaging. Full exploitation of the MPEG7 Video-Objects description format and adaptation of the distributed environment to the Virtual Organization model over a computational grid is in course.

Acknowledgements. The precious and skilled contribution of ideas to design and assistance in testing of FESR prototype from M. Tripiciano, G. L. Alfieri is acknowledged.

References

[1] P.Schallauer, A.Pinz, W.Haas, Automatic Restoration Algorithms for 35mm Film, *Videre: Journal of Computer Vision Research*, *1*(3), 1999,60-85,The MIT Press.
[2] A.C. Kokaram, *Motion Picture Restoration* London: Springer Verlag, 1998.
[3] L. Maddalena "Image Sequences Reconstruction of blue scratches" IEEE ICIAP11 Palermo Italy 2001,547-552.
[4] LIMELIGHT Project (1994-1997):
http//iis.joanneum.ac.at/iis/html_eng/projects/limelight/limelight.html
[5] AURORA Project: Automated Restoration of Original film and video Archives (1995-1998)
http://www.inafr/Recherche/Aurora/index.en.html
http://www.infowin.org/ACTS/RUS/PROJECTS/ac072.htm
[6] W. Plaschzug, G. Hejc "ESPRIT Project FRAME Public Final Report" http://www.hpcn-ttn.org/newActivityDetail.cfm? Activities_ID =1.,
http://www.vcpc.univie.ac.at/activities/projects/FRAME
[7] P. Shallauer, G. Thakllinger, M.J.Addis,,DIAMANT Digital Film Manipulation Platform ",http://www.joanneum.at/cms_img/img509.pdf
[8] A. Machì, M Tripiciano "Video Shot Detection and Characterisation in Semi-automatic Digital Video Restoration" IEEE Proceedings 15th ICPR, Barcelona 2000, pp 855-859.
[9] Alberto Machì, Fabio Collura, Filippo Nicotra: "Detection of Irregular Linear Scratches in Aged Motion Picture Frames and Restoration using Adaptive Masks": IASTED SIP02, Kawai USA 2002
[10] Alberto Machì, Fabio Collura "Accurate Spatio-temporal Restoration of Compact Single Frame Defects in Aged Motion Pictures" IAPR Proceedings 12th International Conference on Image Analysis and Processing ICIAP12, Mantova Italy 2003 pp.454-459 IEEE 2003
[11] ISO/IEC JTC1/SC29/WG11 Coding Of Moving Pictures and Audio
http://www.w3.org/2001/05/mpeg7/w4032.doc
[12] M. Vanneschi: The programming model of ASSIST, an environment for parallel and distributed portable applications. Parallel Computing 28(12): 1709-1732 (2002)
[13] The Qt graphics framework http://trolltech.com/poducts/qt.
[14] G. Sardisco, A. Machì "Development of parallel templates for semi-automatic digital film restoration algorithms. ParCo2001 Naples Italy 2001
[15] A. Machì, F. Collura "Skeleton di componenti paralleli riconfigurabili su griglia computazionale map e farm ". TR ICAR-PA-12-03 - Dec 2003.

On Triangulations

Ivana Kolingerová[1]

Centre of Computer Graphics and Data Visualization
Department of Computer Science and Engineering
University of West Bohemia, Plze , Czech Republic
kolinger@kiv.zcu.cz, http://iason.zcu.cz/~kolinger

Abstract. Triangulations in 2D and 3D are a necessary part of many applications. This paper brings a brief survey of main triangulations used in computer graphics and computational geometry, then focuses on Delaunay and greedy triangulations in 2D and 3D and presents their most important properties and algorithms. Then three applications from computer graphics and image processing areas are presented and results discussed.

1 Introduction

Triangulation in 2D and 3D is a very useful tool for many applications. This paper presents a brief survey of those triangulation methods which very developed or are used in computer graphics and image processing applications, with most focus on the most often used – the Delaunay and greedy triangulations and their constrained counterparts. The triangulation problem presented here is to produce a set of triangles or tetrahedra on the given set of points, without possibility to add new points to refine the triangulation, with possibility to incorporate a set of user-defined edges.

The paper presents pros and contras of the triangulations and experience with their use in three computer graphics and image processing applications.

Section 2 describes the best known planar triangulations, section 3 concentrates on 3D case. Section 4 shows several applications. Section 5 concludes the paper.

2 The Best Known Planar Triangulations

Def. 1. A triangulation. A triangulation $T(P)$ of a set P of n points in the Euclidean plane is a set of edges E such that
- no two edges in E intersect at a point not in P
- the edges in E divide the convex hull of P into triangles.

Many different types of triangulations, concentrated on different criteria, have been derived. They can be categorized into the following groups:

1. *Triangulations optimizing angles.* This class is the most important one as triangles without too big and too small angles are an obviously good choice for

---

[1] This work was supported by the Ministry of Education of The Czech Republic – project MSM 23 5200 005.

2. most applications. To achieve this goal, we can either maximize the minimum angles of triangles (the well-known Delaunay triangulation [4, 9, 36, 37, 40, etc.]), minimize the maximum angles (the so-called minmax angle triangulation [17]), minimize the minimum angle [18], maximize the sum of angles [35] or minimize the maximum eccentricity (i.e. the infimum over all distances between c and the vertices of the triangle, where c is the centre of the triangle's circumcircle [7, 8]) and combinations or derivations of these criteria. The most popular triangulation of this class is the Delaunay triangulation.

3. *Triangulations optimizing edge lengths*. This class is less used in practice as their algorithms have usually higher complexity which limits usability of these triangulations to hundreds or maximally several thousands of points – such data sizes are nowadays uncompetitive. Best known representatives are probably the triangulation minimizing the sum of edge lengths (miminum weight triangulation [1, 5, 6, 13, 30, 33]), its local approximation, optimizing the edge length in each pair of neighbouring triangles (the greedy triangulation [12, 31, 40]). Also a triangulation minimizing the maximum edge length [16] and several others can be found in relevant resources.

4. *Multi-criteria optimized triangulations*. In this class of triangulation, several criteria can be combined to be optimized, using local edge swaps and non-deterministic computation, such as genetic optimization. Disadvantage is a slow computation due to genetic algorithms, reducing usability of the triangulations. This type of triangulations was introduced in [27].

5. *Data-dependent triangulations*. Planar triangulations are satisfactory in most cases of 3D data with one exception, a terrain with a steep slope. Such a kind of data needs a special treatment, the so-called data-dependent triangulations, taking into consideration also heights of points (z-coordinates), angles between triangle normals, etc. This class was established in [14].

Each type of triangulation should allow to insert the so-called **constrained edges**, or **constraints** - the edges prescribed to be present in the triangulation. They are needed to incorporate non-convex boundaries and breaks into the triangulation.

It should be pointed out that no matter which type of 2D triangulation is chosen, each contains the same number of triangles on the same point set.

Besides requirements on the triangulation itself, there are also some demands on the algorithms for their computation, such as:

1. low algorithmic complexity in the worst and expected cases (the proved complexity for the triangulation problem is $\Omega(n \log n)$ in the worst case)
2. low memory consumption
3. easy implementation
4. numerical robustness and low sensibility to singular cases
5. usability in 3D
6. handling of new-coming or even out-going points
7. suitability to parallelization

According to these two lists it can be seen that a good choice of a triangulation method and of an algorithm for its computation can substantially influence efficiency and correctness.

We will concentrate now on the two most important triangulations, which are Delaunay and greedy triangulations, plus on their constrained counterparts.

Def. 2. Delaunay triangulation (DT). The triangulation DT(P) of a set of points P in the plane is a Delaunay triangulation of P if and only if the circumcircle of any triangle of DT(P) does not contain a point of P in its interior.

Delaunay triangulation ensures maximizing minimum angle of each triangle as well as of the whole triangulation. Therefore, it produces the most equiangular triangulation of all. Other positive features exist, as described in [4, 9, 36].

Many algorithms for Delaunay triangulation construction have been developed. They can be subdivided into several groups:

- local improvement by edge flipping [32]
- divide & conquer [15]
- incremental insertion [9, 23]
- incremental construction [11]
- sweeping [19]
- high-dimensional embedding [10]

Reviews and comparison can be found in [11, 43, 28]. According to the comparison in [43], the Dwyer's divide & conquer algorithm is the fastest one. It is worst-case optimal and runs in $O(n \log \log n)$ expected time on uniformly distributed points; its behaviour for non-uniform points is also satisfactory.

In real life applications, divide & conquer methods are often complicated to implement and sensitive to numerical inaccuracy. More simple methods, although not asymptotically optimal, can make a better job. According to our experience, the incremental insertion algorithm, simple in its conception, is a good choice.

The algorithm works as follows. At the beginning, a big triangle containing all the points is constructed. Then points are inserted one at a time. First, the triangle containing the point to be inserted is found, then the triangle is subdivided into new triangles. The newly constructed triangles are checked on empty circumcircle property, if not correct, they are flipped together with their neigbours and check continues until all non-Delaunay triangles are removed. When all the points are processed, the extra triangles incident to the originally added triangle are removed.

This solution provides $O(n^2)$ worst-case time complexity, so it is not optimal, however, it can achieve $O(n \log n)$ or nearly $O(n)$ expected-case time complexity with efficient location data structures. As the point location problem forms a substantial part of the computation, methods to accelerate the computation are numerous and efficient, see a survey in [28, 47]. Memory consumption depends on the choice of location data structures, in most cases is linear. It also allows modifications for insertion constraints, for non-Euclidean metrics and for 3D triangulation. If carefully implemented, it is numerically relatively robust – i.e., in the case of wrong algorithmic decision due to numerical inaccuracy some non-Delaunay triangles may appear, however, their number is low and the triangulation is usually still valid. Possibility to handle new-coming and out-going points depends on the data structure used for point location. Although the algorithm looks inherently serial, parallelization is possible using shared memory and threads - at the same moment, more points can be inserted into the triangulation if concurrent modification of the same triangles is avoided [26].

Def. 3. Greedy triangulation (GT). The triangulation GT(P) of a set of points P in the plane is a greedy triangulation if it consists of the shortest possible compatible edges where a compatible edge is defined to be an edge that crosses none of those triangulation edges which are shorter than this edge.

The main problem connected to GT is its time complexity. It can be computed in $O(n^2 \log n)$ time using demanding algorithms. $O(n)$ expected time solution is possible only for uniform data. Effective algorithms can be found in [12, 31]. If only small n is expected (maximally several thousands of points), we can cope with 'brute force' algorithm, it means a computation according to the Def.3.

To get a better feeling how these two triangulations behave, see Fig.1.

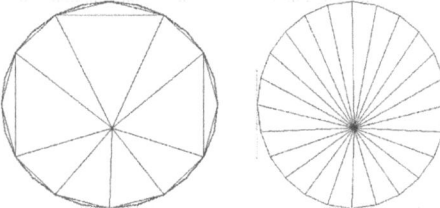

Fig. 1. GT (left) and DT (right) of points sitting on a circle; GT has many small triangles at the convex hull

Let us concentrate on possibilities to include constraints into these two triangulations. DT as defined in Def. 2 is not able to consider constraints - its definition has to be modified: in the constrained Delaunay triangulation (CDT), only non-constrained edges are checked on the empty circumcircle property from Def.2. For the CDT algorithms see [3, 41]. It is easy to use constrained edges in GT: such edges are accepted into the triangulation first, then a usual greedy method continues.

3 3D Triangulations

One of most important problems in a 3D triangulation (tetrahedronization) is that unlike 2D, the number of tetrahedra depends also on the type of triangulation, not only on the point configuration. It means that different tetrahedronizations may differ not only in the quality but also in the number of their tetrahedra. Also, while planar meshes have a linear number of triangles, in tetrahedronization it is necessary to count with $O(n^2)$ in the worst case (although such cases are rare). Also, readability of the generated meshes is low even for low number of points, see Fig.2 for an example.

Most of existing tetrahedronization methods (such as front advancing techniques or octree-based triangulations) originate from FEM areas and their use in computer graphics and image processing is rare – in these applications, 3D Delaunay tetrahedronization reigns. Its definition is a 3D analog of Def.2, having tetrahedra instead of triangles and circumcspheres instead of circumcircles.

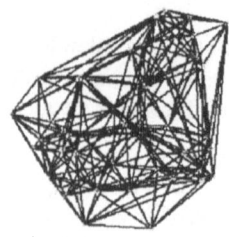

Fig. 2. Delaunay tetrahedronization for 35 points

The main disadvantage of the Delaunay tetrahedronization in comparison with 2D is that it neither maximizes the minimum angles of elements, nor minimizes the maximum radius of a circumsphere. The only shape quality proved for 3D DT is minimization of the maximum radius of the enclosing spheres (i.e. the smallest sphere containing the tetrahedron). To improve the shapes of tetrahedra, some post-optimalization methods for the DT can be used, see [34].

The algorithms for DT 3D computation are also not so many as in the planar case because not all planar approaches converge in 3D (e.g., local improvement for edge flipping can get stuck before reaching the DT [22]. Incremental approaches are prevailing for practical reasons, see [24, 46, 11], also divide & conquer methods are possible [11], even sweeping technique has appeared. Our own experience supports the incremental insertion approach, as well as in 2D. The last piece of sad news is that in 3D, the point location problem forms much smaller part than in 2D, which means that speeding up is much more difficult and less efficient.

4 Triangulations in Computer Graphics and Image Processing

In this section, we will show a usefulness of Delaunay and greedy triangulations in the following applications: contour line computation for GIS systems, surface reconstruction and image representation.

Contour Line Computation

A computation of contour lines on a triangulation is a necessary part of GIS programs. However, automatically computed contours are rarely satisfactory for an experienced user. We tried to compute contour lines on DT triangulation and then to cure some of the problems by manually imposing constraints into the triangulation, followed by local recomputation of triangles and contour lines. More details can be found in [29].

Typical problems appearing on automatically generated contours can be seen in Fig.3. Therefore, we tried to replace the DT by the GT, however neither DT nor GT on itself ensured good behaviour in all cases, therefore, our conclusion was to prefer DT because it is cheaper to compute.

The technique of imposing constraints on the problematic places was successful in our experiments, see Fig.4. The most important task for the future is to derive a criterion to automatically detect the places to be corrected with constraints.

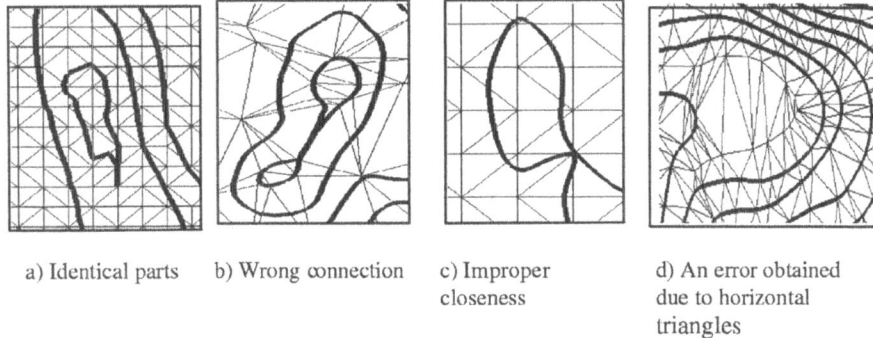

a) Identical parts   b) Wrong connection   c) Improper closeness   d) An error obtained due to horizontal triangles

Fig. 3. Errors detected on contour lines when using DT

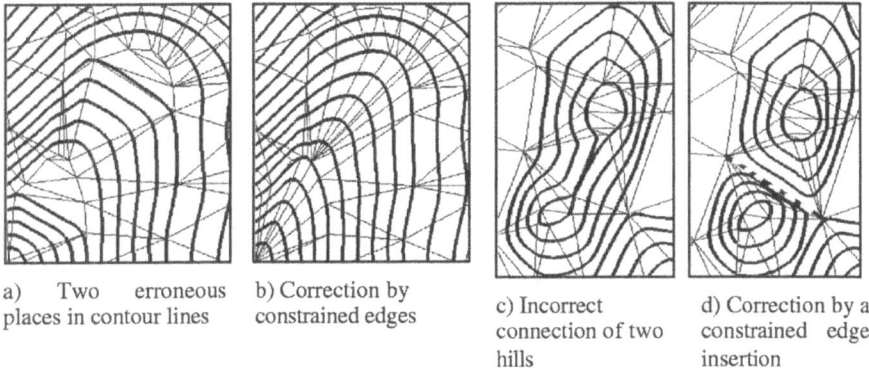

a) Two erroneous places in contour lines   b) Correction by constrained edges   c) Incorrect connection of two hills   d) Correction by a constrained edge insertion

Fig. 4. Two errors in contour lines and their correction with the use of constrained edges

Surface Reconstruction

Having a set of 3D points scanned by special devices from an object surface, we need to reconstruct the object surface. Due to importance of this problem, many algorithms were developed, see survey in [44, 45]. We use the Crust algorithm by N.Amenta [2] as it has a strong theoretical background, it is not too much sensitive to the sampling density and uniformity and also because we had already a fast and robust DT 3D implementation which is a substantial part of the method. We made some smaller modifications to the Crust method, concerning boundary detection and manifold extraction; detailed information can be found in [44, 45].

Let us briefly explain how it works. It can be proved that the surface triangles of the reconstructed object are a subset of DT. The Crust algorithm first computes the Delaunay tetrahedronization of the given point set, then it tries to estimate which of many computed triangles should be kept because they belong to the surface. This estimate is done according to the shape of the Voronoi cells dual to tetrahedra – for a sufficiently sampled surface, they are long in the direction of the estimated surface normal and narrow in the rectangular direction, so we can use them for the surface normal estimation in the given point. After we estimated the normal vectors to the reconstructed surface, we can use this information to find which triangles belong to the surface and which not.

Resulting reconstructed surfaces are not necessarily manifolds and can have problems with correct detection of boundaries, so some post-processing steps to solve these two problems are necessary. If the data are extremely noisy or non-uniform, the solution is not completely correct. Noisy data processing is a big task for our future work. Fig.5 presents examples of the objects reconstructed using the described approach. Data sets presented here belong to standard models for testing these applications, see [42, 20].

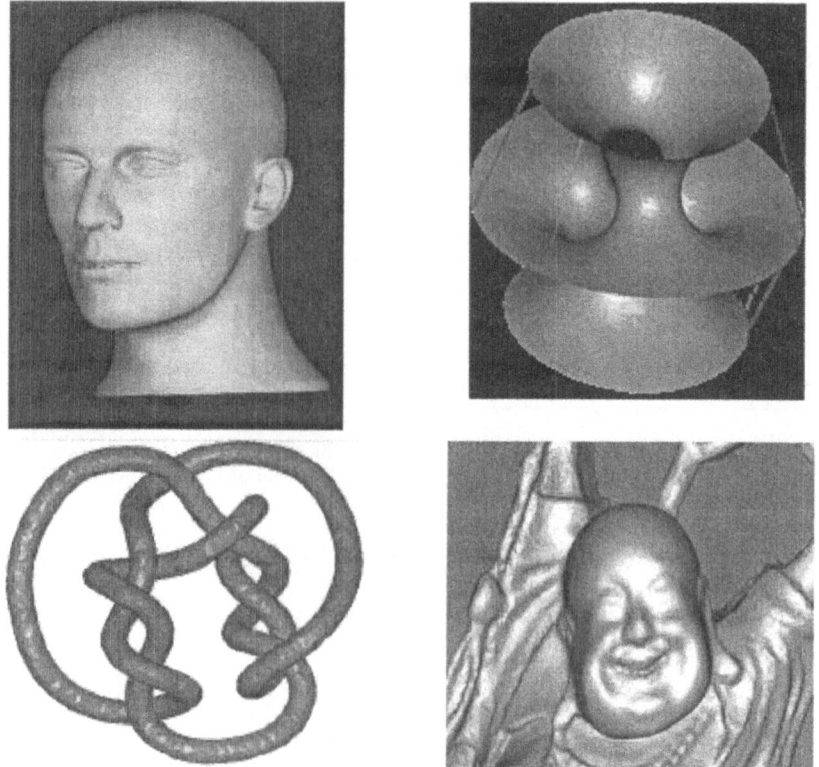

Fig. 5. Successful as well as problematic reconstructions using a modified Crust algorithm

As this method is based on duality between DT and Voronoi diagram, it does not seem possible to use GT for the surface reconstruction. The use of constraints in DT makes more sense and it is our intended future work.

Image Representation

At present, great concern in image processing is devoted to region oriented image representations. In comparison to block-oriented approach, they offer better adaptation to local image characteristics. These arbitrarily-shaped image regions are then coded by a generalized orthogonal transform [21], by iterative cosine transform

[25] or by shape-independent transform with the use of different number of basis functions for each region [39]. If the region boundaries are approximated by a polygon only, it causes a degradation of region boundaries. Therefore, the polygonal regions are triangulated. The triangulation must fit the generally non-convex polygon boundary, so constrained types of triangulations are necessary. Too small triangles are filtered out.

Fig.6a) shows the original 'baboon' image (JPEG), Fig.6b) the CDT using region boundaries as constraints. Fig.6c) presents the image coded on 85 partially triangulated regions. We again experimented with both DT and GT triangulations. Our conclusion is that the results obtained using GT are not that much different to justify higher implementation and computation complexity of the GT, therefore, we again vote for DT in this application. See details in [38].

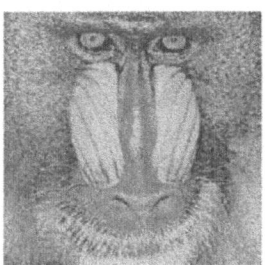
a) JPEG-coded image

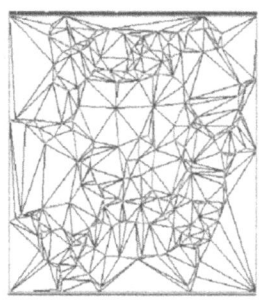
b) CDT on the image region boundaries

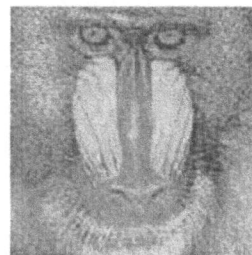
c) CDT-based coding

Fig. 6. Image representation using partially triangulated areas (images with permission of J.Polec, Comenius University, Bratislava, Slovakia)

5 Conclusion

We hope that the paper showed how rich and interesting the triangulation topic is, even if we limit ourselves only to the computer graphics and image processing applications. There is already much work done, however, much challenges still exist, namely in 3D – good and reliable algorithms for well-shaped tetrahedra. Therefore, the triangulation topic provides an interesting and grateful topic for a future research.

Acknowledgements. I would like to express my thanks to Prof. V. Skala from the University of West Bohemia in Pilsen, Czech Republic for providing conditions for this work, to my research partners Prof. B. Žalik from the University of Maribor in Slovenia, Dr. A. Ferko from the Comenius University in Bratislava, Slovak Republic and to Dr. J. Polec from the Slovak Technical University in Bratislava, Slovak Republic for their inspiring cooperation and often the leading role in the above mentioned research topics. My thanks belong also to my PhD students J.Kohout, M.Varnuška, P.Van ek, P. Maur and J. Parus whose work also found a reflection in this paper.

References

1. O. Aichholzer, F. Aurenhammer, R. Hainz (1998) New results on MWT subgraphs, TR Nr. 140, 1998, Institute for Theoretical Computer Science, Graz University of Technology
2. N. Amenta, M. Bern (199) Surface reconstruction by Voronoi filtering, Discrete and Computational Geometry 22(4), pp.481-504
3. M.V. Anglada (1997) An Improved incremental algorithm for constructing restricted Delaunay triangulations, Computers & Graphics 21, pp.215-223
4. F. Aurenhammer (1991) Voronoi diagrams - a survey of a fundamental geometric data structure, ACM Comput Surv, Vol.23, No.3, pp.345-405
5. M. Bartánus, A. Ferko, R. Mag, L. Niepel, T. Plachetka, E. Šikudová (1996) New heuristics for minimum weight triangulation, In : WSCG 96, Conference Proceedings, University of West Bohemia, Pilsen, pp.31-40
6. R. Beirouti, J. Snoeyink (1998) Implementations of the LMT heuristic for minimum weight triangulation, Proc. 14th Annual Symp Comput Geom, Minneapolis, pp.96-105
7. M. Bern, H. Edelsbrunner, D. Eppstein, S. Mitchell, T. S. Tan (1993) Edge insertion for optimal triangulations, Discrete Comput Geom, Vol.10, pp.47-65
8. M. Bern, D. Eppstein (1994) Mesh generation and optimal triangulation, 2nd edn. In: Lecture Notes Series on Computing, Vol.4, World Scientific, pp.47-123
9. M. de Berg, M. van Kreveld, M. Overmars, O. Schwarzkopf (1997) Computational geometry. Algorithms and applications, Springer-Verlag Berlin Heidelberg
10. K.Q. Brown (1979) Voronoi diagrams from convex hulls, Inf. Proc. Letters, Vol. 9, No. 5, pp. 223-228
11. P. Cignoni, C. Montani, R. Perego, R. Scopigno (1993) Parallel 3D Delaunay triangulation, Proc. of Eurographics'93, pp. C129-C142
12. M.T. Dickerson, R.L.S. Drysdale, S.A. McElfresh, E. Welzl, (1994) Fast greedy triangulation algorithms, Proc.10th Annual Symp. on Comp.Geom., pp.211-220
13. M.T. Dickerson, M.H. Montague (1996) A (usually?) connected subgraph of the minimum weight triangulation, Proc. 12th Sym Comput Geom, Philadelphia, pp.204-213
14. N. Dyn, D. Levin, S. Rippa (1990) Data dependent triangulations for piecewise linear interpolation, IMA J Numer Anal 10, pp.137-154
15. R.A. Dwyer (1986) A Simple divide-and-conquer algorithm for constructing Delaunay triangulation in $O(n \log \log n)$ expected time. In Proc. of the 2nd Annual Symposium on Comp. Geom., pp. 276-284
16. H. Edelsbrunner, T.S. Tan (1991) A quadratic time algorithm for the minmax length triangulation, SIAM J Comput 22, pp.527-551
17. H. Edelsbrunner, T. S. Tan, R. Waupotisch (1992) An $O(N^2 \log N)$ time algorithm for the minmax angle triangulation, SIAM J Stat Sci Compu. 13, 1992, pp.994-1008
18. D. Eppstein (1992) The farthest point Delaunay triangulation minimizes angles, Comput. Geom. Theory Appl. 1, pp.143-148
19. S.J. Fortune (1987) A sweepline algorithm for Voronoi diagrams, Algorithmica,Vol. 2, pp. 153-174
20. Georgia Institute of Technology. Large geometric models archive. Available at http://www.cc.gatech.edu/projects/large_models
21. M. Gilge, T. Engelhardt, R. Mehlan R (1989) Coding of arbitrary shaped image segments based on a generalized orthogonal transform, Signal Processing: Image Communication, No.1, pp.153-180
22. J.E. Goodmann, J. O'Rourke,J. [Eds.] (1997) Handbook of Discrete and Computational Geometry, CRC Press
23. L.J. Guibas, D.E. Knuth, M. Sharir (1992) Randomized incremental construction of Delaunay and Voronoi diagrams, Algorithmica, Vol. 7, 1992, pp.381-413

24. B. Joe (1991) Construction of three-dimensional Delaunay triangulations using local transformations. Computer Aided Geometric Design, Vol. 8, pp. 123–142.
25. A. Kaup, T. Aach (1998) Coding of segmented images using shape-independent basis functions, IEEE Transactions on Image Processing, Vol.7, No.7, pp.937-947
26. I. Kolingerová, J. Kohout (2002) Optimistic parallel Delaunay triangulation, The Visual Computer 18, pp.511-529
27. I. Kolingerová, A. Ferko (2001) Multicriteria-optimized triangulations, The Visual Computer, 17, pp.380-395
28. I. Kolingerová, B. Žalik (2002) Improvements to randomized incremental Delaunay insertion, Computers & Graphics 26, pp.477-490
29. I. Kolingerová, V. Strych, V. ada (2004) Using constraints in Delaunay and greedy triangulation for contour lines improvement, accepted for publication on ICCS conference, Krakow, Poland
30. D. Krznaric (1997) Progress in hierarchical clustering & minimum weight triangulation, PhD Thesis, University of Lund, Sweden
31. C. Levcopoulos, A. Lingas, (1992) Fast algorithms for greedy triangulation, BIT 32, pp.280-296
32. C. L. Lawson (1977) Software for C1 interpolation, in: Mathematical Software III (J.R. Rice, Ed.), New York: Academic Press, pp.161-194
33. I. Magová, A. Ferko, L. Niepel (1997) On edges elimination for the shortest mesh, In : WSCG'97 Conference Proceedings, University of West Bohemia, Pilsen, pp.396-403
34. P. Maur, I. Kolingerová(2001). Post-optimization of Delaunay tetrahedrization. In : SCCG IEEE proceedings, ISBN 0-7695-1215-1, Los Alamitos, USA, pp.31-38.
35. T. Midtbo (1993) Spatial modelling by Delaunay networks of two and three dimensions, http://www.iko.unit.no/tmp/term.html
36. A. Okabe, B. Boots, K. Sugihara (1992) Spatial tesselations: concepts and applications of Voronoi diagrams, John Wiley & Sons, Chichester
37. J. O' Rourke (1994) Computational geometry in C, Cambridge University Press, New York
38. M. Partyk, J. Polec, I. Kolingerová, A. B ezina (2003) Triangulations in a hybrid scheme for shape independent transform coding, Proc. ACIVS, Ghent, Belgium, pp. 137-141
39. J. Polec at al. (2002) New scheme for region approximation and coding with shape independent transform, ISPRS Commision III, Proc. Of Symposium Photogrammetric Computer Vision, Graz, Austria, pp.B214-217
40. F. P. Preparata, M. I. Shamos (1985) Computational geometry: an introduction, Springer-Verlag, Berlin Heidelberg New York
41. S.W. Sloan (1993) A fast algorithm for generating constrained Delaunay triangulations, Computers & Structures 47, pp.441-450
42. Stanford Computer Graphics Laboratory. The Stanford data scanning repository. Available at http://graphics.stanford.edu.data/3Dscanrep/
43. P. Su, R.L.S. Drysdale (1995) A comparison of sequential Delaunay triangulation algorithms, In: Proc. 11th Annual Symp. on Comp.Geom., ACM, pp.61-70
44. M. Varnuška, I. Kolingerová (2004) Boundary filtering in surface reconstruction, accepted for publication in ICCA conference, Assissi, Italy
45. M. Varnuška,M., I. Kolingerová, (2004) Manifold extraction for surface reconstruction, accepted for publication in ICCS conference, Krakow, Poland
46. D.F. Watson (1981) Computing the n-dimensional Delaunay tessellation with application to Voronoi polytopes, Comput J 24(2), pp.167-172
47. B. Žalik, I. Kolingerová (2003) An Incremental construction algorithm for Delaunay triangulation using the nearest-point paradigm, Int. J. Geographical Information Science 17, pp.119-138

Probability Distribution of Op-Codes in Edgebreaker

Deok-Soo Kim[1], Cheol-Hyung Cho[1], Youngsong Cho[2], Chang Wook Kang[1], Hyun Chan Lee[3], and Joon Young Park[4]

[1] Department of Industrial Engineering, Hanyang University,
17 Haengdang-Dong, Sungdong-Ku, Seoul, 133-791, South Korea
{dskim,cwkang57}@hanyang.ac.kr, murick@voronoi.hanyang.ac.kr
[2] Voronoi Diagram Research Center,
17 Haengdang-Dong, Sungdong-Ku, Seoul, 133-791, South Korea
ycho@voronoi.hanyang.ac.kr
[3] Department of Industrial Engineering, Hongik University,
72-1, Sangsu-dong, Mapo-gu, Seoul, 121-791, Korea
hclee@wow.hongik.ac.kr
[4] Department of Industrial Engineering, Dongguk University,
26, 3-ga, Pil-dong, Chung-gu, Seoul, 100-715, South Korea
jypark@dgu.edu

Abstract. Rapid transmission of 3D mesh models has become important with the use of Internet and with increased number of studies on the compression of various aspects for mesh models. Despite the extensive studies on Edgebreaker for the compression of topology for meshes, the probability distribution of its five op-codes, C, R, E, S, and L, has not yet been rigorously analyzed. In this paper we present the probability distribution of the op-codes which is useful for both the optimization of the compression performances and a priori estimation of compressed file size.

1 Introduction

Ever since Deering suggested the compression of mesh models for faster transmission of 3-D mesh models on the network, there have been extensive researches on this issue. Since a mesh model usually consists of coordinates of vertices, topology defined by the connectivity among the vertices and faces, and other properties such as colors and normal vectors, the compressions of these aspects of mesh models have been studied extensively [1][2][3][4][5][7][9][10][11][14][15][16].

Edgebreaker has been the most popular compressor for topology of meshes because of its speed in both compression and decompression phases, its simplicity in both its theory and implementation, and availability of rigorous analysis of its performance characteristics[6],[11][12][13]. However, the probability distribution of its five op-codes, C, R, E, S, and L, has never been rigorously analyzed despite of the extensive studies on the various aspects of Edgebreaker.

In this paper we present a mathematical derivation of the probability distribution of five op-codes of Edgebreaker. In particular, we will show the limiting probabilities of these op-codes so that this knowledge can be useful for the estimation of compressed file size of a mesh model as well as the optimization of compression algorithms depending on the probability distribution op-codes in a particular model. In addition, we assume the mesh models are simple meshes.

2 Probability Distribution of Op-Codes in Edgebreaker

Edgebreaker, one of the most popular topology compressor for mesh models, transforms each triangle to one of five alphabets called op-codes, C, R, E, S and L[11]. An appropriate op-code is assigned to a triangular face depending on the states of two triangles next to the current triangle and a vertex shared by these neighboring triangles. Hence, a mesh model is transformed to a string of five alphabets and this string represents the complete topology information. Considering other properties of the sequence such as the frequencies of words in the string, Edgebreaker can even compress the topology information of a mesh model with lower than one bit per triangle[12].

Fig. 1 illustrates a mesh model and a corresponding string produced by Edgebreaker starting from the triangle $v_0v_1v_2$ with the edge v_0v_1 as an initial gate. Note that the mesh is now represented by a string of CCCSERLRRLE.

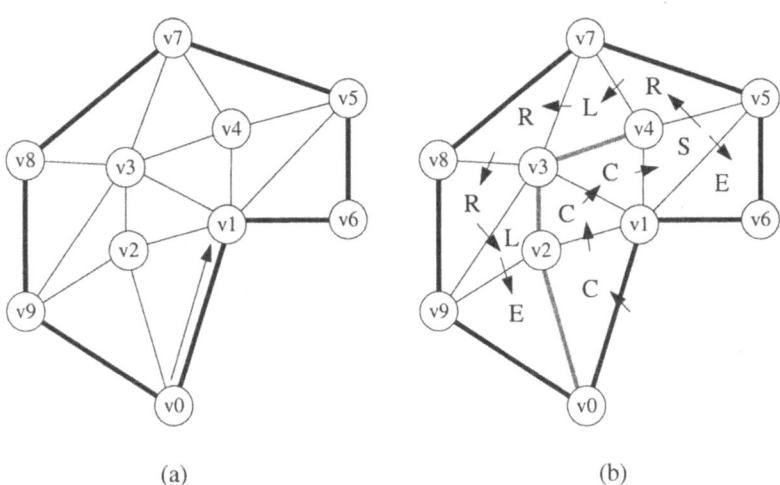

(a)   (b)

Fig. 1. A mesh model and the corresponding string compressed by Edgebreaker: (a) a mesh model (b) the compression sequence by Edgebreaker

For the convenience of presentation, we introduce a few terminologies. Edge v_0v_1 in the above figure is called an incoming door of the triangle $v_0v_1v_2$ since parsing operation of the triangle passes over this edge. Similarly, the edge v_1v_2

is called an outgoing door of the triangle $v_0v_1v_2$. Note that the edge v_1v_2 plays the role of incoming door for triangle $v_2v_1v_3$ and the edge v_1v_3 now becomes an outgoing door of triangle $v_2v_1v_3$. Edges v_0v_2, v_3v_2, and v_4v_3, in this example, are called wall edges since the compression does not proceed over these edges, and the triangle $v_2v_9v_0$ is called a mate of the triangle $v_0v_1v_2$ since both triangles share a wall edge. It is known that the mate of face with op-code C can have neither op-code C nor S.

Shown in Fig. 2 are two popular mesh models: Agrippa and Teeth models defined by 12,660 and 58,300 triangles, respectively. These models show overall features of individual models sufficiently well for ordinary graphical use for most applications.

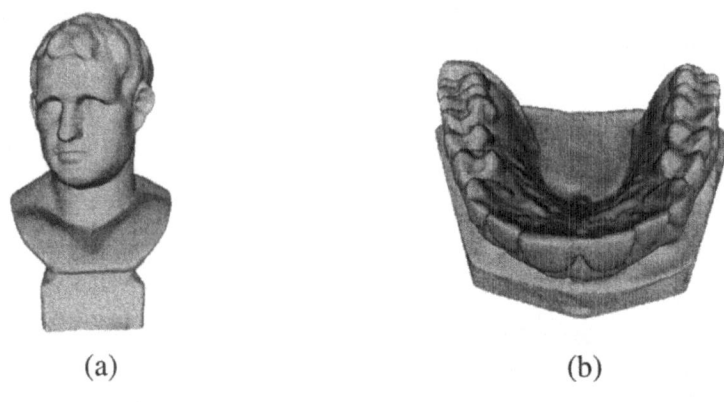

Fig. 2. Simple meshes: (a)Agrippa (b)Teeth

Running Edgebreaker on these models reveals an interesting phenomenon as illustrated in Fig. 3. Shown in Fig. 3 are the probability distributions of four op-codes R, E, S and L for the models shown in Fig. 2. The probability of op-code C is not shown since it is always 0.5[11]. The horizontal axis of each graph represents different initial gates for the start of Edgebreaker for each of the models, and the vertical axis represents the probabilities of each op-code. Even though the graphs show only ten initial gates on the horizontal axis for the convenience of graphical display, we have done all possible experiments using every edge as the initial gate for Edgebreaker and they all show same phenomenon stated as follows.

As illustrated in the graphs, the probabilities of the op-codes, Pr(R), Pr(E), Pr(S), and Pr(L) for the models in Fig. 2 always preserve the following orders: Pr(R) > Pr(E) > Pr(S) > Pr(L). Since Pr(C) is always 0.5, the ordering among five op-codes seems to be complete. Rossignac also reported the relative frequencies of four op-codes R, E, S, and L as 36.3%, 5.6%, 5.6% and 2.5%, respectively, by investigating several large mesh models up

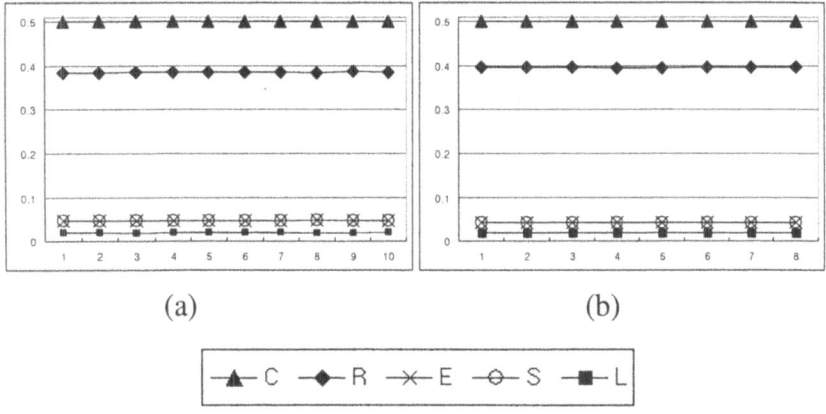

Fig. 3. Op-code distributions of the mesh models in Fig.2: (a) Agrippa (b)Teeth

to 200,000 triangles [11]. Note that the ordering of op-code probabilities is identical to our experiments in the above. Hence, we make a conjecture as follows that the above-mentioned phenomenon may be true for any mesh models.

Conjecture. For a simple mesh model with a sufficient number of triangles, the probabilities of five op-codes produced by Edgebreaker always preserve $Pr(C) > Pr(R) > Pr(E) > Pr(S) > Pr(L)$.

3 Subdivision of Triangles

To investigate the conjecture, we devise a mechanism called subdivision of triangles in the mesh model to facilitate the manipulation of the probability distribution of op-codes without changing the geometries of models. Based on the subdivision of a triangle, we will provide a mechanism to investigate the state changes among the related triangles to derive the limiting probabilities of op-codes.

Suppose that a triangle T be given with three vertices v_1, v_2 and v_3 in a counter-clockwise orientation as shown in Fig.4(a). Let the edge v_1v_2 in T is the incoming door. Then inserting a new vertex v_4 inside T creates three new triangles T_1, T_2, and T_3, while destroys the given triangle T as shown in Fig.4 (b). Note that the subdivision increases the total number of topological entities (such as vertices, edges and faces) while the shape of model does not change if v_4 is placed on the triangle T. Note that this subdivision preserves the vertex conformity with the neighboring triangles in the model.

Once a triangle is subdivided as suggested, we observe the states of op-codes of new triangles and the change of op-code of the mate of destroyed triangle. It turns out that there are a few definite rules governing the subdivision process. Since there is a definite rule corresponding to each op-code, we have

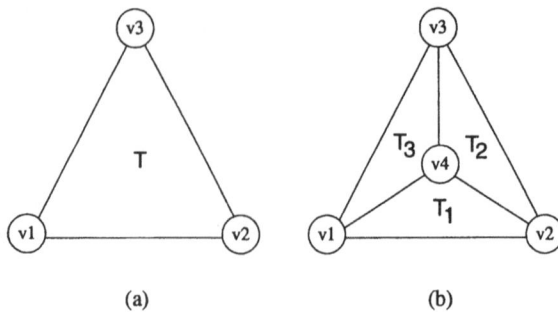

Fig. 4. Proposed subdivision scheme: (a)before subdivision (b)after subdivision

five subdivision rules as a total. Presented in the following are four subdivision rules corresponding to op-codes R, E, S, and L. The rules for op-code C will be discussed separately. Let $Sub(T)$ be an operator subdividing a triangle T, and $\equiv$ be an operator assigning an op-code to a triangle. Then,

S1: $Sub(T\equiv R) \Rightarrow \{T_1\equiv C,\ T_2\equiv R\ \text{and}\ T_3\equiv L\}$
S2: $Sub(T\equiv E) \Rightarrow \{T_1\equiv C,\ T_2\equiv R\ \text{and}\ T_3\equiv E\}$
S3: $Sub(T\equiv S) \Rightarrow \{T_1\equiv C,\ T_2\equiv S\ \text{and}\ T_3\equiv L\}$
S4: $Sub(T\equiv L) \Rightarrow \{T_1\equiv C,\ T_2\equiv S\ \text{and}\ T_3\equiv E\}$

What the subdivision rule S1 does is shown in Fig. 5 and can be explained as follows: Subdividing a triangle with op-code R will result in three triangles such that T_1, T_2 and T_3 have op-codes C, R, and L, respectively. Applying the subdivision rule S2 to a triangle with op-code E results in three triangles T_1, T_2 and T_3 with op-codes C, R, and E, respectively. The meanings of other rules can be interpreted similarly.

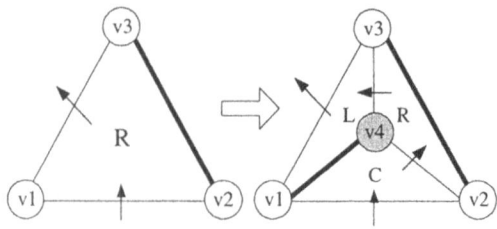

Fig. 5. Subdivision of a triangle with op-code R

On the other hand, subdividing a triangle with op-code C falls into additional four subcases depending on the op-code of the mate of current triangle and the position of incoming door of the mate. Let M be the mate of a current triangle T. Then, the rules are as follows.

S5-1: If T≡C and M≡R, then Sub(T≡C) ⇒
{T_1≡C, T_2≡C and T_3≡E} and M≡S.
S5-2: If T≡C and M≡L, then Sub(T≡C) ⇒
{T_1≡C, T_2≡C and T_3≡E} and M≡S.
S5-3: If T≡C and M≡E with an incoming door X, then
Sub(T≡C) ⇒ {T_1≡C, T_2≡C and T_3≡E} and M≡L.
S5-4: If T≡C and M≡E with an incoming door Y, then
Sub(T≡C) ⇒ {T_1≡C, T_2≡C and T_3≡E} and M≡R.

The op-code for the mate of a triangle with op-code C can be either E, R, or L. Subdividing a triangle of an op-code C always results in T_1≡ C and T_2≡C. However, the op-code of T_3 may vary depending on the op-code of its mate and a little bit of complication may happen. Note that the mate of a triangle with op-code C can be neither C nor S. If the op-code of a mate is R, the compression sequence is as shown in Fig. 6(a). The meanings of S5-2 can be interpreted similarly.

Shown in Fig. 6 (b) is the case that the mate of current triangle has an op-code E. In this case, the process is further categorized into two distinct subcases depending on the location of incoming door. In the case of S5-3 as shown in Fig.6(b), a triangle with op-code C is subdivided into T_1, T_2 and T_3 having op-codes C, C and E, respectively, and the state of mate is changed to L from E. The meanings of S5-4 can be interpreted similarly but its incoming door of mate is other edge.

In an arbitrary mesh model, there should be at least one triangle with op-code R, since the wall edge created by the initial triangle should have a mate of triangle with op-code R. Hence, Pr(R)>0. To get the upper limit of Pr(R), we apply subdivision rules which increase the frequency of R as much as possible. Let $|F_0|$ be the number of triangles and $|R_0|$ be the number of triangle with op-code R of a given mesh model, respectively. Since a subdivision operator increases one, and only one, triangle with op-code R, the number of R faces after N times gives the following upper bound.

$$Pr(R) < \lim_{N \to \infty} \frac{|R_0| + N}{|F_0| + 2N} \approx \frac{1}{2}$$

By similar arguments, the following observations can be made.

$$0 = Pr(L) < \frac{1}{2}, 0 = Pr(S) < \frac{1}{4}$$

4 Test of Conjectures

Shown in Fig. 7 (a) is a tetrahedron and the triangles developed into a plane. Compressing this tetrahedron using Edgebreaker through any edge as initial

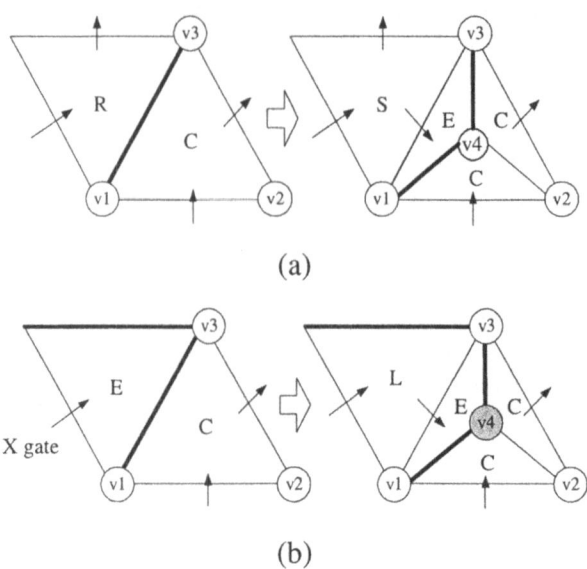

Fig. 6. Subdivision of a triangle with an op-code C: (a) its mate is R type, (b) its mate is E type where the incoming door is X.

gate will produce an identical op code of "CCRE" as shown in Fig. 7 (b). This example, however, does not reject the conjecture yet.

Suppose we subdivide a triangle of the tetrahedron repeatedly as shown in Fig. 8. Fig. 8(a) is a subdivision of triangle with op-code R repeatedly. Since R code operated by S1 keeps alive in T_2, we apply same subdivision on successive T_2 and the process will result in a topology similar to Fig. 8(a). Shown in Fig. 8(b) is a case that triangle with op-code E is subdivided by applying operator S2 successively. Note that triangle with op-code E operated by S2 still alive in T_3. Hence, applying same operator repeatedly will result in a topology similar to Fig.9(b).

Shown in Fig. 9(a) and (b) are the probability distribution of five op-codes for both tetrahedral after sufficient number of subdivisions. Since we started from a tetrahedron and one subdivision increases two triangles in the refined model, applying operators 498 times will create a model with 1000 triangles. Shown in Fig. 8 (a) and (b) are the cases that triangles with op-code R and E are subdivided repeatedly 498 times, respectively. In addition, we call these refined models as TR^{498} and TE^{498}. The horizontal axis in the figures are the locations of initial gates for Edgebreaker and therefore the numbers under the axis is immaterial, but explains different cases and sorted so that the frequencies are in a meaningful order. The vertical axis represents the probabilities of each op-codes.

Probability Distribution of Op-Codes in Edgebreaker   561

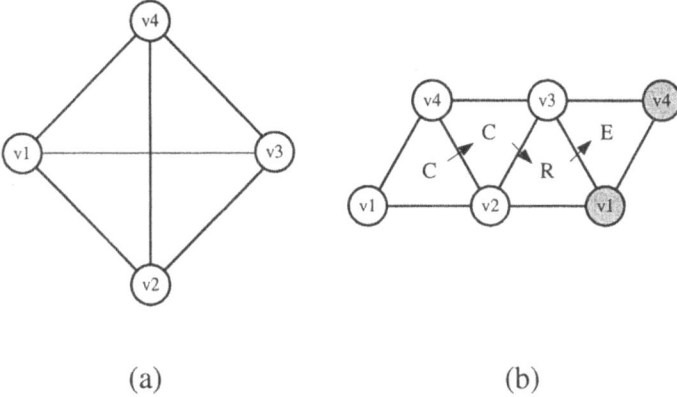

Fig. 7. Tetrahedron and developed figure by Edgebreaker: filled circle is compressed vertex: (a) Tetrahedron (b) Tetrahedron developed by Edgebreaker.

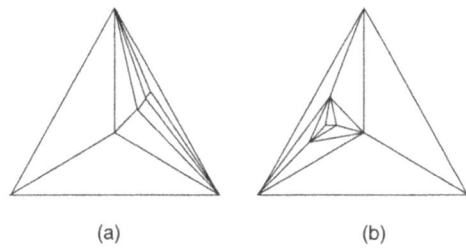

Fig. 8. Topology configuration of experiment model: (a)TR^{498} (b)TE^{498}.

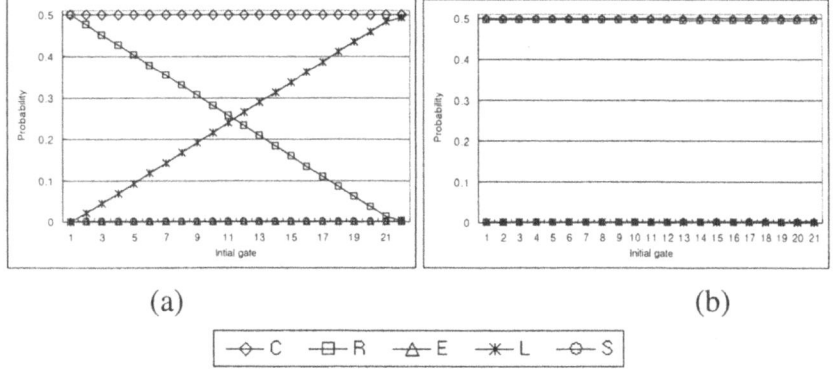

Fig. 9. Op-code probability distribution of (a) TR^{498} and (b) TE^{498}.

Subdividing triangle with op-code R of a tetrahedron repeatedly shows that the ordering of CRESL cannot be preserved. In this model, the relative probability among five op-codes even depends on the initial gate of the compression. On the other hand, Fig.9(b) always preserve the hypothesis very well. Regardless where the Edgebreaker started, the total ordering of CRESL is preserved. Hence, the conjecture is rejected from this experiment as well as the mathematical derivation of limiting probabilities.

5 Conclusion

In this paper we present a mathematical derivation of the probability distribution of five op-codes of Edgebreaker which is most popular compressor for topology of mesh models. In particular, we have shown the probability bounds of these op-codes so that this knowledge can be useful for the estimation of compressed file size of a mesh model as well as the optimization of compression algorithms depending on the probability distribution op-codes in a particular model. Based on usual experimental experiences, we established an hypothesis of probability ordering $Pr(C) > Pr(R) > Pr(E) > Pr(S) > Pr(L)$, and showed in both analytically and experimentally that the hypothesis is not valid. Our result, presented in this paper, can be used to measure performance of mesh models produced by different mesh generation algorithms. If the most part of triangles are regular, then probability of R code is very higher than others and the result of compression in this model completely keeps order of $Pr(C) > Pr(R) > Pr(E) > Pr(S) > Pr(L)$.

Acknowledgments. This research was supported by the research fund of Hanyang University(HY-2001).

References

1. Bajaj, C.L., Pascucci, V., Zhuang, G.: Single resolution compression of arbitrary triangular meshes with properties. Computational Geometry: Theory and Application, Vol. 14. (1999) 167–186.
2. Chow, M.: Optimized Geometry Compression for Real-Time Rendering. Proceedings of IEEE Visualization '97, (1997) 347–354.
3. Deering, M.: Geometry Compression. Proceedings of ACM SIGGRAPH '95, (1995) 13–20.
4. Gumhold, S., Strasser, W.: Real-Time Compression of Triangle Mesh Connectivity. Proceedings of ACM SIGGRAPH '98, (1998) 133–140.
5. Isenburg, M., Snoeyink, J.: Face Fixer: Compression Polygon Meshes with Properties, Proceedings of SIGGRAPH 2000, (2000) 263–270.
6. Isenburg, M., Snoeyink, J.: Spirale Reversi: Reverse decoding of Edgebreaker encoding. Computational Geometry: Theory and Application, Vol. 20. (2001) 39–52.
7. Kim, D.-S., Cho, Y., Kim, D.: The Compression of the Normal Vectors of 3D Mesh Models Using Clustring. Lecture Notes in Computer Science, Vol. 2330. Springer-Verlag, Berlin Heidelberg New York (2002) 275–284.

8. Kim, D-S., Cho, Y., Kim, D., Kim, H.: Probability Distribution of Index Distances in Normal Index Array for Normal Vector Compression, Lecture Notes in Computer Science, Vol. 2657 (2003) 887–896.
9. Kim, Y.-S., Park, D.-G., Jung, H.-Y., Cho, H.-G.: An Improved TIN Compression Using Delaunay Triangulation, Proceedings of Pacific Graphics '99, (1999) 118–125.
10. Lee, E.-S. and Ko, H.-S.: Vertex Data Compression For Triangle Meshes, Eurographics 2000, Vol. 19, No. 3 (2000) 1–10.
11. Rossignac, J.: Edgebreaker: Connectivity Compression for triangle meshes. IEEE Transactions on Visualization and Computer Graphics, Vol. 5. (1999) 47–61.
12. Rossignac, J., Szymczak, A.: Wrap & Zip decompression of the connectivity of triangle meshes compressed with Edgebreaker. Computational Geometry: Theory and Application, Vol. 14. (1999) 119–135.
13. Szymczak, A., King, D., Rossignac, J.: An Edgebreaker-based efficient compression scheme for regular Meshes. Computational Geometry: Theory and Application, Vol. 20. (2001) 53–68.
14. Taubin, G., Horn, W.P., Lazarus, F., Rossignac, J.: Geometric Coding and VRML. Proceedings of the IEEE, Vol. 86. (1998) 1228–1243.
15. Taubin, G., Rossignac, J.: Geometric Compression Through Topological Surgery. ACM Transactions on Graphics, Vol. 17 . (1998) 84–115.
16. Touma, C., Gotsman, C.: Triangle Mesh Compression. Proceedings of Graphics Interface '98. (1998) 26–34.

Polyhedron Splitting Algorithm for 3D Layer Generation

Jaeho Lee[1], JoonYoung Park[1], Deok-Soo Kim[2], and HyunChan Lee[3]

[1] Department of Industrial Engineering, Dongguk University,
26, 3-ga, Pil-dong, Chung-gu, Seoul, Korea
{rapidme, jypark}@dgu.edu
[2] Department of Industrial Engineering, Hanyang University,
17 haengdang-dong, Seong dong-gu, Seoul, Korea
dskim@hanyang.ac.kr
[3] Department of Industrial Engineering, Hongik University,
Sangsu-dong, 72-1. Mapo-gu, Seoul, Korea
hclee@wow.hongik.ac.kr

Abstract. RP(Rapid Prototyping) is often called as Layered Manufacturing because of layer by layer building strategy. Layer building strategy is classified into two methodologies. One is based on the 2D layer and the other is based on the 3D layer. 2D layer is simply created by the intersection between the polyhedron and a slicing plane whereas 3D layer is created with some constraints such as cuttability and manufacturability. We propose a geometric algorithm that uses the feature of individual triangle to create 3D layers.

1 Introduction

RP(Rapid prototyping) has been developed by many vendors and research groups and supplied to market since 1992. RP has a strong point that can usually manufacture a sample within 24 hours. Especially, 3D layer is being used to overcome the limit of 2D based RP technology and handle the composite material. To achieve these goals, 3D models should be divided properly depending on individual RP equipment[6].

Currently, most RP equipments are based on 2D layer. That is, the side of the layer is vertical and this layer is called a vertical layer. Since the vertical layer is basically a 2.5D block, a stair-step effect happens. However, if the height of the layer, h, is small enough, models with complicated shapes can be made with some allowable errors. On the other hand, there are some efforts to make side walls with a curved or sloped layer. This technique is called a 3D layer technology and can be classified into two major approaches. One is a cutting and bonding approach which is to make a large prototype with topologically spherical shape by bonding thick plates. The other is a hybrid approach, which is a combination of 2.5D deposition RP technology and CNC milling operation.

There are some recognizable researches on the cutting and bonding approach by using thick plate. In 1997, Hope *et al.* developed TruSurf(True Surface System), which generates a sloping layer by using a 4-axis water jet machine to the thin plate based RP machine[4]. Imre et al. from Delft University also developed a system based on the foam cutting[2]. In this system, a tool-path is generated for two planes connected side by side and two layers of foams are semi automatically bonded after the machining[5].

This approach has an advantage of making a large prototype which is bigger than one cubic meter in a very short time. Most RP equipments available today can not handle the model with that size. This approach, however, has difficulties for making models with a cavity between the layers. For example, if a pipe with small diameter is passing through a thick layer, making a prototype without partition and cutting is not possible.

There is another approach called SDM(Shape Deposition Manufacturing) developed in Stanford university. In this approach, when a complicated model is to be made, the cutting and deposition are alternatively used after the model is decomposed into the smallest level. Although this approach is not desirable for a large prototype, it can be applied to a geometrical shape with interior cavities or holes where a traditional machining can not be used. Particularly, it has advantages of depositing high quality materials and making a model with embedded components. It can also be used in a rapid tooling which makes prototypes with a high precision. This system has been developed as a hybrid system combining a 5-axis CNC equipment and a powder based RP machine[3, 10].

2 Preliminaries

2.1 3D Layer

The 3D layer is not a common term in the RP industry. However, the concept of the 3D layer is actively adopted in research area. The 3D layer is regarded as a feature in a specific machine whereas a sloped layer in the side cutting system. The machining volume is cutting by the machining tool. We assumed the 3-axis machining. In our work, the 3D layer is defined as the collection of polyhedral surface patches yielding the 3D manufacturable volume.

3 Polyhedron Splitting

3.1 Object Decomposition

In the hybrid RP system, prototypes with different shapes can be made depending on the orientation of decomposition. We use the STL file format, de facto standard format in the RP industry. To create the 3D layer from the STL, we adopt the polyhedron splitting strategy. In Fig. 1, it shows that a valid STL file format is composed of the simple B-rep information which is composed of facets by marking the three vertices and one face normal. Despite of the limitations of the STL file format[1, 7, 8, 9], we use the format because the STL file format is the de facto standard in the RP industry and many input models for RP is converted from RE(Reverse Engineering), CT(Computed Tomography), MRI(Magnetic Resonance Imaging) graphics packages to the STL files. However most hybrid RP systems are adopted by the native solid modeling kernel format[10]. We intend to get the more broader input area rather than the native file format by the using of the STL file format.

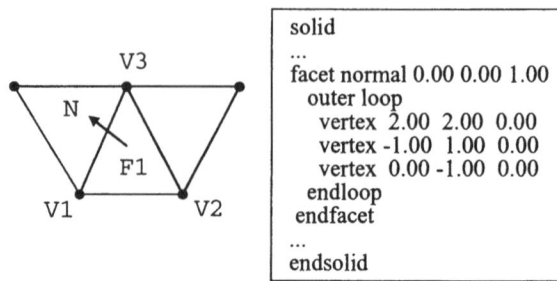

Fig. 1. STL File format(ASCII type)

The polyhedron splitting algorithm is composed of six steps: (1) visibility test of a triangle (2) closed manufacturing volume generation via triangles (3) monotone surface splitting (4) slicing and resulting 3D layer generation (6) support structure generation.

3.2 Triangles Visibility

The manufacturability of 3D layers is verified by testing the visibility for each triangle of 3D layer. Fig 2 shows the ray vector and the MT, NMT, HT, VT of the polyhedron.

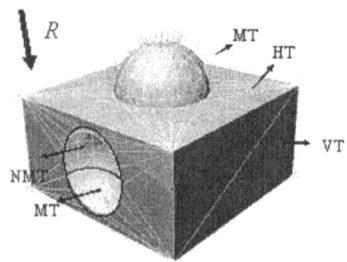

Fig. 2. Polyhedron and its MT, NMT, VT, HT.

The accessibility of a cutting tool is determined by a ray vector. The proposed procedure is as follows.

Procedure 1. Visibility test of triangles
(1) After initializing the ray vector(In this work, we use a fixed value[0,0,1]) and BB(Bounding Box) of object, calculate the inner product between the ray vector and the each triangle T_i with three vertices.

[a] If $0<(N \cdot R)<1$, such triangles are machinable triangles, MT.(N is the normal vector of the triangle)
[b] If $-1<(N \cdot R)<0$, such triangles are non-machinable triangles, NMT.
[c] If $(N \cdot R) = 0$, such triangles are the vertical triangles.
[d] If $(N \cdot R) = 1$, we call that the triangle as the horizontal triangle.
(2) Save the list of 4 categories for triangles(MTlist, NMTlist, HORlist, VERlist).
We classified the triangles into the four types by using the procedure 1.
· MT(Machinable Triangle) : MT means a triangle is machinable by a milling machine.
· VT(Vertical Triangle) : VT means a triangle that composes a side wall(Vertical Wall).
· HT(Horizontal Triangle) : HT means a triangle put on support structure.
· NMT(Non-Machinable Triangle) : NMT means a triangle that needs the creation of support structure(Support structure) as triangles put on curved surface (overhang surface) that need support structure. Triangle classed by NMT is combined by Non-machinable Surface.

Each triangle is classified by the our decomposition module in the initial stage. Fig. 3 shows the each triangle is verified by procedure 1 and saved the list respectively.

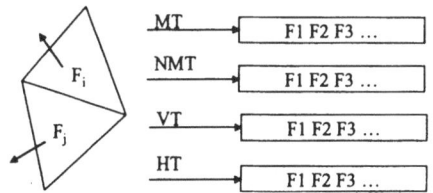

Fig. 3. The classification of MT and NMT triangles

3.3 Closed Manufacturable Volume Generating via Triangles

Most polyhedral model include vertical walls and holes, but information of hole is not stored because of the limitation of STL file format. Thus we adopt the all manufacturable volume by the properties of triangles. For this purpose, three classes are defined. Table 1 shows the three classes. The class one is the Machinable Feature Set of triangles and the class two is the Non-Machinable Feature Set of triangles. Finally, the class three is the collection of the triangles on the self overlapping surface or cavity.

Table 1. Three classes in the combination of four types of triangles and their properties

Class Type	Combination	Properties
Class one	MT-VT-HT	3-axis manufacturing
Class two	NMT-VT-HT	Overhang-surface Need support structure
Class three	NMT-MT-VT-HT	Self-overlapping surface, cavity, etc

568   J. Lee et al.

After the three classes are identified using the procedure 1, class one and class two are saved in the list respectively. However, additive splitting is needed in class three since it contains the cavity or self-overlapping surface. The class one and two, both have the monotonicity in the ray vector.

The monotonic surface is manufactured to use the three axis machining system in the one setup orientation. However, class three is not monotone then is not possible to machining by using the three axis machining. We assume the 3-axis machining system by cutting the 3D layer in our work. This simplifies the formulation of the manufacturability check routine.

3.4 Monotone Surface Splitting

In general, a complex shape cannot be manufactured by one setup orientation. Since the monotonic surface can be machined by one setup orientation, monotonic surface splitting becomes a main operation of dividing a polyhedral object to be machined by a milling operation. Therefore, we should find the monotonic surface set in the polyhedral object. The monotonic surface can be obtained by using the silhouette curve[10]. In fig. 4, a silhouette curve is defined as a curve along with $N \cdot R = 0$, where N is the normal vector at point(u,v) on surface S, and R is the selected ray direction. However, since we use the polyhedron with facets, our silhouette detection procedure is based on triangles.

The procedure of finding the silhouette curve is as follows. First, compute the inner product between the ray vector and the surface normal of the object. If the result is zero, then the curve is generated on the triangle containing the origin of the normal vector. Then the procedure of the monotonic surface splitting is following:

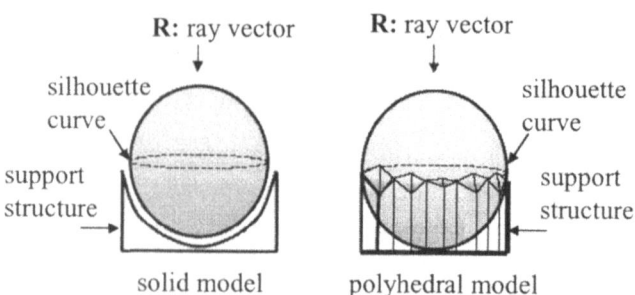

Fig. 4. Silhouette curve with a ray vector in a solid and polyhedral model

Procedure 2. monotonic surface splitting for non-monotonic surface
(1) non-monotonicity check routine.
    [a] scan the triangles in the list defined in procedure 1.
    [b] sort by z coordinate of the centroid for each triangle.
    [c] check weather the sign of the normal vector is changing.
    [d] if it is true, goto (2).
    [e] otherwise, stop.

(2) monotonic surface splitting
  [a] test adjacency among the facets.
  [b] if it is true, then generate a slice plane at that height.

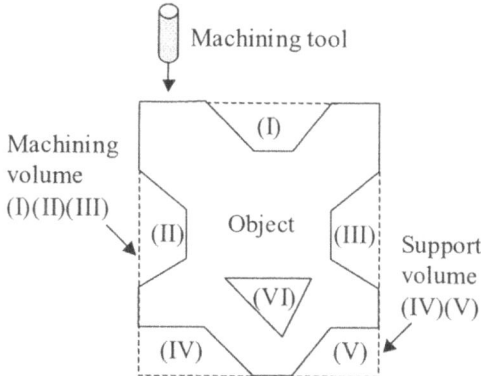

Fig. 5. Machining volume/support volume detection for 3D layer generation

Fig. 5 shows the polyhedron with some machining volumes and support volumes. If the hole or cavity is in the polyhedron, it is necessary to split the object. If the overlapping surface exists, the object also has to be splitted. For this purpose, as shown in Fig. 6, the slicing planes should be generated at the height by checking the adjacency between MT and NMT.

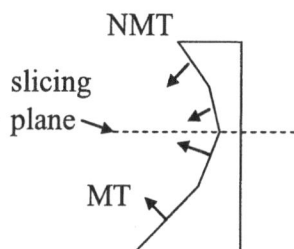

Fig. 6. Slicing plane between MT and NMT

3.5 Slicing and 3D Layer Generation

Manufacturability of 3D layers is verified by testing the monotonicity of splitted surface. Since most manufacturable set of triangles is open rather than a closed volume, it is difficult to calculate deposition paths. Thus we make the closed volume by using the MT-VT-HT-NMT and the slicing plane. This needs a slicing operation between the polyhedron and planar plane with $z = h$. Slicing planes make the triangle as a trapezoid or a smaller triangle. Retrianglulation is our choice to handle that case. Fig. 7 shows the slicing plane entering the triangle, then IP_{in}(entering intersection point) and IP_{out}(go-out intersection point) are generated.

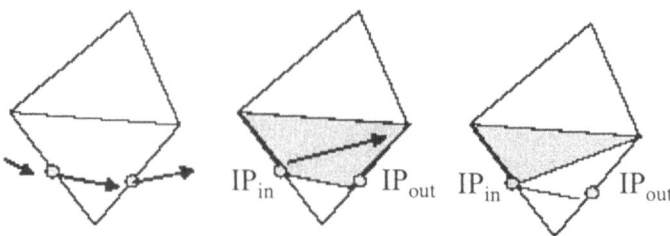

Fig. 7. Slicing and retriangulation with IP_{in} and IP_{out}

3.6 Support Structure Generation

Support structure is needed to support the overhanging surface. The cutting and binding approach do not need the support structure generation. However, the area where the support structure is needed has to be calculate. The area is used to calculate the bonding force and the weight of the layers above it. To generate the support structure, we project NMT to the slice plane. We call the projected triangle as a shadow triangle. If a support structure is needed by the user, a triangular prism is generated by connecting the vertices of the shadow triangle and NMT. The triangular prism can be modified if necessary.

4 Patterns of Polyhedron Splitting

We observed that polyhedron splitting patterns are related to MT and NMT. To demonstrate the effectiveness of polyhedron splitting algorithm, several experiments have been conducted. A cylinder model with three holes is in Fig. 8 (a). First, we make the bounding box and then we make the slicing plane for 3D layer generation.

Based on the classification of three classes of 3D layer, the polyhedron splitting program is generated automatically. Fig. 8 (b) shows the 2D layer using constant layer thickness. Fig. 8 (c) shows MT set and NMT set by using our algorithm. Fig. 8 (d) shows the slicing planes and the resulting 3D layers.

5 Conclusion and Future Works

In this research, 3D layer generation was done based on a polyhedron model. Geometric properties of the individual triangle were used to the generation. We developed a polyhedron splitting algorithm for the 3D layer generation. Although some limitations in the STL file format exit, we effectively extracted the 3D layer from the STL. In the future, we need to study the optimal orientation for the 3D layer generation and support structure.

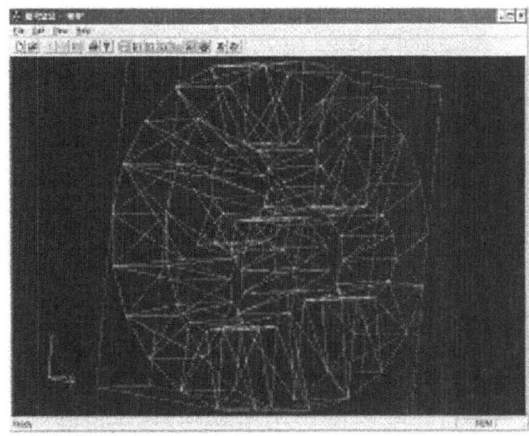

(a)

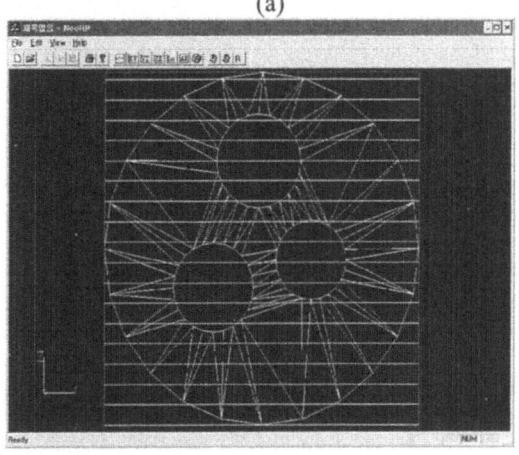

(b)

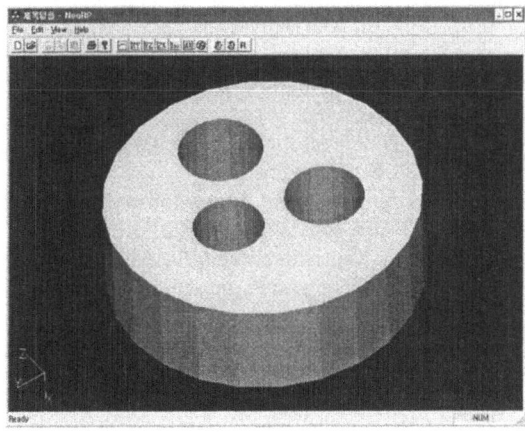

(c)

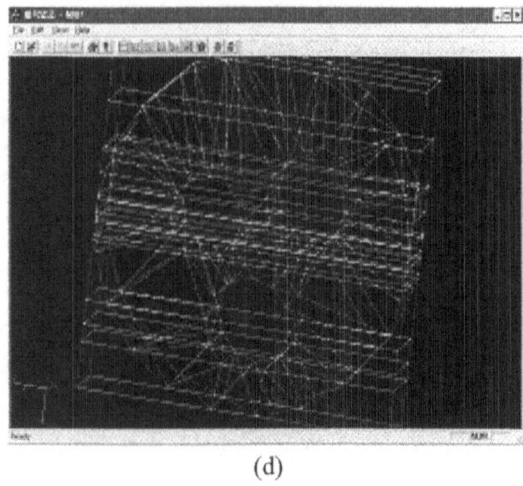
(d)

Fig. 8. Implementation result. (a) Input model (b) 2D layer (c) Shading model in MT and NMT set (d) Slice planes and the 3D layers

References

1. Bechet, E. J., Cuilliereb, -C. and Trochua, F.: Generation of a finite element MESH from stereolithography (STL) files. Computer-Aided Design, Vol. 34, Issue 1, January (2002) 1-17.
2. Broek, J.J., Horvath, I., Smit, B. de, Lennings, A.F., Rusak, Z. and Vergeest, J.S.M.: Freeform thick layer object manufacturing technology for large-sized physical models, Automation in Construction, Vol. 11 No. 3, (2002) 335-347
3. Chang, Y.C., Pinilla, J.M., Kao, J.H., Dong, J., Rawaswami, K. and Prinz, F.B.: Automated Layer Decomposition for Additive/Subreactive Solid Freeform Fabrication, proc. Solid Freeform Fabrication Symposium, (1999) 111-120
4. Hope, R.L., Jacobs, P. A. and Roth, R, N.: Rapid Prototyping with Sloping Surfaces, Rapid Prototyping Journal, Vol. 3, No. 1, (1997) 12-19
5. Horvath, I., Vergeest J.S,M., Broek, J.J., Rusak, Z. and Smit, B. de: Tool profile and tool path calculation for free-form thick-layered fabrication, Computer-Aided Design, Vol. 30, No. 14, (1998) 1097-1110
6. Jacobs, P. F.: Stereolithography and other RP&M Technologies form Rapid Prototyping to Rapid Tooling, ASME Press, 1996
7. Jamieson, R., and Hacker, H.: Direct slicing of CAD models for rapid prototyping, Rapid Prototyping Journal, Vol. 1, No. 2, (1995)
8. Kumar, V. and Dutta, D.: An assessment of data formats for layered manufacturing, Advances in Engineering Software, Vol. 28, No. 3, (1997) 151-164
9. McMains S. and Sequin, A.: Coherent Sweep Plane Slicer for Layered Manufacturing, Proceedings of the Fifth ACM Symposium on Solid Modeling and Applications, (1999) 285-295
10. Ramaswami, K. Yamaguchi, Y. and Prinz, F.P.: Spatial Partitioning of solids for Solid Freeform Fabrication, Proceedings of the Fourth ACM/SIGGRAPH Symposium on Solid Modeling and Applications, (1997)
11. Rock, S. J. and Wozny, M. J.: A Flexible File Format for Solid Freeform Fabrication, Solid Freeform Fabrication Symposium Proceedings, (1991) 1-12

Synthesis of Mechanical Structures Using a Genetic Algorithm

In-Ho Lee[1], Joo-Heon Cha[2], Jay-Jung Kim[1], and M.-W. Park[3]

[1] School of Mechanical Engineering at Hanyang Univ., 17, Haengdang-Dong, Sungdong-Ku, Seoul 133-791, Korea.
[2] School of Mechanical and Automotive Engineering at Kookmin Univ., 861-1, Jungrung-Dong, Sungbuk-Ku, Seoul 136-702, Korea.
cha@kookmin.ac.kr, TEL: +82-2-910-4816, FAX: +82-2-910-4839
[3] CAD/CAM Research Center at Korea Institute of Science and Technology, P.O.Box 131, Cheongryang, Seoul 130-650, Korea.

Abstract. This paper proposes a representation for the embodiment design and a genetic algorithm suited for the representation. Since embodiment design consists of highly interrelated design processes, how to formalize and integrate the processes has become an important issue. In this paper, embodiment design is modeled as simultaneous multi-objective optimization of parametric designs for parts and of layout generation for structures. The study, thus, involves genotypes that are adequate to represent phenotypes of the models for the genetic algorithm to solve the given problems. We demonstrate the implementation of the genetic algorithm with the result applied to the gear equipment design.

1 Introduction

The use of CAD systems for automation of the earlier stages of the embodiment designs that have, to some extent, properties of the conceptual design stages, is currently less common, while CAD systems for highly formalized design phases are much prevalent today [1]. Although the fundamentally important nature of the early stages in design tasks has prompted much work into various design approaches, there has not been much progress on them. Complications may arise when CAD systems for embodiment design problems that involve early stages, which partially are conceptual designs, and latter stages, which partially are detail designs, are to be implemented; large design spaces and high efficiency are equally required [2].

This paper is an attempt to propose a genetic algorithm that evolves designs that are represented with parametric variables and building blocks; it is applied to development of a CAD system for the entire embodiment design phases. The basic concept of the study is that for a practical embodiment design of mechanical structures, the two categories of aspects should be taken into consideration simultaneously, since a design is not accomplished by single processes of modification of parameterized parts or modification of structures of existing parts alone. This is a way to an effective search method and to combine merits of each design aspect.

574   I.-H. Lee et al.

The embodiment design problems modeled after the basic concept involve simultaneous optimization of two different, even competing, sorts of objectives. For the genetic algorithm to effectively solve such problems, they should be regarded as multi-objective optimization problems [3]. In the genetic algorithm, problem-specific sophisticated building blocks with parametric variables are involved as genotypes for the problems and a series of problem-specific evolution processes are involved according to the genotypes too. Additionally, the approach is illustrated with case studies of a gear system design.

2 Modeling Embodiment Design Problems

Designing is a group of modification processes of several properties in existing designs according to given design tasks and conditions. When properties of designs are represented in terms of several variable parameters, they are called parametric designs [4-5]. They are a group of the most prevalent, effective methods especially fit for highly formalized design cases. When properties of designs are represented in terms of several elementary components, they are called building blocks approaches [4-5]. They are another group of effective representation methods especially suited for comparatively less formalized design cases.

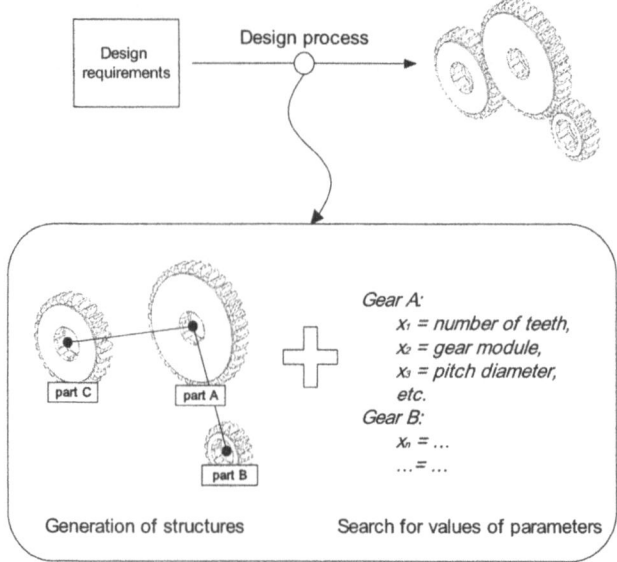

Fig. 1. The two aspects of design processes

Each approach has its own relevant aspect of design, which is only a part of the design processes. The approaches concentrate on representing parameters or organization of parts, but they are not independent aspects; modification on structures in a design process affects their parameterized parts and vice versa. Taking this into consideration, we can suppose that an approach that combines the two aspects of design in a simultaneous manner is a way to improve computer-aided embodiment designs. Now, embodiment design problems, e.g. gear system design, to be solved in this paper can be illustrated as shown in Fig. 1.

Performance of designs depends greatly on how the design problems given are modeled. As discussed previously, the problems that are identified and to be solved are embodiment design problems that have two different types of design aspects. They are inseparable in the course of design processes and they should be dealt with in a simultaneous manner. The design problems are modeled after this concept. The generalized model of embodiment design problems for this study can be described as depicted in Fig. 2. The model is a compound problem of generating a structure with n elements, which are instances of several pre-defined element types, and finding values of m variables involved in the elements.

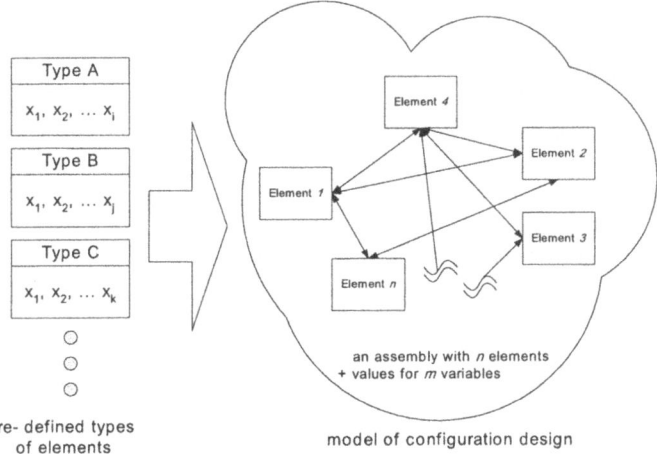

Fig. 2. Generalized embodiment design problem

There can be several advantages of using the model. First, since it is a building blocks approach, it defines large design spaces [4]. Based on this, even when design conditions are changed to some extent, the model can cover them with existing components. Second, since it is also a parametric design, it defines small design spaces at high efficiency [4]. Based on this, designs can involve several important properties such as geometric features or physical rules, which should be determined with accuracy and speed.

Solutions to the problem represented with this model can be achieved with a series of simultaneous computation processes. Since the problem has two independent objectives, it can be dealt with as a multi-objective optimization problem, for which evolutionary computation approaches, genetic algorithms more specifically, are particularly suited [3]. In many cases of multi-objective optimization problems, goals conflict with no preference information available e.g. a ranking of the objectives. Thus, a set of alternative solutions, called Pareto-optimal solutions, is used in exploring the design space [3,7].

3 A Genetic Algorithm for Embodiment Design Problems

A GA (genetic algorithm) is proposed and applied to the problems. Phenotypes of the problem require problem specific genotypes and evolution processes. The GA we propose here should involve several properties different from general GA's as the following:
- Genotypes are designed to support layout generation of structures.
- Genotypes are designed to support optimization of parameterized parts.
- Evolution processes are designed to search the two aspects of space simultaneously.

3.1 Phenotypes and Genotypes

Phenotypes represent mechanical structures that are composed of several parts and parameterized variables. Genotypes for the phenotypes are to combine parametric designs and building block approaches in a genetic algorithm.

Genetic algorithms evolve genotypes that are coded phenotypes. Generalized forms of a phenotype and a genotype are illustrated in Fig. 3. In order for the genetic algorithm to search effectively, well-formed genotypes can be one of the most important requirements. Every genotype proposed in this paper is composed of several pre-defined primitive elements that permit synthetic evolution. Every element is again composed of a set of variables that represent parametric variables and a set of relationships that define feasible assemblies between elements.

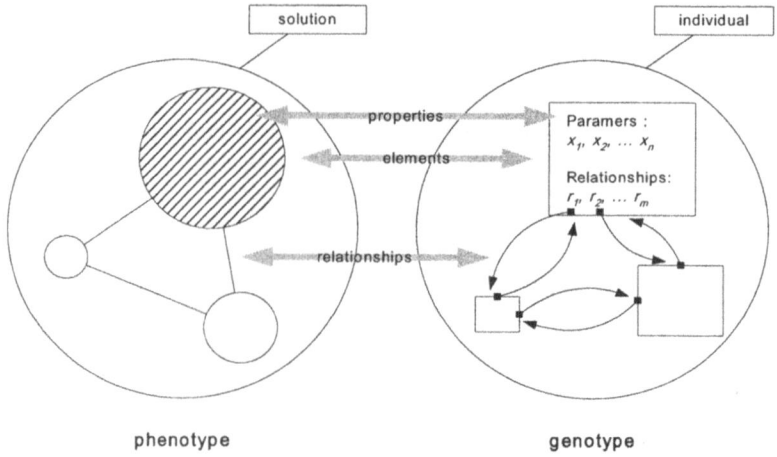

Fig. 3. Phenotypes and genotypes

3.2 Mutation Operators

A series of new mutation operators are required for specified modification of the genotypes because of several differences in genotypes between regular GA's and the GA proposed in this study. There are three mutation operators such as 1) modification on values of parametric variables 2) modification on relationships between given elements and 3) addition or deletion of elements, as shown in Fig. 4.

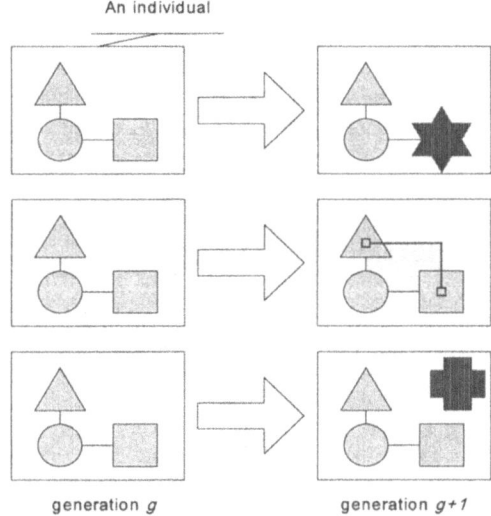

Fig. 4. Mutation operators and modified genotypes

3.3 Evaluation and Selection

A multi-objective optimization problem has a set of fitness functions i.e. a vector function. The fitness functions are categorized into two groups, since evaluation terms are concerned with generated structures or parametric values in genotypes. To evaluate properties of an individual, the two separate groups of fitness are used. Each group of fitness is represented as a summation of ranks for each evaluation term. The summation of ranks S about a specific individual c_i can be calculated with the equation (1).

$$S(c_i) = \sum_{j=1}^{k} R(f(c_i^j)) \cdot \qquad (1)$$

where k number of evaluation terms
         $f(c_i)$ fitness function for each evaluation term
         $R(c_i)$ rank of an individual for the evaluation term

A set of ranks *(Rank₁, Rank₂)* can be achieved with the two summations, S_1 for generated structures and S_2 for parametric values. The two ranks are used in different ways to select fitter individuals: *Rank₁* defines whether the current individual just will reproduce or not; and *Rank2* defines parents' probabilities of reproduction. In this paper, a roulette wheel selection method is used to set the probability of reproduction. The evaluation processes can be regarded as a Pareto-ranking method.

3.4 Test Example

In this chapter, we have taken a simple test example to show how the GA works. There are three types of elements, *A, B* and *C,* in the example. Each element is set be combined with other elements so that it can be a component of a structure. Only several types of assemblies between two elements are allowed as illustrated in Fig. 5; the element *A*, for example, can be combined with an element *A*, an element *B* and an element *C*. Each element has also two parametric variables, x_1 and x_2.

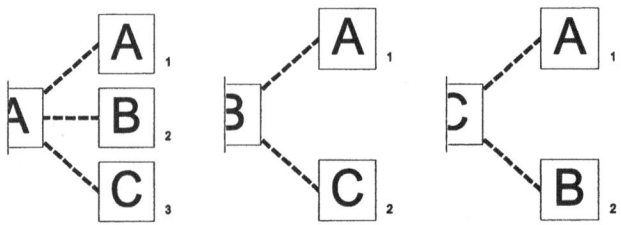

Fig. 5. Allowed assembly types between two elements

Objectives of the example are 1) to generate layouts involving maximum 10 linkages between elements from single-element genotypes and 2) to optimize parametric variables involved in the *n* elements comprising the generated layouts with a simple objective function, which is commonly used to evaluate the effectiveness of different search algorithms [8]. Formally:

$$\text{minimize: } f(\mathbf{x}) = \sum_{i=1}^{n} H(x_{1,i}, x_{2,i}) = \sum_{i=1}^{n} \{(x_{1,i}^2 + x_{2,i} - 11)^2 + (x_{1,i} + x_{2,i}^2 - 7)^2\}$$

subject to: $-5 \leq x_1 \leq 5$, $-5 \leq x_2 \leq 5$. (2)

The fitness value for generated structures is calculated from percentile errors between the number of current linkages and the number of objective linkages. The fitness value for parametric variables is calculated from the given function. In this example, the size of the population is set to 100 and only fitter 80% of them are to reproduce in every generation.

The actual evolution behaviors of the individuals can be shown with several transition graphs. Transition of evolution in structures of genotypes is illustrated as shown in Fig. 6; individuals with simple forms in earlier state of the evolution processes evolve into more complex ones through generations. Fitness values for parameterized parts decrease through generations till optimal values are found.

One of designs produced are shown in Fig. 7. It is a potential solution to the multi-objective problem. They have feasible structures that suit the requirements and optimal parametric values, $x = (3.0, 2.0)$, as shown in the figure. The designs have different topologies and different numbers of elements, but they satisfy the design requirements.

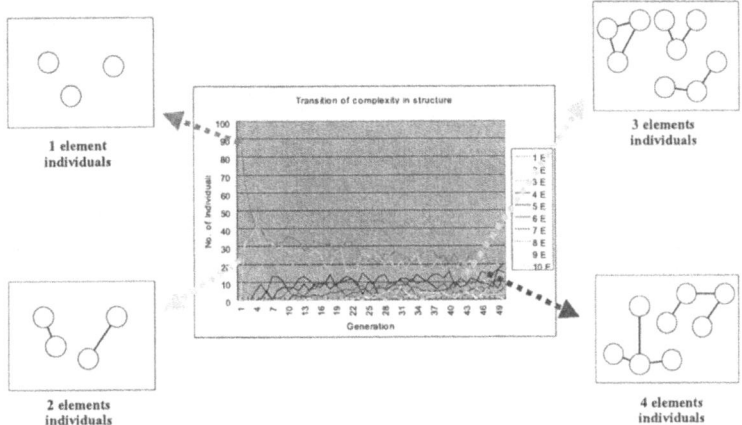

Fig. 6. Transition of evolution: complexity of structures

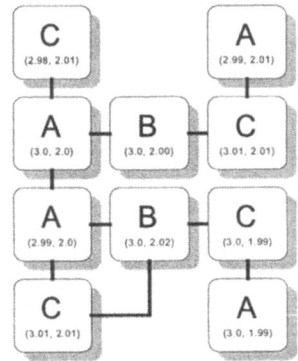

Fig. 7. One of the design results

4 Case Study

The genetic algorithm can also be illustrated with a case study. In this case study, the genetic algorithm is applied to the development of a CAD system for designs of gear

drives that reduce the input rotational speed by a specific reduction ratio. Besides the reduction of speed, the gear drives should satisfy the following requirements:
− The gear equipment is composed of shafts and spur gears.
− Every part of the gear equipment has enough strength to transmit specific power.
− The gear equipment has minimum size.

The genotypes for the case study consist of two types of elements such as gear elements and shaft elements as shown in Table 1. Relationships and parametric variables for the elements are modeled from the physical properties of the parts. The gear element, for example, involves the gear module, the pitch diameter, the number of teeth *etc.* as parametric variables. Possible types of assemblies between elements are defined according to the physical relations between parts. Assemblies can only be made between two gear elements and between a gear element and a shaft element.

Table 1. Elements for genotypes

Name	Gear element	Shaft element
Relationship set	Gear-to-gear	Shaft-to-gear
	Gear-to-shaft	
Parameter set	Gear module	Radius
	Pitch radius	Rotational speed
	Number of teeth	
	Rotational speed	

In the modification processes, constraints also can be involved according to the properties of the parametric variables of the elements. The number of teeth for a gear is integral and it changes from 12 to 150 with consideration of undercuts and practical size [9]. The gear module changes according to the standard gear module data. Pitch diameters of gears and shafts are positive real numbers.

In order to evaluate individuals, two groups of fitness values are used as shown in the test example of the previous chapter. The fitness of generated structures can be measured by comparing output rotational speed of the individuals and the desired one; the percentile error of the difference represents the fitness value.

The fitness of parameterized parts in this case study can be calculated by measuring strength and size of an individual. Strength of a shaft element can be represented with torsional stresses on sections and strength of a gear element can be represented with bending stresses on teeth. The fitness of an individual can be measure by a mean-error of strength of the elements comprising an individual. The following formulas are useful for estimating the errors: the error of strength for shaft parts is defined as the function (3) and the error of strength for gear parts is defined as the function (4).

$$E_{shaft} = abs\left(\frac{d_a - d_c}{d_a}\right). \tag{3}$$

where $d_a = 171\sqrt[3]{\frac{H}{\tau \times n}}$ = allowable diameter of shafts, mm

d_c = current diameter of shaft, mm

H = power, kW
τ = allowable torsional stress, kg/mm
n = rotational speed, rpm

$$E_{gear} = abs\left(\frac{H_a - H_c}{H_a}\right). \qquad (4)$$

where $H_a = f_v \sigma b m y$ = allowable power, kW (from the Lewis formula)
f_v = velocity factor by G. Barth
σ = allowable bending stress, kg/mm
b = width of gear tooth, mm
m = gear module
y = Lewis form factor

To measure fitness of size, the second evaluation term, the mass moments of inertia of gears and shafts are considered. Rank of an individual can be achieved by comparing values of the mass moments of inertia. The mass moments of inertia of gears and shafts are calculated by the general formula for cylinders since sections of gears and shafts are regarded as circles.

Two gear drives are designed by the CAD system implemented in this case study. Each gear drive is designed according to each specification by a single CAD system without any modifications in the codes of the system. One drive is to reduce the input rotation of 3500rpm into the output rotation of 240rpm and the other is to reduce 35000rpm into 240rpm. After successive iterations of evolution, designs were achieved. The designs proposed by the GA are a group of feasible solutions to the given problem. One of the gear drives designed based on the first design specification are graphically represented in Fig. 8 (a) and one of the gear drives designed based on the second design specification in Fig. 8 (b). Parameters of the design for the first case is provided in Table 2, where R is radius of a gear or a shaft, W is width of a gear, m is gear module and z is the number of teeth.

Table 2. Design result: the first case

No.	Element	Rotational speed	Geometry information			
1	Shaft	3500.000	R: 2.5			
2	Gear	3500.000	R: 4.2	W: 4.8	m: 0.6	z: 14
3	Gear	960.784	R: 15.3	W: 4.8	m: 0.6	z: 51
4	Shaft	960.784	R: 3.6			
5	Gear	960.784	R: 5.6	W: 6.4	m: 0.8	z: 14
6	Gear	240.196	R: 22.4	W: 6.4	m: 0.8	z: 56
7	Shaft	240.196	R: 6. 5			

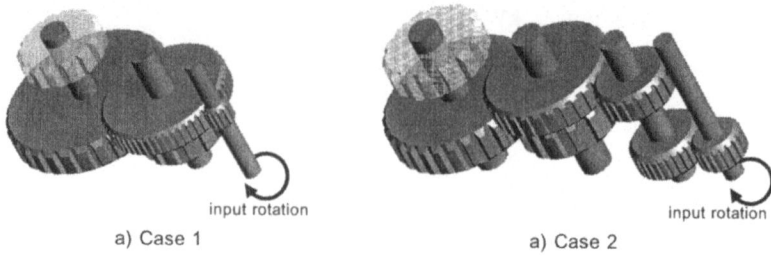

a) Case 1 a) Case 2

Fig. 8. Designed gear drive: the second case

5 Conclusions

With the aim of making a CAD system capable of solving embodiment design problems of mechanical structures, we have described a design approach involving a synthetic representation and a GA that is suited for the representation. In the approach, search for solutions to the given design problems are regarded as a group of simultaneous processes to get a structure and to get a set of parametric values. Genotypes and evolution processes of the GA have also been proposed taken the representation into consideration. The approach has been tested on implementation of a CAD system for gear equipment designs. The purpose of this case study was not to show a more or less complete implementation of the automatic design; rather, we intended to show a general-purpose prototype on which we are going to expand actual design subjects based on the new approach proposed.

References

1. Yokoyama, M., Endo, T. and Cha, J.H., Knowledge-based CAD (1997), Corona publishing co.
2. Pahl, G.and Beitz, W., Engineering Design (1984), The Design Council.
3. Obayashi, S., Sasaki, D., Takeguchi, Y. and Hirose, N., Multiobjective evolutionary computation for supersonic wing-shape optimization, IEEE Trans. on evolutionary computation, Vol.4, No.2 (2000), p.182-187.
4. Gero, J.S. and Kazakov, V.A., Evolving building blocks for design using genetic algorithm, Advances in Formal Design method for CAD - Proceedings of IFIP95 (1995), p.31-50.
5. Coyne, R.D., Rosenman, M.A., Radford, A.D., Balachandran, M. and Gero, J.S., Knowledge based design systems (1990), Addison-Wesley publishing co.
6. Cha, J.H., Lee, I.H. and Kim, J.J., Computer-aided innovative mechanical design framework, Proceedings of FAN Symposium '00 in Tokyo (2000), p.405-410.
7. Bentley, P.J., Genetic evolutionary design of solid objects using a genetic algorithm, Ph.D. thesis, the university of Huddersfield (1996).
8. Whitley, D., Rana, S., Dzubera, J. and Mathias, K.E., Evaluating evolutionary algorithms, Artificial intelligence, Vol.85 (1996), p.245-276.
9. Dudley, D.W., Handbook of Practical Gear Design, McGraw-Hill Book Company (1984).

Optimal Direction for Monotone Chain Decomposition

Hayong Shin[1] and Deok-Soo Kim[2]

[1] KAIST, Dept. of Industrial Engineering, S.Korea, hyshin@kaist.ac.kr
[2] Hanyang Univ., Dept. of Industrial Engineering, S.Korea, dskim@hanyang.ac.kr

Abstract. Monotone chain is an important concept in computational geometry. A general polygon or polygonal chain can be decomposed into monotone chains. Described in this paper are two algorithms to find an optimal direction with respect to which a polygonal chain can be split into the minimal number of monotone chains. The first naive algorithm has $O(n^2)$ time complexity, while the improved algorithm runs in $O(n \log n)$ time, where n is the number of vertices of input chain. The optimal direction can improve the performance of the subsequent geometric processing.

Keywords: monotone chain, monotone decomposition

1 Introduction

A *polygonal chain*, or *chain* for short, is a sequence of connected line segments. A *polygon* is a closed chain. The following three definitions are borrowed and slightly modified from [3].

Definition 1 A polygonal chain P is *strictly monotone* w.r.t. a line L if any line L° orthogonal to L meets P in at most one point, i.e. L° ∩ P is either empty or a single point. L is called the *line of monotony* and the angle from x-axis to L is called the *direction of monotony*.

Definition 2 A polygonal chain P is *monotone* if L° ∩ P has at most one connected component: L° ∩ P is either empty, a single point, a line segment, or connected line segments.

Definition 3 A polygon is *(strictly) monotone* w.r.t. L if it can be split into two chains which are (strictly) monotone w.r.t. L.

The chain P_a in Fig.1 (a) is strictly monotone w.r.t. L_1, and monotone w.r.t. L_2, but not monotone w.r.t. L_3. The polygon P_b in Fig.1 (b) is monotone but not strictly monotone w.r.t. L. The concept of monotone chain plays an important role in computational geometry. As depicted in [3], many problems which are difficult for a general polygon can be efficiently done for monotone polygon, primarily because the vertices of a monotone chain are already sorted w.r.t. the direction of monotony. Triangulation of a polygon is an example.

With a given line L, a general polygonal chain can be split into a number of sub-chains which are monotone w.r.t. L. This process is called *monotone decomposition* of a chain. The number of resulting monotone chains depends on the direction of L.

For example, the polygon P in Fig.2 (a) is decomposed into 4 monotone chains w.r.t. the vertical line L_b as shown in Fig.2 (b), and 2 monotone chains w.r.t. L_c in Fig.2 (c). Many important problems in computational geometry can be solved efficiently by using the monotone decomposition. For example, multiple point in polygon problem (also known as point inclusion problem), which is to classify a set of point against a polygon, can be efficiently done by decomposing the polygon into monotone chains and classifying points against each monotone chain [5]. The performance of those algorithms depends on the number of monotone chains.

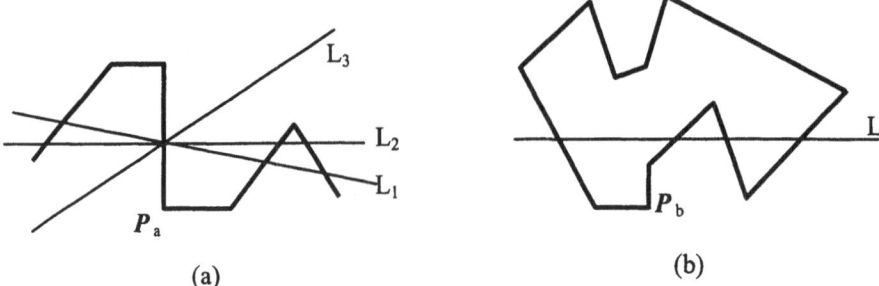

Fig. 1. Monotone chain or polygon

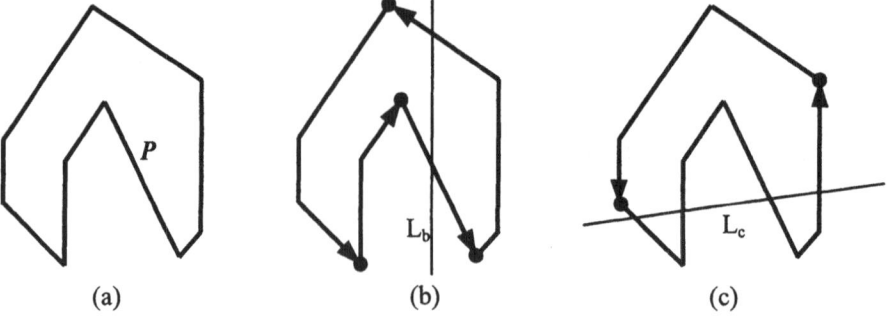

Fig. 2. Monotone decomposition of a polygon

Presented in this paper are algorithms to find an *optimal direction* w.r.t. which a polygonal chain can be split into the minimal number of monotone chains. Surprisingly little works regarding this problem are done. Chandru et al. tackled similar problem in [1]. They presented an algorithm to decompose a polygonal chain into minimal number of monotone chains. In their work, however, the resulting monotone chains have different directions of monotony. A linear time algorithm to test if a simple polygon is monotone w.r.t. some direction can be found in [6]. Recently Park and Shin [4] reported monotone chain decomposition algorithm briefly in the context of finding intersections between polygonal chains.

This paper elaborates the work reported in [4]. Section 2 defines the *extreme region*, which is the key concept of the algorithms presented in this paper. A simple algorithm to find the optimal direction is described in Section 3, and Section 4 contains a faster algorithm. Addressed in Section 5 are issues to be considered for the proposed algorithm, followed by concluding remarks.

2 Extreme Region of a Vertex

Let a polygonal chain P be a sequence of edges, denoted by $P = \{ E_1, \ldots, E_n \}$. P is not necessarily a simple polygon, meaning that it may have self-intersections and it may be an open chain. An edge E_i is the directed line segment joining vertices V_i and V_{i+1}, denoted by $E_i = V_i V_{i+1}$. If P is closed, V_{n+1} is identical to V_1. For an open chain, V_1 and V_{n+1} are called *start vertex* and *end vertex*, respectively. The others are *internal vertices*. All vertices of a closed chain (polygon) are internal. An internal vertex is further classified. An internal vertex V_i is called *right (left) convex vertex* if it is on the right (left) of a directed line passing through V_{i-1} and V_{i+1}. If it is on the line, it is a *collinear vertex*. To denote the *type* of vertex, *SV*, *EV*, *LCV*, *RCV*, and *CV* will be used for a *start vertex*, an *end vertex*, a *left convex vertex*, a *right convex vertex*, and a *collinear vertex*, respectively, as shown in Fig.3. For the brief of explanation, the input polygonal chain is assumed to have no collinear vertex. This assumption can be achieved by merging the consecutive collinear edges into a single edge in preprocessing step.

Now consider the region around an internal vertex V_i as shown in Fig.4. An internal vertex V_i is called an *extreme vertex* w.r.t. L if two incident edges E_{i-1} and E_i are on the same side of L^o which is orthogonal to L and passes through V_i, as shown in Fig.4 (a). The cases where E_{i-1} or E_i is perpendicular to L are delayed until Section 5. A monotone decomposition of a chain can be obtained by splitting it at each extreme vertex.

Let L_i be the undirected infinite line perpendicular to E_i, and A_i be the angle from x-axis to L_i ($0 \leq A_i < \pi$). Here and henceforth an angle is measured in CCW direction. And, let $L(\theta)$ be the line having an angle θ with x-axis. V_i is extreme w.r.t. $L(\theta)$ if $L(\theta)$ is inside of the shaded region, which does not contain L_{i-1} and L_i. The shaded region is called the *extreme region* of V_i, and is denoted by R_i. It is easy to note that a chain $\{ E_{i-1}, E_i \}$ is monotone w.r.t. $L(\theta)$ if $L(\theta)$ is not in R_i. Furthermore, $\{ E_{i-1}, E_i \}$ is monotone, but not strictly monotone, w.r.t. L_{i-1} or L_i. The extreme region R_i can be expressed with A_{i-1} and A_i. Since A_i is a value between 0 and π, extra attention should be paid when extreme region contains x-axis as shown in Fig.5. Let (a, b) denote the interval $\{ x \mid a < x < b \}$, and $[a, b)$ denote the half-closed interval $\{ x \mid a \leq x < b \}$. For a right convex vertex V_i, if $A_{i-1} < A_i$ as in Fig.5 (a), $R_i = (A_{i-1}, A_i)$. If $A_{i-1} > A_i$ as in Fig.5 (b) or (c), $R_i = (A_{i-1}, \pi) \cup [0, A_i)$. Similarly, for a left convex vertex V_i, if $A_i < A_{i-1}$, $R_i = (A_i, A_{i-1})$, else $R_i = (A_i, \pi) \cup [0, A_{i-1})$. For a collinear vertex V_i, $A_i = A_{i-1}$. If E_{i-1} and E_i have the same direction, $R_i = \emptyset$. If they have opposite directions, $R_i = [0, \pi) - \{A_i\}$.

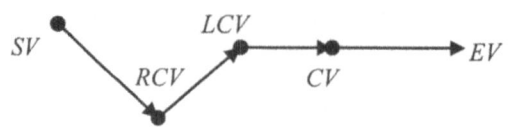

Fig. 3. Types of vertices

SV : Start vertex
EV : End vertex
LCV : Left-concave vertex
RCV : Right-concave vertex
CV : Collinear vertex

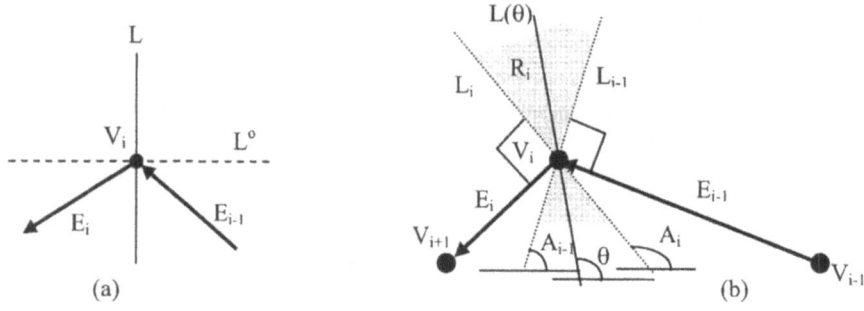

Fig. 4. Extreme vertex and extreme region (a) V_i is extreme w.r.t. L, (b) extreme region of V_i

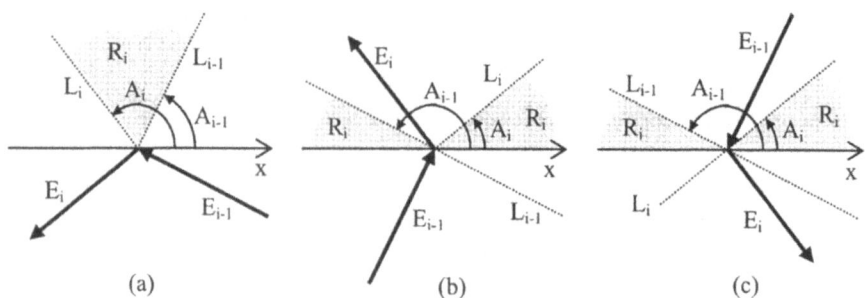

Fig. 5. Cases for the extreme region of an RCV

3 Basic Observation

The entire interval $[0, \pi)$ can be partitioned into a number of cells (sub-intervals) by $\{ A_1, .., A_n \}$. Let $\mathbf{C} = \{ C_0, .., C_m \}$ be the ordered set obtained by sorting $\{ A_1, .., A_n \} \cup \{ 0, \pi \}$. (Note that $\{ A_1, .., A_n \}$ may contain duplicated elements, while $\mathbf{C}$ can not.) For each cell (C_k, C_{k+1}), let N_k be the number of extreme vertices of the input chain w.r.t. $L(\theta)$ when $C_k < \theta < C_{k+1}$. See Fig.6. A brute force approach to optimal direction of monotony is to compute N_k by counting the number of extreme regions containing the cell (C_k, C_{k+1}). This can be done with $O(n^2)$ worst case time complexity, because each R_i may contain $O(n)$ cells. For example, in Fig.6 (a), the cell $(\pi/4, \pi/2)$ is included only in the extreme regions of V_3 and V_6, meaning that V_3 and V_6 are extreme vertices w.r.t. $L(\theta)$ for $\pi/4 < \theta < \pi/2$. Hence N_k for the cell is 2.

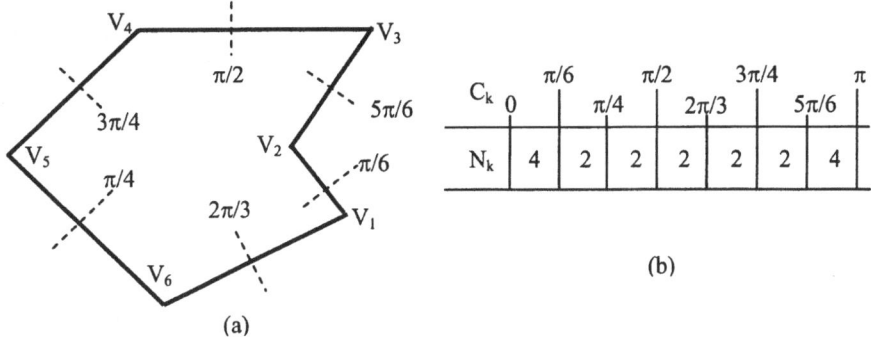

Fig. 6. The number of extreme vertices (a) a polygon (b) table of N_i

By observing the changing behavior of N_k, this can be much improved. In Fig. 6, note the change of extreme vertices w.r.t. $L(\theta)$ as θ increases from the cell $(0, \pi/6)$ to the cell $(\pi/6, \pi/4)$. Let ε be a sufficiently small positive number. When $\theta = \pi/6 - \varepsilon$, V_1, V_2, V_3, and V_5 are extreme vertices w.r.t. $L(\theta)$. As θ increases to $\theta = \pi/6 + \varepsilon$, V_1 and V_2 become non-extreme vertices, and hence N_k decreases by 2. Similarly when θ increases across $5\pi/6$, N_k increases by 2. This observation can be generalized. An edge can be classified by the type of its vertices as shown in Fig.7. Let $E_i=V_iV_{i+1}$ denote the edge joining V_i and V_{i+1}. The type of edge E_i in Fig.7(a) is called by *LL_edge* as it joins two vertices of *LCV* type. Similarly, *LR_edge*, *RR_edge*, *RL_edge*, *SL_edge*, *SR_edge*, *RE_edge*, *LE_edge*, and *SE_edge* denote the types of E_i's for the rest of Fig.7. The nine edge types listed in Fig.7 covers every possible combination of V_i and V_{i+1}.

In Fig.7, let $N(L)$ be the number of extreme vertices w.r.t. L among $\{V_i, V_{i+1}\}$. Note that a vertex of type *SV* or *EV* cannot be an extreme vertex, since they are not internal. When L is perpendicular to $E_i=V_iV_{i+1}$, only one of V_i and V_{i+1} is considered to be an extreme vertex for Fig.7 (a) and (c), and none of V_i and V_{i+1} are extreme for the rest of Fig.7. This means that $N(L_i) = 1$ for Fig.7 (a),(c) and $N(L_i) = 0$ for the rest of Fig.7, where L_i is the line perpendicular to E_i.

Let L^- and L^+ denote the lines having angles $A_i - \varepsilon$ and $A_i + \varepsilon$ with x-axis, where ε is a very small positive value. Now observe the change of $N(L)$ when L rotates from L^- to L^+ through L_i. For Fig.7 (b), as an example, V_i and V_{i+1} are not extreme w.r.t. L^- nor L_i. But they are extreme w.r.t. L^+. Similar analysis can be done for the other cases, and the results are listed in Table 1. The first three columns of Table 1 show the number of extreme vertices w.r.t. $N(L^-)$, $N(L_i)$, and $N(L^+)$, respectively. The last two columns show the change of $N(L)$ as L rotates from L^- to L^+ through L_i. This observation enables us to process the edge locally in computing N_i's. And it also tells us that *RR_edge*, *LL_edge*, and *SE_edge* has no contribution in computing N_i's.

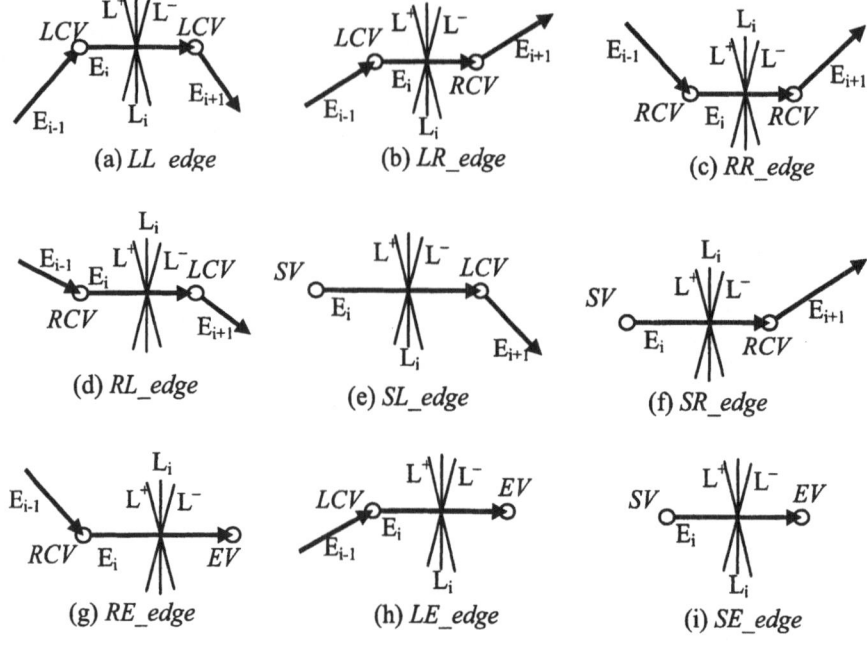

Fig. 7. Types of edges

Table 1. The number of extreme points

Type of E_i	$N(L^-)$	$N(L_i)$	$N(L^+)$	$L^- \to L_I$	$L_i \to L^+$
LL_edge	1	1	1	0	0
LR_edge	0	0	2	0	+2
RR_edge	1	1	1	0	0
RL_edge	2	0	0	-2	0
SL_edge	1	0	0	-1	0
SR_edge	0	0	1	0	+1
RE_edge	1	0	0	-1	0
LE_edge	0	0	1	0	+1
SE_edge	0	0	0	0	0

4 Algorithm to Find Optimal Direction of Monotony

In the following **algorithm-ODM**, $C = \{C_0, .., C_m\}$ is an ordered set of distinct angle values from A_i's, representing the partitioning of the entire range $[0, \pi)$ of direction. C_i and C_{i+1} are the lower and the upper boundaries of the ith cell. A structure e_node is introduced to represent an edge with additional data members, e_node.*angle* and e_node.*type*. For each E_i, an e_node having A_i and the type of E_i will be created. T is the sorted set of e_nodes in an increasing order of associated angle, and T_i denotes the ith e_node in **T**. *n_wrap* represents the number of extreme regions containing x-axis. N_i is the number of extreme vertices w.r.t. $L(\theta)$ when $C_i < \theta < C_{i+1}$. And M_i is the number of extreme vertices w.r.t. $L(C_i)$.

In this algorithm, N_i's and M_i's are computed by using the results in Table 1. To be more detail, while θ increases in discrete-event manner as in plane sweep algorithm, N_i's and M_i's are changed by the rules in the last two columns of Table 1.

algorithm-ODM : algorithm to determine the optimal direction of monotony
    0. input : polygonal chain $P = \{ E_1, .. , E_n \}$, where $E_i = V_iV_{i+1}$
    1. **for** each vertex V_i
        set vertex type as *SV, EV, LCV, RCV,* and *CV*.
    2. $n\_wrap = 0$, $\mathbf{T} = \varnothing$
        if *P* is closed **then** compute A_n
        for i=1 to n {
            A_i = angle from x-axis to the line perpendicular to E_i
            create an e_node T_i = <E_i, A_i, type of E_i> and add it to **T**
            if V_i is internal and R_i contains x-axis **then** $n\_wrap = n\_wrap + 1$
        }
    3. Sort **T** in increasing order of angle.
    4. // Compute N_i and M_i.
        4-1. $N_0 = N_1 = n\_wrap$, $i = 1$, $C_0 = T_n.angle - \pi$
        4-2. **for** k=1 to n {
            switch $T_k.type$ {
                case *LR_edge* : $M_i = N_i$, $N_i = N_i + 2$
                case *RL_edge* : $N_i = N_i - 2$, $M_i = N_i$
                case *LL_edge, RR_edge,* or *SE_edge* : $M_i = N_i$
                case *SL_edge* or *RE_edge* : $N_i = N_i - 1$, $M_i = N_i$
                case *SR_edge* or *LE_edge* : $M_i = N_i$, $N_i = N_i + 1$
            }
            if k = n or $T_k.angle \neq T_{k+1}.angle$ **then** // create a cell
                $C_i = T_k.angle$, $N_{i+1} = N_i$, $i = i+1$
        }
    5. Find the minimum from $\{N_i\} \cup \{M_i\}$. Ties are broken arbitrarily.
        If it is N_i, output $\theta = (C_i + C_{i+1}) / 2.0$
        else if it is M_i, output $\theta = C_i$

Step 1 of **algorithm-2** classifies all vertices. In step 2, an e_node is generated for each edge. After increasingly sorting the set of e_node's by their angle values in step 3, the number of extreme vertices for each cell is computed in step 4 by using the results in Table 1. And finally output is selected in step 5. Some variation of step 5 will be addressed in the next section. Fig.8 shows the result of this algorithm for the polygon in Fig.6 (a). Step 1, 2, and 4 of **algorithm-2** can be done in a linear time, while step 3 requires $O(n \log n)$ time. Hence algorithm-2 has $O(n \log n)$ time complexity and $O(n)$ space requirement.

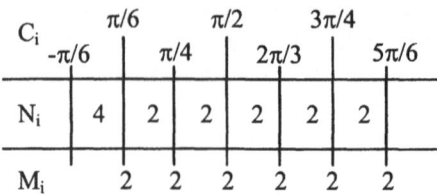

Fig. 8. The result of **algorithm-2** for the polygon in Fig.6 (b)

5 Further Considerations

As depicted in [7], degeneracy is one of the main reasons which makes the practical implementation of an algorithm difficult. Degeneracy requires many exceptional branches of processing in geometric algorithms. A monotone chain which is not strictly monotone can cause degeneracy in a subsequent processing. Hence, the direction which gives strict monotony is often preferred. For example, a plane sweep algorithm to find self-intersections of a polygon can assume that no segment is parallel to the sweep line (in other words, perpendicular to sweep direction) if the sweep direction is carefully taken. (For the details of the plane sweep paradigm, readers are referred to [2,4,5].) This assumption will help the intersection algorithm to be easily implemented. Let $N_p = \min\{N_i\}$ and $M_q = \min\{M_i\}$. Such a direction can be obtained by selecting θ as $(C_p+C_{p+1})/2$ rather than C_q in the step 5 of **algorithm-ODM**.

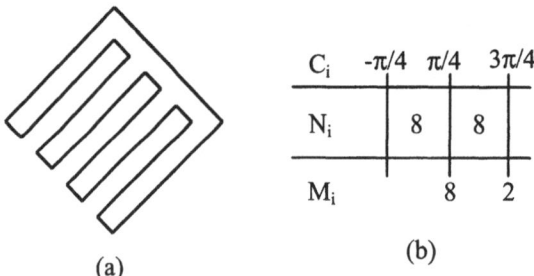

Fig. 9. Degenerate case (a) degenerate polygon (b) the result of algorithm

It is worth to mention that $N_p \geq M_q$ always, and sometimes N_p can be much larger than M_q as shown in Fig.9. The polygon in Fig.9 (a) is monotone w.r.t. L(3π/4) only. Any other direction will split the polygon into 8 monotone chains. To summarize these considerations, the followings can be said :
- Actual implementation of step 5 of **alorithm-ODM** depends on the purpose and assumption of the subsequent process.
- Usually the direction which splits input the polygonal chain into strictly monotone chains is preferred. In that case, select θ as $(C_p+C_{p+1})/2$, which is the median of the cell with $\min\{N_i\}$.
- If strict monotony is not required, compare N_p and M_q. If they are same, select θ as $(C_p+C_{p+1})/2$. If not, select θ as C_q.

6 Concluding Remarks

In this paper, an efficient algorithm is presented to find an optimal direction which can split a polygonal chain into (strictly) monotone chains. The proposed algorithm can handle arbitrary polygonal chains : open or closed, simple polygon or self-intersecting. The possible uses of the algorithms are as follows.

- Sweeping direction of a plane sweeping type of algorithm for polygonal chains can be determined by this algorithm. With the optimal direction of sweep, the performance of a plane sweep algorithm for polygonal chains can be improved.
- It can be applied to several polygonal chains simultaneously without major modification. For example, finding intersection between or within a polygon as shown in [4].
- A direction which split a chain into *strictly monotone* chains can be determined. With this direction, the subsequent algorithms may be simplified.
- It can be used to test if a polygon is (strictly) monotone, though this problem can be done more efficiently by the method in [6].

Acknowledgements. This research was supported in part by Korean government through NRL grant.

References

1. Chandru, V., Rajan, V.T., and Swaminathan, R., Monotone pieces of chains, *ORSA Journal on Computing*, Vol.4, No.4, 1992, pp.439-446.
2. Guibas, L. and Stolfi, J., Ruler, compass and computer: the design and analysis of geometric algorithms in Earnshaw, R. ed., *Theoretical Foundations of Computer Graphics and CAD*, in *NATO ASI*, Vol.F40, Springer-Verlag, pp.111-165, 1988. Also can be found as Research Report 37 of Digital System Research Center.
3. O'Rourke, J., *Computational geometry in C*, Cambridge University Press, 1993.
4. Park, S. and Shin, H., Polygonal chain intersection, *Computers & Graphics*, Vol.26, 2002, pp.341-350.
5. Preparata, F. and Shamos, M., *Computational geometry : An introduction*, Springer-Verlag, 1985.
6. Preparata, F. and Suposit,K.J., Testing a simple polygon for monotonicity, *Information Processing Letters*, Vol.12, 1981, pp.161-164.
7. Sugihara, K., "A simple method for avoiding numerical errors and degeneracy in Voronoi diagram construction" *IEICE Trans. Fundamentals*, Vol. E75-A, No.4 1992, pp.468-477.

GTVIS: Fast and Efficient Rendering System for Real-Time Terrain Visualization

Russel A. Apu and Marina L. Gavrilova

Dept. of Computer Science, University of Calgary
Calgary, AB, Canada T2N1N4
{apu,marina}@cpsc.ucalgary.ca

Abstract. The paper presents an improved scheme for the visualization of 3D terrain in real-time using Digital Elevation Model (DEM). The presented method is primarily based on a Real-time Optimally Adapting Mesh (ROAM), and boosts ROAM performance by improving the rendering speed, enhancing the visual continuity and achieving frame rate constancy. Based on the proposed methodoly, Geo-morph Terrain VIsualization System (GTVIS) was created. Experimental results confirm striking improvement in the rendering efficiency, frame rate constancy and visual continuity, while preserving important features of the terrain (such as sharp peaks and valleys).

1 Introduction

The 3D terrain modeling is a term that often refers to a process of visualizing geographical and spatial data. Although the power of recent graphics hardware has improved significantly due to the development of GPUs, it is still inadequate for rendering a large mesh in a real-time. A viewer-dependent multi-resolution terrain mesh generation technique was developed by Cohen-Or and Levanoni [4] to remedy this situation. The method improves visualization performance by a great magnitude.

High quality rendering and meshing techniques for displaying complex geographical data, such as terrain models, play an important role in the fast growing domain of CAD oriented towards Geographic Information Systems (GIS). Terrain rendering also finds numerous applications in computer graphics, simulation and games industry [2],[7]. Recent advances in a research on a multi-resolution dynamic mesh allow rendering large terrains effectively by taking advantage of viewer-dependent refinement of LOD (Level of Detail) and various culling techniques [1][2][3][14]. Moreover, most applications for real-time rendering involve some sort of walkthrough or flyover, allowing exploitation of a frame-to-frame coherence.

This paper presents the modified ROAM algorithm for terrain visualization, which can be easily implemented and highly practical. It outperforms its counterpart in several key aspects. In particular, it allows to achieve a better *rendering efficiency, frame rate constancy* and *visual continuity* than the presently known methods while preserving important *terrain features*. We also propose a fast and effective scheme for *mesh smoothing* and *geo-morphing*. The result is *Geo-morph Terrain VIsualization System (GTVIS)* for efficient and fast real-time terrain visualization.

2 Related Work

In SIGGRAPH 1996, Lindstorm proposed the real-time continuous level of detail terrain rendering algorithm that focused on a block based mesh understructure [17]. Lindstorm used a bottom up approach to assess the level of each block by means of defining a set of error index. The goal thereafter is as simple as minimizing this error while not violating a set of constraint (i.e. limited triangle count). The primary approach is to arrange entities in a priority queue that exploits the monotonic subdivision refinement principle [14][15].

A variety of research papers sprang up based on the original work of Lindstorm. The work on progressive meshes by Hoppe presents an effective scheme for smooth rendering of adaptive terrain mesh [9]. From the practical point of view, however, the scheme proposed by Duchaineauy seems to be more practical and was adapted in our implementation [5]. It is primarily based on the framework proposed by Lindstorm and presents a highly flexible and adaptable method representing continuous mesh with controlled LOD [16]. Many of the later papers that are based on ROAM (or adaptive mesh in general) dealt with viewer dependent refinements and culling techniques [3][4][12][13][14]. The paper by Hesse in WSCG'2003 provided a comprehensive comparison of various culling and refinement techniques and runtime benchmarks to test and compare the improved scheme [14]. The effective stripping strategy was presented in the papers by Hoppe in SIGGRAPH'97 and Shafae in Pacific Graphics'03 [10][15].

In our work on creating a system for fast, real-time rendering of terrain, we used the ROAM implementation proposed by Duchaineauy, and suggested a number of algorithms to extended and improved methodology significantly. The observation of hardware performance reacting to strip length allowed us to design a fast and effective greedy approach to dynamically maintain triangle stripping. The method resulted in achieving rendering efficiency and frame rate constancy.

We further improved the rendering efficiency while preserving important terrain features by introducing the following procedures:
- Eliminating the priority queue for merge operation,
- Amortizing update cost by allowing degenerate split queue,
- Boosting performance by preprocessing DEM,
- Taking advantage of frame-to-frame coherence for local update.

Since almost all of the recent demonstrations of ROAM in different conferences exhibited visual popping and discontinuity due to the dynamic refinement process (split/merge), we also derived an effective scheme for mesh-smoothing while preserving important terrain features, using *geo-morphing*. As far as we know, this is a first successful application of Geo-morph technique in real-time terrain visualization.

Finally we implemented the proposed methodology in the *GTVIS* software, which boosted the frame rate and constancy of frame rates, which in turn resulted in striking improvement in the rendering efficiency, frame rate constancy and visual continuity, as proven by extensive experimentation.

3 Analyzing the Real-Time Optimally Adapting Mesh

The basic idea in ROAM is to use triangle pairs (diamonds) as a basic geometric element and use a split and a merge priority queue to decide on the next refinement. Although ROAM is a very robust model for adaptive mesh understructure, there are a number of problems that discourages its use in practical application [7],[23]. For example, the constant update of the Split/Merge queues on every frame is somewhat a costly process when we have lots of triangles in the system. Performing this update every frame (which guarantees optimally adapting viewer dependent mesh) inevitably consumes a lot of processing power (for updating and adjusting a heap). This might not be a problem for large visualization projects implemented for high-end multiprocessor machines, but it is certainly a barrier for gaming and simulation applications that run on a single processor desktop computer. Another potential problem is the noticeable visual popping effect due to the split and merge operation. This creates notorious discontinuity in smooth shading and highlights.

4 Enhancing the Performance of ROAM

To overcome the problems associated with ROAM based terrain models, we propose a number of modifications, described in the following sections.

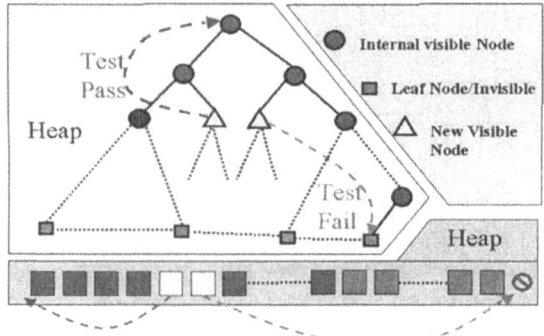

Fig. 1. Algorithm for priority queue update

4.1 Eliminating the Priority Queue

We examined runtime analysis of several flyover and walkthrough demos and discovered that almost all of the merge operation takes place on the obscured region (assuming there is no occlusion testing). Merge occurs within the visible frustum is when we are moving backwards. In this case merge occurs mostly in distant regions. Assuming that we have far clipping beyond a reasonable distance, we can force merge operation only beyond a certain distance. This way, we can make sure that merge does not take place at regions that has high screen pixel contribution. This eliminated the need for a priority queue for merge operation. In order to enhance performance we therefore suggest the following algorithm (see illustration in Fig 1):

Method 1.
1. Replace the prioritized merge queue (which imposes high maintenance cost at every frame) by a simple queue (FIFO).
2. Allow merge operation to occur arbitrarily on the obscured region satisfying specific criteria (i.e. triangle surface area).
3. Remove an element from the queue and test the criteria against an element.
    If the criteria are satisfied, simplify the region.
    Else add the element at the end of the queue (no search is required).

The visibility update operation can also update the content of the queue without additional computation, reducing run time complexity for performing update from $O(\log n)$ to $O(1)$. Experiment show that in the worst case (moving backward all the time without turning) an average of 21.2% (approx) of triangles is unable to simplify for a finite number of frames. During a forward coherent motion no more than 3% of the triangles are unable to simplify (which would be simplified by a priority queue). By providing a reasonable far clipping refinement [1], this effect can be reduced significantly and the overall mesh differs insignificantly from the expected mesh.

4.2 Amortizing Update Cost by Allowing Degenerate Split Queue

In our simulation we identified that mesh update becomes a bottleneck for high triangle counts. Since merge queue is no longer prioritized, the majority of the computation time is spent maintaining the split queue and error index/visibility update operations. Ideally the split queue must be updated every frame, before any diamond could be split. However if we could achieve higher frame rate, it will also ensure us a very good frame-to-frame coherence. Consequently we exploit this coherence to achieve even higher frame rates! On the other hand to maintain visual continuity we cannot allow too many split operations taking place all at once. Therefore we have to ensure that refinements occur at a steady rate. Putting all these together we derived a nice symbiotic solution presented below.

Instead of strictly maintaining a heap (split queue) every frame, we let the queue to degenerate and make sure that it could still perform effectively. In our method we only update error indexes and build the heap after every k frames where k is a constant. We perform a constant number of refinements within these frames. However the problem is that within these k frames, the queue is no longer optimal or even valid. Therefore we take measures to counter this effect. We suggest to achieve it by ensuring that details are added to the newly visible region.

Method 2.
1. Compute the average error e_{avr} of the first few diamonds in the split queue.
2. *If* any invisible block b_i becomes visible, compute the error e_r.
3. *If* $e_r < e_{avr}$ place block b_i at the beginning of the queue,
   Else block b_i at the beginning of the end of the queue.

When a block **b** is split, the error index of the newer blocks are computed and compared to the average value e_{avr}. This will make sure of two things. Firstly it will force coarser obscured region to split when they become visible. Secondly it will refine regions that need attention. Thus in average case of a coherent simulation, the deliberate use of degenerated heaps will greatly boost speed without any noticeable artifacts.

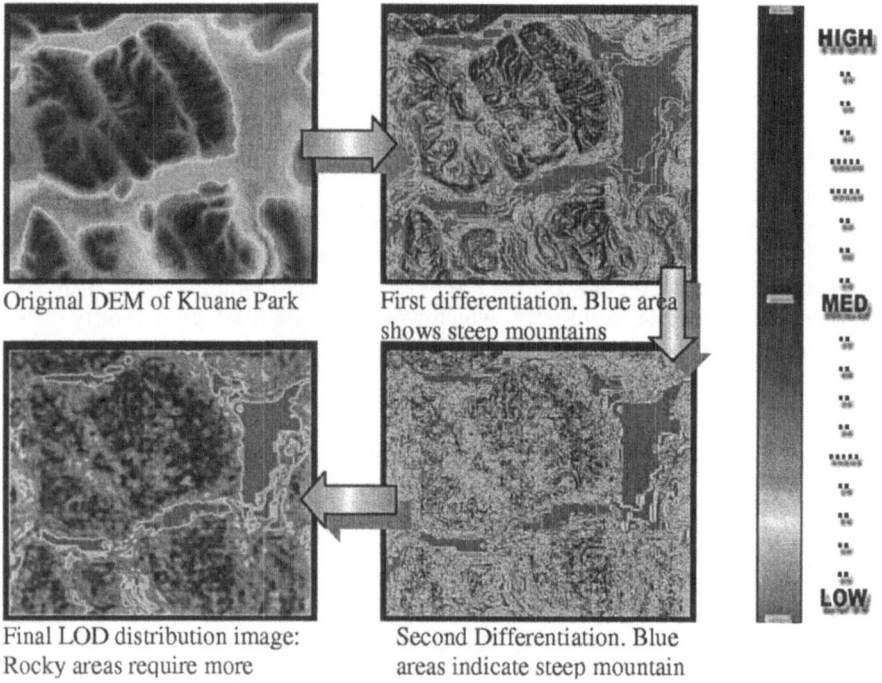

Fig. 2. The processing of DEM data to produce LOD distribution. The colors represent relative values (spectrum shown on the right).

4.3 Boosting Performance by Preprocessing DEM

In order to determine the error index of a face based on the DEM, we evaluate an influence map in the preprocessing stage. Logically more details on steep edges along various curvatures (where slopes are changing faster) are required, and less detail along areas with relatively constant slope are needed. But the question is, how does the curvature relate to DEM?

The answer is simple. In order to find the rate of changes of pixel gradients we take the bitmap and perform a two-pass convolution (differentiate) [6]. The idea is to find the changes between neighboring pixels and estimate a weighted average. Therefore the second difference gives us the rate of gradients for each pixel (as a bitmap). However to avoid sampling during the runtime, we also apply simple multi-pass radial blurring to obtain an average radial sampling of pixel-influence (Fig.2).
Our algorithm for these processing seems quite efficient and has a linear performance in a number of pixels. Although we have generated the maps first, it can nevertheless

be embedded in the application as a pre-render input processing stage. This will allow method to be independent of anything except the DEM bitmap as the input to the application. This usually doesn't delay the loading of the application significantly.

4.4 Taking Advantage of Frame-to-Frame Coherence for Local Update

Ideally, we only compute normal when a split operation occurs. To do so first we compute face normal of the new faces and update the vertex normal of surrounding vertices. Traditionally, vertex smoothing is used to achieve a smooth shaded terrain. The problem that arises, however, is that vertex smoothing diffuses sharp features of the terrain, so it loses details along sharp mountain peaks. We propose a method to recover higher details at the cost of negligibly alleviated computation. For this we propose the *hybrid-shading scheme*, which is based on using the linear combination of face shading and vertex shading. Thus, when each face is rendered the weighted average of the vertex normal and the face normal is computed (Fig. 3). The update is local and practically does not influence the simulation time. As a weight distribution coefficient, the *curvature map* generated during preprocessing is used. The result is preservation of sharp peaks and valleys at the minimal cost, due to the locality of update.

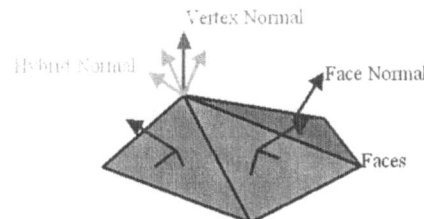

Fig. 3. Procedure for preserving sharp features of the terrain

5 Modifying the Algorithm to Incorporate Geo-morph

Since we have eliminated the merge queue, all the merge operation will occur outside the visible frustum. Therefore there is no need to geo-morph vertices for merge operation. However for a split, we have to interpolate coordinates and other attributes of a vertex (and neighboring faces) over time. There are two problems to resolve in this case: how to deal with vertex split when the original vertex is in transition and how to minimize the overhead of computing smooth transition between refinements. To address this problem, we introduced a new queue to the model called a G-queue. When any diamond is refined, we generate a center vertex, which we insert to the geo-morph queue. The G-Queue is FIFO in nature (since every transition is sequential) and contains only reference to a vertex. When the interpolation is complete, the algorithm removes items from the queue (if a vertex in transition needs to be deleted, its reference is removed from the queue). Finally when a new vertex is created its initial position is placed on the plane (Fig. 4) of the parent triangle pair (taking average). This eradicates the second problem. The high level algorithm for our method is given below:

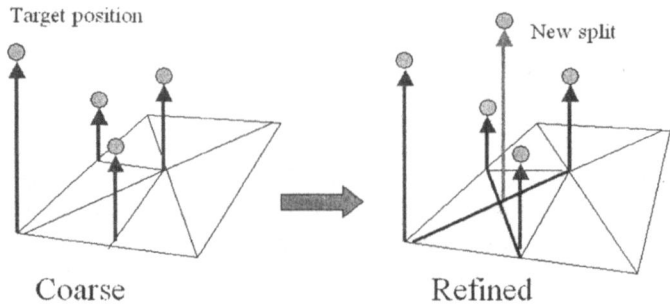

Fig. 4. Procedure for continuity maintenance through grid refinement in Geo-morph

Method for Mesh Refinement:

```
1    Procedure Refine_Mesh
2        Refine visibility and error indexes
3        If current_step_# mod k = 0 then do
4            Update all priorities
5            Re-Build Priority Queues
6        Else
7            Compute Average Error
8            Filter new visible faces for split_queue
9        For I←1 to COARSE_MAX do
10           If Face# < FACE_TARGET then exit for loop
11           F←Dequeue(Merge_Queue)
12           Merge(f)
13       Advance Current_step_#
14       If G-Queue is full or Face# > FACE_TARGET then exit proc
15       For I←1 to REFINE_MAX do
16           f←Dequeue(Split_Queue)
17           Split(f)
```

6 Analysis of the Performance

We conducted trial runs on the newly developed system by importing the DEM of Kluane National Park, Canada (Fig. 5 and Fig. 6). Experiments demonstrated that implementation of the suggested methodology did in fact performed significantly well compared to other recent implementations [13][14][15]. The novelty of our approach is that we have selected the subset of refinement and culling techniques that has been proven to be the most effective. We also suggested a novel technique for amortizing the cost of maintenance over several frames without potential side effects. The following figure shows the terrain rendered by our software.

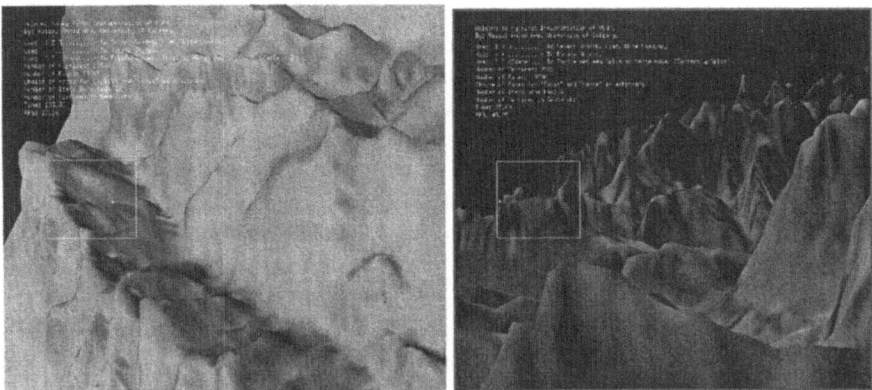

Fig. 5. Rendered Terrain (left) and terrain with Smoothing and Texture features enabled (right) using GVIS software. The level of detail obtained at a frame rate of 47FPS is quite amazing.

The plot of simulation time over rendered frame indicates the runtime behavior of the system in general. We have found that the relationship is almost linear regardless of all the refinements and wild navigation characteristics to tease the system during trial run (Fig. 7). This proves the effectiveness of our scheme in achieving continuity and frame rate constancy.

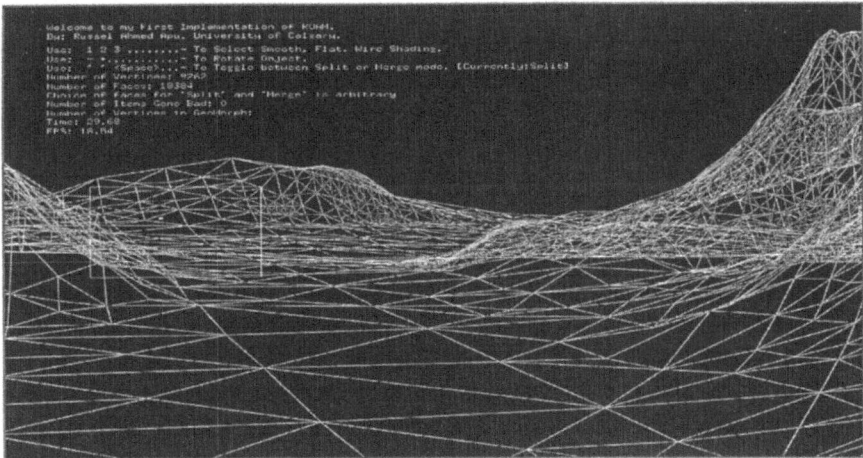

Fig. 6. Underlying data structure for rendered Terrain. The level of detail obtained at a frame rate of 47FPS is quite amazing.

Our next experiment involved measuring Frame Per Second (FPS), since this is the most important concern for real-time terrain rendering. To our surprise we found that the average frame rate was about 47 FPS on a single 1.7GHz P4 processor machine with Nvidia Geforce graphics adaptor. This result combined with the fact that our average number of rendered faces was 19000 and that for a naïve system, rendering a static mesh with 19000 faces cannot be done faster than 55 FPS (using the same configuration and uniform patches) proves that the system is reasonably efficient. We found that our implementation maintains almost a constant frame-rate with slight variation (Fig. 7). The variation however is inevitable since the effect of culling and clipping from different view effects the number of visible triangle (and especially during steep turns of the aircraft). Also the rendering depends on the size of each triangle in the mesh. There are also effects of hardware back-face culling that influence the rendering time of a frame.

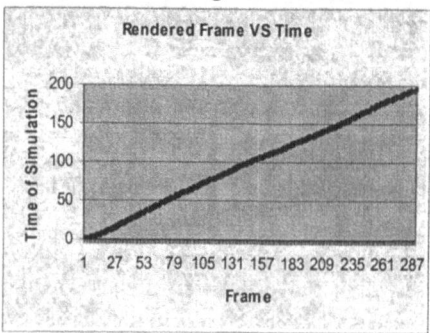 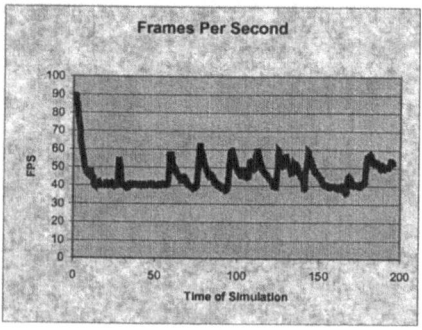

Fig. 7. The linear relationship of frame vs. time (left) and graph showing frame per second rendered over time

On the other hand, the face counts of the mesh provide us some interesting observations. We measured two different counts: the number of faces generated by the ROAM structure and the number of faces rendered by our system after culling is performed. The chart shows that at some points that the two counts converge (Fig. 8). This is because we didn't use occlusion culling. We observed that the effect of occlusion culling is minimal in a typical application. The deltas formed at several region is due to sharp turning but it appears that details are removed from obscured region and added to the visible part, as the mesh adapts.

Finally, we show the behavior of the Geo-Morph sequence (Fig. 8). The chart shows the dynamic nature of the Geo-Morph queue as load factors in terms of active morphing vertices. For our experimental run assuming 30 frames geo-morph and a face count of 20000, The highest ceiling of vertices in Geo-morph was less than 460. On average there were less than 300 vertices each frame involved in morphing. This gives us the notion that with negligible processing load, we were able to achieve greater visual continuity over the simulation by using Geo-morph sequencing.

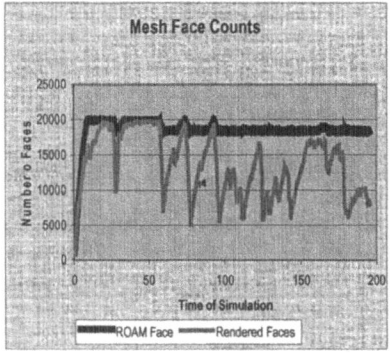 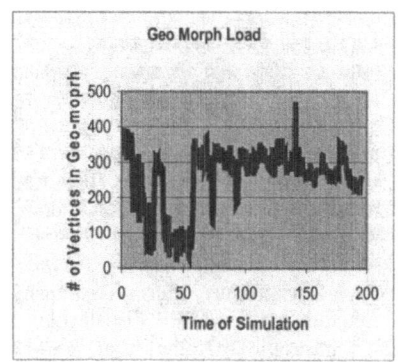

Fig. 8. Face counts generated/rendered by the simulation (left) and vertices involved in active Geo-Morph sequence (right)

7 Conclusion

In this paper we outlined an easy but effective scheme for the efficient implementation of ROAM overcoming the fallacies and bottlenecks. We boosted the performance of ROAM by proposing a modified scheme to achieve relatively constant and higher frame rate and visual continuity. We also incorporated Geo-morph technique integrating with ROAM and resolved some of the problems that came inherently with it. The GTVIS software provides a versatile system for terrain visualization which can be also extended to other applications. The incorporated methodology makes GTVIS system very attractive and justifiable for utilization in other practical real-time graphics applications. Future work will include but not be limited to enhancing visual details at high rendering speed.

The research was partially sponsored by NSERC and GEOIDE granting agencies.

References

[1] Assarsson, U. and Moller, T. *Optimized View Frustum Culling Algorithms for Bounding Boxes,* Journals of Graphics tools, 2000.
[2] C. Silva, J. Mitchell, and A. Kaufman. Automatic generation of triangular irregular networks using greedy cuts. *Visualization '95.* IEEE Press, 1995.
[3] Coorg, S. and Teller, S. *Real-time Occlusion Culling for Models with Large Occludes,* ACM symposium on Interactive 3D Graphics, 1997
[4] Daniel Cohen-Or and Yishay Levanoni. Temporal continuity of levels of detail in delaunay triangulated terrain. *Visualization '96,* pp. 37-42. IEEE. Press, 1996.
[5] Duchaineauy, M. et. al. (1997), ROAMing Terrain: Real-Time Optimally Adapting Meshes, *IEEE Visualization '97 Proceeding,* 81-88.
[6] E. Horowit, S. Sahni, *Fund. of Computer Algorithms,* Galgotia Publ., 1998.
[7] ESRI Inc., Your Internet guide to Geographical Information
[8] Hanan Samet. *Applications of Spatial Data Structures.* Addison-Wesley, 1990.
[9] Hoppe, H. Progressive Meshes. *SIGGRAPH '96* pp. 99-108, 1996.
[10] Hoppe, H. View-Dependent Refinement of Progressive Meshes. *Computer Graphics (SIGGRAPH '97 Proceedings)* pp. 189-198, 1997.

[11] James H. Clark. Hierarchical geometric models for visible surface algorithms. *CACM*, 19(10):547-554, October 1976.
[12] Julie C. Xia and Amitabh Varshney, Dynamic view-dependent simplification for polygonal models,*Visualization '96,* pp. 327-334, IEEE Comp. Soc. Press, 1996.
[13] Li Sheng, Liu Xuehui and Wu Enhau, Feature-Based Visibility-Driven CLOD for Terrain, *In Proc. Pacific Graphics 2003,* pp 313-322, IEEE Press, 2003.
[14] M. Hesse and M. Gavrilova, Quantitative Analysis of Culling Techniques For Real-time Rendering of Digital Elevation Models, *In Proc. WSCG 2003,* 2003.
[15] M. Shafae and R. Pajarola, Dstrips: Dynamic Triangle Strips for Real-Time Mesh Simplification and Rendering, *Pacific Graphics 2003,* pp. 271-280, 2003.
[16] Mark Duchaineauy, Murray Wollinshy, et al. ROAMing Terrain: Real-Time Optimally Adapting Meshes, IEEE Visualization '97 Proceeding, 1997
[17] P. Lindstrom, D. Koller, et al. *Real-time continuous level of detail rendering of height fields,* Computer Graphics (SIGGRAPH 1996 Proceedings), pp. 109-118.

Target Data Projection in Multivariate Visualization – An Application to Mine Planning

Leonardo Soto[1,2], Ricardo Sánchez[2], and Jorge Amaya[3]

[1] Department of Systems Engineering, University of Talca
Casilla 747 - Talca - CHILE
lesoto@utalca.cl
[2] Department of Electrical Engineering, University of Concepción
Casilla 160-C - Concepción 3 - CHILE
leonardosoto@ieee.org, risanchez@telsur.cl
[3] Centre for Mathematical Modeling, University of Chile
Casilla 170 - Santiago 3 - CHILE
jamaya@dim.uchile.cl

Abstract. Visualization is a key issue for multivariate data analysis. Multivariate visualization is an active research topic and many efforts have been made in order to find suitable and meaningful visual representations. In this paper we present a technique for data projection in multivariate datasets, named Target Data Projection (TDP). Through this technique a vector is created for each multivariate data item considering a subset of the available variables. A new scalar variable is generated projecting those vectors over a target vector that defines the *direction of interest* for visual analysis. End-users set up target vectors in order to explore particular relationships by means of application meaningful projections. Hence, it is possible to map a combination of multivariate data into one scalar variable for graphical representation and interaction. This technique has proved to be very flexible and useful in mine planning providing valuable information for decision making.

Keywords: Multivariate visualization, mine planning, multiple visualization workspaces, target data projection, human-computer interaction.

1 Introduction

Data representation and manipulation are key activities in problem solving and knowledge discovery. For this reason many efforts have been made to find suitable and meaningful representations of multivariate data [1]. Visualization supports multivariate data analysis by allowing the generation of interactive visual representations aimed to extract knowledge from data. Interactive graphical depicts encourages the users to explore data to find patterns, relationships, and clusters. However, the graphical display of data is constrained to two or three dimensions but data may have dozens of variables. Furthermore, interacting with such a high number of variables may be frustrating.

In this context, one approach is to project the multivariate data over carefully selected data vectors to reduce the dimensionality of the data representation, and still be able to get some information from it. In the statistics domain, several projection methods for multivariate data analysis have been devised. For instance, principal components analysis, factor analysis, and projection pursuit [2]. In general terms, these methods are used to find optimal transformations considering statistical properties of the dataset. Thus, projected representations retain the most informative features of the data. These projections are useful for clustering purposes, supported by orthogonal components. Usually the first principal components are used to define a two dimensional coordinated system, revealing data clusters. However, resulting principal vectors are likely to have no intrinsic meaning for the application domain. Therefore, the expressiveness of data representations is limited when those vectors are used to create a spatial substrate for visualization. It is worth stressing that spatial encodings are perceptually dominant and usually visualization design take advantage of its metric structure [3,4].

Several methods have been developed for multivariate data visualization, including scatterplot matrices, worlds within worlds, parallel coordinates, icon-based display, hierarchical plotting, and pixel based techniques [5]. However, non trivial projections of multivariate datasets have played just a minor role in visualization. Trivial projections refer to those achieving dimensionality reductions by just discarding some data variables. Multivariate data projections methods have been used in text database applications, where documents are treated as points in a multidimensional conceptual space [4,6,7]. Documents are mapped to multidimensional points through text analysis. The associated points are used to classify documents according to its proximity to keyword queries. Tools for text visualization and visual information retrieval has been developed [6,7]. The first attempt to expand the use of data projection was called *data signature* [8]. Data signatures are aimed to capture the essence of large scientific datasets in a compact data vector and to unify different data types. It seems clear that increasing the use of data projection methods in multivariate visualization would result in enhanced analytical capabilities for visualization tools.

In this paper we propose a new projection mechanism for the analysis of quantitative variables in multivariate datasets. Our technique, named Target Data Projection (TDP), relies not on dataset statistics but on application meaningful data vectors for the projection process. Those vectors are referred to as target data vectors because they define a *direction of interest* for visual analysis in the multidimensional space of the dataset. The user specifies a target data vector either by selecting a determined data item directly from the dataset or by defining a vector in terms of interesting data attributes for the analysis (e.g., a particular combination of values for a set of variables). Then, the entire dataset is projected over the established target vector, resulting in a new scalar variable. Generated scalars convey quantitative information about the direction each data item points to, with respect to the target data vector. This information may be extracted with the use of traditional methods for scalar visualization [9]. Besides, the use of classical interaction techniques [3] like brushing or attribute walk is encouraged. Threshold based filtering over this derived variable allows the users to interact in terms of data direction for exploring attribute relations over the dataset, furthering the classical use of range filtering for each variable separately. Similarly, attribute walk is improved with the use of TDP because it is possible to recognize not only data items with similar attribute values, but with similar attribute relations.

Mine planning is a challenging application domain for visualization. Large multivariate datasets describe the mineral resource and establish the starting point for plan design. Strategic, economical, and technical constraints are used to define an optimization model to find an optimal extraction plan [10]. However, the optimization model is far from complete, and many decisions have to be made considering the opinions of several experts. For this reason, an iterative process involving mine plan optimization and discussions of experts from different disciplines takes place throughout plan development. The results obtained in each step should be evaluated by the experts and new constraints would appear. This process relies in the experts to understand complex relations and constraints over a three-dimensional geometry and multivariate data. Hence, visualization tools are of major importance. Our work on TDP was motivated while we were developing a visualization system to support mine planning activities. The focus was in the evaluation of designed plans and planning heuristics considering the available multivariate data.

The rest of this paper is organized as follows: In section 2 target data projection is formally defined. Section 3 describes visualization challenges in the mine planning application domain. Section 4 refers to design and implementation issues in the development of a visualization system for mine planning. Section 5 is devoted to discuss results obtained with TDP, implemented in a visualization prototype. Conclusions and future work are presented in Section 6.

2 Target Data Projection

Reference models provide the foundations to extend visualization techniques and to develop more effective applications [5]. Although work is still required to state an intuitive reference model and notation, *Data State Reference Model* [11] and Stuart Card's model [3] have been of major importance. In the context of these models, *target data projection* would be described as a data transformation. The analytical abstraction behind TDP is a set of multidimensional data vectors defined from the original data table. A derived scalar variable is generated by projecting the whole data table (i.e., its data vectors) in the direction of a particular *target data vector*. Target data vector is intended to be an application-meaningful vector that allows the user to define a *direction of interest* for visual analysis. Finally, original data and generated scalars are assembled resulting in a transformed data table. Exploiting projected variables in the visualization process would result in an improved visual representation or interaction.

2.1 Definition

Let $C_{n \times m}$ be the corresponding matrix for a given data table $\mathcal{T}$ with n data items and m variables, such that c_{ij} element refers to the value of the j-th variable in the i-th data item, $i = 1, \ldots, n$, $j = 1, \ldots, m$.

Now let matrix $D_{n \times m}$ be defined in terms of elements $d_{ij} \in [-1.0, 1.0]$ determined by:

$$d_{ij} = \begin{cases} -1 + 2\frac{c_{ij} - \alpha_j}{\beta_j - \alpha_j} & \text{if } \alpha_j \neq \beta_j \\ 0 & \text{otherwise} \end{cases} \quad (1)$$

with,

$$\alpha_j = \min_{i=1,\ldots,n} \{c_{ij}\}. \quad (2)$$

$$\beta_j = \max_{i=1,\ldots,n} \{c_{ij}\}. \quad (3)$$

for $i = 1, \ldots, n$, $j = 1, \ldots, m$. Consider the i-th row of D (i.e. the i-th data item) as a m-dimensional vector D_i such that $D_i = [d_{i1}\ d_{i2}\ \ldots\ d_{im}]^t$, $i = 1, \ldots, n$.

Assume a m-dimensional *target data vector* g with coordinates $g_j \in [\alpha_j, \beta_j]$, $j = 1, \ldots, m$. A transformed target vector u is defined with elements u_j given by the expression:

$$u_j = \begin{cases} -1 + 2\frac{g_j - \alpha_j}{\beta_j - \alpha_j} & \text{if } \alpha_j \neq \beta_j \\ 0 & \text{otherwise} \end{cases} \quad j = 1, \ldots, m. \quad (4)$$

Note that u components are such that $u_j \in [-1.0, 1.0]$, $j = 1, \ldots, m$. Furthermore, a constraint must be imposed in the selection of g to prevent a null target vector u.

$$\|u\| \neq 0 \iff \exists j \in [1, m] : u_j \neq 0. \quad (5)$$

$$u_j \neq 0 \iff \left\{ \left(g_j \neq \frac{\alpha_j + \beta_j}{2}\right) \wedge (\alpha_j \neq \beta_j) \right\}. \quad (6)$$

then,

$$\|u\| \neq 0 \iff \exists j \in [1, m] : \left\{ \left(g_j \neq \frac{\alpha_j + \beta_j}{2}\right) \wedge (\alpha_j \neq \beta_j) \right\}. \quad (7)$$

Let a_i be the m-dimensional vector obtained from D_i by projecting to zero every component d_{ij} orthogonal to the target vector u. Thus, a_{ij} coordinates of a_i are determined by:

$$a_{ij} = \begin{cases} d_{ij} & \text{if } u_j \neq 0 \\ 0 & \text{otherwise} \end{cases} \quad i = 1, \ldots, n,\ j = 1, \ldots, m. \quad (8)$$

then, the *target data projection* of the data table $\mathcal{T}$ in the *direction of interest* determined by the *target data vector* g is defined as the n-dimensional vector s, with elements s_i such that:

$$s_i = (Proj_{\hat{u}}(\hat{a}_i)) \cdot \hat{u} = \left(\frac{\langle \hat{a}_i, \hat{u} \rangle}{\|\hat{u}\|^2} \hat{u}\right) \cdot \hat{u} = \hat{a}_i \cdot \hat{u} = \cos\theta. \quad (9)$$

$$\hat{a}_i = \begin{cases} \frac{a_i}{\|a_i\|} & \text{if } \|a_i\| \neq 0 \\ 0 & \text{otherwise} \end{cases} \quad i = 1, \ldots, n . \qquad (10)$$

where $\hat{u} = u/\|u\|$, and θ is the angle between the data item vector a_i and the vector u pointing in the *direction of interest*.

The resulting s vector is used for constructing the transformed data table $\tilde{\mathcal{T}}$, associated with the augmented matrix $\tilde{C}$

$$\tilde{C} = [\, C \mid s \,] . \qquad (11)$$

A slight variant of the method, named *target data projection for unnormalized data vectors* (TDPu) is defined replacing the normalized data vector $\hat{a}_i$ by its unnormalized version a_i. Rewriting equation (6) yields:

$$s_i = (Proj_{\hat{u}}(a_i)) \cdot \hat{u} = \left(\frac{\langle a_i, \hat{u} \rangle}{\|\hat{u}\|^2} \hat{u} \right) \cdot \hat{u} = a_i \cdot \hat{u} \quad i = 1, \ldots, n . \qquad (12)$$

2.2 Constructing Target Data Vectors

As stated in the previous description, target data vectors are intended to be constructed by the user. The first activity the user should perform is to define a k-dimensional space for data to be projected, $1 \leq k \leq m$. Hence, the original set of m variables is separated in two disjoint subsets. The first subset is named *variables of interest* and is specified by selecting the k important variables for the analysis at hand. For each variable of interest a component should be specified in the target data vector, defining its direction in a k-dimensional space. The second subset is denominated *discarded variables* and is composed of $m - k$ variables excluded from the analysis.

The next step is to compose a target data vector with a particular combination of values in each variable. Components of the target data vector may be specified between two different data ranges: $[\alpha_j, \beta_j]$ or $[-1.0, 1.0]$. If the original data range is selected, then values $g_j \in [\alpha_j, \beta_j]$ are mapped to normalized range values u_j through equation (4). This is useful for defining target vectors by direct selection from the dataset and for specifying its components with critical values for the application. When normalized range is used, values $u_j \in [-1.0, 1.0]$ may be transformed to its original range (see Table 1).

Table 1. Specifying target data vector components.

Variable j	Normalized Range	Original Range
Discarded	0	$g_j = \frac{\alpha_j + \beta_j}{2}$
Maximum	1	β_j
Minimum	-1	α_j
g_j	$u_j = -1 + 2\frac{g_j - \alpha_j}{\beta_j - \alpha_j}$	g_j
u_j	u_j	$g_j = \alpha_j + \frac{(u_j+1)(\beta_j - \alpha_j)}{2}$

Furthermore, each component of the target data vector corresponding to a discarded variable is set automatically according to Table 1, in order that no direction is defined for this variable.

3 Visualization Challenges in Mine Planning

Mine planning is devoted to the management of processes involved in exploiting mineral resources. Mine optimization and strategic business planning converge to elaborate strategic mine plans for mining projects. Its objective is to develop a strategic mine plan for a given mining project, such that the *net present value* (NPV) of the project is maximized while risk is controlled. From a visualization standpoint, there are several interesting features of the planning process:

1) Handling, exploring and analyzing large multivariate datasets are essential tasks in order to extract valuable information for decision making in plan design.
2) The design process should integrate geological and grade models, geotechnical and operational considerations, economical factors and strategic objectives.
3) Mine optimization is computational intensive. For this reason, heuristics are used to develop the optimized plan. Adjusting planning heuristics requires for fast assessment.
4) Optimization model is far from complete. Discussions of experts should take place in order to recognize critical constraints and to assess the resulting plans. Hence, an iterative process relating mine optimization and experts review is carried out to develop the mine plan.
5) Multidisciplinary interaction must take place to develop realistic plans, including geologists, geo-statisticians, mining engineers, metallurgists and managers.
6) Although high group interaction is required in order to asses design alternatives, there is a lack of efficient and effective ways to communicate ideas in the planning process. Besides, tools are required to present resource information integrated with mine plans to support decision making and plan assessment.
7) The basis for plan design and project analysis is a mineral resource estimate. Consequently there is an associated uncertainty that should be considered.

4 Design and Implementation Issues

Target Data Projection was implemented in a software prototype named SVPM, employing an object oriented approach. Evolving prototypes were developed under C++ using VTK [9]. Close work with mine planning practitioners and the continuous assessment provided for them were key aspects in developing a useful functional prototype. Design was carried out considering the mine as the central object encapsulating data and its graphical representations. Mine object design was organized in three levels: data table, workspace and visual structure.

It is worth to mention that for resource modeling and planning purposes, a mine is conceptually divided in many blocks. Hence, a mine has an associated block model, with each block containing multivariate data. At data table level, the block model is used to create a multivariate data table. This level works as a data repository, providing data to construct visual structures in different workspaces. Besides, data table abstraction is considered essential to think about data independently of its intrinsic geometrical meaning and to conceptualize data vectors. At this level, TDP is used to generate user defined projections for expanding visualization alternatives.

Workspace level allows the user to set up a concrete dataset by selecting the variables used to define geometrical coordinates. Hence, spatial substrate is established by choosing one variable for each axis in the visual representation. The remaining variables may be used for color mapping over all the visual structures contained in a workspace. Several workspaces may be used in order to explore different aspects of a dataset. Furthermore, TDP might help, in some measure, to overcome the limitations of a three-dimensional workspace for visualizing multivariate data. This is accomplished by first projecting data vectors comprising several variables, in the direction of a target vector. Then, the projected variable may be used for color mapping or to define one axis in the workspace. Finally, each workspace provides standard view transformations in order to explore its content.

Visual structure level stands for the user to create and manipulate elements in the workspace. Visual structures are generated through visualization techniques, including isosurface extraction, cutting planes, and direct representation of point clouds or mine blocks (blocks representation is supported only for workspaces composed with the original x, y, and z coordinates). User interface elements provide access to each visual structure in a given workspace, allowing the user to set up graphical parameters. Visual structures are arranged according to visualization techniques to create tree-like interface widgets. Standard check boxes are provided to enable and disable the visibility of visual structures in the workspace, either individually or by technique. In addition, threshold filtering is performed in a per visual structure basis. For this reason, a determined visualization technique may be used with the same parameters, but different filter thresholds to create diverse elements in the workspace.

5 Results

SVPM has been used for resource exploration, evaluation of mining plans and validation of planning heuristics with real mine data. High user satisfaction has been achieved through flexible visualization tools and interactive rates for large datasets. In this section, TDP is illustrated for resource exploration tasks. The mine used for illustrations comprises 525.000 blocks and ten scalar variables,

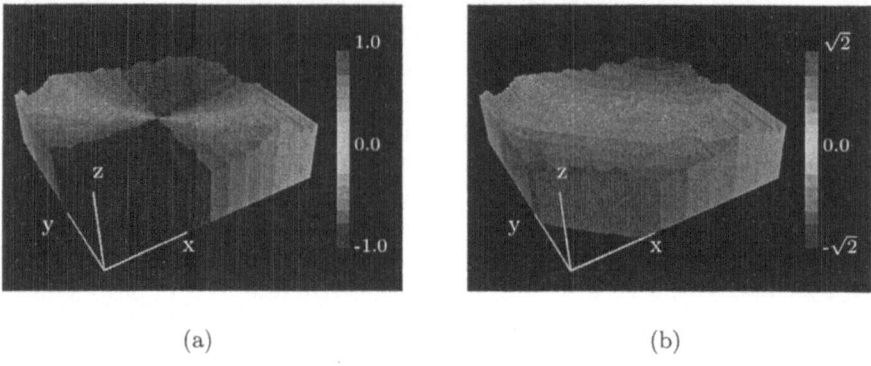

(a) (b)

Fig. 1. Comparing TDP and TDPu in a geographical workspace.

including geographical coordinates, geotechnical parameters, and ore characterization for each block.

Fig. 1 illustrates the differences between TDP (left) and TPDu (right) in a workspace built from geographical coordinates. Classical threshold based filtering is used in order to select only the blocks with projected value in the defined range for the filter. The generated scalars are color mapped over the geometry of the dataset, revealing how to define thresholds for filtering (i.e., how to interact with filters). A target data vector with direction {x-maximum, y-maximum} was specified by end-users. Fig. 1(a) demonstrate angular filtering with respect to the target vector. This task is achieved using the variable derived through TDP for filtering. In the current image, threshold is set in the range $[\cos 170, \cos 0] = [-0.9848, 1.0]$. It should be noted that, because cosine is an even function, the specified value is considered to both sides of the target vector (this issue is inherent to the projection concept). On the other hand, Fig. 1(b) displays the application of linear filtering in the direction defined by the target vector. This is accomplished using TDPu generated scalars for filtering. In the image, blocks in the range $[-1.0, \sqrt{2}]$ for the generated variable are represented.

TDP may be exploited in more attractive tasks. Ore concentration is a fundamental issue for mining activities. Assume we are interested in finding the spatial distribution of the richest cooper concentrations in the mine. Color coding is set to cooper grade in Fig. 2(a,b). Fig. 2(a) shows blocks above half the range of cooper concentration in the mine. High ore concentrations are revealed, but exploiting mineral resources in the profundity of the mine represents higher costs than extracting resources close to the surface. Note that in open pit mining, extracting a given block requires the preceding exploitation of all the blocks comprising its subsidence cone. This means that to reach a block far from the surface many blocks must be extracted before, and this may take as long as decades. Even rich blocks located in the depth of the mine may present low economic interest, provided that discount rate lowers its net present value (NPV)

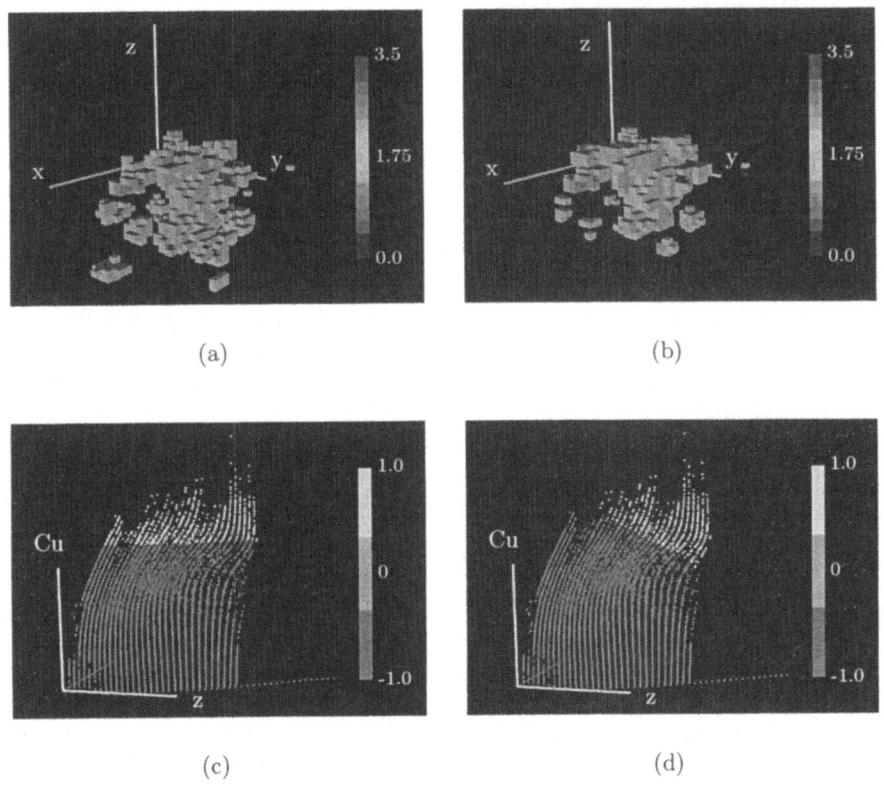

Fig. 2. Using TDP in blocks models and point clouds.

yearly. In this context, TDP may be used to search for blocks in the direction of interesting multivariate combinations. Set a target data vector with direction {z-maximum, cooper-maximum, gold-maximum} and compute TDP for the dataset. Fig. 2(b) shows blocks in the range $[0, 1]$ for the projected variable P. Compare to Fig. 2(a), note that less blocks are considered interesting in the lower part of the mine. Besides, some blocks with low cooper grade may present high projection values because they are closer to the surface or rich in gold content.

An interesting use of TDP is to split the dataset by using one-dimensional spaces for projections. From expressions (8) and (9), if the target vector u has only one component $u_j \neq 0$ all the projected scalars are going to be either -1, 0 or 1. Consider the previous example, and construct a workspace with coordinates z, cooper and P (projected variable for direction z-maximum, cooper-maximum, gold-maximum). Fig.2 (c,d) shows a point cloud representation corresponding to the block models in Fig. 2(a,b). Fig. 2(c), clearly shows that blocks in Fig. 2(a) were selected according to cooper grade. On the other hand, Fig. 2(d) reveals that threshold filtering in Fig. 2(b) was based on a z-cooper relation.

6 Conclusions and Future Work

Target Data Projection is a flexible, yet easy to implement, tool for visualization systems. It is useful for defining a direction of interest for visual analysis through the specification of a target data vector. The primary application for TDP is to analyze data with respect to vectors composed of maximums and minimums in data ranges of critical variables. This allows application meaningful projections to be generated.

Domains like mine planning, and more recently bioinformatics, demand visualization tools designed with an integrated approach considering both, scientific and information visualization. TDP facilitates this convergence and provides a flexible tool for generating scalars to be visualized with existing techniques. Data table manipulation and workspace definition are key issues in achieving such integration.

Collaborative visualization environments could provide improved support to mine planning activities, from resource analysis to exploitation. SVPM was used in face to face meetings to support discussions, but work should be done to develop a collaborative visualization system for decision making.

References

1. Wong, P.C., Bergeron, R.D.: 30 years of multidimensional multivariate visualization. In: Scientific Visualization - Overviews, Methodologies, and Techniques. IEEE Computer Society Press, Los Alamitos, CA (1997) 3–33
2. Hyvärinen, A.: Survey on independent component analysis. Neural Computing Surveys 2 (1999) 94–128
3. Card, S.K., et al.: Readings in Information Visualization: Using Vision to Think. Morgan Kaufmann Publishers, San Francisco, CA (1999)
4. Ware, C.: Information Visualization: Perception for Design. Morgan Kaufmann Publishers, San Francisco, CA (2000)
5. Ferreira de Oliveira, M.C., Levkowitz, H.: From visual data exploration to visual data minning: A survey. IEEE Transactions on Visualization and Computer Graphics 9 (2003) 378–394
6. Wise, J.A., et al.: Visualizing the non-visual: Spatial analysis and interaction with information from text documents. In: Proceedings of InfoVis'95, IEEE Symposium on Information Visualization. (1995) 51–58
7. Spoerri, A.: Infocrystal: A visual tool for information retrieval. In: Proceedings of IEEE Visualization'93 Conference. (1993) 150–157
8. Wong, P.C., et al.: Data signatures and visualization of scientific data sets. IEEE Computer Graphics and Applications 20 (2000) 12–15
9. Schroeder, W., Mmartin, K., Lorensen, B.: The Visualization Toolkit: An Object Oriented Approach to 3D Graphics. Prentice-Hall, Upper Saddle River,NJ (1998)
10. Turek, C.: Strategic mine planning: Beyond npv - understanding the resource. In: Proceedings of the Whittle North American Strategic Mine Planning Conference. (2000)
11. Chi, E.H., Riedl, J.T.: An operator interaction framework for visualization systems. In: Proceedings of InfoVis'98, IEEE Symposium on Information Visualization. (1998) 63–70

Parametric Freehand Sketches

Ferran Naya[1], Manuel Contero[1], Nuria Aleixos[2], and Joaquim Jorge[3]

[1] DEGI - ETSII, Universidad Politécnica de Valencia, Camino de Vera s/n,
46022 Valencia, Spain
{fernasan, mcontero}@degi.upv.es
[2] Departamento de Tecnología, Universidad Jaume I, Avda. Sos Baynat
12071 Castellón, Spain
naleixos@tec.uji.es
[3] Computer Science Department, IST/UTL, Av. Rovisco Pais
1049-001 Lisboa, Portugal
jorgej@acm.org

Abstract. In this paper we present the 2D parametric freehand sketch component of an experimental prototype called GEGROSS (GEsture & Geometric ReconstructiOn based Sketch System). The module implements a gesture alphabet and a calligraphic interface to manage geometric constraints found in 2D sections, that are later used to perform modeling operations. We use different elements to implement this module. The geometric kernel ACIS[1] stores model data. The constraint manager 2D DCM handles restrictions. Finally, we use the CALI library to define gestural interfaces. In this paper we present a strategy for integrating these tools, and a calligraphic interface we developed to provide dimensional controls over freehand sketches. Our system allows users to build simple sketches composed by line segments and arcs, which are automatically tidied and beautified. Proportional and dimensional information over sketched parts is provided by handwriting their corresponding sizes.

1 Introduction

The WIMP (Windows, Icons, Menus and Pointing) interface paradigm dominates user interfaces in most CAD applications. Recently, work on sketch-based modeling aims at a paradigm shift to change the way geometric modeling applications are built, in order to develop user-centric systems rather than systems that are organized around the details of geometry representation. New hardware devices such as Tablet-PCs open new opportunities to experiment with different user interfaces, based on drawing and sketching using a digitizing tablet and a pen, an approach we have termed *calligraphic interfaces*. These rely on interactive input of vector information (pen-strokes) and gestures instead of menu selection and point input as their main organizing principle.

---

[1] http://www.spatial.com/products/3D/modeling/ACIS.html

Our paper describes work on the 2D parametric freehand sketch module of GEGROSS. This module implements a gesture alphabet and a calligraphic interface to manage the geometric constraints found in 2D sections, that later are used to perform geometric modeling operations.

Our sketching system aims at providing an intelligent modeling tool to support visual thinking in the conceptual design stage of product development, where designers usually employ freehand sketches to record different design alternatives. Integrating both gesture and reconstruction modeling capabilities, we offer a very simple interface to create geometric models, with an interesting possibility: "dimensional control". For users who want to explore design alternatives, or adjust a sketch to satisfy a set of dimensional constraints, we provide parametric capabilities to edit 2D freehand sections. In this way, users can dynamically drag a section edge to modify its geometry, or they can impose dimensional constraints by drawing the corresponding dimension label and writing its value. To this end, handwritten text recognition capabilities provide a very simple method to change dimension values. Moreover, geometric constraint manipulation is possible via the gesture alphabet we developed.

Our approach differs from previous sketch systems in the integration of both gestural and reconstruction methods. These define a new kind of systems we have called "hybrid" modelers that we think provide greater flexibility and modeling capabilities over other techniques.

2 Related Work

Form surveying the literature we identify two main approaches to sketch-based modeling. The first is based on the calligraphic interfaces that use gestures and pen-input as commands [1, 2, 3]. These *gestural modeling* systems rely on gestures as commands for generating solids from 2D sections. One good example is Sketch [4], where the geometric model is entered by a sequence of gestures according to a set of conventions regarding the order in which points and lines are entered as well as their spatial relations. Quick-Sketch [5] is a system oriented to mechanical designs that consists of a 2D drawing environment based on constraints. From these it is possible to generate 3D models through modeling gestures. Another system, Teddy [6] is oriented to free-form surface modeling using a very simple interface of sketched curves, pockets and extrusions. Users draw silhouettes through a series of pen strokes and the system automatically proposes a surface using a polygonal mesh whose projection matches the object contour. In [20] the authors present an improved sketch-based mesh extrusion method. GIDeS [7] allows data input from a single-view projection. In addition the dynamic recognition of modeling gestures provides users with contextual menus and icons to allow modeling using a reduced command set. The ISID [21] system can accept strokes, just like drawing on paper with a pen freely. ISID transforms strokes into edit commands or beautifies them into a complete design schematic. Based on the intelligent recognition technique and integrated modeling templates, ISID tries to simplify and optimize sketching to provide a natural, intensive and tangible interface for conceptual design.

The second approach, which we call *geometric reconstruction*, uses computer vision algorithms to reconstruct geometric objects from sketches that best match a two dimensional projection. The systems we surveyed use two main techniques. The first is based on Huffman-Clowes labeling scheme [8], [9]. The second approach handles reconstruction as an optimization problem [11]. This enables us to obtain what is unrealizable, from the point of view of geometry: a three-dimensional model from a single projection. However, it is well known by psychologists that humans can identify 3D models from 2D images by using a simple set of perceptual heuristics [12]. Authors such as Marill [13], Leclerc and Fischler [14], and Lipson and Shpitalni [15] provide interesting references in the development of this field. The Digital Clay system [16], which supports basic polyhedral objects in combination with a calligraphic interface for data input, uses Huffman-Clowes algorithms to derive three-dimensional geometry. Stilton [17] uses a calligraphic interface directly implemented on a VRML environment. Its reconstruction process uses an optimization technique based on genetic algorithms. Finally, CIGRO [19] which supports drawing constrained polyhedral objects (normalon and quasi-normalon ones), and, in combination with a calligraphic interface for data input, uses an axonometric inflation engine to derive three-dimensional geometry.

3 Overview of Operations

In contrast to surveyed work, the GEGROSS application integrates both gestural and reconstruction based approaches. This extends the capabilities of our previous reconstruction modeling system CIGRO [19], in order to support gestural modeling. At its current level of development, system behavior is explicitly controlled by users, who can alternate between gestural or reconstruction operation modes. A novel contribution of this technique is to provide support for defining freehand parametric sketches. Gestural modeling is organized in two stages. First, we must define a 2D profile. Then, using a combination of modeling gestures, we build new geometry. In this paper we will describe the 2D parametric freehand sketch module. This could be extended to implement a stand-alone 2D parametric drawing application.

The main objective that guided the interface design for this module was to provide a familiar drawing strategy that takes into account the way engineers create sketches and technical drawings. For the first version of our system, we imposed some sketch creation rules, to simplify implementation tasks. This sped up the development of the first application prototype, allowing us to test the application with engineering students and engineering companies' technical staff to get their feedback.

The first condition we imposed to define sketching was to use stylus pressure level as the main criterion to distinguish between auxiliary information, such as center lines, construction lines, dimensions, editing gestures, etc. and geometric primitives, such as line segments, arcs and circles. Drawings made applying high pressure on the stylus signify intended for geometry input, while soft pressure conveys auxiliary information.

The second condition is related to sketch order. Some multi-stroke entities must be drawn sequentially in time. For example, dimension elements must be sketched in a "logic" order: first we draw extension lines, then dimension lines and arrows, and finally dimension text is drawn last.

We designed the user interface to minimize interaction with menus or icons in an attempt to emulate the traditional use of pen and paper by resorting to drawing conventions used in technical drawings.

3.1 Gesture Recognition

The current implementation of GEGROSS processes strokes drawn directly on the screen of a Tablet-PC or LCD tablet, as captured by Microsoft Windows XP Tablet-PC Edition SDK. This API allows retrieving additional information such as the pressure applied on the stylus tip at each point of the stroke. Raw strokes are then processed by an enhanced analyzer based on the CALI library [3]. This library recognizes elemental geometric forms and gestural commands in real time, using fuzzy logic. Recognized gestures are inserted into a list, ordered according to a computed degree of certainty. This list is then returned to the application. CALI recognizes the elementary two-dimensional geometric shapes, such as triangles, rectangles, circles, ellipses, lines, arrows, etc. and some gestural commands, such as delete, move, copy, among others.

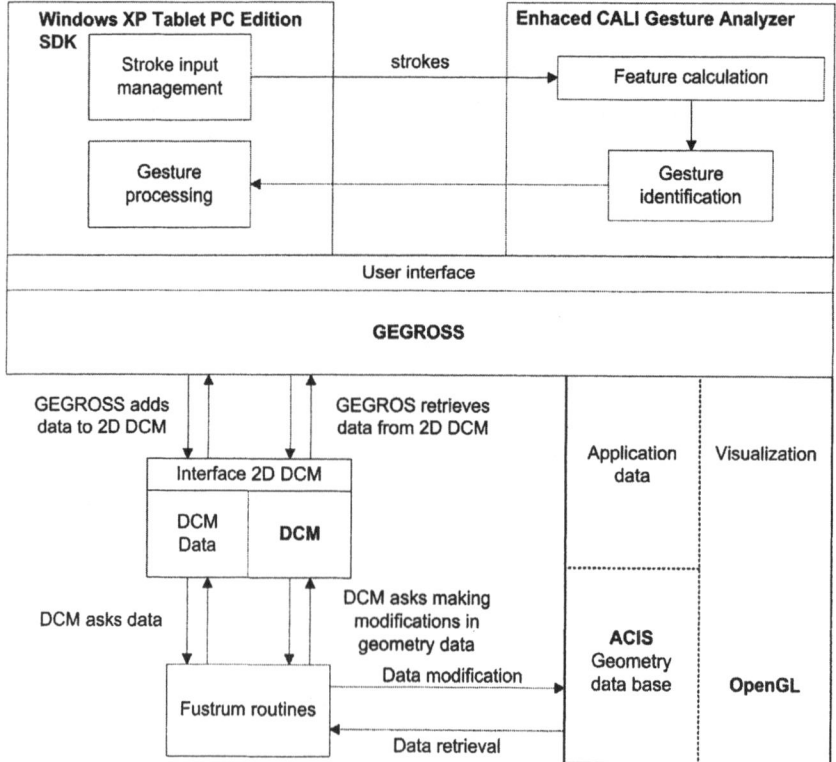

Fig. 1. Structure of the freehand parametric sketch module in GEGROSS.

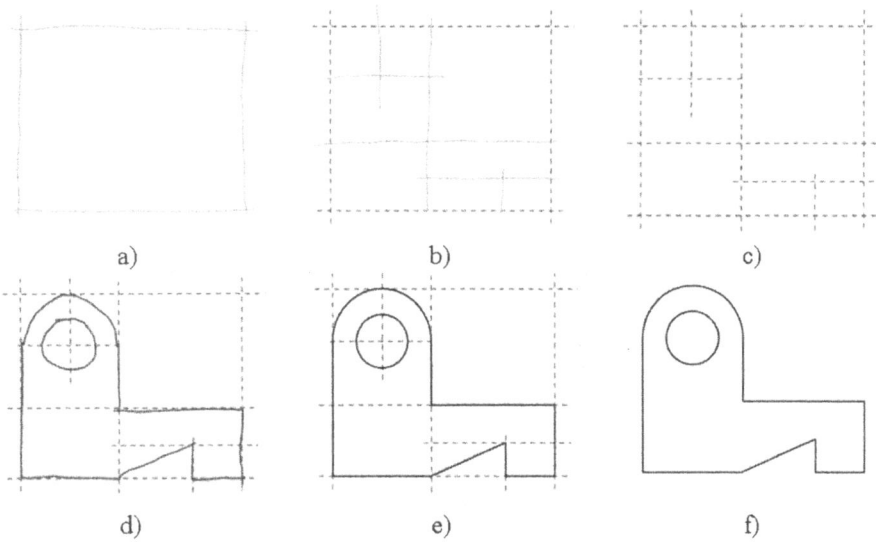

Fig. 2. Sketching sequence in GEGROSS. Grey color represents raw strokes corresponding to auxiliary construction lines. Cyan dashed lines represent adjusted auxiliary lines. Black raw strokes in d correspond to real geometry that is snapped to auxiliary-line skeleton depicted in c.

Currently, GEGROSS only supports strokes which can be recognized as entities of *line, arc or circle* types or as gestural commands of the *delete* or *geometric constraint* classes. Through a user-defined pressure threshold, strokes are classified either as real (geometric) or auxiliary (construction) entities.

3.2 Sketching Procedure

In drafting, most engineers draw a set of auxiliary lines (see Figure 2.a to 2.c) to define main geometric features of a section, using this framework as a drawing template. This sketch is refined by applying pressure with the pencil and drawing over the previous template (see example in Figure 2.d). GEGROSS implements this sketching paradigm by allowing oversketching of real geometry over auxiliary lines, which provide a set of references for snapping purposes.

The user interface supports three visualization modes: rough sketch, beautified sketch and parametric sketch. The user can alternate between views via visualization buttons. Depending on intent, users can choose the adequate visualization mode. If they want only to explore design alternatives, they can do so in rough sketch mode. This is equivalent to drawing a sketch on paper. To work with automatic beautified sketches they can press the corresponding push button. When they want to perform dimensional analysis, they can switch to the parametric sketch mode. In this later mode it is possible to visualize geometric constraints and dimensions found by applying the auto-constraint and auto-dimensioning capabilities of the parametric engine.

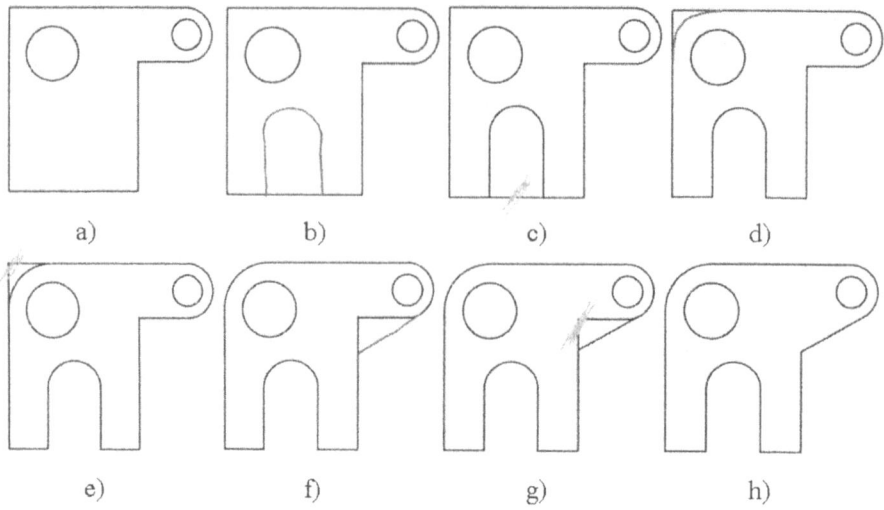

Fig. 3. Refining sketch geometry from simple shapes

Drawing entities can be removed using a scratching gesture. This not only allows errors to be corrected but it also enables more complicated shapes to be drawn by refining "simpler" forms as illustrated in Figure 3. When users draw a scratching gesture, the application identifies the entities that they want to delete as those intersected by the smallest quadrilateral enclosing the gesture.

3.3 Line Drawing Beautification

If users prefer to work with interactively beautified sketches, GEGROSS provides the corresponding visualization mode (see Figure 3). Online beautification provides immediate feedback, since it operates online as a user draws the sketch. This concept of beautification bears some relation to the drawing beautification proposed by Igarashi [6], [18]. Also it exploits the auto-constraining capabilities of the parametric engine in a way transparent to the user. Both parametric and beautified models are continuously updated by the system, regardless of the current visualization mode.

With the help of the 2D DCM parametric engine, from D-cubed [9], we clean up input data in several ways. These include adjusting edges to make sure they meet precisely at common endpoints to get geometrically consistent figures. We also filter all defects and errors of the initial sketches that are inherent to their inaccurate and incomplete nature. In beautification visualization mode, users have immediate feedback (see Figures 2 and 3). To this end we have implemented a number of drawing aids.

Automatic line slope adjustment checks whether newly-created lines are either vertical or horizontal or parallel to other line by considering a slope tolerance.

Vertex point snap looks for vertices close to geometric entities' endpoints, within a vertex proximity tolerance. Should there be several such vertices, we select the one closest to the candidate entity endpoint.

Vertex on entity snap and automatic line breaking applies to endpoints of the new entity which do not lie close to a model vertex. Our system analyzes whether the points are close to an existing edge, taking into account a given edge proximity tolerance. If several edges match this criterion, we select the edge that lies closest to the given endpoint. Then, the selected edge is broken at the contact point; in order to simplify later delete operations, as depicted in figure 3.g.

Snapping and adjustments are performed in real time. Users can see the effect immediately if they are working in beautification or parametric mode. A general tolerance parameter controls beautification actions. This is because some users prefer a less automatic drawing control.

Parallel	Perpendicular
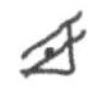	
Vertical	Horizontal
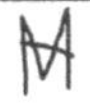	
Tangent	Concentric
Dimension	Symmetric
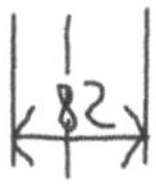	

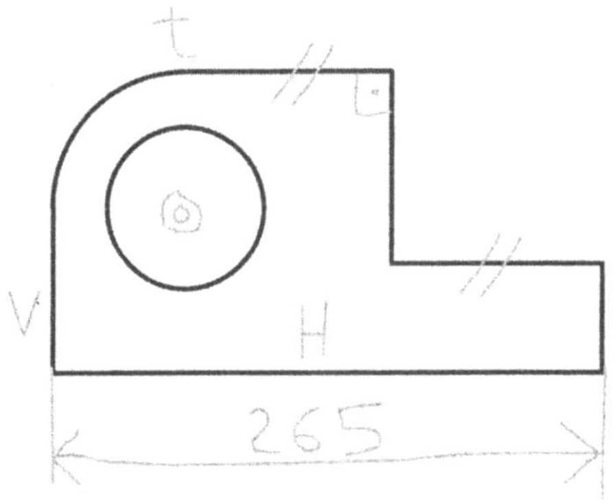

Fig. 4. Constraint alphabet and its use in describing a section.

3.4 Parametric Edition

This optional stage for section definition is intended to provide geometric control over the sketch. For this purpose we adopted the 2D DCM parametric engine. Using the auto-constraint and auto-dimension features provided by the parametric engine we make sure that the section is adequately constrained and dimensioned at any stage of the section creation. Two visualization buttons enable users to visualize constraints and dimensions. Here we distinguish between "automatic" and "user defined" constraints and dimensions. The former are provided by the system and represented in different color. The latter are sketched by user. Geometric constraints are represented according to the symbols presented in Fig. 4. These symbols can be written by the user to impose some desired constraint. Also, they can be deleted using the scratch gesture. User defined dimensions are sketched using a logic order to simplify parsing. Extension lines, dimension line, arrows and dimension value are drawn in sequence using low pencil pressure. A dimension sketched without a value, is considered a measure, which the system provides automatically. If a user wants to change this value, he/she deletes the text and then writes down the new value (see Figure 5 for an example sequence). Handwritten recognition is provided by Windows XP Tablet PC Edition system. In this way, we provide a natural way of specifying the desired dimensions on a sketch. Also, users can drag any geometric entity in a section. The system presents the sketch evolution in real time while preserving geometric constraints.

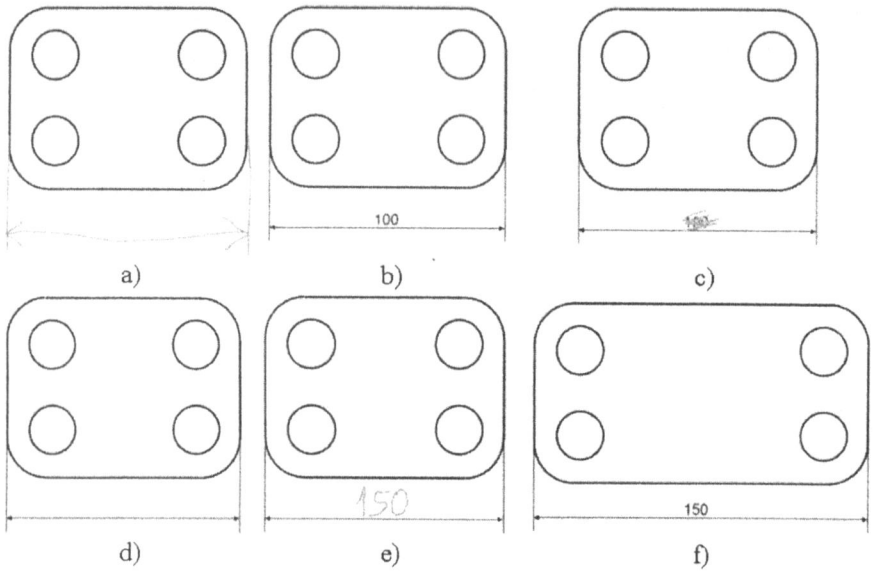

Fig. 5. Example of parametric edition in GEGROSS

4 Conclusions and Future Work

We have described an approach to input two dimensional sections through the gesture modeling capabilities of our GEGROSS application. The main objective of this work has been to provide dimensional control over the sections in a easy and natural way. With this purpose, handwritten dimensions offer a natural method, that is known by any engineer. Preliminary usability tests have shown encouraging results. Users with an engineering background find the system behavior very natural. Also, the learning process to manage the application is both simple and fast, since it matches conventional drafting techniques. In the near future we aim at implementing ISO 10303 capabilities to support parametric sections, as those defined in part 108 "Parameterization and constraints".

Acknowledgments. This work was supported in part by European Commission grant IST-2000-28169 (SmartSketches) and grant CTIDIB/2002/51 from Spanish Generalidad Valenciana.

References

1. Rubine, D.: Combining Gestures and Direct Manipulation. Proceedings ACM CHI'92 Conference Human Factors in Computing Systems (1992) 659–660
2. Long, A.C., Landay, J.A., Rowe, L.A., Michiels, J.: Visual Similarity of Pen Gestures. Proceedings of Human Factors in Computer Systems (SIGCHI), (2000) 360–367
3. Fonseca, M., Jorge, J.: Experimental Evaluation of an On-Line Scribble Recognizer. Pattern Recognition Letters, **22** (12), (2001) 1311–1319
4. Zeleznik, R.C., Herndon, K.P., Hughes, J.F.: SKETCH: An Interface for Sketching 3D Scenes. SIGGRAPH'96 Conference Proceedings (1996) 163–170
5. Eggli, L., Hsu, C., et al.: Inferring 3D Models from Freehand Sketches and Constraints. Computer-Aided Design, **29** (2), (1997) 101–112
6. Igarashi, T., Matsuoka, S., Tanaka, H.: Teddy: A Sketching Interface for 3D Freeform Design. ACM SIGGRAPH99 Conference Proc. (1999) 409–416
7. Pereira, J., Jorge, J., Branco, V., Nunes, F.: Towards Calligraphic Interfaces: Sketching 3D Scenes with Gestures and Context Icons. WSCG'2000 Conf. Proc. Skala V. Ed. (2000)
8. Huffman, D.A.: Impossible Objects as Nonsense Sentences. In: Meltzer B., Michie D. (eds.) Machine Intelligence, No. 6, Edinburgh UK. Edinburgh University Press (1971) 295–323
9. D-Cubed, Cambridge, UK, http://www.d-cubed.co.uk/
10. Clowes, M.B.: On Seeing Things. Artificial Intelligence, **2**, (1971) 79–116
11. Wang, W., Grinstein, G.: A Survey of 3D Solid Reconstruction from 2D Projection Line Drawing. Computer Graphics Forum, **12**(2), (1993) 137–158
12. Hoffman, D.: How We Create What We See. Visual Intelligence, Norton Pub., **2**, (2000)
13. Marill, T.: Emulating the Human Interpretation of Line-Drawings as Three-Dimensional Objects. International Journal of Computer Vision, **6**(2), (1991) 147–161
14. Leclerc, Y., Fischler, M.: An Optimization-Based Approach to the Interpretation of Single Line Drawing as 3D Wire Frames. Int. Journal of Computer Vision, **9**(2), (1992) 113–136
15. Lipson, H., Shpitalni, M.: Optimization-Based Reconstruction of a 3D Object from a Single Freehand Line Drawing. Computer Aided Design, **28**(8), (1996) 651–663
16. Schweikardt, E., Gross, M.D.: Digital Clay: deriving digital models from freehand sketches. Automation in Construction, **9**, (2000) 107–115
17. Turner, A., Chapmann, D., and Penn, A.: Sketching space. Computers & Graphics, **24**, (2000) 869–879
18. Igarashi, T., Matsuoka, S., Kawachiya, S., Tanaka, H.: Interactive Beautification: A Technique for Rapid Geometric Design, UIST'97, (1997) 105–114
19. Contero, M., Naya, F., Jorge, J. and Conesa, J.: CIGRO: A Minimal Instruction Set Calligraphic Interface for Sketch-Based Modeling. Lecture Notes in Computer Science, **2669**, (2003) 549–558
20. Wang, C.C.L. and Yuen, M.M.F.: Freeform Extrusion by Sketched Input. Computers & Graphics **27**, (2003) 255–263
21. Baohua, S., Mingxi, T., Frazer, J. H. and Haicheng, Y.: Stroke-based Intelligent Sketching Interface". Proceeding of the 5th Asia Pacific Conference on Computer Human Interaction APCHI2002, (2002) 500–509

Variable Level of Detail Strips

J.F. Ramos and M. Chover

Departamento de Lenguajes y Sistemas Informaticos
Universistat Jaume I, Campus de Riu Sec, 12071, Castellon, Spain
{Jromero,chover}@uji.es

Abstract. Some multiresolution models offer variable resolution capabilities. This property allows us to display areas with different levels of detail on the same mesh. In this paper we present a continuous multiresolution model that allows variable resolution. An important characteristic of the model is the use of triangle strips both in the data structure and in the rendering stage. The proposed model extends a previous strip-based multiresolution scheme that implements uniform resolution. We present new data structures, algorithms and some experiments that demonstrate the model efficiency.

1 Introduction

The multiresolution modelling techniques [5][9] allow us to represent an object with different levels of detail (LOD). These multiresolution models establish the data structures necessary to store the information, and the algorithms to recover it in the most efficient form. The model design is related with their efficiency and will determine its applicability.

The initial intention of these modelling techniques was to accelerate the visualization process. Nevertheless, as a result of the continuous multiresolution models evolution, new applications were found like progressive transmission, geometric compression, multiresolution edition, etc. A characterization of multiresolution modelling techniques and their applications can be found in the paper by Ribelles et al. [12].

Some of the developed models allow variable resolution whereas others uniform resolution. If the model implements variable resolution it allows to extract different LODs on the mesh that represents an object. Recently have appeared multiresolution models that use primitives with implicit connectivity as triangle strips [2][13][19] or triangles fans [11]. The use of theses primitives allows the models to decrease their storage cost and accelerate their rendering time.

In this article we present an extension of the model presented by Ramos et al.[19] that supports variable resolution. Section 2 begins with an overview of multiresolution models that exploit connectivity primitives. In section 3, a review of the strip-based model with uniform resolution is explained. Later, in section 4 we present the variable resolution model with internal details, data structures and algorithms. In section 5 the results obtained from this model are explained. Finally, in section 6, conclusions and future work are presented.

2 Related Work

In this section we present an overview of the continuous multiresolution models that use primitives with implicit connectivity. It has been distinguished between the models that allow uniform resolution of those with variable resolution.

2.1 Uniform Resolution Models

Ribelles et al. introduced MOM-Fan[11], the first model that uses graphics primitives with implicit connectivity. This model uses the triangle fan primitive both in the data structure and in the rendering stage. The main drawback of this model is the high number of degenerated triangles used in the representation, although they are purged out before the rendering stage. Another drawback to the model is that the average number of triangles in each triangle fan is small.

Following on from this approach, MTS by Belmonte et al. appeared. This is a model that uses the strip primitive in the storage and in the rendering stage [1]. The model is made up of a collection of multiresolution strips. Each multiresolution strip represents a triangle strip at every LOD, and this is coded as a graph. Only the strips that are modified between two consecutive LOD extractions are updated before rendering.

Another approach to the use of triangle strips in a multiresolution model is the work carried out by A. James Stewart [15], and extended by Porcu [8]. This work uses a tunneling algorithm to connect isolated triangle strips, thus obtaining triangle strips with high numbers of triangles while reducing the number of triangle strips in the model as it is simplified. Again, its main drawback is the time consumed by the stripification algorithm.

The last model that uses triangle strips was developed by Ramos et al. [19]. This model has been extended in this paper in order to incorporate a variable resolution capability.

2.2 Variable Resolution Models

One of the first models to use triangle strips is VDPM by Hoppe [7]. After calculating the set of triangles to be rendered, this model performs an on-the-fly determination of the strips to be rendered. This is a time-consuming task but the final rendering time is reduced because triangle strips are faster than triangles.

Later, El-Sana et al. introduces the Skip-Strips model [2]. This is the first model to maintain a data structure to store strips, thus avoiding the need to calculate them on-the-fly.

Recently, some works based on the triangle strip primitive have been presented. These focus on the dynamic simplification of the triangle strips for each demanded LOD. The model by Shafae et al. called DStrips [13][14] manages the triangle strips in such a way that only those triangle strips that are being modified are processed, while the rest of the triangle strips in the model remain unmodified. This updating mechanism reduces the extraction time. However, results published from this work still show a high extraction time.

3 Review of the Strip-Based Model with Uniform Resolution

Recently, Ramos et al. [19] have presented a multiresolution model that uses triangle strips as primitive. This model was developed in order to manage uniform resolution meshes.

The model represents a mesh as a set of multiresolution strips. We denote a triangle strip mesh M as a tuple (V;S), where V is a set of vertices v_i with positions $v_i \in R^3$, and S is a collection of sub-triangulations $s^1,\ldots,s^m$, so each $s^i \in S$ is an ordered vertex sequence (1) also called a strip.

$$\{s_1^i \cdots s_q^i\} \qquad S = \begin{Bmatrix} s_1^1 \cdots s_k^1 \\ \vdots \\ s_1^m \cdots s_r^m \end{Bmatrix} \qquad V = \{v_1 \ldots v_n\} \tag{1}$$

Each row inside the S matrix represents a triangle strip. After some modifications, this matrix will be adapted to become a multiresolution triangle strip data structure in order to be used in our model. In this way, this data structure will change during level of detail transitions.

The model has been built in order to optimize data access as well as the vertices sent to the graphics pipeline. In this way, it manages the triangle strips both in the data structure and in the rendering stage.

A set of vertices with their 3D coordinates and a set of multiresolution strips are needed to support the multiresolution model. What is more, an auxiliary structure is used to improve level of detail transitions.

4 Variable Resolution in LodStrips

The meaning of variable resolution LOD consist of displaying different LODs on the same object. In Figure 1, a sphere interactively located by a user selects the area to be represented in high detail. This area is drawn at the highest resolution LOD, the rest of the object is drawn with the coarsest resolution possible while maintaining the connectivity of the mesh. Thus, in the same object, two sets of faces can be observed: a set of faces with the lowest resolution LOD and another with the highest.

To retrieve a variable resolution LOD a criterion or set of criteria is needed. The criterion is used to decide which part of the object is simplified and which is refined. This decision could be decided by the user interactively, indicating what regions should be displayed with the higher or lower resolution or by the application. Several authors [2][7][20] have defined criteria to represent areas in high detail, which are:

- View frustum, increasing detail in the regions inside the view frustum.
- Silhouette boundaries, increasing detail in the regions where there are edges for which one of the adjacent faces is visible and the other is invisible.
- Orientation surface, increasing detail to the regions oriented near the viewer.

- Screen-space projections, increasing or decreasing detail in the region depending on the length of its screen-space projection.
- Local illumination, increasing detail in a direction perpendicular to, and proportional to, the illumination gradient across the surface.

These criteria can easily be added to our model because our algorithm is independent of the criteria used.

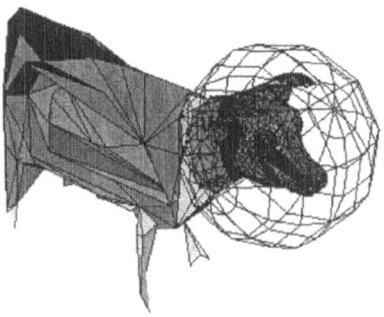

Fig. 1. Variable resolution cow with the strip-based model

4.1 Data Structures

Three data structures are needed to support variable resolution capabilities: lVerts, lStrips and lChanges.

We denote a lVerts structure as a set V which contains an ordered vertex sequence and where each $v_i \in V$ consists of four items (2). The first three items are vertex positions: $(x_i, y_i, z_i) \in R^3$ and the last one, v_{k_i}, is the vertex which v_i collapses in the lowest LOD.

$$V = \{v_1, ..., v_n\} \qquad v_i = \left(x_i, y_i, z_i, v_{k_i}\right) \tag{2}$$

Thus, the lVerts data structure stores 3D coordinates and information about vertex simplification of each vertex in the mesh.

Transitions between levels of detail involve vertex collapses. This information is stored in the lVerts data structure and when a vertex v_i has to be collapsed, it is replaced by v_{k_i} in every strip where it appears.

The multiresolution strip set is stored by the lStrips data structure. It consists of a collection L (3), where each $L^i \in L$ is an ordered vertex sequence, which denotes a multiresolution strip.

$$L = \begin{Bmatrix} v_1^1 & \cdots & v_r^1 \\ \vdots & \vdots & \vdots \\ v_1^m & \cdots & v_t^m \end{Bmatrix}$$

(3)

Each row $L^i \in L$, or each strip in the L collection, changes dynamically with vertex collapses and with strip resizing.

Vertex collapses are performed by replacing vertices in the data structure L by others that simplify them.

The model also incorporates a data structure, lChanges, which allows us to quickly recover the positions of the vertices that are changed in each level of detail transition.

We denote a lChanges data structure as a level of detail ordered set, C (4), where each tuple c_j^i represents a position in L that determines where is the vertex to collapse.

$$C = \begin{Bmatrix} c_1^1 & \cdots & c_s^1 \\ \vdots & \vdots & \vdots \\ c_1^m & \cdots & c_t^m \end{Bmatrix}$$

(4)

This data structure increases model performance because it allows us to quickly apply level of detail changes between transitions. Without this data structure it would be very expensive to apply these changes.

4.2 Algorithms

Multiresolution models need algorithms to be able to support multiresolution capabilities. This model and most multiresolution models have two main algorithms to do these tasks, i.e. a level of detail recovery algorithm and a drawing algorithm. We assume the rendering stage to be a stage that contains these two algorithms, which are applied in a sequential order, first extraction and then drawing.

The level of detail recovery algorithm goes into action when a level of detail change is induced by the application or by the user. Then, data structure C is traversed from $C^{currentLOD}$ to C^{newLOD}, applying changes stored in each tuple Cij ∈ C, where i is in the interval [currentLOD , newLOD], although taking into account that, for each change, a test function is evaluated to decide whether the criterion is achieved. Depending on the result, the vertex included inside the change is simplified or not, obtaining a mesh with different lods.

It is important to notice that, depending on whether the level of detail is bigger or smaller than the current one, splits or collapses will be applied to the lStrips data structure, although the information stored in c_j^i referring to collapses is also used to perform splits. The associated pseudo-code is shown below.

```
for lod=currentLOD to newLOD
   if newLOD>currentLOD   //To a coarser mesh
     for change=lChanges.Begin(lod) to lChanges.End(lod)
       if (test(change)=TRUE) then
         if (change.isCollapse()) then
           lStrips.Collapse(lod,change);
         end if
       end if
     end for
   else  //To a more detailed mesh
     for change=lChanges.Begin(lod) to lChanges.End(lod)
       if (test(change)=TRUE)
         if (change.isSplit()) then
           lStrips.Split(lod,change);
         end if
       end if
     end for
   end if
end for
```

Fig. 2. Level of detail recovery algorithm

After the level of detail recovery algorithm has processed multiresolution strips and the criteria have been applied to them, the drawing algorithm takes over, traversing each strip to obtain their vertices in order to send them to the graphics system. This algorithm is easily implemented, as shown in Figure 3.

```
for strip=lStrips.Begin() to lStrips.End()
   for vertex=strip.Begin() to strip.End()
     Send(vertex);
   end for
end for
```

Fig. 3. Simple drawing algorithm

5 Results

This model has passed two types of tests: plain test and sphere test. Plain test consist of a plain that transverses the object from an extreme to the other in a linear way, thus in front of the plain and behind it the resolution is the highest and the lowest respectively. Sphere test consist of a sphere with origin in a point of the 3D space, then its radius grows in a linear way changing the resolution of the object inside it to the highest, outside the object has the lowest resolution possible.

To carry out the tests, two meshes from the Stanford 3D Scanning Repository were taken as a reference, so as to make it easy to compare this model with other well-developed models. Tests were carried out with a PC with an Intel Xeon 2.4 Ghz processor and 512 Mb of main memory, using an ATI Fire GL E1 64 Mb graphics card.

The aim of the experiments is to observe and compare the visualization times of the proposed algorithm when the area in high detail varies in size. Results from cow and horse objects are shown. The characteristics of these models are shown in Table 1.

Table 1. Characteristics and storage cost

	Cow	Horse
Vertices	2904	48485
Strips	136	1964
LODs	2758	46061
Size	0.45 Mb	5.70 Mb

Figure 4 and 5 show the performance of the algorithm as the plain test is working. Axis X represents the number of vertices sent to the graphics system whereas axis Y represents the time in milliseconds. In the beginning, the objects are shown in the lowest LOD possible.

Let be the rendering time the sum of the LOD extraction time and the drawing time, then it is possible to observe how the extraction is almost constant and, logically, the drawing time grows proportionally when the area with higher LOD is bigger or, what is the same, when more vertices are sent to the graphics system.

As a result of these experiments, we conclude that the algorithm increases its performance when the area in high detail is small compared with the size of the model. The smaller the size of the area, the better the performance of the algorithm and vice versa, the performance decreases as the size of the area increases.

6 Conclusions and Future Work

A strip-based model to retrieve variable resolution LODs has been presented. Based on [19], this model has been modified to support variable resolution capabilities, although new data structures has no needed. Thus, it encodes enough information to correctly retrieve any variable resolution LOD.

The experimental results show an improvement in performance when the area in high detail is small compared with the whole surface, which is logic because the vertices sent to the graphics system decrease in the same way as the extension of the area with the highest LOD.

Further improvement can be achieved by using data structures more efficient to retrieve variable level of detail, moreover by using acceleration techniques we can diminish the time required to render.

Acknowledgments. This work was supported by the Spanish Ministry of Science and Technology grants TIC2001-2416-C03-02 and TIC2002-04166-C03-02, and FEDER funds.

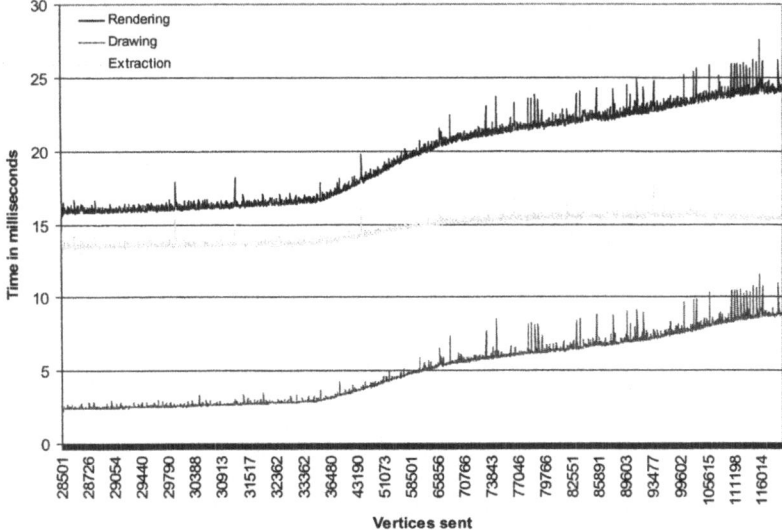

Fig. 4. Plain test from the horse object

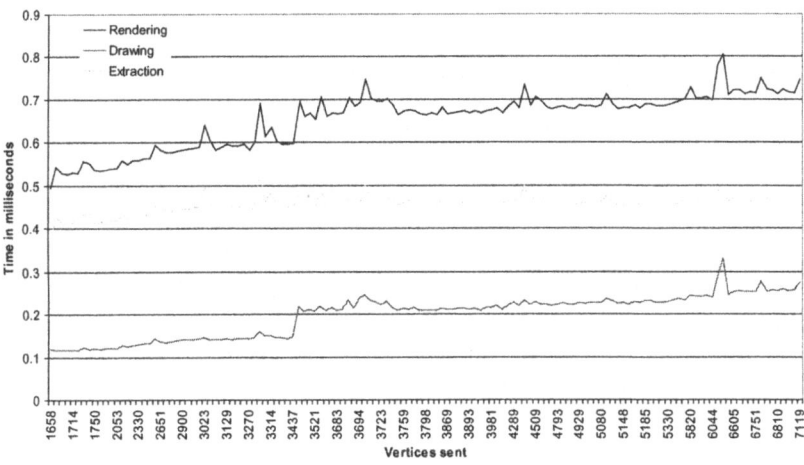

Fig. 5. Plain test from the cow object

References

1. O. Belmonte, I. Remolar, J. Ribelles, M. Chover, M. Fernández. Efficient Use Connectivity Information between Triangles in a Mesh for Real-Time Rendering, Future Generation Computer Systems, Special issue on Computer Graphics and Geometric Modeling, 2003. ISSN 0167-739X.

2. El-Sana J, Azanli E, Varshney A. Skip strips: maintaining triangle strips for view-dependent rendering. In: Proceedings of Visualization 99, 1999. p.131-7.
3. F. Evans, S. Skiena and A. Varshney, Optimising Triangle Strips for Fast Rendering, IEEE Visualization '96, 319-326, 1996. http://www.cs.sunysb.edu/~stripe
4. M. Garland, and P. Heckbert, Surface Simplification Using Quadratic Error Metrics, Proceeding of SIGGRAPH'97, 209-216, 1997.
5. M. Garland, Multiresolution Modelling: Survey & Future Opportunities. State of the Art Reports of EUROGRAPHICS'99, 111-131, 1999.
6. Hoppe H. Progressive Meshes. Computer Graphics (SIGGRAPH), 30:99-108, 1996.
7. Hoppe H. View-dependent refinement of progressive meshes. In: Procedeeding of SIGGRAPH 97, 1997. P.189-98.
8. Massimiliano B. Porcu, Riccardo Scateni. An Iterative Stripification Algorithm Based on Dual Graph Operations. EUROGRAPHICS 03.
9. Puppo E, Scopigno R. Simplification, LOD and multiresolution-principles and applications. Tutorial notes of EUROGRAPHICS 99, vol. 16, no. 3, 1997.
10. J. Ribelles, M. Chover, A. Lopez and J. Huerta. A First Step to Evaluate and Compare Multirresolution Models, Short Papers and Demos EUROGRAPHICS'99, 230-232, 1999.
11. J. Ribelles, A. López, I. Remolar, O. Belmonte, M. Chover. Multiresolution Modelling of Polygonal Surface Meshes Using Triangle Fans. Proc.of 9th DGCI 2000, 431-442, 2000. ISBN 3-540-41396-0.
12. J. Ribelles, A. López, Ó. Belmonte, I. Remolar, M. Chover, Multiresolution modeling of arbitrary polygonal surfaces: a characterization, Computers & Graphics, ISBN/ISSN 0097-8493, vol. 26, num. 3, pp. 449-462, 2002.
13. Michael Shafae, Renato Pajarola. DStrips: Dynamic Triangle Strips for Real-Time Mesh Simplification and Rendering. Proceedings Pacific Graphics Conference, pp. -, 2003.
14. Michael Shafae, Renato Pajarola. DStrips: Dynamic Triangle Strips for Real-Time Mesh Simplification and Rendering. Slides and Videos, Eurographics 2003.
15. A. James Stewart: Tunneling for Triangle Strips in Continuous Level--of--Detail Meshes. Graphics Interface 2001: 91-100.
16. L. Velho, L.H. de Figueiredo, and J. Gomes. Hierarchical Generalized Triangle Strips. The Visual Computer, 15(1):21-35, 1999.
17. A. Bogomjakov, C. Gostman. Universal Rendering Sequences for Transparent Vertex Caching of Progressive Meshes. Proceedings of Graphics Interface 2001.
18. Leif P. Kobbelt, Thilo Bareuther. Hans-Peter Seidel. Multiresolution Shape Deformations for Meshes with Dynamic Vertex Connectivity. Computer Graphics Forum
19. Ramos, J.F., Chover M. An approach to improve strip-based multiresolution schemes, in proceeding of WSCG 2004, Vol. 12, N°2, P. 349-354. ISSN 1213-6972.
20. Luebke, D. and Erikson, C.: View-Dependent Simplification of Arbitrary Polygonal Environments, Proc. of SIGGRAPH'97, pp. 199-208, 1997.
21. Puppo, E.: Variable resolution terrain surfaces. Proc. of Eight Canadian Conference on Computational Geometry, pp 202-210, 1996
22. Puppo, E., Scopigno R.: Variable resolution triangulations, Computational Geometry, Vol.11(3-4), pp. 219-238, 1998.
23. Ramos, J. F., Chover, M. LodStrips: Level of detail strips. Submitted and accepted in CGGM 2004..

Bézier Solutions of the Wave Equation

J.V. Beltran and J. Monterde

Dep. de Geometria i Topologia, Universitat de València,
Burjassot (València), Spain. {beltranv,monterde}@uv.es,

Abstract. We study polynomial solutions in the Bézier form of the wave equation in dimensions one and two. We explicitly determine which control points of the Bézier solution at two different times fix the solution.

1 Introduction

The wave equation describes a lot of vibrating processes in mathematical physics, for example, a vibrating string (1 dimensional) or membrane (2 dimensional). It is one of the simplest hyperbolic partial differential equation.

There are some papers dealing with Bézier or B-splines techniques applied to waves ([3], [2]), but, to our knowledge, there is not a study of the polynomial solutions in Bézier form of the wave equation.

The usual approach to the search of solutions to the wave equation is to state a Cauchy problem of the kind: To find a function verifying the wave equation with initial conditions the value of the function at time $t = 0$ and the value at time $t = 0$ of its partial derivative with respect to the time.

Along this paper the approach will be different. We shall try to find solutions of the wave equation with the value of the function at time $t = 0$ as an initial condition, and with the value of the function in another time $t = t_0$ as a final condition.

For the 1D-wave equation this can be done using Bézier curves. Given an initial Bézier curve at time $t = 0$ and another Bézier curve at another time, for the sake of simplicity, $t = 1$, there is a unique Bézier solution of the 1D-wave solution verifying such initial and final conditions. Moreover, the control points of the Bézier solution are computed as the solution of a linear system from the control points of the initial and final Bézier curves.

This result can be useful to reproduce the motion of a string which evolves according to the wave equation. If snapshots of the string are taken at different times, then it is possible to interpolate the motion of the string by approximating first the string by a Bézier curve and finding then the Bézier solutions of the 1D-wave equation according to our results.

For the 2D-wave equation the results are similar. Given an initial Bézier surface at time $t = 0$ and another Bézier surface at time $t = 1$, there is a unique Bézier solution of the 2D-wave solution verifying such initial and final conditions. The only difference is that the initial and final Bézier surfaces must verify some conditions. Such conditions can be expressed in terms of their control points and then some of the control points of the initial Bézier surface can be determined from the other control points. The same for the final Bézier surface.

Usually, the wave equation is stated on functions. Nevertheless, here we are considering parametrized surfaces whose coordinate functions verify the wave equation. Due to the fact that we shall follow a Bézier approach, the amount of work needed is the same if we deal with parametrized surfaces or with functions.

Anyway, notice that the wave operator is not intrinsic to the surface in the same way that the usual Laplace operator is not intrinsic. In that case, the intrinsic operator is the Laplace-Beltrami operator. In spite of this, we shall study the non intrinsic wave operator as a first step towards the study of the corresponding intrinsic operator.

Moreover, if the first two coordinates are chosen appropriately, this is $P_{ijk} = (\frac{i}{n}, \frac{j}{n}, c_{ijk})$ for the 2D-wave equation, then the third coordinate function of the Bézier surface can be understood as a function depending on the first two variables, and then, the wave equation has full sense.

One of the advantages of this approach is that we only need as data the initial and final state of the string (1D) or membrane (2D) in order to simulate the vibration. Moreover, the initial an final position are described using Bézier curves or surfaces, and therefore we only need a few control points to fully determine the model. The degree of the Bézier curve or surface will be determined by the accuracy that we want to obtain.

Finally, let us remark that our intention in a future work is to compare our method with other methods simulating waves and/or with true waves.

2 Bézier Solutions of the 1D-Wave Equation

If a string of uniform linear density is stretched to a uniform tension and if, in the equilibrium position, the string coincides with the x-axis, then when the string is disturbed slightly from its equilibrium position, the transverse displacement $u(x,t)$ satisfies an equation in which the second partial derivative of u with respect to x is proportional to the second partial derivative with respect to t,

$$\frac{\partial^2 u}{\partial x^2} = c^2 \frac{\partial^2 u}{\partial t^2}. \qquad (1)$$

This is known as the one-dimensional wave equation. The usual approach is to state a Cauchy problem of the kind: To find a function $u(x,t)$ verifying Eq. 1 with initial conditions $u(x,0) = u_0(x)$ and $\frac{\partial u}{\partial t}(x,0) = v_0(x)$.

Along this section we shall take $c = 1$ as the constant of proportionality. Then, the 1D-wave equation can be written as

$$\frac{\partial^2 u}{\partial x^2} - \frac{\partial^2 u}{\partial t^2} = 0. \qquad (2)$$

The only difference with the harmonic equation is a sign. So we can borrow the same study we did for the harmonic Bézier surfaces in [1] and [5]. Moreover, we shall look for solutions of Eq. 2 where not only for each t_0 the string $u(x,t_0)$ is a Bézier curve, but even the evolution along time of the string is of Bézier kind:

$$u(x,t) = \sum_{i,j=0}^{m,n} B_i^m(x) B_j^n(t) P_{ij}, \quad x,t \in [0,1].$$

2.1 The 1D-Wave Equation in Terms of the Control Points

A simple change of sign transforms the harmonic conditions (see [1], [4]) into the 1D-wave conditions.

Theorem 1. *Given a control net in $\mathbb{R}^3$, $\{P_{ij}\}_{i,j=0}^{m,n}$, the associated Bézier surface, $u : [0,1] \times [0,1] \to \mathbb{R}^3$, is a solution of the 1D-wave equation if and only if for any $i \in \{1, \ldots, m\}$ and $j \in \{1, \ldots, n\}$*

$$P_{i+2,j}a_{im} + P_{i+1,j}(b_{i-1,m} - 2a_{im}) + P_{i-1,j}(b_{i-1,m} - 2c_{i-2,m}) + P_{i-2,j}c_{i-2,m}$$
$$-P_{i,j+2}a_{jn} - P_{i,j+1}(b_{j-1,n} - 2a_{jn}) - P_{i,j-1}(b_{j-1,n} - 2c_{j-2,n}) - P_{i,j-2}c_{j-2,n}$$
$$+P_{ij}\left((a_{im} - 2b_{i-1,m} + c_{i-2,m}) - (a_{jn} - 2b_{j-1,n} + c_{j-2,n})\right) = 0,$$

where, for $i \in \{0, \ldots, m-2\}$

$$a_{in} = (n-i)(n-i-1), \quad b_{in} = 2(i+1)(n-i-1), \quad c_{in} = (i+1)(i+2),$$

and $a_{in} = b_{in} = c_{in} = 0$ otherwise.

Note that the equations in the statement of the Theorem can be understood as generated by an adaptable mask.

The results obtained for harmonic Bézier surfaces indicate that given the first and last rows of control points, all the other control points are determined. The same is true for Bézier solutions of the 1D-wave equation.

For example, for $m = n = 3$, given $P_{00}, P_{10}, P_{20}, P_{30}$ and $P_{03}, P_{13}, P_{23}, P_{33}$, the other control points are determined by

$$P_{01} = \tfrac{1}{3}(4P_{10} + 2P_{13} - 2P_{20} - P_{23}),$$
$$P_{31} = \tfrac{1}{3}(-2P_{10} - P_{13} + 4P_{20} + 2P_{23}),$$
$$P_{02} = \tfrac{1}{3}(2P_{10} + 4P_{13} - P_{20} - 2P_{23}),$$
$$P_{32} = \tfrac{1}{3}(-P_{10} - 2P_{13} + 2P_{20} + 4P_{23}),$$
$$P_{11} = \tfrac{1}{9}(-4P_{00} - 2P_{03} + 12P_{10} + 6P_{13} - 2P_{30} - P_{33}),$$
$$P_{12} = \tfrac{1}{9}(-2P_{00} - 4P_{03} + 6P_{10} + 12P_{13} - P_{30} - 2P_{33}),$$
$$P_{21} = \tfrac{1}{9}(-2P_{00} - P_{03} + 12P_{20} + 6P_{23} - 4P_{30} - 2P_{33}),$$
$$P_{22} = \tfrac{1}{9}(-P_{00} - 2P_{03} + 6P_{20} + 12P_{23} - 2P_{30} - 4P_{33}).$$

Let us recall that $\{P_{i0}\}_{i=0}^{m}$ are the control points of the Bézier string at time $t = 0$ and $\{P_{in}\}_{i=0}^{m}$ are the control points of the Bézier string at time $t = 1$. This gives us a non usual approach to the wave equation. Instead of fixing the initial state of the string and its first partial derivative with respect to the time we are fixing the initial and final states of the string.

The next figure shows the solution of the 1D-wave equation as a two-dimensional surface.

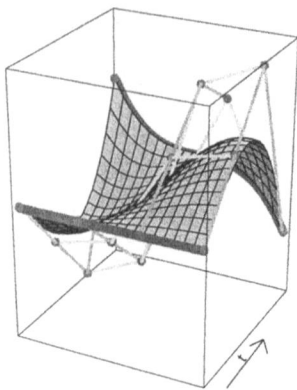

Fig. 1. A solution of the 1D-wave equation as a two-dimensional surface. The initial and final states are plotted thicker.

2.2 Computing the Solutions of the 1D-Wave Conditions

We have seen that in low dimensions, the 1D-wave conditions imply that some of the control points can be expressed as linear combinations of the other control points. This is true for any dimension, i.e., the initial and final states fully determine the Bézier solution of the 1D-wave equation. We prove that for a square net for simplicity.

Proposition 1. *Let $u(x,t) = \sum_{i,j=0}^{n} B_i^n(x) B_j^n(t) P_{ij}$ be a Bézier solution of degree $n \geq 2$ of the 1D-wave equation with control net $\{P_{ij}\}_{i,j=0}^{n}$, then*

1. *If n is odd, control points in the inner rows $\{P_{ij}\}_{i=1,j=0}^{n-1,n}$ are determined by the control points in the first and last rows, $\{P_{0j}\}_{j=0}^{n}$ and $\{P_{nj}\}_{j=0}^{n}$.*

2. *If n is even, control points in the inner rows $\{P_{ij}\}_{i=1,j=0}^{n-1,n}$ and also the corner control point P_{nn} are determined by the control points in the first and last rows, $\{P_{0j}\}_{j=0}^{n}$ and $\{P_{nj}\}_{j=0}^{n-1}$.*

The analogous statement and proof for the harmonic case can be seen in [5]. We shall include here the proof for the 1D-wave case for the sake of completeness and because it gives some clues for what we will do later in the 2D case.

Proof: Let us consider the case n odd. Let us write the Bézier chart in the usual basis of polynomials

$$u(x,t) = \sum_{i,j=0}^{n} \frac{A_{ij}}{i!j!} x^i t^j,$$

where $A_{ij} \in \mathbb{R}^3$.

The 1D-wave equation $(\frac{\partial^2}{\partial x^2} - \frac{\partial^2}{\partial t^2})u = 0$ can be translated into a system of linear equations in terms of the coefficients $\{A_{ij}\}_{i,j=0}^n$

$$A_{i+2,j} - A_{i,j+2} = 0, \quad i,j = 0, \ldots, n,$$

but with the convention $A_{n+1,j} = A_{n+2,j} = A_{i,n+2} = A_{i,n+1} = 0$.

This means that any coefficient A_{ij} with $i > 1$ can be related with $A_{i+2,j-2}$ and so on until the second subindex is 0 or 1, or until the first subindex is greater than n. In this second case, A_{ij} is directly 0. Indeed, if $i + 2j > n$ then $A_{i,2j} = A_{i,2j+1} = 0$, otherwise

$$A_{i,2j} = A_{i+2j,0}, \quad A_{i,2j+1} = A_{i+2j,1}. \tag{3}$$

Note that the first and last rows of control points determine the starting and final curves $u(x,0), u(x,1)$, $x \in [0,1]$. The first border curve is

$$u(x,0) = \sum_{i=0}^n \frac{A_{i0}}{i!} x^i, \tag{4}$$

and the second one is

$$u(x,1) = \sum_{i=0}^n \left(\sum_{j=0}^n \frac{A_{ij}}{i!j!} \right) x^i. \tag{5}$$

From Eq. 4 we can obtain coefficients A_{i0} for $i = 0, \ldots, n$. Having in mind the expression of $u(x,t)$ in terms of Bernstein polynomials the formula of the change of basis from the basis of Bernstein polynomials to the usual basis says us that

$$\frac{A_{i0}}{i!} = \binom{n}{i} \Delta^{i0} P_{00}, \quad i = 0, 1, \ldots, n,$$

where Δ^{i0} is the incremental operator.

Thanks to Eq. 3, we can reduce Eq. 5

$$\sum_{j=0}^n \frac{A_{ij}}{i!j!} = \binom{n}{i} \Delta^{i0} P_{0n}, \quad i = 0, 1, \ldots, n,$$

to just a system of linear equations involving the coefficients A_{i1}.

Moreover, the matrix of coefficients of this system is triangular and with the unit in the diagonal entries. Therefore, the knowledge of the first and last rows of control points, implies the knowledge of the coefficients A_{i0} and A_{i1} and then, the knowledge of all the coefficients, i.e., of the whole solution $u(x,t)$, or equivalently, of the whole control net.

For the even case, the arguments are similar. □

3 Bézier Solutions of the 2D-Wave Equation

If a thin elastic membrane of uniform areal density is stretched to a uniform tension and if, in the equilibrium position, the membrane coincides with the xy-plane, then the small transverse vibration $u(x, y, t)$ of the point (x, y) of the membrane satisfies an equation in which the two dimensional Laplacian of u is proportional to the second partial derivative with respect to t,

$$(\frac{\partial^2}{\partial x^2} + \frac{\partial^2}{\partial y^2})u = \Delta u = c^2 \frac{\partial^2 u}{\partial t^2}.$$

This equation is called the two-dimensional wave equation.

As before, we shall assume that the constant of proportionality is $c = 1$. Then, the 2D-wave equation can be written as

$$(\Delta - \frac{\partial^2}{\partial t^2})u = 0. \tag{6}$$

The operator $\Box = \Delta - \frac{\partial^2}{\partial t^2}$ is sometimes called the D'Alembertian operator. It is the typical example of hyperbolic operator.

As in the previous section, we shall look for solutions of Eq. 6 where not only for each t_0 the membrane $u(x, y, t_0)$ is a Bézier surface, but even the evolution along time of the membrane is of Bézier kind:

$$u(x, y, t) = \sum_{i,j,k=0}^{\ell,m,n} B_i^\ell(x) B_j^m(y) B_k^n(t) P_{ijk}, \quad x, y, t \in [0, 1]. \tag{7}$$

We shall call the set of control points $\{P_{ijk}\}_{i,j,k=0}^{\ell,m,n}$ the control web.

3.1 The 2D-Wave Equation in Terms of the Control Points

Theorem 2. *Given a control web in $\mathbb{R}^3$, $\{P_{ijk}\}_{i,j,k=0}^{\ell,m,n}$, the associated Bézier membrane, $u : [0,1] \times [0,1] \times [0,1] \to \mathbb{R}^3$, is a solution of the 2D-wave equation if and only if for any $i \in \{1, \ldots, \ell\}$, $j \in \{1, \ldots, m\}$ and $k \in \{1, \ldots, n\}$*

$$a_{i\ell} \Delta^{200} P_{ijk} + b_{i-1,\ell} \Delta^{200} P_{i-1,jk} + c_{i-2,\ell} \Delta^{200} P_{i-2,jk}$$
$$+ a_{jm} \Delta^{020} P_{ijk} + b_{j-1,m} \Delta^{020} P_{i,j-1,k} + c_{j-2,m} \Delta^{020} P_{i,j-2,k} \tag{8}$$
$$- a_{kn} \Delta^{002} P_{ijk} - b_{k-1,n} \Delta^{002} P_{ij,k-1} - c_{k-2,n} \Delta^{002} P_{ij,k-2} = 0,$$

where, for $i \in \{0, \ldots, \ell-2\}$

$$a_{in} = (n-i)(n-i-1), \quad b_{in} = 2(i+1)(n-i-1), \quad c_{in} = (i+1)(i+2),$$

and $a_{in} = b_{in} = c_{in} = 0$ otherwise.

The proof is analogous to that of harmonic Bézier surfaces (see [1]). To obtain a statement similar to that of Th. 1 all it lacks to do is to expand the incremental operators.

3.2 A Tricubical Example

A study of Eqs. 8 for $\ell = m = n = 3$ shows that given the border control points of the initial and final states (2×12) of the Bézier membrane, the other control points ($4 + 16 + 16 + 4$) of the whole control web are determined. In the next Figure the independent border control points of the initial and final states are plotted thicker.

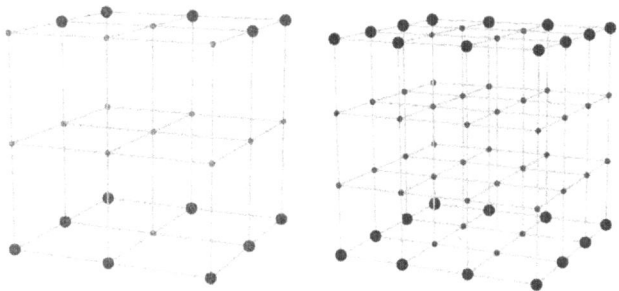

Fig. 2. Left $n = 2$. Right $n = 3$.

The Figure IV shows six states of the evolution of a solution for the 2D-wave equation with border control points (the black ones) in the initial and final states as follows in Figure III,

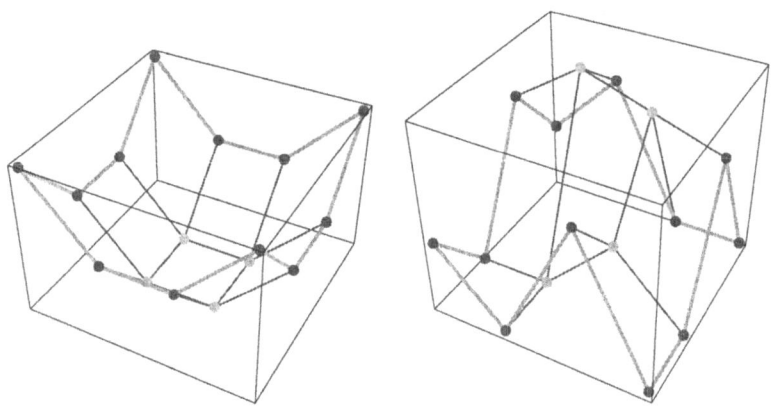

Fig. 3. Initial state   Final state

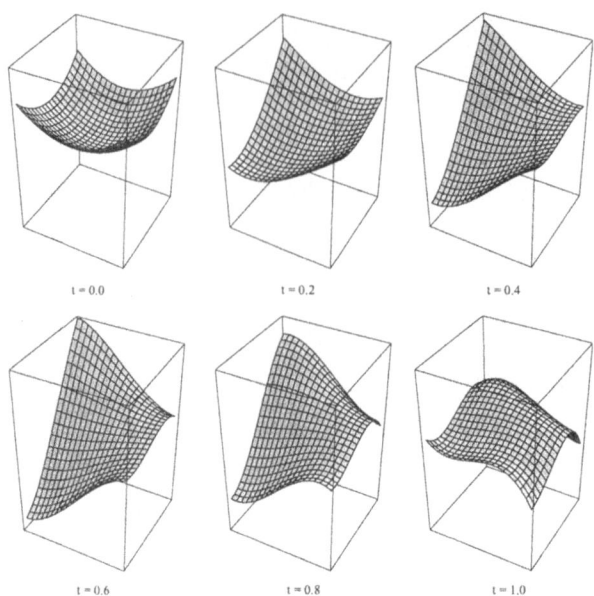

Fig. 4.

3.3 Computing the Solutions of the 2D-Wave Conditions

To solve in general Eqs. (8) for degrees other than 2 and 3 is a difficult task even for a symbolic computation program. In order to do that, it is better to come back to the usual basis of polynomials.

Lemma 1. *Let $f(x,y,t) = \sum_{i,j,k}^{n} \frac{a_{i,j,k}}{i!j!k!} x^i y^j t^k$ be a polynomial solution of degree $n \geq 2$ of the 2D-wave equation, then, all coefficients $\{a_{ijk}\}_{i,j=0,k=2}^{n}$ are totally determined by the coefficients $\{a_{ij0}, a_{ij1}\}_{i,j=0}^{n}$ and, moreover these two sets of coefficients verify if $n = 2p$ that*

$$\sum_{r=0}^{p} \binom{p}{r} a_{i+2r,j+2(p-r),1} = 0 \quad , \quad \sum_{r=0}^{p+1} \binom{p+1}{r} a_{i+2r,j+2(p+1-r),0} = 0, \quad (9)$$

and if $n = 2p - 1$ that

$$\sum_{r=0}^{p} \binom{p}{r} a_{i+2r,j+2(p-r),0} = 0 \quad , \quad \sum_{r=0}^{p} \binom{p}{r} a_{i+2r,j+2(p-r),1} = 0, \quad (10)$$

for $i + j < n$ in both cases.

Proof: The 2D-wave condition $\Box f = 0$ can be translated into a system of linear equations in terms of the coefficients $\{a_{i,j,k}\}_{i,j,k=0}^{n}$

$$a_{i+2,j,k} + a_{i,j+2,k} - a_{i,j,k+2} = 0, \quad (11)$$

for $i,j,k = 0,\ldots,n$, but with the convention $a_{ijk} = 0$ if any of the indexes is greater than n.

This means that any coefficient a_{ijk} with $k > 1$ can be written as $a_{ijk} = a_{i+2,j,k-2} + a_{i,j+2,k-2}$ and so on until the last subindex is 0 or 1, or until one of the first two subindexes is greater than n. Indeed,

$$a_{i,j,2k} = \sum_{r=0}^{k} \binom{k}{r} a_{i+2r,j+2(k-r),0},$$
$$a_{i,j,2k+1} = \sum_{r=0}^{k} \binom{k}{r} a_{i+2r,j+2(k-r),1}. \tag{12}$$

Note that using relations (12) we can solve all the Eqs. 11 in terms of $\{a_{i,j,0}, a_{i,j,1}\}$ for $k = 0,\ldots,n-2$, but not for $k = n-1, n$. For these two values, Eqs. 11 are

$$a_{i+2,j,n-1} + a_{i,j+2,n-1} = 0,$$

$$a_{i+2,j,n} + a_{i,j+2,n} = 0,$$

for $i, j = 0,\ldots,n$, or, equivalently

$$a_{i,j,n+1} = a_{i,j,n+2} = 0,$$

and with relations (12) we get the expressions (9) for n even and (10) when n is odd. Those are a set of relations between coefficients with last subindex 0 or 1. Moreover, it can be seen that if $i + j \geq n$ then the corresponding equations identically vanishes. □

Now, let us translate this result in terms of the control web.

Proposition 2. *Let $u(x,y,t) = \sum_{i,j,k=0}^{n} B_i^n(x) B_j^n(y) B_k^n(t) P_{ijk}$ be a Bézier solution of degree n of the 2D-wave equation with $\{P_{ijk}\}_{i,j,k=0}^{n}$ as control web, then the control points in the inner levels $\{P_{ijk}\}_{i,j=0,k=1}^{n,n-1}$ are determined by the control points in the first and last rows, $\{P_{ij0}\}_{i,j=0}^{n}$ and $\{P_{ijn}\}_{i,j=0}^{n}$. Moreover, there are also central control points of the initial and final states which are determined by the other control points.*

Sketch of proof: Let us write the Bézier solution in the usual basis of polynomials

$$u(x,y,t) = \sum_{i,j,k}^{n} A_{ijk} x^i y^j t^k,$$

where $A_{ijk} \in \mathbb{R}^3$. Coefficients with $k = 0$ are determined by $u(x,y,0)$, or, equivalently, by control points $\{P_{ij0}\}_{i,j}^{n}$. Indeed, by the formula of the change of basis from the basis of Bernstein polynomials to the usual basis

$$A_{ij0} = \binom{n}{i}\binom{n}{j} \Delta^{ij0} P_{000},$$

where Δ^{ij0} is the incremental operator.

Now, Eqs. (9) and (10) can be translated and solved in terms of the control points at time $t = 0$.

The change $t \to 1-t$ in $u(x,y,t)$ allows to compute the analogous relations of the control points at time $t = 1$. So, the same dependence scheme is valid for the initial and the final states.

Finally, as it happens for harmonic surfaces, the computation of the coefficients A_{ij1} from the control points of the initial and final states is reduced to a triangular linear system.

Once the first two levels of coefficients A_{ijk} are known, then, by Lemma 1, the solution $u(x,y,t)$ is completely known and, therefore, all its control points. This uniquely solves Eqs. (8) for $u(x,y,t)$ as in (7). □

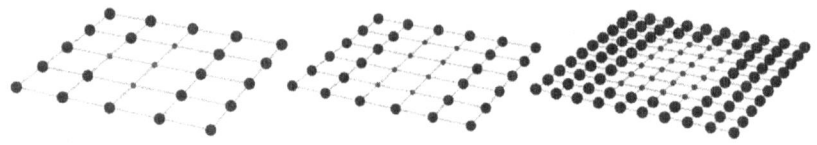

Fig. 5. Representation for $n = 4$ (left), $n = 5$ (center) and $n = 9$ (right) of the independent control points (the bigger ones) in the initial and final states. Compare with $n = 2$ and $n = 3$, Fig. II.

Note that for higher degrees, the number of control points near to the boundary increases. This means that we can fix not only the boundary of the initial and final states, but also, some of the partial derivatives of the initial and final states at the boundary.

Acknowledgements. This work has been partially supported by a Spanish MCyT grant BFM2002-00770.

References

1. C. Cosín, J. Monterde, *Bézier surfaces of minimal area*, Proceedings of the International Conference of Computational Science, ICCS'2002, Amsterdam, eds. Sloot, Kenneth Tan, Dongarra, Hoekstra, Lecture Notes in Computer Science, 2330, vol II, pages 72–81, Springer-Verlag, (2002).
2. B. Lang, *The synthesis of waveforms using Bézier curves with control point modulation*, The Second CEMS Research Student Conference.
3. A. T. Layton, M. van dePanne, *A numerically efficient and stable algorithm for animating water waves*, The visual Computer, 18, 41-53 (2002).
4. J. Monterde, *Bézier surfaces of minimal area: The Dirichlet approach*, to appear in CAGD, (2004).
5. J. Monterde, *The Plateau-Bézier problem*, in Mathematics of Surfaces X, M.J. Wilson and R.R. Martin (eds.), LNCS 2768, Springer-Verlag, 262-273, (2003).

Matlab Toolbox for a First Computer Graphics Course for Engineers

A. Gálvez[1], A. Iglesias[1]*, C. Otero[2], and R. Togores[2]

[1] Department of Applied Mathematics and Computational Sciences
[2] Department of Geographical Engineering and Graphical Expression Techniques
University of Cantabria, Avda. de los Castros, s/n, E-39005, Santander, Spain
iglesias@unican.es
http://personales.unican.es/iglesias

Abstract. This paper describes a new Matlab toolbox designed for teaching a first Computer Graphics course for Engineering students. The aim of the toolbox is to provide the students with a tool for a comprehensive overview on the fundamentals of Computer Graphics in terms of basic algorithms and techniques. Firstly, the paper discusses the design of such a course taking into account its contents, goals, duration and other constraints. Then, the toolbox architecture is described. Finally the application of this toolbox to educational issues is discussed through some illustrative examples. The system has been successfully used by the authors during the last three years at the University of Cantabria. The main conclusions from this experience are also reported.

1 Introduction

Nowadays, Computer Graphics (CG) is not longer the domain of a small group of professional experts. Conversely, during the last recent years people from different fields and with very different background are increasingly demanding CG courses "a la carte", adapted to suit different requirements and oriented to particular interests and objectives. Many universities have included Computer Graphics courses in their curricula, in response to the new students' needs and interests. Although this is particularly the case for Computer Science students, it is still true for other scientific studies, such as Engineering.

Since 1999 the authors have taught a first Computer Graphics course for Engineering students at the University of Cantabria, Spain. The course, with a total of 60 hours, is a part of a set of elective courses designed to provide students with a deeper knowledge on Computer Science and related fields. Those courses are open to almost everyone from Engineering studies. This implies that our audience is extremely diverse, from sophomore to senior students (freshmen are still not allowed to choose the course), from Electrical, Electronical, Mechanical, Civil and Chemical Engineering. Therefore, these students exhibit a wide range of different background and skills.

---

* Corresponding author

Another important issue is that this is the first course on Computer Graphics for our students. This fact restricts dramatically the course's goals to simply offer a comprehensive overview on the fundamentals of Computer Graphics as well as a gentle introduction to the main topics and techniques. Consequently, the tools to be used in the course must be carefully chosen in order to prevent students from boredom or discouragement.

In this context, the present paper describes a new computer tool for teaching a first Computer Graphics course for Engineering students. The program is actually a Matlab toolbox implemented by the authors and aimed to be complemented with both theoretical and practical sessions and homework. Furthermore, the toolbox does assume neither familiarity with Matlab or its environment nor previous background on CG techniques. Once implemented, we have applied the toolbox to our CG courses during the last three years. The students' results and their feedback have been so successful that convinced us to share our experience with others. This is actually the core of the present work.

The structure of this paper is as follows: in Section 2 we describe the framework of our development. Section 3 describes the process of designing a first CG course for Engineering students. A discussion on how the course's contents and goals do determine the toolbox architecture is also given in this section. The architecture of the Matlab toolbox itself and some remarkable features are described in Section 4. Section 5 shows some illustrative examples on its application to a CG course. Finally, Section 6 closes with the main conclusions and some further remarks.

2 Framework

In the last few years, Computer Graphics has become one of the most important fields in Computer Science, with outstanding applications to many areas: Science, Engineering, Medecine, advertising, entertainment, etc. and the list is expanding rapidly. Current Computer Graphics hardware is orders of magnitude faster and cheaper, and it is much more robust and powerful than earlier technology. Today, we are used to have powerful graphical cards in our desktop computers and laptops. These graphical cards are especially designed for high performance and often incorporate OpenGL or other APIs in hardware. In addition, we have witnessed extraordinary advances in Computer Graphics software, such as:

- the appearance and acceptance of standardized APIs (Application Programmer Interfaces), such as OpenGL, Direct3D, QuickDraw3D, Java3D, etc.
- extensive general-purpose libraries providing high level graphical capabilities and simplified GUIs (Graphical User Interfaces)
- powerful graphical programs (for instance, 3D Studio Max, LightWave or RenderMan) that provide the users with a large and very powerful collection of tools for design, rendering and animation
- the Web and its wealth of free demos, software tools (POVray, rayshade, VRML), data sets and examples, etc.

Another point that deserves consideration from a Computer Graphics education standpoint is that current students are quite different from those in earlier years. Although less skilled in mathematical reasoning and tools, they have access to a considerable amount of technology: computer tools (such as the web, video games, graphical interfaces), video and TV devices, etc. Consequently, they are quite familiar with concepts like RGB color, texture, navigation, etc.

All these new features have introduced dramatic changes in both the methodology to teach Computer Graphics and the tools used in CG courses. While in the past those courses were mostly intended for Computer Graphics experts or people with a solid background and outstanding skills in programming and Computer Science, today there is an increasing demand of CG courses adapted to a wide range of different skills and interests.

Most of these courses are organized on the basis of a graphics library such as Open GL or Java3D [6]. In fact, as remarked in [13], OpenGL appears to be the standard tool in many CG courses. However, these libraries require both a mathematical background and certain programming skills that either have not been acquired yet (for example, freshman or sophomore university students still lack many advanced programming concepts) or have already been forgotten. Graphical programs (3D Studio Max, LightWave, RenderMan) do not represent a feasible alternative, due to their costs and the fact that they are best suited for modeling and rendering rather than for learning the CG principles and techniques. On the other hand, the material available at the Web is disperse among many different resources and often oriented to specific goals (such as POVray for ray tracing techniques or VRML for virtual reality and navigation).

An interesting (and not exhaustively explored yet) alternative is given by the general-purpose symbolic/numerical computer algebra systems (CAS). Originally intended for scientific purposes, these systems currently integrate very powerful symbolic, numerical and graphical capabilities within a unified framework. For several reasons, these features are specially valuable: on one hand, they provide the students with the required mathematical tools and functions. In addition, the mathematical algorithms are optimized for high performance. On the other hand, the CAS include many graphical commands and options. Especially remarkable are the graphical functions of Matlab, a scientific program that has emerged as one of the most popular computer algebra systems. It is a very powerful system used by hundreds of thousands of industrial, government and academic users around the world. It provides versions at affordable price for the most important platforms while keeping the external appearance (the user interface) as similar as possible, thus reducing the span to become familiar with the environment. Furthermore, since Matlab is based on C, it runs very fast and its memory requirements are very reasonable.

In a previous paper [4], we explored the strenghts and weaknesses of Matlab for Computer Graphics and Geometric Modeling tasks. Although that paper was mostly oriented towards visualization issues, we realized that Matlab might also be an excellent program for Computer Graphics courses for engineers, provided that some additional libraries are effectively incorporated into the system. As a consequence, we decided to go through this possibility by implementing a set of numerical, symbolic and graphical libraries to be linked with the Matlab kernel. Details on the architecture of the system will be described in Section 4.1. Since the program is intended for educational purposes, a powerful user interface has also been designed and implemented. Of course, the user interface's design depends strongly on the course's contents and goals, which are discussed in the next section.

3 Designing a First Computer Graphics Course for Engineers

Based on the students' profile and limitations described in the previous sections, a first Computer Graphics course for engineers should restrict its objectives to:

- offer a comprehensive overview on the fundamentals of Computer Graphics in terms of basic algorithms and techniques
- develop the students' visual sense, which is not usually acquired from most of the traditional courses.

On the other hand, the curriculum of the course depends strongly on the number of hours and their structure. In our case, the course consists of 40 lectures (of one hour and half each), of which the half are given at the classroom and the other half at the computer labs. Under these constraints, the topics for the course have been organized into ten chapters:

1. *Introduction*: basic definitions, history of CG, goals and principles.
2. *Basic geometrical concepts*: 2D and 3D transformations, perspectives (including some mathematical review on vectors and matrices).
3. *Geometric Modeling*: parametric and implicit curves and surfaces. Free-form schemes (including NURBS) and their basic algorithms: manipulation of control points, knot vectors and weights, derivatives, etc.
4. *Basic algorithms for Geometric Modeling*: subdivision, degree raising, knot insertion, knot removal, convex hull property, etc.
5. *Advanced algorithms for Geometric Modeling*: intersection, blending, offsetting, implicitization. Approximation and interpolation techniques.
6. *Basic Differential Geometry techniques*: tangent vectors and planes, curvatures (Gaussian, mean), curvature lines, fundamental forms, etc.
7. *Illumination models*: ambient, specular and diffuse illumination terms. Flat, Gouraud and Phong shading. Illumination techniques: ray-tracing, radiosity.
8. *Textures*: texture mapping, bump mapping, environment map, solid texture, fractal texture.
9. *Industrial files*: IGES format. Visualization of industrial files.
10. *Principles of animation*: basic concepts. Animation of natural phenomena, virtual characters and avatars.

Each chapter (typically from 2 to 6 lectures, depending on the complexity of the topics involved) consists of a first visual introduction to the subject, a theoretical explanation of the concepts and techniques, some computer work at the computer labs and homework. At the end of the course, a final project on some of the topics of the course is also required. As the reader can easily realize, the list of topics is quite large so our students are required to perform some tasks by themselves, either as homework or as computer training at our labs. To this aim, essential ingredients are (1) the bibliography, (2) additional documentation and material and (3) the software.

Regarding the bibliography used in this course, students are suggested to use the books in [3,11,12] for a gentle and general introduction to Computer Graphics. These books also cover most of the course's topics at a general (but still enough) level. Some topics, however, require additional bibliography. For Geometric Modeling we recommend the excellent books in [1,2,7,9,11], while illumination models are also analyzed in [8] and in [5] for ray-tracing techniques.

Additional material comprises ACM Siggraph videotapes, freeware/shareware programs and libraries available from the Web (such as POVray, Open GL or VRML) that are shown just for general knowledge (but they are basic tools for a subsequent Computer Graphics course not described here), journals (such as "Computers and Graphics", "ACM Transactions on Graphics", "Computer Graphics Forum" or "Computer Graphics World" and others available at our library) and additional documentation (manuals, presentations, slides, etc.). Concerning the software, in addition to some specific material designed for elementary CG courses such as the program TERA (Tool for Exploring Rendering Algorithms) and its complementary book in [14] or the software described in the previous paragraph, we use a specific tool designed by ourselves to fulfill our requirements and needs. The tool is described in the following sections.

4 The Matlab Toolbox

4.1 System's Architecture

The two course's objectives described in Section 3 imply that the software to be used therein must be fundamentally both *interactive* and *visual*. In addition, we looked for an easy-to-use system with good graphical capabilities, a powerful user interface and a C++-like object oriented programming language.

After a careful analysis of our needs, goals and constraints, we decided to use Matlab. Reasons for this choice have already been given in Section 2. We implemented a Matlab toolbox with three main components:

1. **A system kernel**: it comprises an extension of the Matlab kernel with an exhaustive set of numerical and symbolic libraries for Computer Graphics and Geometric Modeling. These libraries contain hundreds of commands for: the accurate manipulation and visualization of geometric entities, symbolic computation routines, efficient rendering, illumination techniques and animation. Of course, the libraries must be continuously updated, so the system must be flexible enough to allow the programmer to improve the algorithms and codes in an efficient, quick and easy way.

2. **A user interface**: Matlab provides a framework, called GUI (Graphical User Interface) for developing powerful user-friendly interfaces. In particular, Matlab offers a set of routines for generating a graphical tool easily: by a simple drag-and-drop procedure buttons, menus, dialog boxes and many other interactivity tools are automatically incorporated into the project. The GUI generates some parts of the code, the gaps being filled in by the programmer so that the interface can be effectively linked with the kernel.
3. **Matlab graphics library**: this module includes the graphical commands needed for our setup. For instance, in the case of PC platforms, it is a set of DLLs (Dynamic Link Libraries). In general, they are not included in the Matlab standard version and should be purchased separately.

4.2 Remarkable Features

Some remarkable features of this system are modularity and the possibility of code conversion into a standalone application. The first requirement is given by the fact that the course's contents are extremely diverse, ranging from geometric modeling to illumination or visualization of industrial files. Consequently, the topics of the course have been organized in different chapters, as described in Section 3. Then, the system has been splitted up into several independent windows associated with those topics. Currently, we have designed six different windows, devoted to curves, surfaces, illumination models, industrial files, visualization and differential geometry topics. Figures 1 and 2 are examples of these windows. As the reader can see, their external appearance is very similar and simple, so students can become familiar with the system in only a few minutes, even although they had never used a computer at all. On the other hand, the system allows the programmer to build a standalone application through a two-step process: (1) the code is converted to either C or C++ via a Matlab compiler and (2) the new C/C++ code can be subsequently compiled and linked with the graphics libraries. By this simple procedure a new, independent application can be obtained.

Table 1. Minimum hardware and software requirements for the system

Hardware/Software	Requirement
Operating System	Windows (9x, 2000, NT, Me, XP), MacOS, UNIX
RAM	128 MB
Disk storage	25 MB (60 MB recommended)
Monitor	Super VGA monitor
Screen resolution	640 × 480

The program, whose minimum hardware and software requirements are listed in Table 1, supports many different platforms, such as PCs (with Windows 9x,

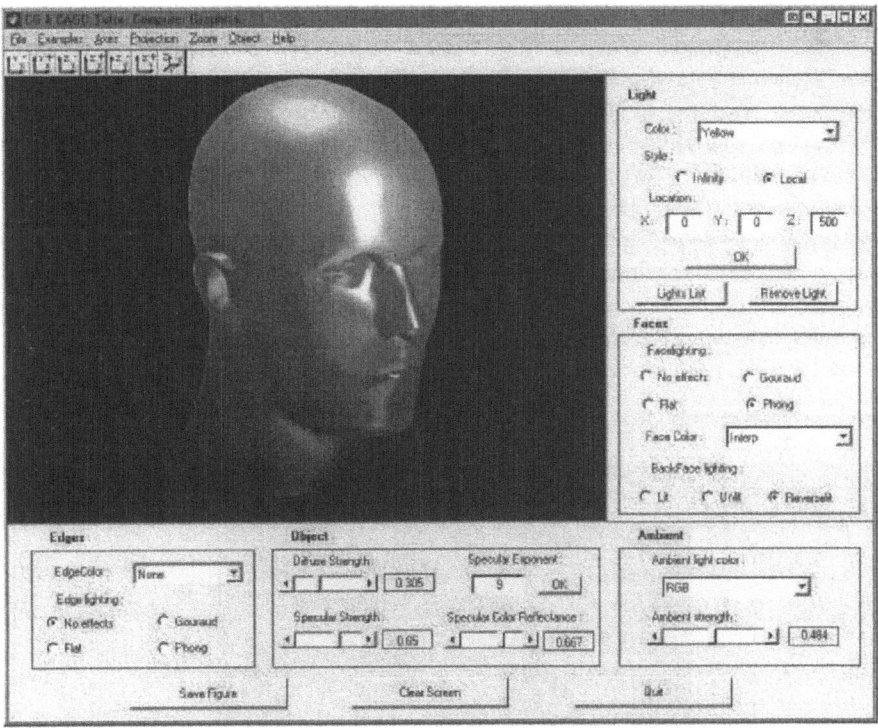

Fig. 1. Window for exploring illumination models

2000, NT, Millenium or XP), Macintosh (Power Mac and 68040) and UNIX workstations from Sun, Hewlett-Packard, IBM, Silicon Graphics and Digital. The figures of this paper correspond to the PC version. It should be remarked that the codes for numerical and symbolic routines are basically the same in all versions, while the graphical interface varies (depending on the platform you are working on) and so do the Matlab graphical libraries.

5 Some Illustrative Examples

Figure 1 shows the window devoted to illumination models. In addition to the powerful based-on-OpenGL graphical commands and functions provided by Matlab, some graphical libraries for rendering are to be incorporated into the system. These libraries include different surface algorithms (wireframe, hidden-line removal or z-buffer), shaders (constant, faceted, Gouraud, Phong, etc.), lighting techniques (ambient, diffuse, specular, etc.), texture mapping, one or several light sources, interactive manipulation (coloring, zooming, rotation), etc. Many of these algorithms have been applied to the shape in Fig. 1 in order to obtain a realistic illumination model.

Because of our students' profile and interests, we have included some topics especially oriented to Engineering. A good example is the chapter on industrial files, in which our students can easily learn how the geometric information is generated, stored, transferred and visualized as a prior step to manufacturing. To this end, real industrial files corresponding to different parts of cars or planes are analyzed. In previous chapters, the *parametric surfaces* window allows the students to learn about the fundamentals of computer design, the basic tool for generating those industrial files. Then, real shapes are analyzed by using the *industrial files* window, shown in Figure 2. In this figure several thousands of different geometric entities (arcs, lines, spline curves, NURBS curves and surfaces, trimmed NURBS surfaces) representing the front seat door of a car body in IGES format are displayed.

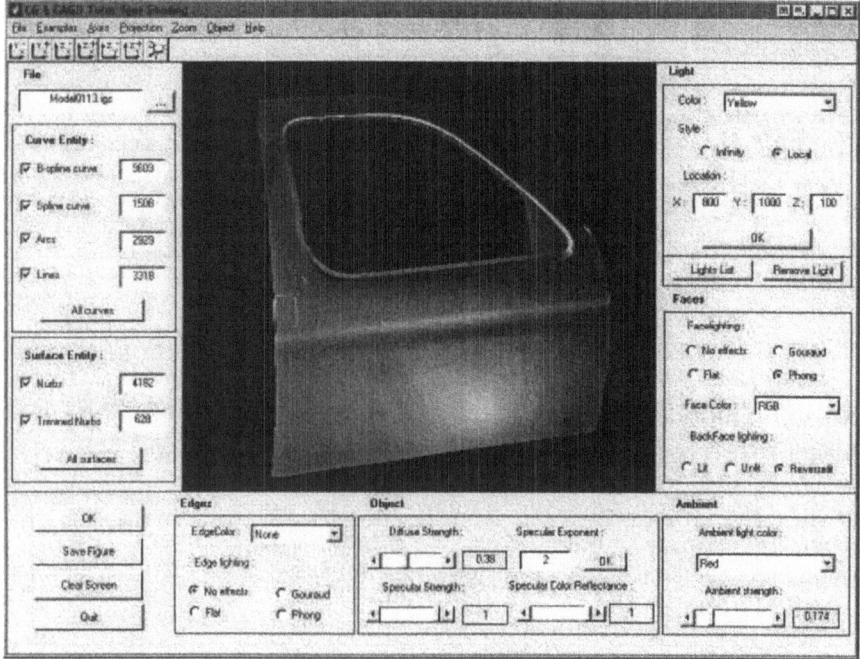

Fig. 2. Window for analyzing industrial files

Several illumination effects similar to those in Fig. 1 contribute to simulate that image by computer with a high level of quality. In addition to their aesthetic effect, the illumination models are important to detect irregularities on surfaces, for example, by using some illumination lines called *isophotes* [10]. This procedure can simulate the real situation of looking for the different reflection lines on the surface of a car body.

6 Conclusions and Further Remarks

In this paper a new Matlab toolbox designed for helping Engineering students to get a visual insight into Computer Graphics techniques is described. The program is mostly intended for computer training on the topics of the course and does assume neither familiarity with Matlab or its environment nor previous background on Computer Graphics techniques.

The program has been applied during the last three years for teaching a first Computer Graphics course for Engineering students at the University of Cantabria (Spain). In general, the students were enthusiastic about the use of this tool and several took the subsequent course on Computer Graphics that was taught as a second course in sequence. The results of this second course evidenced that students did effectively learn many concepts on the subject from the first one, and were able to follow the new material quicker than students from previous years that did not use the program at all. However, this last statement must be filtered by the fact that current students are more accustomed to concepts related to CG that might have been acquired from the real world.

In order to get a more reliable information about students' opinion they were required to fill a questionnaire. It showed that the software was found to be intuitive and easy to use, and generally satisfied users. Some students proposed different ways to improve the system (particularly, the addition of a set of interactive exercises for each module, which are currently in progress), demonstrating the good comprehension level of the program. Moreover, the students recognized the utility of the system as a learning tool and appreciated the possibility of using it in order to gain certain visual insight into the effects of certain algorithms and techniques.

In spite of these encouraging results, there is still a long way to walk. In particular, further experiments must be done to investigate more about the impact of this software in the classroom environment and in the learning process. Of course, we had to face some problems as well. Perhaps the most important one is related to the dissimilarity of the background knowledge, since it was an elective course for sophomore, junior and senior students from five different (Mechanical, Chemical, Electrical, Electronical and Civil) Engineering degrees. A natural solution could be to implement different curricula adapted to each student's profile. But because the number of students never excedeed 30, we decided to keep a similar curriculum for all the attendees. Finally, some students found the syntax to enter equations symbolically to be quite confuse. Additional warning dialogs should fix the problem.

Another interesting test is given by the oral and written project to be presented by the students at the end of the course. Some students put a lot of work and effort into the project, and presented interesting material. However, a few students were not so stimulated for this assignment and presented poor material, in spite of the fact they were allowed to propose a subject of their own interest. In the questionnaire, some pointed out that they do not like to make an oral presentation and become upset by this duty (this was actually the first thing they suggested to remove from the course). Other students claimed that, being an elective course, the project is perhaps too demanding. These opinions were carefully analyzed and the course's methodology and evaluation were modified accordingly. For example, last year we gave the possibility to replace the oral presentation by a standard written examination.

It is our hope that sharing our positive experience of teaching Computer Graphics for enginners by using this toolbox will be of benefit to other educators. We would be delighted to receive the feedback and comments from others with similar experiences.

References

1. Anand, V.B.: Computer Graphics and Geometric Modeling for Engineers. John Wiley and Sons, New York (1993)
2. Farin, G.: Curves and Surfaces for Computer Aided Geometric Design. Academic Press, Orlando Fl (1988)
3. Foley, J.D., van Dam, A., Feiner, S.K., Hughes, J.F.: Computer Graphics. Principles and Practice. (2nd. Ed.) Addison Wesley, MA (1990)
4. Gálvez, A., Iglesias, A., Gutiérrez, F.: Applying Matlab to computer graphics and CAGD. Application to a visualization problem in the automotive industry. Proc. of the IX International Conference on Computer Graphics and Vision, Graphicon'99. Moscow (1999) 214-221
5. Glassner, A.: An Introduction to Ray Tracing. Academic Press, San Diego (1989)
6. Hitchner, L.E., Sowizral, H.A.: Adapting computer graphics curricula to changes in graphics technology. Proc. of the Eurographics/Siggraph Workshop on Computer Graphics and Visualization Education, CGE'99, Coimbra (1999) 23-29
7. Hoschek, J., Lasser, D.: Fundamentals of Computer Aided Geometric Design. A.K. Peters, MA (1993)
8. Hall, R.: Illumination and Color in Computer Generated Imagery. Springer-Verlag, New York (1989)
9. Piegl, L., Tiller, W.: The NURBS Book (2nd. Ed.) Springer-Verlag, Berlin Heidelberg (1997)
10. Poeschl, T.: Detecting surface irregularities using isophotes. Computer Aided Geometric Design 1 (1989) 163-168
11. Rogers, D.F., Adams, J.A.: Mathematical Elements for Computer Graphics (2nd. Ed.) Mc Graw-Hill, New York Boston (1989)
12. Rogers, D.F.: Procedural Elements for Computer Graphics (2nd. Ed.) Mc Graw-Hill, New York Boston (1998)
13. Wolfe, R.J.: OpenGL: Agent of change or sign of the times? Computer Graphics November-98 (1998) 29-31
14. Wolfe, R.J.: 3D Graphics. A Visual Approach. Oxford University Press, New York (2000)

A Differential Method for Parametric Surface Intersection

A. Gálvez, J. Puig-Pey*, and A. Iglesias

Department of Applied Mathematics and Computational Sciences, University of Cantabria, Avda. de los Castros, s/n, E-39005, Santander, Spain
{puigpeyj,galveza,iglesias}@unican.es

Abstract. In this paper, a new method for computing the intersection of parametric surfaces is proposed. In our approach, this issue is formulated in terms of an initial value problem of first-order ordinary differential equations (ODEs), which are to be numerically integrated. In order to determine the initial value for this system, a simple procedure based on the vector field associated with the gradient of the distance function between points lying on each of the parametric surfaces is described. Such a procedure yields a starting point on the nearest branch of the intersection curve. The performance of the presented method is analyzed by means of some illustrative examples that contain many of the most common features found in parametric surface intersection problems.

1 Introduction

The intersection of surfaces is one of the most outstanding problems in many fields, such as computational geometry, solid modeling, geometric processing, computer aided design (CAD), computer aided geometric design (CAGD), manufacturing, numerical-controlled machining, visualization, robotics, etc. (see, for example, [5,7,8,11,12,16,19]).

During the last few years, a number of different methods to compute the intersection of surfaces have been described in the literature[1]. Among them, those methods intended to calculate the intersection of parametric surfaces have received much attention, since the parametric representation is the most used mathematical representation in industry. Most of them [12,16] are based on the idea of implicitizing the equations of one of the parametric surfaces so that the problem is reduced to a simpler parametric-implicit intersection problem. A remarkable fact is that this parametric-implicit conversion is always possible. However, a tensor-product parametric patch of degree (m,n) has an implicit equation of degree $2mn$. In the simple case of bicubic patches, this means an

---

* Corresponding author
[1] See, for example, the excellent reviews on this topic in [12] (Chapter 12) or [16] (Chapter 5) and the references therein. We also recommend the survey in [19]. Additional references can be found in [8]. An up-to-date survey on surface intersection can also be found in [9].

equation of degree 18 and 1330 terms. In addition, it is known that two surface patches intersect in a curve whose degree is much higher than the parametric degree of the two patches. Thus, two bicubic patches intersect in a curve of degree 324. Because of these facts, many alternative techniques for parametric surface intersection have been described, including the subdivision methods, the discretization methods, the hybrid methods, etc. (see [9,13,16] and references within for recent surveys on computational methods for surface intersection).

Another interesting approach that is gaining interest during the last few years is the differential scheme, such as the second order boundary algebraic-differential approach in [10,11]. There is also a family of methods known as marching methods based on generating a sequence of points of an intersection curve branch by stepping from a point on such a curve in a direction determined by some local differential geometry analysis [3,4,15]. These methods are globally incomplete since they require starting points for every branch of the solution. Motivated by this, great effort has been devoted to the determination of such starting points by using hodographs [21], elimination methods [6], iterative optimization and Moore-Penrose pseudo-inverse local methods [1]. It should be remarked, however, that there has been no known algorithm that can compute the intersection curve of two arbitrary rational surfaces accurately, robustly and efficiently [9,12,16].

This paper focuses on the problem of computing the intersection of two parametric surfaces[2] $\mathbf{S}_1(u_1, v_1)$ and $\mathbf{S}_2(u_2, v_2)$. The method presented in this paper belongs to the family of differential approaches and follows up a previous method for computing the intersection of two surfaces given in parametric and implicit form, respectively [20]. In our approach, the parametric surface intersection problem is formulated in terms of an initial value problem of first-order ordinary differential equations (ODEs). The resulting system of ODEs is then numerically integrated through an adaptive 4-5-order Runge-Kutta method [18]. From this point of view, the present method can also be classified as a marching method. As pointed out above, one of the main shortcomings of the marching methods is the need for a starting point on the intersection curve. In this paper, a procedure to overcome this limitation is proposed. Starting at an arbitrary point on each of the parametric surfaces, we trace a path on those surfaces by following the direction indicated by the gradient of the distance between both points. As it will be shown later, this simple procedure yields a starting point on the nearest branch of the intersection curve.

The structure of this paper is as follows: in Section 2 we introduce some basic concepts and the terminology to be used throughout the paper. Then, a new differential method for computing the intersection between two parametric surfaces is presented in Section 3. The determination of a starting point on the intersection curve for the marching method of this paper is also discussed in this section. In order to show the good performance of the method, some illustrative examples are analyzed in Section 4. All the examples have been carefully chosen so that they contain several problems similar to those usually found in many

---

[2] In this paper vectors will be denoted in bold.

parametric surface intersection scenarios. Finally, the conclusions and future work are discussed in Section 5.

2 Basic Concepts and Terminology

In this paper we restrict ourselves to parametric surfaces, which are assumed to be differentiable at any point. The parametric surfaces are described by a vector-valued function of two variables:

$$\mathbf{S}(u,v) = (x(u,v), y(u,v), z(u,v)), \qquad u,v \in \Omega \subset \mathbb{R}^2 \qquad (1)$$

where u and v are the surface parameters. Expression (1) is called a parameterization of the surface $\mathbf{S}$. At regular points, the partial derivatives $\mathbf{S}_u(u,v)$ and $\mathbf{S}_v(u,v)$ do not vanish simultaneously. These partial derivatives define the unit normal vector $\mathbf{N}$ to the surface at $\mathbf{S}(u_0, v_0)$ as:

$$\mathbf{N} = \frac{\mathbf{S}_u \times \mathbf{S}_v}{|\mathbf{S}_u \times \mathbf{S}_v|} \qquad (2)$$

where "×" denotes the cross product.

A curve in the domain Ω can be described by means of its parametric representation $\{u = u(t), v = v(t)\}$. This expression defines a three-dimensional curve $\mathbf{C}(t)$ on the surface $\mathbf{S}$ given by $\mathbf{C}(t) = \mathbf{S}(u(t), v(t))$. Applying the chain rule, the tangent vector $\mathbf{C}'(t)$ of this curve at a point $\mathbf{C}(t)$ becomes:

$$\mathbf{C}'(t) = \mathbf{S}_u \, u'(t) + \mathbf{S}_v \, v'(t) \qquad (3)$$

In this work the curve $\mathbf{C}$ will usually be parameterized by the arc-length s on the surface. Its geometric interpretation is that a constant step s traces a constant distance along an arc-length parameterized curve. Since some industrial operations require an uniform parameterization, this property has several practical applications. For example, in computer controlled milling operations, the curve path followed by the milling machine must be parameterized such that the cutter neither speeds up nor slows down along the path. Consequently, the optimal path is that parameterized by the arc-length. In this case:

$$E \left(\frac{du}{ds}\right)^2 + 2F \frac{du}{ds}\frac{dv}{ds} + G \left(\frac{dv}{ds}\right)^2 = 1 \qquad (4)$$

where E, F and G are the coefficients of the First Fundamental Form of the surface given by:

$$E = \mathbf{S}_u \cdot \mathbf{S}_u \quad , \quad F = \mathbf{S}_u \cdot \mathbf{S}_v \quad , \quad G = \mathbf{S}_v \cdot \mathbf{S}_v \qquad (5)$$

and "." indicates the dot product.

3 The Parametric Surface Intersection Method

In this section a new differential method for computing the intersection between two parametric surfaces is described. This problem is formulated in terms of an initial value problem of first-order ordinary differential equations (ODEs). However, for this initial problem to be well defined, we need a starting point on each branch of the intersection curve. A procedure for obtaining such a point based on the gradient of the difference between points of both surfaces is also discussed in this section. Then, the initial value ODE system will be integrated numerically by using a Runge-Kutta method.

3.1 Obtaining the ODE System

Let us consider two non-tangent surfaces $S_1(u_1, v_1)$ and $S_2(u_2, v_2)$ given in parametric form whose unit normal vectors at a point on the intersection curve $C(s)$ between S_1 and S_2 are N_1 and N_2 respectively. Following (3), the unit tangent vector $T_1(s)$ ($T_2(s)$ resp.) to the curve $C(s)$ considered as belonging to the surface S_1 (S_2 resp.) is given by:

$$\begin{cases} T_1(s) = \dfrac{dC(s)}{ds} = \dfrac{\partial S_1}{\partial u_1}\dfrac{du_1}{ds} + \dfrac{\partial S_1}{\partial v_1}\dfrac{dv_1}{ds} \\ \\ T_2(s) = \dfrac{dC(s)}{ds} = \dfrac{\partial S_2}{\partial u_2}\dfrac{du_2}{ds} + \dfrac{\partial S_2}{\partial v_2}\dfrac{dv_2}{ds} \end{cases} \quad (6)$$

Because T_1 and N_2 are orthogonal and so are T_2 and N_1 we have

$$\begin{cases} \left(\dfrac{\partial S_1}{\partial u_1}.N_2\right)\dfrac{du_1}{ds} + \left(\dfrac{\partial S_1}{\partial v_1}.N_2\right)\dfrac{dv_1}{ds} = 0 \\ \\ \left(\dfrac{\partial S_2}{\partial u_2}.N_1\right)\dfrac{du_2}{ds} + \left(\dfrac{\partial S_2}{\partial v_2}.N_1\right)\dfrac{dv_2}{ds} = 0 \end{cases} \quad (7)$$

On the other hand, since the curve $C(s)$ belongs to both S_1 and S_2, from (4) we have

$$\begin{cases} E_1 \left(\dfrac{du_1}{ds}\right)^2 + 2 F_1 \dfrac{du_1}{ds}\dfrac{dv_1}{ds} + G_1 \left(\dfrac{dv_1}{ds}\right)^2 = 1 \\ \\ E_2 \left(\dfrac{du_2}{ds}\right)^2 + 2 F_2 \dfrac{du_2}{ds}\dfrac{dv_2}{ds} + G_2 \left(\dfrac{dv_2}{ds}\right)^2 = 1 \end{cases} \quad (8)$$

Solving (7) and (8) for $\dfrac{du_1}{ds}$, $\dfrac{dv_1}{ds}$, $\dfrac{du_2}{ds}$ and $\dfrac{dv_2}{ds}$ we obtain:

$$\begin{cases} \dfrac{du_1}{ds} = \pm \dfrac{\dfrac{\partial \mathbf{S}_1}{\partial v_1}.\mathbf{N}_2}{\sqrt{E_1 \left(\dfrac{\partial \mathbf{S}_1}{\partial v_1}.\mathbf{N}_2\right)^2 - 2F_1 \left(\dfrac{\partial \mathbf{S}_1}{\partial v_1}.\mathbf{N}_2\right)\left(\dfrac{\partial \mathbf{S}_1}{\partial u_1}.\mathbf{N}_2\right) + G_1 \left(\dfrac{\partial \mathbf{S}_1}{\partial u_1}.\mathbf{N}_2\right)^2}} \\[2ex] \dfrac{dv_1}{ds} = \mp \dfrac{\dfrac{\partial \mathbf{S}_1}{\partial u_1}.\mathbf{N}_2}{\sqrt{E_1 \left(\dfrac{\partial \mathbf{S}_1}{\partial v_1}.\mathbf{N}_2\right)^2 - 2F_1 \left(\dfrac{\partial \mathbf{S}_1}{\partial v_1}.\mathbf{N}_2\right)\left(\dfrac{\partial \mathbf{S}_1}{\partial u_1}.\mathbf{N}_2\right) + G_1 \left(\dfrac{\partial \mathbf{S}_1}{\partial u_1}.\mathbf{N}_2\right)^2}} \\[2ex] \dfrac{du_2}{ds} = \pm \dfrac{\dfrac{\partial \mathbf{S}_2}{\partial v_2}.\mathbf{N}_1}{\sqrt{E_2 \left(\dfrac{\partial \mathbf{S}_2}{\partial v_2}.\mathbf{N}_1\right)^2 - 2F_2 \left(\dfrac{\partial \mathbf{S}_2}{\partial v_2}.\mathbf{N}_1\right)\left(\dfrac{\partial \mathbf{S}_2}{\partial u_2}.\mathbf{N}_1\right) + G_2 \left(\dfrac{\partial \mathbf{S}_2}{\partial u_2}.\mathbf{N}_1\right)^2}} \\[2ex] \dfrac{dv_2}{ds} = \mp \dfrac{\dfrac{\partial \mathbf{S}_2}{\partial u_2}.\mathbf{N}_1}{\sqrt{E_2 \left(\dfrac{\partial \mathbf{S}_2}{\partial v_2}.\mathbf{N}_1\right)^2 - 2F_2 \left(\dfrac{\partial \mathbf{S}_2}{\partial v_2}.\mathbf{N}_1\right)\left(\dfrac{\partial \mathbf{S}_2}{\partial u_2}.\mathbf{N}_1\right) + G_2 \left(\dfrac{\partial \mathbf{S}_2}{\partial u_2}.\mathbf{N}_1\right)^2}} \end{cases} \quad (9)$$

which together with an initial point of the intersection curve:

$$(u_1(0), v_1(0), u_2(0), v_2(0)) = (u_0^1, v_0^1, u_0^2, v_0^2) \quad (10)$$

constitutes an initial value problem for this system of four explicit first-order ordinary differential equations. The signs $\pm$ and $\mp$ in (9) mean that there are two arcs of curve starting at $(u_0^1, v_0^1, u_0^2, v_0^2)$ associated with the two possible opposite directions of the tangent vectors $\mathbf{T}_1(s)$ and $\mathbf{T}_2(s)$.

3.2 Determining the Starting Point on the Intersection Curve

In the previous section an initial value system of ODEs associated with an intersection curve between two parametric surfaces has been determined. For such a system to be unambiguously defined, we must provide a starting point on that intersection curve. Several methods have been described to solve this problem (see Section 1 for details).

In [5,16] the function d given by

$$d = (\mathbf{S}_1(u_1, v_1) - \mathbf{S}_2(u_2, v_2)) \cdot (\mathbf{S}_1(u_1, v_1) - \mathbf{S}_2(u_2, v_2)), \quad (11)$$

that represents the square of the distance between two arbitrary points of the surfaces $\mathbf{S}_1$ and $\mathbf{S}_2$, has been proposed in the framework of distance problems between parametric surfaces. Note that the gradient of the function d, given by

$$\nabla d(u_1, v_1, u_2, v_2) = \left(\frac{\partial d}{\partial u_1}, \frac{\partial d}{\partial v_1}, \frac{\partial d}{\partial u_2}, \frac{\partial d}{\partial v_2}\right)$$

can be considered as a vector field in $\mathbb{R}^4$ that can be successfully applied to determine a starting point on the intersection curve. The basic idea can be summarized as follows: firstly we start at an arbitrary point on the set $\Lambda \subset \mathbb{R}^4$ defined by the parametric domain of the variables (u_1, v_1, u_2, v_2). Then, we move on that domain from this initial point following the direction of the gradient vector field $-\nabla d(u_1, v_1, u_2, v_2)$ that minimizes the distance between $\mathbf{S}_1$ and $\mathbf{S}_2$. The corresponding trajectory is a gradient curve obtained by numerical integration of the system of first-order explicit ODEs

$$\left(\frac{du_1}{ds}, \frac{dv_1}{ds}, \frac{du_2}{ds}, \frac{dv_2}{ds}\right) = \frac{-\nabla d(u_1, v_1, u_2, v_2)}{||\nabla d(u_1, v_1, u_2, v_2)||} \quad (12)$$

where s is the euclidean arc-length in $\mathbb{R}^4 = (u_1, v_1, u_2, v_2)$. This numerical integration is accomplished until we reach a point $\mathbf{P}^* = \mathbf{S}_1(u_1^*, v_1^*) = \mathbf{S}_2(u_2^*, v_2^*)$ such that $d(u_1^*, v_1^*, u_2^*, v_2^*) = 0$, that is, a point $\mathbf{P}^*$ that lies on the intersection curve of surfaces $\mathbf{S}_1$ and $\mathbf{S}_2$.

4 Some Illustrative Examples

In this section, the performance of the proposed method is analyzed. Although our discussion has been restricted to just a few examples because of limitations of space, they have been carefully chosen so that they contain several problems similar to those usually found in many parametric surface intersection scenarios.

In general, the method works well for any pair of parametric surfaces. However, because of their outstanding advantages in industrial environments, their flexibility and the fact that they can represent well a wide variety of shapes, our examples will focus on NURBS surfaces. In this case, the corresponding derivatives in Eq. (9) have been obtained by following the procedure described in [17].

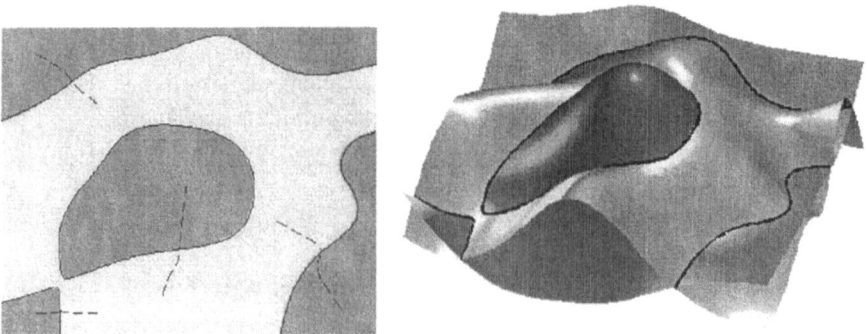

Fig. 1. Intersection of two NURBS surfaces: (left) ground view; (right) 3D view

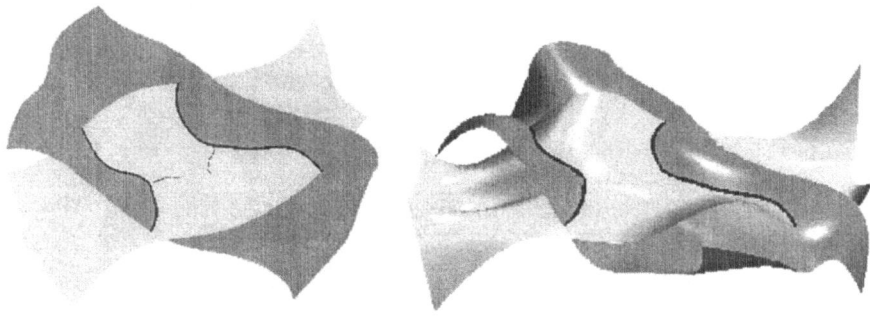

Fig. 2. Intersection of two NURBS surfaces: (left) ground view; (right) 3D view

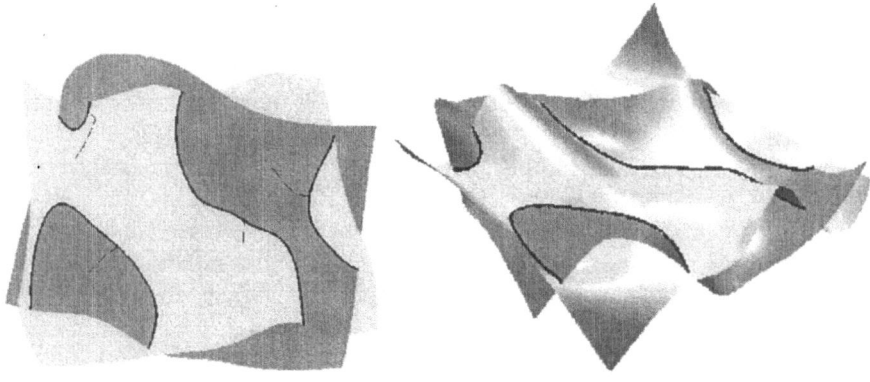

Fig. 3. Intersection of two NURBS surfaces: (left) ground view; (right) 3D view

Fig. 1 shows the intersection curves between two $(3,4)$-order NURBS surfaces, each defined by a grid of 6×6 control points (with weights $w_{i,j} \neq 1$ except for the corner points) and nonperiodic knot vectors for both variables (according to the classification used in [2]). Figs. 2-4 show additional examples of the intersection of NURBS surfaces. For example, in Figs. 2 and 3 the intersection curves of a $(5,3)$-order surface given by a mesh of 8×5 control points and a $(5,4)$-order surface given by a mesh of 8×5 control points are displayed. Finally, Fig. 4 shows the intersection curves of a $(4,4)$-order surface with 6×8 control points and a $(4,3)$-order surface with 8×4 control points.

As explained above, the method starts with the determination of the starting points for each of the intersection branches. To this end, we consider arbitrary points on each of the parametric surfaces, and apply the algorithm described in Section 3.2 to minimize their distance by integrating numerically Eq. (12) until the distance between these points vanishes. The dashed lines in Figs. 1-4(left) show the corresponding trajectories from this integration process. As it can be seen, the procedure will yield a starting point on the nearest branch

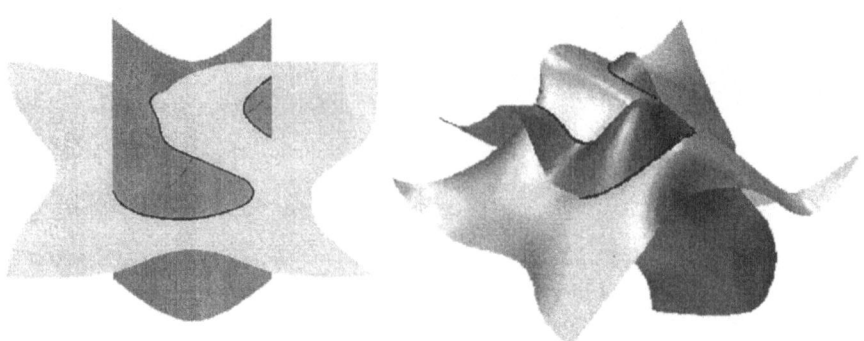

Fig. 4. Intersection of two NURBS surfaces: (left) ground view; (right) 3D view

of the intersection curve between the two surfaces. Of course, the intersection between two parametric surfaces may contain several unconnected branches. In this case, an additional analysis must be performed to determine all branches for the surface intersection problem. Although such analysis is usually carried out by hand, the algorithm still exhibits a very good performance. On the other hand, the domain of the involved surfaces must be taken into account during the integration process. For instance, the integration to find a starting point on the intersection curve will stop if the trajectory reaches: (a) a singular point, (b) the limits of the domain of the parametric surface or (c) a zero value for d, meaning that such a point on the intersection curve has been already obtained.

Once the starting point for each intersection curve between the two surfaces is obtained, such a curve is determined by using the system (9)-(10) with that starting point as initial value. However, the calculation of the analytical solutions for Eqs. (9)-(10) is not possible in general. Fortunately, the system (9)-(10) does involve first-order explicit ODEs only and hence many standard numerical techniques can be applied instead. In particular, all the numerical work in this paper has been performed by using an adaptive 4-5-order Runge-Kutta algorithm [18] with control of absolute and relative error tolerance. In our examples, the absolute tolerance error (a threshold below which the value of the ith solution component is unimportant) was 10^{-8} while the relative tolerance error (a measure of the error relative to the size of each solution component) was 10^{-5}.

Figs. 1-4 illustrate the good performance of the method. The continuous lines in these figures correspond to the intersection curves between the NURBS surfaces. For example, in Fig. 1 four intersection curves have been found. In three cases the integration process stops when the trajectory of the intersection curve reaches the boundaries of the parametric surfaces. In the fourth case, a closed intersection curve has been found, meaning that additional tests for self-intersection had to be incorporated into the method as well. In addition, when the surfaces to be intersected consist of several patches some kind of continuity conditions are required to assure that the differential model is still valid in the neighbourhood of the boundaries between patches. For example, each parametric surface in Figs. 1-3 consists of 12 patches connected with at least C^1 continuity for both u and v directions, while the surfaces in Fig. 4 are comprised of 15 and 10 patches, respectively.

We would like to remark that the method is quite fast. Although it is obvious that the computation time depends on the complexity of the problem, it typically ranges from 10^{-2} to 10^{-1} seconds for the examples reported in this paper and many others with complexity levels similar to those of this paper. All results have been obtained by using Matlab release 11 [14] on a PC Pentium IV at 1.8 Ghz. with 256 Mb. of RAM.

5 Conclusions and Future Work

In this paper, a new method for calculating the intersection curves of parametric surfaces is proposed. In this approach, the problem is formulated in terms of an initial value problem of first-order ordinary differential equations (ODEs), which are numerically integrated through an adaptive 4-5-order Runge-Kutta method. In order to determine the initial value for this system, a simple procedure based on the vector field associated with the gradient of the distance function between points lying on each of the parametric surfaces is described. Such a procedure yields a starting point on the nearest branch of the intersection curve.

The method is very general, it can be applied to any pair of parametric surfaces and it has shown a very good performance in the examples described in this paper as well as in many others not reported here. However, further research is still required in order to improve some features of the method. The automatic determination of different branches for the intersection curves and the analysis of singularities on those curves are examples of unsolved problems. These and other questions are currently under analysis and the future results will be reported elsewhere.

The authors are grateful to the CICYT of the Spanish Ministry of Science and Technology (project DPI2001-1288), the European Union (GAIA II, Project IST-2002-35512) and the University of Cantabria (Spain) for their partial support of this work.

References

1. Abdel-Malek, K. Yeh, H.J.: On the determination of starting points for parametric surface intersections, Computer Aided Design **28**(1) (1997) 21–35
2. Anand, V.: Computer Graphics and Geometric Modeling for Engineers. John Wiley and Sons, New York (1993)
3. Bajaj, C., Hoffmann, C. M., Hopcroft, J. E. H., Lynch, R. E.: Tracing surface intersections, Computer Aided Geometric Design **5** (1988) 285–307
4. Barnhill, R.E., Kersey, S.N.: A marching method for parametric surface/surface intersection, Computer Aided Geometric Design **7** (1990) 257–280
5. Barnhill, R.E. (Ed.): Geometry Processing for Design and Manufacturing, SIAM, Philadelphia (1992)

6. Chandru, V., Dutta, D, Hoffmann, C.M.: On the geometry of Dupin cyclides, The Visual Computer **5** (1989) 277–290
7. Choi, B.K., Jerard, R.B: Sculptured Surface Machining. Theory and Applications. Kluwer Academic Publishers, Dordrecht/Boston/London (1998)
8. Farin, G.: An SSI bibliography. In: Barnhill, R. (ed.) Geometry Processing for Design and Manufacturing, SIAM, Philadelphia (1992) 205–207
9. Farin, G., Hoschek, J., Kim, M.S. (eds.): The Handbook of Computer-Aided Geometric Design, North-Holland (2002)
10. Grandine, T.A., Klein, F.W.: A new approach to the surface intersection problem, Computer Aided Geometric Design **14** (1997) 111–134
11. Grandine, T.A.: Applications of contouring, SIAM Review **42** (2000) 297–316
12. Hoschek, J., Lasser, D.: Computer-Aided Geometric Design, A.K. Peters, Wellesley, MA (1993)
13. Iglesias, A.: Computational methods for geometric processing of surfaces: blending, offsetting, intersection, implicitization. In: Sarfraz, M. (Ed.) Advances in Geometric Modeling, John Wiley and Sons, Chichester, England (2003) Chapter 5
14. The Mathworks Inc: Using Matlab. Natick, MA (1999)
15. Kriezis, G.A., Patrikalakis, N.M., Wolters, F.E.: Topological and differential-equation methods for surface reconstructions, Computer Aided Design **24**(1) (1992) 41–55
16. Patrikalakis, N.M., Maekawa, T.: Shape Interrogation for Computer Aided Design and Manufacturing. Springer-Verlag, New York, Berlin Heidelberg (2002)
17. Piegl, L., Tiller, W.: The NURBS Book, Springer Verlag, Berlin Heidelberg (1997)
18. Press, W.H., Teukolsky, S.A., Vetterling, W.T., Flannery, B.P.: Numerical Recipes (2nd edition), Cambridge University Press, Cambridge (1992)
19. Pratt, M.J., Geisow, A.D.: Surface-surface intersection problems. In: Gregory, J.A. (ed.) The Mathematics of Surfaces, Clarendon Press, Oxford (1986) 117–142
20. Puig-Pey, J., Gálvez, A., Iglesias, A.: A new differential approach for parametric-implicit surface intersection. In: Computational Science-ICCS'2003. Sloot, P.M.A., Abramson, D., Bogdanov, A.V., Dongarra, J.J., Zomaya, A.Y., Gorbachev, Y.E. (Eds.) Lectures Notes in Computer Science, Vol. 2657. Springer-Verlag, Berlin Heidelberg (2003) 897–906
21. Sederberg, T.W., Meyers, R.J.: Loop detection in surface patch intersections, Computer Aided Geometric Design **5** (1988) 161–171

A Comparison Study of Metaheuristic Techniques for Providing QoS to Avatars in DVE Systems *

P. Morillo[1], J.M. Orduña[2], M. Fernández[1], and J. Duato[3]

[1] Instituto de Robótica. Universidad de Valencia. SPAIN.
[2] Departamento de Informática. Universidad de Valencia. SPAIN
[3] DISCA. Universidad Politécnica de Valencia. SPAIN
{Pedro.Morillo,Juan.Orduna}@uv.es

Abstract. Network-server architecture has become a de-facto standard for Distributed Virtual Environment (DVE) systems. In these systems, a large set of remote users share a 3D virtual scene. In order to design scalable DVE systems, different approaches have been proposed to maintain the DVE system working under its saturation point, maximizing system throughput. Also, in order to provide *quality of service* to avatars in a DVE systems, avatars should be assigned to servers taking into account, among other factors, system throughput and system latency. This highly complex problem is called *quality of service (QoS) problem* in DVE systems. This paper proposes two different approaches for solving the QoS problem, based on modern heuristics (simulated annealing and GRASP). Performance evaluation results show that the proposed strategies are able no only to provide quality of service to avatars in a DVE system, but also to keep the system away from the saturation point.

1 Introduction

Distributed Virtual Environments (DVE) are systems where many users can connect their client computers through different networks and interact in the same 3D virtual scene [16]. Each user of the DVE appears in the virtual world as an entity, usually humanoid, called *avatar*. Avatars are controlled by the users, and each avatar offer a different point of view of the scene. DVE systems are currently used in many different applications such as collaborative design [15], civil and military distributed training [7], e-learning [8] or multi-player games [5]. Nowadays, most of current DVE systems have a network-server architecture. In this architecture (also denoted as mirrored-server) each user of the system is assigned to a server, so that when a client performs a movement in the virtual scene it sends updating messages to the server where it is assigned to. This server will be responsible for distributing this message to the rest of clients and servers of the system, in order to maintain a consistent view of the virtual world for all the avatars. In order to avoid a message outburst when the number of clients increases, concepts like areas of influence (AOI) [16] have been defined. This concept describes a neighborhood area for avatars, in such a way that a given avatar must notify his movements (by sending an updating message) only to those avatars located in that neighborhood. These destination avatars are denoted as neighbor avatars.

---
* Supported by the MCYT under grants DPI2002-04438-C02-02 and TIC2003-08154-C06-04

Lui and Chan have shown the key role of finding a good assignment of avatars (clients) to servers in order to ensure both a good frame rate and a minimum network traffic in DVE systems [6]. The problem of efficiently assigning avatars to the different servers of the system is called the *partitioning problem* [6], and several approaches have been proposed for solving it. Lui and Chan model the problem as a formal numerical optimization problem and obtain the solution using an ad hoc algorithm [6]. Keeping the same specifications of the problem, the results obtained by this technique have been improved by using metaheuristic techniques [9]. However, none of these approaches takes into account the non-linear behavior of DVE systems with the number of avatars in the system, shown in [10]. This work shows that the main purpose of any partitioning method should be keeping all servers in the system away from reaching 99% of CPU utilization. Otherwise, the entire DVE system enters saturation and system latency greatly increases. Recently, an adaptive strategy that takes into account this non-linear behavior of DVE systems has been proposed for solving the partitioning problem [12].

However, once the partitioning method has ensured that the system is under its saturation point (it has provided a partition where the estimated percentage of CPU utilization in all the DVE servers is under 99%), then the computing resources can still be used to decrease the average system time response provided to avatars. This improvement should be carried out also by the partitioning method, since it is really a trade-off between system throughput and system latency. The problem of solving the partitioning problem ensuring that the system is under its saturation point and at the same time the average latency provided to avatars is minimized is known as the *quality of service problem (QoS problem)*. This problem can be modelled as finding a partition minimizing a new quality function.

In this paper, we propose a comparative study of two different metaheuristics for solving the QoS problem. One of them is Simulated Annealing (SA), a stochastic metaheuristic. The other one is GRASP, a constructive metaheuristic. Performance evaluation results show that the proposed metaheuristics are both valid methods for solving the QoS problem, simultaneously providing quality of service to a large set of avatars and also maintaining the DVE system under the saturation point. Therefore, they can be used as a valid partitioning methods to provide QoS to avatars in DVE systems. The rest of the paper is organized as follows. Section 2 details the problem of providing QoS to avatars and how it has been addressed in DVE systems. Also, we propose in this section a method to provide QoS through the partitioning problem. In Section 3, we describe the tuning of two different heuristics when applied to the solving of this problem. Next, Section 4 presents the performance evaluation of the proposed heuristics. Finally, Section 5 presents some concluding remarks and future work to be done.

2 The Quality of Service Problem in DVE Systems

The Quality of Service problem (QoS problem) has been already described in DVE systems, and some strategies have been proposed for solving it [1,17]. Approaches like [17] use latency compensating methods in order to repair the effects of high network jitter. Adaptative rendering strategies like [1] or [16] modify the resolution of the 3-D models depending on the client connection speed. However, none of these strategies takes into account the non-linear behavior of DVE systems with the workload assigned to each server, as described in [10]. Therefore, these strategies cannot guarantee that the performance provided to avatars will not degrade beyond any threshold value.

QoS problem can be expressed in DVE systems as latency constraints. In order to fulfill these constraints, and taking into account the non-linear behavior of these systems described in [10], a trade-off among server saturation, clients' interactivity and system stability must be reached. A DVE system can only offer QoS to clients if it is working under its saturation point and at the same time the average round-trip delay for the messages sent by each avatar (denoted as ASR, for *average system response*) is lower than 250 ms. [17]. However, the ASR provided to a given avatar i depends on where avatars located in the AOI of i are assigned to. If avatar i is assigned to server s then the ASR for avatar i linearly decreases with the number of avatars in the AOI of i that are migrated from other servers to server s. Therefore, the problem of offering QoS to avatars can be expressed as a new partitioning problem. A partitioning method that provides QoS to avatars will have to maximize the number of neighbor avatars assigned to the same server and at the same time it will have to keep the system away from saturation. Additionally, since this strategy is a global load balancing scheme it must not migrate more than 30% of avatars in the system [4]. Therefore, the partitioning problem will consists of finding a partition complying with all these three requirements.

In order to solve this partitioning problem, we propose a quality function that takes into account all these requirements. Equation 1 represents the proposed evaluation function, composed of three terms. This quality function measures the quality of each partition (assignment of n avatars to s servers).

$$f_{QoS} = \sum_{i=0}^{s} h_{cpu}(i) + \sum_{j=0}^{n} h_{asr}(j) + n_m \qquad (1)$$

The term $h_{cpu}(i)$ is a function of the percentage of CPU utilization in server i. The behavior of this function is exponential, as shown in Figure 1(a). While the percentage of CPU utilization in server i is under 80%, this function provides a low value. However, as the percentage of CPU utilization goes beyond this threshold value, function $h_{cpu}(i)$ greatly increases. In this way, this function rejects any partition where at least one of the DVE servers is close to saturation.

The term $h_{asr}(j)$ is a function of the ASR provided to avatars by the systems, and it is composed of two sections. The section from that zero value to an ASR of 25 ms. is has an inverse exponential behavior, as shown in Figure 1(b). From 250 up this function shows a parabolic behavior. Therefore, function $h_{asr}(j)$ penalizes partitions where the ASR of avatars higher than 250 ms.

Finally, the term $n_m(i)$ is a function of the number of avatars that should be migrated in order to obtain a given partition. This function is also composed of two sections. Section from the zero value to one third of the existing avatars shows a linear behavior, as shown in Figure 1(c). From one third up, this function also show a parabolic behavior. In this way, this function avoids partitions that only can be obtained by migrating more than 30% of avatars. are migrated.

Thus, the QoS problem in DVE system is reduced to find the minimum value of f_{QoS}. Because of the high complexity of this problem, labelled as NP-hard in other systems

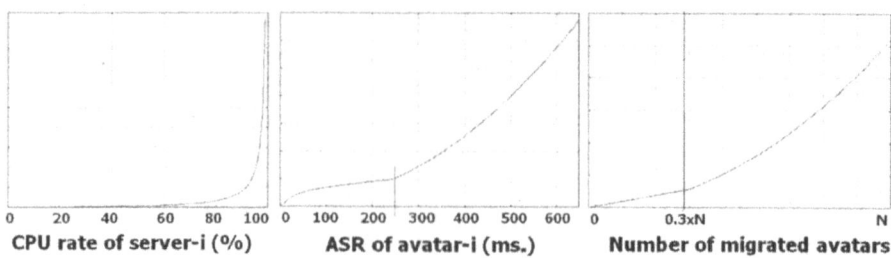

Fig. 1. Behavior of constraints in f_{QoS} for a) CPU utilization b) ASR and and c) migrations

[18], we propose two different approaches based on metaheuristic procedures. These approaches solve the problem in a domain composed by n^s different feasible solutions.

3 Heuristic Adaptation for QoS Problem in DVE Systems

Metaheuristic strategies are widely considered as one of the most practical approaches for highly complex problems. A wide range of different problems have been solved using these strategies. Moreover, metaheuristics have been used in DVE systems for solving the partitioning problem [9]. In this section, we present the implementation and tuning of two different heuristics, based on simulated annealing and GRASP, for solving the QoS problem in DVE systems.

3.1 Simulated Annealing

Simulated Annealing (SA) is a stochastic metaheuristic applicable to arbitrary combinatorial optimization problems [3]. SA is a randomized local search strategy which is able to perform climbing moves. In this sense, it can also escape from local minima and find solutions which are much better than those of pure local search.

SA has been used in a wide range of problems [3,9,13]. This method models system temperature as the probability of accepting a worsening result. SA starts with a high system temperature, and in each iteration system temperature is decreased. In this way, SA can leave local minima by accepting worsening results at intermediate stages. The search method ends when either the number of iterations finishes or system temperature is so low that accepting worsening results is practically impossible.

The proposed implementation of SA method when applied to solving the QoS problem in DVE systems deals with *boarder avatars*. Two avatars (a_i and a_j) are boarder avatars if they are assigned to different servers (s_r and s_x) and their AOIs intersect. The assignment of boarder avatars is critical, and it allows to obtain partitions with low levels of h_{asr}. An iteration of SA consists of randomly selecting two different boarder avatars and randomly performing one of these three actions: exchanging the assignment of servers, both avatars server s_x or assigning both avatars to server s_r. Once this action is performed, the quality function for the resulting partition is computed. If the resulting value of f_{QoS} is higher than the previous one plus a threshold T, that change is rejected. Otherwise, it is accepted (the search method must decrease the value of the quality function f_{QoS} associated with each assignment). The threshold T used in each iteration i of the search depends on the rate of temperature decreasing R, and it is defined as

$$T = R - \left(\frac{R \times i}{N}\right) \quad (2)$$

where N determines the finishing condition of the search. When N iterations are performed without decreasing the quality function f_{QoS}, then the search finishes.

As literature shows ([3], [9]) the two key issues for properly tuning this heuristic search method are the number of iterations N and the temperature decreasing rate R. Figure 2 shows the tuning of SA method. The graphic on the left shows the f_{QoS} values obtained with SA method when different number of iterations are performed, and the graphic on the right shows the f_{QoS} values obtained with SA method when different cooling rates are considered. These results have been obtained for a DVE system composed of 700 avatars assigned to 10 servers.

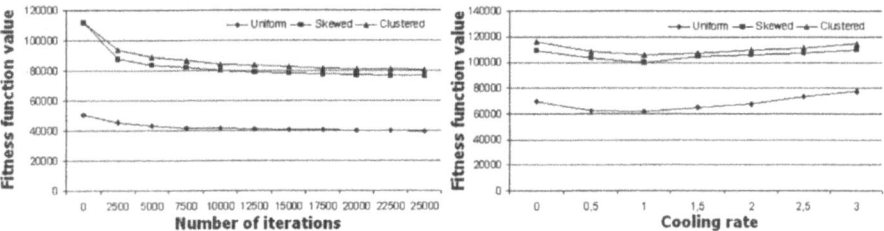

Fig. 2. Variation of quality function f_{QoS} for different number of iterations and cooling rates

Figure 2-left shows that f_{QoS} decreases as the number of iterations increases, until value of 7500 iterations is reached. From this point this value remains constant or decreases very slightly. This behavior is due to the impossibility of finding better search paths even when more iterations are performed. On other hand, Figure 2-right shows that f_{QoS} decreases until a value of 1% is reached. From this point quality function also increases, since cooling rate is too high and the search method accepts too many worsening solutions. Therefore, for this DVE configuration, the number of iterations and cooling rate selected for SA method has been 7500 and 1% respectively.

3.2 Greedy Randomized Adaptive Search (GRASP)

GRASP is a constructive technique designed as a multi-start heuristic for combinatorial problems [2]. It has been shown to quickly produce good quality solutions for a wide variety of problems [14],[9].

The proposed implementation of GRASP method for solving the QoS problem in DVE systems starts with an initial partition. This initial partition is provided by a load balancing technique [12]. Therefore, we ensure that the initial partition is well-balanced and all servers have a percentage of CPU utilization as low as possible. At this point, GRASP method will be used for searching a near optimal partition that provides QoS to the maximum number of avatars migrating the minimum number of avatars.

The first step in our GRASP implementation consists of sorting the avatars whose messages show a round-trip delay higher than 250 ms. (those avatars not provided with QoS) by their presence factor f_p. We define the *presence factor* ($f_p(i)$) of avatar i [11] as the number of avatars in whose AOI avatar i appears. The idea is to provide QoS to those avatars that require the least system efforts. The avatars with the higher presence factor should receive updating messages from a lot of avatars, and it will send messages to a lot of avatars. Therefore, migrating these avatars to the proper server can decrease the round trip-delay for the messages sent by a lot of avatars (can provide QoS with the least effort). Moreover, if GRASP method focuses only on this kind of avatars then it will significantly decrease the term $\sum_{j=0}^{n} h_{asr}(j)$ in f_{QoS} function without significantly increasing the term n_m.

The first c elements in the sorted list of avatars (from a population of n avatars) are denoted as *critical avatars*. GRASP method considers critical avatars as non-assigned avatars, and they will be assigned by GRASP method to a server in such a way that QoS is provided to them. The rest of the n avatars (denoted as the e *easy avatars*, where $n = c + e$) will not be re-assigned, and they will remain assigned to the same server where they were initially assigned to. The assignment of each of the c critical avatars is obtained in each of the iterations of the GRASP method. The number of iterations (the number of re-assigned avatars) is the only parameter to be tuned for GRASP method. If c value is set too low, then only a few avatars will be provided with QoS. Also, if c value is set too high (trying to provide too many avatars with QoS) then GRASP method will not be able to find a partition fulfilling all the requirements for all avatars, and it will take a long time for providing bad partitions.

Each iteration of GRASP method consists of two steps: construction and local search. The *construction* phase builds a feasible solution choosing one critical avatar by iteration, and the *local search* derives this temporal solution following a neighborhood criterion. Concretely, we propose an implementation where each iteration i performs the next steps:

- Constructive phase:
    1. The first avatar in the sorted lists of critical avatar is randomly assigned to a server, and the quality function f_{QoS} is computed for the partition composed of the e easy avatars plus this new avatar assigned to that server.
    2. The previous step is repeated with the remaining $c - 1$ avatars in the sorted list of critical avatars. This step will provide a list of $c - i + 1$ different critical avatars each of them randomly assigned to a server and each one with a f_{QoS} value. This list will be denoted as the *list of candidate avatars for iteration i (LCA(i))*. LCA(i) will have a size equal to $c - i + 1$, that is, the number of non-assigned avatars for iteration i. Each element in LCA(i) will have the form (*non-assigned border avatar, server, resulting f_{QoS}*.
- Local search phase:

1. LCA(i)list is sorted (using Quick-sort algorithm) by the resulting cost f_{QoS} in ascendent order, and then is reduced to its top quartile. An avatar j in this reduced LCA(i) list is randomly selected for local search.
2. Local search on avatar j consists of looking of non-assigned critical avatars that are neighbors of avatar j. If any avatar k exists in LCA(i) and k is a neighbor avatar of j, then all possible assignments of avatar k to the different servers are considered. For each of them, the quality function f_{QoS} is computed for the partition composed of the $e + i - 1$ assigned avatars plus this new avatar k assigned to that server.
3. If there not exists any avatar k that is a neighbor avatar of avatar j and at the same time appears in LCA(i), then the solution of iteration i consists of assigning avatar j to the server where avatar j is assigned in LCA(i).
4. If any avatar k exists, then all assignments of avatar k (and any other existing avatar in the previous step) to the different servers and their resulting f_{QoS} are sorted in ascending order to form the *list of local search assignments LLSA(i)*, together with avatar j and its assignment in LCA(i). The first element in LLSA(i) (the assignment of avatar j or its neighbors with the minor f_{QoS} value) is selected as the solution of GRASP for iteration i

The main parameter to be tuned in GRASP method is the number of avatars c that the initial partition must leave unassigned. Figure 3 shows the tuning of GRASP method when this value is varied. The graphic on the left shows the values of f_{QoS} obtained for different values of this parameter, and the graphic on the right shows different execution times required for performing the search when this parameter is varied. The results shown in these figures have been obtained from a DVE system composed of 700 avatars and 10 servers.

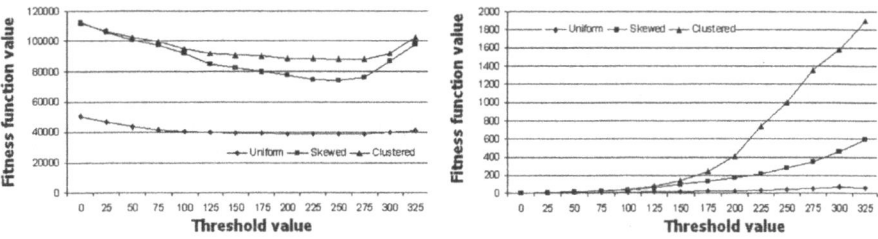

Fig. 3. Variation of quality function f_{QoS} and execution times for different values of c

This figure shows that as the number of critical avatars increases the quality of the provided partitions increases and so does the required execution time. In the case of this DVE configuration, 125 iterations has been chosen in order to obtain good quality solutions without spending too much execution time. A higher number of iterations would require too much execution time and it would not provide significantly better solutions. Finally, if this value is excessively increased (more than 275 iterations) then f_{QoS} even increases. This behavior is produced by the greedy component of the algorithm that offers suboptimal solutions within the construction phase.

4 Performance Evaluation

In this section, we present the performance evaluation of the heuristics described in the previous section when they are used for solving the QoS problem in DVE system. We have empirically tuned SA and GRASP search methods in two different DVE configurations, denoted as MEDIUM1 and MEDIUM2. Empirical results have been obtained from our DVE simulation tool described in [10], [11] and [12]. This tool models the behavior of a generic DVE system with a network server architecture on a real network of heterogeneous computers. MEDIUM1 is composed by 250 avatars and 3 servers, and MEDIUM 2 is composed by 700 and 10 servers. In both configurations uniform, skewed and clustered distribution of avatars have been simulated. However, due to space limitations, we present here the result for MEDIUM2 configuration. The results obtained for MEDIUM1 configuration were very similar.

Table1 shows the performance evaluation results obtained for MEDIUM2 configuration when the proposed method is simulated under different initial distributions of avatars in the virtual world. For comparison purposes, we have evaluated the DVE performance obtained with ALB method [12] and the performance obtained with the proposed methods, SA and GRASP. For each distribution of avatars, table 1 contains three columns, one for each partitioning method. These columns show the performance provided by each method. The first nine rows, labelled with Sx, show the percentage of CPU utilization reached in each DVE server with each partitioning method. Last but two row shows the number of avatars in the partition provided by each partitioning method whose messages showed an average round trip-delay lower than 250ms. That is, this row shows the number of avatars provided with QoS by each partitioning method. Next row shows the number of migrated avatars $\Gamma(P_0)$ in the whole simulation. Finally, last row shows the execution time required by each partitioning method to provide the final partition tested in the simulation.

Table 1 shows that for large DVE configurations the proposed methods allows to provide QoS to a significatively higher amount of avatars. Thus, for example, for a uniform distribution of avatars in the virtual world GRASP method is able to increase about a 22% the amount of avatars provided with QoS with respect of ALB method, while maintaining all servers far from reaching 95% of CPU utilization and also migrating less than one third of the population of avatars. In the case of a clustered distribution of avatars, GRASP is able to increase in about 50% the amount of avatars provided with QoS, in relation to ALB method. In the case of an skewed distribution of avatars, the increasing is about a 300%. These results fully validate the proposed method as a valid approach for providing QoS to the highest number of avatars as possible.

It is also worth mention the great differences in absolute terms that the same method provides for the different distributions of avatars, particularly between the uniform distribution of avatars and the other two distributions. This difference is due to the differences in the presence factor of avatars between the distributions. Since most of the avatars are very close each other in both skewed and clustered distributions, then most of avatars are

Table 1. Results for a MEDIUM2 DVE configuration

	Uniform distribution			Skewed distribution			Clustered distribution		
	ALB	SA	GRASP	ALB	SA	GRASP	ALB	ALB	GRASP
S0 (%)	25	23	21	67	85	42	88	75	87
S1 (%)	28	30	29	58	46	78	71	69	61
S2 (%)	18	19	35	64	69	56	93	74	76
S3 (%)	17	14	14	63	48	58	80	75	76
S4 (%)	20	20	8	66	71	67	69	77	77
S5 (%)	17	18	12	89	67	51	74	76	74
S6 (%)	16	13	41	36	67	58	77	78	61
S7 (%)	22	23	9	55	62	84	71	79	84
S8 (%)	17	21	16	78	59	87	74	77	85
S9 (%)	17	16	12	65	66	59	67	81	80
QoS	544	629	663	92	232	270	256	363	388
$\Gamma(P_0)$	-	97	101	-	209	204	-	201	184
$T_{exe}(s.)$	-	7.6	5.3	-	20.1	18.8	-	38.6	34.3

highly connected with other avatars. As a consequence, it is more difficult to distribute these avatars between more servers while still providing QoS to all of them.

Finally, table 1 also shows similar execution times and number of migrated avatars for both proposed methods. However, for all three distributions of avatars GRASP method obtains partitions with less numbers of avatars without QoS than SA method. The reason of this behavior is the fast and powerful mechanism of GRASP approach in order to explore huge domains of solutions.

5 Conclusions and Future Work

Traditionally, DVE(Distributed Virtual Environments) systems have addressed the QoS of clients with graphical approaches. However these approaches, based on multiresolution models or compensation mechanisms, do not take into account the non-linear behavior of DVE systems with the workload they support.

In this paper, we have proposed the implementation, tuning and comparison study of two different search methods, based on Simulated Annealing (SA) and GRASP, in order to solve the QoS problem in DVE systems. These approaches models the QoS problem as an evaluation function to be minimized when solving the partitioning problem.

Performance evaluation results shows that the proposed methods can be considered as a good mechanism in order to offer QoS to avatars in a DVE system. These results show similar performance, in terms of quality of the provided solutions and execution times, for both methods and for a small DVE configuration (MEDIUM1). For large DVE configurations (MEDIUM2) GRASP method manages to provide more avatars with QoS than SA method.

As future work to be done, we plan to design a parallel implementation of GRASP method that can take advantage of the DVE servers where it will be performed. This new design will be based on a master-slave configuration and will be implemented in conjunction with a post-optimization path-relinking procedure.

References

1. Z. Choukair, D. Retailleau, M. Hellstrom. "Environment for Performing Collaborative Distributed Virtual Environments with QoS", in Proceedings of the IEEE Conference *ICPADS'00*, pp.111-118, Iwate, Japan. July, 2000.
2. T. A. Feo and M. G. C. Resende, "Greedy Randomized Adaptive Search Procedures", *Journal of Global Optimization*, No.1, pp.109-136 1995.
3. P.V. Laarhoven and E. Aarts, "Simulated annealing: Theory and applications", *Reidel Publication*, Dordrecht, Holland, 1987.
4. K. Lee and D., "A Scalable Dynamic Load Distribution Scheme for Multi-Server Distributed Virtual Environment Systems With Highly-Skewed User Distribution", *Proceedings of the 10th ACM Conference VRST-2003*, October 2003, pp.160-168, Osaka, Japan.
5. M. Lewis and J. Jacboson, "Game Engines in Scientific Research", in *Communications of the ACM*, Vol. 45, No. 1, pp. 27-31, January 2002.
6. Jonh C.S. Lui and M.F. Chan, "An Efficient Partitioning Algorithm for Distributed Virtual Environment Systems", *IEEE Transactions TPDS*, Vol.13, No.3, pp. 193-211. March 2002.
7. D.C.Miller and J.A. Thorpe, "SIMNET: The advent of simulator networking", in *Proceedings of the IEEE*, Vol. 83, No.8, pp. 1114-1123. August, 1995.
8. T. Nitta, K. Fujita and S. Cono, "An Application Of Distributed Virtual Environment To Foreign Language", in *Proceedings of FIE'2000. IEEE Education Society*. Kansas City, Missouri, October 2000.
9. P. Morillo, M. Fernández and J. M. Orduña, "A Comparison Study of Modern Heuristics for Solving the Partitioning Problem in Distributed Virtual Environment Systems", in *International Conference in Computational Science and Its Applications (ICCSA' 2003)*, volume 2669 of Springer LNCS, pp. 458-467, Montreal, Canada. May 2003.
10. P. Morillo, J.M. Orduña, M. Fernández and J. Duato, "On the Characterization of Distributed Virtual Environment Systems", *Proceedings of European Conference on Parallel Processing (Euro-Par' 2003)*, Klagenfurt, Austria. August, 2003.
11. P. Morillo, M. Fernández and J. M. Orduña, "On the Characterization of Avatars in Distributed Virtual Worlds", in *Annual Conference of the European Association for Computer Graphics (EUROGRAPHICS' 2003)*, pp. 215-129, Granada, Spain. September, 2003.
12. P. Morillo, J.M. Orduña, M. Fernández and J. Duato, "An Adaptive Load Balancing Technique for Distributed Virtual Environment Systems", *Proceedings of the 15th IASTED International PDCS-03*, pp.256-261, California, USA. November, 2003.
13. C.A.S. Oliveira, D. Paolini and P.M. Pardalos, "A Randomized Algorithm for Minimizing User Disturbance Due to Changes in Cellular Technology", *Proceedings of the International Conference CCCT-03*, Vol. 5, pp. 45-50, Orlando, Florida. July, 2003.
14. M.G.C. Resende and C.C. Ribeiro, "A GRASP with path-relinking for privatevirtual circuit routing", *Networks*, No. 41, pp. 104–114, 2003.
15. J.M. Salles, Ricardo Galli, A. C. Almeida et al, "mWorld: A Multiuser 3D Virtual Environment", in *IEEE Computer Graphics*, Vol. 17, No. 2, pp.55-65. March-April 1997.
16. S. Singhal and M. Zyda, *Networked Virtual Environments* (ACM Press, New York, 1999).
17. T. Henderson and S. Bhatti, "Networked games: a QoS-sensitive application for QoS-insensitive users?", in *Proceedings of the ACM Conference SIGCOMM 2003*, pp. 141-147, Karlsruhe, Germany. August 2003.
18. X. Yuan, "Heuristic Algorithms for Multi-Constrained Quality of Service Routing", *IEEE/ACM Transactions on Networking*, Vol. 10, No. 2, pp. 244-256. April 2002.

Visualization of Large Terrain Using Non-restricted Quadtree Triangulations

Mariano Pérez, Ricardo Olanda, and Marcos Fernández

Instituto de Robótica. University of Valencia.
Polígono de la Coma, s/n.
Aptdo. 22085, 46071-Paterna, Spain. {Mariano.Perez, Ricardo.Olanda, Marcos.Fernandez}@uv.es

Abstract. This paper presents a set of new techniques oriented towards the real-time visualization of large terrains. These techniques are mainly focused on semi-regular triangulations of non-restricted quadtree terrain representations. Despite the fact that the paper shows that triangulations based on non-restricted quadtrees are as simple and efficient as those based on restricted quadtrees, the new triangulations avoid discontinuity problems among the boundaries of different patches without the need for tree balancing and extra triangles addition. Another important feature of the proposed triangulation is that it incorporates an efficient method for building triangle strips and triangle fans for the efficient rendering of the final triangle mesh.

1 Introduction

Terrain representation usually involves the management of large databases to represent realistic virtual models on the screen at interactive frame rates. The huge amount of data needed for large terrain representations exceed in most of the cases the current graphic board capacities and even the main memory of visualization platforms. In order to solve these problems, terrain Databases have to be segmented into several square tiles, and dynamic paging algorithms have to be used in order to use only those tiles really needed at each frame chosen according to position, orientation and movement of the virtual camera.

However, even the amount of data needed at each frame is too large to be used in a raw way at interactive frame rates. To take the maximum performance from visualization platforms, algorithms to compute and generate the appropriate level of detail dynamically in real-time, have to be used. This paper address this problem focused on terrain meshes representation.

To solve the problem of terrain meshes generation efficiently, a new triangulation is proposed. Such triangulation is based on Non-Restricted Quadtrees Triangulations (NRQT) which prevent possible cracks in the terrain surface, and uses a lower number of triangles than restricted hierarchical triangulations. Moreover, the proposed triangulation offers the possibility of building optimized primitives such as triangle strips and triangle fans. Additionally, the triangulation presented provides the usual features included in triangulation methods:

- *Continuous Level Of Detail (CLOD)*. The terrain representation is built using a continuous multiresolution approach. This approach mixes different levels of detail maintaining the continuity over the whole terrain surface.
- *View Dependent Level of Detail control*. Using a top-down approach, the level of detail is selected based on position and geometric distortion in screen space, from a first elemental version of the model, at each frame and for each region.
- *View-Culling*. Polygons that are outside of the viewing frustum are excluded from the drawing process. The proposed triangulation eliminates these triangles in opposition to other methods, such as [9] o [6], which only reduce mesh complexity.
- *Optimized Primitives Generation*. The final set of triangles are grouped into optimized rendering primitives (triangle strips and triangle fans), which increase the performance of current graphic systems, avoiding unnecessary overhead in the drawing process.

2 Hierarchical Triangulations

Due to the regularity of digital elevation terrain models, quadtree or binary tree based on hierarchical triangulations have demonstrated to be the most efficient methods in terms of performance rendering. The most commonly used models are:

- *Restricted Quadtree Triangulations (RQT)*. Restricted Quadtrees, also known as balanced quadtrees, limit the maximum level difference among neighboring nodes. The original RQT model was proposed by Von Herzen [4] (fig. 1), who applied it to elevation maps. Regarding to terrain representation, some of the most interesting works belongs to Lindstrom [8] and Pajarola [10].
- *Restricted Binary Tree Triangulation*. This is a hierarchical triangulation scheme closely related to the RQT. It is based on a top-down approach which divides triangles based on longest edge bisection. This triangulation offers a continuous surface without cracks between patches of different levels of detail [1,2,3,9]. The results obtained with this method are quite similar to RQT (fig. 1).

2.1 Point of View Based Refinement

Lindstrom is one of the pioneers in the development of view dependent multiresolution models and their application to real time terrain representations [8]. His algorithm uses a bottom-up scheme in order to generate a semi-regular simplified version of the original mesh at each frame. Some other authors use a more efficient top-down approach to generate the same kind of meshes, such as Duchaineau (ROAM) [2], Pajarola [10], or even Lindstrom [9].

At the same time, Hoppe [6,7] developed a view dependent multiresolution model based on progressive meshes [5]. The main drawback is the considerable memory requirements of his method, which reduces the immediate available information, and increases the on-demand loading time. Additional drawbacks are: fully implementation is a harder task and a worse spatial access.

From the above mentioned works, the Lindstrom approach [9] is one of the most noticeable, because it is simple, efficient and introduces a monotonic error metric on the screen space, which avoids cracking without introducing costly relations among neighboring nodes in order to balance the tree. In this paper, this method is used in order to make a comparative set of tests with the new approach proposed here.

3 Non-restricted Quadtree Triangulation (NRQT)

The main advantage of the NRQT model is that it builds simplified terrain meshes with fewer triangles than the RQT, because no constrains exist between quadtree nodes. While in NRQT the node level is independent from the rest, in RQT the split of a node produces a cascade effect over its neighboring nodes, which implies an increase in the total number of triangles in the terrain representation. (fig. 1).

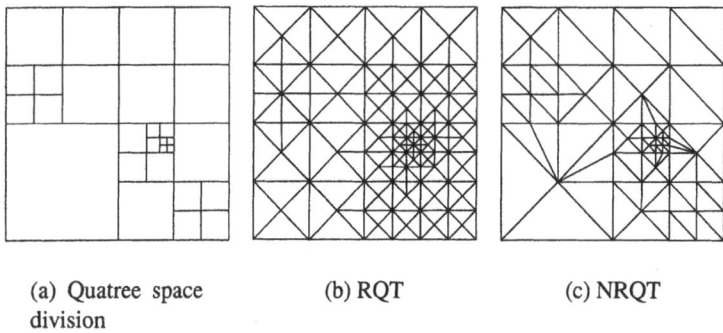

(a) Quatree space division   (b) RQT   (c) NRQT

Fig. 1. RQT and NRQT triangulations of the same space division

3.1 Vertices Extraction Algorithm

The NRQT is based on the hierarchical quadtree non-restricted division of the space in "quadtree regions" driven by view dependent error metrics.

Quadtree generation. The space division is generated following a top-down approach. The process starts from the root node and adds new nodes in a recursive way. The process is driven by the error metric and the *view-culling* algorithm.

The NRQT algorithm is recursive and repeats the following steps for each node (algorithm 1). First, it tests whether or not the region is inside the viewing frustum. When it is fully outside, the quadtree growth is stopped for this branch. When it is inside, two cases are possible: the view dependent error is superior to a certain threshold, then the node is divided into its four children and recursion continues; or the error is below the threshold value, then recursion stops. In the last case, the quadtree node becomes a node leaf and the corners of the region are inserted in order into the vertex lists.

Algorithm 1 Drawing Quatree Generation

```
createQuadtree(node)
    if (insideViewPyramid(node))
        error ← getVDError(node);
        if ( (node.level < tile.maxlevel) and (error > thresholdError) )
            createQuadtree(node.childSE());
            createQuadtree(node.childSW());
            createQuadtree(node.childNE());
            createQuadtree(node.childNW());
            if ( error < getVDError(tile.nearestCollapseNode) )
                tile.nearestCollapseNode ← node;
        else
            if ( error > getVDError(tile.nearestSplitNode) )
                tile.nearestSplitNode ← node;
            vertex ← node.getPatchCornerNW();
            ListsX[vertex.getRow()].addInHeader(vertex);
            ListsY[vertex.getCol()].addInHeader(vertex);
            if ( node.getPatchCornerNE() ≠ vertex.nextX )
                vertexRight ← node.getPatchCornerNE();
                addVertexList(X, vertex, vertexRight);
                vertexAux ← lookForPreviousY(VertexRight); // O(1)
                addVertexList(Y, vertexAux , vertexRight);
            if ( node.getPatchCornerSW() ≠ vertex.nextY )
                vertexDown ← node.getPatchCornerSW();
                addVertexList(Y, vertex, vertexDown);
                vertexAux ← lookForPreviousX(VertexDown); // O(1)
                addVertexList(X, vertexAux , vertexDown));
```

Vertex List. When quatree construction is finished, patches associated to leaf nodes must be broken into compact sequences of triangles (triangle strips, and triangle fans). In order to carry out this task, a set of vertex lists are generated from the patches corners, following directions x e y. The vertices are inserted in the proper order according to their spatial ordering (fig. 2).

There are two sets (arrays) of lists: the first contains vertices which share the same y coordinate, sorted by x coordinate; this set is named ListsX. The other set of lists consists of vertices which share its x coordinate ordered by y coordinate; this set is denoted by ListsY. Each element of ListsX represents a different vertex list (there is one list for each possible y coordinate). ListsY behaves in the same way (fig. 2). Usually, most of these lists are empty because the number of vertices of the simplified version of the mesh is much lower than that of the original mesh.

The insertion process of vertices into lists is done at the same time that quadtree generation, every time a leaf node is inserted. This operation is represented in algorithm 1 by the "AddVertexList" function.

The cost of an insertion operation in a ordered list is linear (O(N)); however, this cost can be reduced to constant (O(1)) if the quadtree is built in the order described in algorithm 1, visiting first the branches located south and east. Following this order, it is only neccesary to add vertices at the beginning of the lists to keep them sorted (algorithm 2).

Functions "lookForPreviousX" and "lookForPreviousY" determine the position in which the vertex has to be inserted into ListsX and ListsY, respectively. The cost of these functions is also constant (O(1)), because the order followed in the quadtree generation assures that the position is always at the beginning of the lists.

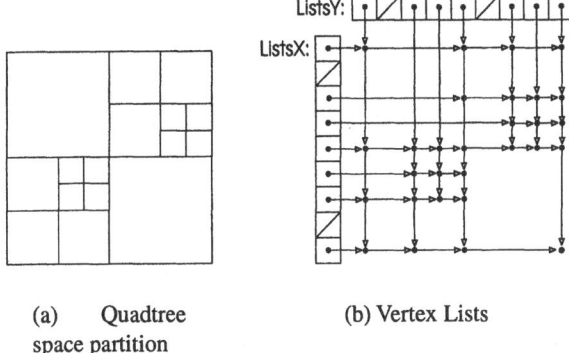

(a) Quadtree space partition

(b) Vertex Lists

Fig. 2. Set of linked Vertex Lists associated to the quadtree space partition

---

Algorithm 2 Sorted Insertion of vertex into a list.

AddVertexList(*orientation, previousNode, newNode*)
  if (*orientation* = X)
    newNode.nextX ← *previousNode.nextX*;
    previousNode.nextX ← *newNode*;
  else
    newNode.nextY ← *previousNode.nextY*;
    previousNode.nextY ← *newNode*;

3.2 Triangle Arrangement Algorithm

The basic elements in the NRQT are the patches or quadtree regions. To keep continuity among neighboring regions with different resolution, patch is split into several triangles but not neighboring nodes (in contrast to other semi-regular triangulations).

Types of Patches. Patches belong to one of the following categories

1. Patches without neighbors of higher resolution.
2. Patches with neighbors of higher resolution.

The process of triangle splitting is different for each kind of patches. Patches belonging to the first category do not present cracks problems, and they are only divided into two triangles (fig. 1). However, the second type of patches has to be split into several triangles which guaranties the boundaries continuity. To perform the splitting process, an extra vertex is added in the center of the quadtree region. The last part of the triangulation process arranges the generated triangles in optimized primitives (triangle strips and triangle fans).

Strips. Triangle strips are generated from triangles belonging to the patches of the first category, which satisfy the following conditions:

1. All the patches are contiguous following x direction (all strip are oriented following the same direction, usually the x direction).
2. All the patches have the same resolution.
3. All top and bottom neighbors have the same or higher resolution (fig. 3).

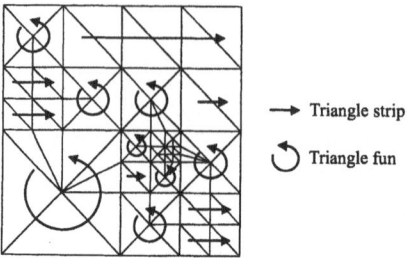

Fig. 3. Triangle arrangement in strips and fans

Algorithm 3 describes the steps followed in the process of triangle strip creation. The process begins from the top-left corner of the first patch; from this position, the algorithm visits vertices from left to right following their spatial ordering.

Due to the fact that the first kind of patches only have vertices at its corners, it is possible to figure out with ease if the next patch satisfies the set of constrains previously indicated (line 2, algorithm 3).

Algorithm 3 Triangle strips generation.

drawTriangleStrip(*vertex*)
    level ← *vertex*.getLevel();
    while ((*vertex.nextX.nextY* = *vertex.nextY.nextX*)
        and *level* = *vertex.nextX*.getLevel())
        TriangleStrip ← *vertex*;
        TriangleStrip ← *vertex.nextY*;
        vertex ← *vertex.nextX*;
    TriangleStrip ← *vertex*;
    TriangleStrip ← *vertex.nextY*;
    return *vertex*;

The function "drawTriangleStrip" (algorithm 3) renders the triangle strip and returns the last vertex of the strip (top-right corner). This vertex starts the next primitive. If the first vertex corresponds to a patch that does not satisfy the inclusion constraints, no strip is generated and the function returns the same initial vertex. From this vertex a triangle fan is created.

Fans. The second type of patches are split into several triangles, joining the center vertex with vertices in the middles of edges in order to avoid cracking (fig. 1). These triangles are arranged in circle around the center vertex generating a triangle fan. The center vertex was not included in the vertex lists, so it is added when the fan is generated (figure 3).

The algorithm 4 describes the process followed in the triangle fan generation. The first step evaluates the coordinates for the center vertex and the limits of the quatree region (corners). Next, the rest of fan vertices are located.

Algorithm 4 Triangle fan generation.

drawTriangleFun(*vertex*)
    TriangleFun ← getPatchCenter(); // Central vertex
    (*xmax*, *ymax*) ← getPatchCornerSE();
    (*xmin*, *ymin*) ← getPatchCornerNW();
    while (*vertex.x* ≠ *xmax*) //Looking for NE vertex
        TriangleFun ← *vertex*;
        vertex ← *vertex*.getNextX();
    nextPrimitiveVertex ← *vertex*;
    while (*vertex.y* ≠ *ymax*) //Looking for SE vertex
        TriangleFun ← *vertex*;
        vertex ← *vertex*.getNextY();
    while (*vertex.x* ≠ *xmin*) //Looking for SW vertex
        TriangleFun ← *vertex*;
        vertex ← *vertex*.getPreviousX();
    while (*vertex.y* ≠ *ymin*) //Looking for NW vertex
        TriangleFun ← *vertex*;
        vertex ← *vertex*.getPreviousY();
    return *nextPrimitiveVertex*;

Primitives Sequence. Algorithm 4 shows how the final version of the mesh is fully covered by triangle strips and triangle fans in an interlaced way. The algorithm visits ListsY[ ] adding its vertices to the triangle strips or the triangle fans.

Functions "drawTriangleStrip" and "drawTriangleFan", apart from the construction of these primitives, return the last vertex of the generated structure. From this vertex the construction process for the next primitive begins.

Algorithm 5 Triangle strips and triangle fans generation sequence.

groupVertex()
    for *j* ← 0 **to** *t*-1 **do** // *t* = number of elements of ListsX
        vertex ← ListsX[*j*].*header*;
        while (*vertex* ≠ null)
            vertex ← drawTriangleStrip(*vertex*);
            vertex ← drawTriangleFun(*vertex*);

3.3 Error Metric

In order to implement an efficient top-down mesh refinement, it is necessary to define a view-independent object space error metric. This metric can be calculated off-line and stored with the elevation data or can be evaluated at loading time. Later on, object space errors are transformed into screen space.

Object Space Error Metric. NRQT uses a similar metric to ROAM [2]. The metric is based on a wedgie bounding of the patch surface. The associated error ϵ to each quadtree region is defined based on its four children regions errors ($\epsilon_{q_1}, \epsilon_{q_2}, \epsilon_{q_3}, \epsilon_{q_4}$), from the vertical distance between the new vertices of the children nodes and the original edges of the patch ($h_{l1}, h_{l2}, h_{l3}, h_{l4}$), and from the vertical distance between the new vertex added in the middle of the divided quadtree region and the two patch diagonals (h_{d1}, h_{d2}):

$$\epsilon = \max\{\epsilon_{q_1}, \epsilon_{q_2}, \epsilon_{q_3}, \epsilon_{q_4}\} + \max\{h_{l1}, h_{l2}, h_{l3}, h_{l4}, h_{d1}, h_{d2}\} \quad (1)$$

The evaluation of these values is performed using a bottom-up approach, beginning with the lowest detail level patches.

Error Projection. There are two established methods to project ϵ onto the screen space [9]: isotropic error projection, which only takes into account the distance between the polygon position and the viewpoint; and the anisotropic error projection where the view direction is also considered. Metric isotropic projection can be written as:

$$\rho(\epsilon) = \lambda \frac{\epsilon}{d} \quad (2)$$

where λ is a constant value and d is the euclidian distance between the polygon position and the viewpoint. This equation is simpler and more efficient than the expression for the anisotropic projection, while the reduction in mesh complexity is not significant [6,8,9].

3.4 Cracks between Tiles

Database segmentation introduces new cracks on boundaries between meshes belonging to different tiles. Because of the independent load of each tile, the relationship among data of different tiles has to be kept at minimum. This makes the definition of a monotonic metric through several tiles a hard task [1,9], introducing an extra complexity to the evaluation of dependencies among neighboring vertices that belong to different tiles [2, 10]. The determination of these dependencies is the base for cracking avoidance (figure 4). Previous triangulations were usually devised to work with an unique mesh covering the whole terrain and are not well prepared for database segmentation.

The NRTQ avoids discontinuities between tiles easily, without extra dependencies. The process is only based on the fusion of the linked vertex list which corresponds to the boundaries between neighboring tiles (figure 4).

3.5 Frame Coherence

NRQT improves the rendering time performance using frame to frame coherence. While the process of quadtree generation associated to a specific tile takes place, the closest node to be divided and collapsed are stored. For every frame, the tiles where the projected errors of the two nodes do not change their status, the associated quadtree is not rebuilt, and the same vertex lists computed in the previous frame is used. (algorithm 6).

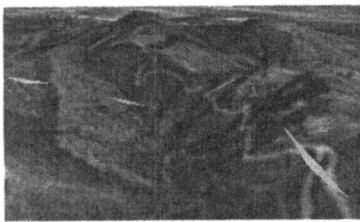 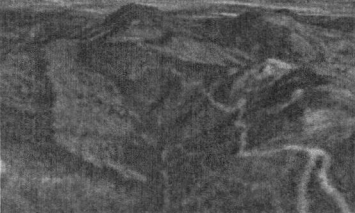

Fig. 4. Left: Cracks between neighboring tiles. Right: NRQT triangulation avoids cracks between neighboring tiles

---

Algorithm 6 Frames Coherence.

calculateQuadtree (*tile*)
    if ((getVDError (*tile.nearestSplitNode*) > *thresholdError*))
    or (getVDError (*tile.nearestCollapseNode*) < *thresholdError*))
        //Adding SE corner
        ListsX[*t*-1] ← *tile*.getCornerSE(); //*t* = number of elements of ListsX
        ListsY[*t*-1] ← *tile*.getCornerSE(); //*t* = number of elements of ListsY
        createQuadtree(*tile.root*);

---

4 Results

In order to compare the performance of the NRQT with Linstrom [9] (which is currently one of the references in this field), a set of comparative tests were performed.

Simplified Meshes Quality. The first comparative test performed has been oriented towards the evaluation of the visual quality of simplified meshes obtained from both approaches. Figure 5 shows the Root Mean-Square Error (RMSE) on meshes projected onto the screen. The test has been carried out using different meshes (full resolution texture mapped) and several points of view. In every case, similar results to those of figure 5 were obtained. In figure 5 it is possible to observe that NRQT offers a better mesh quality with a lower number of triangles. These results are coherent with the fact that in non-restricted quadtrees, mesh resolution is directly driven by the error metric, independently from others aspects such as quadtree balancing. Another issue that contributes to the results is the fact that Lindstrom inflates the error metric to ensure a monotonic behavior on the screen space (for cracking avoidance), producing a poor control of the resolution in each surface region. Meanwhile in NRQT, the cracking avoidance is implicit in the triangulation and it is independent to the error metric, allowing a finer control of the detail required in each surface region.

Triangulation Performance. Figure 6 (a) shows a comparison between computation times used to build the mesh representation at each frame for both approaches (NRQT with and without frame-to-frame coherence). It highlights that Lindstrom approach performs somewhat better than NRQT without using frame-to-frame coherence. However, when this feature is included, NRQT is clearly superior to Lindstrom triangulation.

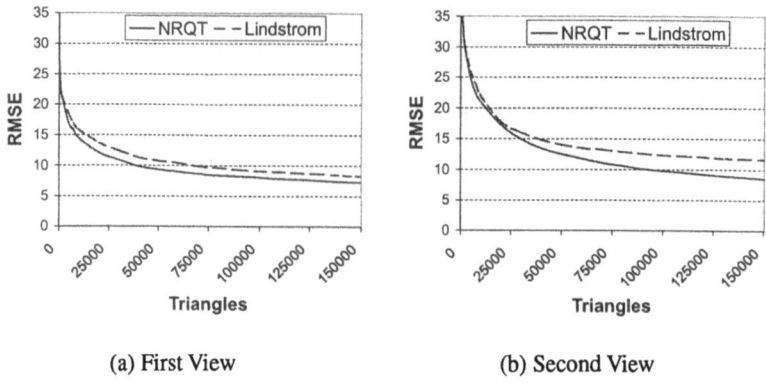

(a) First View   (b) Second View

Fig. 5. Screen space projected errors of Linstrom and NRQT approaches.

Finally, figure 6 (b) compares triangle strip and triangle fran compression rates of both approaches.

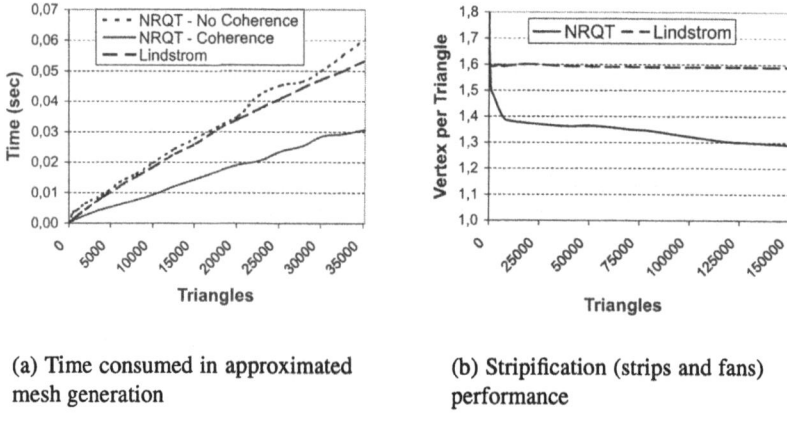

(a) Time consumed in approximated mesh generation

(b) Stripification (strips and fans) performance

Fig. 6. Performance of Lindstrom and NRQT approches

5 Conclusions

In this paper, a new hierarchial triangulation model based on a non-restricted quadtree representation has been presented. NRQT offers the same or higher levels of simplicity than previous hierarchical models, is compatible with database segmentation and with any kind of error metric (even object space or screen space non-monotonic metrics), and offers simple cracking avoidance between contiguous elements (quadtree region and tiles) and optimal view-culling algorithm that really eliminates triangles out of the viewing frustum.

Because of the use of non-restricted quatrees, the visual quality of the simplified mesh versions is superior to the results obtained using restrictive approaches. Another important feature present in NRQT is the use of frame coherence, which allows an important improvement in terms of time performance. Finally, NRQT also improves the stripification process producing long triangle strips and triangle fans.

References

1. Jonathan Blow. "Terrain Rendering at high levels of detail", in *Proceedings of the 2000 Game Developers Conference*, 2000.
2. M. Duchaineau, M. Wolinsky, D. E. Sigeti, M. C. Miller, C. Aldrich and M. B. Minnev-Weinstein. "ROAMing Terrain: Real-Time Optinally Adapting Meshes", in *Proceedings of the IEEE Visualitation'97*, pp.81–88, 1997.
3. W. Evans, D. Kirkpatrick and G. Townsend. "Rigth-triangulated irregular networks", in *Algorithmica*, Vol. 30, No. 2, pp. 264–286, 2001.
4. H. Von Herezen and A. Barr. "Accurate triangulations of deformed intersecting surfaces", in *SIGGRAPH'87, Computer Graphics Proceedings*, pp. 103–110, 1987.
5. H. Hoppe. "Progressive Meshes", in *SIGGRAPH'96 Conf. Proc.*, pp. 99–108, 1996.
6. H. Hoppe. "View-Dependent Refinement of Progresive Meshes", in *SIGGRAPH'97 Conf. Proc.*, pp. 189–198, 1997.
7. H. Hoppe. "Smooth view-dependent level-of-detail control and its applications to terrain rendering", in *Proceedings, IEEE Visualization'98*, pp. 35–42, 1998.
8. P. Lindstrom, D. Koller, W. Ribarsky, L. F. Hodges, N. Faust and G. A. Turner. "Real-Time Level of Detail Rendering of Height Fields", in *Proceedings of the 23rd annual conference on Computer Graphics, SIGGRAPH'96*, pp. 109–118, 1996.
9. Peter Lindstrom and Valerio Pascucci. "Terrain Simplification Simplified: A General Framework for View-Dependent Out-of-Core Visualization", in *IEEE Transactions on Visualization and Computer Graphics*, Vol. 8, No. 3, pp. 239–254, 2002.
10. Renato Pajarola. "Large Scale Terrain Visualization Using The Restricted Quadtree Triangulation", in *Proceedings of the conference on Visualization'98*, pp. 19–26, 1998.

Boundary Filtering in Surface Reconstruction

Michal Varnuška[1] and Ivana Kolingerová[2]

Centre of Computer Graphics and Data Visualization
Department of Computer Science and Engineering
University of West Bohemia, Pilsen, Czech Republic
{miva,kolinger}@kiv.zcu.cz

Abstract. One of the methods for 3D data model acquisition is the real object digitization followed by the surface reconstruction. Many algorithms have been developed during past years, each of them with its own advantages and disadvantages. We use for the reconstruction a CRUST algorithm by Nina Amenta, which selects surface triangles from the Delaunay tetrahedronization using information from the dual Voronoi diagram. Unfortunately, these candidate surface triangles do not form a manifold, so it is necessary to perform some other steps for manifold extraction. In this paper we present some improvements and observations for this step.

1 Introduction

Surface reconstruction is a common problem in modern computer graphics. There exist many applications which need to work with a piecewise linear approximation of the existing real 3D objects. One of the methods for acquiring these models is the digitization of the real 3D object using many types of devices (haptic probes, MRT or laser scanners, etc.) followed by the point cloud reconstruction. We use for the reconstruction the CRUST algorithm [19, 20, 21] that has two parts: selection of the surface triangles and manifold extraction from these triangles. We concentrate to the step of manifold extraction in this paper.

The paper is organized as follows. Paragraphs 1.1 - 1.2 give a short description of the existing methods, data input and effects of the sampling. Section 2 presents the CRUST algorithm by Nina Amenta which we use for the reconstruction and shows the differences between onepass and twopass version. Section 3 explains the manifold extraction step, section 4 our improvements. Results are presented in section 5 and the section 6 concludes the paper.

1.1 State of the Art

Many algorithms have been developed to deal with this problem. The 2D version of this problem, curve reconstruction, has been addressed by many authors, such as

---

[1] The author was supported by the project FRVŠ G1/1349 2004.
[2] The author was supported by the project ACTION 36p9.

Amenta [1], Dey [2] or Atalli [3]. Their algorithms have strong theoretical background and are extendable to 3D.

The 3D problem is more important and more difficult to solve. Methods can be divided into four groups [4] (division is not strict; some methods can belong to more groups):

- warping
- incremental surface construction
- distance function methods
- spatial subdivision

Warping works on the basic idea that we deform some starting surface to the surface that forms the object. The idea of warping is relatively old and is used in Müller's approach [5] or by Muraki [6].

The incremental surface reconstruction is the second huge group of algorithms. Boissonat [7] presented an approach, which begins on the shortest edge from all possible edges between points and incrementally appends the edges to create a triangle mesh. Similar algorithm was developed by Mencl and Müller [8, 9] which creates an extended minimum spanning tree, identify and extract typical features.

Other group of algorithms is based on a distance function, which describes the shortest distance from the point to the surface. The sign of the function for closed surfaces depends on whether the point is inside or outside the object. This function is computed for each point using the tangent plane estimated from k nearest neighbors (points). Hoppe [10, 11] presented an algorithm, where the surface is represented by the zero set of a signed distance function. Curless and Levoy [12, 13] gave an effective algorithm using the signed distance function on a voxel grid which is able to reconstruct eventual holes in a post-processing.

The fundamental property of the methods based on spatial subdivision is that the boundary hull (convex hull, box around points, etc.) of the point input set is divided into independent areas. A typical example is division by a regular grid, adaptive by an octree or irregular tetrahedronization. Algorri and Schmitt [14] gave an effective algorithm in which the space is divided into voxels. Then the voxels are chosen which contain points from the input set and the surface is extracted. Edelsbrunner and Mücke [15, 16] developed a program for uniform sample set surface reconstruction using α-shape algorithm. Bernardini and Bajaj [17] developed the algorithm that gets the surface subcomplex from Delaunay tetrahedronization. This algorithm extends the idea of alpha shapes and it uses the binary search on the parameter alpha to find this subcomplex. Bernardini [18] presented the ball pivoting algorithm.

Amenta introduced the concept of CRUST [19, 20, 21] (details will be explained in section 2). Dey extended the ideas of Amenta, gave an effective COCONE algorithm and presented a way how to handle large data [22], which is the common problem of Delaunay based algorithms, and what to do with boundaries [23], undersampling and oversampling [24]. These methods are based on the observation that the places with a changing of point density can be detected using shape of Voronoi cells in these places. Authors gave then algorithms for watertight surface creation, Amenta's POWERCRUST [25] is built on medial axis transformation (MAT) and Dey [26] is based on COCONE algorithm extension.

1.2 Data Input and Sampling

The output of the scan devices is usually only a point cloud. Some of these devices provide additional information, such as surface normal vectors, which can simplify the reconstruction algorithm. Input to our algorithm is just a points cloud in 3D:
$$\forall p \in P, P \subset S : p = [x, y, z]; x, y, z \in R$$
S is the original surface of the object, P is the discrete set of points lying on the surface S and p are points from P. The surface (triangle mesh) obtained after reconstruction from the set P will be only a surface approximation of the object because some information is lost during the process of digitization.

There exist many influences which may affect the success of the reconstruction; the question of the points cloud sampling is the most important. Data uniformity belongs to the most requested sampling property, but even if the data are sampled uniformly, it does not guarantee the success.

The surface properties are also important, the reconstruction of the smooth closed surfaces is easier than the reconstruction of the surfaces with sharp edges, boundaries or outliers and the algorithms for closed smooth surfaces do not need to be so robust.

The algorithms have some criteria under which they work. We use in our approach the latter criterion based on *local feature size* presented by Amenta [19]. *LFS(s)* of the point $s \in S$ is a function which assigns to each point s the closest distance to the medial axis, a real value *LFS(s): S → R*. Successful reconstruction depends on the ε-sampling. The point set $P \in S$ is called ε-sampled, if every point $s \in S$ has a sample $p \in S$ within the distance of $\varepsilon * LFS(s)$. According to [19, 20] for $\varepsilon < 0.06$ the reconstruction is homeomorphic to surface S, in practice it is able to achieve successful reconstruction for $\varepsilon < 0.5$.

The big advantage of this sampling is that on the planar part of surface, with regards to the distance to the medial axis, it is not necessary to have many sampled points, but on the places with small details the sampling density should increase. Due to this sampling condition, the CRUST algorithm, which we use, is not sensitive to big changes in sampling density, the data need not to be uniformly sampled, but it has problems with outliers and sharp edges.

2 CRUST Algorithm

The planar version of the CRUST algorithm was first presented in [1] and followed by the 3D version in [19, 20]. It works on spatial subdivision, which is achieved by Delaunay tetrahedronization (DT). Auxiliary subdivision, Voronoi diagram (VD), is obtained by dualization from DT. There exist two versions of the algorithm based on onepass or twopass tetrahedronization. We have chosen these methods because they have strong theoretical background, are not so much sensitive to the sampling errors and we have fast and robust implementation of DT.

The details of these methods can be found in [1, 19, 20, 21], we concentrate now only to the definition necessary for later understanding. The algorithms introduce the concept of poles. The positive pole $p+$ is the farthest Voronoi vertex (VV) of the Voronoi cell around some point p, the negative pole $p-$ is the farthest VV on the "second side", so the dot product of the vectors $(p-, p)$ and $(p+, p)$ is negative. For successfully sampled surface all Voronoi cells are thin and long, so the poles lay on the medial axis and vectors to the poles approximate the normal vectors.

The first shared step of both versions is the creation of the DT followed by its dualization to VD. Then the versions differ, especially in the meaning what poles are. The twopass algorithm takes the poles as an approximation of the medial axis while the onepass version takes the vectors from point to the poles as the approximation of the normal vectors. We have implemented both versions but onepass version is more then three times faster and can triangulate more points than twopass version, so we use the onepass version in our approach.

In the onepass version three conditions must hold for the surface triangles: their dual Voronoi edges must intersect the surface, the radius of the circumcircle around the surface triangle is much smaller than the distance to the medial axis at its vertices and the normals of surface triangles make small angles with the surface normals at the vertices.

We can compute the set of surface triangles as follows. For each point p we have an approximation of its normal vector $n = p+ - p$. Each triangle in DT has an edge e dual in VD. For the triangles on the surface, this edge has to pass through the surface P. Let us denote the vertices of the edge e as w_1, w_2, the angles $\alpha = \angle(w_1 p, n)$ and $\beta = \angle(w_2 p, n)$. When the interval $<\alpha, \beta>$ intersects the interval $<\pi/2 - \theta, \pi/2 + \theta>$ and this condition holds for each vertex p of the triangle, then the triangle is on the surface. The parameter θ is the input parameter of the method.

The problem with the algorithm occurs when the surface is badly sampled with noise or on sharp edges. Then the surface normals estimated using poles point to bad directions and the surface reconstruction fails. A lot of bad triangles will appear and the manifold extraction due to them will not be very successful. Every reconstruction algorithm has problem with the data which do not pass its sampling criterion.

3 Manifold Extraction

The result of the CRUST algorithm is the set of surface triangles T (we call it the primary surface). These triangles passed conditions of the CRUST algorithm but they do not form the manifold yet. There can be more than two triangles incident on some edges or some triangles may miss on the places of local discontinuity. For example, very flat tetrahedra in the smooth part of the surface (Fig. 1a) or the tetrahedra on the surface edge (Fig. 1b) may have all faces marked as surface triangles. The number of overlapped triangles differs from model to model and depends on the surface smoothness. For smooth surface it is in tens percent and when the surface is roughly sampled, the rate decreases.

That is why the surface extraction step must be followed by a manifold extraction step. The input to the manifold extraction step is just the set of triangles. Manifold extraction step is independent of the reconstruction method, therefore it could be combined with other algorithms than CRUST. We have developed our own algorithm. The reason was that the manifold extraction methods were explained very briefly in the papers, however, this step is important. Our approach uses breadth-first search for appending triangles on free edges and has a linear time complexity. The algorithm is presented in [27], for clarity of the text we will briefly explain it.

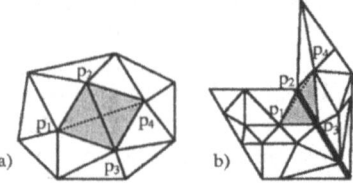

Fig. 1. a) Flat part of the surface b) the part with the sharp edge (bold line). One pair of triangles is $(p_1p_2p_3)$ and $(p_2p_3p_4)$, the second pair is $(p_1p_2p_4)$ and $(p_2p_3p_4)$.

The preprocessing step of the extraction is creation of two structures, a list of incident triangles for each point and a multiple neighbors mesh containing for each triangle the list of incident triangles on the edges. There can exist more then two triangles sharing one edge as the manifold is not ensured yet, see example in Fig. 2. These structures are used for fast search.

First, we have to find (using data structures presented before) starting triangles which will form the root of the searching tree. The starting triangles form a triangle fan; no triangles from the fan overlap when we project them to the plane defined by the point and normal.

Next, we add to the already extracted triangles their neighbors on the non-connected edges. These neighbors form next level of the tree. Because we can have multiple neighbors, we have to find just one triangle of them. We assume that the triangles must be small to form a correct surface, so we take the one, which has the shortest edge length (in Fig. 3, the minimum of $(|p_2p_4p_3|, |p_2p_5p_3|$ and $|p_2p_6p_3|)$.

Although it is a heuristic, we did not find any problem with it.

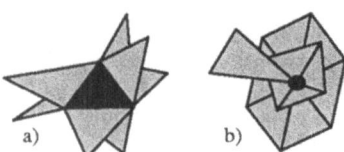

Fig. 2. a) An example of multiple neighbors to some triangle, b) an example of incident triangles to one point.

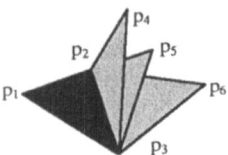

Fig. 3. Extracted triangle $p_1p_2p_3$ and some possible manifold triangles on edge p_2p_3

We need only 2 levels of the tree at one moment, so older levels can be safely deleted. We continue recursively until all edges are processed. More details about the extraction can be found in [27]. The advantage of this manifold extraction method is that it has not big memory requirements and is very fast. The manifold extraction does not depend on the algorithm used for the surface triangles extraction.

4 Boundary Triangle Filtering

When the surface contains the boundary, the CRUST has problems with its recognition and it marks the boundary triangles as surface triangles. This is a problem of all algorithms, there is no chance to find if some place presents a boundary or just a local undersampling and we are left to heuristics. Dey presented a heuristic algorithm [23] for recognition if a point lies on a boundary or not. It is built on categorization of the Voronoi cell. For every cell the width and height are computed and if the rate between width and height exceeds some value and the normals of neighbors satisfy normal condition then the point is marked as boundary.

We present now other heuristic approach based on the observation of the boundary triangles, see Fig. 4. There are two examples of a surface, where boundary triangles appear incorrectly. Fig. 4a) presents the case when the boundary triangles are perpendicular to the surface triangles while Fig. 4b) show the case where boundary triangles are parallel to the surface triangles. When we look closer at the figures, the length of the edges of boundary triangles differ from the length of surface triangle edges. To avoid the situation, when the surface is not uniformly sampled and the length of edges incidenting to a point may differ more, we developed an adaptive criterion based on the edge length and the angle between point and incident triangles.

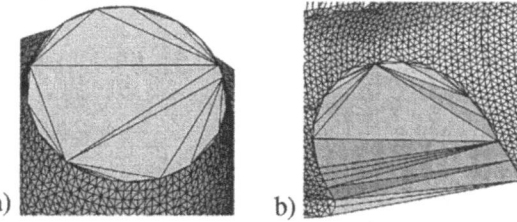

Fig. 4. Two examples of badly triangulated boundary.

The boundary test is computed as follows. For each point p the set E of all incident edges is created. One edge e is chosen as referential. All triangles whose both edges incident to the point p are longer than the length of the referential edge multiplied by some variable m are filtered out. This variable depends on the angle of the triangle and point normals. The multiplicator m is computed :

$$m = c_{be} + c_{se} |n_p \cdot n_t|$$

The constant c_{be} is the maximal allowed length of the triangle edge when the angle between the normal of the triangle n_t and the normal of the point n_p is 90° (then the dot product of these normals is zero). When we compute this multiplicator on a flat smooth part of the surface, the dot product will be one and the multiplicator m is the sum of c_{be} and c_{se}. These constants were set experimentally and almost for every data sufficient results were achieved for c_{be} equal to 2.0 and c_{be} equal to 5.0 (on the boundary where boundary triangles are perpendicular to the surface triangle, m is 2.0 and when the boundary triangles are parallel, m is 7.0).

The most important question is how long should be the referencial edge. First we have tried to take the median of the edges lengths as the refencial edge, but it did not give good results because at one boundary point there can be more boundary triangles than surface triangles (see Fig. 5a) and the boundary triangles were not filtered. Then we have tried to take the smallest edge in the set E (Fig. 5b). But some datasets have due to errors in the scanning process some points very close together, so more triangles were filtered than we wanted. Now the referential edge e is taken as the third smallest edge. It is a heuristic criterion and it is built on the observation that even when there are some points mutually very close, the third shortest edge represents the length of some "normal" triangle.

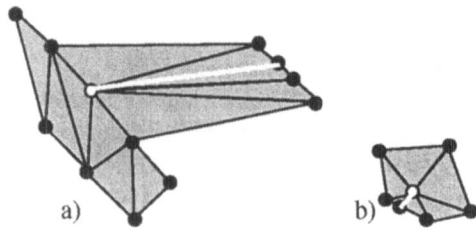

Fig. 5. a) The referential edge is taken as the median of all edges lengths, b) the referential edge is the shortest edge.

5 Results

The implementation of the CRUST algorithm and of our improvement was done in Borland Delphi compiler under the Windows XP system on AMD Athlon XP+ 1500MHz processor with 1GB of memory. The critical part of the algorithm is the computation of the Delaunay tetrahedronization. We have our own library which computes Delaunay tetrahedronization with expected complexity $O(N \log N)$ but practical tests on real data (scanned 3D objects) give the complexity of $O(N)$.

We have tested (testing datasets are in Tab. 1) our implemented algorithm (Amenta's CRUST with our manifold extraction) together with Dey's COCONE algorithm [22, 23, 24], which is conceptually very similar to the CRUST. It contains also the boundaries detection built on Voronoi cell shape so we were able to compare the efficiency of our technique.

Table 1. Number of points in datasets used for testing.

bone	bunny	x2y2	engine	hypshet	14S	mann	nascar	teeth
68537	35947	5000	22888	6752	277228	12772	20621	29166

Let us show now the results of boundary filtering. Fig. 6a) shows the reconstructed surface (by our approach), which has boundary triangles collinear with the surface triangles. The part of the surface in Fig. 6b) has boundary triangles perpendicular to the surface triangles. The boundary filtering works well for this kind of data, too.

For comparison we tried to reconstruct the same objects by COCONE. The reconstruction worked well for the objects in the Fig. 6a) and the reconstructed models were almost the same. The reconstruction of the object as in Fig. 6b) fails and a lot of bad boundary triangles appear there.

Fig. 7 shows the whole reconstruction of the object from Fig. 6a). Fig. 7a) is the reconstruction using COCONE. It is visible that the boundary detection failed. The output of the CRUST followed by our manifold extraction is in Fig. 7b), several incorrect boundary triangles occur there mainly in the windows and the wheel parts of the coachwork of the car. When we apply the boundary filtering, the situation is better, see Fig. 7c). The highlighted parts around the car trace the boundary contour.

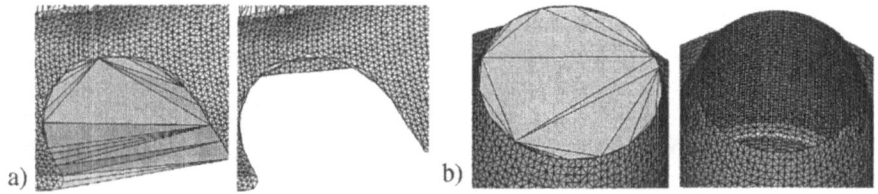

Fig. 6. The reconstruction of the surface with the boundary triangles before and after boundary filtering a) collinear b) perpendicular to the surface triangles.

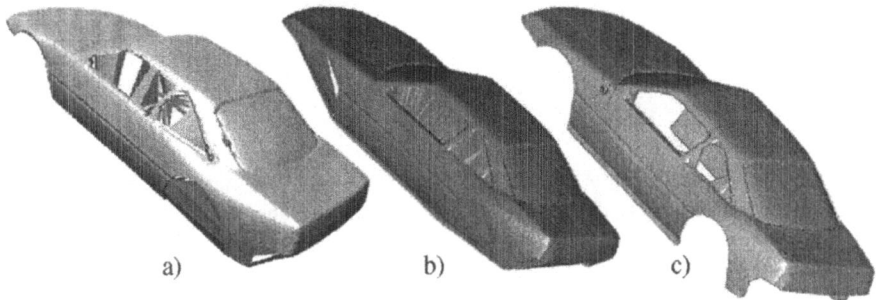

Fig. 7. a) The model reconstructed by the COCONE, detected boundaries are highlighted parts, b) the same model reconstructed using CRUST, c) after applying the boundary filter, highlighted parts are traced boundary edges.

The car model is uniformly sampled, the other test data, some of them nonuniformly sampled, can be seen in Fig. 8 - 10. The hypersheet model was a little problematic for both programs (Fig. 9). The data are not uniformly sampled, sometimes several points are mutually very close and some surface triangles were missing. The boundary triangles were successfully removed except one. COCONE has problem with the dataset knot too, some surface triangles were marked as boundary and holes appear. Other data were successfully reconstructed by both programs.

The bottleneck of the heuristic algorithms is always the choice of parameters settings. Because we use here the adaptive setting depending on the local configuration, the choice described in the Section 4 is useful almost for all data we had. For uniform datasets we also tried successfully to set the constants to $c_{be} = 2.0$ and $c_{se} = 1.0$ but at this moment, we are not able to detect data uniformity automatically.

6 Conclusion

In this paper, we have presented an improvement for the manifold extraction. We gave an efficient heuristic algorithm for the boundary triangles detection. As presented in the Section 4, the boundary detection works well and is fully comparable with the Dey detection algorithm. For some data our results were better, because Dey's approach marks also the places of local undersampling as boundaries.

Our future work is to implement or develop some algorithm for the holes triangulation using existing data structures and big challenge is to develop an algorithm for a noisy datasets because extremely noisy data is impossible to reconstruct with the existing algorithms.

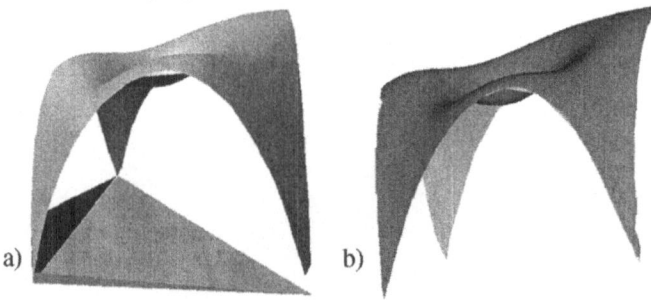

Fig. 8. a) The function x^2y^2 reconstructed by COCONE. The surface is connected by the boundary triangles (incorrectly), b) the reconstruction by our approach.

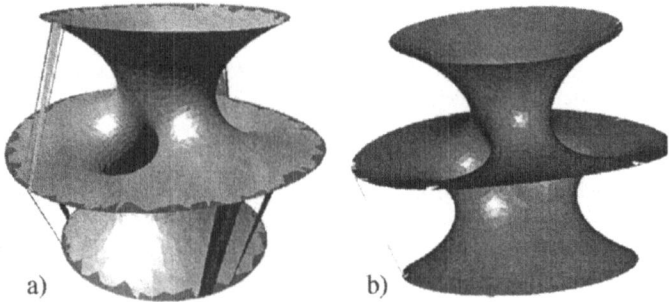

Fig. 9. a) The hypersheet reconstructed by COCONE, the boundary is correctly detected but some incorrect boundary triangles appear, b) reconstruction by our approach, almost all boundary triangles were removed but some of the surfaces triangles too.

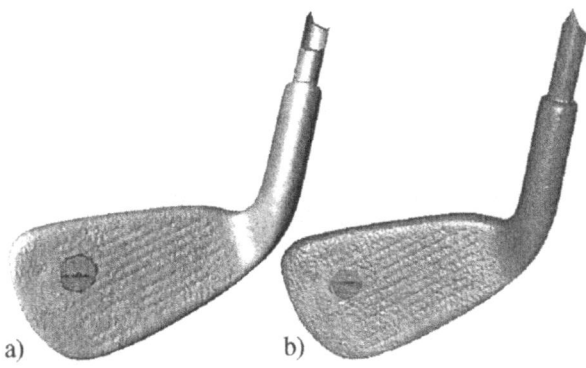

Fig. 10. a) The club dataset reconstructed by COCONE, the reconstruction is correct and boundaries triangles removed, b) the reconstruction by our approach, boundary triangles successfully removed, too.

References

1. N.Amenta, M.Bern, D.Eppstein. *The CRUST and b-skeleton: combinatorical surface reconstruction*. Graph. Models and Image Processing, 1998, p. 125-135.
2. T.K.Dey, P.Kumar. *A simple provable algorithm for curve reconstruction*. Proc. ACM-SIAM Sympos. Discr. Algorithms, 1999, p. 893-894.
3. D.Attali. *R-regular shape reconstruction from unorganized points*. Proc. of 13th ACM Sympos. of Discr. Algorithms, 1997, p. 248-253.
4. R. Mencl, H. Müller. *Interpolation and approximation of surfaces from three-dimensional scattered data points*. Eurographics, 1998.
5. J.V.Miller, D.E.Breen, W.E.Lorenzem, R.M.O'Bara, M.J.Wozny. *Geometrically deformed models: A Method for extracting closed geometric models from volume data*. Proc. SIGGRAPH, 1991, p. 217-226.
6. S.Muraki. *Volumetric shape description of range data using "Blobby model"*. Comp. Graphics, 1991, p. 217-226.
7. J.D.Boissonat. *Geometric structures for three-dimensional shape representation*. ACM Trans. Graphics 3, 1984, p. 266-286.
8. R.Mencl. *A graph based approach to surface reconstruction*. Comp. Graph. forum 14 (3), EUROGRAPHICS 1995, p. 445-456.
9. R.Mencl, H.Müller. *Graph based surface reconstruction using structures in scattered point sets*. Proc. CGI, 1998, p. 298-311.
10. H.Hoppe, T.DeRose, T.Duchamp, M.Halstead, H.Jin, J.McDonald, J.Schweitzer, W.Stuetzle. *Piecewise smooth surface reconstruction*. SIGGRAPH 1994, p. 295-302.
11. H.Hoppe, T.DeRose, T.Duchamp, J.McDonald, W.Stuetzle. *Surface reconstruction from unorganized points*. Computer Graphics 26 (2), 1992, p. 71-78.
12. B.Curless, M.Levoy. A volumetric method for building complex models from range images. SIGGRAPH, 1996, p. 302-312.

13. M.Levoy et all. *The Digital Michelangelo project: 3D scanning of large statues.* In proceedings of the 27th annual ACM conference on Comp. Graphics, p. 131-144.
14. M. E. Algorri, F. Schmitt. *Surface reconstruction from unstructured 3D data.* Computer Graphic Forum, 1996, p. 47-60.
15. H.Edelsbrunner, E.P.Mücke. *Three-dimensional alpha shapes.* ACM Trans. Graphics 13, 1994, p. 43- 72.
16. H.Edelsbrunner. *Weighted alpha shapes.* Technical report UIUCDCS-R92-1760, DCS University of Illinois at Urbana-Champaign, Urbana, Illinois, 1992.
17. F.Bernardini, C.Bajaj. *A triangulation based. Sampling and reconstruction manifolds using a-shapes.* 9th Canad. Conf. on Comput. Geometry, 1997, p. 193-168.
18. F.Bernardini, J.Mittleman, H.Rushmeier, C.Silva, G.Taubin. *The ball pivoting algorithm for surface reconstruction.* IEEE Trans. on Vis. and Comp. Graphics 5 (4), 1999. p. 349-359.
19. N.Amenta, M.Bern, M.Kamvysselis.*A new Voronoi-based surface reconstruction algorithm.* SIGGRAPH, 1998, p. 415-421.
20. N. Amenta, M. Bern. *Surface reconstruction by Voronoi filtering.* Discr. and Comput. Geometry 22 (4), 1999, p. 481-504
21. N. Amenta, S. Choi, T. K. Dey, N. Leekha. *A simple algorithm for homeomorphic surface reconstruction.* 16th. Sympos. Comput. Geometry, 2000, p. 213-222.
22. T.K.Dey, J.Giesen, J.Hudson. *Delaunay Based Shape Reconstruction from Large Data.* Proc. IEEE Sympos. in Parallel and Large Data Visualization and Graphics, 2001, p. 19-27.
23. T.K.Dey, J.Giesen, N.Leekha, R.Wenger. *Detecting boundaries for surface reconstruction using co-cones.* Intl. J. Computer Graphics & CAD/CAM, vol. 16, 2001, p. 141-159.
24. T.K.Dey, J.Giesen. *Detecting undersampling in surface reconstruction.* Proc. of 17th ACM Sympos. Comput. Geometry, 2001, p. 257-263.
25. .Amenta, S.Choi, R.Kolluri. *The Power Crust.* Proc. of 6th ACM Sympos. on Solid Modeling, 2001.
26. T.K.Dey, S.Goswami. *Tight Cocone: A water-tight surface reconstructor.* Proc. 8th ACM Sympos. Solid Modeling application (2003), p. 127-134 [27].
27. M.Varnuška, I.Kolingerová. *Improvements to surface reconstruction by CRUST algorithm.* SCCG 2003 Budmerice, Slovakia, p. 101-109.

Image Coherence Based Adaptive Sampling for Image Synthesis

Qing Xu[1], Roberto Brunelli[2], Stefano Messelodi[2], Jiawan Zhang[1], and Mingchu Li[1]

[1] Institute of Information, Tianjin University,
300072 Tianjin, China
{qingxu, jwzhang, mcli}@tju.edu.cn
[2] ITC-irst, 38050 Povo, Italy
{Brunelli, Messelodi}@itc.it

Abstract. Due to generality, simplicity and robustness of Monte Carlo, as well as the high complexity of the computation of global illumination problem, Monte Carlo is a very good choice for synthesizing image accounting for global illumination effects. However, the well-known problem in Monte Carlo based methods for global illumination is noise. We explore adaptive sampling as a method to reduce noise. We introduce a coherence distance map, which is one kind of formulization for image coherence, to conduct the adaptive sampling scheme. Based on the coherence distance map, we construct an elegant probability density function to drive Monte Carlo importance sampling to adaptively controlling the number of required samples per pixel. The proposed algorithm can not only improve image quality efficiently, but also be implemented easily. In addition, our approach is unbiased and thus superior to mostly earlier adaptive sampling techniques.

1 Introduction

Monte Carlo based methods are quite suitable for the calculation of global illumination problem when highly complicated scenes with very general and difficult reflection models are rendered. Especially, Monte Carlo is applied as a method of last resort when all other analytical or numerical methods fail [1]. However, the synthesized image generated by using Monte Carlo based global illumination algorithms is very noisy. Usually, we can reduce the noise through taking more samples within each pixel to get a visually acceptable image, but a lot of rendering time must be spent because of the slow convergence of the Monte Carlo technique.

Adaptive image sampling tries to use more samples in the difficult region of the image where the sample values vary obviously. For example, the contour of a shadow or a high illumination gradient area should be treated much carefully than a uniformly illuminated part. In this way, each pixel is firstly sampled at a low density, and then more samples are taken smartly for complex part based on the initial sample values. Hence, adaptive sampling avoids the problem of using a fixed and high number of samples per pixel [2].

The research on adaptive sampling is very early, and that can be traced back to the research on anti-aliasing. There exist two main ways for anti-aliasing, namely, deterministic method [3] and stochastic method [4][5][6]. Both of the two kinds of method focus on detecting object edges and placing more samples along edges in the generated image so as to achieve anti-aliasing. However, significance of stochastic noises, which is perceived as spots in image, makes the aliasing or distraction of object edges unimportant for Monte Carlo based global illumination algorithms. Until now, there are not many adaptive sampling algorithms presented for Monte Carlo global illumination. Lee et al. sampled pixel on the basis of variance of pixel value [7]. Purgathofer proposed confidence interval used for pixel sampling resulting in specifying individual error bound for each pixel [8]. Tamstorf et al. did similar work with that of Purgathofer and derived confidence interval with the aid of tone mapping technique [2].

Pixel value is evaluated independently by stochastically sampling the pixel and constructing random walk chains in the framework of Monte Carlo based calculation of global illumination problem, and therefore errors in the resulting images show up as noisy spots [9]. Actually, image coherence, which is essentially the smooth value changing between adjacent pixels, can be applied to stochastic pixel sampling in the context of view dependent Monte Carlo global illumination algorithms such as path tracing [10] and bidirectional path tracing [11][12]. As a result, more samples are taken in the pixel area where the change of global illuminated values is large and on the contrary fewer samples are arranged in the region in which the gradient of pixel values is small. In this paper, we present the concept of Coherence Distance Map (CDM), and the CDM is substantially one kind of formular measurement of image coherence. In particular, a new adaptive sampling algorithm is advanced by making use of the CDM. Based on the CDM, a well-designed probability density function is created to conduct Monte Carlo importance sampling to adaptively deciding the number of required samples per pixel. Our easily implemented method can decrease the noise level of the produced image, and does not introduce a bias in the results. Moreover, the adaptive sampling algorithm presented here has not been found so far from our knowledge.

The following is the organization of this paper. The second part describes the basic idea and normal form of the CDM, which is the basis for the new adaptive sampling. Details of the adaptive sampling scheme are depicted in the third section. The fourth portion discusses the implementation and experimental results. Finally, conclusion and future works are presented in last section.

2 Coherence Distance Map (CDM)

A fine depiction of image coherence is the key to adaptive sampling method proposed by us in this paper. The CDM presented here is the quantitative definition of image coherence.

2.1 Basic Idea of the CDM

The so-called image coherence is the degree to which parts of an image exhibit local similarities, or is actually the continuity for values of adjacent pixels. In fact, image coherence can be studied through analyzing the changing of pixel values for each pixel-centered rectangle region of the image. Thus, the whole of changes of pixel values for all the pixel-centered rectangle regions is the entire image coherence.

Apparently, all the absolute values of differences between the mean of a pixel-centered rectangle region and all the pixel values within this region represent to a certain extent the changing of pixel values for this region. Namely, these absolute values are the description of image coherence for the pixel-centered rectangle region to a certain degree. This kind of image coherence can be exploited by establishing the probability distribution function to direct Monte Carlo importance pixel sampling to achieve the adaptive sampling procedure. As thus, more samples will be used stochastically for pixels whose differences are large and inversely fewer samples will be used for pixels whose differences are small so as to distribute stochastic noises uniformly over the whole region and to avoid spike noises. Nevertheless, the definition of image coherence in this way is not sufficient. For instance, global illumination algorithm can simulate two significantly optical phenomena, soft shadow and caustics. The changing of pixel values around the boundary of either soft shadow or caustics is a little notable. The image details for the boundary of soft shadow and caustics should not be erased in the course of computation of global illumination.

In order to obtain the refined definition of image coherence, we would consider the preservation of image details to treat intensively the differences between the mean of pixel-centered rectangle region and all the pixel values within this region. At first, these differences can be sorted into several subsets by the given thresholds, and all the pixels within the pixel-centered rectangle region can also be divided into several subsets. Then the mean of each pixel subset (called as the mean of pixel subset) and the absolute values of differences between the mean of pixel subset and all the pixel values within this subset (called as the relatively absolute difference) can be gotten. Finally, the image coherence for the pixel-centered rectangle region can be expressed by all the coherence distances for all the pixels within this region. Coherence distance for each pixel can be described by the product of two parts. The first part is related to the division of subsets, and the second part is set as the relatively absolute difference. The first part of coherence distance could have several forms, as long as they can express the meaning of coherence distance correctly. For example, if supposing a pixel is in a subset within a pixel-centered rectangle region and the product of the reciprocal of the number of pixels in this pixel subset and the relatively absolute difference of this pixel is largest in the domain of this pixel-centered rectangle region. And so this pixel is exceptional within the pixel-centered rectangle region because this pixel is the most distant one from others according to the image coherence of this region. Consequently, the first part of coherence distance can be simply defined as the reciprocal of the number of pixels in the pixel subset, or something like that. All the coherence distances for all the pixels within a pixel-centered rectangle region can be utilized as the regional image coherence to build a probability distribution function to guide Monte Carlo importance pixel sampling adaptively.

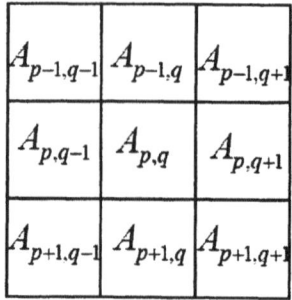

Fig. 1. The rectangle region centered at pixel $A_{p,q}$

For each pixel except the one on the image boundary, we can set a rectangle region centered at the pixel and earn the coherence distances for all the pixels within this region. All the coherence distances for all the pixels of all the pixel-centered rectangle regions compose a CDM for the entire image. The CDM is an illustration for image coherence. Obviously, the CDM is the foundation for the adaptive sampling algorithm proposed in this paper.

2.2 Normal Form of the CDM

In Fig. 1, there is a pixel $A_{p,q}$ centered rectangle region Z that contains $n \times n$ pixels (usually n is odd for computing convenience, and here n is 3). Any pixel within the region Z is expressed as $A_{i,j}(1 \leq i,j \leq n)$, and $M(A_{i,j})$ denotes the illumination value of this pixel. In order to gives out the normal form of the coherence distance, some relative definitions are given out in the following:

Mean of region Z,

$$\overline{M} = \frac{1}{n \times n} \sum_{i=1}^{n} \sum_{j=1}^{n} M(A_{i,j}) . \qquad (1)$$

Difference between mean of region Z and pixel value, and its absolute value,

$$d_{A_{i,j}} = M(A_{i,j}) - \overline{M} \text{ and } |d_{A_{i,j}}| . \qquad (2)$$

Maximum for absolute value of difference between mean and pixel value,

$$d_{\max} = \max(|d_{A_{i,j}}|) . \qquad (3)$$

Normalized difference between mean and pixel value, and its absolute value,

$$D_{A_{i,j}} = \frac{d_{A_{i,j}}}{d_{\max}} \left(-1 \leq D_{A_{i,j}} \leq 1\right) and \left|D_{A_{i,j}}\right|. \quad (4)$$

Please note that the normalized value here is necessary for classifying all the pixels within all the pixel centered rectangle regions by using a unique threshold.

With the k given thresholds $T_1, T_2, \cdots, T_k$ ($0 < T_1 < T_2 < \cdots < T_k < 1$), all the pixels within the region Z can be sorted into $2k+1$ subsets $\{S_l\}$ ($l = 1, 2, \cdots, 2k+1$) according to the situation that the normalized differences between the mean and the pixel values are located at which one of $2k+1$ intervals of threshold $(-\infty, -T_k), [-T_k, -T_{k-1}), \cdots, [-T_1, -T_1], \cdots, (T_{k-1}, T_k], (T_k, +\infty)$. S_l equals to $\{a_{l,m} | m = 1, 2, \ldots, n_{S_l}\}$, and n_{S_l} is the number of pixels within the pixel subset S_l. Here $a_{l,m}$ represents a pixel within the region, and is used for indicating the difference from $A_{i,j}$. Also, we can obtain the mean of pixel subset $\overline{M_{S_l}} = \frac{1}{n_{S_l}} \sum_{m=1}^{n_{S_l}} a_{l,m}$, the relatively absolute difference $RD_{l,m} = |a_{l,m} - \overline{M_{S_l}}|$, and the absolute difference between the mean of pixel subset and the mean of region $DD_{S_l} = |\overline{M_{S_l}} - \overline{M}|$. In the end, we can express the coherence distance for pixel $a_{l,m}$ as $CD = \frac{\frac{1}{n_{S_l}}}{\sum_{l=1}^{2k+1} \frac{1}{n_{S_l}}} \times \frac{RD_{l,m}}{\sum_{m=1}^{n_{S_l}} RD_{l,m}}$.

struct CoherenceDistanceMap {

float CD[n][n];

} CDM[NumberOfPixels]

Fig. 2. Data structure for the CDM

In Fig. 2, we show the data structure CDM[NumberOfPixels] for the CDM. For each inner pixel in image, the $n \times n$ pixel centered rectangle region and the corresponding coherence distances CD[n][n] or CDM[i] ($1 \leq i \leq$ NumberOfPixels) for all the pixels within this region are erected. Here the NumberOfPixels is the number of inner pixels of the image. The CDM[NumberOfPixels] is a sort of formula for image coherence, and the CDM[i] can be taken advantage of to lead to a nice probability density function to accomplish importance based adaptive sampling.

3 Adaptive Sampling Based on the CDM

Our adaptive sampling approach performs the pixel sampling for a pixel A_p depending on the total image coherence of A_p centered rectangle region to pick a

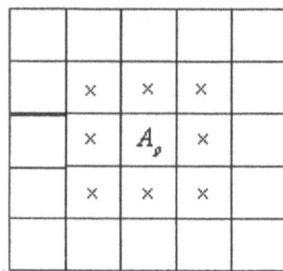

Fig. 3. The regional image coherence

more likely pixel within this region to fulfill the sampling on this pixel, rather than using the fixed number of samples directly. Fig. 3 denotes the image coherence for the pixel A_p centered region by employing the symbol ×. The adaptive pixel sampling determines stochastically the number of samples for each pixel within a pixel centered rectangle region relying on Monte Carlo importance sampling, which is steered by the probability distribution function derived from the coherence distances for all the pixels within the region.

The CDM provides us with the information to establish the required probability distribution function for the adaptive sampling. For each inner pixel, we construct a discrete probability distribution function on a two dimensional $n \times n$ grid by using the corresponding element of the CDM. The $n \times n$ values contained in the CDM element are duplicated respectively to $n \times n$ cells of the grid. If the record for a grid cell is too little and less than a specified value, a small fraction of the overall values stored within the whole grid is added to this grid cell to avoid systematic error. This is our probability distribution function.

In order to construct probability distribution function we create a discrete cumulative probability function from the knowledge in the grid. Fig. 4 demonstrates in the graph by using the dashed line that this function is used to select a cell in the grid based upon a random number ξ. The likelihood that a cell is chosen is proportional to the value recorded in that cell which means that we select a cell (i.e. pixel within the rectangle region) counting on the regional image coherence. Once a pixel is determined, we complete the actual pixel sampling on this pixel.

Unlike other approaches, the adaptive sampling technique presented here does not introduce a bias in the results because it is based on the unbiased Monte Carlo importance sampling procedure.

4 Implementation and Results

We have accomplished the adaptive sampling scheme developed in this paper for a pure Monte Carlo path tracing algorithm on a machine with Pentium IV

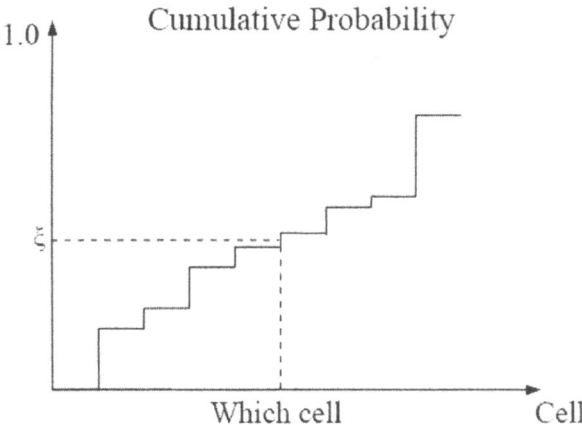

Fig. 4. The cumulative probability function derived from the CDM

2.4GHz processor and 1GB memory running Linux. The dimension of the pixel centered rectangle region used for our adaptive sampling is 3 × 3, and the only one threshold T is set to 0.5.

There are two test scenes used for evaluating the proposed adaptive sampling technique. The test scene 1 is a standard Cornell box with surfaces having diffuse and specular attributes, and with a rectangle light source attached to the ceiling of the box. The reference image of test scene 1, which is computed by way of ordinary Monte Carlo path tracing using 10000 samples per pixel, is used to judge the quantitative magnitude of errors for the images created by our adaptive sampling method. The test scene 2 is a complicated soda shop including 5 spherical light sources, 3 area light sources and 23 glass goblets. This scene is composed of 53134 triangles with both diffuse and specular attributes. Resolution of the generated image for the two scenes is 200 × 170 and 260 × 170 respectively.

We appraise the adaptive sampling scheme by comparing the measurable errors obtained from rendering the test scene 1 using the adaptive sampling scheme and traditional pixel sampling. The error metric is the RMS, $RMS - error = \sum_{all\ pixels\ p} (L_{p,ref} - L_p)^2$, where $L_{p,ref}$ and L_p represent the pixel value for the reference image and the computed image respectively.

Fig. 5 plots the different RMS error and denotes different degrees of improvement on error reduction produced from our adaptive sampling (indicated by Adaptive) and traditional pixel sampling (indicated by Traditional) by rendering the test scene 1 at $10, 20, \ldots, 90, 100, 200, \ldots, 1000$ samples per pixel. As far as the adaptive sampling scheme is concerned, its RMS error is as a function of the average number of samples used per pixel because unfixed samples are used for pixel sampling. The mean RMS error reduction of the adaptive sampling compared with traditional pixel sampling is approximate 9.89%.

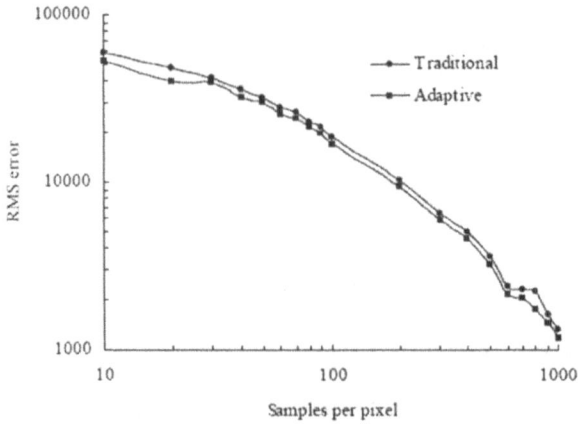

Fig. 5. The RMS error

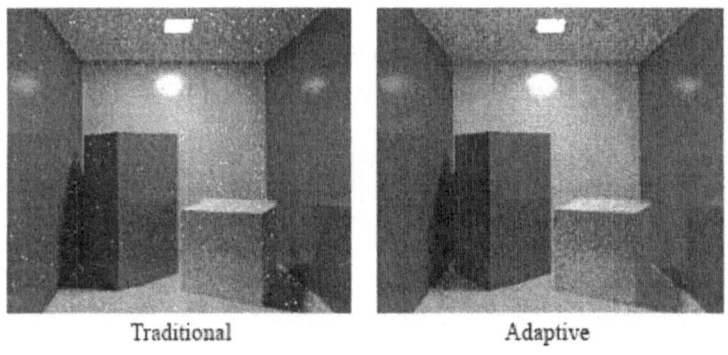

Fig. 6. Results of comparison for the test scene 1

The results for rendering the test scene 1 when using the adaptive sampling scheme and traditional sampling with 20 samples per pixel are shown and compared in Fig. 6. The difference is that the adaptive sampling scheme eliminates almost all the spike spots, and gives a more homogeneous image quality that can be observed from the shadows resulted from the two boxes.

The results for rendering the test scene 2 when using the adaptive sampling scheme and traditional sampling with 20 samples per pixel are shown and compared in Fig. 7. It can be observed that the adaptive sampling scheme leads to a better image quality exhibited from the two grey walls and the pink surfaces of the counter. Also, the reduced noise at the top part of the freezer and at the foot part of the goblet is evident.

Fig. 7. Results of comparison for the test scene 2

Furthermore, memory cost and computational effort are very limited when introducing our adaptive sampling into use. For example, the additional memory is about 800K bytes, which is far less than the memory for the machine, when adopting adaptive sampling to generate the image for the test scene 2. Also, the cumulative distribution function is precomputed and the cost for adaptive sampling is tiny because it is completed only at the additional cost of executing 9 conditions if compared with ordinary pixel sampling. Note that we need 100 samples per pixel to produce a satisfied image by using Monte Carlo path tracing. As a result, the cost for adaptive sampling is negligible when compared with that for producing a visually acceptable image.

5 Conclusion and Future Work

We have presented a new adaptive sampling scheme based on image coherence. The coherence distance map is introduced as a formula to describe image coherence to construct a probability density function and conduct the adaptive sampling scheme. The results obtained by using this method on Monte Carlo path tracing show an improvement on error reduction over the traditional pixel sampling. The proposed method can be implemented easily. Moreover, our approach is unbiased without systematic error.

In the near future, we will study on the more elaborate representation of image coherence and the refinement criterion for processing adaptive sampling falling on some sound theoretical framework, for instance, information theory.

References

1. Bekaert,P., Hierarchical, Stochastic: Algorithms for Radiosity. Ph.D. Dissertation, Katholieke Universiteit Leuven, 1999

2. Tamstorf, R., Jensen, H. W.: Adaptive Sampling and Bias Estimation in Path Tracing. In: Dorsey, J., Slusallek, P.(eds.): Proceedings of Eurographics Workshop on Rendering '97. Springer-Verlag, WienNew York (1997) 285–295
3. Whitted,T.: An Improved Illumination Model for Shaded Display. Communications of the ACM. **32** (1980) 343–349
4. Mark, A. Z., Dippe, Wold, E. H.: Antialiasing through Stochastic Sampling. Computer Graphics. **19** (1985) 69–78
5. Mitchell, D. P.: Generating Antialiased Images at Low Sampling Densities. Computer Graphics. **21** (1987) 65–72
6. Painter,J., Sloan,K.: Antialiased Ray Tracing by Adaptive Progressive Refinement. Computer Graphics. **23** (1989) 281–288
7. Lee, M. E., Redner, R. A., Uselton,S. P.: Statistically Optimized Sampling for Distributed Ray Tracing. Computer Graphics. **19** (1985) 61–65
8. Purgathofer,W.: A Statistical Method for Adaptive Stochastic Sampling.Computers Graphics. **11** (1987) 157–162
9. McCool,Michael D.: Anisotropic Diffusion for Monte Carlo Noise Reduction. ACM Transactions on Graphics. **18** (1999) 171-194
10. Kajiya,J. T.: The Rendering Equation, Computer Graphics. **20** (1986) 143–150
11. Lafortune,E. P., Willems, Y. D.: Bi-directional Path Tracing. In: Santo, H.P.(eds.): Proceedings of CompuGraphics '93. Alvor, Portugal (1993) 145–153
12. Veach,E., Guibas, L.: Bidirectional Estimators for Light Transport. In: Shirley, P., Muller, E. (eds.): Proceedings of Eurographics Workshop on Rendering 1994. Springer-Verlag, Berlin (1995) 147–162

A Comparison of Multiresolution Modelling in Real-Time Terrain Visualisation

C. Rebollo, I. Remolar, M. Chover, and J.F. Ramos

Departamento de Lenguajes y Sistemas Informáticos - Universitat Jaume I.
12071 Castellón, Spain
{rebollo,remolar,chover,jromero}@uji.es

Abstract. One of the most important challenges in Computer Graphics is real-time rendering of terrain. This is due to the huge number of polygons used to represent these surfaces. Up to now, several works have appeared that attempt to reduce terrain visualisation time, some of which are based on multiresolution modelling. This method adapts the number of polygons of an object as a function of particular criteria in order to reduce the information that has to be computed for each frame. The ultimate purpose of this work is to compare the most popular multiresolution algorithms to have appeared in recent years, while at the same time examining some of the fundamental concepts underlying real-time terrain visualisation.

1 Introduction

Terrain relief plays an important role in different applications currently present in our daily lives, such as highway construction, irrigation works, etc. Terrain representation is also necessary in interactive applications like flight and driving simulators, computer games or Geographic Information Systems.

Graphic display of real surfaces requires the elaboration and creation of a *Digital Terrain Model – DTM*. This model represents a terrain's topographical characteristics numerically by the x, y, z coordinates of the critical points defining it. The notation $z = f(x,y)$ characterises the parameter to be modelled and there is a unique z value for each location (x,y) [1]. The most common representation of the DTM is as a regular grid of points.

Digital terrain modelling requires the prior acquisition, conversion and storage of data so that they can later be displayed as graphics on the computer screen. The DTM obtained normally represents large volumes of data, which makes their real-time visualisation practically impossible. Several methods have recently appeared in the literature that offer the possibility of accelerating scene visualisation by using simplified object representations, one of the fundamental methods being multiresolution modelling [2] [3] [4] [5].

The objective of this paper is to compare a number of relevant contributions made over the past few years on the topic of real-time terrain visualisation. In the paper, we introduce the process, followed by the creation of a DTM and its computer representation (Section 2). Subsequently, some fundamental concepts concerning interactive terrain visualisation are reviewed (Section 3) and we also examine how they have been dealt with in recent works in this field (Section 4).

2 Digital Terrain Modelling

DTM is a general term for digitally created three-dimensional terrain representations. Its main advantage lies in the fact that it allows operations to be carried out on a numerical representation of the terrain, which then enables the results of these models to be extrapolated to the original terrain. Its main drawback is the complexity involved in its design and data management.

In a DTM process, the following phases can be distinguished:

- *Data acquisition:* By means of direct methods on terrain or indirect methods from previously elaborated documents.
- *Conversion and storage of data:* It is advisable to preprocess data before structuring and storage to avoid obtaining an excessive number of points derived from the data acquisition process. In this phase, an interpolation function is applied in order to obtain the continuous behaviour of the surface.
- *Grid generation:* Original data acquired are structured in order to simplify management in terrain modelling operations. Most DTMs are adjusted to regular grids or irregular grids.

DTM applications are elaborated on models generated in a regular or irregular grid format. Interactive 3D terrain visualisation is one of these models.

2.1 Regular Grids

A regular grid is a data model that represents surfaces by a polyhedron of rectangular faces, being represented the heights as discrete intervals. Vertices of those polyhedrons are their own sampling points, which are stored in an array. They use a vector of values of regularly spaced height, which provides for easy storage, management and height interpolations. This representation only stores the z value, that is to say, the height. It is difficult to obtain a correct definition of relief by regular geometrical structures. One requirement is the capacity to capture the notable terrain points, as well as the addition of those singular lines that represent changes in slope. Once the resolution and, consequently, the coordinates of each grid point have been defined, an interpolation method is used to calculate the approximate height values. From this information, a grid is generated that represents the surface as faithfully as possible. The two most common representations of regular grids are quadtrees [6], [7], [8], [9] and [13] and bintrees [10], [11] y [12].

2.2 Irregular Grids

An irregular grid (known as Triangulated Irregular Networks – TIN) is a digital model that represents surfaces through a polyhedron with triangular faces [14], [15]. They support variable spacing among vertices and the sampling points are connecting lines forming triangles. Data capture is usually carried out by topographical methods that offer the advantage of employing carefully selected sampling points to model the surface without the need to interpolate the same points. They allow sampling to be greater in rough regions than in planar regions. They also ensure that important morphological information can be considered during grid generation, thus enabling users to model the terrain surface while maintaining their geomorphological features. To generate a grid it is necessary to employ an algorithm that establishes the relations of vicinity among the different points in order to build the triangulation.

2.3 Hybrid Models

Interesting options have been developed for the real-time representation of a terrain by grids. One of these solutions is the construction of hybrid representations that combine the advantages offered by regular grids and TINs [16], [17], [18] and [19].

3 Terrain Visualisation

Due to the huge number of polygons that are used to represent these terrain surfaces, which prevents the visualisation of an adequate number of scenes per second, it is necessary to employ software engineering techniques that adapt this number of polygons to the capacity of the available hardware.

Multiresolution modelling offers the possibility of accelerating scene visualisation by using progressively simpler representations of the objects, that is, by changing the Level of Detail – LoD. This modelling procedure establishes the data structures needed to store the information and the algorithms that provide access to these structures in the most efficient way.

In this paper, the most recent real-time terrain visualisation studies using multiresolution modelling techniques are analysed (Table 1).

3.1 Fundamental Concepts in Interactive Terrain Visualisation

In all the research works studied, we encounter a series of concepts to be borne in mind when it comes to optimising the visualisation process:

- *Data structures used in the representation:* The two main structures used to represent a terrain are regular grids and TINs.
- *Simplification and refinement of the generated meshes:* Simplification means to obtain a new surface that represents the initial one, with a certain amount of error, and that consists of fewer polygons. To this end, triangle union operations are

carried out. Refinement would be the inverse operation to simplification. Beginning with a simple initial surface a representation is obtained with more detail by means of triangle division operations.
- *Processing of cracks and T-junctions*: A crack in a terrain is produced as a result of the existence of adjacent triangles with different LoDs, while a T-junction comes about when the vertex of a triangle of a higher LoD does not share a vertex in the adjacent triangle from a lower level.
- *Frame-to-frame coherence:* Allows the extraction of a new approximation of the model from the last approximation that was required, instead of the base mesh.
- *Mechanism to build triangle strips*: The triangle is the primitive that is most commonly used to visualise meshes of polygons. To represent n triangles, this method sends 3n vertices to the graphic system. Another very popular primitive, and one which accelerates the visualisation of polygonal models, is the triangle strip, which only requires n + 2 to represent the previous n triangles.
- *Geomorphs:* This refers to the transition between approximations by smoothly interpolating vertex geometry, thus resolving the problem of "popping".
- *Out-of-core storage:* To visualise terrain data in real-time, these need to be organised on the disk and to be loaded in the main memory in pieces.
- *View frustum culling*: Eliminates the polygons that are out of the volume of the observer's view.
- *Textures*: The representation of the details on the terrain surface.

Table 1. Recent studies in multiresolution modelling terrain.

Author	Work	Year
Lindstrom	Continuous LoDs Rendering of Height Fields [6].	1996
Duchaineau	ROAM: Real-Time Optimally Adapting Meshes [10].	1997
Röttger	Real-Time Generation of Continuous LoDs [7].	1998
Pajarola	Large Scale Terrain Visualisation using RQT [8].	1998
Hoppe	VDPM: View-Dependent Progressive Meshes [14].	1998
DeFloriani	VARIANT: VAriable Resolution Interactive ANalysis of Terrain [15].	2000
Evans	R.T.I.Ns: Right-Triangulated Irregular Networks [16].	2001
Lindstrom	Visualisation of Large Terrains [11].	2001
Lindstrom	View-Dependent Out-of-Core Visualisation [12].	2002
Pajarola	QuadTIN: Quadtree based TINs [17].	2002
Lario	HB-QTIN: Hyper-Block Quadtree based TINs [18].	2003
Cignoni	BDAM: Batched Dynamic Adaptive Meshes [19].	2003
Bao	LoD-based Clustering Techniques [9].	2003

4 Comparison of Models

In this section, we present a study of how each author deals with these concepts in his algorithm. The result of this analysis appears schematically in Table 2.

4.1 Data Structure Used to Represent the Terrain

Two main structures for terrain representation are regular grids [6], [7], [10], [11], [12], [8] and [9] and TINs [14] and [15]. It must be emphasised that the tendency in recent years is to use hybrid representations that combine both methods [16], [17], [18] and [19]. Evans et al. use a non-arbitrary subset (regular grids) of data of the original triangulated points (TINs) [16]. Pajarola et al. work with TINs that apply a triangularisation based on quadtrees, known as QuadTINs [17]. Lario et al. improve the use of QuadTINs by generating a tree structure of blocks of QuadTINs where the different LoDs are stored [18]. To accomplish the efficient management of textured multiresolution data terrain, Cignoni et al. divide the mesh into small TINs that in turn are made up of triangles [19].

4.2 Simplification and Refinement

There are two ways of building multiresolution models: by simplification or by refinement algorithms. In both methods the meshes are built at run time and each time the point of view changes they are brought up to show the change.
- *Simplification algorithms:* the complete model and simplify zones of interest by means of union operations between triangles. In Lindstrom et al. the process of simplification is carried out in two steps: first, divide the mesh geometry in blocks and then carry out a finer simplification inside each block [6].
- *Refinement algorithms:* These start out with a unit of representation and a more detailed approximation is obtained by means of triangle division operations. In Duchaineau et al. two priority queues are employed to maintain the order of application of the division and union operations that are carried out on pairs of triangles that share their hypotenuses [10]. This order maintains the construction of continuous triangularisations from the preprocessed bintree triangles. In Röttger et al. the refinement of the mesh is carried out depending on the distance to the point of view and on the roughness of the surface [7]. Beginning with a square, Evans et al. generate isosceles right-angled triangles [16]. The model stores the hierarchy of approximations, which, when combined, allow different LoDs to be represented in the same mesh. This enables high LoDs to be obtained in certain regions of the surface. One special case is Pajarola et al. and Lario et al., who recursively subdivide triangles, [17] and [18], but in this case they are no longer isosceles right-angled triangles. An edge itself would not be divided by introducing a vertex in the median point but instead by adding a vertex from the input data.
- *Simplification and refinement algorithms:* Hoppe first breaks down the terrain into square blocks that are simplified by union operations on triangles until a base mesh is obtained [14]. In his model, the base mesh is stored together with the sequence of division operations, which is the opposite to the triangle union operations carried out in the preprocess. A mesh with the desired LoD will be obtained by applying the necessary division operations from the base mesh. It must be emphasised that these processes are carried out by the application of geomorphing. Cignoni et al. employed a refinement algorithm to visualise a continuous terrain surface by merging the small pieces of mesh that go to make up the TINs [19]. The simplification algorithm utilises it out of core in a preprocess for BDAMs construction. Each step in the level of refinement of the texture quadtree corresponds to two refinement steps in the geometry bintree.

4.3 Cracks and T-junctions

Cracks and T-junctions are mainly produced in algorithms that work with blocks or quadtrees. Cracks between vertices in a same block are easily avoided by means of the recursive division of triangles. The complicated part is to avoid cracks between different blocks. To solve this problem Lindstrom et al. suggest that adjacent blocks should share vertices along their edges [6]. There are researchers who avoid cracks by representing the triangularisation as a graph of dependences on the grid [6], [8], [17] and [9]. This method is known as *Restricted Quadtree Triangulation – RQT* [21] and [22]. Each vertex depends on two others in the same or the following level in the hierarchy of the *quadtree*. This means that if a vertex is selected for a particular triangulation of LoD, the vertices that it points at in the graph of dependences have to be selected so as to guarantee a conforming triangulation. These dependences always spread from a fine grid resolution to one which is coarser.

Röttger et al. generate a mesh that forces the edges between two adjacent blocks to have the same number of vertices, although they are at different LoDs [7]. This method works only when the levels of adjacency do not differ by more than one. Hoppe prevents vertex simplification on the edge of a block [14]. Lindstrom et al. implicitly force all the father vertices to be activated with their descendants [11] and [12]. Lario et al. do not work with individual vertices but instead they use blocks of vertices [18]. In this case, to avoid cracks between adjacent hyperblocks, there cannot be more than one level of difference between them. In Cignoni et al., after obtaining the texture the refinement process of the geometry is carried out from the father patches that were selected in the previous step in texture refinement [19].

4.4 Frame-to-Frame Coherence

Only Lindstrom et al. do not support the coherence between consecutive images, and the mesh must be built from scratch for each individual frame [12].

4.5 Mechanisms for Building Triangle Strips

Lindstrom et al. use a recursive algorithm of triangle strip construction which does not take advantage of the coherence between consecutive images [6]. Duchaineau et al. employ an incremental mechanism of triangle strips [10]. Initially, they place an average of 4 or 5 triangles in the mesh strips and they update them accordingly as triangles are merged or split. RQT [8] and [17] obtain results such that the complete mesh of triangles can be represented by just a simple one using generalised triangle strips. Pajarola goes round the centre of the quadtree blocks in an anticlockwise fashion and adds each triangle alternating the orientation each time [8]. Lario et al. build a triangle strip to each block on the level of a hyperblock [18]. Lindstrom et al. build triangle strips at run time [11] and [12] by starting out with the refinement algorithm and following the same method as Lindtrom et al. [6]. Hoppe uses an algorithm to generate triangle strips for each frame; because the connectivity of the mesh is dynamic, it is not possible to use precomputed triangle strips [14]. In Cignoni et al., the small patches of triangles in the mesh can be processed and optimised offline for efficient visualisation thanks to the use of cache coherent triangle strips [19].

4.6 Geomorphing

Authors such as Duchaineau et al. and Hoppe carry out time-based geomorphing techniques [10] and [14]. Duchaineau et al. avoid the production of "popping" effects by tending to carry out few frame-to-frame changes. The algorithm introduced by Hoppe extends the general VDPM framework [20] to provide temporary coherence through the creation of geomorphs in real time. When eliminating or adding a vertex, instead of running the transformation instantly, a process of geomorphing is carried out by gradually changing vertex geometry over several frames. This happens only if the affected surface is visible. Other authors such as Röttger et al., Pajarola and Lindstrom et al. make use of position-based geomorphing techniques [7], [8] and [12]. In Röttger et al. popping is treated when a block is replaced by another belonging to a different LoD. They eliminate the visual appearance of popping by carrying out geomorphing between the two LoDs. Pajarola applies morphing directly to the geometry that is used to visualise. Lindstrom et al. utilise the real space screen error as a morphing parameter and they combine the geometry when this error falls within a value specified by the user. Evans et al. mix detailed and coarse levels of representation [16].

4.7 Out-of-Core Storage

Most authors use data storage outside the core memory. They usually break down the domain into blocks that are subsequently stored in a quick access database [8], [14]. In Pajarola [8], a multiresolution terrain quadtree is maintained in an object-oriented database to enable efficient out-of-core access [8]. In Hoppe [14], terrain data are divided and grouped in square tiles without considering any level of detail (LoD) information and without grouping the data in fixed-size blocks on disk [14]. In Lindstrom et al., we find the most efficient algorithm [11] and [12]. In these studies, the multiresolution hierarchy of terrain triangularisation is linearised in an array and a mechanism for mapping memory files is used (supported by the operating system). In Cignoni et al. the small blocks into which the mesh of triangles is divided are at the same time triangles considered as being TINs [19]. Bao et al. propose a clustering algorithm that introduces the information about the LoD into the organisation of the data terrain for the data mapped out of core [9]. The complete terrain quadtree is divided into a set of small complete quadtrees and this union is then mapped to secondary memory pages.

4.8 View Frustum Culling

All the authors studied carry out culling of the volume against the view. Duchaineau et al. define the view frustum as an intersection of six subspaces [10]. Röttger et al. employ a quadtree for culling [7]. They provide a rectangular bounding box for each node for use in the culling against the view volume. Pajarola uses the concept of windowing to visualise the terrain database [8]. Visualised scenes always represent a window on the world. Partitioning the visible scene in rectangular patches supports efficient updatings of the scene and the assignment of different LoDs to different patches. Lindstrom et al. [11] and [12] make use of the six planes of the view frustum [10] but exploit the nested bounding sphere [23] and [24]. According to Cignoni, the process of culling is done easily as part of a recursive refinement by exploiting the nested bounding volumes [19]. Because a patch bounding volume contains all the geometry of a given subtree, the recursion can stop without visualisation when the bounding volume is detected as being invisible. When it comes to visualising a specific vertex or not, Bao does not only take into account the fact that the vertex will be inside the culling of the selected view volume, but rather that this comes as a side effect determined by a LoD constraint [9].

4.9 Textures

Only Cignoni et al. support the combination of elevation data and texture in the same framework [19], and this is the reason why this concept is not included in Table 2. Their technique is based on a paired tree structure: a tiled quadtree for the texture data and a pair of bintrees of small TINs for the geometry. The terrain textures have similar or greater dimensions than that pertaining to the elevation data. Because of this, they are partitioned in chunks before visualisation using a similar technique to that used for the geometry. If the original image covers the same region as the elevation data, each element of texture quadtree corresponds to a pair of elements from the bintree of adjacent geometry. Descending a level in the texture quadtree corresponds to descending two levels in its pair of bintree elements of associated geometry.

5 Conclusions

Terrain representations based on regular subdivisions of its structure require a greater number of polygons to obtain the same accuracy as TINs, which allow a non-uniform sampling of points to be performed. On the other hand, TINs need more complicated data structures and require greater storage space for the same number of points sampled than is used by regular grids, above all if complementary information about adjacency among triangles is needed. For this reason, the tendency over recent years has been to build hybrid representations that combine the advantages offered by both regular grids and TINs.

At present, although some existing models use triangle strips in their data structure, these have not been employed in terrain visualisation. These primitives take advantage of mesh connectivity, which reduces the number of vertices that are sent to the graphics system in order to visualise an object and thus speeds up the process.

Another aspect to bear in mind is the use of external memory to store these multiresolution models and to be able to handle the large quantities of data.

One of the more active and on-going lines of research currently being undertaken seeks to determine how the advantages of TINs and the use of triangle strips in the level of visualisation, as well in the data structure, can be included in the same model. In this way, it benefits from the acceleration obtained as a result of using these primitives. The model is built from a collection of triangle strips on the original object and of data obtained from a simplification algorithm from the model. All this, together with the use of external memory, will enable more efficient interactive terrain visualisation to be achieved.

Table 2. Results obtained in the comparison.

S/R	Simplification/Refinement.	
RBQ	Regular grid of *Bintree* with *Quadtree* structure.	
RB	Regular grid of *Bintree*.	
RQ	Regular grid of *Quatree*.	
QTIN	TIN with blocks *quadtree*.	
TINR	TIN of *bintrees*	
RTIN	Regular grid of *bintree* with TINs.	

HBQTIN	TIN with Hyperblocks of *Quadtree*.	
Subs	Subspaces.	
Q	Quadtree.	
Win	Windowing.	
NBS	Nested Bounding Sphere.	
NBV	Nested Bounding Volume.	
(*)	Preprocess.	

Ref.	Structure	S/R	Avoid Cracks and T-junctions	Frame-to-frame Coherence	Triangle Strip	Geo-morphing	Out-of-core	View frustum culling
[6]	RBQ	S	Yes	Yes	Yes	Possibility	No	
[10]	RB	R	Yes	Yes	Yes	Yes	No	Subs
[7]	RBQ	R	Yes	Yes		Yes	No	Q
[8]	RQ	S/R	Yes		Yes	Yes	Yes	Win
[11]	RB	R	Yes		Yes	No	Yes	NBS
[12]	RB	R	Yes	No	Yes	Yes	Yes	NBS
[14]	QTIN	S/R	Yes	Yes	Yes	Yes	Yes	Yes
[15]	TIN	S/R	Yes	Yes			Yes	Yes
[16]	RTIN	R	Yes	Yes		Yes	No	Yes
[17]	QTIN	R	Yes	Yes	Yes	No		Yes
[9]	RQ	R	Yes				Yes	Yes
[18]	HBQTIN	R	Yes		Yes			Yes
[19]	TINR	S/R	Yes	Yes	Yes (*)		Yes	NBV

Acknowledgements. This work has been partly financed by project CTIDIB/2002/182 from the *Generalitat Valenciana*, and projects TIC2001-2416-C03-02 and TIC2002-04166-C03 from the Spanish Ministry of Science and Technology.

References

1. M.Ervin, H.H.Hasbrouck. "Landscape Modeling: Digital Techniques for Landscape Visualization", McGraw-Hill Professional, 2001.

2. J.Ribelles, A.López, Ó.Belmonte, I.Remolar, M.Chover. "Multiresolution Modelling of Arbitrary Polygonal Surfaces: A Characterization", C & Gs, 26(3).449-462, 2002.
3. E.Puppo, R.Scopigno. "Simplification, LOD and Multiresolution – principles and applications." Tutorial Notes of EUROGRAPHICS'97. 16(3), 1997.
4. M.Garland. "Multiresolution modelling: survey & future opportunities". State of the Art Reports of EUROGRAPHICS'99, 111-31, 1999:
5. P.SA. Heckbert, M.Garland. "View-dependent refinement of progressive meshes." In Proc. of Graphics Interface ' 94, 1–8, 1994.
6. P.Lindstrom, D.Koller, W.Ribarsky, L.F.Hughes, N.Faust, G.Turne, "Real-Time, continuous level of detail rendering of height fields", Proc. SIGGRAPH'96, 109–118, 1996.
7. S.Röttger, W.Heidrich, P.Slusallek, H.P.Seidel, "Real-Time Generation of Continuous Levels of Detail for Height Fields", Proc. of WSCG '98, 315-322, 1998.
8. R.Pajarola, "Large scale terrain visualization using the restricted quadtree triangulation", Proc. IEEE Visualization'98, 19–26, 1998.
9. Xiaohong Bao, R.Pajarola, "LOD-based Clustering Techniques for Efficient Large-scale Terrain Storage and Visualiztion" Proc. SPIE'03, 225-35, 2003.
10. M.Duchaineau, M.Wolinsky, D.Sigeti, M.Miller, C.Aldrich. "ROAMing terrain: Real-time optimally adapting meshes", Proc. IEEE Visualization'97, 81-88, 1997.
11. P.Lindstrom, V.Pascucci. "Visualization of Large Terrains Made Easy", Proc. Visualization'01, 363-370, 2001.
12. P.Lindstrom, V.Pascucci, "Terrain Simplification Simplified: A General Framework for View-Dependent Out-of-Core Visualization", Proc. IEEE Visualization'02, 1 8(3) 239-254, 2002.
13. H.Samet. "The Quadtree and Related Hierarchical Data Structures". ACM Computing Surveys 16(2), 187–260, June 1984.
14. H.Hoppe, "Smooth view-dependent level-of-detail control and its application to terrain rendering", Proc. IEEE Visualization'98, 35-42, 1998.
15. L.DeFloriani, P.Magillo, E.Puppo, "VARIANT: A System for Terrain Modeling at Variable Resolution", GeoInformatica. 4(3), 287-315, 2000.
16. W.Evans, D.Kirkpatrick, G.Townsend. "Right-triangulated irregular networks", Algorithmica 30(2), 264–286, 2001.
17. R.Pajarola, M. Antonijuan, R.Lario. "QuadTIN: Quadtree based Triangulated Irregular Networks". Proc. IEEE Visualization'02, 395-402, 2002.
18. R.Lario, R.Pajarola, F.Tirado. "HyperBlock-QuadTIN: Hyper- Block Quadtree based Triangulated Irregular Networks". Proc. IASTED VIIP Conference, 2003
19. P.Cignoni, F.Ganovelli, E.Gobbetti, F.Marton, F.Ponchio. "BDAM – Batched Dynamic Adaptative Meshes for High Performance Terrain Visualization", Eurographics 22(3), 2003.
20. H.Hoppe, "Smooth view-dependent level-of-detail control and its application to terrain rendering", Proc. SIGGRAPH '97, 189–198, 1997.
21. E. Puppo. "Variable resolution terrain surfaces". In Proc. of the 8th Canadian Conference on Computational Geometry, pages 202–210, 1996.
22. R. Sivan and H. Samet. "Algorithms for constructing quadtree surface maps". In Proc. 5th Int. Symposium on Spatial Data Handling, pages 361–370, August 1992.
23. S.Rusinkiewicz, M.Levoy. "QSplat: A Multiresolution Point Rendering System for Large Meshes". Proc. SIGGRAPH'00, 349–352, 2000.
24. J. Blow. "Terrain Rendering at High Levels of Detail". Proc. of the 2000 Game. Mar. 2000.

Photo-realistic 3D Head Modeling Using Multi-view Images

Tong-Yee Lee, Ping-Hsien Lin, and Tz-Hsien Yang

Computer Graphics Group/Visual System Lab.
Department of Computer Science and Information Engineering,
National Cheng-Kung University, Tainan, Taiwan, Republic of China
tonylee@mail.ncku.edu.tw

Abstract. In this paper, we present a system for reconstructing photo-realistic 3D head models from multi-view images. In the proposed system, we perform two-pass bundle adjustments to reconstruct a photo-realistic 3D head model. At the first pass it computes several feature points of a target 3D head and then use these features to modify a generic head to obtain a rough head model. Next, we perform the second pass bundle adjustment to compute a detailed 3D head model. The texture of head models is obtained from multi-views. After texturing on the reconstructed head models, several preliminary results of photo-realistic 3D head models are demonstrated to verify the proposed system.

1 Introduction

Three-dimensional face model reconstruction from image sequences is a very challenging problem in computer vision and graphics. Our goal is to reconstruct a photo-realistic 3D face model in a short amount of time by a regular digital camera and personal computer. There are many methods existed for reconstructing 3D head models. Lee et al [1] generate a face model using data generated from laser scanners. In contrast to the laser sensor approach to reconstruct head models, the reconstruction from multi-views has the advantages of inexpensive cost. In the past, there have been many approaches proposed to reconstruct 3D models using multi-view images. The most simple and fast way is to reconstruct 3D head models from two orthogonal views [2]. The selected features on both views are used for 3D feature coordinates (i.e., (x,y) from the front view and (y,z) from the side view). Then, this approach modifies a given generic model to obtain a 3D head model. This approach requires two views are carefully captured so that their directions are orthogonal. Fua et al. [3] propose a regularized bundle-adjustment on every triplet images to reconstruct 3D head models from image sequences. This approach requires dense stereo matching that is always computationally expensive. Later, Liu et al. [4] improve [3] by estimating a small set of features and use a bundle-adjustment only on two the first views instead of all views. Then, the camera motions for other views are estimated using 3D head model. Pighin et al. [5] develop a technique to fit a generic model to a set of photographs of a person's face. This approach create highly realistic face models, but with manually intensive procedure. The texture map is extracted by combining the multi-view photographs into a cylindrical map. Blanz et al. [6] utilize statistics and knowledge of facial variations from large 3D head scans to establish a morphable face model. This morphable modeling scheme reconstructs a face model by a linear combination of geometries and images. In this paper, we propose a two-pass bundle adjustment scheme to reconstruct a 3D face model from multi-views. The preliminary results show that the proposed scheme is very promising and practical. In contrast to previous work, this approach does not require very intensive manual work but it can create very realistic 3D head models. These 3D head models are very useful for many applications and researches such as character design, virtual newscaster and talking head for MPEG-4. The details will be described in the following sections.

2 Reconstruction with Bundle Adjustments

Buddle adjustment [7] is a well-known technique in computer vision to reconstruct 3D points and camera parameters from N images. It can be formally described as follows. Given a set of 3D points X_i ($i = 1..M$) are viewed by a set of cameras with matrices P^j ($j = 1..N$). Let x_i^j be denoted as the corresponding 2D location of X_i visible in the image of P^j. The goal of 3D reconstruction is to find these X_i and P^j such that $P^j X_i = x_i^j$. However, the equation $P^j X_i = x_i^j$ may be not exactly held due to the errors of image measurements. The goal of the bundle adjustment attempts to estimate $\hat{P}^j$ and 3D points $\hat{X}_i$ such that $\hat{P}^j \hat{X}_i = \hat{x}_i^j$, and to minimize the following equation:

$$\min_{\hat{P}^j, \hat{X}_i} \sum_{j,i} d\left(\hat{P}^j \hat{X}_i, x_i^j\right)^2 \qquad (1)$$

where $d(x, y)$ is the image distance between x and y.

The bundle adjustment involves a nonlinear optimization and requires good initialization for proper and quick convergence. It can become an extremely large minimization problem because of the large number of parameters involved and thus can be computationally expensive and not stable. To reduce the complexity and achieve better stability, we perform the following approaches:

1. We adapt Zhang's method [8] to find the internal parameters of cameras and therefore, for each camera, we only compute its external parameters (i.e., 3 rotation angles and a translation vector).
2. We perform two-pass bundle adjustments to reconstruct a detailed 3D head model. At the first pass it computes several feature points of a target 3D head and then use these features to modify a generic head to obtain a rough head model. Finally, we perform the second pass bundle adjustment to find a detailed 3D head model. The details will be described in Section 3.

The bundle adjustment is a non-linear minimization process. We employ a standard Levenberg-Mardquart numerical algorithm to find the solution in an iterative manner. For good initializations for the first pass, we can base on two-view or triple-view point correspondences to estimate 3D corresponding points and external camera parameters using any standard method such as DLT(direct linear transform) [9] method. Figure 1 shows the feature points selected on the images and the corresponding features on the generic model.

Fig. 1. Shows corresponding features on the images and the generic model

In our experiments, we can use three to five images to reconstruct a 3D head. The user is required to manually select the feature correspondences on each image. To specify all features as shown in Fig. 1 is a very tedious work. To reduce human labor, we manually select seven points only on the images (see Fig. 2(a)) and we use these seven 2D locations to compute camera parameters. With these camera parameters, we can rotate the generic model to the same poses of the images and then project other feature points of the generic model onto each image. We can expect the projected features of generic model can not exactly match the corresponding features on the images. However, most feature correspondences can roughly match. At this stage, the user is required to locally adjust the wrong matching points. After this manual adjustment, we determine all features on the input images (see Fig. 2 (c) and (d)). In this manner, we can reduce many human efforts instead of manually selecting all features points on the images. In Section 3, we will describe techniques to further obtain more feature points and to better reconstruct a detailed 3D head model.

3 Detailed Head Reconstruction

After the first pass bundle-adjustment, we obtain 3D points of features as shown in Fig. 1(left). To adjust a 3D model, the radial basis function is used to adjust a 3D model. Since we know the correspondence between generic model and 3D features denoted as m_i ($i = 1..M$) from the first pass bundle adjustment, we can use them to adjust the given generic model. Therefore, we obtain a rough head model. The radial-based model is described as follows.

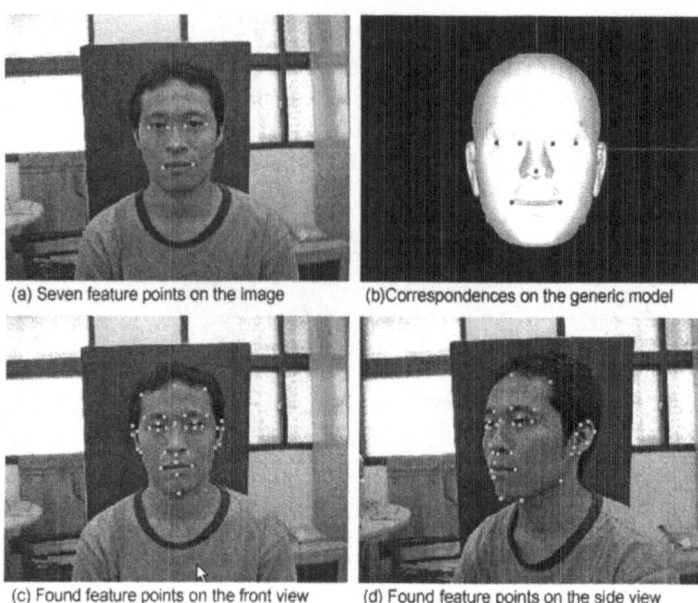

Fig. 2. Shows the determination of the features on the images

$$\hat{m} = \sum_i d_i \phi(\|m - m_i\|) + ax + by + cz + d \qquad (2)$$

where m is a vertex in the generic model, $\hat{m}$ is the corresponding new vertex after adjustment, and m_i is a 3D feature point. This equation includes a radial basis interpolation term: $\phi(r) = \exp(-r/k)$ with a constant k and a low-order polynomial term to model a global shape deformation.

To get a better-look 3D head, we need to obtain more feature points. For this purpose, we project all vertices of the above rough head model on the images as shown in Fig. 3. In Fig. 3 (a) and (b), we know where each vertex X_i of a generic model is projected on (a) and (b), say x_i^a and x_i^b. These two 2D locations x_i^a and x_i^b are potentially matching points on two views (a) and (b). To verify their correctness, we compute the similarity of images in the areas that are centered at x_i^a and x_i^b, evaluated by NCC(normalized correlation coefficient). If NCC value is smaller than a threshold then those corresponding points x_i^a and x_i^b would be invalid and thus be removed. In our experiment, the size of block of NCC is 9*9, and the threshold value is 0.5. After the above process, we obtain many extra matching pairs (x_i^a, x_i^b) on the images Fig. 3 (a) and (b), and next we compute the second pass bundle-adjustment to compute these extra 3D points. Finally, we use these extra feature points to adjust the rough head model using radial basis function to obtain the detailed 3D head model. To obtain better look, we exploit a silhouette-based scheme from our previous work [10] to build a hair model for our face model. This silhouette-based scheme extracts hair silhouettes from images and reconstructs control curves of hairs to modify the head model. See an example in Fig. 4. Finally, Fig. 5 shows the results obtained from the proposed method with hair modeling.

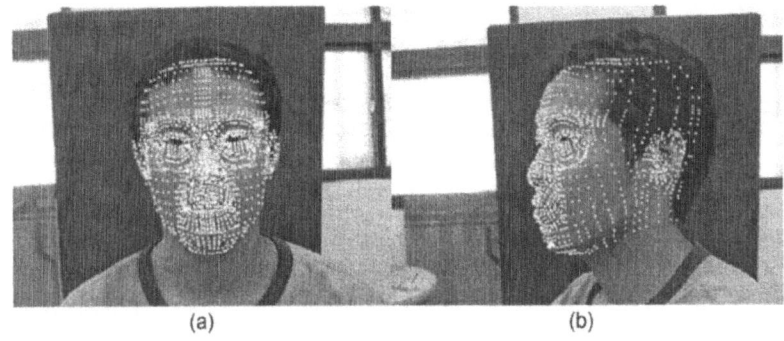

Fig. 3. Shows potential feature correspondences between two views (a) and (b)

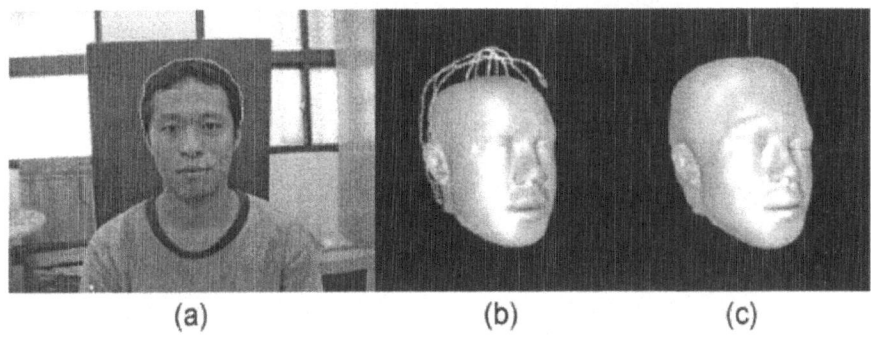

Fig. 4. (a) Silhouettes of hairs from a front view, (b) several 3D control lines from silhouettes, (c) head model modified by control lines with hairs

4 Texture Mapping and Preliminary Results

Using the texture mapping technique, a vivid 3D human face model can be built. We integrate all the images to produce the texture map. The cylinder surface is used as an intermediate object between the 3D head model and images. Any point say (x, y, z) of a head model can be represented as (θ, h) in the cylindrical

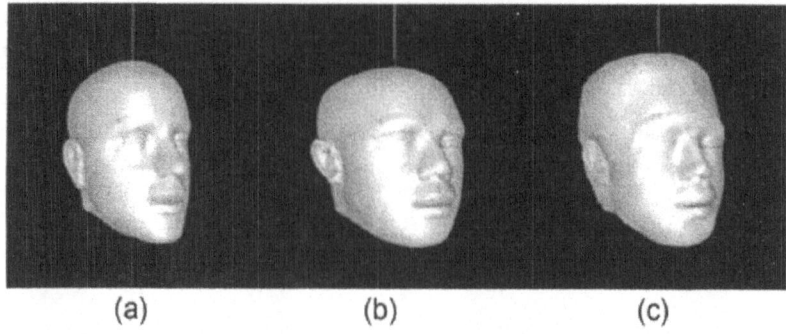

Fig. 5. (a) The generic head model, (b) the rough model after the first pass bundle adjustment and (c) the detailed head model after the second pass of bundle adjustment

coordinate system, where $0 \leq \theta \leq 2\pi$ and $-1 \leq h \leq 1$. In the texture image coordinate, the horizontal axis is the rotation θ and the vertical axis is the height value h. Then, we use a view independent approach [5] to generate a cylindrical texture map. Figure 6 shows a texture map using this technique combining five reference images. Finally, Fig. 7 shows preliminary results of photo-realistic heads and these heads use the same generic model as shown in Fig. 5(a). In this figure, we also show the detailed head models without texture mapping.

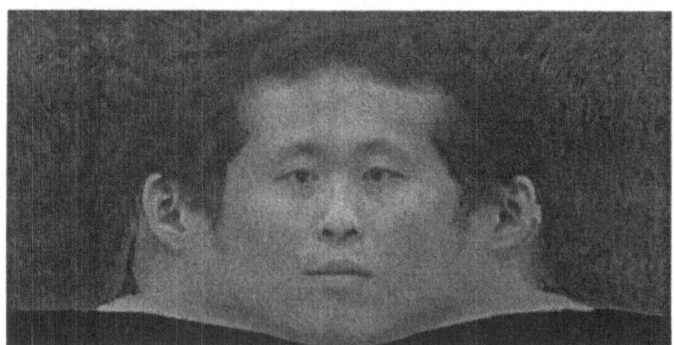

Fig. 6. Shows a texture map by combining multi-views

5 Conclusion and Future Work

In this paper, we have presented a two-pass bundle-adjustment scheme to build photo-realistic 3D head models. In the current implementation, although we have

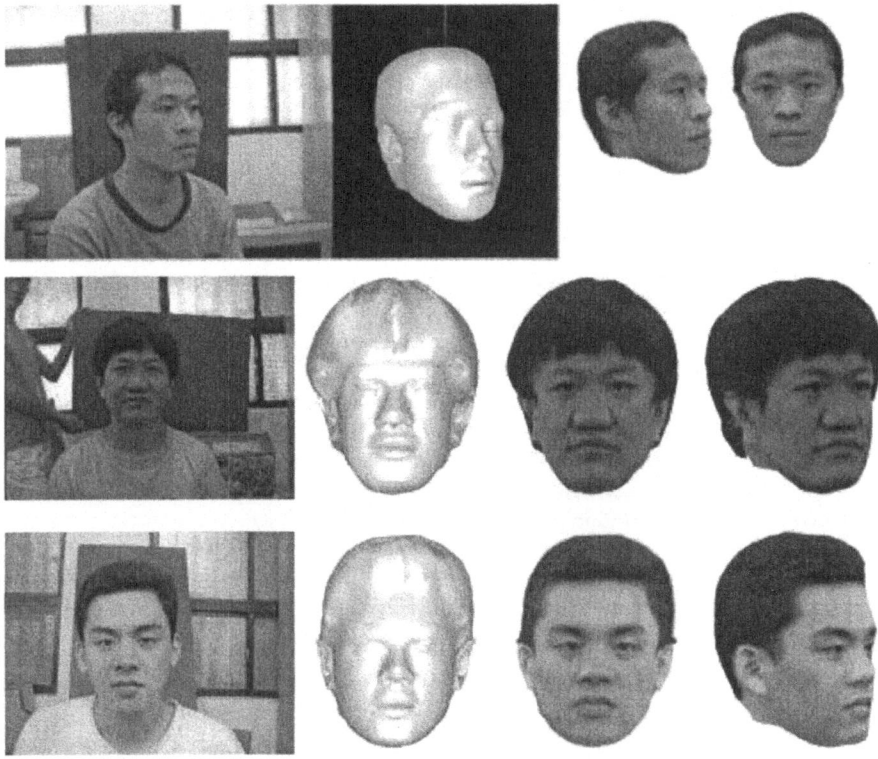

Fig. 7. Shows results of photo-realistic heads using the proposed schemes

attempted to reduce manual efforts, we still need some. In near future, we will improve this issue. In addition, we also would like to extend the current work to build models from video sequence. To make this approach more practical in daily use, we are planning to reduce the execution time.

Acknowledgement. This work is supported by National Science Council, Taiwan, ROC, under Contract No. NSC-92-2213-E-006-066 and NSC-92-2213-E-006-067.

References

1. Y. C. Lee, D. Terzopoulos, and K. Waters, "Constructing physics-based facial models of individuals," in Proceedings of Graphics Interface, pp. 1-8, 1993.
2. W.S. Lee, N.M. Thalmann, "Generating a Population of Animated Faces from Pictures," IEEE International Workshop on Modeling People, 1999, pp. 62-69.
3. P. Fua, "Regularized Bundle-Adjustment to Model Heads from Image Sequences without Calibration Data," International Journal of Computer Vision, 38(2). 153-171, 2000.

4. Z. Liu, Z. Zhang, C. Jacobs, and M. Cohen,"Rapid modeling of animated faces from video," The Journal of Visualization and Computer Animation, Vol. 12, No. 4, pp. 227-240, 2001.
5. F. Pighin, J. Hecker, D. Lischinski, R. Szeliski, D. Salesin, "Synthesizing Realistic Facial Expressions from Photographs," in SIGGRAPH 98 ConferenceProceedings, ACM SIGGRAPH, 1998, pp. 75-84.
6. V. Blanz, T. Vetter, "A Morphable Model For the Synthesis Of 3D Faces," in SIGGRAPH 99 ConferenceProceedings, ACM SIGGRAPH, 1999, pp. 187-194.
7. A. Gruen, H.A. Beyer, "System Calibration through Self-Calibration," in Calibration and Orientation of Cameras in Computer Vision, Washington D.C., Aug., 1992.
8. Z. Zhang, "A flexible new technique for camera calibration," IEEE Transactions on Pattern Analysis and Machine Intelligence, 22(11):1330-1334, 2000.
9. Reinhard Klette, Karsten Schluns, Andreas Koschan, "Computer Vision: Three-Dimensional Data from Images," Springer 1998.
10. Tz-Shian Yang, "Synthesizing Realistic Face From Video Streams," Mater Thesis, Dept. of Computer Science and Information Engineering, National Cheng-Kung University, Taiwan, Republic of China, 2002, July.

Texture Mapping on Arbitrary 3D Surfaces

Tong-Yee Lee and Shaur-Uei Yan

Computer Graphics Group/Visual System Lab(CGVSL)
Department of Computer Science and Information Engineering,
National Cheng-Kung University, Tainan, Taiwan, Republic of China
tonylee@mail.ncku.edu.tw

Abstract. Texture mapping is a common technique in computer graphics to render realistic images. Our goal is to achieve a distortion-less texture mapping on arbitrary 3D surfaces. To texture 3D models, we propose a scheme to flatten 3D surfaces into a 2D parametric domain. Our method does not require the two-dimensional boundary of flattened surfaces to be stationary. It consists of three steps: 1) we find high distortion areas in a 2D parametric domain and find a cutting path over these areas, 2) we add virtual points to adaptively find the better boundary of parametric domain instead of a predefined one and 3) finally, we perform an well-known smoothing technique for better texture mapping. The proposed scheme can be efficiently solved by a linear system and it yields an interactive performance. Finally, several preliminary results are demonstrated to verify the proposed scheme.

1 Introduction

Texture mapping is one of the most important techniques in computer graphics and it can enhance the reality of computer-generated 3D models. The common way to texture 3D surfaces is to map surface into an isomorphic planar representation. This operation is called surface parameterization or surface flattening and this planar representation is called embedding. Except for simple surfaces, such as cylinders, the surface parameterization always creates distortion of surface in 2D parametric domain. If we directly texture map on this 2D parameterization, the texture mapped on 3D surface is also distorted. It is an well-known differential geometry theorem that no isometric parameterization is in the plane for a general surface patch [1]. Hence, there is no easy way to parameterize a general 3D surface over 2D domain without introducing any distortion. In this paper, the propose method attempts to minimize the distortion of surface parameterization and produces a distortion-less texture mapping on arbitrary 3D surfaces.

There have been many texture mapping methods proposed [2,3,5,6,7,8,9,10, 12,13] and most of them discuss the issue of surface parameterization. In general, the topology of 3D surface meshes always are the same with a disk, since the connection of 3D surface meshes will have no change and can be mapped to a planar during the process of surface parameterization. Maillot et al. [2] group the facets by their normals and attempt to reduce distortion by minimizing a norm of the Green-Lagrange deformation tensor. Eck et al. [4] use harmonic maps to minimize the energy of distortion. Float [5] maps a disk-like open mesh to the plane. This approach can be solved efficiently by solving a linear system, but the boundary vertices are required to be on a convex 2D polygon. Hormann and Greiner [6] derive their deformation function from the ratio of singular values about mapping function. This method does not require the boundary vertices to be fixed on the convex 2D polygon. However, its computational cost is expensive. Sander et al. [3] define a non-linear texture stretch metric for surface parameterization. This approach uses relaxation approach to iteratively flatten 3D surfaces. Their texture stretch metric produces results better than those of several previous methods [2,4,6], but this technique may need lots of iterations for computing the optimal mapping. Sorkine et al. [7] introduce a bounded-distortion parameterization to guarantee that each triangle's distortion will strictly be below a user-defined threshold. This method also does not need the surfaces be mapped into a convex boundary and it would automatically map the boundary to the plane. For complicated 3D models, it would create many patches and would be more discontinuous for the seams of neighbor patches. Zigelman et al. [8] map the open mesh to a plane by a multi-dimensional scaling method that tries to preserve the geodesic distances criterion. This method does not ensure no self-intersections between the triangles and the computation cost is expensive due to the calculation about the geodesic distances.

Among the previous work, the proposed scheme is most related to [9,10]. Piponi et al. [9] cut the closed surface and then use the cut seams to be the boundaries for parameterization, and they also blend the seams to reduce the effect of the discontinuity. The method may need lots of user-interactions to define boundaries to the 2D plane. In contrast to [9], our proposed method would avoid iterations by using a standard numerical method to quickly solve the linear system. Gu et al. [10] propose a method for cutting through high distortion area in parameteric domain and they use an iteration procedure that finds only one path to cut at a time until the distortion is below some threshold value. In contrast this method, our proposed method can cut over all high distortion regions at a time by using the clustering strategy.

2 Methodology

Generally, the techniques of parameterization can be classified into two types: 1) one is with fixed boundaries in the parametric domain and 2) another one is with non-fixed boundaries. The methods with fixed boundaries often have large distortion, because the boundary shapes of the original 3D models are very different from those of their flattened surfaces in 2D domain. In this paper, the proposed method can fast flatten arbitrary 3D surfaces without the constraint of fixed boundaries.

2.1 The Initial Setup

Figure 1 is the flow chart of the proposed scheme. We will explain in detail in later sections. We require the 3D models to be consistent with the topology of

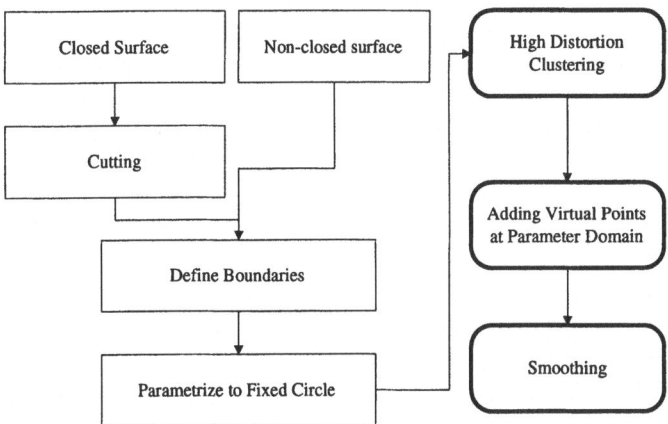

Fig. 1. Flow chart of the proposed scheme

an open disk. If the 3D model is a closed surface, we must cut the model first to make it to be topology equivalent to an open disk. The cutting procedure can be executed interactively by the user. We would cut along model's features with high curvature to reduce the distortion of parameterization. If the closed surface is not genus 0, it can be cut automatically to be a topological disk by the method [10]. Then, we utilize a relaxation-based scheme from our previous work [12] to parameterize the 3D model into a 2D circular embedding. This surface parameterization can be solved by a sparse linear system. The relaxation equation is shown in (1) and we select w to be the inverse of neighboring edge length in 3D.

$$p'_i = (1-\lambda)p_i + \lambda \frac{\sum_{j=1}^{k_i}(w_j p_j)}{\sum_{j=1}^{k_i} w_j} \quad (1)$$

where w:weight, λ:the speed of relaxation, and p_j:1-ring neighboring vertices of p_i.

2.2 Reducing Distortion Using Clustering and Cutting

After the surface is parameterized into a circular embedding using the relaxation-based scheme, we find many vertices are clustered in some areas of the embedding. These clustered areas always have the higher distortion of the surface parameterization. Gu et al. [10] suggest to use cutting for reducing the distortion.

Their approach applies parameterization iteratively to find regions of maximal distortion and connect the regions to the boundary one at a time. However, it is very computationally expensive to repetitively execute parameterization and to find each region with maximal distortion. In contrast, the proposed method does not require repetitive parameterization of surface. We propose a clustering algorithm to find a cutting path going through distorted areas. Then, we connect the cutting path to the boundary, cut the model and then execute the surface parameterization. The proposed clustering algorithm is not a K-mean-like clustering algorithm. The K-mean clustering algorithm needs to know the number of clusters in advance. This is not suitable for our case because we don't know the actual number of clusters occurring in the embedding. Our clustering scheme is described in following:

Clustering Algorithm

1. Determine vertices which would join to the computation of clustering algorithm. In our implementation, we calculate texture stretch [3] of all vertices. If its texture stretch value is larger than the mean of all vertices, it would be added into a queue for clustering calculation.
2. Define a Guassian function G_R that indicates the effective region of a cluster.

$$G_R(C - P_i) = e^{-\|C - P_i\|^2 / R^2} \qquad (2)$$

C:vertex uv position on the parameter domain
P_i:the position of i-th cluster's center
R:the radius of cluster

3. Initialize the first cluster. We choose the vertex with the maximum texture stretch value to be the center of the first cluster.
4. Calculate the centers of other clusters

   repeat
   $$\arg_j \max\left(N\left(C_j; P_1, ..., P_M, j = 1...n\right)\right)$$
   until $N(C_j; P_1, ..., P_M) \leq \varepsilon$

P_i: clusters which have already been determined
C_j: uv position of the vertex j
ε: defined value for the termination of repeat loop

The definition of function N is below:

$$N(C_j; P_1, ..., P_M) = \prod_{k=1}^{M} [1 - G_R(C - P_k)] \qquad (3)$$

Where the N function can be thought as the probability that vertex C_j is not belonging to other clusters $P_1, ..., P_M$.

From the above procedure, we determine all cluster centers. For each loop, the proposed clustering algorithm will find out the vertex that is the most irrelevant to the other clusters. If the function N value is less than the user-defined threshold ε, the clustering procedure would be terminated. Figure 2 shows an example of clustering and three centers are found.

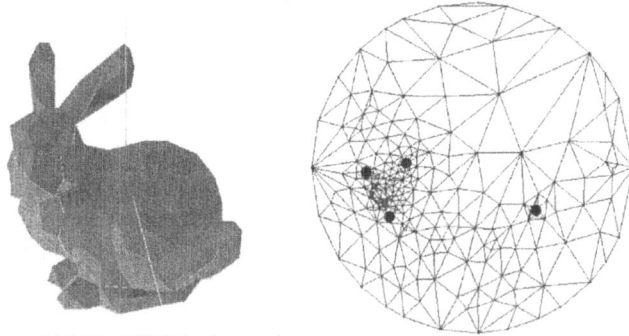

Fig. 2. A Rabbit Model and its result of clustering(solid circle are the centers of the clusters)

Minimal Spanning Tree Cutting. After executing clustering algorithm, we need to get a path to connect all clusters and connect this path to the boundary of the embedding. For this purpose, we find a tree to connect these cluster centers. In our implementation, we adopt the standard Kruskal's minimal spanning tree algorithm. To assign the weight of a graph edge, we combine both edge length and texture stretch together. We attempt to find a minimal spanning tree that is over high distortion areas as well as is with the minimal path length. The cutting path is founded using the following steps:

1. For each pair of clusters, we compute the shortest path. The edge weight is set using the following equation.

$$\text{Edge Weight} = 3D \text{ length} + 1/(\text{texture stretch}). \qquad (4)$$

2. Then, we obtain a complete graph for connecting all clusters.
3. Calculate the minimal spanning tree of this complete graph.

2.3 Virtual Points

Once the cutting path is found, we cut the model and re-parameterize the model to a circular embedding again. At this stage, the boundary of the embedding is still fixed. In this section, we will add virtual points to pull the boundary of this circular embedding and to adaptively modify the boundary for reducing distortion.

The idea of the virtual points is very simple. The circular embedding can be seen as a spring-based system. Each vertex is connected to its 1-ring neighbors through springs. The spring constant for each spring is in the inverse proportion to the 3D length of each edge connecting two vertices. In the outside of the circular embedding, for each boundary vertex, we put a corresponding vertex called virtual point along the direction from the center of the circle to the boundary vertex. See an example in Fig. 3. We add a virtual spring between a boundary vertex and a virtual point. The spring constant of each virtual point is in the inverse proportion to the 3D length of a shortest path from a boundary vertex to the center. Then, we apply an equal force to pull outward each virtual point along the direction of the embedding center to each virtual point. With adding these extra virtual points, we have the flexibility to re-define the positions of boundaries and can potentially reduce the distortion of parameterization.

For calculating the spring constants of boundaries, we can directly apply the Dijkstra's single source shortest path algorithm to compute the 3D surface length. Because the path computation is limited on triangle edges, this method may not always produce good and smooth results (Fig. 4: left). For this problem, we can subdivide the 3D mesh to increase the accuracy of the 3D path length, but it is computationally expensive. Alternatively, we find each path directly from the embedding by a line connecting the center to each boundary vertex (see an example in Fig. 4: right). Then, we find line intersections and compute their corresponding 3D positions. In this manner, we can obtain an approximately shortest path in 3D. Finally, similar to [12], we can organize this spring-based system as a sparse linear system and we can solve it numerically.

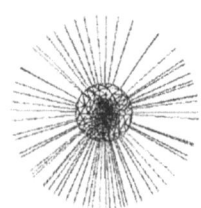

Fig. 3. the vertices of outer circle is virtual points

2.4 Smoothing

Smoothing is the last step of our procedure. The main concept of smoothing is to warp the texture used for texture mapping and the distortion of mapping can be further improved. Figure 5 illustrates the effects of smoothing for texture

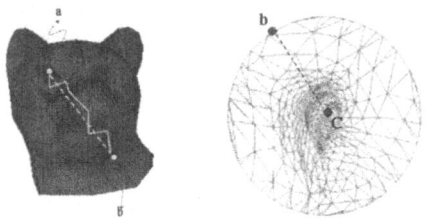

Fig. 4. The line between a boundary vertex and the center of the embedding

mapping. In the current implementation, we do not actually warp the texture. Alternatively, we further smooth the embedding by the method in [13]. However, we adopt texture stretch [3] to be our distortion ratio instead of the edge ratio of 2D edge length and 3D edge length in [13].

Fig. 5. Left: texture mapping without smoothing. Right: texture mapping with smoothing

3 Experimental Results

In this section, we demonstrate three preliminary examples to verify the proposed scheme. Figure 6, 7 and 8 are these three examples. In these examples, we show 1) original model, 2) circular embedding and cutting path from the proposed clustering scheme, 3) embedding after pulling by virtual points, and finally 4) model with texture after smoothing. To better visualize the cutting, we color the embedding. The region with a warm color has larger distortion than that of the region with a cool color. From these figures, the clustering centers are always located in the regions with a warm color. Furthermore, the paths are also going through the warm color regions. In addition, we show quantity information in Table 1. In this table, we measure the quality of texture mapping using mean texture stretch ratio [3]. The ideal mean stretch ratio is equal to one. The experimental measurement shows this stretch metric can be further improved by the consecutive steps of the proposed method. Finally, we should note that each step of the proposed method is executed in few seconds on a PC with Pentium III 1GHz and 256 MB RAM. Therefore, we can interactively perform texture mapping on arbitrary 3D surfaces. We calculate the texture stretch value at each step, so the user can optionally bypass some step of the proposed method if the current texture stretch are good enough. For example, we can bypass the smoothing or even the cutting if satisfactory results are obtained without these steps.

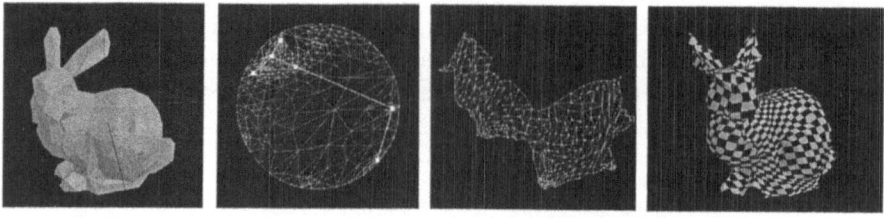

Fig. 6. Left: Rabbit model, Middle Left: clusters and the connected minimal spanning tree, Middle Right: Embedding without a fixed boundary, Right: Rabbit with texture

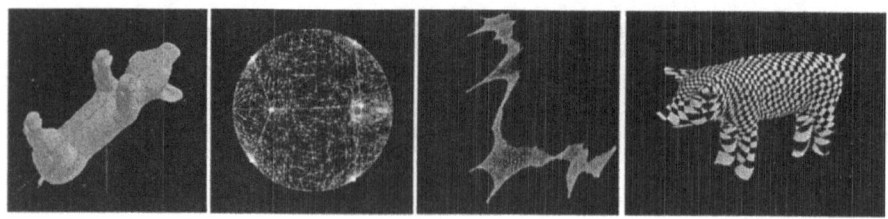

Fig. 7. Left: Pig model, Middle Left: clusters and the connected minimal spanning tree, Middle Right: Embedding without a fixed boundary, Right: Pig with texture

4 Conclusion and Future Work

In this paper, we proposed a texture-mapping scheme on arbitrary 3D surfaces. This new scheme can be solved very efficiently. All experiment results are executed on a PC with Pentium III 1GHz and 256 MB RAM at an interactive performance. In contrast to [9,10], the computational performance is very promising. The proposed method consists of several steps. These steps consecutively

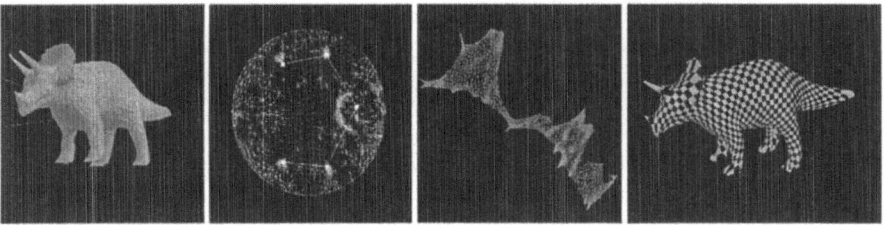

Fig. 8. Left: Dinosaur model, Middle Left: clusters and the connected minimal spanning tree, Middle Right: Embedding without a fixed boundary, Right: Dinosaur with texture

Table 1. Mean texture stretch ratio statistics for three models

Model	Rabbit	Pig	Dinosaur
# of vertices	296	3684	2917
# of triangles	522	7164	5660
Stretch ratio before clustering	2.071	3.256	18.557
Stretch ratio after clustering (without pulling by virtual points)	1.707	2.539	2.833
Stretch ratio after clustering (with pulling by virtual points)	1.314	2.031	1.851
Stretch ratio after smoothing	1.256	1.731	1.12

improve the quality of embedding. From the experimental results, the stretch ratio of texture mapping on the embedding is consecutively improved by these steps. In near future, we plan to extend the current work to the texture mapping with hard constraint problem. Or we are going to apply the proposed method to model re-meshing application.

Acknowledgement. This project is supported by National Science Council, Taiwan, ROC under Contract No. NSC-92-2213-E-006-066 and NSC-92-2213-E-006-067.

References

1. L. V. Ahlfors and L. Sario; *Riemann Surfaces* ; Princeton University Press, Princeton, New Jersey, 1960
2. J. Maillot, H. Yahia, and A. Verroust; *Interactive texture mapping*; Proceedings of SIGGRAPH, 1993, pp. 27-34
3. P. Sander, J. Snyder, S. Gortler and H. Hoppe; *Texture mapping progressive meshes*; Proceedings of SIGGRAPH, 2001, pp. 409-416
4. M. Eck, T. DeRose, T. Duchamp, H. Hoppe, M. Lounsbery, and W. Stuetzle; *Multiresolution analysis of arbitrary meshes*; Proceedings of SIGGRAPH, 1995, pp. 173-182

5. M. S. Floater; *Parametrization and smooth approximation of surface triangulations*; Computer Aided Geometric Design, 14(3):231-250, 1997
6. K. Hormann and G. Greiner; *Mips: an efficient global parameterization method*; Curve and Surface Design: St. Malo 1999, pages 153-162, Vanderbilt University Press, 2000
7. O. Sorkine, D. Cohen-Or, R. Goldenthal, and D. Lischinski; *Bounded-distortion Piecewise Mesh Parameterization*; IEEE Visualization, 2002, pp. 355-362
8. G. Zigelman, R. Kimmel, and N. Kiryati; *Texture mapping using surface flattening via multidimensional scaling*; IEEE Transactions on Visualization and Computer Graphics, Vol. 8, No. 2, pp. 198-207, 2002
9. G. Piponi and D. Borshukov; *Seamless Texture Mapping of Subdivision Surfaces by Model Pelting and Texture Blending*; Proceedings of SIGGRAPH, 2000, pp. 471-478
10. X. Gu, S. J. Gortler, and H. Hoppe; *Geometry Images*; Proceedings of SIGGRAPH, 2002, pp. 355-361
11. C. Gotsman, X. Gu, and A. Sheffer; *Fundamentals of Spherical Parameterization for 3D Meshes*; Proceedings of SIGGRAPH, 2003, pp. 358-363
12. Tong-Yee Lee and P. H. Huang; *Fast and Institutive Polyhedra Morphing Using SMCC Mesh Merging Scheme*; IEEE Transactions on Visualization and Computer Graphics, Vol. 9, No. 1, pp. 85-98, 2003
13. A. Sheffer and E. D. Sturler; *Smoothing an Overlay Grid to Minimize Linear Distortion in Texture Mapping*; ACM Transactions on Graphics, Vol. 21, Issue 4, pp. 874-890, 2002

Segmentation-Based Interpolation of 3D Medical Images

Zhigeng Pan[1,2], Xuesong Yin[1], and Guohua Wu[1]

[1] Institute of Virtual Reality and Multimedia HZIEE Hangzhou, 310018, China
[2] VRDM, State Key Lab of Novel Software Technology Nanjing University 210008, China
zgpan@cad.zju.edu.cn , yxs11@163.com , wugh@hziee.edu.cn

Abstract. This paper introduces a new interpolation algorithm based on images segmentation. Firstly, the algorithm obtains the regions of air, soft tissue and skeleton through segmenting images. Secondly, the algorithm uses matching interpolation in the same density regions, and scales the size of region as the interpolation data to interpolate image in the different density regions. The new image basically satisfies the requirements of medical image interpolation. Compared with linear interpolation, the new algorithm greatly improves the quality of image. The interpolation can be effectively used to construct 3D volume models.

Keywords: segmentation-based interpolation; 3D image; threshold segmentation; matching point pair

1 Introduction

Image interpolation is widely used in computer vision, especially in biomedical image processing, visualization, and analysis. Most 3D biomedical volume images are sampled anisotropically, with the distance between consecutive slices significantly greater than the in-plane pixel size. Either prior to display and measurement or during these manipulations, the volume image must be transformed in order to compensate for this anisotropy. This is done by creating a number of new slices between two known slices using image interpolation. Typically, there are several interpolation methods such as linear interpolation, spline interpolation, Kriging interpolation, matching interpolation and so on. These methods essentially are weighted average which is computed by the gray-level of sampling points and the relative distance between interpolation point and sampling point [1, 2, 3].

YU Wen-xue *et al* [4] presented an improved linear interpolation method. In this scheme, the optimal matching points were obtained from points of intersection of two contours through shifting the one slice which had its center of mass overlapped with the other slice's center of mass, and then matching points were interpolated. The shortcoming of the method is that points of intersection of two contours don't correctly describe the shape of interpolated contours when the actual contour was different from the sampled contour. Therefore the discrepancy is produced between the new slice and the actual slice. Supposed that the gray-level was changed according to the trend change of spline curve on three normal directions, spline interpolation [5] used three variables and three orders tensor multiple to estimate the gray-level of interpolated points. Due to supposing that data were changed according to spline curve in advance, spline interpolation brought the intolerable error when the change rule of the actual data was very different to the change rule of supposed data. Kriging interpolation [6, 7] based on the spatial distribution rule of data, and processed the optimal agonic estimation on statistical viewpoint. The typical matching interpolation manually selected matching points from two slices, and then interpolation was decided by the template which was composed by the distribution of matching points. Another automatic matching method automatically obtained the contour of target as template to use [8]. When interpolation was performed by the template, interpolation efficiency was very low because of many manual interventions or required higher algorithm.

Linear interpolation, spline interpolation, Kriging interpolation and matching interpolation were gray-based interpolation methods, so that they only are suitable for gray-level interpolation, instead of interpolating the satisfactory image shape. The problem is solved by shape-based interpolation method [9, 10, 11]. The method transforms 2D gray-level image to 3D binary image, then the shape is interpolated. Shape-based interpolation only obtains the shape of appointed object, namely a tissue only interpolated one time. Thus, the method loses its merit when we want to obtain the data of the whole image.

In order to overcome these shortcomings of gray-based interpolation and shape-based interpolation, we develop a new method called Segmentation-Based Interpolation of 3D Medical Images. It uses matching interpolation to interpolate images in the same matter, and interpolates the images after regarding interpolation distance as scale factor to scale the region of an image in the different matter. The advantage of this scheme not only interpolates the image gray-level, but also takes into account the image shape. Therefore, the new image can lead to dramatic improvement for 3D volume rendering in visualization.

The remainder of the paper is organized as follows. In Section 2, we first give a threshold segmentation of the proposed method. Then, the theoretical basis and the detailed techniques of our approach are described. Experiments and results are presented in Section 3, followed in Section 4 by conclusions and suggestions for future work.

2 The Proposed Method

2.1 Image Segmentation

Gray-level between the different tissues isn't equal in cross-sectional slices, for example gray-level is dramatically changed in the boundary of soft tissue and skeleton. According to the different tissues, several density regions can be obtained which is the key of proposed algorithm. Experiment shows the gray-level distribution of air, soft tissue and skeleton as is shown in Figure 1. The image is divided into air, fattiness, soft tissue and skeleton by the threshold segmentation in the light of the discrepancy of gray-level value, at the same time four regions marked correspondingly in order to meet the requirement of interpolation later.

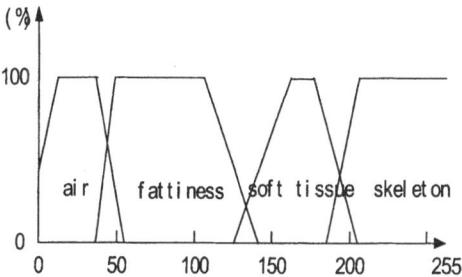

Fig. 1. Gray-level histogram of CT knee image

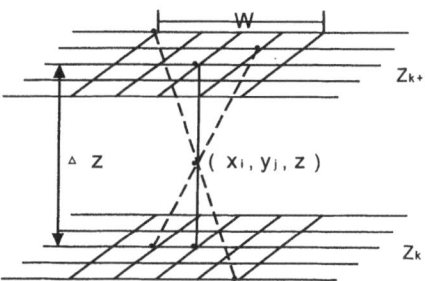

Fig. 2. Sketch map of matching points

2.2 Matching Interpolation

The new slice $V(x_i, y_j, z)$ is obtained by interpolation between the slice $V(x_i, y_j, z_k)$ and the slice $V(x_i, y_j, z_{k+1})$, as shown in Figure 2. To guarantee universality, we assume the distance between the Kth layer slice and the new slice is greater than that between the next layer slice and the new slice. The point (x_i, y_j, z_k) is considered as the center to get the window width of W×W on the Kth layer slice. Then, within this window the connecting lines between the each point on $V(x_n, y_m, z_k)$ and the point (x_i, y_j, z) intersect with the slice $V(x_n, y_m, z_{k+1})$ [12]. Consequently, we can obtain the matching point pairs of W×W.

(1) Computing the position of matching points

The size of matching window is decided by the number of matching point pairs, since matching point pairs locate the selected window. The width of window generally is the fixed value and is set the odd number, such as 3×3, 5×5 in order to guarantee the symmetry. The experimental results show that the proposed method not only incurs extremely time consuming, but also will result in poor approximations and a jagged staircase artifact which is clearly visible in renderings of these volumes if the width of the window is selected greater.

In the Kth layer slice, the coordinates of matching points are as follows:

$$x_m(k) = x_i + \frac{2m+1-W}{2} \qquad (1)$$

$$y_n(k) = y_i + \frac{2n+1-W}{2} \qquad (2)$$

$$m, n = 0, 1, 2, \cdots, W-1$$

Relative to the matching points of the Kth layer slice, the coordinates are as follow in the next layer slice.

$$x_m(k+1) = x_i + \text{int}\{\frac{d_2}{d_1}(x_{W-1-m}(k) - x_i)\} \qquad (3)$$

$$y_m(k+1) = y_i + \text{int}\{\frac{d_2}{d_1}(y_{W-1-m}(k) - y_i)\} \qquad (4)$$

$$m, n = 0, 1, 2, \cdots, W-1$$

where d_1 is the distance from (x_i, y_j, z) to $V(x_i, y_j, z_k)$ ($d_1 = z - z_k$) and d_2 is the distance from (x_i, y_j, z) to $V(x_n, y_m, z_{k+1})$ ($d_2 = z_{k+1} - z$). According to the above four formulas, we can set up the matching point pairs in the appointed window.

(2) Confirmation of the optimal matching points

The key problem of matching interpolation is that the optimal matching point pair should be established from lots of matching point pairs. In this paper, we argue that the optimal matching point pair should satisfy four conditions, as follows:

(a) The connecting line of matching point pairs is as small as possible.

(b) Matching point pairs would possess the same direction of rotation which makes the angle composed with Z axis smaller.

(c) The gray-level value of matching point pairs is equal or near.

(d) The gray-level gradient is near.

We present the vector function below to indicate the matching degree of a pair of matching points (x, y, z) and (x', y', z').

$$\begin{aligned}R(x, y, z, x', y', z') = &[V(x, y, z) - V(x', y', z')]\mathbf{i} \\&+ [D(x, y, z) - D(x', y', z')]\mathbf{j} \\&+ [\theta(x, y, z) - \theta(x', y', z')]\mathbf{k} \\&+ [\sqrt{(x-x')^2 + (y-y')^2}]\mathbf{l}\end{aligned} \qquad (5)$$

where $V(x,y,z)$, $D(x,y,z)$ and $\theta(x,y,z)$ are the gray-level of the point, the gray-level gradient of the point and the direction of the gradient respectively. Also, $V(x',y',z')$, $D(x',y',z')$ and $\theta(x',y',z')$ are the gray-level of the point, the gray-level gradient of the point and the direction of the gradient respectively. $\sqrt{(x-x')^2+(y-y')^2}$ is the distance which is projected on the horizontal plane by matching point pair (x,y,z) and (x',y',z'). Here, the gray-level gradient $D(x,y,z)$ comes from the module of $\nabla D(x,y,z)$.

$$\nabla D(x,y,z) = \frac{1}{2}[V(x+1,y,z)-V(x-1,y,z)]\mathbf{i}$$
$$+\frac{1}{2}[V(x,y+1,z)-V(x,y-1,z)]\mathbf{j}$$
$$+\frac{1}{2}[V(x,y,z+1)-V(x,y,z-1)]\mathbf{k} \qquad (6)$$

We can obtain the formula below according to Formula 5.

$$|R(x_m(k),y_n(k),z_k,x_m(k+1),y_n(k+1),z_{k+1})|=$$
$$(V(x_m(k),y_n(k),z_k)-V(x_m(k+1),y_n(k+1),z_{k+1}))^2$$
$$+(D(x_m(k),y_n(k),z_k)-D(x_m(k+1),y_n(k+1),z_{k+1}))^2$$
$$+(\theta(x_m(k),y_n(k),z_k)-\theta(x_m(k+1),y_n(k+1),z_{k+1}))^2$$
$$+(x_m(k)-x_m(k+1))^2+(y_n(k)-y_n(k+1))^2 \qquad (7)$$

According to the above four conditions that the optimal matching point pair need to achieve, if the point $(x_p(k),y_q(k),z_k)$ and the point $(x_p(k+1),y_q(k+1),z_{k+1})$ consist of the optimal matching point pair, two points must satisfy Formula 8.

$$|R(x_p(k),y_q(k),z_k,x_p(k+1),y_q(k+1),z_{k+1})|$$
$$= \min_{m,n=0,1,\cdots,W-1} |R(x_m(k),y_n(k),z_k,x_m(k+1),y_n(k+1),z_{k+1})| \qquad (8)$$

After the optimal matching point pair is obtained, gray-level value of the target points (x_i,y_j,z) can be obtained using the linear interpolation to the optimal matching pair, as shown in Formula 9.

$$V(x_i,y_j,z) = \frac{1}{d_1+d_2}(V(x_p(k),y_q(k),z_k) \times d_2 + V(x_p(k+1),y_q(k+1),z_{k+1}) \times d_1)$$

(9)

2.3 Segmentation-Based Interpolation

According to the above analysis, the proposed segmentation-based method is applied to interpolate the new slice between the slice $V(x_i,y_j,z_k)$ and the slice $V(x_i,y_j,z_{k+1})$. The steps of the new method are as follows:

Step 1. The slice $V(x_i,y_j,z_k)$ and the slice $V(x_i,y_j,z_{k+1})$ are segmented the four regions, such as air, fattiness, soft tissue and skeleton which are marked correspondingly.

Step 2. The distances between the new slice and the two interpolated slice are obtained.

Step 3. Find out the optimal matching point pair of $(x_p(k),y_q(k),z_k)$ and $(x_p(k+1),y_q(k+1),z_{k+1})$.

Step 4. Matching interpolation is performed when the optimal matching point pair is in the same regions.

Step 5. Present the size of the two interpolated slice region where the target points locate in when the optimal matching point pair is in the different regions.

Step 6. When the size of the above layer region is greater than the size of the low layer region $d_1/(d_1+d_2)$ is regarded as scale factor to scale the above layer region as the interpolated data, or else $(d_1+d_2)/d_1$ is regarded as scale factor to scale the above layer region as the interpolated data. Using the interpolated data, gray-level value of target point can be obtained.

Step 7. Repeat Step 3, Step 4, Step 5 and Step 6 until the new slice is produced.

The most important part of the proposed method is that the position of the optimal matching point pair is ascertained. If the optimal matching point pair is in the same density region, the matching interpolation is used, or taking advantage of the scale factor, the corresponding regions are scaled to obtain the interpolated data so that the new slice is produced. Thus, the error brought by the single matching interpolation is avoided and the image quality is greatly improved too.

3 Experimental Results and Discussion

We employ a group of the knee slices of human being to test the proposed method. The three consecutive gray-level slices (512×512) are acquired and marked I_1, I_2 and I_3 respectively, as illustrated in Figure 3. The new second slice is obtained using the two slices I_1 and I_3. According to the original second slice, we have compared the interpolated result with the result of the simple linear interpolation.

The experimental results of the two interpolation methods are shown in Figure 4. In Figure 4(a), the image is of the referenced image, Figure (b) and Figure (c) are the results of the linear interpolation method and the proposed method respectively. The contour of slice which is interpolated by the linear interpolation method blurs from the experimental result. The cause of the error is that the linear interpolation method (LIM) based on gray-level produces the artifact when the interpolated matching points locate the different regions of the two slices. Segmentation based interpolation method (SIM) takes account of not only the change of gray-level, but also the change of contour. Therefore, the new slice is of the clarity and the artifact is decreased greatly.

Mean-squared difference (MSD) is used to measure the interpolated results. The formula is as follows:

$$\sigma^2 = \sum (v' - v)^2 / N \qquad (10)$$

where v' is the estimation value and v is the actual value. N is the number of points. MSD of two kinds of methods is shown in Table 1.

Table 1. MSD σ^2 ($greylevel^2$)

	LIM	SIM
Knee slices of human being (Figure 4)	31.3794	17.8924
Brain slices of human being (Figure 7)	38.2715	24.6299

4 Conclusions and Future Work

A novel segmentation based interpolation method is presented in this paper in order to avoid the shortcoming of the gray based interpolation method and the shape based interpolation method. The proposed method interpolates not only gray-level, but also the shape of the object, so that the contour blur which is produced by the traditional interpolation method is further solved. Moreover, the new slice is more similar to the original medical slice. The experimental results show that the proposed method improves the interpolating quality substantially.

As to the future study, we will perform further validation for segmentation based interpolation method, including sensitivity to noise tests, more qualitative and quantitative comparisons with the shape-based method on both synthetic and real images. We are also interested in applying segmentation based interpolation method to the dynamic objects. Finally, since good visualization is the major advantage of segmentation based interpolation method, we will also work on combining volume rendering with interpolation to develop a 3D volumetric segmentation and visualization algorithm.

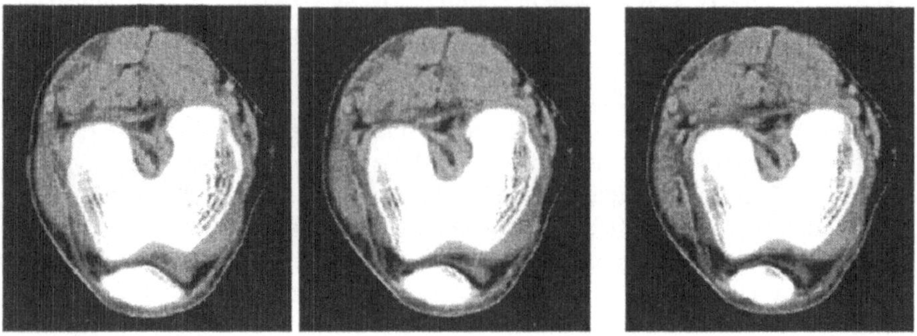

(a) I_1 (b) I_2 (c) I_3

Fig. 3. Original slices

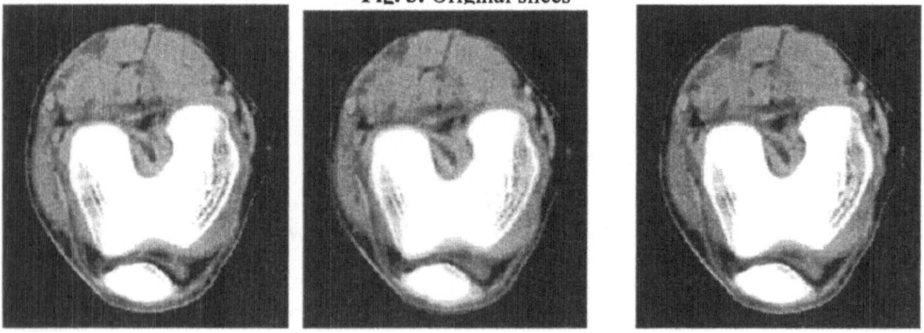

(a) I_2 (b) Linear interpolation   (c) Segmentation interpolation

Fig. 4. Comparison for LIM and SIM

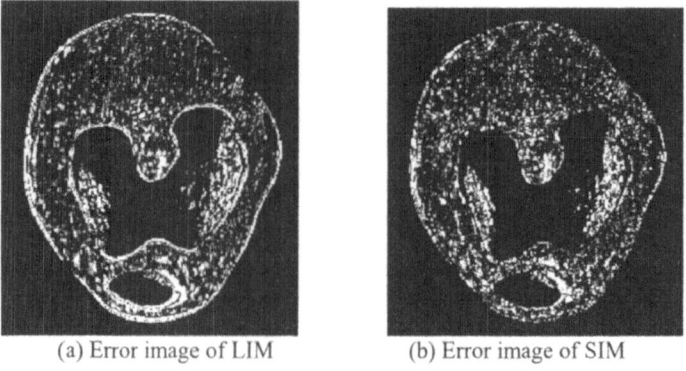

(a) Error image of LIM   (b) Error image of SIM

Fig. 5. Error image of two interpolation methods

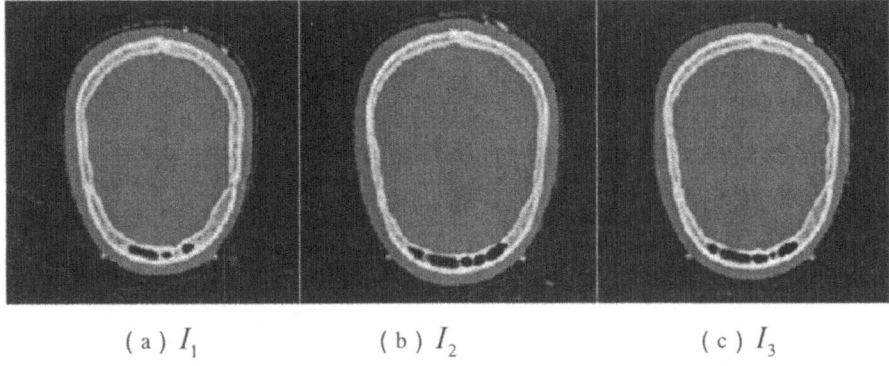

(a) I_1 (b) I_2 (c) I_3

Fig. 6. Original slices

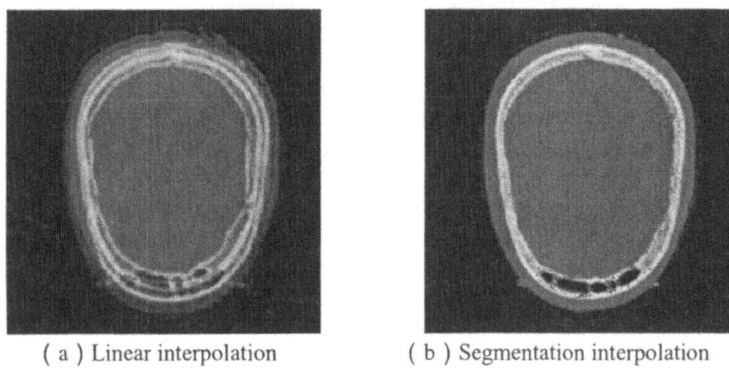

(a) Linear interpolation   (b) Segmentation interpolation

Fig. 7. Comparison for LIM and SIM

Acknowledgements. This research work is co-supported by Excellent Youth NSF of Zhejiang Province(RC 40008) and TRAPOYT Program in Higher Education Institutions of MOE, P.R.C.. The authors would like to express thanks to Dr. Yigang Wang, M.S. Qian Zhang, M.S. Yunfang Ge, M.S. Pengzheng Zou, M.S. Ting Shu who give help on the preparation of the paper.

References

[1] Grevera G J, Udupa J K. *An objective comparison of 3-D image interpolation methods.* IEEE Trans. on Medical Imaging, 1998, 17(4) : 642 - 652
[2] Lehmann T M, Gonner C, Spitzer K. *Survey:interpolation methods in medical image processing.* IEEE Trans Med Imag, 1999, 18(11) : 1049-1075
[3] HUANG Hai-yun, QI Fei-hu CHEN Jian, et al. A wavelet-based interpolation of medical images. ACTA AUTOMATICA SINICA, 2002, 28(5): 722-728

[4] YU Wen-xue, LUO Li-min, FU Yao, et al. *Intermediate contour interpolation for computed altas surface reconstruction of 3-D human brain*. Journal of Electronics, 2000, 28 (2) : 52 - 54

[5] Kokaram A C, Morris R D, Fitzgerald W J et al. Interpolation of missing data in image sequence. IEEE Trans. On Image Processing, 1995, 4(11) : 1509 - 1519

[6] Stylz M R, Parrott R W. *Using kriging for 3-D medical imaging. Computerized Medical Imaging and Graphics* , 1993, 17(6) : 421 - 442

[7] Matechik S M., Stytz M R. Using kriging to interpolate MRI data. Proceedings of the 16th Annual International Conference of the IEEE, 1994, 5(1): 576 -577

[8] JIA Chun-guang, TAN Ou, DUAN Hui-long, et al. *Medical image registration based on deformable contour. Journal of Computer-Aided Design & Computer Graphics,* 1999, 11 (2) : 115 - 119[9]   Grevera G J, Udupa J K. *Shape-based interpolation of multidimensional grey-level images*. IEEE Trans. on Medical Imaging, 1996, 15(6) : 881 - 892

[10] Grevera G J, Udupa J K. *A Task-specific evaluation of three-dimensional image interpolation techniques*. IEEE Trans. on Medical Imaging, 1999, 18(2) : 137 - 143

[11] Bor A G, Kechagias L, Pitas I. *Binary morphological shape-based interpolation applied to 3-D tooth reconstruction*[J]. IEEE Trans Med Imaging 2002 21(2): 100 - 108

[12] MIAO Bin-he, DENG Yuan-mu, HUANG Fei-zeng, et al. *Interpolation of 3-D images based on point matching*. Chinese Journal of medical Physics, 2000, 17(1) : 14 - 16

A Bandwidth Reduction Scheme for 3D Texture-Based Volume Rendering on Commodity Graphics Hardware

Won-Jong Lee[1], Woo-Chan Park[2], Jung-Woo Kim[3], Tack-Don Han[1], Sung-Bong Yang[1], and Francis Neelamkavil[1]

[1] Media System Laboratory, Department of Computer Science,
Yonsei University, Seoul 120-749 Korea,
{airtight, hantack}@kurene.yonsei.ac.kr
{yang, francis}@cs.yonsei.ac.kr
[2] School of Computer Engineering, Department of Internet Engineering,
Sejong University, Seoul 143-747, Korea,
pwchan@sejong.ac.kr
[3] Digital Media R&D Center, Samsung Electronics, 416,
Maetan-3Dong, Paldal-Gu, Suwon City 442-742, Korea,
jwoo.kim@samsung.com

Abstract. In this paper, we propose a bandwidth-effective volume rendering scheme which subdivides the volume into the sub-volumes and transmits them to the texture units in visibility order. Each sub-volume is rendered in the same manner as the original volume on the graphics hardware and the corresponding sub-image is blended in the alpha blending unit. The sub-volume oriented processing improves the cache efficiency and allows empty space skipping. Moreover, it is capable of rendering volume datasets that do not fit into the texture memory. Simulations show that the proposed scheme is effective for 3D texture-based volume rendering on commodity graphics hardware by reducing memory bandwidth up to 30 times when compared with the traditional method.

1 Introduction

Direct volume rendering is one of the popular methods to visualize volumetric data in various areas such as medicine, science, and engineering. Due to the rapid advances in graphics processing unit (GPU), 3D texture-based volume rendering on commodity graphics hardware receives great attention in these days [1, 2, 3]. Although using the commodity graphics hardware for volume rendering allows us to perform the 3D texture mapping with a low-cost, there are several disadvantages as mentioned in [4, 12, 16]. First, a large amount of data traffic causes bottlenecks in memory bandwidth both for 3D texture mapping and pixel processing. In 3D texture mapping, tri-linear interpolation may degrade the temporal locality of the texture memory accesses. Thus, we cannot achieve a satisfiable hit rate with the texture cache optimized only for 2D textures. In pixel processing, one depth test and one read/write operation must be done for each pixel and accesses to the same screen location are separated by too many accesses to other pixels. As a result, using the commodity graphics hardware causes poor pixel cache utilization, which results in excessive frame buffer traffic in the blending operation. Second, empty space skipping is also problematic due to the slice-oriented processing order. Thus unnecessary computation for meaningless space cannot be avoided. Third, the size of a dataset that can be processed on the commodity graphics hardware is very limited. A realtime rendering of a large dataset (512^3 or larger) is infeasible on the current graphics hardware [19, 20] in general.

This paper proposes a bandwidth-effective rendering scheme by improving the cache utilization on the commodity graphics hardware. The proposed scheme subdivides the original volume into uniform sized sub-volumes logically. Each sub-volume is tested if it is an empty space or not, and then it is rendered on GPU separately. Finally each rendered sub-image is blended in the alpha blending unit for generating frame image (see Fig. 3). The advantages of the proposed rendering scheme are as follows. First, sub-volume ordered processing increases the locality of the memory access and hence improves cache efficiency, which results in a dramatic reduction of the memory bandwidth. Second, empty space testing for each sub-volume is possible in the preprocessing step so that we can avoid computation for meaningless space by skipping these empty sub-volumes. Third, a volume dataset that does not fit into texture memory can be rendered, because the original volume is subdivided into much smaller sub-volumes. A drawback of the proposed rendering scheme is that the subdivision operation generates additional vertices for each sub-volume, which causes overhead on a hardware T&L engine. However, according to the specification of a recent commodity graphic hardware [19, 20], the T&L engine has a capability of processing more than 350×10^6 vertices per second. On the other hand, in spite of the worst case of the proposed rendering scheme (the case of 16^3 sub-volume), as shown in the simulation results in Section 4, the required bandwidth for rendering at a rate of 30 frames per second is about 230×10^6 vertices per second. Thus, the additional vertices do not causes any problem in practice.

We have built a simulator to evaluate achievable performance. Memory access traces has been generated during the benchmark volume datasets of size 512^3 were rendered on this simulator. The performance has been evaluated with the trace-driven cache simulator, DineroIII [5]. Simulation results show that the proposed rendering scheme reduces memory bandwidth from 2 to 30 times compared with a non-subdivided method with the cache size varying from 16Kbytes to 128Kbytes. The rest of the paper is organized as follows. Section 2 reviews the related work, Section 3 describes the proposed sub-volume rendering scheme, Section 4 provides simulation results.

2 Related Work

The subdivision method has been adopted in a variety of researches for processing large scaled volume datasets. This method was used for load balancing on parallel rendering machines [6] early in the research. Dedicated hardware for ray-casting [7, 8] partitioned the volume data suitable for their own memory organization. In particular, RaceEngine [7] proposed a new method to keep the visibility order of the sub-volumes. Recently, a networked cluster of PCs each of which is equipped with a fast graphics accelerator to render each sub-volume has been proposed [9]. However, parallel volume rendering is a quite expensive approach, because it requires dedicated hardware or a massive parallel rendering machine.

Nowadays the performance of commodity graphics hardware has been improved dramatically. High-level GPU-programming environment and relevant graphic APIs such as OpenGL and Direct3D are fully supported. Especially various 3D texture-based volume rendering methods using the commodity graphics hardware have been being proposed [2, 3, 4]. TriangleCaster [4] employed the extension units to divide the image plane and blend the corresponding sub-planes. However, the additional hardware units such as the composition buffer and the triangle generator should be implemented to extend the existing graphics hardware.

In order to render large scaled volume datasets on the commodity graphic hardware, the subdivision method can be applied to multi-resolution and compression. Multi-resolution volume rendering uses a spatial hierarchy to adapt the resolution to project onto the screen with hierarchical structures such as an octree. A variety of techniques have been applied to volume rendering [10, 11], since multi-resolution technique was first proposed in polygonal rendering. Moreover, volume data compression for reducing the size of such data as vector quantization [12], fractal compression [13], and wavelet transform [14, 15] utilizes hierarchical division method.

The goal of the proposed scheme is to resolve the memory bandwidth problem in 3D texture mapping by subdividing original volume into optimal sized sub-volumes to increase cache locality and by rendering in visibility order. The proposed scheme can support the rendering of the volume datasets that do not have fit into the texture memory. Also volume data compression in [15] can be easily adopted in the proposed scheme.

3 Bandwidth-Effective Sub-volume Rendering

The proposed rendering scheme is composed of three steps; the preprocessing step for subdivision of the volume data, the rendering step including the classification and the 3D texture mapping, and the composition step for blending each corresponding sub-images. This section describes each step in detail.

3.1 The Preprocessing Step

In the preprocessing, we divide the original volume into uniform sized sub-volumes after determining the sub-volume size. For each sub-volume, vertices and texture coordinates of slices are generated with using the bound cube in [1,

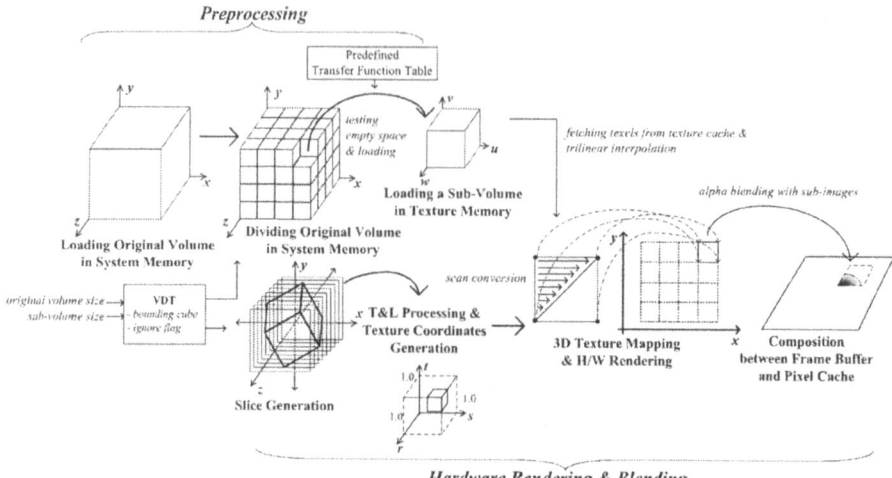

Fig. 1. Overall processing flow of the bandwidth-effective sub-volume rendering scheme

11]. We define a new data structure, called the *volume division table* (VDT) for managing the information of empty spaces and the geometry (min/max bounds of positions in the texture and the object space) of sub-volumes.

Fig. 1 illustrates the overall processing flow of the bandwidth-effective sub-volume rendering scheme. In the preprocessing step, the transfer function table is created for classification and the VDT is generated with the information about the sizes of the original volume and sub-volume. During loading the original volume to the system memory, each sub-volume can be tagged with either empty space or non-empty space after each voxel is compared with a user-defined threshold. If a sub-volume is tagged with empty, a single bit is stored into the corresponding position in the VDT. After the preprocessing step, the visibility order of each sub-volume is determined and each sub-volume is transmitted to GPU in this order which will be described in Section 3.2.

To increase cache locality, the size of a sub-volume is smaller than that of the texture cache, and the size of the slice to be mapped for this sub-volume is smaller than that of the pixel cache. This leads to minimize the bandwidth between the rendering processor and the graphics memory.

3.2 Sub-volume Rendering

In the sub-volume rendering step, each sub-volume is rendered in the same manner as the original volume is done and the corresponding sub-images are blended in the alpha blending unit to generate final frame image as shown in Fig. 3. The sub-volumes should be rendered according to the visibility order, because the volume data is not fully transparent in general and there are overlapped parts between adjacent sub-images. Whenever the viewpoint (for perspective projection) and the view direction (for parallel projection) are changed, this order needs to be sorted again. We adopt the ordering method in [7]. In this method, the sub-volumes are ordered based on three types of classifications. First, viewpoint is parallel with a face of the volume in parallel projection. Second, looking at an edge or at a point of the volume in parallel projection, and the last type is perspective projection. In case of the parallel projection, the order is set either by height or by width based on the oriented position in the volume. In case of the perspective projection, the nearest sub-volume from the viewpoint precedes other sub-volumes.

One of the advantages of the sub-volume based rendering is that we can skip empty space easily. Because a significant portion of the volume data (more than 50% on the average in general) is empty space [18], empty space skipping is a common method to accelerate rendering for ray-casting based volume rendering. However, empty space skipping is difficult to be applied to the conventional 3D texture based volume rendering, because the whole voxels to be mapped onto slices should be fetched [16]. In contrast, the sub-volume based rendering [11, 17] has a capability of checking whether the current sub-volume is empty or not in the preprocessing step. As a result, rendering can be skipped for the empty sub-volume by referring the VDT. According to the simulation results of Section 4, the average empty space is more than 50% of the volume data. Such results can also be found in [7].

3.3 Texture and Pixel Cache Efficiency

Recent commodity graphics accelerators employ both the texture cache and pixel (color and depth) caches to reduce the bandwidth between the rendering processor and the graphics memory. But they are designed to optimize accesses only to process the polygonal data. Thus we cannot achieve satisfiable performance with them for volume rendering applications. We now describe the effective cache utilization issue of sub-volume rendering.

The tri-linear interpolation for 3D texture mapping requires accesses to the neighborhoods of voxels in the x-, y-, z-dimensions. For the volume stored in memory such as the neighboring voxels along the x-axis are adjacent, the neighboring voxels along the y-axis are one row of the x-axis voxels apart. The locality of the neighboring voxels along the z-axis is even worse. They are one row of the x-axis voxels times one column of the y-axis voxels apart. Thus a serious problem occurs in the texture cache as mentioned in [16, 18]. Due to this weak locality, access the neighboring voxel along the z-axis causes subsequent cache misses, which results in frequent replacement of cache blocks. Although some dedicated hardware accelerators for ray-casting [7, 8] are optimized for the three-dimensional data access to solve this problem, the texture cache of a commodity graphic accelerator is optimized only for 2D texture mapping. In this case, the caching is almost useless for volume rendering.

However, we can resolve this problem by subdividing and reorganizing the volume data, which make rendering of the sub-volumes more manageable. An 8bit-16^3 sub-volume consumes 4Kbytes (when 16bit-16^3 sub-volume, 8Kbytes) and a 8bit-32^3 sub-volume consumes 32Kbytes (when 16bit-32^3 sub-volume,

64Kbytes) of memory space. This chunk of data fits easily within the texture cache of a typical graphics accelerator available today. Thus the locality of memory accesses can be improved. Hence we can significantly increase cache hit rate.

There is a similar problem in the pixel cache. The size of each slice is the same as that of a slice in the view-plane in general. Each slice covers the entire screen. Moreover, the blending operation requires read/write operation for both the depth test and alpha-blending per each fragment of a slice. Thus the entire screen must be processed before a particular pixel is visited again. As a result, the locality of the frame buffer accesses is decreased and cache miss rate can be increased enormously. On the other hand, if the size of a sub-image of a sub-volume is smaller than the pixel cache size, cache hit can be guaranteed for all slices except the cache misses (compulsory miss) on the first slice. Therefore, we can improve cache utilization significantly.

4 Experimental Results

We have built a simulator to evaluate cache efficiency, memory traffic, and achievable performance. We model a standard graphics pipeline composed of the T&L engine, the Goraud shader, the 3D texture mapping unit, the blending unit, the framebuffer (color and depth buffer), and the texture memory. Memory access traces have been generated during the benchmark volume datasets are rendered on the simulator. The performance has been evaluated with the trace-driven cache simulator, DineroIII [5]. Because the pixel cache consists of the color cache and the depth cache in most of commodity graphics accelerators [19, 20], a separate simulation has been performed for each cache. All caches are configured with direct-mapped, a block size of 32Bytes. We have experimented varying the cache size from 16K to 128Kbytes.

We have used three datasets, the HeadMR, the Foot, and the Skull (see Fig. 4) from the volren web-page (http://www.volren.org). Each of these datasets has 8bit-intensity and the size of 256^3, but has been re-sampled to 512^3 (128Mbytes) to test the large volume data. Rendering is directed to a 512x512 viewport, with 100 slices, and the back-to-front blending method is used. 16^3 and 32^3 have been chosen as the size of sub-volume. Ten frames have been rendered for each case. For fair generation of memory access patterns, the volume is rotated by width and height for rendering. We describe the simulation results of the proposed rendering scheme in this section.

4.1 Texture and Pixel Cache Efficiency

Fig. 2 shows the comparison of each cache miss rate for our subdivided rendering scheme (SD) and the traditional non-subdivided rendering scheme (NSD). More than 50% of the miss rate of the texture cache with NSD was occurred due to frequent cache thrash, which means the cache could not perform its own duty. On the other hand, the miss rate was reduced sharply in the case of rendering with SD due to locality improvement. As shown in Fig. 2, all the datasets recorded

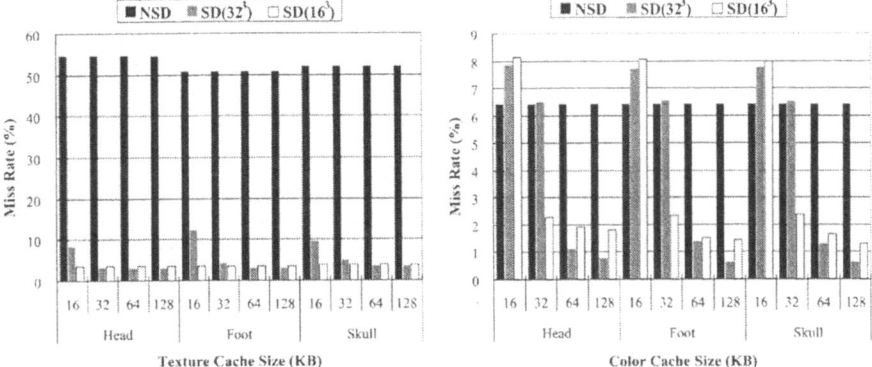

Fig. 2. Comparison of the cache miss rate for our subdivided (SD) rendering and the traditional non-subdivided (NSD) rendering

less than 5% of the miss rate of the texture cache when the cache size is larger than 32Kbytes. Even though the size is larger than 32Kbytes, the miss rate is hardly reduced for all the cases. This result shows that the miss rate had no relation with the cache size when the size is over a critical point as the case of 2D texture cache [21]. The miss rate of SD when the size of a sub-volume is 16^3 is little higher than that of SD when the size is 32^3 from the 32Kbytes for the HeadMR and the 64Kbytes for the Foot and the Skull, because more overlapped voxels are generated as the sub-volume size is decreased.

Similarly the miss rates in the pixel cache for SD have also been reduced due to high cache locality. For simplicity, we show the results for the color cache in the Fig. 2, because the number of accesses to the depth buffer for depth test and accesses to the color buffer for the blending are the same. In case of NSD, the miss rate is almost the same, over 6%, regardless of the cache size, because the size of a slice is larger than that of the pixel cache. In contrast, because the sub-image of a sub-volume is smaller than that of the pixel cache for SD, the locality is increased and the miss rate is reduced. The miss rate is sharply decreased from 64Kbytes for the 32^3 sub-volume and from 32Kbytes for the 16^3 sub-volume. When the size is 16Kbytes, on the contrary, each sub-image could not fit into the pixel cache, and hence the miss rate of SD was higher than that of NSD.

4.2 Total Bandwidth Comparison

Table 1 shows the comparison of the total bandwidth during 10 frames rendering. The total bandwidth is calculated by summing up the amount of fetched data from the texture memory due to texture cache misses, the amount of the read/write data from/to the frame buffer due to pixel cache misses, and the amount of vertices.

Table 1. The comparison of the required bandwidth between SD and NSD

Dataset	Rendering Method	Sub-volumes	Empty-Space Skipped Sub-volumes	Vertices	Texture Cache Size(KB)	Read Data from Texture Memory Due to Cache Miss(Bytes)	Pixel Cache Size(KB)	Read/Write Data from/to Frame Buffer Due to Cache Miss(Bytes)	Total Bandwidth Needed for 30f/s(GB/s)
HeadMR	NSD	-	-	4,000	16	15,025,144,064	16	1,926,144,000	47.37
					32	15,024,721,472	32		47.37
					64	15,023,112,832	64		47.36
					128	15,016,700,416	128		47.34
	SD(32)	4,096	2,042	572,120	16	1,764,230,976	16	2,927,880,896	13.16
					32	680,485,568	32	2,418,177,920	8.71
					64	630,633,824	64	417,452,352	2.98
					128	630,477,568	128	289,171,136	2.62
	SD(16)	32,768	18,894	2,219,840	16	642,788,320	16	2,572,624,960	9.19
					32		32	726,369,856	4.03
					64		64	610,183,616	3.71
					128		128	575,279,040	3.61
Foot	NSD	-	-	4,000	16	13,478,043,488	16	1,926,144,000	43.04
					32	13,477,901,216	32		43.04
					64	13,477,758,976	64		43.04
					128	13,477,473,600	128		43.04
	SD(32)	4,096	1,859	626,360	16	2,772,522,176	16	3,456,836,352	17.47
					32	969,392,416	32	2,935,609,984	10.98
					64	684,198,304	64	616,759,424	3.7
					128	684,198,304	128	215,540,160	2.58
	SD(16)	32,768	17,735	2,405,280	16	880,501,952	16	4,000,987,136	13.86
					32		32	1,155,852,416	5.91
					64		64	748,096,256	4.78
					128		128	710,950,016	4.67
Skull	NSD	-	-	4,000	16	14,249,069,696	16	1,926,144,000	45.2
					32	14,249,069,696	32		45.2
					64	14,249,069,696	64		45.2
					128	14,249,069,696	128		45.2
	SD(32)	4,096	2,935	325,080	16	2,281,828,608	16	1,745,188,160	11.29
					32	578,412,608	32	1,465,006,080	5.74
					64	426,176,352	64	291,141,632	2.04
					128	426,176,352	128	143,732,992	1.63
	SD(16)	32,768	26,585	989,280	16	321,912,992	16	1,193,654,720	4.33
					32		32	358,563,008	2.0
					64		64	247,057,408	1.68
					128		128	193,965,568	1.54

In case of NSD, over 14Gbytes of data was transmitted from the texture memory on the average. If the frame buffer bandwidth is considered, the total required bandwidth for rendering 30 frames per second was more than 40Gbytes. Because the internal memory bandwidth of recent commodity graphics hardware [19, 20] is about 30Gbytes, it is difficult to render at a real-time frame rate.

When rendering the datasets with SD, the total required bandwidth for rendering 30 frames per second is only in the range between 1.5Gbytes and 17Gbytes. Such result was possible because SD could skip more than 50% of empty space on the average and SD improved cache hit rate on both the texture and the pixel caches. As we mentioned before, the miss rate was sharply decreased when the cache size is over the 64Kbytes for the 32^3 sub-volume and when the cache size is over the 32Kbytes for the 16^3 sub-volume. Thus, the required bandwidth for rendering 30 frames per second was from 2 to 3Gbytes in this case. The bandwidth was reduced up to 30 times, compared with NSD in the best case (when the cache size is 128Kbytes for Skull).

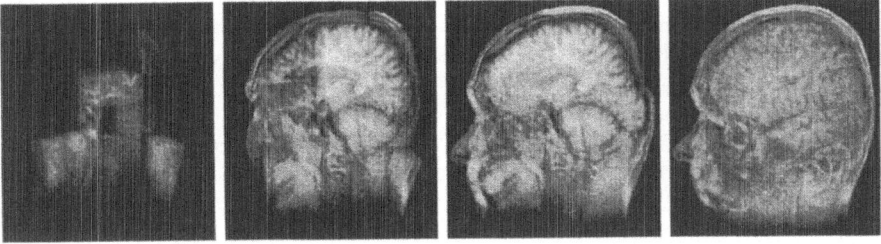

Fig. 3. Some snapshot of our rendering scheme

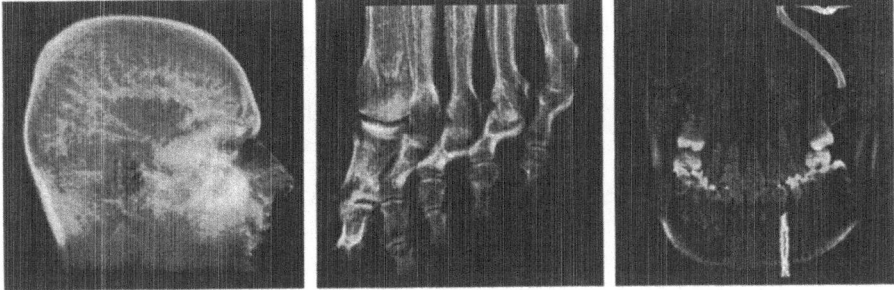

Fig. 4. Datasets rendered images with our simulator. HeadMR, Foot, Skull

5 Conclusion

In this paper, we proposed a bandwidth-effective volume rendering scheme which divides the volume into uniform sized sub-volumes and transmits them to the texture mapping units in visibility order. Simulation results showed that the proposed scheme is effective for 3D texture-based volume rendering on commodity graphics hardware by reducing internal memory bandwidth substantially. Because this scheme manages the sub-volumes, it is expected to be applied to three dimensional volumetric effects such as volumetric lighting, fire, and clouds.

References

1. Gelder, A.V., Kim, K.: Direct Volume Rendering with Shading via Three-Dimensional Textures. In Proc. of ACM Symposium on Volume Visualization (1996) 23-30
2. Engel, K., Kraus, M., Ertl, T.: High-Quality Pre-Integrated Volume Rendering Using Hardware-Accelerated Pixel Shading. In Proc. of Eurographics/SIGGRAPH Workshop on Graphics Hardware (2001) 9-16
3. Kniss, J., Kindelmann, G., Hansen, C.: Interactive Volume Rendering Using Multi-Dimensional Transfer Functions and Direct Manipulation Widgets. In Proc. of IEEE Visualization (2001) 255-262

4. Knittel, G.: TriangleCaster: extensions to 3D-texturing units for accelerated volume rendering. In Proc. of Eurographics/SIGGRAPH Workshop on Graphics Hardware (1999) 25-34
5. Dinero III, http://www.cs.wisc.edu/ larus/warts.html
6. Silva, C.T., Kaufman, A.E., Pavlakos, C.: PVR: High-Performance Volume Rendering. IEEE Computating in Science and Engineering, Vol. 3, No. 4 (1996) 18-28
7. Ray, H., Silver, D.: The Race II Engine for Real-Time Volume Rendering. In Proc. of Eurographics/SIGGRAPH Workshop on Graphics Hardware (2000) 129-136
8. Meiner, M., Kanus, U., Wetekam, G., Hirche, J., Ehlert, A., Straer, W., Doggett, M., Forthmann, P., Proksa, R.: VIZARD II: A Reconfigurable Interactive Volume Rendering System. In Proc. of Eurographics/SIGGRAPH Workshop on Graphics Hardware (2002) 137-146
9. Kniss, J., McCormick, P., McPherson, A., Ahrens, J., Painter, J., Keahey, A., Hansen, C.: TRex: Interactive Texture Based Volume Rendering for Extremely Large Datasets. IEEE Computer Graphics & Applications, Vol. 21, No. 4, July (2001) 52-61
10. Ertl, T., Westermann, R., Grosso, R.: Multiresolution and Hierarchical Methods for the Visualization of Volume Data. Future Generation Computer Systems, Vol. 15, No. 1 (1999) 31-42
11. Boada, I., Navazo, I., Scopigno, R.: Multiresolution Volume Visualization with a Texture-Based Octree. The Visual Computer, Vol. 17, No. 3 (2001) 185-197
12. Schneider, J., Westermann, R.: Compression Domain Volume Rendering. In Proc. of IEEE Visualization (2003) 293-300
13. Cochran, W.O., Hart, J.C., Flynn, P.J.: Fractal Volume Compression. IEEE Trans. Visualization and Computer Graphics, (1996) December 313-322
14. Nguyen, K.G., Saupe, D.: Rapid High Quality Compression of Volume Data for Visualization. Computer Graphics Forum, Vol. 20, No. 3 (2001)
15. Guthe, S., Wand, M., Gonser, J., Straer, W.: Interactive Rendering of Large Volume Data Sets. In Proc. of IEEE Visualization (2002) 53-60
16. Hadwiger, M., Kniss, J.M., Engel, K., Rezk-Salama, C.: High-Quality Volume Graphics on Consumer PC Hardware. In Proc. of Eurographics/SIGGRAPH Workshop on Graphics Hardware, Course Note (2002)
17. Li, W., Mueller, K., Kaufman, A.: Empty Space Skipping and Occlusion Clipping for Texture-Based Volume Rendering, In Proc. of IEEE Visualization (2003) 317-324
18. Lichtenbelt, B., Crane, R., Naqvi, S., Introduction to Volume Rendering, Prentice Hall PTR (1998)
19. GeForce FX 5900, http://www.nvidia.com/page/fx_5900.html
20. ATI's RADEON 9800 Pro, http://mirror.ati.com/products/pc/radeon9800pro/index.html
21. Hakura, Z.S., Goopta, A.: The design and Analysis of a Cache Architecture for Texture Mapping, In Proc. of 24th International Symposium on Computer Architecture (1997) 108-120

An Efficient Image-Based 3D Reconstruction Algorithm for Plants

Zhigeng Pan[1,3], Weixi Hu[2,3], Xinyu Guo[2], and Chunjiang Zhao[2]

[1] Institute of Virtual Reality and Multimedia, HZIEE, Hangzhou, China
zgpan@cad.zju.edu.cn
[2] National Engineering Research Center for Information Technology in Agriculture, Beijing, China
guo-xinyu@263.net
[3] State Key Lab of CAD&CG, Zhejiang University, Hangzhou, China
huweixi@cad.zju.edu.cn

Abstract. Image-based 3D reconstruction technique can build the model quickly and easily. In this paper, we propose an efficient image-based method for plant model reconstruction with modified KLT image matching method. The modification makes the new algorithm search the matching points more accurately and helps users to add matching points manually. We present the method to get 3D information from the matching images. An algorithm to unify different coordinate systems has also been introduced in this paper. Finally we discuss an easy texture-mapping way to the 3D vertex model.

Keywords: Image-based modeling, modified KLT image matching method

1 Introduction

3D models are very popular nowadays. Natural looking 3D models of real objects are basic components in a variety of virtual reality applications, such as virtual shopping, virtual studio production, 3D teleconferencing, 3D information systems and 3D archiving.

In particular, in Virtual Agriculture, 3D information of plants is necessary for further study of growing status of plants. Traditionally, people use a 3D scanner to measure this information. But this manual creation is quite complicated, time-consuming, and dependent on the hardware's stability.Because of the above reasons, techniques are under investigation, which allow the automatic reconstruction of 3D objects. The presented approach is an image-based method, which requires less equipment and can reconstruct the model quickly and easily.

There have been many studies in image-based method already. Some use coded light for reconstruction [1]. Some use "shape from silhouettes" method to build the 3D objects [2], [3] and [4]. Some take advantage of camera model and image matching techniques to obtain 3D coordinates [5], [6], [7] and [13]. "Shape from silhouettes" method usually needs to have more view spots than image matching technique does, and usually this method needs to know the accurate camera's relative place to the real object. In the research field of plants reconstruction, Prusinkiewicz [14] has done a lot of work. But their reconstruction cannot rebuild the model based on the exact data of the plants, such as the accurate height. For this reason, we prefer to use image matching technique, which forces no constraint on the number of view spots, does not need to know the specific position of the camera, and can provide the comparatively accurate data. And as far as we know, it is the first time to use image-based modeling method to plants reconstruction.

In our new system, a modified pyramidal KLT image matching method is presented, which not only can detect huge quantities of matching points from two projective images, but also can enhance the accuracy of original technique greatly. And we will give our method to unify different coordinate systems, which are caused by multiple view points.

In section 2 the developed steps are described. Section 3 illustrates the modified KLT image matching method. Section 4 presents the processing pipeline from calculation of 3D coordinates to texture mapping. Finally, in section 5, we will give our results and make discussion.

2 System Description

Considering the simplicity of system as well as the convenience of users, we design following steps, see Figure 1.

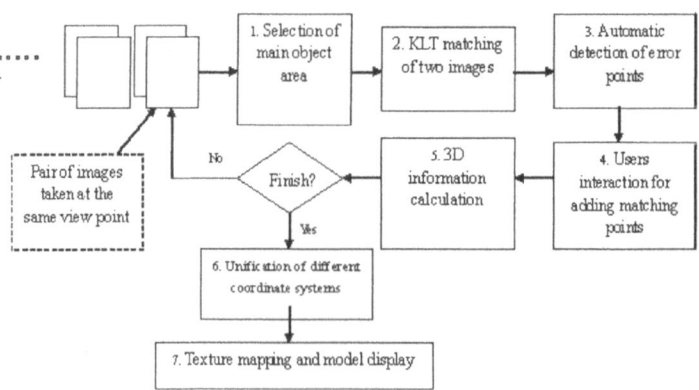

Fig. 1. The pipeline of our system

The first step is shown in Figure 2. In every view direction, the digital camera needs to move a little to take two pictures, so that the whole object can be still in the two pictures but there yet exists a little displacement, which is very crucial to our further work. And the camera parameters nearly don't change at all when we take two pictures in the same direction, which helps us avoid the complicated steps of camera calibration. We can read the parameters from generated digital images directly, such as focus length.

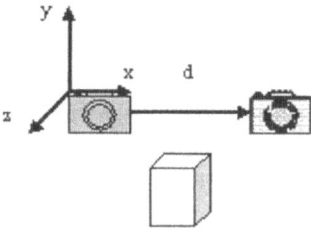

Fig. 2. Take two images in nearly the same view direction

The second step is to use pyramidal Kanade-Lucas-Tomasi Feature Tracker (KLT) [8] and [9] to do the first matching tasks. The third and the fourth steps are our modification on the KLT, which make the result more accurate.

In the fifth step, we use the knowledge of camera projective model. In the situation we have designed, 3D information of every point on the object can be calculated by the x-axis displacement of their corresponding projective points in two images.

We keep on doing from the first step to the fifth step until there is no more image, and we unify all the coordinate systems and display the model.

There are detailed descriptions about the steps in the following sections.

3 Modified KLT Image Matching Method

Image matching technique is very essential to our study. In Figure 3, the two pictures are taken almost in the same direction. We need to find out an effective way to automatically obtain accurate matched points in large quantity in these two images.

Fig. 3. Two pictures are taken almost in the same direction but with a little horizontal displacement of the camera

3.1 Pyramidal Kanade-Lucas-Tomasi Feature Tracker

Among many matching methods, Kanade-Lucas-Tomasi Feature Tracker (KLT) is proved to be a classic and effective method. KLT was proposed by Lucas and Kanade [8] and was developed fully by Tomasi and Kanade. A pyramid implementation of KLT algorithm was presented by Jean-Yves Bouguet [9]. We applied pyramidal KLT and made some modification.

Given two similar pictures I and J, the pyramidal KLT first makes the pyramidal image representation of the two images I and J: $\{I^L\}_{L=1...L_m}, \{J^L\}_{L=1...L_m}$,

Then, for every good feature point in image I, pyramidal KLT algorithm tries to find the matching point in image J. It iterative computes optical flow in both of the two images with the aim to minimize the matching function:

$$\epsilon^L(d^L) = \epsilon^L(d_x^L, d_x^L) = \sum_{x=u_x^L-w_x}^{u_x^L+w_x} \sum_{y=u_y^L-w_y}^{u_y^L+w_y} (I^L(x,y) - J^L(x+d_x, y+d_y))^2. \quad (1)$$

w_x, w_y stand for the width and the height of the calculation neighborhood size of every tracking point respectively.

The computation is taken from the highest image level L, and the every result is propagated to the lower level until it reaches the zero level, and the matching points in image J have been found, see Figure 4.

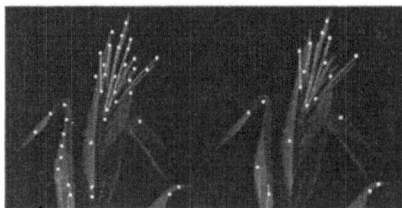

Fig. 4. (a) shows the points that are good to track in one image, and (b) shows the corresponding points the computer finds in the next image (part of the plant)

In order to enhance the effectiveness of matching process, we have made some modification to the original algorithm.

3.2 A New Step before Image Processing

The picture we have taken is usually large and contains not only the object but also much of background scene. This will unexpectedly influence the result of image matching. Many tracked points are often of the background scene, which will be useless in the future reconstruction work. And the experiment shows that the larger the picture is, the longer time the algorithm needs to take. In order to avoid data and time waste, we propose a step that allows users to select the tracking area, in which the target object occupies most of the space. The example is shown in Figure 5, and the result shows that 500 matched points tracked in the left picture mainly lie on the spot light place, while in the right picture, they mainly are on the corn plant. And the speed has been doubled after this step has been taken.

Fig. 5. (a) is the first of original pictures, with the size of 1280 x 960. (b) is the one after area selection, with the size of 612 x 784, taken from the point (442, 46) in the original picture. (c) is the second of original pictures, with the size of 1280 x 960. (d) is the one after area selection, with the size of 612 x 784, taken from the point (402, 46) in the original picture

3.3 Removal of Error Points

Though computer can find out hundreds of matching points from two images, yet it will make mistakes if we don't give any constraint.

We have proposed an algorithm to remove the error points. The steps are as following: 1. Use a proper "moving window" with the size of w_x and w_y to go through the second image, where the matching points have been found;

2. Find out the approximate maximum density area of existing matching points and calculate the average pointtopoint distance d in this density-high area;

3. For every matching point p in the second image, we find out its nearest points (10 points are enough) within a distance of 5xd;

4. Calculate both the x-axis and y-axis displacement of these nearest points in the first image and their matching points in the second image, and obtain the average x-axis displacement and y-axis displacement;

5. Considering the object continuity, points in a small continuous area usually have the same displacement. We will remove those points whose displacement is deviated from the average with a distance t.

Using this algorithm, "error points" mostly vanish. For the few error points that cannot be removed by the algorithm, we enable users to remove them manually.

3.4 Automatically Adding Matching Points with Users' Interaction

In general, the computer can find most of the matching points, but it is probable to miss some important points. The reason for point missing is that for a whole picture, KLT algorithm cannot pay much attention onto certain specific points, but to the global condition. So the calculation window for every point is usually the same, which is bigger for those missing points and thus the problem rises. We need to manually add some matching points.

In order to take full use of KLT algorithm to solve the point-missing problem, we propose an idea to adjust the size of the KLT calculation neighborhood window according to our different needs. During the global points matching process, we keep the original window size as original pyramidal KLT algorithm proposes, but when we want to add points, we will take the following steps.

1. For every manually-added point in the first image, we will focus its corresponding part in the second image;

2. We will do the KLT algorithm in a very small neighborhood of this point with the size of 40 x 40 and change the calculation neighborhood window size smaller than the original one. The experiment proves that this method can give a good result, see Figure 6.

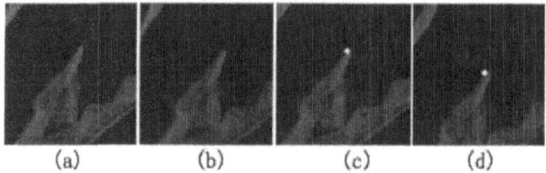

(a)   (b)   (c)   (d)

Fig. 6. (a) and (b) are part of the input images. After automatic selection of matching points, the computer misses the ending point of the leaf. We adjust the calculation window and focus the matching area in a small neighborhood of the given point, shown in (c) the manual-added point is colored in white, and the computer deduces the matching point in another image automatically. (d) gives the result

4 Calculation of 3D Coordinates

4.1 Basic Knowledge of 3D Data Acquirement

With the matching points we obtain from above steps and with the knowledge of camera model, we can calculate 3D coordinates [11].

If the internal parameter of the camera does not change during picturing, and the camera projective model can be considered as "pin-hole" model, taking the horizontal line as x-axis, the vertical line as y-axis, and the direction of picturing as positive z-axis, the camera takes the first picture of the object at the position (0) and takes the second at (0). After point-matching processing, we can calculate 3D coordinates from these two pictures. Figure ?? gives the illustration of picture-taking step.

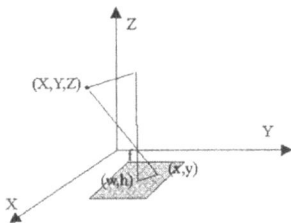

Fig. 7. It is the picture-taking process illustration. The horizontal line is x-axis, the vertical line is y-axis, and the direction of picturing is positive z-axis. The camera is at the position (w, h, 0) to take the first picture, and moves to (-w, h, 0) to take another. f is the camera focus length

The projective points on the first picture can be calculated with the formula:

$$\begin{cases} \frac{w-x}{X-w} = \frac{f}{Z-f} \\ \frac{h-y}{Y-h} = \frac{f}{Z-f}, \end{cases} \quad (2)$$

which can be transformed as

$$\begin{cases} x = w - \frac{f(X-w)}{Z-f} \\ y = h - \frac{f(Y-h)}{Z-f} \end{cases} \quad (3)$$

If the width pixel and the height pixel of the picture are represented by W_{pixel} and H_{pixel}, and the width and the height in measurement of meter are represented by width and height, we finally have

$$\begin{cases} X = w + \frac{w(W_{pixel} - 2x_{pixel})}{x_{pixel} - x'_{pixel}} \\ Y = h + \frac{h(H_{pixel} - 2y_{pixel})}{x_{pixel} - x'_{pixel}} \\ Z = \frac{2Kw}{x_{pixel} - x'_{pixel}} \end{cases} \quad (4)$$

where $K = \frac{f}{C}$.

Points found in the first image and their corresponding points in the second image are actually the projective points on different images of the same spatial point, therefore, with the matching-point information, we can calculate their corresponding spatial point coordinate.

After we get sample points, we can use Marching Cubes Algorithm to interpolate more points easily. Here, we use vtkContourFilter Class of VTK [10] to finish the interpolation.

4.2 Unification of Different Coordinate Systems

Because we take the pictures from different directions of view, after we calculate the 3D coordinates in the different coordinate systems, we should transform them into one unique coordinate system. We take one of coordinate systems S as the final unique system. The steps are as following:

1. Manually pick out the projective 2D points of the same spatial 3D point on two different directions of view. Picking three pairs is the least requirement, while the more pairs are picked, the more accurate the transformation will be.

2. Let the 3D coordinate values in these two different systems be (x, y, z) and (x', y', z') respectively, which (x, y, z) belongs to the system S and (x', y', z') belongs to the system S'.

3. With the transform equation (4),

$$s \begin{pmatrix} x \\ y \\ 1 \end{pmatrix} = R \begin{pmatrix} X \\ Y \\ Z \\ 1 \end{pmatrix} \qquad (5)$$

where

$$R = \begin{pmatrix} r_{11} & r_{12} & r_{13} & r_{14} \\ r_{21} & r_{22} & r_{23} & r_{24} \\ r_{31} & r_{32} & r_{33} & r_{34} \end{pmatrix}$$

we can calculate the every value for the units in the matrix R

4.3 Texture Mapping

If we transform the equation (4), we can have

$$\begin{cases} \Delta x = x_{pixel} - x'_{pixel} = 2Kw/Z \\ x_{pixel} = (W_{pixel} - (X - w) * \Delta x/w) * 2 \\ y_{pixel} = (H_{pixel} - (Y - h) * \Delta x/w) * 2 \end{cases} \qquad (6)$$

5 Results

We have designed a system to reconstruct the vertex model by unifying every view coordinate system. But considering texture mapping, we have just finished in one view coordinate system.

In Figure 8, the images are selected from the original corn plant pictures. And Figure 9 shows the reconstruction result. The size of both images is 612

Fig. 8. The preprocessed images are of the size 612 x 784

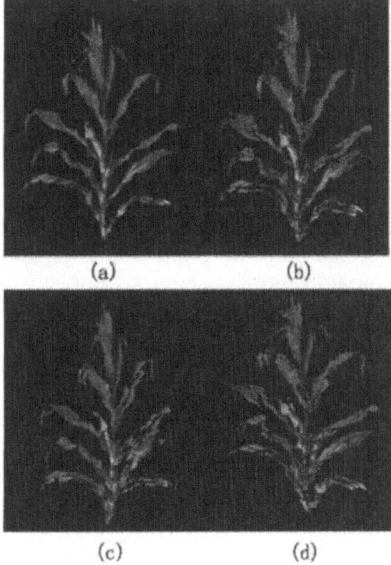

Fig. 9. The texture-mapped 3D model of corn plant. (a) is vertical to the view direction, (b) is the one with 10 degree deviation from the top, (c) is the one with 20 degree deviation from the left, and (d) is the one with 20 degree deviation from the right

x 784. We set 300 points to be matched, and the final matched points are 270, with 30 points removed.

The holes on the leaves are caused by occlusion from one view. This problem will be solved with multi-view texture mapping [12]. From the result, we can see that image-based reconstruction shows a good 3D result. Our algorithm also can be extended to the reconstruction of other models. With the increasing demand for 3D models, this technique can be a very easy and quick method.

As our future work, we will finish texture-mapping in multiple views. And to widen the models' usage, we will export 3D data into VRML file.

Acknowledgement. This research work is co-supported by 973 project(project no.: 2002G3312100), Excellent Youth NSF of Zhejiang Province(RC 40008) and TRAPOYT Program in Higher Education Institutions of MOE, P.R.C.

References

1. F. Wahl, A coded light approach for depth map acquisition, in: Proceedings 8th DAGM-Symposioum Mustererkennung, G. Hartmann, Paderborn, (1986) 12-17.
2. C.H. Chien, J.K. Aggarwal, Identification of 3D objects from multiple silhouettes using quadtrees/octrees, Comp. Vision Graphics and Image Processing 36 (1986) 256-273.
3. W. Niem, Robust and fast modelling of 3D natural objects from multiple views, SPIE Proceedings "Image and Video Processing II" 2182 (1994) 388-397.
4. J.Y. Zheng, Acquiring 3-D models from sequences of contours, IEEE Transactions on Pattern Analysis and Machine Intelligence 16 (2) (1994) 163-178.
5. Z. Zhang, Q. T. Luong and O. Faugeras, Motion of an uncalibrated stereo rig: self-calibration and metric reconstruction, IEEE Trans. Robotics and Automation 12(1), (1996) 103-113.
6. Z. Zhang, R. Deriche, Q. T. Luong, and Faugeras, O., A robust approach to image matching: Recovery of the epipolar geometry, Proc. International Symposium of Young Investigators on information computer control. Beijing, China, (1994) 7-29.
7. Z. Zhang, R. Deriche, Faugeras, O. and Q. T. Luong, A robust technique for matching two uncalibrated images through the recovery of the unknown epipolar geometry, Arti.cial Intelligence Journal 78, (1995) 87-119.
8. Bruce D. Lucas and Takeo Kanade, An Iterative Image Registration Technique with an Application to Stereo Vision. International Joint Conference on Artificial Intelligence, (1981) 674-679.
9. J.Y. Bouguet, Pyramidal Implementation of the Lucas Kanade Feature Tracker Description of the Algorithm. Intel Corporation, Microprocessor Research Labs. OpenCV Documents, 1999.
10. The Visualization Toolkit, http://www.kitware.com/vtk.html
11. Yujin Zhang, Image Project — Image understanding and computer vision, Tsing Hua University press, 2000.
12. W. Niem, H. Broszio, Mapping texture from multiple camera views onto 3d object models for computer animation, Proceedings of International Workshop on Stereoscopic and Three Dimensional Imaging, Eylül 1995, 99-105.
13. Z.Y. Zhang, Image-Based Modeling of Objects and Human Faces, SPIE Conference Videometrics and Optical Methods for 3D Shape Measurement Invited Paper, San Jose, CA, USA. Proceedings of SPIE Vol.4309, January 2001.
14. P. Prusinkiewicz, L. Muendermann, R. Karwowski, and B. Lane, The use of positional information in the modeling of plants. Proceedings of SIGGRAPH 2001 (Los Angeles, California, August 12-17, 2001), 289-300

Where the Truth Lies (in Automatic Theorem Proving in Elementary Geometry)

T. Recio* and F. Botana

Departamento de Matemáticas, Estadística y Computación, Universidad de Cantabria, Avda. de los Castros, 39071 Santander, Spain, http://www.recio.tk
Departamento de Matemática Aplicada I, Universidad de Vigo, Campus A Xunqueira, 36005 Pontevedra, Spain, http://rosalia.uvigo.es/fbotana

Abstract. In this paper we use a new integrated theorem prover (GDI), codeveloped by the second author, to discuss a geometric result due to Maclane, the 8_3 theorem, which has been declared to be true, according some authors, while other claim it is false. Our approach is based in Gröbner bases computations and illustrates the controversial concept of truth in the algebraic automatic theorem proving model, as well as some of the new features provided by GDI. The potential applications to computer graphics of the idea behind these rather unique features, are also briefly discussed.

1 Introduction

This note exhibits, through a well-known example, i) the difficulties of modeling some basic reasoning on simple geometric objects such as points and lines on a plane, and ii) the advantages of using GDI [2,3], an integrated (dynamic geometry screen+algebraic engine) prover prototype created by the second author (with J.L. Valcarce), to exemplify the situation.

We will consider (Section 2) the well established algebraic approach to automatic theorem proving in elementary geometry (we refer to [21] for an exhaustive repository of related papers). Within this approach, a theorem due to Maclane [15] (the so called theorem 8_3) has been declared to be true, according to some authors ("if we adhere to our definition of (algebraic) truth, the theorem should be considered true: an example of an obviously false theorem that is true" [8]); but which has also been reputed by some other authors to be false ("it holds in the real plane but fails in the complex one" [9]). As it has been already remarked [8], what lies behind this controversy is the (intricate) concept of truth in the algebraic automatic theorem proving model. We will discuss briefly this issue (Section 4) and, basically, show (in Section 5) what (and how) GDI has to conclude about this paradigmatic example, as well as (in Section 6) what could become in the future, for computer graphics, of the idea behind this software.

---

* First author supported by grant BFM2002-04402-C02-02 from the Spanish MCyT and by the University of Vigo, as a visiting scholar during October 2003

2 The Algebraic Approach to Automatic Theorem Proving

For the last 20 years, symbolic algebraic techniques have been successfully used for automatically proving theorems in elementary geometry. The practical interest of this goal (i.e to automate, through the algebraic translation of hypotheses and theses, theorem proving in euclidian geometry) could be related, for instance, to its potential applications in geometric constraint solving and parameterized CAD [13,10,11], as well as in computer assisted learning, etc. In [4,19] one can navigate on many interesting details concerning the mentioned topics and their mutual relationship, as well as providing a source for further references.

The standard algebraic approach proceeds, roughly, as follows (see [6] for some basic details). A geometric statement (a finite set of hypotheses and a thesis) is translated into two multivariate polynomial systems, H, T. The statement is declared to be true if the hypotheses variety (where all polynomials in H are zero) is contained in the thesis variety, $Var(H) \subseteq Var(T)$. Different approaches exist to compute whether $Var(H) \subseteq Var(T)$, after performing a suitable algebraic translation (based on Hilbert's Nullstellensatz) of the set inclusion, mainly Wu–Ritt characteristic sets and Gröbner bases. As this algebra/geometry translation usually assumes working over an algebraically closed field, the overall decision procedure about the truth of a geometric statement involves not only real solutions, but complex ones as well.

3 Integrated Theorem Provers

The history of the algebraic approach to elementary geometry theorem proving, very roughly stated, begins with an initial phase devoted to the mathematical modeling of proving (in this specific context), then followed by an era of extensive experimentation (such as in [6]) on different built–on–purpose automatic provers. Thousands of theorems, mostly classic but some also quite new, have been successfully proved in this way.

Moreover, in the last years a handful of these provers [20,12,22,17] integrate, jointly with the traditional computer algebra engine, a dynamic geometry environment, allowing the user to sketch geometric figures, to mark distinguished elements on them, and to graphically state (ie. by selecting with a computer device on the screen some of these geometric elements) different hypotheses and theses, thus avoiding the cumbersome task of explicitly writing the corresponding equations as an input for the algebraic prover.

3.1 GDI

GDI, a Spanish acronym of Intelligent Dynamic Geometry, is a prototype of these last generation provers. Besides offering the user a friendly interface for stating, in a graphical mode, hypotheses and theses, GDI has a distinguished feature, the Discovery option, allowing a user to specify some extra constraints over a geometric construction and then finding necessary conditions on the free elements so

that the given constraints will hold in the resulting modified construction. This
option can also be invoked through the internet (using the webDiscovery feature
of GDI [1]). The algebraic machinery behind this comes from [16] and uses the
symbolic computation packages CoCoA [5] or Mathematica in an automatic way,
at user's choice, who has just to graphically sketch a construction and then start
a dialog (deciding what he/she wants to test about the construction) with some
pre-build menus options.

Let us illustrate it with an elementary example. Consider an arbitrary triangle ABC and an arbitrary point X on its plane and construct the symmetrical
points X', X'', X''' of X with respect to the sides of ABC; then drag X. One
can observe that the triple of symmetrical points X', X'', X''' becomes sometimes
collinear. Asking about the necessary conditions for this alignment GDI discovers two conditions: one is a degenerate case (*The points A, B, C are aligned*),
while the other is the cocircularity of points A, B, C and X (Fig. 1).

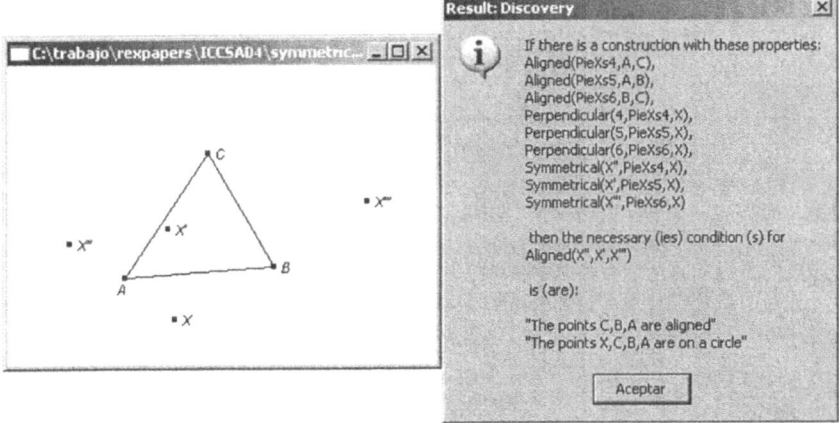

Fig. 1. Necessary conditions for the alignment of the symmetrical points of X

4 Maclane 8_3: Difficult to Prove...

This theorem of MacLane can be stated as follows: *Consider eight points A,
B,..., H such that the following (also eight) triples are collinear ABD, BCE,
CDF, DEG, EFH, FGA, GHB, HAC. Then all eight points lie on a line.* As
noted in the Introduction, Dolzmann, Sturm and Weispfenning [9] point out,
without further discussion, that this theorem holds in the real plane, but fails in
the complex one, while Conti and Traverso [8] consider it as "an example of an
obviously false theorem that is true". Leaving aside reality issues, i.e. considering
just the complex case (as it is usual up to now in most theorem provers), we
have attempted to verify with GDI the truth/falsity of this result.

To begin with, the first difficulty we face is that the GDI prover only accepts geometric statements of constructive type (for a definition of constructibility, see [6, p. 16]) and this one is not of this kind. In fact, using the graphical interface of GDI, one is tempted to proceed as follows: first, freely place the points $A(u_1, u_2)$ and $B(u_3, u_4)$ defining a line, then point D is constructed as a point on it (hence having a free coordinate u_5 and a bounded one x_1). Similarly, point $B(u_3, u_4)$ and a new free point $C(u_6, u_7)$ define another line, and point $E(u_8, x_2)$ lies on it, and so on .... But the coordinates of A have got to be redefined when imposing the alignment of F, G and A (since A must lie on line FG), and the same problem appears with B and C for the last two conditions of alignment. Thus, as remarked in [8, p. 93], "no construction can contain an 8_3 condition, since every point is involved in three alignment conditions".

Of course, one can write, by hand, the set of equations defining in this case the hypothesis variety H (eight alignment conditions, each yielding a second degree equation) and the thesis variety T (for instance, the single equation stating that C is aligned with A, B, hence all the points are aligned). In order to check if $Var(H) \subseteq Var(T)$, one can compute (for instance, with Co-CoA) the saturation [14] of the ideal $Ideal(H)$ by $Ideal(T)$. It yields a proper ideal, and thus it means that the inclusion of varieties does not hold (because $1 \neq Sat(Ideal(H), Ideal(T))$ is equivalent to saying T is not in the radical of $Ideal(H)$). In particular, we can quote Conti and Traverso [7], who explicitly give the collection of points $(0,0), (1,0), (0,\alpha), (\infty, 0), (\alpha+1, -1), (-2\alpha, \alpha), (2, -1)$, $(0,1)$ where $\alpha = i/\sqrt{(3)}$, verifying the hypothesis but not the conclusion of Maclane's theorem. Furthermore the same authors explicitly compute (in a highly non-automatic, but clever way) nine eight dimensional components of the hypothesis variety where the theorem fails [8], and parts of them represent eight–tuples of different points verifying the hypotheses but not the thesis.

So the theorem is not true all over the points of $Var(H)$, although, for each element $G \in Sat(Ideal(H), Ideal(T))$, the theorem holds over the points $Var(H) \cap \{G \neq 0\}$, because for some power m, $G * T^m \in Ideal(H)$, by definition of saturation. Next one could show that, actually, there exists some $G \in Sat(Ideal(H), Ideal(T))$ such that there are points where $Var(H) \cap \{G \neq 0\}$; in fact it is enough to check among the elements of a base of $Sat(Ideal(H), Ideal(T))$ in order to find one of these G's with the required property; computationally speaking, one just verifies that for any base there is some element G such that $Sat(Ideal(H), Ideal(G))$ is not 1.

So the actual issue is, how relevant is this subset of $Var(H)$ where the theorem is true? For some, if this set includes an open subset in every higher dimensional component of the hypothesis variety (and this is the case here as we will see in the next section), it can be said the theorem holds (in general), while for others, the fact that there are components (such as the nine eight dimensional ones of Conti and Traverso, although the variety is 10–dimensional) where the theorem fails is enough to label the theorem as *intuitively false*. Dependent and independent variables play as well a precise role in this discussion... All this is a subtle issue, which has been thoroughly analyzed in the cited papers, but it is not the topic of this short and much less ambitious paper.

5 Maclane 8_3: ...Easier to Discover with GDI

In sharp contrast, it seems much more feasible to study this theorem using the automatic Discovery feature of GDI. Since this approach consists of imposing some extra, hypothetical, conditions on a construction, we can sketch the construction as above, namely, we construct two free points $A(u_1, u_2)$ and $B(u_3, u_4)$ which define a line containing the point $D(u_5, x_1)$. A new free point $C(u_6, u_7)$ is used to define line BC containing the point $E(u_8, x_2)$. Analogously, the line CD contains the point $F(u_9, x_3)$, the line DE contains the point $G(u_{10}, x_4)$, and the line EF contains the point $H(u_{11}, x_5)$ (Fig. 2).

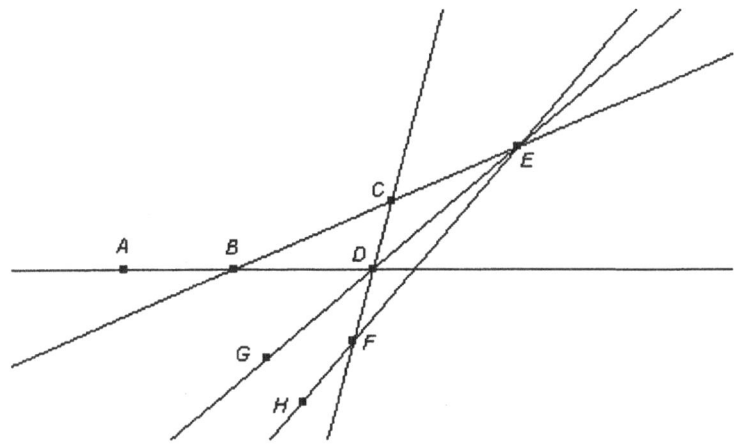

Fig. 2. The construction for discovering 8_3

There are eleven free variables in the above construction: A, B and C each contribute with two new variables, while the other five points give only one free coordinate each. Then the last three alignment conditions are imposed on the sketch by selecting the corresponding triple of points and declaring them as collinear (Fig. 3). Thus the alignment of F, G and A should (in general) reduce by one the number of free variables (as it imposes a new relation between them), and the same should happen for the remaining two conditions. Finally, it seems that there are just eight free variables (except for the case that some of these constraints were locally or globally redundant).

For symbolic purposes, the sketch is internally (in GDI) translated as follows, where A and B have been declared in the construction as the origin and the x-axis unit point, without loss of generality and in order to reduce the number of involved variables,

```
Points
A(0,0)
B(1,0)
C(u[6],u[7])
```

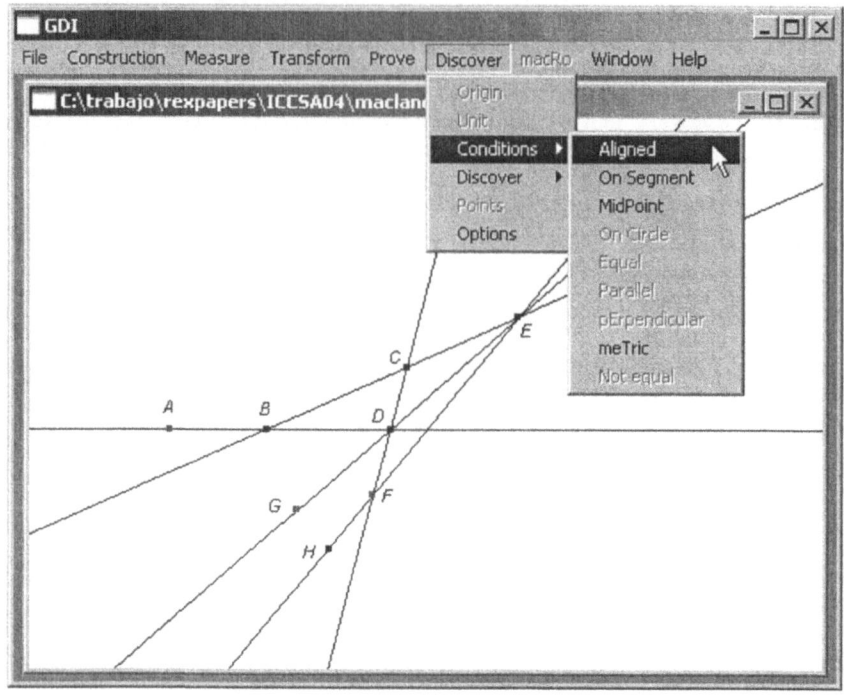

Fig. 3. Imposing the alignment of F, G and A in GDI

```
D(u[5],x[1])
E(u[8],x[2])
F(u[9],x[3])
G(u[10],x[4])
H(u[11],x[5])

Properties
Aligned(D,A,B)  (0-x[1])*(1-0)-(0-0)*(0-u[5])
Aligned(E,B,C)  (0-x[2])*(u[6]-1)-(u[7]-0)*(1-u[8])
Aligned(F,D,C)  (x[1]-x[3])*(u[6]-u[5])-(u[7]-x[1])*(u[5]-u[9])
Aligned(G,D,E)  (x[1]-x[4])*(u[8]-u[5])-(x[2]-x[1])*(u[5]-u[10])
Aligned(H,E,F)  (x[2]-x[5])*(u[9]-u[8])-(x[3]-x[2])*(u[8]-u[11])

Conditions
Aligned(F,G,A)  (x[4]-x[3])*(0-u[10])-(0-x[4])*(u[10]-u[9])
Aligned(G,H,B)  (x[5]-x[4])*(1-u[11])-(0-x[5])*(u[11]-u[10])
Aligned(H,A,C)  (0-x[5])*(u[6]-0)-(u[7]-0)*(0-u[11])
```

Let I be the ideal automatically generated by the eight polynomials (properties + conditions) in the ring $R = Q[x_1, \ldots, x_5, u_5, \ldots, u_{11}]$. The dimension of R/I, computed with CoCoA, turns out to be 6 (it yields dimension 10 if A and

B are not declared as origin and unit, respectively), thus showing that the alignment conditions are not totally independent of each other, since one would have expected dimension 4 (because we have seven apparently independent variables $u_5, \ldots, u_{11}$ in the construction and three extra conditions over them); likewise, one would have expected dimension 8, as counted above, if A and B would have been taken as free points. So there is a dimension increase from 4 to 6 (or from 8 to 10 in the latter case). This agrees with Conti and Traverso [8], who state without further details that there is a highest dimension component in $Var(H)$ of dimension 10 (since they work without fixing any points a priori; and, by the way, it happens that the theorem holds over this component, ie. it consists of points where the eight points are collinear).

Anyway, without taking into consideration this non-automatic observation, GDI proceeds in a mechanic way (at user's request) towards the discovery of properties induced on the eight points configuration after imposing the last three conditions. This is (roughly) done by eliminating the (construction declared) bounded variables x_i in order to find necessary conditions expressed by means of the supposedly free variables. This elimination ideal, computed with CoCoA in a few seconds, is generated by 28 polynomials, say $p_i, i = 1, \ldots, 28$.

The vanishing of any of these polynomials, involving just the variables $u_5, \ldots, u_{11}$, is a mere consequence of the proposed hypotheses and gives some necessary conditions for the configuration to verify the construction and the extra conditions. Now the surprise comes when we realize that the gcd of these polynomials is u_7. But $u_7 = 0$ gives the collinearity of the eight given points! In fact, $u_7 = 0$ expresses that point C is aligned with points A, B. Let us call $q_i, i = 1, \ldots, 28$ the polynomials such that $u_7 * q_i = p_i$. Therefore, if all the polynomials in I vanish (ie. if we are on the hypothesis variety) and some $q_i \neq 0$, then $u_7 = 0$, that is, the theorem is true over the open set of the hypothesis variety described by the union of $q_i \neq 0, i = 1, \ldots, 28$.

Finally, as a personal and non–automatic remark, we might observe that $u_5, u_6, u_8, \ldots, u_{11}$ is a maximal set of 6 independent variables modulo I, and that $Sat(I, Ideal(u_7)) \cap Q[u_5, u_6, u_8, \ldots, u_{11}]$ is not zero (for instance, it contains $q_1 = -u_5 u_6 u_8 u_{11} + u_5 u_6 u_{10} u_{11} + u_5 u_6 u_8 - u_6 u_8 u_{10} + u_6 u_8 u_{11} - u_5 u_{10} u_{11} - u_6 u_{10} u_{11} + u_8 u_{10} u_{11} - u_5 u_6 + u_6 u_{10} + u_5 u_{11} - u_8 u_{11}$) which implies [18] that the theorem holds over all components of the hypothesis variety where these variables (of number equal to the dimension of the variety) remain independent; this fact (for some of us) supports calling this a *generally true theorem*.

6 Conclusion

As mentioned in the introduction, the main point of this paper is showing how such a paradigmatic and debated problem, which has concealed non–automatic efforts from many researchers, can be approached in a simple way via GDI, a mixture of a dynamic geometry software and an automatic theorem prover and discoverer (by algebraic means), built up by the second author (Botana).

Such program, for this particular problem of Maclane, starting from a simple graphic construction of a configuration satisfying some of the hypothesis and then imposing the remaining ones as further conditions on the geometric objects, without further human intervention, automatically (except for a GCD calculation, that could be in the future included as a final check in the program) *discovers* the thesis and provides the user with some solid arguments to conclude that the theorem of Maclane 8_3 might be generally true (after all, it is even guessed by a machine!).

We think that the mere fact that GDI is capable of suggesting the thesis in Maclane's theorem in a purely automatic way and with little effort from the user and the machine, is already interesting on its own (leaving aside the discussion about the truth/falsity of the whole statement). After all, automatic theorem provers should be *tools* for experimentation, not oracles for deciding. And GDI shows in this example its capability for bringing up non-trivial conjectures, such as *if you make such construction and impose such conditions it might happen that ... the eight points are collinear.*

We think that the idea of finding automatically the (sometimes unexpected) geometric consequences of a given sequence of steps in a construction with geometric entities, has a natural repercussion in Computer Aided Geometric Design. It is easy to imagine a situation in which one is imposing different properties to some objects on a sketch, without noticing it will end up with an (undesirable or, perhaps, lucky) configuration.... Maclane's could be an example of this situation. Or, on the other way, it might be the case that one wants to find out what extra conditions should be imposed to a collection of objects on a construction so that a certain desirable property holds: for instance, as in the (toy) example in 3.1.

Both situations will require automatic discovery and could be treated in a unified way: in the second case, one is lead to *discover* the consequence of imposing some concrete condition, such as the alignment of the symmetrical points; in the first case, of Maclane's theorem, one talks about *discovering* the consequence(s) of imposing no extra condition, (i.e. a trivial condition such as 1=1, in algebraic terms) on the set of already given ones, i.e. on the given construction.

GDI automatic discovery feature is quite unique and, if not adequately explained by showing, as in the preceding sections, some of the involved methods and, also, its inherent limitations (concerning, for instance, the choice of independent variables), could be considered as miraculous. This is clearly not the case, but even so, the idea behind it (of automatically discovering) deserves, in our opinion, further research in view of its potential application to various fields in computer graphics, as a helper feature on CAD programs, providing them with an (apparently) omniscent ability to answer queries about a given sketch.

Automatic proving has had in the past an important connection to computational geometric design; automatic discovering should have, in the future, a closer connection. After all, when working with geometric entities, only a few statements we can imagine about them will hold true... most will be false. Thus, in some sense, proving refers just to a small portion of reality, while discovering deals with wide, open, everyday situations...

References

1. Botana, F.: A Web-based intelligent system for geometric discovery. In P.M.A. Sloot et al, editors, *Proc. 2nd International Conference on Computational Science* (ICCS 2003), LNCS, 2657, 801–810 (2003)
2. Botana, F., Valcarce, J.L.: A software tool for the investigation of plane loci. *Mathematics and Computers in Simulation*, 61(2), 141–154 (2003)
3. Botana, F., Valcarce, J.L.: Automatic determination of envelopes and other derived curves within a graphic environment. *Mathematics and Computers in Simulation*, to appear
4. Bouma, W., Chen, X., Fudos, I., Hoffmann, C., Vermeer, P.J.: An electronic primer on geometric constraint solving. http://www.cs.purdue.edu/homes/cmh/electrobook/into.html
5. A. Capani, G. Niesi and L. Robbiano, CoCoA, a system for doing Computations in Commutative Algebra. Available via anonymous ftp from: cocoa.dima.unige.it
6. Chou, S.C.: *Mechanical geometry theorem proving*. D. Reidel Publishing Company, Dordrecht, Boston (1988)
7. Conti, P., Traverso, C.: A case study of semiautomatic proving: the Maclane 8_3 theorem. In G. Cohen, M. Giusti and T. Mora, editors, *Proc. 11th International Symposium in Applied Algebra, Algebraic Algorithms and Error–Correcting Codes* (AAECC-11), LNCS, 948, 183–193 (1995)
8. Conti, P., Traverso, C.: Algebraic and semialgebraic proofs: Methods and paradoxes. In J. Richter–Gebert and D. Wang, editors, *Proc. 3rd International Workshop on Automated Deduction in Geometry* (ADG 2000), LNAI, 2061, 83–103 (2000)
9. Dolzmann, A., Sturm, A., Weispfenning, V.: A new approach for automatic theorem proving in real geometry. *Journal of Automated Reasoning*, 21, 357–380 (1998)
10. Gao, X.S., Chou, S.C.: Solving geometric constraint systems. I. A global propagation approach. *Computer–Aided Design*, 30, 47–54 (1998)
11. Gao, X.S., Chou, S.C.: Solving geometric constraint systems. II. A symbolic approach and decision of Rc–constructibility. *Computer–Aided Design*, 30, 115-122 (1998)
12. Gao, X.S., Zhang, J.Z., Chou, S.C.: *Geometry Expert*. Nine Chapters, Taiwan (1998)
13. Hoffmann, C., Bouma, W., Fudos, I., Cai, J., Paige, R.: A geometric constraint solver. *Computer–Aided Design*, 27, 487–501 (1995)
14. Kreuzer, M., Robbiano, L.: *Computational Commutative Algebra I*. Springer, Berlin (2000)
15. MacLane, S.: Some interpretation of abstract linear dependence in terms of projective geometry. *American Journal of Mathematics*, 58, 236–240 (1936)
16. Recio, T., Vélez, M. P.: Automatic discovery of theorems in elementary geometry. *Journal of Automated Reasoning*, 23, 63–82 (1999)
17. Richter–Gebert, J., Kortenkamp, U.: *The Interactive Geometry Software Cinderella*. Springer, Berlin (1999)
18. Vélez, P., Recio, T., Sterk, H.: Project: Automated Geometry theorem proving. In A.M. Cohen, H. Cuypers and H. Sterk, editors, *Some Tapas of Computer Algebra* (Algorithms and Computations in Mathematics, Vol. 4), Springer, Berlin, 276–296 (1999)
19. Vona, M.: Annotated links on geometry and geometric design. http://www.mit.edu/~vona/links/links-geometry-only.html

20. Wang, D.: GEOTHER: A geometry theorem prover. In M. McRobbie and J. Slaney, editors, *Proc. 13th International Conference on Automated Deduction* (CADE 1996), LNCS, 1104, 166–170 (1996)
21. http://www-calfor.lip6.fr/~wang/GRBib/ Welcome.html
22. http://www.acailabcom/english/mathxp.htm

Helical Curves on Surfaces for Computer-Aided Geometric Design and Manufacturing

J. Puig-Pey*, A. Gálvez, and A. Iglesias

Department of Applied Mathematics and Computational Sciences, University of Cantabria, Avda. de los Castros, s/n, E-39005, Santander, Spain
{puigpeyj,galveza,iglesias}@unican.es

Abstract. This paper introduces a new method for generating the helical tool-paths for both implicit and parametric surfaces. The basic idea is to describe the helical curves as the solutions of an initial-value problem of ordinary differential equations. This system can be obtained from the fact that the helical curve exhibits a constant angle ϕ with an arbitrary given vector **D**, which is assumed to be the axis of the helical curve. The resulting system of differential equations is then integrated by applying standard numerical techniques. The performance of the proposed method is discussed by means of some illustrative examples of helical curves on parametric and implicit surfaces.

1 Introduction

An important and widely researched issue in CAGD (Computer-Aided Geometric Design) is that of *surface interrogation*. Roughly speaking, it consists of questioning geometrically understandable characteristics of already constructed surfaces. Of course, many different methods can be applied to this purpose [2, 5,6,7,9,16]. Among them, a very popular and interesting body of research is the so-called *characteristic curves* on a surface. These are curves reflecting either the visual or the geometric properties of the surface. For the visual properties we can use the *reflection lines* [8] and the *isophotes* [13], which help to evaluate the behavior and aesthetics of the surface under illumination models. On the other hand, the geometric properties can be accurately analyzed through the *contour lines* [2], the *lines of curvature* [2], *geodesic paths* [2], *asymptotic lines* [16], etc.

On the other hand, the generation of numerically controlled tool paths to produce a smooth surface is a central issue in any sculptured surface machining process. There are several types of tool-path topology patterns (see [4], Chapter 22): serial, radial, strip and contour patterns. While the serial and radial patterns are mostly used for machining an area, the contour type is applied to cut a vertical or slant wall. In particular, the helical topology is widely used in high-speed machining [3]. A classical procedure for generating the helical tool-paths is based on the idea of using a set of z-parallel tool-paths that are closed curves wich become the boundary curves of a "surface strip" formed by a pair adjacent

---
* Corresponding author

curves. Such a surface strip is constructed as a ruled surface. A ramp curve is traced on this surface by linking one of the z-parallel curves with the next one. This ramp guide curve is projected onto the working surface to obtain the corresponding path segment. The operation is sequentially repeated for all the pairs of adjacent z-parallel paths [3].

In this paper a new method for generating the helical tool-paths for both implicit and parametric surfaces is introduced. The helical curves are described as the solutions of an initial-value problem of ordinary differential equations (ODEs). To this aim, we take advantage of the fact that the helical curve, given in parametric form[1] by $\mathbf{C}(t) = (x(t), y(t), z(t))$ exhibits a constant angle ϕ with an arbitrary given vector $\mathbf{D}$ (the axis of the helical curve). The resulting initial-value system of ODEs is then numerically integrated through an adaptive Runge-Kutta method [14]. As we will show later, this simple procedure yields a helical curve on the surface to be manufactured. The performance of our approach will be shown by means of some illustrative examples.

The structure of the article is as follows. Some mathematical preliminaries on parametric and implicit surfaces are given in Section 2. Then Section 3 describes our method for computing helical curves lying on a surface given in parametric and implicit forms. Section 4 gives some brief comments about the numerical procedures used in this work. In Section 5 some illustrative examples to show the good performance of the proposed method are presented. Finally, the conclusions are discussed in Section 6.

2 Mathematical Preliminaries

In this paper we will consider differentiable surfaces given in both parametric and implicit forms. In the first case, they are described by a vector-valued function of two variables:

$$\mathbf{S}(u,v) = (x(u,v), y(u,v), z(u,v)), \qquad u, v \in \Omega \subset \mathbb{R}^2 \qquad (1)$$

where u and v are the surface parameters and Ω represents the surface domain. Expression (1) is called a parameterization of the surface $\mathbf{S}$. At regular points, the partial derivatives $\mathbf{S}_u(u,v)$ and $\mathbf{S}_v(u,v)$ do not vanish simultaneously. For $\{u = u_0, v = v_0\}$, $\mathbf{S}_u$ and $\mathbf{S}_v$ are vectors on the tangent plane to the surface at the point $\mathbf{S}(u_0, v_0)$, each being tangent to the parametric or coordinate curve $v = v_0$ and $u = u_0$, respectively.

On the other hand, any arbitrary curve on the surface can be described in parametric form by $\{u = u(t), v = v(t)\}$. This expression defines a three-dimensional curve on the surface $\mathbf{S}$ given by $\mathbf{C}(t) = \mathbf{S}(u(t), v(t))$. Applying the chain rule, the tangent vector of the curve $\mathbf{C}$ at a point $\mathbf{C}(t)$ becomes:

$$\frac{d\mathbf{C}(t)}{dt} = \mathbf{S}_u \frac{du}{dt} + \mathbf{S}_v \frac{dv}{dt} \qquad (2)$$

---

[1] In this paper vectors will be denoted in bold.

It is useful to consider the case in which the curve **C** is parameterized by the arc-length. Its geometric interpretation is that a constant step traces a constant distance along an arc-length parameterized curve. Since some industrial operations require an uniform parameterization, this property has several practical applications. For example, in computer controlled milling operations, the curve path followed by the milling machine must be parameterized such that the cutter neither speeds up nor slows down along the path. Consequently, the optimal path is that parameterized by the arc-length s on the surface **S** given by:

$$E\left(\frac{du}{ds}\right)^2 + 2F\frac{du}{ds}\frac{dv}{ds} + G\left(\frac{dv}{ds}\right)^2 = 1 \qquad (3)$$

where E, F and G are the coefficients of the First Fundamental Form of the surface:

$$E = \mathbf{S}_u.\mathbf{S}_u \;,\; F = \mathbf{S}_u.\mathbf{S}_v \;,\; G = \mathbf{S}_v.\mathbf{S}_v \qquad (4)$$

and "." is used to indicate the *dot* product (see [15] for details). For the sake of clarity, in the following we shall refer exclusively to the parameter s to account for a curve parameterized by the arc-length on the surface.

On the other hand, a parametric curve $\mathbf{C}(t) = (x(t), y(t), z(t))$ lying on an implicit surface given by $f(x, y, z) = 0$ satisfies:

$$f_x\frac{dx}{dt} + f_y\frac{dy}{dt} + f_z\frac{dz}{dt} = 0 \qquad (5)$$

This relation is obvious by simply noticing that $\dfrac{d\mathbf{C}}{dt}$ is orthogonal at each point of $\mathbf{C}(t)$ to the normal vector (f_x, f_y, f_z) to the surface. This relationship also holds in the case of t being the arc-length parameter s. In addition, the parameterization $\mathbf{C}(s) = (x(s), y(s), z(s))$ of the curve by the arc-length on the surface verifies:

$$\left(\frac{dx}{ds}\right)^2 + \left(\frac{dy}{ds}\right)^2 + \left(\frac{dz}{ds}\right)^2 = 1 \qquad (6)$$

3 Helical Curves

In this section the proposed method for obtaining helical curves on a surface is introduced. We firstly describe the case of parametric surfaces in Section 3.1. Then, the case of implicit surfaces is discussed in Section 3.2.

3.1 The Parametric Case

Let us consider a curve $\mathbf{C}(s) = (x(s), y(s), z(s))$ lying on a parametric surface defined as in Eq. (1). For $\mathbf{C}(s)$ to be a helical curve, it must exhibit a constant angle ϕ with an arbitrary given vector **D**, which will be the axis of such ahelical curve. Without loss of generality, we can assume that $\left\|\dfrac{d\mathbf{C}}{ds}\right\| = \|\mathbf{D}\| = 1$. In this case we have:

$$\frac{d\mathbf{C}}{ds}.\mathbf{D} = cos(\phi) \tag{7}$$

Inserting Eq. (2) in (7) we get:

$$(\mathbf{S}_u.\mathbf{D})\frac{du}{ds} + (\mathbf{S}_v.\mathbf{D})\frac{dv}{ds} = cos(\phi) \tag{8}$$

Combining Eqs. (3) and (8) and making some calculations, we obtain:

$$\begin{cases} \dfrac{du}{ds} = \dfrac{-B \pm 2(\mathbf{S}_v.\mathbf{D})\sqrt{(F^2 - EG)\cos^2(\phi) + A}}{2A} \\ \\ \dfrac{dv}{ds} = \dfrac{-B^* \mp 2(\mathbf{S}_u.\mathbf{D})\sqrt{(F^2 - EG)\cos^2(\phi) + A}}{2A} \end{cases} \tag{9}$$

where the values A, B and C are given by

$$A = E\,(\mathbf{S}_v.\mathbf{D})^2 - 2F\,(\mathbf{S}_u.\mathbf{D})\,(\mathbf{S}_v.\mathbf{D}) + G\,(\mathbf{S}_u.\mathbf{D})^2$$
$$B = 2\cos(\phi)\,[F(\mathbf{S}_v.\mathbf{D}) - G(\mathbf{S}_u.\mathbf{D})]$$
$$B^* = 2\cos(\phi)\,[F\,(\mathbf{S}_u.\mathbf{D}) - E\,(\mathbf{S}_v.\mathbf{D})]$$

and E, F and G are defined by (4). For system (9) to be completely determined we need to consider an initial point:

$$\begin{cases} u(0) = u_0 \\ v(0) = v_0 \end{cases} \tag{10}$$

on the surface domain. System (9)-(10) constitutes an initial-value problem of explicit first-order ordinary differential equations. The signs $\pm$ and $\mp$ in (9) mean that there are two helical curves starting at (u_0, v_0) associated with the two possible ϕ directions at this point. The expression $(F^2 - EG)\cos^2(\phi) + A$ under the square root in (9) must be positive for the existence at a given point on $\mathbf{S}(u,v)$ of the two helical curves having the angle ϕ with $\mathbf{D}$. Once system (9)-(10) is solved for u and v, the x, y and z coordinates of the helical curve can be automatically determined by a simple substitution of u and v into the surface equation $\mathbf{S}(u,v)$.

3.2 The Implicit Case

Let $\mathbf{S}$ be a surface given in implicit form by $f(x,y,z) = 0$. Let ϕ be the angle between an arbitrary given vector $\mathbf{D}$ and the helical curve $\mathbf{C}(s)$, given in parametric form by $\mathbf{C}(s) = (x(s), y(s), z(s))$, s being the arc-length parameter on the surface $\mathbf{S}$. Therefore, we have:

$$\frac{d\mathbf{C}}{ds}.\mathbf{D} = cos(\phi) \tag{11}$$

Without loss of generality, we can assume that $\mathbf{D} = (0,0,1)$ (the discussion is quite similar for any other choice of the unit vector $\mathbf{D}$), so from (11) we get:

$$\frac{dz}{ds} = cos(\phi) \qquad (12)$$

Substituting (12) into (6), combining the resulting equation with (5) and making some calculations, we finally obtain:

$$\begin{cases} \dfrac{dx}{ds} = \dfrac{-f_x\, f_z cos(\phi) \pm f_y\sqrt{(f_x^2 + f_y^2)\, sin^2(\phi) - f_z^2\, cos^2(\phi)}}{f_x^2 + f_y^2} \\[2ex] \dfrac{dy}{ds} = \dfrac{-f_y\, f_z cos(\phi) \mp f_x\sqrt{(f_x^2 + f_y^2)\, sin^2(\phi) - f_z^2\, cos^2(\phi)}}{f_x^2 + f_y^2} \\[2ex] \dfrac{dz}{ds} = cos(\phi) \end{cases} \qquad (13)$$

Equations (13) together with an initial point:

$$\begin{cases} x(0) = u_0 \\ y(0) = v_0 \\ z(0) = z_0 \end{cases} \qquad (14)$$

constitute an initial-value problem for an explicit first-order system of ordinary differential equations. Note that there are two different systems of differential equations, associated with the signs $\pm$ and $\mp$ in (13), respectively. They correspond to the two different helical curves passing through a given point on the surface $\mathbf{S}$ for which the value of the expression $\Delta = \left(f_x^2 + f_y^2\right) sin^2(\phi) - f_z^2\, cos^2(\phi)$ is positive, $\Delta > 0$. On the contrary, a value $\Delta < 0$ at a point $\mathbf{P}$ means that there is no helical curve at that point for the chosen direction $\mathbf{D}$ and the angle ϕ.

4 Numerical Procedures

The analytical solution of systems of first-order ODEs presented in (9)-(10) and (13)-(14) is not possible in general, but this kind of problems can be solved very efficiently by using standard numerical techniques. The integration of the previous systems has been performed by using an adaptive 4-5-order Runge-Kutta algorithm [14]. In particular, we have used the Matlab routine ode45 [10] that provides the users with a control of absolute and relative error tolerance. In all our examples, the absolute tolerance error (a threshold below which the value of the estimated error of the ith solution component is unimportant) was set to 10^{-7} while we used 10^{-4} for the relative tolerance error (a measure of the error relative to the size of each solution component).

In general, the method works well for any parametric or implicit surface. It should be noticed, however, that some situations require a careful analysis: for example, when the surface consists of several patches some kind of continuity conditions must be imposed to assure that the differential model is still valid in the neighbourhood of each patch boundary. On the other hand, the domain of the involved surface should be taken into account during the integration process. In particular, we must check whether the trajectories reach the boundaries of the surface domain. Finally, in the case of working with surfaces which are partially or completely closed, additional tests for self-intersection had to be performed.

5 Examples

In this section, the performance of the method described in the previous paragraphs is discussed by means of some illustrative examples. Following the structure of previous sections, we analyze firstly the case of parametric surfaces and subsequently the case of implicit surfaces.

In general, the method works well for any parametric surface. However, because of their outstanding advantages in industrial environments, their flexibility and the fact that they can represent well a wide variety of shapes, our examples will focus on NURBS surfaces. In this case, the corresponding derivatives in Eq. (9) have been obtained by following the procedure described in [12].

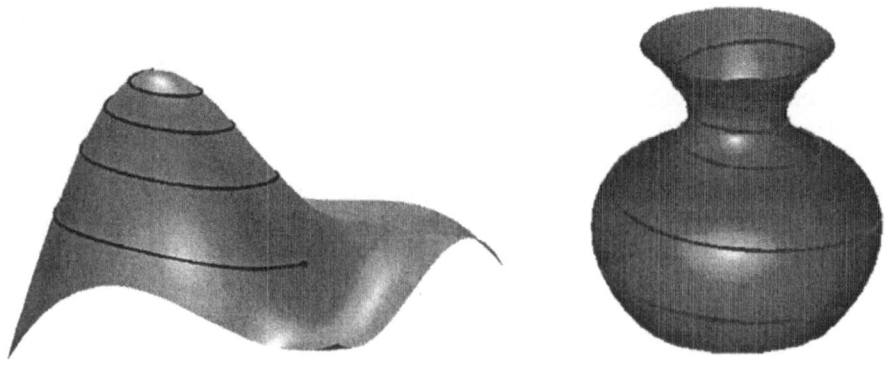

Fig. 1. Helical curves on parametric surfaces: (left) a Bézier surface; (right) a $(3,3)$-order NURBS surface

Figure 1 shows two examples of helical curves on parametric surfaces. On the left the case of a Bézier surface of degree 4 for both variables u and v is displayed. The surface on the right is a $(3,3)$-order NURBS surface defined by a mesh of 6×9 control points and knot vectors $\mathcal{U} = \{0,0,0,1,1,2,2,3,3,4,4,4\}$ and $\mathcal{V} = \{0,0,0,1,2,3,4,4,4\}$. Note that while the surface in Fig. 1(left) consists of a single patch, the surface in Fig. 1(right) is comprised of 16 parametric patches, connected with C^1 continuity. Note also that this last example corresponds to a closed surface in the direction of the variable v.

The case of implicit surfaces is illustrated in Figure 2. The corresponding surface equations in this figure are $x^4 + 2x^2 z^2 - 0.36x^2 - y^4 + \frac{1}{4}y^2 + z^4 = 0$ (left) and $(x^2 + y^2) z^2 + \frac{1}{4}(x^2 + y^2) - \frac{1}{4} = 0$ (right). Especially interesting is the second example, in which the surface is comprised of two vertical cones joined by two branches connected each other. As shown in Fig. 2(right), the helical curve on that surface evolves on the lower cone, goes up on one of the branches and suddenly crosses to the another one via the connected link, then keeps going up on the upper cone until it reaches the upper boundary of the surface. We remark that this behavior depends on the initial condition of system (13)-(14) and hence other initial conditions may eventually lead to different behaviors than that of this picture.

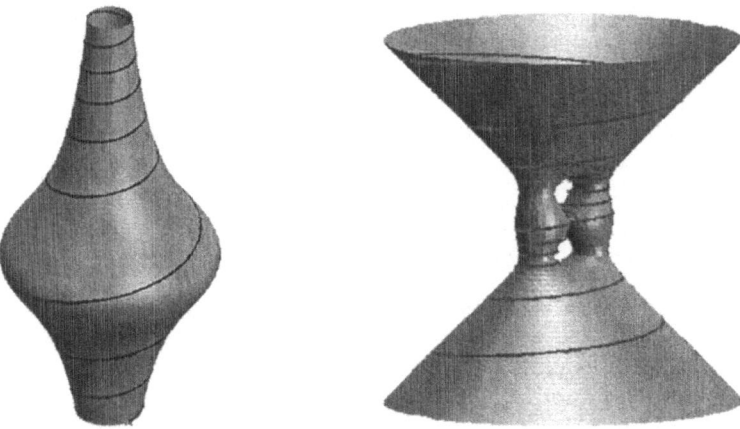

Fig. 2. Helical curves on implicit surfaces: (left) surface $x^4 + 2x^2 z^2 - 0.36x^2 - y^4 + \frac{1}{4}y^2 + z^4 = 0$; (right) surface $(x^2 + y^2) z^2 + \frac{1}{4}(x^2 + y^2) - \frac{1}{4} = 0$

6 Conclusions

This paper describes a new method for generating the helical tool-paths for both implicit and parametric surfaces. The basic idea is to describe the helical curves as the solutions of an initial-value problem of ordinary differential equations. This system is obtained from the fact that the helical curve exhibits a constant angle ϕ with an arbitrary given vector $\mathbf{D}$, which is assumed to be the axis of the helical curve. The resulting system of differential equations is then integrated by applying standard numerical techniques. As the reader can see, the method exhibits a very good performance in the examples shown in this paper and in many others not included here because of limitations of space.

Potential applications of the present method include the generation of numerically controlled tool-paths for high-speed machining. Possible applications also appear in other design problems (roads, railways) in which paths with prescribed slopes and angles are also required.

Acknowledgements. The authors are grateful to the CICYT of the Spanish Ministry of Science and Technology (project DPI2001-1288), the European Union (GAIA II, Project IST-2002-35512) and the University of Cantabria (Spain) for their partial support of this work.

References

1. R.E. Barnhill (Ed.): Geometry Procesing for Design and Manufacturing, SIAM, Philadelphia (1992)
2. Beck, J.M., Farouki, R.T., Hinds, J.K.: Surface analysis methods. IEEE Computer Graphics and Applications, Dec. (1986) 18-36
3. Choi, B.K., Jerard, R.B.: Sculptured Surface Machining. Theory and Applications. Kluwer Academic Publishers, Dordrecht Boston London (1998)
4. Farin, G., Hoschek, J., Kim, M.S.: Handbook of Computer Aided Geometric Design. Elsevier Science, Amsterdam (2002)
5. Hagen, H., Schreiber, T., Gschwind, E.: Methods for surface interrogation. Proc. Visualization'90, IEEE Computer Society Press, Los Alamitos, Calif. (1990) 187-193
6. Hagen, H., Hahman, S., Schreiber, T., Nakajima, Y., Wördenweber, B., Hollemann-Grundstedt, P.: Surface interrogation algorithms. IEEE Computer Graphics and Applications, Sept. (1992) 53-60
7. Hoschek, J., Lasser, D.: Computer-Aided Geometric Design, A.K. Peters, Wellesley, MA (1993)
8. Klass, R.: Correction of local surface irregularities using reflection lines. CAD, **12**(2) (1980) 73-77
9. Koenderink, J.J., van Doorn, A.J.: Surface shapes and curvature scales. Image and Vision Computing, **8**(2) (1992) 557-565
10. The Mathworks Inc: Using Matlab. Natick, MA (1999)
11. Patrikalakis, N.M., Maekawa, T.: Shape Interrogation for Computer Aided Design and Manufacturing. Springer Verlag, 2002.
12. Piegl, L., Tiller, W.: The NURBS Book, Springer Verlag, Berlin Heidelberg (1997)
13. Poeschl, T.: Detecting surface irregularities using isophotes. CAGD **1**(2) (1984) 163-168
14. Press, W.H., Teukolsky, S.A., Vetterling, W.T., Flannery, B.P.: Numerical Recipes (2nd edition), Cambridge University Press, Cambridge (1992)
15. Struik, D.J.: Lectures on Classical Differential Geometry, 2nd ed., Dover Publications, New York (1988)
16. Theisel, H., Farin, G.E.: The curvature of characteristic curves on surfaces. IEEE Computer Graphics and Applications, Nov./Dec. (1997) 88-96

An Application of Computer Graphics for Landscape Impact Assessment

César Otero[1], Viola Bruschi[2], Antonio Cendrero[2], Akemi Gálvez[3], Miguel Lázaro[1], and Reinaldo Togores[1]

[1] Dpmt. of Geographical and Graphical Engineering. EGI-CAD Group. Civil Engineering Faculty. University of Cantabria. Spain.
oteroc@unican.es
[2] Dpmt. CITYMAC. Geomorphology Group. Faculty of Sciences. University of Cantabria. Spain
[3] Dpmt. Applied Mathematics and Computer Science. Computer Graphics Group. University of Cantabria. Spain

Abstract. This communication shows a procedure aimed at assisting in the assessment of the landscape impact caused by new lineal infrastructure services, such as motorways. The computer simulation has demanded the idealisation of the phenomenon to be graphically represented, its theoretical model and the functional analysis of the computational tool that can aid in finding a solution. The core of the problem is that there is a non finite set of landscapes to be considered (and analysed); however, according to the model proposed, if some suitable Computer Graphics libraries are used, it is possible to automatically produce a reduced catalogue of realistic images simulating, each of them, the effect that the new infrastructure will cause on the environment, just at the sites where the landscape results to be more vulnerable.

1 Introduction: Nature of Landscape Impacts and Their Indicators

Visual quality (intrinsic merit of a unit from a perceptual point of view) and fragility (sensitivity to visual intrusion from human activities) of landscape [8] are determined mainly by three groups of factors: geomorphology (relief, shape, and rock type), vegetation, and land use (especially impacting elements such as prominent constructions); presence of water is also important. The introduction of new large structures (Fig. 1) such as a motorway represents visual intrusions which may reduce visual quality. This reduction (*intensity* of the impact) is related to the degree of modification; that is, contrast in size, shape, colour and texture between the structure and the pre-existing landscape. The *magnitude* of the impact can be considered to depend on the number of people and/or extent of the area affected.

Indicators of magnitude can be the total area from which the new structure (or individual sectors of it) can be seen or the number of people affected. A measure of intensity can be obtained by defining levels of visual intrusion in sensitive units (those with high visual quality or fragility) on the basis of different degrees of contrast with the new structure, presented by means of photographs or images.

But measuring the visual impact magnitude and intensity of linear infrastructure facilities turns out to be exceedingly complex because infinite views exist that potentially must be considered. How can we reduce that infinite number of views to a discrete and manageable set? A model of treatment and its corresponding implementation are the subject of this contribution. On the other hand, dealing with the problem at the design level, this tool can add more criteria (environmental in this case) for decision making when deciding the optimal solution among several alternative layouts.

2 A Model for the Simulation of Landscape Impact

2.1 Landscape Units

The procedure to be followed in assessing visual impacts starts with the definition of landscape units directly affected by the new road (units crossed by or adjacent to the motorway). Two possible methods can be used to identify and map landscape units: a) determination of *viewsheds* directly in the field or through the use of Computer Graphics and GIS; b) definition of *units* according to geomorphological criteria (mainly on the basis of type of unit, relief and shape). Visual quality of landscape units [1] can be assessed on the basis of: *Relief* (difference between maximum and minimum altitude); *General shape* (convexity, concavity, etc); *Landform* (irregularity, ruggedness); *Geomorphological diversity* (Number of geomorphological elements in unit); *Vegetation* (rank of vegetation type); *Land use* (intensity of human modification) and *Water bodies/courses*.

2.2 Data Needed

The following initial data [2], [3] are required:
DTM of the zone; NMDM: Layout of the new motorway as DM on top of the one belonging to the area; Entity SIG: Scenic units; Geomorphologic map of the zone; Land-use map; CAD base cartography: Roads, population nuclei, rivers; Geographic DB with population data; Data on the size of another relevant infrastructures of the zone; Examples of known visual intrusion, for comparison.

2.3 Operative Procedure

From the stated data capture, the following layers of graphic information are configured [4]:
DTM: Digital Terrain Model of the zone; GM: Geometry of the Motorway; PD: Areas in the zone, classified according to Population Density; RB: Zone's Areas, classified according to their proximity to other roads (Road Buffers); LU: Landscape Units.
In this first process, notice that the DTM and GM data are obtained directly, PD requires a population analysis of the base cartography through the application of a trivial filter for the selection of population nuclei, RB can be solved by means of a

spatial operation of the type "Area of Influence (Buffer)" upon the roads shown in the Base cartography and the LU arises from the weighting process described in paragraph 2.1.

A second processing leads to the following intermediate results [5]:

GVE: Areas of Greater incidence or Visual Effect, obtained from PD and RB, combining their geometries and establishing the sum of their relative weights for both criteria (See figure 2). MSLU: Motorway's sections in Landscape Units, obtained from GM and LU, by means of the inclusion operator. CMSLU: Centroids of the MSLU sections. VS: Viewsheds for each CMSLU centroid. In figure 2 only the viewshed for one centroid is represented; In general, the viewsheds for several centroids can overlap. SVI: Areas of great Visual Impact in Viewsheds, obtained by the relation of inclusion of GVE within VS. In figure 3, the SVI resulting from only one centroid is shown.

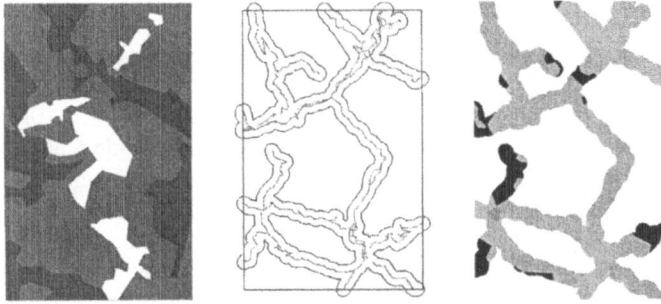

Fig. 1. The spatial intersection of the PD (left) and RB (center) layers yields GVE (right).

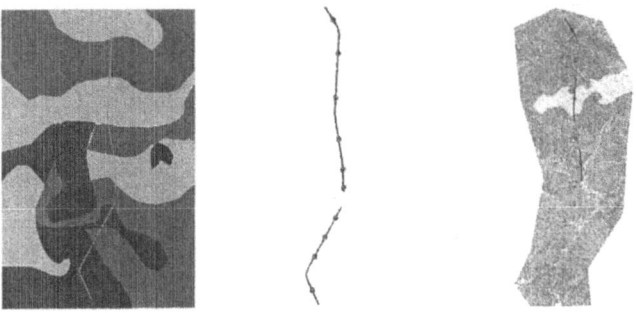

Fig. 2. Each LU polygon (left) yields a MSLU span (center). Each MSLU centroid generates a VS viewshed.

782   C. Otero et al.

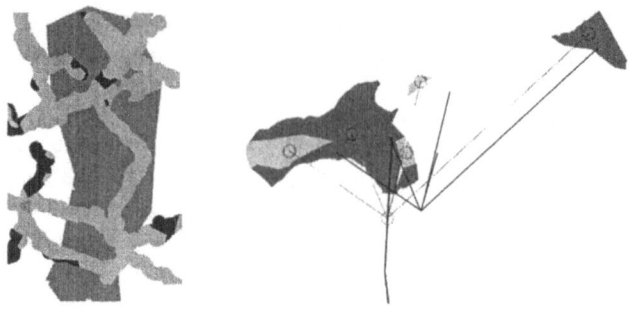

Fig. 3. Each VS viewshed for a CMSLU centroid is used to establish, in association with the GVE polygons (left) the SVI maximum visual impact zones (right).

Let us consider the CSVI (SVI's centroids); These points can be considered as representative VIEWPOINTS of the zones from which the observation of the motorway proves to be more traumatic on the landscape; in fact, CSVI centroids belong to zones of higher population or passerby density, while CMSLU centroids represent stretches where the motorway passes through zones of high scenic value. In the process described, each centroid PT_i of the CMSLU has given origin to a more or less reduced set of PC_j centroids in CSVI. In brief, a simulation of results can be attained automatically considering all the realistic visualizations that arise from considering each PT_i centroid in MSLU as a Target Point and all PC_j centroids in SVI as Viewpoints from which the corresponding visualizations are generated.

Fig. 4. CSVI Centroids are viewpoints and CMSLU Centroids are target points. The images obtained, RV, simulate the visual impact produced by the motorway on the landscape.

The result obtained is RV: a catalog of Realistic Visualizations obtained from considering all of the PTi as Target Points and, for each one of them, all of the corresponding PCj like Viewpoints. This catalog simulates the worst effects of the motorway on the existing landscape.

2.4 Postprocessing and Decision Making

From this outcome we pass on to the analysis and postprocessing phase that allows us to take actions on the two variables in which decision is more involved:
1. On a quantification of the Visual Impact Magnitude, because the determination of the percentage of landscape area affected by the implantation of the motorway in each of the visualizations of the RV catalog implicates an easily measurable 2D problem of overlapping areas.
2. On a qualitative treatment of Intensity, because each one of the visualizations in the RV catalog can be evaluated and compared by experts.

The referred procedure [6] favors decision-making; take notice that, from among the layers of data processed, the majority is independent of the road's layout; in fact, this is the case with DTM, PD, RB and LU, among those proposed in the first processing stage, so that varying GM (that is, only and exclusively the layout for the Motorway project being studied) allows the analysis of new proposals. Otherwise, in the case of GM's variation, the rest of the described procedure can be computer-automated, arriving this way again to the phase of analysis and post-process, which marks an end to a complete cycle of evaluation of simulated results. In short, it is only necessary to modify the proposed layout of the road to ascertain any possible improvement regarding scenic impact, an operation that allows the motorway designer complete freedom to propose technical alternatives.

3 Complexity of the Model

The described procedure succeeds in making manageable and tangible a problem that, in principle, is not so. The main idea consists in that detecting the adequate viewpoint and target-point sets within a zone is the only requisite needed for the generation of a catalog of realistic representations; that is, for *simulating the visual impact*. A priori, that forces us to consider infinite viewpoints and infinite target-points, to which the observer directs his vision from each of the former. However, the visual perception of landscape does not differ notably on looking at a same place from two nearby viewpoints or if, from a same viewpoint, he looks at two nearby places. If we assume that perspectives generated from the Centroids of the greater Visual Impact areas in viewsheds (CSVI) towards the centroids of the motorway sections lying in the most valued Landscape Units (CMSLU) are the most representative visualizations of the panorama (photos with which a native or a visitor would describe or would remember that landscape) it then proves to be acceptable to say that a problem of unmanageable combinatorial magnitude has been reduced to one limited to a reduced set of combinations

4 The Application

The modeling program has been developed for the Windows environment, as an .EXE type application. As follows from the precedent description, once the data has been input, the program's execution demands but a reduced number of settings and carries out the whole processing without any need for user interaction. The critical GIS type procedures have been solved [7] with the aid of Intergraph®'s Geomedia Professional Object Library, for reasons of an economic more than a technological nature; in fact, we have tried other freeware or evaluation GIS Libraries with comparable success.

The main difficulty during the GIS programming phase is that on generating the Recordsets resulting from the spatial analysis (crossing, intersection or buffers), we have found that in many occasions the resulting polygons overlap (which is especially frequent in buffer operations) making it necessary to traverse the recordset for a second time in order to clean any improper overlap detected.

Since to accomplish the measure of the intensity and magnitude of the visual impact we must have a complete access to the DTM, we opted for generating our own DLL in C++ for the calculation of VS viewsheds and handling the DTM triangles' topologic relations. Its fundamental characteristics are described next.

4.1 Independent Functions from the DLL

The DLL exports, in order to be used from any other program, the following language-independent functions, declared in C++:

void Inicializa (void) – Initializes the library. This should be called before any other function. *void Descarga (void)* – Unloads the library, freeing memory and deleting any loaded DTM. *void PonTriang (double Ax, double Ay, double Az, double Bx, double By, double Bz, double Cx, double Cy, double Cz)* – Adds a triangle to the library's internal DTM. It must be called for each triangle in the database's DTM, specifying its vertices. *short ValidaMDT (double ignoreHoles)* – Returns 0, 1 or 2 depending on the DTM's area type, meaning respectively: Connected with no holes, Not connected, Connected but with holes. *void SetTP (double TPx, double TPy, double TPz, double altura=0)* – Sets the TP, specifying its location and optionally its height. *signed long ProcesaTriang (double ignoraAgujeros=0)* – Processes the data (DTM and TP) obtaining internally the areas that see the TP (the viewshed as a collection of connected areas). It will just return the number of connected areas found for the TP set. If a parameter is specified, it will ignore in each connected area (the DTM will be split in connected areas) every hole where: Hole's Area < ignoreHoles · External Polygon's Area. *signed long Cuenta (void) and signed long CuentaSinAmigos (void)* – Return respectively the number of loaded triangles (of which the library's DTM consists) and how many of these have less than three contiguous triangles, this is, how many are placed in the DTM's edge.

4.2 Internal Classes of the Library

Now we will list and briefly comment the basic classes that we have built, together with their more important methods and properties. It will be seen that all of them can be handled by a single object of class TriangSet, which encapsulates the whole functionality of the DLL.

Classes:
P3D or V3D - 3D Point or Vector. Both are the same class. *Poligono* - Polygons, specially qualified for our purposes. *Triang* - Triangles with spatial connectivity capabilities. *TriangSet* - Set of triangles. It will store one DTM and has specific methods.

Detail on important methods and properties:

P3D or V3D
x, y, z - Point or vector coordinates. Objects pertaining to this class can be linked among them. This property will be used by objects of class Poligono to store information about its vertices.

Polígono
PrimVertice - First vertex, the rest of them will be linked after this one.
Add (Vertex) - Adds a vertex, detecting complex polygons and splitting them. *BorraVertices ( )* - Deletes all the vertices in a polygon. *Nvertices ( )* - Counts vertices number in a polygon. *SetMasExterno ( ), SetUltimoIsla ( )* - Sets the polygon as most external and last of the boundary collection, respectively.
SetPrimVertice (Vertex) - Sets the first vertex of a polygon. *SonConsec (Vertex1, Vertex2)* - Returns 'True' if two vertices belong to the same edge of the polygon, this is, if they are consecutive in any sense (clockwise or counter clockwise). *ContienePunto (Point)* - Returns 'True' if the point is inside the polygon. *Area ( )* - Returns the area enclosed by a polygon. *DamePol (px, py, pz)* - Works the same way as the one we have already explained. It returns the points that will define the polygons which in turn will bound the Viewsheds.

Triang
A, B, C – Vertices. *Baricentro ( )* - Returns the baricenter of the given triangle. *YoVeo (P)* - Returns "True" if the triangle can see the TP from point P (P is inside the triangle). *HazAmigos (OtherTriang)* - Checks vicinity with other triangles. If they result to be contiguous, both triangles will remember.

TriangSet
PrimerTriang - First triangle in the DTM. The rest of them are linked to this. *Add (Ax, Ay, Az, Bx, By, Bz, Cx, Cy, Cz,)* - Adds a triangle to the DTM. *EncuentraRecinto (pT, Recinto)* - Selects the triangles that comprise a connected area. *BordeAPoligono (FirstPolyg, ignoreHoles)* - Finds the connected area edges. *ProcesaTriang (ignoreHoles)* - *Main process*: Obtains the viewsheds. *CuentaConexos (fromTriang)* - Counts the triangles that form a connected area with the given. *Cuenta ( )* - Counts all triangles in this set. *CuentaSinAmig ( )* - Counts all triangles in this set with less than 3 neighbors. *ValidaMDT ( )* - Checks DTM according to the already explained code. *DamePol (px, py, pz)* - Works the same way as the one we have already explained. It returns the points that will define the polygons which in turn will bound the Viewsheds.

5 The User Interface

It is assumed we already have a database with all the required data. Once connected to it, two identical lists will appear, with all the classes in the database. We will then select the DTM and the Target as Fig. 5, right, shows. As we select the classes they will appear in the frame titled *Vista de la Geometría* if the checkbox *Vista* (see Fig. 5) is checked. If it isn't, we can know which classes are selected by looking at the legend.

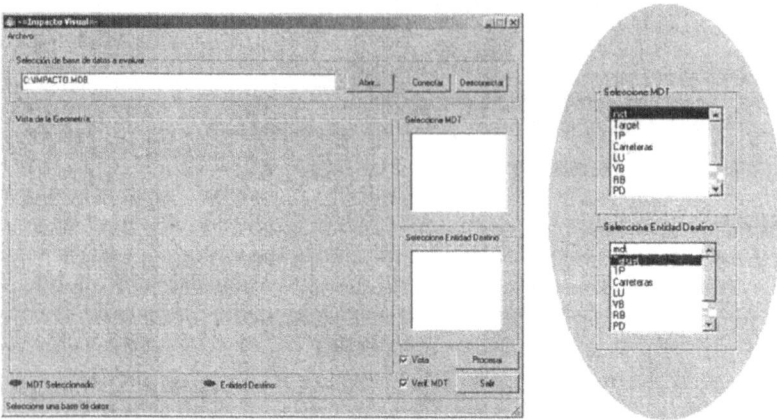

Fig. 5. User interface (left) and detail of the catalog of visible layers

Now, after making sure that we have selected the appropriate classes we click on the button labelled *Procesar*. A dialog box asking for the following data will then pop up (Fig. 6).

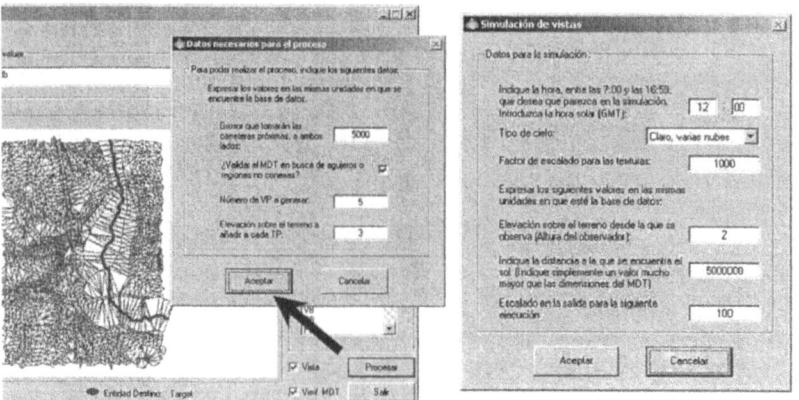

Fig. 6. Settings (left) and last steps of the simulation (right).

Road width: We must specify the road width, in the database units. *Check MDT for holes*: This option defaults to the one in the main screen and has the same meaning.

Number of VPs to find: this setting establishes the amount of pictures of the simulation. *Height over the terrain*: Once we calculate the TPs, some height must be added, no matter how small it is, to avoid them to lay exactly in the terrain. This height should be small compared with the sizes we are working with, but not null. (A few centimeters will be enough). When everything is set, press *Accept* in the dialog box, and the process will begin.

6 Output of the Simulation

Once we have finished this process successfully, some blue spots will appear on screen. We are then ready for the next step: the generation of the realistic simulations of the views. The program will try to obtain several images (one for each VP) representing as faithfully as possible what we would see if we were at each VP looking at its TP. Fig. 7 shows the final result.

Fig. 7. Last steps of the simulation (it shows the landscape before the motorway is constructed).

Acknowledgements. This task was fulfilled as part of the GETS Project (Contract No. FMRX-CT98-0162, DG XII-SLJE-TMR PROGRAM, European Commission).

Some parts of this contribution have been developed under the RRL Program [9], promoted by Intergraph Co. through Geomedia Team International. The GIS Libraries used in part of the application belong to Geomedia Professional®.

Summarizing

The present contribution is relevant mainly in two domains:
In Environmental Impact Assessment, because the Analysis Model and the Simulation Tool described succeed in making the problem manageable: a reduced set of realistic representations characterizes the environment's most vulnerable points. In general, it can be said that it is makes possible an undertaking whose treatment is, a priori, diffuse on account of the unembraceable set of combinations and possibilities implied.
In the purely graphical domain, because it presents a computational procedure that arises from 2D data and considerations, but that is capable of applying and transforming them automatically into 3D graphic output; this step leading from 2D to 3D is extremely valuable in terms of the capabilities of a Geographic Environment Modeling System.

References

[1] Claver, I. et al. Guía para la Elaboración de Estudios del Medio Físico: Contenido y Metodología. MOPT, Madrid, 1991.
[2] Bonachea, J.; Bruschi, V.; et al. An Approach for the Incorporation of Geomorphologic Factors into EIA of Transportation Infrastructures; a Case Study in Northern Spain. Geomorphology. 2003.
[3] González-Díez, A.; Remondo, J.; Bruschi, V.M; et al. Computer Assisted Methods for the Assessment of Impacts on Geomorphological Resources; Application to a Case Study in Northern Spain. ISPRS XIX Congress. Amsterdam. Julio 2000.
[4] Otero C, Bruschi V, Cendrero A., Rivas V. Assessment of visual impacts on landscape. Gets 4TH Workshop. Söelden. Austria. January 2000.
[5] Otero, C.; Togores, R; De la Pedraja, A.; Bruschi, V.; González, A. Métodos Gráficos en la Modelización, simulación y evaluación del impacto ambiental. XII Congreso Internacional de Ingeniería Gráfica. Valladolid. June 2000.
[6] C. Otero; R. Togores; et al. Assessment of Visual Impact on Landscape: Graphical Methods. RRL Paper.
http://www.teamgeomedia.com/orl/rrldetails.asp?track=IMGS404.8341|201631.
[7] C. Otero; R. Togores; et al. Visual Impact Objective Evaluation. RRL Paper.
http://www.teamgeomedia.com/orl/rrldetails.asp?track=IMGS404.8341|201631
[8] Rivas, V. et al. Geomorphological indicators for environmental impact assessment; consumable and non consumable geomorphological resources. Geomorphology 18(3-4), pp. 169.182.
[9] RRL: Registered Research Laboratory. http://www.teamgeomedia.com/

Fast Stereo Matching Using Block Similarity*

Han-Suh Koo and Chang-Sung Jeong**

Department of Electronics Engineering, Korea University
1-5ka, Anam-dong, Sungbuk-ku, Seoul 136-701, Korea
{esprit@snoopy,csjeong@charlie}.korea.ac.kr

Abstract. A lot of engineering applications related to computer vision use stereo vision algorithms and they usually require fast processing capability near to real time. In this paper, we present a new technique of area-based stereo matching algorithm which provides faster processing capability by using block-based matching. Our algorithm employs block-based matching followed by pixel-based matching. By applying block-based matching hierarchically, representative disparity values are assigned to each block. With these rough results, dense disparity map is acquired under very short search range. This procedure can reduce processing time greatly. We test our matching algorithms for various types of images, and shall show good performance of our stereo matching algorithm in both running time and correctness' aspect.

1 Introduction

The stereo matching algorithm is a technique that investigates two or more images captured at diverse view points in order to find positions in real 3D space for the pixels of 2D image. For some vision systems, such as optical motion capture, stereo image coding, and autonomous vehicle navigation, the stereo matching methods can be applied and they are requested to get good performance in both processing time and correctness.

Stereo matching algorithms can be generally classified into two methods: feature-based and area-based ones. feature-based methods match the feature elements between two images, and use interpolation techniques to obtain disparity for each pixel in the image, while area-based methods perform matching between pixels in two images by calculating correlations of the pixels residing in a search window. Area-based methods cannot match feature element with more accuracy than feature-based methods even though they can make dense disparity map[1, 2,3]. Moreover, they have more possibility of error in the area of insufficient texture information or depth discontinuities.

In this paper we present a new technique for area-based stereo matching algorithm which provides faster processing capability by using block-based matching in a coarse-to-fine manner. More accurate and error-prone dense disparity map

---

* This work was partially supported by the Brain Korea 21 Project and KIPA-Information Technology Research Center.
** corresponding author

can be acquired fast through pieces of information from block-based matching. We test our matching algorithm with various types of images including real and synthetic images, and shall show good performance of our stereo matching algorithm in both running time and correctness' aspect.

The paper is organized as follows. Section 2 briefly describes the existing stereo matching algorithms and compares each approach. Section 3 presents our fast stereo matching algorithm using block similarity. Section 4 shows the experimental results of our matching algorithm. Finally, section 5 gives conclusions.

2 Stereo Matching Methods

Stereo matching methods are classified into area-based and feature-based matching. Area-based method uses the correlation of intensity between patches of two images to find, for each pixel in one image which is called reference image, its corresponding pixel in the other image which is called target image based on the assumption that disparity values for the pixels in the neighborhood region of one pixel are nearly constant. It can match each pixel in the reference image to its corresponding pixel in the target image with maximum correlation, minimum sum of squared difference(SSD), or similar measures between two windows in the left and right images[4,5,6]. They can generate dense disparity map directly. However, they are sensitive to noise and variation between images, and have a high probability of error in the area where there are insufficient texture information or depth discontinuities.

Feature-based method extracts primitives such as corners, junctions, edges, or curve segments as candidate features for matching by finding zero-crossings, gradient peaks, and so on[7,8,9,10]. Feature-based method is more accurate and less sensitive to photometric variations than area-based method since matching results of detected feature elements are more precise. However, it requires additional interpolation for sparse feature elements, and has some difficulties in deriving a general algorithm applicable to diverse types of images.

Area-based methods are known to be slow because they require window calculation for each pixel. Feature-based methods are not fast either, considering additional calculation such as feature detection and interpolation for acquiring a dense disparity map. Although some constraints such as epipolar, continuity, or ordering are used to reduce calculation, the performance of existing methods is not so fast sufficiently as to be used for real-time applications[11]. So specially designed algorithm is required to get a dense disparity map in a short time.

3 Fast Stereo Matching Using Block Similarity

Our algorithm employs block-based matching followed by pixel-based matching as shown in figure 1. At the block-based matching stage, a representative disparity value per each block is acquired roughly in short time by using a block

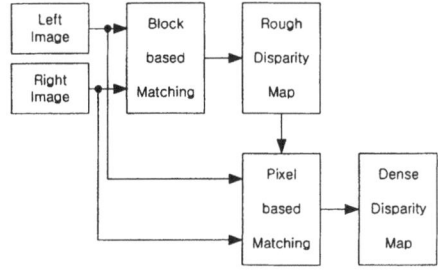

Fig. 1. The overall procedure of our algorithm

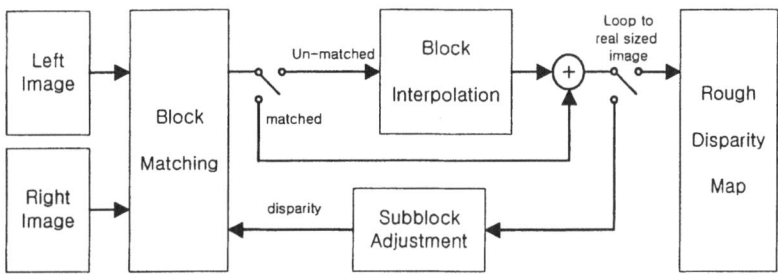

Fig. 2. Diagram of Block-based Matching

matching in a hierarchical framework. At the pixel-based matching stage, a subsequent area-based matching per each pixel gives a reliable dense disparity map with less computation using pieces of information from block-based matching.

3.1 Block-Based Matching

The overall procedure of block-based matching section is shown in figure 2. For a hierarchical matching, input stereo pair is down-sampled to the size of $2n \times 2(n + \alpha)$ where n is the size of a side of a block for matching and α is an arbitrary number for fitting blocks into an image. Then the down-sampled reference image which is one of the stereo pair is divided into non-overlapping $n \times n$ blocks. Each divided block in the reference image is tested with same sized areas in the target image which is the other of stereo pair. For each block in the reference image, a representative disparity $d_{l,i,j}$ which is a distance between correct matching blocks for (i,j)th block of level l is assigned where i and j represent row and column indices of divided image, respectively.

On the assumption that the stereo pair is rectified already, disparity value is surveyed along the identical horizontal scan line by epipolar constraint. At the lowest level, the search range in the target image is confined to $[0,+w]$ and disparities of the rightmost blocks in the reference image $d_{0,1,j}$ are assigned to the same value of the left side block without search. Now that corresponding pixels are shifted consistently in the lowest level image. In other higher levels,

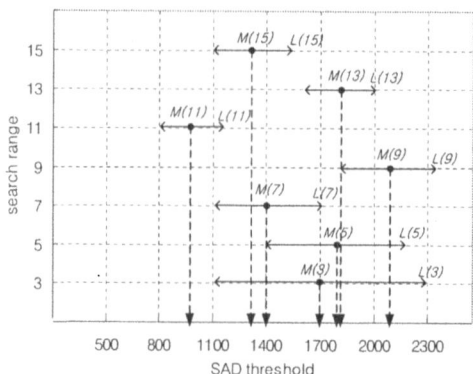

Fig. 3. Selection of Optimal threshold and search range

the search range is set to $[-w,+w]$. However, by the ordering constraint, left search range is constrained not to exceed the matched position of previous left block. When a block is an occlusion or horizontally adjacent blocks do not belong to the same object, the ordering constraint may not apply. In this case, the block is labelled as an *un-matched* block and processed in subblock adjustment step in figure 2. The center of search range is selected as $2D_{l-1,\lfloor i/2 \rfloor,\lfloor j/2 \rfloor}$ which is double of the subtotal disparity from previous image levels. The subtotal disparity of current level $D_{l,i,j}$ is can be written as follows: $D_{l,i,j} = 2D_{l-1,\lfloor i/2 \rfloor,\lfloor j/2 \rfloor} + d_{l,i,j}$.

Among the candidate blocks in the target image, one block which shows maximum correlation is matched with a block in the reference image. We used the sum of absolute differences(SAD) estimate as a formula for area comparison because it has been reported that the SAD estimate is superior to other measures in both execution time and quality of the output disparity map[12]. The formula of SAD is defined as:

$$SAD(i,j,d) = \sum_{m=i-K}^{i+K} \sum_{n=j-L}^{j+L} |r_{m,n} - t_{m+d,n}| \qquad (1)$$

where $r_{m,n}$ and $t_{m,n}$ are the intensity values of reference image r and target image t at position (m,n), respectively. K and L define the size of matching window and d is the amount of window shift in the target image along a scan line. In our algorithm, if correlations in all search range are not greater than selected threshold or they have same values, the block in the reference image is labelled as *un-matched*. Usually, First situation is regarded as an occlusion, and second is usually come from insufficient texture information.

Concerning SAD threshold and search range w, we exploited a statistical strategy. When SAD threshold is varied under a certain camera condition, an interval where the variation of maximum and minimum disparity outputs are constant can be found. The optimal threshold is similarly approximated in the stable interval because error-free results are made in these conditions. Threshold and search range are changed discretely for time efficiency, and maximum and minimum disparities are recorded for each case. From this record, optimal SAD threshold and search range are calculated from the following formulas:

$$Optimal\ threshold \simeq \frac{\sum_{i \in search\ range} L(i)M(i)}{\sum_{i \in search\ range} L(i)} \qquad (2)$$

$$Optimal\ search\ range \simeq \lceil \frac{\sum_{i \in search\ range} iL(i)}{\sum_{i \in search\ range} L(i)} \rceil \qquad (3)$$

where $L(i)$ and $M(i)$ are the length and the center value of the major(longest) stable interval for each search range i, respectively. Figure 3 shows the example of selection of optimal threshold and search range. Bars in graph represent major stable interval for each search range. If sampling step for SAD threshold is 100, $L(3)$ is 13 and $M(3)$ is 1700. In the same manner, optimal values are selected with all measurements.

After disparity assignment for all blocks in current level is completed, disparity values of un-matched blocks are interpolated using matched left and right neighbor blocks. For fast estimation, linear interpolation is used. If one or more neighbors do not exist because of image boundary, double value of the subtotal disparity of previous level $2D_{l-1,\lfloor i/2 \rfloor, \lfloor j/2 \rfloor}$ is used as a disparity of corresponding position.

If the level of stereo images is not reached to the original size yet, block-based matching procedures after up-sampling are repeated in a hierarchical manner. Before entering block matching for next level, subblock adjustment is required because the results from block matching and block interpolation are so blocky that it may produce errors. By continuity constraint, disparity is smoothed by:

$$D_{l,i,j} = \begin{cases} (D_{l,i,j} + D_{l,i+1,j})/2, & \text{if even column} \\ (D_{l,i-1,j} + D_{l,i,j})/2, & \text{if odd column} \end{cases} \qquad (4)$$

Block-based matching produces a rough disparity map which has one representative disparity value in a $n \times n$ sized block. As depth information can be estimated fast, block-based matching can be applied usefully to such applications as need real time processing. Our algorithm omits reverse order matching for fast estimation, however, it also produces reasonable results.

3.2 Pixel-Based Matching

This stage is similar to conventional area-based matching. The original stereo pair is used for pixel-based matching. However, by help of the rough disparity map from block-based matching, very short search range is enough to get correct a dense disparity map. Prior to pixel-based matching, blocks residing in the interior of object which has smooth surface or containing plain texture are selected to prevent unnecessary matching and reduce calculation. In case of sequence of blocks having similar disparity along the horizontal order, inner

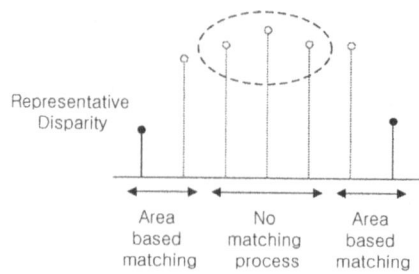

Fig. 4. Object Search

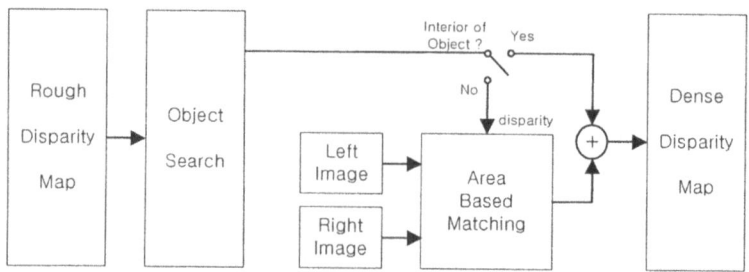

Fig. 5. Diagram of Pixel-based Matching

blocks are excluded from pixel-based matching and the representative disparity values from block-based matching are settled as disparities for every pixel in the corresponding blocks. As shown in figure 4, blocks in dashed line are excluded from area-based matching. They are interior of grayed sequence considered as an object. For pixels in the rest blocks, area-based matching is processed. The size of matching window adopted in this step is $n \times n$ and SAD estimate is used for area comparison also. Because disparity from block-based matching is close to correct value, even the search range which does not exceed the length of side of block n is enough for pixel-based matching. The same measure with block-based matching and very short search range help reducing calculation. The overall procedure of pixel-based matching section is shown in figure 5.

4 Experimental Results

Performance evaluations with some real and synthetic images for our stereo matching algorithm are presented. 512×512 sized images are used for test and 8×8 sized block is used in block-based matching and same sized block is used as a correlation window in pixel-based matching. The quality of disparity map is visually evaluated because there are few standard results to my knowledge. The processing time has been evaluated on a P4-2GHz system.

Figure 6 shows stereo matching results for a pentagon stereo pair. Figure 6(a) is the right image which is used as a reference image. The left image which

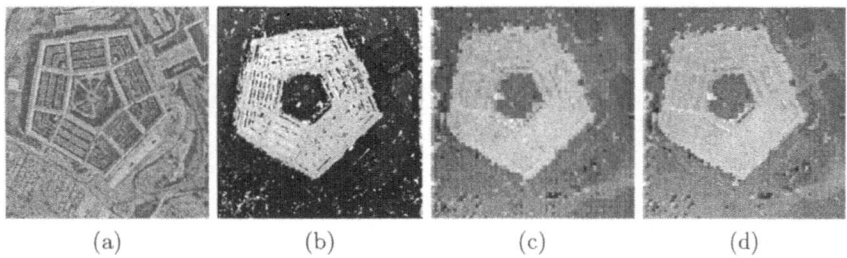

Fig. 6. Stereo matching results for pentagon pair (real image) (a) Right image (b) Disparity map of traditional area-based matching algorithm (c) Disparity map of our block-based matching without pixel-based matching (d) Disparity map of our algorithm

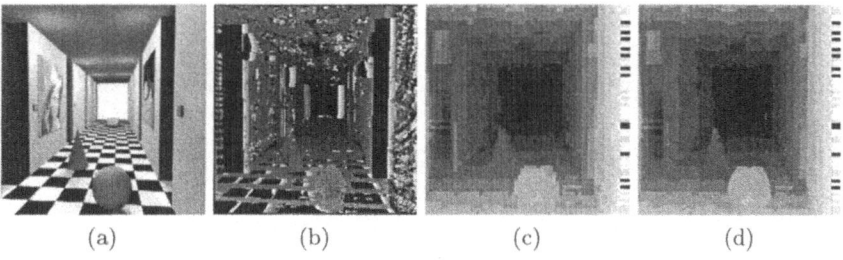

Fig. 7. Stereo matching results for corridor pair (synthetic image) (a) Right image (b) Disparity map of traditional area-based matching algorithm (c) Disparity map of our block-based matching without pixel-based matching (d) Disparity map of our algorithm

is similar but slightly different compared with the right image is used as a target image. Figure 6(b) is a disparity map obtained by the conventional area-based matching algorithm with search range and search window fixed to 30 pixels and 7×7, respectively. The brighter pixel in a disparity map means the closer one to the camera. SAD estimate is used in conventional algorithm to compare with ours. Figure 6(c) shows the disparity map obtained by our block-based matching without pixel-based one. In case block-based matching is used only, matching can be done very fast compared with traditional approaches. In real time applications which are sufficient with rough disparity information, block-based matching can be used only. When there is a need for refined dense disparity map, pixel-based matching is applied. In our experiments, the rough disparity map shown in figure 6(c) was obtained in 0.32 sec and the dense disparity map shown in figure 6(d) was obtained in 1.38 sec. However it took the conventional method 6.58 sec to get a dense disparity map. These results show that time efficiency of our algorithm is excellent. According to our strategy, 9 pixels are chosen as search range in each level for block matching and 1777 is selected as the threshold of SAD for determining matched or un-marched block in block-based matching. Search range for pixel matching is set to 9 pixels to cover the size of block.

Fig. 8. Stereo matching results for apple circuit pair (real image) (a) Right image (b) Disparity map of traditional area-based matching algorithm (c) Disparity map of our block-based matching without pixel-based matching (d) Disparity map of our algorithm

Fig. 9. Stereo matching results for cone pair (synthetic image) (a) Right image (b) Disparity map of traditional area-based matching algorithm (c) Disparity map of our block-based matching without pixel-based matching (d) Disparity map of our algorithm

Figure 7(a) is a right image of a corridor stereo pair. In case there is short texture information like these synthetic images, large plain part is susceptible to serious mismatch as shown in figure 7(b) which is a disparity map from traditional area-based algorithm. This type of error can be corrected by using a hierarchical property during the step of determining disparity. Figure 7(c) is a disparity of our block-based matching and figure 7(d) is a result of our whole algorithm. Correction effect for plain texture is shown well in these two results. Conditions applied in this experiment are same with those of figure 6 except that the SAD threshold is 1600.

Various types of stereo images were tested as shown in Figure 8 to 11. Because other constraints are not applied in the traditional method during matching, disparity map is corrupted seriously. Short search range in each level of our algorithm suppresses a mismatch problem in complex texture as shown in result images. Search range and threshold for block matching are described in table 1 and other conditions are same with previous experiments.

Finally, our algorithm is compared with conventional area-based matching algorithm in view of processing time. Three real and three synthetic stereo image pairs are tested under previously mentioned condition. As shown in table 1, our

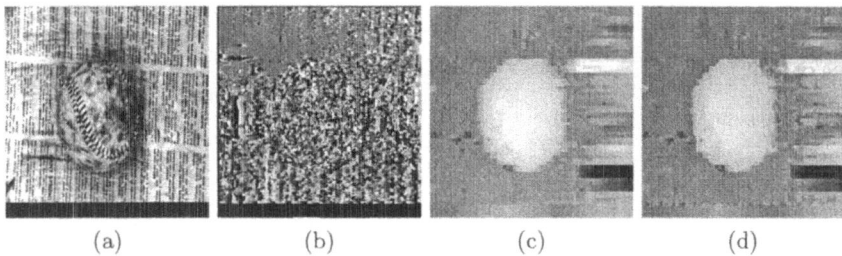

Fig. 10. Stereo matching results for ball pair (real image) (a) Right image (b) Disparity map of traditional area-based matching algorithm (c) Disparity map of our block-based matching without pixel-based matching (d) Disparity map of our algorithm

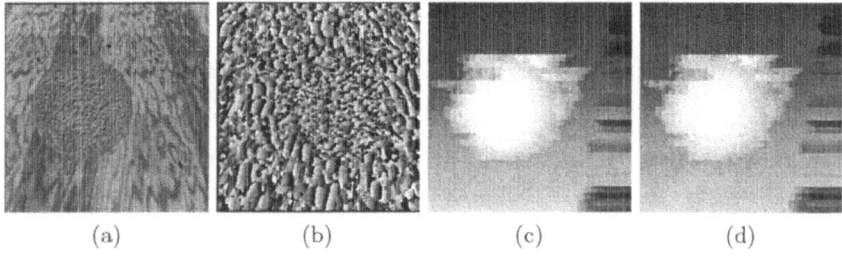

Fig. 11. Stereo matching results for sphere pair (synthetic image) (a) Right image (b) Disparity map of traditional area-based matching algorithm (c) Disparity map of our block-based matching without pixel-based matching (d) Disparity map of our algorithm

algorithm has very good capability in speed. Although tested images are pretty large, the average processing time in block-based matching is about 0.32 sec and that in whole matching is about 1.24 sec.

5 Conclusions

In this paper, we have proposed a fast stereo matching algorithm by finding a representative disparity value to each divided block. Rough disparity map from block-based matching helps reducing search range in pixel-based matching. We have tested our matching algorithms for real and synthetic images and good performance in processing time has been obtained with reasonable disparity map. Using of regular blocks enables our algorithm to be expanded to a parallel and distributed environment easily. Also it can be combined with other applications which require fast matching time such as optical motion recognition, autonomous vehicle navigation, and stereo image coding[13], etc. The applicable areas of our algorithm are very critical for future improvements in stereo applications. As a future work, we are going to upgrade the quality of depth results and adapt our algorithm to parallel and distributed computing environment.

Table 1. Processing Time for 512×512 sized image

Image name	Search range w	SAD threshold	Conventional	Block	Block+Pixel
Apple	9	1766	6.46 sec	0.30 sec	1.64 sec
Pentagon	9	1777	6.58 sec	0.32 sec	1.38 sec
Ball	9	1811	6.50 sec	0.32 sec	1.15 sec
Cone	11	1333	6.62 sec	0.33 sec	1.00 sec
Corridor	9	1600	6.50 sec	0.32 sec	1.28 sec
Sphere	11	1924	6.51 sec	0.33 sec	1.01 sec
Average time			6.53 sec	0.32 sec	1.24 sec

References

1. S. D. Cochran, G. Medioni, "3-D Surface Description From Binocular Stereo," IEEE Transactions on Pattern Analysis and Machine Intelligence, vol. 14, no. 10, pp. 981-994, Oct. 1992.
2. G. Wei, W. Brauer, and G. Hirzinger, "Intensity- and Gradient-Based Stereo Matching Using Hierarchical Gaussian Basis Functions," IEEE Trans. PAMI, vol. 20, no. 11, pp. 1143-1160, Nov. 1998.
3. H. S. Koo and C. S. Jeong, "An Area-Based Stereo Matching Using Adaptive Search Range and Window Size," Lecture Notes in Computer Science, vol. 2074, pp. 44-53, May. 2001.
4. M. Okutomi and T. Kanade, "A Multiple-Baseline Stereo," IEEE Trans. PAMI, vol. 15, no. 4, pp. 353-363, Apr. 1993.
5. T. Kanade and M. Okutomi, "A Stereo Matching Algorithm with an Adaptive Window: Theory and Experiment," IEEE Trans. PAMI, vol. 16, no. 9, pp. 920-932, Sep. 1994.
6. C. Sun, "Fast Stereo Matching Using Rectangular Subregioning and 3D Maximum-Surface Techniques," International Journal of Computer Vision, vol. 47, pp. 99-117, 2002.
7. K. L. Boyer and A. C. Kak, "Structural Stereopsis for 3-D Vision," IEEE Trans. PAMI, vol. 10, no. 2, pp. 144-166, Mar. 1988.
8. W. Hoff and N. Ahuja, "Surface From Stereo: Integrating Feature Matching, Disparity Estimation and Contour Detection," IEEE Trans. PAMI, vol. 11, no. 2, pp. 121-136, Feb. 1989.
9. Shing-Huan Lee and Jin-Jang Leou, "A dynamic programming approach to line segment matching in stereo vision," Pattern Recognition, vol. 27, no. 8, pp. 961-986, 1994.
10. S. B. Maripane and M. M. Trivedi, "Multi-Primitive Hierarchical(MPH) Stereo Analysis," IEEE Trans. PAMI, vol. 16, no. 3, pp. 227-240, Mar. 1994.
11. O. Faugeras, "Three-Dimensional Computer Vision: A Geometric Viewpoint," pp. 169-188, Massachusetts Institute of Technology, 1993.
12. B. Cyganek and J. Borgosz, "A Comparative Study of Performance and Improvementation of Some Area-Based Stereo Algorithms," LNCS, vol. 2124, pp. 709-716, 2001.
13. H. S. Koo and C. S. Jeong, "A Stereo Image Coding using Hierarchical Basis," LNCS, vol. 2532, pp. 493-501, Dec. 2002.

View Morphing Based on Auto-calibration for Generation of In-between Views

Jin-Young Song, Yong-Ho Hwang, and Hyun-Ki Hong

Dept. of Image Eng., Graduate School of Advanced Imaging Science, Multimedia and Film, Chung-Ang Univ., 221 Huksuk-dong, Dongjak-ku, Seoul, 156-756, KOREA
Song74jy@hanmail.net , hwangyongho@hotmail.com,
honghk@cau.ac.kr

Abstract. Since image morphing methods do not account for changes in viewpoint or object pose, it may be difficult even to express simple 3D transformations. Although previous view morphing can synthesize change in viewpoint, it requires camera viewpoints for automatic generation of in-between views and control points by user input for post-warping. This paper presents a new morphing algorithm that can generate automatically in-between scenes by using auto-calibration. Our method rectifies two images based on the fundamental matrix, and computes a morph that is linear interpolation with a bilinear disparity map, then generates in-between views. The proposed method has an advantage that knowledge of 3D shape and camera settings is unnecessary.

1 Introduction

Many studies have been proposed to build up 3D world for virtual reality systems up to present. Since real images are used as input, image-based approaches have an advantage in producing photorealistic scenes and independency of scene complexity. Specifically, Image-Based Rendering (IBR) aims at arbitrary view rendering with real images instead of reconstructing 3D structure [1].

Although previous image morphing methods can generate 2D transitions between images, they do not ensure that the resulting transitions appear natural. In order to overcome the limitation, Seitz presented view morphing to create the illusion that the object moves rigidly between its positions in the two images [2]. However, it requires camera viewpoints for automatic generation of in-between views and control points by user input for post-warping. In addition, the generated in-between views are much affected by self-occlusion.

This paper presents an automatic morphing method that can generate in-between scenes from un-calibrated images. Our algorithm rectifies images based on the fundamental matrix, which is followed by linear interpolation with bilinear disparity map. In final, in-between views are generated by inverse mapping of homography of the rectified images. The proposed method can be applied to photographs and drawings, because knowledge of 3D shape and camera settings is unnecessary. The in-between views may be used in various application areas such as virtual environment, image communication. Fig. 1 shows block diagram for automatic view morphing algorithm.

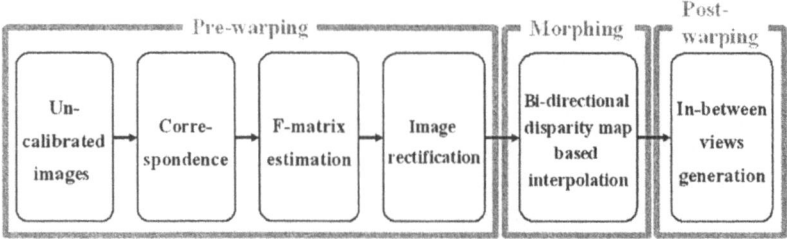

Fig. 1. Automatic view morphing algorithm

The remainder of this paper is structured as follows: Sec. 2 reviews previous methods to generate an arbitrary view, and Sec. 3 describes the proposed algorithm with three steps. After comparisons of experimental results on synthetic and real scenes are given in Sec. 4, the conclusion is described in Sec. 5.

2 Previous Methods

Table 1 summarizes previous methods to generate an arbitrary view, which are classified into two approaches. One is view synthesis with real images, and the other uses additional information such as 3D depth data, camera parameters, etc. As shown in Table 1, most of studies using images generate generally views with limited viewpoints or have assumption about the scenes. In addition, researches for an arbitrary view require prior information of the depth or scene structure.

Even though existing image morphing techniques can generate 2D transitions between images, differences in object pose or viewpoint often cause unnatural distortions. On the contrary, view morphing proposed by Seitz allows current image morphing methods to easily synthesize changes in viewpoint without 3D information. View morphing consists of pre-warping two images, computing a morph between the pre-warped images, and then post-warping each in-between image produced by the morph.

This paper presents a new view morphing algorithm based on auto-calibration from un-calibrated sequences. Our method can generate automatically in-between images without additional information such as camera parameters, scene structures, and user input. Existing morphing methods warp the two original images and average the pixel colors of the warped images, so unwanted smoothing effects may be occurred. Because a bilinear disparity map from dense correspondence is used, we can determine the pixel value of the in-between view effectively and alleviate the effects by occlusion.

Table 1. Previous studies for views generation

Input		Methods	Requirements
Images	S. E. Chen [1]	- mosaics by registering images - views generated by panning	- difficulty making an arbitrary view
	S. M. Seitz [2]	- view morphing for in-between images	- user input - sensitivity to changes in visibility
	N. L. Chang [3]	- 3D representation for uncalibrated sequence	- set-up of camera motion and positions for dense depth maps
	D. Scharstein [4]	- warping image by stereo system	- calibration for stereo system - occlusion problem
	X. Sun [5]	- linear interpolation & rectification based on epipoles	- monotonicity constraints in the scene
	C. R. Dyer [6]	- view morphing for sequence having a moving rigid object	- user input for layering of the moving object - assumption about object motion
Images & 3D information	H. Huang [7]	- forward/backward morphing in 3D space for occlusion problems	- disparity map & camera parameters
	S. E. Chen [8]	- in-between views generation from rendered images	- camera parameters & 3D datum for correspondence
	W. R. Mark [9]	- views generation by 3D warping synthesis & reference images	- a prior information of 3D warping & rendering set-up

3 Automatic View Morphing

3.1 Pre-warping Based on Fundamental Matrix

Epipolar geometry is a fundamental constraint used whenever two images of a static scene are to be registered. Given a point in one image, corresponding point in the second image is constrained to lie on the epipolar line. As shown in Fig. 2, epipoles are the points at which the line through the centers of projection intersects the image planes. All the epipolar geometry is contained in the fundamental matrix (F) and the epipolar constraint can be written as:

$$x'Fx = 0, \qquad (1)$$

where x and x' are the homogeneous coordinates of two corresponding points in the two images. The fundamental matrix that has rank 2, and since it is defined up to a scale factor, there are 7 independent parameters. The fundamental matrix contains the intrinsic parameters and the rigid transformation between both cameras [10].

Given a pair of stereo images, prewarping called as rectification, determines a transformation of each image such that pairs of conjugate epipolar lines become collinear and parallel to the horizontal axis of the image. By rotating cameras around their optical centers until the baseline is located on a plane (Π_R), we can define projection matrices, H_1 and H_2, which satisfy the relation: $\Pi_1 = [H_1 | -H_c C_1]$ and $\Pi_2 = [H_2 | -H_2 C_2]$. Epipoles of the rectified images are on vanishing plane, and a point (i) on the vanishing space satisfies $i = [1 \ 0 \ 0]^T$. The fundamental matrix of the rectified image can be defined:

$$\bar{F} = [i]_x = \begin{bmatrix} 0 & 0 & 0 \\ 0 & 0 & -1 \\ 0 & 1 & 0 \end{bmatrix}, \tag{2}$$

where i_x is un-symmetric matrix.

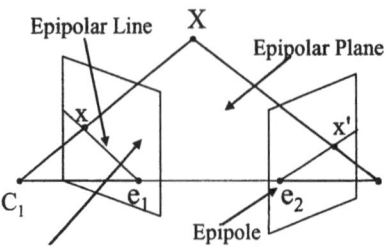

Fig. 2. Epipolar geometry

Corresponding points (x_1 and x_2) on input images, I_1 and I_2, are transformed into $\bar{x}_1$ and $\bar{x}_2$ on the rectified images, $\bar{I}_1$ and $\bar{I}_2$, by Eq. (3). By combining these equations, we can derive Eq. (4) as follows:

$$\bar{x}_1 = H_1 x_1, \quad \bar{x}_2 = H_2 x_2. \tag{3}$$

$$\bar{x}_2^T \bar{F} \bar{x}_1 = 0, \quad x_2^T H_2^T \bar{F} H_1 x_1 = 0,$$

$$F = H_2^T [i]_x H_1. \tag{4}$$

A plane Π_R is parallel to the baseline of C_1 and C_2 as shown in Fig. 3. H_1 and H_2 are obtained by using minimization methods because there is no unique solution, and then we can rectify two images [11].

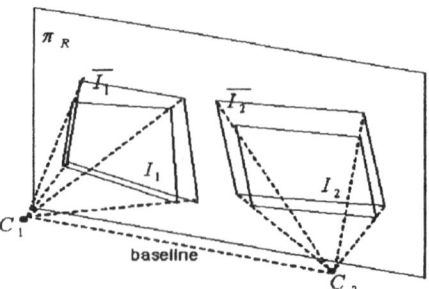

Fig. 3. Rectification plane

3.2 Interpolation by Using Bi-directional Map

Let a point P on 3D space be projected into p_1 and p_2 on two images, respectively. If a dense correspondence on two images is established [12,13], we can use a point disparity for in-between views generation without 3D information. Since pre-warping brings the image planes into alignment, corresponding points are on the same scanline. The disparities on the rectified images have just horizontal direction as,

$$p_2 = \begin{bmatrix} x_2 \\ y_2 \end{bmatrix} = \begin{bmatrix} x_1 + d_{L \to R}(x_1, y_1) \\ y_1 \end{bmatrix} = p_1 + \begin{bmatrix} d_{L \to R}(x_1, y_1) \\ y_1 \end{bmatrix}, \quad (5)$$

where $d_{L \to R}(x_1, y_1)$ is the disparity of corresponding points from the left image to the right.

We normalize the distance between two corresponding points into 1, and generate in-between views as a distance weight function of $s \in [0,1]$. Eq. (6) represents a disparity function in-between views from the left image, and the disparity from the right is obtained in a similar way. Since the point corresponding to (x_1, y_1) on the left image can be found along the scanline ($y = y_1$) in the right, their y-axis coordinates are same as,

$$p_s = \begin{bmatrix} x_s \\ y_s \end{bmatrix} = \begin{bmatrix} x_1 + s \cdot d_{L \to R}(x_1, y_1) \\ y_1 \end{bmatrix}. \quad (6)$$

Changes in visibility may result in folds or holes on the interpolated image. A fold occurs in an in-between image ($\bar{I}_s$) when a visible surface in an input image becomes occluded in $\bar{I}_s$. More specifically, multiple pixels of the input image map to the same point in $\bar{I}_s$, causing an ambiguity. For the rectified images, Z-buffer techniques are used to solve occlusion problem [2]. On the contrary, when an occluded surface suddenly becomes visible, a hole is found. In this case, we can not establish correspondences of $\bar{I}_s$ and the input image, so the hole is interpolated generally with neighboring pixels. However, it is difficult to interpolate the hole having a large disparity. Bi-directional map is used to alleviate the unwanted hole effects [14]. The pixel values of the hole on the image are found on the other image, so we can determine those as following:

$$IP_s(x,y) = \begin{bmatrix} (1-s) \cdot IP_{s,L \to R}(x,y) + s \cdot IP_{s,R \to L}(x,y) & \text{if } IP_{s,L \to R} \neq 0, IP_{s,R \to L} \neq 0 \\ IP_{s,L \to R}(x,y) & \text{if } IP_{s,L \to R} \neq 0, IP_{s,R \to L} = 0 \\ IP_{s,R \to L}(x,y) & \text{if } IP_{s,L \to R} = 0, IP_{s,R \to L} \neq 0 \\ \text{nearest neighbor pixel} & \text{if } IP_{s,L \to R} = 0, IP_{s,R \to L} = 0 \end{bmatrix}, \quad (7)$$

where $IP_s(x,y)$ is the dissolved pixel value, $IP_{s,L \to R}(x,y)$ and $IP_{s,R \to L}(x,y)$ are the interpolated pixel value from the left image to the right and vice-versa, respectively. When all of the interpolated values are non-zero, Eq. (7) computes an average by using morphing weight (s). In the case that one of the interpolated is zero, we determine non-zero value as $IP_s(x,y)$. In addition, when all of the interpolated are zero, the hole is interpolated with neighboring pixels. Fig. 4 shows the interpolated image by one-directional and that by bi-directional disparity map. Comparison of the magnified images of (a) and (b) represents that the use of bi-directional map can achieves better performances.

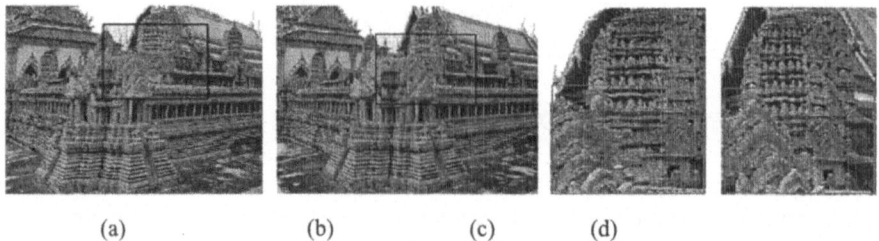

Fig. 4. Interpolation images by (a) one-directional and (b) bi-directional disparity map. (c) and (d) magnified images of (a) and (b)

3.3 Post-warping

Since the rectified image is rotated around optical centers of cameras, we can obtain the homography matrix. The viewpoint of the interpolated image ($\bar{I}_s$) is different from that of the original images. Post-warping transforms the image plane of a new view to its desired position and orientation, so we can generate in-between views ($I_{\text{in-between}}$). Eq. (8) provides the rectified plane (Π_S) and the projective matrix for in-between views from two images. The interpolated images are transformed into in-between views along a natural camera path by Eq. (9).

$$\Pi_s = (1-s)\cdot\Pi_1 + s\cdot\Pi_2, \quad H_s = (1-s)\cdot H_1 + s\cdot H_2 . \tag{8}$$

$$I_{in-between} = H_s^T \bar{I}_s . \tag{9}$$

4 Experimental Results

We have experimented on the synthetic images to evaluate accuracy of in-between views. After setting up camera projective parameters and 3D object on simulation, we generate an image sequence of 3D data by MAYA Ver. 4.5. Fig. 5 shows rendered images and in-between view by using one-directional and that by bi-directional disparity map, respectively. PSNR (Peak Signal/Noise Ratio) of in-between views by two methods and the rendered image are compared in Fig. 6. PSNR is an objective quality measurement between two images, and is given by:

$$PSNR = 10\log_{10}(\frac{255^2}{MSE})[dB],$$

$$MSE = \frac{1}{N^2}\sum_{x=1}^{N}\sum_{y=1}^{N}[f_{original}(x,y) - f_{in-between}(x,y)]^2. \qquad (10)$$

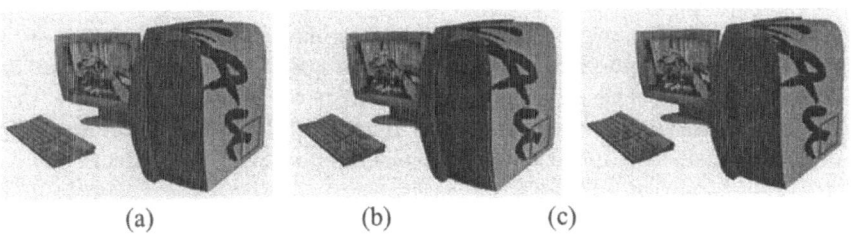

Fig. 5. PSNR comparison (a) original image, (b) and (c) in-between images by using one-directional and bi-directional disparity map

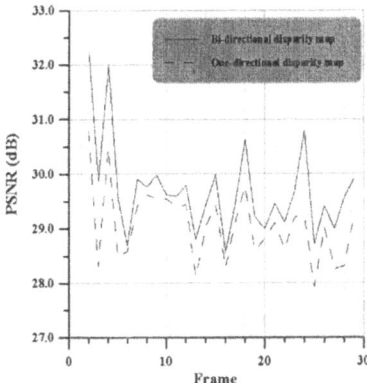

Fig. 6. PSNR comparison on image sequence

These results represent that in-between view by using bi-directional disparity map is more similar to the original image than one-directional disparity map in qualitative as wall as quantitative measure. Although the frame rate of the input sequence is lower than that of image sequences used generally in image communication, PSNR comparison showed that our algorithm achieve relatively good results. In other words, a wide baseline between viewpoints causes it to fail to establish correspondence. Therefore, the proposed method can achieve better performance in the case of the sequence with higher frame rate. In Fig. 6, we can see that PSNR results are affected by the self-occlusion between frames, and the use of the bi-directional map can much more alleviate these effects.

We have simulated on the real images (Fig. 7) captured by Cannon G-2, and the size of each image is 800×600. Fig. 8 shows pre-warped images that are rectified by using the fundamental matrix. Since the correspondence may not be one-to-one, we use Z-buffer techniques to solve occlusion problem at the scanline for parallel views, and alleviate the unwanted hole effects by using bi-directional disparity map. Post-warped 10 frames, in-between views, are represented in Fig. 9.

In order to check accuracy of the generated view, we have experimented on the calibrated turn-table. Fig. 10 presents comparison of real images and the in-between view. (a) and (b) are two input images captured by rotating the turn-table. In addition, (c) and (d) are the real image captured from the middle of two viewpoints and the generated in-between view, respectively. The experimental results show that we can generate a precise in-between view without prior information of rotation angle of turntable and camera parameters. The proposed algorithm can generate considerably many realistic in-between views from just a few frames, which are applicable to various areas such as virtual environment, image communication, and digital effects.

5 Conclusions

This paper presents an automatic view morphing algorithm that can generate inbetween scenes based on auto-calibration. The proposed method requires neither knowledge of 3D shape nor camera settings, and alleviates the hole effects by using bi-directional disparity map. Experimental results showed that our method can generate precise in-between views from two frames in qualitative as wall as quantitative measure. Further study will include real-time imaging system development for virtual environment and various applications to image communication.

Fig. 7. Input images

Fig. 8. Pre-warped images

Fig. 9. Comparison or real images and the in-between view (a) left and (b) right input images (c) real image by camera (d) the generated in-between view

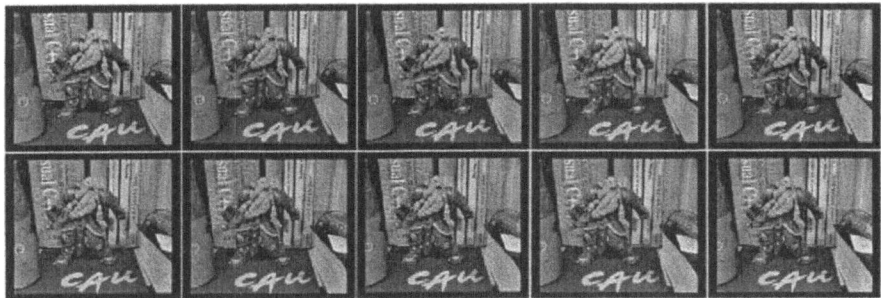

Fig. 10. In-between views (10 frames)

Acknowledgment. This research was supported by the Ministry of Education, Korea, and under the BK21 project, and the Ministry of Science and Technology, Korea, under the NRL(2000-N-NL-01-C-285) project.

References

1. S. E. Chen, "Quicktime VR — An image-based approach to virtual environment navigation," *Proc. SIGGRAPH*, (1995) 29-38
2. S. M. Seitz and Charles R. Dye, "View morphing," *Proc. SIGGRAPH*. (1996) 21-30
3. N L. Chang and A. Zakhor, "View generation for three-dimensional scenes from video sequences," *IEEE Transactions on Image Processing*, vol. 6, no. 4, April (1997) 584 -598
4. D. Scharstein, "Stereo vision for view synthesis," *Proc. IEEE Conference on Computer Vision and Pattern Recognition*, San Francisco, June (1996) 852-858
5. X. Sun and E. Dubois, "A method for the synthesis of intermediate views in image-based rendering using image rectification," *Proc. of the Canadian Conference on Electrical & Computer Engineering*, (2002) 991-994
6. R. A. Manning and C. R. Dyer, "Interpolating view and scene motion by dynamic view morphing," *Proc. IEEE Conference on Computer Vision and Pattern Recognition*, Fort Collins, Colorado, June (1999) 125-131

7. H. Huang, S. Nain, Y. Hung, and T. Cheng, "Disparity-based view morphing — a new techniques for image-based rendering," *Proc. SIGGRAPH,* (1998) 9-16
8. S. E. Chen and L. Williams, "View interpolation for image synthesis," *Proc. SIGGRAPH 93,* (1993) 279-288
9. W. R. Mark, L. McMillan, and G. Bishop, "Post-rendering 3D warping," *Proc. Symposium on Interactive 3D Graphics,* Providence, RI, April (1997) 7-16
10. R. Hartley, "In defense of the eight-point algorithm," *IEEE Transactions on Pattern Analysis and Machine Intelligence,* vol. 19, June (1997) 580 -593
11. Z. Zhang, "Computing rectifying homographies for stereo vision," *Proc. IEEE Conference on Computer Vision and Pattern Recognition,* Fort Collins, Colorado, June (1999) 125-131
12. C. Tsai and A. K. Katsaggelos, "Dense disparity estimation with a divide-and-conquer disparity space image technique," *IEEE Transactions on Multimedia,* vol. 1, no. 1, (1999) 18-29
13. S. S. Intille and A. F. Bobick, "Disparity-space images and large occlusion stereo," *2nd European Conference on Computer Vision,* Stockholm Sweden, (1994) 179-186
14. D. Kim and J. Choi, "Bidirectional disparity morphing considering view ambiguity," *Proc. Frontiers of Computer Vision,* (2003) 31-36

Immersive Displays Based on a Multi-channel PC Clustered System

Hunjoo Lee[1] and Kijong Byun[2]

[1]Electronics and Telecommunications Research Institute,
161 Gajeong-Dong, Yuseong-Gu, Daejeon, 305-350, Republic of Korea
hjoo@etri.re.kr
[2]Samsung Electronics
416 Maetan-Dong, Yeongtong-Gu, Suwon, 443-370, Republic of Korea

Abstract. As virtual reality technologies have become feasible in the market, new arcade games tend to provide a user with an immersive game environment. This paper proposes a method to generate a wide field of view game display on an arbitrary display surface. This method consists of two parts. One is a multi-channel clustered system and the other is a multi-projection display system. The multi-channel clustered system is a scalable system which generates a wide field of view game display in real time. The multi-projection display system projects this wide field of view game display on an arbitrary display surface, eliminating a distortion caused by the shape of the display surface.

1 Introduction

A conventional arcade game has typically used a single monitor as its primary display device. While a single monitor display is a cost-effective and simple solution for an arcade game, it failed to convey a feeling of immersion to its users. An introduction of virtual reality(VR) games made other display devices available such as projectors and head-mounted displays, and these devices were very effective in providing their users with immersive VR game environments.

A multi-projection display provides a wide field of view(FOV) for a user, and, unlike a head-mounted display, it is easy to use. However, applications using multi-projectors, such as a cave system and a dome system, require a huge display surface, and the shape of the display surface must be known. In order to fix this drawback and to project a wide FOV game display on an arbitrary display surface, we propose a method to render a wide FOV game display using a multi-channel clustered system of which each PC renders its uniquely assigned part of the game display and a method to project the rendered game content on an arbitrary display surface.

1.1 Multi-channel Clustered System

Since PC-based 3D accelerators have been improved their performance and their costs have become cheaper, a cluster of PCs with the accelerators is an attractive approach to build a scalable graphics system. PC graphics systems have been improved at an astounding rate over the last few years, and their price-to-performance ratios far exceed those of conventional high-end rendering systems. Therefore a multi-computer display has become increasingly important for applications such as a collaborative computer-aided design, interactive entertainment and medical visualization. A high-resolution display can visualize graphical data sets in detail, and a wide FOV display allows users to interact with virtual objects. These technical enhancements are critical to user's perception of 3D objects in an immersive display environment.

A multi-channel clustered system provides a real-time wide FOV game display to render the very complex 3D graphics objects. This system is composed of a multi-channel distribution server and visualization Clients. The multi-channel distribution server can control any number of the clients to make a wide FOV game environment at runtime. All visualization clients are synchronized to ensure visual uniformity. However, a tightly synchronized PC cluster may cause a drawback in its rendering speed[1]. So, we propose a dual scene graph management method to synchronize these PCs. A conventional scene graph is organized, and it controls its rendering process of constituent objects. And the other scene graph controls its rendering process of objects found at the edges of two adjacent channels.

1.2 Multi-projection Display System

A multi-projection display consists of an arbitrary display surface, a camera and projector pair, and a laser tracker. The camera and projector pair is used to extract an arbitrarily shaped display surface model based on a structured light technique[2], [4]. The same projector, used to extract the display surface model, is used to project a part of a game content on the display surface. The extracted display surface model is used to project a game content without a distortion caused by the shape of the display surface. The laser tracker is used to track user's position and orientation to render the game content on the user's viewing point.

2 Multi-channel PC Cluster Configuration

Our target content is a VR online game with multi-user participants. Our system is composed of multi–channel clustered PCs based on an adaptive synchronization algorithm, and a multi-monitor or a multi-projector display system. Fig. 1 shows our target system.

2.1 Multi-channel PC Cluster

A multi-channel PC cluster displays a VR online game with a wide FOV. Every PC in the cluster is dedicated to render their uniquely predefined view point area of the game scene, and these PCs are synchronized on TCP/IP or IPX network protocol. The clients and server are connected via an Ethernet network.

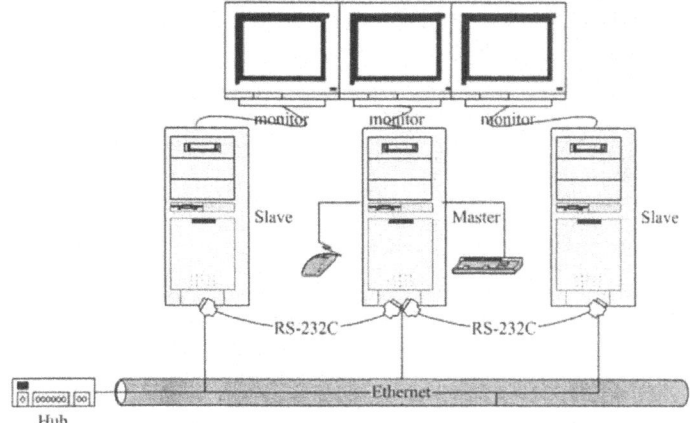

Fig. 1. System overview

In our system, VisMaster is a multi-channel distributor, and VisClient is a visualization client. VisMaster runs on all participating VisClients. This configuration is set based on each PC's predefined view point and FOV. VisMaster configures the cluster at runtime to span any number of VisClients to make a wide FOV display and to manage game events. VisMaster synchronizes the game displays rendered by VisClients. All VisClients are synchronized to ensure visual uniformity and are composed of graphic hardware and software to display the server's messages.

VisMaster controls the user event, such as changes of view point and status changes of the game objects, by using message packets. When VisMaster sends the Nth frame packet to all VisClients, they receive the packet, set the cameras and view port regions to match the Nth frame information sent in the message, and send ready messages for Nth frame back to VisMaster. Then VisMaster receives the ready messages. If the number of the messages matches to that of VisClients, VisMaster broadcasts a render ready message to all VisClients. Then VisClients display the (N-2)th frame and render the (N-1)th frames in frame buffers. Finally, at the end of the frames, VisClients exchange synchronized messages with VisMaster to swap the frame buffers. Fig. 2 shows the rendering states of the VisMaster and VisClients.

2.2 Adaptive Synchronization Algorithm

In a multi-channel PC clustered system, if one system has low-level rendering power, the overall system speed becomes very slow. Nowadays, the 3D Game provides users with more and more realistic virtual environments, resulting in an explosive growth of graphical data. A solution to this problem is to develop an efficient data management mechanism coherently supporting diverse data representations and access patterns of typical visualization calculations. Therefore, we deliberately choose an adaptive synchronization mechanism for our application.

Our target application is a multi-player VR online game. There are so many user interactions and special effects such as explosions and natural phenomena. In order to process these effectively, we could consider a rigid mechanism or an adaptive mechanism. A rigid mechanism, which is based on worst-case assumptions, is not applicable for a 3D game content. During a synchronization process of a multi-channel PC cluster, the user interaction may not be processed in real time. An adaptive mechanism monitors an underlying rendering power and is able to adapt the cluster to changes of user conditions.

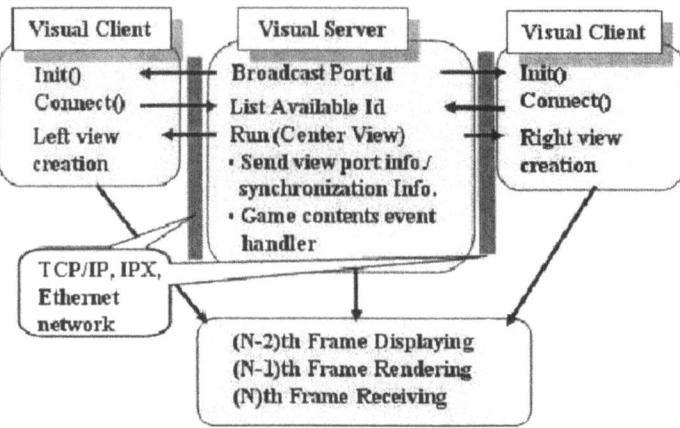

Fig. 2. Synchronization method on multi-channel clustered PCs

The adaptive synchronization protocol supports a dual scene graph to process states and events of game objects. A conventional scene graph organizes and controls a rendering process of objects found solely at one channel, and the other scene graph organizes and controls objects found at the edge of two adjacent channels. The former objects are called local objects and the latter objects are called global objects. To decide whether an object is local or global, we introduce a virtual plane across the visual systems. The dual scene graph periodically detects the collision between a virtual plane and all objects. If an object collides against a virtual plane, it becomes global and thus tightly synchronized. However, local objects are coarsely synchronized. Our results are well suited for immersing in 3D game displays such as virtual soccer and long distance race.

3 Immersive Display Environment

3.1 Multi-projector System

PCs, in the Multi-channel PC Cluster, are connected to their corresponding projectors. These projectors display their own portions of game display, and thus, when they display simultaneously, a user can enjoy an immersive game display with a wide FOV. However, since the display surface for the projectors are shaped arbitrarily, the user must endure watching the distorted game display affected by the shape of the display surface. Furthermore, if there are cases where a display area for a projector is overlapped with another, these will cause an irregular game display in terms of its intensity. These problems can be fixed by using structured light technique[2], [4] and intensity blending[3]. In addition, a laser tracker is used to track user's position to render the game content on the user's viewing point.

3.2 Structured Light Technique

A multi-projector display consists of an arbitrary display surface, camera and projector pairs, and a laser tracker. Each camera and projector pair is used to extract a shape of its corresponding portion of display surface based on a structured light technique. The same projector, used to extract the display surface model, is also used to project a game content on the display surface.

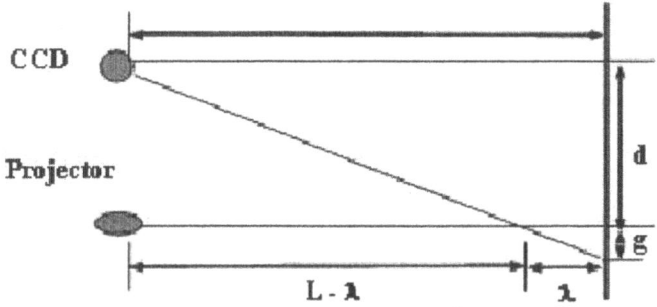

Fig. 3. Setting for extraction of display surface model

In order to carry out the extraction of display surface model, we used the setting in Fig. 3. A projector is used to display an interference pattern on its display area and a camera is used to capture the pattern on the area[2], [4].

The structured light technique that we used to extract the shape of the arbitrary display surface is a well-known two-wavelength phase shifting interferometry. The basic concept of phase-shifting method is to extract heights of the object surface from an interference pattern created by a phase-shift of interferometer. 4-frame phase-shifting method is normally used, and, from this method, we can acquire four interference patterns by shifting a phase of $\pi/2$ four times. The formula (1) expresses the intensity of acquired interference pattern at (x,y).

$$I_p(x,y) = I_0(x,y)\{1 + \gamma(x,y)\cos[\phi(x,y) + \Delta]\} \tag{1}$$

$I_0(x,y)$ is an average moiré intensity at (x,y), $\gamma(x,y)$ is a standardized visibility of moiré interference pattern, Δ is a degree of phase-shifting, and $\phi(x,y)$ is a height of the surface at (x,y).

By the 4 frame phase-shifting method, we can get following four different formulas from formula (1).

$$I_1(x,y) = I_0(x,y)\{1+\gamma(x,y)\cos[\phi(x,y)]\},$$
$$I_2(x,y) = I_0(x,y)\{1+\gamma(x,y)\cos[\phi(x,y)+\pi/2]\},$$
$$I_3(x,y) = I_0(x,y)\{1+\gamma(x,y)\cos[\phi(x,y)+\pi]\},$$
$$I_4(x,y) = I_0(x,y)\{1+\gamma(x,y)\cos[\phi(x,y)+3\pi/2]\}$$
(2)

The four different formulas can simplified to the following formula (3).

$$\frac{I_4(x,y)-I_2(x,y)}{I_1(x,y)-I_3(x,y)} = \frac{\sin\{\phi(x,y)\}}{\cos\{\phi(x,y)\}} = \tan\{\phi(x,y)\},$$
$$\phi(x,y) = \tan^{-1}[\frac{I_4(x,y)-I_2(x,y)}{I_1(x,y)-I_3(x,y)}]$$
(3)

Therefore, we can acquire the height map of the object surface from $\phi(x,y)$. However, since these heights are calculated from arctangents, there can be ambiguities of height in every 2π.

3.3 Two Rendering Algorithm

Based on the structured light technique, an arbitrary display surface area of each projector is acquired. However, user won't be able to enjoy the distortion free display unless this model is taken into account. In order to fix this problem, we use a following real-time rendering algorithm[5], [6], [7].

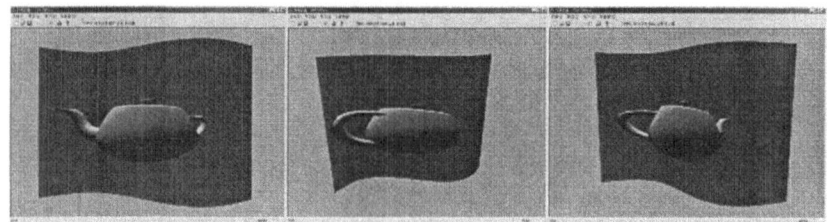

Fig. 4. Projection display without rendering algorithm

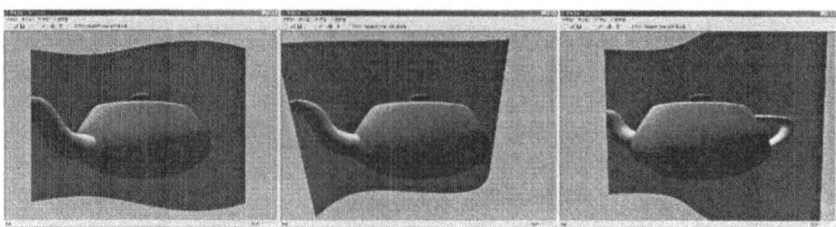

Fig. 5. Projection display with rendering algorithm

The game content is rendered on the user's view point and saved as a series of textures. These textures are projected from user's view point on the display area model extracted using the structured light technique. Then the display area model, with the projective texture mapped onto it, is rendered from the projector's viewing point, and this rendered content is projected onto the real display area[8].

According to the algorithm described above, the user's view point must be known in real time. This can be done utilizing a laser tracker. The tracker's body is attached on the ceiling and its sensor is attached on the user's head. This is required to achieve a completely distortion-free display on the user's view point. However, if the shape of the display surface is not radically curved, the user's view point can be set to an appropriate position and angle, and the user won't notice much difference between the display with the user's view point tracked and that without it. Fig. 4 is the pictures without this rendering algorithm, and Fig. 5 is the pictures with it.

3.4 Intensity Blending

Unlike a Cave which each projector is dedicated to a certain part of the display surface, this multi-projector display doesn't rely on an assumption that each projector is dedicated to its unique part of the display surface. Thus, there may be an overlapped display surface area by projectors. Fig. 6 is an example which this overlapped area occurs.

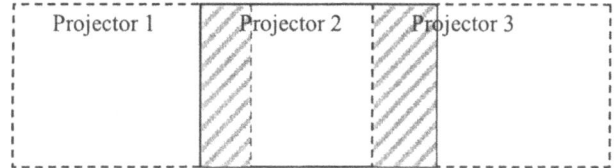

Fig. 6. Overlapped display surface areas by projectors are indicated with stripe

When this overlapped area occurs, the intensity of this area is the sum of projector's intensities which these projectors overlap. Thus, this produces an irregularity of brightness to the multi-projection display[5], [6].

There are several possible solutions available to solve this problem, but they all depend on projector's calibration parameters. These calibration parameters of projector include its focal point and FOV, and are used to calculate an overlapped area. First solution is using a filter for a projector. In order to eliminate the overlapped area by multiple projectors, one of the projectors for this area is fully displayed and the other projectors filter the overlapped area to cull. Second method is to set the alpha value of the overlapped area to 1 divided by number of projectors overlapping the area. The third method is to assign each vertex of display surfaces in overlapped area different alpha value depending on the distance to its nearest edge of the display surface. The vertices of an overlapped area in a display surface are same with its corresponding ones in the other display surfaces. The alpha value is expressed as (4).

$$\alpha_i(x,y) = \frac{d_i(x,y)}{\sum_i d_i(x,y)} \qquad (4)$$

$\alpha_i(x,y)$ is an alpha value of a vertex at (x,y), which is in the display surface i.
$d_i(x,y)$ is the distance between the vertex and the nearest edge of the surface i.

The second and the third methods require that all vertices of an overlapped area in a display surface are the vertices in the overlapped area of the other display surfaces. Thus, these vertices are shared by the display surfaces which overlap one another. This can be achieved by mesh unification.

4 Conclusions and Future Work

We've presented a multi-channel system on a PC cluster. We've developed optimizations that have enhanced the frame rates as our adaptive synchronizations that have dual scene graph with global and local 3D objects. We described a rendering system for scalable display in order to generate high-resolution images at real-time frame rates. And also we introduced our prototype projection system to develop a multi-projector display system which adjusts to the arbitrary shape of display surface and which provides a user with a seamless game display. These systems as a whole could provide a real-time wide FOV game display to render the very complex 3D graphics objects.

Finally, our current research includes further work on the VR game event process, projector calibration process, and extension of the current ideas to handle multiple projectors automatically.

References

1. W. Brian, P. Constantine, L. Vasily.: Scalable Rendering on PC Clusters. IEEE CG&A (July/August 2001) 62-70
2. J. Choi.: A Study on the Projection Moire Topography for Shape Measurement of 3D Object. Dept. of Mechanical Design, Chonbuk National University
3. F. Funkhouser, K. Li.: Large-format Displays, IEEE CG&A (July/August 2000) 20-21
4. W. Ryu.: A Study on the Measurement of Three Dimensional Object Shape by Using Phase Shifting Shadow Moire Method. Dept. of Mechanical Design, Chonbuk National University
5. R. Rasker. (et. al.): The Office of the Future: A Unified Approach to Image-Based Modelling and Spatially Immersive Displays. SIGGRAPH Conference Proceedings (July 1998) 19-24
6. R. Rasker. (et. al.): Multi-Projector Displays Using Camera-based Registration. IEEE Visualization (1999)
7. R. Rasker, B. Paul.: A Self Correcting Projector. TR-2000-46 (January 2002)
8. M. Segal. (et. al.): Fast Shadows and Lighting Effects Using Texture Mapping. Computer Graphics, 26 (July 1992)
9. Z. Zhang.: Flexible Camera Calibration by Viewing a Plane from Unknown Orientations. Computer Vision (1999) 666-673
10. P. H. Smith, Van Rosendale J. (eds.): Data and Visualization Corridors(DVC) in report CACR-164 (September 1998)

Virtual Reality Technology Applied to Simulate Construction Processes

Alcínia Zita Sampaio[1], Pedro Gameiro Henriques[2], and Pedro Studer[3]

[1] Assistant Professor, Dep. of Civil Engineering and Architecture, IST/ICIST, Technical University of Lisbon
Av. Rovisco Pais 1096-001 Lisbon, Portugal
`zita@civil.ist.utl.pt`
[2] Assistant Professor, Dep. of Civil Engineering and Architecture, IST/ICIST, Technical University of Lisbon
Av. Rovisco Pais 1096-001 Lisbon, Portugal
`pgameiro@civil.ist.utl.pt`
[3] Research fellowship in Civil Engineering, Dep. of Civil Engineering and Architecture, IST/ICIST, Technical University of Lisbon,
`studer@netc.pt`

Abstract. This paper describes a didactic application that is part of a research project whose main aim is to develop a computer-aided system which will assist design and construction processes. It is based on the visual simulation of construction activities. Geometric Modeling and Virtual Reality techniques are used to visualize the design process and to define user-friendly interfaces in order to access construction information, which could prove useful to Civil Engineering professionals. As a first step, it was developed a prototype that serves as a didactic tool for Civil Engineering students of disciplines concerned with building construction. The construction of a double brick wall is the case studied. Using the virtual three-dimensional model of the wall it is possible to show, in an interactive way, the sequence of the construction process and observe from any point of view the configurations in detail of the building components.

1 Introduction

Normally, academic and commercial applications of computer-aided design in construction, provide a visual presentation of the final state of the project, that is, the three-dimensional (3D) representation of the building with an animated walk-through, allowing observation of both its interior and exterior. The current computer tools and models are unable to follow changes in the geometry of the building or structure during the construction process.

The visual simulation of the construction process needs to be able to produce changes to the geometry of the project dynamically. Is then important to extend the usefulness of design information to the construction planning and construction phases [1]. The integration of geometrical representations of the building together with scheduling data is the bases of 4D (3D + time) models in construction domain. 4D models combine 3D models with the project timeline [2]. VTT Building Technology has been developing and implementing applications based on Virtual Reality (VR) technology and 4D to improve construction management practice [3].

The virtual model presented here was developed within the activities of a research work: *Virtual Reality in optimization of construction project planning* - POCTI/1999/ECM/36300, ICIST/FCT [4] now in progress at the Department of Civil Engineering and Architecture of the Technical University of Lisbon. The main aim of the research project is to develop interactive 3D models where students can learn about planning construction activities. The innovative contribution lies in the application of VR techniques to the representation of information concerning construction, of practical use to civil construction professionals [5] [6].

As a first step, a prototype serving as a didactic tool for civil engineering students of disciplines concerned with building construction was developed. The study case is a common external wall composed with two brick panels. The wall's virtual model, developed along this work, allows the user to visualize the construction progression, in particular, the following actions:

- The interaction with the construction sequence by means of the production of 3D models of the building in parallel with the phases of construction;
- The accessing of qualitative and quantitative information on the status of the evolution of the construction;
- The visualization of any geometric aspect presented by the several components of the wall and the way they connect together to form the complete wall.

Using the developed virtual model, allows students to learn about construction planning of the specific situation presented. This communication is oriented to teaching construction techniques by means of virtual environments. It is expected this model will be able to contribute to support teaching disciplines concerned with Civil Engineering. Another objective in creating this kind of virtual applications is to show in which way new technologies afford fresh perspectives for the development of new tools in the training of construction processes. The virtual models can be very useful in distance learning using *e-learning* technology.

2 Virtual Reality Technology

Virtual Reality can be described as a set of technologies, which, based on the use of computers, simulates an existing reality or a projected reality [7]. This new tool allows computer-users to be placed in 3D worlds, making it possible for them to interact with virtual objects at levels until now unknown in information technology: turning handles to open doors; switching lights on and off; driving a prototype car or moving objects in a house. To achieve this, elements of video, audio and 3D modeling are integrated in order to generate reality, initially through specific peripheral devices (handles, helmets and gloves) and at present, through the Internet.

Its origin is attributed to flight simulators developed about fifty years ago by the United States Army. The beginning of VR is attributed to Ivan Sutherland, with the introduction in 1965 of the first 3D immersion helmet, which was later divulged to the peripheral device industry with the designation, head mounted display. A precursor, Nicholas Negroponte (and collaborators), in the seventies, produced a virtual map of guided walks using a model of the city of Aspen, Colorado. In 1989, Jaron Lanier, an important driving force behind this new technology, designated it Virtual Reality [8]. In the 90s, with the upsurge of the Internet, a specific programming language was defined, the Virtual Reality Modeling Language (VRML). It is a three-dimensional interactive language devised for the purpose of modeling and visually representing objects, situations and virtual worlds on the Internet, which makes use of mathematical co-ordinates to create objects in space.

Interaction and immersion can be considered the most important characteristics of VR [8]:
- The immersion sensation is obtained by means of special physical devices, that allows the user to have the sensation of finding himself physically present in a world imagined and modeled by the system;
- The interactive characteristic is assumed because the VR technology is not limited to passive visual representation of this world but enables the user to interact with it (touching or moving objects, for example).

What is more, the virtual world responds to such actions in real time. In the developed application only the interactive property was explored.

Technically, the active participation of the user in a virtual environment, or, in other words, the sensation of immersion or presence in that environment, is achieved on the basis of two factors:
- The integration of information technology techniques (algorithms) used to obtain images of the highest degree of visual realism (ray-tracing, luminosity, application of textures etc.);
- The integration of a series of physical devices resulting from specific technologies like visual technology, sensorial technology (sensors of force and positioning) and mechanical technology (for transmitting movement such as a 3D mouse or gloves).

One of the areas in which the incorporation of VR technology as a means of geometric modeling and visual presentation of 3D animated models is most often applied is Architecture and Engineering. However, VR does not merely constitute a good interface but presents applications that provide the possibility of finding solutions to real problems in such diverse fields as Medicine or Psychology.

3 Developing the Virtual Model

As a case study, in the building construction field, an external wall with double brick panels was selected. The developed virtual model allows the student to learn about the construction evolution concerning to an important part of a typical building. The selected construction component focuses different aspect of the construction process: the structural part, the vertical panels and the opening elements. The 3D geometric model of the wall was defined using the AutoCAD system, a computer-aided drawing system common in Civil Engineering offices. Next, the wall model was transposed to a VR system based on a programming language oriented to objects, the EON Studio [9].

3.1 Creating the 3D Geometric Model

First, all building elements of the wall must be identified and defined as 3D models. Structural elements (demarking the brick panels), vertical panels of the wall and two standard opening elements, were modeled. In order to provide, later in a virtual space, the simulation of the geometric evolution of a wall in construction, the 3D model must be defined as a set of individual objects, each one representing a wall component.

Structural elements of the wall. Foundations, columns and beams, were considered as structural elements. The concrete blocks are defined as *box* graphic elements (available in the AutoCAD system) and the steel reinforcements as *cylinder* and *torus* graphic elements. The Fig. 1 shows some details of these components.

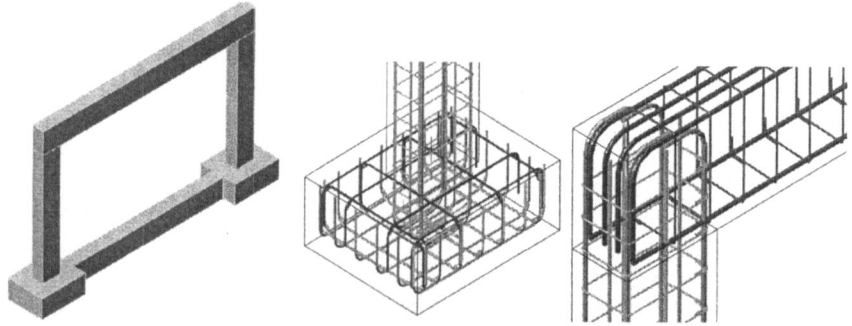

Fig. 1. Details of the structural elements modeled in 3D

In the image, it is possible to observe how to accommodate the steel reinforcements inside the structural elements. This is a real problem that is solved for each case in the work *in loco*. These elements were modeled taking account this kind of difficulty. So, it is an illustrative example.

Vertical panels of the wall. Confined by the structural elements there are two brick panels and a heating proof layer. Over all there are two rendering coats and two painted surfaces. Initially, all these elements were modeled as boxes with different thickness. The selection of thickness values for each panel is made according to the usual practice in similar real cases (Fig. 2).

Next, there were defined openings in the panels to place the window and the door elements.

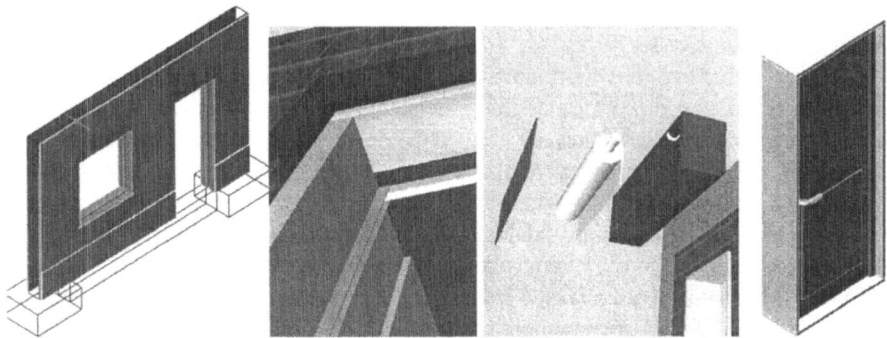

Fig. 2. Details of the wall's vertical and opening components

Opening components of the wall. Finally, the components of two common opening elements, a window and a door, were modeled (Fig. 2). The pieces of the window's and the door's frames were created as individual blocks. Each element was modeled taking in consideration the real configuration that such type of elements must present in real situations. By this, at the virtual animation of the wall construction, it is possible to observe each one separately and analyze conveniently the configuration details of those frames.

3.2 Creating the Virtual Environment of the Construction Process

One by one every part of each element considered as a building component of the wall was modeled. The Fig. 3 shows the complete 3D model of the double brick wall. Next, the geometric model was exported as a 3DStudio-drawing file (with the file extension .3ds) to the VR system used, EON Studio [8].

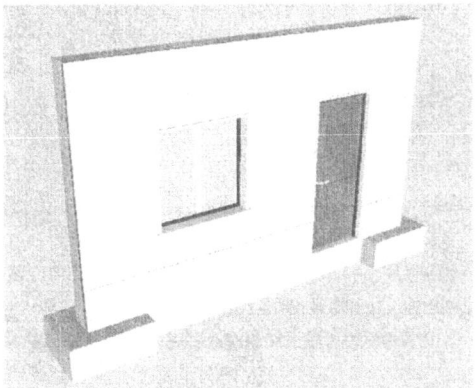

Fig. 3. The complete 3D model of the wall

The VR system should allow the manipulation of the elements of the wall model according to the planning prescribed for the carrying out of the construction. Supporting that, a range of nodes or function is available in the system to build up convenient virtual animations.

The Fig. 4 presents the work ambience of the EON system. In the left side of the image, is the nodes window (containing the virtual functions available in the system), in the central zone is the simulation tree (is the place where the drawing blocks hierarchy is defined and the actions are imposed to blocks) and, in the right side, the routes simulation window (where the nodes are linked).

When the 3D model is inserted into the RV system, the drawing blocks of the model are identified in the central window. To define an animated presentation the nodes or actions needed are picked from the nodes window and put into the simulation tree. Here, those nodes are associated to the blocks to be affected by the programmed animation.

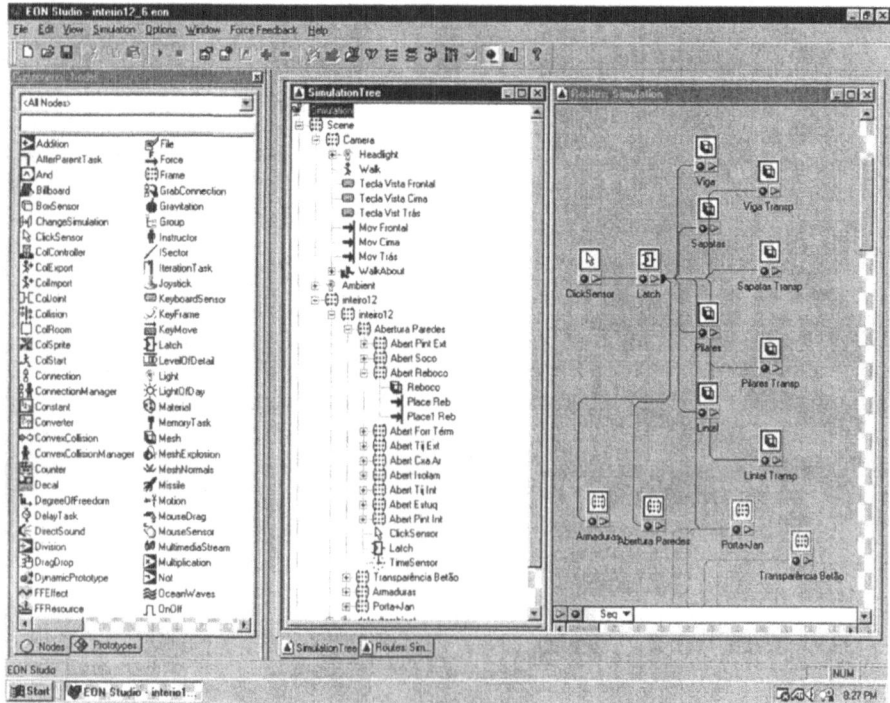

Fig. 4. The work ambience of the EON Studio

For instance, to impose a translation to the external rendering coat panel (named "reboco" in the simulation window of Fig. 5), two *place* nodes are needed (one to define the place where to go and the other to bring back to its original position). The *ClickSensor* and *Latch* nodes allow the initialization of a programmed action. For that the user interacts with the virtual scenario pressing on a mouse button.

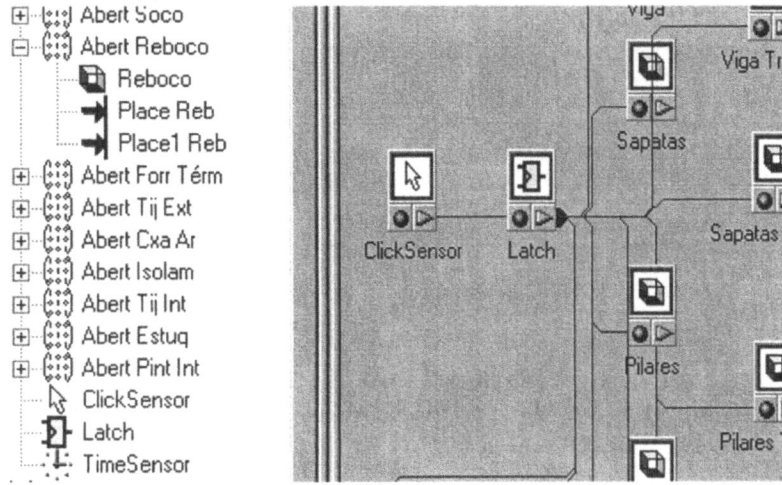

Fig. 5. Aspects of the simulation tree and the routes windows

Vertical panels presented in explosion. The exhibition of the several vertical panels of the wall presented in explosion is a kind of animation with a great didactic interest. Fig. 6 includes two steps of this presentation, the opened and closed situations. The translation displacement value attributed to each panel was distinct from each other in order to obtain an adequate explosion presentation. This type of animation allows the student to understand the correct sequence of the vertical panels in a wall and to observe the different thickness of each one.

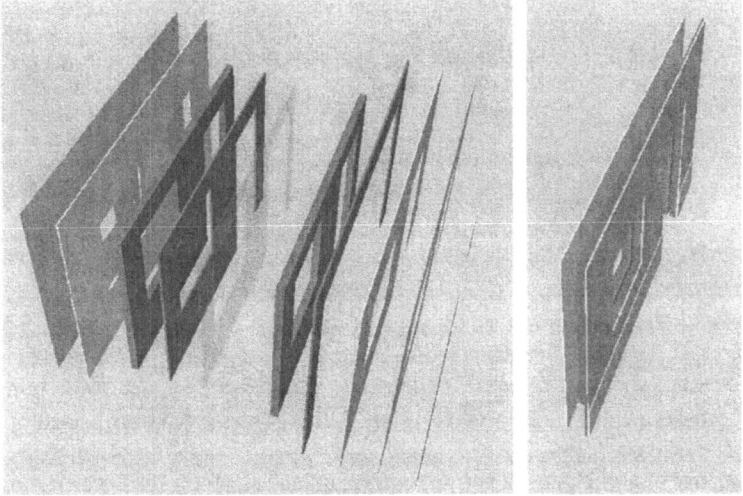

Fig. 6. Vertical panels presented in explosion

Animation of the wall's construction. The construction process was decomposed in 23 phases following the real execution of this kind of building component in the work *in loco*. The first element to become visible, at the virtual scenario, is the steel reinforcement of the foundation (Fig. 7) and the last is the door pull. The programmed animation simulates the progression of the wall construction. In each step the correspondent 3D geometric model is shown (Fig. 7). In this way, the virtual model simulates the changes that really occur while the wall is in construction in a real work place.

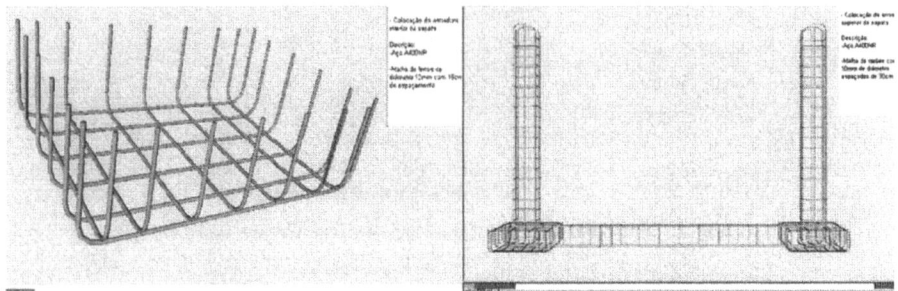

Fig. 7. Presentations of two steps of the virtual construction progress

For each new wall component becoming visible in a construction phase, the virtual model allows the user to pick the element and to manipulate the camera around it (Fig. 8). The user can then observe the element (displaced from the global model of the wall) from any point of view. Then all configuration details that the components of a real wall must present can be observed and analyzed. This capacity is important in construction process training.

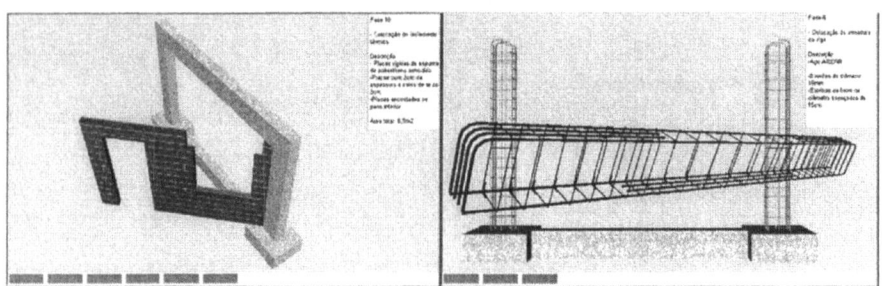

Fig. 8. Pictures presenting elements displaced from the global model of the wall and the box text construction data

While the animation is in progress, a box text is presented, fixed at the upper right corner of the display (Fig. 8). It contains construction information about the step in exhibition. The text includes the order number in the sequence, the activity description and the material specification and quantification concerned to each phase. The visualization of this type of data following the virtual construction evolution is useful to students.

The virtual animation presents, below the visualization area, a toolbar (Fig. 9). The set of small rectangles included in it shows the percentage, in relation to the wall fully constructed, up to the step visualized. To exhibit the next phase the user must click in any part of the model. To go back to an anterior step the user must click over the pretended rectangle in that progression toolbar.

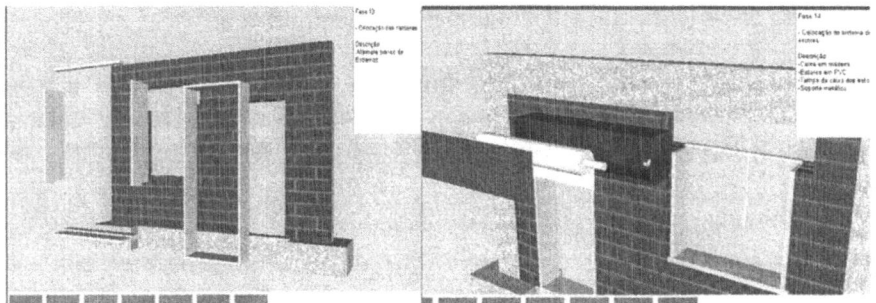

Fig. 9. Images showing the progression toolbar and wall components presented in explosion

Finally, the animation allows the user to visualize the pieces of wall elements in an exploded exhibition. The images included in Fig. 9 shows two elements presented in explosion. This type of presentation allows the student to know how the different parts connect each other to form wall components and can observe the configuration of those parts with detail. This capacity provided by the virtual model is also of great interest in construction domain instruction.

This type of didactic material can be used in face-to-face classes and in distance learning based on *e-learning* technology. The insertion of the model in the site of disciplines concerning construction is now in preparation.

4 Future Perspectives

Other type of building components can be modeled and manipulated in a virtual scenario for construction learning proposes.

Two other applications, also in the construction area, are now in progress. One concerns to the construction of a bridge by the segmental free cantilever method and the other to the construction of different cases of roofs. With this type of virtual models students can learn about construction technologies and analyses the sequence of construction, the steps required along the correspondent planned execution process and the configuration detail of each element.

5 Conclusions

The VR technology applied to construction field made possible to represent a three-dimensional space realistically. The visual simulation of the construction evolution of a common case was achieved. The user can interact with the virtual model of the wall and impose any sequence time in the construction process, select from the wall's model any component or parts of element and manipulate the camera as desire in order to observe conveniently any detail of the components configuration. While the animation is in process, the construction information associated to each step is listed. The use of these capacities, allowed by the developed virtual model, is beneficial to Civil Engineering student in construction process subjects.

Acknowledgement. This work was developed within the research program POCTI/1999/ECM/36300 - *Virtual reality in optimization of construction project planning* [4] supported by the Foundation for Science and Technology (Fundação para a Ciência e Tecnologia - FCT, Portugal).

References

1. Fischer, M. 4D CAD - 3D Models Incorporated with Time Schedule". *CIFE in Finland, VTT-TEKES*, Helsinki, Finland. (2000).
2. Retik, A. Planning and monitoring of construction projects using virtual reality projects, *Project Management Journal*, n° 97 (3), (1997), 28–31.
3. Leinonen, J., Kähkönen, K., Retik, A. New construction management practice based on the virtual reality technology, in book *4D CAD and Visualization in Construction: Developments and Applications*, editors Raja R.A. Issa, Ian Flood William J. O'Brien. Ed. A.A. Balkema Publishers, (2003), 75-100.
4. Henriques, P.G., Sampaio, A.Z., Project Program: Virtual Reality in Optimization of Construction Project Planning. POCTI/1999/ECM/36300, ICIST/FCT, Lisbon, Portugal (1999)
5. Henriques, P.G., Sampaio, A.Z., Visual Simulation in Building Construction Planning, 4th European Conference on Product and Process Modelling, Portoroz, Slovenia (2002) 209-212
6. Henriques, P.G., Sampaio, A.Z., Interactive project planning in construction based on virtual reality technology, IKM – 16th International Conference on the Applications of Computer Science and Mathematics in Architecture and Civil Engineering, Bauhaus-Universität, Weimar, Alemanha (2003), abstract 29 (CDROM 6 pgs)
7. Burdea, G., Coiffet, P., *Virtual Reality technology*. 2nd Edition, John Wesley & Sons, (2003)
8. Vince, J., Virtual Reality Systems, ACM SIGGRAPH Books series, Addison-Wesley, (1998)
9. Introduction to Working in EON Studio, EON Reality, Inc.(2003).

Virtual Reality Applied to Molecular Sciences

Osvaldo Gervasi[1], Antonio Riganelli[2], and Antonio Laganà[2]

[1] Department of Mathematics and Computer Science, University of Perugia,
via Vanvitelli, 1, I-06123 Perugia, Italy
ogervasi@computer.org
[2] Department of Chemistry, University of Perugia,
via Elce di Sotto, 8, I-06123 Perugia, Italy
{auto,lag}@dyn.unipg.it

Abstract. The paper describes the developments of the prototype implementation of a portal[1] related to a Virtual Reality application to Molecular Science on a Grid environment, called **VMSLab-G**, and will describe two running virtual experiments. Some aspects related to the applications of **VMSLab-G** to research and education are also outlined.

1 Introduction

The rapid development undergone by computer and information technologies is having a tremendous impact on research, education and, in more general terms, on knowledge and information management. In particular, the continuous growth of network bandwidth and middleware components to manage the Grid [1] is enabling the use of virtual reality approaches (VR) as the building blocks of the construction of realistic simulation environments. This work is part of an ongoing project [2,3] aimed at exploiting Virtual Reality, at both meter (HVR, Human Virtual Reality) and nanometer (MVR, Molecular Virtual Reality) level to integrate virtual elements in chemical laboratory.

Existing experiences [4,5] clearly show that the use of VR tools does not only facilitate the approach of the user to the experiment by offering an effective way of virtually exploring its components but also offers him/her an environment in which studying, learning, training and teaching are available at any time and everywhere. The additional particularly valuable feature of **VMSLab-G** is the unique combination of HVR and MVR to support the simulation of the chemical laboratory with an immersion in the microscopic world of the molecules and their laws. To this end well established information technologies (such as JAVA and XML) offer a new generation of simulation and knowledge management facilities coupling a realistic representation of the events with the management of more traditional textual type information.

Our work originates from the activities of both the D23 Action (*METACHEM: Metalaboratories for complex computational applications in Chemistry* [6] and from the **MuTALC** (*Multimedia in Teaching And Learning in Chemistry*) working group of the European Chemistry Thematic Network

---

[1] **VMSLab-G** portal URL: http://vmslab.org

(ECTN). In particular within the D23 Action several working groups have developed useful tools for MVR. Among them of particular relevance is the activity of **SIMBEX** [7,8].

SIMBEX is aimed at constructing a portal for the simulation of molecular beam experiments. The project goals are achieved by combining the expertises of several chemical and computer science laboratories. The chemical groups are competent in carrying out electronic structure and dynamics calculations and in rationalizing molecular beam experiments. The computer science laboratories are competent in Cluster and Grid computing, Portal development, advanced Scientific Visualization and Virtual Reality technologies and collaborate to implement the simulation environment on the Web, to develop related middle-ware and to manage the Metacomputing back-end. The simulator is used to reproduce reactive scattering properties of some gas phase chemical reactions, to rationalize their reaction mechanisms and to pivot experimental measurements.

Another working group of D23 is **ELCHEM** whose goal is to develop HVR for chemical processes and technologies [9]. The **ELCHEM** project is aimed at developing chemistry courseware based on problem solving environments, Metacomputing and WEB technologies. Its goals are going to be achieved by combining the know how of several groups (expert in developing material for computer assisted teaching and evaluation in chemistry, in designing related middleware and managing Metacomputer platforms) with the systemic support of a European large scale computing facility. This group of researchers have established links with the **ECTN** that already develops e-learning tools as a means for carrying assessment and remote training in chemistry Universities, Adult education centres and Industries. The group is also establishing strong ties with Prometeus [10], CEN/ISS Workshop on Learning Technologies [11] and the CAPRI group of the European Organization of the Chemical industries (CEFIC).

On the **MuTALC** side a great deal of work has been made to develop computer assisted protocols to support chemical practice, to develop teaching and learning units, to accumulate demonstration material, etc., for ubiquitous education in chemistry and molecular processes.

Our work is aimed at combining the two experiences by developing a portal for an ubiquitous usage of the outcomes of the above mentioned two lines of research by exploiting the web and the innovative features of the Grid. In particular this paper reports on the work done to build the **VMSLab-G** portal that focuses on the management of a virtual laboratory. The scope of the portal is twofold. First, it represents the main entrance to the virtual world in its HVR and MVR components. Second, it gives an environment aimed at collecting and managing molecular science related knowledge for the specific use of a virtual laboratory for a virtual chemical laboratory.

As a test case we have chosen to simulate a chemical laboratory infrastructure where low-cost, user-driven experiments are realized.

This paper is organized as follows. In section 2 we describe the characteristics of **VMSLab-G** portal. In section 3 the virtual laboratory environment implemented using virtual reality technologies are illustrated. In section 4 two

Fig. 1. Welcome page of **VMSLab-G** portal

examples of the virtual experiments implemented are discussed. In section 5 the main conclusions ares outlined.

2 The VMSLab-G Portal

In Figure 1 the entrance environment of **VMSLab-G** portal is shown. From it the user can access to the specific services of a virtual laboratory. In particular he/she can explore the HVR environment and once selected a virtual room and the related experiment (or another component of **VMSLab-G** portal) use the MVR tools.

In recent papers [2,3] we describe some freshman practice laboratory experiences held in the introductory chemistry courses at the University of Perugia. This project is now progressing into developing a virtual environment for some of the complex organic laboratory protocols (**iChemLab**) of J. Froehlich and collaborates [12] of the Technical University of Vienna.

2.1 The VMSLab-G Portal Toolkit

The portal has been realized by using standard web technologies and following the W3C recommendations. In particular we made use of VRML[13] and X3D[14] for the description of the virtual worlds. Java and Javascript have been used to implement the scripts acting as computational engines for driving the evolution of the virtual worlds and to perform the required calculations.

X3D and VRML are used to represent and model both the virtual laboratory and its components, and the chemical structure of the molecules under study. To this end a library of Java classes has been created to represent the properties of atoms and molecules and to facilitate a standard representation of them on the virtual worlds. We have also made extensive use of PHP and a PHP framework has been implemented in order to facilitate the programmers work.

From the Web programming side, although an important part of the contents has been written in HTML, the usage of XML has been made to deal properly with the representation of the knowledge stemming from the experiments. The management in a distributed environment of scientific knowledge is one of the main challenge of next generation web technologies. For domains such as chemistry, physics and mathematics where much of the knowledge is represented through not standard textual information and symbolic representation plays an important role, the added value of semantic web environments[15] is crucial. Although the future research efforts will be focused on implementing the semantic environment, on the definition of the ontologies and on using the semantic languages for such class of problems, **VMSLab-G** is moving in this direction also by using the Chemistry Markup Language (CML)[16]. An example of such approach is represented by the UV-VIS virtual experiments (see section 4.1) where CML has been used to represent in the Web environment the chemical reaction involved in the experiment.

3 The Virtual Laboratory

The main feature of **VMSLab-G** is to provide the user with tools to manipulate and interact with the traditional instruments of an experimental apparatus and to visualize and represent at the same time molecular processes in a virtual world.

3.1 The Human Components (HVR)

The virtual laboratory was designed to reproduce as close as possible a real laboratory even if a certain degree of idealization needed to be introduced. Virtual experiments were built mainly with the purpose of offering pre and post laboratory assistance to the student. From the design point of view, the laboratory is made up of a main hall and several rooms in which the user can walk-by and enter to run a specific experiment. In this way one can get in contact with specific experimental setups, use the various components and work with them in a reality-like fashion. The virtual experiments are based on program scripts that simulate quantitatively the movement of the objects and produce images emulating the real experiment as dictate by a well established protocol.

The main hall and the rooms where the virtual experiments are run, are made of objects. These objects may be active or inactive. In some cases, when the visitor touches an object, an action is performed (active objects) that causes a modification of the virtual world. The most common active objects are laboratory objects (bottles, nipples, apparatuses, etc) and chemical objects (atoms, molecules, reagents, etc). An active object can be associated with one or more tasks. As already mentioned the core computation associated with the various tasks and laboratory operations are handled on the client side, using a script program.

The application starts by entering the hall of the laboratory from where one can access the various rooms in which the experiments take place. Figure 2 shows the main hall of the virtual laboratory and the doors of the specific experimental laboratories. By entering one of doors shown in the Figure, one enters in an environment equipped with the apparatus necessary for running a particular class of experiments.

Fig. 2. The main hall of the virtual laboratory

3.2 The Molecular Components (MVR)

The goal of combining an experimental protocol with MVR tools is to give the possibility of accompanying the laboratory procedure with a molecular description of chemical events occurring during the experiments undertaken. The simulation of molecular processes in fact benefits from advanced visualization techniques not only in order to provide an illustration of the involved processes

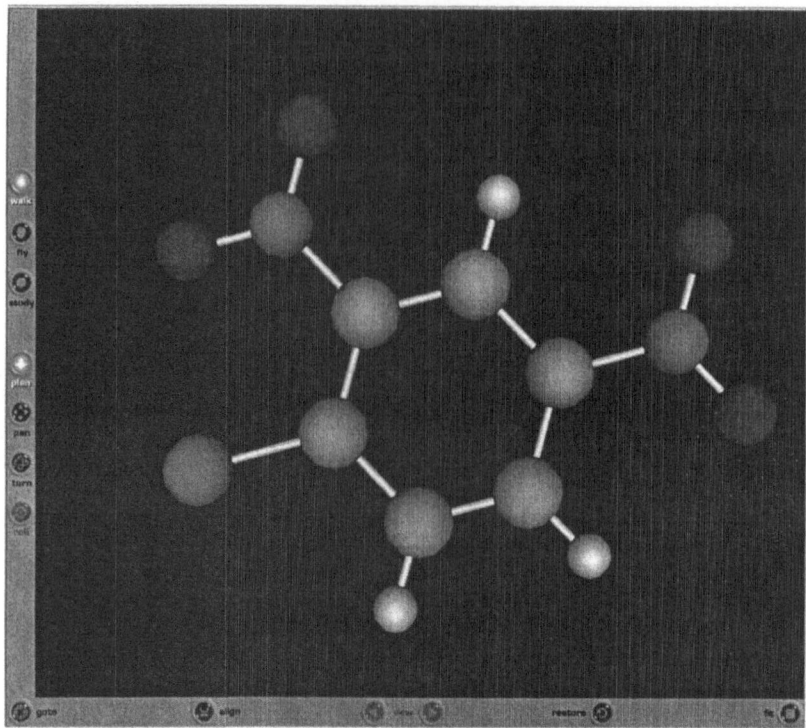

Fig. 3. VR (Ball and Stick) representation of 1-Chloro-2,4-dinitrobenzene obtained from the Sybil MOL2 description file.

but also in order to gain additional insight [17,18] on the nature of the atomic and molecular events taking place during the experiment. Although this is at the moment restricted to some simple cases, in principle, a full a priori description of realistic processes in which theory coupled with computational models can reliably simulate the outcome of an experiment.

The MVR machinery of **VMSLab-G** is that of **SIMBEX**. As already mentioned it makes use of Molecular Dynamics approaches to produce a priori quantitative treatments of the atomic and molecular processes. The complexity and the duration of related calculations makes it necessary to distribute the computational efforts on the Grid and use parallel computing techniques.

At present MVR technologies have been particularly developed as Virtual Monitors of **SIMBEX** in which the outcome of the molecular dynamics calculations are preprocessed with Java or PHP tools. **SIMBEX** Virtual Monitors may also produce VR representations of complex molecules using some legacy description formats (like Sybil MOL2) or CML[16]. In Figure 3 the structure of the molecule 1-Chloro-2,4-dinitrobenzene used in one of the **VMSLab-G** experiments is shown. The MOL2 structure used to generate the VRML representation showed in the figure was obtained by using CORINA software [19]

4 Virtual Experiments Taken from Spectroscopic Domain

As already mentioned the paper focuses on some case studies of chemistry practice laboratories. The virtual practice session is structured so as to provide the user with a work-flow template outlining the sequence of steps to be performed during the experiment. In this work we show two case studies representing the experimental apparatus and standard experiments in the process of being integrated with the **iChemLab** protocols.

The first case study concerned with the measurement of the rate coefficients of an organic reaction using the UV-visible spectroscopy.

The second case study is concerned with the measurement of the rate coefficients of an organic reaction of aromatic nucleophilic substitution using the IR spectroscopy.

4.1 Rate Coefficient Measurements by UV Spectroscopy

In this virtual experiment the goal is to evaluate the rate constant of a chemical reaction by measuring the UV spectrum of the bulk at different times. As a particular case study the rate coefficients of an organic reaction of aromatic nucleophilic substitution is considered. The VRML part of the virtual experiment (see Figure 4) reproduces a UVICON 923 UV/VIS spectrophotometer. The dynamic part of the experiment allows the student to carry out virtually all the operations relevant to the experiment. Three virtual monitors showing respectively a preliminary graphic that allows the user to choose the operative wavelength (in this case it will be $\lambda=365\ nm$), a plot showing the product absorbance versus time, and a third plot (obtained with the data coming from the second one) reporting the $\log[(A_0 - A_\infty)/(A_0 - A_t)]$ versus time (A_0, A_t and A_∞ are respectively the absorbances at time 0, t, and ∞). Working over the information provided by this last plot one can evaluate (this is also performed in the background by the virtual experiment) the rate constant of the reaction. The mathematical transformation involved are showed in this last data-elaboration virtual monitor. In this virtual monitor are also given the results otained for the first order nucleophilic aromatic substitution reaction of 1-Chloro-2,4-dinitrobenzene

4.2 Concentration Measurements by IR Spectroscopy

The IR spectroscopy experiment is shown in Figure 5. It deals with the evaluation of the concentration of a chemical substance in a solution by means of IR spectra measurements. In this experiment the student has to carry out operations at two different levels. At the first level, he/she has to perform a series of mechanical operations such as opening and closing the door slot, insert the sample cells, push a virtual button to produce the vacuum, etc.

The second operational level corresponds to data collection and analysis. These actions are performed in the real experiment by a dedicated personal computer, whereas in **VMSLab-G** this is implemented by means of a virtual monitor. From the virtual monitor the student has to select the peak from the

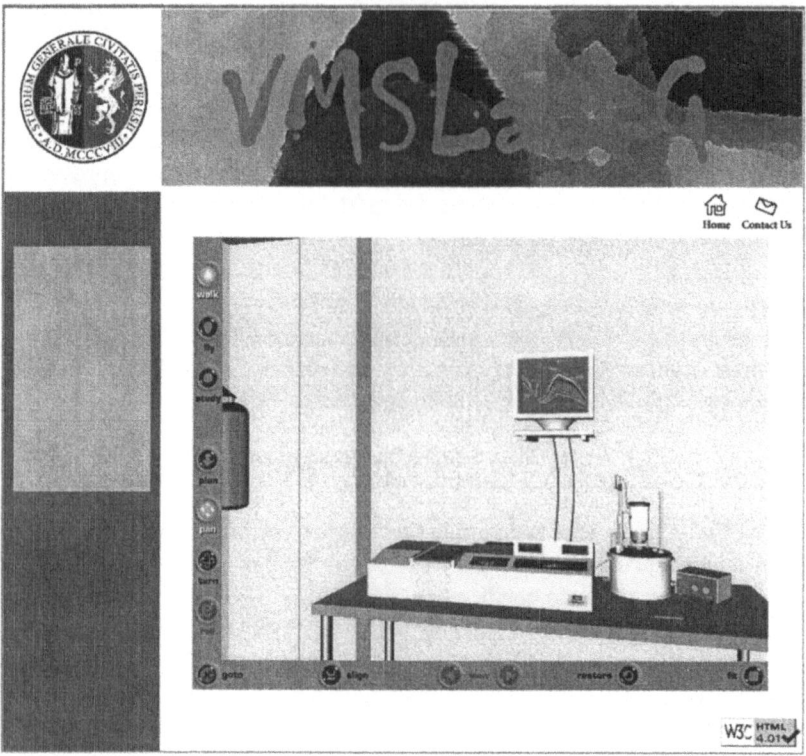

Fig. 4. A detail of Chemical kinetics experiment as implemented in the **VMSLab-G** portal.

registered IR spectrum curve. At this point a background calculation starts (this is done via a Java applet on the client side) in order to evaluate the integral of the area selected. The current concentration of the sample is evaluated from a stored calibration curve.

5 Conclusions

In this work we have described the main features of the **VMSLab-G** portal. The paper shows the modalities for using virtual reality approaches to describe chemical experiments at both human and molecular level. To this end a multi-scale approach has been adopted to deal with the description of the physical environment, at human virtual reality level that can be connected to the molecular virtual reality modules of **SIMBEX**. Two case studies have been discussed implying to a different extent various levels of virtual reality operations.

Acknowledgments. We are deeply grateful to our chemistry colleagues S. Alunni and A. Morresi, who have set up the real experiment and have col-

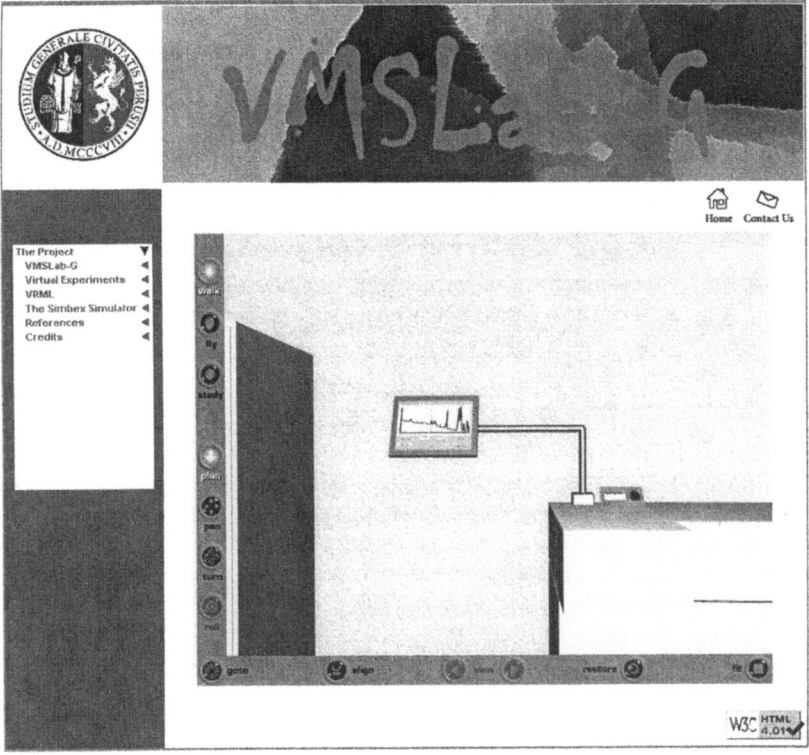

Fig. 5. A detail of IR spectroscopic experiment as implemented in the **VMSLab-G** portal.

laborated to write down their description. This research has been financially supported by MIUR, CNR and ASI as well as by the D23 Action (Metachem) of COST in Chemistry.

References

1. Foster, I., Kesselman, C. (eds.): *The Grid: Blueprint for a Future Computing Infrastructure*. Morgan Kaufmann Publishers, USA (1999); Berman, F., Fox, G., and Hey T. (rds) *Grid Computing Making the Global Infrastructure a Reality*. ISBN 0-470-85319-0 - John Wiley & Sons (2003).
2. Gervasi, O., Riganelli, A., Pacifici, L., Laganà, A.: VMSLab-G: a virtual laboratory prototype for molecular science application, Future Generation Computer System, (in press).
3. Riganelli, A., Gervasi, O., Laganà, A., and Alberti, M.: A Multi scale virtual reality approach to chemical experiments. Lecture Notes on Computer Science, **2658** (2003) 324-330.
4. Casher, O., Leach, C., Page, C.S., and Rezpa, H.S.: Virtual reality modeling language (VRML) in Chemistry. Chemistry in Britain, **34** (1998) 26-31.

5. Ruiz, I.R., Espinosa, E.L., Garcia, G.C., Gómez-Nieto, M.À.: Computer-Assisted Learning of Chemical Experiments through a 3D Virtual Laboratory, Lecture Notes in Computer Science, **2329** (2002) 704-712.
6. A. Laganà, METACHEM: Metalaboratories for cooperative innovative computational chemical applications, METACHEM workshop, Brussels, November (1999) (see also [7]);
   http://costchemistry.epfl.ch/docs/D23/d23-main.htm
7. "**SIMBEX**: *A Metalaboratory for the a priori simulation of crossed molecular beam experiments*" is the COST Project D23/003/001 of the D23 action. "*Metachem*: Meta-laboratories for complex computational applications in chemistry*". the coordinator is Osvaldo Gervasi, Dept. of Mathemathics and Computer Science, University of Perugia.
   http://www.eu.cec.int/.
8. Gervasi, O., and Laganà, A.: **SIMBEX**: a portal for the a priori simulation of crossed beam experiments. Submitted to *Future Generation Computer Systems*, to be published.
9. *ELCHEM: A Metalaboratory to develop e-learning technologies for Chemistry* is the COST Project D23/005/001 under the D23 action, called *Metachem*. The coordinator is Antonio Laganà, Dept. of Chemistry, University of Perugia.
10. http://www.prometeus.org/.
11. http://www.cenorm.be/cenorm/index.htm.
12. http://www.ichemlab.at
13. Ames, A.L., Nadeau, D.R., and Moreland, J.L.: *VRML 2.0 Sourcebook*. Wiley Computer Publishing. New York Tokio (1997)
14. http://web3d.org
15. Lassila, O.: *Web Metadata: A Matter of Semantics*. IEEE Internet Computing, **2**, No. 4, July 1998, 30-37;
    Luke, S., et al.: *Ontology-Based Web Agents*. Proc. First Int. Conf. Autonomous Agents, ACM Press, New York, 1997;
    Heflin, J., and Hendler, J.: *Dynamic Ontologies on the Web*, Proc. 17th Nat. Conf. AI (AAAI-2000), MIT-AAAI Press, Menlo Park, Calif., 2000, 443-449;
    Decker, S., et al.: *Knowledge Representation on the Web*. Proc. 2000 Int. Workshop on Description Logics (DL2000), Sun SITE Central Europe (CEUR), Aachen, Germany, 2000,
    http://sunsite.informatik.rwth-aachen.de/Publications/CEUR-WS/Vol-33/
16. Chemistry Markup Language: http://www.xml-cml.org
17. Wolff, R.S., L. Yaeger: *Visualization of Natural Phenomena*. Springer-Verlag New-York (1993).
18. Zare, N.R.: *Visualizing Chemistry*. Journal of Chemical Education, **79** (2002) 1290-1291.
19. http://www2.chemie.uni-erlangen.de/software/corina/free_struct.html

Design and Implementation of an Online 3D Game Engine

Hunjoo Lee and Taejoon Park

Electronics and Telecommunications Research Institute,
161 Gajeong-Dong, Yuseong-Gu, Daejeon, 305-350, Republic of Korea
{hjoo, ttjjpark}@etri.re.kr

Abstract. This paper proposes Dream3D, an online 3D game engine. We analyze requirements to build MMORPGs together with the techniques to satisfy them. And then, the techniques are classified into four categories: 3D rendering, animation, server, and network techniques. We design and implement Dream3D to provide all the required functionalities. Related with the technique classification, Dream3D consists of four subsystems: rendering engine, animation engine, server engine, and network engine. For each of the subsystems, we propose an implementation model to satisfy the functionality requirements.

1 Introduction

Because of the improvement of graphic hardware technologies, it becomes possible to generate high-quality images even on conventional personal computers in real-time. Without any expensive hardware devices, 3D virtual environments such as virtual shopping malls and virtual museums can be constructed on personal computers. Recently, 3D computer games for conventional personal computers are also developed and announced. Furthermore, since it becomes general for personal computers to employ graphic processing units (GPU) that are faster than CPU, 3D PC games such as Unreal and Quake that utilize such devices are now released and become popular.

Moreover, Due to the advances of network technologies, it becomes possible to build a large distributed virtual environment shared by several client systems. The most typical example of these kinds of applications is online game services, especially massive multi-user online role playing games (MMORPGs). There have been released several MMORPG services around the world, and the number of game players enjoying such kind of games have also been increased.

In this paper, we propose an online 3D game engine called Dream3D for 3D MMORPGs. Software techniques required to build 3D MMORPGs are analyzed and used to make an implementation model of the engine. We subdivide the engine into four subsystems: server engine, network engine, rendering engine, and animation engine. And then the functional requirements for each of the subsystems and software techniques to satisfy them are analyzed. We also implement a prototype MMORPG system to verify the soundness of the proposed game engine.

2 Advantages of Employing Game Engines to Develop MMORPGs

Sever, network, and computer graphics techniques that are used once to develop a commercial MMORPG service are usually reused to produce another game services. Therefore, it is common for a game provider to maintain the developed software techniques as a software library or a set of software modules. We call the set of libraries and modules that are reused during the game development procedure as a game engine [1]. Recently, some game providers even sell their engines.

Typical MMORPG services form very large virtual environments shared by more than ten thousands of client systems. Therefore, game server systems must be stable, fault tolerant, and consistent to every client. And the game server systems must provide functions to control the amount of network traffics between servers and clients. It is not easy to develop all the components of MMORPG services from the beginning.

3 Designing the Game Server Engine

Typically, there are thousands of client systems simultaneously connecting to a certain server system. Therefore it is necessary to develop a server engine structure to share a game world by several servers and reduce the amount of network packet transferred to the server system. In Dream3D server system, the following game world subdivision scheme is used to accomplish the system requirements.

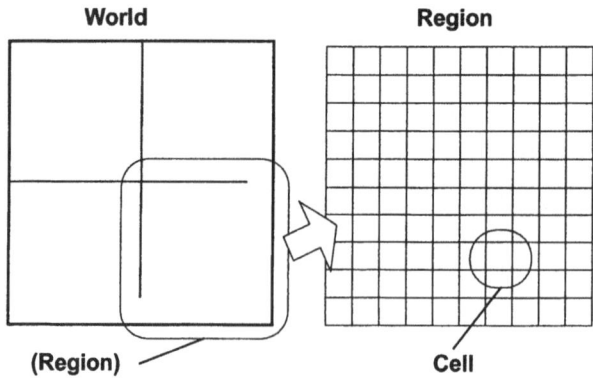

Fig. 1. Game world subdivision scheme

A game world is divided by several regions and each region is also subdivided by regular sized cells (See Fig. 1). The region is an area controlled by a server system. If there is only a server system controlling a given game world, the game world and the region assigned to the server become identical. We made it possible to control the size of each region according to the amount of the network traffic. If there occur lots of traffic around a certain area, more server systems take control around this area and thus the size of regions assigned those server systems become small.

The cell is a fixed sized unit area. The size of a region is controlled by the number of cells included in the region. To accomplish the seamless multi server environment, the cells at the boundary area of each region transfer the occurring game events to the server controlling the region together with those of adjacent regions.

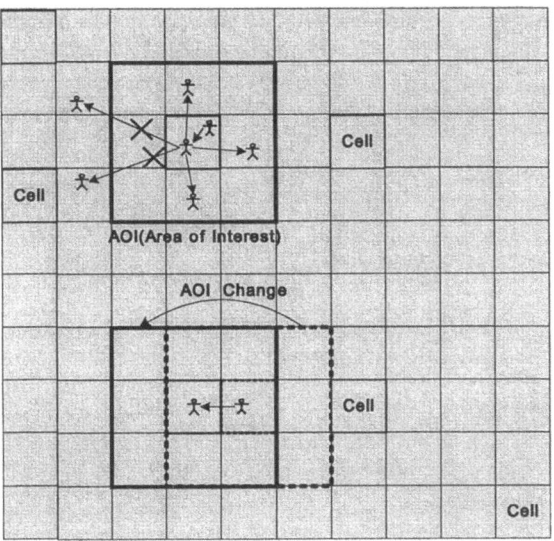

Fig. 2. AOI control based on cell structure

To control the amount of network traffics and maintain consistent game environments, the Dream3D server system provides area of interest (AOI) control based on the cell structure. For a user character controlled by a client system, only the game events occurred in the cell including the character and its neighboring cells are transferred to reduce the network traffic. If the user character moves, the AOI is changed according to the movement.

4 Designing the Network Engine

4.1 Server Side Network Engine

We built the Dream3D server system on Windows NT platforms and thus the server side network engine can utilize the input and output completion port (IOCP) provided by the Windows system. The network engine sends and receives game events as network packets transmitted through the IOCP and then transfers them into the server engine implemented on the network system. The server engine maintains a number of pre-created threads as a thread pool. For each event transmitted through the IOCP, the server system assigns a thread from the tread pool, processes the event by activating the corresponding event processor on the thread, and then sends the processing results through the IOCP again as resulting event packets. The basic structure of the server side network engine is shown in Fig. 3.

4.2 Client Side Network Engine

At the client side, the network engine only transfers the game events from server to client and vice versa. Game events transferred from the server engine are input to the game main loop of the client system, and analyzed by the predefined game logic together with other game events occurred by the user input and certain game situation. The results of the analysis change the game situation by modifying the position of the game objects and activating several game effects. If necessary, the analysis results are transferred back to the server engine as forms of resulting game events (See Fig. 4).

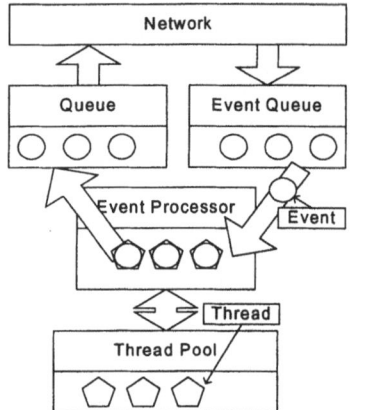
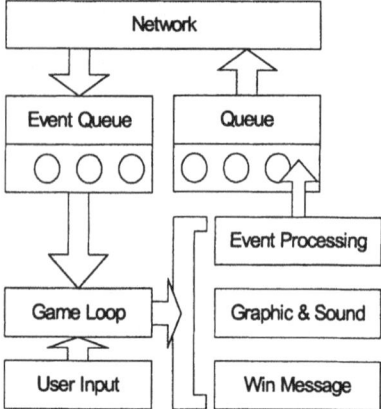

Fig. 3. Server side network engine

Fig. 4. Client side network engine

5 Designing the Rendering Engine

To generate impressive game situation on computer screens, a 3D rendering engine must draw high quality game scenes in real-time. The Dream3D rendering engine is designed to satisfy such requirements. By adapting geometric culling algorithms that cull out a part of polygons that are not visible from the camera among the whole polygons, the rendering engine can generate game scenes rapidly. To perform such kind of culling operation, additional geometric data structures such as the BSP-PVS structure [2, 3] used to choose polygons to be culled out must be constructed at the game design phase. The rendering engine also provides progressive mesh based level of detail (LOD) techniques [4] for static game objects. The rendering engine also utilizes multi-texture techniques that are common features of advanced graphic acceleration hardwares. Light mapping and shadow mapping functions can be provided using those features.

6 Designing the Animation Engine

To produce realistic motions of human-like game characters, the Dream3D animation engine provides functions to handle captured motion data obtained from real human motion. Techniques such as motion blending, motion retargeting, and motion transition are implemented to realistic motions generated from the captured data. Furthermore, to get more flexibility on using the captured data, the animation engine provides inverse kinematics rules to edit and modify the data. And the animation engine also has the physics system that can simulate the game environment physically correctly.

7 Experimental Results: Implementation of a MMORPG

We implemented a prototype MMORPG system to verify the soundness of the proposed game engine. The server system is built on a Windows platform with four Xeon processors. The client system is implemented on a Pentium 3 with an NVIDIA GeForce 3 acceleration board. We also implemented a special purpose plug-in software that converts 3D Studio MAX model data into the implemented game system. It is possible to use all image formats including tga, bmp, and jpg if Direct3D supports them. In Fig. 5, we show screen shots of the implemented MMORPG system as the experimental results.

Fig. 5. Screenshots from the implemented MMORPG service

8 Conclusions

In this paper, we proposed the Dream3D game engine for developing 3D MMORPG services. We analyzed requirements for developing 3D MMORPGs and then designed an implementation model of a game engine that satisfied the requirements.

As future works, we plan to adjust the proposed engine so that it can run on several platforms including game consoles and mobile devices.

References

1. D. H. Eberly.: 3D Game Engine Design: A Practical Approach to Real-Time Computer Graphics. Morgan Kaufmann Publishers (2001)
2. T. Moeller, E. Haines.: Real-Time Rendering. A K Peters (1999)
3. A. Watt, F. Policarpo.: 3D Games: Real-time Rendering and Software Technology. Addison-Wesley (2001)
4. H. Hoppe.: Progressive Mesh. Proceedings of ACM SIGGRAPH (1996) 99-108
5. A. Watt.: 3D Computer Graphics, 3rd edition. Addison-Wesley (2000)
6. H. Hoppe, T. DeRose, T. Duchamp, J. McDonald, and W. Stuetzle.: Mesh Optimization. SIGGRAPH (1993) 19-26
7. Andrew Rollings, Dave Morris.: Game Architecture and Design. Coriolis (2000)
8. P. Lindstorm, D. Koller, W. Ribarsky, L. F. Hidges, N. Faust, and G. A. Turner.: Real-Time, Continuous Level of Detail Rendering of Height Fields. SIGGRAPH (1996) 109-118
9. Mark DeLoura.: Game Programming Gems 3. Charles Rivermedia (2002)

Dynamically Changing Road Networks – Modelling and Visualization in Real Time

Christian Mark[1], Armin Kaußner[1], Martin Grein[1], and Hartmut Noltemeier[2]

[1] Center for Traffic Sciences (IZVW), Germany
http://www.izvw.de
[2] University of Würzburg, Department of Computer Science D-91405 Würzburg, Germany
http://www1.informatik.uni-wuerzburg.de/en/

Abstract. In this paper we present a new concept for a driving simulator database. The database consists of a three-dimensional road network, which is not fixed like in conventional databases. The parts lying outside the viewing range of the driver can be changed during simulation.
Changes in the road network are either initiated interactively by the researcher or automatically by the system based on predefined conditions. Conventional visualization methods are not usable for such a road network. Therefore we present new visualization algorithms for real time graphics. The visualization is based on a layered coarsening of the geometrical representation. The algorithms presented in this paper guarantee a framerate of at least 60fps on standard PCs.
The concept is implemented in a high-fidelity simulator and currently used for psychological research on traffic and driver behaviour.

1 Introduction

It is commonly understood that driving simulation is an efficient tool in driver training as well as in traffic research. The more simulation is used in both areas the higher performance the user asks for. Therefore, new flexible concepts are needed. One of them is dynamic modelling of road networks.

Conventional Driving Simulators
Conventional driving simulators use a part of a virtual or real world as a database for the scenery (see [1]). They define a road network based on the geographical information about the world. Therefore, the road network is fixed and cannot be changed during simulation. A road network of this kind is called globally geometrically consistent and can be represented by a map.

Requirements of the Simulator at the IZVW
The restriction to drive within a fixed database often conflicts with the requirements of experimental research.
The following example - coming from research carried out at the Center for Traffic Sciences - may demonstrate the practical requirements:

To examine the effects of workload on driving, the driver has to handle a specific sequence of situations.

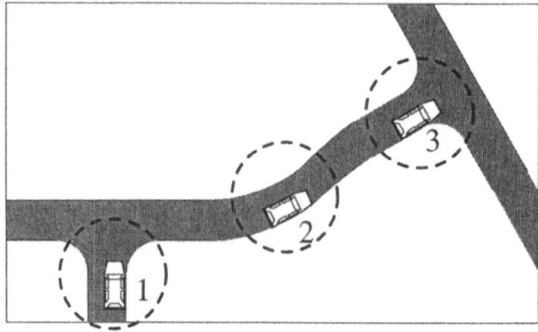

Fig. 1. Locally geom. consistent road network

Figure 1 shows an example. The circles around the driver illustrate his region of visibility. The radius of the circles in the simulation is about two kilometres. The driver is in position 1 and should turn left. Ignoring the instructions of the experimenter, he turns right. If the driver is in position 2 the crossroad of position 1 is moved to position 3 and the requested situation is repeated until he turns left.

The basic principle of the example in Figure 1 is that the road network is adapted to the behaviour of the driver. To make this adaptation possible, the experimenter has to define a topological representation of a road network. The resulting network is no longer globally geometrically consistent but only locally geometrically consistent. "Local" means that only the driver's region of visibility is geometrically consistent. Outside this region the road network can be changed without considering geometric limitations (see [2], [5]).

As a consequence the polygons for visualization have to be computed for every frame and cannot be built offline as for fixed databases. The framerate of this visualization must not drop below 60 fps to meet the requirements of modern driving simulators. In addition it should be possible to run the simulation completely on standard hardware (PCs) instead of specialized graphics computers.

This paper shows an alternative method for design and visualization that is built up at runtime. The main purpose is to find algorithms for visualization that satisfy our needs with as few resources as possible. The saved resources will be needed for changing the scenery at runtime as well as for traffic simulation.

2 Road Network

The road network is the basis of the scenery. It is a graph $RN = (V, E)$. The set of nodes V is divided into two parts: $V = COURSES \cup AREAS$. A $v \in COURSES$

is a continuous road segment of arbitrary length without turn-offs. The profile of the road's cross section (number of lanes, their width, type, ...) does not change. A $v \in AREAS$ is a rectangular area containing lanes that model junctions as well as transitions between $COURSES$. An $AREA$ can have a different cross section at every port in order to connect $COURSES$ with different cross sections e.g. drive-ups to highways.

Each $v \in V$ has ports $v.p_i$ that connect to other nodes of V. A $v \in COURSES$ has the ports $v.BEGIN$ and $v.END$. A $v \in AREAS$ has an arbitrary number of ports $v.p_1, \ldots, v.p_n$. Each port is placed at one side (top, bottom, left, right) of the underlying rectangle. The ports on one side are labeled as $TOP_1, TOP_2, \ldots,$ $BOTTOM_1, \ldots$. An edge out of E always connects ports of two nodes: $E = \{(v.p_i, w.p_j) : v, w \in V\}$.

From a topological view these connections can be ambiguous, as shown in Figure 2. A node can be connected to more than one other node as for $COURSE$ C_1 in the example. If the region "behind" or on the right side of $COURSE$ C_1 gets visible for the driver, one of these connections must be chosen to solve the ambiguity and create a geometrically consistent road network. They are resolved by an algorithm called geometric instantiation (see Algorithm 1).

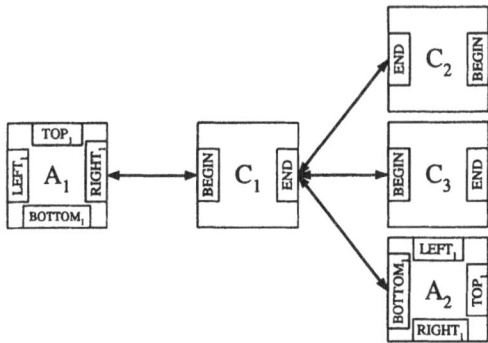

Fig. 2. A simple road network: $C_1, C_2, C_3 \in COURSES$; $A_1, A_2 \in AREAS$

Remark on Figure 2:
The three connections from $C_1.END$ to $C_2.END$, $C_3.BEGIN$ and $A_2.BOTTOM_1$ represent alternatives in the road network. They are selected at runtime, as described in section 3.5.
In this paper we will focus on modelling and visualizing courses.

3 Modelling of *COURSES*

The model of a course consists of the following three parts:

1. Surface of the road
2. Height profile of the surrounding landscape
3. Objects on the surface

3.1 Modelling of the Road

The path a *COURSE* takes through the environment is described by a unit speed curve. Let c be a *COURSE* of length S. The unit speed curve $c.p : [0, S] \to \mathbb{R}^3$ specifies the path of the road centre. A position on this curve is specified by a number $s \in [0; S]$, which indicates the distance in meters from Port *BEGIN*. For the sake of convenience this number will be called "roadposition" from now on. The paths can be restricted to conform to german guidelines for road design (see [6]).

3.2 Height Profile of the Surrounding Landscape

It has to be guaranteed that the simulation will not reach undefined states in the case the driver leaves the road. Therefore, the representation of the surrounding landscape must provide detailed information. For example, the physical vehicle model needs the angles of inclination in each point of the landscape.

Let $c \in COURSES$ be a node of a road network. Each *COURSE* has a landscape on the left and on the right side of the road. On each side of c, a set of so-called height functions $\{F_1(d), \ldots, F_n(d)\}$ running perpendicularly to the unit speed curve is defined. $F_i(d)$ starts at a certain position $s_i \in [0, S]$ on the road. It defines the elevation of a point (relative to the road surface) which has distance d to the road. An example of a height curve can be:

$$F_i(d) = G(d, d_g^{(i)}) \cdot \left(A_1^{(i)} \cdot \sin\left(f_1^{(i)} \cdot d\right) + A_2^{(i)} \cdot \sin\left(f_2^{(i)} \cdot d\right) + o^{(i)} \right) \quad \text{(HC)}$$

Two superposed sine waves and a constant offset $o^{(i)}$ are multiplied by a smoothing function G. The function F_i would then be specified by so-called shape parameters: $A_1^{(i)}, A_2^{(i)}, f_1^{(i)}, f_2^{(i)}, o^{(i)}$.

A smoothing function G has the following properties:

1. $G(0, d_g^{(i)}) = 0$
2. $G(d, d_g^{(i)}) = 1$ if $d \geq d_g^{(i)}$
3. $G'(0, d_g^{(i)}) = G'(d_g^{(i)}, d_g^{(i)}) = 0$

Independently from the slopes of the sine waves and the offset, this function ensures a smooth increase of elevation up to a distance $d_g^{(i)}$ from the road.

The maximum value of d is computed at runtime in a such way that no intersections of two height functions exist on the inner side of a bend. The height of points beyond such an intersection is ambiguous and therefore useless. The surface in such areas is just plane. This convention is sufficient because offroad driving is not intended.

Let s^* be an arbitrary point lying between two pre-defined height curves e.g. $s_i < s^* < s_{i+1}$. Let p^* be a point on the virtual height curve in s^* with a distance d^* to the road (see Figure 3). To determine the exact height of p^*, we must interpolate the height between the two neighbouring height curves. Therefore we create a new height curve by a linear interpolation of the neighbouring parameters. With increasing distance between s_i and s_{i+1} we have to build more than one additional height curve.

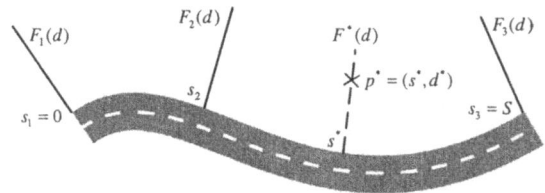

Fig. 3. A COURSE with height functions on its left side

3.3 Objects on the Surface

To design a realistic landscape, various objects obj must be placed onto the surface. Therefore, each $COURSE$ c has a list $c.objList$ of objects obj. Each object consists of a roadposition $obj.s$, a distance $obj.d$ and a 3D graphical representation. The graphical representation of an object is placed at the position specified by $obj.s$ and $obj.d$.

3.4 Automatic Generation of Landscapes

The presented way of modelling a course is very time consuming and in part difficult to handle. One reason is that the experimenter has to specify a lot of height curves. In most cases he is not interested in details of the landscape, he only wants to control the basic appearance by a few parameters. Therefore, the experimenter has the option to specify the following parameters:

1. altitude difference
2. ripple
3. offset

For each parameter the experimenter may specify an interval. During simulation a random number is drawn from the interval and assigned to the appropriate parameter. Altitude difference controls the amplitudes and ripple controls the frequencies of the sine waves defining height curves. Height curves are created every 500m at each side of a course automatically.

Designing a realistic surface also means to distribute many objects within the landscape. To shorten this lengthy operation, the experimenter has the option to select a landscape type. At the moment there are three types: farmed, forested and inhabited. The objects which are characteristic for a specific landscape type are placed on the surface in a reasonable way by using R-Trees (see [3] as a method for spatial indexing)

Regarding the example from the beginning, it is now possible to vary the height profile as well as the objects on the landscape every time a node gets visible for the driver.

3.5 Geometric Instantiation

In this section the subgraph of the road network representing the drivers' region of visibility will be constructed. The first *COURSE* in this subgraph is the one nearest to the driver. Starting with this *COURSE*, the geometric instantiation algorithm traverses the road network along the ports. If there is more than one course connected to a port the algorithm selects one of them according to criteria defined by the experimenter. Furthermore, any *COURSE* in a distance of more than two kilometres to the driver is defined as being outside the region of visibility.

During simulation a global *timer* is counting the number of displayed frames. Every *COURSE* has a variable *tInstantiation* for saving the highest frame number at which the associated *COURSE* was visible. This variable is used for avoiding infinite loops (particularly if areas are also used for creating a road network). In addition, every port has a pointer called *edge* where a reference to the next visible *COURSE* is saved.

Algorithm 1 *COURSE*::Geom_Inst(*COURSE* caller)

    if tInstantiation == time **then** {abortion of recursion}
        return;
    end if
    bool outside = (*COURSE* is completely invisible ?)
5: **if** outside **then**
        Reset edges of port *BEGIN* and port *END*
        return
    end if
    if tInstantiation < time-1 **then** {*COURSE* not treated in last inst. step ?}
10:     generate_height_profile();
        transform(caller); {transform *COURSE* smoothly to caller}
        distribute_objects();
        Port p = Port *BEGIN*;
        while $p \in \{BEGIN, END\}$ **do** {loop over all ports connected to *COURSE*}
15:         p.edge = SelectEdge(); {select edge out of the alternatives of port p}
            p = NextPort();
        end while
    end if
    tInstantiation = time {set instantiation timestamp}
20: (Course connected to *BEGIN*)→Geom_Inst(this)
    (Course connected to *END*)→Geom_Inst(this)

Remarks on the geometric instantiation algorithm:

Line 10:
: Due to the definition of the experimenter a new height profile is generated for *COURSES* getting visible.

Line 11:
: A *COURSE* getting visible has to be connected to the calling *COURSE*. Therefore the path of their roads has to be connected smoothly. In addition the first height curve of the new *COURSE* is set to the same values as the last height curve of the *COURSE* already visible.

Line 12:
: A set of objects is chosen and distributed on the surface generated in the line before.

Line 15:
: If a port is connected to more than one *COURSE*, one of them is chosen due to various measurements (already described in this paper).

Line 20, 21:
: Every neighbouring *COURSE* to the current *COURSE* will be instantiated.

4 Visualization of *COURSES*

This section focuses on the visualization of the subgraph given by algorithm 1. First a visualization of the road surface is generated. Then, the surface of the landscape will be generated and the objects on the terrain will be visualized.

4.1 Surface of Roads

In order to visualize the road of a *COURSE* c, a sequence of quadrilaterals (quadstrips) parallel to the centre of the road has to be generated.

First of all a set of points is chosen on the unit speed curve, at which the vectors of the quadstrips are calculated. These points are distributed along the *COURSE* unevenly: Closer to the driver and in curves the point density is higher in order to ensure smoothness of representation; otherwise, the distances between points are increased to save resources in visualization.

All other elements of a road which are parallel to the unit speed curve are visualized in the same way. Vertical quadstrips can be used to create guardrails or barriers. Examples for these elements are guardrails or noise barriers.

Figure 4 shows the visualization of a *COURSE*. The larger the distance to the driver, the coarser is the subdivision; the higher the curvature, the finer is the subdivision.

4.2 Landscapes of Courses

Starting with the boundary points used for visualizing the road surface, the surface of a landscape is visualized by a series of quadstrips. These quadstrips

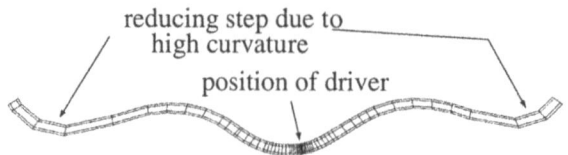

Fig. 4. Visualization of a road

are placed parallel to the road. The width (parallel to the path of a road) of each quad between two boundary points depends on the distance to the driver.

Figure 5 shows a complete subdivision of a landscape which was generated according to these conventions. In the figure the surface is restricted to the areas in which no height curves overlap each other.

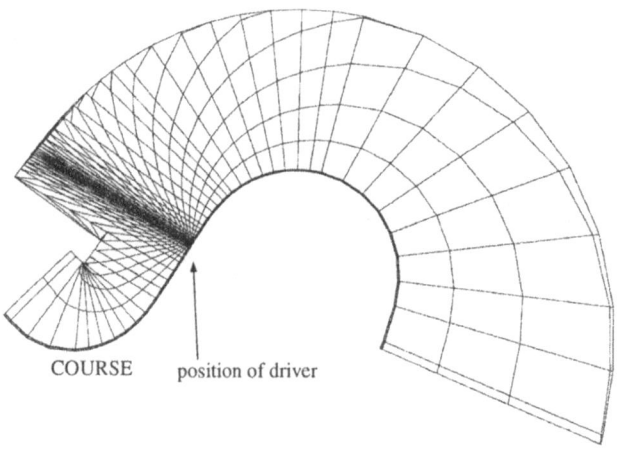

Fig. 5. Visualization of a landscape

The generated subdivision is aligned perpendicularly to the road. Along the road the subdivision of the landscape depends on the subdivision of the road and not on the distance to the driver.
Therefore the length of a quad must be greater than a minimal value. This minimal value depends on the distance to the driver. If the length of a quad is smaller than the value, a boundary point is dropped out. Figure 6 shows the same landscape as Figure 5 using this advanced method.

The presented methods generate a surface consisting of polygons. Local extreme values will be visualized if they coincide with the corner of a polygon. This will not be the case if they lie inside or on the edge of a polygon. This would result in a surface which seems to "flutter". To avoid this effect, a simple heuristic algorithm was developed which detects those points or at least points

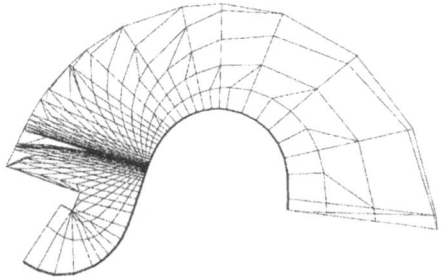

Fig. 6. Visualization of a landscape with subdivision tangential to the road

nearby. In equation (HC) two superposed sine waves are used to build a height curve. Assuming that the first sine wave has a very low amplitude in comparison to the second one, it suffices to calculate the extreme values of the second and to disregard the influence of the first sine wave.

Figure 7 shows the number of vectors used for a surface. The three described methods are compared concerning the number of required vectors.

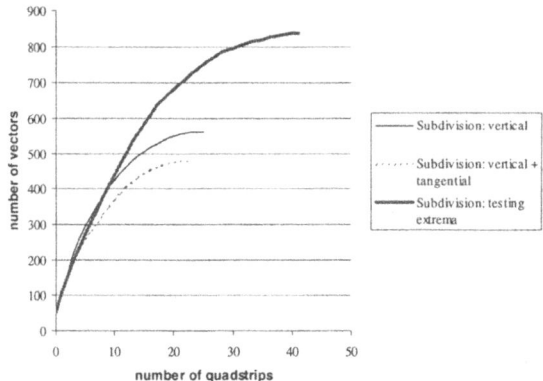

Fig. 7. Comparison of methods generating surfaces of landscapes

The x-axis gives the number of quadstrips parallel to the road which have been used to visualize the surface. We can save lots of vectors using these subdivision techniques.

4.3 Objects on the Surface

The next step after visualizing the surface is to place objects within the landscape. Let obj be an object on a landscape. The associated graphical representation must be transformed in such a way that the reference point has a distance of $obj.d$ to the road at roadposition $obj.s$.

Figure 8 shows an example of a *COURSE* with different height profiles and landscape types on the left vs. right side of the road and various examples of objects on the surface.

Fig. 8. Example of a visualization of a road network

The usage of predefined graphical objects which are transformed to fit to the surface makes it possible to change the appearance of the surface. This means that the driver cannot realize that he is driving on the same course.

5 Conclusion

The presented methods enhance the use of driving simulators in traffic sciences. Dynamic modelling of road networks and automatic generation of landscapes open up new experimental procedures for driving simulation. Simulation becomes fully adaptive to the behaviour of the driver, offering him or her the adequate driving situation throughout the whole simulation.

References

1. M. Desrochers, "Database Structures", Tagungsband 6. Workshop Sichtsysteme - Visualisierung in der Simulationstechnik [Proceedings of the 6.th Workshop for Visual Systems - Visualization for Simulation Technology], Shaker Verlag, Bremen, 1999, pp. 165-190.
2. M. Grein, A. Kaußner, H.-P. Krüger, H. Noltemeier, "A flexible application framework for distributed real time systems with applications in PC based driving simulators", Proceedings of the Driving Simulation Conference, DSC2001, Sophia-Antipolis France, 2001.
3. A. Guttman, "R-TREES: A dynamic index structure for spatial searching", Proc. ACM SIGMOD Int. Conference on Management of Data, Boston, 1984, pp. 47-57.
4. J. Hammes, "Modeling of ecosystems as a data source for real-time terrain rendering", LNCS 2181, Springer, 2001.

5. A. Kaußner, M. Grein, H.-P. Krüger, H. Noltemeier, "An architecture for driving simulator databases with generic and dynamically changing road networks", Proceedings of the Driving Simulation Conference, DSC2001, Sophia-Antipolis France, 2001.
6. Forschungsgesellschaft für Straßen- und Verkehrswesen (Arbeitsgruppe Strassenentwurf), "Richtlinien für die Anlage von Straßen RAS Teil: Linienführung RAS-L und Querschnitte RAS-Q" [German Guidelines for Road Construction], 1995.

EoL: A Web-Based Distance Assessment System

Osvaldo Gervasi[1] and Antonio Laganà[2]

[1] Department of Mathematics and Computer Science, University of Perugia,
via Vanvitelli, 1, I-06123 Perugia, Italy
ogervasi@computer.org
[2] Department of Chemistry, University of Perugia,
via Elce di Sotto, 8, I-06123 Perugia, Italy
lag@unipg.it

Abstract. An implementation of a Web-based distance assessment system, called **EoL**, developed and used at our University is presented. **EoL** is based on the outcome of the DASP project, devoted to the evaluation of the competencies acquired by stagers during in-company placements in the *Leonardo da Vinci* program of the European Commission for lifelong learning. We discuss here the architecture of the system and the related methodological aspects. Results obtained are discussed.

1 Introduction

The software systems designed for the distance assessment of competencies and skills through Internet and Web environments are powerful and ubiquitous evaluation instruments. They have the social added value of facilitating equal opportunities and reducing barriers for disadvantaged people. Their assemblage makes use of combined statistical treatments, hypertexts and networking tools.

In this paper we present an assessment system, called **EoL** [1], developed by our group and adopted as an assessment system for the examination of students of university courses during the last five years.

EoL has been developed starting from the outcomes of the European project *DASP - Distance Assessment System for accreditation of competences and skills acquired trough in company Placements* [2,3,4] (1997-2000) aimed at:

1. implementing a distance assessment system for the evaluation of the competencies and skills acquired by the stagers during the Leonardo da Vinci program activities. The stages performed in small and medium enterprises (SMEs), in Research Centers and Educational Institutions have been a successful complement of their education curricula to optimize their professional training.
2. implementing tools for the self-assessment and Open and Distance Learning, oriented to lifelong learning.
3. facilitating the mapping of the competencies and skills of the stagers, acquired during the professional training on the necessities of the Enterprises (in particular of SMEs).

EoL produces individual tests assembled by selecting a certain set of questions from a bank created by the teacher of the course to be assessed. This makes the assessment process more tailored to the actual teaching contents.

The selection algorithm guarantees that the questions are extracted following the rules defined (or modified) by the teacher (a detailed description of the criteria adopted for the selection of each question will be given in sec.3.2) enabling the teacher to assess in each individual test all arguments of the course and to obtain a homogeneous level of difficulty (averaged over all the questions of the named test).

In each test scores are assigned according to the answers returned by the user in a quantitative form.

EoL may be accessed simply through a Web browser. As discussed in section 3, when **EoL** is being used for formal examinations, some restrictions are adopted in order to guarantee the reliability of the assessment. In particular, security mechanisms have been implemented in the back-office side in addition to the security mechanism implemented on the server side.

The paper is structured as follows: in sec. 2 we present the architecture of the distance assessment system; In sec. 3 the implementation and the functionalities of **EoL** system are presented; in sec. 4 some considerations about the application of **EoL** in the industry are outlined and, finally, in section 5 some conclusions are drawn.

2 The EoL Architecture

EoL is a Web-based distance assessment system entirely based on Opensource components. The main software components of **EoL** are the following:

1. The PHP language parser
2. The Web server Apache, enabled to invoke the PHP language parser
3. The RDBMS Firebird or the RDBMS MySQL.

EoL has been extensively tested and used on two Linux Operating System environments, the RedHat 7.3 and the RedHat 8.1 releases, but the portability of the **EoL** system on other Linux distributions or other Unix systems depends on the availability of the above mentioned software components on the lasting system.

The architecture on the Server side of the distance assessment system **EoL** is shown in figure 1. As shown in the figure the whole information system of **EoL** may be viewed as made of two separated agents: the *front office agent* and the *back office agent*.

2.1 The EoL Front Office Agent

The front office agent is accessed by the Web browser allowing the user to perform the assessment test. Before enabling the user to enter the test, the teacher has to activate the session for the named examination, entering the accounting data

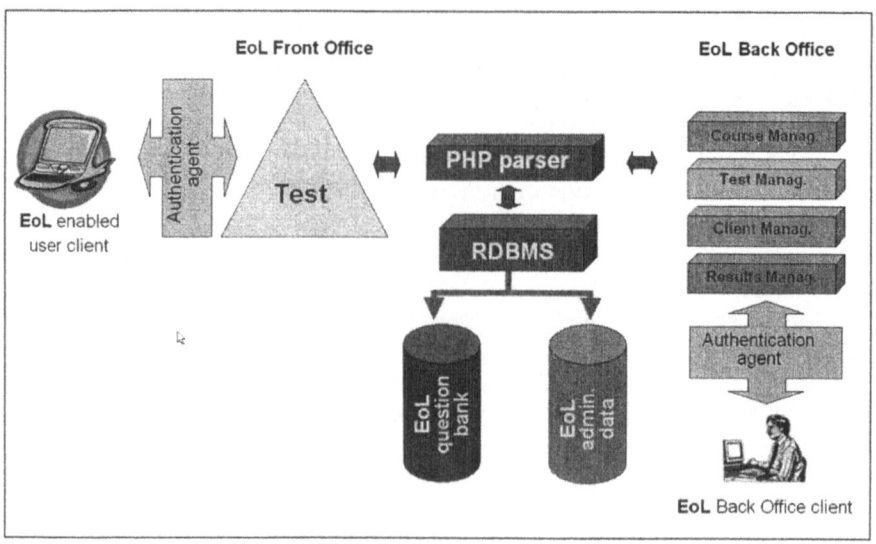

Fig. 1. The architecture of **EoL**

(login and password) specific for the named test session. After **EoL** system has verified the accounting data specific of the current examination session, a check is performed to verify whether the IP address of the client belongs to the set of hosts allowed to connect to the **EoL** server. Once the check has been passed, the user is requested its identification data and can start the test answering to the questions in a strict sequential fashion. The result of the tests will be issued by the teacher after reviewing the answers and having evaluated the performance as a whole.

Each assessment test must be performed under the visual control of an assistant to avoid frauds and cribs. The computers clients must be able to connected to **EoL** server through Internet. A computer room is a good example of a site able to access **EoL** to perform quantitative distance assessment tests.

The output of each test is stored on the **EoL** administrative bank.

2.2 The EoL Back Office Agent

This agent relies to the completely on the work carried out by the teacher as far as the implementing of the question bank is concerned. The question bank is articulated in sections, each related to a specific course. The back office agent manages each course, analyzes the results of the various tests and manages the clients. The access via Web to the management section of this agent is restricted to the teacher's login and password.

The back office is composed by the following sub-agents:

1. Courses management
2. Test sessions management

3. Clients management
4. Test results management

Sub-agents will be described in detail in the next section.

2.3 EoL as a Self-Assessment System

EoL is designed to perform the distance assessment of large numbers of students at the same time. However, a special module of EoL can also be used to carry out individual self-assessment in distance learning environments. In this case a simplified approach has been implemented both for the algorithm selecting the questions and for the choice of the pool from which questions are going to be selected. In fact, to implement the self assessment environment, only a limited number of questions is made available. This leads to a drastic reduction of the complexity of the algorithm while still keeping all the related functionalities. In particular only a small subset of questions per pool is allowed. An *intelligent assessment system* (see for example [5,6]) is being implemented in order to improve the efficiency of EoL.

3 The Management Functions of EoL System

The EoL back office agent is concerned with the main management functions of the system and is composed by the set of sub-agents already mentioned in sec. 2.2. Each sub-agent performs a set of complex functions and a detailed description will be given in this section.

3.1 Courses Management

This section is related to the management of the test of the various courses related to a given teacher. There is a relationship one-to-many between the teacher and each set of course tests since the material of each course may be managed by more than one teacher. A key prerogative of the teacher is the definition of the main characteristics of the tests: the type of test (single or double path, see below for details about this feature), their name, the login needed to enable the clients to perform each test, the total number of questions per test, the scale in which the results are expressed (tenths, thirtieths, etc), the value of the mean difficulty coefficient D_T, the value of the optional bonus[1] the teacher wants to assign to each test.

As said above EoL enables two types of test: the *single path* test and the *double path* test. In the single path test system all questions are selected from the same bank and have an homogeneous scoring system. In the double path test system, after some preliminary questions selected from a *general knowledge (GK) bank* the remaining questions are chosen from a *specific knowledge (SK) bank*. In

---

[1] We have experienced some time the need for shifting the evaluation scale to compensate some biases of the adopted questions bank.

this way it will be possible to activate for the user two different assessment paths, each with a specific scoring system. In fact, depending on the score obtained in the first part of the test during the GK phase, the questions of the SK phase, are selected according to either a low profile path or a high profile path.

A low profile path for the SK phase will be chosen when the score obtained by the user in the GK phase is lower than a given threshold T. On the contrary, the high profile path will be chosen when the score obtained in the GK phase is larger than T.

In the case of the low profile path the questions of the test will be selected in a way that the value of the mean difficulty coefficient D_T is lower than that of the questions selected in the high profile path. The score assigned to a correct answer in the SK assessment in the low profile is accordingly lower.

After having defined the course, the teacher is able to manage (through the functions add/delete/modify) the main topics of the course and, for each topic, the specific questions of the bank. For each question the teacher has to specify the text of the question, optionally including the specification of a multimedial component, like an image, a drawing or a movie, and the text of each answer. The text can be specified in one or more languages (the supported languages are at the moment English, Spanish, French, German and Italian).

Two types of systems are currently supported for the question specification: the multiple choice system (MCS) and multiple response system(MRS). In both cases the teacher has to specify the text of the answers and the score associated with them. In the MCS system one answer is true (score = 1) and the remaining answers are false (score = 0). In the MRS system a set of answers are true and contribute to the whole score; the set of wrong answers give a negative contribution to the score of the specific question though, if the overall score for that question is negative, it is set equal to zero.

3.2 The Algorithm Adopted for Selecting the Questions from the Bank

EoL selects the questions out of the data bank so as to have N_q questions for each of the N_a subjects of the course (accordingly the total number N_T of questions of the test is $N_T = N_a N_q$). Each question i has its own difficulty d_i and the selection of the question is driven by an algorithm that minimizes the deviation δ of the average difficulty (D_s) of the N_s already selected questions from the requested difficulty (D_T) of the test

$$|D_s - D_T| \leq \delta \qquad (1)$$

with D_s being defined as

$$D_s = \sum_{i=1,N_s} \frac{d_i}{N_s} \qquad (2)$$

3.3 Test Sessions Management

This agent is concerned with the management of the tests already performed and the definition of a new test session.

When a test session has been completed (i.e. all tests has been performed and the teacher has defined the final score of each test) the test session may be archived, to inhibit further modifications.

Using this agent a teacher is enabled to define a new test session, choosing the test session's name (selected among those courses for which the teacher is administratively enabled and proposed on a menu by the system), the session's date and time, the expected number of students attending the session (this number may be later varied) and the name of the room where the test will be performed. Finally, the teacher selects from a menu showing all enabled clients, which pool of clients will be used for the test. This option enables the strongest control mechanism for the test management: only requests to the front office agent coming from an IP address poll defined in this section will be granted access to the test environment.

During the test session the teacher will enable each client opening a Web connection to the site http://eol.unipg.it and entering on the welcome page the user and password associated to the named test session as generated during the test session definition phase.

Each user will then be able to enter its identification data and start the test session.

At the end of the test the Web session is destroyed. At this point new authentication data are required and a new test session can be opened as described above.

3.4 Client Pools Management

This agent is related to the management of the client pools enabled to access **EoL** system for the test sessions. Requests coming from the authorized IP address pools will be allowed to enter the welcome page of http://eol.unipg.it. The agent allows to modify an existing entry or to add a new client pool.

The definition of a pool consists simply on the specification of a range of IP addresses (in the form *first IP - last IP*) and a mnemonic name associated to the client pool. When the clients are configured using private IP addresses making usage of the Network Address Translation (NAT), attention must be payed to check which are the public IP addresses from which all Web requests are coming because this are the only values that identify the named client pool to **EoL** system.

3.5 Test Results Management

This agent is accessed by the teacher at the end of the test session in order to review the whole set of tests of the session and to set the final score of each test.

The agent presents a web page extracting the information from the **EoL** administrative bank where the output of each test has been stored by the front office agent. In the web page a table representing the list of all tests performed in the session is shown. In each row of the table the information about the user and the final score are shown as hyperlinks. Once selected the hyperlink the detailed information regarding the named test are shown (administrative data of the user, date, starting and ending time, duration of the test) followed by the detailed summarization of each question (text of the question, answers selected with explicit symbols indicating if the response is exact or not and if an exact response has been omitted). An example of the web page described above is shown in figure 2.

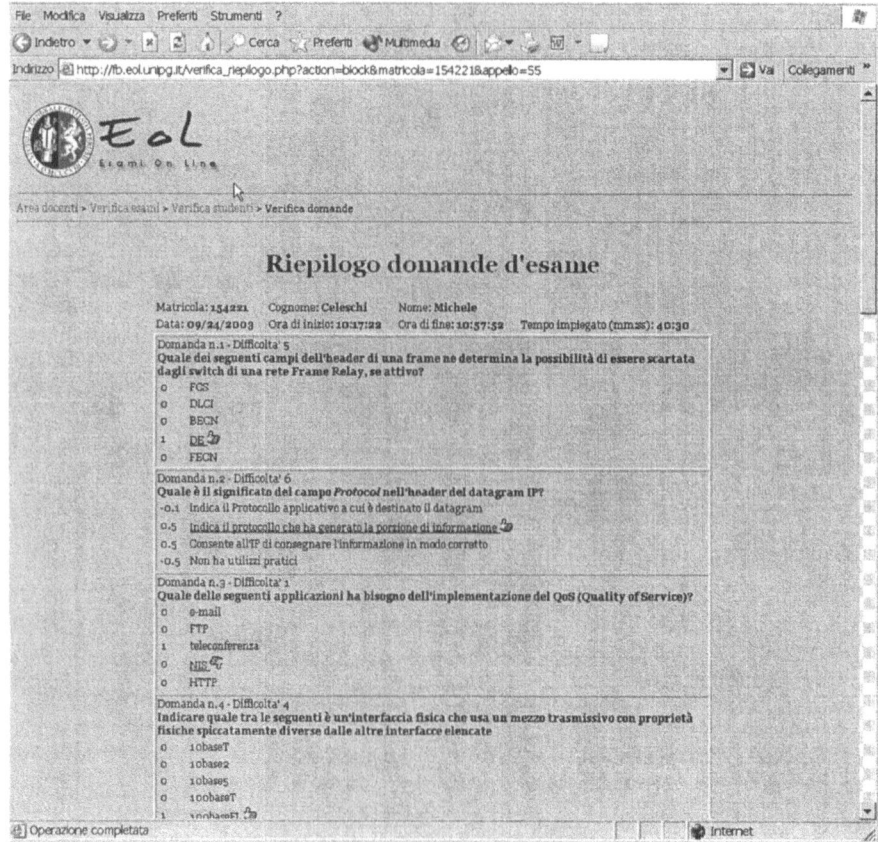

Fig. 2. An example of the test results management environment of **EoL**.

In the bottom of the page the score assigned by **EoL** and a line allowing the teacher to set the final score of the test are shown. The manual intervention of the teacher is necessary because sometimes the text of the question it is unclear or ambiguous. A future development of **EoL** system will be to insert here an intelligent assessment agent that may reduce or even replace the human intervention [5,6].

4 Industrial Applications of EoL

As already mentioned, **EoL** has been already used in an industrial context for the assessment of students' placing. **EoL**, however, could be also used for the selection of new employees or the promotion to a new responsibility or in any other case in which traditional methods (like interviewing and reference checking) are unable to provide a sufficient support for the final decision.

EoL based approach could provide additional evaluation support on

Productivity : people are more suitable and capable to perform the function.
Morale : people have been selected because of their abilities
Quality : the right person is doing the right job
Customer Satisfaction : a more reliable product is made by people skilled and aware of the matter
Training Outcomes : the assessment process allows both to evaluate the goodness of the training performed and its optimization.

The use of products like **EoL** will reduce:

turnover : people is less frustrated for covering inadequate functions and shows more fidelity to the position covered.
inappropriate hiring decisions : the assessment process is more reliable
tardiness : people are involved in a positive way
absenteeism : people are interested on their functions and believe in the role play.
theft : a positive behavior is more likely when people is well inserted in the working environment
fraud : people do not need to do bad actions against a positive environment

5 Conclusions

In this work we have presented a prototype implementation of **EoL**, a distance assessment system available on the Internet and currently used in some university courses with success. We have described the architecture of the system and we have presented the achievements obtained in developing such system on a free-software platform. In particular the back-end system of **EoL** is important because enables the teacher to control very important aspects of the assessment system. The future developments of **EoL** and its possible industrial application have been also outlined.

Acknowledgments. We acknowledge all colleagues that have contributed to the implementation of the system. In particular we thanks Federico Giorgetti, Francesca Tufi, Donatella Ermini for the work done and Sergio Tasso for the precious suggestions and support. A special mention is made to Leonardo Angelini, Agnese Beatrici and Web-Engineering for the developments made in **EoL** and for providing an ad-hoc PHP Framework environment. This research has been partly financially supported by the European Commission under the Leonardo da Vinci program.

References

1. The **EoL** system is available at the URL: http://eol.unipg.it
2. The project *Distance Assessment System for the evaluation of competencies and skills acquired through in-company Placements* (DASP) has been carried out by a group of partners led by Perugia University in the Leonardo 2 initiative of the European Commission; see:
   http://leonardo.cec.eu.int/pdb/Detail_fr.cfm?numero=831&annee=97;
3. Gervasi, O.,Giorgetti, F., and Laganà, A.: Distance Assessment System for Accreditation of Competencies and Skills Acquired Through in-Company Placements(DASP), INET99: The Internet Global Summit, S.Jose, CA (USA), 23-26 June 1999;
   http://www.isoc.org/inet99/proceedings/posters/216/index.htm;
4. Martinez-Rubio, J.M., Dominguez-Montagud, C., Perles-Ivars, A., Albaladejo, J., Hassan, H.: Distance Assessment System for Accreditation of Competencies and Skills Acquired Through in-Company Placements, 2000 Int. Conf. on Engineering Education (ICEE 2000), Taipei, Taiwan, 14-18 August 2000;
   http://www.ineer.org/Events/ICEE2000/Proceedings/papers/WB6-1.pdf
5. Shen, R.M., Tang, Y.Y. Zhang, T.Z.: The intelligent assessment system in Web-based distance learning education, Frontiers in Education Conference, 2001. 31st Annual , 1 , 10-13 Oct. 200, 17-11
6. Patel, A., Kinshuk, K., Russell, D.: A computer-based intelligent assessment system for numeric disciplines, Information Services & Use, 18, n.1-2, (1998) 53–63;
   http://citeseer.nj.nec.com/patel98computer.html

Discovery Knowledge of User Preferences: Ontologies in Fashion Design Recommender Agent System

Kyung-Yong Jung[1], Young-Joo Na[2], Dong-Hyun Park[3], and Jung-Hyun Lee[4]

[1] HCI Laboratory, School of Computer Engineering, Inha University, Incheon, Korea
   kyjung@gcgc.ac.kr
[2] Department of Clothing and Textiles, Inha University, Incheon, Korea
   youngjoo@inha.ac.kr
[3] Department of Industrial Engineering, Inha University, Incheon, Korea
   dhpark@inha.ac.kr
[4] School of Computer Engineering, Inha University, Incheon, Korea
   jhlee@inha.ac.kr

Abstract. Solutions for filtering the WWW exist, but they are hampered by the difficulty of discovery knowledge of user preferences in such a dynamic environment. We explore an ontological approach to discovery knowledge of user preference in fashion design recommender agent system. Information filtering can be used for the discovery knowledge of user preference and is therefore a key-technology for the construction of personalized recommeder system. In this paper, we focus in the application of hybrid collaborative filtering and content-based filtering to improve the performance. And we validate our web based fashion design recommender agent system according to discovery knowledge of user preference in on-line experiments. Design merchandizing may meet the consumer's needs more exactly and easily.

1 Introduction

The WWW hypertext system is a very large distributed digital information space. Some estimates suggested that the Web included about 150 million pages and this number doubled every 4 months. As more information becomes available, it becomes increasingly difficult to search for information without specialized aides. During a browsing session, these recommender systems work collaboratively with a user without the need of an explicit initiation. It has a static or dynamic collection of pages to be recommended. It assists the user by recommending items that match his/her needs [1]. Many users may find articulating what they want hard, but they are very good at recognizing it when they see it. This insight has led to the utilization of relevance feedback, where user rate web pages as interesting or not interesting and the recommender system tries to find pages that match the interesting example and do not match the not interesting examples. With sufficient positive and negative examples, modern machine learning techniques can classify new pages with impressive accuracy. Obtaining sufficient examples is difficult. However, especially when trying to obtain negative examples. The problem with asking user for examples is that the cost, in terms of time and effort, of providing the examples generally outweighs the reward they will eventually receives. Negative examples are particularly unrewarding, since there could be many irrelevant items to any typical query. Hybrid 2-way filtering systems, attempting to combine the advantages of collaborative filtering and content-based filtering, have proved popular to-date [17]. The user feedback required for content-based filtering is shared, allowing collaborative filtering as well. Another important issue for adaptation deals with the impact on the user interface [2,3,14]. Our work focuses on the use of information filtering techniques to construct web based textile design recommeder agent system. In particular, we address the issues of hybrid collaborative filtering and content-based filtering, as well as the user feedback. A hybrid 2-way filtering approach is used by our current prototype recommender system, FDRAS: fashion design recommender agent system.

2 FDRAS: Fashion Design Recommender Agent System

In fabric research, a Web based fabric information system has been developed. Recently, the numbers of textile companies that have their own homepages to advertise their product fabrics for apparel through the Web-based E-commerce web site rapidly increase. Unfortunately, traditional fabric information system based on direct meeting and trust cannot give sufficient information to numerous visitors of the Internet sites including fabric buyer for apparel. To develop product with sensibility, it needs to investigate consumer's sensibility for the product. After this investigation, they could develop the system that connects each sensibility to specific designs of product. There had been several studies, such as in the fields of car interior/exterior design, interior design (for example, HULIS) and apparel design (FAIMS, etc.), those related to school uniform, and wedding dresses [9,12]. These systems deal with the various aspects of structure, designs, appearance, structure, shape, color, and space. Through these systems, consumers can create the product textile designs on the computer screen simply by entering his/her preferences or needs on the product. The human sensibility on textile designs was related to their external aspects, such as motif size, motif striking and color as well as the inner aspects of their performance and comfort properties [12,13]. The purpose of this study is to develop design-supporting system based on consumer's sensibility with adaptive multi users interface. Thus, we investigated the consumers' sensibility and established databases on sensibility adjectives and on user's attributes. We developed programs connecting database and extracting proper designs according to user preference needs. And finally we proposed recommending design system with adaptive user interface window. The purpose of our research is to explore the textile design recommender agent system using hybrid collaborative filtering and content-based filtering based on server-client.

2.1 Web Based Fashion Design Recommender Agent System

FDRAS is a project that aims at studying how the hybrid of collaborative filtering and content-based filtering can be used to build the web based textile design recommender system, where the navigation through information is adapted to each user. Please refer to [12] for the detailed formulas. The chosen application is a recommender system that allows user to browse through of textile designs. In the FDRAS project, we use filtering techniques to develop the web based recommender system, in which the hypertext structure is created for each specific user, based on predictions of what this user should prefer. This basic idea is that the user is asked to provide ratings for the textile designs that he views during his visit. The system then selects textiles similar to textile design with the high rating according to content-based filtering. Collaborative filtering is also used to make predictions based on the ratings that other users have provided during previous visits. These predictions are then used to present textile designs to the user accordingly, so that more relevant textile designs are seen first. Figure 1 shows web-based fashion design recommender agent system.

Fig. 1. FDRAS's Web based Interface: Fashion Design Recommeder Agent System

FDRAS consists of server and client module. The computer uses the PentiumIV, 1.9GHz, 256 MB RAM, and the algorithms use MS-Visual Studio C++ 6.0, ASP, MS-SQL Server 2000. We will open FDRAS's full source code and sensibility database for the more development. We construct the database of design sensibilities and designs made up previously and their connecting modules including collaborative filtering and content-based filtering [9,11,12].

If the proper textile designs are not found, the system goes through information filtering technique [11]. The system lets the client evaluate his/her sensibility on a few designs to predict his/her sensibility on the other textile designs. Also the database on previous users classifies a client into a group according to their socio-information. The server module regarded the client having similar sensibility with the group with based on representative attributes of neighborhood, such as age, gender [11]. And the degree of similarity of the client with the group was calculated. By sample evaluation and user similarity, hybrid 2-way filtering could select the best-fit designs. Filtering techniques recommends the textile design that the user seeks. For filter textile designs, another solution is to make use of metadata that can be made available with digitized textile design recommeder system. Whatever the technique, personalized web based recommender system gives the user a better change at getting to information that interests user.

We developed the design recommender system to coordinate design sensibility and design products, and enhance the communication efficiency between designers and design merchandisers in design industry. The system has a convenient multi-user interface window, by which the user can input their sensibility on three sample designs into a database for collaborative filtering. The system recommend five designs on screen according to client's preferred sensibility, and he/she can select one of them and also manipulate its image with the control bar. The users can control the textile design with view factor control window, this input will change the textile of apparel on visual model to approach the sensibility preferred by him/herself. View factor control includes Illuminant, Flags[None, Negative, Logarithmic Filter, Negative & Logarithmic Filter], Basic[Contrast, Brightness, Colorfulness, RedGreenTint], Reference[Black, White], Gamma[Red, Green, Blue]. This gives the tool controlling in detail on the hue, chroma, value and texture of textile design. Therefore, design recommender system will become the referencing tool on planning sensibility product to the merchandisers or buyers of textiles and apparel industry.

2.2 The System Model and Acquiring Preference

In FDRAS project, we use filtering techniques to create textile recommeder system, in which the navigation is created for each user, based on discovery knowledge of user preference. To allow restructuring of recommder agent system based on collaborative filtering, we introduced some simplification: First, the content is determined by a corpus of textile designs. Each textile design is contained in one page, which display the textile design and further given additional information e.g. title, maker, and date. Second, the pages are organized in adaptive classes, which can be accessed through virtual adaptive classes [5,14]. By changing existing adaptive classes and create new ones, which contain reference to pages showing textile designs, it is possible to restructure the design recommender system in a way adapted to each user. Ontologies are use both to structure the web, as in Yahoo's search space categorization, and to provide a common basis for understanding between systems, such as in the knowledge query modelling language (KQML). In depth ontological representations are also seen in knowledge-based systems, which use relationships between web entities e.g. bookmark, web page, page author etc. to infer facts given situations [17]. We use ontologies to investigate how domain knowledge can help in the acquisition of user preference.

In our FDRAS, users can express preferences by giving ratings to textile designs (five value: -2, -1, 0 +1, +2). This is considered as the bimodal distribution that expresses the total of the distribution, which is biased to right, and distribution to left. As a matter of convenience of calculation and an economy of memory, this study uses the user ratings as mapping values on 1, 2, 3, 4, and 5 instead of –2, -1, 0, +1, and +2.

This is to expresses the vocabulary pair that has an opposite propensity ([−2, −1, 0, +1, +2]→[1, 2, 3, 4, 5] ‖ [−2, -1, 0, +1, +2]→[5, 4, 3, 2, 1]). The user ratings on both −1 and +1 are manifested as mapping value 2. According to the bimodal distribution of the user ratings, we define 2 as the default voting. We found in our experiment that assuming a default value for the correlation between the user rating vectors is helpful when the data set is very small. Therefore, the ratings can be conveniently provided while wandering within FDRAS. We noticed in initial trials, that users are hesitant in giving ratings, because giving a rating demand the inconvenience of having to make a decision. Therefore, we provided that the ratings can be provided with very little effort without disrupting the users chosen tour, and if a textile design is viewed in detail a rating is mandatory so that the user evaluates this textile design or otherwise cannot continue. Other access criteria can be suggested, such as the frequency of motif source, interpretation (realistic, stylized, abstract), arrangement (all-over, 1-way, 2-way, 4-way), hue contrast, chroma contrast, articulation, motif-ground ratio etc. But we did not emphasize this issue since the focus of our work is on filtering technique.

3 Performance Evaluation

We describe the metrics and experimental methodology. We use to compare different prediction algorithm; and present the result of our experiments. In order to evaluate various approaches of filtering, we divided the rating dataset in test-set and training-set. The training-set is used to predict ratings in the test-set using a commonly used error measure. From each user is the test set, ratings for 30% of textiles were withheld. Predictions were computed for the withheld items using each of the different predictors. The qualities of the various prediction algorithms were measured by comparing the predicted values for the withheld ratings to the actual ratings. The metrics for evaluating the accuracy of a prediction algorithm are used mean absolute error (MAE) and rank scoring measure (RSM) [4,7,10]. [4,7,10,11,15,16] used the same metrics to compare their algorithms. MAE is used to evaluate recommendation systems about single textile. RSM is used to evaluate the performance for systems that recommend textiles from ranked lists. MAE between actual ratings and recommender scores is a generally used evaluation measure. MAE treats the absolute error between true user rating value and predicted score by recommender system. Formally, MAE is calculated as follows Equation (1).

$$MAE(u) = \frac{\sum_{i \in I_u / r^{test}} |r_{u,i} - p_{u,i}|}{|I_u / r^{test}|} \qquad I_u / r^{test} = \{i : r_{u,i} \in r^{test}\} \qquad (1)$$

The lower MAE means that the predictive algorithm predicts more accurate numerical rating of users. The MAE for several users is then accumulated as Equation (2). |U| is the number of textiles that have been evaluated by the new user.

$$MAE = \sum_{u \in U} \frac{MAE(u) | I_u / r^{test} |}{|U|} \qquad (2)$$

The RSM of an textile in a ranked list is determined by user evaluation or user visits. RSM is measured under the premise that the probability of choosing a product lower in the list decreases exponentially. Suppose that each product is put in a decreasing order of value j, based on the weight of user preference. Equation (3) calculates the expected utility of user U_a's RSM on the ranked textile design list.

$$R_a = \sum_j \frac{\max(V_{a,j} - d, 0)}{2^{(j-1)/(\alpha-1)}} \quad (3)$$

In Equation (3), d is the mid-average value of the textile design, and α is its halflife. The halflife is the number of textiles in a list that has a 50/50 chance of either review or visit. In the evaluation phase of this paper the halflife value of 5 shall be used.

$$R = \frac{\sum_{i=1}^{u} R_i}{\sum_{i=1}^{u} \max(R_i)} \times 100 \quad (4)$$

In Equation (4), the RSM is used to measure the accuracy of predictions about the new user. If the user has evaluated or visited textile design that ranks highly in a rank list, $max(R_i)$ is the maximum expected utility of the RSM.

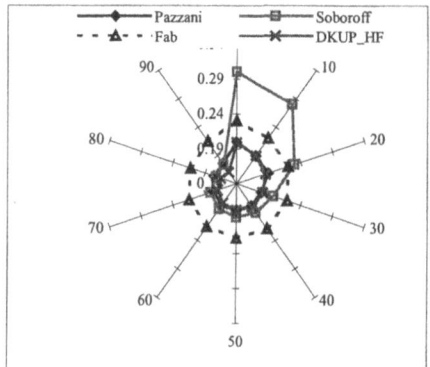

Fig. 2. MAE at nth rating at nth rating

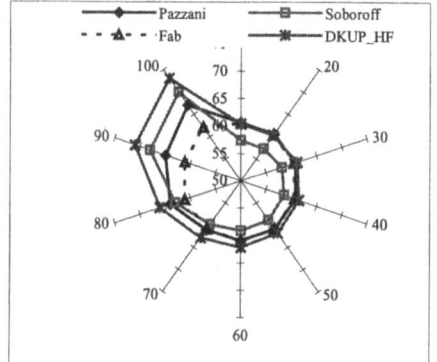

Fig. 3. RSM at nth rating at nth rating

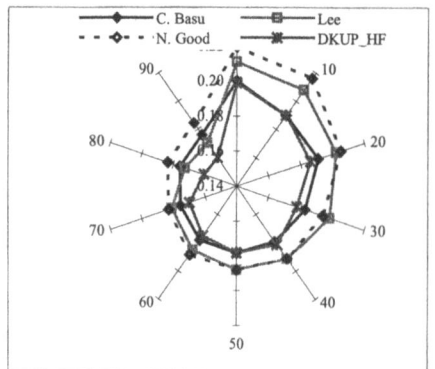

Fig. 4. MAE at nth rating at nth rating

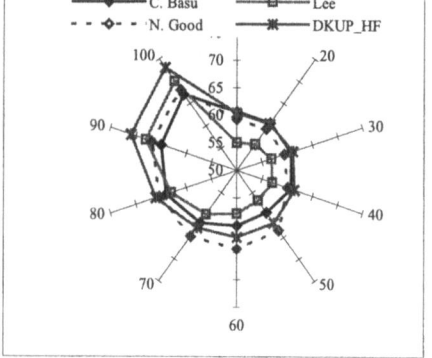

Fig. 5. RSM at nth rating at nth rating

For evaluation, this paper uses the following method: the proposed discovery knowledge of the user preference through hybrid filtering approach (DKUP_HF), Pazzani's approach [18], Fab [6], Soboroff et al. [19], N. Good et at [8], W. S. Lee [15], and C. Basu et al. [3] for the detail formulas at 4 Related Works.

Figure 2 and Figure 4 is used to show the mean absolute error when the number of user's evaluations is increased based on Equation (2). Figure 3 and Figure 5 is used to show the rank scoring measure based on Equation (4) when the number of user's evaluations is increased. In Figure 2 and Figure 3, the Soboroff method shows low performance when there are few evaluations; the other methods outperform the Soboroff method. In Figure 4 and Figure 5, the W. S. Lee method shows low performance when there are few evaluations; the other methods outperform the W. S. Lee method. Although the Pazzani method and C. Basu method along with the DKUP_HF shows high rates of accuracy, the DKUP_HF shows the highest accuracy of all methods.

Table 1. The Result of Our Experiments

Method	MAE	RSM
Content	1.057	64.7
P_Corr	1.004	64.3
N_Com	1.019	61.2
DKUP_HF	0.934	69.5

The results of our experiments are summarized in Table 1. Table 1 lists the measured precision for the previously discussed predictors. Here, it is interesting to note the improvements of discovery knowledge of the user preference through hybrid collaborative filtering and content-based filtering approach (DKUP_HF) compared to collaborative filtering (P_Corr), content-based filtering (Content), and naïve combined approach (N_Com). As can be seen, our DKUP_HF approach performs better than the other algorithms on both metrics. On the MAE metric, DKUP_HF performs 9.4% better than Content, 4.2% better than P_Corr and 4.9% better than N_Com. All the differences in MAE are statistically significant.

On the RSM metric, DKUP_HF performs 5.4% better than Content, 4.6% better than P_Corr and 9.1% better than DKUP_HF. This implies that our discovery knowledge of user preference through hybrid collaborative filtering and content-based filtering, compared to others, a better of job of recommending high quality textiles, while reducing the probability of recommending bad textiles to the user. Therefore, in terms of accuracy of prediction, it is evident that method DKUP_HF, which used the discovery knowledge of the user preference through hybrid collaborative filtering and content-based filtering, is more superior to method N_Com.

4 Related Works

There have been a few other attempts to combine content information with collaborative [6,14,16]. One simple approach is to allow both collaborative filtering and content-based filtering to produce separate ranked lists of recommendations, and then merge their results to produce a final list [7]. There can be several schemes to merging the ranked lists, such as interleaving content and collaborative recommendations or averaging the rank or ranking predicted by the two approaches. This is essentially what our naïve combined approach does.

Pazzani's approach [18] user profiles are represented by a set of weighted words derived from positive training examples using the Winnow algorithm. This collection of user profiles can be thought of as the content-profile matrix. Predictions are made by applying collaborative filtering directly to the content-profile matrix. Billsus and Pazzani use single value decomposition. They use the SVD of the original user-ratings matrix to project user-ratings and rated items into a lower dimensional space. By doing this they eliminate the need for users to have co-rated items to be predictions for each other. The distance between the new vector and every other vector is measured. If the new vector is close enough to other items then it is deemed interesting to the user and added to the short-term model. Any stories that are not close enough to the items of short-term interest are then classified by the long-term model. Billsus and Pazzani's hybrid model outperformed either of the two individual approaches.

Fab [6] is a system that helps web site users to discover new and interesting sites. Fab uses an alternative approach to combining both collaborative filtering and content-base filtering. The system maintains user profiles based on content analysis, and directly compares these profiles to determine similar users for collaborative filtering. The system delivers a number of pages that it thinks would interest the users. Users evaluate the pages and provide explicit relevance feedback to the system. Fab uses the web for collecting pages for recommendation.

Soboroff et al. [19] propose a novel approach to combining collaborative filtering and content-based filtering using latent semantic indexing (LSI). In their approach, first a term-document matrix is created, where each cell is weight related to the frequency of occurrence of a term in a document. The term-document matrix is multiplied by the normalized ratings matrix to give a content-profile matrix. The singular value decomposition (SVD) of this matrix is computed. Term vectors of the user's relevant documents are averaged to produce a representing the user's profile. New documents are ranked against each user's profile in the LSI space.

Good et at [8] use collaborative filtering along with a number of personalized information agents. Predictions for a user were made by applying collaborative filtering on the set of another users and the active user's personalized agents. Our approach differs from this by also using collaborative filtering on the personalized agents of the other users.

W. S. Lee [15] treats the recommending task as the collaborative learning of preference functions that exploits item content by multiple users as well as the ratings of similar users of recommender system. They perform a study of several mixture learning models for this task.

Basu et al. [3] integrate collaborative filtering and content-based filtering in a framework in which they treat recommending as a classification task. They use an inductive logic program to learn a function that takes a user and movie and predicts a label indicating whether the movie will be liked or disliked. They combine collaborative filtering and content information, by creating features such as comedies liked by user and users who liked movies of genre. The movies are categorized into 10 categories: action, animation, art foreign, classic, comedy, drama, family, horror, romance, and thriller, where a movie may belong to more than one category.

5 Conclusion

Unlike collaborative filtering and content-based filtering recommending hold the promise of being able to effectively recommend un-rated personalized textile designs and to prove quality recommendations to users with unique, individual tastes by exposing a personalized structure. We implemented a current prototype, the web based FDRAS: fashion design recommender agent system, which has been test by a number of users. We focus in FDRAS of hybrid collaborative filtering and content-based filtering to improve the performance. It is useful to hybrid these independent filtering approaches to achieve better filtering results and therefore better recommeder systems. Privacy has been one of the central topics when talking about the problems with collaborative filtering. Party the problem has already been solved as general web site standard have been originated. Still, the privacy will be one main area of discussion when developing the construction of personalized FDRAS. Future work in FDRAS is to make virtual environments and tours around the FDRAS available online. VRML (Virtual Reality Modeling Language) can be used to create 3-dimentional environments. It is possible to represent translation, rotation, shearing, and scaling in a matrix using perspective projection. The picture below shows a part of this 3D fitting simulation (www.dnmco.com).

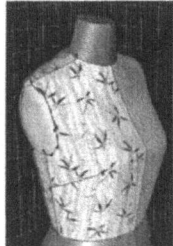

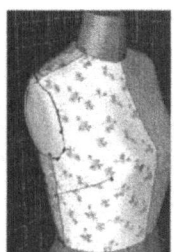
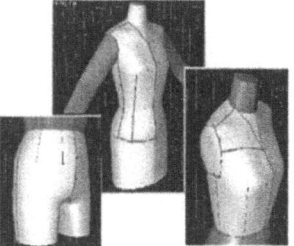

Acknowledgements. The author would like to thank the anonymous reviewers whose comments helped improve the paper. This work was supported by grant No. R04-2003-000-10177-0 from the Basic Research Program of the Korea Science & Engineering Foundation. Sincere thanks go to A. Kohrs, B. Mérialdo who provided the idea for this thesis.

References

1. M. Ahmad Wasfi, "Collecting User Access Pattern for Building User Profiles and Collaborative Filtering," In Proc. of the International Conf. on Intelligent User Interfaces, pp. 57-64, 1999.
2. C. Avery, P. Resnick, R. Zeckhauser, "The Market for Evaluation," American Economic Review, Vol. 89, No. 3, pp. 564-584, 1999.

3. C. Basu, H. Hirsh, W. Cohen, "Recommendation as Classification: Using Social and Content-based Information in Recommendation," In Proc. of the 15th National Conf. on Artificial Intelligence, pp. 714-720, 1998.
4. J. S. Breese, D. Heckerman, C. Kadie, "Empirical Analysis of Predictive Algorithms for Collaborative Filtering," In Proc. of the 14th Conf. on Uncertainty in Artificial Intelligence, 1998.
5. M. Balabanovic, "An Interface for Learning Multi-Topic User Profiles from Implicit Feedback," Recommender System: AAAI Workshop, 1998.
6. M. Balabanovic, Y. Shoham, "Fab: Content-based, Collaborative Recommendation," Communications of the ACM, Vol. 40, No. 3, pp. 66-72, 1997.
7. P. Cotter, B. Smyth, "PTV: Intelligent Personalized TV Guides," In Proc. of the 20th Conf. on Innovative Applications of Artificial Intelligence, pp. 957-964, 2000.
8. N. Good, et al., "Combining Collaborative Filtering with Personal Agents for Better Recommendations," In Proc. of the 16th National Conf. on Artificial Intelligence, pp. 439-446, 1999.
9. S. Han, et al., "A Study on Design factor Transfer Supporting System based on Sensibility Science," In Proc. of the Ergonomics Society of Korea, pp. 129-135, 1996.
10. J. Herlocker, et al., "An Algorithm Framework for Performing Collaborative Filtering," In Proc. of ACM SIGIR'99, 1999.
11. K. Y. Jung, J. H. Lee, "Prediction of User Preference in Recommendation System using Association User Clustering and Bayesian Estimated Value," In Proc. of 15th Australian Joint Conf. on Artificial Intelligence, pp. 284-296, LNAI 2557, Springer-Verlag, 2002.
12. K. Y. Jung, et al., "Development of Design Recommender System using Collaborative Filtering," In Proc. of 6th International Conference of Asian Digital Libraries, pp.100-110, LNCS 2911, Springer-Verlag, 2003.
13. M. Kim, "A Study on Sensibility Science Technology of Texture and Colour coordination", Hanyang University, Ph D. Dissertation, 1996.
14. A. Kohrs, et al. "Improving Collaborative Filtering with Multimedia Indexing Techniques to Create User-Adapting Web Sites," In Proc. of the 7th ACM International Conf. on Multimedia, pp. 27-36, 1999.
15. W. S. Lee, "Collaborative Learning for Recommender Systems," In Proc. of the 18th International Conf. on Machine Learning, pp. 314-321, 2001.
16. P. Melville, et al., "Content-Boosted Collaborative Filtering for Improved Recommendations," In Proc. of the National Conf. on Artificial Intelligence, pp. 187-192, 2002.
17. S. E. Middleton, et al, "Capturing Knowledge of User Preferences: Ontologies in Recommender Systems," In Proc. of the First International Conf. on Knowledge Capture, pp. 100-107, 2001.
18. M. J. Pazzani, "A Framework for Collaborative, Content-based, and Demographic Filtering," Artificial Intelligence Review, Vol. 13, No. 5-6, pp. 393-408, 1999.
19. I. Soboroff, C. Nicholas, "Combining Content and Collaboration in Text Filtering," In Proc. of the IJCAI'99 Workshop on Machine Learning in Information Filtering, pp. 86-91, 1999.

When an Ivy League University Puts Its Courses Online, Who's Going to Need a Local University?

Matthew C.F. Lau[1] and Rebecca B.N. Tan[2]

[1] Gippsland School of Computing & IT, Monash University, Victoria 3842, Australia
Matthew.Lau@infotech.monash.edu.au
[2] School of Business Systems, Monash University, Victoria 3800, Australia
Rebecca.Tan@infotech.monash.edu.au

Abstract. In recent years there has been a prolific growth in on-line courses primarily provided by "less traditional" university providers in order to tap into the vast potential local and overseas markets. These universities are trying to overcome depleted government funding by taking advantage of the ease of Internet access. Traditional internationally renowned elite universities are cautiously monitoring this trend but have not jumped into the fray, preferring to rely on their own unique success factors in providing quality on-campus education. However, they have not dismissed this mode of delivery. The provision of both on-campus and on-line courses by internationally renowned elite universities does not necessarily sound the death knell for other providers, especially the elite on-line provider/universities in other countries. This paper will address the fact that there are important survival factors which are intrinsic to home universities within a country - the traditional on-campus providers, whereby the international universities taking a share in this expanding market will not have such a significant impact. The onerous and challenging implications in implementing quality on-line delivery for teaching and learning in education and training are considered in this paper.

1 Introduction

In recent years there has been a prolific growth in on-line courses primarily provided by "less traditional" universities to tap into the vast potential local and overseas market. Governments are providing less educational and research funding, and this has impacted significantly on the less research oriented and less traditional universities. In order to survive, these universities have had to overcome depleted government funding by taking advantage of the ease of Internet access and doing "more with less". Elite traditional universities are cautiously monitoring this trend to on-line delivery but have not jumped into the fray, preferring to rely on their own unique success factors in providing quality on-campus education. However, they have not dismissed this delivery mode.

The provision of both on-campus face-to-face and on-line courses does not necessarily sound the death knell for other providers, especially the elite on-campus face-to-face provider universities in their home countries. There are a number of intrinsic factors for these home universities within a country that ensure their survival and success. The competitive pressures on universities to provide and adopt flexible learning or on-line delivery are intense. There is room for quality face-to-face providers to service the local and international community, and are quietly taking measures to address this challenge.

In this paper, we have used two elite universities, one which is internationally renowned, Harvard University in the United States of America (Harvard), while the University of Melbourne (Melbourne) in Australia has an established reputation in its home country, as examples to highlight our argument. Both these universities are traditional providers of on-campus education. Harvard has an international reputation, while Melbourne, one of the group of eight, is among the top universities in Australia. No untoward intentions are implied.

2 Place for Both on-Campus and Online Delivery Modes

In this section, we will present the following scenarios:
- Harvard on-campus courses have not affected Melbourne enrolments,
- Harvard on-line
- Melbourne on-line
- If Harvard goes on-line, who will go to Melbourne?
- Fees implications

Harvard has traditionally provided on-campus face-to-face learning. Its courses and reputation are recognized not only in their home country but also internationally. The high cost of study at Harvard has not deterred those seeking the quality education that is uniquely Harvard. Their critical success factors include its reputation, quality faculty members and students, quality learning materials and the unique Harvard university experience. These factors have been zealously guarded. [1]. Although Harvard is considered to be among the top elite universities in the world, its on-campus courses have not impacted on the reputation and enrolment at Melbourne. There is room for Melbourne which has its own unique features, as one of the elite universities in Australia and in the Asia Pacific region as well as for Australian citizens living overseas who intend to enrol in on-line course. Melbourne has always been a popular choice for both Asian and Southeast Asian students [2].

2.1 Harvard – Online

Even if Harvard provides its courses on-line for international students including Australian students, Melbourne will continue to thrive, as each one has its own niche market. The common critical success factors of both universities include their reputation, the quality of both faculty and of learning materials and the uniqueness of their university experience. Students are becoming discriminating about their courses and so would be inclined to select a particular course from a specific university to fit their individual requirements. Another significant consideration is the cost of the course [3, 4].

Both Harvard and Melbourne recognize the dichotomy of the need for an on-line component and its potential effect on their market share of their traditional on-campus delivery, and are taking measures to address this on-line assault by less research-oriented universities. Though Harvard has not gone completely on-line, it is cautiously viewing and monitoring this trend, and has been offering some limited courses [5].

2.2 Melbourne – Online

Melbourne is cautiously monitoring the on-line market with interest and has started to offer some of its courses on-line. Its Education Faculty [6] offers a number of courses and subjects through the external mode of delivery either by: -correspondence, on-line via the Internet, or face-to-face teaching in approved off-campus or offshore locations.

While a university might have the expertise and experience in conducting on-campus delivery effectively, the on-line mode of delivery may pose very different and unique issues and hurdles. This is discussed in section 3. However, most universities appreciate the need to have a small component of on-line delivery to complement but not replace their expertise in traditional on-campus delivery to cater for the diverse market.

2.3 If Harvard Goes Online, Who Will Go to Melbourne?

Harvard cannot service the needs of all students globally, both logistically and resource-wise. However, it would be reasonable to suggest that Harvard on-line courses would be considered favourably by many international students. The experience of studying at Harvard on-campus is not the same as studying through Harvard on-line. Face-to-face lectures are so different from on-line videos even though on-line programs are supported by email, newsgroups, chat-rooms, etc. Students go to Harvard or Melbourne not just for their reputation, but also for the unique experience of studying on campus at that university including the opportunity of face-to-face contact with the "wisdom" of the faculty members.

On-line learning cannot replace the campus experience. Even though on-line lecture notes can be fascinating and studying on-line have many advantages, face-to-face discussion on the lecture topics can be much more stimulating and meaningful. Many on-line students wish they had the opportunity to attend but are unable to do so because of work, financial or family commitments. On-campus students in some respects have the best of both worlds; they not only get the arm waving and the contact with other students, whose influence could be substantial, but can also look at the lectures on the Web and participate in all the on-line forums or discussions, though not many do so. Apart from activities in the lecture theatres and laboratories, there are ongoing activities around the student union, interactions with classmates, on clubs and sports teams, and through interaction with the community in volunteer activities, not forgetting the brown bag talks by visiting academics. These valuable experiences can add to the experience of university study.

Melbourne on-campus will continue to thrive even if Harvard and other local Australian universities go on-line. Local Australian students, especially those from Victoria, who meet tertiary entrance requirements, will continue to choose Melbourne as their first preference because of its prestige. Every year, the demand for the number of available university places outstripped the supply. In 2003, about 40 per cent of eligible applicants in Australia missed out on a university place, an increase of 9.7 per cent from last year [7]. Despite this the federal government is not funding more university places. The demand for places among the top local universities will be more intense.

2.4 Fees Implications

Melbourne on-campus courses will continue to attract its own cohort of local and international students because of its reputation, prestige, uniqueness, convenience/location, and cost, which remains relatively high, in the Australian context, despite the fluctuations in the exchange rate of the Australian dollar. International students will continue to choose Melbourne because of its relatively low fees and cost of living expenses when compared to that of studying in the United States or the United Kingdom [4]. A comparison of the tuition fees (approximate) is given in Table 1.

There is no sign that demand from Asia will abate. A recent IDP survey report [2] predicts that, by 2025, Asia will represent 70% of total worldwide demand for places at Australian universities. The article continued to indicate that there would be a tertiary boom down under. Foreign students represented 20.4% of the total number of students enrolled in Australian higher education courses in 2002. The total demand for Australian education is forecast to increase more than nine-fold between 2000 and 2025, to nearly one million students.

Cost will increasingly be the major factor influencing student choice. Administrators would like to think that if they get enough students the fixed costs will fall, unfortunately they do not. This is one reason for the move to on-line courses. The quality, reputation and interaction with quality faculty members or tutors will remain the centerpiece of education at Melbourne, no matter what the medium. This will be its key asset.

3 Implications for Online Delivery

In this paper, we will present some salient points regarding the implications of implementing on-line delivery for teaching and learning in education and training. Our aim is to show that on-line delivery is by no means without its own issues and demands. As new technologies become available, the motivations of educational innovators must be supported and sustained. Innovation must be made compatible with current work practices through provision of adequate time and resources [8]. Rogers's model [9] offers a useful lens through which the efforts of those who foster innovation may be viewed. Teachers cannot be relegated to just the "guide on the side", a motivation facilitator or instruction designer/developer, to supplement learning materials on the Internet or on CD ROMs [10]. Rather, education and learning, be it on-campus or online, should be delivered by living teachers, fully qualified and interested in doing so. Technologies should be used to enhance the mode of delivery. On the other hand online courses are beneficial to students who feel intimidated by peers in classroom situations as they allow them to find a voice and those who are unable to attend on-campus courses.

Table 1. Comparative tuition fees in Australia, United States, and United Kingdom (All fees are in US dollars. A$1.00 is approximately equal to £0.42 or US$0.72 as on Nov 24, 2003).

	Local fees (US$)		International (private) fees (US$)		
	Australia Melbourne University	UK Imperial College	Australia Melbourne University	US (Harvard University)	UK (Oxford University)
University courses	2500	5000	10000	25000	18000
Clinical courses	4200	17000	23000	50000	35000

3.1 Instructional Design

The on-line environment introduces new demands on teachers, trainers, students, and methods of instructional design and curriculum development. However, the potential freedom of the new technologies that accompanies the on-line environment means that the more interactive, navigationally focused and communication-hungry technologies will also require thoughtful implementation with scaffolded support for students, and teachers and trainers who are both confident and comfortable with this new way of working [11]. Lebow [12] explores the implications of on-line instructional design and identifies a number of factors as being important.

The McKavanagh NREC study [13] found that the new technologies have not yet been matched with an equivalent change in the instructional design of the learning materials. Most importantly the flexibility and encouragement of interaction has not been 'designed' for, and traditional materials and methodologies still dominate the materials prepared for on-line learning. However, imaginative teachers are developing applications software such as WebCT [14], virtual labs and expert systems that have the potential for developing their new and higher order skills [15].

The focus on information design dominates commercial Web publishing. Whereas this is motivated by the need to make material attractive, scalable, convincing, informative, stylish and easy to navigate, instructional design has a very different charter. Apart from providing information in educationally well-organized ways to maximize student learning, instructional design needs to provide the 'scaffolds' and communication tools for independent learning and contact, structuring information to optimize its usefulness. Brennan [10] developed a profile for successful on-line learning, which includes these features based on [16]. Ikegulu & Ikegulu [17] have identified some of these technical limitations of design.

3.2 Curriculum Design

The exactness or specificity of the curriculum, which determines the content of delivery, will have an influence on the curriculum design itself. Tight curriculum and assessment methods usually focus on a controlled and formalised teaching, which will not foster learning. The medium and the message are equally important. A broad view of the curriculum should take into consideration and builds on the different backgrounds, predispositions and experiences of the learners and puts them in touch with both materials and strategies that fit with their prior learning.

3.3 Teachers

Like any changes, on-line teaching and learning will not just happen automatically and magically for teachers. The changes brought about by on-line teaching and learning has created considerable new demands for teachers and trainers. The contexts of their work have changed and the skills demanded have altered accordingly often amidst very disjointed staff development. Teachers and trainers not only have to adapt to the new work environment and transfer their skills from one context to another, but still have to perform their traditional roles, where their performances continued to be monitored and reviewed. In a world where the economic imperative seems to be "to do more with less", teachers and trainers are confronted by increasing workloads to an already crowded and demanding schedule of research and administration work.

Technology is changing so rapidly that many teachers are wondering how they can cope without sacrificing the pedagogical aspect of teaching and education. Another important consideration is that this may require many laborious hours of unpaid work. Teachers should not be relegated from the 'sage on the stage" to the "guide on the side" to supplement materials on CD-ROMs or on the Internet. Students need 'scaffolding' through the new ways of learning. Tony Bates [18, 19] used this phrase in an attempt to make students more responsible for their own learning practices. Not all learners can study independently. There are a variety of student profiles. Some are confident with new technologies while others are not. Assessment strategies must complement the process and not detract from it. Again on-line design and delivery materials need to take these into consideration. This needs to be balanced against those students who are on campus but who do not participate in class but who do participate in on-line discussion as more than one person can input their ideas at one time.

With the commoditization of instruction, teachers as labour are drawn into a production process designed for the efficient creation of instructional commodities. They therefore become subject to all the pressures that other production workers in other industries have encountered when undergoing rapid technological transformation from above. The reality is that teachers have to "face the music", adapt to change, and most importantly, to exploit change in order to do more with less, and yet have a satisfying career; else they become dinosaurs. And Jurassic Park scenarios are unlikely to occur in the new future, not yet anyway.

3.4 Students

Although virtual learning sites have no walls and no "nagging" teachers, but while this might appear to be liberating for some learners, for others it is a deepening divide. The culture of face-to-face classes and the contact and rapport that tends to build up between teacher/trainer and students in a traditional delivery environment is impossible to replicate on screen, no matter how interactively busy the site might be [20]. This is especially so as the semester comes to a close. The emerging focus on interactivity and communication via email, newsgroups, etc. in on-line delivery tends to address the other extreme of the faceless distance education learning. The assumption that the responsibility for learning lies with the student is a difficult issue [21].

Access is certainly one of the selling points of the new technologies. However, the creation of opportunities has to be enhanced by other forms of support [16]. The quantity and quality of information, which learners can access, rather than impressing people with liberating powers, may simply represent another layer of threat to their own sense of themselves as learners [22]. For some students the move to online simply is seen as a cost shifting exercise with the cost of downloading and printing information being transferred from the university to the student.

4 Discussion

Both Harvard and Melbourne continue to rely on their own critical success factors even though both are fully aware of the threat of on-line delivery to their traditional role. Even though Harvard may opt to put its courses on-line, there is still a need for Melbourne on-campus. With the increased demand for quality education the size of the cake continues to grow. Harvard on-line is not the same as Harvard on-campus. To address some of these issues posed by on-line providers from other universities, both have complemented their traditional delivery mode with the new enhanced on-line delivery as well as other means, to maintain or increase their market share.

Any drop or threat to demand for on-campus places in Melbourne, whether it be due to competition from on-line delivery at Harvard or other local universities, is likely to be short-lived. There will certainly be a place for on-line delivery of higher education, but the fundamental place of prestigious universities like Melbourne is not under threat.

The implications of implementing on-line delivery for teaching and learning in education and training are rather onerous and challenging. However, new virtual universities or virtual incarnations of established universities will not dominate higher education across the globe. If information technology alone could make this sort of difference, then surely the VCR and TV would have already done so. Neither would interactive TV or iTV have much impact to this line of argument. Tradition, faculty members, quality courses, location, cost, and the campus educational experience will continue to matter. The relative cost-effectiveness of technology–assisted delivery would allow more resources to be spent on the campus experiences that matter most to students.

5 Conclusion and Future Research

In this paper, the important factors have been identified, which are intrinsic to elite on-campus providers within a country that will ensure international universities taking a share in an expanding market will not have a significant impact. The onerous and challenging implications in implementing on-line delivery for teaching and learning in education and training were presented and discussed.

However, other education providers that are not among the elite universities in Australia or US have not been considered. Also we have not considered the number of eligible students missing out on university places of their choice, nor how overseas students make their choice. There is a need to consider the degree of importance placed on financial, cultural and social factors in influencing students' choice as well as the advantages and disadvantages of studying on-campus as compared to on-line.

References

1. Fineberg, H V, "Fineberg sees tradition amid change", Harvard gazette, February 2001
2. Bagwell, S., "Edu.: Europe's students head Down Under", BRW vol. 25 (2003)
3. Higher Education Contribution Scheme (HECS) Loans and Fees Manual, 2003
4. Higher Education Strategy, In Depth HE Reaction 2003
5. Distance Education Program at Harvard Extension School 2002-3
6. Melbourne Education – The University of Melbourne
7. Dunn, A. "Tertiary offers ", The Age (2003).
8. Housego, S. & Freeman, M. "Case studies: Integrating the use of web based learning systems into student learning." *Aust. Journal of Edu. Tech.*, 16(3), (2000) pp. 258-282
9. Rogers, E. M. "Diffusion of Innovations." 4th ed. New York: Free Press (1995).

10. Brennan, R. "An evaluation of the effectiveness of on-line delivery in education and training: multiple discourses". Charles Sturt University, Wagga (2000).
11. Holzl, A. & Khurana, R. "Flexible Learning: can we really please everyone?" (On-line). *Educational Technology & Society,* Vol. 2 (4) (2000).
12. Lebow, D. "Constructivist values for instructional systems design: five principles towards a new mindset". *Edu. Tech. Research and Development,* 41 (3), (1993) pp. 4-16.
13. McKavanagh, C., Kanes, C., Bevan, F., Cunningham, A. & Choy, S. "Evaluation of WEB-based flexible learning in the Australian VET sector", Griffith University (1999).
14. Goldberg, M., Salari, S., and Swoboda, P. "WWW – Course Tool: An Environment for Building WWW-Based Courses", Fifth International WWW Conference (1996)
15. Bates, A.W. "The impact of technological change on Open and Distance Learning", Queensland Open Learning Network Conference (1996)
16. Alexander, S. & McKenzie, J. "An evaluation of information technology projects for university learning". AGPS: Canberra (1998).
17. Ikegulu, P. & Ikegulu, T. "The effectiveness of window presentation strategy and cognitive style" Centre for Statistical Consulting, Ruston, L.A. (1999).
18. Bates, A.W. "Crossing Boundaries: Making Global Distance Education a Reality", Journal of Distance Education, Vol. XII, No.1/2, (1997a) pp49-66.
19. Bates, A.W. "Restructuring the university for technological change", The Carnegie Foundation, (1997b)
20. Palfreeman, A. "The Internet and distance learning: interaction and information". Literacy Broadsheet, 49, pp. 11-15 (1998)
21. University of Illinois Faculty Seminar, "Teaching at an internet distance: the pedagogy of On-line teaching and learning." (2000)
22. Ross, J. & Schulz, R., "Can CAI accommodate all learners equally?" British Journal of Educational Technology, 30, 1, (1999) pp. 5-24.

Threads in an Undergraduate Course: A Java Example Illuminating Different Multithreading Approaches

H. Martin Bücker[1], Bruno Lang[2], Hans-Joachim Pflug[3], and Andre Vehreschild[1]

[1] Institute for Scientific Computing, RWTH Aachen University,
D-52056 Aachen, Germany
[2] Applied Computer Science Group, University of Wuppertal,
D-42097 Wuppertal, Germany
[3] Center for Computing and Communication, RWTH Aachen University,
D-52056 Aachen, Germany

Abstract. Multithreading is a fundamental approach to expressing parallelism in programs. Since Java is emerging as the de facto standard language for platform independent software development in higher education, there is need for teaching multithreading in the context of Java. We use a simple problem from scientific computing to explain two different multithreading approaches to second-year students. More precisely, a simple boundary value problem is considered, which makes visible the differences between native Java threads and the OpenMP interface. So, the students are able to appreciate the respective merits and drawbacks of a thread package that is integrated into the Java programming language and an approach combining compiler support with library calls.

1 Java and Parallel Computing

Java is increasingly becoming popular among young students and is emerging as the de facto standard language for platform independent software development in higher education. In the context of computational science, the Java programming language is also rapidly gaining ground; see [1,2,3,4,5] and the references therein. While Java offers ease of programming through a high level of abstraction, large simulations in various areas of computational science typically require computational power and memory, often to an extent that only parallel computers can accommodate. We feel that, to benefit from the merits of both Java and parallelism, students should be introduced to parallel computing in Java. Given the importance of parallelism, this should be done near the beginning of a curriculum rather than at its end. For this purpose, we have designed a hands-on practical laboratory in parallel computing for students of computer science to be taken in their second year. One aim of this course is to introduce three different parallel programming paradigms: native threads, OpenMP, and message passing. The course is unique in using Java as a common umbrella to teach these parallel programming models. A detailed description of the complete course is given in [6].

In the basic version of the course, each paradigm is applied to solve one problem which maps particularly well to that paradigm. To complement the students' experience, we also recommend to have them solve one of the problems using both shared-memory paradigms, native threads and OpenMP. In addition, both paradigms allow for several different parallelization techniques. By comparing the different approaches, the students are able to appreciate the respective merits and drawbacks of the shared-memory paradigms. This optional extension of the course is the focus of the present note. It should be straight forward to use our example as a building block in any related course.

In Sect. 2, we briefly present the problem to be solved, a three-body system arising in putting a satellite onto a closed trajectory. This problem is carefully chosen to represent an illustrative, yet simple example from scientific computing involving a moderate degree of parallelism.

In Sect. 3, we examine two different designs of parallel Java programs to solve the sample problem using native Java threads [7,8]. By its feature of native threads, Java offers a simple and tightly-integrated support for multiple tasks. One situation where threads are an effective programming model is when a program must respond to asynchronous requests because, using different threads, an object does not have to wait for the completion of unrelated actions. However, we will restrict the discussion of threads to parallelism where the computing power of multiple processors is exploited to improve performance.

In Sect. 4, we solve the same computational problem with a different methodology. The OpenMP interface [9,10] is a programming model specifically designed to support high-performance parallel programs on shared-memory computers. In OpenMP, a combination of compiler directives and library calls is used to express parallelism in a program. The OpenMP directives instruct the compiler to create threads, perform synchronization, and manage the shared address space. We use a package called JOMP [11,12] providing an OpenMP-like set of directives and library routines for Java heavily based on the existing OpenMP C/C++ specification. JOMP enables a high-level approach to shared-memory programming in Java. A prototype compiler and a runtime library are available from the University of Edinburgh.

Though we will not consider parallel processing on distributed-memory architectures in the following, we mention that, for Java, a preliminary prototype interface to the Message Passing Interface (MPI) standard [13] is available. The Java Grande Forum [5] proposed a reference message passing interface for Java called MPJ [14]. Unfortunately, there is currently no complete implementation of MPJ.

2 A Sample Problem from Scientific Computing

We consider the trajectory of a satellite under the gravitational influence of earth and moon (i.e., a simple three-body system); cf. [15]. Prescribing the initial position $\mathbf{x}_0$ of the satellite, we seek an initial velocity $\mathbf{v}_0$ that leads to the satellite flying in a closed orbit with a given periodicity T. This requirement results in a nonlinear system of equations,

$$\mathbf{g}(\mathbf{v}_0) = \mathbf{0} ,$$

where $\mathbf{g}(\mathbf{v}_0) = \mathbf{x}(t, \mathbf{v}_0) - \mathbf{x}_0 \in \mathbb{R}^2$ is the deviation of the satellite's position at time $t = t_0 + T$, given the initial velocity $\mathbf{v}_0 \in \mathbb{R}^2$ at t_0, from its initial position $\mathbf{x}_0$.

Newton's method is used to solve the nonlinear system, and Gaussian elimination for solving the linear systems occurring in each Newton step. Note that the function $\mathbf{g}$ is not given by an explicit formula. Instead, since the motion of the satellite is described by an ordinary differential equation, the nonlinear system in fact describes a boundary value problem (BVP), and an evaluation of $\mathbf{g}$ amounts to the numerical solution of an initial value problem (IVP). As a consequence, the Jacobi matrix $\mathbf{g}'(\mathbf{v}_0)$, which is needed in Newton's method, cannot be determined via symbolic differentiation. Therefore, forward finite differences are used to approximate the 2×2 Jacobian matrix, thus requiring the solution of two more IVPs whose initial velocities $\mathbf{v}_0 + \varepsilon \mathbf{e}_i$ are perturbed by ε in the ith direction.

To summarize, one step of Newton's method involves three calls to a numerical IVP solver. At this point of the course, the students already have implemented a Runge-Kutta-Fehlberg (RKF) initial value problem solver with step-size control. The three IVP solves dominate the execution time and, furthermore, are independent of each other so that they can be executed in parallel. This observation directly leads to the concept of task parallelism with three tasks, each consisting of the numerical solution of an IVP.

Two approaches to executing the three tasks in parallel are considered: on the one hand Java threads [7,8] and on the other hand OpenMP [9,10]. Both approaches use Java threads to parallelize the tasks, the difference being that the first approach has to be explicitly coded by the user and the second one is done by adding a few lines (in fact, comments) to the serial program and running it through a preprocessor.

3 Native Java Threads

Since each task involves one call to the Runge-Kutta-Fehlberg solver RKF, the latter is wrapped in a class IVP, which in turn is derived from the Java Thread class. This wrapping enables an easy exchange of the RKF solver with another initial value problem solver. Now we must decide how many tasks are executed within one thread. In the single-task approach, we create, start, and terminate the threads in each iteration of the main loop in Newton's method. Thus, each thread performs a single task. In the alternative server approach, we start all three threads just before the loop and keep them alive until the iteration terminates. In this case, each thread must wait until a new IVP is set up and the data are available, then solve the IVP and finally notify the caller that the solution is available. Then the thread waits for the next IVP or a termination signal. This way, each thread acts as a server performing as many tasks as there are iterations in Newton's method, thus removing the overhead of repeatedly creating and destroying threads. Both designs will be discussed in the following.

3.1 The Single-Task Approach

In the single-task approach the Newton nonlinear solver, which is sketched in Fig. 1, creates three IVP threads ivp_g, ivp_dg1, and ivp_dg2 within each iteration of its main loop. Upon creation of each thread, the constructor IVP of the wrapper class initializes all necessary data (e.g., the initial velocity v0) and starts the thread at the end of the constructor call. The run method of the thread calls the IVP solver RKF to determine the final position x of the satellite, and then terminates. The object x remains in memory and is accessible via the thread's getX() method as long as the thread object itself exists.

```
/** Newton's method in the context of a BVP */
class Newton {

  /* solve system, starting with approximation v0 */
  public RealVector solve( RealVector v0 ) {

    RealVector g, dg[] ;          // function value, Jacobi matrix
    // ... declare more local variables here ...

    do {                           // set up and start IVP threads
      IVP ivp_g = new IVP( Satellite, RKF, v0, ... ) ;
        // v1 is v0, with 1st component perturbed by +eps
      IVP ivp_dg1 = new IVP( Satellite, RKF, v1, ... ) ;
        // v2 is v0, with 2nd component perturbed by +eps
      IVP ivp_dg2 = new IVP( Satellite, RKF, v2, ... ) ;

      try {                        // wait for their completion
        ivp_g.join() ;
        ivp_dg1.join() ;  ivp_dg2.join() ;
      } catch ( InterruptedException e ) { ... } ;

      g = ivp_g.getX() - x0 ;      // assemble results
      dg[1] = ( ( ivp_dg1.getX() - x0 ) - g ) / eps ;
      dg[2] = ( ( ivp_dg2.getX() - x0 ) - g ) / eps ;

      // ... update v0 according to Newton's method ...
      // ... check convergence and number of iterations ...
    } while ( ! finished ) ;
    return v0 ;
  }
}
```

Fig. 1. Sketch of the single-task-style Java-threaded code.

In the meantime, the master thread calls the join() method of each IVP thread to wait for its termination. Finally, the position x is extracted from the IVP thread object by calling its getX() selector. After that the value of the initial velocity stored in v0 is updated according to Newton's formula, and the convergence of the iteration is checked. The loop terminates if the current value v_0 is considered "good enough" (meaning, e.g., that the function value $g(v_0)$ is small or that v_0 has changed only slightly during the last iteration), or if some prescribed maximum number of iterations has been reached.

This approach is easily designed and implemented because no explicit communication is needed. If a serial version of the problem exists then the parallelization takes just about half an hour.

3.2 The Server Approach

The design and implementation of the server approach is more complicated because it uses explicit communication and therefore synchronization. Java's language specification offers constructs (such as keywords and classes) for synchronization, which is rarely seen in other programming languages. Nevertheless, the programmer needs some experience to design a correct multithreaded Java program with explicit synchronization.

The server-style parallel Newton solver is sketched in Fig. 2. Again, the objects ivp_g, ivp_dg1, and ivp_dg2 are threads, but in contrast to the single-task approach they are created and started before the iteration loop is entered.

The overall activity of the threads is shown in Fig. 3. The run() method consists of an infinite loop. Once the thread is started, in each pass of the loop it first checks if a value for the initial velocity v_0 is available, which is indicated by the flag data_available. If no data is available yet, then the thread will wait. When new data is available, then the RKF initial value problem solver is run, and after that the master thread is notified about the availability of the result.

The master thread synchronizes with the IVP threads by calling two data transfer methods setV() and getX(), which are part of the IVP wrapper class.

First, a call to the setV() method (do-loop in Fig. 2) equips the IVP thread with new data and thus enables the RKF initial value solver to be run. The setV() method is defined as follows:

```
synchronized public void setV( RealVector v0 ) {
   this.v0 = v0 ;
   data_available = true ;
   notify() ;
}
```

Note that the method is declared synchronized. This means that at most one thread may be within this method at any given time and ensures that both the data_available flag is set *and* the signal is sent to the IVP thread. According to Fig. 3 this guarantees that the RKF solver can be started no matter if the IVP thread has been waiting or not. In the former case, the notify()

```
class Newton {

   public RealVector solve( RealVector v0 ) {
      RealVector g, dg[] ;            // function value, Jacobi matrix

                                      // create and start IVP threads
      IVP ivp_g   = new IVP( Satellite, RKF, ... ) ;
      IVP ivp_dg1 = new IVP( Satellite, RKF, ... ) ;
      IVP ivp_dg2 = new IVP( Satellite, RKF, ... ) ;

      do {
         ivp_g.setV( v0 ) ;           // provide data for threads
         ivp_dg1.setV( v1 ) ;         // v1 and v2 are as in Fig. 1
         ivp_dg2.setV( v2 ) ;

         g     = ivp_g.getX() - x0 ;        // assemble results
         dg[1] = ( ( ivp_dg1.getX() - x0 ) - g ) / eps ;
         dg[2] = ( ( ivp_dg2.getX() - x0 ) - g ) / eps ;

         // ... update v0 according to Newton's method ...
         // ... check convergence and number of iterations ...
      } while ( ! finished ) ;
      return v0 ;
   }
}
```

Fig. 2. Sketch of the server-style Java-threaded code.

ends the wait(), whereas in the second case no waiting is necessary since the data_available flag is set.

After the setV calls, the main thread collects the results from the IVP threads by calling their getX() method, which is shown below. In contrast to its serial counterpart which only extracts the results, this version of the getX() method forces the main thread to wait until the result is available, which in turn is indicated by the thread's run() method setting the result_available flag and calling notify().

```
synchronized public RealVector getX() {
   if ( ! result_available ) {
      try {
         wait() ;
      } catch ( InterruptedException e ) { ... } ;
   }
   result_available = false ;
   return x ;
}
```

The design and implementation of the server approach takes more than one hour even for an experienced programmer. Particular care has to be taken about data consistency and possible deadlocks.

```java
public void run() {
   while ( true ) {
      synchronized( this ) {      // wait for data
         if ( ! data_available ) {
            try {
               wait() ;
            } catch ( InterruptedException e ) { ... }
         }
         data_available = false ;
      }

      // ... call RKF solver to compute final position x ...

      synchronized( this ) {      // report completion
         result_available = true ;
         notify() ;
      }
   }
}
```

Fig. 3. Sketch of the run() method in the server-style Java-threaded code.

4 The OpenMP Interface

The OpenMP interface offers an alternative way to parallelize a program using threads. OpenMP provides several work sharing constructs. Parallel sections are used for executing (possibly different) tasks concurrently, and the for construct distributes the passes of a loop over several threads. The OpenMP constructs automatically perform all the low-level work for creating, synchronizing, and terminating the threads. The sample problem is parallelized using each construct in turn.

4.1 OpenMP Solution with Sections

Using the OpenMP sections work sharing construct, an existing serial Newton solver is easily parallelized by adding four directives (and a pair of braces for each directive) and running the modified Newton class through the OpenMP compiler for Java, called JOMP [11,12]. Figure 4 shows a sketch of the OpenMP version of the Newton-solver. Since OpenMP handles the thread management, now the class IVP needs no longer be derived from the Java class Thread, and no explicit call to the start() method is present anymore. The IVP class is exactly the same as in the serial program providing a wrapper to the call of the Runge-Kutta-Fehlberg IVP solver as well as an access method to the results, getX().

In our example, the three tasks of computing the function value g and the partial derivatives correspond to three sections within the parallel region that is defined by the parallel sections directive. The latter directive creates as many threads as there are section directives within its scope (i.e., three), assigns

```
class Newton {

  public RealVector solve( RealVector v0 ) {

    RealVector g, dg[] ;           // function value, Jacobi matrix
    // ... declare more local variables here ...

    do {
      // omp parallel sections
      {                                  // set up and start threads
        // omp section
        {                                // task for first thread
          IVP ivp_g = new IVP( Satellite, RKF, v0, ... ) ;
          g = ivp_g.getX() - x0 ;
        }
        // omp section
        {                                // task for second thread
          IVP ivp_dg1 = new IVP( Satellite, RKF, v1, ... ) ;
          g1 = ivp_dg1.getX() - x0 ;
        }
        // omp section
        {                                // task for third thread
          IVP ivp_dg2 = new IVP( Satellite, RKF, v2, ... ) ;
          g2 = ivp_dg2.getX() - x0 ;
        }
      }
      dg[1] = ( g1 - g ) / eps ; dg[2] = ( g2 - g ) / eps ;

      // ... update v0 according to Newton's method ...
      // ... check convergence and number of iterations ...
    } while ( ! finished ) ;
    return v0 ;
  }
}
```

Fig. 4. Sketch of the OpenMP sections code.

one task to each thread, runs the threads in parallel, and terminates them at the closing brace following the sections.

Inserting and preprocessing this code with the JOMP compiler takes less than half an hour even for a less experienced programmer. With more experience it can be done in a few minutes.

4.2 OpenMP Solution with for

Using the OpenMP for work sharing construct requires the existence of a loop. The three sections shown in Fig. 4 are redesigned to be executed within the body of a for loop. Figure 5 shows the sketch of the resulting algorithm. To enable the use of a for loop, all three initial values v0, v1, and v2 have to be copied into an array, and the three results also must be buffered in an array.

```
class Newton {

   public RealVector solve( RealVector v0 ) {

      RealVector g, dg[] ;           // function value, Jacobi matrix
      RealVector v[], x[] ;          // for initial values and results
      // ... declare more local variables here ...

      do {
         v[0] = v0 ; v[1] = v1 ; v[2] = v2 ;

         // omp parallel for
         for ( int i = 0 ; i < 3 ; i++ )
            x[i] = ( new IVP( Satellite, RKF, v[i] ) ).getX() ;

         g = x[0] - x0 ;
         dg[1] = ( ( x[1] - x0 ) - g ) / eps ;
         dg[2] = ( ( x[2] - x0 ) - g ) / eps ;

         // ... update v0 according to Newton's method ...
         // ... check convergence and number of iterations ...
      } while ( ! finished ) ;
      return v0 ;
   }
}
```

Fig. 5. Sketch of the OpenMP `for` code.

5 Concluding Remarks

We have presented four thread-based parallel versions of Newton's method in the context of a boundary value problem. All four programs were run on a Sun-Fire 6800 with 24 UltraSparcIII processors (900 MHz) and a total of 24 GB RAM. All approaches need approximately the same execution time. However, they differ in the time needed to implement these approaches. So, the implementation times which are given in Tab. 1 should be the prime criterion for choosing an approach. The implementation times are rough estimates rather than precise measurements, essentially reflecting the differences in handling the thread management explicitly by native Java threads or implicitly by OpenMP.

By comparing the four approaches, the students are able to assess the relative merits and drawbacks of implementations based on Java threads and on the

Table 1. Implementation times of the four parallel programs.

	Java threads single-task	Java threads server	OpenMP sections	OpenMP for
implementation time	30min	120min	5min	15min

OpenMP interface. Using OpenMP restricts the programmer to a set of common (regular) synchronization patterns, whereas directly relying on Java threads offers the potential of (almost) arbitrary patterns. On the other hand, if OpenMP approaches are viable then their implementation is *much* more convenient and less error-prone than a design based on Java threads. Except for the OpenMP sections technique, all approaches easily generalize to variable dimensions.

Some students also will notice that the OpenMP paradigm would also allow a server-type design by letting the parallel region comprise the whole do loop of the Newton solver, but currently this optimization is hard to achieve by the JOMP compiler because of its prototype character.

References

1. Philippsen, M., et al.: JavaGrande – Work and Results of the JavaGrande Forum. In Sørevik, T., et al., eds.: Applied Parallel Computing: New Paradigms for HPC in Industry and Academia, Proc. 5th Intern. Workshop, PARA 2000, Bergen, Norway, June 2000. Volume 1947 of LNCS., Berlin, Springer (2001) 20–36
2. Boisvert, R.F., Moreira, J., Philippsen, M., Pozo, R.: Java and numerical computing. IEEE Computing in Science and Engineering **3** (2001) 18–24
3. Moreira, J.E., Midkiff, S.P., Gupta, M., Wu, P., Almasi, G., Artigas, P.: NINJA: Java for high performance numerical computing. Scientific Programming **10** (2002) 19–33
4. Moreira, J.E., Midkiff, S.P., Gupta, M.: From flop to megaflops: Java for technical computing. ACM Transactions on Programming Languages and Systems **22** (2002) 265–295
5. Java Grande Forum: http://www.javagrande.org (2003)
6. Bücker, H.M., Lang, B., Bischof, C.H.: Parallel programming in computational science: an introductory practical training course for computer science undergraduates at Aachen University. Future Generation Computer Systems **19** (2003) 1309–1319
7. Lea, D.: Concurrent Programming in Java: Design Principles and Patterns. Addison-Wesley, Reading (1996)
8. Oaks, S., Wong, H.: Java Threads. O'Reilly (1997)
9. OpenMP Architecture Review Board: OpenMP C and C++ Application Program Interface, Version 1.0 (1998)
10. OpenMP Architecture Review Board: OpenMP Fortran Application Program Interface, Version 2.0 (1999)
11. Bull, J.M., Kambites, M.E.: JOMP – an OpenMP-like interface for Java. In: Proceedings of the ACM 2000 Java Grande Conference. (2000) 44–53
12. Kambites, M.E., Obdrzalek, J., Bull, J.M.: An OpenMP-like interface for parallel programming in Java. Concurrency and Computation: Practice and Experience **13** (2001) 793–814
13. Message Passing Interface Forum: MPI: A message-passing interface standard. International Journal of Supercomputer Applications **8** (1994) 165–414
14. Baker, M., Carpenter, B.: MPJ: A proposed Java message passing API and environment for high performance computing. In Rolim, J., et al., eds.: Parallel and Distributed Processing: Proc. 15 IPDPS 2000 Workshops, Cancun, Mexico, May 2000. Volume 1800 of LNCS., Berlin, Springer (2000) 552–559
15. Stoer, J., Bulirsch, R.: Introduction to Numerical Analysis. 2nd edn. Springer, New York (1993)

A Comparison of Web Searching Strategies According to Cognitive Styles of Elementary Students

Hanil Kim[1], Miso Yun[2], and Pankoo Kim[3]

[1] Department of Computer Education, Cheju National University, Jeju, 690-756, Korea
hikim@cheju.ac.kr
http://educom.cheju.ac.kr/~hikim
[2] Jeju Ara Elementary School, Jeju, 690-121, Korea
miso123456@hanmail.net
[3] Department of Computer Science, Chosun University, Gwangju, 501-759, Korea
pkkim@chosun.ac.kr

Abstract. The popularization of the Internet brought about easy access to a huge amount of information, yet it is not easy for one to find information in need. Therefore information users should have appropriate information search competencies to gather, analyze, and utilize it efficiently. In general people have their unique information search strategies, and consequently the retrieved results also significantly vary from one to another. This paper describes an experimental study on web searching strategies of elementary students and analyzes the variation in their searching strategies by examining their individual personalities, specifically their cognitive styles.

1 Introduction

The development of information technology has contributed to an exponential growth of the amount of information available. Despite fast advances in making information accessible to vast number of users, it is not easy for us to retrieve information that best fits our needs and to make use of them ultimately. Therefore the ability to search information efficiently becomes highly valued to a great extent. The efficient and effective information search allows ones to search and manipulate their required information and applying it for their uses.

To search a certain information, we usually set a specific goal according to their experience, objective, knowledge, personality, and cognitive style, and then perform all required tasks by feeding various searching phrases into selected search engines. Setting up the information search strategy is often affected by the personality of each retriever. In other words an information retriever has his/her own information retrieval strategy which is an individual preference [7].

Previous research efforts in web searching have mainly focused on improving the efficiency of search engines, which includes designing and developing the information retrieval system and storing and indexing information. However not many works have researched on the relationship of the personality in web searching.

In this paper, we present how elementary students' individual cognitive styles affect their web searching strategies and the ratio of correctness, and propose a suggestion for improving the web searching competency. The rest of the paper is organized as follows. In section 2, we describe some basic concepts covered in related works. In section 3, we explain our methods and procedures, and results are analyzed in section 4. Finally we conclude our paper in section 5 with a set of suggestions on educational requirements for improving the competency and effectiveness of information searching on the web.

2 Related Works

2.1 Information Searching on the Web

Computers, the Internet and the World Wide Web have made vast amounts of information available for our daily life. The term "Information Search" means a series of tasks done in order to locate the expected information using specific web search tools and utilize them [2]. These tasks are accomplished through a cyclic procedure as shown in Figure 1, which is adapted from the prior effort by [7].

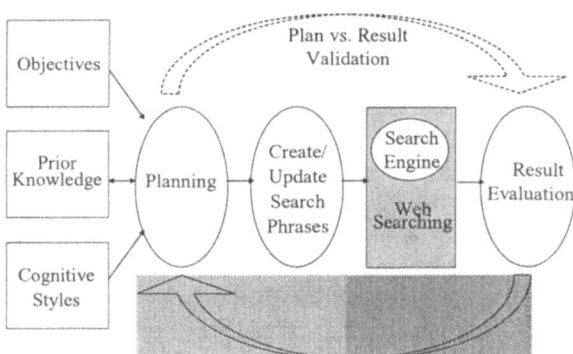

Fig. 1. Information Search Procedure

Prior to web searching, specific objectives need to be set up. Various types of prior knowledge have a potent influence on understanding given objectives and solving problems within the cycle, and each retrieval strategy can vary with individual cognitive styles as well. Steps in information search procedure are represented in circles in Figure 1. Firstly, understanding the objectives should be assured from the given search problems and then a plan is set up as a result. Secondly, that plan helps proper search keywords or conditions to be created or updated and fed into the designated search tools. Subsequently the search result is analyzed and its correctness and closeness are evaluated. Those steps are repeated until the web search tasks are completed and the search keywords and conditions become more accurate as each cycle is iterated. In this procedure information retrievers may have their own information search strategies when writing their own specific search keywords and conditions according to the characteristics of given problems and information search engines, and their cognitive styles.

2.2 Cognitive Styles

Each individual has a unique approach to learning activities and problem solving. This means one recognizes external environments according to his/her own preferred or habitual patterns of recognition, and memorizing. Those patterns are often called "Cognitive Styles" and sometimes described as styles of collecting and processing information [9].

Reflective-Impulsive. Reflective-impulsive cognitive style, which was introduced by Kagan(1964), is a cognitive style known as one directly related with learning and problem solving activities. In general this type of cognitive style is used to classify learners based on their decision making speed and its differences can be explained by speed, accuracy, and reflectiveness shown for searching and manipulating the information [5].

[1][6] show how these two groups, one is reflective and the other is impulsive, are different in the way of thinking and problem solving. Impulsive learners have a strong tendency to stick to the very first idea which they have encountered and be less efficient for controlling their first reaction expressed in an uncertain problem solving situation. To the contrary reflective learners usually consider several alternatives at the same time and can control their first reactions accordingly to reach a better solution. Moreover they typically work on the given tasks carefully in no hurry since they value the accuracy of work above their reaction speed [4].

Field independence - Field dependence. Field independence-Field dependence, a concept firstly introduced by Herman Witkin (1961), is defined as the degree of ability to distinguish insignificant background details from truly significant details [8]. Field-dependent individuals tend to be synthetic people and are easily affected by insignificant details provided within the given field. On the other hand, field-independent individuals are analytic learners who have selective attention noticing important aspects.

3 Methods and Procedure

This section describes how participants are selected, which materials and instruments are used to collect data, and through which procedure the experiment is performed.

3.1 Participants

Our research experiment was conducted at an elementary school located in Jeju-si, South Korea. Fifty 6th-grade students participated in the experiment. The cognitive style was determined for each participant using the Matching Familiar Figures Test (MFFT) and the Group Embedded Figures Test (GEFT). The MFFT and GEFT are tests that are widely accepted and used for determining an individual cognitive style. The MFFT was used to measure the dimension of reflectivity-impulsivity. According to the result based on the response time and the number of errors, 15 reflective students and 18 impulsive students were identified as participants. 10 students in fast-accurate group and 7 in slow-inaccurate group were excluded from the experiment. By using the GEFT, 15 students in the higher 30% group, who showed the high degree of field-independence, were classified as a field-dependent group, while ones in the lower 30% group were identified as a field-dependent group.

3.2 Materials and Instruments

Matching Familiar Figures Test (MFFT). The MFFT, which was developed by Kagan *et al.* in 1964, is used to measure the dimension of reflectivity- impulsivity. The test consists of 14 items including two preview problems, each of which requires the student to match a sample figure to one of six test figures that best matches the one given as a sample.
Group Embedded Figures Test (GEFT). To classify the participants into field-independent and field-dependent groups, we used the GEFT which was developed by Witkin *et al.* in 1971. Students are asked to identify simple forms hidden within complex figures for 10 minutes. The more figures found, the better the student is at the separation tasks and is said to be more field-independent.
Search Problems. Five search problems with various difficulty levels are developed for this experiment. They are designed so that the information can be retrieved using only search phrases in Korean in order to avoid any influence of English fluency on the experiment result. Search steps anticipated for each problem were pre-performed by research staffs prior to the experiment.
Search Engines. A web search engine was pre-selected for this experiment, because allowing ones to use different search engines can be an undesirable to the result. We chose a Korean search engine, which name is *simmani* (http://www.simmani.com), since it is one of search engines showing the superiority in web searching on information in Korean. Students were taught the basic instructions of using search engine in advance.

3.3 Procedure

Our research experiment was conducted at the computer laboratory of "*A*" Elementary School during the period of March/April in 2003. A total of 50 students were given two types of tests to categorize their cognitive styles. The data for 33 participants were retained for analysis on reflective-impulsive groups, and another set of data from 30 participants were used for analysis on field independence-field dependence as mentioned in Section 3.1.

Figure 2 shows the detail of the experiment procedure. Five web searching problems are posted, which are developed by analyzing questions previously used in the information scavenger hunts on the web. All students are requested to solve 5 problems within 40 minutes using a recommended Korean search engine, *simmani*.

Fig. 2. Research Experiment Procedure

All keyboard and mouse operations performed by each student are recorded using the automatic screen capture program and are stored in avi file format. A total of 50 avi files were acquired from this experiment and the running time of each file was 30-40 minutes. Search behaviors captured in avi files are evaluated according to the following factors adapted from ones in [3]:

- the number of total search attempts,
- the number of repeated search attempts using the same phrases,
- the number of AND operators used in search phrases,
- the number of browsed documents,
- the tendency to depend on search options provided by the given search system to increase the search precision/recall ratio, instead of considering variations in search keyword (option-based search style),
- the tendency to provide variations in search keywords to increase the search precision/recall ratio rather than depending on operating features of information search system (keyword-based search style).

Statistical t-tests were conducted to investigate the differences between two groups. Experiment data were collected to a Microsoft Excel worksheet and a principal factor analysis was done by the SPSS 10.1 for Windows.

4 Results and Analysis

Attained results were analyzed to show the relevance between cognitive styles and web search strategies. Two aspects of results were evaluated; search strategies and search achievement level.

4.1 Search Strategy Analysis

Relevance to Reflective-Impulsive Cognitive Styles. Table 1 shows the analysis on search behaviors of reflective group and impulsive group. Students in impulsive group expose higher scores over all factors except the number of browed documents. This explains people in impulsive group tend to be inattentive and catch and apply their idea instantly without any careful consideration. On the other hand ones in reflective group show the deliberate attempt to understand and solve problems.

Table 1. Search strategy differences for reflective group vs. impulsive group

Analysis Factors	Reflective Mean (SD)	Impulsive Mean (SD)	t	p
No. of repeated search attempts using the same phrases	1.512 (1.151)	2.889 (2.602)	-2.020	0.054
No. of total search attempts	4.182 (1.983)	5.724 (3.199)	-1.623	0.115
No. of used AND operators	4.684 (3.618)	5.638 (4.742)	-0.663	0.512
No. of browsed documents	3.1841 (0.399)	2.9031 (0.573)	0.538	0.594
option-based search style	0.559 (0.811)	1.6391 (0.675)	-2.417	0.023*
keyword-based search style	0.571 (0.736)	1.035 (0.751)	-1.784	0.084
No. of students: Reflective-15, Impulsive-18				* p<0.05

Students with impulsive cognitive style made frequent use of both option-based and keyword-based search styles without any significant preference to one style. As illustrated in Figure 3, their numbers of search attempts reach to three times as many as reflective group's attempts in case of the option-based search style, and twice in case of the keyword-based style. This phenomenon is exposed since impulsive students usually take many actions within a short time as instant reactions with no considerate thoughts. According to the t-test analysis on 6 factors, the group difference in option-based search style was significant (p=0.023, i.e., p<0.05).

Relevance to Field independence - Field dependence Cognitive Styles. As shown in Table 2 and Fig.4, the field-dependent group marks the maximum three times as many as field-independent group in the following factors: the number of repeated search attempts, the number of total search attempts, the number of search operators, and option-based search style. However for keyword-based search style, the field-independent group showed more search attempts by retrials with changed keywords.

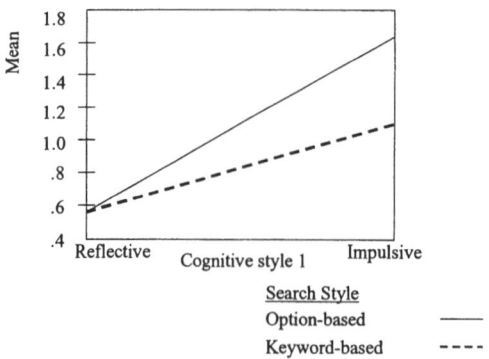

Fig. 3. No. of total search attempts according to reflective-impulsive cognitive style

Table 2. Search strategy differences for field-independent group vs. field-dependent group

Analysis Factors	Field-independence Mean (SD)	Field-dependence Mean (SD)	t	p
No. of repeated search attempts using the same phrases	1.192 (1.160)	3.890 (2.535)	-3.748	0.001*
No. of total search attempts	4.109 (1.632)	6.330 (3.469)	-2.244	0.036*
No. of used AND operators	2.933 (2.564)	6.516 (5.138)	-2.416	0.022*
No. of browsed documents	3.001 (2.428)	1.001 (1.581)	1.183	0.247
option-based search style	0.613 (0.905)	1.618 (1.796)	-1.934	0.067
keyword-based search style	0.944 (0.689)	0.754 (0.790)	0.702	0.049
No. of students: Field-independence-15, Field-dependence-15				* p<0.05

It clearly exhibits the characteristics of field-dependent group who are easily influenced by the field environment. Especially field-dependent students browse less number of documents than field independent students. They tend to strongly depend on the document summaries provided by the search engines, instead of browsing and reading the content of documents. This tendency may result in missing target documents which have solutions within the part not shown in a summary. It happens because they accept the structure of given summary with no restructuring it at all.

The field-independent group prefer the keyword-based style when they perform the alternative searches to improve their precision/recall ratio, while the field-dependent group have a preference for changing options and attempting the alternative option features provided by search engines. These tendencies are understood in that field-independent group have a tendency to make the given structure (search problems) be restructured by giving new search phrases. The relevance analysis on field independence-field dependence groups shows p=0.001(p<0.05) for the number of repeated searches, p=0.036(p<0.05) for the number of total search attempts, p=0.022(p<0.05) for the number of used operators. In other words, significant differences were found in those three factors.

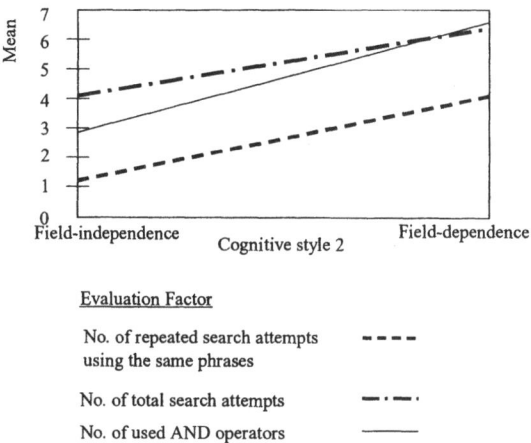

Fig. 4. No. of total search attempts according to field independence-field dependence cognitive style

4.2 Search Correctness Level Analysis

The correctness level of web searching results according to cognitive styles is illustrated in Table3. The impulsive group succeeded in a total number of 1.56 problems out of 5 and the reflective group got 2.53 problems out of 5. The percentage of number of people according to scores they got is shown in Figure 5. The t-tests analysis explains there is a significant difference in reflective and impulsive groups, i.e., p=0.036 (p<0.05). With results of analysis of each cognitive style, the reflective group found out one more problem than the impulsive group, even though impulsive people worked with all higher factors related with the number of searches. This implies that reflective people perform their web searching tasks more effectively.

Table 3. Search correctness level according to cognitive styles

Cognitive Styles	Mean	SD	t	p
Reflective	2.533	1.245	2.198	0.036*
Impulsive	1.565	1.293		
Field-independence	3.400	0.910	5.196	0.000*
Field-dependence	1.600	0.985		

p<0.05

Table3 also proclaims that the field-independent group got 3.4 correct answers out of 5 problems and 1.6 answers were correct in the field-dependent group. Figure 6 clearly describes the field-independent group are showing the higher correctness ratio than the field-dependent group.

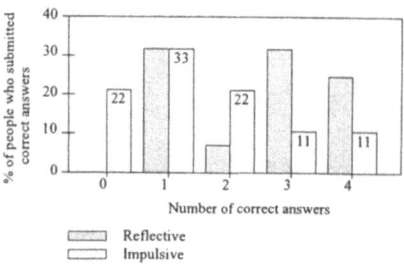

Fig. 5. Search Correctness of Reflective-Impulsive

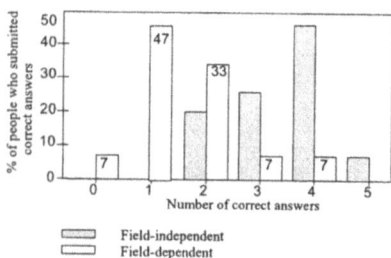

Fig. 6. Search Correctness of Field independence- field dependence

According to t-test analysis, there turns out to be a significant difference between field independence and field-dependence, $p=0.000 (p<0.05)$. The field-independent group perform more efficient search by finding 1.8 more correct answers, though field-dependent group use both option-based and keyword-based search styles with high factors related with the number of searches, operators, etc.

5 Conclusion

This research aims to compare and examine various individual web search strategies of elementary students according to their cognitive styles. Data collected from the developed experiment were analyzed to show the relevancy of cognitive styles to search strategies.

The web search correctness ratio attained from the experiment shows the result varied by cognitive styles. Especially reflective group and field-independent group show much higher search efficiencies than the impulsive group and field-dependent group accordingly. This result proves the search correctness is affected by cognitive styles as well as prior search experiences or competency in using computer itself.

Search strategies are also affected by cognitive styles. There were significant differences between reflective-impulsive groups in option-based search styles and between field independent-field dependent groups in the number of repeatedly used search phrases, the number of search attempts, and the number of used operators.

By understanding the effects of different individual cognitive styles on web search strategies, we can provide several suggestions. Firstly, we can improve the effectiveness of web searching education by varying educational methods according to cognitive styles. Secondly, the design of search systems can be advanced when the cognitive styles of their users are considered. Thirdly, other relevant factors which can affect search strategies should be studied as well, though only cognitive styles are concerned in this paper.

References

1. Goldstein, K.M., & Blackman, S: Cognitive Style: Five Approaches and Relevant Research. New York: John Wiley (1978)
2. Kim, H.: The Comparative Research of Domestic Information Search Engines Depending On Internet Information Searching Technology from the Viewpoint of Elementary Students. Master Thesis, Information & Communication Department, Pai-chai University, Korea (2001)
3. Kim, J.E.: Information Search Styles according to Individual Personality in Searching Web-based Educational Information Database. Master Thesis, Educational Engineering department, Ehwa Women's University, Korea (2000)
4. Kim, Sung Duk: Relationships among Cognitive Style, Problem Solving Processes and Achievements, Major in Curriculum Graduate School of Korea National University of Education, Chung-Buk, Korea (1990)
5. Lee S. Y.: The Effects of the Types of Concept Map on the Academic Achievement by Learner's Cognitive Styles under the Hypermedia Learning Environment, Major in Curriculum, Graduate School of Korea National University of Education, Chung-Buk, Korea (2000)
6. Oh, Young-soon: The Effects of Inductive & Deductive Teaching on Academic Achievement by Reflective & Impulsive Cognitive Style, Major in Curriculum Graduate School of Korea National University of Education Chung-Buk, Korea (1997)
7. Park, C.-H., Kim, Y.-J., Kwak, H.-W., Lee, J.-K., Sung, K.-J., & Lee, J.-M.: Exploratory learning and user strategy in search on the Internet. Paper presented at the 1997 Annual Meeting of the Korean Psychological Association (1997) 483-494
8. akar, Asım: Exploring the Relationships among Learning Styles, Annotation Use and Reading Comprehension for Foreign Language Reading in a Hypermedia Environment, Master Thesis, English Language Teaching, Bo aziçi University, Istanbul, Turkey (2003) Retrieved on January 12, 2004 from,
http://www.sfli.bahcesehir.edu.tr/teacherpages/asim/Thesis/
9. Yoo, J. O.: A Theoretical Study of Personal Characteristics of Online Searchers, Journal of the Korean Society for Library and Information Science 30(4) (1996) 39-60

The Development and Application of a Web-Based Information Communication Ethics Education System

Suk-Ki Hong[1] and Woochun Jun[2]

[1] Division of Business & Economics, Dankook University, Seoul, Korea
skhong017@dankook.ac.kr
[2] Dept. of Computer Education, Seoul National University of Education, Seoul, Korea
wocjun@ns.snue.ac.kr

Abstract. This paper introduces a Web-based information communication ethics education system. The system is designed and implemented specifically for elementary school students. It is distinguished by the following characteristics: First, the system allows students to learn information communication ethics by discussion, rather than providing information for them to memorize. For this purpose, the system's design is based on the project-based model (PBM). The second feature of this education system is its support of various types of interactions. Students can exchange their ideas with other students, teachers, as well as experts in the field. This encourages students to organize small groups for themselves. Third, the system permits students to find materials that they can apply to real life.

1 Introduction

Advanced computer and communication technologies have made human life more efficient and convenient than ever before. However, along with the positive aspects of a society brought together by computer networks are certain disadvantages. Some examples of these side effects include hacking, the use of violent language, the spread of harmful computer viruses and lewd materials, and the invasion of intellectual property or personal privacy [1].

As a way of reducing the harmful effects, strict rules and technical measures have been put in place. These steps, however, are neither preventive nor effective. For this reason, people have become more concerned with ethics education, considering it a fundamental and effective way to lessen the appearance of improper online behavior. A solid education in ethics builds upright values and promotes the right code of conduct [2]. Furthermore, to achieve greater results, the information communication ethics education should be conducted early, particularly before students learn technical computer skills. The ethics program should also be included in the curriculum of each school. Traditional computer education that focuses on the technical aspects of computers has been proven to be of little help. Information communication ethics programs that provide examples of real world problems and their practical solutions have exhibited better educational results. This paper, therefore, presents a new teaching method related to the latest information communication systems [3, 4].

This research proposed an information communication ethics education through the Web. Specifically, the program suggests that a teacher divide class students into small groups to teach problem solving methods. Students then exchange opinions with one another, and learn proper ethics including basic communication manners. With the support of a teacher as a facilitator, this web-based ethics program was determined to be an effective way to build students' values and etiquette.

In this paper, we designed and implemented a Web-based information communication ethics education system for elementary school students. This system has the following advantages. First, this system allows students to learn information communication ethics by discussion rather than by traditional forms of learning, like memorizing. For this purpose, the system's design is based on PBM. Second, the system supports various types of interactions. That is, it enables students to exchange ideas with other students, with teachers as well as experts. In addition, the system encourages students to organize small groups for themselves. Third, the system stresses that the real world context offers a variety of materials and examples from which they can learn proper forms of behavior. These materials and examples will build the students' ability to apply ideas to real life.

This paper is organized as follows: Section 2 presents the basic concepts and principles dealing with information communication ethics. Theoretical backgrounds for this system are introduced in Section 3. The design and application of this system is explained in Section 4 and 5, respectively. Section 6 outlines a case study related to an information communication ethics issue. Finally, our work and suggestions for further study are summarized in Section 7.

2 Information Communication Ethics

2.1 The Definition and Principles of Information Communication Ethics

The Moral Philosophy in Computer Science is the theory analyzing social roles and the effectiveness of computer technology, specifically identifying and justifying regulations for its ethical usage [5]. The Moral Philosophy in Computer Science is equally known as information communication ethics, or just information ethics.

From previous study [6], four principles were proposed for the information communication ethics with an explanation on its application: (i) Respect for intellectual property rights, (ii) Respect for one's privacy, (iii) Legal language, and (iv) Not to cause harmful effects. The application of these principles follows the four steps below [6]:

1) Collect facts accurately.
2) Confirm the moral dilemma.
3) Evaluate the moral dilemma using the principles of the information communication ethics to decide which side guarantees major ethical support.
4) Verify own solution based on the rule of generalization possibility.

The principles and application steps above provide a good guideline for ethics formation in modern society. However, in spite of their effectiveness, they do not provide a detailed structure for the teaching of information communication ethics. This is especially the case with elementary school students who are not familiar with using a computer. Thus, personal experiences must be considered during the teaching-learning processes rather than simply providing principles and application methods.

2.2 Netiquette as Educational Content

In cyberspace, people often make mistakes like hurting another person's feelings due to anonymity and two-way communication. It is, therefore, necessary to learn the right "netiquette," or manners required for cyber space, as it is to learn etiquette in real life. Although a definite standard for netiquette has not yet been established, some basic rules have been introduced [7, 8]:

1) Remember the human
2) Adhere to the same standards of behavior online that you follow in real life
3) Know where you are in cyberspace
4) Respect other people's time and bandwidth
5) Make yourself look good online
6) Share expert knowledge
7) Help keep flame wars under control
8) Respect other people's privacy
9) Don't abuse your power
10) Be forgiving of other people's mistakes

3 Theoretical Backgrounds

3.1 Internet Application for Information Communication Ethics Education

The Internet is widely considered one of the most useful tools for the information communication ethics program. The Web has many advantages. Students have access to the class subject along with multimedia materials, which can be collected through websites and various online sources. In addition, students can conduct projects or discuss with people in distant locations through electronic mail, online chat rooms, bulletin boards or news groups. Experts outside classes can also provide advice and offer suggestions upon request.

3.2 Small Group Discussions Focusing On Real Cases

The small group discussion method can be far more effective than traditional means of instruction, especially in classes covering social issues like information communication ethics [7]. As the subject relates to diverse attitudes and behaviors in cyberspace, the instruction method needs to be both specific and adaptable. Most important, however, is that the instruction method enables students to make their own decisions through critical reasoning processes.

In small group discussions, the application of real cases offers valuable guidelines to the students, who have already experienced or who will experience similar experiences in the future. Teachers provide discussion topics for the class in small group discussions, and students study cases related to the given topics. Students debate on the subject together, and further evaluate the debating process, which leads to the appropriate moral codes in that subject area.

The most typical procedure in small group discussion classes is to assign class students into small groups, and provide discussion time for a common topic or different topics for each group. After the discussion, small groups present their discussion results in the class. Students can also make use of the Internet as a discussion tool even during regular class hours. If the discussion topic requires a relatively long completion period, debate over the Internet would prove to be more advantageous.

3.3 Project-Based Model

Project-Based means that study is conducted through relevant case examples. Students lead each step of the study, address topics and possible problems, identify issues in dispute, and then come to their own conclusion [9]. The primary characteristics of PBM are as follows [10, 11, 12]:

1) *Learning to learn*
PBM encourages students to work on a problem in depth, rather than cover many topics superficially. Students learn what is required to solve a problem or complete a project in a "just-in-time" manner, rather than in a preset curriculum sequence.

2) *Life-long learning*
PBM emphasizes that learning experiences happen throughout life, rather than just learning in formal educational institutions only for a certain period of time. For this purpose, PBM offers diverse tools for acquiring information beyond traditional textbooks. For example, students can find relevant information and people on the Internet.

3) *Active learning*
People usually learn best by "doing." In a well-designed Web project, students can conveniently gather information and data; explore, create, experiment, manipulate items; and reorganize information. They have complete access to different people and information, and understand issues in the real-world context. Interactions with accurate situations promote students' learning process.

4) *Cooperative learning*
PBM gives students an expanded audience and the opportunity for cooperative learning through communication with diverse people. Students can collaborate with other students, teachers, and experts whenever they wish.

The PBM program makes students become independent learners and active information users. In this regard, PBM is one of the ideal models to teach information communication ethics. The model is also useful in motivating students in their study activities, which usually involve ICT (Information Communication Technology) application classes [13].

4 The Design and Implementation of a Web-Based Information Communication Ethics Education System

This section presents a system for Web-based information communication ethics education. The teaching model of this system is based on PBM, reflecting the three stages of PBM for small group discussions. The title of each stage is revised for small group discussion. The system's learning tools include information communication technology, like the Internet, and these are available for use inside as well as outside of the classroom. It is important to notify teachers of the information communication technology applicable for their courses, and to discuss with them their application results. Figure 1 shows the overall structure of the teaching diagram. From this diagram we are able to derive a new teaching model, which consists of 5 stages. Detailed processes for each stage of this model are shown in Figure 2.

4.1 Project Setup

Teachers suggest current issues as topics of interest to students and use news articles, online materials, and CD-ROM or videos to encourage further study. A teacher can provide references like reading materials, relevant websites and a list of reference books to delve deeper into the topics, which should be appropriate for the students' intellectual capabilities and experiences. After students look through those references, they are then instructed to organize their ideas. After, students come up with the subject of their project. The subject should be selected carefully, because it will be the main theme until the end of the project. To ensure that each student comes up with a proper theme, the teacher can help form possible research and discussion subjects with the students. The teacher then provides background information on the project once that subject has been determined.

4.2 Planning the Project

The first step of the project plan is to organize small groups for research and discussion activities. The right number of each group ranges from 4 to 6 students. After groups have been assigned, each student must handle a specific role. Within each group, students share information and plan together. Detailed plans are recommended to determine the ways to conduct research, develop discussions, and arrange student outputs. Project strategies can be divided into two parts: the research activity and the discussion activity.

4.3 Conducting the Project

Project terms vary depending on the characteristics of the project and planning strategies. The project consists of two parts. The first part is related to research activities like the collection of information and materials related to research themes. Students conduct various activities. Examples of these include the collection of materials by searching on the Internet, getting opinions from domain experts via e-mail, conducting surveys through websites, and information sharing with group members online.

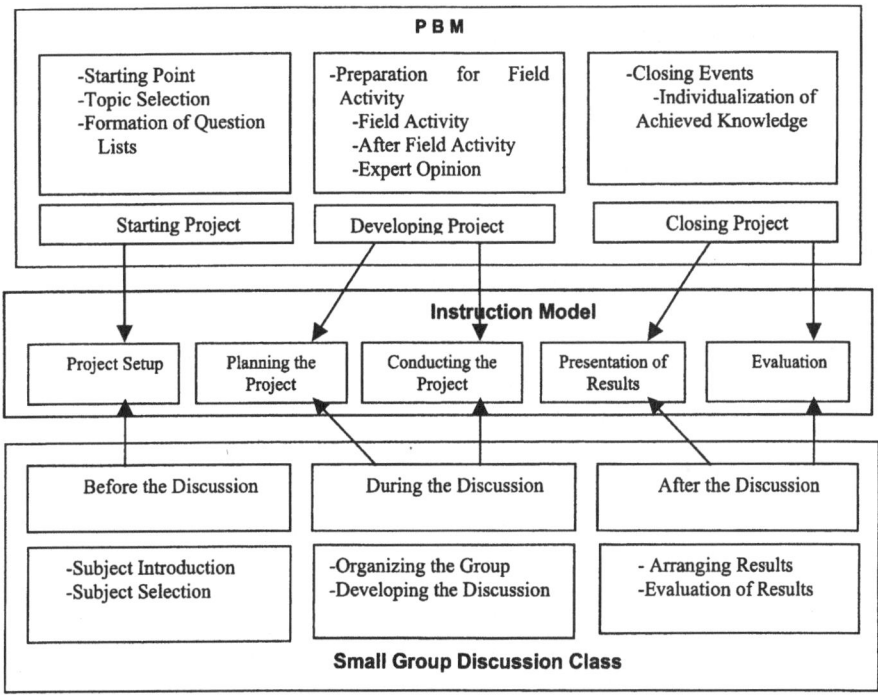

Fig. 1. The overall structure of the teaching diagram

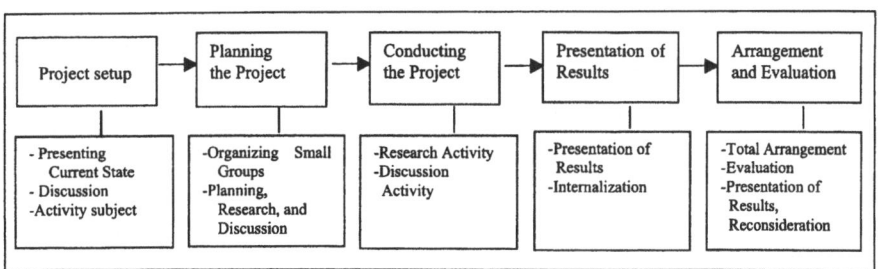

Fig. 2. The proposed Web-based, project-based teaching model

The second part involves discussion activities, and group members talk about their subject in a limited period of time. After exchanging ideas and opinions through Internet chat rooms or online bulletin boards, students can organize the results of their discussion.

The teacher will encourage these research and discussion activities, and also offer *advice and suggestions*. For example, teachers may provide tips for overcoming obstacles to discussion or amend wrong directions taken during debates. Either the research or the discussion can start first, or both of them can be conducted at the same time.

4.4 Presentation of Project Results

After completion of the project, students can present the project results. All the small groups organize their final output, and present to the other class members the project process and their final conclusions.

Results of the project can be presented in various creative ways like publishing a newspaper, posting information on the Web, or with a research paper. Group members are allowed to ask questions about the results. At this stage, students should realize the appropriate behavior necessary for the situation, the basis for this behavior, and the proper perspectives they should maintain. The experience gained from this project, as well as the project's results, should leave a positive impression on each student.

4.5 Arrangements and Evaluation

Teachers may assist students in organizing and evaluating their project results. Examples of teacher assistance include the identification of errors, negative or positive attitudes, etc. Results are then posted on the Web so that all those concerned, including the school, students' families and the local community, can view project results. This sharing of information will promote further cooperative efforts in new projects. Self-evaluation during the entire activity process will enable participants to improve their capabilities for the next project.

5 Application of the Teaching Model

The following website (http://user.chollian.net/~est0718/icee) proposes the implementation of a teaching model. Figure 3 shows the overall menu structure of the system.

5.1 Project Setup

Figure 4 shows the *Hot Issue* screen of the system. *Hot Issue* is a room used for diverse purposes, and moral conflict-related issues are introduced here. In addition to these, there are also reading materials, website information, and reference books. After reading carefully what is posted in this section, students should reconsider their own thoughts or opinions, and relevant cases can be used as examples.

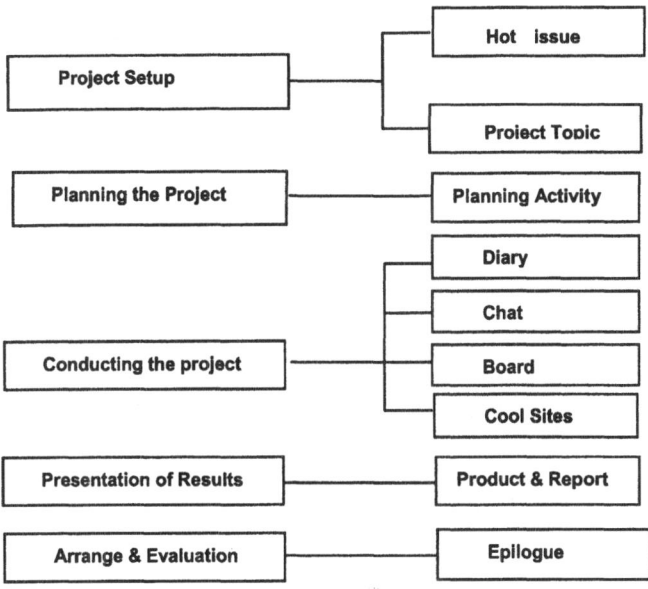

Fig. 3. The overall menu structures of the system

Project Topic is the section that introduces the project's research subject and discussion subject. Research subjects are first common themes for all the students, and then within each theme small groups choose a different topic. Both the common subject and different themes can be given to the students either simultaneously, or one by one. Figure 5 shows the *Project Topic* screen of our system.

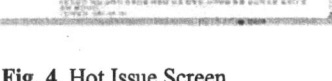

Fig. 4. Hot Issue Screen **Fig. 5.** Project Topic Screen

5.2 Planning the Project

The design of the *Planning Activity* enables students to submit activity plans for research, and to discuss the secondary themes after being organized into small groups. The plan should outline the activities and time schedules for the activities. Students also present ways of generating output and give a detailed sequence of activities.

5.3 Conducting the Project

The *Diary* is a type of free bulletin board for students to record their own thoughts and feelings. Content examples are related to the use of e-mail, a dialogue list with other people joining chat room or Internet bulletin boards, Web information, the total usage time on the computer, and the reasons for using it to conduct the project.

Chat is designed for simultaneous discussion during research activities. Group members can make use of the *Board* when they cannot complete discussions within given class hours. Use of this *Board*, however, must follow netiquette standards and teachers should try to enforce these rules. In the *Cool Site*, students can recommend useful websites and other information for others as they go about their project. Figures 6 and 7 show *Diary* and *Cool Site* screens, respectively.

Fig. 6. Diary Screen

Fig. 7. Cool Site Screen

5.4 Presenting Project Results

The *Product & Report* screen offers a room for students to organize what they have conducted and discussed before, to produce and post the results of their activities, and finally to notify others of the results. Project results can be presented via different means, like in the form of a newspaper, or through a Web presentation, and other documents. Students should make note of discussion results.

5.5 Arrangements and Evaluation

The *Epilogue* screen expresses students' conclusions after finishing their project. When students have a theme to study or discuss later on, they can further write the detailed contents on this board.

6 A Case Study of the Proposed System

In this section, we selected the topic *Online Computer Game Addiction*, and presented how our model can be applied to this subject.

6.1 Project Outline

Table 1 below gives the entire project outline.

Table 1. Project Outline

Title	Online Computer Game Addiction						
Study Subjective	* To explain the advantages and disadvantages of online games. * To explain the reason behind and consequences of online game addiction. * To understand healthy online game culture and set practical attitudes.						
Project Subject	**0Research Subject**			**1Discussion Subject**			
	[Common Research Subject] After testing the level of online game addiction, people should conduct a self-examination of their own addiction levels. [Group Research Subject] 1. Conduct a survey on the actual condition of students playing online games and their parents' responses to the situation. 2. Research online game addiction symptoms and provide an expert's opinion about the problems. 3. Research and analyze real cases of online game addiction, and identify reasons behind it and possible consequences. 4. Research the reasons behind online game addiction, and identify possible solutions and precautionary measures.			[Common Discussion Subject] What are the advantages and disadvantages of online games? [Group Discussion Subject] 1. What's the best way to cope with online game addiction? 2. What's the proper use of online games? 3. How can we help our friends who are in this situation?			
Related School Subject	Ethics, Computer	Grade Level	4th~6th	Project Period	4 weeks	Required Items	Internet-Accessible Computers

6.2 The Process of Conducting the Project

The project process is summarized in Table 2.

Table 2. The Process of Conducting the Project

Steps	Detailed Steps	Activity Details
Project Setup (1st week)	Case Suggestion	Present the abuses of online game addiction and gather opinions from students.
	Giving the Activity Subject and Discussion	Present general information about online game addiction and allow each group to choose a subject for further research and discussion.
Planning the Project (1st week)	Organizing Small Groups	After organizing six students into one group, discuss each person's role and share tasks.
	Planning Research and Discussion Activity	After making plans, make students hand in an outline of the project's stages.
Conducting the Project (2nd week)	Research Activity	[Common Research Activity] After individual tasks, let students hand in the results of their research in the form of paper work. [Group Activity Subject] Each member conducts his or her own role depending on the assigned subject. [Research and share materials through e-mail, Web search engines, chat and other online bulletin boards.]
	Discussion Activity	[Common Discussion Subject] Supervise the entire class discussion based on the common subject. [Group Discussion Subject] Conduct group discussions and let students submit results on paper.
Presentation of Results (3rd week)	Presentation of Results	Present reports or final conclusions, and record these on the online bulletin board to share materials.
	Internalization	To identify the advantages and disadvantages of online games and to discuss proper ways to enjoy them.
Arrangement and Evaluation (4th week)	Total Arrangement	Arrange results by the making of mistakes, proper attitudes, and the required steps to take into practical action when playing online games.
	Evaluation	Evaluate how students conduct their project and estimate their final report and output. (This includes self-evaluation and reviewing other members' performances.)
	Posting the Results	Post the arranged results on the Web to share with others.
	Reflection and Self-Examination	Reflect on the project process with discussion on the next project theme.

7 Conclusions and Further Work

The objective of the information communication ethics education is to reduce the negative consequences that may arise from online exposure at an early age. In this paper, we developed and implemented a new teaching model for the teaching of information communication ethics of elementary school students. Our model has several advantages. First, our system allows students to control their computer use by recording Internet access time. Second, students learn to respect other people's opinions, especially in cyberspace, and identify appropriate ways to express their own views. Third, students become more active and curious during class hours. This is because discussions are based on real cases, which students have either already experienced or will experience later on. Fourth, the program emphasizes small group discussions, and students learn how to interact actively with their classmates. Well-facilitated discussion enables them to utilize advanced information technologies with the proper netiquette.

In order to achieve a successful information communication ethics education, relevant parties should make greater efforts to motivate students as well as to develop good study environments. In the near future, cyberspace will play a central role in our lives and in the learning field. As members of our society, we must, therefore, support, promote, and develop the positive utilization of the Internet. This can only be achieved with the information communication ethics education.

References

1. Study on preventing the reverse effect of information-oriented educational institution, Korea Education and Research information Service (2000)
2. http://www.kedi.re.kr/Exec/Pds/Cnt/128-16.htm
3. Couger, J. D.: Preparing IS students to deal with ethical issues. MIS Quarterly, 13(2) (1989)
4. Johnson, D. G.: Education toward ethical responsibility, Ethics and the management of computer technology. Cambridge, MA: Oelegschlager, Gunn&Hain Publisher (1981)
5. Edgar, S. L.: Morality and Machines, Johns and Bartlett (1997)
6. Severson, R. J.: The principles of information ethics, ME Sharpe (1997)
7. How will instructor teach the cyber ethics? (2001), Seoul Metropolitan Office of Education (2001)
8. http://www.albion.com/netiquette/corerules.html
9. Katz, L.G.: The Project Approach, ERIC Digest: EDO-PS-94-6, [Online] http://ericps.crc.uiuc.edu/eece/pubs/digests/1994/lk-pro94.html (1994)
10. http://24.16.119.79:1444/web/pbl/reform.htm
11. http://www.globalschoolhouse.org/pbl/finding.html
12. http://gsh.lightspan.com/web/lib/ed/pbl.htm
13. Jung, H., Jun, W., Gruenwald, L., and Hong, S.: "The Design and Implementation of a Web-based Teaching-Learning Model for Information Communication Technology Application Education", The 1st International Conference on Web-based learning, Hong Kong, Aug. (2002) 56-68.

An Interaction Model for Web-Based Learning: Cooperative Project

Eunhee Choi[1], Woochun Jun[2], Suk-Ki Hong[3], and Young-Cheol Bang[4]

[1] Seoul Dongil Elementary School, Seoul, Korea
[2] Dept. of Computer Education, Seoul National University of Education, Seoul, Korea
[3] Division of Business & Economics, Dankook University, Seoul, Korea
[4] Dept. of Computer Engineering, Korea Polytechnic University, Siheung, Korea
ghachee@naver.com, wocjun@ns.snue.ac.kr,
skhong017@dankook.ac.kr, ybang@kpu.ac.kr

Abstract. The purpose of this paper is to propose an interaction model, *Cooperative Project*, and develop the Web-based education system based on this model. Interaction is essential to education, and specifically it is emphasized on the Web-based instruction. *Cooperative Project* is devised to promote interaction. With *Cooperative Project*, learners can cooperate smoothly, and complete a project under the Web-based environment where the cooperation between learners can be effectively achieved. The major advantage of this model is that it can motivate learners to process a project with the positive interdependence until the project is completed. Further, this research also shows that *Cooperative Project* can be applied through the web-based instruction to maximize the educational benefits.

Keywords: Web-based Instruction, Interaction

1 Introduction

Many people say that a class is a teaching for a little less knowledgeable people by a little more knowledgeable people. Obviously, the former is a student and the latter is a teacher. Since the Industrial Revolution, the traditional class is that only one person (a teacher) educates many persons (students). However, the traditional class has been changed with the social change. Interaction among the learning process has been more emphasized than ever.

Nowadays, the world becomes a global village, and the whole world is close to each other. International connection is possible for anyone, anytime, and anywhere. People can use the Internet to exchange information. The Web-based instruction is to apply the characteristics of the Web to education. A web-based instruction becomes an effective tool for self-driven learning. It can overcome space and time limitation in an education process, where communication can be activated [8].

Communication is a key term in a Web-based instruction. In Web-based learning environments, communication is done in the form of interaction. The interaction that is essential to an instruction is important to teachers, students, and the developers of education systems. The education under one-side instruction by a teacher usually cannot be effective. It is an interaction that solves this problem.

914   E. Choi et al.

In the traditional class, most of the cases, only one teacher educate many learners, which is "1:n" class. However, Web-based instruction enables a class to have "n:n" relationship. For the meaningful learning, the cooperative learning is emphasized in the "1:n" class.

The purpose of the research is to propose a model to promote interaction, and to build the Web-based instruction system based on this model. This paper is organized as follows. In Section 2, we present the theoretical background of interaction and cooperative learning. In Section 3, we propose an interaction model. In Section 4, we describe the implementation of the Web-based education system. Finally, In Section 5, we give conclusions and further research issues.

2 Theoretical Backgrounds

2.1 Interaction of a Web-Based Instruction

Khan defines the Web-based instruction is the hypermedia-based instruction program which uses the data and the usability of Web in order to create the meaningful instruction [11]. Na defines that the Web-based instruction is the thing that uses the abundant information in the open, integrated environment, which the Web provides. It is a kind of distance learning. In addition to the definition of Khan, Na also argued that the Web-based instruction has brought a new prospect for the education in the form of the integrated environments, which the traditional instruction has not been given yet [2].

The Web-based instruction is clearly different from that of the traditional classroom instruction. Specially, the Web-based instruction can promote self-driven learning, and the learner-directed study is possible. Crossman explained the characteristics of the Web-based instruction as follows [10].

First, a Web-based instruction enlarges the physical location of the education. That is, the education can be performed at not only a classroom, but also home or workplace. Second, a Web-based instruction promotes learning by experience. Learners have the interactions among others, experience through graphic data, and the practical information on the study subject. Third, it provides a way of social interactions and new perspectives through instruction. Fourth, it can present dynamic contents information. Fifth, the learner can control learning progress for himself or herself through a hypertext format. Sixth, it can get individual feedback about an assignment in manner of flexible time and contents. Finally, learner can select various media according to the contents, time, resources, feedback, understanding, and presentation.

Interaction means that more than two subjects affect each other. It does not mean that only one affects the other. That is, a learner becomes the object as well as the subject of an influence in the relation of interaction at the same time. Interactions have been applied to the development of education systems in the learner-centered instruction. The meanings of interactions are follows. Na and Jung mentioned that a learner has control on self-learning assignment [3]. Kim explained that interaction is learner's capability, which can communicate bi-directionally and dynamically, in self-driven manner for obtaining information and knowledge [1].

Interactions must be environment making multiple communications possible. With the development of information communication technology, the interaction environments overcome a local limit, provide new various types of data, and promote information sharing and exchange through a network. According to Yu, interactions must be a process to search, select, reconstruct, and share data through communications between learners in order to solve learners' concrete target or an assignment [4]. In the mean while, a model of interactions in the Web-based instruction has been studied. In [1], it is argued that interaction is the function of six factors such as the degree of satisfaction, the degree of effective control of a learner, dynamics, smooth communication, usability, and effective delivery of contents.

Three models of Moore have been widely used in a model of interaction [12]. The models include learner-instructor interaction, learner-learner interaction, and learner-content interaction as an interaction model. Learner-instructor interaction means that interactions happen between the learners and the instructors, who are contents specialist providing learning contents. As for the learner-learner interaction, it was appeared with the development of various media after 1990's. Especially, an emergence of a computer and network can make learners act in a group. That is, it is suitable for group activities, cooperative learning, and problem solving learning. It will be more important in the future. Learner-contents interaction is the most essential element. Only contents are able to change learners' viewpoints and cognition schema. However, only interactions between a learner and contents have been emphasized in the traditional correspondence education [12].

As a result, the Web-based instruction should be the learner-oriented, and the quality and quantity of interactions on Web-based instruction are essential to the effective learning.

2.2 Cooperative Learning

According to the Slavin, cooperative learning, different from individual learning, is the way of learning that members who have different abilities belong to the same group to act together toward learning target. It gives learners the attitude called all-for-one, one-for all. Members of group help each other for successful learning [13]. Jang argued that cooperative learning is a general class operating principle, which combines a teacher and a student, as well as students [6].

Interaction is a key of success in cooperative learning. The basic principle of the cooperative learning is to maximize the instruction by promoting the quantity and quality of interactions. If the principle is not kept, the following problem may happen. That is, some lazy members share the results from other members' hard work. Therefore, an excellent member is no longer work hard, which may reduce members' interactions. There are some basic principles such as positive interdependence, individual accountability, simultaneous interaction, equal participation in order to solve this problem, and to promote positive interaction between learners [7]. As stimulating an efficient level of stress, positive and constructive feedback among members can induce the target accomplishment [5].

From the literature review, the critical success factors of cooperative learning can be summarized as follows. First, cooperative learning is more effective method than traditional learning in promoting academic records. Second, members can learn how to adjust their own actions with the other members through interaction. Third, the participant in cooperative learning shows higher pride and positive values. Finally, the cooperative learning induces the positive attitude [9].

3 The Proposed Interaction Model

3.1 The Definition of Cooperative Project

Based on the discussions so far, we argue that the most effective way to increase interaction is to provide appropriate problems to be solved and close dependency among students. The following figure 1 shows the overview of the proposed interaction model called Cooperative Project.

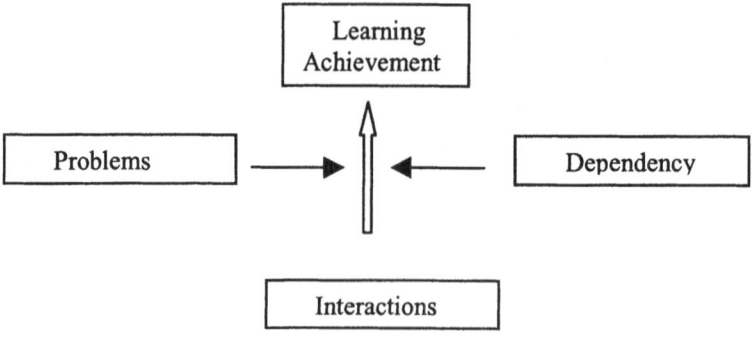

Fig. 1. The overview of Cooperative Project Model

The purpose of the Cooperative Project is to let students complete a project successfully on Web-based environments, in which members can interact among them smoothly. A project is divided into smaller projects, which a group takes charge of. Without the completion of the smaller projects, the whole project cannot be accomplished. The Web provides the environments to help members solve their own projects in interactive and self-directed manner.

3.2 The Procedures of Cooperative Project

Figure 2 shows the procedures of the cooperative project.

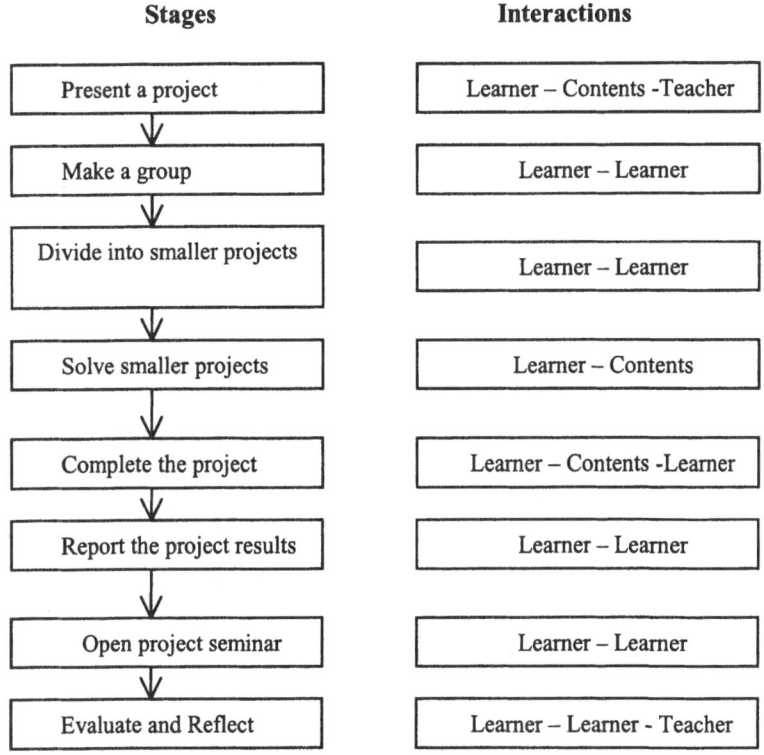

Fig. 2. The Procedures of Cooperative Project

Each procedure is described further as follows.
Procedure 1: Present a project
- Present various projects on the Web.
- Project must be very specific.
- Present simple guidance about the project contents, configuration personnel and a period.
- Instruct necessary netiquettes for smooth communication.

Procedure 2: Make a group
- Each learner can participate in a group.
- Each group consists of 3-4 students.

Procedure 3: Divide into smaller projects
- Group members use electronic communication and exchange their opinions.
- Classify a project into small projects.
- Every member needs to determine the small project to be solved.

Procedure 4: Solve smaller projects
- Each member needs to solve small project on the Web.
- Put results on the Web in the form of a personal report.
- The each member refers to other member's contents and each member solves own small project.
Procedure 5: Complete the project
- Complete a whole project by combining results of smaller projects
Procedure 6: Report the project results
- After the whole project is finished, each group must report the project results.
Procedure 7: Open project seminar
- Each group should report the project results for other project group and share ideas.
Procedure 8: Evaluate and reflect
- Based on the evaluation and reflection on the project, each member prepares better future project.

3.3 The Merits of a Cooperative Project Model

The major advantage of the cooperative project is to increase efficiency by matching groups to predefined smaller projects in advance, which saves the total project completion time. Especially, our model has the following merits.
- The positive high-quality interaction is possible.
  Each member belongs to both a group project and a small project at the same time. This kind of positive dependency maximizes interaction among members. It also brings the better results into each member.
- Being a member of group provides high motivation to students
Since each smaller project is assigned to an each group and a learner can participate in a smaller project of his choice, the learner can have high motivation from the beginning.
- There is no wasting time to discuss project topics after group configuration.
This is because each member has already decided a topic of his/her choice after group configuration.
- A learner can have positive interdependence.
Members in a group are tied up together since they select the same topic. Also, members need to help each other for the successful completion of the project.
- Each learner has individual accountability.
Only after each member finishes his/her own project assigned, the whole group project can be completed. Therefore, each member must have personal duty and responsibility on other member.
- Heterogeneous group configuration is naturally achieved.
This is because that a group can be configured by members' free will. When members select a project, they use information on only project itself regardless of members' personal relationship or history.

4 The Implementation of the Proposed Cooperative Project

Several hardware/software tools were used to implement the proposed cooperative project. Some of them are JavaScript and PHP 4.0 as script languages, MySQL as database, DreamWeaver 4.0 as an authoring tool, and Apache as a Web server. The URL for the implementation of our cooperative project is http://my.netian.com/~ghachee/.

In the system, some representative screens such as main screen, introduction screen, project screen, the first project group screen, the second project group screen, the third project group screen, work screen, discussion screen, evaluation screen, and advice screen are shown in figure 3 to figure 12, respectively.

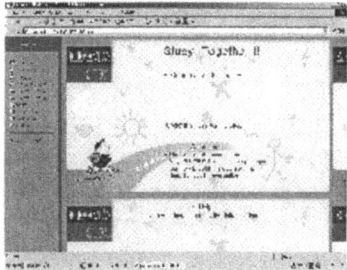

Fig. 3. Main Screen

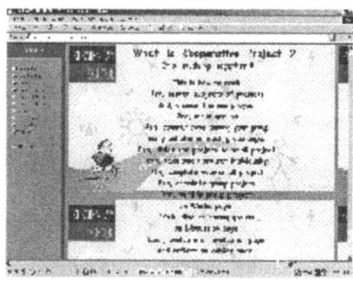

Fig. 4. Introduction Screen

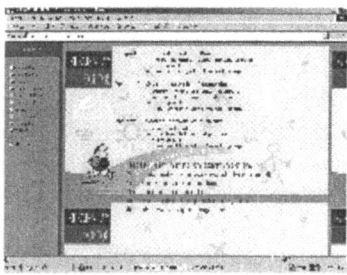

Fig. 5. Project Screen

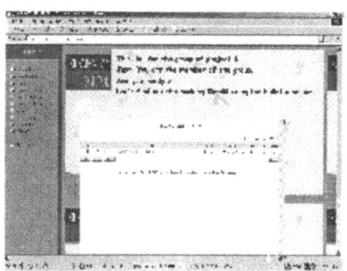

Fig. 6. 1st Project Group Screen

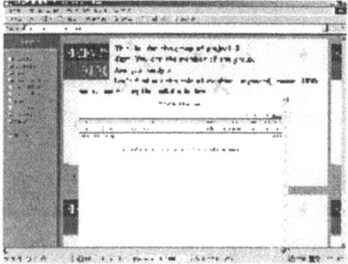

Fig. 7. 2nd Project Group Screen

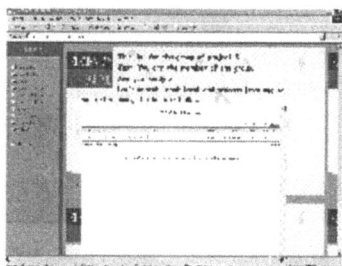

Fig. 8. 3rd Project Group Screen

Fig. 9. Work Screen

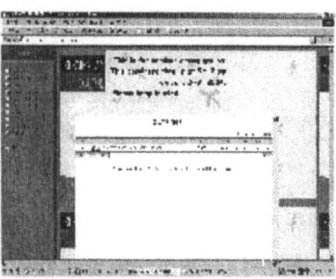

Fig. 10. Discussion Screen

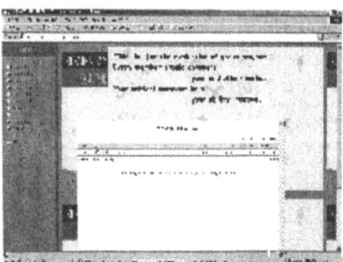

Fig. 11. Evaluation Screen

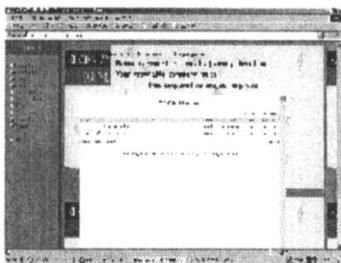

Fig. 12. Advice Screen

5 Conclusions and Further Work

The Web provides better means as a learning tool supporting interaction than any other tool. Many works have been focused on the Web application in distance learning. However, currently works on interaction and learner-directed study are more emphasized. This means that the subject of learning becomes learners, neither teachers nor contents.

In our research, we proposed the cooperative project as an exemplary model of interactions through the Web-based instruction system. Achieving a whole project through the completion of smaller but predefined projects requires more interactions under positive interdependence. Therefore, learning effects can be maximized.

The advantages of this model are as follows: (1) providing active interaction, (2) providing learners' motivation, (3) saving time to complete a project, (4) supporting positive interdependence, (5) allowing learners to have individual accountability, etc.

Although we have finished the implementation of this model, further research is required in some areas. For instance, the evaluation of interactions is sometimes subjective, which makes the results obscure. Therefore, evaluation criteria should be developed in diverse manner. Quantitative measures like click through rates and visiting numbers are not sufficient, and some other qualitative ones should be developed.

References

[1] Kim, M.: A study on an Interaction Promotion Strategy in a Hypertext Education System, Doctoral Dissertation, Seoul National University, Seoul, Korea (1998)
[2] Na, I.: Web-based Education, Educational Technology Press, Seoul, Korea (1999)
[3] Na,I. and Jeong, I.: The Understanding of Educational Technology, HakJiSa, Seoul, Korea (1996)
[4] Yu, S.: A Study of Interaction in Web-based Instruction Using the Character of a Learner. Master Thesis, Seoul National University, Seoul, Korea (1999)
[5] Lee, B.: Educational Meaning of Cooperative Learning – Centered on Structure and Function of Brain, Master Thesis, Seoul National University of Education, Seoul, Korea (2002)
[6] Jang, J.: The Application Case Study 1 of Cooperative Learning Structure of Kagen. Cooperative Learning Training Material, Seoul, Korea (2001)
[7] Jung, M. and Kim, D.: Theory and Practice of Cooperative Learning for Open-Education, edited by. Na, I., Hyeongsul Publication, Seoul, Korea, (1998)
[8] Jung, I.: "Effect Factor Analysis of Web-based Education" in Web-based Education, Science Education Publication, Seoul, Korea (1999) 299-330
[9] Jo, G.: "Activation of the Elementary School Humanity Education t Using Cooperative Learning", Elementary Education Study, 15(1) (2000) 8-9
[10] Crossman, D. M.: "The Evolution of the World Wide Web as an Emerging Instructional Technology Tool", in Web-based Instruction, edited by B. H. Khan, Educational Technology Publications, Englewood Cliffs, NJ, USA (1997)
[11] Khan, B. H.: Web-based Instruction, Englewood Cliffs, NJ: Educational, Technology Publications (1997)
[12] Moore, M. G.: "Three Types of Interactions", in Distance Education: New Perspectives, edited by Harry, K., John, M. and Keegan, D., London, United Kingdom (1993)
[13] Slavin, R.: Cooperative Learning: The Theory, Research and Practice, Allyn and Bacon, Upper Saddle River, NJ, USA (1995)

Observing Standards for Web-Based Learning from the Web

Luis Anido, Judith Rodríguez, Manuel Caeiro, and Juan Santos

Departamento de Ingeniería Telemática, Universidade de Vigo
Campus Universitario S/N – E-36200 Vigo, Spain
lanido@det.uvigo.es, Fax: +34 986 812116

Abstract. Learning technology standardization is a lively process that will last for years until a clear and precise set of recommendations are identified. So far, there is only one official standard and some final specifications in different areas. Most documents are working drafts that are still in their development process before being approved and, most important, accepted by the global e-learning community. Currently, many institutions collaborate to produce new updates almost every week. In this situation it is really hard to get familiar with the learning technology standardization process and, even worse, to keep updated. The Workshop for Learning Technologies, within the European Committee for Standardization (CEN/ISSS WS-LT), produced the Learning Technology Standards Observatory, a web portal where those interested in this process can get familiar with it, learn the main differences between related specifications, follow what is going on, what is planned for the near future and furthermore participate in the process itself. This paper presents the main functionalities and content areas within this site and how users can benefit from its use.

1 Introduction

Application of information and communication technologies to the learning domain has been very fruitful. A lot of e-learning systems and resources have been developed, and as usual, problems of reusability and interoperability appear.

As a consequence, a standardization process was initiated. Currently, this effort has produced some proposals for e-learning standards. Although most of them are in draft form, and therefore exhibit varying degrees of stability, eventually, some of them will become generally accepted standards that people involved in the e-learning business should be aware about. This process has produced an important number of working drafts, specifications, and even working prototypes. Within the same standardization area, there are different specifications from different institutions that deal with the same problems and, therefore, are related with each other. Nevertheless, there is no reference source where the main differences among them are identified, which their advantages/disadvantages are. In short, learning technologists would need a huge effort to get familiar with this standardization process: many related and non-related technical specifications, minutes from working groups and other documents produced by standardization bodies would need to be reviewed.

This paper presents an important effort taken by the CEN/ISSS WS-LT [1] to set up a Web-based observatory that could serve as a reference point to follow the learning technology standardization process. This site has been called the "Learning Technology Standards Observatory" (LTSO) [2] (see Figure 1):

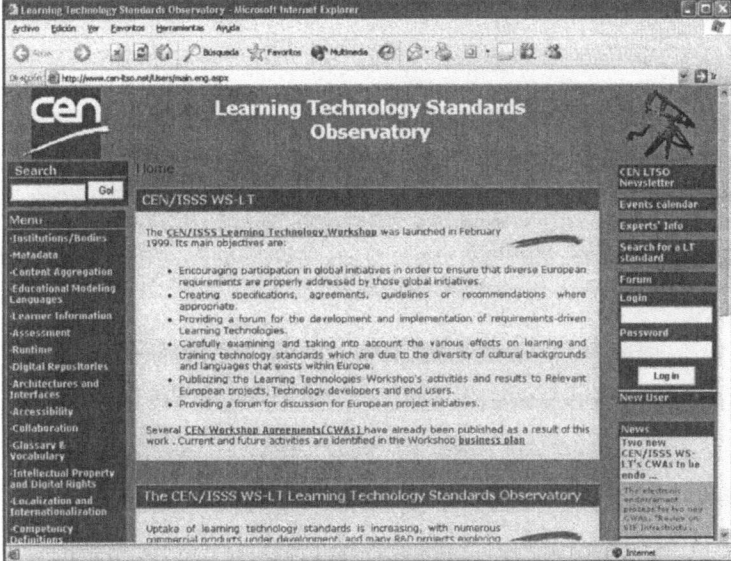

Fig. 1. LTSO English home page

Information available at the LTSO includes but is not be restricted to:

- Summary of each standard including key data in order to allow the user to grasp the gist of the specification. The main aim is to prevent users from needing to read the technical details of the whole specification.
- Tracking data on the evolution of the standard through the different drafts.
- Relationship and main difference among proposed specifications and standards for the same category (e.g. AICC CMI and ADL SCORM, LTSC LOM and IMS metadata, IMS LIP and LTSC PAPI, etc.).
- Clarification of the relationships between formal standardization bodies (ISO, CEN, IEEE), specification development consortia (such as ARIADNE and IMS) and profiling bodies (such as CANCORE, ALIC, and others)
- Links to the complete set of specifications (if available on-line) or to information on where to obtain them.
- Information on the actual uptake of specifications and standards.
- Information about relevant events, activities and organizations.

The LTSO is multilingual-capable. Contents can be published in any of the 11 official languages in the European Union (Danish, German, Greek, English, Spanish, French, Italian, Dutch, Portuguese, Finnish, and Swedish). Contents are published using a proprietary web-based authoring tool. This means that both contents and structure are completely dynamic and can be updated automatically using a web-based interface. Section 2 of this paper introduces the main sections available at the LTSO today.

Section 3 shows other additional functionalities oriented to create a virtual community for those interested in e-learning standardization (experts' info system, subscriptions to news and events, events calendar, etc.) and ease the access to the LTSO contents (e.g. downloading of a PDF version of the whole LTSO contents).

2 LTSO Contents

LTSO contents offer up-to-date information on the main institutions and bodies involved in e-learning standardization and their outcomes. Currently, there is information about the following institutions (references to each institution can be obtained from the LTSO itself): ISO/IEC JTC1 SC36, IEEE LTSC, CEN/ISSS WS-LT, ADL, AICC, IMS, European Schoolnet, ARIADNE, PROMETEUS, GEM, EdNA, ALIC, and DCMI.

For each institution the following subsections are maintained:

- *Working groups*: Data of each working group within the institution.

- *Liaisons*: Liaison policies, active agreements and their results.

- *Meetings*: It gathers information on the next meetings scheduled.

- *News*: Last or most relevant piece of news.

- *Future activities*: Summary of future activities.

The gist of the LTSO is structured around the different areas within the learning technology standardization process. For each group of specifications/standards the LTSO provides a classification for it and information on its scope. Within each group, the LTSO provides key information on every standard/specification like the institution that produced it, version tracking, its present-day status, references to related sources, etc. In addition to this basic information, several sub-categories are also included:

- *Summary*: An overall description containing the minimum information needed to understand the gist of the standard.

- *Current version*: Name and reference to the currently available version as well as its main updates.

- *Previous versions*: Tracking data on the evolution of the standard.

- *Relations*: Similarities and main differences with related specifications.

- *Future work*: Main working lines and tasks being developed. It may include the schedule.

- *Conformance*: Requirements to be compliant with this standard.

- *News*: News related to this standard.
- *Meetings*: Next meetings scheduled that may affect this specification.

The next paragraphs offer an overview on the different current standardization fields within the LTSO. Further information and references to the actual specifications can be obtained from the LTSO itself.

Metadata. Metadata standards allow creators of documents and managers of resource collections to describe resources in a detailed manner facilitating targeted queries by search engines and the exchange of descriptive information about the resources themselves. In particular, educational metadata standards extend the scope of description that can be included in a metadata record with information that has particular educational relevance. This is done by either defining education-specific elements, element refinements or encoding schemes.

For this category the LTSO addresses the next specifications: *LOM, ARIADNE Metadata, IMS Metadata, ADL Metadata, DC, Cancore, GEM metadata,* and *EdNA metadata*.

Content Aggregation. The need for educational resource sharing among learning systems and authoring tools has motivated the development of common formats and procedures for encapsulating learning resources into cohesive units of instruction, applying structure and associating learning taxonomies so that the structure and intended behavior (sequencing of activities) can be uniformly represented, communicated and reproduced across heterogeneous environments.

The LTSO provides information about the next content aggregation specifications: *SCORM CAM, IMS Content Packaging, IMS Simple Sequencing, AICC Course Structure,* and *IEEE P1484.6 Course Sequencing*.

Educational Modeling Languages. An Educational Modelling Language (EML) is a semantic notation to create units of learning to support the reuse of pedagogical entities like learning designs, learning objectives, learning activities, etc. EMLs are used to create highly-structured course material.

The next specifications are included in the LTSO: *OUNL EML, IMS Learning Design, PALO, CDF, LMML, Targeteam,* and *TML / Netquest*.

Learner Information. Learner Information specifications are devoted to support the exchange of learner information between different systems. They provide data models, including the syntax and the semantics, to describe both the characteristics of a learner and his or her knowledge/abilities.

More specifically, the specifications addressed in this section by the LTSO are: the formerly known *LTSC PAPI, IMS LIP, IMS Enterprise,* works by the *ISO/IEC JTC1 SC36 WG3,* and *European Diploma Supplement*.

Assessment. Assessment specifications address the need of defining common formats and procedures for the exchange of evaluation material among different e-learning tools. The most outstanding proposal in this field is the *Question & Test Interoperability* (QTI) specification from the IMS consortium.

Runtime. The basic tasks of runtime environments are content delivery to the student, support of the interaction between the content and the Learning Management System, and to decide the content to be delivered next depending on the static and dynamic course structure, and previous student actions. To facilitate reuse, the logic needed to provide this functionality should be clearly separated from other educational resources like multimedia elements, or even software modules responsible for other tasks (e.g. content transfer, communication among students, etc.).

The specifications addressed by the LTSO are: *AICC CMI/Lesson Communication, SCORM Runtime*, and the works by the *IEEE P1484.11*.

Digital Repositories. Repository systems provide key infrastructure for the development, storage, management, discovery and delivery of all types of electronic content. Digital repositories store both assets – and the metadata describing them – which can be retrieved through the network without a previous knowledge of the collection structure. At present, the most outstanding proposal addressing interoperability issues among digital repositories is the *Digital Repositories Interoperability* (DRI) by the IMS Global Learning Consortium.

Architectures and Interfaces. The purpose of developing system architectures is to discover high-level frameworks for understanding certain kinds of systems, their subsystems, and their interactions with related systems, i.e., more than one architecture is possible. An architecture is not a blue print for designing a single system, but a framework for designing a range of systems over time, and for the analysis and comparison of these systems, i.e., an architecture is used for analysis and communication.

The LTSO includes information about the following specifications: *IEEE LTSA, OKI, LSAL, SIF* and *OASIS*.

Accessibility. As with many types of products and technologies, people with disabilities may be inadvertently excluded if accessibility is not considered and incorporated into products and technologies. However, accessibility is not only of concern to those with disabilities. The potential for online distributed learning expands when developers embrace the widest possible range of individual learning styles, preferences and abilities. Specifications in this are cover these needs.

The LTSO includes an analysis of the next proposals: *CEN works, IMS Accessibility*, and *PROMETEUS works*.

Collaboration. Collaborative learning is a form of learning concerned with the collaboration among users. It is designed for faculty and groups of learners to fulfill learning objectives of groups and learners, through sharing resources or interaction.

Currently there is only one group, *ISO/IEC JTC1 SC36 WG2*, devoted to this standardization area, but other proposals (e.g. EMLs) also consider collaboration issues in their proposals.

Glossary and Vocabulary. Specifications and technical reports on glossaries and vocabularies provide definitions of terms, acronyms, abbreviations, and notations used in the standards and appropriate mechanisms to exchange them. They also provide a set of explanations of terms, which complement the definitions by specifying conceptual links within groups of related terms.

The LTSO provides information about *AICC works*, proposals by the *ISO/IEC JTC1 SC36 WG1*, and *IMS VDEX*.

Intellectual Property and Digital Rights. Intellectual Properties and Digital Rights specifications are concerned with the syntax and semantics needed to specify rights expressions on how digital content may be distributed or used.

The LTSO includes information about *CEN* works, and *IEEE DREL*.

Localization and Internationalization. Localization & Internationalization initiatives aim to ensure that standards consider language and cultural diversity in order to improve the provision of technology-based learning experiences.

CEN works, *LTSC Localization*, and *ISO/IEC JTC1 SC36 CLFA Rapporteur Group* are the proposals included in the LTSO.

Competency Definitions. Competency Definition data objects are used to specify competencies, skills and learning objectives in a Learning Management System and in Competency Profiles. These specifications provide the means to create common understandings of competencies that appear as part of a learning or career plan, as learning pre-requisites, or as learning outcomes.

The LTSO provides information about the next proposals: *LTSC Competency Definitions*, and *IMS RDCEO*.

User Interfaces. User interfaces standardization is related to the specification of widget and graphical metaphors.

Only the *AICC* has issued some proposals in this field.

Platform and Media. The generic category "Platform & Media" groups a set of documents that provide recommendations, guidelines and best practices in order to achieve interoperability on the different media formats used by courseware systems.

The LTSO includes information about the *AICC* whose proposals are the most outstanding in this field so far.

Quality. Quality specifications are concerned with the specification of metrics, guidelines, taxonomies, etc. for the development of technology-based learning systems. Allowing end users to specify their specific quality needs and providers to declare their specific quality provisions in corresponding formats will be a highly important instrument for a global market in this field.

The LTSO provides information about the *ISO/IEC JTC1 SC36 WG5* proposal.

3 LTSO Additional Functionality

In addition to the actual data on learning technology standards, the CEN/ISSS WS-LT's LTSO provides a set of additional functionality:

- *News system.* The home page of the observatory includes a news display area where they are shown from more to less recent publication date. A RSS [3] interface is provided to allow external systems to access the LTSO news system.
- *Events system.* LTSO authors are allowed to enter data on e-learning standardization events. The home page of the observatory includes an events display area where they are shown according to the event dates (nearer events first). Past events will be automatically deleted from this area. An events calendar is also provided. Dates within any event' celebration period are highlighted and events' data can be accessed clicking on them.
- *Notification mechanism.* There is a notification mechanism to allow users to subscribe to the news and/or events information system. Different topics may be selected to filter those data to be received. News and information on events are sent by e-mail.
- *Newsletter.* Gathering data from news and events, the repository creates a newsletter including news published and events celebrated during that month. This bulleting is created in PDF format and can be downloaded from an "observatory bulleting" area, where links are provided for each produced bulletin. Every day of the month a new draft version of the newsletter is created automatically. This version becomes final the last day of the month.
- *Experts' information system.* This mechanism provides access and contact information for those individuals actively involved in the learning technology standardization process. Approval for being included in the agenda is needed from the LTSO general administrator.
- *Open discussion forum.* Registered users (anyone can register) and LTSO experts may use a shared space to post comments, questions, replies, etc.
- *Internal search engine.* To locate concrete information on both the LTSO contents area, news and/or events.
- *Standards description.* The LTSO uses the "Standards Metadata Element Set version 3.0" [4] to describe learning technology standards. This specification provides a metadata schema to describe any kind of standard. In this way, LTSO users are allowed to easily locate a given specification or access to further information.
- *LTSO PDF version.* This version is a printable book that is created automatically every day with an updated version of contents and can be downloaded from the LTSO home page.
- *LTSO IMS package.* LTSO contents are also packaged following the IMS Content Packaging specification [5]. In this way, it is possible to extract LTSO contents into an external repository where this information can be downloaded from.

4 LTSO Maintenance

The LTSO is a fully-dynamic web-site where contents of each page can be edited and customized from the distance using its own authoring tool. Also, the whole structure of the web is customizable allowing for a flexible and easy maintenance of the site. There is no tight to any initial particular structure for the information. On the contrary, both contents and structure could be adapted to future modifications or new fields of interest or activities inside the dynamic learning technology standardization process. Therefore, in addition to the current contents, it could be included in the observatory any new data about standardization bodies or standard specifications or new relationship that may appear as an important information in the future. They are automatically displayed to learning technology standards "observers", see Figure 2:

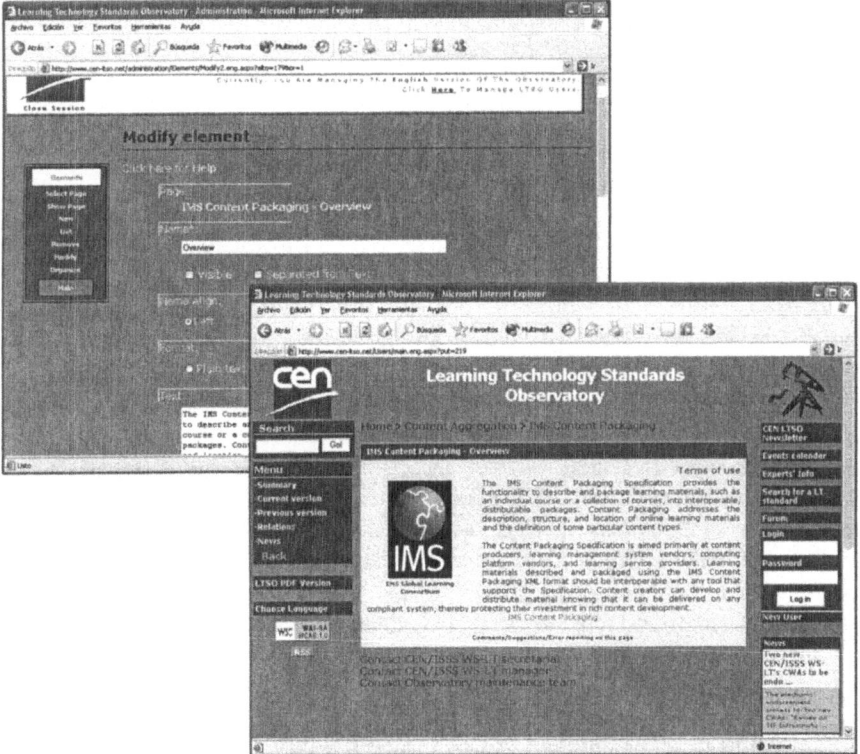

Fig. 2. LTSO authoring tool (left side) and produced contents (right side)

There are different roles for those managing the site. In this way, a general administrator can appoint individuals as responsible for entering information about a particular standardization body or standard. Access is provided to the main standardization bodies at present to let them update the observatory and publicize their work through it.

This dynamic approach is also used to provide a multilingual user interface. Those responsible for entering information about standards or standardization bodies are provided with a suitable mechanism to enter information in the different EU languages. For this, the general administrator defines the different language versions of the repository that each user is allowed to develop and update.

LTSO maintenance is sponsored since 2004 by two outstanding institutions within the educational domain in Europe: CEDEFOP [6] and the European SchoolNet [7].

5 Summary

Uptake of learning technology standards is increasing, with numerous commercial products under development, and many R&D projects exploring the issues in this area. However, there is widespread confusion and misunderstanding about the relationships between the relevant standards and specifications, as well as between the organisations that develop, define, profile or implement them.

Especially in Europe, it is crucial that communication on these aspects increases both in quality as well as in accessibility, as there is a danger that only an U.S. centric point of view will be widely disseminated otherwise.

This is the rationale that lead the European Committee of Standardization's Workshop on Learning Technologies (CEN/ISSS WS-LT) to establish an accessible and sustainable web based repository that acts as a focal access point to projects, results, activities and organizations that are relevant to the development and adoption of e-learning technology standards: the Learning Technology Standards Observatory (LTSO).

The LTSO can be used by those unfamiliar with the learning technology standardization process: it includes summarized information describing the main characteristics of outstanding specifications, the most important differences among related specifications, and an access points to the main standardization bodies. Also, it can be used by experts who want to share their knowledge or learn more about a specific subject: detailed information, like versioning tracking data or relationships with other specifications, is also available at the LTSO.

Available authoring tools allow us to distribute the maintenance work among a set of e-learning standardization experts in the different areas managed by the LTSO. Current contents cover 17 standardization areas – with more than 50 standards, specifications and proposals described – 13 institutions and bodies and more than 80 news and events have been published since June 2003. Also, some of the most relevant experts, in some cases editors of well-known specifications/standards and chairmen of the most important working groups, have subscribed to the LTSO experts' info system.

Activities within 2004 are focused on the dissemination of this information service worldwide. The most important institutions and bodies within the learning technology standardization process are revised every day (more than 24 web sites) to make sure the LTSO contents are daily updated. So far, contents are only available in the English version. It is expected to fill other lingual versions in the near future.

Acknowledgements. The LTSO development has been funded by the CEN/ISSS Workshop on Learning Technologies through contract No. CEN/VIGO/2002/094. LTSO maintenance is supported since 2004 by CEDEFOP and the European SchoolNet.

References

[1] CEN: Learning Technologies Workshop (CEN/ISSS WS-LT). Official website address at http://www.cenorm.be/cenorm/businessdomains/businessdomains/informationsocietystandardizationsystem/elearning/learning+technologies+workshop/index.asp
[2] Learning Technology Standards Observatory (LTSO). Website at http://www.cen-ltso.com
[3] Miller, E., et al.: W3C RSS 1.0 News Feed Creation How-To (2001). Electronic version available at http://www.w3.org/2001/10/glance/doc/howto
[4] ANSI Public Document Library: Standards Metadata Element Set, v3.0 (2003). Electronic version available at http://public.ansi.org/ansionline/Documents/Other Services/Standards Registry Committee/public_comments_metadata_specv3.htm
[5] Smythe, C.: IMS Content Packaging Information Model, version 1.1.3, Final Specification. IMS Global Learning Consortium (2003). Electronic version available at http://www.imsglobal.org/content/packaging/cpv1p1p3/imscp_infov1p1p3.html
[6] CEDEFOP: European Centre for the Development of Vocational Training. Official website address at http://www.cedefop.eu.int/
[7] European Schoolnet. Official web site at http://www.eun.org

On Computing the Spectral Decomposition of Symmetric Arrowhead Matrices

Fasma Diele[1], Nicola Mastronardi[1], Marc Van Barel[2], and Ellen Van Camp[2]

[1] Istituto per le Applicazoni del Calcolo "M. Picone", sez. Bari, via Amendola
122/D, I-70126 Bari, Italy
{f.diele, n.mastronardi}@area.ba.cnr.it
[2] Department of Computer Science, Katholieke Universiteit Leuven, Celestijnenlaan
200A, B-3001 Leuven (Heverlee), Belgium
{Ellen.VanCamp,Marc.VanBarel}@cs.kuleuven.ac.be

Abstract. The computation of the spectral decomposition of a symmetric arrowhead matrix is an important problem in applied mathematics [10]. It is also the kernel of divide and conquer algorithms for computing the Schur decomposition of symmetric tridiagonal matrices [2,7,8] and diagonal–plus–semiseparable matrices [3,9]. The eigenvalues of symmetric arrowhead matrices are the zeros of a secular equation [5] and some iterative algorithms have been proposed for their computation [2,7,8]. An important issue of these algorithms is the choice of the initial guess. Let $\alpha_1 \leq \alpha_2 \leq \ldots \leq \alpha_{n-1}$ be the entries of the main diagonal of a symmetric arrowhead matrix of order n. Denoted by $\lambda_i, i = 1, \ldots, n$, the corresponding eigenvalues, it is well know that $\alpha_i \leq \lambda_{i+1} \leq \alpha_{i+1}, i = 1, \ldots, n-2$. An algorithm for computing each eigenvalue $\lambda_i, i = 1, \ldots, n$, of a symmetric arrowhead matrix with monotonic quadratic convergence, independent of the choice of the initial guess in the interval $]\alpha_{i-1}, \alpha_i[$ is proposed in this paper. Although the eigenvalues of a symmetric arrowhead matrix can be computed efficiently, a loss of orthogonality can occur in the computed matrix of eigenvectors [2,7,8].In this paper we propose also a simple, stable and efficient way to compute the eigenvectors of arrowhead matrices.

1 Introduction

The computation of the spectral decomposition of a symmetric arrowhead matrix is an important problem in applied mathematics [10]. It is also the kernel of a divide and conquer algorithm to compute the Schur decomposition of symmetric tridiagonal matrices [2,7,8] and diagonal–plus–semiseparable matrices [3,9].

The eigenvalues of symmetric arrowhead matrices are the zeros of a secular equation [5] and some algorithms have been proposed for their computation [2, 7,8]. An important issue of these algorithms is the choice of the initial guess.

Let $\alpha_1 \leq \alpha_2 \leq \ldots \leq \alpha_{n-1}$ be the entries of the main diagonal of a symmetric arrowhead matrix of order n. Denoted by λ_i, $i = 1, \ldots, n$, the corresponding eigenvalues, it is well know that $\alpha_i \leq \lambda_{i+1} \leq \alpha_{i+1}, i = 1, \ldots, n-2$. An algorithm for computing each eigenvalue $\lambda_i, i = 1, \ldots, n$, of a symmetric arrowhead matrix with monotonic quadratic convergence, whatever the choice of the initial guess in the interval $]\alpha_{i-1}, \alpha_i[$, is proposed in this paper.

Although the eigenvalues of an arrowhead matrix can be computed efficiently, a loss of orthogonality can occur in the computed matrix of eigenvectors [2,7,8]. Recently a stable method for computing the eigenvectors has been proposed, that, computed accurately the approximate eigenvalues, requires to compute the eigenvectors of a new arrowhead matrix whose exact eigenvalues are the computed ones [2,7,8].

In this paper a simple, stable and efficient way of computing the eigenvectors of symmetric arrowhead matrices is described.

The paper is organized as follows. The spectral properties of symmetric arrowhead matrices are shortly introduced in § 2. In § 3 an algorithm for computing the eigenvalues of a symmetric arrowhead matrix is proposed. A simple and efficient way to compute the eigenvectors of arrowhead matrices is described in § 4 followed by the numerical results in § 5.

2 Eigenvalues and Eigenvectors of a Symmetric Arrowhead Matrix

Let
$$A = \begin{bmatrix} D & \mathbf{b} \\ \mathbf{b}^T & \gamma \end{bmatrix}$$

be a real symmetric arrowhead matrix, where $A \in \mathbb{R}^{n \times n}$, $D = \text{diag}(\alpha_1, \ldots, \alpha_{n-1})$, $\alpha_1 \leq \alpha_2 \leq \ldots \leq \alpha_{n-1}$, $\mathbf{b} = [\beta_1, \ldots, \beta_{n-1}]^T$. Let λ_i, $i = 1, \ldots, n$, be the eigenvalues of A. The following interlacing property holds [2,8,6,7]

$$\lambda_1 \leq \alpha_1 \leq \lambda_2 \leq \alpha_2 \leq \ldots \leq \lambda_{n-1} \leq \alpha_{n-1} \leq \lambda_n. \tag{1}$$

In the following cases some eigenvalues of A are explicitely known [2,7,8].

1. If $\beta_j = 0$, then $\lambda_j = \alpha_j$, with corresponding eigenvector $\mathbf{e}_j$, where $\mathbf{e}_j$ is the j-th vector of the canonical basis of $\mathbb{R}^n$.
2. If $\alpha_j = \alpha_{j+1}$, for some j, then $\lambda_j = \alpha_j$. In this case, a 2×2 orthogonal similarity transformation is applied to the arrowhead matrix in order to make $\beta_{j+1} = 0$.

From a computational point of view, we consider α_j eigenvalue of A if $|\beta_j| \leq 3(n+6)\varepsilon \|\mathbf{b}\|_2$, where ε is the machine precision [2]. The other eigenvalues can be computed deleting the j-th row and the j-th column from A. This process is called *deflation* [4].

After the deflation, the problem is to compute the Schur form of an *irreducible* arrowhead matrix A, with $\alpha_1 < \alpha_2 < \ldots < \alpha_{n-1}$, and $|\beta_j| > 3(n+6)\varepsilon \|\mathbf{b}\|_2$, for $j = 1, \ldots, n-1$.

It is well known [5,10] that the eigenvalues of A are the zeros of the *secular equation*

$$f(\lambda) = \lambda - \gamma + \sum_{j=1}^{n-1} \frac{\beta_j^2}{\alpha_j - \lambda}.$$

3 An Algorithm for Computing the Eigenvalues of Symmetric Arrowhead Matrices

In this section we construct an algorithm to compute the eigenvalues of the symmetric irreducible arrowhead matrix

$$A = \begin{bmatrix} \alpha_1 & & & \beta_1 \\ & \ddots & & \vdots \\ & & \alpha_{n-1} & \beta_{n-1} \\ \beta_1 & \cdots & \beta_{n-1} & \gamma \end{bmatrix} \qquad (2)$$

with

$$\alpha_1 < \alpha_2 < \cdots < \alpha_{n-1}, \quad \text{and} \quad \beta_i \neq 0, \quad i = 1, \ldots, n-1.$$

Let $\lambda_1 < \lambda_2 < \cdots < \lambda_n$ be the eigenvalues of (2). Then λ_i, $i = 1, \ldots, n$ are the roots of

$$p_n(\lambda) = (\gamma - \lambda) \prod_{j=1}^{n-1}(\alpha_j - \lambda) - \sum_{k=1}^{n-1} \beta_k^2 \prod_{\substack{j=1 \\ j \neq k}}^{n-1}(\alpha_j - \lambda).$$

Since the arrowhead matrix is irreducible, $\lambda_i \neq \alpha_j$, $i = 1, \ldots, n$, $j = 2, \ldots, n$. Let

$$\alpha_0 = \min \left\{ \gamma - \sum_{i=1}^{n-1} |\beta_i|, \alpha_1 - |\beta_1|, \ldots, \alpha_{n-1} - |\beta_{n-1}| \right\},$$

$$\alpha_n = \max \left\{ \gamma + \sum_{i=1}^{n-1} |\beta_i|, \alpha_1 + |\beta_1|, \ldots, \alpha_{n-1} + |\beta_{n-1}| \right\},$$

$$q_i(\lambda) = \prod_{\substack{j=1 \\ j \neq i}}^{n-1}(\alpha_j - \lambda), \quad i = 0, \ldots, n.$$

The eigenvalues of (2) are the zeros of the following rational functions:

$$\phi_i(\lambda) = \frac{p_n(\lambda)}{q_i(\lambda)} = (\gamma - \lambda)(\alpha_i - \lambda) - \sum_{k=1}^{n-1} \beta_k^2 \frac{\alpha_i - \lambda}{\alpha_k - \lambda}$$
$$= (\gamma - \lambda)(\alpha_i - \lambda) - \sum_{k=1}^{n-1} \beta_k^2 + \sum_{k=1}^{n-1} \beta_k^2 \frac{\alpha_k - \alpha_i}{\alpha_k - \lambda}$$

To derive the algorithm we need the following theorem.

Theorem 1. *Let $\tilde{\lambda} \in]\alpha_i, \alpha_{i+1}[$. Define the modified secular functions*

$$\psi_i(\lambda) = \begin{cases} \phi_i(\lambda) & \text{if } \phi_i(\tilde{\lambda}) \geq 0 \\ \phi_{i+1}(\lambda) & \text{if } \phi_i(\tilde{\lambda}) < 0. \end{cases}$$

The sequence $\left\{\lambda_i^{(j)}\right\}_{j \in N}$ generated by the Newton method applied to the function ψ_i, quadratically converges to λ_i, with initial guess $\lambda_i^{(0)} = \tilde{\lambda}$.

Proof. To prove the theorem, we study the functions $\phi_i(\lambda)$ and $\phi_{i+1}(\lambda)$ in $[\alpha_i, \alpha_{i+1}]$ (see Figure 1):

$$\phi_i(\alpha_i) = -\beta_i^2 < 0, \quad \phi_{i+1}(\alpha_{i+1}) = -\beta_{i+1}^2 < 0, \quad \lim_{\lambda \to \alpha_{i+1}^-} \phi_i(\lambda) = \lim_{\lambda \to \alpha_i^+} \phi_{i+1}(\lambda) = +\infty.$$

Moreover

$$\phi_i''(\lambda) = 2 + 2 \sum_{\substack{j=1 \\ j \neq i}}^{n-1} \beta_j^2 \frac{\alpha_j - \alpha_i}{(\alpha_j - \lambda)^3} > 0, \quad \lambda \in [\alpha_i, \alpha_{i+1}[, \tag{3}$$

$$\phi_{i+1}''(\lambda) = 2 + 2 \sum_{\substack{j=1 \\ j \neq i+1}}^{n-1} \beta_j^2 \frac{\alpha_j - \alpha_{i+1}}{(\alpha_j - \lambda)^3} > 0, \quad \lambda \in]\alpha_i, \alpha_{i+1}]. \tag{4}$$

Therefore

$$\begin{cases} \phi_i(\lambda) < 0, \ \phi_{i+1}(\lambda) > 0, \text{ if } \lambda \in]\alpha_i, \lambda_i[\\ \phi_i(\lambda) > 0, \ \phi_{i+1}(\lambda) < 0, \text{ if } \lambda \in]\lambda_i, \alpha_{i+1}[. \end{cases} \tag{5}$$

From (3) and (5), the Fourier conditions for the convergence of the sequence obtained by applying the Newton method to ϕ_i are satisfied in the interval $[\lambda_i, \alpha_{i+1}[$. In the same way, (4) and (5) assure the convergence of the sequence obtained by applying the Newton method to ϕ_{i+1} in the interval $]\alpha_i, \lambda_i]$. The theorem follows from the definition of ψ_i. □

Taking the previous theorem into account, if $\tilde{\lambda} \in]\alpha_i, \alpha_{i+1}[$, the sequence obtained by applying the Newton method to the function ψ_i, with $\tilde{\lambda}$ as initial guess, quadratically converges to λ_i. A possible choice for the initial guess is $(\alpha_i + \alpha_{i+1})/2$, the middle point of the interval $[\alpha_i, \alpha_{i+1}]$.

Newton Algorithm

```
for i= 1..., n
    k = 0
    λ_i^(k) = (α_i + α_{i+1})/2
    if φ_i(λ_i^(k)) ≥ 0 then
        while | φ_i(λ_i^(k)) |> tol
            λ_i^(k+1) = λ_i^(k) - φ_i(λ_i^(k))/φ'_i(λ_i^(k))
            k = k + 1
        end while
    else
        while | φ_{i+1}(λ_i^(k)) |> tol
            λ_i^(k+1) = λ_i^(k) - φ_{i+1}(λ_i^(k))/φ'_{i+1}(λ_i^(k))
            k = k + 1
        end while
    end if
    λ_i = λ_i^(k)
end for
```

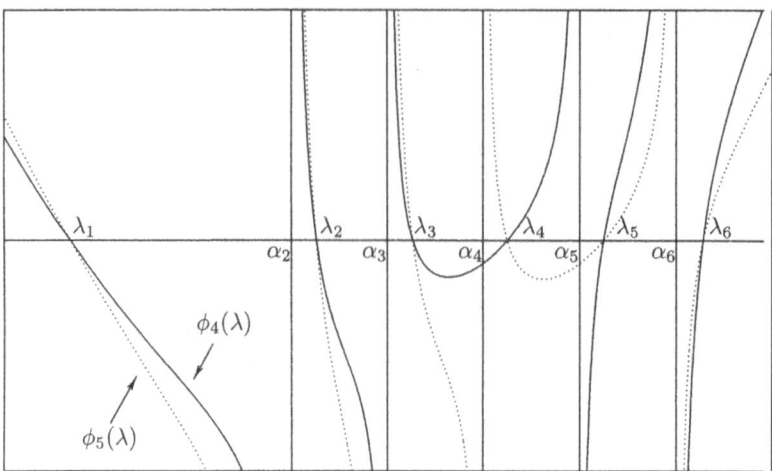

Fig. 1. Modified secular functions ϕ_4 and ϕ_5 associated to the arrowhead matrix with $D = \mathrm{diag}(2,3,4,5,6)$, $\gamma = 1$, $\beta = (1,1,1,1,1)$.

4 Computing the Eigenvectors of an Arrowhead Matrix

Let λ_i be an eigenvalue of the symmetric irreducible arrowhead matrix (2). The eigenvector associated to λ_i is given by

$$v(\lambda_i) = \begin{bmatrix} \frac{\beta_1}{\lambda_i - \alpha_1} \\ \vdots \\ \frac{\beta_{n-1}}{\lambda_i - \alpha_{n-1}} \\ 1 \end{bmatrix},$$

and the normalized eigenvector, with the last entry positive, is

$$u(\lambda_i) = \frac{v(\lambda_i)}{\sqrt{f'(\lambda_i)}},$$

since $f'(\lambda_i) = \|v(\lambda_i)\|_2^2 = 1 + \sum_{j=1}^{n-1} \frac{\beta_j^2}{(\alpha_j - \lambda_i)^2}$.

In practice, we can compute only an approximation $\tilde{\lambda}_i$ of λ_i. Although $\tilde{\lambda}_i$ could be close to λ_i, the approximate ratio $\beta_j/(\tilde{\lambda}_i - \alpha_j)$ can be very different from the exact ratio $\beta_j/(\lambda_i - \alpha_j)$ and a loss of orthogonality in the computed matrix of eigenvectors can occur.

To overcome this problem, in [2,6] the following technique is considered: computed accurately the approximations $\tilde{\lambda}_i$ of the exact eigenvalues λ_i of A, a new arrowhead matrix $\tilde{A}$ is constructed with exact eigenvalues $\tilde{\lambda}_i$, $i = 1, \ldots, n$. Then the computed normalized eigenvectors $u(\tilde{\lambda}_i)$ of $\tilde{A}$ are taken as the normalized eigenvectors of A.

Now we propose a more simple and efficient way to compute a *numerically orthogonal* matrix of eigenvectors of A.

We suppose, without loss of generality, that $\alpha_i - \lambda_i < \lambda_i - \alpha_{i-1}$. Then $\lambda_i - \alpha_i = \min_j |\lambda_i - \alpha_j|$, since, from (1), $\alpha_{i-1} < \lambda_i < \alpha_i$, $i = 2, \ldots, n-1$. Since an eigenvector defines a one–dimensional subspace, any multiple of $v(\lambda_i)$ is also an eigenvector associated to λ_i. Then we consider the modified eigenvector associated to λ_i

$$\tilde{v}(\lambda_i) = (\lambda_i - \alpha_i)v(\lambda_i) = \begin{bmatrix} \beta_1 \frac{\lambda_i - \alpha_i}{\lambda_i - \alpha_1} \\ \vdots \\ \beta_{i-1} \frac{\lambda_i - \alpha_i}{\lambda_i - \alpha_{i-1}} \\ \beta_i \\ \beta_{i+1} \frac{\lambda_i - \alpha_i}{\lambda_i - \alpha_{i+1}} \\ \vdots \\ \beta_{n-1} \frac{\lambda_i - \alpha_i}{\lambda_i - \alpha_{n-1}} \\ \lambda_i - \alpha_i \end{bmatrix} \qquad (6)$$

and, the corresponding orthonormalized modified eigenvector

$$\tilde{u}(\lambda_i) = \frac{\tilde{v}(\lambda_i)}{\|\tilde{v}(\lambda_i)\|_2}. \qquad (7)$$

On the other hand, if $\lambda_i - \alpha_i > \alpha_{i-1} - \lambda_i$, we choose as modified eigenvector associated to λ_i

$$\tilde{v}(\lambda_i) = (\lambda_i - \alpha_{i-1})v(\lambda_i) = \begin{bmatrix} \beta_1 \frac{\lambda_i - \alpha_{i-1}}{\lambda_i - \alpha_1} \\ \vdots \\ \beta_{i-2} \frac{\lambda_i - \alpha_{i-1}}{\lambda_i - \alpha_{i-2}} \\ \beta_{i-1} \\ \beta_i \frac{\lambda_i - \alpha_{i-1}}{\lambda_i - \alpha_i} \\ \vdots \\ \beta_{n-1} \frac{\lambda_i - \alpha_{i-1}}{\lambda_i - \alpha_{n-1}} \\ \lambda_i - \alpha_{i-1} \end{bmatrix}, \qquad (8)$$

and as orthonormalized modified eigenvector

$$\tilde{u}(\lambda_i) = \frac{\tilde{v}(\lambda_i)}{\|\tilde{v}(\lambda_i)\|_2}. \qquad (9)$$

We observe that each difference and each ratio in (6) can be computed to componentwise high relative accuracy and the matrix of the computed orthonormalized modified eigenvectors will be numerically orthogonal.

The following theorem shows that the orthogonality of the computed eigenvectors can be assured whenever it is possible to provide small relative errors in computing the differences $\alpha_j - \lambda$, $j = 1, \ldots, n - 1$.

Theorem 2. *Let λ and μ, $\lambda \neq \mu$, be eigenvalues of the symmetric irreducible arrowhead matrix (2). Let $\tilde{\lambda}$ and $\tilde{\mu}$ be the computed approximations of λ and μ, respectively, such that*

$$\alpha_j - \lambda = (\alpha_j - \tilde{\lambda})(1 + \delta_j), \quad \alpha_j - \mu = (\alpha_j - \tilde{\mu})(1 + \delta'_j), \quad j = 1, \ldots, n-1, \quad (10)$$

and $\delta = \max\{\max_j |\delta_j|, \max_j |\delta'_j|\} \ll 1$. Let $\tilde{u}(\tilde{\lambda}), \tilde{u}(\tilde{\mu})$ be the associated orthonormalized modified eigenvectors, computed either by (6)-(7) or by (8)-(9). Then

$$|\tilde{u}^T(\tilde{\lambda})\tilde{u}(\tilde{\mu})| \leq 2\delta(2 + \delta)\left(\frac{1+\delta}{(1-\delta)^2}\right)^2. \quad (11)$$

Proof. Without loss of generality, we suppose that $|\alpha_i - \lambda| = \min_j |\alpha_j - \lambda|$ and $|\alpha_k - \mu| = \min_j |\alpha_j - \mu|$. We note that (10) is equivalent to

$$\frac{\alpha_j - \lambda}{\alpha_j - \tilde{\lambda}} = 1 + \delta_j, \quad \frac{\alpha_j - \lambda}{\alpha_j - \tilde{\mu}} = 1 + \delta'_j, \quad j = 1, \ldots, n-1.$$

Let $\tau = \|\tilde{v}(\tilde{\lambda})\|_2 \|\tilde{v}(\tilde{\mu})\|_2$. Then

$$-\tau \tilde{u}^T(\tilde{\lambda})\tilde{u}(\tilde{\mu}) = -\tilde{v}^T(\tilde{\lambda})\tilde{v}(\tilde{\mu})$$

$$= -\sum_{\substack{j=1 \\ j \neq i,k}}^{n-1} \beta_j^2 \frac{\alpha_i - \tilde{\lambda}}{\alpha_j - \tilde{\lambda}} \frac{\alpha_k - \tilde{\mu}}{\alpha_j - \tilde{\mu}} - \beta_i^2 \frac{\alpha_k - \tilde{\mu}}{\alpha_i - \tilde{\mu}} - \beta_k^2 \frac{\alpha_i - \tilde{\lambda}}{\alpha_k - \tilde{\lambda}} - (\alpha_i - \tilde{\lambda})(\alpha_k - \tilde{\mu})$$

For the orthogonality of the exact eigenvectors, we have

$$-\tau \tilde{u}^T(\tilde{\lambda})\tilde{u}(\tilde{\mu}) = \tilde{v}^T(\tilde{\lambda})\tilde{v}(\tilde{\mu}) - \tilde{v}^T(\lambda)\tilde{v}(\mu)$$

$$= \sum_{\substack{j=1 \\ j \neq i,k}}^{n-1} \beta_j^2 \frac{\alpha_i - \lambda}{\alpha_j - \lambda} \frac{\alpha_k - \mu}{\alpha_j - \mu} \left(1 - \frac{\alpha_i - \tilde{\lambda}}{\alpha_i - \lambda} \frac{\alpha_k - \tilde{\mu}}{\alpha_k - \mu} \frac{\alpha_j - \lambda}{\alpha_j - \tilde{\lambda}} \frac{\alpha_j - \mu}{\alpha_j - \tilde{\mu}}\right)$$

$$+ \beta_i^2 \frac{\alpha_k - \mu}{\alpha_i - \mu}\left(1 - \frac{\alpha_k - \tilde{\mu}}{\alpha_k - \mu} \frac{\alpha_i - \mu}{\alpha_i - \tilde{\mu}}\right) + \beta_k^2 \frac{\alpha_i - \lambda}{\alpha_k - \lambda}\left(1 - \frac{\alpha_i - \tilde{\lambda}}{\alpha_i - \lambda} \frac{\alpha_k - \lambda}{\alpha_k - \tilde{\lambda}}\right)$$

$$+ (\alpha_i - \lambda)(\alpha_k - \mu)\left(1 - \frac{\alpha_i - \tilde{\lambda}}{\alpha_i - \lambda} \frac{\alpha_k - \tilde{\mu}}{\alpha_k - \mu}\right)$$

$$= \sum_{\substack{j=1 \\ j \neq i,k}}^{n-1} \beta_j^2 \frac{\alpha_i - \lambda}{\alpha_j - \lambda} \frac{\alpha_k - \mu}{\alpha_j - \mu}\left(1 - \frac{1+\delta_j}{1+\delta_i} \frac{1+\delta'_j}{1+\delta'_k}\right)$$

$$+ \beta_i^2 \frac{\alpha_k - \mu}{\alpha_i - \mu}\left(1 - \frac{1+\delta'_i}{1+\delta'_k}\right) + \beta_k^2 \frac{\alpha_i - \lambda}{\alpha_k - \lambda}\left(1 - \frac{1+\delta_k}{1+\delta_i}\right)$$

$$+ (\alpha_i - \lambda)(\alpha_k - \mu)\left(1 - \frac{1}{(1+\delta_i)(1+\delta'_k)}\right).$$

It turns out,

$$\frac{1}{\tau} = \frac{1}{\|\tilde{v}(\tilde{\lambda})\|_2 \|\tilde{v}(\tilde{\mu})\|_2} = \frac{1}{\|\tilde{v}(\lambda)\|_2 \|\tilde{v}(\mu)\|_2} \frac{(1+\delta)^2}{(1-\delta)^2},$$

Then (11) follows, since

$$|\tilde{u}^T(\tilde{\lambda})\tilde{u}(\tilde{\mu})| \leq \frac{2\delta(2+\delta)}{(1-\delta)^2} \frac{(1+\delta)^2}{(1-\delta)^2}.$$

□

5 Numerical Results

The computation of the eigendecomposition of symmetric arrowhead matrices is the kernel of divide and conquer algorithms to compute the Schur form of symmetric tridiagonal matrices and diagonal–plus–semiseparable matrices [2,3,6,9].

In particular, a stable and efficient divide and conquer algorithm to compute the eigendecomposition of symmetric tridiagonal matrices was presented in [8]. At each *conquer step*, the spectral decomposition of a symmetric arrowhead matrix needs to be computed. In this section we compare the algorithm for computing the eigendecomposition of symmetric tridiagonal matrices described in [8] (we refer to this algorithm as tdc) with the same divide and conquer algorithm, where the conquer step, i.e. the computation of the spectral decomposition of a symmetric arrowhead matrix, is done by means of the algorithm described in § 3 and § 4 (we refer to the latter algorithm as tdc1). Both algorithms are used to compute the eigendecomposition of various classes of tridiagonal matrices. All codes are written in matlab with machine precision $\varepsilon \sim 2.22 \times 10^{-16}$.

Let us denote by

$$[\mathbf{v}, \mathbf{w}, \mathbf{v}] \equiv \begin{bmatrix} \omega_1 & v_1 & & \\ v_1 & \omega_2 & \ddots & \\ & \ddots & \ddots & v_{N-1} \\ & & v_{N-1} & \omega_N \end{bmatrix}$$

the involved symmetric tridiagonal test matrices of order N.

- a random matrix, generated by the matlab function rand;
- the Wilkinson matrix W_N^+ with $\omega_i = |(N+1)/2 - i|$ and $v_i = 1$;
- a glued Wilkinson matrix

$$W_{25,g}^+ = \begin{bmatrix} W_{25}^+ & g\mathbf{e}_{25}\mathbf{e}_1^T & & \\ g\mathbf{e}_1\mathbf{e}_{25}^T & W_{25}^+ & \ddots & \\ & \ddots & \ddots & g\mathbf{e}_{25}\mathbf{e}_1^T \\ & & g\mathbf{e}_1\mathbf{e}_{25}^T & W_{25}^+ \end{bmatrix}_{k \cdot 25 \times k \cdot 25}$$

Table 1. Relative residuals (3rd and 4th column) and orthogonality among the eigenvectors (5th and 6th column) obtained computing the spectral decomposition of the involved matrices by tdc and tdc1.

Matrix type	Order N	$\frac{\max_i \|T\tilde{x}_i - \tilde{\lambda}_i \tilde{x}_i\|_2}{N\epsilon\|T\|_2}$		$\frac{\max_i \|X^T \tilde{x}_i - e_i\|_2}{N\epsilon\|T\|_2}$	
		tdc	tdc1	tdc	tdc1
Random	64	4.00×10^{-1}	4.00×10^{-1}	4.75×10^{-2}	1.48×10^{-1}
	128	8.46×10^{-1}	8.46×10^{-1}	4.16×10^{-2}	1.01×10^{-1}
	256	1.17×10^{0}	1.17×10^{0}	2.14×10^{-2}	8.91×10^{-2}
W_N^+	65	1.02×10^{-1}	1.02×10^{-1}	6.49×10^{-2}	7.02×10^{-2}
	129	2.61×10^{-2}	2.61×10^{-2}	3.32×10^{-2}	2.42×10^{-2}
	257	6.61×10^{-3}	6.60×10^{-3}	1.66×10^{-2}	1.25×10^{-2}
$W_{25,10^{-14}}^+$	50	2.93×10^{-2}	2.95×10^{-2}	6.96×10^{-2}	6.15×10^{-2}
	125	5.97×10^{-2}	5.97×10^{-2}	3.03×10^{-2}	3.30×10^{-2}
	275	2.91×10^{-2}	2.91×10^{-2}	1.46×10^{-2}	1.25×10^{-2}
$[1,2,1]$	64	1.51×10^{-2}	2.66×10^{-2}	8.28×10^{-2}	3.12×10^{-1}
	128	9.75×10^{-3}	1.75×10^{-2}	6.64×10^{-2}	2.67×10^{-1}
	256	9.97×10^{-3}	1.49×10^{-2}	4.55×10^{-2}	3.54×10^{-1}
$[1/100, 1+\gamma_i, 1/100]$	64	1.36×10^{-2}	1.45×10^{-2}	1.00×10^{-1}	4.44×10^{-1}
	128	7.43×10^{-3}	7.04×10^{-3}	6.61×10^{-2}	3.88×10^{-1}
	256	3.69×10^{-3}	3.75×10^{-3}	4.53×10^{-2}	4.01×10^{-1}
$[1, \gamma_i, 1]$	64	2.78×10^{-2}	5.27×10^{-2}	1.15×10^{-1}	3.07×10^{-1}
	128	1.96×10^{-2}	3.69×10^{-2}	7.94×10^{-2}	2.89×10^{-1}
	256	1.55×10^{-2}	3.21×10^{-2}	4.68×10^{-2}	3.53×10^{-1}

where e_1 and e_{25} are the first and the 25-th vectors of the canonical basis of $\mathbb{R}^{25}$, respectively, and $g = 10^{-14}$ for $k = 2, 5, 11$;
- the tridiagonal Toeplitz matrix $[1, 2, 1]$ with $\omega_i = 2$ and $v_i = 1$;
- the tridiagonal matrix $[1/100, 1 + \gamma_i, 1/100]$, with $\gamma_i = i \times 10^{-6}$ and $v_i = 1$.
- the tridiagonal matrix $[1, \gamma_i, 1]$, with $\gamma_i = i \times 10^{-6}$ and $v_i = 1$;

W_N^+ has pairs of close eigenvalues, $W_{25,10^{-14}}^+$ has clusters of 50 close eigenvalues, $[1, 2, 1]$ has no close eigenvalues, $[1, \gamma_i, 1]$ and $[1/100, \gamma_i, 1/100]$ do not deflate.

Table 1 presents the numerical results. In columns 3 and 4 the relative residuals obtained computing the spectral decomposition by tdc and tdc1, respectively, are reported.

In columns 5 and 6 the orthogonality among the eigenvectors computed by tdc and tdc1, respectively, is shown. We observe that the results obtained by the two algorithms are comparable.

References

1. Arbenz, P., Golub, G.H.: QR-like algorithms for symmetric arrow matrices. SIAM J. Matrix Anal. Appl. **13** (1992) 655–658

2. Borges, C.F., Gragg, W.B.: A parallel divide and conquer algorithm for the generalized real symmetric definite tridiagonal eigenproblem. In Numerical Linear Algebra and Scientific Computing, L. Reichel, A. Ruttan, and R. S. Varga, eds., de Gruyter, Berlin (1993) 10–28
3. Chandrasekaran, S., Gu, M.: A divide-and-conquer algorithm for the eigendecomposition of symmetric block-diagonal plus semiseparable matrices. Numerische Mathematik (to appear)
4. Dongarra, J.J., Sorensen, D.C.: A fully parallel algorithm for the symmetric eigenvalue problem. SIAM J. Sci. Stat. Comput. **8** (1987) 139–154
5. Golub, G.H.: Some modified matrix eigenvalue problems. SIAM Rev. **15** (1973) 328–334
6. Gu, M., Eisenstat, S.C.: A stable and efficient algorithm for the rank–one modification of the symmetric eigenproblem. SIAM J. Matrix Anal. Appl. **15** (1994) 1266–1276.
7. Gu, M., Eisenstat, S.C.: A divide and conquer algorithm for the symmetric tridiagonal eigenproblem, SIAM J. Matrix Anal. Appl. **16** (1995) 172–191
8. Gragg, W.B., Thornton, J.R., Warner, D.D., Parallel divide and conquer algorithms for the symmetric tridiagonal eigenproblem and bidiagonal singular value problem. In Modelling and Simulation, W.G Vogt and M. H. Mickle, eds., **23** University of Pittsburg School of Engineering, Pittsburg, (1992) 49–56
9. Mastronardi, N., Van Camp, E., Van Barel, M.: Divide and Conquer algorithms for computing the eigendecomposition of symmetric diagonal–plus–semiseparable matrices. IAC–CNR Tech. Rep. **7** (2003) submitted to Numer. Alg.
10. O'Leary, D.P., Stewart, G.W.: Computing the eigenvalues and eigenvectors of symmetric arrowhead matrices. J. Comp. Phys. **90** (1990) 497–505

Relevance Feedback for Content-Based Image Retrieval Using Proximal Support Vector Machine

YoungSik Choi and JiSung Noh

Department of Computer Engineering, Hankuk Aviation University,
200-1 Hwajun-Dong Dukyang-Gu
Kyonggi-Do, Koyang-City, South Korea
{choimail, jisung}@hau.ac.kr

Abstract. In this paper, we present a novel relevance feedback algorithm for content-based image retrieval using the PSVM (Proximal Support Vector Machine). The PSVM seeks to find the optimal separating hyperplane by "regularized least squares". The obtained hyperplane comprises the positive and negative "proximal planes". We interpret the proximal vectors on the proximal planes as the representatives among training samples, and propose to use the distance from the positive proximal plane as a measure of image dissimilarity. In order to reduce computational time for relevance feedback, we introduce the "expanded sets" derived from the pre-computed dissimilarity matrix, and apply the feedback algorithm to these expanded sets rather than the entire image database, while preserving the comparable precision rate. We demonstrate the efficacy of the proposed scheme using unconstrained image databases that were obtained from the Web.

1 Introduction

The rapid growth of digital images has brought about the need for CBIR (Content-Based Image Retrieval) [2][4][8]. The CBIR system represents images as low-level visual features such as color, texture, and retrieves relevant images in order of the similarities with respect to the visual features [3][4]. The researches in CBIR have been focusing on how to describe images as a concise set of low-level visual features, and how to measure the similarities between the low-level features [10][17].
Despite of such endeavors, there is still a "big gap" between the low-level similarity measure and the human perception of image similarity [4][6][10][17]. In order to bridge this semantic gap, the research efforts on relevance feedback for CBIR have been rigorously made during the last half a decade [2]-[4][8][10][15]-[17]. In relevance feedback, a user submits his/her own perceptual judgements on the first-round retrieval results to the CBIR system so that the system can retrieve more relevant images on the next round.
Among such attempts to capture human perceptual measures, the SVM-based (Support Vector Machine) relevance feedback methods have gained much attention [5][14], and have shown several promising results [8][11][15][17]. The SVM seeks to find the optimal separating hyperplane that bisects the positive and negative samples

in a high dimensional feature space. Although the SVM has been widely used in many applications due to its powerful classification capabilities, one cannot directly apply the SVM algorithm to relevance feedback. In relevance feedback, the goal is not to classify the images but to rank the relevance of images according to users' feedback. Therefore, the additional ranking methods should be provided into the SVM-based relevance feedback. The simplest way is to use the decision function, the optimal separating hyperplane as in [17]. This approach is, however, lacking the rational for the use of the distance from the hyperplane as the similarity measure. Addressing this problem in [8], they used the color-based similarity measure for the images in the positive side of the decision boundary after classifying images into two classes. However, they used the distance from the decision hyperplane as the distance measure for the images in the negative side [9]. These problems naturally arise from the nature of the SVM, where the support vectors are located on the decision boundary.

In order to overcome this limit of the SVM-based approach to the ranking problems, we propose to adopt the PSVM (Proximal Support Vector Machine) [7] for image relevance feedback. The PSVM seeks to find two hyperplanes, so called "proximal planes" that best represent the positive and negative training samples, respectively, while maximizing the distance between those two proximal planes. The PSVM can be viewed as "regularized least squares". The PSVM shows a near optimal classification performance compared with that of the SVM, and is much faster than the SVM [7]. The proximal planes in the PSVM give more weights to more typical samples among training samples whereas the support vectors in the SVM represent the separating boundary. Taking into this aspect of the PSVM, we propose to use the distance from the positive proximal plane as a dissimilarity measure. The positive proximal plane describes the representatives among positive samples. As a consequence, the distance from those representatives can be used as the dissimilarity measure.

The computational time is as critical as the similarity learning scheme for the relevance feedback system. Since the similarity measures are only available after learning the relevance, one cannot compute the similarities among images in advance for fast retrieval. In order to reduce the time for searching the entire database, we introduce "expanded sets", which are the subsets of the database. In the relevance feedback rounds, we only compute the dissimilarities among the images in the expanded set. In other words, the proposed method retrieves the relevant images from the expanded set so that the computation time can be significantly reduced. The experimental results show that this approach performs comparable with the case of searching a whole database.

In the following Sections 2 and 3, we describe the proposed relevance feedback scheme. We illustrate the experimental results on several image databases in Section 4, and provide a discussion of the approach as well as conclusions in Section 5.

2 Relevance Feedback Using PSVM

2.1 Proximal Support Vector Machine (PSVM)

The linear PSVM (Proximal Support Vector Machine) seeks to find the "proximal planes" that best represent training samples while maximizing the distance between the two proximal planes. The obtained two proximal planes comprise the optimal separating hyperplane [7].

The training samples are represented as a matrix $A \in R^{m \times n}$, where m denotes the number of the training samples and n represents the dimensionality. Matrix D denotes an $m \times m$ diagonal matrix with the values $\{-1, +1\}$ of each element and y represents a column vector whose element represents an error. The identity matrix will be denoted as I.

In linearly separable cases, the optimal separating hyperplane can be represented as $x^T \omega - \gamma = 0$, where $x \in R^n$. The objective function for the PSVM can be stated as

$$\min_{(\omega,\gamma,y) \in R^{n+1+m}} v \frac{1}{2}\|y\|^2 + \frac{1}{2}(\omega^T \omega + \gamma^2) \tag{1}$$

$$\text{subject to} : D(A\omega - e\gamma) + y = e$$

Note that v is a non-zero constant and e is the column vector of ones. The objective function in equation (1) can be restated as the "regularized least squares". Figure 1 shows that the positive proximal plane ($x\omega$-γ=1) and the negative proximal plane ($x\omega$-γ=-1) are located in the middle of positive and negative training samples, respectively.

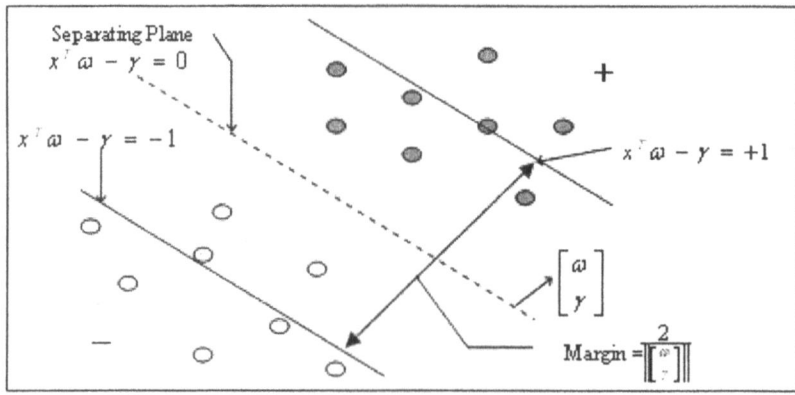

Fig. 1. Proximal Support Vector Machines: positive and negative proximal planes

One can solve the optimization problem in equation (1) using the Lagrangian multipliers $\alpha \in R^m$ as follows.

$$L(\omega, \gamma, y, \alpha) = \frac{v}{2}\|y\|^2 + \frac{1}{2}\left\|\begin{bmatrix}\omega\\\gamma\end{bmatrix}\right\|^2 - \alpha^T(D(A\omega - e\gamma) + y - e) \tag{2}$$

Taking the derivative of equation (2) with respect to ω, γ, y and setting them to zero, the following equations can be obtained.

$$\begin{aligned}\omega - A^T D\alpha &= 0\\\gamma + e^T D\alpha &= 0\\vy - \alpha &= 0\\D(A\omega - e\gamma) + y - e &= 0\end{aligned} \tag{3}$$

$$\omega = A^T D\alpha, \quad \gamma = -e^T D\alpha, \quad y = \frac{\alpha}{v} \qquad (4)$$

Substituting equation (4) into equation (3), one can obtain α as follows.

$$\alpha = \left(\frac{I}{v} + D(AA^T + ee^T)D\right)^{-1} e = \left(\frac{I}{v} + HH^T\right)^{-1} e, \qquad (5)$$

$$H = D[A \ -e]$$

One can derive the PSVM for non-linear cases by using kernel function as follows.

$$\min_{(u,\gamma,y) \in R^{m+1+m}} v\frac{1}{2}\|y\|^2 + \frac{1}{2}(u^T u + \gamma^2) \qquad (6)$$

$$\text{subject to}: D(KDu - e\gamma) + y = e$$

, where u is a feature vector in a kernel space corresponding to a vector x in a data space. In equation (6), K is the abridged form of matrix $K(A, A^T)$ whose elements are dot products of two samples in a given kernel space. Using the Lagrangian multipliers $v \in R^m$, the following objective function can be obtained.

$$L(u,\gamma,y,v) = \frac{v}{2}\|y\|^2 + \frac{1}{2}\left\|\begin{matrix}u\\ \gamma\end{matrix}\right\|^2 - v^T(D(KDu - e\gamma) + y - e) \qquad (7)$$

Setting the partial derivative of equation (7) to zero leads to

$$u - DK^T Dv = 0$$
$$\gamma + e^T Dv = 0$$
$$vy - v = 0 \qquad (8)$$
$$D(KDu - e\gamma) + y - e = 0$$

$$u = DK^T Dv, \quad \gamma = -e^T Dv, \quad y = \frac{v}{v} \qquad (9)$$

Substitution of equation (9) into equation (8) leads to

$$v = \left(\frac{I}{v} + D(KK^T + ee^T)D\right)^{-1} e = \left(\frac{I}{v} + GG^T\right)^{-1} e, \qquad (10)$$

$$G = D[K \ -e]$$

The kernel induced decision function can be obtained from the linear decision function.

$$x^T \omega - \gamma = x^T A^T D\alpha - \gamma = 0 \qquad (11)$$

Replacing $x^T A^T$, α, and γ by the corresponding kernel representations, and substituting them into equation (11), one can obtain the following kernel induced PSVM decision function as follows.

$$f(x) = K(x^T, A^T)Du - \gamma$$
$$= K(x^T, A^T)DDK(A, A^T)^T Dv + e^T Dv \qquad (12)$$
$$= (K(x^T, A^T)K(A, A^T)^T + e^T)Dv = 0$$

In (10), it was shown that the performance of the PSVM should be comparable with that of the SVM with fast computational time.

2.2 Dissimilarity Measure for Image Retrieval

As indicated in Section 2.1, the PSVM finds the optimal separating hyperplane $f(x)$ and the corresponding proximal hyperplanes are located in $f(x) = \pm 1$ as illustrated in Figure 1. Note that the two proximal planes represent the positive and negative training samples. Taking into account these observations, we propose the following similarity measure $S(x)$.

$$S(x) = |1 - f(x)|. \tag{13}$$

The similarity measure $S(x)$ computes how close to the positive proximal plane a data point x lies. Since the positive and negative proximal planes lie in the region where most of positive and negative samples reside, respectively, the positive proximal plane captures what users want to retrieve and keeps away from what users want not to retrieve.

In this paper, we used the following Gaussian kernel function to deal with linearly non-separable data sets.

$$K(x, z) = \exp\left(-\frac{\|x - z\|^2}{2\sigma^2}\right) \tag{14}$$

The scaling factor σ significantly affects the performance of the kernel based SVMs. In general, if σ is too small, then the SVMs tend to over-fit the training samples. Particularly in the case of small size of training samples like the relevance feedback, this over-fitting effect might severely degrade the performance of the SVMs. In order to avoid the over-fitting effect, we begin with a larger value of scaling factor. As the feedback rounds go, we use a smaller value of σ. In this way, we can reduce the chances of over-fitting and enhance the retrieval results.

3 Expanded Sets for Fast Retrieval

In a typical CBIR paradigm, users input an example image as a query (Query By Example), and the CBIR system computes the similarities according to the pre-defined similarity measure, and retrieve a set of relevant images. In order to reduce the similarity computation time, most of CBIR systems deploy high dimensional indexing or clustering methods. One cannot directly use these schemes for relevance feedback since the similarity measure can be available only after learning the relevance. In order to reduce such computation time, we introduce the expanded sets inspired by "link analysis" in the Web document retrievals [1][12].

We denote the set of retrieved images at the first round as "root set". The root set contains k-nearest neighbors to the query image. From this root set, we define two expanded sets, DFS (Double Forward Set) and FBS (Forward and Backward Set). The DFS denotes a union set of the root set itself and k-nearest neighbors of each element in the root set. The FBS denotes a union of the DFS and the images whose k-nearest neighbors are in the root set. Figure 2 illustrates the two expanded sets. We also define the FS (Forward Single) as an extension of the root set. The FS simply contains the root set and the additional nearest neighbors to the query image. The size of FS is set to the size of the DFS.

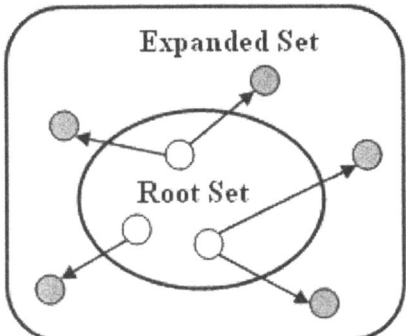 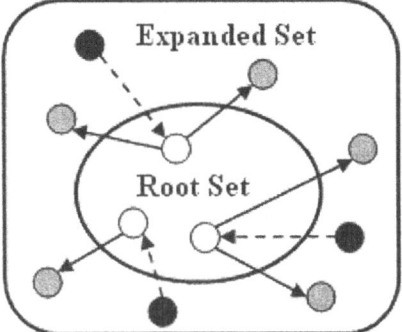

Fig. 2. Expanded sets: (a) Double Froward Set (DFS), (b) Forward and Backward Set (FBS)

The proposed system shows the root set to users as an initial retrieval result, then users select negative and positive examples from the root set. After obtaining the new dissimilarity measure, i.e. $S(x)$ as described in Section 2, we only compute the dissimilarities of the images in the expanded set DFS, FBS or FS. Since we can pre-compute the DFS, FBS, and FS by using pre-defined similarity measure just as in the similar document retrieval, this approach requires a small amount of computational cost. This approach assumes that the group of relevant images should be located in their local areas. It is noteworthy that complex non-linear regions can be represented in the kernel based PSVM.

4 Experiments

For the experiments, we collected the images from Corel Photo Images and the Web images and constructed 3 image databases. Database 1, 2, and 3 contain 1000, 3248, and 6295 images, respectively. Database 1 contains 10 different classes, each of which comprises about 100 images. Database 2 and 3 contain 18 and 49 classes. In Database 1, the memberships to classes are manifest. However, in Database 2 and 3, the memberships are somewhat vague and subjective since we collected images partly from the Web.

We used several visual features, including color moments (10 dimension), normalized wavelet moments (20 dimension), region color moments (10 dimension) [13]. For initial retrieval, we used the Euclidean distant measure.

To demonstrate the performance of the proposed approach, we first experimented without the expanded sets. That is, we computed the dissimilarities over the entire database. We compared the performance of the PSVM-based approach with that of the SVM-based algorithm [17] with various scaling factors σ in equation (14). As a performance measure, we used the precision as follows.

$$precision = \frac{n}{k}, \qquad (15)$$

where k is the size of the root set, and n is the number of relevant images out of k images. In this experiment, we set k to 40. We selected 5 positive, 5 negative samples after the first round. Figure 3 shows the experimental results on database 1 with various scaling factors. Note that the PSVM-based method is favorably comparable with the SVM-based method when the value of σ is small. Moreover, the proposed PSVM-based algorithm performs better when the value of σ gets larger.

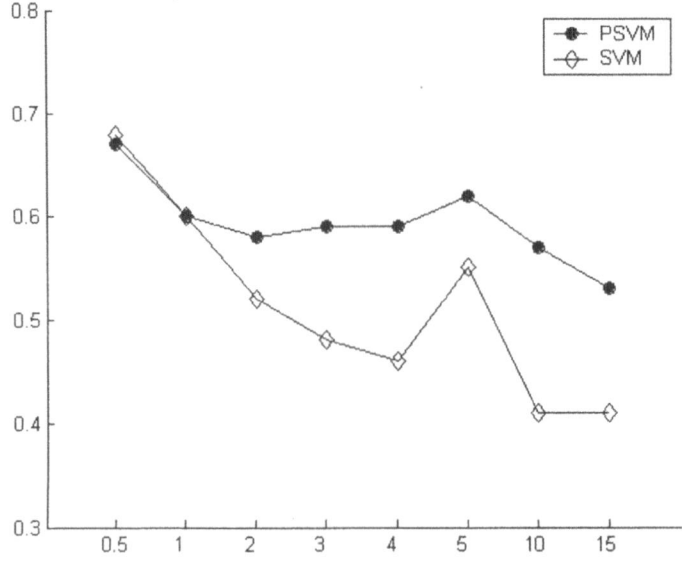

Fig. 3. Precision of the PSVM and SVM with respect to various σ

In the next experiment, we tested the performance with the expanded sets, DFS, FS and FBS. We varied the size of the root set from 20 to 80 and iterated the feedback process 7 times. We showed the experimental results in Figures 4, 5 and 6. In the Figures, the Total denotes the case of searching the entire databases. Figure 4, 5, and 6 show the average precision rates with various sizes of root sets after 7 iterations. One can see that the precision rates are comparable with that of the Total case. As expected, the precision rates decrease as the recall rates increase.

5 Conclusions

In this paper, we proposed a novel relevance feedback algorithm for content-based image retrieval using the PSVM. The SVM, well known for the powerful classification capabilities, has been successfully used for learning image similarities. However, one cannot directly apply the SVM algorithm to relevance feedback. In relevance feedback, the goal is not to classify the images but to rank the relevance of images according to users' feedback. Therefore, the additional ranking methods should be provided into the SVM-based relevance feedback.

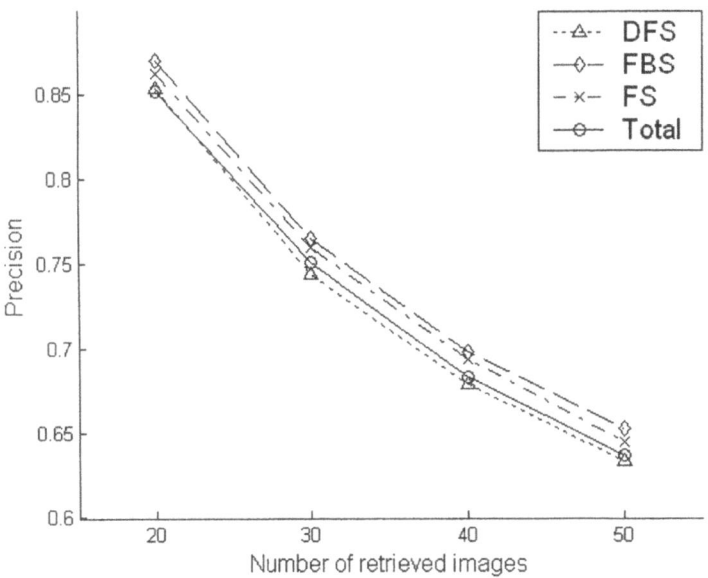

Fig. 4. Experimental result with varying the size of root set in Database 1

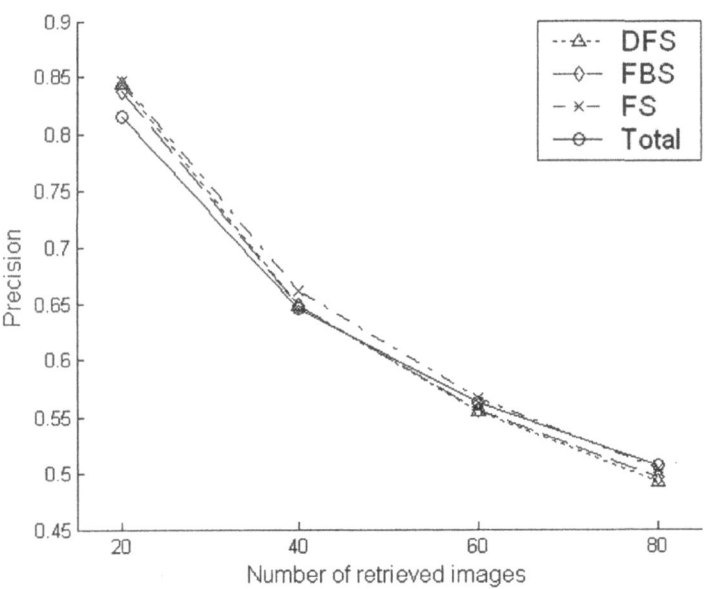

Fig. 5. Experimental result with varying the size of root set in Database 2

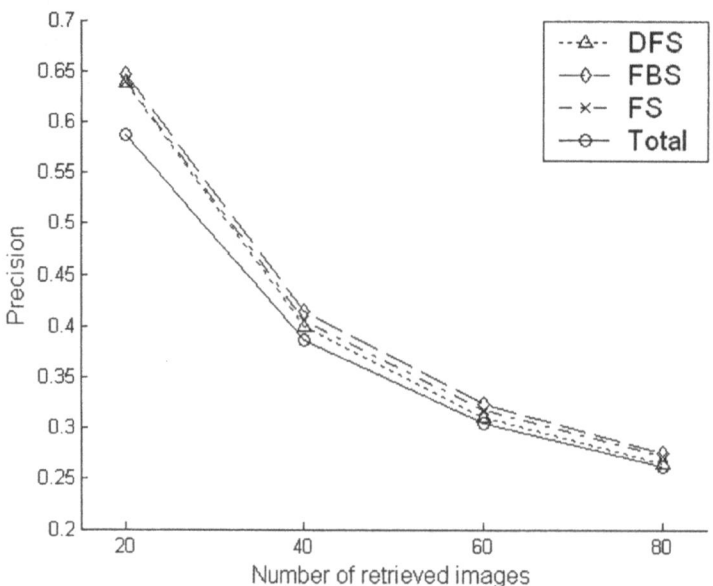

Fig. 6. Experimental result with varying the size of root set in Database 3

In order to overcome this limit, we adopted the PSVM for the image relevance feedback. The PSVM seeks to find the optimal separating hyperplane that comprises the positive and negative "proximal planes". We interpreted the proximal plane as the representative among the training samples, and proposed to use the distance from the positive proximal plane as a measure of dissimilarity. We also presented a scheme to reduce the computational time by introducing the expanded sets. Experiments on real image databases demonstrated the efficacy of the proposed method.

Acknowledgement. This research was supported by IRC (Internet Information Retrieval Research Center) in Hankuk Aviation University. IRC is a Kyounggi Province Regional Research Center designated by Korea Science and Engineering Foundation and Ministry of Science & Technology.

References

1. S. Brin and L. Page, "The Anatomy of a Large-Scale Hypertextual Web Search Engine," 7th WWW Conference on Computer Networks and ISDN Systems, Vol 30, pp. 107-117, April 1998
2. Y. Chen, J.Z. Wang, R. Krovetz, "An unsupervised learning approach to content-based image retrieval," Seventh International Symposium on Signal Processing and its Applications (ISSPA 2003), Paris, Vol. 1, pp. 197-200, July 2003
3. Y. Chen, X.S. Zhou, T.S. Huang, "One-class svm for learning in image retrieval," IEEE Conference on Image Processing, Vol. 1, pp. 34-37, Oct. 2001

4. YoungSik Choi, Daewon Kim, Raghu Krishnapuram, "Relevance feedback for content-based image retrieval using the Choquet integral," IEEE International Conference on Multimedia and Expo, pp. 1207–1210, 2000
5. C. Cortes, V. Vapnik, "Support vector network," Machine learning, 20, pp.273-297, 1995
6. P. Enser, C. Sandom, "Towards a comprehensive survey of the semantic gap in visual image retrieval," CIVR 2003, LNCS 2728, pp. 291-299, 2003
7. G. Fung, O. L. Mangasarian, "Proximal support vector machine classifiers," In Knowledge Discovery and Data Mining, pp. 77-86, 2001
8. G.D. Guo, A.K. Jain, W.Y. Ma, H.J. Zhang, "Learning similarity measure for natural image retrieval with relevance feedback," IEEE Transactions on Neural Networks, Vol. 13, Issue: 4 , pp. 811-820, July 2002
9. G.D. Guo, H.J. Zhang, S.Z. Li, "Distance from boundary as a metric for texture image retrieval," International Conference on Acoustics, Speech, and Signal Processing, v. 3, May 2001
10. P. Hong, Q. Tian, T.S. Huang, "Incorporate support vector machines to content-based image retrieval with relevant feedback," IEEE International Conference on Image Processing, Vol 3, pp. 750-753, Sept 2000
11. D. Heesch, A. Yavlinsky, S. Ruger, "Performance comparison of different similarity models for CBIR with relevance feedback," International Conference on Image and Video Retrieval (CIVR, Urbana-Champaign, Il, Jul 2003), LNCS 2728, pp. 456-466, Springer-Verlag, 2003
12. J. Kleinberg, "Authoritative sources in a hyperlinked environment," 9th Annual ACM-SIAM Symposium on Discrete Algorithms, pp. 668--677, January 1998
13. B.S. Manjunath, J.R. Ohm, V.V. Vasudevan, A. Yamada, "Color and texture descriptors," IEEE Transactions on Circuits and Systems for Video Technology, Vol 11, No. 6, pp. 716-719, 2001
14. J.C. Platt, "Sequential minimal optimization: a fast algorithm for training support vector machines," Technical Report MSR-TR-98-14, April 21, 1998
15. Q. Tian, P. Hong, T. S. Huang, "Update relevant image weights for content-based image retrieval using support vector machines," IEEE International Conference on Multimedia and Expo, Vol. 2, pp. 1199-1202, June 2000
16. S. Tong, E. Chang, "Support vector machine active learning for image retrieval," ACM International Conference on Multimedia, pp.107-118, Ottawa, October 2001
17. L. Zhang, F. lin, B. Zhang, "Support vector machine learning for image retrieval," IEEE International Conference on Image Processing, Vol. 2, pp. 721-724, October 2001
18. X.S. Zhou, T.S. Haung, "Comparing discriminate transformations and SVM for learning during multimedia retrieval," ACM Multimedia2001, Sept. 30-Oct 5, 2001, Ottawa, Ontario, Canada, 2001

Orthonormality-Constrained INDSCAL with Nonnegative Saliences

Nickolay T. Trendafilov

Faculty of Computing, Engineering and Mathematical Sciences,
University of the West of England, Bristol BS16 1QY, UK
Nickolay.Trendafilov@uwe.ac.uk

Abstract. INDSCAL is a specific model for simultaneous metric multi-dimensional scaling (MDS) of several data matrices. In the present work the INDSCAL problem is reformulated and studied as a dynamical system on the product manifold of orthonormal and diagonal matrices. The problem for fitting of the INDSCAL model to the data is solved. The resulting algorithms are *globally* convergent. Numerical examples illustrate their application.

1 Introduction

Carroll and Chang (1970) (see also Cox and Cox, 1995) introduced the INDSCAL (**IN**dividual **D**ifferences **SCAL**ing) model to consider a three-way array consisting of m symmetric $n \times n$ slices S_i. In the context of classical MDS (Cox and Cox, 1995) these slices S_i are composed by the doubly centered dissimilarity matrices. The INDSCAL model decomposes each slice as

$$S_i = QD_iQ^T + E_i, \tag{1}$$

where Q be $n \times r$ matrix (assumed of full column rank), D_i-diagonal matrix, and E_i be $n \times n$ matrix, containing the errors of the model fit. That means all slices share a common loading matrix Q and differ each other only by the diagonal elements of D_i called idiosyncratic saliences. The INDSCAL model, appeared initially in psychometric literature, is now widely used in social sciences, marketing research, etc. More recent applications are in new areas as chemometrics (e.g. Courcoux et al., 2002) and signal processing (for CANDECOMP-like algorithm for independent components analysis) (e.g. De Lathauwer, 1997). The INDSCAL model can be quite useful for modification of the standard eigenfaces approach for face recognition (e.g. Turk and Pentland, 1991) and will be considered in detail elsewhere.

The INDSCAL problem seeks for $(Q, D_1, D_2, \ldots, D_m)$ such that the model fits the data in least squares sense, i.e. for given and fixed $n \times n$ symmetric matrices $S_i, i = 1, 2, \ldots, m$, minimizes the function:

$$F(Q, D_1, D_2, \ldots, D_m) = \sum_{i=1}^{m} ||S_i - QD_iQ^T||^2, \tag{2}$$

where the unknowns Q and D_i should have the form described above. This optimization problem has no direct analytical solution. The standard numerical solution of the problem as unconstrained one is given by an alternating least squares algorithm, called CANDECOMP. The two appearances of Q in (2) are represented by different matrices, say Q and P; then (2) is called CANDECOMP function. Optimization is carried out on Q and P independently (Carroll and Chang, 1970; Cox and Cox, 1995). The belief is that after convergence of CANDECOMP eventually $Q = P$. This is known as "symmetry requirement" and seems to be satisfied in practice, although there is no general proof of the claim. On the contrary, it has been shown in (ten Berge and Kiers, 1991) that CANDECOMP algorithm can produce asymmetric INDSCAL solutions ($Q \neq P$) even for positive semi-definite (p.s.d.) data S_i. However, for an orthonormality-constrained INDSCAL model, i.e. $Q^T Q = I_r$ in (2), ten Berge and Kiers (1991) proved that CANDECOMP yields solution with $Q = P$ for p.s.d. S_i, with at least one of them strictly positive. In fact, this is the strongest result known yet concerning the CANDECOMP convergence. In our opinion, this is a good reason to pay more attention on the orthonormality-constrained INDSCAL model. Note that in the recent INDSCAL applications (e.g. Courcoux et al., 2002; De Lathauwer, 1997) the orthonormality-constrained INDSCAL model is used.

So far, there are very few papers studying this problem. The minimization of the INDSCAL loss function (2) subject to the constraint *square orthogonal Q* was considered in (Kroonenberg, 1983). Later this problem was treated in (ten Berge et al., 1988), see also (Chu and Trendafilov, 1998). The orthonormality-constrained INDSCAL has been also considered in (Kiers, 1991). The need of such a modified INDSCAL model has been motivated as being useful in a technical way and simplifying the interpretation of the results (Kiers, 1991). There are important implications from the orthonormality-constrained reformulation of the INDSCAL model. In case S_i are correlation matrices, the INDSCAL can be seen as simultaneous principal component analysis of several correlation matrices. More generally, the INDSCAL (with the nonnegativity of D_i removed) can be seen as a simultaneous eigenvalue decomposition of several symmetric matrices (e.g. Chu, 1991; Chu and Trendafilov, 1998). The problem for simultaneous PCA of several covariance matrices is considered in detail in (Flury, 1988). Under normality assumptions, likelihood function of the model is constructed and fitted, and asymptotic results are derived. Recently in (Boik, 2003), the study is extended to simultaneous PCA of several correlation matrices.

Another serious problem is that, in the most applications, a solution with nonnegative D_i makes sense only. Indeed, while Q represents the group stimulus space and $D_i, i = 1, 2, \ldots, m$ – the subject space, the individual stimulus space for ith individual is presented by $QD_i^{1/2}$, i.e. the square root must be real. INDSCAL solutions with nonnegative D_i can not be guaranteed by the original CANDECOMP algorithm. Negative saliences can occur, even when the data matrices S_i are p.s.d. (ten Berge and Kiers, 1991). Moreover in (ten Berge et al., 1993), an example is provided where three symetric matrices (two strictly positive definite and one p.s.d.) are subject to INDSCAL analysis and the CAN-

DECOMP solution has a negative salience. The nonnegativity problem can be handled by ALSCAL program (Takane et al., 1977) for the non-metric version of INDSCAL, and by MULTISCALE program (Ramsay, 1977) – for its maximum likelihood version respectively. The problem with negative saliences for the metric (scalar-product) version of the INDSCAL has been solved simultaneously in (ten Berge et al., 1993) and (De Soete et al., 1993), subject to diag$(Q^TQ) = I$.

In the present paper we consider an orthonormality-constrained INDSCAL with nonnegative saliences, i.e. the loss function (2) is minimized subject to the constraints $Q^TQ = I_r$ and D_i be $n \times n$ p.s.d. diagonal matrices.

First, we show that the unconstrained INDSCAL model with nonnegative saliences can be defined with orthonormal Q, without loss of generality. We start with the more general IDIOSCAL model (e.g. Cox and Cox, 1995):

$$F(Q, W_1, W_2, \ldots, W_m) = \sum_{i=1}^{m} \|S_i - QW_iQ^T\|^2, \qquad (3)$$

where Q is a common $n \times r$ loading matrix (assumed of full column rank) and W_i are south p.s.d. Let $Q = VU$ read the QR decomposition, where V is $n \times r$ orthonormal matrix and U is $r \times r$ upper triangular matrix with positive main diagonal. Then the IDIOSCAL problem (3) is transformed into

$$F(V, U, W_1, W_2, \ldots, W_m) = \sum_{i=1}^{m} \|S_i - V(UW_iU^T)V^T\|^2, \qquad (4)$$

where V is $n \times r$ orthonormal and UW_iU^T is p.s.d. because U is non-singular. The equation (4) is identical to (3) with $Q := V$ and $W_i := UW_iU^T$. Thus, without loss of generality, the IDIOSCAL problem can be considered with orthonormally constrained Q.

The INDSCAL case is a little bit more complicated. Substituting the QR decomposition of Q in (2) gives $QD_iQ^T = VUD_iU^TV^T$ which makes the INDSCAL problem (with nonnegative saliences) equivalent to the IDIOSCAL problem (4) with $Q := V$ and $W_i := UD_iU^T$. If all $W_i(= UD_iU^T)$ are orthogonally similar, i.e. they can be diagonalized by a common $r \times r$ orthogonal P ($W_i = PD_i^*P^T$), then the INDSCAL model (2) with $Q := VP$ and $D_i := D_i^*$ takes place. Alternatively, if all W_i are *not* orthogonally similar, then the problem is considered as an orthonormality-constrained IDIOSCAL problem.

There is no available algorithms for solving the orthonormality-constrained INDSCAL with nonnegative saliences. Here we propose a method that solves the problem in an unified way regardless of the form of the input symmetric data S_i (p.s.d. or not) and produces solution with orthonormality constrained Q and nonnegative saliences D_i.

We show that the INDSCAL solution can be presented as a steepest descent flow on the product manifold of orthonormal and diagonal matrices. This approach has been applied successfully in the analysis of number of discrete numerical algorithms (Helmke and Moore, 1994). A formulation of the INDSCAL problem is given as an initial value problem for matrix ordinary differential equation(ODE) of first order. In the present approach the constrained manifold is considered embedded into a high-dimensional Euclidean space. Thus the standard ODE integrators can be used to solve the problem which may ease the interested user. Particularly, the INDSCAL solution in this work is found by the MATLAB ODE numerical integrator (Shampine and Reichelt, 1997), which is widely available. Alternative, but mathematically more involved approaches, are under intensive research: (Edelman et al., 1998), (Lippert and Edelman, 1998) – differential geometry and topology; (Iserles et al., 2000), (Engø et al., 1997), (Owren and Welfert, 1996) – Runge-Kutta methods on Lie groups; (Diele et al., 1998) – Cayley transform. The projected gradient method leads to globally convergent algorithm, i.e. the convergence is reached *independently* of the starting (initial) point. Note, that the currently available INDSCAL solutions lack this feature. At the end of the work, numerical results illustrating the algorithm proposed are commented.

The INDSCAL problem considered in this work is concerned with the following equality constrained optimization problem: for given fixed $n \times n$ arbitrary symmetric matrices $S_i, i = 1, 2, \ldots, m$

$$\text{Minimize } \sum_{i=1}^{m} \|S_i - QD_i^2 Q^T\|^2 \tag{5}$$

$$\text{Subject to } (Q, D_1, D_2, \ldots, D_m) \in \text{St}(n, r) \times \mathcal{D}(r)^m, \tag{6}$$

where $\text{St}(n, r)$ denotes the set of all $n \times r$ column-wise orthonormal matrices known as the Stiefel manifold (Stiefel, 1935–1936), i.e.:

$$\text{St}(n, r) := \{Q \in R^{n \times r} | \ Q^T Q = I_r\} \tag{7}$$

and $\mathcal{D}(r)^m = \underbrace{\mathcal{D}(r) \times \ldots \times \mathcal{D}(r)}_{m}$ and $\mathcal{D}(r)$ denotes the set of all $r \times r$ diagonal matrices.

2 Fitting the INDSCAL Model

In this work we reconsider the INDSCAL problem (5) – (6) in terms of the projected gradient approach. This is a specific continuous-time method based on the classical gradient approach (Hirsch and Smale, 1974) for function optimization and modified for analyzing and solving constrained optimization problems (Helmke and Moore, 1994).

If $X(t)$ is restricted to move on a certain feasible set the gradient $\nabla F(X(t))$ of the objective $F(X(t))$ may move the flow $X(t)$ out the feasible set because it is determined by the function F only but not at all by the constraints imposed. The idea of the projected gradient method is to keep the flow $X(t)$ "nailed" on the constrained manifold. The projected gradient is concerned with the following dynamical system:

$$\frac{dX(t)}{dt} = -g(X(t)), \tag{8}$$

where $g(X(t))$ is the projection of the gradient $\nabla F(X(t))$ onto the tangent space of the feasible set. It is shown in (Chu and Driessel, 1990; Chu and Trendafilov, 1998a) that the flow $X(t)$ defined by (8) also defines a steepest descent flow for the function F on the feasible set.

In order to apply the projected gradient approach to the INDSCAL problem (5) – (6) we need to know the projection of the gradient of the objective function (5) onto the feasible set (6), i.e. onto its tangent space (11).

We shall regard $St(n,r)$ as embedded in the nr dimensional Euclidean space $R^{n\times r}$ equipped with the Frobenius inner product:

$$\langle X, Y \rangle := \text{trace}(XY^T) \tag{9}$$

for any $X, Y \in R^{n\times r}$. Hereafter we suppose that Q depends on the real parameter t, such that for all $t \in R$ the matrix $Q(t)$ is orthonormal, i.e. we have $Q(t)^T Q(t) = I_r$. For short, we write simply Q. By definition, a tangent vector H of $St(n,r)$ at Q is the velocity of the smooth path $Q(t) \in St(n,r)$ at $t=0$.

The feasible set (6) of the problem is a product set. In the product space $R^{n\times r} \times (R^{r\times r})^m$, we shall use the induced Frobenius inner product:

$$\langle (A_0, \ldots, A_m), (B_0, \ldots, B_m) \rangle = \langle A_0, B_0 \rangle + \ldots + \langle A_m, B_m \rangle , \tag{10}$$

where the inner products $\langle \, , \, \rangle$ on the right hand side are defined in (9).

The feasible set $St(n,r) \times \mathcal{D}(r)^m$ of the problem is a smooth manifold. It is not difficult to show (Chu and Driessel, 1990) that the tangent space to $St(n,r) \times \mathcal{D}(r)^m$ at a point $(Q, D_1, D_2, \ldots, D_m) \in St(n,r) \times \mathcal{D}(r)^m$ is given by

$$\mathcal{T}_{(Q,D_1,D_2,\ldots,D_m)}(\mathcal{O}(n,r) \times \mathcal{D}(r)^m) = \mathcal{T}_Q St(n,r) \times \underbrace{\mathcal{T}_{D_1}\mathcal{D}(r) \times \ldots \times \mathcal{T}_{D_m}\mathcal{D}(r)}_{m}$$

$$= \mathcal{T}_Q St(n,r) \times \mathcal{D}(r)^m , \tag{11}$$

which follows from the fact that $\mathcal{D}(r)$ is a vector subspace in $r \times r$ and it is identical to its tangent subspace. Thus, the projection onto $\mathcal{T}_Q St(n,r) \times \mathcal{D}(r)^m$ is easy to be obtained if the projection onto $\mathcal{T}_Q St(n,r)$ is known. A formula for this projection is given in Trendafilov (1998), Edelman et al. (1999). Let $Z \in R^{n\times r}$, then

$$\pi_{\mathcal{T}_Q St(n,r)}(Z) := Q\frac{Q^T Z - Z^T Q}{2} + (I_n - QQ^T)Z \tag{12}$$

defines the projection of Z onto the the tangent space $\mathcal{T}_Q St(n,r)$.

The gradient ∇F of the objective function F in (5) at $(Q, D_1, \ldots, D_m) \in St(n,r) \times \mathcal{D}(r)^m$ with respect to the induced Frobenius inner product (10) can be interpreted as $(m+1)$-tuple of matrices, i.e.:

$$\nabla F(Q, D_1, \ldots, D_m) = (\nabla_Q F, \nabla_{D_1} F, \ldots, \nabla_{D_m} F) .$$

Thus the differential equations

$$\frac{dQ}{dt} = \frac{Q}{2}\left[\sum_{i=1}^{m} Q^T S_i Q D_i^2 - \sum_{i=1}^{m} D_i^2 Q^T S_i Q\right]$$

$$+(I - QQ^T) \sum_{i=1}^{m} S_i Q D_i^2 \qquad (13)$$

$$\frac{dD_i}{dt} = (Q^T S_i Q) \odot D_i - D_i^3 \qquad (14)$$

for $i = 1, 2, \ldots, m$, define $m + 1$ simultaneous steepest descent flows on $\text{St}(n, r)$ and m copies of $\mathcal{D}(r)$ respectively for the objective function F. The standard elementwise (Hadamard) matrix product is denoted by $\odot$. Starting with an initial point for each of them, we may use these flows to approximate a solution to the INDSCAL.

At the end of this Section, we briefly outline the INDSCAL problem with *arbitrary* diagonal weights (idiosyncratic saliences), i.e. the nonnegative constraints *removed*:

$$\text{Minimize} \quad \sum_{i=1}^{m} ||S_i - QD_iQ^T||^2 \qquad (15)$$

$$\text{Subject to} \quad (Q, D_1, D_2, \ldots, D_m) \in \mathcal{O}(n,r) \times \mathcal{D}(r)^m . \qquad (16)$$

The corresponding descent flows that approximate its solution are governed by the following ODEs:

$$\frac{dQ}{dt} = Q \left[\sum_{i=1}^{m} Q^T S_i Q D_i - \sum_{i=1}^{m} D_i Q^T S_i Q \right]$$

$$+2(I - QQ^T) \sum_{i=1}^{m} S_i Q D_i \qquad (17)$$

$$\frac{dD_i}{dt} = \text{diag}(Q^T S_i Q) - D_i . \qquad (18)$$

We mention that projected gradient matrix flows can be derived also for the case with oblique loadings Q, i.e. $\text{diag}(Q^T Q) = I$ (Ten Berge et al., 1993). One simply needs to replace (12) with the corresponding projection formula for oblique Q, but this is out of the scope of the paper.

3 Numerical Results

In this section, we report some of our numerical experiments with equations (13) – (14). The computations are carried out in MATLAB 6.3. We have taken the solver **ode15s** from the MATLAB ODE SUITE Shampine and Reichelt (1997) as integrator of the initial value problems. The code **ode15s** is a quasi-constant step size implementation of the Klopfenstein-Shampine family of numerical differential formulas (implicit) for stiff systems. More details of these codes can be found in Shampine and Reichelt (1997).

In our experiments, the tolerance for the absolute error is set at 10^{-6}, and for relative error at 10^{-3} that applies to all components of the solution vector. This criterion is used to control the accuracy following the solution path. Experiments show that higher accuracy does not influence the typical dynamics of the underlying vector field. Lower accuracy requirements in the calculation saves CPU time. The output values at time interval $[0, 100]$ are examined. The integration terminates automatically when the absolute improvement of the objective function between two consecutive output points is less than 10^{-4}, indicating a local minimizer has been found. This stopping criterion can be modified if necessary.

The initial values $(Q_{in}, D_{1,in}, \ldots, D_{m,in})$ for INDSCAL algorithm based on solving (13) – (14) can be taken either rationally or randomly. The following rational starts seem reasonable. Let $P\Lambda P^T$ read the EVD of the data $\sum_{i=1}^{m} S_i$, where the eigenvalues in Λ are arranged in descending order. Let Λ_r be the $r \times r$ diagonal matrix containing the r largest eigenvalues. Then $Q_{in} = P_r$, i.e. an $n \times r$ matrix containing those eigenvectors corresponding to the r largest eigenvalues. Then the starting values for $D_{i,in}$ are computed as follows:

$$D_{i,in}^2 = \max(\text{diag}(Q_{in}^T S_i Q_{in}), 0). \tag{19}$$

If the r largest original eigenvalues of S_i for $i = 1, 2, \ldots, m$, are all positive than they can be taken as alternative good start.

The sensitivity of the algorithms to hit false local minima for different starting values $(Q_{in}, D_{1,in}, \ldots, D_{m,in})$ has been studied by applying the algorithms to 50 data sets and for each data set 20 times with randomly generated initial values for Q and D_i. It has been found that the algorithm is not quite sensitive. In about 45 of all 50 runs, the sample variance of the minimal value of the objective function (5) obtained from 20 different starting values is of order or less than 10^{-4}. This indicates that practically an identical minima have been found for the corresponding data set for all 20 different starting values.

Example 1. We reconsider the small artificial example given in (ten Berge and Kiers, 1991). The following p.s.d. matrices, one of which strictly positive definite (with eigenvalues 4, 2, and 0, and 4, 2, and 1, respectively):

$$S_1 = \begin{bmatrix} 3 & 1 & 0 \\ 1 & 3 & 0 \\ 0 & 0 & 0 \end{bmatrix} \text{ and } S_2 = \begin{bmatrix} 3 & -1 & 0 \\ -1 & 3 & 0 \\ 0 & 0 & 1 \end{bmatrix},$$

are subject to INDSCAL analysis with $r = 1$. In (ten Berge and Kiers, 1991) it is reported that the CANDECOMP solution is asymmetric and is a local minimizer of the problem. This local minimum of the loss function is 39. The global solutions are symmetric: with $Q = (\sqrt{.5}, \sqrt{.5}, 0)^T$ or $Q = (\sqrt{.5}, -\sqrt{.5}, 0)^T$ and D_i^2 reduced to scalars: 4 and 2, and 2 and 4, respectively (ten Berge and Kiers, 1991). Note, the global solutions are orthonormal, as well as the asymmetric local CANDECOMP solution is. The global minimum of the INDSCAL loss function is 21.

First, we integrate the INDSCAL ODEs (13) – (14) starting with a *rational* initial value for Q as described above and the original eigenvalues of S_1 and S_2 – for $D_{i,in}^2$, i.e. $D_{1,in}^2 = D_{2,in}^2 = 4$. The solution found is $Q = (0, 1, 0)^T$ and $D_1^2 = D_2^2 = 3$. This is a local minimizer and the minimum is 23.

Next, we integrate the INDSCAL ODEs (13) – (14) starting with 20 *random* initial values for (Q, D_1, D_2). In each of the 20 runs a *global* minimizer of the problem is fond. The overall results are: the mean of the 20 minima of the loss function (5) found is 21 with variance 4.6447 10^{-9}, i.e. all 20 minima found are identical. There are four global minimizers (Q, D_1, D_2) of the INDSCAL loss function (5). They are:

$$Q_I = \begin{bmatrix} .7071 \\ .7071 \\ 0 \end{bmatrix}, Q_{II} = \begin{bmatrix} -.7071 \\ -.7071 \\ 0 \end{bmatrix}, Q_{III} = \begin{bmatrix} .7071 \\ -.7071 \\ 0 \end{bmatrix} \text{ and } Q_{IV} = \begin{bmatrix} -.7071 \\ .7071 \\ 0 \end{bmatrix},$$

The first two go with $D_1^2 = 4$ and $D_2^2 = 2$; the second two – with $D_1^2 = 2$ and $D_2^2 = 4$. Among all 20 runs, we have convergence to Q_I in 6 runs, to Q_{II} – in 2, to Q_{III} – in 3, and to Q_{IV} – 9. Q_I and Q_{III} are reported global minimisers in (ten Berge and Kiers, 1991). Indeed, Q_{II} and Q_{IV} can be derived from Q_I and Q_{III} by multiplication with -1.

References

1. Boik, R. J.: Principal component models for correlation matrices. Biometrika **90** (2003) 679–701
2. Carroll, J. D., Chang, J. J.: Analysis of individual differences in multidimensional scaling via an n–way generalization of "Eckart-Young" decomposition. Psychometrika **35** (1970) 283–319
3. Chu, M. T.: A continuous Jacobi-like approach to the simultaneous reduction of real matrices. Linear Algebra and its Applications **147** (1991) 75–96
4. Chu, M. T., Driessel, K. R.: The projected gradient method for least squares matrix approximations with spectral constraints. SIAM J. Numer. Anal. **27** (1990) 1050–1060
5. Chu, M. T., Trendafilov, N. T.: ORTHOMAX rotation problem. A differential equation approach. Behaviormetrika **25** (1998) 13–23
6. Chu, M. T., Trendafilov, N. T.: On a differential equation approach to the weighted orthogonal Procrustes problem. Statistics and Computing **8** (1998a) 125–133
7. Cox, T.F., Cox, M.A.A.: Multidimensional Scaling. Chapman & Hall, London (1995)
8. De Lathauwer, L.: Signal Processing Based on Multilinear Algebra, PhD Thesis, Katholieke Universiteit Leuven, http://www.esat.kuleuven.ac.be/sista/members/delathau.html (1997)
9. De Soete, G., Carroll, J. D., Chaturvedi, A. D.: A modified CANDECOMP method for fitting the extended INDSCAL model. Journal of Classification **10** (1993) 75–92
10. Diele, F., L. Lopez, Peluso, R.: The Cayley transform in the numerical solution of unitary differential systems. Advances in Computational Mathematics **8** (1998) 317–334
11. Edelman, A., T. Arias, Smith, S. T.: The geometry of algorithms with orthogonality constraints. SIAM J. Matrix Anal. and Appl. **20** (1998) 303–353
12. Engø, K., A. Marthinsen, Munthe-Kaas, H.: DiffMan–an object oriented MATLAB toolbox for solving differential equations on manifolds (User's Guide). http://www.math.ntnu.no/num/synode/ (1997)
13. Flury, B.: Common Principal Components and Related Multivariate Models, John Wiley & Sons, New York (1988)

14. Helmke, U., Moore, J. B.: Optimization and Dynamical Systems, Springer, London (1994)
15. Hirsch, M. W., Smale, S.: Differential Equations, Dynamical Systems, and Linear Algebra. Academic Press, London (1974)
16. Iserles, A., Munte-Kaas, H., Norset, S. P., Zanna, A.: Lie group methods, Acta Numerika **9** (2000) 1-151
17. Kiers, H. A. L.: Simple structure in component analysis techniques for mixture of qualitative and quantitative variables. Psychometrika **56** (1991) 197-212
18. Kroonenberg, P. M.: Three Mode Principal Component Analysis: Theory and Applications, DSWO Press, Leiden (1983)
19. Lippert, R.A., Edelman, A.: Nonlinear eigenvalue problems. http://www.mit.edu/people/ripper/Template/template.html (1998)
20. Magnus, J. R., Neudecker, H.: Matrix Differential Calculus with Application in Statistics and Econometrics, Wiley, New York (1988)
21. Courcoux, Ph., Devaux, M-F., Bouchet, B.: Simultaneous decomposition of multivariate images using three-way data analysis. Application to the comparison of cereal grains by confocal laser scanning microscopy. Chemometrics and Intelligent Laboratory Systems **62** (2002) 103-113 Massart, D.L., to be changed!!!!!
22. Owren, B., Welfert, B.: The Newton iteration on Lie groups. BIT **40** (2000) 121-145.
23. Ramsay, J. O.: Maximum likelihood estimation in multidimensional scaling. Psychometrika **42** (1977) 241-266
24. Shampine, L. F., Reichelt, M. W.: The MATLAB ODE suite. SIAM Journal on Scientific Computing **18** (1997) 1-22
25. Stiefel, E.: Richtungsfelder und fernparallelismus in n-dimensionalel manning faltigkeiten. Commentarii Mathematici Helvetici **8** (1935-1936) 305-353
26. Takane, Y., Young, F.W., De Leeuw, J.: Nonmetric individual differences multidimensional scaling: an alternating least squares method with optimal scaling features. Psychometrika **42** (1977) 7-67
27. ten Berge, J. M. F., Knol, D. L., and Kiers, H. A. L.: A treatment of the ORTHOMAX rotation family in terms of diagonalization, and a re-examination of a singular value approach to VARIMAX rotation. Computational Statistics Quarterly **3** (1988) 207-217
28. ten Berge, J. M. F., Kiers, H. A. L.: Some clarifications of the CANDECOMP algorithm applied to INDSCAL. Psychometrika **56** (1991) 317-326
29. ten Berge, J. M. F., Kiers, H. A. L., Krijnen, W. P.: Computational solutions for the problem of negative saliences and nonsymmetry in INDSCAL. Journal of Classification **10** (1993) 115-124
30. Turk, M., Pentland, A.: Face recognition using eigenfaces, Proc. IEEE Conf. on Computer Vision and Pattern Recognition (1991) 586-591.

Optical Flow Estimation via Neural Singular Value Decomposition Learning

Simone Fiori[1], Nicoletta Del Buono[2], and Tiziano Politi[3]

[1] Facoltà di Ingegneria, Università di Perugia, Polo Didattico e Scientifico del Ternano, Loc. Pentima bassa, 21, I-05100 Terni, Italy. fiori@unipg.it
[2] Dipartimento di Matematica, Università di Bari, Via Orabona 4, I-70125 Bari, Italy. delbuono@dm.uniba.it
[3] Dipartimento di Matematica, Politecnico di Bari, Via Amendola 126/B, I-70126, Bari, Italy. politi@poliba.it

Abstract. In the recent contribution [9], it was given a unified view of four neural-network-learning-based singular-value-decomposition algorithms, along with some analytical results that characterize their behavior. In the mentioned paper, no attention was paid to the specific integration of the learning equations which appear under the form of first-order matrix-type ordinary differential equations on the orthogonal group or on the Stiefel manifold. The aim of the present paper is to consider a suitable integration method, based on mathematical *geometric integration* theory. The obtained algorithm is applied to optical flow computation for motion estimation in image sequences.

1 Introduction

The computation of the singular value decomposition (SVD) of a non-square matrix [11,12,19], plays a central role in several signal/data automatic processing. It has found widespread applications e.g. in automatic control [15], digital circuit design [14], time-series prediction [18] and image processing [3,16]. Recently, some efforts have been devoted to SVD computation by neural networks in the neural community [2,17,21]. The aim of this paper is to present some notes on neural SVD computation with application to optical flow estimation. Optical-flow (OF) estimation is a well-known image-processing operation that allows estimating the motion of portions of an image over an image-sequence. It is closely related to motion estimation. Most of the OF estimation algorithms used in video encoding belong to either Block Matching Algorithms (BMAs) or Pixel Recursive Algorithms (PRAs) [13]: The majority of the current estimation methods employ a block-matching algorithm.

The BMA methods are based on the concept of template-matching: It is supposed that a single block in a time-frame has moved solidly to another location in a subsequent time-frame, so the image-block is regarded as a template to be searched for in the subsequent frame. The BMA methods try to achieve motion computation by reducing the number of search locations in the search range and/or by reducing the number of computations at each search location. Such algorithms are either *ad hoc* or are based on the assumption that the error increases monotonically from the best-match location. However, typically the error surface may exhibit local minima and the majority of the OF estimation methods get trapped in one of the local minima depending on the starting point and the search direction. Also, the matching algorithms aim at finding the best match with respect to some selected mismatch (error) measure, but the best match may not represent the true motion [4].

Conversely, the standard PRAs try to estimate the motion at each pixel. In the method for OF estimation considered here, based on paper [4], a methodology similar to the PRA is employed but we operate on a pixel-block basis and find a single motion-vector for each block. In order to make the method robust in a noisy environment, we use the total least squares (TLS) estimation approach.

Through the paper we use the following notation. Symbol $I_{m,n}$ denotes the pseudo-identity matrix of size $m \times n$ and $I_m = I_{m,m}$ while symbol $0_{m,n}$ denotes a $m \times n$ all-zero matrix. Symbol X' denotes the transposition of the matrix X while X^* denotes Hermitian-transposition; symbol $\mathrm{tr}(X)$ denotes the trace of the square matrix X, i.e. the sum of its in-diagonal entries. The orthogonal group $\mathrm{O}(m, \mathbb{K}) \stackrel{\text{def}}{=} \{X \in \mathbb{K}^{m \times m} | X^* X = I_m\}$, where the field $\mathbb{K}$ may be either $\mathbb{R}$ or $\mathbb{C}$. For details on this Lie group see e.g. [8]. Also, the Frobenius norm of a matrix $X \in \mathbb{K}^{n \times n}$ is defined as $\|X\|_\mathrm{F} \stackrel{\text{def}}{=} \sqrt{\mathrm{tr}(X^*X)}$.

2 Optical Flow Estimation by Total-Least-Squares

Let us consider a sequence of gray-level images $\{\mathcal{I}(x,y,t)\}_t$, where $\mathcal{I}$ denotes the scalar image intensity, the pair (x,y) denotes the coordinate-pair of any pixel, and t denotes the frame index.

During motion, any pixel moves from frame t and position (x,y) to frame $t+\Delta t$ at position $(x + \Delta x, y + \Delta y)$. The fact that the pixel-intensity has moved over the image support may be formally expressed with the *optical-flow conservation* equation, namely:

$$\mathcal{I}(x,y,t) = \mathcal{I}(x + \Delta x, y + \Delta y, t + \Delta t) \ . \tag{1}$$

On the basis of the above conservation equation and on the knowledge of a sequence of two subsequent frames it is possible to estimate the motion of any pixel within the sequence $\{\mathcal{I}(x,y,t)\}_t$.

In fact, let us define $\Delta\mathcal{I}(x,y,t) \stackrel{\text{def}}{=} \mathcal{I}(x,y,t+\Delta t) - \mathcal{I}(x,y,t)$. For this quantity we have:

$$\Delta\mathcal{I}(x,y,t) = \mathcal{I}(x - \Delta x, y - \Delta y, t) - \mathcal{I}(x,y,t) \ ,$$
$$= -\mathcal{I}_x(x,y,t)\Delta x - \mathcal{I}_y(x,y,t)\Delta y + \mathrm{h.o.t.} \ ,$$

where $\mathcal{I}_x$ and $\mathcal{I}_y$ denote the partial derivates of the image function, h.o.t. denotes the sum of higher-order terms in the Taylor expansion of the image function, and the vertical and horizontal displacements $(\Delta x, \Delta y)$ have been supposed small enough for the Taylor series to represent accurately the optical flow change. The latter hypothesis is equivalent to assuming slow motion or sufficiently high-rate image sampling.

As mentioned, we make the solid-block-motion assumption, thus the above equation holds true, with the same values of displacements, for a set of pixels located within the square described by $x \in [x_1, x_{N_x}]$ and $y \in [y_1, y_{N_y}]$, where constants N_x and N_y denote the block-size. On the basis of these considerations, it is possible to write the resolving system for any block between frames t and $t + \Delta t$, that is:

$$\mathcal{I}(x_1, y_1, t+\Delta t) - \mathcal{I}(x_1, y_1, t) = -\mathcal{I}_x(x_1, y_1, t)\Delta x - \mathcal{I}_y(x_1, y_1, t)\Delta y ,$$
$$\mathcal{I}(x_2, y_1, t+\Delta t) - \mathcal{I}(x_2, y_1, t) = -\mathcal{I}_x(x_2, y_1, t)\Delta x - \mathcal{I}_y(x_2, y_1, t)\Delta y ,$$
$$\mathcal{I}(x_3, y_1, t+\Delta t) - \mathcal{I}(x_3, y_1, t) = -\mathcal{I}_x(x_3, y_1, t)\Delta x - \mathcal{I}_y(x_3, y_1, t)\Delta y ,$$
$$\vdots$$
$$\mathcal{I}(x_1, y_2, t+\Delta t) - \mathcal{I}(x_1, y_2, t) = -\mathcal{I}_x(x_1, y_2, t)\Delta x - \mathcal{I}_y(x_1, y_2, t)\Delta y ,$$
$$\vdots$$
$$\mathcal{I}(x_{N_x}, y_{N_y}, t+\Delta t) - \mathcal{I}(x_{N_x}, y_{N_y}, t) = -\mathcal{I}_x(x_{N_x}, y_{N_y}, t)\Delta x$$
$$-\mathcal{I}_y(x_{N_x}, y_{N_y}, t)\Delta y ,$$

where high-order terms have been neglected.

By defining the unknowns vector $v \stackrel{\text{def}}{=} [\Delta x \; \Delta y]'$ and properly defining a matrix $L \in \mathbb{R}^{N_x N_y \times 2}$ and a vector $c \in \mathbb{R}^{N_x N_y}$, the above system casts into $Lv = c$. This is an over-determined linear system of $N_x \cdot N_y$ equations in two unknowns which may be solved by the help of a total least squares technique [10]. The resulting algorithm is as follows:

1. Define the $N_x N_y \times 3$ matrix $Z = [L \; c]$;
2. Compute the SVD (U, D, V) of Z, with $V \in O(3, \mathbb{R})$;
3. Define the partition $V = \begin{bmatrix} V_{11} & V_{12} \\ V_{21} & V_{22} \end{bmatrix}$ with $V_{12} \in \mathbb{R}^{2 \times 1}$;
4. Estimate v as $-V_{12}V_{22}^{-1}$.

3 Neural SVD Learning Algorithm

Denoting as $Z \in \mathbb{C}^{m \times n}$ the matrix whose SVD is to be computed and as $r \leq \min\{m, n\}$ the rank of Z, its singular value decomposition writes $Z = UDV^*$, where $U \in \mathbb{C}^{m \times m}$ and $V \in \mathbb{C}^{n \times n}$ are unitary matrices and D is a pseudo-diagonal matrix that has all-zero values except for the first r diagonal entries, termed singular values. It is easily checked that the columns of U coincide with the eigenvectors of ZZ^* while V contains the eigenvectors of Z^*Z with the same eigenvalues.

Here we consider the Helmke and Moore (HM) algorithm [11], which was studied in details in [9]. This algorithm is utilized to train in an unsupervised way a neural network.

3.1 Learning Differential Equations

The HM dynamics arises from the maximization of a specific criterion Φ_W : $O(m, \mathbb{C}) \times O(n, \mathbb{C}) \to \mathbb{R}$ defined as:

$$\Phi_W(A, B) \stackrel{\text{def}}{=} 2\operatorname{Re}\operatorname{tr}(WA^*ZB), \tag{2}$$

where $W \in \mathbb{R}^{n \times m}$ is a weighting kernel and $Z \in \mathbb{C}^{m \times n}$ is the matrix whose (complex-valued) SVD is looked for, in the hypothesis that $m \geq n$. The dynamical system, derived as a Riemannian gradient flow on $O(m, \mathbb{C}) \times O(n, \mathbb{C})$, reads:

$$\begin{cases} \dot{A} = A(W^*B^*Z^*A - A^*ZBW), & A(0) = A_0 \in O(m, \mathbb{C}), \\ \dot{B} = B(WA^*ZB - B^*Z^*AW^*), & B(0) = B_0 \in O(n, \mathbb{C}). \end{cases} \tag{3}$$

By construction it holds $A(t) \in O(m, \mathbb{C})$ as well as $B(t) \in O(n, \mathbb{C})$. The weighting matrix W has the structure $[W_1 \; 0_{n,m-n}]$, where $W_1 \in \mathbb{R}^{n \times n}$ must be diagonal with, in general, unequal entries on the diagonal [9].

3.2 Learning Algorithm through Integration

Whereas the continuous-time versions of the learning algorithms leave the orthogonal-group invariant, this is not true for their discrete-time counterparts, which are obtained by employing a numerical integration scheme, unless a suitable integration method is selected.

In the present case, we may employ a convenient Lie integration method drawn from geometric integration theory (see e.g. the recent contribution [1] and the previous reviews in [5,6,7]).

First, in the present case, we only consider learning over the real orthogonal group, therefore the learning equations for computing the SVD of the matrix P simply write as:

$$\begin{cases} H = A'PB, \\ \dot{A} = A(W'H' - HW), & A(0) = A_0 \in O(N_x N_y, \mathbb{R}), \\ \dot{B} = B(WH - H'W'), & B(0) = B_0 \in O(3, \mathbb{R}). \end{cases} \tag{4}$$

Also, we note that the product WH may be simplified as $W_1 H(1:3,:)$ (using standard MATLAB notation), while the product HW may be computed as the composite matrix $[HW_1 \; 0_{N_x N_y, N_x N_y - 3}]$.

If we denote by s the generic learning step index, the learning algorithm may thus be written as:

$$\begin{cases} H_s = A'_s P B_s, \\ (HW)_s \stackrel{\text{def}}{=} [H_s W_1 \; 0_{N_x N_y, N_x N_y - 3}], & (T_a)_s \stackrel{\text{def}}{=} (HW)'_s - (HW)_s, \\ A_{s+1} = A_s \exp(\eta(T_a)_s), & A_0 \in O(N_x N_y, \mathbb{R}), \\ (WH)_s \stackrel{\text{def}}{=} W_1 H_s(1:3,:), & (T_b)_s \stackrel{\text{def}}{=} (WH)_s - (WH)'_s, \\ B_{s+1} = B_s \exp(\eta(T_b)_s), & B_0 \in O(3, \mathbb{R}), \end{cases} \tag{5}$$

with η being a convenient learning step-size (or integration step).

The exponential map 'exp' in this case lifts the Lie algebra of the skew-symmetric matrices to the associated orthogonal group and the above expressions ensure that the state-matrices A and B keep within the respective orthogonal groups up to machine precision. The efficient computation of the matrix exponential is a sensitive point in geometric integration and there exist many different ways of performing exponentiation, which differ by their computational burden [1,7]. The selection of an efficient exponentiation method, tailored to the considered problem, is an interesting topic, that, however, falls outside the scope of the present contribution.

In the present contribution, as it is desirable that the matrix B is computed accurately for optical flow estimation purposes, for the second of equations (5) we relied on MATLAB's expm primitive; conversely, as the matrix A do not need to be computed accurately and the matrix T_a is generally very large, in the first of equations (5) we invoked the (rather coarse) Taylor approximation truncated to first order, namely we used:

$$\exp(\eta(T_a)_s) \approx I_{Nx \cdot Ny} + \eta(T_a)_s \ . \tag{6}$$

4 Numerical Experiments

A useful performance measure for the numerical algorithm just explained is the norm of the off-diagonal part of the argument-matrix of the criterion (2), particularized to the problem at hand. Namely, we may define an index as:

$$\delta(A,B) \stackrel{\text{def}}{=} \|\text{offdiag}(WA'ZB)\|_F \ , \tag{7}$$

which may be computed at any step s. It is interesting to note that the above-defined measure is 'blind' in the sense that it is able to measure how far the algorithm is from the sought solution without actually knowing it.

Furthermore, the course of the learning criterion function is by itself a good indicator of the network's internal state of activation.

Also, as a general-purpose quality measure we may consider two indices that take into account the loss of orthogonality of the matrices A and B, namely:

$$n(A) \stackrel{\text{def}}{=} \|A'A - I_m\|_F \ , \ n(B) \stackrel{\text{def}}{=} \|B'B - I_n\|_F \ . \tag{8}$$

As a toy example, which mainly aims at verifying the effect of the considered numerical integration method, we considered a real-valued randomly-generated matrix Z having $m = 100$ and $n = 3$. In this case we know in advance the true SVD-pair (U, V) and may therefore compare the results provided by the neural network with the exact result. In this special case, if A_n, B_n, U_n and V_n denote the sub-matrices formed by the first n columns of the SVD and network matrices, it is known that the columns of A_n should tend to the columns of U_n, while the columns of B_n should tend to the columns of B_n, ordered in the same way but with a possible sign switch for every column; therefore, a proper measure of (A, B) convergence is:

$$\epsilon(A_n) \stackrel{\text{def}}{=} \||U_n| - |A_n|\|_F \ , \ \epsilon(B_n) \stackrel{\text{def}}{=} \||V_n| - |B_n|\|_F \ , \tag{9}$$

where $|X|$ stands for component-wise absolute-value extraction. These are termed 'subspace errors'.

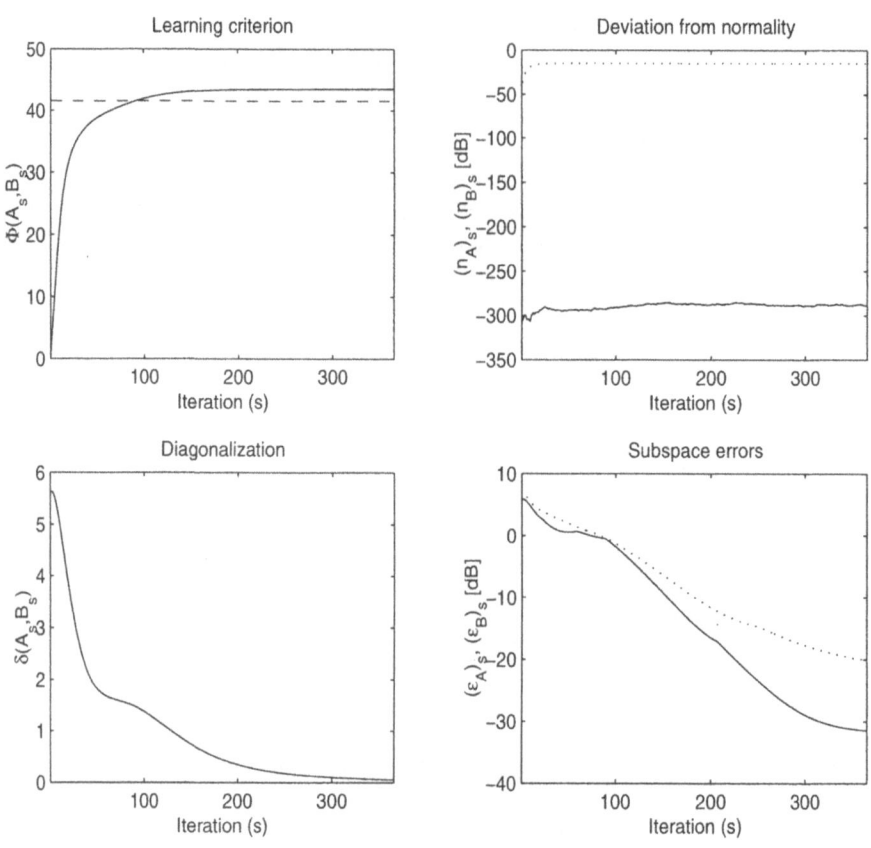

Fig. 1. Results of toy experiment with a randomly-generated 100×3 matrix. (The values on the vertical axis of the right-hand panels are expressed in decibels, that is $20 \log_{10}(\cdot)$.)

The result of this toy experiment is illustrated in the Figure 1. As it is readily seen, the singular values of Z are slightly over-estimated. However, the matrix B is rather well-adherent to the true value and orthogonal to machine precision. The iteration stopped when the ratio $\delta(A_s, B_s)/\delta(A_0, B_0) < 0.01$, namely when the residual off-diagonality is less than 1% of the initial off-diagonality.

In the experiment with real-world data, we consider a sequence of two images, drawn from the public database [20]. The image support has dimension 256×256. The images are represented with $0 \div 255$-range integer numbers capturing the local optical intensity, which was preventively scaled down of a factor 50 for numerical purposes. The sequence pertains to Walter Cronkite's screenplay in which he is rotating his head to his right. Note that the images were digitized

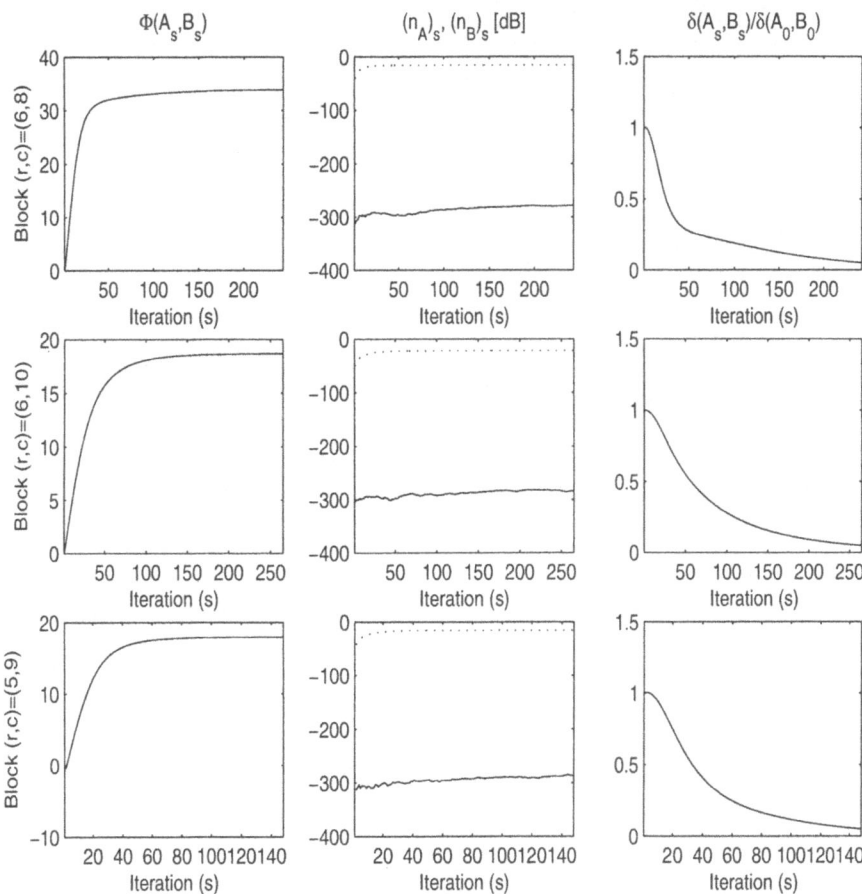

Fig. 2. Courses of the SVD-neural-computation task performance indices on 'Walter Cronkite' sequence for three selected blocks. The indicated blocks' coordinates (r,c) denote the centers of the blocks in N_x and N_y units. (The values on the vertical axis of the central panels are expressed in decibels, that is $20\log_{10}(\cdot)$.)

from an original low-quality support, therefore they are quite noisy: This makes the optical flow estimation a rather difficult task.

Here we illustrate the behavior of the algorithm by showing two subsequent images with superimposed motion vectors on the first one, and the off-diagonality index pertaining to integrated HM-run illustrated for two image-blocks; the block-size was $N_x = N_y = 16$ pixels.

The Figure 2 shows the courses of the performance indices $\delta(A_s, B_s)$, $n(A_s)$ and $n(B_s)$ as well as $\Phi_W(A_s, B_s)$ at every iteration-step s. The results pertain to the step-size value $\eta = -0.01$. In this case, the iteration stop criterion was $\delta(A_s, B_s)/\delta(A_0, B_0) < 0.05$.

Fig. 3. Two subsequent frames on a sequence of images. The arrows on the earlier image denote the estimated motion-vectors.

The Figure 3 shows two considered frames and the estimated motion vectors on the three blocks. (The arrows have been scaled to exhibit the same length for illustrative purposes only.)

The obtained results are interesting and confirm the good SVD-based OF-estimation ability of the discussed PRA approach: The estimated displacement vectors look locally consistent with the platform motion, and the learning algorithm behaved properly, as confirmed by the relatively low values of the off-diagonality measure.

5 Conclusions

In the contribution [9], a unified view of four neural-network-learning-based singular-value-decomposition algorithms was considered. No attention was paid to the specific integration method of the involved learning equations, which appear under the form of first-order matrix-type ordinary differential equations on the orthogonal group or on the Stiefel manifold. In the present paper, we considered a suitable integration method, based on mathematical geometric integration theory.

The discussed algorithm was applied to optical flow computation for motion estimation in image sequences. The obtained results confirmed the quality of the mentioned approach.

Future work could be directed along a) the search of an integration method specifically tailored to the solution of the learning differential equations defined over the Cartesian product of two orthogonal groups of different dimensions and b) the search of a partial learning algorithm that allows to extend an already learnt solution for a frame-pair to the subsequent frame(s) through partial updating.

Acknowledgment. The present work has been completed while the author SF was a short-term visitor of Professor Amari's Laboratory for Mathematical Neuroscience at the Brain Science Institute (RIKEN, Japan).

References

1. Celledoni, E., Fiori, S.: Neural Learning by Geometric Integration of Reduced 'Rigid-Body' Equations. J. Comp. Appl. Math. (to appear)
2. Cichocki, A., Unbehauen, R.: Neural networks for computing eigenvalues and eigenvectors. Biological Cybernetics **68** (1992) 155–164
3. Costa, S., Fiori, S., Image Compression Using Principal Component Neural Networks. Image and Vision Computing Journal (special issue on "Artificial Neural Network for Image Analysis and Computer Vision") **19** (9-10) (2001) 649–668
4. Deshpande, S.G., Hwang, J.-N.: Fast Motion Estimation Based on Total Least Squares for Video Encoding. Proc. International Symposium on Circuits and Systems, Vol. 4 (1998) 114–117
5. Fiori, S.: A theory for learning by weight flow on Stiefel-Grassman manifold. Neural Computation **13** (7) (2001) 1625–1647
6. Fiori, S.: A Theory for Learning Based on Rigid Bodies Dynamics. IEEE Trans. on Neural Networks **13** (3) (2002) 521–531
7. Fiori, S.: Fixed-Point Neural Independent Component Analysis Algorithms on the Orthogonal Group. Fut. Gen. Comp. Syst. (to appear)
8. Fiori, S.: Unsupervised Neural Learning on Lie Group. International Journal of Neural Systems **12** (3-4) (2002) 219–246
9. Fiori, S.: Singular Value Decomposition Learning on Double Stiefel Manifold. International Journal of Neural Systems **13** (2) (2003) 155–170
10. Golub, G.H., Van Loan, C.F.: *Matrix Computations*. The John Hopkins University Press, Third Edition, 1996
11. Helmke, U., Moore, J.B.: Singular value decomposition via gradient and self-equivalent flows. Lin. Alg. Appl. **169** (1992) 223–248
12. Hori, G.: A general framework for SVD flows and joint SVD flows. Proc. International Conference on Acoustics, Speech and Signal Processing, Vol. II (2003) 693–696
13. Liu, L.K., Feig, E.: A Block-Based Gradient Descent Search Algorithm for Block Motion Estimation in Video Coding, IEEE Trans. Circuits and Systems for Video Technology **6** (4) (1996) 419–422
14. Lu, W.S., Wang, H.P., Antoniou, A.: Design of two-dimensional FIR digital filters by using the singular value decomposition. IEEE Trans. on Circuits and Systems **CAS-37** (1990) 35–46
15. Moore, B.C.: Principal component analysis in linear systems: Controllability, observability and model reduction, IEEE Trans. on Automatic Control **AC-26** (1) (1981) 17–31
16. Nestares, O., Navarro, R.: Probabilistic estimation of optical flow in multiple bandpass directional channels. Image and Vision Computing journal **19** (6) (2001) 339–351
17. Sanger, T.D.: Two iterative algorithms for computing the singular value decomposition from input/output samples. In J.D. Cowan, G. Tesauro and J. Alspector (Ed.s), Advances in Neural Processing Systems, Vol. 6, (1994) 141–151, Morgan-Kauffman Publishers

18. Salmeron, M., Ortega, J., Puntonet, C.G., Prieto, A.: Improved RAN sequential prediction using orthogonal techniques, Neurocomputing **41** (1-4) (2001) 153–172
19. Smith, S.T.: Dynamic system that perform the SVD, Systems and Control Letters **15** (1991) 319–327
20. USC-SPC Image Database, Sequences: Walter Cronkite, Signal & Image Processing Institute, Electrical Engineering Department, University of Southern California. URL: http://sipi.usc.edu/service/database/Database.html
21. Weingessel, A.: An Analysis of Learning Algorithms in PCA and SVD Neural Networks. Ph.D. Dissertation, Technical University of Wien, Austria, (1999)

Numerical Methods Based on Gaussian Quadrature and Continuous Runge-Kutta Integration for Optimal Control Problems

Fasma Diele[1], Carmela Marangi[1], and Stefania Ragni[2]

[1] Istituto per le Applicazioni del Calcolo M. Picone, CNR, Via Amendola 122, 70126 Bari, Italy {f.diele,c.marangi}@area.ba.cnr.it
[2] Facoltà di Economia, Università di Bari, Via Camillo Rosalba 56, 70100 Bari, Italy
s.ragni@area.ba.cnr.it

Abstract. This paper provides a numerical approach for solving optimal control problems governed by ordinary differential equations. Continuous extension of an explicit, fixed step-size Runge-Kutta scheme is used in order to approximate state variables; moreover, the objective function is discretized by means of Gaussian quadrature rules. The resulting scheme represents a nonlinear programming problem, which can be solved by optimization algorithms. With the aim to test the proposed method, it is applied to different problems.

1 Introduction

Many real-life phenomena in biology, medicine, engineering, economics involve optimal decisions in a multiperiod framework. Such a kind of problems can be handled through numerical optimal control techniques which conjugate functional optimization with numerical integration of the state equations. So far, the best choice to get a consistent approximation of control problems, has been represented by high order explicit Runge-Kutta methods (see [6]). Nevertheless, a further improvement can be obtained by exploiting the advantages of continuous extension of RK integration, a technique formerly introduced in the field of delay differential equations (see e.g. [1], [7]). The main advantage of introducing the continuous method is that it provides a dense output aimed to enhance and refine the ODE solvers. Since the continuous scheme is obtained by interpolating the weights without using extra stages, continuous RK methods allow to achieve an efficient and accurate approximation of the dynamics of the system of interest with a reduced number of variable evaluations. In this paper we focus on applications of continuous Runge-Kutta schemes to a subclass of optimal control problems where the functional to be optimized appears in integral form. In this case, a further refinement of the solutions, is achieved by coupling the ODE integrators with high-order Gaussian quadrature rules (see [2]) for the discretization of the functional to be optimized. Once that the continuous optimal control problem has been converted into a discrete optimization one, it can be solved by classical algorithms.

The continuous problem we are concerned with, is described in Section 2, hence the numerical approach is provided in Section 3; finally in Section 4 we apply our schemes on some test problems.

2 The Optimal Control Problem

We deal with a class of optimal control problems given by

$$\min_{\mathbf{u}} \left\{ J(\mathbf{u}, \mathbf{x}) := \int_0^T g(\mathbf{u}(t), \mathbf{x}(t), t) \, dt \right\} \quad (1)$$

where control parameters $\mathbf{u}(t)$ belong to a feasible set $\mathcal{U} \subset \mathbb{R}^m$ for every time t in a fixed and finite interval $[0, T]$ (i.e. $\mathbf{u}(t) \in \mathcal{U}$) and state variables $\mathbf{x}(t) \in \mathbb{R}^n$ satisfy an ordinary differential equation

$$\frac{d\mathbf{x}}{dt}(t) = \mathbf{f}(\mathbf{x}(t), \mathbf{u}(t), t), \quad t \in [0, T] \quad (2)$$

with initial condition $\mathbf{x}(0) = \mathbf{x}_0$, being $g : \mathbb{R}^m \times \mathbb{R}^n \times \mathbb{R} \to \mathbb{R}$ and $\mathbf{f} : \mathbb{R}^n \times \mathbb{R}^m \times \mathbb{R} \to \mathbb{R}^n$ suitable functions. Moreover, supposing that $\mathcal{U} \subset V$ where

$$V := \{\mathbf{v} : [0, +\infty) \longrightarrow \mathbb{R}^m \mid \mathbf{v} \text{ is piecewise continuous}\},$$

we notice that feasible set $\mathcal{U}$ accounts for any constraint on control parameters such as inequality conditions.

3 Numerical Discretization

We focus on the numerical discretization of the previous optimal control problem by using Gaussian quadrature rules and continuous extensions of Runge-Kutta schemes.

Let us consider a discretization $0 = t_0 < t_1 < \ldots < t_N = T$ of the time horizon $[0, T]$ with constant step length h, thus $t_i = ih$ for each $i = 0, \ldots N$. Then, the objective function in (1) can be written as

$$J(\mathbf{u}, \mathbf{x}) = \int_0^T g(\mathbf{u}, \mathbf{x}, t) \, dt = \sum_{i=0}^{N-1} \int_{t_i}^{t_{i+1}} g(\mathbf{u}, \mathbf{x}, t) \, dt$$
$$= \frac{h}{2} \sum_{i=0}^{N-1} \int_{-1}^{1} g\left(\mathbf{u}\left(t_i + h\tfrac{s+1}{2}\right), \mathbf{x}\left(t_i + h\tfrac{s+1}{2}\right), t_i + h\tfrac{s+1}{2}\right) ds. \quad (3)$$

In order to compute an approximation for every integral, we apply a Gaussian rule; therefore, functional $J(\mathbf{u}, \mathbf{x})$ in (3) is discretized as

$$J(\mathbf{u}, \mathbf{x}) \approx \frac{h}{2} \sum_{i=0}^{N-1} \left(\sum_{k=0}^{1} \overline{w}_k \, g\left(\mathbf{u}(t_i + hk), \mathbf{x}(t_i + hk), t_i + hk\right) \right.$$
$$\left. + \sum_{l=0}^{L} w_l \, g\left(\mathbf{u}\left(t_i + h\tfrac{s_l+1}{2}\right), \mathbf{x}\left(t_i + h\tfrac{s_l+1}{2}\right), t_i + h\tfrac{s_l+1}{2}\right) \right)$$

where w_k, $\overline{w}_k$ and s_k represent the weights and nodes of the chosen quadrature formula (for more details, see [2]). Setting

$$\xi_l = \frac{s_l + 1}{2}, \quad l = 0, \ldots, L,$$

we obtain

$$\frac{h}{2} \sum_{i=0}^{N-1} \left(\sum_{k=0}^{1} \overline{w}_k \, g\left(\mathbf{u}(t_i + hk), \mathbf{x}(t_i + hk), t_i + hk\right) \right. \\ \left. + \sum_{l=0}^{L} w_l \, g\left(\mathbf{u}\left(t_i + h\xi_l\right), \mathbf{x}\left(t_i + h\xi_l\right), t_i + h\xi_l\right) \right). \tag{4}$$

Remark 1. In the sequel, we will account for different Gauss formulae. In particular, we consider the so-called Gauss-Legendre rule

$$\frac{h}{2} \sum_{i=0}^{N-1} \sum_{l=0}^{L} w_l \, g\left(\mathbf{u}(t_i + h\xi_l), \mathbf{x}(t_i + h\xi_l), t_i + h\xi_l\right) \tag{5}$$

where we have set $\overline{w}_0 = \overline{w}_1 = 0$, ξ_l's are related to the zeros s_l of Legendre polynomial of order $L+1$ and w_l's represent the coefficients of the chosen scheme. Furthermore, we also perform a Gauss-Lobatto integration

$$\frac{h}{2} \sum_{i=0}^{N-1} \left(\sum_{k=0}^{1} \overline{w}_k \, g\left(\mathbf{u}(t_i + hk), \mathbf{x}(t_i + hk), t_i + hk\right) \right. \\ \left. + \sum_{l=0}^{L} w_l \, g\left(\mathbf{u}\left(t_i + h\xi_l\right), \mathbf{x}\left(t_i + h\xi_l\right), t_i + h\xi_l\right) \right) \tag{6}$$

where $\overline{w}_0 = \overline{w}_1 = \frac{2}{(L+2)(L+3)}$, w_l represent the coefficients and each ξ_l is related to every node s_l of the given quadrature rule.

We notice that, in order to evaluate both (5) and (6), control parameters $\mathbf{u}$ and state variables $\mathbf{x}$ sampled at every inner instant $t_i + h\xi_l$ are needed. We assume to approximate control parameters in the class of piecewise linear functions

$$\mathcal{V} = \{\mathbf{v} : [0, T] \to \mathbb{R}^m | \mathbf{v}|_{[t_i, t_{i+1}]} \text{ is linear}\},$$

thus we have that

$$\mathbf{u}(t_i + h\xi_l) = (1 - \xi_l)\, \mathbf{u}(t_i) + \xi_l\, \mathbf{u}(t_{i+1}), \quad l = 0, \ldots, L.$$

Moreover, aiming to compute $\mathbf{x}(t_i + h\xi_l)$'s, state equation (2) is discretized by applying continuous extensions of a Runge-Kutta scheme. Therefore, we consider an explicit s-stage Runge-Kutta method with Butcher array

$$\begin{array}{c|c} \mathbf{c} & A \\ \hline & \mathbf{b}^T \end{array}$$

and its continuous extension related to coefficients $\tilde{b}_j(\theta)$ ($j = 1, \ldots, s$) which are suitable polynomials in $\theta \in [0, 1]$ (see [1], [7]). The resulting scheme consists of

$$\mathbf{x}(t_i + h\xi_l) = \mathbf{x}(t_i) + h \sum_{j=1}^{s} \tilde{b}_j(\xi_l) \mathbf{f}(\mathbf{X}_j, \mathbf{u}(t_i + hc_j), t_i + hc_j),$$

$$\mathbf{x}(t_{i+1}) = \mathbf{x}(t_i) + h \sum_{j=1}^{s} \tilde{b}_j(1) \mathbf{f}(\mathbf{X}_j, \mathbf{u}(t_i + hc_j), t_i + hc_j), \qquad (7)$$

$$\mathbf{X}_j = \mathbf{x}(t_i) + h \sum_{v=1}^{j-1} a_{jv} \mathbf{f}(\mathbf{X}_v, \mathbf{u}(t_i + hc_v), t_i + hc_v), \quad j = 1, \ldots, s$$

starting from $\mathbf{x}(t_0) = \mathbf{x}_0$ and evaluating

$$\mathbf{u}(t_i + hc_j) = (1 - c_j) \mathbf{u}(t_i) + c_j \mathbf{u}(t_{i+1}), \qquad j = 0, \ldots, s.$$

Set

$$\mathbf{U} = \begin{pmatrix} \mathbf{u}(t_0) \\ \mathbf{u}(t_1) \\ \vdots \\ \mathbf{u}(t_N) \end{pmatrix} \in \mathbb{R}^{m(N+1)}.$$

Using this notation, the sum in (4) can be written as

$$\tilde{J}(\mathbf{U}) = \frac{h}{2} \sum_{i=0}^{N-1} \left(\sum_{k=0}^{1} \overline{w}_k\, g\left(\mathbf{u}(t_{i+k}), \mathbf{x}(t_{i+k}), t_{i+k}\right) \right. \\ \left. + \sum_{l=0}^{L} w_l\, g\left((1 - \xi_l)\mathbf{u}(t_i) + \xi_l \mathbf{u}(t_{i+1}), \mathbf{x}(t_i + h\xi_l), t_i + h\xi_l\right) \right) \qquad (8)$$

where the values $\mathbf{x}(t_i + hk)$ and $\mathbf{x}(t_i + h\xi_l)$, depending on $\mathbf{U}$, are computed by recursive formula (7). Therefore, the numerical approximation of the given optimal control model reduces to the nonlinear programming problem which consists of

$$\min_{\mathbf{U}} \tilde{J}(\mathbf{U})$$

where function $\tilde{J}(\mathbf{U})$ is defined in (8).

4 Numerical Examples

We have implemented different schemes in order to test the effectiveness of the proposed numerical procedure. In each of them differential equation (2) has been discretized by using the continuous extension of the 2-stage *Improved Euler* method (see [4]) where

$$b_1(\theta) = -\frac{1}{2}\theta^2 + \theta \quad \text{and} \quad b_2(\theta) = \frac{1}{2}\theta^2.$$

Moreover, by different choices of the weights and knots in (5) and (6) three schemes have been considered:

- Leg0: $L = 0$, $\overline{w}_0 = \overline{w}_1 = 0$, $w_0 = w_1 = 1$, $\xi_0 = \frac{3-\sqrt{3}}{6}$, $\xi_1 = \frac{3+\sqrt{3}}{6}$;
- Lob0: $L = 0$, $\overline{w}_0 = \overline{w}_1 = \frac{1}{3}$, $w_0 = \frac{4}{3}$, $\xi_0 = \frac{1}{2}$;
- Lob1: $L = 1$, $\overline{w}_0 = \overline{w}_1 = \frac{1}{6}$, $w_0 = w_1 = \frac{5}{6}$, $\xi_0 = \frac{5-\sqrt{5}}{10}$, $\xi_1 = \frac{5+\sqrt{5}}{10}$.

Finally, in order to solve the obtained nonlinear programming problem, we have used the fminunc and fmincon routines in Matlab environment.

Problem 1. (see [3]) Find

$$\min_u J(u,x) := \int_0^1 (0.625\ x^2(t) + 0.5\ x(t)u(t) + 0.5\ u^2(t))\,dt$$

subject to

$$\frac{dx}{dt}(t) = \frac{1}{2}x + u, \quad t \in [0,1], \quad x(0) = 1.$$

This optimal control problem has an analytic solution $u^*(t)$ given by

$$u^*(t) = -(\tanh(1-t) + 0.5)\cosh(1-t)/\cosh(1), \quad t \in [0,1]$$

with optimal cost $J^* = e^2 \sinh(2)/(1+e^2)^2 \approx 0.380797$. In Figure 1, we plot the approximate time evolution of the solution obtained by the different schemes with stepsize $h = 0.05$. All of them give the same value for the approximated functional that is $\widetilde{J}^* = 0.3807381$; hence, numerical results are in agreement with theoretical ones.

Problem 2. (see [5]) Find

$$\min_u J(u,x) := \int_0^{2.5} (x_1^2(t) + u^2(t))\,dt$$

subject to

$$\frac{dx_1}{dt}(t) = x_2(t), \qquad\qquad x_1(0) = -5,$$

$$\frac{dx_2}{dt}(t) = -x_1(t) + (1.4 - 0.14x_2^2(t))x_2(t) + 4u(t), \qquad x_2(0) = -5.$$

In Figure 2, we plot the behaviour of the discrete solution obtained by means of the three different methods when applied with stepsize $h = 0.05$.

Problem 3. Find

$$\max_u J(u,x) := \int_0^1 c\,(1 - u(t))x(t)\,dt$$

subject to

$$\frac{dx}{dt}(t) = (bu - \mu)x, \quad t \in [0,1], \quad x(0) = 1$$

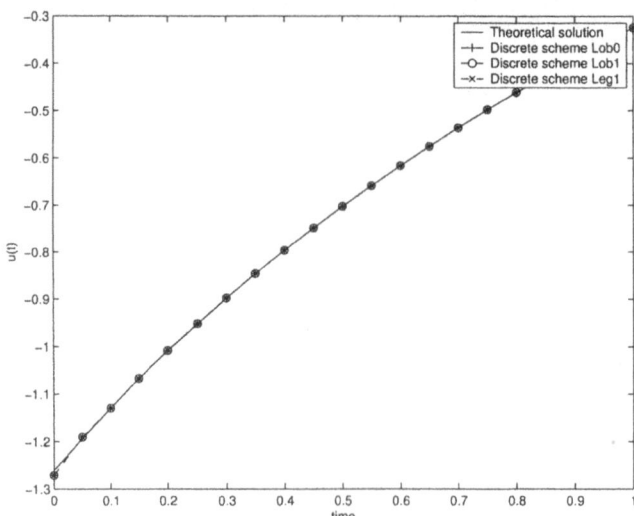

Fig. 1. Theoretical and discrete solutions to Problem 1.

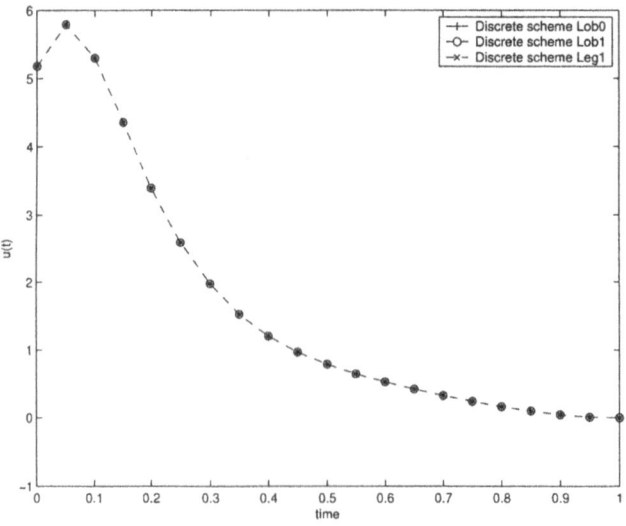

Fig. 2. Discrete solutions to Problem 2.

and

$$0 \leq u(t) \leq 1, \quad t \in [0,1],$$

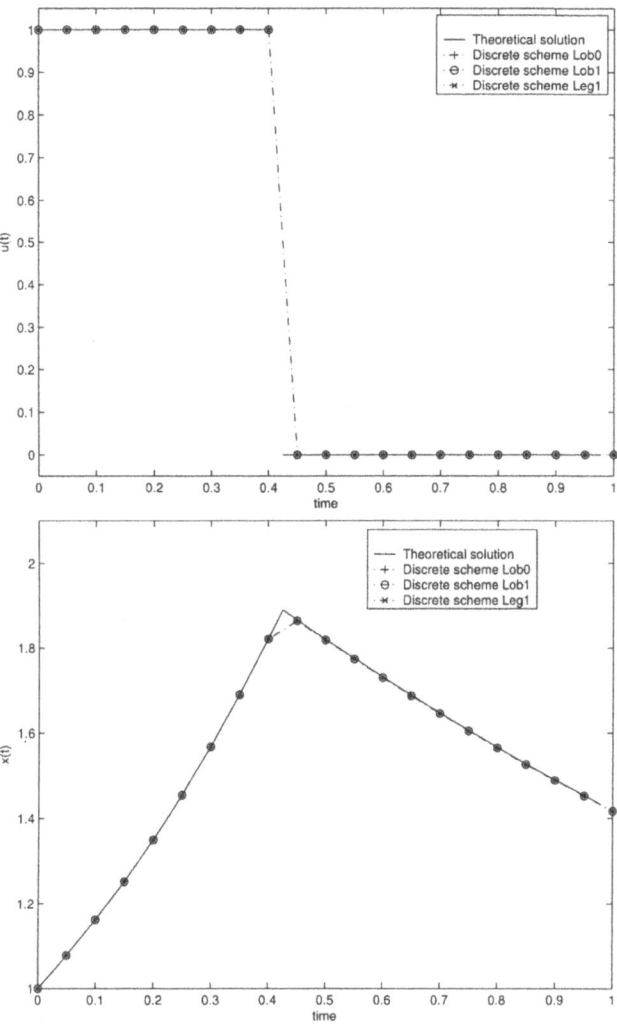

Fig. 3. Problem 3: Theoretical and discrete control parameter (top) and state variable (bottom).

being c, b, μ positive constants with $b > \mu$. Since the Hamiltonian is linear with respect to the control parameter, the optimal solution $u^*(t)$ is a *bang-bang* function with switching time $\tau = \frac{1}{\mu} log(1 - \frac{\mu}{b}) + 1$ that is

$$u^*(t) = \begin{cases} 1 & \text{if } t \in [0, \tau], \\ 0 & \text{if } t \in (\tau, 1]. \end{cases}$$

Moreover the corresponding optimal state variable is given by

$$x^*(t) = \exp\left(\int_0^T (bu^*(s) - \mu)\, ds\right).$$

In Figure 3, we plot the theoretical and discrete solutions obtained for $c = 1$, $b = 2$, $\mu = 0.5$ by means of the proposed schemes with stepsize $h = 0.05$.

5 Conclusions

This paper introduces the use of continuous extensions of Runge-Kutta methods in the field of optimal control problems governed by ordinary differential equations. These methods are used to discretize state variables and they are coupled with Gaussian quadrature rules, which approximate the objective function. The numerical procedure we obtain reduces the continuous problem to a nonlinear programming one that can be solved by optimization algorithms. The main advantage of this approach is that it allows the reduction of nodal variable values necessary for the chosen discretization.

The proposed scheme has been implemented in order to solve test problems; the numerical results are in agreement with the theoretical ones.

References

1. Bellen, A., Zennaro, M.: Numerical methods for delay differential equations. Clarendon Press, Oxford (2003).
2. Davis, P.J., Rabinowitz, P.: Methods of numerical integration, 2nd Ed., Academic, New York (1984).
3. Hager, W.W.: Rates of convergence for discrete approximations to unconstrained control problems. SIAM J.Numer.Anal. **Vol. 13**, No. 4, (1976) 449–472.
4. Lambert, J. D.: Numerical methods for ordinary differential systems. Wiley & Sons, Chichester (1997).
5. Nedeljkovic', N.B.: New algorithms for unconstrained nonlinear optimal control problems. IEEE Trans. Autom. Cntrl., **Vol. 26**, No. 4,(1981) 868-884.
6. Schwartz, A.: Theory and implementation of numerical methods based on Runge-Kutta integration for solving optimal control problems. PhD Thesis, U.C. Berkeley (1996).
7. Zennaro, M.: Natural continuous extensions of Runge-Kutta methods. Mathematics of Computation, **Vol. 46**, No. 173 (1986) 119–133.

Graph Adjacency Matrix Associated with a Data Partition

Giuseppe Acciani, Girolamo Fornarelli, and Luciano Liturri

Dipartimento di Elettrotecnica ed Elettronica, Politecnico di Bari, Via E. Orabona n° 4, 70125 Bari, Italy
{acciani,fornarelli,liturri}@deemail.poliba.it

Abstract. A frequently recurring problem in several applications is to compare two or more data sets and evaluate the level of similarity. In this paper we describe a technique to compare two data partitions of different data sets. The comparison is obtained by means of matrices called Graph Adjacency Matrices which represent the data sets. Then, a match coefficient returns an estimation of the level of similarity between the data sets.

1 Introduction

In several applications, such as detecting flaws or image retrieval, comparing two or more data sets is a very important task. Moreover the comparison methods must evaluate level of similarity between the data sets. The most common data-comparing methods are based on the search of eigenvalues of appropriate matrices, the search of recurrences by the use of filters and the definition of proper metrics on the whole data set (e.g. single pixel classification in comparing images) [1,2,3]. In methods performing the comparison by data filtering is very difficult to set the optimal filter parameters. Moreover, the search eigenvalues-based methods and the ones operating on the whole data sets are both very expensive. In fact, the inversion of large matrices or the two-by-two comparisons of each single datum of large sets are needed. To overcome these drawbacks we propose a technique that requires, in the comparison step, simple operations on small size matrices. It is based on a Graph Adjacency Matrix (GAM) associated with a partition of data set, an isomorphism operation on the graphs which represent the GAM and a match coefficient (MC) to evaluate the level of similarity between the two graph adjacency matrices.

This paper is organized as follows. In Sections 2 and 3 we define the graph adjacency matrix associated with the data set and describe the comparing method introducing the match coefficient. In Section 4 the testing procedure is shown. The results of a test performed on a reference data set and some data sets affected by increasing Gaussian noise are shown in Section 5. Finally we summarize our findings and outline future works.

2 Graph Adjacency Matrix

Let us consider an $n \times p$ dimensional data set X ($X_i=x_{i1}, x_{i2}......, x_{ip}$, $i=1,2,...n$) and k prototypes C ($C_j=c_{j1}, c_{j2}......, c_{jp}$ $j=1,2,...k$) which can represent the data X splitted into clusters. Let $\Pi(X)$ be a partition of the data set X represented by the k prototypes C, where a partition is a way to split into clusters a data set. Let $U(X,C)$ a $k \times n$ dimension fuzzy membership function matrix, which measures the membership of each data point in X with respect to the k prototypes and let $W(C,U)$ be the following weight function which returns a real nonnegative value for each pair of prototypes:

$$W = w_{ij}(C_i, C_j) = \frac{1}{n}\sum_{k=1}^{n} u_{ik} \cdot u_{jk} \quad \forall C_i, C_j \in C \quad (1)$$

Now we define a $k \times k$ dimensional matrix $A(C,W)$, called graph adjacency matrix (GAM) associated with the partition $\Pi(X)$, whose entries are given by the values of the weight function W. In this way the graph adjacency matrix $A(C,W)$ is:

$$A(C,W) = [a_{ij}] = w_{ij}(C_i, C_j). \quad (2)$$

The choice for $W(C,U)$ allows to build a graph adjacency matrix $A(C,W)$ whose entries depend on the reciprocal position of data points with respect to prototypes. If a data point X_k is near to the center C_i but far from the center C_j the contribution to the value of the entry is poor, according to the definition of a membership function $U(X,C)$ whose value decreases when the distance of a data point X_k from the center C_i increases. In fact at least one between u_{ik} and u_{jk} has a very small value because the values of membership function u_{ik} is as higher as the data point X_k is near to the center C_i. On the contrary if it is near to both prototypes the contribution is high. In other words, the built graph adjacency matrix $A(C,W)$ represent the internal structure of data set and can be useful to compare two or more data sets.

3 Comparing the Data Partitions

Let $\Pi_1(X_1)$ be the partition of the $n_1 \times p$ dimensional data X_1 represented by the k_1 prototypes C_1 and the fuzzy membership matrix U_1 ($k_1 \times n_1$) and $\Pi_2(X_2)$ the partition of the $n_2 \times p$ dimensional data X_2 represented by the k_2 prototypes C_2 and the fuzzy membership matrix U_2 ($k_2 \times n_2$). Let $A_1(C_1,W_1)$ be the GAM associated with the partition $\Pi_1(X_1)$ and $A_2(C_2,W_2)$ be the GAM associated with the partition $\Pi_2(X_2)$. To compare the two data sets by means of the graph adjacency matrices we need two matrices with the same dimensions and with rows or columns (the graph adjacency matrices are square matrices) arranged in same way. In fact, if the entry (i,j) of $A_1(C_1,W_1)$ is referred to prototypes C_i and C_j of partition $\Pi_1(X_1)$ the analogous entry (i,j) of $A_2(C_2,W_2)$ must be referred to the analogous prototypes C_i and C_j of the partition $\Pi_2(X_2)$. To reach this target we can use an isomorphism, borrowing concepts of graph matching operations. The graph matching operation is a method to match two isomorphic graphs. We recall that two graphs are isomorphic if there exists a one to one correspondence between their vertex sets which preserves adjacency. In the same way, we can define a matrix matching operation to match two isomorphic matrices and we can define the isomorphic matrices. Two matrices are isomorphic if there exists a one to one correspondence in the way that are arranged its rows or/and columns. As matter of fact two matrices are isomorphic if the entry a_{ij} is obtained in the same way in the two matrices according the idea seen before.

We can visualize each graph adjacency matrix as a complete undirected weighted graph with a number of vertexes equal to the order of the graph adjacency matrix and with branch weights equal to the entries of graph adjacency matrix. In Fig. 1 there is a simple graph of bi-dimensional case that represents a graph adjacency matrix with five prototypes. For this reason we call the matrix $A(C,W)$ graph adjacency matrix.

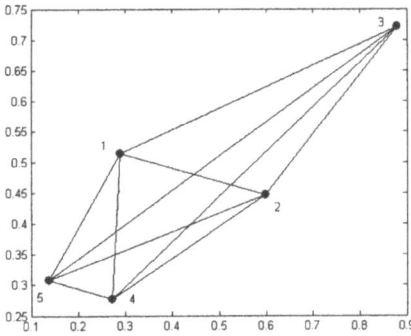

Fig. 1. A simple example of a graph representing a graph adjacency matrix with five vertexes (prototypes)

After this graph representation of the GAM, we can generate an isomorphism between two graph adjacency matrices through the realization of an isomorphism of two graphs that represent the matrix.

Matrices $A_1(C_1,W_1)$ and $A_2(C_2,W)$ can have a different number of rows. To build an isomorphism we have to associate uniquely each prototype of the set C_1, or each entry of $A_1(C_1,W_1)$, with each prototype of the set C_2 or each entry of $A_2(C_2,W_2)$. For this purpose we adopt a hierarchical cluster tree algorithm [4].

It consists of some steps. The first step is to find the similarity between every pair of objects in the two sets. In this step, we calculate the Euclidean distances between the prototypes C_1 and C_2. The second step is to group the objects into a binary, hierarchical cluster tree. In this step, we link together pairs of prototypes that are in close proximity, that is the distance calculated in step 1. As prototypes are paired into binary clusters, the newly formed clusters are grouped into larger clusters until a hierarchical tree is formed. Finally, it is determined where to divide the hierarchical tree into clusters. In this step, we divide the prototypes in the hierarchical tree into clusters. The cluster can be created by detecting natural groupings in the hierarchical tree or by cutting off the hierarchical tree at an arbitrary point. We "cut" the tree at the first level of association as we want a binary association.

Let n_a be the number of associated prototypes of $A_1(C_1,W_1)$ and $A_2(C_2,W_2)$. The number of unassociated prototypes is $n_{u1}= k_1-n_a$ for $A_1(C_1,W_1)$ and $n_{u2}=k_2-n_a$ for $A_2(C_2,W_2)$. To match the dimensions of graph adjacency matrices we must introduce some prototypes that will be associated with the unassociated prototypes, they are $n_d = n_{u1}+ n_{u2}$ prototypes called dummy [5]. The i-th prototype is dummy if $w_{ij}=0$ for each j. It is interesting to note that the introduction of dummy prototypes generates n_{u1} rows and columns in graph adjacency matrix of $A_2(C_2,W_2)$ and n_{u2} rows and columns in graph adjacency matrix of $A_1(C_1,W_1)$. Therefore the GAMs associated with the data partitions $\Pi_1(X_1)$ and $\Pi_2(X_2)$ become $n \times n$ matrices where:

$$n = k_1 + n_{u2} = k_1 + k_2 - n_a = k_2 + n_{u1}. \tag{3}$$

The correspondences find above constitute the isomorphism and it can be represented through an orthogonal permutation matrix P [6]. In fact, the matrix P performs a matrix operation which arranges the matrix $A_2(C_2,W_2)$ to be isomorphic with $A_1(C_1,W_1)$ as follows:

$$A_2(\Phi) = P A_2 P^T. \tag{4}$$

An orthogonal permutation matrix P, is a square matrix defined as:

$$P \cdot P^T = P^T \cdot P = I, \tag{5}$$

where P is a square matrix ($n \times n$) that defines uniquely the isomorphism Φ between the pair of matrices. Each row and column of the matrix P has only one non zero element, $P(i,j)$ equal to 1 if the i-th row of the prototype set C_1 is associated with the j-th row of the prototype set C_2, and vice versa, while the unassociated prototypes of $A_1(C_1,W_1)$ and $A_2(C_2,W_2)$ are associated with dummy prototypes. In Fig. 2 an example of association of prototypes of two graph adjacency matrices is shown. The prototype number 2 is an unassociated prototype so it will be associated to a dummy prototype. The prototypes with apex in Fig. 2 are the prototypes of graph adjacency matrix $A_2(C_2,W_2)$, the prototypes without apex are the prototypes of graph adjacency matrix $A_1(C_1,W_1)$.

SEQTo evaluate the level of similarity between the two GAMs we need to define a match coefficient (MC).

Let Φ be the isomorphism between graph adjacency matrices $A_1(C_1,W_1)$ and $A_2(C_2,W_2)$ as in (2). We define the Matching Coefficient (MC) as [6]:

$$MC = \|A_1 - A_2(\Phi)\|, \tag{6}$$

where $\| X \|$ is the Frobenius norm defined as:

$$\|X\| = \sqrt{\sum_{i=1}^{n} X_1(i,i)} \quad \text{where} \quad X_1 = X^T \cdot X. \tag{7}$$

If P is the permutation matrix defining the isomorphism Φ as in (4), the Matching Coefficient is:

$$MC = \|A_1 - P A_2 P^T\|. \tag{8}$$

The coefficient *MC* evaluates the level of similarity of two isomorphic matrices and for this reason it is a measure of similarity between the two data sets represented by their graph adjacency matrices. In the following section an example of application of these concepts is shown.

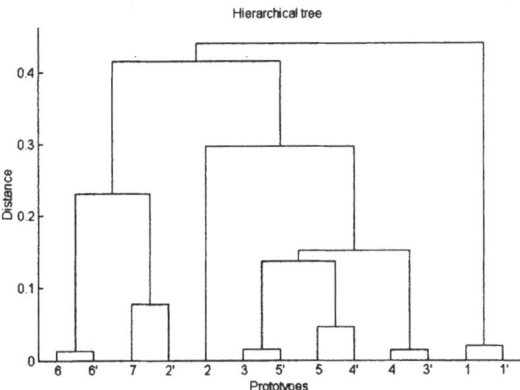

Fig. 2. The prototypes of A1(C1,W1) are associated with every prototype of A2(C2,W2). The fifth prototype is associated with the dummy prototype.

4 Testing Procedure

The method has been performed on nine data sets obtained from nine images. Each image is 128 x 192 pixel sized in RGB color space and is represented by a three dimensional matrix (128 x 192 x 3). The first image is the reference data set, while the other eight images are obtained overlapping an increasing value of the variance of zero-mean Gaussian noise.
In Fig. 3 we see the reference image, while Fig. 4 shows the image with the higher value of noise variance. The comparison is made between the reference data set and each data set originated from noisy images.
To compare the two data sets we need the graph adjacency matrices associated with each data set, that is the number of prototypes C and the fuzzy membership function $U(X,C)$ since the weight function W is defined in (1).

Fig. 3. Reference image (reference data set).

Fig. 4. Noisy image with Gaussian noise of variance 5×10^{-1}.

An alternative method to the neural networks is to use a fuzzy partition algorithm like Fuzzy C-Means (FCM) [4], or Gustafson Kessel (GK) [10] since they give both the prototypes C and the fuzzy membership function $U(X,C)$. For sake of simplicity, in the following the comparisons between the different data sets are performed with the FCM.

The FCM clustering algorithm is based on the minimization of the following objective function with respect to the fuzzy membership u_{ij} and the cluster center C_i:

$$J_m(U,C) = \sum_{j=1}^{n}\sum_{i=1}^{k}(u_{ij})^m d^2(X_j,C_i), \qquad (9)$$

where $d(X_j, C_i)$ is the distance, in a generic metric, between the data point X_j and the center C_i, k is the number of centers, n is the number of data points and $m>1$ is the fuzziness index (in this work the Euclidean distance and $m=2$ have been used). The FCM algorithm is executed in the following steps:

1) Initialize membership u_{ij} of X_j belonging to cluster i so that:

$$\sum_{i=1}^{k} u_{ij} = 1, \qquad (10)$$

2) Compute the fuzzy center C_i for $i=1,2....,k$, using:

$$C_i = \frac{\sum_{j=1}^{n}(u_{ij})^m X_j}{\sum_{j=1}^{n}(u_{ij})^m}, \qquad (11)$$

3) Update the fuzzy membership u_{ij} using:

$$u_{ij} = \frac{\left[\dfrac{1}{d^2(X_j,C_i)}\right]^{1/(m-1)}}{\sum_{i=1}^{k}\left[\dfrac{1}{d^2(X_j,C_i)}\right]^{1/(m-1)}} . \qquad (12)$$

Repeat steps 2) and 3) until the value of J_m is no longer decreasing.

In this case is very easy to understand that the centers in (11) represent a partition of data set X, therefore they can be used like the prototypes described in Section 2, while the values of the fuzzy membership function $U(X,C)$ are given in (12).

FCM needs an a priori knowledge about the number of clusters, since only if the number of clusters is correct it works well. To overcome this problem is possible to use a validation index. In this work Xie-Beni validation index (XB) [11] is adopted. It has been chosen because of its more stable behavior with respect to other indexes which search the best fuzzy partition. The XB is defined as:

$$XB(f) = \frac{\sum_{x \in X}\sum_{k \in K} u_{ij}^2 \cdot d^2(X_i,C_j)}{n \min_{i,j} d^2(C_i,C_j)} . \qquad (13)$$

For the FCM algorithm with $m=2$, XB is:

$$XB = \frac{J_2}{n \min_{i,j} d^2(C_i,C_j)} . \qquad (14)$$

The optimal number of clusters to initialize the FCM algorithm can be found with the following procedure:

1) Run the FCM with $k^*=2$;
2) Evaluate Xie-Beni index, XB*, as in (14) using the objective function computed in step 1;
3) Run the FCM with k=k*+1;
4) Evaluate Xie-Beni index, XB, using the objective function calculated in step 3. If XB>XB*, k* is the optimal partition else go to step 3 with k*=k.

This procedure minimizes the Xie-Beni index. Minimizing XB corresponds to minimizing J_2. The additional factor $\min_{i,j} d^2(C_i, C_j)$ in XB, is the separation measurement between clusters i and j. The more separated the clusters are, the larger $\min_{i,j} d^2(C_i, C_j)$, and the smaller XB are. Thus, the smallest XB indicates a valid optimal partition with well distinguishable clusters.

Now it's clear why the graph adjacency matrices associated with two data sets could have different order, FCM could find different number of prototypes according to the Xie-Beni index.

5 Numerical Results

The measure of the comparison, that is the level of similarity, between data sets can be immediately pointed out by the coefficient MC shown in Table 1. In this table, the results of the test are reported. Each row is related to a different data sets. The first and second columns indicate the values of the variance of the Gaussian noise added to the reference image and the number of centers found by FCM respectively. In the third column the number of the associated prototypes between the reference image and the noisy image are provided.

Table 1. Testing procedure result.

Noise Variance (10^{-2})	No. of centers	No. of associated centers (n_a)	Total No. of vertex (n)	MC (10^{-3})
0	3	3	3(0 dummy)	0
0.01	3	3	3(0dummy)	2.9
0.05	3	3	3(0dummy)	13.6
0.1	3	3	3(0dummy)	26.9
0.5	3	3	3(0dummy)	130.3
0.7	3	3	3(0dummy)	180.5
1	2	1	4(3dummy)	272.8
3	2	1	4(3dummy)	539.5
5	2	1	4(3dummy)	554.8

The size of the graph adjacency matrix and, in brackets, the number of dummy prototypes are given in the fourth column. Last column reports the value of MC. As expected, the value of the matching coefficient increases as the value of the variance increases. As matter of fact a small value of MC indicates a good match between two data sets. In fact, in the first row MC is zero because the data set matches perfectly itself. It can be pointed out that, as the noise increases, some dummy prototypes appear, since they are a qualitative measure for the comparison.

The appearing of the dummy prototypes and the variations of value of the entries in the graph adjacency matrix increase the matching coefficient MC.

6 Conclusions

In this paper we present a graph adjacency matrix associated with a data set able to represent it in a compact way. Moreover a match coefficient to evaluate the level of similarity between the two graph adjacency matrices is provided. Therefore a technique to compare more data set and based on the graph adjacency matrices is developed. The results performed by means of the Fuzzy C-Means clustering algorithm in the testing procedure show that this technique is very useful to compare two, or more similar data sets. In fact, the value of the matching coefficient provides immediately the level of similarity between the analyzed data. Moreover, in the comparison operation this method uses only the information content of entries of graph adjacency matrix so it is easy to realize that the method proposed is a very simple and computationally light instrument. It does not need to work with the whole data set or to search the eigenvalues in a matrix with large size. The aim of our future work is to perform some additional experiments to find the best algorithm to generate the prototypes and the entries of graph adjacency matrix.

References

1. Bodnarova, A., Bennamoun, M., Latham, S.: Optimal Gabor Filter for Textile Flaw Detection. Pattern recognition, **35** (2002) 2973-2991.
2. Aksoy, S., Haralick, R. M.: Graph Theoretic Clustering for Image Grouping and Retrieval. IEEE, Conference on Computer Vision and Pattern Recognition (1999) 63-68.
3. Sahasrabudhe, N., West, J.E., Machiraju, R., Janus, M: Structured Spatial Domain Image and Data Comparison Metrics. Visualization '99, Proceedings, October (1999), 97-105.
4. Hoppner, F., Klawonn, F., Kruse, R., Runkler, T.: Fuzzy Cluster Analysis. John Wiley & Sons, LTD, Chichester, New York, Weinheim, Brisbane, Singapore, Toronto, 2000.
5. Papadimitiou, C. H., Steiglitz, K.: Combinatorial Optimization: Algorithm and Complexity. Englewood Cliffs, NJ: Prentice-Hall 1982.
6. Umeyama, S.: An Eigendecomposition Approach to Weighted Graph Matching Problems. IEEE, Transaction on Pattern Analysis and Machine Intelligence, **10** (5) (1988) 695-703.
7. Xu, L.: Best Harmony, Unified RPCL and Automated Model Selection for Unsupervised and Supervised Learning on Gaussian Mixtures, Three-Layer Nets and ME-RBF-SVM Models, International Journal of Neural Systems **11** (2001) 43-69.
8. Chiarantoni, E., Acciani, G., Fornarelli, G., Vergura, S.: Robust Unsupervised Competitive Neural Network by Local Competitive Signals, Proceedings of International Conference On Artificial Neural Networks, (2002) 963-968.
9. Kohonen T.: Self-organizing maps, 3rd ed., Ed. Springer-Verlag, 2001.
10. Gustafson, D.E., Kessel, W.C.: Fuzzy clustering with a fuzzy covariance matrix. Proc. IEEE CDC (1979) 761-766.
11. Xie, X. L., Beni, G.: A Validity Measure for Fuzzy Clustering. IEEE, Transaction on Pattern Analysis and Machine Intelligence, **13** (8) (1991) 841-847.

A Continuous Technique for the Weighted Low-Rank Approximation Problem

Nicoletta Del Buono[1] and Tiziano Politi[2]

[1] Dipartimento di Matematica, Università degli Studi di Bari, Via E. Orabona 4, I-70125 Bari, Italy. delbuono@dm.uniba.it
[2] Dipartimento di Matematica, Politecnico di Bari, Via Amendola 126/B, I-70126 Bari, Italy. politi@poliba.it

Abstract. This paper concerns with the problem of approximating a target matrix with a matrix of lower rank with respect to a weighted norm. Weighted norms can arise in several situations: when some of the entries of the matrix are not observed or need not to be treated equally. A gradient flow approach for solving weighted low rank approximation problems is provided. This approach allows the treatment of both real and complex matrices and exploits some important features of the approximation matrix that optimization techniques do not use. Finally, some numerical examples are provided.

1 Introduction

Low rank matrix approximation with respect to the Frobenius and the Euclidean norm can be easily solved using the Singular Value Decomposition (SVD). The SVD decomposition has become a successful tool for numerous scientific and engineering applications, such as data compression, automatic control, image processing, digital signal reduction techniques, filter design, perceptual systems ([4,6,7,13]). An important property of the SVD of a matrix $A \in \mathbb{R}^{n \times m}$ is that it offers a series of optimal low rank approximation of A in both Euclidean and Frobenius norm sense. In fact, it is well known that if

$$A = USV^\top = \sum_{i=1}^{r} \sigma_i \mathbf{u}_i \mathbf{v}_i^\top$$

is the SVD of the rank r ($r \leq \min\{m, n\}$) matrix A, being $U \in \mathbb{R}^{n \times r}$, $V \in \mathbb{R}^{m \times r}$ and $S \in \mathbb{R}^{r \times r}$ a diagonal matrix with nonnegative entries $\sigma_1 \geq \sigma_2 \geq \cdots \geq \sigma_r > 0$. Then for any k between 1 and r:

$$\min_{\mathrm{rank}(X)=k} \|A - X\| = \|A - A_k\|$$

where A_k is the k−th truncated SVD of A:

$$A_k = \sum_{i=1}^{k} \sigma_i \mathbf{u}_i \mathbf{v}_i^\top$$

and $\|\cdot\|$ denotes the Euclidean or the Frobenius norm.

Despite of the importance that the SVD tool has in the low rank approximation problem, it presents the drawback of treating all the entries of the target matrix A equally. In fact, many applications require to discriminate between portions of a matrix and to locate the important and unimportant ones ([12]). In other cases, entries in the target matrix represents aggregates of many samples so that while the unweighted low rank approximation considers a uniform number of samples for each entry, the incorporation of weights allows to account for varying numbers of samples ([13]). Additionally, when some of the entries in the target A are not observed a zero/one weighted norm has to be introduced in order to tackle the approximation problem ([11,9]). All the previous examples highlight the necessity of finding a weighted low rank approximation of a target matrix.

More precisely, given a target matrix $A \in \mathbb{R}^{n \times m}$, a corresponding non negative weight matrix $W \in \mathbb{R}_+^{n \times m}$ and a desired rank k, we would like to find a matrix X_k of rank k, that minimizes the weighted Frobenius distance

$$\min_{\text{rank}(X)=k} \|W \circ (A - X)\|_F = \|W \circ (A - X_k)\|_F = \left(\sum_{i=k+1}^{n} \sigma_i^2 \right)^{1/2} \quad (1)$$

where $W \circ B$ denotes the entrywise multiplication of W with B (known also as the Hadamard matrix product). The matrix X_k in (1) is called weighted rank k approximation of A.

The weighted low rank approximation (WLRA) problem appeared early in the context of factor analysis ([14]), but its numerical treatment has been only considered recently. Numerical techniques for the WLRA problem derived from the design of two dimensional digital filters have been considered in [6,7,10]. In [10] the WLRA problem is treated as a numerical minimization problem so that alternating optimization techniques can be used to find a solution. The approach developed in [6] reviewed the WLRA as an extension of the conventional singular value decomposition of a complex matrix to include nontrivial weighting matrix in the approximation error measure to be minimize. The rank k weighted approximation is then obtained computing iteratively k generalized Schmidt pairs via the solution of nonlinear equations. Most recently in [11] and [12] the general WLRA problem has been viewed as a maximum likelihood problem with missing values and an efficient expectation-maximization (EM) procedure has been provided for its numerical solution.

In this paper we suggest a projected gradient flow approach to tackle the WLRA problem. Working with ordinary differential systems derived from the minimization problem (1) allows us to preserve the original dimension of the problem (which could be eventually increased if an optimization tool is directly applied) and furthermore, permits to deal with both the real and the complex case and to exploit some important features of the approximation matrix that optimization techniques do not take advantage of.

This paper is organized as follows. In the next section, we show how to derive the projected gradient flow system to be solved. Then we applied our procedure to a specific problem and finally, in the last section, we report some numerical test showing the behavior of our approach.

2 The Continuous Approach

This section is devoted to the derivation of the projected gradient flow whose limiting solutions solve the WLRA problem (1) in interest. Introducing the Frobenius inner product of two matrices M and N in $\mathbb{R}^{n \times m}$, defined as

$$\langle M, N \rangle = \text{trace}(MN^T) = \sum_{i,j} m_{ij} n_{ij} \qquad (2)$$

the WLRA problem is equivalent to

$$\min_{\text{rank}(X)=k} \|W \circ (A - X)\|_F^2 = \min_{\text{rank}(X)=k} \langle W \circ (A - X), W \circ (A - X) \rangle.$$

In order to ensure the low rank condition, we represent X via the parameters (U, S, V) where

$$X = USV^\top \qquad (3)$$

is the singular value decomposition of X. The task now is equivalent to minimizing the functional

$$F(U, S, V) := \langle W \circ (A - USV^\top), W \circ (A - USV^\top) \rangle, \qquad (4)$$

subject to the conditions that $U \in St(n, k)$, $S \in \mathbb{R}^k$, and $V \in St(m, k)$, where $St(n, k)$ and $St(m, k)$ denote the Stiefel manifolds, e.g. the groups of all real matrices of dimension $n \times k$ and $m \times k$, respectively, with orthonormal columns. For compatibility, we shall use the same notation $S \in \mathbb{R}^k$ for the diagonal matrix in $\mathbb{R}^{n \times m}$ whose first k diagonal entries are those of S. Observe that by limiting S to $\mathbb{R}^k$, the rank of the matrix $X = USV^\top \in \mathbb{R}^{n \times m}$ is guaranteed to be at most k, and exactly k if S contains no zero entry. The Fréchet derivative of F at (U, S, V) acting on any $(H, D, K) \in \mathbb{R}^{n \times k} \times \mathbb{R}^k \times \mathbb{R}^{m \times k}$ can be considered as

$$F'(U, S, V).(H, D, K) = \frac{\partial F}{\partial U}.H + \frac{\partial F}{\partial S}.D + \frac{\partial F}{\partial V}.K, \qquad (5)$$

where $\Lambda.\eta$ denotes the result of the action by the linear operator Λ on η. We now calculate each action in (5) as follows. First

$$\frac{\partial F}{\partial V}.K = \langle -W \circ (USK^\top), W \circ (A - USV^\top) \rangle =$$

$$= \langle -W \circ (W \circ (A - USV^\top)), USK^\top \rangle =$$

$$= \langle -SU^\top (W \circ (W \circ (A - USV^\top))), K^\top \rangle =$$

$$= \langle -(W^\top \circ (W^\top \circ (A^\top - VSU^\top)))US, K \rangle.$$

Using the property that

$$\langle MN, P \rangle = \langle M, PN^\top \rangle = \langle N, M^\top P \rangle$$

for any matrices M, N and P of compatible sizes, it follows from the Riesz representation theorem that with respect to the Frobenius inner product the partial gradient can be represented as

$$\frac{\partial F}{\partial V} = -(\widehat{W}^\top \circ (A^\top - VSU^\top))US, \qquad (6)$$

where we have put

$$\widehat{W} = W \circ W.$$

Observe next that

$$\frac{\partial F}{\partial S} \cdot D = \langle -W \circ (UDV^\top), W \circ (A - USV^\top) \rangle =$$

$$= \langle -W \circ (W \circ (A - USV^\top)), UDV^\top \rangle =$$

$$= \langle -U^\top(\widehat{W} \circ (A - USV^\top))V, D \rangle.$$

We thus conclude that

$$\frac{\partial F}{\partial S} = -\text{diag}_k \left(U^\top(\widehat{W} \circ (A - USV^\top))V \right), \qquad (7)$$

where, for clarity, we use diag_k to denote the first k diagonal elements and to emphasize that $\frac{\partial F}{\partial S}$ is a vector in $\mathbb{R}^k$. Finally,

$$\frac{\partial F}{\partial U} \cdot H = \langle -W \circ (HSV^\top), W \circ (A - USV^\top) \rangle =$$

$$= \langle -W \circ (W \circ (A - USV^\top)), HSV^\top \rangle =$$

$$= \langle -(\widehat{W} \circ (A - USV^\top))VS, H \rangle.$$

and hence

$$\frac{\partial F}{\partial U} = -(\widehat{W} \circ (A - USV^\top))VS. \qquad (8)$$

Formulas (6), (7), and (8) constitute the gradient of the objection function F, that is,

$$\nabla F(U, S, V) = \left\langle \frac{\partial F}{\partial U}, \frac{\partial F}{\partial S}, \frac{\partial F}{\partial V} \right\rangle$$

in the general ambient space $\mathbb{R}^{n \times k} \times \mathbb{R}^k \times \mathbb{R}^{m \times k}$.

Now it remains to compute the projected gradient. Embedding $St(n, k)$ in the Euclidean space $\mathbb{R}^{n \times k}$ equipped with the Frobenius inner product, it is easy to see that any vector H in the tangent space $T_Q St(n, k)$ is of the form (see [2,3])

$$H = QM + (I_n - QQ^\top)Z$$

where $M \in \mathbb{R}^{k \times k}$, and $Z \in \mathbb{R}^{n \times k}$ are arbitrary, and M is skew-symmetric. The space $\mathbb{R}^{n \times k}$ can be rewritten as the direct sum of three mutually perpendicular subspaces:
$$\mathbb{R}^{n \times k} = Q\mathcal{S}(n) \oplus \mathcal{N}(Q^\top) \oplus Q\mathcal{S}(n)^\perp$$
where $\mathcal{S}(n)$ is the subspace of $k \times k$ symmetric matrices, $\mathcal{S}(n)^\perp$ is the subspace of $k \times k$ skew-symmetric matrices, and
$$\mathcal{N}(Q^\top) = \{X \in \mathbb{R}^{n \times k} | Q^\top X = 0\}.$$
Hence any real $n \times k$ matrix B can be uniquely splitted as
$$B = Q\frac{Q^\top B + B^\top Q}{2} + (I_n - QQ^\top)B + Q\frac{Q^\top B - B^\top Q}{2}.$$
Hence it follows that the projection $\mathcal{P}_{St(n,k)}(B)$ of $B \in \mathbb{R}^{n \times k}$ onto the tangent space $\mathcal{T}_Q St(n,k)$ is given by
$$\mathcal{P}_{St(n,k)}(B) = Q\frac{Q^\top B - B^\top Q}{2} + (I_n - QQ^\top)B. \tag{9}$$
Replacing B by $\frac{\partial F}{\partial U}$ in (9) and repeating the same computations for $\frac{\partial F}{\partial V}$ in $St(m, k)$ we obtain the dynamical system:
$$\frac{dU}{dt} = -\mathcal{P}_{St(n,k)}\left(\frac{\partial F}{\partial U}\right) \tag{10}$$
$$\frac{dS}{dt} = -\frac{\partial F}{\partial S} \tag{11}$$
$$\frac{dV}{dt} = -\mathcal{P}_{St(m,k)}\left(\frac{\partial F}{\partial V}\right). \tag{12}$$

3 WLRA with Missing Data

In this section we sketch a further application of the proposed approach. Consider the case where only some of the elements of a target $m \times n$ matrix A are observed while others are missing and a best rank k approximation of A is needed ([9]). To tackle this problem, we can define the nonempty subset $\emptyset \neq N \subset \{(i,j) \in \mathbb{N} \times \mathbb{N} | 1 \leq i \leq m, 1 \leq j \leq n\}$ of indexes such that for any $(i,j) \in N$, the corresponding elements a_{ij} of the target matrix A is known. Defining the pseudonorm:
$$\|A\|_* = \sum_{(i,j) \in N} a_{ij}^2. \tag{13}$$
the problem drawn above is equivalent to find a rank k matrix X_k such that
$$\|A - X_k\|_* = \min_{\text{rank}(X)=k} \|A - X\|_*. \tag{14}$$

If a weight matrix $W \in \mathbb{R}^{m \times n}$ with only zero/one entries is used to weight with ones the known elements of A and with zeros the unknowns, that is:

$$w_{ij} = \begin{cases} 1 & (i,j) \in N \\ 0 & (i,j) \notin N \end{cases} \tag{15}$$

then (13) can be rewritten as

$$\|A\|_* = \|W \circ A\|_F$$

and (14) corresponds to

$$\|A - X_k\|_* = \|W \circ (A - X_k)\|_F = \min_{\text{rank}(X)=k} \|W \circ (A - X)\|_F.$$

Now to solve (14) one can use the projected gradient flow derived in the previous section considering that for the weight matrix W, defined by (15), it results that

$$\widehat{W} = W \circ W = W.$$

4 Numerical Tests

In this section we show a numerical example obtained applying the projected gradient flow approach to the WLRA problem. Here we have considered the rank 5 approximation of a square random matrix A of order 8 and a weight matrix W sketched in Figure 1. The differential systems (10)–(12) have been solved numerically in the interval $[0, 20]$. In Figure 2 we display the behaviour of the objective function (4), while in Figure 3 we reported the dynamics of the singular values of the numerical solution X obtained solving the system (11). Matrices U_n and V_n are the numerical approximations of the solutions $U(t)$ and $V(t)$, computed at the instant $t = t_n$, firstly integrating the differential systems (10) and (12) with the MatLab ode routine ode113 and then projecting the numerical solutions on the Stiefel manifold. This projection has been computed taking the orthogonal factor of the Gram-Schmidt decomposition. The initial conditions for (10) and (12) are both random matrices in the Stiefel manifold. Figures 4 and 5 illustrate how the use of the projection of the numerical solution on the manifolds is necessary. In these pictures solid lines denote the manifold errors of the numerical solutions for $U(t)$ and $V(t)$ using the projection, while dashed lines denote the errors given by the MatLab integrator. The solution is computed by routine ode113 with a relative tolerance set to 10^{-6}, but it departs from the manifold very soon. Figure 6 plots the difference between the values of the objective function computed without the projection and those computed using the projection. We observe that the difference is almost 10^{-5}, hence although the projection is necessary to obtain numerical approximations for $U(t)$ and $V(t)$ belonging to the Stiefel manifold to machine precision it causes only a negligible improvement in the value of the objective function.

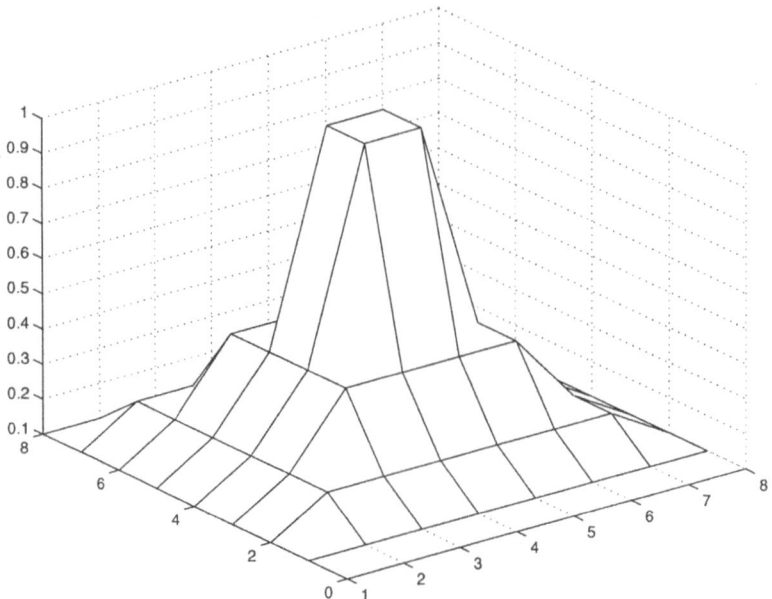

Fig. 1. The weight matrix W.

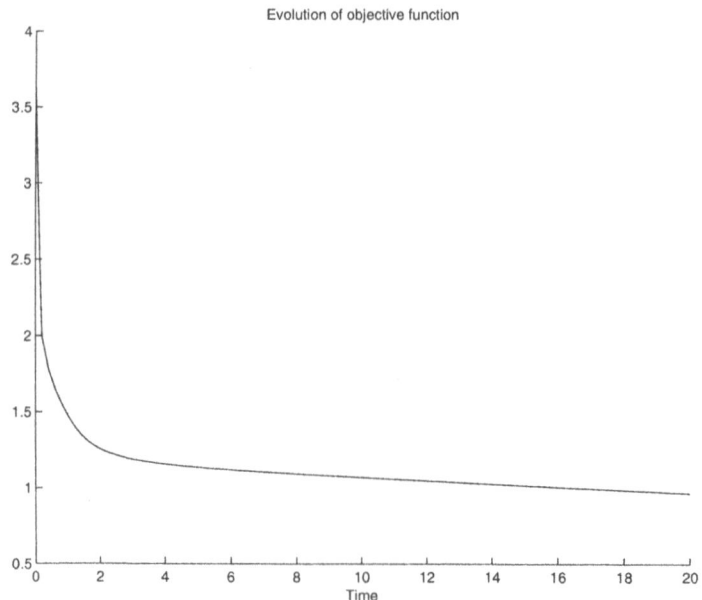

Fig. 2. Evolution of the objective function.

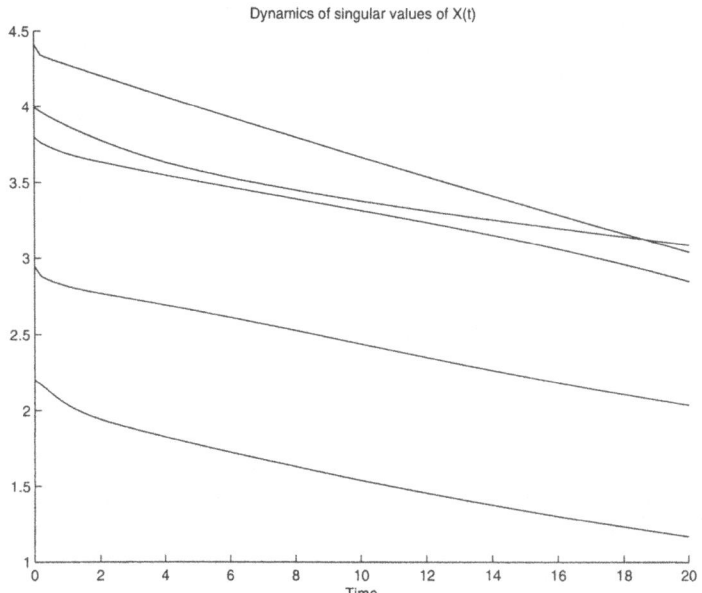

Fig. 3. Dynamics of the Singular Values of $X(t)$.

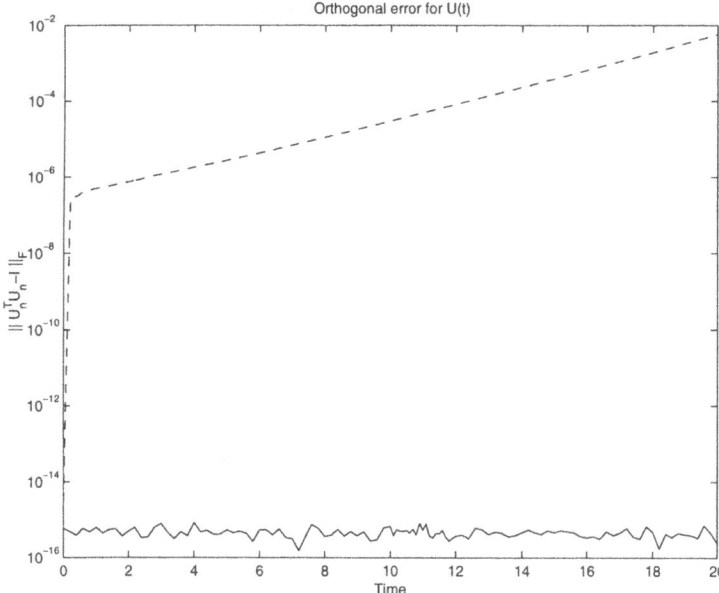

Fig. 4. Orthogonal error for the numerical solution of $U(t)$.

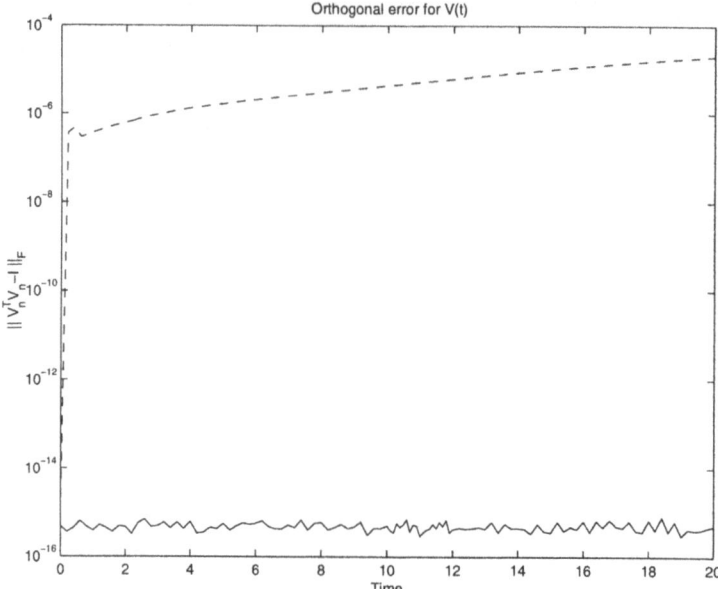

Fig. 5. Orthogonal error for the numerical solution of $V(t)$.

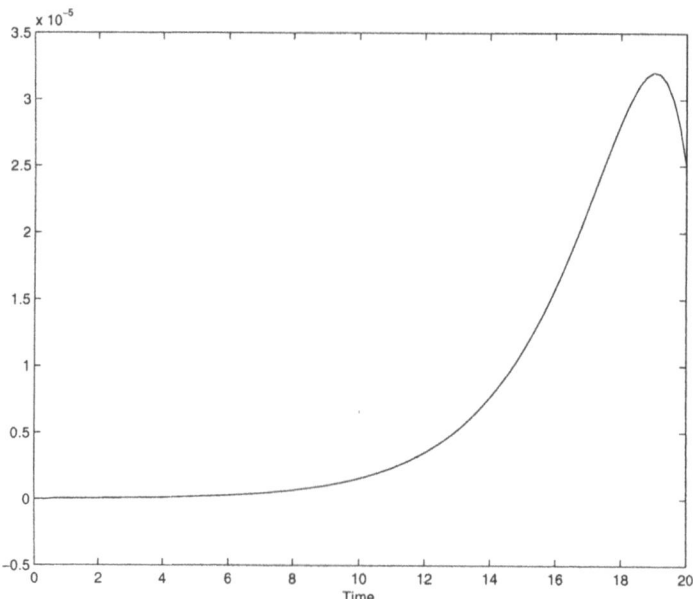

Fig. 6. Difference between the values of the objective functions.

References

1. Chu, M.T., Del Buono, N., Lopez, L., Politi, T.: On the Low Rank Approximation of Data on the Unit Sphere. Submitted for publication
2. Del Buono, N., Lopez, L.: Runge-Kutta Type Methods Based on Geodesics for Systems of ODEs on the Stiefel Manifold. BIT **41** (5) (2001) 912–923
3. Edelman, A., Arias, T.A., Smith, S.T.: The Geometry of Algorithms with Orthogonality Constraints. SIAM J. Matr. Anal. Appl. **20** (1998) 303–353
4. Klema, V.C., Laub, A.J.: The singular value decomposition: its computation and some applications. IEEE Trans. Automat. Contr, **25** (1980) 164–176
5. Lu, W.S.: Is the SVD-Based Low-Rank Approximation Optimal? Proc. 38th Midwest Symp. CAS, Rio, Brazil, (1995) 470–473
6. Lu, W.S., Pei, S.C., Wang, P.H.: Weighted Low-Rank Approximation of General Complex Matrices and its Application in the Design of 2-D Digital Filters. IEEE Trans. Circuits Syst. I **44** (7) (1997) 650–655
7. Lu, W.S., Pei, S.C.: On Optimal Low-Rank Approximation of Multidimensional Discrete Signals. IEEE Trans. Circuits Syst. II **45** (3) (1998) 417–422
8. Lu, W.S., Antoniou, A.: New Method for Weighted Low-Rank Approximation of Complex-Valued Matrices and its Application for the Design of 2-D Digital Filters. ISCAS Vol. III (2003) 694–697
9. Simonsson, L., Eldén, L.: Computing a Partial SVD of a Matrix with Missing Data. Numerical Linear Algebra and its Applications: XXI International School and Workshop, Porto Giardino, Monopoli, Italy, 2003.
10. Shpak, D.: A weighted-least-squares matrix decomposition method with applications to the design of two-dimensional digital filters. IEEE Thirty Third Midwest Symposium on Circuits and Systems (1990).
11. Srebro, N., Jaakkola, T.: Generalized Low-Rank Approximations. AI Memo 2003-001, MIT Artificial Intelligence Laboratory (2003).
12. Srebro, N., Jaakkola, T.: Weighted Low-Rank Approximations. Proceedings of Twentieth International Conference on Machine Learning (ICML-2003), Washington DC (2003).
13. Tenenbaum, J.B., Freeman, W.T.: Separating style and content with bilinear models. Neural Computation **12** (2000) 1247-1283.
14. Young, G.: Maximum likelihood estimation and factor anlysis. Psychometrika **6** (1940) 49–53.

A Spatial Multivariate Approach to the Analysis of Accessibility to Health Care Facilities in Canada

Stefania Bertazzon

Department of Geography, University of Calgary,
2500 University Dr. NW, T2N 1N4Calgary, AB, Canada,
bertazzs@ucalgary.ca

Abstract. The paper discusses an analytical framework for the specification of a multivariate spatial model of the accessibility of health care facilities to the Canadian population. The model draws from simple economic theory to define an optimum level of service provision, based on the demand-supply equilibrium. The focus is on the demand side model: the demand is represented by the population requiring the service, and since the population is spatially distributed over the study region, accessibility of the service becomes the key variable. A set of modules forms the entire model, the main feature of which are the determination of the potential origin region and the specification of an efficient spatial model that explicitly incorporates local factors.

1 Introduction

Accessibility can be described as a capacity measure; with reference to a health care facility, it can be more specifically defined as the number of persons that can theoretically access that service, or in other words, that are capable of traveling to that facility within a predefined time limit. Like in many other cases (e.g. carrying capacity) this measure represents a threshold, characteristic of the facility in question, that must be related to the capacity of that facility of providing health services. Applying simple economic theory, accessibility can be represented as a demand function, and the facility capacity as the supply function. In classical economic theory, the demand-supply optimum is ruled by the price variable; in this model it is represented by the number of persons that can travel to the facility in an adequate time and receive adequate care. The most important aspect, in a spatial perspective, is the analysis of the demand: the optimum problem can be re-stated in the following terms: given the location of the facility, what is the number of persons that have access to it? The focus of this paper is on the measurement of accessibility, and the discussion of the most appropriate spatial models to achieve this goal.

1.1 Background

The importance of accessibility to health care facilities is topical in the Canadian context. The specifically Canadian problem is a sparsely distributed population over a very large geographic area: this spatial characteristic of the population distribution induces specific needs particularly for the rural residents and the native population living in reserves. The counterpart of this sparse population distribution is a relatively small number of health care facilities, typically clustered in major urban areas. As the Canadian population, like the population of most industrialized countries, ages, the need for such services becomes increasingly acute. The problem of supplying an aging population with adequate health care services is not uncommon in western societies, and adequate policy measures are typically taken by governments. The specifically Canadian problem is the spatial distribution of its population: providing an adequate level of service (supply) may not be an adequate policy response in itself, unless those services are effectively accessible to all the interested population. Hence, the problem of an optimum service supply assumes a specifically spatial dimension that should be addressed by explicitly spatial models: the first goal of such models is to measure, or quantify, accessibility itself.

The interest of geographers in the concept of accessibility dates back to the time of the quantitative revolution in geography (Hagerstrand, 1967; 1970) and significant work appears in the context of spatial interaction models (Haynes and Fotheringham, 1984; Wilson and Bennett, 1985; Lowe and Moryadis, 1975). More recently, particularly in the context of GIS, the interest in accessibility evolves beyond the need for measurement to practical ways of including it in GIS models: to this end several analytical tools have been explored, including map algebras and geo-algebras, local statistics and fuzzy set theory (Wu and Miller, 2001; Hanson, 1998; Janelle, 1995). One of the essential issues in the traditional and current debate is the operational difference between access and accessibility, i.e. between potential and realized contact (Harris, 2001). In this paper we make use of such difference, as access and accessibility take up very different roles in the spatial models discussed.

Access will be defined here as the number of persons who do receive services at the facility within a specified time unit; accessibility will be the number of persons who can access the facility within a pre-specified travel time. Accessibility is thus a theoretical capacity measure that does not relate directly to the time unit of fruition of the services: it does provide a hypothetical number of persons who are physically able to reach the facility within the predefined travel time. Based on this definition, accessibility can be more formally defined as a function of several variables. The main variable is distance from the health care facility, which defines a geographic area, inhabited by a number of residents. The definition of the geographic area is based on travel time, which, in turn, depends on the transportation mode, and the characteristics of the available route network: the mode of transportation determines the travel speed, hence, given a pre-defined travel time, can define a threshold distance between the heath care facility and the place of residence. Consideration of the route network (typically road network) provides a more accurate definition of such threshold, including, for example, main route axis and physical barriers; additional factors affecting travel speed depend on the time when travel occurs: rush hour vs. non-rush hour; weekday vs. weekend; day vs. night; and season, i.e. winter vs. non-winter driving conditions.

The second component of the definition of accessibility deals with the number of residents of the region within the threshold, and with their likelihood of needing to access the facility. Based on the number of residents, defined by the travel-time threshold, such likelihood can be approximated by a probability, assigned on the basis of demographic and socio-economic factors. Defining such likelihood in probabilistic terms is a common method in spatial epidemiology (Elliott et al., 2000): we propose that relevant individual characteristics be weighted by the known incidence of a specific health condition associated with that characteristic in the given geographic area and at the time of the study. Individual characteristics include demographic attribute, i.e. gender and age; complementary socio-economic variables, such as level of income and education; ethnic origin, where it is relevant to the condition in question and, where these are available, additional health information, such as drinking and smoking habits and the presence of other related conditions. In general, all the above attributes are not available for analysis at the individual level, but reasonably detailed spatial databases can be accessed: these include data at the postal code level for health databases, and at the census dissemination level for census (demographic and socio-economic) variables. The definition of such likelihood function depends heavily upon the type of health condition and health care facility under consideration: we will refer to one particular condition as an example, but the framework proposed can be adapted to other conditions likewise. Assigning a probability value to the resident population provides a measure of the number of individuals who may effectively access the health care facility within the given time unit, thus providing an indicator of access: this can be viewed as a final step of the model, relating the theoretical accessibility function to an estimated access value, that can be used in a predictive model, an important tool for analysis and simulation in the provision of accessible health care. This final step will be achieved in the demand-supply framework.

The two fundamental aspects of the accessibility function, i.e. distance threshold and individual probability, can be formally represented by multivariate models: the first one is clearly configured as an explicitly spatial model, the second one deals with a series of essentially aspatial factors, which are, however, attributes of a spatially distributed population: including such factors in the spatial model will provide a more accurate representation of the population.

2 Methods

2.1 Case Study and Study Area

The definition of a probabilistic model on the population of a specific geographic region requires a number of attribute data and assumptions that depend on the specific health condition being analyzed. In the current application we will consider cardiac catheterization, a procedure performed in after the occurrence of myocardial infarction; for this application we refer to the APPROACH database, covering the 1995-2002 period for the city of Calgary (Alberta, Canada).

The spatial resolution of the database is the postal code, roughly coinciding with a city block (Ghali and Knudtson, 2000). The database is completed by census variables for the demographic and socio-economic variables: the 1996 - 2001 censuses are considered, at the finest spatial resolution available, i.e. census enumeration areas (1996 census) and census dissemination areas (2001 census). The APPROACH database also contains detailed information about the day and time when the procedure was performed. The latter information is useful to refine the travel time model. Fig. 1 represents the density of cardiac catheterization cases and the ratio of catheterization cases over the population. The database can be considered a set of panel data, representing the basis for spatio-temporal analysis; the spatial resolution is high, the temporal database is a rather narrow basis for time series analysis, particularly for multivariate temporal analysis; the latter is aggravated by the lack of contemporary census data, as the time interval is covered by only two census.

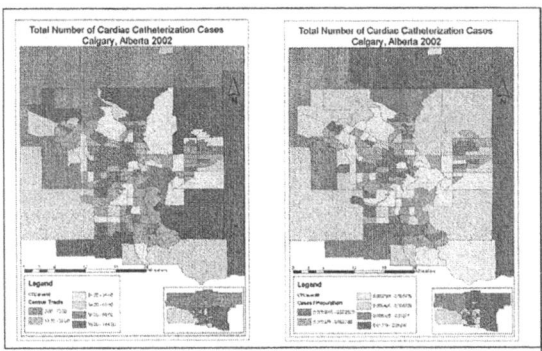

Fig. 1. 2002 Cardiac Catheterization Cases (a); over population (b).

2.2 The Demand – Supply Framework

The simplest analytical framework for the analysis of the demand and supply of a commodity or service an a-spatial framework where the equilibrium, or optimum quantity is determined as a function of price. We refer to this basic model to analyze the demand (D) and supply (S) for the provision of health care services. The general form of the model is represented in Fig. 2: the quantity of service provided is a function of a generic price[1].

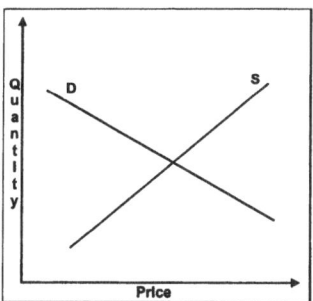

Fig. 2. The demand-supply model for health service provision

The health care service (S) is provided by a specific, localized facility, whereas the demand (D) is represented by the population residing within a threshold distance from the facility and requiring that service. The problem of optimum service provision can thus be analyzed using the tools of simple economic theory, representing the

---

[1] Note that unlike the traditional economic approach of representing the dependent variable on the x axis, in Fig. 4 and all of the following discussion, we revert to the general approach of representing the independent variable on the x axis and the dependent variable on the y axis.

population requiring health services as the demand variable, and the health care facilities, providing such services, as the supply variable. The problem can be put in a simple scheme, analyzing a single facility and its pertinent population, or, more interestingly, as a system, where all the facilities providing a specific service can be jointly analyzed with reference to their pertinent population, thus providing an effective model, or planning tool, for an entire city, or province.

The advantage of this approach is that the generic price variable can be considered as a transportation price, or the price of reaching the facility, which allows for the specification of the spatial models for the demand. At present, the supply curve is represented by a generic positive function of price, which might be subsequently refined to include more appropriate models of supply. This paper focuses on a demand-side model, and specifically on a spatial model for the estimation of the demand for health care services as a function of accessibility of that facility to the resident population. The demand function, D, is thus a function of the threshold distance, based on travel time, and of the resident population, weighed by the probability that each individual resident requires the health care service in question.

The variable accessibility is initially represented by the variable access, which can be thought of as the number of cardiac catheterization cases for each temporal unit at the facility over the entire study period (e.g. number of cardiac catheterization cases per week at the Foothills Hospital in Calgary between 1995 and 2002). The variable access is thus one particular value of the accessibility function. We can formally specify the accessibility function as a multivariate function (A), and access as the value (a) of that function, a (A). Conceptually, the variable access can be viewed as a set of realizations of the random variable accessibility, and the two variables can be related in using a maximum likelihood (ML) approach: the assumption that the observed values (access) have the highest probability of occurrence among all the possible outcomes of the random variable accessibility, allows to derive a set of functional parameters, based on the probability density of A, linking the potential with the realized variable.

$A = f(v_1, v_2, ..., v_k)$

Equation 1

$\mathbf{a} = a_{ML}(A)$

Equation 2

Equation 1 expresses a generic multivariate function *Accessibility*, and Equation 2 expresses the variable access as a function of Maximum Likelihood parameters of Accessibility.

$[a_{ML}(A)_1, a_{ML}(A)_2, ..., a_{ML}(A)_n]$

Equation 3

This analytical framework allows for the specification of spatial, temporal, and spatio-temporal a regression models, based on the observed values of the accessibility function. The entire set of access values, Equation 3, represents a vector of observed values of the variable, which can be regressed on a set of independent (explanatory) variables. The simple Demand curve in Fig. 2 can now be expressed by Equation 1, where the generic variables and functional form will be replaced by specific factors and appropriate functional form for the multivariate spatial model. The model follows a stepwise procedure, which is detailed in the following paragraphs; the core of the model is spatial. Some temporal analyses, as well as some a-spatial analyses, are necessary steps for the completion of the model.

2.3 Distance/Travel Time Model

At the core of the multivariate accessibility function is the hypothesis that distance from the health care facility is the crucial explanatory variable: based on a predefined threshold travel time, such distance defines the boundaries of the potential origin region, and consequently not just the parameters, but even the spatial sample of all the other explanatory variables.

The definition of an appropriate distance function includes the definition of an appropriate metric that takes into consideration the characteristics of the road network, the transportation means, contains a model of travel time. The latter two components – distance and travel time- are strongly related, as a travel time measurement is necessarily based on some measurement of distance. Bertazzon (2003) and Gavrilova and Bertazzon (2003) have proposed the comparison of alternative distance metrics to enhance the calculation of spatial autocorrelation. Extensions of that model (manuscript currently in preparation) include the definition of different metrics within a single study region to better approximate distance on a network of variable regularity and size. The proposed method encompasses the definition of a flexible grid based on the same principle, aimed at addressing the spatial heterogeneity problem arising from the variability of shape and size of the spatial units used by census Canada.

Proposed for the study of the first and second order properties of spatial patterns, the variable grid method presents important advantages for the estimation of the accessibility function, mainly because the explanatory variables of the A function (demographic, socio-economic, and medical) are sampled at the traditional spatial units (census units for the demographic and socio-economic variables) and released at the postal code level (medical variables). The distance metric and varying grid model can also be used as a model of distance along the network – thus replacing empirical measurements along the network itself. The advantage of the use of such model is that it can be refined to include weights that represent the inflation of travel time at specific times, e.g. during rush hours or during particular seasonal or other place-specific conditions: snow storms and harsh winter driving conditions are common events in Calgary, but heavy fogs, seasonal rains and floods, or important local events, e.g. trade-shows, local fares, religious or popular holidays or tourist season are but a few examples. The degree of variability of such conditions is specific to the element considered: for example, the inflation of travel time due to rush hours in week days is a more deterministic process than assigning a probability of heavy snow storm to a winter day and the calculation of the consequent inflation of travel time and duration of the winter driving conditions. The provision of inflating weights applies to both a model of distance, such as the one proposed by Bertazzon and Gavrilova, and an alternative model based on empirical calculations of travel time on a road network – i.e. distance by speed limit. In both cases, the whole distance and travel time model is used on the side of the main model, in order to provide the spatial borders within which the population variables should be considered.

2.4 Multivariate Specification and Functional Form

Based on the distance/travel time model, the spatial extent of the potential origin region is defined, and the variables sampled and weighed accordingly. For each specific health condition a different functional form will be the most appropriate. Our proposal is to consider the population residing in the potential origin region as the main explanatory variable (the variable which will determine the total number of patients who will use the health care service). Such variable can be weighed directly, or multiplied by a number of other factors, to express the probability that any individual in that population will need the health care service within the hypothesized time unit. The proposed model contains a set of specific attributes of the population, weighed by the known incidence of the condition on that population segment or a model of its probability. The model thus parts from a simple linear specification and is likely to contain some multiplicative effect on the base variable. The functional form is not completely explicit at this point: the multivariate specification on the explanatory variable is conditional to the distance/travel time model. The entire set of explanatory variables of the accessibility function, consists thus of three separate groups of factors:

- **Distance/Location**: Travel time; transportation mode; road network; time of travel (day, hour, season).
- **Demographic and Socio-Economic Factors**: Gender, age; income, education.
- **Health Factors**: Smoking and drinking habits; ethnic origin; co-presence of other health conditions.

The first group, *distance* or *location* variables, does not enter directly in the multivariate specification of the Accessibility function, but represents the crucial set of variables of the distance/travel time model. In a comprehensive specification of the model, this part of the function can be represented as a conditioning factor, because it determines the spatial extent of the potential origin region. In this perspective, the individual probability of requiring the health care service can be view as a conditional probability, i.e. a probability conditional to residing within the threshold distance from the health care facility, as determined by the distance/travel time model.

The next two groups of variables contain demographic and socio-economic (census) variables, and medical variables, respectively. None of these groups of variables is intrinsically spatial, but they are attributes of variables sampled at because they are sampled at specific spatial units. The leading variable in the first part of the equation is the population value. All the other variables in this model are weighed by a weighing factor which assigns the probability of occurrence of the disease to each category in the variable. In order to perform this step, the explanatory variables must be categorized so that a specific weight can be assigned at each category. In the case of cardiac infarction, gender and age are relevant factors. The variable gender will be assigned the incidence on male and on female, respectively.

2.5 Spatial Models

We propose a spatial model that expands on the traditional geographical analysis techniques, to provide an effective management and planning tool for the provision of adequate health care access to the Canadian population. The accessibility function comprises a locational component, i.e. distance factors, and factors, that are not locational in nature, but are spatially distributed and nested in the population distribution. The main framework is a multivariate spatially autoregressive model (Anselin, 1988), where access to health care services can be modeled as a function of locational factors (i.e. distance from clinic, measured by travel time) and non-locational factors (e.g. income, etc.). Important local factors should be included, such as discrimination between local and rural populations, and explicit consideration of rural vs. urban, as well as residence in reserves. To this end, geographically weighted regression (Fotheringham et al., 2000) will be combined within the spatial autoregressive framework.

The novelty of this approach is the explicit consideration of local and specific factors by means of geographically weighted regression, while at the same time considering the effect of spatial autocorrelation, by means of spatial regression. The resulting model possesses two important advantages: it gives adequate consideration to local factors, and it is efficient (statistical efficiency enhances the reliability of the estimated parameters.)

2.6 System Models

The accessibility model discussed so far is defined in the perspective of a single facility and the accessibility function is discussed with reference to a single facility. Particularly in a planning perspective, at the provincial or national level, this model can be extended, to incorporate the entire set of facilities of the same type in the study region, each of them characterized by a specific accessibility function, whose parameters will be estimated with the same procedure of the single facility model. Even though accessibility can be analyzed independently for each facility, a joint approach, or a system model, presents definite advantages. The Distance/Travel time model would immediately identify over- or under-serviced areas, i.e. areas which have access to more than one facility, and areas that have access to none, respectively.

The problem of over-served areas is best addressed in the S-D (supply-demand) model, which can specifically address the unbalance between required and provided service, linking such unbalance to the associated cost. Further application of simple economic theory can estimate the cost associated with such inefficiency. The problem of under-serviced areas is best address by the step-wise procedure proposed in this paper: the probability model for the likelihood of requiring the service would assess the effective need of covering such areas.

The main advantage of the system approach is the joint estimation of all parameters, including a joint consideration of all the errors. The main disadvantage is the need for extensive datasets, and the associated loss of degrees of freedom.

3 Application and Discussion

The discussion of the proposed model focuses on the selected dataset, study region, and time period. The limitation of the available data, as discussed in Section 2.1, prevent us from estimating the model at this point, but the application remains interesting in itself and the case study raises some interesting points.

Module 1: Specification of a Demand-Supply model as a function of a generic price. Goal of this module is to provide a framework to determine an optimum level of health care services, based on the population's requirements. The model is applied to the provision of cardiac catheterization, and while the demand (D) remains a generic function of price, the supply (S) is modeled by a stepwise multivariate spatial model. Step 1 can be summarized as the determination of an equilibrium, or optimum, level of health care service.

Module 2: Definition of access as a proxy variable for accessibility. The spatial, temporal, and spatio-temporal dimensions of the APPROACH database provide as many samples of the variable access, a. Considering the limitations of the database, only spatial models are considered: only these models can reliably be estimated using the available data.

Module 3 defines the distance/travel time model to determine the potential region of origin to the health care facility. The variable grid method proposed by Gavrilova and Bertazzon encompasses all these issues and provides a flexible method that can be applied to other case studies.

Module 4: assignment of probability values to the individual characteristics of the resident population and explicit formulation of the functional form. We propose a multiplicative form, where each secondary variable (demographic, socio-economic, and health factors) represents a fraction of the leading variable (population). Each secondary variable must be further refined to include the incidence of the cardiac infarction on that specific segment of the population. The best mathematical formalism to represent the relationships among all the factors is one of the most crucial elements of the model, yet the lack of a theoretical guidance and the potential complexities of some specifications suggest the use of a relatively simple, but computable solution. We choose to assign the weighs as exponents to each secondary variable: in a multiplicative function, a logarithmic computation of the basic model returns a linear form that can easily be estimated. The need for a computationally simple model relates directly to the need to avoid an increased complication of the spatially autoregressive and geographically weighted models.

Module 5 is the final step and consists in the specification of a spatial structure for the model. We propose the use of a combination of geographically weighted and spatially autoregressive models. The variable grid method proposed for the distance/travel time model will be employed in this stage to compute the spatial autocorrelation structure within the spatially autoregressive model.

Applying the proposed framework to the APPROACH data is a difficult process, due to the limitations of the database. The specification of 5 independent modules and the proposition of a functional form for the multivariate model are operational results, that can further the application to this as well as other case studies.

4 Conclusion and Future Research

Goal of the paper was to provide an analytical framework for the specification of spatial analytical models of the accessibility of health care facility. The staring point of the model is a supply-demand framework. The model expands into a series of modules that specifically address the spatial issues involved in the specification of a realistic accessibility model. The focus is on the demand side, the spatial part of the system: the combination of geographically weighted and spatially autoregressive models provides an efficient estimation of the parameters, while allowing for specific consideration of local factors. The proposed model bears great potemtial for applications in planning of health care provision.

The demand side model requires some refinement, and a major topic for future research is the specification of a complete demand-side model.

Acknowledgements. This paper is part of a larger project funded by the GEOIDE network: I would like to acknowledge the GEOIDE network for supporting our project DEC#BER - Multivariate Spatial Regression in the Social Sciences: Alternative Computational Approaches for Estimating Spatial Dependence. I would also like to acknowledge the APPROACH project for providing us with data and support on our work.

References

1. Anselin L (1988) *Spatial Econometrics: Methods and Models*, New York: Kluwer Academic Publisher.
2. Bertazzon S (2003) A Definition of Contiguity for Spatial Regression Analysis in GISc: Conceptual and Computational Aspects of Spatial Dependence. *Rivista Geografica Italiana*, Vol. 2, No. CX, June 2003.
3. Elliott P Wakefield J C Best N G Briggs D J (eds) (2000) *Spatial Epidemiology. Methods and Applications*. Oxford: University Press.
4. Fotheringham A S, Brundson C, Charlton M, 2000, *Quantitative geography. Perspectives on Spatial Data Analysis*. London: Sage.
5. Gavrilova M, Bertazzon S (2003) *L_P Nearest-Neighbor Distance Computation in Spatial Pattern Analysis for Spatial Epidemiology*. AAG 2003 Annual Meeting. New Orleans, March 4-8, 2003. Abstracts, p.130-131.
6. Ghali MD, William A., and Knudtson MD, Merril L., "Overview of the Alberta Provincial Project for Outcome Assessment in Coronary Heart Disease." *Canadian Journal of Cardiology* 16, no. 10 (October 2000): 1225-1230.
7. Haynes, K A and Fotheringham, A S, 1984. *Gravity and Spatial Interaction Models*. Sage Publications, Beverly Hills, California.
8. Hagerstrand, T, 1967, Innovation Diffusion as a Spatial Process. Chicago, The University of Chicago Press.
9. Hagerstrand, T, 1970, What about People in Regional Science? Papers of the Regional Science Association, 24, 7-21.
10. Haggett P, Cliff A D, Frey A, 1977, *Locational Analysis in Human Geography*, London: Edward Arnold.

11. Hanson, S. 1998. Reconceptualizing Accessibility. *Paper presented at the Varenius meeting on Measuring and Representing Accessibility in the Information Age.* 19-22nd November 1998 (Asilomar, California).
12. Harris B, 2001, Accessibility: Concepts and Applications, *Journal of Transportation Statistics* Vol. 4, No 2/3. September/December 2001.
13. Janelle, D G, 1995 Metropolitan Expansion, Telecommuting and Transportation. In Hanson, S. *The Geography of Urban Transportation, 2nd Ed.* Guilford Press, pp. 407-434.
14. Lowe, J. C. and Moryadis, S. (1975), *The Geography of Movement.* Waveland Press, Prospect Heights, Illinois.
15. Wilson, A. G. and Bennett, R. J. (1985), *Mathematical Models in Human Geography*, John Wiley & Sons, New York.
16. Wu Y-H, Miller HJ, 2001, Computational Tools for Measuring Space-Time Accessibility within Transportation Networks with Dynamic Flow. *Journal of Transportation Statistics* Vol. 4, No 2/3. September/December 2001.

Density Analysis on Large Geographical Databases. Search for an Index of Centrality of Services at Urban Scale

Giuseppe Borruso[1] and Gabriella Schoier[2]

[1] Universitá di Trieste, Centro d'ecellenza in Telegeomatica-GeoNetLab, Universitá di Trieste, Piazzale Europa 1, 34127 Trieste, Italia,
`giuseppe.borruso@econ.units.it`
[2] Universitá di Trieste, Dipartimento di Scienze Economiche e Statistiche,Piazzale Europa 1, 34127 Trieste, Italia

Abstract. Geographical databases are available to date containing detailed and georeferenced data on population, commercial activities, business, transport and services at urban level. Such data allow examining urban phenomena at very detailed scale but also require new methods for analysis, comprehension and visualization of the spatial phenomena. In this paper a density-based method for extracting spatial information from large geographical databases is examined and first results of its application at the urban scale are presented. Kernel Density Estimation is used as a density based technique to detect clusters in spatial data distributions. GIS and spatial analytical methods are examined to detect areas of high services' supply in an urban environment. The analysis aims at identifying clusters of services in the urban environment and at verifying the correspondence between urban centres and high levels of service.

1 Managing Geographical Databases to Detect Clusters at Urban Level

Geographical databases [1] become today available containing a large amount of data referred to spatial features. Such databases allow storage, retrieval and management of geographically-related information concerning different topics and aspects of the real world. When dealing with inhabited areas geographical databases become very important to analyse settlements and urban environments as both population and other human activities, facilities, landmarks, infrastructures and services that can be located on the land and related to each other. Spatial database systems ([6]), consisting of database systems for managing spatial data, combine 'traditional', non spatial data components with

---

[1] Although the general structure of the paper reflects the common aims of the two authors, Dr Borruso is mainly responsible for paragraph 2, 3 and 4, while Prof. Schoier for paragraph 1

geographical, spatial ones. Coupling these two aspects involve a wealth of opportunities in information retrieval as well as problems in the organization and representation of knowledge from such data storage devices.

Exploratory Spatial Data Analysis (ESDA - [1]) is important in order to assist in the process of discovering patterns to identify relationships and detect clusters. Spatial clustering represents therefore the process of grouping sets of objects into clusters so that similarity within a class is maintained while dissimilarity exist between different classes ([7]). Spatial clustering constitutes also an important branch of spatial data-mining, that consists of the application of data analysis and discovery algorithms that, under acceptable computational efficiency limitations, produce a particular enumeration of patterns over the data ([4]).

In the geographical literature clusters are very important subjects of study and research. Following Tobler's 'First Law of Geography', close features are more related with each other than distant ones. The study of spatial clusters could reveal information about the underlying geographical process that generates the spatial pattern. Different methods exist to identify and measure clusters in spatial databases ([8]), corresponding to different aspects of clusters questions (i.e. whether there exist clusters, where are they, what are their characteristics) or because based on different philosophies (i.e. scale, measurement, separation). One of the main issues within clustering techniques and algorithms applied to geographical databases involves the scale of observations and distribution of the dataset. Spatial clusters at a certain scale may not appear at a different scale, as well as the distance between objects can influence the formation of clusters.

2 Density Estimation for the Analysis of Spatial Clusters

Density based techniques can be used to analyze spatial databases.

A technique known as Kernel Density Estimation (KDE) can overtake some of the drawbacks of other clustering techniques. It allows in particular to examine the overall dataset and derive information at both local and global scales. The problem of scale of observation and analysis, that is a very important factor in all spatial analyses, is therefore limited in such an application. KDE allows representing spatial phenomena, expressed as point data, as a continuous surface, that means obtaining a uniform estimate of a density distribution starting from a sample of observations. The kernel consists of 'moving three dimensional functions that weights events within its sphere of influence according to their distance from the point at which the intensity is being estimated' ([5]). The general form of a kernel estimator is

$$\hat{\lambda}(s) = \sum_{i=1}^{n} \left[\frac{1}{\tau^2} k(\frac{s-s_i}{\tau}) \right] . \qquad (1)$$

where $\hat{\lambda}(s)$ is the estimate of the intensity of the spatial point pattern measured at location s, s_i the observed i-th event, k( ) represents the kernel weighting

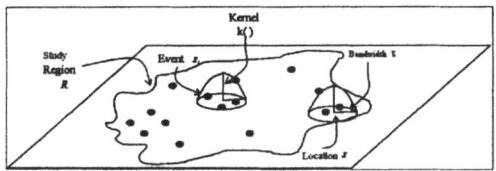

Fig. 1. Kernel Density Estimator

function and τ is the bandwidth. For two-dimensional data the estimate of the intensity is given by

$$\hat{\lambda}(s) = \sum_{d_i \leq \tau}^{n} \left[\frac{3}{\pi \tau^2} (1 - \frac{d_i^2}{\tau^2})^2 \right] . \qquad (2)$$

where d_i is the distance between the location s and the observed event point s_i. The kernel values therefore span from $\frac{3}{\pi \tau^2}$ at the location s to zero at distance τ ([2]).

It presents considerable advantages in studying point patterns if compared with other techniques, as for instance the simple observation of the point pattern and the quadrat count analysis. It allows to estimate the intensity of a point pattern and to represent it by means of a smoothed three-dimensional continuous surface on the study region. It is therefore possible to highlight the presence of clusters or regularity in the distribution (first order properties). The only arbitrary variable in the KDE is represented by the bandwidth ([5]). Different bandwidths produce different patterns. A wider bandwidth produces smoothing of the spatial variation of the phenomenon. On the contrary a narrow bandwidth highlights more peaks in the distribution (Figure 1).

Bandwidth becomes therefore the main parameter for the density distribution estimate. KDE can be however applied over homogeneous distributions of spatial phenomena (i.e. population, crimes, car accidents, etc.).

An issue refers therefore to large and complex databases containing different and articulated data. The density algorithm aimed at detecting patterns and presence of clusters over a spatial distribution should therefore consider a variable at a time within a spatial database. Variables can be weighted so that not only coordinates are considered but also a value of the spatial phenomena at that location.

In order to detect patterns in the whole database, KDE can be integrated in a GIS environment and different fields or variables of the database considered as different layers. Doing this can help detecting patterns into the database while considering more variables at a time. Standard spatial operations between different layers in a GIS environment can therefore allow representing the combined effect of the different attributes of phenomena at different locations.

3 Search for an Index of Centrality of Urban Services

The application involved analyzing a geographical database for the Municipality of Trieste containing information on the different services available. The aim was to obtain some exploratory information for an index of the services' supply at urban level. The method is based on some of the characteristics of services that can be found in urban areas. Some first indicators to derive an index of urban services are examined here. The indicators as well as the resulting index are expressed as a 3-dimensional continuous surface expressing a density function that represents areas where different services clusters within the urban environment. The main aim of the application is to obtain an image of the city as a centre (or as a set of centres) of supply of urban services for more in depth analyses on different urban phenomena. The starting point for the analysis was the distribution of services at urban level georeferenced using address points. Such detail allowed to transform services in Trieste Municipality as a spatial distribution of point data. The density analysis was performed from subset of the overall database of the urban services.

The analysis was organized in different steps: 1) organization of services; 2) density analysis on intermediate indicators; 3) standardization of results; 4) sum of standardized values, overlay and representation of synthetic index.

1) *Organization*. Services were organized according to their characteristics and weight the different services have in the urban environment. Services were therefore grouped in different categories in order to obtain homogenous intermediate indicators. In particular elements as public administration offices, transport services, leisure, emergency, banks and insurance companies, education, etc. where considered.

2) *Density analysis*. The second step involved the density analysis on subsets of the overall services' distribution to obtain the intermediate indicators. KDE was performed on the different subsets of point data using a 500 m bandwidth. Such value was chosen following other researches carried out on density distribution of urban-related phenomena ([3]). Such distance allows representing efficiently the density of urban phenomena under exam. In other researches ([9]) a 300 m bandwidth was used for urban phenomena. Only some of the services were weighted in the density analysis as only some of the intermediate indicators required it. It was experimented that the use of weights in the distribution did not affect considerably the overall shape of the 3-dimensional surface. Figures 2 and 3 shows some examples of density analysis performed on intermediate indicators. It is worth noting that with reference to the density of bank and insurance companies within the Municipality of Trieste we can highlight a 'financial cluster' in the central area of the city.

KDE produced density surfaces from the initial distributions of point data for subset of services. Such density functions are organized as grid of cells with a resolution of 100 m weighted by their density deriving from the analysis.

3) *Standardization of intermediate indicators*. Given the existing differences in data and density values, we tried to keep intermediate values homogeneously

Density Analysis on Large Geographical Databases 1013

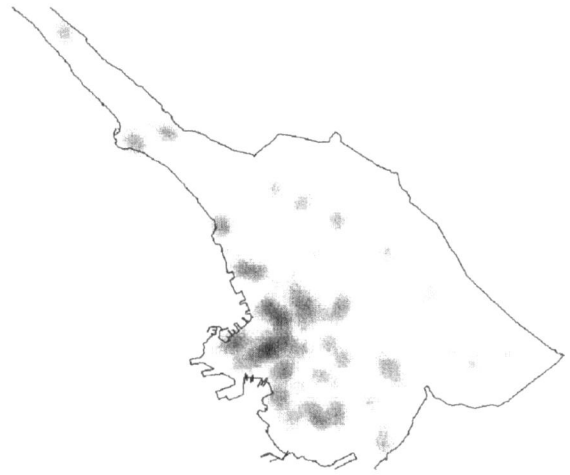

Fig. 2. KDE on intermediate indicators: Education

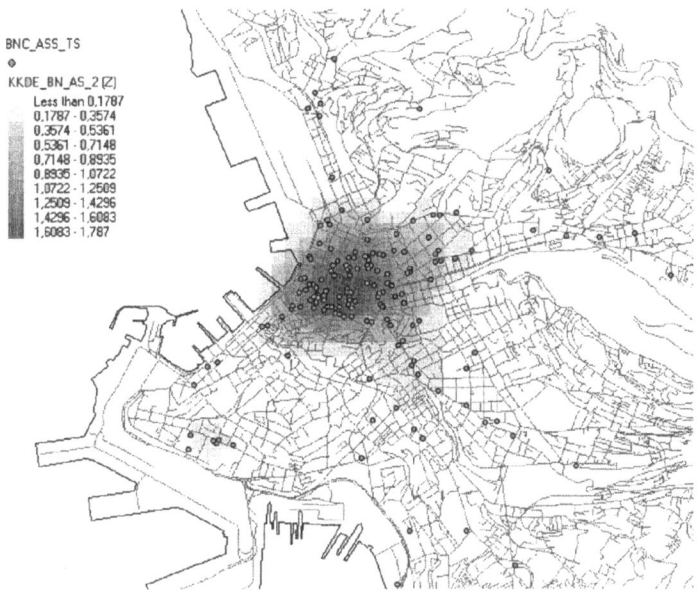

Fig. 3. KDE on intermediate indicators: Banks and Insurance Companies (Financial district)

in order to efficiently insert them into a final index. All data were normalized using the z - score function ([9]).

$$Z = \frac{X - \mu}{\sigma}.\qquad(3)$$

where X is the value to be transformed, μ the average and σ the standard deviation. Doing this allow to compare and combine the different density distributions into a single index.

4) *Overlay.* Standardized values were than combined by means of an overlay in a GIS environment, where density surfaces for the intermediate indicators were organized as layers, summing the density values of overlaying cells. In figure 4 the 2-dimensional representation of the index of services' supply is portrayed.

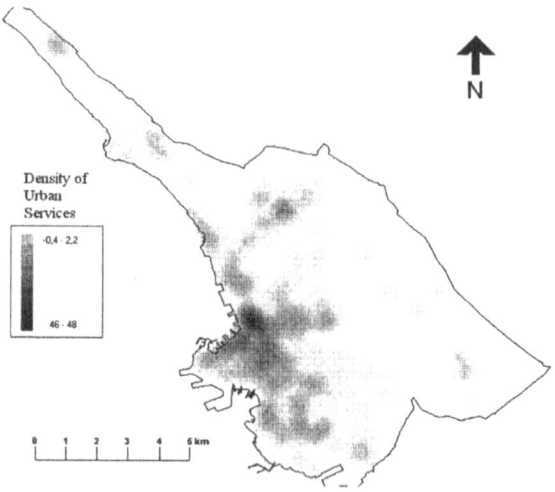

Fig. 4. Index of services' supply in the Municipality of Trieste

The index found present clusters located mainly in the central part of the city of Trieste, as could be expected. We notice a decreasing level of density of the index when moving out from the city centre but also some other 'lower order' clusters in other parts of the city, where services are present but not so numerous as in the city centre. Such peaks can be noticed both in the urban area of Trieste and in the minor settlements surrounding the city.

4 Conclusions

The research has analysed the possibility of using density estimation techniques to detect and examining spatial clusters at urban level from geographical databases. Kernel Density Estimator was used to obtain an index of centrality of services' supply in an urban environment starting from a dataset referred to spatial distribution of services. The density analysis performed allowed to obtain surfaces of density from subsets of the original dataset, each one of them referred to a particular class of services. Peaks or clusters in the point data distributions were detected representing intermediate indicators of services' supply that were eventually combined in a final index. This approach allowed both detecting clusters in the subsets and in the overall index that summarize the entire database of services. The advantages in using KDE are that it depends only on bandwidth as the principal variable, and therefore allowing examining phenomena at both global and local scale, and also in representing immediately the pattern drawn by the phenomena under exam. Drawbacks are mainly due to the limit of KDE functions performed over a variable or field in the database at a time, however this latter problem can be resolved by appropriately combining the different density surfaces obtained. Intermediate indicators as well as the final index allowed obtaining a first image of the services' supply, highlighting centres of different order linked to a higher or lower supply. Places with higher density of services are located in the central areas of the city and also in minor settlements.

References

1. Anselin, L.: The Moran scatterplot as an ESDA tool to assess local instability in spatial association, in Fisher M., Scholten H., Unwin D. (eds), Spatial Analytical Perspectives on GIS. London: Taylor and Francis (1996) 111–125
2. Bailey, T. C., Gatrell, A. C.:Interactive spatial data analysis. Addison Wesley Longman Edinburgh, UK.(1995)
3. Borruso G. : Network density and the delimitation of urban areas, in Transactions in GIS. (2003) **7 (2)** 177–191
4. Fayyad U., Piatesky -Shapiro G.,Smyth P.: From Data Mining to Knowledge Discovery in Databases. (1996) $http://www.kdnuggets.com/gpspubs/aimag-kdd-overview-1996-fayyad.pdf$
5. Gatrell A., Bailey T, Diggle P. Rowlingson B.: Spatial Point Pattern Analysis and its Application in Geographical Epidemiology. Transactions of the Institute of British Geographers. (1996) **(2)** 1256–74.
6. Gueting R. H.: An Introduction to Spatial Database Systems. The VLDB Journal 3(4)(1994) 357-399.
7. Han J., Kamber M., Tung A.K.H.: Spatial Clutering Methods in Data Mining: A Survey. (2001) $ftp://ftp.fas.sfu.ca/pub/cs/han/pdf/gkdbk01.pdf$.
8. Lu ,Y.: Spatial Cluster Analysis for Point Data: Location Quotients Versus Kernel Density, Department of Geography, State University of New York at Buffalo.(1998) $http://www.ucgis.org/oregon/papers/lu.htm$
9. Thurstain-Goodwin M., Unwin D. J.: Defining and Delimiting the Central Areas of Towns for Statistical Modelling Using Continuous Surface Representations. In Transactions in GIS, **4** (2000) 305–317

An Exploratory Spatial Data Analysis (ESDA) Toolkit for the Analysis of Activity/Travel Data

Ronald N. Buliung and Pavlos S. Kanaroglou

School of Geography and Geology, McMaster University, Hamilton, Ontario, Canada
{buliungr, pavlos}@mcmaster.ca

Abstract. Recent developments in geographic information systems (GIS) and increasing availability of micro-data for cities present an environment that encourages innovations in approaches to visualizing and exploring outcomes of urban processes. We describe an object-oriented, GIS-based environment that facilitates visualization, exploration, and description of household level activity-travel behaviour. The activity pattern of a sample household from the 1994/1995 Portland Household Activity and Travel Behaviour Survey is used in the discussion to demonstrate the current capabilities of our system.

1 Introduction

In this paper we discuss the development of a prototype, GIS-based, object-oriented, system that facilitates spatial analysis of activity/travel behaviour micro-data. The paper documents the development of this system and demonstrates its use with applications that reveal its potential for enhancing our understanding of household level activity/travel behaviour. The system has been developed using ESRI's Geodatabase with additional programming of ArcObjects using Visual Basic for Applications (VBA). The software has been prototyped, in the ArcGIS Desktop environment, using data from Portland Metro's 1994/1995 Household Activity and Travel Behaviour survey. These data have been organized and stored in an object-relational spatial database. Formal design of this database resulted in an object-oriented data model of contextually relevant portions from the Portland survey. This process conformed to the principles of object-oriented design and benefited from application of the Unified Modeling Language (UML).

The data model was implemented as an ArcGIS Geodatabase and acts as an information foundation upon which several Exploratory Spatial Data Analysis (ESDA) tools have been constructed. These tools operate on household activity location geocodes that are retrievable from the underlying database. Analysts can (1) develop measures of central tendency and dispersion from the spatial properties of household activity patterns, (2) construct convex hull polygons from the spatial pattern of household activity geocodes, and (3) assemble individual and household level three-dimensional space-time paths. While some of the analytical concepts that have been implemented are not new the implementation methodology is innovative.

The system provides enhanced capabilities for data visualization and exploration while minimizing user interaction with the underlying database. In addition, the development environment supports further extension and refinement of existing tools as the need arises.

The paper has four sections including the introduction. In the second section of the paper we outline the general characteristics of the data used to develop our prototype. The discussion elaborates on the formal approach to database modeling that we have adopted. The third section of the paper details the specific analytic tools that we have developed. We use the two-day activity pattern of a sample household to illustrate the visualization and exploratory capabilities of our system. Ongoing research and conclusions are addressed in the fourth and final section of the paper.

2 Data Modeling and Integration

The prototype system has been built and tested using data from Portland Metro's 1994/1995 Household Activity and Travel Behaviour survey [1], Metro's Regional Land Information System (RLIS) and the 1990 U.S. Census. The Portland survey has emerged as a rich data source that continues to support activity/travel behavior research [2][3][4]. The final survey includes comprehensive description of the activity/travel behaviour of 4,451 households (63% of 7,090 recruited). At the individual level, 9,471 respondents reported a total of 122,348 activities and 67,891 trips [1]. The survey was completed in phases with phases one and two occurring during the spring and autumn of 1994 and phase three during the winter of 1995.

RLIS is an integrated, detailed regional geographic database that includes data from 24 cities and 3 counties. The system, was developed during the 1980's, and is managed by Metro's Data Resource Centre (DRC). We have incorporated several geographic data layers from RLIS including regional traffic analysis zones, and political boundaries. Both the Activity and Travel Behavior survey and a subset of the RLIS data can be acquired from Metro's web space. We have also incorporated data from the 1990 U.S. Census. Specifically, we have included the 1990 Urbanized Area (UA) layer for the purpose of assigning households to either urban or rural space depending upon the location of the place-of-residence. The UA cartographic boundary file was acquired from the Census Bureau's Internet based data repository. The UA concept is used by the Census Bureau to tabulate data on urban and rural populations and housing characteristics. According to the Census Bureau, UAs are considered dense settlements that contain at least 50,000 individuals.

Data from these sources have been examined and integrated using an object-oriented analysis and design (OOAD) approach. The process resulted in construction of a formal database design model using the Unified Modeling Language (UML) notation and the Microsoft Visio visual modeling tool. The development of this database has been described in detail elsewhere and will not be repeated here [2]. The database contents and relationships between stored classes are shown in Figure 1. We have organized the data model in a manner that logically separates classes that have geographical content (points, lines or polygons) from those containing tabular data only. All classes have attributes defined for them that have been suppressed in the figure to economize on space.

The figure serves as an example of a UML *class diagram*. This type of diagram communicates information concerning the structure of objects and specified relationships. As an example, consider the Household and Person classes. The structure of each class is given by a set of attributes or properties that have been programmed into each class. The Household class is described by properties such as the number of persons, vehicles, and income. The Person class is described by properties that include gender, age, and employment status. The relationship defined between these classes indicates that a Household consists of zero to many persons. In fact, the relationship programmed into our data model is a special type of *association* known as a *composite*. This has conceptual and practical implications. At the conceptual

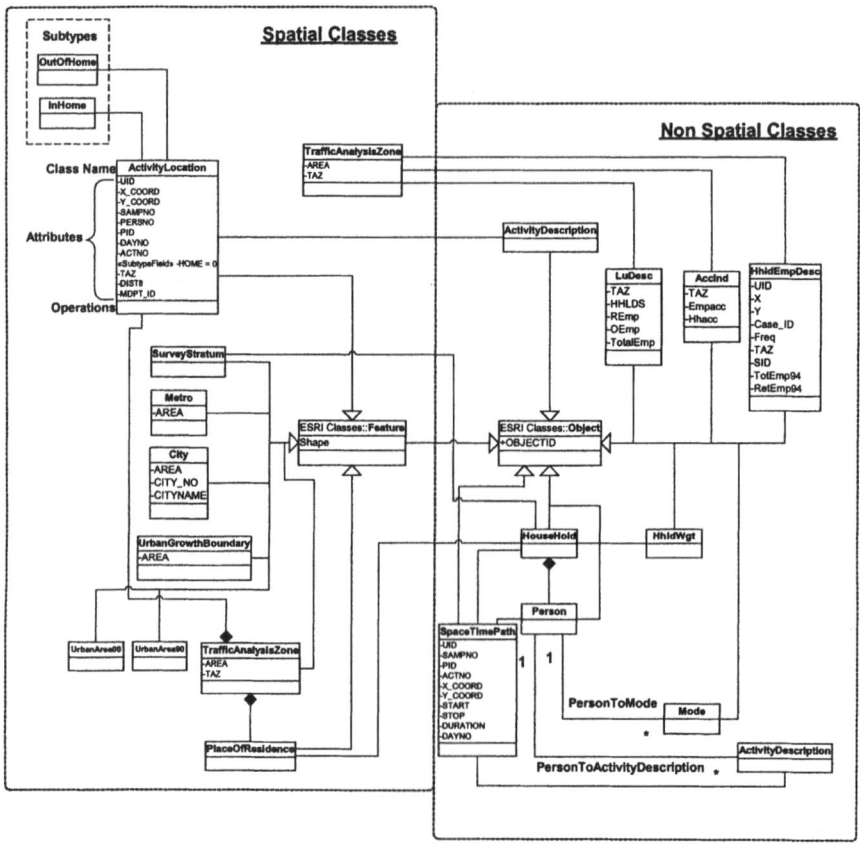

Fig. 1. Activity/Travel Database Schema

tual level we are suggesting that a Household is a collection of persons. At a practical level, the objects of the Household class control the existence of objects in the Person class. If a Household object is deleted, related Person objects will automatically be removed from the database. This helps to maintain consistency within the database.

One advantage of the approach is that we have been able to incorporate detailed descriptions of the database directly into the component classes. These descriptions include the spatial properties of the data, information sources and contacts, and detailed field descriptions. For example, the metadata available on the Census Bureau website has been programmed into the UA class (UrbanArea00 and UrbanArea90) descriptions in our data model. Another advantage of the approach, relates to the development of our ESDA toolkit. We use several relationships between programmed classes to automate extraction of spatial data that are then used in the assembly of structures and measures that describe household activity patterns. In this sense the approach to data management that we have adopted supports rapid development of analytical tools.

The design model has been implemented as an object-relational spatial database using tools from ESRI's ArcGIS suite and its Geodatabase data model. The database implementation is considered an object-relational hybrid because, while the data modeling exercise is completely object-oriented, the physical implementation relies on relational database technology. That is, classes in the data model become relations in a relational database, and objects become rows or tuples in these relations. Resulting classes have been populated with corresponding data from the sources described earlier. The database is accessible within the ArcGIS software environment both programmatically and through the various applications that are part of the ArcGIS suite. The resulting Geodatabase serves as the information foundation upon which the ESDA tools discussed in the remainder of this paper have been developed.

3 ESDA Tools Development

Recent advances in computer science, geocomputing, and the increasing availability of micro-data contribute to an environment that encourages innovations in data processing and the analysis of the activity/travel behaviour of individuals or households. We now have an opportunity to explore complex behaviour, using detailed information, with greater computational efficiency than in the past. While several software environments exist that support visualization, exploration, and modeling of geographic processes in general [5] [6], there are few examples that have been developed specifically to accommodate study of activity/travel behaviour.

Frihida et al. [7] discuss a prototype system for exploring individual level activity-based travel behaviour using data from a 1991 origin-destination survey conducted in the Québec City region. Schwarze and Schönfelder [8] have developed the VISAR (Visualisierung von Aktionsräumen) extension for ArcView. The tool has been used to characterize the activity spaces of respondents from the Mobi*drive*, six-week travel diary survey conducted in two German cities [9]. Shaw and Xin [10] have developed a prototype GIS-based application for exploring transport/land use interactions in space and time. The system was made operational using transportation and land use data for Dade County, Florida. These systems are related in terms of substantive focus, but also have common ground in the context of the developmental paradigm. With varying degrees of compliance, all of these researchers have adopted an object-oriented approach and have incorporated the analytic and data management capabilities of GIS.

Contributing to this area of research, we have developed a prototype set of tools that facilitate visualization and exploration of activity-based micro-data in space and time. Our primary behavioural unit of interest is the household although it is possible, using our system, to extract the behavioural data for individual household members. Our system has been developed as an analytic interface to the activity-based object-relational spatial database discussed earlier in the paper. Development of these tools involved programming of ESRI ArcObjects using Visual Basic for Applications (VBA) (Environmental Systems Research Institute, Redlands CA). Our application is an extension of the core capabilities of the ArcGIS desktop and in this sense provides an example of the *encompassing framework* to coupling spatial analysis capabilities and GIS [11] [6]. Typically, customizations of this sort will rely on scripting or macro languages that are native to the GIS software [6]. In the past, this presented certain disadvantages partially related to the proprietary nature of many GIS-based scripting languages. Convergence on VBA makes the approach that we have followed more accessible to analysts and programmers not completely familiar with ArcGIS but sufficiently familiar with object-oriented development and VB.

Exploratory analysis with our system begins with creation of a layer containing activity geocodes for a survey household. To do this, the analyst spatially selects a place-of-residence geocode. These locations have been associated with survey households in the underlying database. The application then searches through the database to retrieve appropriate activity objects. Household activity layers are then built for assigned survey days. A third layer is built that contains the joint distribution of activity sites visited on both survey days. Analysts can then explore these layers using the custom tools programmed for the system. Spatial structures generated by the tools emerge through collaborations between objects within the database and associated methods. These methods have their foundations in spatial statistics, computational geometry, and time-geography.

3.1 Spatial Description of Household Activity Patterns

In the remainder of this section we describe the set of tools that have been developed so far, and discuss examples that illustrate their application. We focus the discussion on the two-day activity pattern of a sample household (Figure 2). The selected household contains two adults ($20 \leq$ age ≤ 59) with paid employment (full or part-time), and a young child (age ≤ 12). This household recorded their activity/travel behaviour for two consecutive days. On the first survey day, Wednesday, they recorded 7 in-home and 7 out-of-home activities for a total of 14 activities. On the second survey day, Thursday, they recorded 10 in-home and 11 out-of-home activities for a total of 21 activities. When studying the examples in Figure 2 it is useful to keep in mind that multiple activities can occur at a single location.

Layers of extracted household activities for corresponding survey days are equivalent to spatial point patterns. The spatial statistical literature provides several approaches to describing the spatial distribution of point events [12] [13]. When describing the characteristics of a univariate distribution we rely on measures of central tendency and dispersion. We have similar measures available when describing a spatial distribution that arises from a particular spatial process. For any set of observed activity events our application can generate a bivariate, unweighted, mean center as a measure of central tendency. The dispersion of these activity patterns can then be explored using a measure of *standard deviational distance* (standard distance) and by constructing what is known as a *standard deviational ellipse*. In our implementation we have adopted the computational formulae from [12] with appropriate modifications reported in [14] to produce unbiased estimates with appropriate geometries.

The *standard distance* is the standard deviation of the distance of each point in a point pattern from the mean center [15]. A general criticism of standard distance is that while it provides a decent single measure of dispersion around the mean center it cannot be used to diagnose the presence of anisotropy or a directional trend in the spread of points around the mean [16] [12] [14]. Our tool generates an unweighted

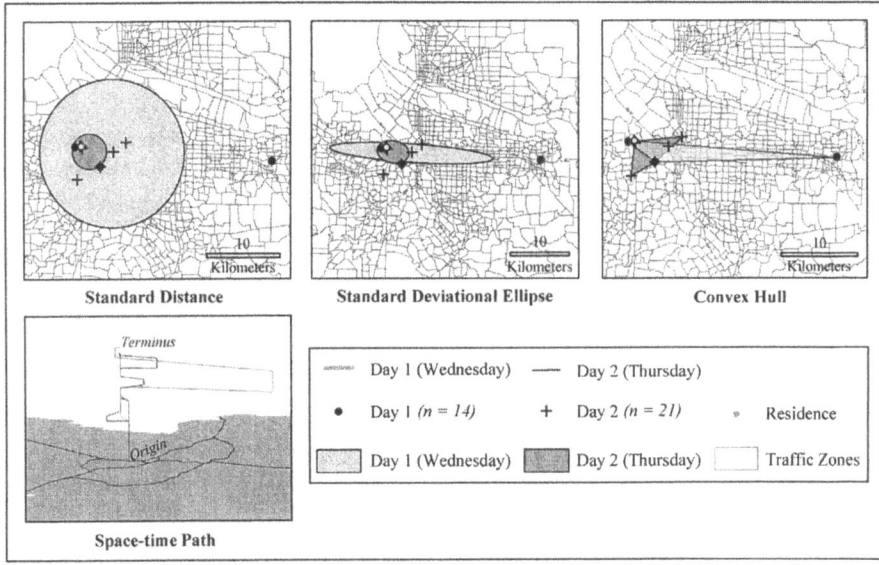

Fig. 2. Household Activity/Travel Behaviour

mean center from the spatial distribution of household activity events and then estimates the standard deviational distance around that mean. The tool can generate the mean center, standard distance, and a circle with the radius given by the standard distance (Figure 3).

The standard distance polygon for our sample household suggests that the dispersion of consecutive daily patterns is unequal (Figure 2). In this particular instance, we are examining the entire pattern of daily household activities, including those that occur in the home. The presence of multiple in-home activity events tends to bias the mean centre toward the home location and similarly impacts the standard distance calculation. The reason for this is that the coordinates associated with the household location enter the mean centre and standard distance calculations for each recorded in-home activity. The finding suggests that the home location is important to the overall spatial structure of a household's daily activity pattern. We can also use this environment to independently describe the spatial pattern of out-of-home activities for the same household.

The *standard deviational ellipse* represents an alternative to standard distance deviation and has several useful applications. Perhaps the most useful of these, is that the approach can be used to study the directional dispersion of a spatial point pattern. In our implementation the ellipse is focused on the mean center of the household activity pattern. The long or major axis is oriented in the direction of maximum dispersion and the short or minor axis in the direction of minimum dispersion. Using our application, the analyst can extract numerous properties that can be used to characterize the statistical and geometric properties of an ellipse. These include the mean center, length of the major and minor axes, eccentricity, angle of rotation, area, perimeter, and several others (Figure 3).

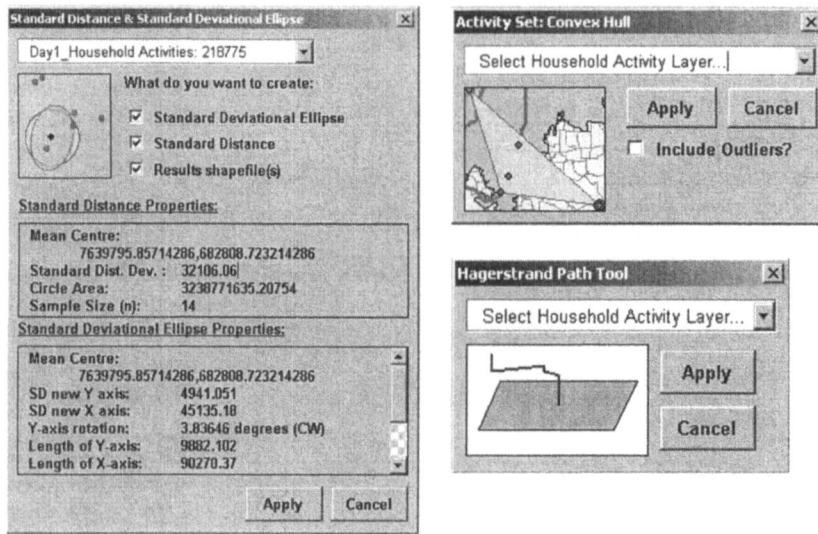

Fig. 3. Exploratory Tools

The use of a standard deviational ellipse to describe patterns of events in geographical space is not new. Initial methodological developments and applications occurred in the 1920's [17] [18]. Yuill [16] seems to be responsible for introducing the concept to the discipline of Geography. More recently, ellipses defined using other methods have been used to explore the action spaces of households [19] and to investigate the concept of social exclusion [9]. As our example suggests, the mean center and dispersion of the spatial pattern of activity events is not constant across survey days (Figure 2). In addition, by applying the standard deviational ellipse we observe that for the selected household, the direction of dispersion of activity events is not uniform across survey days (Figure 2). Similar to the mean centre and standard distance calculations, the ellipse is also biased toward the home location due to the presence of multiple in-home activities. We expect that the size and orientation of the ellipse will change with an independent investigation of out-of-home activities.

3.2 Convex Hull and Urban/Rural Assignment

A convex hull of a discrete set of points S is given by the intersection of all convex sets that contain S. A convex set occurs when line segments connecting any pair of points fall completely within the area occupied by the set S. Another way to understand these concepts is to consider a convex hull as a polygon that is created by connecting the extreme points in a point pattern. In our application, we treat the set of activity sites visited by a household as the set S and use a convex hull method to generate a polygon that contains all activity locations. Convex hull polygons are constructed by applying an ArcObjects hull method to a geometry object containing household activity location objects. The ESRI hull method applies a variation of the Quickhull algorithm [20]. The analyst has the flexibility to develop hull-based measures of household activity spaces for each survey day or both survey days in combination (Figure 3).

The tool generates spatial datasets that contain hull geometries and a table of hull-based measures. These include area, perimeter, and a hull generation code that is used to identify reasons for hull creation failure. We are also generating measures that attempt to characterize the "urbanity" of households. Using a point-in-polygon routine, the tool assigns a binary urban/rural code to a household of interest. The criteria for this assignment is based on a determination of the household's geocoded place-of-residence being located inside or outside the 1990 U.S. Census Bureau's Urban Area (UA) boundary. In addition, through a topological intersection routine, we are able to generate the respective proportions of a household's hull polygon falling inside and outside the UA boundary. With respect to the multi-day activity/travel behaviour of the sample household we again observe differences in the geographic extent of household activities across survey days (Figure 2).

3.3 Implementing the Space-Time Path Concept

While conceptualization of human activity patterns as three-dimensional paths in space and time is not new [21] [22] implementation of the concept using GIS has only recently been achieved [23] [3]. Kwan [3] has focused on visualization and exploration of individual space-time paths and has also used the 1994/1995 Portland survey for this purpose. Our innovation involves simultaneous assembly of household level space-time paths as collections of the paths of individual household members. This approach also enables extraction of individual paths from the household structures. A second innovation is that our implementation facilitates rapid creation of these spatiotemporal structures through a user interface that minimizes interaction of the analyst with the underlying survey database (Figure 3).

The tool can automatically generate separate paths for assigned survey days given the activity geocodes and timing information for all household members. Path vertices are given by the location of activity events in space and time. The origin of each space-time path is at the point of planar intersection (Figure 2). The terminus of each path coincides with the timing of a household's last recorded activity and is given by the maximum coordinate location along the vertical axis (Figure 2). The household space-time path is essentially a collection of individual paths that, when viewed in two or three-dimensional space, allows the analyst to visualize and explore the complexity of activity/travel behaviour at the household level. The paths of individual members can also be extracted from this data structure. This can be useful for examining interactions between individual household members.

Two space-time paths have been developed for the sample household to visualize activities performed on consecutive survey days (Figure 2). In this particular example the path, on both days, begins and ends at the home location. The vertical axis represents time spent at a particular location, or activity duration. Changes in the path location in planar space are attributed to activity locations visited by household members. We can observe that, for this household, the Thursday path is more focused around the home location than the Wednesday path. In addition, the household is accessing more distant locations on Wednesday suggesting that the spatial and temporal structure of activity/travel behaviour is not uniform across survey days for this particular household.

4 Conclusion

We have developed an environment that facilitates exploration of household level activity/travel behaviour. The prototype has been implemented using the 1994/1995 Portland Household Activity and Travel Behaviour survey and other regional data. The tools provide different options for visualizing and exploring household activity/travel behaviour. Area-based measures of daily household activity patterns have been operationalized and can be extracted from the software for further exploratory analysis using other statistical software. The development environment promotes extension and refinement of existing tools as the need arises.

The current system has been developed closely with the underlying Portland survey database. We will be working toward improving the flexibility of the system to accommodate similar activity-based micro-data stored in different formats. We will also be exploring the transition of the system from its current prototype implementation within the ArcGIS environment to a standalone system.

While the area-based measures discussed in this paper are useful and interesting conceptualizations of household activity/travel behaviour they are not universally applicable. That is, for certain households in the survey, it is not possible to generate area-based measures. Examples include households with activity events occurring at a single location and households with activity events that are collinear in planar space. In response, we are currently developing a link-based measure of daily household kilometers traveled for performing activities.

Acknowledgements. The authors wish to thank Metro for the provision of data and supporting documentation. The second author gratefully acknowledges the financial assistance of the Social Sciences and Humanities Research Council (SSHRC) Canada Research Chairs (CRC) program.

References

1. Cambridge Systematics: Data Collection in the Portland, Oregon Metropolitan Area: Case Study. Report DOT-T-97-09. U.S. Department of Transportation (1996)
2. Buliung, R.N., Kanaroglou, P.S.: On Design and Implementation of an Object-Relational Spatial Database for Activity/Travel Behaviour Research. Submitted to the *Journal of Geographical Systems*. (2003)

3. Kwan, M-P.: Interactive Geovisualization of Activity-travel Patterns Using Three-dimensional Geographical Information Systems: A Methodological Exploration with a Large Data Set. *Transportation Research C*, Vol. 8. (2000) 185-203
4. Buliung, R.N.: Spatiotemporal Patterns of Employment and Non-Work Activities in Portland, Oregon. In *Proceedings of the 2001 ESRI International User Conference*, San Diego, CA, (2001)
5. Levine, N.: Spatial Statistics and GIS: Software Tools to Quantify Spatial Patterns. *Journal of the American Planning Association*, Vol. 62(3). (1996) 381-391
6. Anselin, L.: Computing Environments for Spatial Data Analysis. *Journal of Geographical Systems*, Vol. 2. (2000) 201-220
7. Frihida, A., Marceau, D.J., Theriault, M.: Spatio-Temporal Object-Oriented Data Model for Disaggregate Travel Behavior. *Transactions in GIS*, Vol. 6. (2002) 277-294
8. Schwarze, B., Schönfelder, S.: ArcView-Extension VISAR - Visualisierung von Aktionsräumen. Version 1.6, Arbeitsbericht Verkehrs - und Raumplanung, 95, Institut für Verkehrsplanung, Transporttechnik, Strassen- und Eisenbahnbau, ETH, Zürich, (2001)
9. Schönfelder, S., Axhausen, K.W.: Activity Spaces: Measures of Social Exclusion? *Transport Policy*, Vol. 10. (2003) 273-286
10. Shaw S-L., Xin X.: Integrated Land use and Transportation Interaction: A Temporal GIS Exploratory Data Analysis Approach. *Journal of Transport Geography*, Vol. 11. (2003) 103-115
11. Anselin, L., Getis, A.: Spatial Statistical Analysis and Geographic Information Systems. *Annals of Regional Science*, Vol. 26. (1992) 19-33
12. Ebdon, D: Statistics in Geography. 2nd edn. Blackwell, Oxford (1988)
13. Bailey, T.C., Gatrell, A.C.: Interactive Spatial Data Analysis. Addison Wesley Longman, Essex (1995)
14. Levine, N.: CrimeStatII User Manual. Levine and Associates, URL at http://www.icpsr.umich.edu/NACJD/crimestat.html. (2002)
15. Bachi, R.: Standard Distance Measures and Related Methods for Spatial Analysis. *Papers of the Regional Science Association*, Vol. 10. (1963) 83-132
16. Yuill, R.S.: The Standard Deviational Ellipse; An Updated Tool for Spatial Description. *Geografiska Annaler Series B Human Geography*, Vol. 53(1). (1971) 28-39
17. Lefever, D.W.: Measuring Geographic Concentration by Means of the Standard Deviational Ellipse. *The American Journal of Sociology*, Vol. 32(1). (1926) 88-94
18. Furfey, P.H.: A Note on Lefever's "Standard Deviational Ellipse". *The American Journal of Sociology*, Vol. 33(1) (1927) 94-98
19. Dijst, M.: Two-earner Families and their Action Spaces: A Case Study of two Dutch Communities. *Geojournal*, Vol. 48. (1999) 195-206
20. Barber C., Dobkin, D., Huhdanpaa, H.: The Quickhull Algorithm for Convex Hulls. *ACM Transactions on Mathematical Software*, Vol. 22 (1997) 469-483
21. Hägerstrand, T.: What About People in Regional Science? *Papers of the Regional Science Association*, Vol. 24. (1970) 7-21
22. Lenntorp, B.: A Time-Geographic Simulation Model of Individual Activity Programmes. In: Carlstein, T., Parkes, D., Thrift, N. (eds.): Human Activity and Time Geography. Wiley and Sons, New York (1978)
23. Kwan, M-P.: Gender, the Home-work Link and Space-time Patterns of Non-employment Activities. *Economic Geography*, Vol. 75(4). (1999) 370-394

Using Formal Ontology for Integrated Spatial Data Mining

Sungsoon Hwang

Department of Geography
State University of New York at Buffalo
105 Wilkeson Quad, Buffalo, NY 14261, U.S.A.
shwang5@buffalo.edu

Abstract. With increasingly available amount of data on a geographic space, spatial data mining has attracted much attention in a geographic information system (GIS). In contrast to the prevalent research efforts of developing new algorithms, there has been a lack of effort to re-use existing algorithms for varying domain and task. Researchers have not been quite attentive to controlling factors that guide the modification of algorithms suited to differing problems. In this study, ontology is examined as a means to customize algorithms for different purposes. We also propose the conceptual framework for a spatial data mining (system) driven by formal ontology. The case study demonstrated that formal ontology enabled algorithms to reflect concepts implicit in domain, and to adapt to users' view, not to mention unburdened efforts to develop new algorithms repetitively.

1 Introduction

No single spatial data mining method is best suited to all research purposes and application domains. However, determining which algorithm is suited to a certain problem, and how to set the values of parameters is not a straightforward task. Rather it requires an explicit specification of domain-specific knowledge as well as of task-oriented knowledge. Given this problem, ontology, which is defined as "the active component of information system" [1] in addition to "the explicit specification of conceptualization" [2], can play an important role in organizing the mechanism underlying the spatial data mining phenomenon. Spatial data mining can be thought of as an information system where different kinds of ontologies serve as active components. In this context, spatial data mining system is driven by formal ontology.

This study is concerned with endowing algorithms with semantics and adapting algorithms to users view rather than developing new algorithms for different domains and problems. The focus is placed on the role of ontology in customizing existing algorithms. Thus, the purpose of this study is to illustrate how formal ontology can be used to re-use existing algorithms suited to varying domain and task. To make this clear, we propose a conceptual framework for spatial data mining system driven by formal ontology. The framework will clearly show how ontologies can be incorporated into spatial data mining algorithms. The conceptual framework is implemented in finding hot spots of traffic accidents. We evaluate whether using ontology is beneficial in spatial data mining by comparing ontology-based method and existing methods. A case study shows that spatial data mining methods using ontology can take into account domain-specific concepts and users view that have not been handled well before. In short, results (i.e. pattern discovered) are both natural and usable.

The rest of this paper is organized as follows. It begins by describing the conceptual framework for ontology-based spatial data mining in Sect. 2. In Sect. 3, the case study using the proposed method is illustrated, and results are analyzed. We conclude by summarizing the study.

2 Conceptual Framework for Ontology-Based Spatial Data Mining

2.1 Relation between Data Mining and Ontology Construction

Data mining enhances the level of understanding by extracting a high-level knowledge from a low-level data [3]. Ontology construction makes implicit meaning explicit by formalizing how the knowledge is conceptualized. Here we first discuss different notion of knowledge between data mining and ontology construction. Second, we discuss how they are related. Third, we examine how data mining can be used for ontology construction, and vice versa.

In the context of data mining, knowledge is *discovered*. In the context of ontology construction [4], knowledge is *acquired*. Wherein different notion of knowledge can be noted; the knowledge discovered from data mining is, to a large extent, data-specific whereas the knowledge acquired for ontology construction is data-independent. This fact arises from different approaches taken. Data mining is a bottom-up (i.e. data-drive) approach while ontology construction is a top-down approach. In terms of cognitive principle, data mining is similar to induction while ontology construction is similar to deduction. The knowledge discovered from data mining is applied in small scope with high level of detail whereas the knowledge acquired for ontology construction is applied in large scope with low level of detail.

The next question is, how are they interrelated? Data mining and ontology construction is bridged by the varying level of abstraction. Fig.1 shows that two kinds of knowledge are at the continuum that is manipulated by the level of abstraction. Where the knowledge is the result of data mining, and becomes the source of ontology construction. Not surprisingly, high level of abstraction in ontology construction entails human intervention such as expert knowledge or concept hierarchy. In contrast, data mining can be, to some extent, automated by machine learning programs.

The potential role of ontology in data mining is (1) to guide algorithms such that they can be suitable for domain-specific and task-oriented concepts (2) to provide the context in which the information or knowledge extracted from data is interpreted and evaluated [5]. Conversely, the knowledge confirmed in the knowledge discovery process can be seen as candidates for ontology in the long term. In such a way, the interaction between induction (knowledge discovery) and deduction (ontology construction) process can enrich our knowledge base.

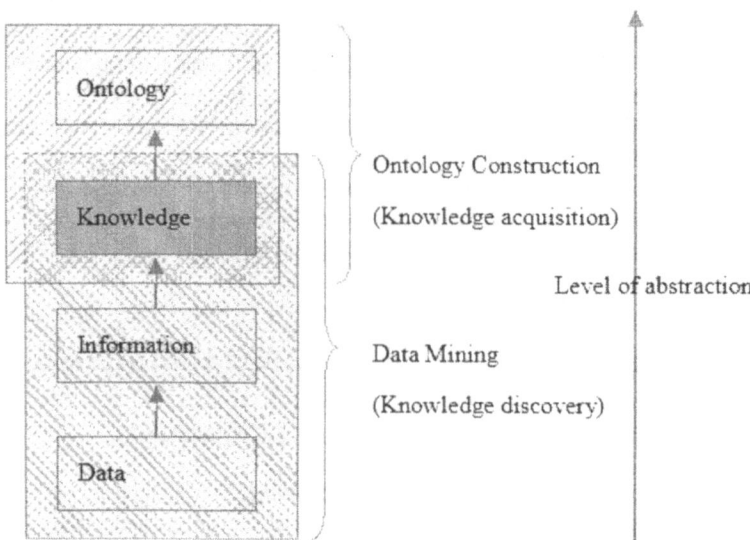

Fig. 1. The relation between data mining and ontology construction seen from varying level of abstraction

Along this line, examining the role of ontology in data mining is significant in explicating the interaction between knowledge discovery and ontology construction. Null hypothesis can be stated as follows: ontology-based data mining method does not improve the quality of knowledge discovery as compared to data mining without using ontology. The study is focused on the first role of ontology. The study is restricted to the spatial data mining. In sum, research questions we attempt to address here are, (1)"Can ontology really enhance spatial data mining?" (2)"If yes, how can it be done?"

2.2 Rationale of Using Ontology for Spatial Data Mining

The same spatial data mining algorithms [6] can yield results inconsistent with fact without considering the domain knowledge. The same data may have to be mined in different ways depending on users' goals. In sum, the domain and users' need associated with input data are the major factors that control the mechanism underlying the spatial data mining [7]. Therefore, we need to build a new spatial data mining algorithm that is dictated by domain and task model (Fig.2) [8] [9]. The focus is placed on customizing existing algorithms suited to a certain domain and problem rather than developing new algorithms.

To illustrate the domain model incorporated into spatial data mining algorithms, suppose we have two different geospatial data for the purpose of detecting their hot spots: one is traffic accident data, and the other is the location of supermarket. Wherein, traffic accident *is-an* event whereas a store *is-a* physical object. Thus we can conclude they should be handled differently. Moreover traffic accidents can only occur on the road whereas a store is located in the pedestrian blocks. As a result, the topological relation to road network is different: traffic accident occurs *in* road network, but store is located *outside of* road network. To build clustering algorithms, we need to define features to compose similarity measures (or distance function). Different features of similarity measures should be used due to different domain model (or conceptualization) associated with the input data.

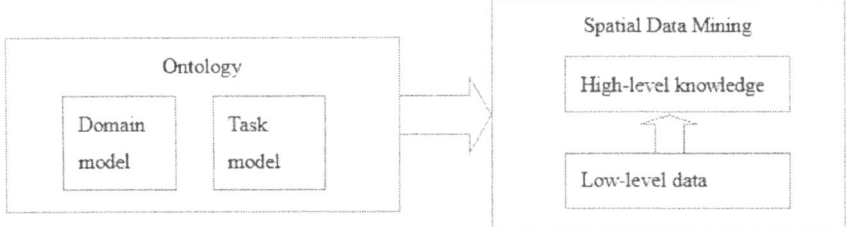

Fig. 2. Using ontology for integrated spatial data mining

To illustrate the task model incorporated into spatial data mining algorithms, let us consider two different tasks of spatial clustering (see [10] for the survey of spatial clustering algorithms): One is to detect hot spots, and the other is to assign customers to market areas. The number of clusters, k can be derived from overall data distribution for the former task. In contrast, the number of clusters, k will be preliminarily given as a resource constraint for the latter task. In addition to different goals, the number of cluster, k will vary with the level of details in which users want to examine. Therefore, different notion of arguments (e.g. k) should be adopted depending on task model.

Users are often prompted to specify input parameters without understanding the mechanics of parameters when they use spatial data mining tools. For example, classic k-means clustering algorithms prompt users to specify the number of clusters. But the best number of clusters should not be chosen arbitrarily. Rather, the number of clusters should be obtained as a result of learning the data or underlying phenomenon. Likewise, the desired level in hierarchical clustering (i.e. cutting the link in dendrogram) should not be chosen arbitrarily, but rather should be chosen depending on the level of details in which users intend to examine the problem in hand.

Best spatial data mining methods will be achieved by taking into account the factors such as users view, goal, domain-specific concepts, characteristics of data, and available tools. However, if considerations were given to those factors in an arbitrary manner, it would not meet our desire. Given this, ontology can provide the systematic way of organizing those factors. The need for users to specify the input parameters will be reduced if we are able to utilize available ontology in a generic level (e.g. top-level ontology of space and time, domain ontology) [11] and to construct ontology in a specific level (e.g. application ontology) [1]. Therefore, ontology-based spatial data mining can overcome the shortcomings of existing spatial data mining methods.

2.3 Conceptual Framework for Ontology-Based Spatial Data Mining

The component of ontology-based spatial data mining systems can be divided into three parts: Input, Ontology-based spatial data mining method, and Output (Fig.3).

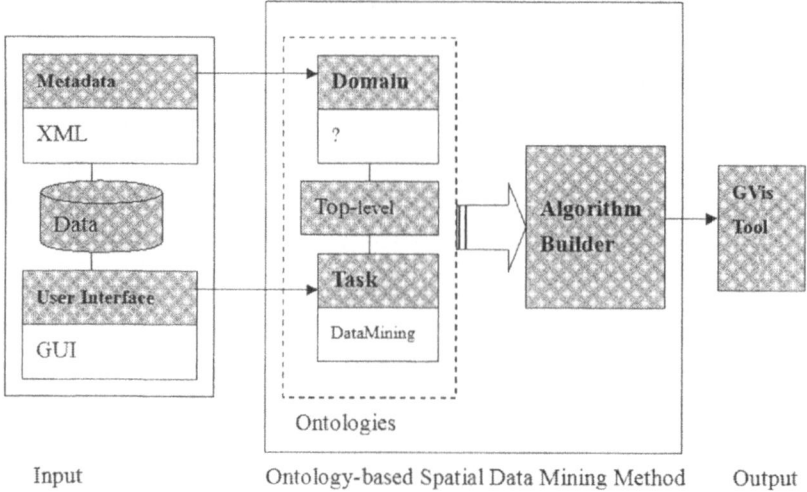

Fig. 3. Conceptual Framework for Ontology-based Spatial Data Mining System

Input component is composed of metadata and user interface that are linked to data. Metadata contains the information on data content. User interface allows users to select a goal to be achieved and related parameters wherever necessary. Ontology-based spatial data mining method is composed of ontologies and an algorithm builder. The natural language given by metadata (domain) and users (task) is translated into corresponding components (domain/task ontologies). Task ontologies define methods adequate to the goal specified. Domain ontologies specify domain-specific concepts, relation, function, and properties. The generic characteristics inherit from the top-level ontologies [12]. Task ontologies interact with domain ontologies to filter the relevant information to activate operations defined in the method. In a spatial data mining algorithm builder, algorithms are dynamically built from the items (method, concept inherent in domain) derived from ontologies. Output component presents results through geographic visualization (GVis) tools. The remainders of this section describe each component in more detail.

Metadata. Metadata is a summary document providing content, quality, type, creation, and spatial information about a data set. A parser program can easily retrieve the theme of data in hand utilizing the tag structure in XML. In such a way, metadata informs domain ontologies of the semantics of data.

User Interface. User interface allows users to effectively explore and select information relevant to a task in hand. Task-centered user interface prompts users to select their goals. Moreover, users can select the level of detail (e.g. jurisdiction level), and the geographic area of interest.

Domain Ontologies. Terms within the "theme" tag in the metadata are used as a token to locate the appropriate domain ontologies. Domain ontologies specify their definition, class (e.g. Accident is a Subclass-Of Temporal-Thing) and properties (e.g. Road has a Geographic-Region as a Value-Type) (see [13] for the ontology of geography). Properties of class inherit from upper-level ontologies.

Task Ontologies. The goal selected by users in the user interface is translated into task ontologies. The method, requirement, and constraints adequate to the user-supplied goal are specified in task ontologies. A certain class requires the interaction with domain ontologies (also known as interaction problem). For example, in order to detect the hot spots of traffic accidents, algorithms need to get the information from domain ontologies, such as spatial constraints (e.g. traffic accident occurs along the road segments) [14] [15].

Algorithm Builder. The method (hierarchical) suited to user-supplied goal ("find hot spots"), requirements (data, scale, detail level), and constraint (road) has been already supplied from the interaction between input components (metadata, user interface) and ontologies (domain, task, top-level). Algorithm builder puts together these items to build the best algorithm. That way, ontology provides the arguments necessary for running algorithms without a need to prompt users to select them. More specifically, the algorithm builder filters the data content through domain ontology and users' requirement through task ontology.

GVis Tool. The geographic visualization tool displays the resulting clusters. Results will be displayed differently depending on goal and scale. Well-designed GVis tool facilitates hypothesis formulation, pattern identification (that was the purpose of spatial clustering task also) and decision-making.

3 Case Study

3.1 Ontology-Based Spatial Clustering Algorithm

We used 7413 geocoded fatal accident cases that have been reported in New York State from year 1996 to 2001 (see [16] for data description; see [17] for georeferencing procedures). Fig.4 shows (a) input data and (b) output clusters in Buffalo, NY, where output clusters are generated by ontology-based spatial clustering (OBSC) algorithms. All components (with a focus on spatial clustering) discussed in the preceding section are combined to find the hot spots of traffic accident in the case study.

To evaluate the benefit of using ontologies, we compare a control algorithm (i.e. without using ontologies) [18] with a test algorithm (OBSC) in terms of Geographic-Scale and Spatial-Constraint with other factors controlled. Scale is implicit in task specification, and constraint is given in domain ontologies.

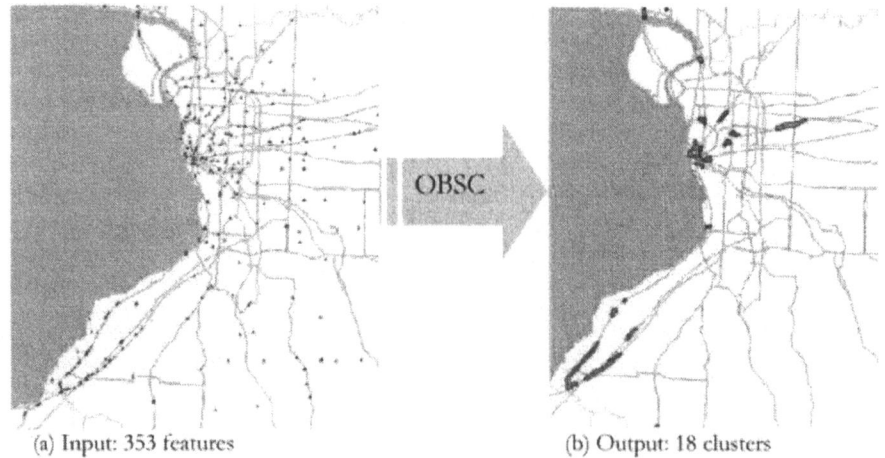

(a) Input: 353 features   (b) Output: 18 clusters

Fig. 4. The Result of Ontology-Based Spatial Clustering Algorithm

3.2 Analyses of Results

Effect of Scale (Task Ontologies). To illustrate the point, suppose a user wants to pinpoint the spot where traffic accidents occur with higher frequency in Manhattan, not other localities. In Fig.5, map (a) results from a control algorithm, and map (b) results from an ontology-based algorithm. Two algorithms are the same except that an ontology-based algorithm prompts a user to choose geographic scale of his interest. Two maps show clusters in different level of detail. With a control algorithm, traffic accident cases are lumped into three large clusters, which mask the detail. It is mainly due to the averaging effect of fixed scale. On the other hand, an ontology-based algorithm discovers hot spots of traffic accidents in the desired level of detail because the necessary information is conveyed to task ontologies through a user interface. To recapitulate, the result of an ontology-based algorithm reflects spatial distribution specific to the scale of users' interest, thereby resulting in *usable* clusters.

Effect of Constraint (Domain Ontologies). In Fig.6 (a), no consideration of spatial constraint (i.e. the occurrence of accidents is spatially constrained on the road) produces a large cluster spanning both sides of New York harbor. The control algorithm overlooks the existence of a body of water between Manhattan and Brooklyn. On the other hand, an ontology-based algorithm separates clusters because domain ontologies inform the algorithm that an accident cannot be on the body of water. It shows that domain ontologies enable clustering algorithms to embed domain knowledge, thereby resulting in *natural* clusters.

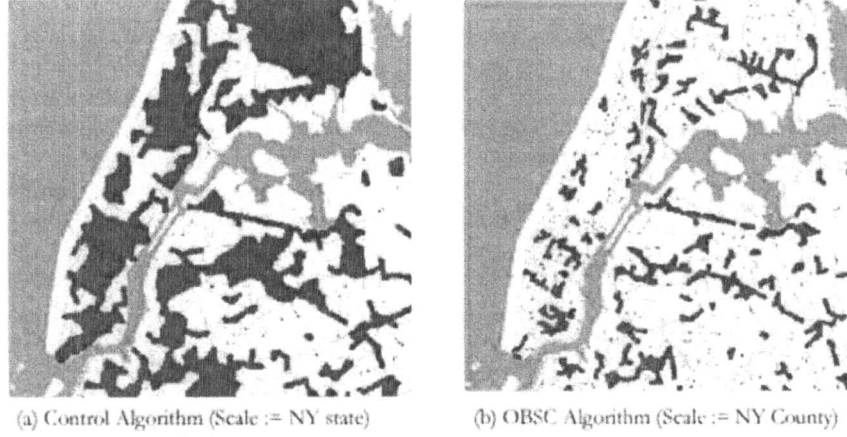

(a) Control Algorithm (Scale := NY state)   (b) OBSC Algorithm (Scale := NY County)

Fig. 5. OBSC clusters reflect spatial distribution specific to the scale of users' interest

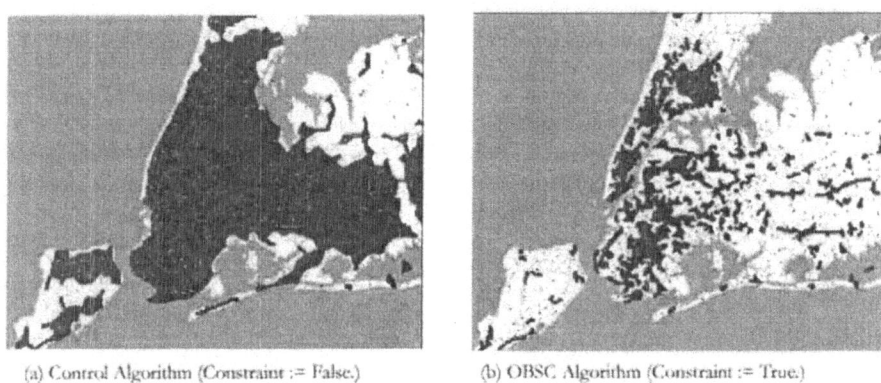

(a) Control Algorithm (Constraint := False.)   (b) OBSC Algorithm (Constraint := True.)

Fig. 6. OBSC clusters identify the physical barrier due to concept implicit in domain

4 Conclusion

This study demonstrates that it is worthwhile using ontology in a spatial data mining task in several respects: (a) Ontology provides the systematic way of organizing various features that consist of mechanism underlying the data mining phenomenon. (b) Ontology-based methods produce more intuitive results. (c) The need to specify input parameters arbitrarily is reduced. (d) Ontology provides the semantically plausible way to re-use existing algorithms.

Findings can be summarized as follows: First, in ontology-based method data mining mechanisms are dictated by concepts implicit in domain. For instance, the resulting clusters of traffic accidents are concentrated along road network because a spatial constraint is a priori implicit in domain. Second, ontology-based method is responsive to users view. The user-supplied task requirements make a cut-off value depend on the distribution specific to the scale of users' interest. To sum up, ontology-based methods make the result of spatial data mining natural and usable.

This study can advance the field of geographic knowledge discovery by introducing a novel approach to data mining methods based on ontology. This study is important because the attempt has been made to present the mechanism that ontologies are incorporated in spatial data mining algorithms in a way that algorithms can be built at a semantic level. Formalizing knowledge (i.e. ontology construction) is not a focus of this study, but the semantic linkage between ontologies and algorithms through the parameterization process has been emphasized.

References

1 Gruber, T. R., 1993, Formal Principles for the Design of Ontologies Used for Knowledge Sharing, *Formal Ontology in Conceptual Analysis and Knowledge Representation*, Kluwer Academic Publishers
2 Guarino, N., 1998, Formal Ontology and Information Systems, *Proceedings of FOIS'98*, Trento, Italy, 6-8 June 1998, Amsterdam, IOS Press, pp. 3-15
3 Fayyad, U. M, Piatetsky-Shapiro, G., and Smyth, P., 1996, From Data Mining to Knowledge Discovery: An Overview. *Advances in Knowledge Discovery and Data Mining*, AAAI/MIT Press, pp.1-34
4 van Heijst, G., Schreiber, A., and Wielinga, B., 1997. Using explicit ontologies for KBS development, *International Journal of Human-Computer Studies*, 46(2/3): 183--292.
5 Visser, U., Stuckenschmidt, H., Schuster, G., and Vogele, T., 2002, Ontologies for Geographic Information Processing, *Computers & Geosciences* 28: 103-117
6 Koperski, K, Adhikary, J., and Han, J., 1996, Knowledge Discovery in Spatial Databases: Progress and Challenges, *Proceedings of Workshop on Research Issues on Data Mining and Knowledge Discovery*, pp. 55-70, Montreal, QB, June 1996.
7 Witten, I. H. and Frank, E., 2000, *Data Mining: practical machine learning tools and techniques with java implementations*, Morgan Kaufmann
8 Newell, A., 1982, The Knowledge Level. *Artificial Intelligence*, 18, pp. 87-127
9 van de Velde, W., 1993, Issues in Knowledge Level Modeling, in David, J-M., Krivine, J-P., and Simmons, R. (Eds.) *Second Generation Expert Systems*, pp.211-231. Springer Verlag, Berlin
10 Han, J., Kamber, M., and Tung, A.K H., 2001, Spatial Clustering Methods in Data Mining: A Survey, in H. Miller, J. Han, (Eds.) *Geographic Data Mining and Knowledge Discovery*, Research Monographs in Geographic Information Systems, Taylor and Francis
11 Russell, S., and Norvig, P., 1995, *Artificial Intelligence: A Modern Approach*, Prentice Hall
12 Fonseca, F. T., Egenhofer, M. J., Agouris, P, and Gamara, G, 2002, Using Ontologies for Integrated Geographic Information Systems, *Transactions in GIS*, 6(3): 231-57
13 Smith, B. and Mark, D. M., 2001, Geographic Categories: an Ontological Investigation, *International Journal of Geographical Information Science*, 15(7): 591-612
14 Tung, A K H, Hou, J., and Han, J., 2001, Spatial Clustering in the Presence of Obstacles, *17th International Conference on Data Engineering* April 02 - 06, 2001 Heidelberg, Germany
15 Estivill-Castro, V and Lee, I., 2001, AUTOCLUST+: Automatic Clustering of Point-Data Sets in the Presence of Obstacles, in Roddick, J. F., and Hornsby K. (Eds.) *Temporal, Spatial, and Spatio-Temporal Data Mining*, Springer-Verlag pp.133-146

16 NHTSA, 1995, FARS 1996 Coding and Validation Manual, National Center for Statistics and Analysis, National Highway Traffic Safety Administration, Department of Transportation, Washington, D.C.
17 Hwang, S., and Thill, J-C, 2003, Georeferencing FARS accident data: Preliminary Report, NCGIA and Department of Geography, State University of New York at Buffalo, Unpublished document
18 Kang, I., Kim, T., and Li, K., 1998, A Spatial Data Mining Method by Delaunay Triangulation, *Proceedings of the 6th ACM GIS Symposium*, pp.157-158, November 1998

G.I.S. and Fuzzy Sets for the Land Suitability Analysis

Beniamino Murgante and Giuseppe Las Casas

University of Basilicata, Via S.Caterina, 85100 Potenza, Italy
murgante@unibas.it

Abstract. This paper reports about uncertainty in defining boundaries, which assume an institutional significance when transposed in planning prescription. Every discipline involved in environmental planning uses different approaches to represent its own vision of reality. Geological sciences or hydraulics evaluate risks by consistent mathematical models which are relevantly different to non linear models emploied in the field of ecology, and at the same time information about significance and value of cultural heritage in a given environment does not easily correspond to a value attribution. These questions represent an interesting field of research, related with the different character of information deriving from different disciplinary approaches, and with the more appropriate way of combining the same information. Different ways of managing values correspond to different ways of giving information. The result is a set of discrete representations of the physical space which correspond to a set of different values referring to areas which are considered homogeneous according to each disciplinary point of view, but very difficult to combine to create landscape units according to the whole of disciplines. The present paper illustrates a reflection on a G.I.S. application in a land suitability study on a sub-regional area of Southern Italy. Emerging questions are related to the need to combine contributions of all environmental information which are represented at different scales, with different interpretative models, with different precision of identification of landscape unit, etc.

1 Introduction

Urbanisation and the diffusion of haphazard settlements belong to the most important features of the spatial development in many industrialized countries. Current land use instruments are not enough to steer the environmental protection for several reasons: scattering of the extra urban settlement system, ongoing soil sealing by construction, considerable abandonment of potentially agricultural soils, increase in hydro-geologic hazard, pauperization of environmental and landscape valuable components are often conceived as a decrease of life quality in rural areas.

In urban planning, the attempt to change this tendency needs to consider many aspects as demographic dynamics, economic factors, environmental issues, protection of soils, morphological features, agronomic evaluations.

During the last three decades, after the fundamental contribution by Mcharg [1], several authors in the world developed evaluation procedures of land capability or suitability, based on overlay mapping techniques. One of the aims was the integration of all the aspects mentioned above. In fact, the evaluation of land use attitudes is a multidisciplinary question and can be supported by a multicriterial approach.

A good example of intersection between fuzzy aspects of multidisciplinary approaches in geographical classification and multicriterial evaluation is represented by the identification of boundaries of areas suitable for the realisation of settlements. In that case the amount of features to be considered in defining urbanised areas (e.g. by the definition of a combinatory rule of criteria such as density of settlements, land economic value, degree of urbanisation, degree of naturality and so on) has some characteristics of a fuzzy multicriterial evaluation problem [2]; on the other hand, the intrinsic fuzzy characteristics of geographical elements to be identified (e.g. by the identification of thresholds of settlement density, land economic value, degree of urbanisation, degree of naturality and so on) link to the concept of fuzzy boundary of a region.

This paper reports about G.I.S. application in the experience of a rural planning carried out in Italy. The plan proposes a reorganization project for both insediative and infrastructural systems, to thicken the existing settlements with disadvantage for development of new residential areas in open territory; it favours actions of building completion and new transformation, inside the centres, or of intensifying the centres themselves.

To achieve this purpose a mechanism of transfer of development rights has been adopted in transformation areas from zones in which promoting the export of volumetric rights was suitable to guarantee the safeguard of environmental, naturalistic and agricultural characteristics, in agreement with the Administration.

This paper aims to compare the classical approach to land suitability through G.I.S. with an approach based on G.I.S. and fuzzy evaluation.

2 G.I.S. and Classification by Sharp Boundaries

In Italy recent developments in regional planning systems give great emphasis to the production of maps, which represent both actual and more suitable attitudes in land use, as a support for developing local planning instruments. This operation can be carried on only by the support of G.I.S. application, and evaluation routines. In the present case of study the classical approach to land suitability was adopted, using sharp boundaries and definite thresholds. According to this method the following inclusive rules were developed:
1. better accessible areas: after the classification of road network relative to its functional role, 100 m buffers starting from the upper level road network were created. In this way areas with a better accessibility, defined areas close to the road network, were identified;
2. small rural settlements:
    - 137 small rural settlements were defined through a simple geometric rule (at least 20 buildings and a distance among them not less than 25 m);

- a proximity area for each rural settlement was defined by means of the representation of a 250 m buffer around each of them, according to administrative directives;
- areas close to the road network were intersected with the rural settlement buffers to determine areas close to both road network and small rural settlements;
- the remaining part of centre buffers, outside the small rural settlement perimeter and not comprised in road network buffers, were defined as areas close to small rural settlements.

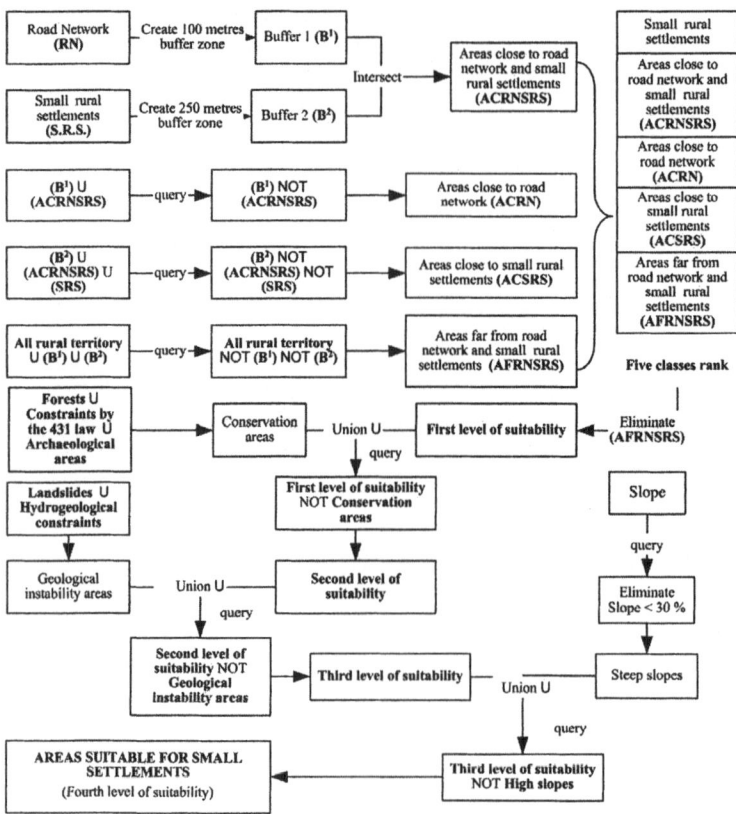

Fig. 1. Flow chart of land suitability procedure

These rules determine the first level of suitability, and are schematized in figure 1. The degree of suitability characterizes a first level of attitude to the urbanisation. The last level of suitability for settlements was determined by means of a methodology similar to the one used by Mcharg [1] having the remarkable advantage of using the G.I.S. (figures 1, 2, show more clearly this methodology). In all rural territories the four classes described above were taken in account, by subtracting all factors considered poorly compatible with settlement aims. Forests, areas close to rivers and higher than 1200 m a.s.l., archaeological sites, landslides, hydro-geological constraints and high slopes were not considered as suitable areas. The reduction can be seen clearly in geographic components (see figure 2): a part of the previous four classes has been eliminated in all the steps among the levels of suitability. When looking at the buffer of small rural settlements (figure 2, lower right corner) this kind of reduction appears obvious.

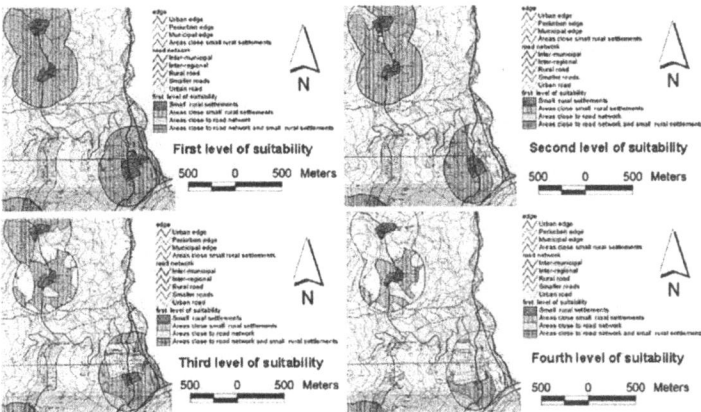

Fig. 2. Layout of the four-steps of the progressive selection process

3 The Fuzzy Classification

The planning context creates further constraints in representing spatial informations. The difficulty of identifying certain boundaries in spatial analysis is not considered in the institutional planning system, especially in the field of land use planning, where physical limits of spatial categories correspond unequivocally to the limits of spatial prescriptions. The geographical classification and the estimate of land use attitudes show operational and evaluative issues which are interesting. The geographical classification of land use deals with problems of uncertainty, which typically are the topic of G.I.S. literature.

The uncertainty of analysis needs to be solved necessarily within the land-use plan, characterised by "institutional certainty" of boundaries and land use rights. A possible perspective is to transform the dichotomy $(0,1)$ = (precise, imprecise) in a passage to a ownership function (by the use of fuzzy logic), which varies for each category of spatial entities. The identification of boundaries appears often as a "fuzzy" question. In fact, Couclelis [3] refers to ill-boundaries, Leung [4] to core areas with maximum degree of certainty, and so on. The accuracy in definition of "border" varies along a line. The integration of elements of fuzzy set theory in geographical database applications can give a measure of the variation of accuracy along the border line of areas identified in land use maps. The question of boundaries is usually treated as a question of threshold definition. A hill-shaped ownership function, for instance, can be used to represent a spatial attribute which varies from a core area to a background area. The threshold is a horizontal line which defines the passage from an unacceptable ownership value to an acceptable one. But the hill-shaped ownership functions vary by each entity, as well as the threshold which distinguishes core and background areas [5]. Consequently we can find easy thresholds for hill-defined contiguity of entities when existing a symmetric transition from one attribute to another, or very difficult conditions, when contiguous attributes vary by very different hill-shaped ownership functions.

The second situation is quite frequent in spatial analysis. The characteristic of a fuzzy information depends on several factors. An information can be intrinsically fuzzy or not. An information can appear fuzzy in some conditions, and can become crisp in some other conditions. In this case the information is not intrinsically fuzzy (fig.3).

The causes of fuzziness in information can be roughly considered as follows:
- **lack of information:** the nature of the information/phenomenon is quantitative, but there are not enough data and the expression of the information is anyway qualitative (*e.g.* the information/phenomenon "employment rate" can be expressed by a percentage, but when data are missing, it can be expressed by expert's judgement, such as "high level of employment" or "low level of employment");
- **complexity:** due to the multiplicity of factors, the information is so complex that it is impossible to be completely expressed by any kind of reduction to quantitative data. In this case each subject expresses a judgement according to complex implicit mental relationships between various factors (*e.g.* the aesthetic value of a landscape can be considered as a subjective value, or as the synthesis of several judgements, expressed by the use of qualitative terms).

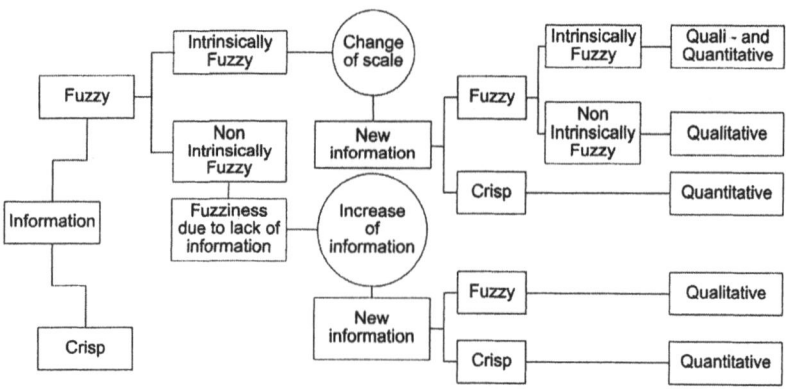

Fig. 3. Fuzzy and crisp information

The intrinsically fuzzy information is expressed by linguistic variables, which are not referable to quantitative information. A high level of precision can be obtained only when studying simple spatial systems.

As Zadeh [6] postulated, precision and complexity are often based on conflict. The principle of incompatibility affirms that when system complexity increases, the possibility of describing the behaviour of the same system in a precise way decreases until the complexity leads to the impossibility of description. In planning questions the character of fuzziness is also related to the spatial character of information.

Starting from traditional objects with sharp boundaries defined as Crisp-Crisp Object (CC-Object) Cheng et al. [7] classified fuzzy objects according to the following three patterns: Crisp-Fuzzy Object (CF-Object), with well defined boundaries and uncertain content; Fuzzy-Crisp Object (FC-Object), with precise thematic content and undefined spatial edge; Fuzzy-Fuzzy Object (FF-Object), with uncertainty in both contents and boundaries.

Two different ways of representing spatial data in G.I.S. application can be often conceptualized differently under field and object views. The first approach allows to assimilate the space to a grid, where each element is considered homogeneous. The space is therefore subdivided in finite elements, to be classified according to a multidimensional analysis. Each spatial information represents one dimension of the analysis. The boundary of an object is defined *a priori*, corresponding with the border of the element of the grid; the fuzziness factor can concern the attribution of each element to a cluster or to another. The fuzziness is managed by substituting the traditional way of classification with a fuzzy clustering. In this case a threshold defines the minimum value of acceptability of the membership function, which expresses the belonging to each cluster. The spatial partition is represented by the expression in figure 4 (left panel) and equation 1:

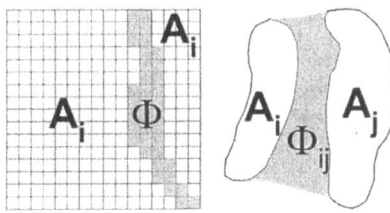

Fig. 4. Field fuzzyness (left) and Object fuzzyness (right)

$$S = U_{i=1}^{n} A_i + \Phi. \tag{1}$$

$$S = U_{i=1}^{n} A_i + U_{ij=1}^{n} \Phi_{ij}. \tag{2}$$

where Ai represents all finite elements of the grid which belong certainly to a cluster, and Φ represents the group of elements without certain membership to any cluster.

On the contrary the second approach is characterised by punctual, linear and polygonal elements. The density of these elements defines the belonging to a cluster. In this case the fuzziness factor of clustering is related to the threshold value for density of each type of punctual elements; the traditional way can be substituted by the construction of a set of complex fuzzy rules to identify different fuzzy clusters. In figure 4 (right panel) and equation 2 Ai represents all finite elements, which have certain spatial definition, and Φij represents the element ij which lies between the elements Ai and Aj without certain spatial definition. The spatial partition is therefore composed by geometric elements which can be identified by a core (certain) and a background (uncertain) area.

In the case of study we reasoned about the variability of the borderline situation. The boundary of Ai and Aj in fact, can depend on several conditions. These conditions can coexist or cannot occur along the borderline of Ai and Aj. This is the reason of the variability of uncertainty of boundary definition along the borderline of Ai and Aj. This means also that Φij represents a combination of conditions which can appear fuzzy along some partitions of the boundary, but can appear crisp along other partitions of the same boundary.

The function Φ in both models represents the grey zone, *i.e.* the transition between two core areas. It can be considered as an ownership function which varies form 0 to 1. Analyzing the previous land suitability case, we can consider the function Φ as a combination of the inclusive rules (areas close to the road network and areas close to small rural settlements), defined by the following membership functions, respectively:

$$\mu_A = \begin{cases} \dfrac{x}{10}, & 0 > x \leq 10 \\ 1, & 10 < x \leq 100 \\ 0, & x > 100 \end{cases} \quad \mu_B = \begin{cases} 1, & x \leq 100 \\ \dfrac{250-x}{150}, & 100 < x \leq 250 \\ 0, & x > 250 \end{cases}. \quad (3)$$

These functions can have many shapes, according to different requirements. For instance the areas close to the road network and the areas close to small rural settlements (see figure 5) are represented with two different trapeziums.

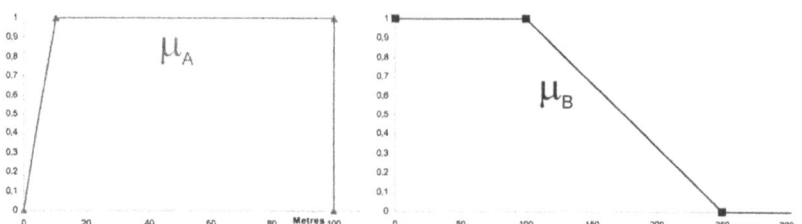

Fig. 5. Areas close to road network (left) and Areas close to small rural settlements (right)

It is possible to combine the last two images following the rule of the fundamental operation of fuzzy set. In this way we can identify the same classes of suitability achieved with sharp boundaries. For instance, areas close to both the road network and small rural settlements, obtained as the intersection of areas close to small rural settlements and areas close to road network, can be achieved as the intersection (fig.6):

$$\mu_A \cap \mu_B(x) = \min[\mu_A(x), \mu_B(x)]. \quad (4)$$

G.I.S. and Fuzzy Sets for the Land Suitability Analysis   1043

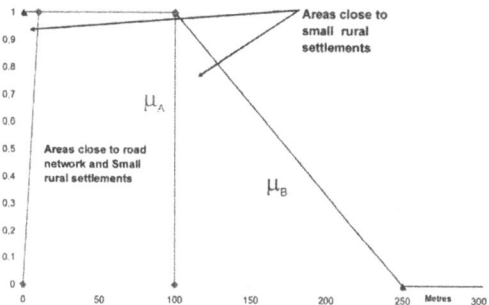

Fig. 6. Areas close to road network and small rural settlements

In the same way it is possible to consider the exclusive rules taking in account all factors poorly compatible with settlement aims, such as areas in geological hazard, on steep slopes or in forests. For the sake of clearness the exclusive rules can be grouped in two different classes. The first includes all the features that can be classified with sharp boundaries which identify an abrupt transition; for instance areas close to the river, upper than 1200 m a.s.l., archaeological sites, landslides and areas in hydro-geological hazard can be considered without a gradual transition.

The second class considers all the features that present a certain degree of fuzziness as the slopes and the forests; in these two cases it is possible to realize two ownership functions µ, related to the function Φ previously defined.

$$\mu_c = \begin{cases} 0, x = 0 \\ \dfrac{x}{30}, 0 < x \le 30 \\ 1, x > 30 \end{cases} \quad \mu_d = \begin{cases} 0, x \le 10 \\ \dfrac{x-10}{150}, 10 < x \le 150 \\ 1, x > 150 \end{cases}. \tag{5}$$

The function μ_c represents the slope suitable for settlement aims; its value cannot exceed 30 % (see figure 7 left panel).

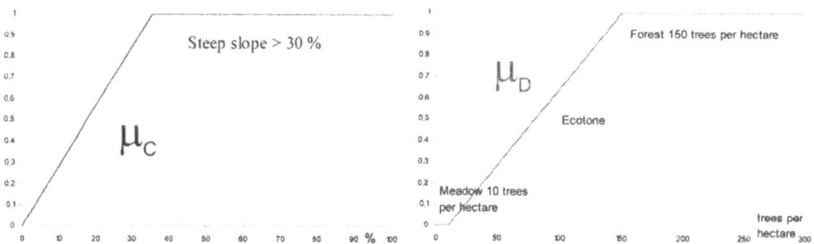

Fig. 7. Areas in steep slopes (left) and Areas in forest (right)

In general it is possible to define a "forest" if stand density is greater than 150 trees per hectare. The function μ_d represents the transition among meadow, ecotone and forest (see figure 7 right panel).

4 Final Remarks

The relationship among fuzziness and classical planning theory has been described by Leung [4] through four classes: going from a typical Rational Comprehensive kind of plan to a Strategic one.

The first class is an ideal case where it is possible to build a plan with certain objectives based on certain analyses. The result is a set of discrete representations of the physical space which correspond to a set of different values, referring to areas considered homogeneous according to each disciplinary point of view. This situation involves the production of a huge number of polygons, which are difficult to combine for creating landscape units according to the whole of disciplines. The second class is the case of a plan with crisp boundaries, based on fuzzy analyses. If data are imprecise, human evaluation and experience are employed. A typical example is the need to give an answer to a specific problem with certain economic resources. The third class is very frequent in sector plans (*e.g.* transport) where the need to achieve widely flexible results and the possibility to obtain several alternatives correspond to a certainty of data. The fourth class is the case of the strategic plan which identifies the fundamental issues and purposes driving the update process. This is a typical cyclic process where all original choices can be modified and the policy is continually made and re-made, avoiding errors related to radical changes in policy. In this case we have the combination of a fuzzy objective based on a fuzzy analysis.

The planner has his own patterns, both classificatory and decisional, often far from the data model processing schematized in three steps by Molenaar [8]: acquisition oriented data model, query oriented data model and output oriented data model.

Each discipline involved in environmental planning uses a different approach to represent its own vision of reality. Geological sciences or hydraulics evaluate risks by consistent mathematical models which are relevantly different to non linear models utilised in the field of ecology, and at the same time information about significance and value of cultural heritage in a given environment does not easily correspond to value attribution. Different ways of managing values correspond to each method of giving information.

The growing importance of the relationship between interdisciplinarity and technological innovation in environmental analyses connected to plans leads to a dangerous consciousness about the possibility of managing great amounts of data, by the use of Geographic Information Systems.

Normally, planners face with incomplete, heterogeneous and multiscale spatial information which constitutes a weak support to the construction of prescriptive land use plan. Normally there is a difference of scale between each kind of analyses (*e.g.* settlement system and environmental system); if data with a different accuracy are combined adopting a common scale there is the risk of an error propagation and the results can be meaningless or potentially dangerous [9].

This situation, widely diffuse among planners, is related to the attitude of managing data and objects in order to build complex spatial analyses, starting from mono-thematic studies or simple spatial data. In this case, planners often forget that the original sources of data have different degrees of precision and that geographic objects have different spatial characters. Synthetically, planners using G.I.S. developed the attitude to cross and overlay, without taking in account some relevant differences between objects to be mixed. This attitude becomes more significant, when the approach to G.I.S. tends to build continuous spatial representations, closer to the real world [10]. Entities become objects in the planner's perception. Often objects do not have crisp boundaries or do not have boundaries at all [11], [12]. In many cases geographic entities are compressed in a crisp boundary described by a single attribute; this representation often is very far from the reality. Contextually a new difficulty rises to the fore, regarding the need of traducing G.I.S. supported complex analyses and evaluative routines in planning instruments which have only crisp definition of zoning, due to their normative issues in land-use regulation. In fact, as Couclelis [3] highlights, spatial analyses in planning context are characterised by uncertain nature of entities and uncertain mode of observing real world, even if the user's purpose is certain (land-use zoning).

References

1. McHarg I. L.: Design with Nature. The Natural History Press, Garden City, New York (1969)
2. Malczewski J.: Gis and Multicriteria Decision Analysis. John Willey & Sons Incorporated, Toronto (1999)
3. Couclelis H.: Towards an Operational Typology of Geographic entities with ill-defined boundaries. In Burrough P. A., Frank A. U. (eds): Geographic Objects With Indeterminate Boundaries. Taylor and Francis, London (1996) 45-55
4. Leung Y.: Spatial analysis and planning under imprecision. Elsevier Science Publishers, Amsterdam (1988)
5. Jawahar C.V., Biswas P.K. and Ray A.K.: Investigation on fuzzy tresholding based on fuzzy clustering. Pattern Recognition Vol. 30, No.10. Elsevier Science Ltd., Great Britain (1997) 1605-1613
6. Zadeh L. A.: Outline of a new approach to the analysis of complex systems and decision processes. IEEE Trans. On Systems, Man, Cybernet, Vol. 3 No.1 (1973) 28-44
7. Cheng T. Molenaar M. and Lin H.: Formalizing fuzzy object uncertain classification results. International Journal of Geographical Science, Vol. 15, No. 1. Taylor and Francis, London (2001) 27-42
8. Molenaar M.: An Introduction to the Theory of Spatial Object Modelling for GIS. Taylor & Francis, London, (1998)
9. Zhang J., Goodchild M.F.: Uncertainty in geographic information. Taylor and Francis, London, (2002)
10. Johnston K.M.: Geoprocessing and Geographic Information System Hardware and Software: Looking towards the 1990's. In H.J. Sholten and J.C.H. Stillwell (eds): Geographical Information System for Urban and Regional Planners, Kluwer Academic Publisher, Dordrecht (1990) 215-227
11. Couclelis H.: People manipulate objects (but cultivate fields): beyond the raster-vector debate in GIS. In Frank A.U., Campari I., Formentini U. (eds): Theories and Methods of Spatio-temporal Reasoning in Geographic Space. Computer Science, Springer, Berlin (1992) 45-55
12. Burrough Peter A., Frank Andrew U. (eds): Geographic Objects With Indeterminate Boundaries. Taylor and Francis, London (1996)

Intelligent Gis and Retail Location Dynamics: A Multi Agent System Integrated with ArcGis

S. Lombardo, M. Petri, and D. Zotta

University of Pisa, Department of Civil Engineering, Via Diotisalvi, 2
56126 Pisa, Italy
{s.lombardo, m.petri, d.zotta} @ing.unipi.it

Abstract. The main step towards building "intelligent" Gis is to connect them with spatial simulation models. Multi Agent Systems (Mas) allow to represent, through a computer code, the behaviour of entities operating in a given environment and the system dynamics that derive from the interactions of such agents. We integrated (through the VBA programming language) a Mas into a Gis, where, through a friendly interface, it is possible, during the simulation, to modify the model by adding individual behavioural rules and/or new typologies of agents. In our urban Mas, two typologies of agents are defined: retail users and retail entrepreneurs, interacting according to a spatial demand-supply matching mechanism. We describe a prototypal application of a model whose aim is simulating, in an urban system, the dynamics of retailing location.

1 The System Structure

The main step towards building "intelligent" Gis is to connect them with spatial models. Multi Agent Systems (Mas) allow to represent, through a computer code, the behaviour of entities operating in a given environment and the system dynamics that derive from the interactions of such agents. A Mas is a system which allows the emerging of a kind of synergy that could be unforeseeable by considering the sum of effects of single agents actions: this synergy is an emergent property of the system as a whole.

We integrated (through the VBA programming language) a Mas into a Gis where, through a friendly interface, it is possible, during the simulation, to modify the model by adding individual behavioural rules and/or new typologies of agents. The system structure is shown in figure no.1. The input data of the GIS are provided by DBMS and directly by the user; the data are processed by the GIS. At the end of each cycle of the dynamic simulation, the outputs, the new distribution of the retail services and the structure of the transport flows are recorded in the DBMS. When the simulation stops, that is to say when the system reaches the equilibrium or the given maximum number of iterations, the final outputs, processed by GIS, are visualized using thematic maps. A relevant component of this tool is the friendly interface which were designed to make easier the data entry and the interpretation of the outputs.

The simulation starts through a button added to the ArcGis 8.1 interface, and a window appears which allows to select the input data. They appear summarised in a window and then some "forms" ask what are the changes the user wants to introduce in the system during its dynamics: it gives the opportunity to introduce exogenous intervening changes in the properties both of agents and of environment.

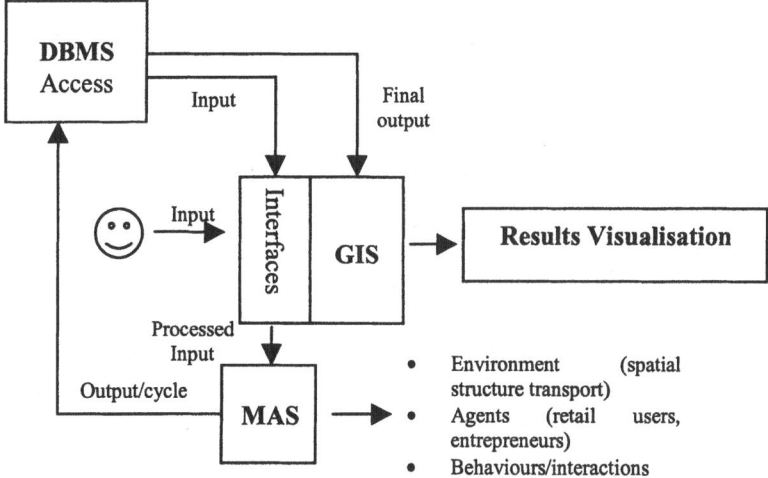

Fig. 1. The system structure

In the following section we describe the framework of this model, whose aim is simulating, in an urban system, the dynamics of retailing location. In our Mas, two typologies of agents are defined: retail users and retail entrepreneurs, interacting according to a spatial demand-supply matching mechanism.

2 The Framework of the Model

The framework of our Mas is characterized by three fundamental components: the environment in which the agents operate, the agents themselves (including their aims and their choice processes), and the interaction mechanism which defines the evolution of the system dynamics.

2.1 The Environment

In the prototypal application the environment is characterised by two components:
- the spatial framework, cell space: a grid of 8 cells with different sizes;
- the transport infrastructure. The transport network is represented by a graph and its quality and efficiency is measured by the travel time from cell x to cell y of the spatial grid.

2.2 The Agents

An agent, which represents the elementary entity of a Multi Agent System can be defined as a "real or virtual entity which is capable of acting in an environment, which can communicate directly with other agents, which is driven by a set of tendencies (in the form of individual objectives or of a satisfaction/survival function which it tries to optimize), which possesses resources of its own, which is capable of perceiving its environment (but to a limited extent), which has only a partial representation of its environment (and perhaps none at all), which possesses skills and can offer services, which may be possibly be able to reproduce itself, whose behaviour tends towards satisfying its objectives, taking account of the resources and skills available to it and depending on its perception, its representation and the communications it receive." [11]

In our Mas, two typologies of agents are defined: retail users and retail entrepreneurs. Both of them are spatial agents which are represented as aggregated entities whose behaviour is described by the evolution of the variables used to characterize the entities themselves. In this model, "aggregated entity" means that each retail user is characterized by the number of the consumers in the cell where the agent is located; and that the retail entrepreneur is represented by the size of the services located in the cell (this size is measured by the number of the service employments).

Agents are able to undertake actions that can modify the environment and, subsequently, the other agents decisions. For example, during the simulation of our Mas, congestion can appear in some transport network routes; this phenomenon modifies the accessibility of some cells (measured by travel time), and, consequently, it modifies the environment in which the agents act. Therefore, the agents choices will be affected by the new accessibility, that is to say new behaviour patterns of the agents, retail users and retail entrepreneurs, will appear.

Agents have their own objectives and act in order to reach these aims which, sometimes, can be represented by functions to be optimized. In our model, the retail user aim is to reach the services trying to minimize costs and disadvantages; the aim of the retail entrepreneur is to locate and dimension their enterprises trying to get the largest profit.

Agents have a partial representation of the environment: the retail user knows the state of each retail entrepreneurs, but they do not know the choices of the other retail users (they see "signals" of such choices, e.g. through congestion phenomena), the retail entrepreneur knows only the retail users, which make use of that service, but it does not know anything about the state of the other retail users, and, consequently, it does not know the state of the other retail entrepreneurs.

Retail Users. The aim of these agents is to reach the services trying to minimize costs and disadvantages. In order to choose, they assess:
- the quantity of services in each cell of the system
- the travel time to go to the services located in each cell of the system
- the service agglomeration level in each cell of the system
- the "comfort" of travelling to reach the services located in the different cells

Because the above variables levels change in function of the retail users and entrepreneurs choices, the assessment and choice process is continuous and triggers the system dynamics.

Retail Entrepreneurs. The aim of these agents is to locate and dimension their enterprises trying to maximize their profit. In order to choose, they assess:
- the choice behaviours of the service users
- the installation and management costs in each cell of the system

The system would tend to the equilibrium between the demand and supply, but it is "disturbed" by the changing preferences of the agents.

2.3 The Behaviour Rules and the Interaction Mechanism

System dynamics can be described as follows:
- a grid that includes 8 cells (variable)
- an initial spatial distribution of the retail users (not variable)
- an initial spatial distribution of the retail services (variable)
- an initial accessibility for each cell (variable)
- a preference level of the retail users for the agglomeration of the services (variable)
- a "comfort" level of the retail users to travel from one cell to another cell
- an installation and management aggregated cost of the service in each cell (variable)
- the hypothesis that the system tends to a local equilibrium between demand and supply of services.

Retail users located in a cell "x" of the spatial grid "see" and assess the initial distribution of the retail services, the initial travel time among the cells, the initial "comfort" level to travel from one cell to another cell, their own preference level for the agglomeration of the retail services, and then they choice the cell where to go.

Retail entrepreneurs react to the retail users choices by deciding to open or close or increase or reduce their activities. In order to make a decision, they assess the installation and management costs of the service in the cells and compare them with the revenue they can get there. In this way, a new spatial distribution of the retail services appears, and this new distribution produces changes as to choices of the retail users which can stimulate the retail entrepreneurs to make new location choices.

3 The Experiments

In this section we discuss the prototypal application results of the model in an urban system. The aims of the experimentations are, on one hand, to test the model technical strenght, and, on the other hand, to test the model ability to simulate possible scenarios. To this end, the urban system built for the experiment is much smaller and simpler than a real one, in order to be able to interpret easily the results and then evaluate the model's simulation capabilities.

The dynamics of our Mas include three different speeds of change:
- high speed for the changes of the agents-retail users preferences as a reaction to environmental changes

- medium speed for the changes of the entrepreneurs choices
- much slower speed for the environmental changes (i.e. a new residential area).

In our experiments, we assumed that the system speed of change is proportional to the speed for the changes of the choices of the entrepreneurs, while that of the agents-retail users preferences is, in comparison, instantaneous. The environmental changes, being much slower, are, for the moment, treated as exogenous.

3.1 The Urban System

The urban system (fig. no. 2) is a cell space: a grid of 8 cells with different sizes, in which are located 80.000 retail users and 12.000 retail entrepreneurs (table no. 1). The introduction of the cell no. 9 (fig. no. 2) at a given step of the simulation is considered as an environmental exogenous change which depends on local administrators and/or firms decisions.

Table 1. The distribution of retail users and retail entrepreneurs at time t=0

	CELL								
	1	2	3	4	5	6	7	8	9
USERS	12.000	20.000	15.000	5.000	10.000	7.000	4.000	7.000	4.000
SERVICES	3.000	3.000	1.500	600	1.500	1.000	500	900	600

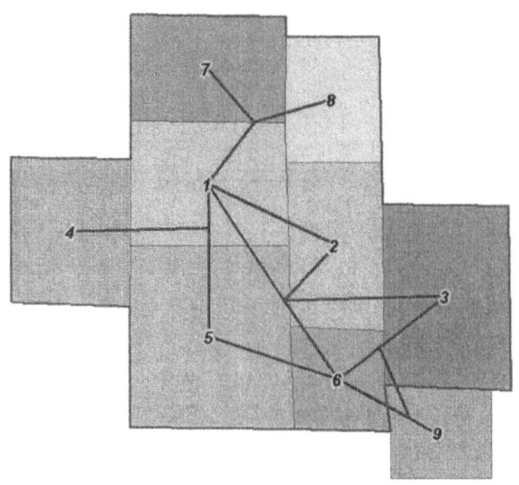

Fig. 2. The urban system

The retail users and entrepreneurs initial distribution and the transport network structure are based on a general and spread idea of an urban development; the centre of the system (cell no.1) is well endowed with retail services; the development of the system in south-east direction generates an area (cell no. 2) which is characterized by relatively good level of retail services and accessibility, and a peripheral area (cell no. 3) which is densely populated, but is lacking in retail services and accessibility.

3.2 The Simulation of Scenarios

In this section we briefly discuss some results of the simulated scenarios. The different scenarios related to both the changes of retail users preferences, and environmental changes which occur adding other agents, particularly policy maker and firms who decide to carry out a new residential area (cell no. 9, fig. no. 2).
The simulated scenarios are:
scenario I: "natural" evolution of the system, without changes in the preferences and not adding new agents;
scenario II: introduction of users sensitivity to the transport network congestion that can take place on some routes during the system's evolution;
scenario III: introduction of new actors by the addition of a new cell (i.e. an environmental change)
scenario IV: the combination of scenario II and scenario III

Scenario I. This scenario simulates a natural evolution of the system, there are not changes in the preferences of the service users and entrepreneurs and no new typology of agents is added.
The simulation results show the centrality role of the cell no. 1 (fig. no. 2). The initial condition of this cell is characterized by a high level of accessibility from the other cells of the system, and by a high number of retail services.
At the end of the simulation, the cell no. 1 is again the cell which has the highest concentration of services. The services increase (they double, fig no. 3) and the flow of service users from other cells increases at the same time (fig. no. 4). The peripheral area, cell no. 3, shows a different dynamics from the other ones: the service users located in this cell generate an increase of flows increase in the cell itself. This event makes the cell almost self-contained. In this scenario, the main factor which influences the agents choices is the accessibility. Under the same conditions, a good level of accessibility makes the central cell more suitable than the others.

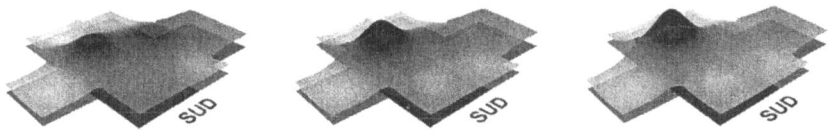

Fig. 3. The progressive growth in zone no. 1 (the decrease in zone no.2 can also be seen). Starting from the left, the services at time t=0, t=14, t=22

Scenario II. In this scenario, we introduce the sensitivity of the service users to a given level of transport network congestion.
The system dynamics is the same as the scenario I until a level of transport network congestion is reached at about 11th cycle. Starting from this point, the system dynamics totally changes: the central area (cell no. 1) loses its supremacy, and it generates flows towards other system areas, mainly towards cell no. 5. The above mentioned area that in scenario I was a satellite of cell no. 1 becomes an important zone endowed with retail services which increase more and more after the 11th cycle.

The peripheral area (cell no. 3) is characterized by the same dynamics of the scenario I and the bad accessibility towards the central areas makes the service users located in this zone not affected by the transport network properties of the whole system. The presence of the transport network congestion makes cell no. 2 self-contained. Moreover, the good physical accessibility is not sufficient to ensure the development of a given area; the "comfort" level of the retail users to travel from one cell to another one plays an important role in the choice of the users.

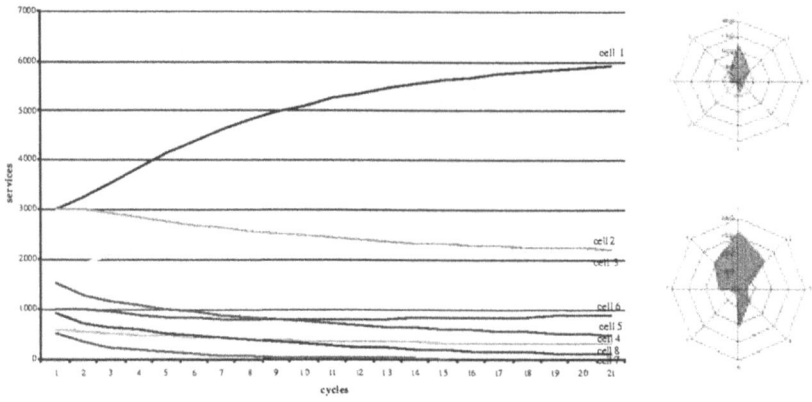

Fig. 4. On the left, the variation of the services during the simulation. On the right, flows entering in cell no. 1: on top, the initial flows; below, the final flows

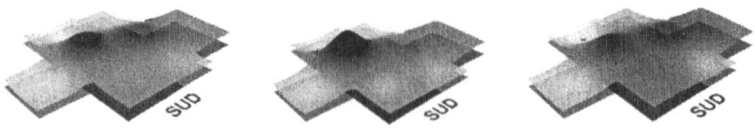

Fig. 5. The progressive growth and decrease in zone no. 1. Starting from the left: the services at time t=0, t=14, t=22

Scenario III. In this scenario we introduce new actors adding a new cell to the urban system (cell no. 9, fig. 2). We suppose that this cell is a new residential area which could represents, for example, the intervention of a plan. The new area is added at about 15th cycle, it is located in the south-east of the urban system and is lacking in accessibility towards the other areas. The system dynamics is the same as the scenario I, except for cell no. 6 whose retail services increase more than in the scenario I. It is reasonable to claim that also in this case, the good infrastructural system is the most important factor in the users choices. The bad accessibility forces the users to stay in the area where they are located (see for example the dynamics of the cell no. 3) or to move towards the nearest area endowed with services, as the dynamics of the cell no. 9 in this scenario.

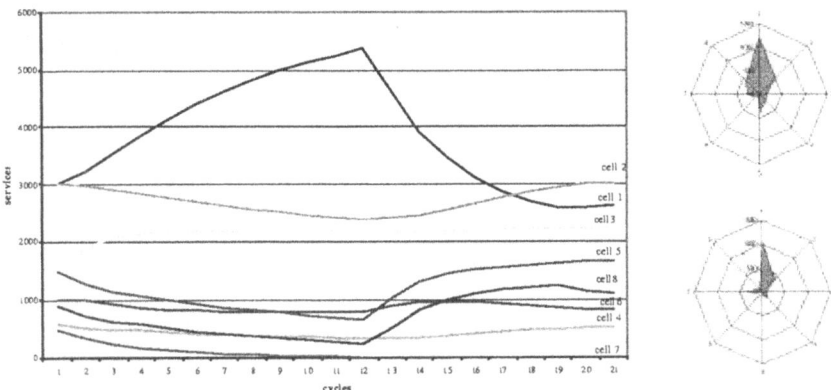

Fig. 6. On the left, the variation of the services during the simulation. On the right, flows entering in cell no. 1: on top, the initial flows; below, the final flows

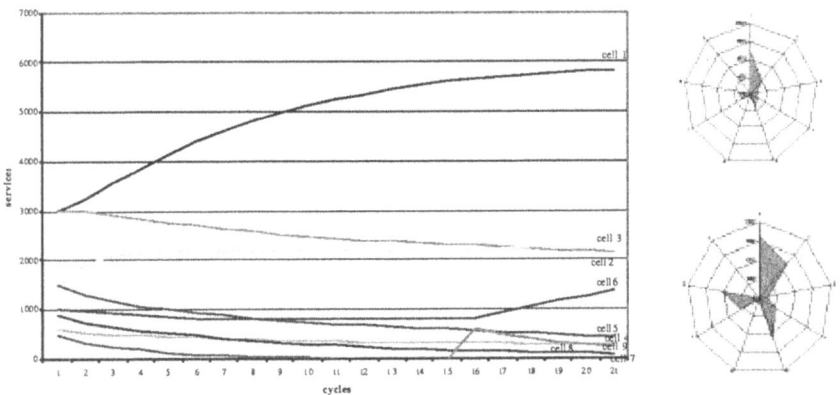

Fig. 7. On the left, the variation of the services during the simulation. On the right, flows entering in cell no. 1: on top, the initial flows; below, the final flows

Fig. 8. The progressive growth in zone no. 1. Starting from the left: the services at time t=0, t=15, t=22

Scenario IV. In this scenario we introduce both the sensitivity of the service users to a given level of transport network congestion and new actors adding a new cell to the urban system (cell no. 9, fig. 2). The system dynamics is the same as the scenario II, except for cell no. 6. The retail services located in this area increase more and more after the 15th cycle, until the 25th cycle. At this point of simulation transport network congestion appears near cell no. 6. This phenomenon makes retail services in cell no. 6 decreasing, and, in turn, retail services located in cell no. 9 (new residential area) increase more.

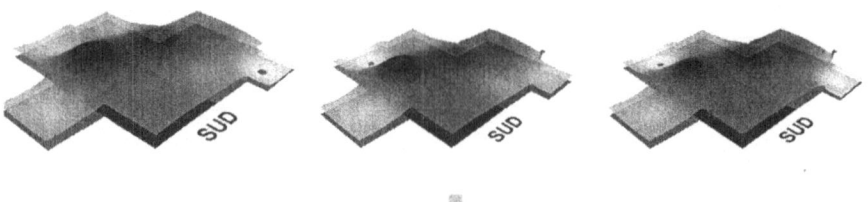

Fig. 9. The progressive growth and decrease in zone no. 1 (the growth in zone no.9 can also be seen). Starting from the left: the services at time t=0, t=14, t=22

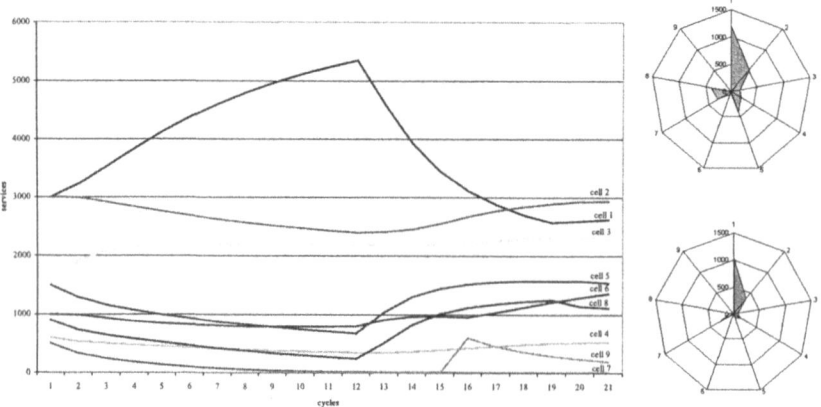

Fig. 10. On the left, the variation of the services during the simulation. On the right, flows entering in cell no. 1: on top, the initial flows; below, the final flows

4 Conclusions

As said before, Mas allow macro level patterns of an urban system emerge from micro level interactions among the actors of the system and then represent a powerful tool to try to understand the mechanisms that underlie the evolution of spatial urban patterns. Moreover, by building an appropriate implementation, it is possible to overcome the more traditional "one shot" dynamic simulation: indeed, in our model, it is possible, during the simulation, to modify endogenously and/or exogenously individual behavioural rules and/or to add new typologies of agents.

The results of the developed simulations are satisfactory enough, as they reproduce our simple urban system evolution in likely way. The research agenda now includes:
- experiments on a real urban system (the data base for a part of Tuscany region is under costruction);
- introduction of institutional of agents, whose behaviour is not exogenous for the system as it is strongly influenced by the choice behaviour of the other agents;
- improvement of the representation and simulation of the decision making processes which underlie the urban systems dynamics.

References

1. Allen P.: Spatial models of evolutionary systems: subjectivity, learning and ignorance.In Pumain D. (eds): Spatial Analysis and Population Dynamics. John Libbey-INED, Congresses and Colloquia, 6, Paris (1991) 147-160
2. Axtell R.: Why agents? On the varied motivations for agent computing in the social sciences. Center on Social and Economic Dynamics, working paper no. 17, November 2000
3. Batty M., Jiang B: Multi-agent simulation: new approaches to exploring space-time dynamics within gis. Paper presented at the Annual Meeting of GISRUK '99 (Geographical Information Systems Research - UK), University of Southampton, 14-16 April, 1999
4. Benenson I. Multi-agent simulations of residential dynamics in the city. Computers, Environment and Urban Systems vol. 22, no.1, (1998) 25-42
5. Benenson I.: Modelling population dynamics in the city: from a regional to a multi-agent approach Discrete Dynamics. In Nature and Society 3, (1999) 149-170
6. Bregt A. K., Lammeren van R., Ligtenberg A.: Multi-actor-based land use modelling:spatial planning using agents. In Landscape and Urban Planning, vol. 56, Elsevier, (2001) 21-33
7. Bura S., Guérin-Pace F., Mathian H., Pumain D., Sanders L.; Multi-agents system and the dynamics of a settlement system. In Geographical Analysis, vol 28, no.2, (1996) 161-178
8. Bura S., Guérin-Pace F., Mathian H., Pumain D., Sanders L.: SIMPOP: a multi-agents system for the study of urbanism. In Environment and Planning B, vol 24, (1997) 287-305
9. Bura S., Guérin-Pace F., Mathian H., Pumain D., Sanders L.: SIMPOP, A Multi-Agents Model for the Urban Transition. In Sikos T. Bassa L. Fischer M. (eds): Recent Developments in Spatial Information, Modelling and Processing. Studies in Geography in Hungary, Budapest (1995)
10. Davidsson P.: Agent Based Social Simulation: A Computer Science View. In Journal of Artificial Societies and Social Simulation vol. 5, no. 1, (2002)
11. Ferber J.: Multi-Agent Systems: An Introduction to Distributed Artificial Intelligence. Addison-Wesley, Harlow (1999)
12. Ferrand N.: Modelling and Supporting Multi-Actor Spatial Planning using Multi-Agent Systems. Paper presented at 3rd National Center for Geographic Information and Analysis (NCGIA) Conference on "GIS and Environmental Modelling", Santa Fe (1996)
13. Gilbert N., Terna P.: How to build and use agent-based models in social science. In Mind & Society, no. 1, (2000) 57-72
14. Gimblett R. (edited by): Integrating Geographic Information Systems and Agent-based Modeling Techniques for Simulating Social and Ecological Processes. Oxford University Press, (2002)
15. Leung Y.: Intelligent spatial decisions support systems. Springer-Verlag, Berlin (1997)
16. Lombardo S. T., Rabino G. A.: First Experiments on Non-Linear DynamicModels of Metropolitan Areas: The Roma Case Study. In Atti delle Giornate di lavoro Airo, Guida, Napoli, (1983a) 575-592

17. Lombardo S. T., Rabino G. A.: Dinamiche non lineari dell'interazione spaziale e processi di polarizzazione urbana. In Atti della IV Conferenza italiana di scienze regionali, vol. 2, Firenze (1983b) 470-495
18. Lombardo S. T., Rabino G. A.: Un modello della localizzazione dei servizi commerciali e della mobilità dei consumatori. In Atti delle Giornate di lavoro Airo, Tecnoprint, Bologna (1985) 339-358
19. Maes P.: Modeling adaptive autonomous agents. In Journal of Artificial Life, vol. 1, (1994) 135-162
20. Mentges E.: Concepts for an agent-based framework for interdisciplinary social science simulation. In Journal of Artificial Societies and Social Simulation vol. 2, no. 2, (1999)
21. Nicolis G., Prigogine I.: La complessità. Esplorazioni nei nuovi campi della scienza. Giulio Einaudi editore s.p.a, Torino (1991)
22. Ostrom T.: Computer simulation: the third symbol system. In Journal of Experimental Social Psychology, 24 (1988) 381-392
23. Otter H. S., van der Veen A., de Vriend H. J.: ABLOoM: Location behaviour, spatial patterns, and agent-based modelling. in Journal of Artificial Societies and Social Simulation vol. 4, no. 4, (2001)
24. Paolucci M., Pistolesi G.: Simulazione Sociale e Sistemi Multi-Agente. Serie su Simulazione Sociale basata su Agenti, ThinkinGolem pscrl, Novembre 2000, www.thinkingolem.com
25. Sandip S.:, Multiagent systems: milestones and new horizons. In Trends in Cognitive Sciences, vol. 1, no. 9 (1997)
26. Terna P.: Simulation Tools for Social Scientists: Building Agent Based Models with SWARM. In Journal of Artificial Societies and Social Simulation, vol. 1, no.2, (1998)
27. Terna P.: Simulazione ad agenti in contesti di impresa. In Sistemi Intelligenti, 1, XVI, (2002) 33-51
28. Wilson A. G.: Complex Spatial Systems: The modelling foundations of urban and regional analysis. Prentice Hall, (2000)

ArcObjects Development in Zone Design Using Visual Basic for Applications

Sergio Palladini

Department of Geography, University of Calgary, Canada
spalladini@enghouse.com

Abstract. The Modifiable Areal Unit Problem (MAUP) is commonplace in the realm of geographical analysis. The MAUP presents issues concerning scale and configuration of geographic partitions that are representative of finer resolution spatial data. To date, these problems cannot be resolved completely. They can only be manipulated. This "manipulation" can be accomplished using an Automated Zoning Procedure (AZP). In theory, this algorithm amalgamates spatial partitions into a set of zones that are optimized for a function of interest, thus an optimal result is achieved regardless of spatial configuration. Using a Geographic Information Systems (GIS) software environment, an Automated Zoning Procedure was constructed for ArcGIS 8.3 desktop software using ArcObjects development with Visual Basic for Applications (VBA). The fully functional program represents a deterministic, global optimization approach that aggregates n number of zones into a smaller number of m zones where the Moran's I statistic of spatial autocorrelation is the primary function under optimization.

1 Introduction

Space, being it a continuous phenomenon, is difficult to study directly. Therefore, geographers and spatial analysts alike try to simplify space by classifying it into discrete entities of study in the hopes of achieving a better understanding of space and how it relates to other abstractions within the real world. Unfortunately, this approach presents a problem. How does one divide space into these discrete entities? Most researches have reasons, rules, criteria, and an appreciation of artistic design in order to answer such a question, but the harsh reality is; there is no universal standard that instructs one to quantitatively divide space into discrete entities. The scale at which space partitioning is done and the geometric representation of these discrete entities proves the basis of what is known as the Modifiable Areal Unit Problem (MAUP). This ongoing concern of scale and aggregation in geographical research and its relationship with descriptive statistics is less than complimentary. "How this problem may be resolved cannot be foreseen (choice of areal units). But it appears that men trying to develop cogent theories of areal structure will have to reckon with it for some time to come." (Duncan, Cuzzort, and Duncan, 1961). An instance of foreshadowing indeed, for forty-two years later the MAUP still plagues the realm of geographical thought and practice.

It is true, there is no way to completely solve for the MAUP. However, some work has been done in the area of zone design in which classifying space can be accomplished while "controlling" the modifiable nature of areal data. Openshaw (1976) and Openshaw (1984) suggests a method in which to design spatial partitions that represent a phenomenon of interest in an optimal manner. Following the advancement of computer aided mapping, Openshaw and Rao (1995) implemented this zone design methodology into a GIS software package. The concept behind this optimal zoning approach in combination with a GIS platform provides the basis of this document in which a version of this optimal zoning procedure is constructed in an ArcGIS 8.3 software environment using Visual Basic for Applications programming language.

The effectiveness of space partitioning and zone design algorithms has proven an excellent tool as a method of developing census geography for purposes of policy design and socio-economic study. The utility of automated zone procedures is evident as these algorithms have become an integral part in the construction of the United Kingdom's 2001 Census geography (Martin, Nolan, and Tranmer, 2001). The combination of GIS with zone design extensions is an invaluable resource for those involved with census activities.

The widespread classification and simplification of space into discrete entities for spatial analysis is commonplace in the realm of geographical research. This aspect of geographical analysis is not frowned upon, but ignoring the influence of the MAUP inherent with this approach is (Openshaw, 1984). The level of scale and choice of geographical partitions will ultimately affect the results of any spatially explicit analysis unless precautionary measures are otherwise taken. This choice of scale and aggregation is in essence a matter of personal preference influenced directly by study specific goals and so there is no universal solution. There are ways of defining space to overcome the MAUP, one of which revolves around the use of zoning procedures. These methods have proven to be quite effective when implemented. The MAUP will undoubtedly persist to be a "thorn" in the side of geographers for some time to come, but simply ignoring this phenomenon is intolerable and if done so conclusions based on areally constructed data will be anything but reliable. The effect of the MAUP in geographical theory and practice is not to be addressed lightly and appropriate methods should be implemented to control this problem inherent in spatially explicit research using aggregated data.

2 The Modifiable Areal Unit Problem

The modifiable areal unit problem, coined the MAUP, is a simple concept to understand, and yet from this simplicity arises one of the most difficult problems in spatial analysis to date. The effect of the MAUP on geographical inquiry is the result of two components; the problem of modifiable units and the problem of scale (areal aggregation), hence the name "Modifiable Areal Unit Problem". Changing the scale of a map from n number of zones to m number of zones and then following through with some statistical calculations will produce discrepancies between the two map scales even though the study area is left unaltered. As a result, changing the scale of areal units towards coarser or finer scales of resolution will definitely have an effect on any subsequent statistics performed. The problem of modifiable units concerns the geometric orientation of aggregated data at any one particular level of map scale. A map at scale n can have an infinite number of possibilities in which areal units can be spatially arranged. Altering the orientation of these areal units while keeping the scale constant can suffer from statistical discrepancies following any succeeding analysis on the final map as well as visual variations of space. A solid understanding of this phenomenon is established by Svancara, Garton, Chang, Scott, Zager, and Gratson (2002) using a theoretical dataset for simplistic purposes.

Clearly, the MAUP is worthy of mention and study as it pertains to the realm of analytical geography. Making inferences about geographic phenomenon recorded at the micro level based on aggregated data analysis is a dangerous practice that can yield misleading results, unwanted bias, and an overall misconception of the "true" phenomenon under examination.

3   A Geographical Solution

It seems as though the MAUP is here to stay. To date there is no universal solution to the MAUP due to the complexity of modifiable units and scale of analysis, which are quite often study dependent. However, not all hope is lost for a spatial analyst. An enormous amount of work has been done within this field of interest by Stan Openshaw beginning as early as 1976 in which, he proposes a geographical solution to the MAUP. Openshaw (1976) argues that values assigned to the independent parameters of a model are fashioned so that the model itself will reflect a phenomenon of interest as best as possible. Why not devise a geographical zoning system that follows the same mentality? Openshaw (1976) suggests a creation of partitions to represent underlying spatial data in such a way that the function of interest is optimally represented by such zones. Optimization in this context refers to the maximization or minimization of a chosen attribute towards a target value. From this idea spawned the Automated Zoning Procedure.

The AZP algorithm, as defined by Openshaw (1976) and again in by Openshaw (1984), can be summarized into a series of basic steps. Step one requires the generation of a set of random zones. These zones are then assigned to the individual data in question. The second step is the beginning of the actual optimization process. The algorithm should at this point randomly select a zone and examine the calculated result on the objective function following the combination of the initially randomly selected zone with one of its neighbors. This is done for each adjacent zone to that of the original. Once this is complete, the algorithm should merge the two zones that illustrate the greatest degree of change in the statistical function while maintaining contiguity between zones following a merge process. This is done for each original zone in the dataset. After each zone has been analyzed the algorithm should check to see whether or not any moves have been made. If so, then return to the second step of randomly choosing a zone and begin the amalgamation process again by searching for an improvement in the objective function. Otherwise, no more moves have been made and the algorithm should stop. Upon completion, the resulting configuration and number of zones represents the objective function in an optimal way.

The application of this heuristic procedure defines a set of geographically oriented zones that are of interest for a particular study and function under inspection. As a result, the MAUP does not have a significant effect on such zones due to the fact that these zones in question have been re-engineered to highlight a statistical function in an optimal manner, thus avoiding discrepancies in results as attributed to the characteristics of discretely represented space. It is important to realize that the design of such partitions have meaning, however, any one set of zones constructed using the AZP is quite often function specific and does not necessarily mean that a set of zones for one statistical parameter is the same for another even though the study data remains the same. More importantly, the AZP algorithm handles the unwanted effects of the MAUP in a geographic manner making it possible to execute spatial analysis on aggregated data regardless of scale.

3.1 A Version of the Automated Zoning Procedure

The concept of establishing a zone design algorithm for use with a common GIS software package represents the primary objective of the program associated with this document. More specifically, a version of the Automated Zoning Procedure algorithm as defined by Openshaw (1984) is developed for use in ESRI's ArcGIS 8.3 environment.

The program is designed to run from the core ArcMap application within ArcGIS 8.3. The algorithm is instructed to manipulate a map of classified zones into a smaller set of regions as defined by the user. The program will receive a polygon input file, and only a polygon input file, from the user as stored in a geodatabase file format. The user will then have the choice of an input variable for optimization, a function for statistical analysis (Moran's I for spatial autocorrelation), a target number of polygons in which to aggregate (a smaller number than the original set), and a file destination for the creation of a summary report. Once the user has satisfied all the input parameters the program can follow through with the execution. Upon execution the program will aggregate neighboring zones of all possible combinations in which the statistical function of interest shows the greatest amount of improvement and will continue to do so until the specified number of zones is reached or until there are no better moves that best optimizes the statistical parameter. The final product of the AZP will be a map output in which the original polygons are manipulated in such a manner that represents the variable in an optimal way with respect to the chosen statistical function. As a result, regardless of the scale (number of zones) required for a particular study, the areal units will be modified with respect to the statistical parameter of interest, thus controlling discrepancies associated with statistical output as influenced by the MAUP. The statistical parameter in this case represents the zone engineering attribute being optimized towards a target value. This can be done for any particular study in which information is tied to zones or regions represented as polygon data.

The Graphical User Interface. The graphical user interface was designed using Visual Basic for Applications. Since the AZP program relies heavily on user input, the form contains a series of combo boxes in order to assist the user in choosing valid parameters.

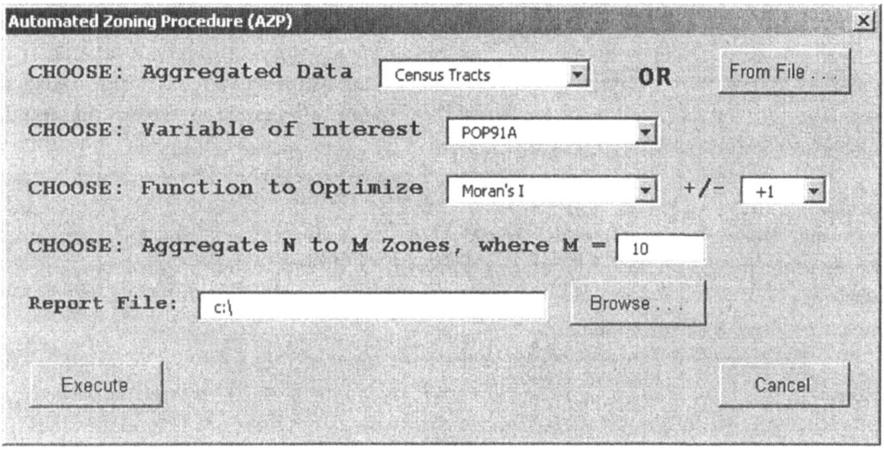

Fig. 1. The Automated Zoning Procedure graphical user interface.

The "Aggregated Data" combo box allows the user to choose an input layer file. This list is populated only when there are layers available in the ArcMap table of contents. If there are no layers present in the table of contents the "From File" button allows the user to browse for a specific feature class and add this feature class to the map.

The "Variable of Interest" combo box allows the user to choose one variable that he or she wishes to construct a zone system on. The variable list is only populated after the user selects a feature class as use for the input file. Next, the variable list will be populated with all the attributes associated with the feature class in question.

The "Function to Optimize" combo box allows the user to choose from a selected list of possible statistical measures to execute on the variable of interest. Currently, the AZP built for ArcGIS 8.3 only offers one statistical measure as a basis of zone engineering criteria. The function available is the Moran's I statistic which is a measure of spatial autocorrelation.

$$I = \frac{\sum_i \sum_j w_{ij} c_{ij}}{s^2 \sum_i \sum_j w_{ij}} \quad (1)$$

The variable w represents the weight given to each geographic entity reflecting its spatial relationship to the initial entity under examination. With regards to zone partitioning, w is given a value of one if a zone is adjacent to the initial zone in the dataset and a value of zero otherwise. The variable c represents the actual numerical value of the entity at that spatial location. For example, $zone_{ij}$ is equal to 244. The variable s^2 represents the standard deviation of the entire dataset. As a result, this statistic is standardized to provide values between positive one and negative one, approximately. Therefore, a "positive/minus" combo box is provided on the user form giving the option of defining a zoning system that can be optimized in either direction for spatial autocorrelation or towards a neutral value of zero. A value close to positive one indicates that neighboring polygons are very similar to one another in a spatial extent with respect to the variable under examination. A value close to negative one is the exact opposite and a value of zero represents no spatial autocorrelation between geographic entities.

The "Aggregate N to M Zones" parameter box allows the user to specify the target number of zones in which to amalgamate the original input layer file of N zones into an output layer file of M zones. In this case, M is less than N for the purpose of aggregation into a smaller number of partitions. A default value of ten resides in the input text box mainly for aesthetics and to provide an indication to the user as to the purpose of this AZP parameter.

The final required parameter before the execution of the AZP can continue is the "Report File" input text box. This text box requires the user to enter a file path name that exists on the client or server they are operating on in order to save a summary of the AZP process. The summary output file is saved as an ASCII text file. The user can specify the file path manually or they can browse for a destination folder using the "Browse" button directly to the left of the text box.

Once all these parameters are satisfied according to the rules of the AZP, the algorithm can be processed by pressing the "Execute" button in the lower left corner of the user form. Otherwise the user can choose to quit the program simply by pressing the "Cancel" button in the lower right corner.

Sample Code. The code was written using Visual Basic for Applications within the ArcGIS 8.3 ArcMap environment. The AZP was designed using the ArcObjects scripting platform. ArcObjects is a COM compliant, object-oreinted programming structure developed by Environmental Systems Research Institute Inc (ESRI). An example of ArcObjects scripting can be found below. This sub-routine is part of the AZP program.

```
Private Sub UserForm_Activate()

'the following code populates the aggregation combo box
with the 'available layers in the map's table of
contents (TOC) and the 'function of interest combo box

        Dim intLayerCount As Integer, intI As Integer

        Dim strLayerName As String

        Dim pMxDoc As IMxDocument

        Dim pMap As IMap
```

```
            Set pMxDoc = Application.Document

            Set pMap = pMxDoc.FocusMap

            Dim pLayer As ILayer

            intLayerCount = pMap.LayerCount

            If intLayerCount > 0 Then

              For intI = 0 to intLayerCount - 1

              Set pLayer = pMap.Layer(intI)

              strLayerName = pLayer.Name

              Cmbo_Layer.AddItem strLayerName, intI

              Next intI

            End If

            If Cmbo_Function.ListCount = 0 Then

              Cmbo_Function.AddItem "Moran's I"

              If Cmbo_Function = "Moran's I" Then

                Cmbo_PlusMinus.AddItem "+1"

                Cmbo_PlusMinus.AddItem "-1"

                Cmbo_PlusMinus.AddItem "0"

              End If

            End If

       End Sub
```

4 Results

The two most prominent components of the AZP are the calculation of the function statistic and the actual execution of the zone construction process. The debugging portion of the program development cycle was done in situ with the actual development of the AZP, whereas the testing of the calculations associated with the Moran's I statistic was done external of program development for comparisons with the AZP calculation. The calculation of the Moran's I statistic was one of the most difficult components of the AZP. The second portion of the algorithm simply calls the Moran I function for each subsequent calculation on a "merge" move and then iterates through each calculation to choose the test statistic value that exhibits the greatest improvement over the original value. Therefore, testing of the second component was done after the execution of the AZP by examining the summary output to make certain that the value of the test function was increasing to an optimal target over the course of the program's lifespan. A small collection of census tracts from the city of Calgary, Alberta Canada were used to test the optimization ability of the AZP. The AZP program for the following sets of test examples was run on a DELL desktop computer operating with a Pentium 4 processor, stepping at approximately 2.4 Ghz, and equipped with 512 Mb of SDRAM.

The AZP was executed on the test area for optimization of zone configuration towards a positive Moran's I value from $N = 17$ zones to $M = 14$ zones. This was done on the variable, POP91A, a census variable that records the total population per census tract.

Original 17 Zone
Configuration

Final 14 Zone
Configuration

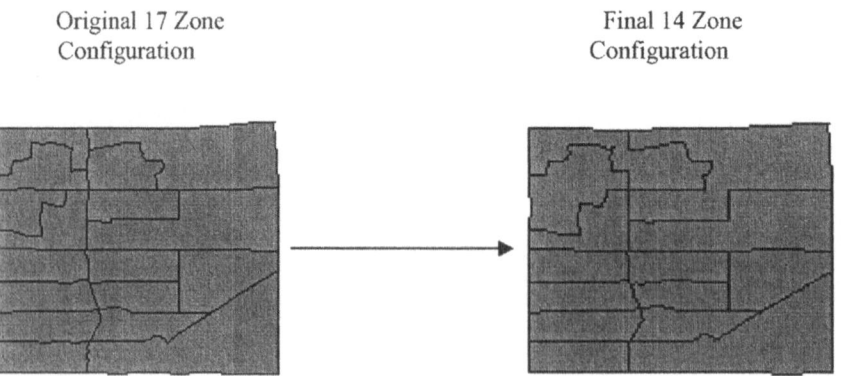

Fig. 2. Automated Zoning Procedure example for a positive Moran's I target.

The algorithm continued until it reached 14 zones. In figure 3 the test statistic improves with each iteration of the AZP moving from -0.1613 to 0.3888 after aggregation from 17 zones to 14 zones. The AZP for this test took approximately four minutes and 45 seconds to complete.

The AZP was tested for optimization towards a negative value of the Moran's I statistic and towards a neutral value of zero. In all three cases, the objective function improved towards a target goal thus indicating that the AZP executed properly. Upon completion of the testing phase it has been determined that the AZP algorithm calculates the Moran's I statistic correctly and recognizes this improvement of the objective function validating the reliability of the AZP as a whole.

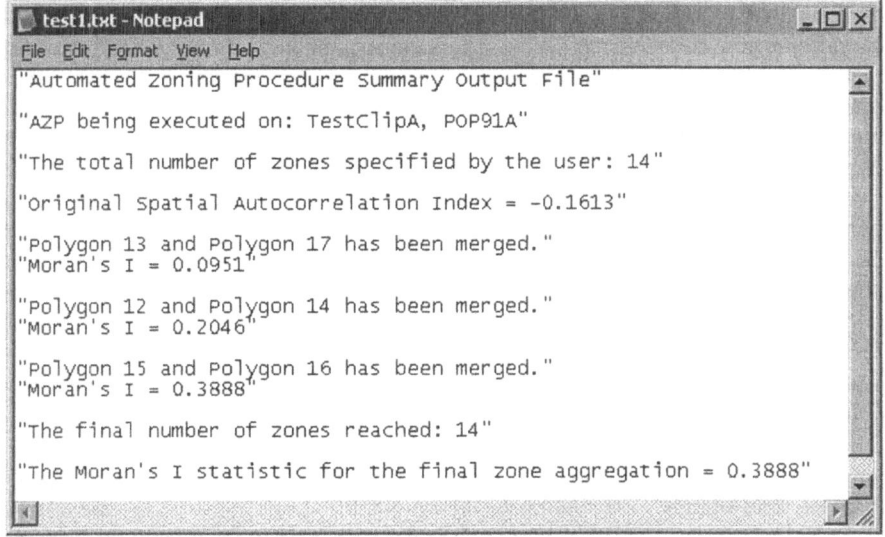

Fig. 3. Automated Zoning Procedure output summary for optimization of Moran's I towards a value of +1.

5 Discussion

The Automated Zoning Procedure built for ArcGIS 8.3, which will be referred to as the "new" AZP for the remainder of this discussion, is designed in order to mimic the general steps of the AZP algorithm implemented by Openshaw (1984) and again by Openshaw and Rao (1995), but with two slight variations to that of the original. Firstly, in step two of the original procedure by Openshaw (1984), he suggests the creation of a random set of zones to apply to the underlying spatial data in which to execute the AZP on. The new AZP assumes that the input polygon feature class is a random aggregation. The input file is somewhat a random aggregation keeping in mind that zones were probably created for another purpose other than the study at hand. The construction of zones is often done haphazardly and formulated based on criteria that are, more times than none, irrelevant to that of the study in question. Therefore, this variation of the AZP assumes that the input polygon feature is a random aggregation.

Secondly, the original AZP algorithm is considered a mildly steepest descent algorithm (Openshaw and Rao, 1995). This means that the algorithm aggregates zones when there is a *local* improvement in the overall function. In this case, the algorithm randomly chooses a zone of all possible zones and searches for an improvement in the objective function based on contiguity with respect to the zone under examination. In contrast, the new AZP is a steepest descent algorithm that continues to search for a *global* improvement in the objective function for all possible combinations of n zones into m regions. Searching for a global improvement in the overall objective function can be a time consuming process. The Moran's I statistic requires the determination of contiguity for each polygon present on the map and is done so for every possible merge combination within the feature class. Using the 17 zone test area as an example, there is a possible forty-two merge combinations right from the start that must be evaluated by the algorithm. Adjacency in this case refers to any zones that share a border or are connected at a vertex. Calculating the Moran's I statistic on its own is done quickly taking only five seconds of process time, but the calculation is done for the each of the forty-two initial moves taking approximately 210 seconds to complete or three minutes and thirty seconds. This is done until the requested number of zones as defined by the user is reached. The process time will decrease with every iteration of the algorithm, but the new AZP will inevitably require a bit of time to complete. The length of processing time for the program will definitely be a side effect of the function used for optimization, but another aspect of time consumption can be attributed to the fact that the algorithm is being executed in interpreted mode rather than in compiled mode as a stand-alone executable in a dynamic link library. Depending on the geometric complexity of the dataset, the relative size of the dataset, and the parameters specified by the user (including statistical function of interest), the AZP operation can be a time consuming operation.

Differences aside, the new AZP for ArcGIS 8.3 exhibits parallel characteristics to that of the ZDES program discussed by Openshaw and Rao (1995). The AZP can be executed with spatial autocorrelation as the function of interest and this algorithm allows the user to choose the number of zones in which to amalgamate to. The new AZP outputs a final map product in which zones are constructed based on the optimization of an objective function. Unfortunately, this version of the AZP does not have as many options as that of the ZDES with respect to statistical parameters, but the program is versatile enough to incorporate these options easily as additional functions within the AZP program structure. The resulting program is a version of the Automated Zoning Procedure defined by Openshaw (1984) implemented within a modern GIS environment. The new AZP is a slight variation to that of the original, but the logistical steps behind the algorithm remain the same.

6 Conclusion

In the realm of spatial research, the capability to categorize continuous space into discrete entities is common practice. Unfortunately, this simplification of space is problematic for the simple fact that there is no universal standard in which to define space using discrete entities. Creating these entities of space introduces questions that are difficult to answer. At what scale do we partition space and how are these partitions oriented geometrically? Once defined, undergoing any form of statistical inference on a classification of areal units is undoubtedly influenced by the MAUP. This geographically explicit phenomenon may introduce bias and uncertainty into statistical analysis for the simple fact that alternate scales of analysis and the ability to modify zones freely will produce varying results even though the underlying spatial information is left unaltered. To date, there is no way of solving this problem in its entirety. Mathematical models have been developed in order to deal with the MAUP (Steel and Holt, 1996), but such models remove the geographical component that makes a study distinctive in a geographic sense. Fortunately, there is a uniquely geographical approach to handling the MAUP as suggested by Openshaw (1984). In this case, areal units are defined for a study specific purpose in which a function of interest is optimized. As a result, a statistical parameter is relatively consistent regardless of the scale and aggregation used. At the same time this procedure is able to identify the limits of the MAUP. From this idea and the increasing use of computational geography Openshaw and Rao (1995) built an Automated Zoning Procedure that takes advantage of GIS environments in order to implement the concept of zone engineering as a pseudo-solution to the MAUP. The AZP has proved its usefulness in the area of census design and with its popularity has worked its way into standard exploratory policy with the Office for National Statistics in England and Wales (Martin et al., 2001). As geographers realize the importance of zone development and the implications it can have with respect to the MAUP, the AZP, or a version thereof, is more than likely to become a required application for many GIS statistical extensions in the near future.

References

1. Amrhein, C.G.: Searching for the elusive aggregation effect: evidence from statistical simulations. Environment and Planning A. 27 (1995) 105-119
2. Burt, J.E., Barber, G.M.: Elementary Statistics for Geographers. 2nd edn. The Guilford Press, New York (1996)
3. Clark, W.A.V., Avery, K.L.: The effects of data aggregation in statistical analysis. Geographical Analysis. 8 (1976) 428-438
4. Duncan, O.D., Cuzzort, R.P., Duncan, B.: Statistical Geography: Problems in Analyzing Areal Data. The Free Press, Glencoe Illinois (1961)
5. Fotheringham, A.S., Wong, D.W.S.: The modifiable areal unit problem in multivariate statistical analysis. Environment and Planning A. 23 (1991) 1025-1044
6. Gehlke, C.E., Biehl, K.: Certain effects of grouping upon the size of the correlation coefficient in census tract material. Journal of the American Statistical Association Supplement. 29 (1934) 169-170
7. Goodchild, M.F.: The aggregation problem in location-allocation. Geographical Analysis. 11 (1979) 240-255
8. Goodchild, M.F.: Spatial Autocorrelation. Geo Books, Norwich (1987)
9. Haining, R.S., Wise, S., Ma, J.: Designing and implementing software for spatial statistical analysis in a GIS environment. Journal of Geographical Systems. 2 (2000) 257-286
10. Macmillan, W.: Redistricting in a GIS environment: An optimization algorithm using switching points. Journal of Geographical Systems. 3 (2001) 167-180
11. Martin, D.: Optimizing census geography: The separation of and collection of output geographies. International Journal of Geographical Information Science. 12 (1998) 673-685
12. Martin, D., Nolan, A., Tranmer, M.: The application of zone-design methodology in the 2001 UK census. Environment and Planning A. 33 (2001) 1949-1962
13. Martin, D.: Extending the automated zoning procedure to reconcile incompatible zoning systems. International Journal of Geographical Information Science. 17 (2003) 181-196

14. Openshaw, S.: A geographical solution to scale and aggregation problems in region-building, partitioning and spatial modeling. Transactions of the Institute of British Geographers. 2 (1976) 459-472
15. Openshaw, S.: Optimal zoning systems for spatial interaction models. Environment and Planning A. 9 (1977) 169-184
16. Openshaw, S.: The Modifiable Areal Unit Problem. Geo Books, Norwich (1984)
17. Openshaw, S., Rao, L.: Algorithms for reengineering 1991 census geography. Environment and Planning A. 27 (1995) 425-446
18. Robinson, A.H.: Ecological correlation and the behavior of individuals. American Sociological Review. 15 (1950) 351-357
19. Steel, D.G., Holt, D.: Analysing and adjusting aggregation effects: The ecological fallacy revisited. International Statistical Review. 64 (1996) 39-60
20. Svancara, L.K., Garton, E.O., Chang, K.T., Scott, J.M., Zager, P., Gratson, M.: The inherent aggravation of aggregation: An example with elk aerial survey data. Journal of Wildlife Management. 66 (2002) 776-787
21. Wise, S., Haining, R., Ma, J.: Providing spatial statistical data analysis functionality for the GIS user: The SAGE project. International Journal of Geographical Information Science. 15 (2001) 239-254

Searching for 2D Spatial Network Holes

Femke Reitsma and Shane Engel

Geography Department, 2181 LeFrak Hall, University of Maryland College Park, MD
20742, femke@geog.umd.edu, engeljs@umd.edu

Abstract. Research involving different forms of networks, such as internet networks, social networks, and cellular networks, has increasingly become an important field of study. From this work, a variety of different scaling laws have been discovered. However, these aspatial laws, stemming from graph theory, often do not apply to spatial networks. When searching for network holes, results from graph theory frequently do not correlate with 2D spatial holes that enforce planarity. We present a general approach for finding holes in a 2D spatial network, and in particular for a network representing street centrelines of an area in Washington, D.C. This methodology involves finding graph holes that can be restricted to 2D spatial holes by examining topological relationships between network features. These spatial network holes gain significance as the number of edges encompassing the hole, and the length of these edges increase. For this reason, our approach is designed to classify these holes into different sets based on the number of edges found and the length of those edges. The results of this application provide valuable insights in the nature of the network, highlighting areas that we know from experience are poorly connected and thus suffer from low accessibility.

Keywords: topological analysis, street network, urban morphology, network

1 Introduction

A network is a valuable representation of data. This data can be observably network in form, such as internet and street networks, or can be interpreted as a network, such as social and other relational networks. The benefit of representing data as a network lies in the graph theoretical measures available for network analysis. Such benefits have resulted in an explosion of research towards the development of new network measures, and the representation of old data newly dressed as a network [8, 7].

Instances like the New York power outage in mid August 2003, and a similar Italian crisis in late September 2003, have highlighted holes that exist in infrastructural networks. Similar holes can also be found in our social networks, giving power to certain players, and limiting information flow to others [1]. Holes have also formed the topic of substantial philosophical debate [3]. The significance of holes lies in their ability to obstruct the flow of information or substance through or over a network. These areas tend to act as barriers to movement, where things are forced to travel around these holes in order to traverse the network. Whether that network represents social networks, transport networks, or any other data that can be represented as an adjacency matrix of relations, the hole forms a virtual object by affecting the rest of the network. Therefore, detecting such holes within networks provides us with information about parts of the network that are poorly connected. The automated discovery of 2D (two-dimensional) spatial network holes allows us to explore the impact of these holes on the 2D spatial network, thereby evaluating their importance to the processes operating upon this underlying fabric of structural relations [1].

This paper focuses on the algorithm for finding 2D spatial network holes, that is, those network holes that are found in planar-enforced topologies. This algorithm involves a number of steps, beginning with a search for graph holes. If these graph holes are established to remain holes under planar enforcement, they are graphically represented for visual analysis and measurement. This methodology is then applied to a case study of an urban street network in Washington D.C. The underlying thesis is that large holes are more important because they identify parts of the network that need to be connected. Hence, filling a hole is likely to be of more value for larger holes, however "larger" is defined [1].

The remainder of this paper is structured as follows; Section 2 discusses background issues regarding the search for holes. Section 3 describes the general methodology that forms the basis of a search for 2D spatial network holes, which is then applied to the case study of street networks in Section 4. Section 5 concludes the paper.

2 Background

Because space is inherent in the structure of spatial networks, these unique networks cannot be fully understood with common graphic theoretic measurements. Space confounds deterministic measures as it does stochastic measures through spatial autocorrelation. For example, research by [8] has found that scale-free patterns that are found in a large range of networks, such as the world wide web and social networks, do not exist in geographic networks, such as internet routers and power grid structures. Similarly, spatial holes do not conform to traditional metrics.

Loosely defined (see Section 3 for a formal definition), a network hole is a part of the network formed by four or more nodes that are connected by links, forming a non-repeating path where the first node equals the last. As a consequence of planar enforcement, graph holes do not equate to 2D spatial holes. In figure 1 below, the nodes (A, B, C, D) and the edges linking them form a graph hole. However, it does not form a spatial 2D hole as there is a crossing set of nodes and vertices, that is, nodes (D, E, B) and the edges connecting them. Hence a general search for graph holes will include holes that are not 2D spatial holes.

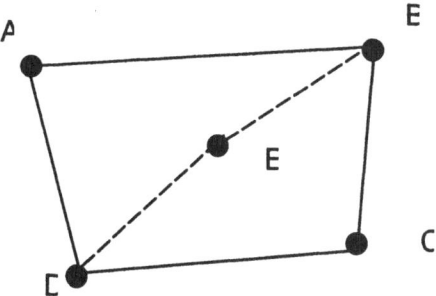

Fig. 1. Graph hole

Furthermore, the problem of detecting spatial network holes is not merely a problem of shortest path analysis from a node back to itself. As expressed in Figure 2, a spatially internal path (D, E, F, G, H, I, B) could be longer than the external path (A, B, C, D). In a shortest path analysis from a node such as A, this would result in (A, B, C, D) being selected.

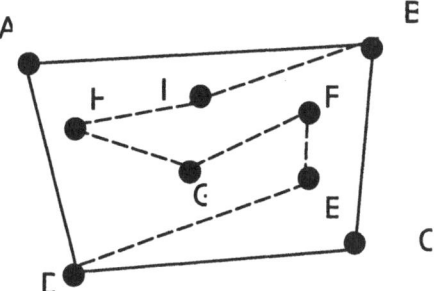

Fig. 2. Graph holes versus spatial holes

3 General Methodology

In overview, the methodology for finding 2D network holes involves three steps which couple GIS with a program implemented in Java. First, the network, represented as an adjacency matrix, is used to find graph holes. Second, the results of the graph hole search are spatially filtered for 2D spatial network holes. And finally, the output is visually represented for analysis.

3.1 Graph Hole Search

The problem of searching for holes in a network translates to a graph theoretical problem of searching for chordless cycles in a graph, also called graph holes. Given a set of vertices V and edges E, we form a graph $G(V,E)$. A graph cycle C is a subset of the set of graph edges that forms a path, a sequence of non-repeating edges, where the first edge is also the last. A chord of a cycle is an edge that joins two non-consecutive vertices of the cycle, that is, an edge not in the edge set of the cycle whose endpoints lie in the vertex set of C (Figure 3). A chordless cycle of graph G is a graph cycle of length at least four in G that has no cycle chord [4].

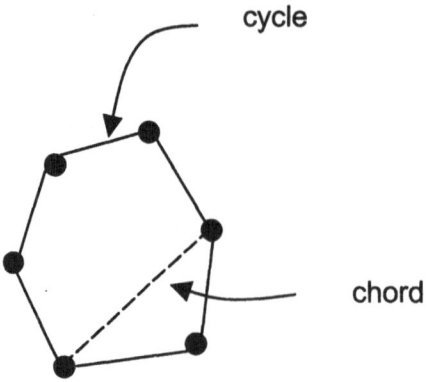

Fig. 3. Graph cyle and chord

The network to be analyzed must be represented as an adjacency matrix. An adjacency matrix is a matrix with rows and columns that are labeled with the network's nodes, or in terms of graph theory, graph vertices. The adjacency matrix $A = A(G) = [a_{ij}]$ is a $n \times n$ matrix in which the $a_{ij} = 1$ if v_i and v_j, where $v_i, v_j \in V$, are adjacent and $a_{ij} = 0$ otherwise. This adjacency matrix forms the input for the detection of chordless cycles or holes.

The method for detecting chordless cycles was informed by the work of [5]. It was implemented in the Java programming language using the JAMA matrix library (available online at: http://math.nist.gov/javanumerics/jama/). The algorithm is structured as follows:

```
for (each edge {x, y} of a matrix) do {

    for (each chordless path ≥ 4) do {

        if (the first vertex is connected to the last vertex)
    then{

            chordless path = chordless cycle

            write chordless path to array

    }

}

}
```

3.2 Spatial Hole Search

Searching for spatial holes is based on the output of the graph hole search. Each graph hole is checked to see whether it spatially contains a subgraph that connects at least two nodes of the hole. The JTS Topology Suite, which is an API of 2D spatial predicates and functions, was utilized for searching for the spatial holes (available online at http://www.vividsolutions.com/jts/jtshome.htm).

The general algorithm for searching for 2D spatial holes is as follows:

```
for (each graph hole) do {
   for (each node in the network) do {
      if (there is no node spatially within the hole that
   connects
         two unique nodes of hole) then {
            write hole to spatial holes array
      }
   }
}
```

The result of this algorithm is an array of 2D spatial network holes that can be used for visualization and analysis.

3.3 Spatial Hole Visualization

The array of 2D spatial network holes is transferred into ESRI's ArcGIS environment, and is manipulated with Visual Basic for Applications programming language. Each node from the vector street network is coordinated with its corresponding node in the array of spatial holes.

The basic algorithm for transferring the array of holes into graphic polygons is as follows:

```
for (each graph hole) do {
   for (each arc in the network) do {
      if (the edge contains two nodes that make up the hole)
   then{
         add edge vertices to point collection
      }
   }
   create new polygon from point collection
}
```

A resulting polygon geometry file is created to visualize the relationships between each hole. This output also provides the means to calculate some basic measurements, such as the area and perimeter of each hole.

4 Street Network Application

A subset street network from Washington D.C. was selected in order to test the algorithm and explore its potential in the analysis of street network structure. The street network is displayed below in Figure 4. It is already visually evident that parts of the network will be more difficult to traverse, in particular where there are large areas without any crossing streets.

Fig. 4. Selected street network in Washington D.C.

4.1 Method and Results

The method implemented for finding 2D spatial network holes in street networks involved a number of steps beyond the general methodology described in Section 3 above. A street centerline network provided by the Washington, D.C. government was used for the analysis. Since these networks can vary in representation of real-world features, it is important to define some basic rules for the study area. The data represents the theoretical centerline of the actual ground feature; with no more than one edge for each route (divided highways and boulevards are often represented with more than one edge). Nodes define the start and end of each edge, and need to be placed where centerlines intersect each other. Overpasses need to be represented so that the edges that meet at this area do not intersect; in other words, without a node present at the intersection. For this example, private drives and alleyways were excluded, as this study is only concerned with main thoroughfares. Finally, any cul-de-sacs that are represented in a "lollipop" configuration should not be included, but should instead be shown with an ending edge.

As mentioned earlier, the vector street network is transferred into an adjacency matrix by describing the relationships between each pair of nodes in the network. Each node in this network has a relationship with one another through a connecting edge. From this understanding, it was necessary to use the general algorithm below to build the adjacency matrix:

```
create matrix consisting of columns and rows for each unique
node
    for (each node in the network) {
        for (each edge in the network) {
            if (from-node or to-node of the edge matches the
            coordinates of node) {
                get the opposite node of either the from-node or
to-
                node and write a 1 in the node row, opposite node
                column {x, y} value of the matrix
            }
        }
    }
}
```

Once holes were found from the adjacency matrix produced from the algorithm above, holes defined with four edges were removed from the array, as they do not greatly impact on accessibility of the network.

The results of the analysis are presented in Figure 5 below. The holes detected range in size from 5 edges to 15. Clearly hole number 3 provides the largest barrier to traffic crossing from one side of the street network surrounding the hole to another.

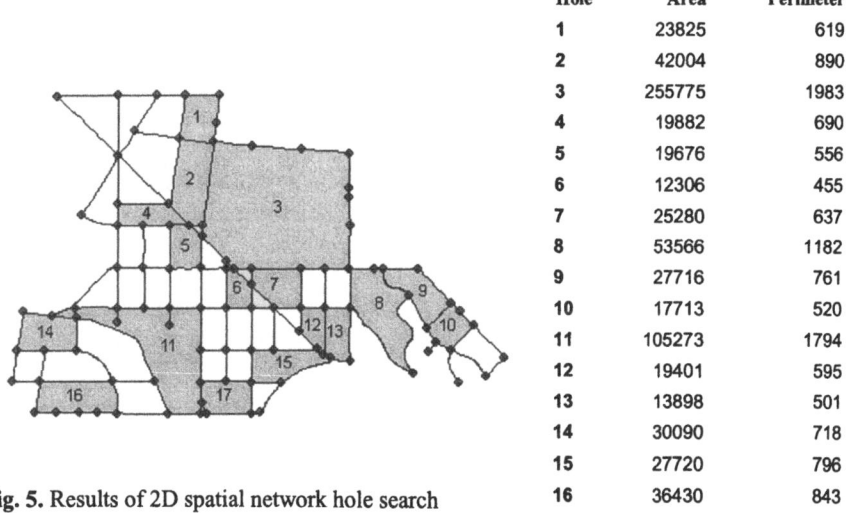

Fig. 5. Results of 2D spatial network hole search

Table 1. Measurements of spatial network holes (meters)

Hole	Area	Perimeter
1	23825	619
2	42004	890
3	255775	1983
4	19882	690
5	19676	556
6	12306	455
7	25280	637
8	53566	1182
9	27716	761
10	17713	520
11	105273	1794
12	19401	595
13	13898	501
14	30090	718
15	27720	796
16	36430	843
17	23173	618

Furthermore, we can see from examining the measurements in Table 1 that hole 3 also has the largest perimeter and area, correlating with the relatively large number of nodes that make up this region. We can therefore conclude that the network would benefit the most by having this hole crossed by an edge that divides the area and number of nodes the best. Unfortunately, the presence of the rather large and historic *National Cathedral* may make this slightly difficult.

4.2 Discussion

Finding holes within the street networks gains significance by providing an indication of barriers to accessibility. Street network holes can increase the cost of travel for individuals traversing the network from two locations on opposing sides of the hole. This is a purely structural approach to accessibility, defining the potential of the network for impeding the flow of traffic, rather than more traditional approaches which incorporate patterns of land use as activity elements [6]. It provides a measure of accessibility that is inherent in the network structure. Furthermore, holes identify areas that would be best suited for the addition of new roads, in order to benefit the entire network structure.

Street network holes cannot be inferred from other transportation network metrics as there are no graph theoretic measures that capture the spatial hole either directly or indirectly. The nodes and edges that form the hole do not give themselves away by local measures such as through lower node degree, that is, by fewer connections to those nodes composing a hole. Nor can holes be detected through global measures such as the diameter of the network, which is the length of the shortest path between the most distant nodes in the graph, or indices such as the alpha index, which is a measure of connectivity comparing the number of cycles in a network with the maximum number of cycles.

4.3 Limitations and Extensions

Although this application is simple, it presents the first findings and analysis of spatial network holes. Yet there are limitations to the example presented in this paper, which directs us to extending this work. A limitation not recognized in example above may arise when the size of a hole in terms of the number of edges and nodes that describe it does not correlate with the area of the hole or perimeter of the hole. In other words, we could have a hole that has a large number of edges and nodes, but a rather small area and perimeter (or vice-versa). For example, roundabouts are prevalent in the D.C. area, resulting from the large number of streets that converge on these small areas, and can produce smaller holes that are densely populated with nodes. In these cases, naturally occurring in larger study areas, a ranking calculation would need to be developed incorporating the length of the edges in each hole.

Furthermore, in certain cases, holes defined with less than four edges could turn out to be large areas that act as greater barriers to the network than normal. An extreme scenario could present itself with the occurrence of a large area that is only connected through two edges, as in Figure 6. This is likely to be the case in mountainous terrain. This area would be excluded from the results of the methodology discussed above, as holes are defined as having equal to or more than four edges. A solution to this problem would be to increase the density of the nodes in the network by creating pseudo nodes if an edge exceeds a certain distance. These pseudo nodes would not define intersections but produce the same results in the analysis.

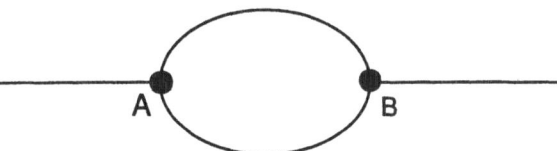

Fig. 6. Large hole consisting of only two nodes and edges

A logical addition would be to weight the links based on factors such as the number of lanes, usage, or other factors that can affect the impedance of a route. The results would then rank the holes according to importance based on these weights. For example, holes can be found at various levels of connectivity through slicing those levels at specific weights. In graph theory, the weighted adjacency matrix forms a graph with multiple edges connecting vertices which can be represented as a network together with a function which maps the edge set into the set of integers. The resultant graph is termed a multigraph or hypergraph [2, 5]. Given a graph S is a subgraph of multigraph $M(V,E)$, where $V(S) \subseteq V(M)$ and $E(S) \subseteq E(M)$, the level of connectivity $i \in I$ of S is thus defined as the union of all edges with a value greater than or equal to i, and their connected vertices. That is, if

$$E_i \neq \emptyset$$
$$\bigcup_{i=I} E_i = E(S)$$

The general algorithm discussed above could then be applied at particular levels of connectivity by specifying the value to be found in the entry of the matrix.

Another approach to determining the importance of the hole is to analyze its impact on transportation paths. The hole could be weighted according to the number of times it impedes the accessibility of an individual trip. For example, a shortest path analysis could be executed for a specified origin-destination matrix, where if half of the nodes of a hole are used in the path, then that hole is impeding the flow. Such an approach integrates this analysis method to traditional measures typically based on Origin-Destination matrices. The hole structure could also be incorporated into other current metrics of accessibility, particularly gravity-based measures, where the hole may add to the impedance function [4].

5 Conclusion and Future Work

The search for 2D spatial network holes presents a new perspective on network structure, where the absence of network structure itself plays an important role in the nature of the network. Finding these holes allows us to evaluate the thesis that filling big and important holes. The application example of the street network provides some first insights into how these holes may impact on the network and its use for transportation.

Beyond the extensions proposed for street networks, this method has significant implications for networks in general, whether they are explicitly spatial or not. As expressed in the introduction, spatial networks, such as infrastructural networks, are susceptible to failure. An analysis of spatial holes may provide insights into the reasons behind these breakdowns. Furthermore, extending the dimensionality of these holes to three dimensions may provide insights into visual hollows that may have an impact on wayfinding research.

References

1. Atkin, R. H. (1977). Combinatorial Connectivities in Social Systems: an application of simplicial complex structures to the study of large organizations. Basel, Birkhauser Verlag.
2. Berge, C. (1970). Graphs and Hypergraphs. Amsterdam, North-Holland Publishing Company.
3. Casati, R. and A. C. Varzi (1994). Holes and Other Superficialities. Cambridge, MA, MIT Press.
4. Chandrasekharan, N., V. S. Lakshmanan and M. Medidi (1993). Efficient Parallel Algorithms for Finding Chordless Cycles in Graphs. Parallel Processing Letters 3(2): 165-170.
5. Chartrand, G. (1977). Introducutory Graph Theory. New York, Dover Publications.
6. Handy, S. L. and D. A. Niemeier (1997). Measuring Accessibility: an Exploration of Issues and Alternatives. Environment and Planning A 29: 1175-1194.
7. Jeong, H., Z. Néda and A. L. Barabási (2003). Measuring Preferential Attachment for Evolving Networks. Europhysics Letters 61: 567-572.
8. Ravasz, E. and A. Barabasi (2003). Hierarchical Organization in Complex Networks. Physical Review E 67(026112).
9. Watts, D. J. and S. H. Strogatz (1998). Collective Dynamics of 'Small-World' Networks. Nature 393: 440-442.

Extension of Geography Markup Language (GML) for Mobile and Location-Based Applications[1]

Young Soo Ahn, Soon-Young Park, Sang Bong Yoo, and Hae-Young Bae

School of Computer Science, Inha University
Incheon, Korea
fax: 82-32-874-1435
{owlet76, sunny}@dblab.inha.ac.kr, {syoo, hybae}@inha.ac.kr

Abstract. Geography Markup Language (GML) has been extended for mobile and location-based applications. Three schemas are included in the extension, i.e., voice schema, tracking schema, and POI (Point Of Interest) schema. In order to handle effectively the extension, we also have designed a Spatial XQuery language and its processing modules. Because our work is based on a spatial database system, the Spatial XQuery statements are first translated into SQL statements. By working on an existing spatial database system, we can use of the existing amenities of database systems such as query optimization, spatial indexes, concurrency control, and crash recovery. Translation of the Spatial XQuery into SQL has been explained using examples. Because the results from the spatial database system are in the form of tables, we need to translate again the results into XML statements. A working example of the proposed system as an Emergency Support System is also presented. Prospected application areas of the proposed system are almost all mobile and location-based systems such as LBS (Location-based System), POI systems, tracking systems, ubiquitous systems, and distributed control and management systems.

1 Introduction

As the uses of mobile and location-based applications are popularized and extended recently, requirements to geographic information systems also have been changed. Geography Markup Language (GML) [15] is the active standard for exchanging spatial data on the Internet and mobile terminals, but it is limited only for text and graphic data of static geographic information. Location-based Systems (LBS) can provide more attractive services when they utilize tracking information of moving objects and path information of Point-of-Interest (POI) objects such as hospitals and gas stations. It would be very convenient if a car navigation system provide voice mode as well as the basic screen mode. In this paper we propose S-XML (Spatial-eXtensible Markup Language), which extends the GML by adding three schemas: voice schema, tracking

---

[1] This research was supported by University IT Research Center Project in Korea.

schema, and POI schema. The characteristics of each schema can be summarized as follows:

- **Voice Schema**: A grammar of voice information of geographic direction has been designed. The syntax of voice information has been designed as simple as possible because it should be brief and concise. The direction information will enable the user to move to the destination through the shortest path.
- **Tracking Schema**: This schema models the dynamic status of moving objects. It includes data type of moving objects, specific coordinate information and the time when the object pass the point, direction information, and velocity information.
- **POI Schema**: This schema has been designed to provide the user with the position and path information to interesting objects such as public offices, schools, hospitals, restaurants, and gas stations. It can refer the voice schema in order to provide voice mode interface.

In order to handle the S-XML data effectively, a spatial XQuery language [20] has been designed. In this research, the geographic data are stored in a spatial database management system and the spatial XQuery constructs submitted by users are translated into spatial SQL programs and evaluated by the spatial database management system. The spatial database management system supports the query optimization methods and concurrency control for processing multiple queries simultaneously. The spatial XQuery supports basic operations (e.g., =, <, and >), spatial operations (e.g., sp_disjoint, sp_intersects, and sp_contains), and operations for moving objects (e.g., mo_after, mo_before, mo_first, and mo_last). With these spatial operations and the operations for moving objects, the spatial XQuery can be effectively used on LBS (Location-Based Service) platforms. The spatial XQuery also supports the VOICE tag, using which the direction information can be provided by voice mode interface.

This paper is organized as follows. Section 2 summarizes the various approaches to handle XML queries. There are two basic approaches, i.e., database approaches and native approaches. Section 3 presents the information model of S-XML and the syntax and semantics of spatial XQuery. In Section 4, we describe the overall processing of spatial XQuery. Spatial XQueries are first translated into equivalent SQL statements and then evaluated by a spatial database system. In Section 5, we provide a working scenario of using the spatial XQuery. In this scenario, the user first locates the nearest hospital from an accident and the driver of an ambulance from the hospital gets the direction to the accident site when the car approaches crossroads. Section 6 summarizes the contributions of the paper.

2 Related Work

Two research groups have contributed to the developments of XML query languages. First, the document research group has their experience in designing languages and tools for processing structured documents. For example, XQuery uses XPath in order to support operations on document path expressions. Second, the database research group has their experience in designing query language and processors for data-oriented computations. XQuery also uses the basic structure of SQL language for relational databases. Consequently, there are two major approaches to XML data management and query evaluation, i.e., database approaches and native approaches.

In database approaches, XML documents are stored in preexisting database systems [2, 8, 16] and the XML queries are translated into SQL or other database languages [10, 19]. The problems of updating XML data stored in tables [18] and indexing XML data [7, 13] have also been presented. Prototypes such as SilkRoute [9] and XPERANTO [4] have proposed algorithms for efficiently encoding relational data as XML documents. Commercial products such as SQL Server, Oracle, and DB2 also support encoding relational data as XML documents. XQuery has been designed based on Quilt[5], which is based on other XML query languages such as XML-QL, XPath, and XQL.

Native approaches design new management systems for XML data. Some examples of native approaches to XML document processing include Natix [11], Tamino [17], and TIMBER [12]. Most work on native approaches has focused on efficient evaluation of XPath expressions using structural join [1] and holistic join [3], and optimal ordering of structural joins for pattern matching [21]. TIMBER extensively uses structural joins for pattern match, as does the Niagara system [14]. A structure called generalized tree pattern (GPT) is proposed for concise representation of a whole XQuery expression [6]. Using the GPT evaluation plans are generated for XQuery expressions that includes join, quantifiers, grouping, aggregation, and nesting.

In general, database approaches make good use of the existing amenities of database systems, which include query optimization, indexes, concurrency control, and crash recovery. On the other hand, native approaches are more customized for XML documents. Because huge amount of geographic information is already stored in spatial database systems, we take the database approach. By using the spatial XQuery language, the user can access not only the extended spatial information but also the ordinary text and graphic information in terms of XML forms.

3 S-XML and Spatial XQuery

In this section, the structure of S-XML is introduced first and then the syntax and semantics of Spatial XQuery are discussed with some examples.

3.1 S-XML

S-XML is based on GML 3.0 and is designed to provide effective location based service to the user in mobile environments. S-XML consists of three main schemas, i.e., voice schema, tracking schema, and POI schema.

(a) Voice Schema

In mobile environments, voice information is often more useful than text or graphic information especially when the user is driving a car. To be effective the grammar of the voice information is designed as simple as possible (see Figure 1). The voice information can be either location information or turn information. When the location information is chosen, the distance and direction to the target is provided. For example, "Central Hospital is located 10 miles away on the East." The turn information can be used when the user is driving a car or walking. An example of turn information is "turn left in 30 meters toward City Hall."

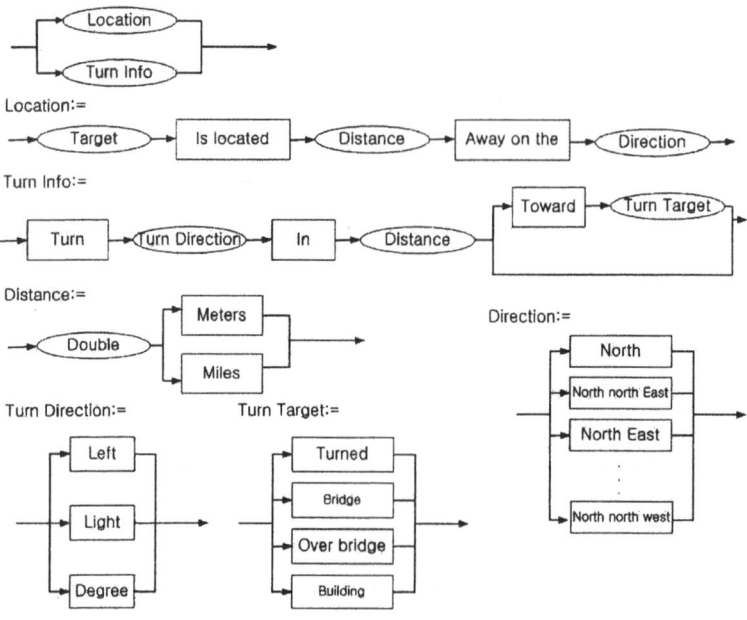

Fig. 1. Grammar of voice schema.

(b) POI Schema

POI schema has been designed based on POIX(Point of Interest eXchange Language Specification) [14] of W3C and references Feature.xsd, Geometry.xsd, Topology.xsd, DynamicFeature.xsd, and Direction.xsd of GML. POI schema includes the location information of objects in specific area, the optimal path to destination, the type of moving objects (e.g., car, human, subway, ...), information of the source and the destination, information of representative objects near the destination, and additional information about the destination (e.g., phone number and email address). Figure 2 shows a part of POI schema, which include a starting point, a target point, and a route.

```
<!-- access -->
<xs:complexType name="accessType">
   <xs:sequence>
       <xs:element ref="method"/>
       <xs:element name="ipoint" type="ipointType"/>
       <xs:element name="tpoint" type="tpointType"/>
       <xs:element name="route" type="routeType" minOccurs="0"/>
   </xs:sequence>
</xs:complexType>
<!-- access element(ipoint, tpoint, route) -->
<!-- ipoint -->
<xs:complexType name="ipointType">
   <xs:sequence>
       <xs:element ref="iclass"/>
       <xs:element ref="gml:location"/>
       <xs:element name="name" type="nameType" minOccurs="0"/>
   </xs:sequence>
</xs:complexType>
<!-- tpoint-->
<xs:complexType name="tpointType">
   <xs:sequence>
       <xs:element ref="tclass"/>
       <xs:element ref="gml:location"/>
       <xs:element name="name" type="nameType" minOccurs="0"/>
   </xs:sequence>
</xs:complexType>
```

Fig. 2. Part of POI schema that includes the specification of starting and target points.

(c) Tracking Schema

Tracking schema references DynamicFeature.xsd, Feature.xsd, Geometry.xsd, Temporal.xsd, Direction.xsd, and Observation.xsd. It specifies the changes of status of an specific object as time changes. The information model includes the type of moving objects, information of specific points and the time when the point passed, and information of direction and velocity. Figure 3 shows a part of Tracking schema, which specifies the locus type. A locus consists of a series of points, time, location, and direction.

```
<!-- locus Type definition-->
 <xs:complexType name="locusType">
     <xs:sequence>
         <xs:element name="point" type="sxml:pointType"/>
         <xs:element name="dateTime" type="xs:dateTime"/>
         <xs:element name="location" type="xs:string"/>
         <xs:element name="dir" type="xs:unsignedShort"/>
     </xs:sequence>
 </xs:complexType>
```

Fig. 3. Part of Tracking schema that includes the specification of locus.

3.2 Spatial XQuery

Spatial XQuery follows basic FLWR (For Let Where Return) structure of XQuery with path expressions so that SQL users are easy to use and the elements and attributes of S-XML can be effectively accessed. Spatial XQuery also supports the spatial data, spatial relational operators, and spatial functions that recommended by OGC specifications [4]. In Spatial XQuery, spatial operators and spatial functions have 'sp' as a prefix in their names. Functions of moving objects have 'mo' as a prefix in their names. Due to the space constraints, details of XQuery syntax are not discussed in this paper. The meaning of Spatial XQuery will be clear with the following examples.

Example 1. This query retrieves the names of cities that located within 5 km from any river in rivers.xml.

```
for    $a in document("cities.xml")//city
       $b in document("rivers.xml") //river
where  sp_overlap($a/obj, sp_buffer($b/obj, 5000)) eq true
return <Within5000CityName>$a/name</Within5000CityName>
```

Example 2. This is a nested query. It retrieves the names of cities and the names of rivers that run through the cities.

```
for    $a in document("cities.xml")//city
return <City>
         <name>$a/name/text()</name>
       <river>
                for    $b in document("rivers.xml")//river
                where  sp_intersects($a//obj, $b//obj) eq true
                return <name>$b/name/text()</name>
       </river>
       </City>
```

4 Processing of Spatial XQeury

The query processor of Spatial XQuery consists of Parser, SQL Info Generator, Result Info Generator, and SQL Translator (see Figure 4). The queries from the user interface are first parsed to generate parse trees, which are sent to SQL Info Generator and Result Info Generator. Spatial XQueries finally translated into SQL based on the information generated by SQL Info Generator and Result Info Generator. SQL Info Generator analyzes the clauses FOR, LET, and WHERE from the Spatial XQuery statements and generates a mapping table for query translation. The RETURN clause of Spatial XQuery is analyzed by Result Info Generator and a linked list of tags and elements is generated.

Extension of Geography Markup Language (GML)

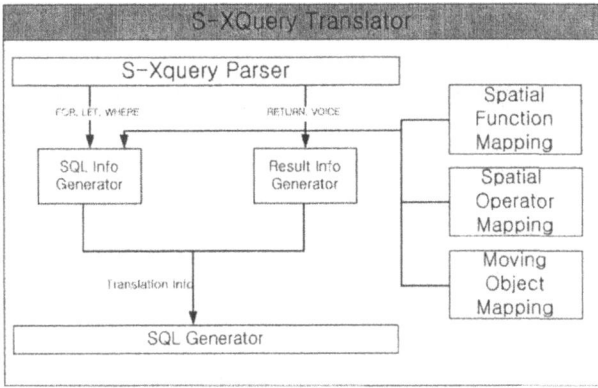

Fig. 4. Structure of the query processor of Spatial XQuery.

Details of the procedure of translating Spatial XQuery are presented using an example. Consider the query in Example 1. By analyzing the FOR and WHERE clauses, we can generate the mapping table as in Table 1. In Table 1, tag names and variable names are translated into corresponding Table names in spatial database. Operators in Spatial XQuery are also mapped into the corresponding ones in spatial database. Such mapping information is provided by the Mapping Module of Spatial Operators. In this example, the spatial operator of Spatial XQuery *sp_overlap* in mapped into operator *overlap* in spatial database.

Table 1. Mapping from Spatial XQuery to SQL.

	Spatial XQuery	SQL
From	Cities.xml//City Rivers.xml//River	City River
Select	$a/name	City.name
Where	sp_overlap($a/obj, Buffer($b/obj, 5000)) eq true	overlap(City.obj, Buffer(River.obj, 5000)) = true
Order by	NULL	NULL

Based on the mapping table in Table 1, an SQL query can be generated as follows.

    Select City.name
    From City , River
    Where Overlap(City.obj, Buffer(River.obj, 5000)) = true;

Using the table returned as a result for the SQL statement, the following XML document will be generated for this example.

    <Within5000CityName>
     <name> Han River </name>
    < /Within5000CityName>

5 Working Scenario

A scenario to apply Spatial XQuery in LBS environment is discussed in this section. This is an example of Emergency Support System that includes POI Service and ShortestPath Service. Suppose that a car accident takes place and there are some injuries (see Figure 5). The user of this system is looking for the nearest hospital and wants an ambulance for injured people. Once the nearest hospital is identified by the Emergency Support System, the hospital is requested to dispatch an ambulance to the accident place. Then the driver of the ambulance asks the direction to the accident place while he or she drives. The Spatial XQueries made for the requests, SQL queries that translated, and XML results generated from the system are explained in this Section.

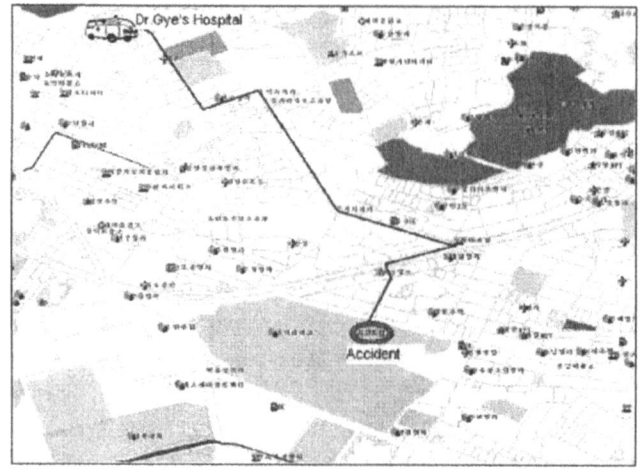

Fig. 5. Car accident and ambulance from the nearest hospital.

Figure 6 is a Spatial XQuery that looks for the nearest hospital from the accident place. The coordinates of the accident place are available from a GPS.

6 Conclusion

Geography Markup Language (GML) has been extended for mobile and location-based applications. Three schemas are included in the extension, i.e., voice schema, tracking schema, and POI schema. In order to handle effectively the extension, we also have designed a Spatial XQuery language and its processing modules. Because our work is based on a spatial database system, the Spatial XQuery statements are first translated into SQL statements. By working on existing spatial database system, we can use of the existing amenities of database systems such as query optimization, spatial indexes, concurrency control, and crash recovery.

```
for    $a in document("textshape.xml")//TextShape
let    $b := min(sp_distance($a/obj, point(4559734.910290 1348143.837131)))
where  contains($a/tex(), "%hospital%") and text_kind eq '50'
return <POI>
       {
              for $c in document("textshape.xml")//TextShape
              where contains($a/tex(), "%hospital%") and sp_distance($a/obj,
                           point(4559734.910290 1348143.837131)) <= $b and text_kind eq '50'
              return
                     {
                            <Label> $c/text() </Label>
                            <Location> $c/obj</Location>
                            <Distance>
                                   sp_distance($c/obj, point(4559734.910290 1348143.837131))
                            </Distance>
                     <voice>
                            <Direction>POIDirection(point(4559734.910290
                                                         1348143.837131))</Direction>
                     </voice>
                     }
       }
       </POI>
```

Fig. 6. Spatial XQuery looking for the nearest hostpital.

Translation of the Spatial XQuery into SQL has been explained using an example. Because the results from the spatial database system are in the form of tables, we need to translate again the results into XML statements. A working example of the proposed system as an Emergency Support System is also presented. The contribution of this paper can be summarized as follows.

- Geography Markup Language (GML) has been extended for mobile and location-based applications.
- A query language and its processing system have been proposed and a prototype is presented.
- Simple form of voice interface for mobile systems has been proposed and implemented.

Prospected application areas of the proposed system are almost all mobile and location-based systems such as LBS, POI systems, Tracking systems, Ubiquitous systems, and distributed control and management systems. In order to be more powerful, it needs to be extended with various development tools. Novice users may have difficulties to program their queries using Spatial XQuery. Some user-friendly designed GUI tools could make the system more effective. Even current execution model allows multiple queries; we need to devise more sophisticated optimization strategies for multiple queries.

References

[1] S. Al-Khalifa et al., "Structural joins: A primitive for efficient XML query pattern matching," *Proceedings of IEEE International Conference on Data Engineering (ICDE)*, 2002.

[2] P. Bohannon, J. Freire, P. Roy, and J. Simeon, "From XML schema to relations: A cost-based approach to XML storage," *Proceedings of IEEE International Conference on Data Engineering (ICDE)*, 2002.

[3] N. Bruno et al., "Holistic twig joins: Optimal XML pattern matching," pp.310-321, *Proceedings of ACM SIGMOD International Conference on Management of Data*, 2002.

[4] M. Carey et al., "XPERANTO: Publishing Object-Relational Data as XML," *Proceedings of Workshops on Web and Databases (WebDB)*, 2000.

[5] Don Chamberlin, Jonathan Robie, Daniela Florescu, "Quilt: an XML Query language for heterogeneous data sources" LNCS, 1997.

[6] Z. Chen, H. Jagadish, L. Lakshmanan, and S. Paparizos, "From Tree Patterns to Generalized Tree Patterns: On Efficient Evaluation of XQuery," *Proceedings of International Conference on Very Large Databases (VLDB)*, pp. 237-248, Berlin, Germany, September 2003.

[7] B. Cooper et al., "A Fast Index for Semistructured Data," *Proceedings of International Conference on Very Large Databases (VLDB)*, 2001.

[8] A. Deutsch, M. Fernandez, and D. Suciu, "Storing semistructured data with STORED," *Proceedings of the ACM SIGMOD International Conference on Management of Data*, pp431-442, Philadelphia, Pennsylvania, June 1999.

[9] M. Fernandez et al., "Publishing Relational Data as XML: The SilkRoute Approach," *IEEE Data Engineering Bulletin*, 24(2), 2001.

[10] M. Fernandex, Y. Kadiyska, A. Morishima, D. Suciu, and W. Tan, "SilkRoute: A framework for publishing relational data in XML," *ACM Transactions on Database Systems*, 2002.

[11] T. Fiebig et al., "Anatomy of a native XML base management system," *VLDB Journal*, 11(4):292-314, 2002.

[12] H. Jagadish et al., "TIMBER: A native XML database," *VLDB Journal*, 11(4):274-291, 2002.

[13] D. Kha, M. Yoshikawa, and S. Uemura, "An XML Indexing Structure with Relative Region Coordinate," *Proceedings of IEEE International Conference on Data Engineering (ICDE)*, 2001.

[14] J. Naughton et al., "The Niagara Internet Query System," http://www.cs.wisc.edu/niagara/papers/NIAGARAVLDB00.v4.pdf.

[15] OGC, Geography Markup Language (GML) Implementation Specification 3.0, 2003.

[16] S. Park, J. Lee, and H. Bae, "Easily Accessible GML-based Geographic Information System for Multiple Data Server over the Web," *Proceedings of 2nd International Conference on Information System Technology and its Applications*, June 2003.

[17] H. Schoning, "Tamino – A DBMS designed for XML," *Proceedings of IEEE International Conference on Data Engineering (ICDE)*, pp. 149-154, 2001.

[18] I. Tatarinov, Z. Ives, A. Halevy, and D. Weld, "Updating XML," *Proceedings of ACM SIGMOD International Conference on Management of Data*, 2001.

[19] I. Tatarinov, S. Viglas, K. Beyer, J. Shanmugasundaram, E. Shekita, and C. Zhang, "Storing and querying ordered XML using a relational database system," *Proceedings of ACM SIGMOD International Conference on Management of Data*, 2002.

[20] W3C, XQuery 1.0: An XML Query Language, 2002.

[21] Y. Wu et al., "Structural Join Order Selection for XML Query Optimization," *Proceedings of IEEE International Conference on Data Engineering (ICDE)*, 2003.

A Clustering Method for Large Spatial Databases

Gabriella Schoier[1] and Giuseppe Borruso[2]

[1] Universitá di Trieste, Dipartimento di Scienze Economiche e Statistiche, Piazzale Europa 1, 34127 Trieste, Italia,
gabriella.schoier@econ.units.it

[2] Universitá di Trieste, Centro d'ecellenza in Telegeomatica-GeoNetLab,Dipartimento di Scienze Geografiche e Storiche, Piazzale Europa 1, 34127 Trieste, Italia

Abstract. The rapid developments in the availability and access to spatially referenced information in a variety of areas, has induced the need for better analysis techniques to understand the various phenomena. In particular spatial clustering algorithms which groups similar spatial objects into classes can be used for the identification of areas sharing common characteristics. The aim of this paper is to present a density-based algorithm for the discover of clusters in large spatial data set which is a modification of a recently proposed algorithm.This is applied to a real data set related to homogeneous agricultural environments.

1 Introduction

The paper[1] explores the spatial issues related to the application of clustering methods to geographically relevant phenomena.

The development of new techniques and tools that support the human in transforming data into useful knowledge has been the focus of the relatively new and interdisciplinary research area: knowledge discovery in databases (KDD) (see e.g. [1]), term coined to describe the process for finding relations among observed data (see e.g. [1]). His heart is Data Mining which consists in the process of selection, modelling and application of algorithms to discover relations among large quantities of data and Web Mining which is the application of Data Mining techniques to the w.w.w. [8]. In particular Spatial Data Mining can be used for browsing spatial databases, understanding spatial data, discovering spatial relationships, optimizing spatial queries. Recently, clustering techniques have been recognized as primary Data Mining methods for knowledge discovery in spatial databases, i.e. databases managing 2D or 3D points, polygons etc. or points in some d-dimensional feature space. The well-known clustering algorithms, however, have some drawbacks when applied to large spatial databases (see e.g. [4]). On one side traditional algorithms seems to be inefficient when managing spatial

---

[1] Although the general structure of the paper reflects the common aims of the two authors, Prof. Schoier is mainly responsible for paragraph 2, 3 and 4, while Dr Borruso for paragraph 1

data; on the other side problems arise when considering spatial and non-spatial data together for clustering. Suited for purpose algorithms for spatial data detects clusters in the geographical distribution of data but not always seem to be suited for considering also their attributes, as intensity, frequency or other characteristics of the phenomena observed.

In this paper we present a new algorithm which is a modification of the $DBSCAN$ algorithm proposed in [2] which take into consideration both the spatial aspect and the non spatial variables relevant for the phenomenon that has to be analyzed. Homogeneous agricultural environments are used as a case study for our analysis.

2 The Proposed Spatial Clustering Algorithm

Spatial clustering is a method of spatial data analysis, which has been used widely in fields like medicine, ecology, urban studies etc.. Its aim is to create groups of units so that there is a high degree of similarity inside the group but a high degree of dissimilarity among elements of different groups.

There are different types of clustering algorithms: the partitioning methods,the hierarchical methods,the density-based methods and the grid methods [2].

Many factors are involved in the choice of the algorithm in the application to large databases, the tradeoff between quality and speed, the capacity of discovering clusters of arbitrary shape and the characteristics of the data.

In this paper we will consider clustering methods based on the notion of density. These regard clusters as dense regions of units which are separated by regions of low density (representing noise), they may be used to discover clusters of arbitrary shape. Among these the $DBSCAN$ [2] judge the density around the neighborhood on an unit to be sufficiently dense if the number of points within a distance $EpsCoord$ of an unit is greater than $MinPts$, in this case the unit is a core point otherwise is a border point [2]. This algorithm has been generalized in [7] by considering the $GDBSCAN$ that can cluster points units such as spatially extended units.

Our proposed generalization "the Modified Density-Based Spatial Clustering of Applications with Noise" $MDBSCAN$ considers an approach density based that take into account at the same time the spatial variables and the non spatial variables.

It has a similar structure of the $DBSCAN$ but introduce a notion of proximity not only for spatial characteristics but also for non spatial characteristics.

The key idea is that the cardinality of the neighborhood of an unit is given not only by counting the number of units that have distance from it less than the radius $EpsCoord$ but by the points that have distance less than $EpsCoord$ and that are "sufficiently" similar as regards non spatial attributes. In order to have a sufficient homogeneity for the non spatial attributes another radius Eps that represent the threshold for the distances calculated on the bases of the non spatial variables is evaluated.

In so doing we want to find clusters of elements which are spatially close to each other and homogeneous as regards other observed variables.The elements of such a clusters may be interpreted as elements similars as regards some variables and that belong to the same spatial area.

In the following we present the main steps of the algorithm:

Algorithm 1 (MDBSCAN)

Step 1. *Insert the values of the parameters: SetOfPoints representing the matrix with the values of the non spatial variables,Coordinates representing the spatial variables ,Eps the limiting distance value for the non spatial variables ,EpsCoord the limiting distance value for the spatial variables,MinPts the minimum number of points to consider a point as a core point.*

Step 2. *Chose an arbitrary point i in the database if i is a core point built the cluster by choosing all the points which are density-reachable from i else if it is a border point the algorithm pass to the point i+1*

Step 3. *Classify the points which are not density-reachable as noise*

The value of the parameter *MinPts* is fixed as in [2].

In order to determine the parameter *EpsCoord*, regarding the spatial variables, and the parameter *Eps*, regarding the non spatial variables that are considered important for explaining the phenomenon of interest, we consider the algorithms *SorteKdist* and respectively *SorteKdist2* . The former evaluates, for every point of the database represented by the matrix *Coordinates*, the distances from the nearest k-points (belonging to the same database),orders them in decreasing way and gives a graphical representation of these distances. The latter is similar, it evaluates, for every point of the database represented by the matrix *SetOfPoints*, the distances from the nearest k-points (belonging to the same database) and gives the mean value.

The algorithm *SorteKdist* uses the spatial variables; it is implemented following [2]; it evaluate for every unit of the input database the distances from the k nearest points (for the choice of k see [7]), it orders the distances and gives a graphical representation. The algorithm *SorteKdist2* uses the non spatial variables, it evaluate for every unit of the input database the distances from the k nearest points and gives the mean values.

3 An Application on a Real Dataset

In order to test the algorithm and the chosen procedure, we have examined a spatial database containing municipalities in the Friuli Venezia Giulia Region in Northeastern Italy for the year 1990.

The aim is relating the spatial position of municipalities with a particular set of indices of performance of the agricultural sector in order to obtain a zoning system of municipalities in terms of both location and structural and economical characteristics. Such zoning should therefore allow a classification of agricultural

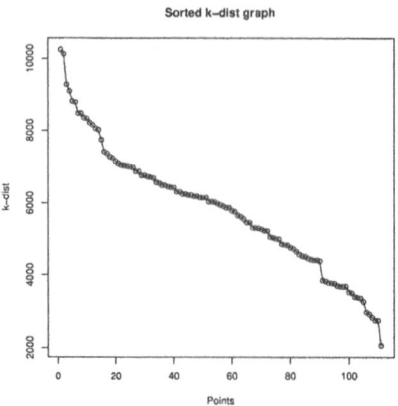

Fig. 1. Determination of the parameter $EpsCoord$

zones according to their characteristics of production and structure to supply public authorities with a tool for examining performances in different parts of the Region ([6]). Different indicators have been used to classify the different municipalities according to their characteristics. The method involved examining the productive structure of the agricultural sector referred to the smallest area.

Municipalities represent the smaller spatial unit for which data are available. Both static and dynamic indices have been used in order to evaluate the weight of agricultural in the demographic environment; the diffusion of agricultural entrepreneurship; the presence and diffusion of small-size forms (pulverization); the diffusion of entrepreneur as main actor in the firm. Agricultural land use has also been considered as well as density of bovine stock-farm.

Moreover the coordinates have also been used as variables for the spatial clustering analysis performed. In this stage we considered the coordinates of municipalities that is the *centroids*. The database obtained contained therefore the following data referred to municipalities in the Region. Only lowland municipalities were considered so we examined 111 municipalities over a total of 219.

The $MDBSCAN$ algorithm implemented involved choosing different thresholds for the coordinates and for the other indices. As regards the spatial variables the algorithm $SorteKdist$ has been implemented, the result is represented in Fig. 1 .

Looking at the graph we have chosen $EpsCoord = 7741$ that is the first point in the first valley [2].

As regards the non spatial variables the algorithm $SorteKdist2$ has been implemented the result is $Eps = 1.71$

At this point we have applied the $MDBSCAN$. The variables used are: S_j (agricultural land use), R_j (weight of agriculture in the economical system), z_j (agricultural entrepreneurship), d_j (agricultural firms' pulverization (number of

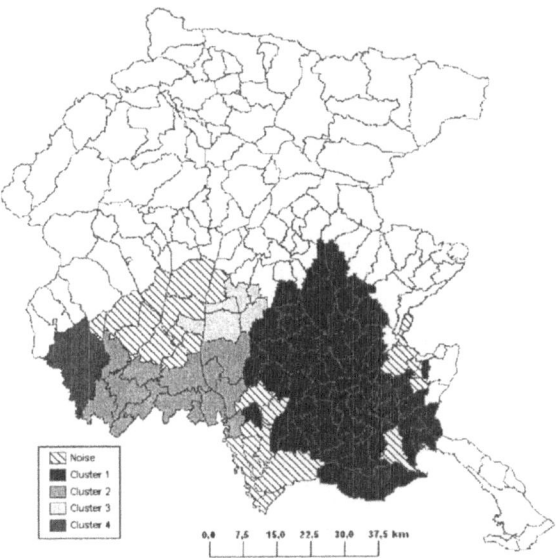

Fig. 2. Spatial clusters of municipalities of the region F.V.G.

firms)); d_{1j} (agricultural firms' pulverization (surface of firms)), t_j (diffusion of entrepreneur as main actor), a_j (agricultural surface used for sown), doj (agricultural surface used for permanent cultivations), p_j (density of bovine stock-farm), g_j (dynamic of firms lead by working firm); $XCOORD$ (X coordinate (Eastings) in UTM 33 ED50 reference system), $YCOORD$ (Y coordinate (Northings) in UTM 33 ED50 reference system).

The distance used for running our algorithm is the euclidean one.

Interesting results arise after performing the cluster analysis on the dataset. Fig. 2 shows the results of the analysis.

The algorithm has detected four clusters and classified 24 municipalities as 'noise', therefore not belonging to none of the clusters detected. If compared to other traditional, non-spatial techniques applied in the past, spatial clustering allow a more homogenous zoning of municipalities, with the most of them belonging to a cluster contiguous and close to each other. From this point of view therefore the algorithm seems to operate proficiently in order to obtain a spatial zoning of the municipalities considered. As threshold plays a key role in determining if a municipality belongs or not to a cluster, different threshold could be tested in order to allocate all the municipalities in a cluster. At this stage it is anyway interesting that most of the municipalities are allocated and continuous.

When moving to examining the different non-spatial indices we can observe interesting dynamics taking place over the municipalities in the lowland area.

The exam of Fig. 3 can help in examining the characteristics of the different clusters in terms of the indices used.

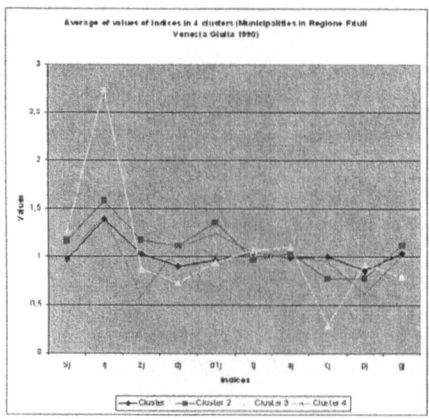

Fig. 3. Average values of the indices.

The graph was built calculating mean values of each index for each cluster. A selection of indices has been done as pivot variable to characterize the clusters. In particular, S_j, z_j d_j, t_j a_j have been analyzed. Only the first three ones however present some diversities in their figures, while the other ones present values very close to each other.

As regarding a general analysis on the composition of the clusters *Cluster 2* seems to be the more proficient one in terms of agricultural performance. This can be said after examining the general performance over the entire set of indicators. It present higher values than the other ones in the most of the indices observed, with particular reference to the agricultural entrepreneurship, weight of agriculture in the economical system, agricultural firms' pulverization (surface of firms) and dynamic of firms lead by capitalistic firm. *Cluster 3* follows, characterized mainly by high values of agricultural land use and low values of agricultural surface used for permanent cultivations. *Cluster 1* and *Cluster 4* follow in the lower end of the class, displaying quite close values in the most of the indices, apart for the agricultural entrepreneurship, where Cluster 4 present lower values.

As a general impression on the results obtained from the analysis of indices we can however notice that different in mean values between different clusters are not so high: that can mean that the general situation of agriculture portrayed in the Region is quite a homogeneous one and differences can be attributed to different shape of small areas used (municipalities) as well as to micro-economical characteristics of firms.

4 Conclusions

The spatial clustering analysis allowed in any case to group together municipalities according to the values offered by the indices and therefore also small differences in values allow to discriminate between clusters.

In this paper we present a new algorithm which is a modification of the *DBSCAN* algorithm proposed in [2] which take into consideration both the spatial aspect and the non spatial variables relevant for the phenomena that has to be analyzed. Homogeneous agricultural environments are used as a case study for our analysis.

Further research is however necessary with particular reference to the spatial component of the data. At this stage we considered centroids of municipalities as identifiers of geographical location. In geographical terms, municipalities are represented by means of irregularly-shaped polygons, with very different characteristics of surface, perimeter and attributes that can be related to them. There is the risk that, given a certain threshold, only small-area municipalities clusters while larger ones tend to be left out or considered as belonging to a different cluster. The topics to tackle in the future involve therefore considering polygons for the cluster analysis instead of their centroids' coordinates.

References

1. Bailey, T. C., Gatrell, A. C.:Interactive spatial data analysis. Addison Wesley Longman Edinburgh, UK.(1995)
2. Ester, M., Kriegel, H.,P.,Sander, J.,Xiaowei, X.: A Density-Based Algorithm for Discovering Clusters in Large Spatial Databases with Noise.Proceeding of the 2nd International Confererence on Knowledge Discovery and Data Mining.(1996) 94–99
3. Fayyad, U., Piatesky -Shapiro, G.,Smyth, P.: From Data Mining to Knowledge Discovery in Databases. (1996) $http : //www.kdnuggets.com/gpspubs/aimag - kdd - overview - 1996 - fayyad.pdf$
4. Han, J., Kamber, M., Tung, A.K.H.: Spatial Clutering Methods in Data Mining: A Survey. (2001) $ftp : //ftp.fas.sfu.ca/pub/cs/han/pdf/gkdbk01.pdf$.
5. Koperski K., Han J.,Adhikary J.: Mining Knowledge in Geographical Data.(1998) $ftp : //ftp.fas.sfu.ca/pubcs/han/pdf/geo_survey98.pdf$.
6. Prestamburgo M.:La classificazionedegli ambiti agricoli: una proposta metodologica. (1981)
7. Sander, J.,Ester, M., Kriegel, H.,P.,Xiaowei, X.: Density-Based Clustering in Spatial Databases: The Algorithm GDBSCAN and its applications.(1999) $http : //www.dbs.informatik.uni - muenchen.de/Publikationen/$
8. Schoier, G.: Blockmodeling techniques for Web Mining, in Procedings in Computational Statistics,COMPSTAT 2002,Springer & Verlag, Berlin. pp.201-206 .(2002)

GeoSurveillance: Software for Monitoring Change in Geographic Patterns

Peter Rogerson[1,2] and Ikuho Yamada[1]

[1]Department of Geography and National Center for Geographic Information and Analysis,
University at Buffalo, Buffalo, NY 14261 USA
[2]Department of Biostatistics, University at Buffalo, Buffalo, NY 14261 USA
{rogerson,iyamada}@buffalo.edu

Abstract. In this paper we first describe how statistical process control methods in general, and cumulative sum methods in particular, may be used to monitor changes in geographic patterns. We give examples for several different possible data scenarios, including those where regional data consist of very small counts each time period. In addition, we describe software we have developed for this purpose. The paper describes both the development of the statistical methods and the software in detail. The authors intend to make the software freely available and downloadable so that researchers interested in a wide variety of applications may potentially make use of the methods.

1 Introduction

There are many situations where it is important to detect changes in geographic patterns as quickly as possible, and to discriminate between "typical" random variation in map patterns, and significant, meaningful spatial change. For example, crime analysts and epidemiologists are interested in finding new clusters of cases as quickly as possible, without raising public fears by declaring a new cluster when it could in fact be attributable to random variation.

In the next section of the paper, we describe how traditional cumulative sum (cusum) methods may be used to monitor variables for each of a set of regions, and how statistical adjustment may be made for the multiple cusums that are maintained. The objective is to find, as quickly as possible, any deviation from a region's baseline expectation, given an acceptable rate of false alarms.

In the third section, we describe the GeoSurveillance software. By linking the regional cusums to a map, the user may visualize the geographic changes that are taking place. The software utilizes Visual Basic 6.0 and MapObjects to portray cumulative sums of the deviations between observed and expected values for each region over time. The user provides a .shp file containing base map information, along with the regional values that are expected and observed during the monitoring period. The user also sets the false alarm rate, and from this the software calculates the critical threshold for the cumulative sum. Separate windows depict a map of the study region (divided into subareas), the data table, and a cusum chart for a specified region. The windows are linked so that, e.g., clicking on a region of the map will display the cumulative sum for that region. Clicking on a point of the cusum chart will highlight that particular data point in the data table.

2 Cumulative Sum Methods for Detecting Changes

Cumulative sum (or cusum) methods are designed to detect changes in the mean value of a quantity of interest [1, 2]. They are widely used in industrial process control to monitor the quality of production characteristics. It is assumed that the quantity being monitored is a normally distributed variable that exhibits no serial autocorrelation.

The monitoring process begins by converting the variable to a z-score with mean 0 and variance 1. Then the cumulative sum, following observation t, is defined as

$$S_t = \max(0, S_{t-1} + X_t - k), \qquad (1)$$

where k is a parameter. A change in mean is signaled if $S_t \geq h$, where h is a predefined critical threshold. Thus values of z in excess of k are cumulated. The parameter k (where a standardized variable is being monitored) is often chosen to be equal to ½; in the more general case, k is often chosen to be equal to ½ the standard deviation associated with the variable being monitored.

The parameter h is chosen in conjunction with an acceptable rate of "false alarms"; high values of h will lead to a low probability of a false alarm, but also a lower probability of detecting a real change.

ARL is used to denote the "average run length" between false alarms. When $k=1/2$, an approximation for the in-control average run length (ARL_0) may be derived from Siegmund's approximation [3]

$$ARL_0 = 2(e^b - b - 1), \qquad (2)$$

where $b=h+1.166$. But this can not be solved directly for h. Rogerson [4] shows that the LambertW function may be used to find the following approximate solution for the threshold, as a function of the desired ARL:

$$h \approx \ln(1 + \frac{ARL_0}{2}) - 1.166. \qquad (3)$$

More generally, for other values of k,

$$h \approx \frac{\ln(1 + 2k^2 ARL_0)}{2k} - 1.166. \qquad (4)$$

Rossi et al. [5] suggest using the following transformation for small counts:

$$y = \frac{x - 3\lambda + 2\sqrt{\lambda x}}{2\sqrt{\lambda}}, \qquad (5)$$

where x is the observed count and λ is the expected count. This works well when the expectation is sufficiently large, but will not be acceptable for small expectations.

2.1 Cusum Method for Poisson Variables

Let λ_0 be the mean value of the in-control Poisson parameter. The Poisson cusum is
$$S_t = \max(0, S_{t-1} + X_t - k), \tag{6}$$
where
$$k = \frac{\lambda_1 - \lambda_0}{\ln \lambda_1 - \ln \lambda_0} \tag{7}$$
and where the objective is to minimize the time to detecting a change from λ_0 to λ_1. The appropriate threshold h can be found from the values of the parameter k and the desired ARL_0 by using either a table [6], Monte Carlo simulation, or an algorithm such as the one provided by White and Keats [7], which makes use of a Markov chain approximation.

Often the expected, in-control mean is varies over time. For example, there may be seasonal fluctuations in either crime rates or disease rates that are being monitored. In this case, the Poisson cusum may be written as
$$S_t = \max(0, S_{t-1} + c_t(X_t - k_t)), \tag{8}$$
where k_t is determined as in (7), from the time-specific values of the in- and out-of-control means [8]. The quantity c_t is equal to h_t / h, where h is the threshold associated with the mean of the time-varying Poisson parameter (and the associated value of k), and where h_t is the threshold associated with k_t and the desired ARL_0.

2.2 Spatial Considerations

There has been increasing interest in monitoring spatial patterns – particularly in the field of public health, where it is desirable to detect emergent clusters of disease as quickly as possible [9]-[16]. Applications also have been made to crime analysis, where regional crime rates are monitored [17, 18]. To implement either the cusum for normal or Poisson distributed variables in a regional context, one option is to maintain separate cusums for each region. Suppose it is desired that the average time until a false alarm in the system is ARL_0. In this case the threshold can not be determined for each region by simply using ARL_0; an adjustment is necessary to the threshold. Without an adjustment, the system of cusums would signal *somewhere* in the multiregional system more frequently than desired. To yield a systemwide false alarm of ARL_0, the ARL used in each region (designated $ARL_{0,r}$) should solve the following:
$$ARL_0 = [1 - (1 - ARL_{0,r}^{-1})^r]^{-1}, \tag{9}$$
where r is the number of regions in the system. This may be solved approximately as $ARL_{0,r} \approx r(ARL_0)$.

The implementation of the cumulative sum approach in a regional setting has, to this point, been rather uninteresting from a spatial context – each region is simply monitored separately, and there is no explicitly spatial connection between the evolution of regions that are near to one another. An extension of the above is to construct "local statistics" in association with each geographic unit. These are defined as a weighted sum of the region's observation and surrounding observations, where the weights could potentially decline with increasing distance from the region. Cumulative sums associated with these local statistics may be monitored. Because the local statistics are spatially autocorrelated, a Bonferroni adjustment would result in too high a value of h, making it difficult to detect change when it actually occurs. Monte Carlo simulation of the null hypothesis, where there is no deviation from expected values of the Poisson or normal parameters, can instead be carried out to determine appropriate thresholds for the cumulative sums.

One can monitor the quantities $y_{i,t} = \sum_j w_{ij} x_{j,t}$, where $x_{j,t}$ is the observed count in region j at time t, and w_{ij} is a weight associated with, for example, the distance from region i to region j. These observed quantities are then compared with their corresponding expectations, $\sum_j w_{ij} \lambda_{0,tj}$ (where the subscript tj refers to region j at time t), and used in a cumulative sum for region i.

To determine appropriate critical thresholds for each region, Monte Carlo simulation of the null hypothesis may be used (where observed counts are realizations from Poisson or normal distributions with parameters set equal to the corresponding expectations). In particular, with a desired average run length of ARL_0, the critical thresholds should be determined using $s(ARL_0)$; the value of s is less than the number of regions (r), and is determined via simulation to lead to the desired average run length. The greater the correlation between the local regional statistics, the lower s will be relative to r.

3 GeoSurveillance ~ Software for Monitoring Changes in Geographic Patterns

GeoSurveillance is GIS-based multi-window software designed to implement cumulative sum methods in a visual and interactive manner in the context of spatial surveillance. It is intended to provide a user-friendly interface to those methods using a dynamically linked multi-window system and GIS-based map representation. GeoSurveillance is developed with Microsoft Visual Basic® and ESRI's MapObjects component, and it runs under the Microsoft Windows operating systems.

GeoSurveillance provides the user with several types of cusum methods and a spatial autocorrelation analysis method. The cusum methods here assume that time-series data are available on observed and expected numbers of incidences for multiple spatial units (for example, counties and census blocks), together with a shapefile containing the units. The software, its user's manual, and sample datasets may be found at http://wings.buffalo.edu/~rogerson.

Figure 1 shows the user interface of GeoSurveillance, which consists of a main window displaying a map and several other windows that are opened when a particular analytical method is performed. The main window consists of a map frame to show shapefiles, a menu bar to select an analytical method to perform, and a button bar to activate basic map manipulation tools such as zoom-in and pan tools. When the user selects an analytical method from the menu bar in the main window, the corresponding window opens up to let him/her set up the analysis (for example, to specify files to be analyzed and input parameters to be used).

GeoSurveillance has several options for monitoring regional variables. For large frequencies or counts, normality may be assumed, or a transformation to normality may be made. Then the user may choose either multiple univariate analysis, where individuals regions are monitored separately and simultaneously, or they may choose multivariate cusum analysis, where input of the matrix of covariances between regional variables is required. For situations where counts are low, Poisson cusum analysis is the appropriate option. This option also allows for truly spatial analysis, where the small counts may be aggregated into neighborhoods consisting of the region in question and all regions that are immediately adjacent.

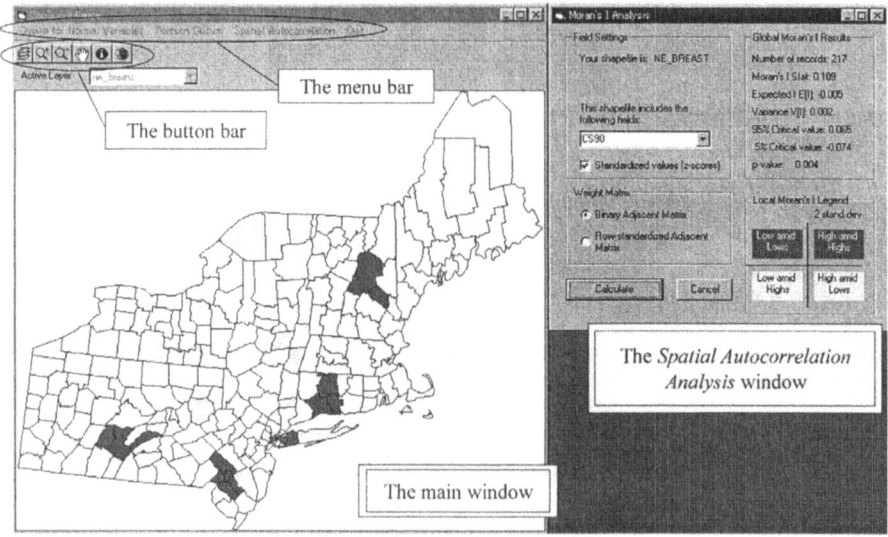

Fig. 1. GeoSurveillance window structure

4 Analysis of Breast Cancer Mortality in the Northeastern U.S. Using GeoSurveillance

4.1 Breast Cancer Mortality and Data

Previous studies of breast cancer mortality in the northeastern United States have either focused upon a cross-sectional assessment of geographic clustering for one period of time [19], or have examined temporal change from a retrospective view [20]. In the latter case, annual data for the period 1968-1998 is described and statistically significant changepoints are found, where the geographic pattern of breast cancer shifts. In this section we examine what would have happened had we began monitoring the pattern of breast cancer in 1978, using the period 1968-1977 as a baseline. Specifically, we seek to find deviations from the baseline geographical pattern as quickly as possible.

From the Compressed Mortality File (National Center for Health Statistics [21]), annual, county-level data on breast cancer mortality were extracted for the 217 counties comprising the northeastern United States for the period 1968-1998. The observed and expected numbers of cases are stored in comma-delimited text files, *Obs.txt* and *Exp.txt*, respectively. The expectations were age-adjusted, and were computed based on the nationwide breast cancer rate and the age-specific county populations for each year provided on the file. Therefore, changes in population size and age structure were taken into account. In the following analysis, data for the first ten years will be used to find the mean annual relative risk (O/E) for each county, which will serve as the "baseline risk" of each county. Then, the expectations for the subsequent years (i.e., 1978-1998) will be multiplied by the mean annual relative risk of each county to determine expectations during each of the monitored years, beginning in 1978. The purpose is to monitor the observed, county-specific frequencies of breast cancer mortality cases, and to determine quickly when significant deviations from the baseline (1968-1977), while allowing for any regional variation that pre-existed during the baseline period. As a final step, the observations and the adjusted expectations will be converted into z-scores. Another data element that is required is a shapefile, *ne_breast.shp*, containing the 217 counties.

4.2 Results, Summary, and Future Work

Figure 2 shows the display of GeoSurveillance at the end of the cusum analysis explained in the previous section. The window in the lower left shows the cusum chart for Philadelphia County, PA, which signaled first among the 217 counties in 1988, the 11th year since the monitoring started in 1978. Its cusum value started increasing around 1983 and reached 8.35 in 1988, exceeding the threshold parameter h, and sounding a signal. The signal indicates that the level of breast cancer mortality in this county has risen significantly above its baseline level that held during the period 1968-1977. The complete list of the counties that signaled during the monitoring period (1978-1998) is given in Table 1.

The detailed steps that are required to carry out this analysis using GeoSurveillance are provided in the appendix. It should be stressed that in this example, signals indicate a deviation from the baseline mortality rates. It is possible that the baseline itself was "unusual" in some way. Perhaps, for example, rates were unusually low during the baseline – a signal might simply indicate a return to more typical mortality rates.

This form of analysis is of course exploratory – it suggests regions of interest to explore in more detail. It does not, by itself, offer explanations for why change may have occurred.

There are several extensions to cumulative sum methods that could provide valuable additions to the initial steps described here. For example, the fast initial response (FIR) feature starts the cumulative sum at one-half of the threshold value ($h/2$) instead of at zero; the intuition is that if the process is already out of control, a signal will be given more quickly, and if it is not out-of-control, the cusum will likely return to zero [22]. Another difficulty is the heavy reliance on the baseline mortality levels. One possible solution is to estimate the baseline as the cusum surveillance is started. This is referred to as a self-starting cusum; Hawkins and Olwell [22] provide details on how this may be implemented. Finally, current work is aimed at monitoring the maximum value observed on a map; the distribution of the maximum is typically not normal, and modifications to the usual cusum procedure must therefore be made. In related work, we are also developing methods for monitoring the maxima of smoothed surfaces, where the surface height represents a variable of interest (such as a crime rate), and the smoothing occurs using a weighted sum of surrounding values to emphasize that one is monitoring the variable in a geographic neighborhood of the region of interest.

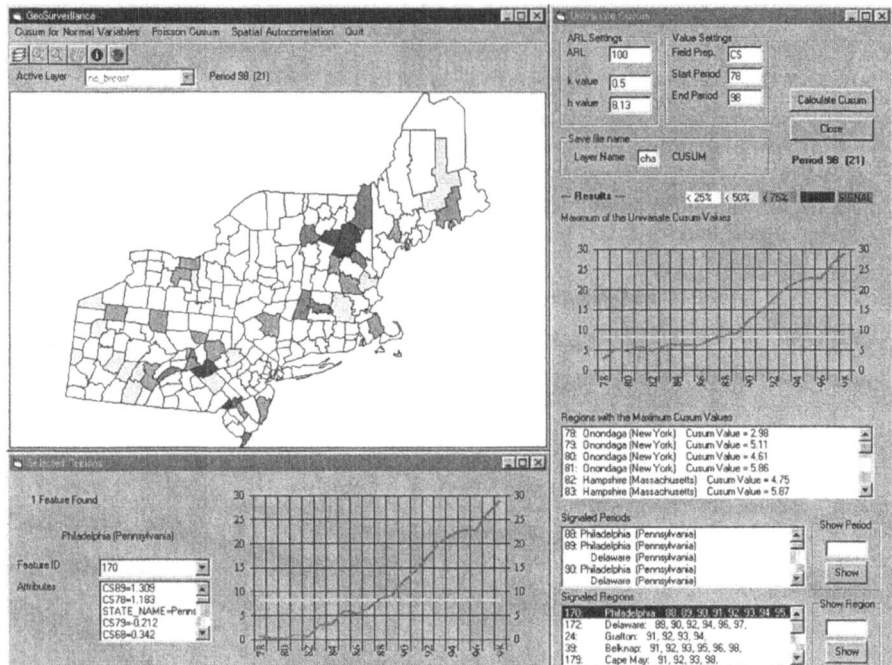

Fig. 2. GeoSurveillance display for the univariate cusum analysis

Table 1. Counties that signaled in the cusum analysis for normal variables

County	Signaling years	Number of signaling years
Philadelphia, PA	1988 - 98	11
Delaware, PA	1989 - 90, 92, 94, 96 - 97	6
Grafton, NH	1991 - 94	4
Belknap, NH	1991 - 93, 95 -96, 98	6
Cape May, NJ	1991 - 93, 98	4
Coos, NH	1992 - 98	7
Huntingdon, PA	1992 - 93, 95	3
Montour, PA	1993 - 98	6
Juniata, PA	1994, 98	2
Berkshire, MA	1995 - 98	4
Schuylkill, PA	1995	1
Addison, VT	1997 - 98	2
Hampshire, MA	1998	1

References

[1] Wetherill, G.W. and Brown, D.W.: Statistical Process Control: Theory and Practice. Chapman and Hall, New York (1991)
[2] Montgomery, D.: Introduction to Statistical Quality Control. Wiley, New York (1996)
[3] Siegmund, D.O.: Sequential Analysis: Test and Confidence Intervals. Springer-Verlag, New York (1985)
[4] Rogerson, P.: Formulas for the Design of CUSUM Quality Control Charts. Submitted for publication (2004).
[5] Rossi, G., Lampugnani, L., and Marchi, M.: An Approximate Cusum Procedure for Surveillance of Health Events. Stat. Med. **18** (1999) 2111-2122
[6] Lucas, J. M.: Counted Data Cusums. Technometrics **27** (1985) 129-144
[7] White, C.H. and Keats, J.B.: ARLs and Higher Order Run Length Moments for Poisson CUSUM. J. Qual. Tech. **28** (1996) 363-369
[8] Rogerson, P. and Yamada, I.: Approaches to Syndromic Surveillance When Data Consist of Small Regional Counts. Morbidity and Mortality Weekly Report (2004) forthcoming
[9] Farrington, P, Beale, A.D.: The Detection of Outbreaks of Infectious Disease. In: Lierl, L., Cliff, A.D., Valleron, A., Farrington, P., and Bull, M. (eds.): Geomed '97. BG Teubner, Stuttgart (1999)
[10] Kleinman, K., Lazarus, R., Platt, R.: A Generalized Linear Mixed Models Approach for Detecting Incident Clusters of Disease: Biological Terrorism and Other Surveillance. Am. J. Epi. **156** (2004) 217-24
[11] Kulldorff, M.: Prospective Time Periodic Geographical Disease Surveillance Using a Scan Statistic. J. Roy. Stat. Soc. A **164** (2001) 61-72
[12] Lawson, A.: Statistical Methods in Spatial Epidemiology. Wiley, New York (2001)
[13] Raubertas, R.F.: An Analysis of Disease Surveillance Data that Uses the Geographic Locations of the Reporting Units. Stat. Med. **8** (1989) 267-71
[14] Rogerson, P.: Surveillance Systems for Monitoring the Development of Spatial Patterns. Stat. Med. **16** (1997) 2081-93
[15] Rogerson P.: Monitoring Point Patterns for the Development of Space-Time Clusters. J. Roy. Stat. Soc. A **164** (2001) 87-96
[16] Rogerson, P. and Yamada, I.: Monitoring Change in Spatial Patterns of Disease: Comparing Univariate and Multivariate Cumulative Sum Approaches. Stat. Med. (2004) forthcoming
[17] Rogerson, P.: Geographic Surveillance of Crime Frequencies in Small Areas. In: Wang, F. (ed.): Geographical Information Systems and Crime Analysis. Idea Group Publishing, Hershey, Pennsylvania (2004) forthcoming
[18] Rogerson, P. and Sun, Y.: Spatial Monitoring of Geographic Patterns: an Application to Crime Analysis. Computers, Environment, and Urban Systems, **25/6** (2001) 539-56
[19] Han, D. and Rogerson, P.: Application of a GIS-Based Statistical Method to Assess Spatio-temporal Changes in Breast Cancer Clustering in the Northeastern United States. In: Khan, P. (ed.): Geographic Information Systems and Health Applications. Idea Group Publishing, Hershey, Pennsylvania (2003) 114-38
[20] Kulldorff, M., Feuer, E.J., Miller, B.A., and Freedman, L.S.: Breast Cancer Clusters in the United States: a Geographic Analysis. Am. J. Epi. **146** 161-70
[21] National Center for Health Statistics. Compressed Mortality File. CD-ROM Series 20 No. 1A. Atlanta, GA: Centers for Disease Control and Prevention.
[22] Hawkins, D.M. and Olwell, D.H.: Cumulative Sum Charts and Charting for Quality Improvement. Springer, New York, (1998)

Appendix

This section explains how to analyze the data described above with GeoSurveillance assuming that the regional counts follow the normal distribution.

1. Start GeoSurveillance by double-clicking the *GeoSurveillance1.exe* icon.
2. In the *Data Saving Directory* window, navigate to a drive and directory where you want to save files containing results of your analysis. Then, press the *OK* button to open the main window.
3. Click on the 🗃 icon; the *Layer Management* window will pop up.
4. Click on the *Add Layer* button and select *ne_breast.shp*. Then click on the *Open* button. Close the *Layer Management* window.
5. In the *Active Layer* combo box, select the *ne_breast* layer.
6. In the *Select Name Fields* window opened, set *Name field 1* to *NAME* and *Name field 2* to *STATE_NANE*. Click on the *OK* button.
7. Select the *Calculate z-scores* menu from the *Cusum for Normal Variables* menu in the main window. Specify the observation and expectation files, that is, *Obs.txt* and *Exp.txt*, respectively, when prompted. This will open the *Calculate z-scores* window.
8. Input settings such as starting and ending years of the data, in this case, 1968 and 1998, respectively. In the following, it is assumed that the *AreaKey Name* and the *Field Prep.* are set to *AREAKEY* and Z, respectively. In the *Adjustment* frame, select both options and set R and X to 1.186 and 10, respectively. The first option adjusts the expectations so that difference between the nationwide incidence rate and that of the study region is taken into account. The second option takes care of regional variations in the "baseline risks." In the *Transformation Type* frame, select the *Standard method* option. Then, click on the *Create Score File* button.
9. Open the created score file by spreadsheet software. Make sure that the AREAKEY column is formatted as text even though the AREAKEY is actually a number such as FIPS code. Save the file in the dBASE format.
10. Select the *Relate data file* menu from the *Cusum for Normal Variables* menu in the main window and, when prompted, specify the dBASE file you have just created. This will open the *Relate DBF file to Active Layer* window.
11. Input the AREAKEY names for the dBASE file and the shapefile; in this example, *AREAKEY* and *FIPS*, respectively. Click on the *Relate* button. When the process is successfully completed, you will get a message, "Relate Created." Click on the *OK* button to close the window.
12. Select the *Calculate h value* menu from the *Cusum for Normal Variables* menu; this will open the *Calculate h value* window.
13. In the *Settings* frame, input *100* to the *target ARL* box and select *0.5* for the k value. The *Number of regions* box will automatically have been set to *217*.
14. Click on the *Calculate univariate h value* button; you will be given *8.13* in the *h value* box. Close the *Calculate h value* window.
15. Select the *Univariate Cusum Analysis* menu from the *Cusum for Normal Variables* menu; this will open the *Univariate Cusum* window.

GeoSurveillance: Software for Monitoring Change in Geographic Patterns   1105

16. In the *ARL Settings* frame, input *100*, *0.5*, and *8.13* into the *ARL*, *k value*, and *h value* boxes, respectively.
17. In the *Value Settings* frame, input Z into the *Field Prep*. box, *78* and *98* into the *Start Period* and *End Period* boxes, respectively.
18. GeoSurveillance will create a comma-delimited text file that contains results of the cusum analysis. The file name will be *ne_breast* {?} *cusum.txt*, and you can set the part of {?} in the *Save file name* frame as you like. (Note: if a file with the specified name already exists, the application will overwrite the file without warning.) The file will include cusum values for the individual counties for each year and items in the *Signaled Periods* and *Signaled Regions* lists explained later, together with the parameter settings.
19. After finishing the settings above, click on the *Calculate Cusum* button.
20. GeoSurveillance will calculate cusum values year by year and report the results as below.
    - Map in the main window:
        Each county is colored with respect to the ratio of its current cusum value to the threshold parameter h. When the ratio is less than 25%, the region will be white, for example. If the ratio is greater than or equal to 100%, i.e., the cusum value exceeds the threshold h, the county will be represented by red, indicating a signal.
    - *Regions with the Maximum Cusum Values* list:
        For each year, the name of the county with the maximum cusum value will be shown with its cusum value.
    - *Signaled Periods* list:
        When at least one county has a cusum value exceeding the threshold h, the year will be shown in this list box together with the name of the counties that sound signals.
    - *Signaled Regions* list:
        When a county first sounds a signal, its ID and name will be shown in this list together with the current year. After the second signal for a particular county, only the current year will be added at the end of the row for the county.
21. After the cusum analysis for all the years specified has been finished, you can examine the results in detail in the following ways.
    - *Identify* button in the main window:
        - Activate the *Identify* tool by clicking on the 🛈 icon in the main window; then select a county (or counties) on the map by either clicking on the county or drawing a rectangle to cover the counties by the mouse.
        - The *Selected Regions* window will pop up and the outline(s) of the selected county (or counties) will be drawn with a thick line.
        - If multiple counties are selected, specify one in the *Feature ID* combo box; then corresponding attributes stored in the shapefile and the file related in Steps 10 and 11 will be listed in the *Attributes* list and the univariate cusum value for the county will be shown by a red line in the line chart, along with the value of threshold h represented by a green line. When you click on the red line, the z-score of the corresponding year will be highlighted in the *Attributes* list.

- You can also open the *Selected Regions* window by selecting a county in the *Signaled Regions* list or by specifying the county name in the *Show Region* frame and clicking on the *Show* button.
- *Signaled Periods* list:
  - Select a year in this list, and you can see the map in the main window restored to the cusum value pattern for the corresponding year.
  - This can also be done using the *Show Period* frame; input in the box a year that you want to examine, and click on the *Show* button.
22. To close the window, click on the *Close* button.
23. To close GeoSurveillance, click on the *Quit* menu in the main window.

From Axial Maps to Mark Point Parameter Analysis (Ma.P.P.A.) – A GIS Implemented Method to Automate Configurational Analysis

V. Cutini, M. Petri, and A. Santucci

Università di Pisa – Dipartimento di Ingegneria Civile
L.I.S.T.A. – Laboratorio di Ingegneria dei Sistemi Territoriali e Ambientali
Via Diotisalvi, 2 – 56126 Pisa – tel. 050.553502 – fax 050.553495
v.cutini@ing.unipi.it, maxines@katamail.com,
alsantucc@tiscali.it

Abstract. Methods based on configurational theory were proved to be highly effective in urban space analysis, in order to highlight the distribution of the levels of attractiveness towards activities. Nevertheless, both Axial Analysis and Visibility Graph Analysis (the most prominent analytic techniques) appear affected by some evident faults, that somehow do limit their actual use: in particular, those faults affect the definition of the axial system, due to the arbitrariness in drawing the lines that cover the urban grid and the lacking of correspondence between them and the streets of the settlement. In VGA, that studies the configurational features of the internal points of the grid, the problem concerns the heaviness of the data (thousands of numerical values) and the difficulty in referring them to a single urban space (either a street or a square).
In other words, on a one hand we can't but appreciate the useful capability of configurational variables in reproducing the distribution of attractiveness towards activities; on the other hand we complain about their actual use, because of the lacking of automation in defining the system, as well as the poor correspondence between it and the observed urban structure.
On such basis, our research aims at enriching the configurational approach with ArcGIS tools: our GIS application works both as the data source for the configurational model (so as to automatically and objectively construct the system) and as its receiver (so as to recipe its outputs, to elaborate them and to graphically render them, with reference to the actual elements of the grid). Besides, our GIS implementation allows an easy interface of configurational analysis with the most widespread urban models.

1 Introduction

An extended description of configurational theory and of its referred techniques is here assumed as hardly appropriate. Since the last nineties, in fact, the methods related with the configurational approach are broadly well known among the most interesting and useful tools for analysing urban settlements and supporting the decision making around their transformations. Nonetheless, a brief sketch of configurational approach will be here provided, so as to highlight some of the limits and the faults that affect its actual use, as well as to show how our research tries to overcome them, allowing a significant "step ahead" towards a reliable urban approach.

Bill Hillier introduced the configurational theory in 1984, publishing with Julienne Hanson "The social logic of space" [9]. Basis of the new approach, that was called Space Syntax, is the assumption of the urban grid (that is the set of all the open spaces of a settlement: streets, squares and so on) as the primary element in all urban phenomena and, in particular, in the generation of movement along its streets and in the location of activities on their sides.

At that time, such an assumption was to appear somehow revolutionary, since it seemed inverting the traditional approach to urban analysis: classic urban modelling was based on the spatial interaction principle, assuming the location of activities as an independent variable (that is the actual input of models processing) and the movement as the direct result of mutual relations between activities.

On the basis of such an impressive assumption is the hypothesis of the existence of the so-called "natural movement", defined as the portion of movement which depends on the configuration of the grid of urban paths and is not influenced by the presence of located activities.

On this basis, activities will follow the distribution of natural movement (and hence the grid configuration), aiming at taking benefit of higher traffic flows. Their location will obviously produce further movement (an attracted movement, according to an interactional vision), and this movement on its turn will induce the location of other activities. And so on, in an exponential process that identifies the activities as a mere multiplier effect of the primary element of urban genesis, that is the grid.

According to configurational theory, the spatial structure of a settlement (i.e. the way its streets and squares are disposed and mutually related) is therefore the actual key for the comprehension of urban phenomenal, both material and immaterial: pedestrian and vehicular movement, activities location, land values distribution, social behaviour, and so on [10].

The full comprehension of the grid configuration appears hence fundamental in order to understand and to manage its effects. And here configurational analysis comes into play. Configurational analysis is aimed at studying the grid of a settlement as a spatial system, by means of its decomposition in a set of elements, of the analysis of the relation mutually connecting them and of the attribution to each of them of a set of numerical parameter which are called configurational indices. The configurational indices are hence assumed as the state variables of the system, suitable for describing its spatial consistence, that is the configuration of the urban grid.

Since now on, the several configurational techniques (all sharing this conceptual foundation) do separate.

The first method, proposed by Hillier himself, is now generally called Axial Analysis, and provides the reduction of the urban grid into a set of linear segments, the axial lines, which compose the axial map. Belonging condition to the system is the mutual connection of the lines; the notion of depth is then introduced, so as to define the distance between a couple of lines, topologically measured as the number of interposed lines along the shortest path which actually connect them.

The axial analysis provides the processing of the axial map so as to provide each single line with its own configurational indices. Several indices are proposed; here we only mention the most important, the integration value, defined as the mean depth of a line from all the other lines of the axial map, normalised with respect to the number of the lines of the whole system by means of the following formal expression:

$$I = 2 (Md - 1) / k - 2 \qquad (1)$$

where Md is the mead depth of the observed line and k is the total number of the lines.

It could be easily shown that the integration index resulting from that expression will vary from a minimum threshold of 0 (corresponding to the most integrated line one could imagine) to a maximum value of 1 (which is assigned to the most segregated ones).

In order to show such an evidence, a small example will be here shown in figure 1, representing two axial maps. In the A map, we mark in red the most integrated line, whose integration value will result:

$$I = 2 (1 - 1) / (7 - 2) = 0 \qquad (2)$$

In the B map, we mark in blue the least integrated line (that is the most segregated), characterised by the following integration value:

$$I = 2 (3,5 - 1) / (7 - 2) = 1 \qquad (3)$$

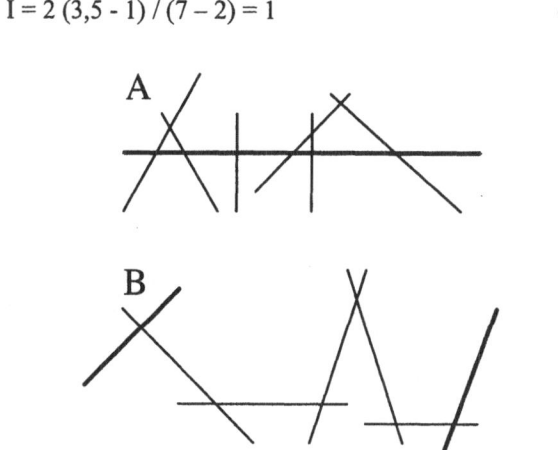

Fig. 1. Two different axial maps; the most integrated line and the less integrated ones are respectively marked in red (A) and in blue (B)

A further approach was developed on the basis of axial analysis, taking into account not only the number of interposed lines but also their intersection angle: some Gestaltpsychology studies, in fact, seem to prove that urban paths consisting of wide angle turns are actually appreciated as more spatially impeding than orthogonal (or nearly orthogonal) ones. On such basis, the so-called Angular Analysis determines the integration values computing the weight of each connection with reference to its own intersection angle [13].

If angular analysis can hardly seen as more than a slight specification of the axial one, the Visibility Graph Analysis proposes a significant modification of the original approach; although it is based on the same configurational foundation, nonetheless

VGA achieves the reduction of the urban grid into a system by means of its complete coverage with a homogeneous mesh of points, that is called visibility graph. Each point (vertex) will then result connected with all the other vertices by means of a sequence of viewsheds, so as to be divided from them by different depth values, which obviously are computed along the shortest paths [1]. On the basis of those visual connections, each single vertex can hence be provided with several configurational parameters; as we did above, also dealing with VGA we here restrict them to the most important that, once again, is integration. And integration is again defined as the mean depth of the vertex from all the other vertices of the visibility graph [14].

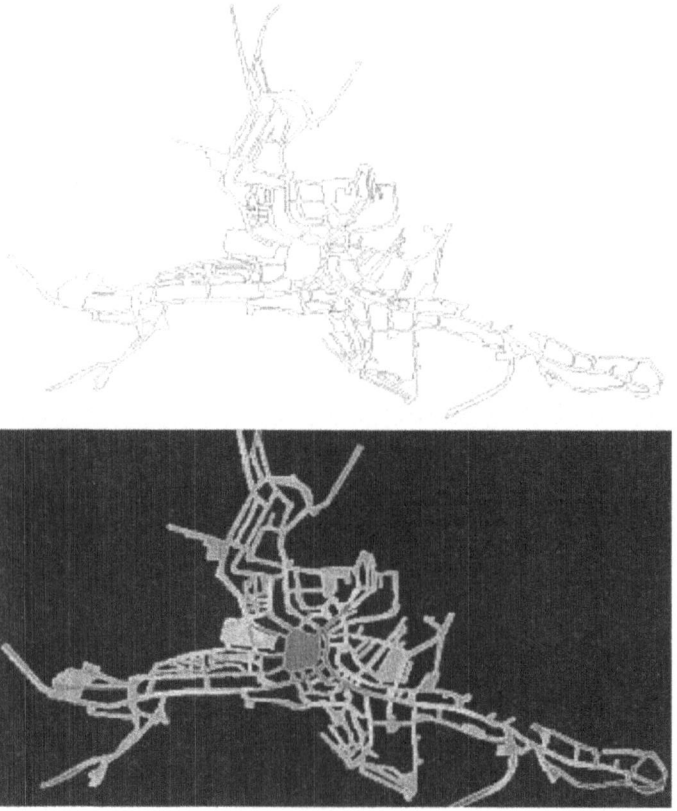

Fig. 2. The processed visibility graph of Siena; the distribution of integration value is here described with a chromatic representation, from red (most integrated vertices) to blue (most segregated ones)

The assessment of integration as the most significant configurational index (either in axial analysis, in angular analysis and in visibility graph analysis) does here appear worth explaining. We observe that integration, defined as the mean depth value of an *element* (line or vertex) from all the others, can be seen as an accessibility index, that is a parameter suitable for measuring the potential of a place with reference to the spatial impedance of its surroundings. It's worth noticing that here we're talking of a "pure" accessibility, that is not influenced by the presence of the located activities (as, instead, accessibility is classically appreciated), but is only determined by the spatial configuration of the grid.

Nonetheless, such a significant notion would remain unfruitful, if integration itself were not proved to narrowly reproduce the distribution of activities along the streets, and then their respective attractiveness level. On this regard, several studies were aimed at verifying this correspondence, and hence to attest the reliability of the methods of configurational analysis both as knowledge techniques and town planning support tool. Integration was proved as narrowly related with pedestrian flows distribution [3] [4], as well as a highly reliable indicator of the distribution of attractiveness towards activities [5] [2].

The distribution of the integration values can help to understand the genesis of a settlement, accounting for the actual location of activities and for the shifting of centrality we are frequently used to observe: especially in the last few decades, in fact, we often notice activities leaving the ancient historic cores seeking for a location in more attractive areas. Those new development areas are therefore getting more and more "central", despite their lacking in historical importance, in functional pregnancy and in architectural and morphological appeal. Configurational analysis can hence help to understand what those activities are actually seeking for, or, from a different point of view, the actual reasons of the appeal of new development areas [6].

But configurational analysis can also be used as a planning tool, suitable for supporting the town planner both in the location of activities and in the transformation of the urban grid; its use will allow to avoid the discrepancy between the provided position of the activities and its actual attractiveness, the activities, in fact, would not accept and follow the planner's indication in case of an unappealing location [7].

Despite this outstanding capability, which is proved by several published researches, yet all the above sketched methods are undoubtedly affected by some evident faults, that somehow do limit their actual use.

While referring to axial analysis, we can notice as its main fault the poor correspondence between the street that actually form the urban space and the lines of the respective analysed system. It won't hence be possible, by axial analysis, to exactly provide each street with its own integration value: in many cases, a line is associated with several streets, as well as sometimes several lines cover the space of a single street. The same problem also affects the case of long straight streets, whose internal differences cannot be appreciated: despite their internal variety, in fact, they result characterised by a single integration value.

Such a relevant fault gets ever increased in the angular version of configurational analysis, that only takes into account the angle of intersection between the connected lines. A street that is nearly straight, covered by lines which intersect with very open angles, will result characterised by slightly different integration indices, as its correspondent lines are mutually divided by very small depth values.

In other words, neither axial nor angular analysis can't account for the variation of integration (that is attractiveness towards activities, or, what's the same, appeal) along a single straight (or nearly straight) urban path, whatever its length can actually be. Visibility graph analysis can overcome this fault, since it discretises the space of the urban grid by means of a mesh of (as many as useful) vertices: the street does hence disappears, as transformed in a set of points. On a one hand, such a different approach allows to observe the (nearly) continuous variation of the integration value along the same urban path, even if straight, highlighting its more appealing parts and the less attractive ones. On the other hand, such a capillary analysis is, on its turn, affected by two significant limits. The first is the actual heaviness of the processing: notice that the analysis of the visibility graph of a small-medium size town such as Siena (about 50,000 inhabitants) involves the introduction and the processing of a very high number of vertices (at least around 100.000). obviously, that provides the need of a strong computing power as well as some troubles in collecting, rendering and using the resulting outputs (a complete set of configurational indices per each vertex). A second fault derives from the uncertain correspondence between the artificially introduced vertices (which the analytical indices are referred to) and the actual urban spaces (streets, roads, squares and so on), in order to define their configurational attractiveness towards activities. In other words, a street, as well as each part of it, will be associated to a set of vertices, and each of them will be characterised by a set of configurational indices: it will be not easy, nor certain to provide that street with a single parameter, suitable for properly representing its own configurational value.

2 The Evolution of the Space Syntax Theory

In last years, a lot of users of the "Space Syntax Anlysis" drew attention to the need of a best method of analysis. In fact, it doesn't exist an unequivocal solution to the research's problem, inside of the convex map, of the low number of segments with more lengths and especially, doesn't automatically calculate.

The developed methodology is based on the fundamental rule of the configuration analysis, that is, it takes on the urban grid as the first matrix of the generation of the "natural movement" and of the location of activities: however, in respect with the Axial Line methodology, we have developed a new method to calculate authomatically urban features, called "Characteristic Points". These points, derived from urban street maps, are representative of the network structure in the sense that, within an urban enviroment, people make a navigation decision on where to head next when they reach these characteristic points. These points include road junctions and turning points (i.e. a turning point is defined as a peak of a curve).

The new algorithm has been developed within GIS software ArcGIS ESRI, using programming language "Visual Basic for Application": in this way it possible authomatically generate Characteristic Points starting from urban street network and built enviroment. For each of these Points it is also possible to calculate Space Syntax indexes: *connectivity*, *control value*, *mean depth* and *global* and *local integration*.

In the first step of the authomatical derivation, we have employed a parameter to establish if a new point have to be insert in the set of Characteristic Points or not; this parameter has been calculated on the basis of a threshold value considering the ratio between the *n-point* distance from the directly linked *n-1* and *n+1* points straight line and the total length of the straight line itself (figure 3).

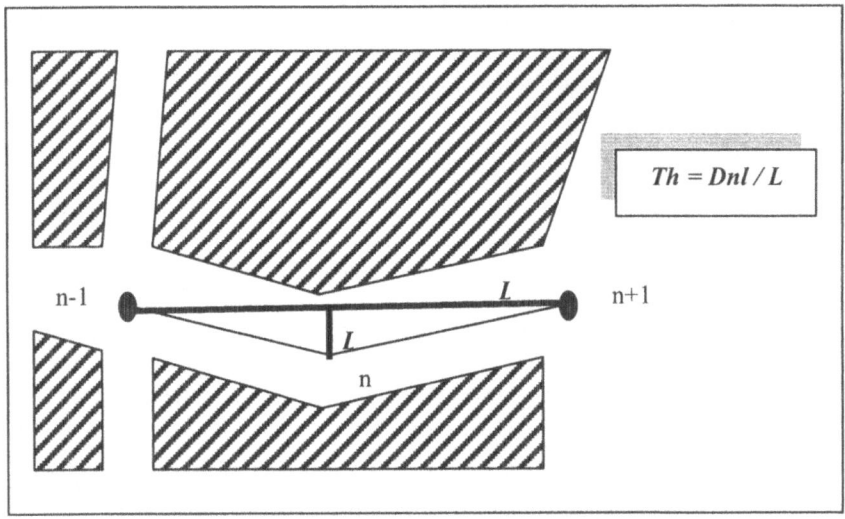

Fig. 3. Example of the derivation af a turning point

In order to examine how feasible the proposed point-based approach is for modelling an urban space, and how the point-based analysis results conform to that of line-based analysis, we conduce two cases study. We chose the well-known mid-sized Tuscany towns of Pisa and Siena. The two experiments have been conducted with two different approaches:
- raster approach, performed by Map Algebra calculations and Viewshed analysis;
- vector approach, performed by vector map analysis.

The first step of the analysis is to derive the city landmarks (borders) to encompass an urban enviroment and to make a closed system. To do this, we have considered the medieval urban walls (in the expermiment regarding the city of Pisa) and we have selected all the city entities (building, streets, etc.) falling within these walls.

2.1 Viewshed Analysis

For each city, we have made the Digital Terrain Model (DTM), starting from quoted points and contour lines. Then, we have derive the grid of the built enviroment, considering the heigth of each building: the two grids have been added by means of the gis function analysis, Map Algebra calculation, to make an "Integrated Digital Terrain Model". The second step, is to authomatically derive characteristic points; this procedure has been implemented within the new Esri software realase, ArcGIS, by using language programming Visual Basic for Application.

These points, as specified above, are representative of the network structure in the sense that, within an urban enviroment, people make a navigation decision on where to head next when they reach these characteristic points.

By using the well-known Gis analysis function, viewshed, it is possible to calculate the visibility of each characteristic points in each town.
In this way, we calculate the connectivity of each point, so that the biggest value correspond to more integrated points.

2.2 Vector Analysis

The configurational analysis, as broadly enunciated already, foresees the presence, to the analytical aims, of the bodies of structure (buildings) and of the "characteristic points"; these last will be individuated by means of various phases of spatial analysis.
In fact such points have been individuated through three different procedures:
- singling out of the intersections derived by the database of the axes of the street;
- calculation of the centroids of the squares, of the courts and of the urban road widens;
- calculation of the turning points of the curves of every road arc

For the derivation of the intersections it is necessary to build the topology, that is the whole spatial relationships that characterizes every geographical database and, by means of the topology, to extract the layer concerning the nodes. It is opportune, in this phase, to perform a control analysis to recognize the possible presence of pseudo-nodes, that is points that join only two arcs and that, therefore, cannot be considered as real intersections.

The operation that allows the drawing of the centroids of every squares is certainly more complex and somehow more hard-working than the previous. This calculation needs, at first, the selection through place names of everything that is not a street (squares, courts, road widens, etc..) and the relative check. This selection is made by means of an SQL expression (Structured Query Language) and particularly through the use of "joker characters" it have been selected all that place names in which the first part of the name was "Square", "Courts" or "Wide". The database related to the road arcs represents these elements through their contour. Therefore, in the moment in which they are recognized, it has been simple to transform theirs geometric primitive from polyline to polygon.

Every polygon has been encoded subsequently with a following numerical and with the place name derived from the road axes theme. Now it's possible to calculate their centroids, defined by the average of the coordinates of the vertexes that delimit every polygon, by means of an automatized function.

The central point of the single square has been connected with fictitious arcs to every road that is leaned out on the considered square.

To operate such connection has been used an extension of the software GIS ArcView, besides created already for other researches. This extension, through the interpretation of an origin-destination matrix (From_Node-To_Node), produces in automatic the fictitious arcs that connect the centers of every squares.

The next step is the authomatic calculation, for each mark point, of the Space Syntax configurational indexes.

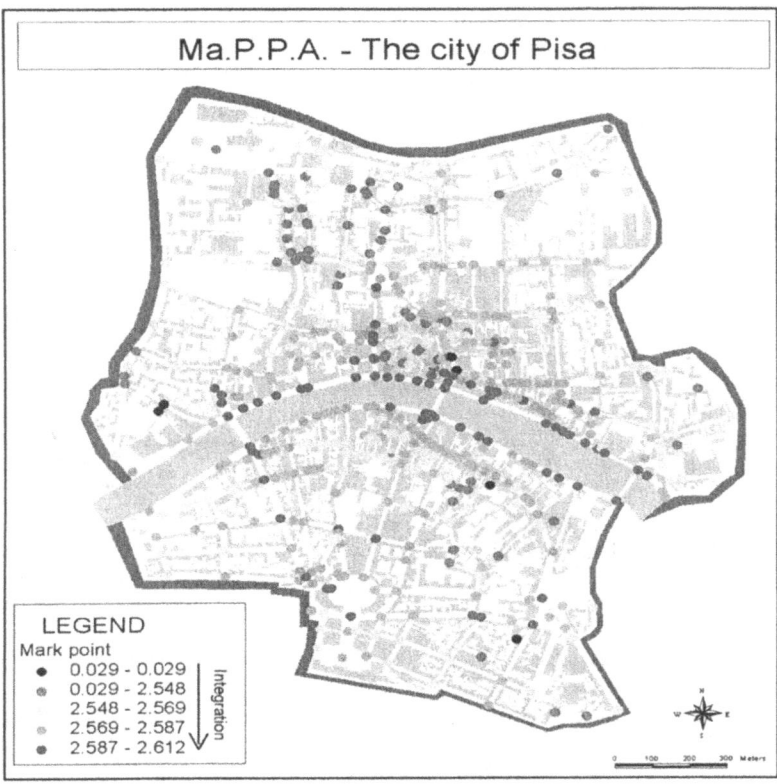

Fig. 4. Integration index of mark points (City of Pisa)

3 Conclusions

The two described experiments represent a first step towards a new approach to configurational urban analysis. Our research aim is to optimize the mark points analysis extending it to larger and more complex cities

References

1. Batty M. (2001) Exploring isovists fields: space and shape in architectural and urban morphology, in *Environment and Planning B: Planning and Design*, volume 28, pagg. 123-150-
2. Bortoli M., Cutini V. (2001) *Centralità e uso del suolo urbano. L'analisi configurazionale del centro storico di Volterra*, ETS, Pisa.

3. Cutini V. (1999a) Configuration and Movement. A Case Study on the Relation between Movement and the Configuration of the Urban Grid, in AA.VV. (1999) *CUPUM '99 Computers in Urban Planning and Urban Management on the Edge of the Millennium*. Proceedings of the 6th International Conference, Franco Angeli, Milano.
4. Cutini V. (1999b) Urban Space and Pedestrian Movement – A Study on the Configurational Hypothesis, in *Cybergeo, Revue Européenne de Geographie*, n° 111, 26 Octobre 1999.
5. Cutini V. (2000) Configuration and Urban Activities Location. A Decision Making Support Tool, in *Proceedings of the 2nd International Conference on Decision Making in Urban and Civil Engineering*, Lyon (F) 20-22 novembre 2000, pagine 151-162.
6. Cutini V. (2001a) Centrality and Land Use: Three Case Studies on the Configurational Hypothesis, in *Cybergeo, Revue Européenne de Geographie*, n° 188, 26 mars 2001.
7. Cutini V. (2001b) Configuration and Centrality. Some Evidence from two Italian Case Studies, in *Proceedings of the Space Syntax 3rd International Symposium*, Atlanta (USA) 7-11 maggio 2001, Alfred Tauban College of Architecture and Urban Planning, University of Michigan, pagine 32.1-32.11.
8. Cutini V. (2003) Lines and squares. Towards a Configurational Approach to the Analysis of the Open Spaces, in *Proceedings of the Space Syntax 4rd International Symposium*, University College of London, London (UK), pagine 32.1-32.11.
9. Hillier B., Hanson J. (1984) *The Social Logic of Space*
10. Hillier B. (1996) *Space is the Machin*
11. *Hillier B., Hanson J., Penn A.*
12. Jiang, Claramunt, Klaqvist (2000) *An integration of space syntax into GIS for modelling urban spaces*
13. Turner A. (2001a) Angular Analysis, in *Proceedings of the 3rd Space Syntax Symposium*, Atlanta, 7-11 May 2001, Alfred Tauban College of Architecture, University of Michigan.
14. Turner A. (2001b) Depthmap. A program to perform visibility graph analysis, in *Proceedings of the 3rd Space Syntax Symposium*, Atlanta, 7-11 May 2001, Alfred Tauban College of Archotecture, University of Michigan.
15. Turner A. Doxa M., O'Sullivan D., Penn A. (2001) From isovists to visibility graphs: a methodology for the analysis of architectural space, *in Environment and Planning B: Planning and Design*, volume 28, pagg. 103-121

Computing Foraging Paths for Shore-Birds Using Fractal Dimensions and Pecking Success from Footprint Surveys on Mudflats: An Application for Red-Necked Stints in the Moroshechnaya River Estuary, Kamchatka-Russian Far East

Falk Huettmann

Department of Wildlife Biology; Institute of Arctic Biology, EWHALE lab, University of Alaska-Fairbanks, Fairbanks 99775, U.S.A.
fffh@uaf.edu

Abstract. Foraging strategies for Red-necked Stints (*Calidris ruficollis*) at migration staging sites are virtually unknown, nor exist any methods to achieve such crucial knowledge. Here, for the first time a non-invasive and quantitative computing method of foraging paths is presented from a study carried out during fall migration 1999 for fine-sediment mudflats of the Moroshechnaya River Estuary (56° 50' N, 156° 10' E), eastern Sea of Okhotsk, Russian Far East. Footprint surveys on a mudflat were used in order to compute the distances and angles between foraging patches, including pecking success information. It was then analyzed how the pecking success at a foraging patch affects the selected distance between foraging patches and turning angles; no correlations were found. However, with increasing scale (size of divider to measure the foraging path) the fractal dimension of the foraging path generally increases, peaking at intermediate scales and then decreases at larger scales. This indicates for the working scale (grain size 1cm, extend 40cm) that Red-necked Stints make their foraging decisions using a static approach (< 20cm step-length, and 10-40° absolute change of angle). On a smaller scale, and once prey was located, they forage in food patches along 'cracks' of the mudflat. It is suggested that fractal dimensions of foraging paths can serve as a basic and quantitative description for the arrangement and distribution of prey patches in mudflats relevant to shorebirds. The presented approach has large potential to describe the efficiency of foraging shorebirds when analyzing pecking success at mudflats along their entire flyways.

1 Introduction

Shorebirds make extensive use of mudflats during their migration at 'stop over' sites where they feed on small prey to fuel up for their migration. At these locations foraging efficiency is believed to be of major importance to make their short stay and the migration successful, not losing energy and time, and not being exposed to predators. For these periods, the spatial distribution of shorebirds within the mudflat is generally expected to reflect the spatial distribution of their prey. However, spatial distribution of prey and shorebirds on mudflats are difficult to describe [1], nor are any methods known which quantify foraging paths of shorebirds in relation to prey distribution ([2] for general concepts and radio-tracking of larger shorebirds) and which are non-invasive. Being 'non-invasive' refers to leaving birds undisturbed during foraging, and carrying out the analysis after the birds have left. Estimating and analysing the fractal dimension of a foraging path allows to investigate how animals perceive the landscape [3], what role scale plays in their ecology [4], [5], [6] and finally how animals forage. This method is now considered a 'classic' [7], [8], and was already applied to insects [6], [9], [10], [11], mammals [12], [13] (see [14], [15] for snowtracking), woodpeckers (A. Desrochers pers. com.), seabirds [16], plant communities [17] and general landscape studies [18]. The fractal dimension of a line describes its 'crookedness' also called tortuosity, or how often angles are turned along this line. It is ranging between 1 for a straight line (straight movement) and 2 for an area fully covered (very intense searching). If the fractal dimension of a foraging path is 'fractal', or scale invariant, this would mean that it does not change across scales (= independent of divider size to measure fractal dimension of a path) [12]. If the fractal dimension of a foraging path increases with scale, then there are relatively more angle-turns for larger scales than for smaller ones (Fig 1 A), and vice versa if the fractal dimension of a foraging path decreases with scale (Fig 1 B) (e.g. see [7], [12], [15], [17] for more details on theoretical background).

This project was carried out in the larger framework of the long-term investigation of fall migration of shorebirds in the northeastern Sea of Okhotsk shore, Moroshechnaya River Estuary (56° 50' N, 156° 10' E), during a study in August 1999 [19], [20], [21], [22], [23]. Here an approach is shown how shorebird foraging processes can easily be measured in the field for quantification. It is also shown how these findings can be analyzed and brought into relation to the distribution of prey and prey patchiness within a landscape scale of a mudflat by using the fractal dimension of the foraging path and the pecking success for Red-necked Stints (*Calidris ruficollis*). The exact foraging strategy of this long-distance migrant is still unknown [24]. Due to the entire lack of investigations and knowledge about the fractal dimension of foraging paths from small shorebirds, it was hypothesized that these birds forage with equal intensity across scales. The overall applicability of this method, and also the gain on how migratory Red-necked Stints make use of mudflats along the flyway, is assessed.

2 Methods

During August 1999 the mudflats of the Moroshechnaya River Estuary, eastern Sea of Okhotsk [19] were surveyed. Surveys for footprints and pecking success were carried out during daylight and calm weather (low wind, no rain, sunny) for foraging Red-

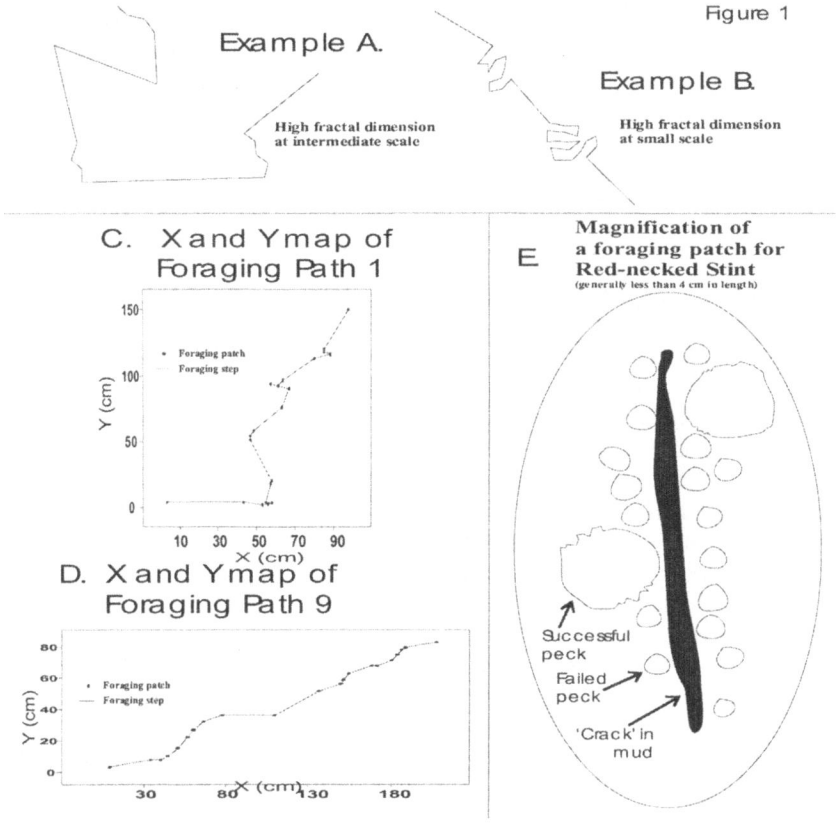

Fig. 1. Concepts of fractal dimensions applied to foraging paths derived from footprint surveys on mudflats of migratory small shorebirds.

necked Stints during fall migration. Footprints can easily be seen in some of the mudflats of the Moroshechnaya River Estuary due to its very fine sediments. Initially, over 80 foraging paths were followed from which it was possible to use data for 10 individual birds having 'readable' and complete foraging paths accompanied by acceptable data on pecking success. For each of these 10 foraging paths the distances (>1 cm) and bearings (degrees from North) to the net foraging location were measured for 22 foraging step changes. The 'absolute change of angle to next location' (turning angle; see also [6], [25], [26] for definition) was calculated by subtracting the new 'bearing of line to the next location' from the previous 'bearing of line to the next location'. For each pecking location, the percent of successful peckings was computed. The actual 'bearing of line to next location' cannot be used in a meaningful analysis here since foraging paths were collected randomly within the mudflat for representativeness and thus have only relative directions.

The overall length of the foraging path was determined by the ability to completely 'read' the entire foraging path and pecking information for as long as possible. Using 22 foraging step changes allowed for the use of 10 complete foraging paths from the initially collected data pool of 80 foraging paths. The time that a bird spent on the foraging path or at a foraging patch was not known. Foraging path measurements were only taken for birds which were truly found foraging on the ground, as confirmed by pecking marks in the mud. Almost all changes of bearings within the identified foraging path had associated pecking marks on the ground (=foraging patch). Backed up by observations, this indicates that Red-necked Stints move directly from prey patch to prey patch (=defined as 'step size') to forage, but not foraging while moving between these patches, e.g. Fig. 1 C and D. Based on observations from foraging birds it was concluded that successful peckings had a larger pecking opening in the mud than failed peckings from probing bills only (Fig. 1 E). This assumption was confirmed with observations for over 20 individual foraging birds and using a scope (20* magnification) on a tripod. Successful pecks and failed pecks were counted for each foraging patch along the foraging path tracked in the mudflat; as well, the percentage of successful pecks with each patch was calculated. Following [14] and Nams (pers. com.) the distances and directions of the foraging steps were used as input variables to obtain cartesian x and y coordinates for construction and analyzing the complete foraging path. The following formulas were used to obtain x and y values from these measurements:

$x = $ previous $x + $ length $*$ sin (angle) $\hspace{2cm}$ (1)
$y = $ previous $y + $ length $*$ cos (angle) $\hspace{2cm}$ (2)

These x and y values then were imported as ASCII format into the software program FRACTAL 3 [8], [12] available for free from the WWW. Fractal V is determining the 'crookedness', search intensity, of a bird along the foraging path measured at a range of scales by computing the fractal dimension using 'dividers' of increasing lengths. The foraging paths were analysed separately for individuals using the 'Fractal V' as outlined in [12], and also for all 10 foraging paths pooled together. Spatial heterogeneity [15] was not addressed in this project. 'Fractal D' (e.g. see [7], 14], [17]) was not used since this would not allow to obtain 95 % confidence intervals for reliable conclusions.

3 Results

Locations where directions of foraging steps changed were all characterized by pecking marks along small openings in the mudflat ('cracks', e.g. Fig 1 E). These were caused by shrinking soil while drying out when the tide withdraws. Over 80% of the distances between foraging patches were less than 20 cm long (Fig. 2 A); the shortest was 1 cm, the longest was 90 cm. Over 70 % of all angle turns were made between 10° and 40° (Fig. 2 B); the change of directions varied between 0° and 180°.

The distance between foraging patches is not a function of pecking success (successful pecks, failed pecks or successful pecking percent) neither and all R^2 values for Fig.4 are as well almost 0 (therefore, further details of linear regression are not presented here). Instead, the distance of foraging steps between foraging patches is constant and not changing for the scales investigated (Fig. 4).

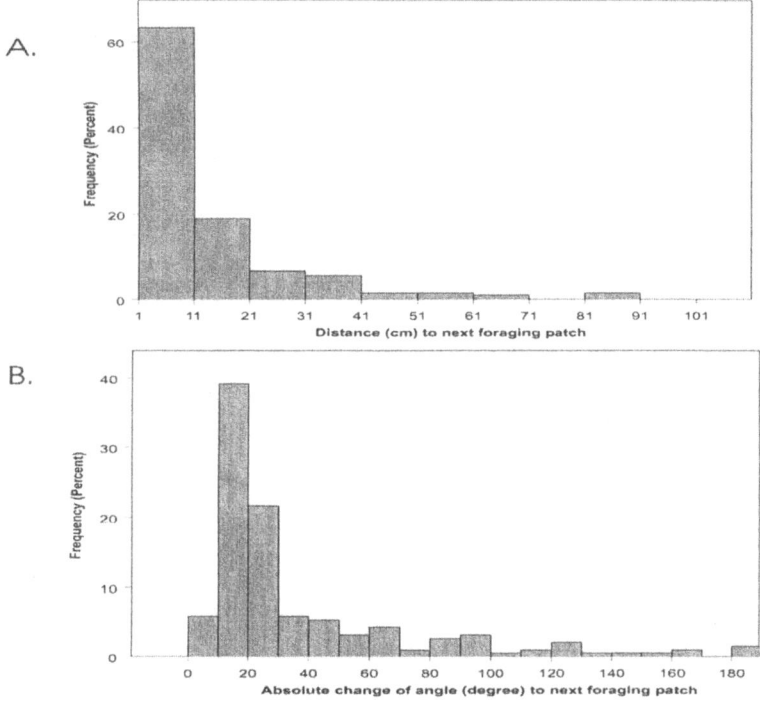

Fig. 2. Frequency distributions of distances to next foraging patch, and absolute change of angles to next foraging patch, using all 10 foraging paths for migratory Red-necked Stints in Moroshechnaya River Estuary.

The absolute change of angle between foraging steps that connect foraging patches is not a function of pecking success, and all R^2 values in Fig. 3 are almost 0 (details of linear regression are not shown). Instead, the linear regression is constant and not changing for the scales investigated (Fig. 3).

Fig. 5 and 6 give evidence that these foraging patterns are driven by the arrangement and distribution of local prey patch size and prey distribution on the mudflat only.
It was found for all individuals that Fractal V changes across scales along their foraging paths. It is generally increasing with scale, peaking at intermediate scales and usually decreasing from there on. However, none of these changes are outside of the 95 % confidence limits (Fig. 5, only means are shown).

Changes of Fractal V across scale can vary strongly in their magnitude, but trends of changes are somewhat similar (Fig. 5). Put in quantitative terms, Fractal V changes across scales and generally increases for the scale up to 13 cm, peaks at intermediate scale and decreases for larger scales > 20m of the foraging path; this trend still holds when all 10 foraging paths are pooled together (Fig. 6). Sample sizes are small, but findings would indicate that Red-necked Stints search more intense (judged by the tortuosity) at the intermediate scale, but usually not at the lower (<4 cm) and higher scales (>20 cm).

1122 F. Huettmann

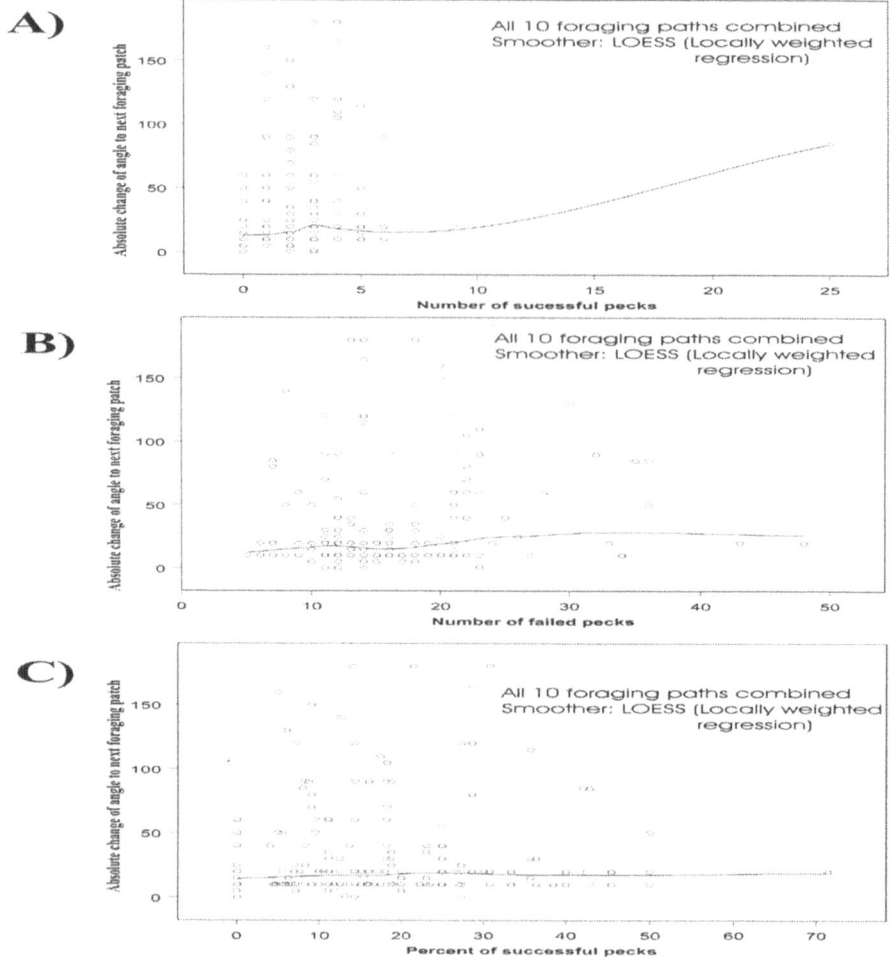

Fig. 3. Absolute change of angle to next foraging patch by foraging success, using all 10 foraging paths for migratory Red-necked Stints in Moroshechnaya River Estuary.

Due to the descriptive nature of this non-invasive study, sound evidences and interpretations linking and explaining directly the mechanisms of the underlying foraging process are hard to make and not done, yet; here the groundwork is laid.

4 Discussion

Using fractal dimensions to describe foraging paths quantitatively and in concert with pecking success was never applied to shorebirds as well as to mudflats in Russia.

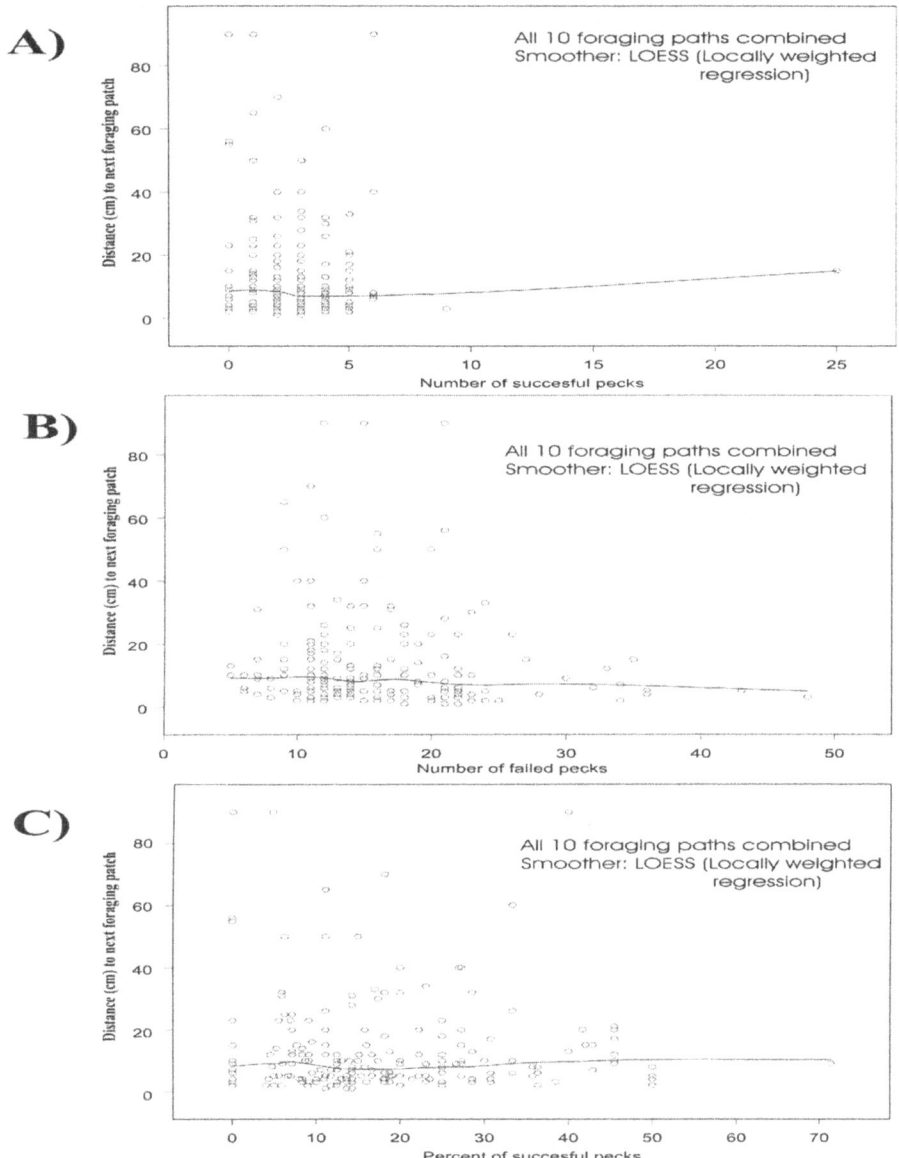

Fig. 4. Distance (cm) to next foraging patch by foraging success, using all 10 for aging paths for migratory Red-necked Stints in Moroshechnaya River Estuary.

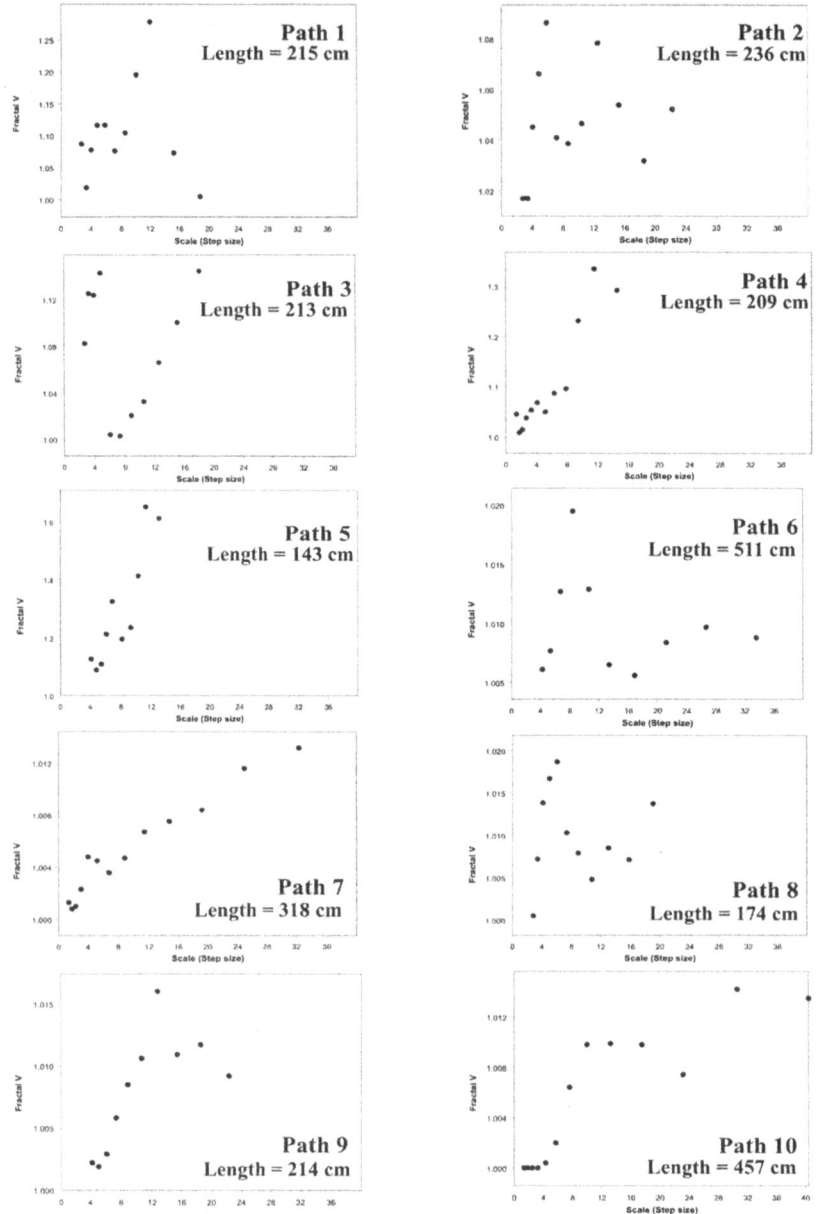

Note: Scale of the Y-axis differs across graphs due to varying sizes of Fractal V by individual foraging paths

Fig. 5. Fractal V across scales for each foraging path, using all 10 foraging paths for migratory Red-necked Stints in Moroshechnaya River Estuary.

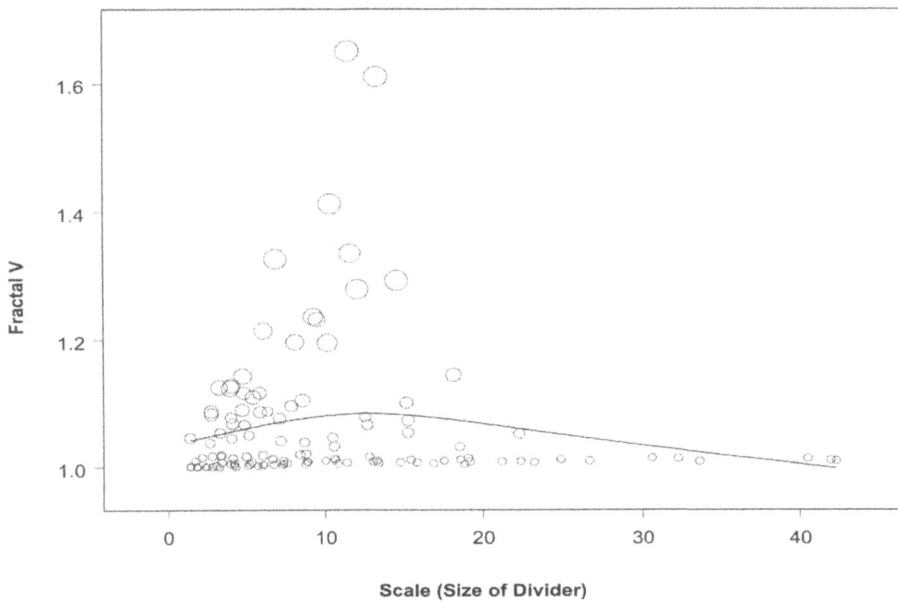

Fig. 6. Fractal V for all 10 Red-necked Stint foraging paths vs. scale for the Moroshechnaya River Estuary; the size of the circle indicates the standard deviation of Fractal V. Smoother: Spline.

The migration strategies of Red-necked Stints are still unknown. Here a convenient method is shown to be carried out in the field during daylight hours (for nocturnal feeding investigations of shorebirds see [27]). It uses a freely available computer algorithm to describe and to improve our knowledge how shorebirds forage during migration. All foraging paths used in this investigation showed intense pecking marks in the mud at cracks (foraging patches). Therefore, at the working scale (maximum extent of 40 cm, grain size of >1 cm for foraging steps) the fractal dimension of a foraging path is in response to prey patch distribution within the mudflat. The results suggested that the prey for Red-necked Stints in the studied mudflat is patchily distributed. The suggested patchiness of prey would be characterized by the average moving distance between foraging processes (foraging patch with peckings in the mud), which were found to be small patches along 'soft sand cracks' appr. <20 cm away from each other (intermediate scale; Fig 1 E). First, these cracks in the soil get exploited by Red-necked Stints foraging without major turning angles and by simply following these cracks along resulting into small fractal dimensions. Moving from crack to crack requires greater turning angles, which results overall into an increasing fractal dimension at the intermediate scale. On a larger scale the fractal dimension declines again promptly, which indicates direct movements without intense searching. These investigations did not take into account time spend at a foraging patch, or time it took to complete the entire foraging path (e.g. [28]). Due to its disturbing nature, parallel prey abundance sampling cannot be carried out while shorebirds are foraging, nor can it be done for potential and neighbouring patches in order to address spatial prey availability and preferences for cracks in the mudflat. Therefore, currently one cannot fully conclude whether the cracks investigated would be the only prey patches available to foraging shorebirds, if they are exploited 'ad lib' [29] and whether the measured distances among the prey patches reflect their real spacing patterns. Constraints may be put onto the bird's foraging decision by energetic limitations [1], competitors [30] and predators [31], weather [32] and body conditions [1].

Mudflats can differ in their environmental set-ups, e.g. spatial extent, configuration, soil types, microclimate, prey abundance, prey availability and prey diversity. It is suggested that fractal dimensions of foraging paths can serve as a non-invasive method and basic quantitative description for the arrangement and distribution of prey patches in mudflats relevant to shorebirds; it can be used to describe the efficiency of foraging shorebirds by analyzing pecking success. Further, it can allow to evaluate the 'quality' of entire mudflats along the flyway, across and within seasons. More research is to be done on these topics. 'Reading' foraging paths of shorebirds and estimating pecking the success from size of pecking holes is only possible in mudflats with very fine sediment, when tides withdraw and preferably when it is very calm (lack of wind); otherwise holes could get filled with sand or get destroyed. These assumptions, and if applied elsewhere along the flyway, could introduce methodological errors and biases in regards to sample coverage, sample selection and conclusions, e.g. mudflats with very fine sediment could present a very specific, non-representative, habitat (see [22] for overview of coastal habitats in the Sea of Okhotsk region). This study only investigated whether the angle and the length of the foraging step is related to the pecking success at the previous location. However, it could be probable that foraging patches do not get fully exploited, that neighbouring ones get ignored, and that the foraging direction is driven by the bird's expectation of foraging success for the next, or neighbouring, foraging patch [28]. This could be indicated to the shorebird for instance by smell or by other clues of potential feeding patches, e.g. feeding behaviour of other individuals [1]. Pecking locations along foraging paths are probably not independent of each other. This needs more study, too. An additional concept worthwhile to pursue in future studies for foraging paths and based on methods described here presents 'Graph Theory' [33]. Graph Theory can address how individual foraging patches are connected for a certain mudflat location (patch), and what their specific 'degree of connectivity' would be to reach the patches best.

The methodological approaches presented here could be applied for other wildlife species and similar habitats (compare [8], [9], [10], [13], [14], [15], [16]), and might even allow to analyse foraging strategies quantitatively among coexisting and competing shorebirds [30]. More samples need to be collected and analysed on different mudflat types along the entire flyway. Results from foraging path work need to be evaluated using complete prey surveys and prey sampling, e.g. complimented by analyzing droppings [34], stomach contents [19] or by indirect measures such as detailed behaviour studies [1] and video taping [35].

References

1. Piersma, T. Close to the edge: energetic bottlenecks and the evolution of migratory pathways in Knots. - PhD Thesis. University of Groningen. (1994)

2. Blackwell, P. G. Random diffusion models for animal movement. Ecol. Mod. (1997) 100: 87-102.
3. With, K.A. Using fractal analysis to assess how species perceive landscape structure. Landscape Ecol. (1994) 9: 25 - 36.
4. Wiens, J. A. Spatial scaling in ecology. Funct. Ecol. (1989) 3:385-397.
5. Levin, S.A. The problem of pattern and scale in ecology. Ecology (1992) 73: 1943 –1967.
6. Wiens, J.A., Crist, T., Milne, B.T. On quantifying insect movements. Env. Entomol.(1993) 22: 709 - 715.
7. Mandelbrot, B. How long is the coast of Britain? Statistical self-similarity and fractional dimension. Science (1967) 156:636-638.
8. Nams, V. O. The VFractal: a new estimator for fractal dimension of animal movement paths. Landscape Ecol. (1996) 11:289-297.
9. Kareiva, P. M., Shigesada, N. Analyzing insect movement as a correlated random walk. Oecologia (Berl.) (1983) 56: 234-238.
10. Christ, T. O., Guertin, D.S., Wiens, J.A, Milne, T.A. Animal movement in heterogeneous landscapes: an experiment with *Eleodes* beetles in shortgrass prairie. Funct. Ecol. (1992) 6:536-544.
11. Wiens, A., Schooley, R.L., Weeks, R.D. Patchy landscapes and animal movements: do beetles percolate ? OIKOS (1997) 78: 257-264
12. Nams, V.O. Fractal 3.0. Software and The Manual, http://www.nsac.ns.ca/envsci/staff/vnams/fractal.htm (2000)
13. Benhamou, S. An analysis of movements of the wood mouse *Apodemus sylvaticus* in its home range. Behav. Proc. (1990) 22:235-250.
14. Edwards, M., Bowman, J. Relationship of weasel track turtosity to prey abundance and habitat structure. In: Huettmann, F., Bowman,J. (eds.): Investigation of Animal Movement. Workshop Proceedings, 12 – 14 November 1998 Fredericton. Supported by the Sir James Dunn Wildlife Research Centre, Atlantic Cooperative Wildlife Ecology Research Network, Biology Dept. of the University of New Brunswick NB. (1999) 22-26.
15. Nams, V.O., Bourgeois, M. Using fractal analysis to measure habitat use at different spatial scales: an example with marten. Oecologica (in press)
16. Viswanathan, G.M, Afansyev, V., Buldrev, S.V., Murphy, E.J., Prince, P.A., Stanley, H.E. Levy flight search patterns of Wandering Albatrosses. Nature (1996) 381: 413 - 415.
17. Palmer, M. W. Fractal geometry: a tool for describing spatial patterns of plan communities. Vegetatio (1988) 75: 91-102.
18. Milne, B.T. (1991) : Lessons from applying fractal models to landscape patterns. In: Turner, M. G., Gardner, R.H. (eds.): Quantitative methods in landscape ecology. Springer, New York. (1991) 199-235.
19. Huettmann, F. Final Report, Pilot Study 1999: Fall migration of waders in the Sea of Okhotsk region. - Wetlands International Oceania, Canberra. (1999)
20. Huettmann, F. Summary of Sea of Okhotsk studies investigating migration of shorebirds May 2000 on southern Sakhalin Island, and August 2000 on western Kamchatka and Magadan region. Stilt (2001) 35: 21-26.
21. Huettmann, F. Summary: Shorebird migration in early fall 2002 on northern Sakhalin Island. Stilt (2002) 43: 34-36.
22. Huettmann, F. Literature Review: Shorebird migration in the Sea of Okhotsk region, Russian Far East, along the East Asian Australasian flyway for selected species (Great Knot *Calidris tenuirostris*, Red Knot *Calidris canutus*, Bar-tailed Godwit *Limosa lapponica*). Report for Environment Australia, Wetlands International. (2003)
23. Huettmann, F. Findings from the 'Southward Shorebird Migration' Expedition in Aniva Bay (Sakhalin Island) and Iturup (Kurile Islands), Russian Far East, during August 2003. Stilt. (in review)

24. Higgins, P.J., Davies, S.J.J. (eds). Handbook of Australian, New Zealand and Antarctic birds. Volume 3: Snipe to Pigeons. Oxford University Press, Melbourne. (1996)
25. Batschelet, E. Circular statistics in biology. Academic Press, New York. (1981)
26. Turchin, P. Translating foraging movements in heterogenous environments into the spatial distribution of foragers. Ecology (1991) 72: 1253-1266.
27. Mouritsen, K.N. Diurnal and nocturnal prey detection by Dunlins. Bird Study (1993) 40:212-218.
28. Stillmann, R.A., Goss-Custard, J.D, Alexander, M.J. Predator search pattern and the strength of interference through prey depression. Behav. Ecol. (200) 11: 597-605.
29. Manly, B. J., McDonald, L.L, Thomas, D.L., McDonald, T.L., Erickson, W.P. Resource Selection by Animals. Kluwer Academic Publishers, Dordrecht, Holland. (2002)
30. David, C.A., Smith, L.M. Foraging strategies and niche dynamics of coexisting shorebirds at stopover sites in the southern Great Plains. Auk (2001) 118: 484-495.
31. Ydenberg, R.C., Butler, R.W., Lank, D.B., Guglielmo, C.G., Lemon, M., Wolf, N. Trade offs, condition dependence and stopover site selection by migrating sandpipers. J. Avian Biol. (2002) 33: 45-55
32. Butler, R.W., Williams, T.D, Warnock, N., Bishop, M.A.. Wind assistance: a requirement of migration of shorebirds ? Auk (1997) 114: 456-466.
33. Bunn, A.G., Urban, D.L., Keith, T.H. Landscape connectivity: a conservation application of graph theory. J. Env. Manag. (2000) 59: 265-278.
34. Dekinga, A., Piersma, T. Reconstructing diet composition on the basis of faeces in a mollusc-eating wader, the Knot *Calidris canutus*. Bird study (1993) 40: 144-156.
35. Huettmann, F., Battley, P., Rogers, D., Dorogoy, D. On abdominal profiles and biometrics of Great Knots (*Calidris tenuirostris*) during southward migration near the breeding grounds, northern shores of the Sea of Okhotsk. Stilt (in prep).

Author Index

Abawajy, J.H. II-107
Abawajy, Jemal II-87
Abdullah, Azizol II-146
Abellanas, Manuel III-1, III-22
Acciani, Giuseppe II-979
Acosta-Elías, Jesús IV-177
Aggarwal, J.K. IV-311
Ahmad, Muhammad Bilal IV-877, IV-940, IV-948
Ahn, Byoungchul III-566, III-993
Ahn, Byungjun I-1125
Ahn, In-Mo IV-896
Ahn, Jaemin III-847
Ahn, JinHo III-376, IV-233
Ahn, Kiok I-1044
Ahn, ManKi I-517
Ahn, Seongjin I-142, I-1078
Ahn, Sung IV-489
Ahn, Yonghak I-1044
Ahn, Young Soo II-1079
Albertí, Margarita II-328, II-374
Albrecht, Andreas A. III-405
Alcaide, Almudena I-851
Alegre, David III-857
Aleixos, Nuria II-613
Alinchenko, M.G. III-217
Amaya, Jorge II-603
An, Beongku IV-974
An, Changho I-25
An, Ping IV-243
Anido, Luis II-922
Anikeenko, A.V. III-217
Annibali, Antonio III-722
Apu, Russel A. II-592
Asano, Tetsuo III-11
Atiqullah, Mir M. III-396
Attiya, Gamal II-97
Aung, Khin Mi Mi IV-574

Bachhiesl, Peter III-538
Bae, Hae-Young I-222, II-1079
Bae, Ihn-Han I-617
Bae, Sang-Hyun I-310, II-186, IV-359
Baik, Kwang-ho I-988

Baik, Ran III-425
Baik, Sung III-425, IV-206, IV-489
Bajuelos, António Leslie III-117, III-127
Bala, Jerzy IV-206, IV-489
Bang, Young-Cheol I-1125, II-913, IV-56
Bang, Young-Hwan I-491
Barel, Marc Van II-932
Barenco Abbas, Clàudia Jacy I-868
Barua, Sajib III-686
Becucci, M. II-374
Bekker, Henk III-32
Bellini, Francesco III-722
Beltran, J.V. II-631
Bencsura, Ákos II-290
Bertazzon, Stefania II-998
Bhatt, Mehul III-508
Bollman, Dorothy III-481, III-736
Boluda, Jose A. IV-887
Bonetto, Paola II-505
Bonitz, M. II-402
Borgosz, Jan III-715, IV-261
Borruso, Giuseppe II-1009, II-1089
Bose, Prosenjit III-22
Botana, F. II-761
Brass, Peter III-11
Brink, Axel III-32
Broeckhove, Jan IV-514
Brunelli, Roberto II-693
Bruno, D. II-383
Bruschi, Viola II-779
Bu, Jiajun III-886, IV-406
Bücker, H. Martin II-882
Buliung, Ronald N. II-1016
Buono, Nicoletta Del II-961, II-988
Buyya, Rajkumar IV-147
Byun, Kijong II-809

Cacciatore, M. II-366
Caeiro, Manuel II-922
Camp, Ellen Van II-932
Campa, S. II-206
Campos-Canton, Isaac IV-177
Capitelli, Francesco II-338
Capitelli, M. II-383

Carbonell, Mildrey I-903
Carretero, Jesús IV-496
Carvalho, Sílvia II-168
Casas, Giuseppe Las II-1036
Cendrero, Antonio II-779
Čermák, Martin III-325
Cha, Eui-Young II-486, IV-421
Cha, JeongHee I-17, I-41
Cha, Joo-Heon II-573
Chae, Jongwoo III-965, IV-983
Chae, Kijoon I-673
Chae, Oksam I-1044
Chambers, Desmond II-136
Chang, Beom H. I-191, I-693, IV-681
Chang, Byeong-Mo I-106
Chang, Hoon I-73
Chang, Min Hyuk IV-877
Chang, Yongseok IV-251
Chelli, R. II-374
Chen, Chun III-886, IV-406
Chen, Deren II-158
Chen, Tzu-Yi IV-20
Chen, Yen Hung III-355
Chen, Zhenming III-277
Cheng, Min III-729
Cheung, Chong-Soo I-310
Cheung, Wai-Leung II-246
Chi, Changkyun IV-647
Cho, Cheol-Hyung II-554, III-53
Cho, Chung-Ki III-847, III-926
Cho, Dong-Sub III-558
Cho, Haengrae III-548, III-696
Cho, Hanjin I-1007
Cho, Jae-Hyun II-486, IV-421
Cho, Jeong-Hyun IV-359
Cho, Jung-Hyun IV-251
Cho, Kyungsan I-167
Cho, Mi Gyung I-33
Cho, Seokhyang I-645
Cho, SungEon I-402
Cho, TaeHo I-567
Cho, We-Duke I-207, I-394
Cho, Yongsun I-426
Cho, Yookun I-547, I-978, IV-799
Cho, Youngjoo IV-647
Cho, Youngsong II-554, III-62
Choi, Chang-Gyu IV-251
Choi, Chang-Won I-302

Choi, Changyeol I-207
Choi, Dong-Hwan III-288
Choi, Doo Ho I-1151
Choi, Eun-Jung I-683
Choi, Eunhee II-913
Choi, Hoo-Kyun IV-11
Choi, Hoon II-196
Choi, HyungIl I-17, I-41
Choi, Joonsoo III-837
Choi, Kee-Hyun I-434
Choi, SangHo IV-29
Choi, Sung Jin IV-637
Choi, Tae-Sun IV-271, IV-291, IV-338, IV-348, IV-877
Choi, Uk-Chul IV-271
Choi, Won-Hyuck IV-321, IV-451
Choi, Yong-Soo I-386
Choi, Yoon-Hee IV-271, IV-338, IV-348
Choi, YoungSik I-49, II-942
Choi, Yumi I-663
Choirat, Christine III-298
Chong, Kiwon I-426
Choo, Hyunseung I-360, I-663, I-765, III-315, IV-56, IV-431
Choo, Kyonam III-585
Chover, M. II-622, II-703
Choy, Yoon-Chul IV-743, IV-772
Chu, Jie II-126
Chun, Jong Hun IV-940
Chun, Junchul I-25
Chun, Myung Geun I-635, IV-828, IV-924
Chung, Chin Hyun I-1, I-655, IV-964
Chung, Ilyong II-178, IV-647
Chung, Jin Wook I-142, I-1078
Chung, Min Young I-1159, IV-46
Chung, Mokdong I-537, III-965, IV-983
Chung, Tai-Myung I-183, I-191, I-238, I-693, IV-681
Cintra, Marcelo III-188
Clifford, Gari I-352
Collura, F. II-536
Contero, Manuel II-613
Costa Sousa, Mario III-247
Crane, Martin III-473
Crocchianti, Stefano II-422
Crothers, D.S.F. II-321
Cruz R., Laura III-415, IV-77

Cruz-Chavez, Marco Antonio IV-553
Cutini, V. II-1107
Cyganek, Bogusław III-715, IV-261

D'Amore, L. II-515
Dağ, Hasan III-795
Daly, Olena IV-543
Daněk, J. II-456
Danelutto, M. II-206
Das, Sandip III-42
Datta, Amitava IV-479
Datta, Debasish IV-994
Delaitre, T. II-30
Demidenko, Eugene IV-933
Denk, F. II-456
Díaz, José Andrés III-158
Díaz-Báñez, Jose Miguel III-99, III-207
Díaz-Verdejo, Jesús E. I-841
Diele, Fasma II-932, II-971
Discepoli, Monia III-745, IV-379
Djemame, Karim II-66
Dong, Zhi II-126
Dózsa, Gábor II-10
Duato, J. II-661
Durán, Alfonso I-949, III-857

Effantin, Brice III-648
Eick, Christoph F. IV-185
Engel, Shane II-1069
Eom, Sung-Kyun IV-754
Ercan, M. Fikret II-246
Erciyes, Kayhan III-518, III-528
Esposito, Fabrizio II-300
Estévez-Tapiador, Juan M. I-841
Estrada, Hugo IV-506, IV-783
Eun, Hye-Jue I-122

Fan, Kaiqin II-126
Farias, Cléver R.G. de II-168
Faudot, Dominique III-267
Feng, Yu III-498
Fernández, Marcos II-661, II-671
Fernández-Medina, Eduardo I-968
Ferrer-Gomila, Josep Lluís I-831, I-924, IV-223
Filinov, V. II-402
Fiori, Simone II-961
Flahive, Andrew III-508
Formiconi, Andreas Robert II-495

Fornarelli, Girolamo II-979
Fortov, V. II-402
Foster, Kevin III-247
Fragoso Diaz, Olivia G. IV-534, IV-808
Fraire H., Héctor III-415, IV-77
Frausto-Solís, Juan III-415, III-755, IV-77, IV-553
Fung, Yu-Fai II-246

Galpert, Deborah I-903
Gálvez, Akemi II-641, II-651, II-771, II-779
Gameiro Henriques, Pedro II-817
García, Alfredo III-22
Garcia, Ernesto II-328
García, Félix IV-496
García, Inmaculada III-877
García, José Daniel IV-496
García-Teodoro, Pedro I-841
Gardner, Henry III-776
Gavrilova, Marina L. II-592, III-217
Gerace, Ivan III-745, IV-379
Gerardo, Bobby D. I-97
Gervasi, Osvaldo II-827, II-854
Giansanti, Roberto III-575
Go, Hyoun-Joo IV-924
Gola, Mariusz III-611
Gómez, Francisco III-207
González Serna, Juan G. IV-137
Gourlay, Iain II-66
Goyeneche, A. II-30
Gregori, Stefano II-437
Grein, Martin II-843
Guan, Jian III-706
Guarracino, Mario R. II-505, II-515
Gulbag, Ali IV-389
Guo, Wanwu IV-471, IV-956
Guo, Xinyu II-751
Gupta, Sudhir IV-791
Gutiérrez, Carlos I-968
Gutiérrez, Miguel III-857

Ha, Eun-Ju IV-818
Ha, JaeCheol I-150
Ha, Jong-Eun IV-896, IV-906, IV-915
Ha, Kyeoung Ju IV-196
Ha, Yan I-337
Hackman, Mikael I-821
Hahn, Kwang-Soo III-837

Hamam, Yskandar II-97
Hamdani, Ajmal H. II-350
Han, Dongsoo IV-97
Han, Jongsu III-955
Han, Qianqian II-272
Han, Seok-Woo I-122
Han, Seung Jo IV-948
Han, Sunyoung I-1115
Han, Tack-Don II-741
Han, Young J. I-191, I-693, IV-681
Haron, Fazilah IV-147
Healey, Jennifer I-352
Heo, Joon I-755
Herges, Thomas III-454
Hernández, Julio César I-812, I-851, I-960
Hiyoshi, Hisamoto III-71
HlaváII-ček, I. II-456
Hlavaty, Tomas III-81
Hoffmann, Kenneth R. III-277
Hong, Choong Seon I-755, I-792, I-915, I-1134
Hong, Chun Pyo III-656, IV-106
Hong, Dong Kwon I-134
Hong, Hyun-Ki II-799
Hong, Inki I-1125
Hong, Kwang-Seok I-89, IV-754
Hong, Man-Pyo IV-611
Hong, Manpyo III-867, IV-708
Hong, Maria I-57
Hong, Seong-sik I-1060
Hong, Suk-Ki II-902, II-913
Hong, Youn-Sik III-1002
Hosseini, Mohammad Mahdi III-676
Hruschka, Eduardo R. II-168
Hu, Hualiang II-158
Hu, Weixi II-751
Huang, Changqin II-158
Huettmann, Falk II-1117
Huguet-Rotger, Llorenç I-831, IV-223
Huh, Eui-Nam I-370, I-738, I-746
Hur, Hye-Sun III-1002
Hurtado, Ferran III-22
Hwang, Byong-Won III-386, IV-281
Hwang, Chan-Sik III-288
Hwang, Chong-Sun I-286, III-945, IV-233, IV-584
Hwang, EenJun IV-838, IV-859

Hwang, Ha Jin I-577
Hwang, Jun I-1, I-655, I-746
Hwang, Seong Oun II-46
Hwang, Sun-Myung I-481
Hwang, Sungsoon II-1026
Hwang, Yong Ho I-442
Hwang, Yong-Ho II-799
Hwang, YoungHa IV-460

Ibrahim, Hamidah II-146
Iglesias, A. II-641, II-651, II-771
Im, Chaetae I-246
Im, Jae-Yuel IV-655
In, Chi Hyung I-792
Inguglia, Fabrizio II-505
Izquierdo, Antonio I-812

Jabbari, Arash II-432
Jacobs, Gwen III-257
Jang, HyoJong I-41
Jang, Jong-Soo I-988, IV-594
Jang, Jongsu I-776
Jang, Kyung-Soo I-434
Jang, Min-Soo III-489
Jang, Sang-Dong II-216
Jang, Seok-Woo I-9
Jang, Tae-Won I-386
Je, Sung-Kwan IV-421, II-486
Jedlovszky, P. III-217
Jeon, Hoseong I-765
Jeon, Jaeeun III-566
Jeong, Chang Yun I-337
Jeong, Chang-Sung I-319, II-789
Jeong, Eunjoo I-418
Jeong, Hae-Duck J. III-827
Jeong, Ok-Ran III-558
Jeong, Sam Jin IV-213
Jiang, Minghui III-90
Jin, Guiyue III-993
Jin, Hai II-116, II-126
Jin, Min IV-763, IV-849
Jin, Zhou II-272
Jo, Hea Suk I-711, III-1010
Jo, Jang-Wu I-106
Jo, Sun-Moon IV-524
Jonsson, Erland I-821
Jonsson, Håkan III-168
Joo, Pan-Yuh I-394
Jorge, Joaquim II-613

Jun, Woochun II-902, II-913
Jung, Changryul I-294
Jung, Il-Hong I-451
Jung, Kyung-Yong II-863
Jung, Yoon-Jung I-491

Kacsuk, Péter II-10, II-37, II-226
Kanaroglou, Pavlos S. II-1016
Kang, Chang Wook II-554
Kang, Dong-Joong IV-896, IV-906, IV-915
Kang, Euisun I-57
Kang, HeeGok I-402
Kang, Ho-Kyung III-602
Kang, Ho-Seok I-1105
Kang, Hyunchul I-345
Kang, Kyung-Pyo IV-348
Kang, KyungWoo I-65
Kang, Min-Goo I-302, I-386, I-394
Kang, SeokHoon I-270, III-585
Kang, Seung-Shik IV-735
Kang, Sunbu III-926
Kang, Sung Kwan IV-940
Kang, Sungkwon III-847, IV-11
Kang, Tae-Ha IV-281
Kang, Won-Seok IV-167
Kasahara, Yoshiaki I-915
Kasprzak, Andrzej III-611
Kaußner, Armin II-843
Kelz, Markus III-538
Kheddouci, Hamamache III-267
Kim, Backhyun I-345
Kim, Bonghan I-1007
Kim, Byoung-Koo I-998, IV-594
Kim, Byunggi I-418
Kim, Byungkyu III-489
Kim, Chang Hoon III-656, IV-106
Kim, Chang-Soo I-410
Kim, ChangKyun I-150
Kim, Changnam I-738
Kim, ChaYoung IV-233
Kim, Cholmin III-867
Kim, D.S. I-183
Kim, Dae Sun I-1134
Kim, Dae-Chul IV-271
Kim, Daeho I-1078
Kim, Deok-Soo II-554, II-564, II-583, III-53, III-62
Kim, Dohyeon IV-974

Kim, Dong S. I-693, IV-681
Kim, Dong-Hoi I-81
Kim, Dong-Kyoo III-896, III-906, IV-611
Kim, Dongho I-57
Kim, Donguk III-62
Kim, Duckki I-378
Kim, Gwang-Hyun I-1035
Kim, Gyeyoung I-9, I-17, I-41
Kim, Haeng-Kon I-461
Kim, Haeng-kon IV-717
Kim, Hak-Ju I-238
Kim, Hak-Keun IV-772
Kim, Hangkon I-587
Kim, Hanil II-892
Kim, Hie-Cheol II-20
Kim, Hiecheol III-656
Kim, Ho J. IV-791
Kim, Hyeong-Ju I-998
Kim, Hyun Gon I-1151
Kim, Hyun-Sung IV-617
Kim, Hyuncheol I-1078
Kim, Hyung-Jong I-567, I-683
Kim, Ik-Kyun I-998, IV-594
Kim, Iksoo I-270, I-345
Kim, Injung I-491
Kim, Jae-Kyung IV-743
Kim, Jaehyoun I-360
Kim, Jay-Jung II-573
Kim, Jeeyeon I-895
Kim, Jeom Goo I-1026
Kim, Jin I-81
Kim, Jin Geol IV-29
Kim, Jin Ok I-1, I-655, IV-964
Kim, Jin Soo IV-964
Kim, Jong G. II-1
Kim, Jong-bu IV-725
Kim, Jong-Woo I-410
Kim, Joo-Young IV-338
Kim, JoonMo I-567
Kim, Jung-Sun I-175, III-985, IV-321, IV-451
Kim, Jung-Woo II-741
Kim, Kee-Won IV-603, IV-672
Kim, Keecheon I-1115
Kim, Ki-Hyung IV-167
Kim, Ki-Tae IV-524
Kim, Ki-Young I-988, IV-594

Kim, KiIl IV-460
Kim, KiJoo I-49
Kim, Kweon Yang I-134
Kim, Kyungsoo II-467
Kim, Mansoo I-537
Kim, Mi-Ae I-159, I-722
Kim, Mi-Jeong I-394
Kim, Mihui I-673
Kim, Min-Su I-1159
Kim, Minsoo I-175, I-230
Kim, Misun I-199, I-262
Kim, Miyoung I-199, I-262
Kim, MoonJoon I-73
Kim, Moonseong IV-56
Kim, Myuhng-Joo I-683
Kim, Nam-Chang I-1105
Kim, Nam-Yeun IV-87
Kim, Pan Koo IV-940
Kim, Pankoo II-892
Kim, Pyung Soo III-975, IV-301
Kim, Sang Ho I-608, I-1069
Kim, SangHa IV-460
Kim, Sangkyun I-597
Kim, Seokyu I-150
Kim, Seong-Cheol III-837
Kim, Seonho I-328
Kim, Seungjoo I-645, I-895
Kim, Shin-Dug II-20
Kim, Soon Seok I-215
Kim, Soon-Dong IV-611
Kim, Soung Won I-577
Kim, Su-Hyun I-1035
Kim, Sung Jo I-278
Kim, Sung Ki I-246
Kim, Sung Kwon I-215
Kim, Sung-Ho IV-251
Kim, Sung-Hyun I-150
Kim, Sung-Min III-602
Kim, Sung-Ryul III-367
Kim, Sung-Suk IV-924
Kim, Sunghae I-1078
Kim, Sungsoo I-207
Kim, SungSuk I-286
Kim, Tae-Kyung I-238
Kim, Taekkeun III-926
Kim, Tai-Hoon I-451, I-461, I-1052, IV-717
Kim, Won I-17

Kim, Wonil III-896, III-906
Kim, Woo-Hun IV-617
Kim, Wu Woan II-216, II-262
Kim, Yong-Guk III-489
Kim, Yong-Sung I-122, I-337
Kim, Yoon Hyuk II-467
Kim, Young Kuen III-975
Kim, Young-Chon IV-994
Kim, Young-Sin I-738, I-746
Kim, YounSoo II-196
Kiss, T. II-30
Kizilova, Natalya II-476
Ko, Myeong-Cheol IV-772
Ko, Younghun I-360
Kóczy, László T. I-122
Koh, JinGwang I-294, I-310, I-402
Koh, Kwang-Won II-20
Kolingerová, Ivana II-544, II-682, III-198
Koo, Han-Suh II-789
Kouadri Mostéfaoui, Ghita I-537, III-965, IV-983
Kouh, Hoon-Joon IV-524
Ku, Kyo Min IV-196
Kulikov, Gennady Yu. III-345, III-667
Kwak, JaeMin I-402
Kwak, Jin I-895, III-955
Kwak, Keun Chang I-635, IV-828, IV-924
Kwon, Chang-Hee I-310
Kwon, Ki Jin I-1159
Kwon, Kyohyeok I-142
Kwon, Soonhak III-656, IV-106
Kwon, Taekyoung I-728
Kwon, Yong-Won I-319
Kwon, YongHoon III-847, III-926

Laccetti, G. II-515, II-525
Laganà, Antonio II-328, II-357, II-374, II-422, II-437, II-827, II-854
Lagzi, István II-226
Lang, Bruno II-882
Lara, Sheila L. Delfín IV-808
Lau, Matthew C.F. II-873
Lázaro, Miguel II-779
Lee, Bo-Hyeong IV-46
Lee, Bong Hwan I-352
Lee, Bum Ro IV-964
Lee, Byong Gul I-134

Lee, Byong-Lyol   I-663
Lee, Byung Kwan   I-33
Lee, Byung-Wook   I-746
Lee, Byunghoon   III-53
Lee, Dae Jong   I-635, IV-828
Lee, Dea Hwan   I-915
Lee, Deok-Gyu   IV-66
Lee, Dong Chun   I-1052, I-1097
Lee, Dongkeun   I-1115
Lee, Dongryeol   I-510
Lee, Eun-ser   I-451
Lee, Gang-Soo   I-491
Lee, Gunhee   III-906
Lee, Gunhoon   III-566
Lee, Hae-Joung   IV-994
Lee, Hae-ki   IV-725
Lee, Han-Ki   I-159
Lee, Ho-Dong   III-489
Lee, HongSub   I-567
Lee, HoonJae   I-517
Lee, Hunjoo   II-809, II-837
Lee, Hwang-Jik   II-20
Lee, Hyon-Gu   I-89
Lee, Hyun Chang   II-186
Lee, HyunChan   II-554, II-564
Lee, Hyung-Woo   I-302, I-386
Lee, HyungHyo   I-701
Lee, Im-Yeong   I-557, III-1020, IV-66
Lee, In Hwa   I-278
Lee, In-Ho   II-573
Lee, Jae Kwang   I-254, I-1007
Lee, Jae-il   I-728
Lee, Jaeheung   I-547
Lee, Jaeho   II-564
Lee, Jaewan   I-97
Lee, Jong Sik   III-621, III-630
Lee, Jong-Suk Ruth   III-827
Lee, Joongjae   I-17
Lee, Ju-Hyun   IV-11
Lee, Jung-Hyun   II-863
Lee, Jungsik   I-97
Lee, KangShin   I-567
Lee, Keon-Jik   III-638
Lee, Key Seo   IV-964
Lee, Ki Dong   III-566, III-993
Lee, Kwan H.   III-178
Lee, Kwang-Ok   I-310
Lee, Kwnag-Jae   IV-451

Lee, Kyong-Ho   IV-743
Lee, Kyung Whan   I-451
Lee, Malrey   I-97
Lee, Myung Eui   III-975, IV-301
Lee, Myung-Sub   IV-441
Lee, Namhoon   I-491
Lee, Okbin   II-178
Lee, Ou-Seb   I-394
Lee, Pil Joong   I-442, I-471, I-802
Lee, Sang Hyo   IV-964
Lee, Sang-Hak   III-288
Lee, Sang-Ho   IV-689
Lee, Sangkeon   I-1017, I-1088
Lee, SangKeun   I-286
Lee, Seok-Joo   III-489
Lee, Seung   IV-725
Lee, SeungYong   I-701
Lee, Soo-Gi   I-625
Lee, SooCheol   IV-838, IV-859
Lee, Soung-uck   III-867
Lee, Sung-Woon   IV-617
Lee, Sungchang   I-1125
Lee, Sungkeun   I-294
Lee, Tae-Jin   I-1159, IV-46
Lee, Tae-Seung   III-386, IV-281
Lee, Taehoon   II-178
Lee, Tong-Yee   II-713, II-721
Lee, Won Goo   I-254
Lee, Won-Ho   III-638
Lee, Won-Hyung   I-159, I-722
Lee, Won-Jong   II-741
Lee, Woojin   I-426
Lee, Woongjae   I-1, I-655
Lee, YangKyoo   IV-838, IV-859
Lee, Yeijin   II-178
Lee, YoungSeok   II-196
Lee, Yugyung   I-410
Leem, Choon Seong   I-597, I-608, I-1069
Lendvay, György   II-290
Levashov, P.   II-402
Lho, Tae-Jung   IV-906
Li, Chunlin   IV-117
Li, Gang   II-252
Li, Layuan   IV-117
Li, Mingchu   II-693
Li, Shengli   II-116
Li, Xiaotu   II-252
Li, Xueyao   IV-414

Li, Yufu II-116
Lim, Heeran IV-708
Lim, Hwa-Seop I-386
Lim, Hyung-Jin I-238
Lim, Joon S. IV-791
Lim, SeonGan I-517
Lim, Soon-Bum IV-772
Lim, Younghwan I-57
Lin, Hai II-236
Lin, Ping-Hsien II-713
Lin, Wenhao III-257
Lindskog, Stefan I-821
Lísal, Martin II-392
Liturri, Luciano II-979
Liu, Da-xin III-706
Liu, Yongle III-498
Llanos, Diego R. III-188
Lombardo, S. II-1046
Longo, S. II-383
Lopez, Javier I-903
López, Mario Alberto III-99
Lovas, Róbert II-10, II-226
Lu, Chaohui IV-243
Lu, Jianfeng III-308
Lu, Yilong III-729
Lu, Yinghua IV-956
Lu, Zhengding IV-117
Luna-Rivera, Jose Martin IV-177
Luo, Yingwei III-335

Ma, Zhiqiang IV-471
Machì, A. II-536
Maddalena, L. II-525
Maponi, Pierluigi III-575
Marangi, Carmela II-971
Mariani, Riccardo III-745
Marinelli, Maria III-575
Mark, Christian II-843
Marques, Fábio III-127
Marshall, Geoffrey III-528
Martínez, Alicia IV-506, IV-783
Martoyan, Gagik A. II-313
Mastronardi, Nicola II-932
Matsuhisa, Takashi III-915
Maur, Pavel III-198
Medvedev, N.N. III-217
Mejri, Mohamed I-938
Melnik, Roderick V.N. III-817
Ménegaux, David III-267

Merkulov, Arkadi I. III-667
Merlitz, Holger III-465
Messelodi, Stefano II-693
Miguez, Xochitl Landa IV-137
Milani, Alfredo III-433, IV-563
Min, Byoung Joon I-246
Min, Hongki III-585
Min, Jun Oh I-635, IV-828
Min, Young Soo IV-869
Minelli, P. II-383
Ming, Zeng IV-127
Mitrani, I. II-76
Moh, Sangman IV-97
Molina, Ana I. III-786
Mollá, Ramón III-877
Monterde, J. II-631
Moon, Aekyung III-696
Moon, Kiyoung I-776
Moon, SangJae I-150, I-517
Moon, Young-Jun I-1088
Mora, Graciela IV-77
Moradi, Shahram II-432
Moreno, Oscar III-481
Moreno-Jiménez, Carlos III-1
Morici, Chiara III-433
Morillo, P. II-661
Mukherjee, Biswanath IV-994
Mumey, Brendan III-90
Mun, Youngsong I-199, I-262, I-378, I-738, I-1144
Murgante, Beniamino II-1036
Murli, A. II-515
Murri, Roberto III-575
Muzaffar, Tanzeem IV-291

Na, Jung C. I-191, I-693, IV-681
Na, Won Shik I-1026
Na, Young-Joo II-863
Nam, Dong Su I-352
Nam, Junghyun I-645
Nandy, Subhas C. III-42
Navarro-Moldes, Leandro IV-177
Naya, Ferran II-613
Nedoma, Jiří II-445, II-456
Neelamkavil, Francis II-741, IV-743
Németh, Csaba II-10
Nguyen, Thai T. IV-791
Nicotra, F. II-536

Nielsen, Frank III-147
Niewiadomski, Radoslaw III-433
Nishida, Tetsushi III-227
Nock, Richard III-147
Noh, Bong-Nam I-175, I-230
Noh, BongNam I-701
Noh, JiSung II-942
Noh, SungKee IV-460
Noltemeier, Hartmut II-843

O'Loughlin, Finbarr II-136
O'Rourke, S.F.C. II-321
Oh, Am Sok I-33
Oh, ByeongKyun I-527, IV-698
Oh, Jai-Ho I-765
Oh, Kyu-Tae III-985
Oh, Soohyun III-955
Oh, Sun-Jin I-617
Oh, Wongeun I-294
Oh, Young-Hwan I-222
Ohn, Kyungoh III-548
Olanda, Ricardo II-671
Oliveira Albuquerque, Robson de I-868
Onieva, Jose A. I-903
Orduña, J.M. II-661
Orozco, Edusmildo III-481, III-736
Orser, Gary III-257
Ortega, Manuel III-786
Otero, César II-641, II-779, III-158
Othman, Abdulla II-66
Othman, Abu Talib II-146
Othman, Mohamed II-146
Ouyang, Jinsong I-345
Ozturk, Zafer Ziya IV-398

Pacifici, Leonardo II-357
Pakdel, Hamid-Reza III-237
Palladini, Sergio II-1057
Palmer, J. II-76
Palmieri, Francesco I-882
Palop, Belén III-188
Pan, Zhigeng II-236, II-731, II-751, III-308
Pardo, Fernando IV-887
Park, Chang Won IV-627
Park, Chang-Hyeon IV-441
Park, Dong-Hyun II-863
Park, Goorack I-25
Park, Gwi-Tae III-489

Park, Gyung-Leen I-114
Park, Hee-Un I-557
Park, Hong Jin I-215
Park, Hyoung-Woo I-319, II-1, III-827
Park, Hyunpung III-178
Park, IkSu I-527, IV-698
Park, JaeHeung I-73
Park, Jaehyung I-1159
Park, Jihun IV-311, IV-369
Park, Jong An IV-877, IV-940, IV-948
Park, Jong Sou IV-574
Park, Jongjin I-1144
Park, Joo-Chul I-9
Park, Joon Young II-554, II-564
Park, Jun-Hyung I-230
Park, Ki heon IV-29
Park, Kyeongmo I-500
Park, Kyung-Lang II-20
Park, M.-W. II-573
Park, Mingi I-97
Park, Namje I-776
Park, Sangjoon I-418
Park, Seong-Seok I-410
Park, Seung Jin IV-877, IV-948
Park, SeungBae I-527, IV-698
Park, Sihn-hye III-896
Park, Soohong III-975
Park, Soon-Young II-1079
Park, Sunghun IV-311, IV-369
Park, Taehyung I-1017, I-1088
Park, Taejoon II-837
Park, Woo-Chan II-741
Park, Yongsu I-547, I-978, IV-799
Pastor, Oscar IV-506, IV-783
Payeras-Capella, Magdalena I-831, IV-223
Pazos R., Rodolfo A. III-415, IV-77
Pedlow, R.T. II-321
Peña, José M. II-87
Pérez O., Joaquín III-415, IV-77
Pérez, José María IV-496
Pérez, María S. II-87
Pérez, Mariano II-671
Petri, M. II-1046, II-1107
Petrosino, A. II-525
Pfarrhofer, Roman III-538
Pflug, Hans-Joachim II-882
Piantanelli, Anna III-575

Piattini, Mario I-968
Pieretti, A. II-366
Piermarini, Valentina II-422
Pierro, Cinzia II-338
Pietraperzia, G. II-374
Pineda, Ulises IV-177
Ping, Tan Tien IV-147
Pişkin, Şenol III-795
Podesta, Karl III-473
Poggioni, Valentina IV-563
Politi, Tiziano II-961, II-988
Ponce, Eva I-949
Porschen, Stefan III-137
Puchala, Edward IV-39
Pugliese, Andrea II-55
Puig-Pey, J. II-651, II-771
Puigserver, Macià Mut I-924
Puttini, Ricardo S. I-868

Qi, Zhaohui II-252
Qin, Zhongping III-90

Ra, In-Ho I-310, IV-359
Radulovic, Nenad III-817
Ragni, Stefania II-971
Rahayu, Wenny III-443, III-508
Ramos, J.F. II-622, II-703
Ramos, Pedro III-22
Rebollo, C. II-703
Recio, T. II-761
Redondo, Miguel A. III-786
Reitsma, Femke II-1069
Remigi, Andrea III-745
Remolar, I. II-703
Rho, SeungMin IV-859
Ribagorda, Arturo I-812
Riganelli, Antonio II-374, II-827
Rivera-Campo, Eduardo III-22
Ro, Yong Man III-602
Robinson, Andrew III-443
Robles, Víctor II-87
Rodionov, Alexey S. III-315, IV-431
Rodionova, Olga K. III-315, IV-431
Rodríguez O., Guillermo III-415, IV-77
Rodríguez, Judith II-922
Rogerson, Peter II-1096
Roh, Sun-Sik I-1035
Roh, Yong-Wan I-89
Rosi, Marzio II-412

Rotger, Llorenç Huguet i I-924
Roy, Sasanka III-42
Rui, Zhao IV-127
Ruskin, Heather J. III-473, III-498
Rutigliano, M. II-366
Ryoo, Intae I-1026
Ryou, Hwang-bin I-1060
Ryou, Jaecheol I-776
Ryu, Eun-Kyung IV-603, IV-655, IV-665, IV-672
Ryu, So-Hyun I-319
Ryu, Tae W. IV-185, IV-791

Safouhi, Hassan II-280
Samavati, Faramarz F. III-237, III-247
Sampaio, Alcínia Zita II-817
Sánchez, Alberto II-87
Sánchez, Carlos II-328
Sánchez, Ricardo II-603
Sánchez, Teresa I-949
Sanna, N. II-366
Santaolaya Salgado, René IV-534, IV-808
Santos, Juan II-922
Santucci, A. II-1107
Sanvicente-Sánchez, Héctor III-755
Sasahara, Shinji III-11
Sastrón, Francisco III-857
Schoier, Gabriella II-1009, II-1089
Schug, Alexander III-454
Sellarès, Joan Antoni III-99
Senger, Hermes II-168
Seo, Dae-Hee I-557, III-1020
Seo, Heekyung III-837
Seo, Kyong Sok I-655
Seo, Seung-Hyun IV-689
Seo, Sung Jin I-1
Seo, Young Ro IV-964
Seong, Yeong Kyeong IV-338
Seri, Raffaello III-298
Seung-Hak, Rhee IV-948
Seznec, Andre I-960
Sgamellotti, Antonio II-412
Shahdin, S. II-350
Shen, Liran IV-414
Shen, Weidong IV-1
Shim, Hye-jin IV-321
Shim, Jae-sun IV-725
Shim, Jeong Min IV-869

Shim, Young-Chul I-1105
Shin, Byung-Joo IV-763, IV-849
Shin, Dong-Ryeol I-434
Shin, Hayong II-583
Shin, Ho-Jun I-625
Shin, Jeong-Hoon IV-754
Shin, Seung-won I-988
Shin, Yongtae I-328
Shindin, Sergey K. III-345
Sierra, José María I-851, I-812, I-960
Silva, Fabrício A.B. da II-168
Silva, Tamer Américo da I-868
Sim, Sang Gyoo I-442
Singh, Gujit II-246
Sipos, Gergely II-37
Skala, Václav III-81, III-325
Skouteris, Dimitris II-357
Slim, Chokri III-935
Smith, William R. II-392
So, Won-Ho IV-994
Sodhy, Gian Chand IV-147
Sohn, Sungwon I-776
Sohn, Won-Sung IV-743, IV-772
Sohn, Young-Ho IV-441
Song, Geun-Sil I-159, I-722
Song, Hyoung-Kyu I-386, I-394
Song, Il Gyu I-792
Song, Jin-Young II-799
Song, Kyu-Yeop IV-994
Song, Mingli III-886, IV-406
Song, Myunghyun I-294
Song, Seok Il IV-869
Song, Sung Keun IV-627
Song, Teuk-Seob IV-743
Sosa, Víctor J. Sosa IV-137
Soto, Leonardo II-603
Sousa Jr., Rafael T. de I-868
Soykan, Gürkan III-795
Stefano, Marco Di II-412
Stehlík, J. II-456
Stevens-Navarro, Enrique IV-177
Stögner, Herbert III-538
Strandbergh, Johan I-821
Studer, Pedro II-817
Sturm, Patrick III-109
Sug, Hyontai IV-158
Sugihara, Kokichi III-53, III-71, III-227
Sulaiman, Md Nasir II-146

Sun, Jizhou II-252, II-272

Tae, Kang Soo I-114
Talia, Domenico II-55
Tan, Rebecca B.N. II-873
Tang, Chuan Yi III-355
Taniar, David III-508, IV-543
Tasaltin, Cihat IV-398
Tasso, Sergio II-437
Tavadyan, Levon A. II-313
Techapichetvanich, Kesaraporn IV-479
Tejel, Javier III-22
Temurtas, Fevzullah IV-389, IV-398
Temurtas, Hasan IV-398
Thanh, Nguyen N. III-602
Thulasiram, Ruppa K. III-686
Thulasiraman, Parimala III-686
Togores, Reinaldo II-641, II-779, III-158
Tomás, Ana Paula III-117, III-127
Tomascak, Andrew III-90
Torres, Joaquín I-851
Torres-Jimenez, Jose IV-506
Trendafilov, Nickolay T. II-952
Turányi, Tamás II-226

Uhl, Andreas III-538
Uhmn, Saangyong I-81
Um, Sungmin I-57

Valdés Marrero, Manuel A. IV-137, IV-534, IV-808
Vanmechelen, Kurt IV-514
Vanzi, Eleonora II-495
Varnuška, Michal II-682
Vásquez Mendez, Isaac M. IV-534, IV-808
Vavřík, P. II-456
Vehreschild, Andre II-882
Ventura, Immaculada III-207
Verduzco Medina, Francisco IV-137
Ves, Esther De IV-887
Villalba, Luis Javier García I-859, I-868
Voloshin, V.P. III-217

Wang, Huiqiang IV-414
Wang, Tong III-706
Wang, Xiaolin III-335
Watson, Anthony IV-471
Wenzel, Wolfgang III-454, III-465
Willatzen, Morten III-817

Winter, S.C.   II-30
Won, Dongho   I-645, I-895, III-955
Woo, Yoseop   I-270, I-345, III-585
Wouters, Carlo   III-508
Wozniak, Michal   III-593
Wu, Bang Ye   III-355
Wu, Guohua   II-731
Wyvill, Brian   III-247

Xinyu, Yang   IV-127
Xu, Guang   III-277
Xu, Jinhui   III-277
Xu, Qing   II-693
Xu, Zhuoqun   III-335

Yamada, Ikuho   II-1096
Yan, Shaur-Uei   II-721
Yang, Bailin   II-236
Yang, Jin S.   I-191, I-693, IV-681
Yang, Jong-Un   IV-359
Yang, Shulin   IV-1
Yang, Sun Ok   I-286
Yang, Sung-Bong   II-741, IV-743
Yang, SunWoong   I-73
Yang, Tz-Hsien   II-713
Yang, Zhiling   II-126
Yao, Zhenhua   III-729
Yap, Chee   III-62
Yaşar, Osman   III-795, III-807
Yavari, Issa   II-432
Yen, Sung-Ming   I-150
Yi, Myung-Kyu   III-945, IV-584
Yi, Shi   IV-127
Yim, Wha Young   IV-964
Yin, Xuesong   II-731
Yoe, Hyun   I-294, I-402
Yong, Chan Huah   IV-147
Yoo, Hyeong Seon   I-510
Yoo, Jae Soo   IV-869
Yoo, Kee-Young   III-638, IV-87, IV-196, IV-603, IV-617, IV-655, IV-665, IV-672

Yoo, Kil-Sang   I-159
Yoo, Kook-yeol   IV-329
Yoo, Sang Bong   II-1079
Yoo, Weon-Hee   IV-524
Yoo, Wi Hyun   IV-196
Yoon, Eun-Jun   IV-665
Yoon, Hyung-Wook   IV-46
Yoon, Jin-Sung   I-9
Yoon, Ki Song   II-46
Yoon, Miyoun   I-328
You, Il-Sun   I-167
You, Mingyu   III-886, IV-406
You, Young-Hwan   I-386, I-394
Youn, Chan-Hyun   I-352
Youn, Hee Yong   I-114, I-711, III-1010, IV-627, IV-637
Yu, Chansu   IV-97
Yu, Kwangseok   III-62
Yu, Qizhi   II-236
Yum, Dae Hyun   I-471, I-802
Yumusak, Nejat   IV-389, IV-398
Yun, Byeong-Soo   IV-818
Yun, Miso   II-892

Zaia, Annamaria   III-575
Zeng, Qinghuai   II-158
Zhang, Hu   III-764
Zhang, Jiawan   II-252, II-272, II-693
Zhang, Jing   IV-994
Zhang, Mingmin   II-236
Zhang, Minming   III-308
Zhang, Qin   II-116
Zhang, Rubo   IV-414
Zhang, Yi   II-272
Zhang, Zhaoyang   IV-243
Zhao, Chunjiang   II-751
Zhou, Jianying   I-903
Zhu, Binhai   III-90, III-257
Zotta, D.   II-1046

Lecture Notes in Computer Science

For information about Vols. 1–2951

please contact your bookseller or Springer-Verlag

Vol. 3060: A.Y. Tawfik, S.D. Goodwin (Eds.), Advances in Artificial Intelligence. XIII, 582 pages. 2004. (Subseries LNAI).

Vol. 3059: C.C. Ribeiro, S.L. Martins (Eds.), Experimental and Efficient Algorithms. X, 586 pages. 2004.

Vol. 3058: N. Sebe, M.S. Lew, T.S. Huang (Eds.), Computer Vision in Human-Computer Interaction. X, 233 pages. 2004.

Vol. 3056: H. Dai, R. Srikant, C. Zhang (Eds.), Advances in Knowledge Discovery and Data Mining. XIX, 713 pages. 2004. (Subseries LNAI).

Vol. 3054: I. Crnkovic, J.A. Stafford, H.W. Schmidt, K. Wallnau (Eds.), Component-Based Software Engineering. XI, 311 pages. 2004.

Vol. 3053: C. Bussler, J. Davies, D. Fensel, R. Studer (Eds.), The Semantic Web: Research and Applications. XIII, 490 pages. 2004.

Vol. 3046: A. Laganà, M.L. Gavrilova, V. Kumar, Y. Mun, C.J.K. Tan, O. Gervasi (Eds.), Computational Science and Its Applications - ICCSA 2004. Part IV. LIII, 1016 pages. 2004.

Vol. 3045: A. Laganà, M.L. Gavrilova, V. Kumar, Y. Mun, C.J.K. Tan, O. Gervasi (Eds.), Computational Science and Its Applications – ICCSA 2004. Part III. LIII, 1040 pages. 2004.

Vol. 3044: A. Laganà, M.L. Gavrilova, V. Kumar, Y. Mun, C.J.K. Tan, O. Gervasi (Eds.), Computational Science and Its Applications – ICCSA 2004. Part II. LIII, 1140 pages. 2004.

Vol. 3043: A. Laganà, M.L. Gavrilova, V. Kumar, Y. Mun, C.J.K. Tan, O. Gervasi (Eds.), Computational Science and Its Applications – ICCSA 2004. Part I. LIII, 1180 pages. 2004.

Vol. 3042: N. Mitrou, K. Kontovasilis, G.N. Rouskas, I. Iliadis, L. Merakos (Eds.), NETWORKING 2004, Networking Technologies, Services, and Protocols; Performance of Computer and Communication Networks; Mobile and Wireless Communications. XXXIII, 1519 pages. 2004.

Vol. 3035: M.A. Wimmer, Knowledge Management in Electronic Government. XII, 342 pages. 2004. (Subseries LNAI).

Vol. 3034: J. Favela, E. Menasalvas, E. Chávez (Eds.), Advances in Web Intelligence. XIII, 227 pages. 2004. (Subseries LNAI).

Vol. 3033: M. Li, X.-H. Sun, Q. Deng, J. Ni (Eds.), Grid and Cooperative Computing. XXXVIII, 1076 pages. 2004.

Vol. 3032: M. Li, X.-H. Sun, Q. Deng, J. Ni (Eds.), Grid and Cooperative Computing. XXXVII, 1112 pages. 2004.

Vol. 3031: A. Butz, A. Krüger, P. Olivier (Eds.), Smart Graphics. X, 165 pages. 2004.

Vol. 3028: D. Neuenschwander, Probabilistic and Statistical Methods in Cryptology. X, 158 pages. 2004.

Vol. 3027: C. Cachin, J. Camenisch (Eds.), Advances in Cryptology - EUROCRYPT 2004. XI, 628 pages. 2004.

Vol. 3026: C. Ramamoorthy, R. Lee, K.W. Lee (Eds.), Software Engineering Research and Applications. XV, 377 pages. 2004.

Vol. 3025: G.A. Vouros, T. Panayiotopoulos (Eds.), Methods and Applications of Artificial Intelligence. XV, 546 pages. 2004. (Subseries LNAI).

Vol. 3024: T. Pajdla, J. Matas (Eds.), Computer Vision - ECCV 2004. XXVIII, 621 pages. 2004.

Vol. 3023: T. Pajdla, J. Matas (Eds.), Computer Vision - ECCV 2004. XXVIII, 611 pages. 2004.

Vol. 3022: T. Pajdla, J. Matas (Eds.), Computer Vision - ECCV 2004. XXVIII, 621 pages. 2004.

Vol. 3021: T. Pajdla, J. Matas (Eds.), Computer Vision - ECCV 2004. XXVIII, 633 pages. 2004.

Vol. 3019: R. Wyrzykowski, J. Dongarra, M. Paprzycki, J. Wasniewski (Eds.), Parallel Processing and Applied Mathematics. XIX, 1174 pages. 2004.

Vol. 3015: C. Barakat, I. Pratt (Eds.), Passive and Active Network Measurement. XI, 300 pages. 2004.

Vol. 3012: K. Kurumatani, S.-H. Chen, A. Ohuchi (Eds.), Multi-Agnets for Mass User Support. X, 217 pages. 2004. (Subseries LNAI).

Vol. 3011: J.-C. Régin, M. Rueher (Eds.), Integration of AI and OR Techniques in Constraint Programming for Combinatorial Optimization Problems. XI, 415 pages. 2004.

Vol. 3010: K.R. Apt, F. Fages, F. Rossi, P. Szeredi, J. Váncza (Eds.), Recent Advances in Constraints. VIII, 285 pages. 2004. (Subseries LNAI).

Vol. 3009: F. Bomarius, H. Iida (Eds.), Product Focused Software Process Improvement. XIV, 584 pages. 2004.

Vol. 3008: S. Heuel, Uncertain Projective Geometry. XVII, 205 pages. 2004.

Vol. 3007: J.X. Yu, X. Lin, H. Lu, Y. Zhang (Eds.), Advanced Web Technologies and Applications. XXII, 936 pages. 2004.

Vol. 3006: M. Matsui, R. Zuccherato (Eds.), Selected Areas in Cryptography. XI, 361 pages. 2004.

Vol. 3005: G.R. Raidl, S. Cagnoni, J. Branke, D.W. Corne, R. Drechsler, Y. Jin, C.G. Johnson, P. Machado, E. Marchiori, F. Rothlauf, G.D. Smith, G. Squillero (Eds.), Applications of Evolutionary Computing. XVII, 562 pages. 2004.

Vol. 3004: J. Gottlieb, G.R. Raidl (Eds.), Evolutionary Computation in Combinatorial Optimization. X, 241 pages. 2004.

Vol. 3003: M. Keijzer, U.-M. O'Reilly, S.M. Lucas, E. Costa, T. Soule (Eds.), Genetic Programming. XI, 410 pages. 2004.

Vol. 3002: D.L. Hicks (Ed.), Metainformatics. X, 213 pages. 2004.

Vol. 3001: A. Ferscha, F. Mattern (Eds.), Pervasive Computing. XVII, 358 pages. 2004.

Vol. 2999: E.A. Boiten, J. Derrick, G. Smith (Eds.), Integrated Formal Methods. XI, 541 pages. 2004.

Vol. 2998: Y. Kameyama, P.J. Stuckey (Eds.), Functional and Logic Programming. X, 307 pages. 2004.

Vol. 2997: S. McDonald, J. Tait (Eds.), Advances in Information Retrieval. XIII, 427 pages. 2004.

Vol. 2996: V. Diekert, M. Habib (Eds.), STACS 2004. XVI, 658 pages. 2004.

Vol. 2995: C. Jensen, S. Poslad, T. Dimitrakos (Eds.), Trust Management. XIII, 377 pages. 2004.

Vol. 2994: E. Rahm (Ed.), Data Integration in the Life Sciences. X, 221 pages. 2004. (Subseries LNBI).

Vol. 2993: R. Alur, G.J. Pappas (Eds.), Hybrid Systems: Computation and Control. XII, 674 pages. 2004.

Vol. 2992: E. Bertino, S. Christodoulakis, D. Plexousakis, V. Christophides, M. Koubarakis, K. Böhm, E. Ferrari (Eds.), Advances in Database Technology - EDBT 2004. XVIII, 877 pages. 2004.

Vol. 2991: R. Alt, A. Frommer, R.B. Kearfott, W. Luther (Eds.), Numerical Software with Result Verification. X, 315 pages. 2004.

Vol. 2989: S. Graf, L. Mounier (Eds.), Model Checking Software. X, 309 pages. 2004.

Vol. 2988: K. Jensen, A. Podelski (Eds.), Tools and Algorithms for the Construction and Analysis of Systems. XIV, 608 pages. 2004.

Vol. 2987: I. Walukiewicz (Ed.), Foundations of Software Science and Computation Structures. XIII, 529 pages. 2004.

Vol. 2986: D. Schmidt (Ed.), Programming Languages and Systems. XII, 417 pages. 2004.

Vol. 2985: E. Duesterwald (Ed.), Compiler Construction. X, 313 pages. 2004.

Vol. 2984: M. Wermelinger, T. Margaria-Steffen (Eds.), Fundamental Approaches to Software Engineering. XII, 389 pages. 2004.

Vol. 2983: S. Istrail, M.S. Waterman, A. Clark (Eds.), Computational Methods for SNPs and Haplotype Inference. IX, 153 pages. 2004. (Subseries LNBI).

Vol. 2982: N. Wakamiya, M. Solarski, J. Sterbenz (Eds.), Active Networks. XI, 308 pages. 2004.

Vol. 2981: C. Müller-Schloer, T. Ungerer, B. Bauer (Eds.), Organic and Pervasive Computing – ARCS 2004. XI, 339 pages. 2004.

Vol. 2980: A. Blackwell, K. Marriott, A. Shimojima (Eds.), Diagrammatic Representation and Inference. XV, 448 pages. 2004. (Subseries LNAI).

Vol. 2979: I. Stoica, Stateless Core: A Scalable Approach for Quality of Service in the Internet. XVI, 219 pages. 2004.

Vol. 2978: R. Groz, R.M. Hierons (Eds.), Testing of Communicating Systems. XII, 225 pages. 2004.

Vol. 2977: G. Di Marzo Serugendo, A. Karageorgos, O.F. Rana, F. Zambonelli (Eds.), Engineering Self-Organising Systems. X, 299 pages. 2004. (Subseries LNAI).

Vol. 2976: M. Farach-Colton (Ed.), LATIN 2004: Theoretical Informatics. XV, 626 pages. 2004.

Vol. 2973: Y. Lee, J. Li, K.-Y. Whang, D. Lee (Eds.), Database Systems for Advanced Applications. XXIV, 925 pages. 2004.

Vol. 2972: R. Monroy, G. Arroyo-Figueroa, L.E. Sucar, H. Sossa (Eds.), MICAI 2004: Advances in Artificial Intelligence. XVII, 923 pages. 2004. (Subseries LNAI).

Vol. 2971: J.I. Lim, D.H. Lee (Eds.), Information Security and Cryptology -ICISC 2003. XI, 458 pages. 2004.

Vol. 2970: F. Fernández Rivera, M. Bubak, A. Gómez Tato, R. Doallo (Eds.), Grid Computing. XI, 328 pages. 2004.

Vol. 2968: J. Chen, S. Hong (Eds.), Real-Time and Embedded Computing Systems and Applications. XIV, 620 pages. 2004.

Vol. 2967: S. Melnik, Generic Model Management. XX, 238 pages. 2004.

Vol. 2966: F.B. Sachse, Computational Cardiology. XVIII, 322 pages. 2004.

Vol. 2965: M.C. Calzarossa, E. Gelenbe, Performance Tools and Applications to Networked Systems. VIII, 385 pages. 2004.

Vol. 2964: T. Okamoto (Ed.), Topics in Cryptology – CT-RSA 2004. XI, 387 pages. 2004.

Vol. 2963: R. Sharp, Higher Level Hardware Synthesis. XVI, 195 pages. 2004.

Vol. 2962: S. Bistarelli, Semirings for Soft Constraint Solving and Programming. XII, 279 pages. 2004.

Vol. 2961: P. Eklund (Ed.), Concept Lattices. IX, 411 pages. 2004. (Subseries LNAI).

Vol. 2960: P.D. Mosses (Ed.), CASL Reference Manual. XVII, 528 pages. 2004.

Vol. 2959: R. Kazman, D. Port (Eds.), COTS-Based Software Systems. XIV, 219 pages. 2004.

Vol. 2958: L. Rauchwerger (Ed.), Languages and Compilers for Parallel Computing. XI, 556 pages. 2004.

Vol. 2957: P. Langendoerfer, M. Liu, I. Matta, V. Tsaoussidis (Eds.), Wired/Wireless Internet Communications. XI, 307 pages. 2004.

Vol. 2956: A. Dengel, M. Junker, A. Weisbecker (Eds.), Reading and Learning. XII, 355 pages. 2004.

Vol. 2954: F. Crestani, M. Dunlop, S. Mizzaro (Eds.), Mobile and Ubiquitous Information Access. X, 299 pages. 2004.

Vol. 2953: K. Konrad, Model Generation for Natural Language Interpretation and Analysis. XIII, 166 pages. 2004. (Subseries LNAI).

Vol. 2952: N. Guelfi, E. Astesiano, G. Reggio (Eds.), Scientific Engineering of Distributed Java Applications. X, 157 pages. 2004.